BIOLOGY

for a Changing World

with Physiology

BIOLOGY for a Changing World with Physiology

third edition

Michèle Shuster / Janet Vigna / Matthew Tontonoz

Vice President, STEM Ben Roberts
Editorial Program Director Brooke Suchomel
Senior Development Editor Susan Moran
Development Editor Erica Frost
Senior Editor, Teaching and Learning Strategies Elaine Palucki
Assistant Editor Kevin Davidson
Executive Marketing Manager Will Moore
Marketing Assistant Cate McCaffery
Director of Content, STEM Clairissa Simmons
Media Editor Jennifer Compton
Content Development Manager, Biology Amber Jonker
Lead Content Developer, Biology Mary Tibbets
STEM Project Manager Erin Inks
Director, Content Management Enhancement Tracey Kuehn
Managing Editor Lisa Kinne
Senior Content Project Manager Liz Geller
Media Project Manager Daniel Comstock
Copyeditor Nancy Brooks
Director of Design, Content Management Diana Blume
Senior Design Manager Blake Logan
Art Manager Matthew McAdams
Artwork Eli Ensor
Media Permissions Manager Christine Buese
Photo Researcher Teri Stratford
Senior Content Workflow Manager Paul Rohloff
Composition Lumina Datamatics, Inc.
Printing and Binding LSC Communications
Cover Photo ZenShui/Michele Constantini/Getty Images

Library of Congress Control Number: 2017958526

ISBN-13: 978-1-319-05058-0
ISBN-10: 1-319-05058-1

Printed in the United States of America

First printing

W. H. Freeman and Company
One New York Plaza
Suite 4500
New York, NY 10004-1562
www.macmillanlearning.com

TO OUR TEACHERS AND STUDENTS

You are our inspiration

About the Authors

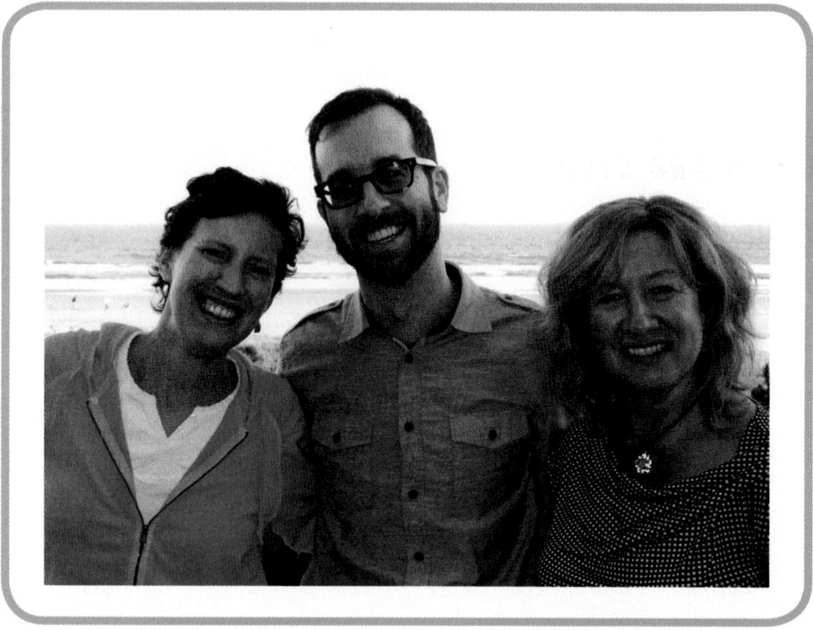

Michèle Shuster, Matthew Tontonoz, and Janet Vigna

MICHÈLE SHUSTER, PH.D., is an associate professor in the biology department at New Mexico State University in Las Cruces, New Mexico. She focuses on the scholarship of teaching and learning and teaches introductory biology, microbiology, and cancer biology classes at the undergraduate level, as well as working on several K–12 science education programs. Michèle is involved in mentoring graduate students and postdoctoral fellows in effective teaching, preparing the next generation of undergraduate educators. She is the recipient of numerous teaching awards, including the Westhafer Award for Teaching Excellence at NMSU. Michèle received her Ph.D. from the Sackler School of Graduate Biomedical Sciences at Tufts University School of Medicine, where she studied meiotic chromosome segregation in yeast.

JANET VIGNA, PH.D., is a professor of biology and chair of the Biology Department at Grand Valley State University in Allendale, Michigan. She is a science education specialist in the Integrated Science Program, with a passion for training and mentoring K–12 science teachers and teaching nonmajors biology. She has recently been recognized as the Michigan Science Teachers Association College Science Teacher of the Year. Her scholarly interests include biology curriculum development, retention strategies for academically at-risk students, and research on the effects of biological pesticides on amphibian communities. She received her B.S. in biology from the University of Michigan and her Ph.D. in microbiology from the University of Iowa.

MATTHEW TONTONOZ is a science writer and independent scholar living in Brooklyn, New York. For ten years, he was a development editor for textbooks in biology before shifting his focus to writing. He is currently senior science writer at Memorial Sloan Kettering Cancer Center, where he covers advances in basic science and clinical cancer research. Matt received his B.A. in biology from Wesleyan University and his M.A. in the history and sociology of science from the University of Pennsylvania.

Dear Student,

Thank you for opening this book! We hope that your journey through it will be as rewarding for you as our journey in writing it has been. When we first came together to collaborate on the development of this text, our biggest overarching goal was to get students interested in biology by showing its relevance to daily life. We wanted to create a textbook that students would actually want to read. Our model and partner in this process has been *Scientific American*, a visually stunning magazine that's been successfully bringing science to the public for more than 150 years. The result is a unique textbook that takes a novel approach to teaching biology, one that we think has the potential to greatly improve learning. We hope that this brief introduction will serve as a road map of the book, so that you can get the most out of your experience with it and be as captivated by the wonders of life as we are.

The main approach of each chapter is the presentation of key science concepts within the context of a relevant and engaging story—a story of discovery, of determination, of human interest, of adventure. From the search for life on Mars to the problem of antibiotic-resistant bacteria, we use stories to bring science to life and to show scientists in action. After all, science is not just a collection of facts, so why would we present it that way? We ask you, our students, to study biology so you can use knowledge to make choices in the real world. We value those stories that will lead you to ask questions about life and how it works and to see the relevance of biology to daily activity. We have seen how stories engage students in our classrooms, and we hope you will be similarly intrigued.

While gripped by a story, you may not even realize how much you are learning. To reinforce the basic learning process, we rely on several strategies:

- Each story is prefaced by a set of **Driving Questions**. By keeping these in mind as you navigate the story, you will have a good framework for learning the key science concepts.
- Eye-catching **Infographics** highlight and drill down into the science of each story. The set of Infographics in a chapter provides a science storyboard for that chapter, illustrating the key scientific concepts and linking them to the story.
- Each Infographic has a **question** to help you ensure that you have grasped the concept illustrated.
- **Key terms** are defined in the margins, making it easy to check a definition without having to leave the story.
- **Chapter summaries** provide a concise set of bullet points that distill the key scientific concepts.
- **Test Your Knowledge** questions at the end of each chapter reinforce basic facts and allow you to apply these facts through data interpretation and mini cases.

By taking full advantage of these resources, you will be better able to appreciate how biology affects each and every one of us as well as our close and distant relatives on this planet. We hope that you will talk about biology with your friends and family, and that what you learn here will be applicable to your life. We hope that you will think as critically about choices you make outside the classroom as we will ask you to do here in these pages.

Welcome to *Biology for a Changing World*. We hope that you enjoy your journey, and complete it more prepared for your life in a changing world.

Michèle Shuster

Janet Vigna

Matthew Tontonoz

Science through Stories

Each chapter of ***Biology for a Changing World*** is written in the style of a ***Scientific American*** article. This story-based approach captures student interest immediately and teaches not only the fundamental concepts of the discipline but also why understanding those concepts matters to students' lives and the world in general.

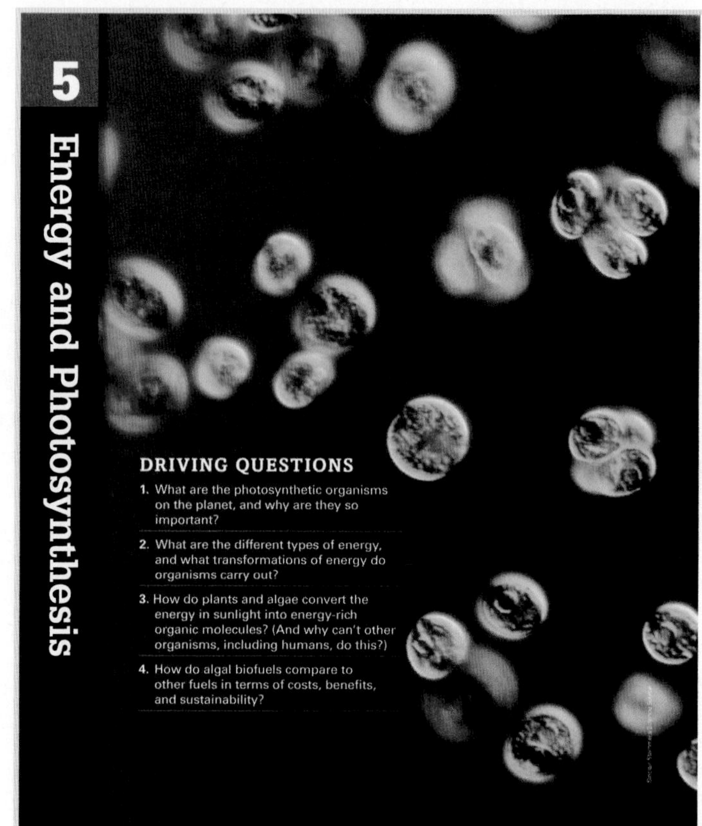

5
Energy and Photosynthesis
DRIVING QUESTIONS
1. What are the photosynthetic organisms on the planet, and why are they so important?
2. What are the different types of energy, and what transformations of energy do organisms carry out?
3. How do plants and algae convert the energy in sunlight into energy-rich organic molecules? (And why can't other organisms, including humans, do this?)
4. How do algal biofuels compare to other fuels in terms of costs, benefits, and sustainability?

the FUTURE of *FUEL*?

Scientists seek to make algae the next alternative energy source

AS AN ENGINEER WORKING FOR the Navy Seals in 1978, Jim Sears took a nighttime scuba dive off the coast of Panama City, Florida, one of many he took to do underwater research. The dive started out routinely, but then, suddenly, glowing phosphorescent algae appeared as if out of nowhere. When Sears put his hands out in front of him, sparkling streamers of microbes trickled off his fingertips. "It was magical," he recalls.

Sears is an inventor with many and varied devices to his credit. In the 1970s and 1980s, he built an underwater speech descrambler and a portable mine detector, among other gadgets. Later, he moved on to more creative technologies, including a "hump-o-meter" that could tell farmers when their animals were in heat or mating.

But the seeds of his real claim to fame weren't sown until 2004, when Sears was working in agricultural electronics. That's when he turned his attention toward what he felt was the world's biggest problem: dwindling fossil fuel reserves. After he did some thinking and a little research, the tiny, glowing organisms that had wowed him during his nighttime dive more than two decades earlier came to mind. He realized suddenly that they might be able to help.

Algae are perhaps best known as the layer of green scum coating the surfaces of ponds and swimming pools, but they have other claims to fame as well. Like plants, algae have the impressive ability to capture the **energy** of sunlight and convert it into a form that other organisms can use. Even more remarkable, algae trap much of this energy in the form of oils ideally suited to making fuel. The oil that

97

NEW CHAPTER STORIES

The Sitting Disease: Understanding the causes and consequences of obesity
(Ch. 6, Dietary Energy and Cellular Respiration)

Bulletproof: Scientists hope to spin spider silk into the next indestructible superfiber
(Ch. 8, Genes to Proteins)

Can Rubber Save the Rain Forest? A small state in Brazil aims to find out
(Ch. 18, Eukaryotic Diversity)

Plants 2.0: Is genetic engineering the solution to world hunger?
(Ch. 24, Plant Growth and Reproduction)

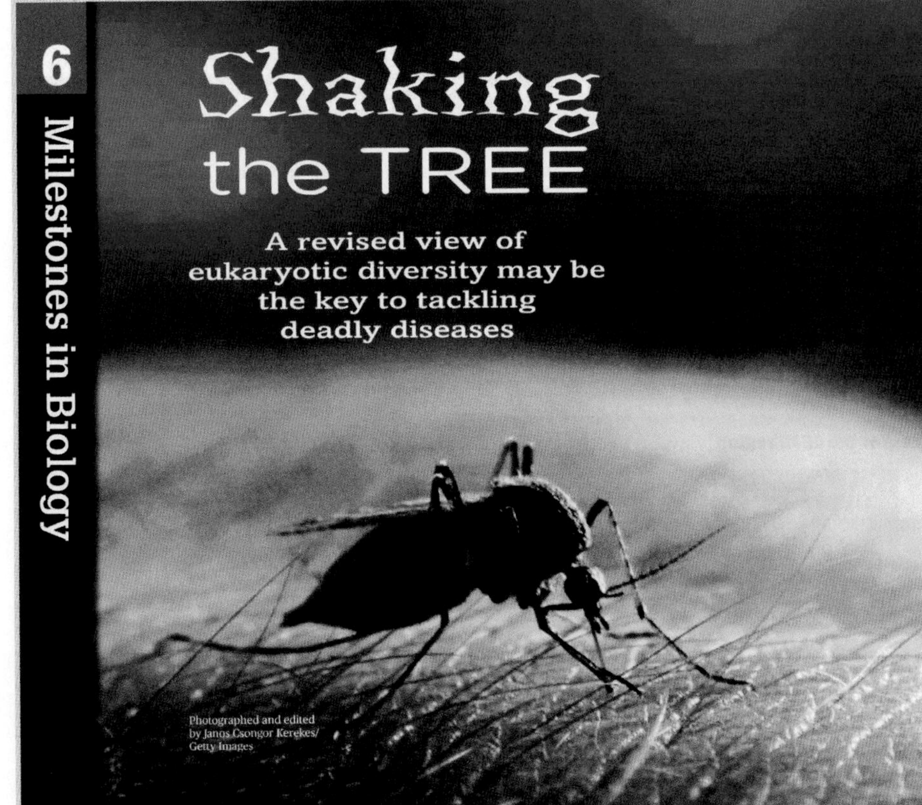
6 Milestones in Biology

Shaking the TREE

A revised view of eukaryotic diversity may be the key to tackling deadly diseases

Photographed and edited by Janos Csongor Kerekes/ Getty Images

DRIVING QUESTIONS

1. How has genetic evidence transformed the classification of protists?
2. Why is it important to classify protists accurately?
3. What are current challenges in preventing and treating malaria?

As the war in Vietnam raged, soldiers on both sides of the conflict faced an unrelenting enemy: malaria. This parasite-caused illness–transmitted through the bite of a mosquito–produces devastating fever, headache, chills, and vomiting in victims. If not treated within 24 hours, the condition can be fatal.

Malaria had traditionally been treated with drugs such as quinine, derived from the bark of the cinchona tree, and its synthetic derivative chloroquine. But increasingly, these drugs were unable to stem the tide of infection; the parasites had begun to evolve resistance.

Worried that a malaria-weakened army would be unable to fight off U.S.-backed military forces, the prime minister of North Vietnam, Ho Chi Minh, turned to Communist China for help. Recognizing their common interest in defeating a shared set of enemies, China's leader, Mao Zedong, launched a secret mission to find a malaria cure.

Project 523, as it was called–it was launched on May 23, 1967–enlisted hundreds of Chinese scientists and traditional Chinese healers. They were charged with screening thousands of known plant compounds for antimalarial effects and scouring traditional sources of Chinese medicine for leads on promising new medicines.

Taking their cue from an ancient medical text, the scientists homed in on one particular herb, called qinghao (known in the West as *Artemisia annua* or sweet wormword). Qinghao had been used for centuries in treating "intermittent fevers," a description that aptly describes a symptom of malaria. The scientists tested the herb against malaria-infected mice but the results proved inconsistent.

ALSO IN THIS EDITION

Enhanced Plant and Diversity Coverage

New two-chapter plant unit:

- Plant Growth and Reproduction (new chapter)
- Plant Physiology (new to the "without Physiology" version)

Expanded coverage of eukaryotic diversity in Chapter 18 and new Milestone 6.

MILESTONES

These **mini-chapters on historically important discoveries** in biology teach students how we know what we know, prompting them to consider how future research will expand our understanding of biology.

NEW MILESTONE! Shaking the Tree: A revised view of eukaryotic diversity may be the key to tackling deadly diseases

Infographics

Engaging and informative **Infographics** are used throughout the book to provide careful and complete explanations of the key scientific concepts. These powerful pieces of art teach students how to learn from charts, graphs, and images and add visual appeal to the science. Animated Infographics in LaunchPad are accompanied by assignable quiz questions.

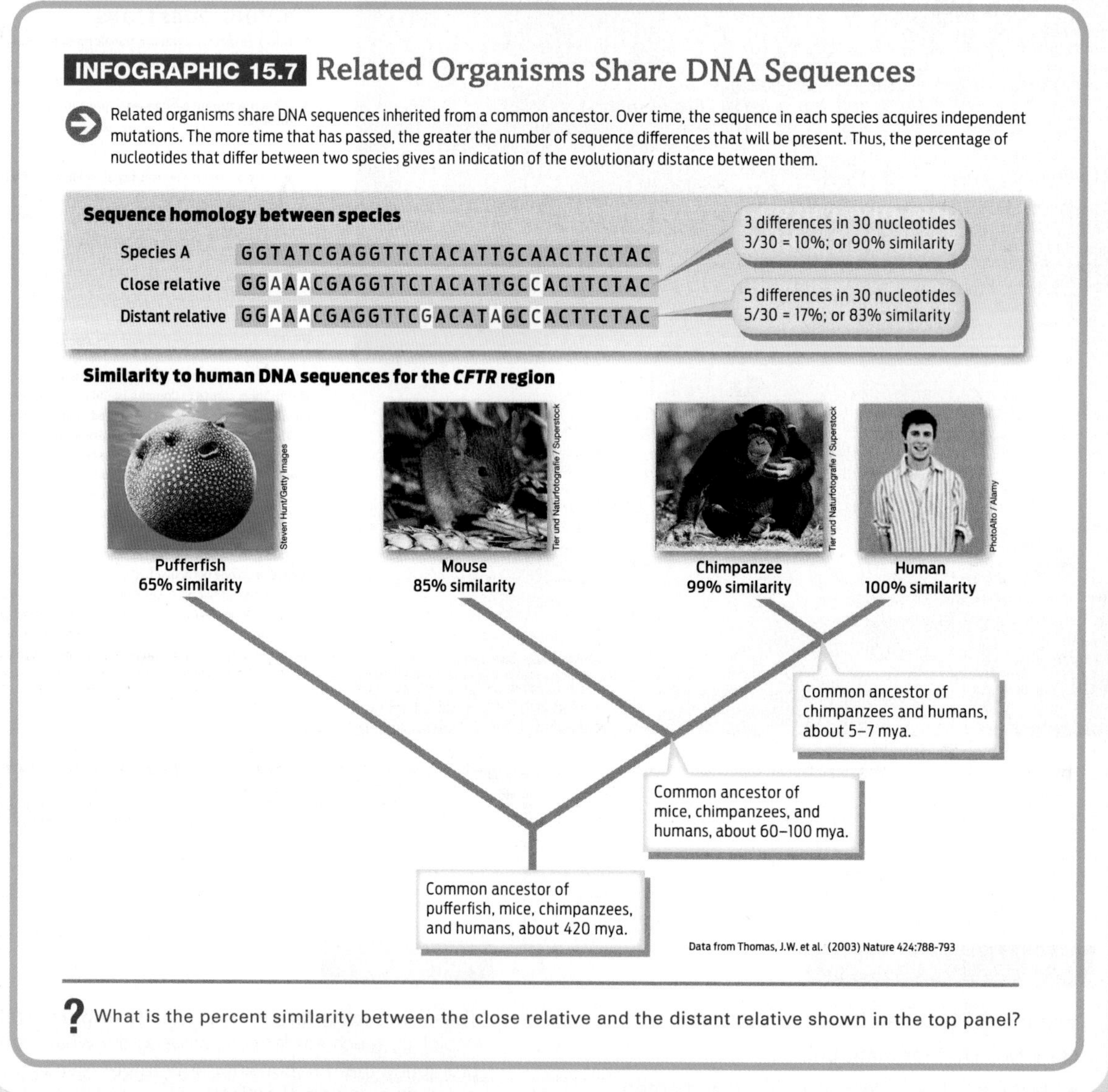

NEW INFOGRAPHIC QUESTIONS

Each Infographic now includes a thought-provoking question at the end to encourage students to think critically about the information presented in the figure.

INFOGRAPHIC 6.6 Aerobic Respiration Transfers Food Energy to ATP

During aerobic respiration, our cells use the oxygen we inhale to help extract energy from food. Cells convert the energy stored in food molecules into the bonds of ATP, the cell's energy currency.

1. Blood delivers oxygen from the lungs and food-derived subunits from the small intestines to the body's cells.

2. Cells break the chemical bonds of food molecules to release energy, which is used to make ATP. Water and carbon dioxide are produced as waste products.

3. Carbon dioxide exits, travels via the blood to the lungs, and is exhaled as waste.

4. ATP is used to power cell functions.

Blood vessel

Glucose (from the intestine)

Oxygen (from the lungs)

Mitochondrion

CO_2

ATP

Water

Animal cell

Carbon dioxide (to the lungs)

Inputs Glucose $C_6H_{12}O_6$ + Oxygen O_2 → **Outputs** Energy ATP + Carbon dioxide CO_2 + Water H_2O

? What is the source of the glucose and the oxygen used in aerobic respiration?

ADDITIONAL PEDAGOGICAL FEATURES

Driving Questions provide the pedagogical framework for the chapter content by prompting students to consider the questions they need to be able to answer to have a full understanding of the material.

End-of-chapter questions, written by Michèle Shuster, are framed around the chapter's **Driving Questions**. Each question set includes **Interpreting Data**, **Mini Case**, and **Bring It Home** questions to help students develop higher-order thinking skills. Selected questions are also assignable online through LaunchPad.

DRIVING QUESTION 2 What features make *Tiktaalik* a transitional fossil, and what role do these types of fossils play in the fossil record?

INTERPRETING DATA

apply YOUR KNOWLEDGE

20 The gene responsible for hairlessness in Mexican hairless dogs is called corneodesmosin (*CDSN*). This gene is present in other organisms. Look at the sequence of a portion of the *CDSN* gene from pairs of different species, given below. For each pair, determine the number of differences. From the variations in this sequence, which organism appears to be most closely related to humans? Which organism appears to be least closely related to humans?

Species	Sequence
Homo sapiens (human)	ACTCCGGCCCCTACATCCCCAGCTCCCA
Canis lupus familiaris (dog)	ATTCTGGCTCCTACATTTCCAGCTCCCA
Homo sapiens (human)	ACTCCGGCCCCTACATCCCCAGCTCCCA
Pan troglodytes (chimpanzee)	ACTCCGGCCCCTACATCCCCAGCTCCCA
Homo sapiens (human)	ACTCCGGCCCCTACATCCCCAGCTCCCA
Sus scrofa (pig)	AGTCTGGCTCCTACATCTCCAGCTCCCA
Homo sapiens (human)	ACTCCGGCCCCTACATCCCCAGCTCCCA
Macaca mulatta (rhesus monkey)	ACTCTGGCCCCTACATCCCCAGCTCCCA

MINI CASE

apply YOUR KNOWLEDGE

21 Fossils allow us to understand the evolution of many lineages of plants and animals. They therefore represent a valuable scientific resource. What if *Tiktaalik* (or an equally important transitional fossil) had been found by amateur fossil hunters and sold to a private collector? Do you think there should be any regulation of fossil hunting to prevent the loss of valuable scientific information from the public domain?

BRING IT HOME

apply YOUR KNOWLEDGE

7 Do an Internet search to find out about fossils discovered in your home state. Determine what kinds of organisms they represent, how old they are, and where in your state you would need to go in order to have a chance of finding fossils in the field.

Expanded Media Program in LaunchPad

Built to address the biggest classroom issues instructors face, **LaunchPad** gives students everything they need to prepare for class and exams, while giving instructors everything they need to quickly set up a course, shape the content to their syllabus, craft presentations and lectures, assign and assess homework, and guide the progress of individual students and the class as a whole.

Simple Inheritance Worksheet

Activity Guide

Chapter 11: Simple Inheritance and Meiosis

Introduction

This activity gives students the opportunity to practice genetics problems in class, either individually or with a partner/group.

Learning Objectives

After completing this activity, students should be able to:

- Determine the genotypes and phenotypes of individuals from information given
- Determine possible gametes
- Predict genotypes and phenotypes of individuals

Materials

- Activity guide and Answer Key
- Student Handout
- Activity PowerPoint presentation with introductory slides and clicker questions

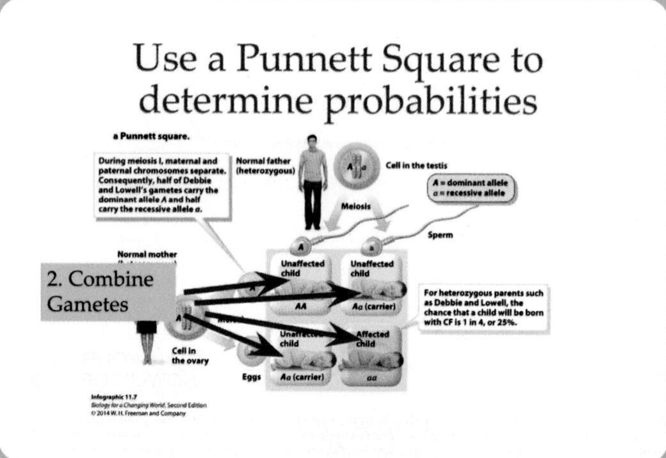

Simple Inheritance Worksheet

Because we know that one copy of each allele is inherited from each parent, we can figure out the possible genotypes and phenotypes of offspring. Working individually or with a partner, answer the following questions.

1. In humans, normal skin pigmentation (B) is a dominant trait and albinism (b) is recessive.
   a. Two people with normal pigmentation have an albino child. Use a Punnett square to demonstrate how the child could have a different phenotype than the parents.
   b. If the mom had normal pigmentation and the dad was albino, how many of each phenotype would you expect out of 4 children? What are the phenotypic and genotypic ratios of the offspring? (hint: there may be more than one answer).

Simple Inheritance Worksheet

Answer Key

1. In humans, normal skin pigmentation (B) is a dominant trait and albinism (b) is recessive.
   a. Two people with normal pigmentation have an albino child. Use a Punnett square to demonstrate how the child could have a different phenotype than the parents.

   Both parents have to be heterozygous for any offspring to be white.

	B	b
B	BB	Bb
b	Bb	bb

   a. If the mom had normal pigmentation and the dad was albino, how many of each phenotype would you expect out of 4 children? What are the phenotypic and genotypic ratios of the offspring? (hint: there may be more than one answer).

   Either 2 normal and 2 albino (the mother is heterozygous); 1:1

	B	b
b	Bb	bb
b	Bb	bb

Active Learning Lesson Plans

Curated and customizable, these sets of student activities provide instructors with an easy-to-use framework for including active learning in their lectures, deepening student engagement with the text. Provided with lecture slides, instructor guides, and worksheets, they are available for all chapters and organized by each chapter's Driving Questions.

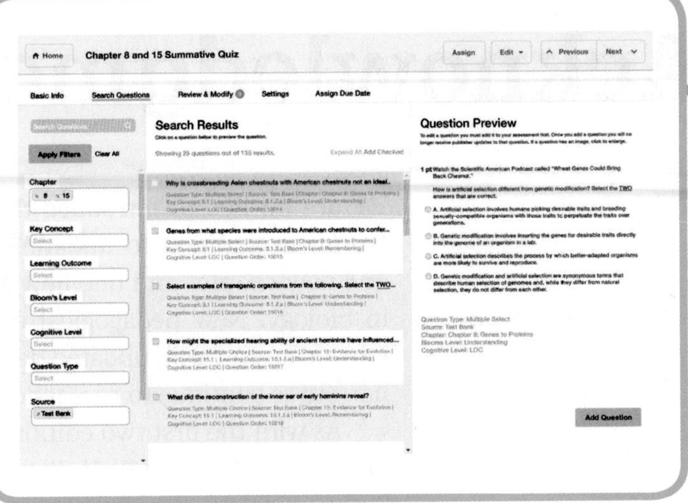

Assignable End-of-Chapter Questions

Selected from Michèle Shuster's question sets, these include Interpreting Data activities that ask students to analyze data in tables, charts, or graphs and draw their own conclusions about their meaning.

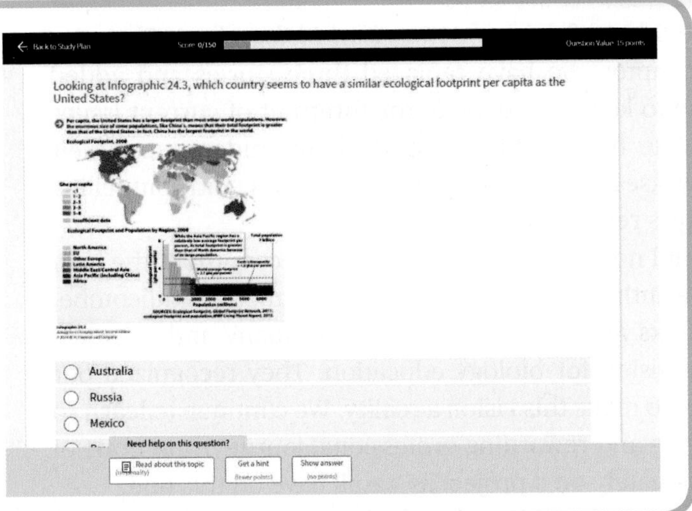

LearningCurve

Putting "testing to learn" into action, LearningCurve is the perfect tool to get students to engage before class and review after. With game-like quizzing, it creates individualized activities for each student, selecting questions—by difficulty and topic—according to the student's performance.

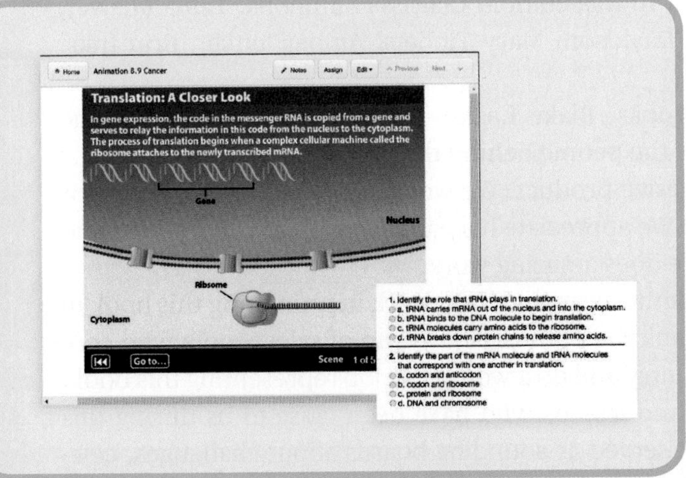

Infographic Animations with Questions

Animated Infographics include assignable questions that encourage students to think critically about the information presented in the Infographic.

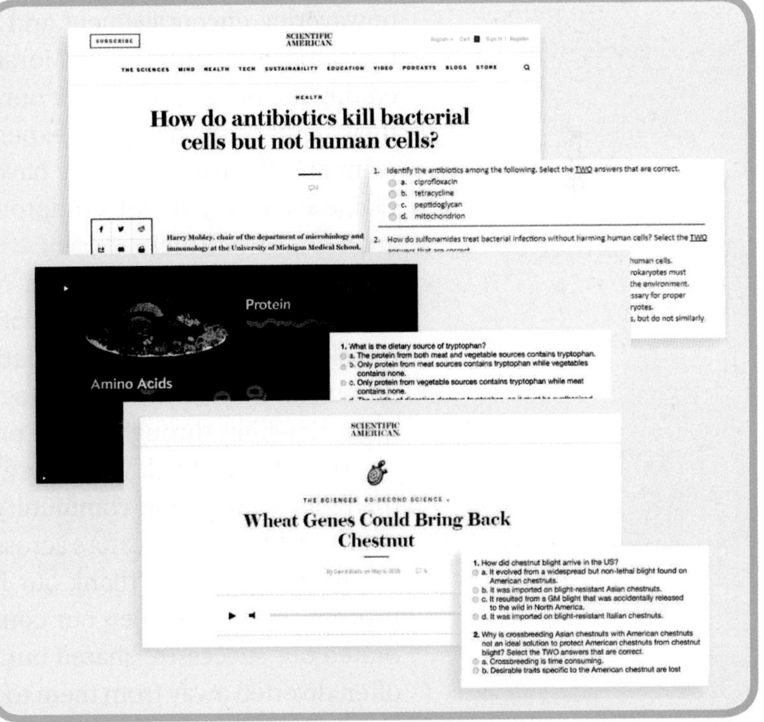

Content from *Scientific American*

Assignable activities that integrate *Scientific American* content—including articles, podcasts, and videos—are available for each chapter in LaunchPad.

Acknowledgments

We are thrilled to introduce the third edition of *Scientific American Biology for a Changing World*. In addition to updating every chapter, we have replaced three stories and added one new chapter and one new Milestone to keep this book at the forefront of current issues in biology. New pedagogical elements like Infographic questions help students test their knowledge. A fresh, updated design showcases our unparalleled Infographics, and compelling narratives continue to reveal how biology is relevant to daily life.

As with the first two editions, we could not have completed this project without the help of a fantastic team at W. H. Freeman. The authors would like to thank Elizabeth Widdicombe, Susan Winslow, Ben Roberts, and the folks at W. H. Freeman and Company and *Scientific American* for continuing to support this vision for biology education. They recognized our diverse strengths and brought us together to make this vision a reality. We continue to learn so much from one another on this challenging and rewarding professional journey, and none of us has likely worked so hard and so passionately on a project as we all have on this one.

We would like to thank all those who were interviewed and who generously contributed information for the chapters in this edition. Their stories are central to the impact that this book will have on the students we teach. They are authentic examples of biology in a changing world, and they bring this book to life.

A special thank you is required for our Program Director, Brooke Suchomel, for her unwavering encouragement and ability to bring stable direction and support to the project. Development Editors Susan Moran and Erica Frost spent many hours in the pages of this book, editing the details, managing our chaos, and smoothing our rough edges. We thank them for their dedication, patience, experience, and expertise. We are grateful to the skilled media team and the expertise they have brought to this edition: Clairissa Simmons, Tania Mlawer, Elaine Palucki, Jennifer Compton, Kevin Davidson, Mary Tibbets, Amber Jonker, Erin Inks, and Daniel Comstock.

Many thanks to Liz Geller, Nancy Brooks, Blake Logan, Matthew McAdams, Christine Buese, Teri Stratford, Paul Rohloff, and all the people behind the scenes at W. H. Freeman for translating our ideas into a beautiful, cohesive product. We would like to thank Eli Ensor for his outstanding work on the Infographics. We appreciate his patience with the many edits and quick timelines throughout the project. He does amazing work.

We'd like to thank Will Moore for his enthusiasm and hard work in promoting this book in the biology education community. We thank the enthusiastic group of salespeople who connect with biology educators across the country and do a wonderful job representing this book.

We would like to thank our families and friends who have been close to us during this process. They have been our consultants, served as sounding boards about challenges, celebrated our successes, shared our passions, and supported the extended time and energy we often diverted away from them to this project. We are grateful for their patience and unending support.

And finally, a sincere thank you to our many teachers, mentors, and students over the years who have shaped our views of biology and the world, and how best to teach about one in the context of the other. You are our inspiration.

Media and Assessment Authors

Ann Aguanno, Marymount Manhattan College, *Chapter Quizzes, Lecture PowerPoint Slides*
Michelle Cawthorn, Georgia Southern University, *Study Guide*
Beth Collins, Iowa Central Community College, *Chapter Quizzes, Clicker Questions, LearningCurve Activities*
Jennifer Cymbola, Grand Valley State University, *Lecture PowerPoint Slides*
Jodi DenUyl, Grand Valley State University, *Chapter Quizzes*
Elizabeth Geils, Polk State College, *Animation Quizzes, Scientific American Activities*
John Harley, Eastern Kentucky University, *LearningCurve Activities*
Megan Litster, University of Wisconsin-La Crosse, *Active Learning Resources*
Brett Macmillan, McDaniel College, *Student Study Guide*
Cindy Malone, California State University, Northridge, *Lecture PowerPoint Slides*
Lisa Maranto, Prince George's Community College, *Learning Outcomes Tagging*
Crystal McAlvin, The University of Tennessee, Knoxville, *Milestone Quizzes*
Teresa Mika, University of Wisconsin-La Crosse, *Test Bank*
Margaret Oliver, Carthage College, *Chapter Quizzes, LearningCurve Activities*
Barb Salvo, Carthage College, *Active Learning Resources, Scientific American Activities*
Mark Sarvary, Cornell University, *Lecture PowerPoint Slides*
Ryan Shanks, North Georgia College & State University, *Test Bank*
Melissa Walsh, University of Texas at Arlington, *Active Learning Resources*
Carolyn Wetzel, Holyoke Community College, *Learning Objectives*
Satya Witt, University of New Mexico, *Clicker Questions, Test Bank*

Reviewers

We would like to extend our deepest thanks to the following instructors who reviewed, tested, and advised on the book manuscript at various stages of its development.

Mari Aanenson, *Western Illinois University*
Amy Abdulovic-Cui, *Augusta University*
Mark Ainsworth, *Seattle Central Community College*
Corrie L. Andries, *Central New Mexico Community College*
Bobby R. Baldridge, *Asbury University*
Liz Balko, *Cornell University*
Randall Barley, *University of Lethbridge*
Tonya C. Bates, *University of North Carolina, Charlotte*
Diane B. Beechinor, *Northeast Lakeview College*
Stacy Bennetts, *Augusta University*
Tiffany Bensen, *University of Mississippi*
Neelima Bhavsar, *Saint Louis Community College*
Donna H. Bivans, *Pitt Community College*
Mark W. Bland, *University of Central Arkansas*
Lisa Ann Blankinship, *University of North Alabama*
Barbara I. Blonder, *Flagler College*
Lanh Bloodworth, *Florida State College at Jacksonville*
Lisa L. Boggs, *Southwestern Oklahoma State University*
Mary E. Bonds, *Northwest Mississippi Community College*
Kelsey Bowlin, *Northwest Missouri State University*
Jett Bradley, *Oklahoma Baptist University*
Ben Brammell, *Asbury University*
Susan E. Brantley, *University of North Georgia*
Randy Brewton, *University of Tennessee, Knoxville*
Joshua M. Brokaw, *Abilene Christian University*
Julio E. Budde, *Youngstown State University*
Diep Burbridge, *Long Beach City College*
Stephanie Burdett, *Brigham Young University*
Rebecca Burt, *Southeast Community College*
Aimee K. Busalacki, *Lincoln University*
Nancy M. Butler, *Kutztown University*
P. Byrd-Williams, *Los Angeles Valley College*
David Byres, *Florida State College at Jacksonville*
William Caire, *University of Central Oklahoma*
Kelli Carter, *Pasco Hernando State College*
Kelly S. Cartwright, *College of Lake County*
Jocelyn Cash, *Central Piedmont Community College*
Joe N. Caudell, *Murray State University*
Michelle Cawthorn, *Georgia Southern University*
Carol Cleveland, *Northwest Mississippi Community College*

Kimberly Cline-Brown, *University of Northern Iowa*
Curt Coffman, *Vincennes University*
Glenn M. Cohen, *Troy University*
Yvonne Cole, *Florissant Valley Community College*
Elizabeth Collins, *Iowa Central Community College*
David T. Corey, *Midlands Technical College*
Andrew Corless, *Vincennes University*
Angela J. Costanzo, *Hawaii Pacific University*
Dave Cox, *Lincoln Land Community College*
Kathleen Curran, *Wesley College*
Gregory A. Dahlem, *Northern Kentucky University*
Don C. Dailey, *Austin Peay State University*
Kristy Daniel, *Texas State University*
Deborah Dardis, *Southeastern Louisiana University*
Douglas Darnowski, *Indiana University Southeast*
James Dawson, *Pittsburg State University*
Jonathan Davis, *Dona Ana Community College*
Carolyn A. Dehlinger, *Keiser University*
Tom D'Elia, *Indian River State College*
Jodi Denuyl, *Grand Valley State University*
Sandra G. Devenny, *Delaware County Community College*
Gregg Dieringer, *Northwest Missouri State University*
Danielle Dodenhoff, *California State University, Bakersfield*
Dani DuCharme, *Waubonsee Community College*
Jeannette Dumas, *Rosemont College*
Stephanie Edelmann, *Bluegrass Community and Technical College*
Eden Effert-Fanta, *Eastern Illinois University*
Julie Ehresmann, *Iowa Central Community College*
Faye Ellis, *University of Wisconsin–La Crosse*
Ray Emmett, *Daytona State College*
Michele Engel, *California State University, Bakersfield*
Marirose T. Ethington, *Genesee Community College*
Kathy McCann Evans, *Reading Area Community College*
Craig Fenn, *Reading Area Community College*
Eugene J. Fenster, *Metropolitan Community College, Longview*
Geneen Fitchett, *Central Piedmont Community College*
Linda Flora, *Delaware County Community College*
Samantha R. Fowler, *Florida Institute of Technology*
Sarah Gall, *Ashford University*
J. Yvette Gardner, *Clayton State University*
Richard D. Gardner, *Southern Virginia University*
Antonio Garza, *East Los Angeles College*
Kellie M. Gauvin, *Finger Lakes Community College*
Solomon Gebru, *Howard University*
Rebecca Gehringer, *Ozarks Technical Community College*
Carrie L. Geisbauer, *Moorpark College*
Bagie George, *Georgia Gwinnett College*
P. Gerard, *University of Maine, Rockland College*
Laci Gerhart-Barley, *University of Hawai'i, West O'ahu*
Richard Gill, *Brigham Young University*
Amy Marie Glaser, *Erie Community College*
David Goldstein, *Wright State University*
Tammy Goulet, *University of Mississippi*
Jen Grant, *University of Wisconsin-Stout*
A. E. Gray, *Eastern Washington University*
Neil Greenberg, *University of Tennessee, Knoxville*
Sara Gremillion, *Armstrong State University*
Christine Griffiths, *Nova Southeastern University*
Cheryl Hackworth, *West Valley College*
Charles T. Hanifin, *Utah State University, Uintah Basin Campus*
Margaret S. Harris, *Faulkner University*
Janet M. Harouse, *New York University*
Joyce Phillips Hardy, *Chadron State College*
Olivia Harriott, *Fairfield University*
Timothy Henkel, *Valdosta State University*
Jay Hodgson, *Armstrong State University*
Kirsten Hokeness, *Bryant University*
Tina T. Hopper, *Missouri State University*
K. M. Horn, *DeVry University*
Laura Houston, *Northeast Lakeview College*
Carina Endres Howell, *Lock Haven University*
Ching-Yu Huang, *University of North Georgia*
C. Hurlbut, *Florida State College at Jacksonville*
Barbekka Hurtt, *University of Denver*
Joseph D. Husband, *Florida State College, Jacksonville*
Evelyn F. Jackson, *University of Mississippi*
Charles W. Jacobs, *Henry Ford College*
Wendy Jamison, *Chadron State College*
Jamie Jensen, *Brigham Young University*
Carl Johansson, *Fresno City College*
Kristy Y. Johnson, *The Citadel*
Robert Johnson, *Pierce College*
Gregory Jones, *Santa Fe College*
Jacqueline Jordan, *Clayton State University*
Anthony Ippolito, *DePaul University*
Hinrich Kaiser, *Victor Valley College*
Joanna Kazmierczak, *Community College of Allegheny County*
Karen Kendall-Fite, *Columbia State Community College*
Michele C. Kieke, *Concordia University*
Sarah J. Krajewski, *Grand Rapids Community College*
Dana Robert Kurpius, *Elgin Community College*
Anne Macek Lachelt, *Iowa Central Community College*
Diane M. Lahaise, *Perimeter College at Georgia State University*
Archana Lal, *Independence Community College*
Kirk Land, *University of the Pacific*
Nancy Lane, *Bellevue College*
Emily Schmitt Lavin, *Nova Southeastern University*
Brenda Leady, *University of Toledo*
Lee Ann Lippincott, *Bucks County Community College*
Kathryn L. Lipson, *Western New England University*
K. J. Lodrigue, Jr., *Baton Rouge Community College*
Suzanne Long, *Monroe Community College*
David A. Luther, *George Mason University*
Rebecca Maas, *Rock Valley College*
William J. Mackay, *Edinboro University of Pennsylvania*
Margaret Major, *Perimeter College at Georgia State University*
Mark Manteuffel, *St. Louis Community College*
Lisa Maranto, *Prince George's Community College*
Mary Charlotte Martin, *Northern Michigan University*
Michael S. Martin, *Ozarks Technical Community College*
Debra M. Mayers, *Missouri State University–West Plains*
Crystal McAlvin, *University of Tennessee*
Renee McFarlane, *Clayton State University*
Paul Guy Melvin, *Clayton State University*
Paige Mettler-Cherry, *Lindenwood University–Belleville*
Kiran Misra, *Edinboro University of Pennsylvania*
Jeanelle Morgan, *University of North Georgia*
Winnie Mukami, *DeVry University*
Maria Neuwirth, *DeVry University*
Lori Nicholas, *New York University*
James Nichols, *Abilene Christian University*
A. Nickens, *Northwest Mississippi Community College*
Sue Nightingale, *Bellevue College*
Fran Norflus, *Clayton State University*
Margaret Oliver, *Carthage College*
Margaret Olney, *St. Martin's University*
Christopher J. Osovitz, *University of South Florida*
Mary O'Sullivan, *Elgin Community College*
Brent D. Palmer, *University of Kentucky*
Jessica Pamment, *DePaul University*
Murali Panen, *Luzerne County Community College*
Monica Parker, *Florida State College at Jacksonville*
Rhonda Patterson, *Western Kentucky University*
Louis L. Pech, *University of Wisconsin-Marathon County*

Mark B. Pellegrino, *Finger Lakes Community College*
Krista Peppers, *University of Central Arkansas*
Angel C. Pimentel, *The University of Arizona*
Mary Poffenroth, *San Jose State University*
Kristen Przyborski, *University of New Haven*
Amber A. Qureshi, *University of Wisconsin–River Falls*
Asha Rao, *Texas A&M University*
Wes Rogers, *Georgia Gwinnett College*
Peggy Rolfsen, *Cincinnati State Technical and Community College*
Barbara J. Salvo, *Carthage College*
Helen D. Sarantopoulos, *East Los Angeles College*
Mark A. Sarvary, *Cornell University*
Georgianna Saunders, *Missouri State University*
Arleen Sawitzke, *Salt Lake Community College*
Gary Schultz, *Marshall University*
Fayla Schwartz, *Everett Community College*
Steve Schwendemann, *Iowa Central Community College*
Jennifer J. Scoby, *Illinois Central College*
Ryan A. Shanks, *University of North Georgia*
Nilesh Sharma, *Western Kentucky University*
Allan M. Showalter, *Ohio University*
Ann M. Showalter, *Clayton State University*
Jack Shurley, *Idaho State University*
Daniel Sigmon, *Alamance Community College*
Donald F. Slish, *SUNY Plattsburgh*
Anna Bess Sorin, *University of Memphis*
John P. Sotolongo, *Miami Dade College–Hialeah*
Salvatore A. Sparace, *Clemson University*
William G. Sproat, *Jr., Walters State Community College*
John H. Starnes, *Somerset Community College*
Richard T. Stevens, *Monroe Community College*
Andrew Stoehr, *Butler University*
Bethany B. Stone, *University of Missouri, Columbia*
Anna Strimaitis, *Florida State University*
Ignatius Tan, *New York University*
Susan E. Tappen, *Central New Mexico Community College*
Diane Tata, *Dayton State College*
Jamey Thompson, *Hudson Valley Community College*
Rita A. Thrasher, *Pensacola State College*
Jonathan Titus, *SUNY–Fredonia*
Kurt A. Toenjes, *Montana State University, Billings*
M. N. Tremblay, *Georgia Southern University*
Chris Trzepacz, *Murray State University*
Marsha Turell, *Houston Community College*
Brittany Linn Twibell, *Ozarks Technical Community College*
Muatasem Ubeidat, *Southwestern Oklahoma State University*
Steven M. Uyeda, *Pima Community College*
Pete van Dyke, *Walla Walla Community College*
Melinda Verdone, *Rock Valley College*
Jonathan Visick, *North Central College*
Jennifer Vlk, *Elgin Community College*
R. Steve Wagner, *Central Washington University*
Stephen Wagner, *Stephen F. Austin State University*
Rebekah L. Waikel, *Eastern Kentucky University*
Melissa J. Walsh, *University of Texas at Arlington*
Kristen L. Walton, *Missouri Western State University*
Jennifer M. Warner, *University of North Carolina, Charlotte*
Lisa Weasel, *Portland State University*
Kathy Webb, *Bucks County Community College*
K. Derek Weber, *Raritan Valley Community College*
Amy Wernette, *Hazard Community at Technical College*
Alexander J. Werth, *Hampden-Sydney College*
Jeremy Whisenhunt, *NorthWest Arkansas Community College*
Robert S. Whyte, *California University of Pennsylvania*
Helen Wiersma-Koch, *Indian River State College*
Emily Williamson, *Mississippi State University*
Leslie Kelso Winemiller, *Texas A&M University*
Satya Maliakal Witt, *University of New Mexico*
Janet Wolkenstein, *Hudson Valley Community College*
Michael J. Wolyniak, *Hampden-Sydney College*
Edwin M. Wong, *Western Connecticut State University*
Holly Woodruff, *Central Piedmont Community College*
Dwight Wray, *Brigham Young University-Idaho*
Gary T. ZeRuth, *Murray State University*

About Scientific American:

Scientific American is the authority on science and technology for a general audience, with coverage that explains how research changes our understanding of the world and shapes our lives. First published in 1845, *Scientific American* is the longest continuously published magazine in the US. The magazine has published articles by more than 150 Nobel Prize-winning scientists and built a loyal following of influential and forward-thinking readers. With daily coverage in digital media, 12 issues per year of *Scientific American*, 6 issues of *Scientific American Mind* and more than 170 years of archives, the magazine continues to be the leading source for business and policy leaders, education professionals, and science enthusiasts of all kinds.

Scientific American is published by Springer Nature, a leading global research, educational, and professional publisher, home to an array of respected and trusted brands providing quality content through a range of innovative products and services.

www.ScientificAmerican.com

Contents

Unit 1: What Is Life Made Of? Chemistry, Cells, Energy

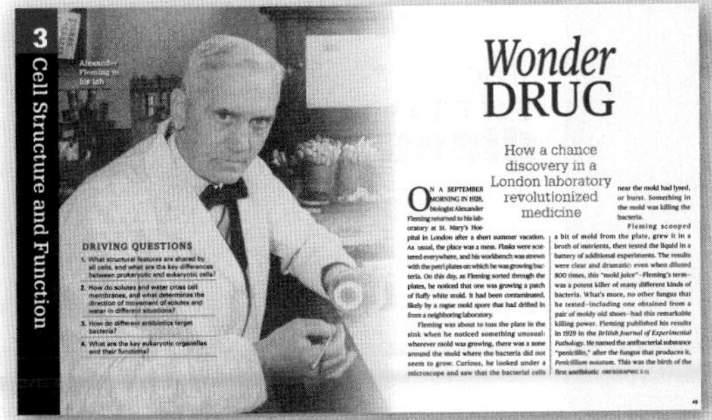

CHAPTER 3

Cell Structure and Function 44

WONDER DRUG

How a chance discovery in a London laboratory revolutionized medicine

M1 Milestones in Biology

SCIENTIFIC REBEL 66

Lynn Margulis and the theory of endosymbiosis

CHAPTER 4

Nutrition, Enzymes, Metabolism 74

THE PEANUT BUTTER PROJECT

One doctor's crusade to end malnutrition in Africa, a spoonful at a time

"I had not the slightest suspicion that I was at the beginning of something extraordinary."

— Alexander Fleming

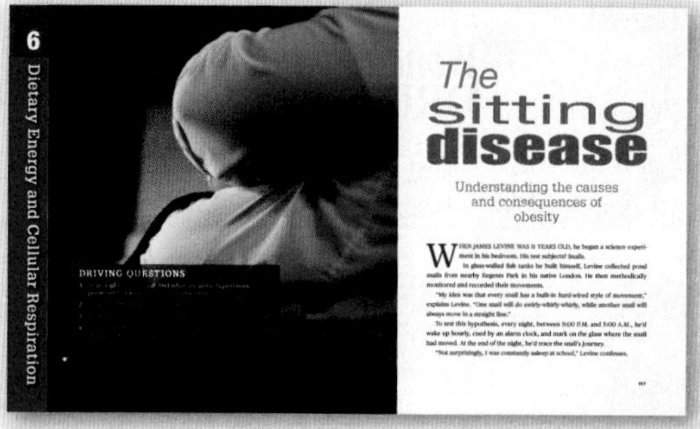

Our waistlines obey the principle of conservation of energy: energy is neither created nor destroyed but merely converted from one form into another.

Unit 2: How Does Life Reproduce? Cell Division and Inheritance

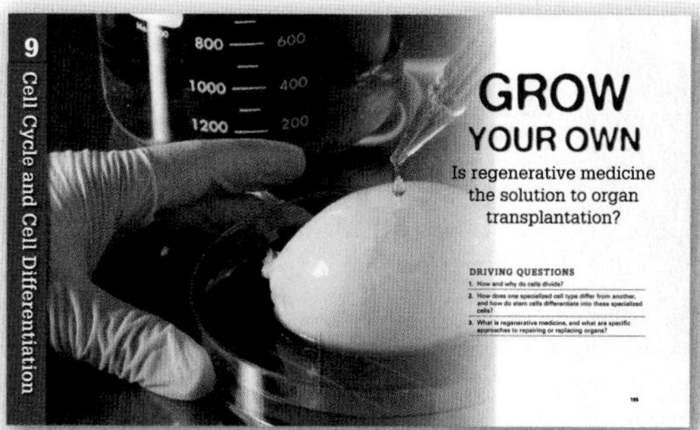
9
Cell Cycle and Cell Differentiation
GROW
YOUR OWN
Is regenerative medicine the solution to organ transplantation?
DRIVING QUESTIONS

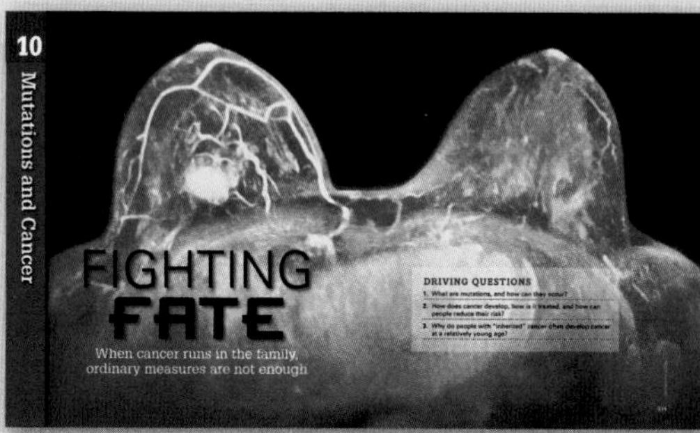
10
Mutations and Cancer
FIGHTING
FATE
When cancer runs in the family, ordinary measures are not enough
DRIVING QUESTIONS

11
Simple Inheritance and Meiosis
CATCHING
BREATH
One woman's mission to outrun a genetic disease
DRIVING QUESTIONS

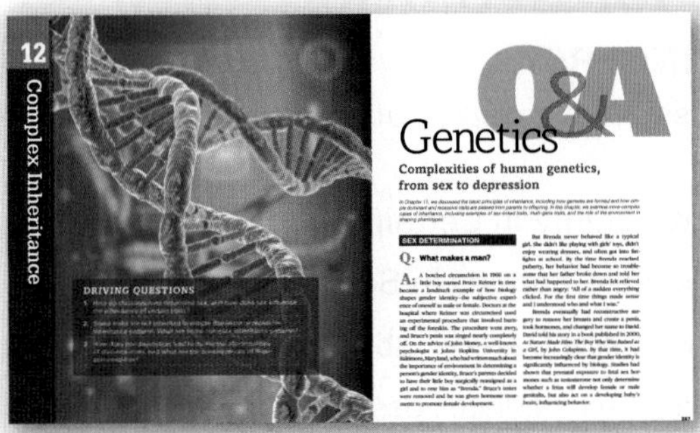

Unit 3: How Does Life Change over Time? Evolution and Diversity

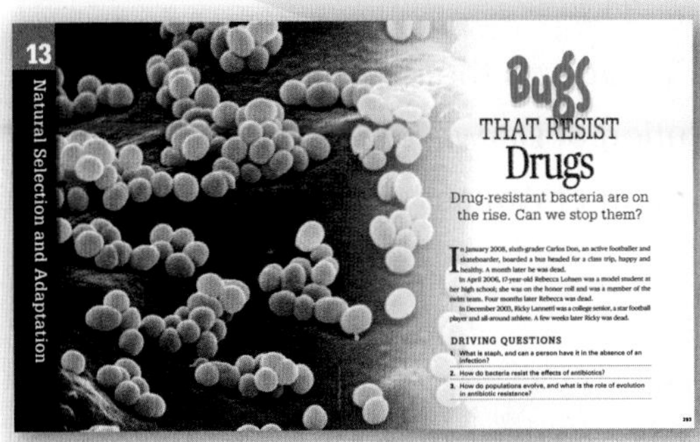

Scientists are now seeing bacterial infections that don't respond to any known antibiotics, leading many to fear the day when we run out of treatment options altogether.

For products that truly must not fail—like airplane tires, surgical gloves, and condoms—natural rubber is the go-to choice.

M6 Milestones in Biology

SHAKING THE TREE 440

A revised view of eukaryotic diversity may be the key to tackling deadly diseases

CHAPTER 19

Human Evolution 450

SKIN DEEP

Science redefines the meaning of racial categories

Unit 4: How Do Organisms Interact? Ecology

CHAPTER 20

Population Ecology 470

ON THE TRACKS OF WOLVES AND MOOSE

Ecologists learn big lessons from a small island

CHAPTER 21
Community Ecology 490
WHAT'S HAPPENING TO HONEY BEES?
A mysterious ailment threatens a vital link in the food chain

CHAPTER 22
Ecosystem Ecology 510
THE HEAT IS ON
From migrating maples to shrinking sea ice, signs of a warming planet

M7 Milestones in Biology
PROGRESS OR POISON? 536
Rachel Carson, pesticides, and the birth of the environmental movement

CHAPTER 23
Sustainability 546
THE MAKINGS OF A GREEN CITY
A Kansas town reinvents itself sustainable

Unit 5: What Makes Plants Unique? Plant Biology

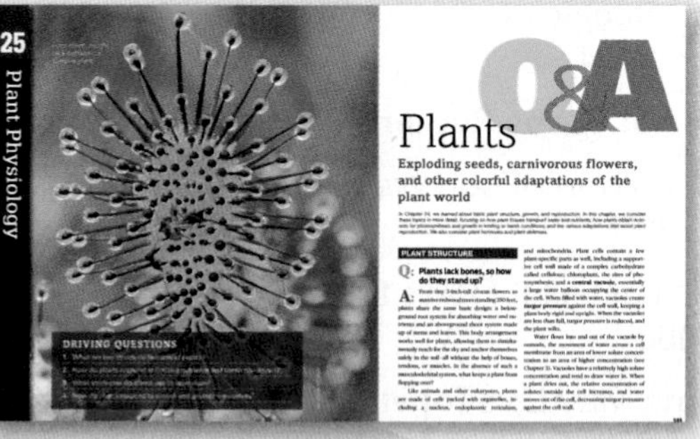

"CRISPR is one of those things I feel like I've been waiting for my entire career."

— Joyce Van Eck

Unit 6: How Do Animals Work? Physiology

The 1918 flu pandemic killed more people in 1 year than the Black Death killed in the entire 14th century; it killed more people in 25 weeks than AIDS killed in 25 years.

1 Process of Science

JavaReport

Making sense of the latest buzz in health-related news

DRIVING QUESTIONS

1. How is the scientific method used to test hypotheses?
2. What factors influence the strength of scientific studies and determine whether the results of any given study are applicable to a particular population?
3. How can you evaluate the evidence in media reports of scientific studies?
4. How does the scientific method apply in clinical trials designed to investigate important issues in human health?

IN 1981, A STUDY IN THE PROMINENT *New England Journal of Medicine* made headlines when it reported that drinking two cups of coffee a day doubled a person's risk of getting pancreatic cancer. Drinking five cups a day supposedly tripled the risk. "Study Links Coffee Use to Pancreas Cancer," trumpeted the *New York Times*. "Is there cancer in the cup?" asked *Time* magazine. The lead author of the study, Brian MacMahon of the Harvard School of Public Health, appeared on the *Today* show to warn of the dangers of coffee.

"I will tell you that I myself have stopped drinking coffee," said MacMahon, who had previously drunk three cups a day.

Just 5 years later, MacMahon's research group was back in the news, reporting in the same journal that a second study had found *no* link between coffee and pancreatic cancer. Subsequent studies by other researchers also failed to reproduce the original findings.

Today, more than 30 years later, coffee is once again making headlines. Recent studies have suggested that, far from causing disease, coffee may actually help *prevent* a number of conditions—everything from Parkinson's disease and diabetes to cancer and tooth decay. "Java Junkies Less Likely to Get Tumors," announced a CBS News headline. "How Coffee Can Help You Live Longer," read an article in *Time*. *Prevention* magazine ran a piece titled "4 Surprising Coffee Cures," which touted coffee's protective effects on Alzheimer's disease, diabetes, heart disease, and skin cancer.

Andrew Burton/Getty Images

Coffee is popular around the world. But is it good for you?

Not everyone is buying the coffee cure, however. Public health advocates are increasingly alarmed by our love affair with—some might say, addiction to—caffeine. Emergency rooms are reporting more caffeine-related admissions, and poison control centers are receiving more calls related to caffeine overdoses. In response, politicians are pressuring the Food and Drug Administration (FDA) to force manufacturers to place warning labels on energy drinks. Nevertheless, caffeine's "energizing" effect is advertised on nearly every street corner, where, increasingly, you're also likely to find a coffee shop.

Why the mixed messages about caffeine? Are researchers making mistakes? Are journalists getting their facts wrong? While both of these possibilities may be true at times, the bigger problem is widespread confusion over the nature of science and the meaning of scientific evidence.

"Consumers are flooded with a firehose of health information every day from various media sources," says Gary Schwitzer, publisher of the consumer watchdog site HealthNewsReview.org and former director of health journalism at the University of Minnesota. "It can be—and often is—an ugly picture: a bazaar of disinformation."

Why might consuming coffee or caffeine be associated with such dramatically different results? There are many possibilities. The risks or benefits of a caffeinated beverage may depend on the amount a person drinks—one cup versus a whole pot. Or it may matter *who* is drinking the beverage. The *New England Journal of Medicine* study, for example, looked at hospitalized patients only. Would the same results have been seen in people who weren't already sick? Sometimes to properly evaluate a scientific claim, we need to look more closely at how the science was done (**INFOGRAPHIC 1.1**).

Science Is a Process

When many people think about science, they think of a body of facts to be memorized: water boils at 100°C; the nucleus is a part of a cell. But that's not the whole story. **Science** is a *way* of knowing, a *method* of seeking answers to questions on the basis of observation and experiment. Scientists draw conclusions from the best evidence they have at any one time, but the process is not always easy or straightforward. Conclusions based on today's evidence may be modified in the future as scientists uncover additional data or make better observations. Improved technology may support more-refined data gathering. This new information can cast a different light on old conclusions. Science is a never-ending process.

Perhaps the best way to understand science is to do it. Let's say you want to evaluate a commonly heard claim about coffee—that it boosts energy. How might you go about investigating this claim scientifically? A logical place to start would be your own experience. You may notice that you feel more awake when you drink coffee or that coffee helps you concentrate. Such informal personal observations are called **anecdotal evidence.** This is a type of evidence that may be interesting but is often unreliable, since it isn't based on systematic study. Asking your friends how they experience coffee would

SCIENCE
The process of using observations and experiments to draw conclusions based on evidence.

ANECDOTAL EVIDENCE
An informal observation that has not been systematically tested.

INFOGRAPHIC 1.1 Conflicting Conclusions

A variety of studies published in peer-reviewed scientific journals report different conclusions about the risks and benefits of coffee. In order for the public to understand and make sense of these varying outcomes, a closer look at the scientific process and the factors that surround coffee drinking is necessary.

Coffee's good for you!

- Fights liver disease
- Lowers the risk of colorectal cancer
- Protects against type 2 diabetes
- Protects against Parkinson's and Alzheimer's
- Reduces risk of depression

Coffee's bad for you!

- Mildly addictive
- Increases anxiety and irritability
- Can increase cholesterol
- Elevates blood pressure
- May increase the risk of heart disease

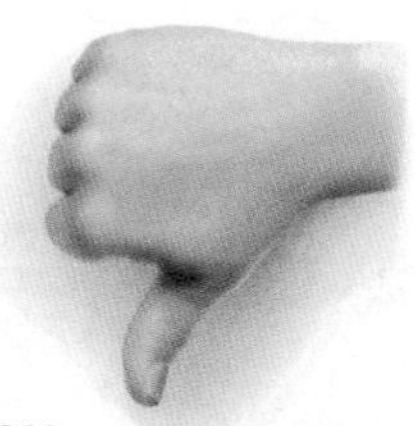

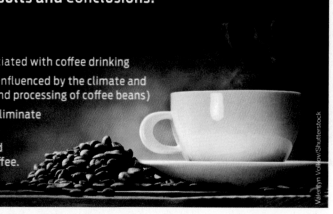

Factors That May Contribute to Conflicting Results and Conclusions:

- The amount of coffee a person drinks
- The gender, age, and general health of an individual
- Social factors such as eating and smoking that are often associated with coffee drinking
- The chemical makeup of the coffee, including caffeine levels (influenced by the climate and soil in which coffee plants are grown, as well as the roasting and processing of coffee beans)
- Flawed study design (lack of appropriate controls, failure to eliminate bias, statistical errors, etc.)
- Other unknown factors may correlate with coffee drinking, and these may be responsible for risks or benefits attributed to coffee.

Valentyn Volkov/Shutterstock

? Give three reasons why one study may find that coffee has health benefits and another find that coffee is associated with health risks.

INFOGRAPHIC 1.2 Science Is a Process: Narrowing Down the Possibilities

Multiple scientists doing multiple experiments narrow down the pool of possible hypotheses. Those that are rigorously tested and supported by other experiments emerge with greatest confidence.

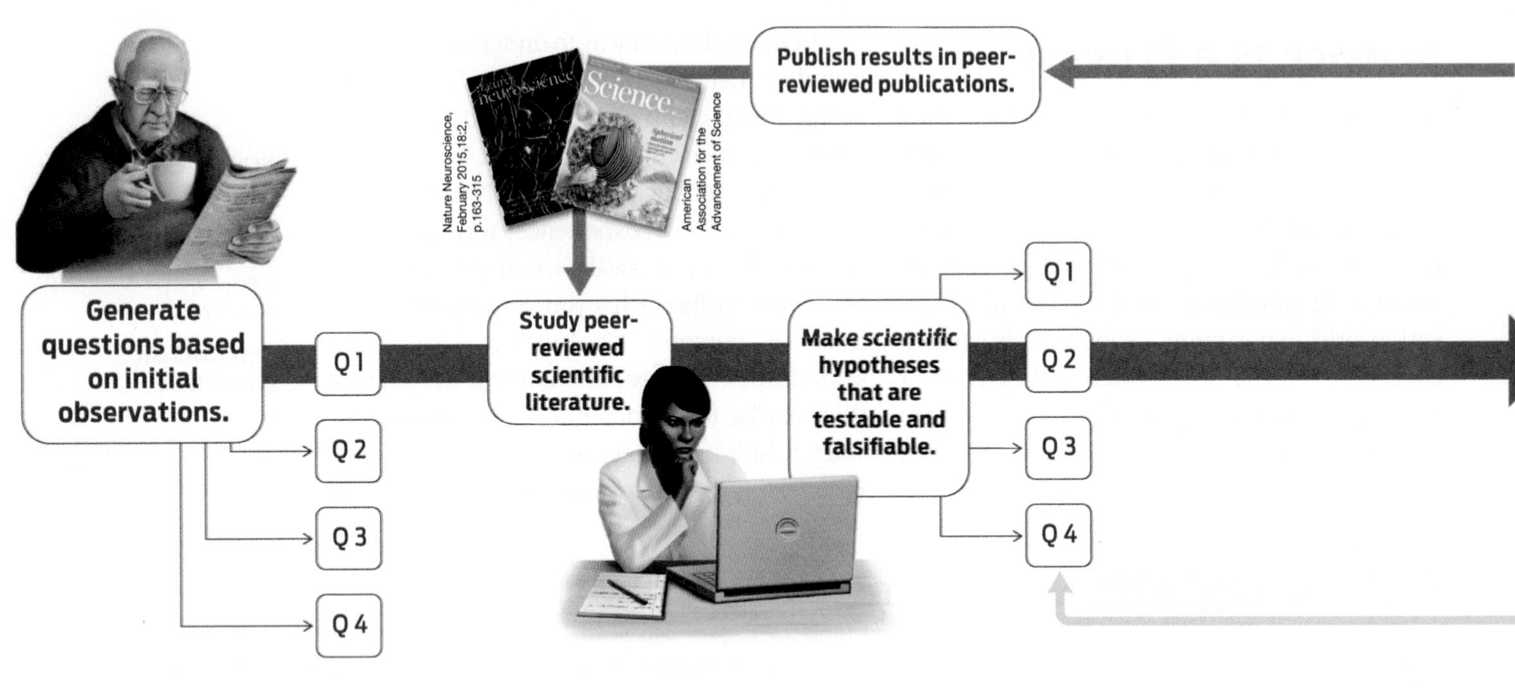
Publish results in peer-reviewed publications.
Nature Neuroscience, February 2015,18:2, p.163-315
American Association for the Advancement of Science
Science
Generate questions based on initial observations.
Q1
Q2
Q3
Q4
Study peer-reviewed scientific literature.
Make scientific hypotheses that are testable and falsifiable.
Q1
Q2
Q3
Q4

? If the data from an experiment appear to support a hypothesis, what are the next steps?

Psychological SCIENCE
VOLUME 27 NUMBER 11 NOVEMBER 2016
aps

PSYCHOLOGICAL SCIENCE

Research Report

CAFFEINE REDUCES TIME-OF-DAY EFFECTS ON MEMORY PERFORMANCE IN OLDER ADULTS

Lee Ryan, Colleen Hatfield, and Melissa Hofstetter

University of Arizona

The studies reported in scientific journals are reviewed by experts before publication to ensure accuracy.

also be anecdotal evidence, since your friends might not be typical or representative of the larger group.

Nevertheless, this anecdotal evidence might lead you to formulate a question that you want to investigate further: does coffee boost energy? To find out what information currently exists on the relationship between coffee and energy, you could read relevant studies that have already been conducted, available in online databases of journal articles or in university libraries. Generally, the information in scientific journals is reliable because it has been subject to **peer review,** a process in which experts in the same field as the investigator review an article before it is published**.** The aim of peer review is to weed out sloppy research as well as overstated claims, and thus ensure the integrity of the journal and its scientific findings. To further reduce the chance of bias, authors must declare any possible conflicts of interest and name all funding sources (for example, pharmaceutical or biotechnology companies). With this information, reviewers and readers can view the study with a more critical eye.

From your perusal of the scientific literature, you would soon discover (if you didn't know it already) that coffee contains caffeine, and that caffeine is a chemical known to stimulate the brain. Brain regions controlling

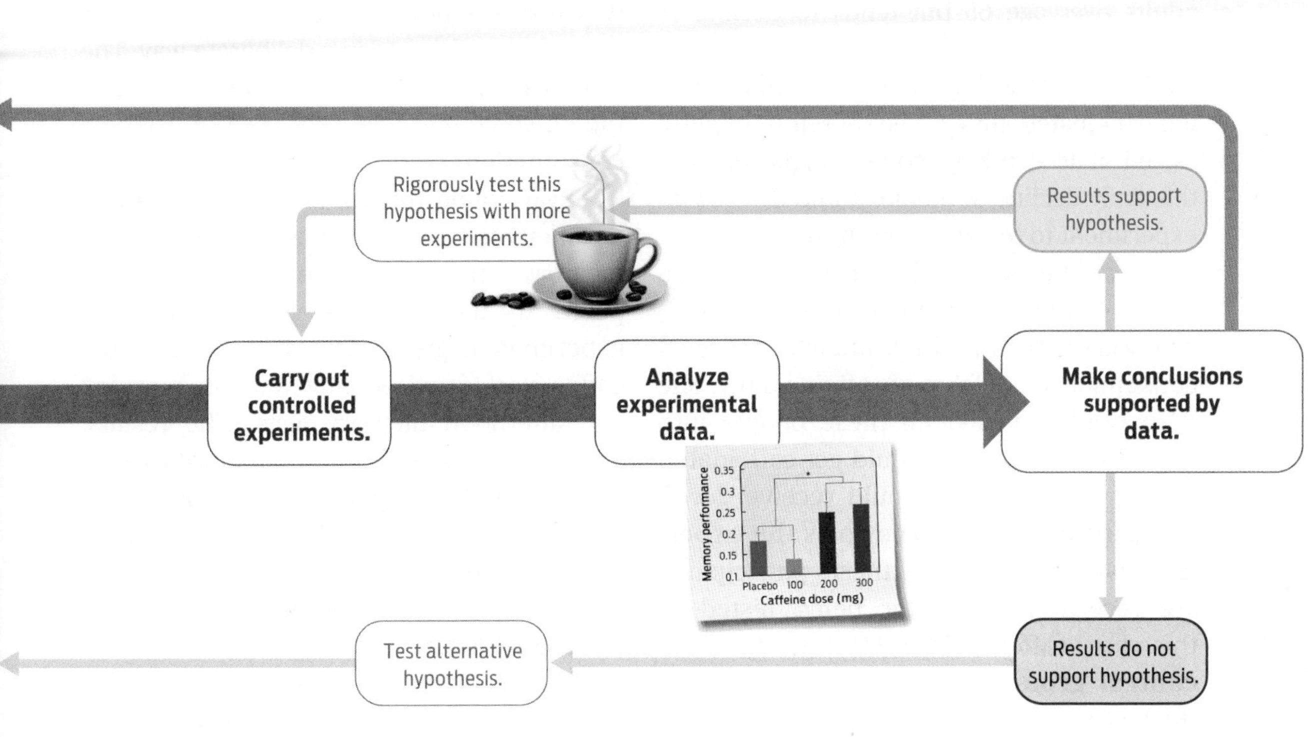

sleep, mood, memory, and concentration are especially sensitive to this chemical, and some researchers believe that caffeine accounts for the general feeling of alertness that many coffee drinkers experience.

Armed with this information, you could go one step further and formulate a specific hypothesis about coffee and energy. A **hypothesis** is a possible answer to the question under investigation. For example, one hypothesis about coffee might be that consuming caffeinated coffee would increase alertness, but drinking decaf would not. Another might be that caffeine boosts levels of blood flow to the brain. Not all possible explanations will be *scientific* hypotheses, though. A scientific hypothesis must be **testable** and **falsifiable**–that is, it can be established or rejected by experiment; and if it is false, it can be proved wrong. Explanations that depend on supernatural forces are not scientific hypotheses because they cannot be tested or refuted.

With a clear scientific hypothesis in hand–"caffeinated coffee improves alertness"–the next step is to test it, generating evidence for or against the idea. If a hypothesis is shown to be false–caffeinated coffee does not improve alertness–the hypothesis can be rejected and removed from the list of possible answers to the original question. As the scientist, you are then forced to consider other hypotheses. On the other hand, if data support the hypothesis it will be accepted, at least until further testing and data show otherwise **(INFOGRAPHIC 1.2)**.

There are a number of ways in which a hypothesis can be supported (or rejected). One is to design a controlled **experiment**, a carefully designed test. In this case, you might measure the effects of coffee drinking

PEER REVIEW
A process in which independent experts read scientific studies before they are published to ensure that the authors have appropriately designed and interpreted the study.

HYPOTHESIS
A tentative explanation for a scientific observation or question.

TESTABLE
Describes a hypothesis that can be supported or rejected by carefully designed experiments or observational studies.

FALSIFIABLE
Describes a hypothesis that can be ruled out by data that show that the hypothesis does not explain the observation.

EXPERIMENT
A carefully designed test, the results of which will either support or rule out a hypothesis.

on a group of participants. In 2002, Lee Ryan, a psychologist at the University of Arizona, decided to do just that. Ryan knew that memory is often optimal early in the morning in adults over age 65 but tends to decline as the day goes on. She also noticed that many adults report feeling more alert after drinking caffeinated coffee. She therefore hypothesized that drinking coffee might prevent this daily decline in memory, and devised an experiment to test her hypothesis.

First she collected a group of participants–40 men and women, all over age 65, who were active, healthy, and who reported consuming some form of caffeine daily. She then randomly divided these people into two groups: one that would get caffeinated coffee, and one that would receive decaf. The caffeine group is known as the **experimental group**, since these participants are receiving the factor being tested–in this case, caffeine. The decaf group is the **control group,** the group that serves as the basis of comparison. Both groups were given memory tests at 8 a.m. and again at 4 p.m. on two nonconsecutive days. The experimental group received a 12-ounce cup of regular coffee containing approximately 220-270 mg of caffeine 30 minutes before each test. The control group received an inactive treatment, or **placebo**: in this experiment, a 12-ounce cup of decaffeinated coffee containing no more than 5-10 mg of caffeine per serving. No participants knew to which group they were assigned–in other words, the study was "blind." In this case, the study was actually "double blind," since neither the investigator nor the participants knew who got what treatment.

By administering a placebo to the control group, Ryan could ensure that any change observed in the experimental group was a result of consuming caffeine and not just any hot beverage. In addition, all participants were forbidden to eat or drink any caffeine-containing foods or drinks–like chocolate, soda, or coffee–for at least 4 hours before each test. Thus, the control group was identical to the experimental group in every way except for the consumption of caffeine.

In this experiment, caffeine consumption was the **independent variable**–the factor that is being changed in a deliberate way. The tests of memory are the **dependent variable**–the outcome that may "depend" on caffeine consumption.

Ryan found that participants who drank decaffeinated coffee did worse on tests of memory function in the afternoon compared to the morning. By contrast, the experimental group that drank caffeinated coffee performed equally well on morning and afternoon memory tests. The results, which were published in the journal *Psychological Science,* support the hypothesis that caffeine, delivered in the form of coffee, prevents the decline of memory over the course of a day–at least in certain people **(INFOGRAPHIC 1.3)**.

Because other factors might possibly explain the link between coffee and mental performance (perhaps coffee drinkers are more active, and their physical activity rather than their coffee consumption explains their mental performance), it's too soon to see these results as proof of coffee's memory-boosting powers. To win our confidence, the experiment must be repeated by other scientists and, if possible, the results strengthened.

Sample Size Matters

Consider the size of Ryan's experiment–40 people, tested on two different days. That's not a very big study. Could the results have simply been due to chance? What if the 20 people who drank caffeinated coffee just happened to have better memory?

One thing that can strengthen our confidence in the results of a scientific study is sample size. **Sample size** is the number of individuals participating in a study, or the number of times an experiment or set of observations is repeated. The larger the

EXPERIMENTAL GROUP
The group in an experiment that experiences the experimental intervention or manipulation.

CONTROL GROUP
The group in an experiment that experiences no experimental intervention or manipulation.

PLACEBO
A fake treatment given to control groups to mimic the experience of the experimental groups.

INDEPENDENT VARIABLE
The variable, or factor, being deliberately changed in the experimental group relative to the control group.

DEPENDENT VARIABLE
The measured result of an experiment, analyzed in both the experimental and control groups.

SAMPLE SIZE
The number of experimental subjects or the number of times an experiment is repeated. In human studies, sample size is the number of participants

INFOGRAPHIC 1.3 Anatomy of an Experiment

There are many ways to approach a scientific problem. Controlled experiments are one way. As illustrated here, controlled experiments have two groups—the control group and the experimental group—that differ only in the independent variable.

Hypothesis: Drinking caffeinated coffee prevents daily memory decline.

Population of 40 men and women over age 65

	Control Group	Experimental Group
Random placement into equivalent groups (with respect to age, gender, health, activity level, etc.)		
Independent variable (the variable that is changed in a systematic way)	Placebo Treatment 12 oz. decaffeinated coffee (30 minutes prior to test)	Test Treatment 12 oz. caffeinated coffee (30 minutes prior to test)
Dependent variable (the variable that is measured in the experiment)	Memory Test given morning and afternoon on different days	Memory Test given morning and afternoon on different days
Results from data	Memory Test Scores Afternoon scores were lower than morning scores	Memory Test Scores Afternoon scores were the same as morning scores
Evidence-based conclusion	**Caffeinated coffee prevents memory decline in this population.**	

? What are the dependent and independent variables in this experiment? Which one is intentionally changed between the control and experimental groups?

sample size, the more likely the results will have **statistical significance**–that is, they will not be due to chance (**INFOGRAPHIC 1.4**).

News reports are full of statistics. On any given day, you might hear that 75% of the American public opposes a piece of legislation. Or that 15% of a group of people taking a medication experienced a certain unpleasant side effect–like nausea or suicidal thoughts–compared to, say, 8% of people taking a placebo. Are these differences significant, or important? Whenever you hear such numbers being tossed around, it's important to keep in mind the sample size. In the case of the side effects, was this a group of 20 patients (15% of 20 patients is 3 people), or was it 2,000 (15% of 2,000 is 300)? Only with a large enough sample size can we be confident that the results of a given study are statistically significant and represent something other than chance. Moreover, it's important to consider the population being studied. For example, do the people reporting their views on a piece of legislation represent

STATISTICAL SIGNIFICANCE
A measure of confidence that the results obtained are "real" and not due to chance.

INFOGRAPHIC 1.4 Sample Size Influences Statistical Significance

The more data collected in an experiment, the more you can trust the conclusions.

Data from a few participants:
While a casual observation of these data suggests a slight positive relationship between caffeine and memory, a statistical test is not likely to be conclusive because of the small sample size.

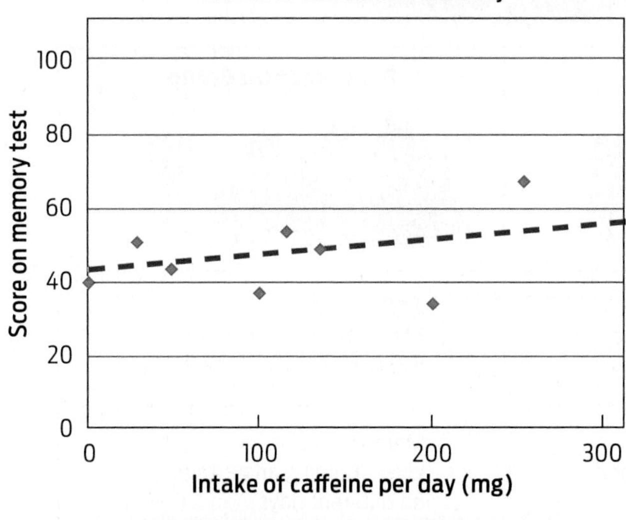

Statistical analysis is inconclusive.
The observed increase may be **due to chance**.

Data from dozens of participants:
Repeating the experiment with a larger sample size may reveal a different relationship between caffeine intake and memory, and the statistical analysis provides more support for the validity of the results.

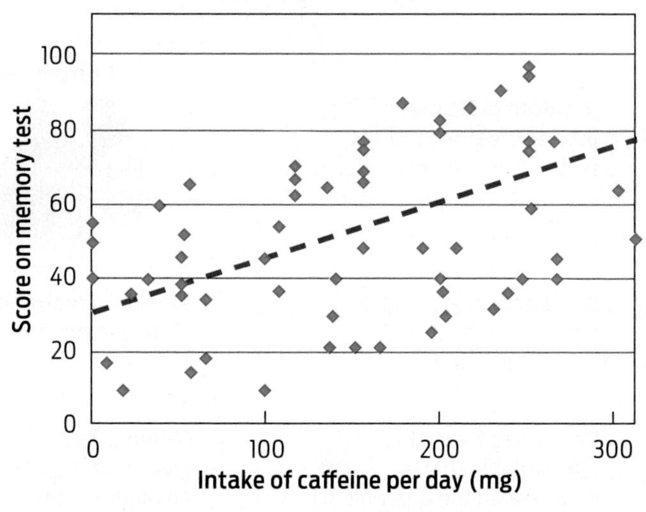

Statistical analysis shows the observed increase is **significant** and therefore **not due to chance**.

? Why are results from a large study likely to be more reliable (and reproducible) than results from a small study?

a broad cross section of the public, or are most of them watchers of the same television network, whose views lie at an extreme? Likewise, in Ryan's study, are the 65-year-old self-described "morning people" who regularly consume coffee representative of the wider population?

If you search for "caffeine and memory" on PubMed.gov (a database of medical research papers), you'll see that the memory-enhancing properties of caffeine is a well-researched topic. Many studies have been conducted, at least some of which tend to support Ryan's results. Generally, the more experiments that support a hypothesis, the more confident we can be that it is true.

SCIENTIFIC THEORY
An explanation of the natural world that is supported by a large body of evidence and has never been disproved.

Truth in science is never final, however. What is accepted as fact today may tomorrow need to be modified or even rejected when more evidence comes to light. Nevertheless, scientific knowledge does progress. The highest point of scientific knowledge is what's called a **scientific theory.** The word "theory" in science means something very different from its everyday meaning. In casual conversation, we may say something is "just a theory," meaning it isn't proved. But in science, a theory is an explanation of the natural world that is supported by a large body of evidence compiled over time by numerous researchers. Far from being a fuzzy or unsubstantiated claim, a theory is

INFOGRAPHIC 1.5 Everyday Theory vs. Scientific Theory

In daily life, people use the word "theory" to refer to an idea that explains an everyday event. In science, a theory is a hypothesis that has never been disproved, even after many years of rigorous testing.

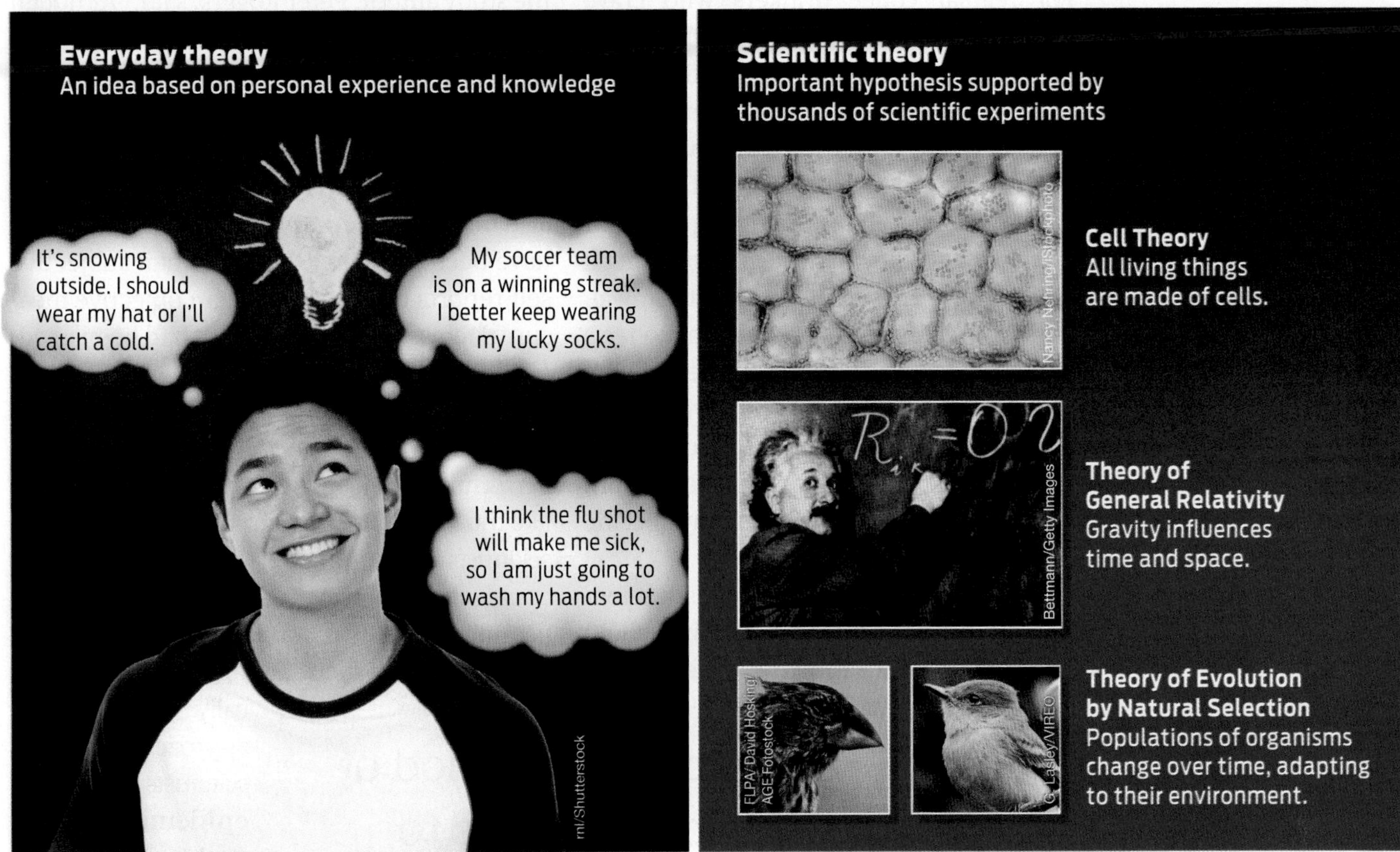

Everyday theory
An idea based on personal experience and knowledge
It's snowing outside. I should wear my hat or I'll catch a cold.
My soccer team is on a winning streak. I better keep wearing my lucky socks.
I think the flu shot will make me sick, so I am just going to wash my hands a lot.
ml/Shutterstock
Scientific theory
Important hypothesis supported by thousands of scientific experiments
Nancy Nehring/iStockphoto
Cell Theory
All living things are made of cells.
Bettmann/Getty Images
Theory of General Relativity
Gravity influences time and space.
FLPA/David Hosking/AGE Fotostock
G. Lasley/VIREO
Theory of Evolution by Natural Selection
Populations of organisms change over time, adapting to their environment.

? In early 2016, a reporter stated that researchers were testing the theory that Zika virus causes birth defects. What is wrong with this statement?

a scientific explanation that has been extensively tested and has never been disproved (**INFOGRAPHIC 1.5**).

This Is Your Brain on Caffeine

Caffeine is a stimulant. It is in the same class of psychoactive drugs as cocaine, amphetamines, and heroin (although less potent than these). While the exact mechanisms are not fully understood, scientists think that caffeine exerts its energizing effect primarily by counteracting the actions of a chemical in the brain called adenosine. Adenosine is the body's natural sleeping pill—its concentration increases in the brain while we are awake and by the end of the day, its high level promotes drowsiness. Caffeine blocks the effect of adenosine in the brain, thereby delaying fatigue and keeping us more alert.

Consumption of caffeinated beverages has skyrocketed in the past 25 years, especially among young people. A 2009 study in the journal *Pediatrics* found that teenagers in a Philadelphia suburb consume anywhere from 23 mg to 1,458 mg of caffeine a day—the equivalent of nearly 10 cups of coffee, and more than three times the recommended safe dose for an adult of 400 mg/day or less. In excess, caffeine can cause anxiety, jitters, heart palpitations, trouble sleeping, dehydration, and more serious symptoms as well—especially in people who are sensitive to it.

For regular coffee drinkers who crave their morning buzz, such side effects are unlikely to persuade them to kick the habit. This may be because, like many other psychoactive substances, caffeine is addictive. Those who drink a significant amount of coffee every day may notice that they don't feel quite right if they skip a day; they may be cranky or get a headache. These are symptoms of withdrawal. In fact, some researchers contend that coffee's mind-boosting effects are an indirect result of the cycle of dependency. Improvement in mood or performance following a cup of coffee, they say, may simply represent relief from withdrawal symptoms rather than any specific beneficial property of coffee.

To test this dependency hypothesis, scientists could conduct an experiment. They could compare the effects of drinking coffee in two groups: one group of regular coffee drinkers who had abstained from coffee for a short period, and another group of non-coffee drinkers. Does coffee give both groups a boost, or only the regular coffee drinkers looking for their fix?

Science is a method of seeking answers to questions on the basis of observation and experiment.

This very experiment was done in 2010 by a group of researchers at the University of Bristol in England. Their study, published in the journal *Neuropsychopharmacology,* looked at caffeine's effect on alertness. Researchers gave caffeine or a placebo to 379 participants and asked them to rate their level of alertness before and after receiving the caffeine or placebo. The study found that caffeine did not boost perceptions of alertness in non-coffee drinkers compared to those drinking a placebo (although it did boost their level of anxiety and headache). Heavy coffee drinkers, on the other hand, experienced a steep drop in alertness when given the placebo.

EPIDEMIOLOGY
The study of patterns of disease in populations, including risk factors.

"What this study does is provide very strong evidence for the idea that we don't gain a benefit in alertness from consuming caffeine," the study author, Peter Rogers, said. "Although we feel alert, that's just caffeine bringing us back to our normal state of alertness."

For those of us who rely on coffee to perk us up, the results of this one study are unlikely to change our minds—or our habits. We might also wonder whether the conclusions would have been the same if, rather than subjective self-reports of participants, an objective measure such as a test of alertness had been used.

Finding Patterns

Performing controlled laboratory experiments like those discussed earlier is one way in which scientists try to answer questions. Another approach is to make careful observations of phenomena that exist in nature, generating data that can then be analyzed systematically. This is the approach taken by scientists who study **epidemiology**—the study of the incidence of disease in populations. It is also the approach of scientists who study topics like the movement of stars or the nature of prehistoric life, phenomena that cannot be directly experimented upon.

For example, an epidemiologist who wanted to learn about the relationship between cigarette smoking and lung cancer could compare the rates of lung cancer in smokers and in nonsmokers. But an experiment in which participants were asked to smoke cigarettes and were observed to see whether or not they developed cancer would be highly unethical.

Although epidemiological studies do not provide the immediate gratification of a laboratory experiment, they do have certain advantages. For one thing, they can be relatively inexpensive to conduct, since often the only procedure involved is a participant questionnaire. And you can study factors that are considered harmful, such as excess alcohol or smoking, that you would be unable to test experimentally. Finally, epidemiological studies have the power of numbers and time. The Framingham Heart Study, for example, is a famous epidemiological study that has tracked rates of cardiovascular disease in a group of people and their descendants in Framingham, Massachusetts, in order to identify common risk factors for the disease. Begun in 1948, the study has been going on for decades and has provided mountains of data for researchers in many fields, from cardiology to neuroscience.

Most of the health studies featured in the news are epidemiological studies. Consider a study on coffee and Parkinson's disease published in the *Journal of the American Medical Association (JAMA)* in 2000. Researchers examined the relationship between coffee drinking and the incidence of Parkinson's disease, a neurodegenerative condition that afflicts more than 1 million people in the United States, affecting men and women of all ethnic groups. There is no known cure, only palliative treatments to lessen symptoms, which include trembling limbs and difficulty coordinating speech and movement.

For more than 30 years, researchers at the Veterans Affairs Medical Center in Honolulu followed more than 8,000 Japanese-American men, gathering all sorts of information about them: their age, diet, health, smoking habits, and other characteristics. Of these men, 102 developed Parkinson's disease. What did these 102 men have in common? Epidemiologists found that most of them did not drink caffeinated beverages—no coffee, soda, or caffeinated tea.

By contrast, coffee drinkers had a lower incidence of Parkinson's disease. In fact, those who drank the most coffee were the least likely to get the disease. Men who drank more than two 12-ounce cups of coffee each day had one-fifth the risk of getting the disease compared to non-coffee drinkers. The same trend was observed for overall caffeine intake that included sources other than coffee.

So does coffee (specifically, caffeine) prevent Parkinson's disease? The occurrence and progression of many diseases are affected by a complex range of factors, including age, sex, diet, genetics, and exposure to bacteria and environmental chemicals, as well as lifestyle factors like drinking, smoking, and exercise.

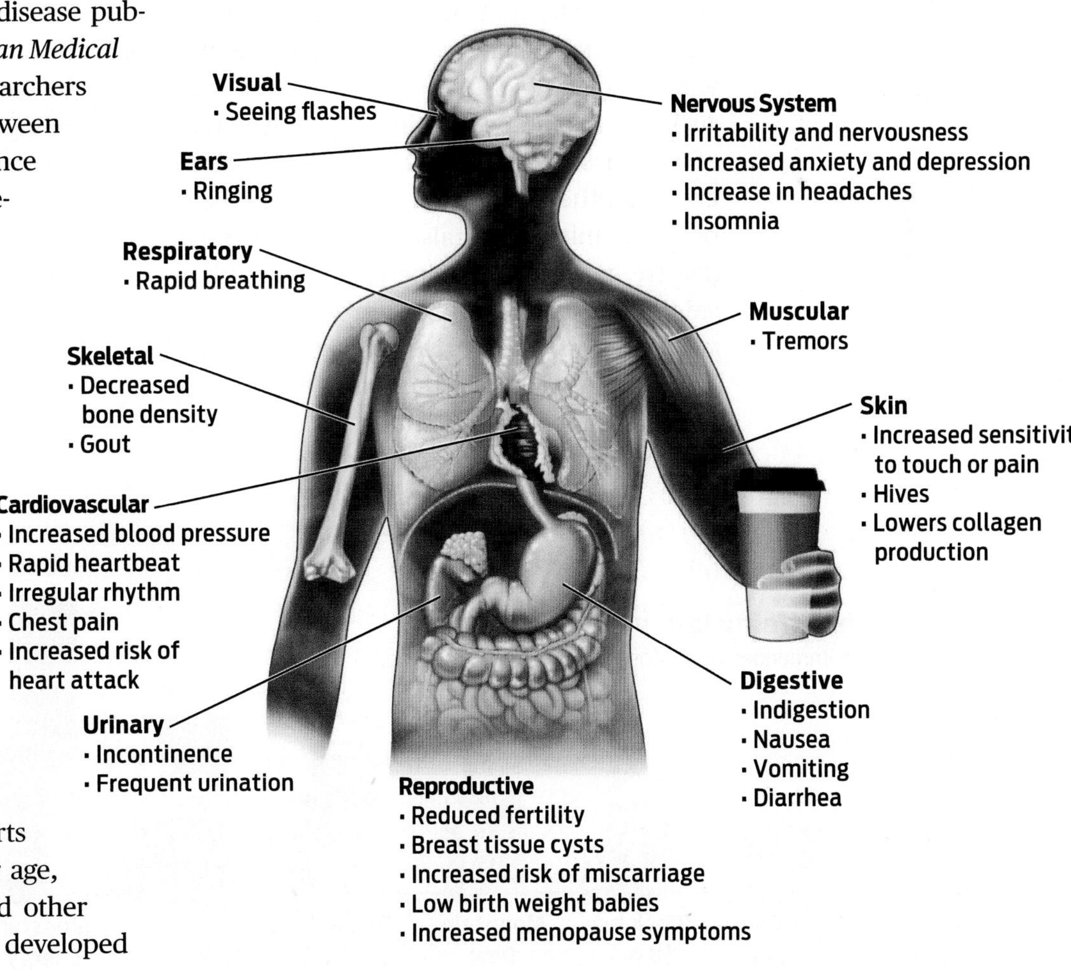

Despite potential benefits as a memory enhancer, the caffeine in coffee has some powerful side effects.

CORRELATION
A consistent relationship between two variables.

Although the study discussed here suggests a link–or **correlation**–between caffeine and lower incidence of Parkinson's disease, it does not necessarily show that caffeine prevents the disease. In other words, correlation is not causation. Perhaps the people who like to drink coffee have different brain chemistry, and it's this different brain chemistry that explains the differing incidence of Parkinson's disease among coffee drinkers **(INFOGRAPHIC 1.6)**.

Indeed, other studies have found that cigarette smoking also correlates with a lower risk of Parkinson's disease. Both coffee drinking and smoking could be considered types of thrill-seeking behavior, observed in people who enjoy the high they get from stimulants such as caffeine or nicotine. The lower risk of Parkinson's disease among coffee drinkers might therefore result from thrill-seeking brain chemistry that also happens to resist disease–rather than being caused by either smoking or drinking coffee per se.

Moreover, the study followed Japanese-American men. Would the same relationship between caffeine and Parkinson's disease be seen in other ethnic groups or in women? Several other epidemiological studies have found a correlation between caffeine consumption and a lower incidence of Parkinson's disease in men of other ethnicities. But in women the results have been inconclusive. All in all, there's still no direct evidence that caffeine actually prevents the disease in either men or women.

> "Consumers are flooded with a firehose of health information every day from various media sources."
>
> — Gary Schwitzer

"While our study found a strong correlation between coffee drinkers and low rates of Parkinson's disease," stated the study's lead author, G. Webster Ross, "we have not identified the exact cause of this effect. I'd like to see these findings used as a basis to help other scientists unravel the mechanisms that underlie Parkinson's onset."

Finding a correlation that is not necessarily a cause is a typical result of epidemiological

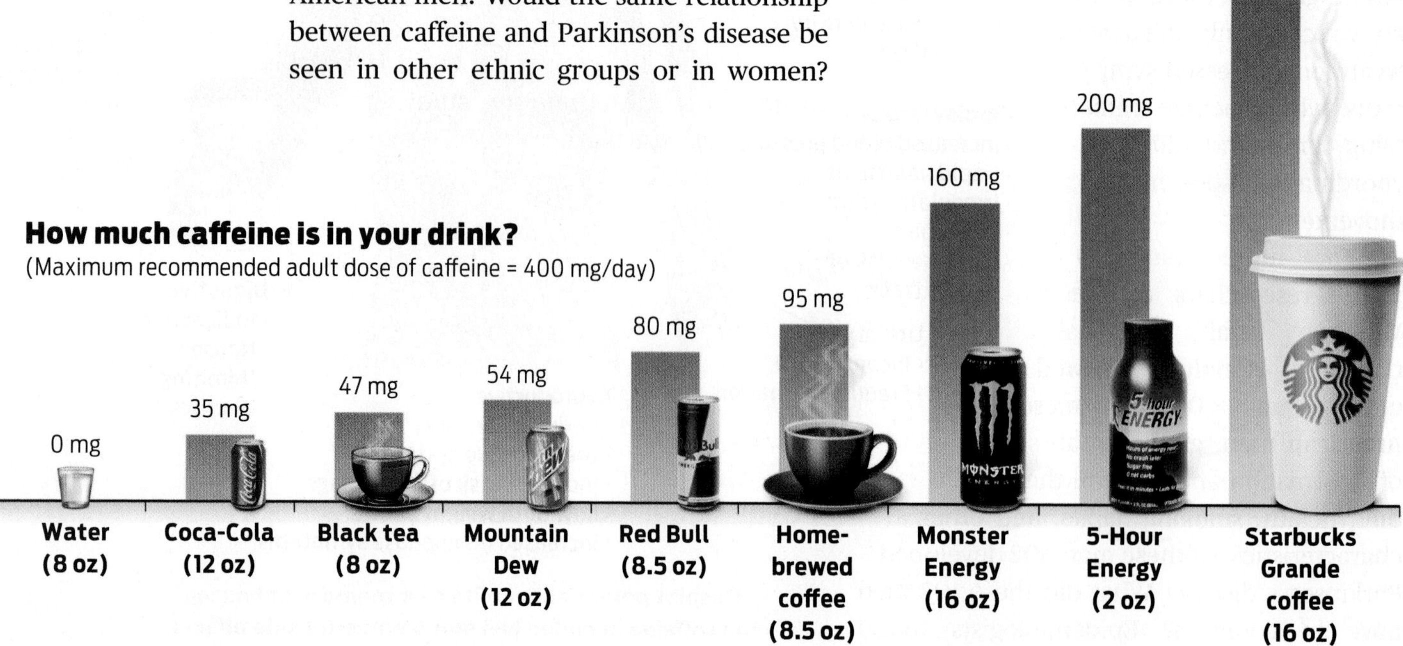

INFOGRAPHIC 1.6 Correlation Does Not Equal Causation

While the data shown below show a convincing **correlation** between reduced caffeine intake and an increased risk of Parkinson's disease, it is impossible to state that less coffee **causes** Parkinson's disease. Other factors that were not tested or controlled for could be causing the increased risk of Parkinson's disease in those with reduced coffee intake.

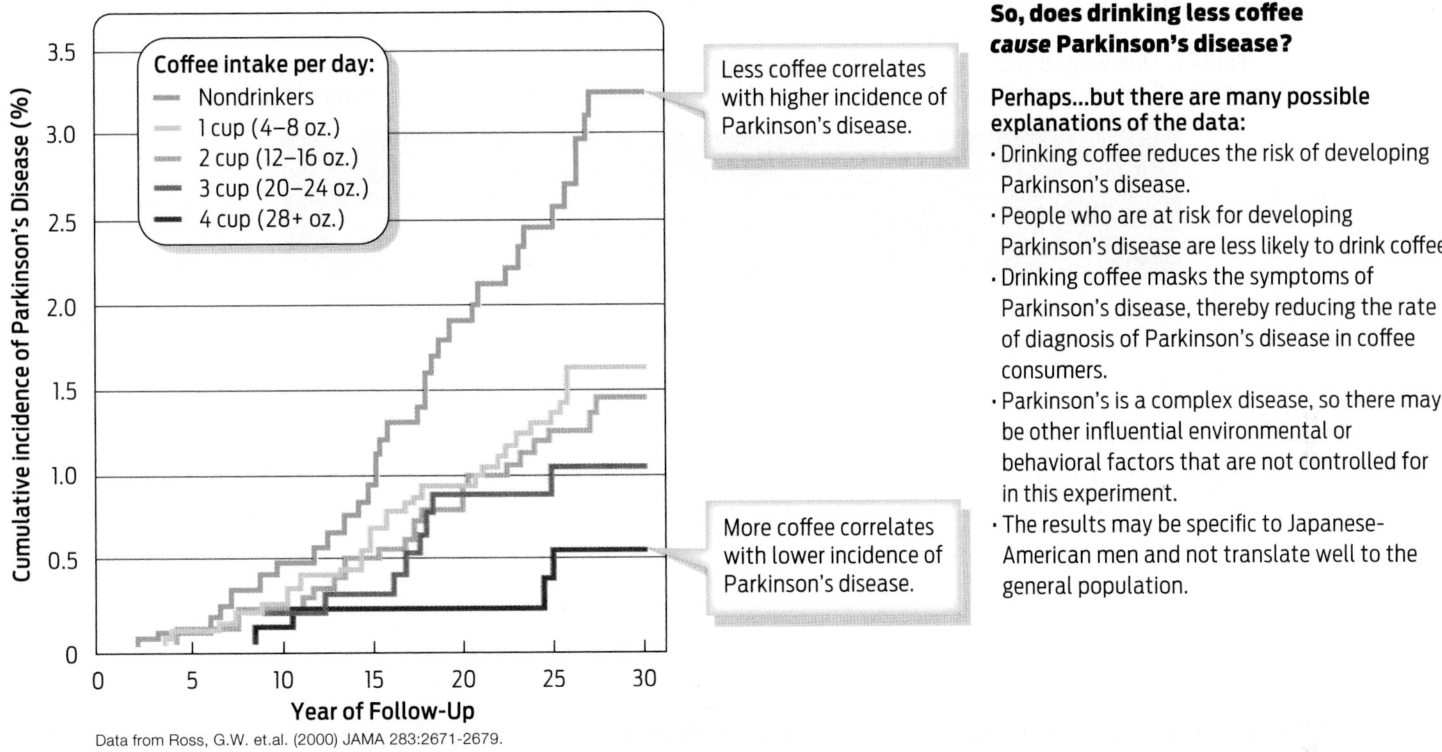

Data from Ross, G.W. et.al. (2000) JAMA 283:2671-2679.

So, does drinking less coffee *cause* Parkinson's disease?

Perhaps...but there are many possible explanations of the data:

- Drinking coffee reduces the risk of developing Parkinson's disease.
- People who are at risk for developing Parkinson's disease are less likely to drink coffee.
- Drinking coffee masks the symptoms of Parkinson's disease, thereby reducing the rate of diagnosis of Parkinson's disease in coffee consumers.
- Parkinson's is a complex disease, so there may be other influential environmental or behavioral factors that are not controlled for in this experiment.
- The results may be specific to Japanese-American men and not translate well to the general population.

? From these data, what can you conclude about the relationship between coffee drinking and Parkinson's disease?

studies, but that does not mean such studies are worthless. In fact, as the author of this study of Japanese-American men indicates, correlations can be a jumping-off point for additional research aimed at identifying root causes. Correlations provide suggestive evidence that merits additional research.

To get a clearer picture of caffeine's role in Parkinson's disease, researchers could conduct a type of experiment known as a **randomized clinical trial,** in which the effects of coffee on randomly chosen participants would be measured directly under controlled conditions. One could divide a population into two groups, put one group on coffee and the other on decaf, and then follow both groups for a number of years to see which one had the higher incidence of disease. The problem with such a study is that it is often logistically challenging to conduct, since it can be difficult to get people to stick to the regimen for the length of the study. Furthermore, such studies are unethical if the experimental treatment is likely to cause harm.

Getting Beyond the Buzz

While a lower risk of Parkinson's disease represents a potential boon to coffee drinkers, the news for caffeine addicts isn't all good. Over the years, epidemiological studies have also linked caffeine consumption to *higher* rates of various diseases, including osteoporosis, fibrocystic breast disease, and bladder cancer. As with the link to Parkinson's disease, however, such correlations do not

RANDOMIZED CLINICAL TRIAL
A controlled medical experiment in which subjects are randomly chosen to receive either an experimental treatment or a standard treatment (or a placebo).

INFOGRAPHIC 1.7 From the Lab to the Media: Lost in Translation

The data as reported in peer-reviewed journals are often very complex. Scientists interpret these data in lengthy discussions, but the public receives them as isolated media headlines.

Data from scientific studies provide a large amount of information:

Table 1. Unadjusted and Age-Adjusted Incidence of Parkinson's Disease (PD) According to Amount of Coffee Consumed per Day (Based on 30 Years of Follow-Up after Examinations from 1965 to 1968)

Coffee intake (oz/day)	No. Cases of PD/No. Subjects at Risk	Incidence Rate/10,000 Person-Years: Unadjusted	Adjusted for Age	Adjusted Relative Hazard (95% Confidence) Compared with Top Category of Coffee Intake**
Nondrinker	32/1286	10.5	10.4	5.1 (1.8 – 14.4)
4 to 8	33/2576	5.5*	5.3*	2.7 (1.0 – 7.8)
12 to 16	24/2149	4.7*	4.7*	2.5 (0.9 – 7.3)
20 to 24	9/1034	3.6*	3.7*	2.0 (0.6 – 6.4)
≥28	4/959	1.7*	1.9*	Reference

**Adjusted for age and pack-years of cigarette smoking.
*Significantly different from nondrinkers.

Data from Ross, G.W. et.al. (2000) JAMA 283:2671-2679.

Data collected only from males

Self-reported by the participants, not measured

The amountof caffeine per cup may vary.

Large sample sizes

People drinking less coffee have a higher risk for PD.

In all groups of coffee-drinkers, some develop PD.

People drinking more coffee have a lower risk for PD.

Media reports simplify and often sensationalize the results for the public:

"Really? A cure for Parkinson's disease? Maybe I should drink more coffee..."

Tony West/Alamy

Key details of the study not reported by the media:

- Some coffee drinkers develop Parkinson's disease, so not everyone will benefit.
- The results reflect a correlation, not a causation. While there is a significant link between coffee drinking and the disease, factors besides coffee may be contributing to the results.
- The study was conducted with a particular male population, so we cannot generalize the results to other populations (e.g., women).

? What key information may be missing in media reports of scientific studies?

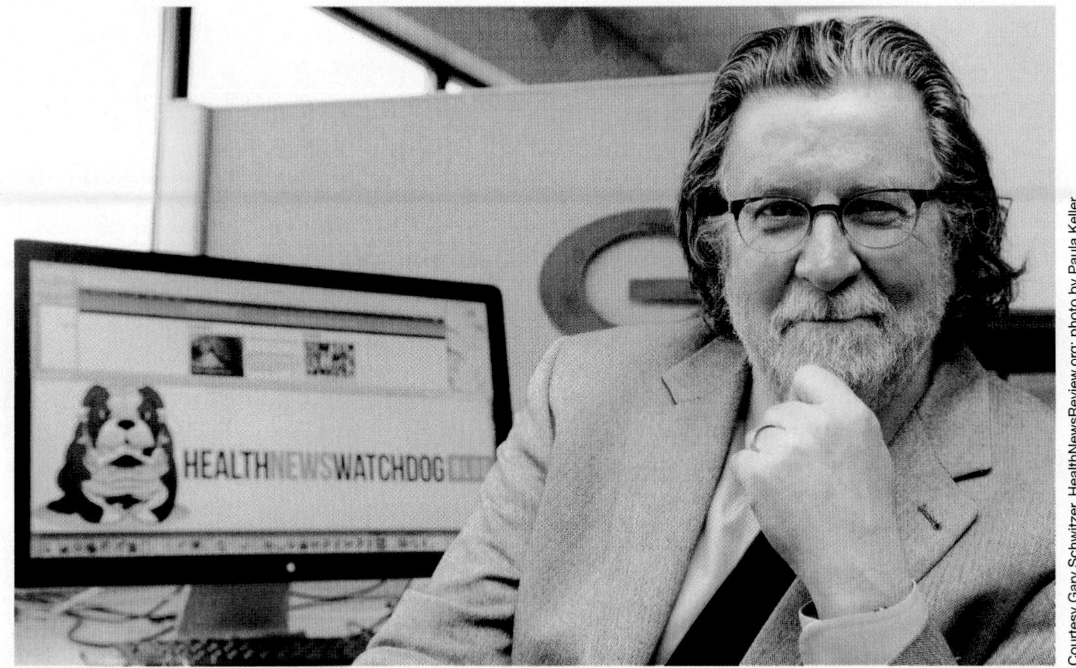

Courtesy Gary Schwitzer, HealthNewsReview.org; photo by Paula Keller.

Gary Schwitzer is the founder and publisher of HealthNewsReview.org, a consumer watchdog website that evaluates health care claims.

necessarily prove that caffeine causes any of these diseases. Nevertheless, such studies are often quite influential and newsworthy–like the apparent correlation between coffee and pancreatic cancer that made headlines in 1981. That study was based on a single, small epidemiological study, which was later shown to be flawed in its design.

Journalists face unique challenges in covering health news, says Gary Schwitzer of HealthNewsReview.org: "They must cover complex topics, do it quickly, creatively, accurately, completely and with balance–and then be sure they don't 'dumb it down' too much for a general news audience," he says. "If they can't do it right, they must realize the *harm* they can do by reporting inaccurately, incompletely, and in an imbalanced way" (**INFOGRAPHIC 1.7**).

Journalists and scientists aren't the only ones who bear the responsibility of determining what information is trustworthy. As consumers and citizens, we can become more knowledgeable about how science is done and which studies deserve to influence our behavior. Whether it's the latest media report linking cell phones to brain tumors or vaccines to autism, the only way to really judge the value of a study is to sift through the evidence ourselves. Of course, to do that, we might first need a cup of coffee. ■

CHAPTER 1 SUMMARY

- Science is a process—a way of seeking answers to questions on the basis on observations and experiments.
- Scientists rely on peer-reviewed scientific reports to learn about new advances in the field. Peer review helps to ensure that the scientific results are valid as well as accurately and fairly presented.
- A potential answer to a scientific question is called a hypothesis. Hypotheses are tested in controlled experiments or in observational studies, the results of which can support or rule out a hypothesis. Hypotheses can be supported by data but cannot be proved absolutely, as future studies may provide new findings.
- Every experiment should have a control—a group that is identical in every way to the experimental group except for one factor: the independent variable.
- The independent variable in an experiment is the one being deliberately changed in the experimental group (e.g., coffee intake). The dependent variable is the measured result of the experiment (e.g., effect of coffee on memory).
- Often a control group takes a placebo, a fake treatment that mimics the experience of the experimental group.
- The strength of the conclusions of a scientific study depends on, among other factors, the type of study carried out and the sample size.
- Scientific theories are different from everyday theories. A scientific theory has undergone extensive testing, is supported by a significant body of evidence, and has never been disproved.
- In epidemiological studies, a relationship between an independent variable (such as caffeine intake) and a dependent variable (such as development of Parkinson's disease) does not necessarily mean one caused the other; in other words, correlation does not equal causation.
- A randomized clinical trial is one in which test participants are randomly chosen to receive either a standard treatment (or a placebo) or an experimental treatment (e.g., caffeine).
- Most of the general public relies on media reports for scientific information. Media reports are not always completely accurate in how they portray the conclusions of scientific studies.
- To understand a study properly, it is often necessary to look at how the study was designed and to analyze the data oneself.

MORE TO EXPLORE

- **PubMed.gov: a collection of more than 25 million biomedical articles maintained by the National Institutes of Health.**
- **HealthNewsReview.org: a consumer watchdog website that helps consumers critically analyze health care claims.**
- **Center for Science in the Public Interest (http://www.cspinet.org): a non-profit watchdog and consumer advocacy group that advocates for safer and healthier foods.**
- **Ryan, L., et al. (2002) Caffeine reduces time-of-day effects on memory performance in older adults. *Psychological Science* 13:68–71.**
- **Ross G. W., et al. (2000) Association of coffee and caffeine intake with the risk of Parkinson disease. *JAMA* 283(20):2674–2679.**

CHAPTER 1 Test Your Knowledge

DRIVING QUESTION 1 How is the scientific method used to test hypotheses?

By answering the questions below and studying Infographics 1.1, 1.2, 1.3, 1.4, 1.5, and 1.6, you should be able to generate an answer for the broader Driving Question above.

KNOW IT

1 When scientists carry out an experiment, they are testing a ________.

- **a.** theory
- **b.** question
- **c.** hypothesis
- **d.** control
- **e.** variable

2 Of the following, which is the earliest step in the scientific process?

- **a.** generating a hypothesis
- **b.** analyzing data
- **c.** conducting an experiment
- **d.** drawing a conclusion
- **e.** asking a question about an observation

3 In a controlled experiment, which group receives a placebo?

- **a.** the experimental group
- **b.** the control group
- **c.** the scientist group
- **d.** the independent group
- **e.** all groups

4 In the studies of coffee and memory discussed in this chapter, the independent variable is ________ and the dependent variable is ________.

- **a.** caffeinated coffee; decaffeinated coffee
- **b.** memory; caffeinated coffee
- **c.** caffeine; memory
- **d.** memory; caffeine
- **e.** decaffeinated coffee; caffeinated coffee

5 Can an epidemiologist who finds a correlation between the use of tanning beds and melanoma (an aggressive form of skin cancer) in college-age women conclude that tanning beds cause skin cancer?

- **a.** yes, as long as the correlation was statistically significant
- **b.** yes, but only for college-age women
- **c.** yes, but only melanoma skin cancer, not other forms of skin cancer
- **d.** no; the study would have to be done with a wider range of participants (males and females of different ages) before it can be concluded that tanning beds cause melanoma
- **e.** no; correlation is not proof of causation

USE IT

6 You carry out a clinical trial to test whether a new drug relieves the symptoms of arthritis better than a placebo. You have four groups of participants, all of whom have mildly painful arthritis (rated 6 on a scale of 1 to 10). Each group receives a daily pill as follows: group 1 (control), placebo; group 2, 15 mg; group 3, 25 mg; group 4, 50 mg. At the end of 2 weeks, participants in each group are asked to rate their pain on a scale of 1 to 10. What is the independent variable in this experiment?

- **a.** the amount of pain experienced at the start of the experiment
- **b.** the amount of pain experienced at the end of the experiment
- **c.** the degree to which pain symptoms changed between the start and the end of the experiment
- **d.** the drug
- **e.** The independent variable could be a, b, or c.

7 You are working on an experiment to test the effect of a specific drug on reducing the risk of breast cancer in postmenopausal women. Describe your control and experimental groups with respect to age, gender, and breast cancer status.

8 Design a randomized clinical trial to test the effects of caffeinated coffee on brain activity. Design your study so that the results will apply to as many people in as many scenarios as possible.

DRIVING QUESTION 2 What factors influence the strength of scientific studies and determine whether the results of any given study are applicable to a particular population?

By answering the questions below and studying Infographics 1.1, 1.3 and 1.4, you should be able to generate an answer for the broader Driving Question above.

KNOW IT

9 In which of the following would you have the most confidence?

- **a.** a randomized clinical trial with 15,000 subjects
- **b.** a randomized clinical trial with 5,000 subjects
- **c.** an epidemiological study with 15,000 subjects
- **d.** an endorsement of a product by a movie star
- **e.** a report on a study presented by a news organization

10 What is the importance of statistical analyses?

- **a.** They can reveal whether or not the data have been fabricated.
- **b.** They can be used to support or reject the hypothesis.
- **c.** They can be used to determine whether any observed differences between two groups are real or a result of chance.
- **d.** all of the above
- **e.** b and c

USE IT

11 You carry out a clinical trial to test whether a new drug relieves the symptoms of arthritis better than a placebo. You have four groups of participants, all of whom have mildly painful arthritis (rated 6 on a scale of 1 to 10). Each group receives a daily pill as follows: group 1 (control), placebo; group 2, 15 mg; group 3, 25 mg; group 4, 50 mg. At the end of 2 weeks, participants in each group are asked to rate their pain on a scale of 1 to 10. The mean pain rating of the participants was 6.5 for the placebo, 6.0 for 15 mg of the drug, 4.5 for 25 mg of the drug, and 4.5 for 50 mg of the drug. What is your next step?

- **a.** Invest in the drug company.
- **b.** Conclude that the drug relieves arthritis pain.
- **c.** Run a statistical analysis to determine if the differences are significant.
- **d.** Conclude that the drug doesn't work very well (even the placebo group went down on the pain scale, and there was no difference in results between doses of 25 mg and 50 mg of the drug).
- **e.** a and b

12 Looking at Infographic 1.4 (Sample Size Matters), you see that both graphs show a positive impact of caffeine on memory. However, the data in the graph on the right carry more weight. Why is that? If you read a study that reported only the data in the left graph, would you find the relationship to be compelling? Why or why not?

13 From what you have read in this chapter, would you say a 21-year-old Caucasian female can count on caffeinated coffee to reduce her risk of Parkinson's disease?

- **a.** yes, because the results of a peer-reviewed study showed that drinking caffeinated beverages reduced the risk of Parkinson's disease
- **b.** no, because participants in that peer-reviewed study were Japanese-American males; it cannot be inferred that the same results would hold for Caucasian females
- **c.** no; she would have to restrict her consumption of coffee to decaffeinated coffee to reduce her risk of Parkinson's disease
- **d.** yes; coffee is known to reverse the symptoms of Parkinson's disease
- **e.** There are no data on the relationship between drinking caffeinated beverages and Parkinson's disease because it would be unethical to conduct such an epidemiological study.

apply YOUR KNOWLEDGE

INTERPRETING DATA

14 Most statistical tests report a *p* value that determines whether or not the results are statistically significant (i.e., not produced by chance). Usually the cutoff for *p* values is 0.05: if the *p* value is less than 0.05, the results are considered to be statistically significant. The graph below shows data from a 2012 study published in the *New England Journal of Medicine*. The study examined the impact of the drug Tofacitinib on ulcerative colitis. From the data shown, what dose(s) of Tofacitinib is/are significantly better than the placebo in treating ulcerative colitis?

- **a.** 0.5 mg
- **b.** 3 mg
- **c.** 10 mg
- d. 15 mg
- e. both 10 mg and 15 mg
- f. All doses are more effective than the placebo.

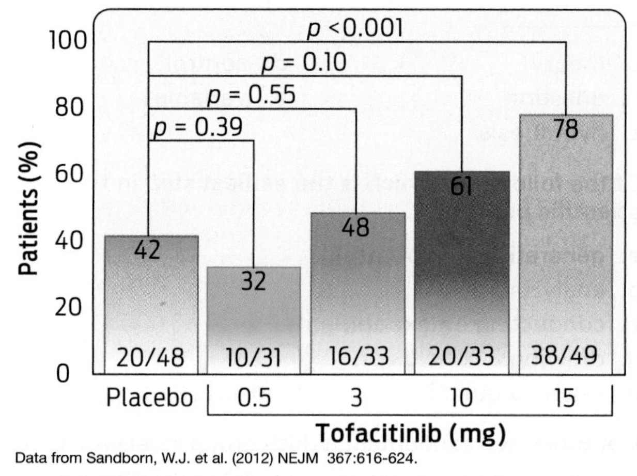

Data from Sandborn, W.J. et al. (2012) NEJM 367:616-624.

DRIVING QUESTION 3 How can you evaluate the evidence in media reports of scientific studies?

By answering the questions below and studying Infographics 1.2, 1.3, 1.4, 1.6, and 1.7, you should be able to generate an answer for the broader Driving Question above.

KNOW IT

15 You hear a news report about a new asthma treatment. What would you want to know before you asked your doctor if this treatment was right for you?

- **a.** Was the drug tested in a randomized clinical trial?
- **b.** How many participants were in the trial?
- **c.** Was there a significant difference between the effect of the new drug and the treatment used in the control group?
- **d.** Did any of the researchers have financial ties to the manufacturer of the new asthma drug?
- **e.** all of the above

16 You are listening to a news report that claims a new study has found convincing evidence that a particular weight-loss product is much more effective than diet and exercise. What can you infer about the "convincing evidence" in this case?

- **a.** It agrees with the hypothesis.
- **b.** Statistical tests showed significantly more weight loss in the participants who used the weight-loss product than those who relied on diet and exercise.
- **c.** All the participants lost at least 10 pounds.
- **d.** Only the participants who used the weight-loss product lost weight.
- **e.** The participants who used the weight-loss product lost an average of 3 pounds, while the participants that used diet and exercise lost an average of 2 pounds.

USE IT

17 How can two different studies investigating the same thing (e.g., the relationship, if any, between caffeinated coffee and memory) come to different conclusions?

- **a.** They may have had different sample sizes.
- **b.** They may have used different types of participants (e.g., participants of different ages or professions).
- **c.** They may have used different amounts of caffeine.
- **d.** They may have evaluated memory differently (e.g., long-term vs. short-term memory).
- **e.** all of the above

18 A scientist who reads an article in a scientific or medical journal can be confident that the report has been peer reviewed.

- **a.** What is a "peer-reviewed" report? Is an article in a daily newspaper a peer-reviewed article?
- **b.** What is the role of a peer reviewer of a scientific article?
- **c.** Why do scientists place so much value on the peer-review process?

19 The mother of a friend is a self-described "coffee addict." She recently received a diagnosis of Parkinson's disease. Does her experience negate the results of the *JAMA* study described in this chapter? Why or why not?

20 You may have seen advertisements on television that show beautiful people with clear skin who claim that a specific skin care product is "scientifically proven" to reduce acne. The product reportedly gave these people glowing, clear skin.

- **a.** Is their testimony alone strong enough evidence for you to act on? Why or why not?
- **b.** What kind of scientific evidence would persuade you to spend money on this product? Explain your answer.

DRIVING QUESTION 4 How does the scientific method apply in clinical trials designed to investigate important issues in human health?

By answering the questions below and studying Infographics 1.2 and 1.3, you should be able to generate an answer for the broader Driving Question above.

21 Following the prompts below, design a clinical trial to test the impact of a particular intervention on a specific aspect of human health. You will need to use everything you have learned in this chapter to do this.

- **a.** From scientific articles or press releases from health organizations you have read, or from your own experiences, what observation(s) can you start with?
- **b.** Do some reading and research to generate a testable hypothesis.
- **c.** Design the trial. Consider sample size, whether or not you will use a placebo, and possible independent and dependent variables.

apply YOUR KNOWLEDGE

BRING IT HOME

22 There are many misconceptions about breast cancer and its causes. In the late 1990s, there were rumors that antiperspirants cause breast cancer. There are still retail sources that offer alternative underarm hygiene products that claim to reduce the risk of breast cancer. One viral e-mail claimed that by blocking perspiration, antiperspirants prevent the body from purging toxins, instead forcing the body to store the toxins in lymph nodes in the underarm area near breast tissue. The e-mail stated that men were less likely to develop breast cancer from antiperspirants because their underarm hair trapped most of the product away from direct contact with skin. And as men are less likely to shave their underarms, they are less likely to have shaving nicks through which antiperspirants can enter the body.

- **a.** Read the abstracts of the two articles for which URLs are provided below.

  Darbre, 2005: http://is.gd/pPLxwZ

  Harvey and Everett, 2004: http://is.gd/ycqDD8

  From the abstracts, and from any other investigation you do, name the components of underarm deodorants and antiperspirants that have been identified as possible culprits in causing breast cancer.
- **b.** Briefly comment on the strengths and weaknesses of each study (consider sample size, control groups, and overall study design).
- **c.** From what you read in the abstracts and from other research you do (cite any additional reliable sources that you consulted), do you think that use of antiperspirants or deodorants or both is a consistent risk factor for breast cancer? Has your opinion about underarm hygiene changed? Explain how and why your opinion has either changed or remained consistent, referring to the abstracts that you have reviewed.

2 Chemistry of Life

Steven Hobbs/Stocktrek Images/Getty Images

Mission to MARS

Prospecting for life on the red planet

DRIVING QUESTIONS

1. How do we define life, and does this definition apply when we search for life on other planets?
2. How is matter organized into the molecules of living organisms?
3. What is the basic structural unit of life, and what are its properties?
4. Why is water so important for life and living organisms?

SEVEN MINUTES OF TERROR. That's how NASA scientists described the anticipated final moments of the 2012 Mars landing. In that harrowing interval, the speeding spacecraft would need to slow from about 13,200 mph to less than 2 mph as it dropped like a stone through the thin Martian atmosphere.

"Those seven minutes are the most challenging part of this entire mission," said Pete Theisinger, project manager at NASA. "For the landing to succeed, hundreds of events will need to go right, many with split-second timing and all controlled autonomously by the spacecraft."

The most frightening part was the last few moments of the descent, when–if all went well–the spacecraft would release its cargo–a 1-ton, SUV-size rover named *Curiosity*–down a floating "sky crane," essentially three nylon cables suspended from a rocket-powered backpack. The maneuver had never been attempted before, and even NASA's own scientists had their doubts about it, dubbing it "rover on a rope."

But it worked. On August 6, 2012, at 1:32 A.M. Eastern Standard Time, NASA's *Mobile Science Laboratory,* aka *Curiosity,* landed successfully on the surface of the red planet. "Touchdown confirmed," announced NASA engineer Allen Chen. "We're safe on Mars!"

NASA/JPL-Caltech

***Curiosity* is lowered to Mars down the floating sky crane.**

News of the successful landing sent NASA's Jet Propulsion Laboratory into loud cheers as people hugged and high-fived one another. Moments later, *Curiosity* began sending back grainy black and white images of its landing site to a planet full of eager witnesses.

This isn't the first time NASA has sent a rover to Mars. Other rovers have explored the red planet, including *Spirit* and *Opportunity* in 2004, but none has done so with as much flair as *Curiosity*. With a $2.5 billion price tag and an entire laboratory built into its sleek frame, *Curiosity* is unquestionably the most technologically advanced rover to date. It's also the most socially connected. Shortly after touching down, *Curiosity* began tweeting news of its progress to its more than 1 million followers, even checking into Mars on Foursquare: "One check-in closer to being Mayor of Mars!" it chirped.

The purpose of NASA's daring trip to Mars is to find out whether the planet could have once supported life–and might support it still. NASA calls the mission the "prospecting" stage of its search for life on Mars. What *Curiosity* discovers will allow scientists to begin to answer some fundamental questions, not only about life on Mars, but also about life on Earth. Questions like: How did life begin? Is there more than one type of life? Could life have arrived here on a meteorite from outer space?

These are big questions, but NASA's new rover is nothing if not curious.

The Search for Martian Life

Answering the question of whether there is life on Mars seems as if it should be pretty straightforward: look and see if anything is growing, or running around, or texting. By these measures, clearly, there is no life on Mars. The earliest pictures of Mars obtained by NASA in 1965 revealed a dry, rocky landscape–more reminiscent of our lifeless moon than the lush, blue marble we call home. But what if Mars harbors microscopic life, invisible to the naked eye? Could life be lurking in the Martian soil?

The first NASA spacecraft to investigate this question was *Viking 1* lander, which touched down on July 20, 1976. Equipped with mechanical arms that could grab and test Martian soil, *Viking* was designed to look for signs of microscopic life. NASA scientists hypothesized that if life were present in the soil, they should be able to measure its activity. For example, was anything in the Martian soil emitting carbon dioxide, as many organisms on Earth do? The scientists added nutrients to Martian soil and waited to see what would happen.

Initially, the results seemed promising: something in the Martian soil did indeed seem to be breaking down the added nutrients and producing carbon dioxide gas. Even more intriguing, when the soil was heated to a very high temperature–a temperature that would kill most life–no carbon dioxide was measured. This result seemed like evidence of life.

But scientists now know that such chemical reactions, or reactivity, can occur even in the absence of life–in dry, lifeless deserts on Earth, for example. "It turns out reactivity is not uniquely biological," says Chris McKay, an astrobiologist with NASA's Ames Research Center in California who studies the evolution of the solar system and the origin of life. Still, he says, the *Viking* experiments were instructive, as they focused attention on a fundamental question in biology: what is life?

Biology is the study of life, so naturally it's important for biologists to know what things fall under that heading. Mosses, for example, are living, but not the rocks they grow on. There are many ways one could define life–on the basis of what it looks like, what it's made of, how it behaves, and so on. And, indeed, scientists and philosophers have offered many definitions of life over the years. But, as the *Viking* experiments show, it can be tricky to search for life on the basis of these definitions.

BRIAN VAN DER BRUG/Newscom/United Press International (UPI)/PASADENA/CA/UNITED STATES

Relief and excitement at NASA's Jet Propulsion Laboratory when Curiosity landed safely.

INFOGRAPHIC 2.1 Properties That Define Life

Living organisms share five properties that define them as "alive."

All living organisms...

Africa Studio/Shutterstock

Grow
For unicellular (one-celled) organisms, growth is an increase in cell size before reproduction. For multicellular organisms, growth refers to an increase in an organism's size as the number of cells making up the organism increases.

JUNIORS BILDARCHIV/AGE Fotostock

Reproduce
Reproduction is the process of producing new organisms. Offspring are similar, but not necessarily identical, to their parents in general structure, function, and properties.

globestock/IStockphoto

Maintain Homeostasis
Organisms maintain a stable internal environment, even when the external environment changes.

© Kazuo Ogawa/ age fotostock

Sense and Respond to Stimuli
Organisms respond to stimuli in many ways. For example, they may move toward a food source or move away from a threatening predator.

Andres Rodriguez/age fotostock

Obtain and Use Energy
All living organisms require an input of energy to power their activities. Organisms obtain energy from food (which they either produce themselves by photosynthesis or consume from the environment). Chemical reactions convert that energy into usable forms. The sum total of all these reactions is metabolism.

? Consider an avocado tree. How does it demonstrate each of the five properties that define it as living?

Biologists generally agree that—on Earth, at least—living things have five life-defining properties. (1) Living things grow—they increase in size or cell number. (2) They reproduce, by producing offspring that are similar if not quite identical to themselves. (3) They maintain a relatively stable internal environment in the face of changing external circumstances—producing heat when they're cold, for example—a phenomenon called **homeostasis.** (4) They sense and respond to their environment, as when a plant grows toward sunlight. (5) And to carry out these various activities, they obtain and use **energy,** the power to do work **(INFOGRAPHIC 2.1)**.

If scientists found an alien being with all five of these properties, they could make a good case for having found life. But what if the alien specimen had only some of these properties? Would it be alive? No one can say for sure, and we won't know until we have a candidate life form to consider.

Even on Earth, our definitions of life don't always hold. For example, a mule (the offspring of a female horse and a male donkey) is clearly alive, but it is sterile and cannot reproduce. Likewise, fire grows, reproduces, and uses energy, but most people would not say fire is alive. By what criteria, then, should NASA evaluate evidence of life, past or present, on Mars? One option, the one that *Curiosity* is using, is chemistry.

Curious about Chemistry

As one of the scientists behind *Curiosity*, Chris McKay's main job is figuring out the best way to look for life on Mars. He was part of the team responsible for packing *Curiosity*'s scientific toolkit. The high-tech rover is equipped with 10 scientific instruments, including a laser that can identify the chemical composition of rocks at a distance, an on-board mass spectrometer, and a gas chromatograph. With these sophisticated instruments, *Curiosity* will be able to analyze the chemical components of the rocks, soil, and atmosphere it encounters on Mars.

NASA set *Curiosity* down in a large canyon called Gale Crater. The rover's ultimate destination, to which it will travel over weeks and months, is the slope of a steep peak named Mount Sharp, located at the crater's center. Geologically, the region is similar to the Grand Canyon, with exposed strata of past eons layered one on top of the other, like a many-layered cake. NASA chose this as the best spot to try to reconstruct the past history of Mars, including whether the ancient atmosphere and soil of the planet could have supported life.

Curiosity will not actually test for living organisms. Instead, it will look for other evidence of life: the chemical building blocks necessary to assemble it.

All life we know of–from amoeba to zebra–uses the same basic chemical recipe: a stew of carbon-based ingredients floating in a broth of water. Carbon (C) is one of 118 known elements in the universe. **Elements** are substances that cannot be broken down by chemical reactions into smaller substances. They are themselves considered the fundamental components of anything that takes up space or has mass–the **matter** in the universe. Elements make up both living and nonliving things. The rocky surface of Mars, for example, appears red because of an abundance of the element iron (Fe) that has long since rusted.

> "For the landing to succeed, hundreds of events will need to go right, many with split-second timing and all controlled autonomously by the spacecraft."
>
> — Pete Theisinger

The smallest unit of an element that still retains the property of that element is an **atom.** Atoms are made up of three types of subatomic particles: positively charged **protons,** negatively charged **electrons,** and neutral **neutrons.** The relatively heavy and large protons and neutrons are packed into the atom's dense core, or **nucleus,** while the light and tiny electrons orbit it in electron shells defined by their energy. The number of subatomic particles in an atom determines its physical and chemical properties–its heaviness, for example, and how it reacts with atoms of other elements. The number of protons in an atom is called the **atomic number**, which is what, by convention, determines the atom's identity. A carbon atom, for example, is carbon because it has six protons. It also has six electrons and six neutrons **(INFOGRAPHIC 2.2)**.

Carbon is the fourth most common element in the universe, and the second most common element in the human body. In fact, just six elements make up the bulk of you: oxygen (65%), carbon (18%), hydrogen (10%), nitrogen (3%), calcium (1.5%), and phosphorus (1%).

When astrobiologists (and science fiction writers) talk about "carbon-based life forms," they are referring to the fact that carbon forms the backbone of nearly every molecule making up living things. Just as humans have a backbone made of interconnected vertebrae, the molecules making up living things have backbones of interconnected carbon atoms. The carbon backbones can be linear, like our spine, or circular–in which case the first carbon in the chain binds to the last carbon in the chain.

"Carbon is very cool, very flexible, very useful," says McKay. "We don't see any other element that has the sort of flexibility and utility that carbon has." That's the main reason scientists are so interested in it.

SAM I Am

In his lab at NASA's Goddard Space Flight Center in Greenbelt, Maryland, chemist Paul Mahaffy monitors a device the size of

HOMEOSTASIS
The maintenance of a relatively constant internal environment.

ENERGY
The ability to do work. Living organisms obtain energy either directly from sunlight (through photosynthesis) or from food they consume.

ELEMENT
A pure substance that cannot be chemically broken down; each element is made up of and defined by a single type of atom.

MATTER
Anything that takes up space and has mass.

ATOM
The smallest unit of an element that still retains the property of the element.

PROTON
A positively charged subatomic particle in the nucleus of an atom.

ELECTRON
A negatively charged subatomic particle with negligible mass.

NEUTRON
An electrically uncharged subatomic particle in the nucleus of an atom.

NUCLEUS
The dense core of an atom.

ATOMIC NUMBER
The number of protons in an atom, which determines the atom's identity.

INFOGRAPHIC 2.2 All Matter Is Made of Elements

Elements are the fundamental building blocks of matter. The periodic table of elements represents all known elements in the universe. Each element is placed in order on the table by its atomic number, the number of protons found in the nucleus of its corresponding atom.

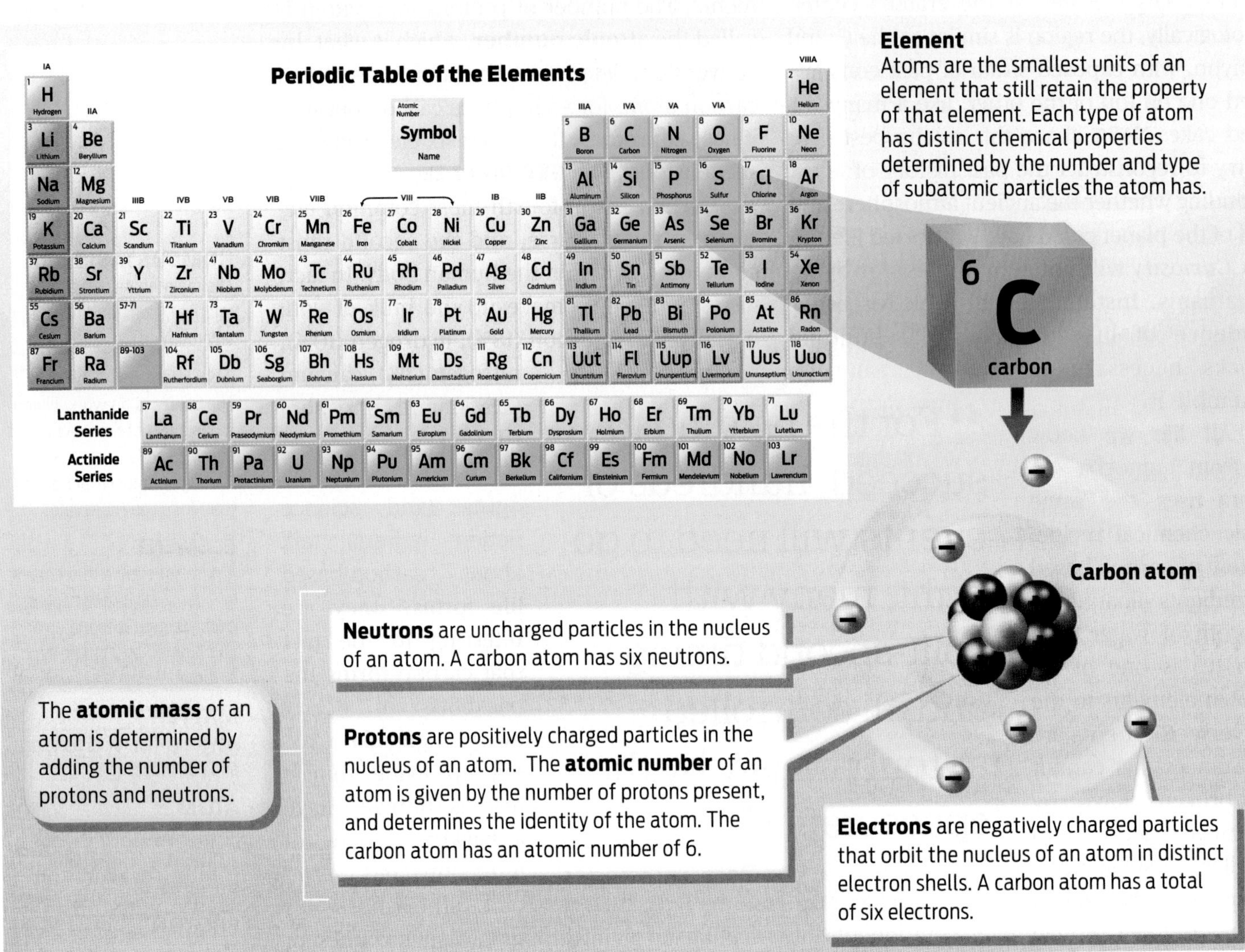

Periodic Table of the Elements
Atomic Number
Symbol
Name
Lanthanide Series
Actinide Series
Element
Atoms are the smallest units of an element that still retain the property of that element. Each type of atom has distinct chemical properties determined by the number and type of subatomic particles the atom has.
6
C
carbon
Carbon atom
Neutrons are uncharged particles in the nucleus of an atom. A carbon atom has six neutrons.
The atomic mass of an atom is determined by adding the number of protons and neutrons.
Protons are positively charged particles in the nucleus of an atom. The atomic number of an atom is given by the number of protons present, and determines the identity of the atom. The carbon atom has an atomic number of 6.
Electrons are negatively charged particles that orbit the nucleus of an atom in distinct electron shells. A carbon atom has a total of six electrons.

? What is the atomic number of magnesium? How many protons does it have?

a microwave oven, hairy with cords and wires. When it boots up, the device declares: "Sam I am, I am Sam!" in playful homage to Dr. Seuss. SAM stands for "Sample Analysis on Mars." It is an exact replica of the tool that *Curiosity* will use to check for life's chemical building blocks. Mahaffy is SAM's principal investigator.

SAM is mounted on *Curiosity* like a backpack. If the powerful cameras on board *Curiosity* are its eyes, then SAM is its extremely sensitive nose. On Mars, says Mahaffy, SAM is doing a lot of sniffing.

The sniffing takes place in a series of interconnected chambers. Once a sample of rock or dirt is scooped up by *Curiosity*, it is loaded into SAM, where it is baked to a high temperature. The gases given off from this high-tech oven are then analyzed to determine their precise chemical makeup. The whole operation takes about a day, but interpreting the results can take much longer.

Curiosity safely landed in Gale Crater

It has been navigating toward and up Mount Sharp, collecting and analyzing soil and rock samples, and capturing important images along the way.

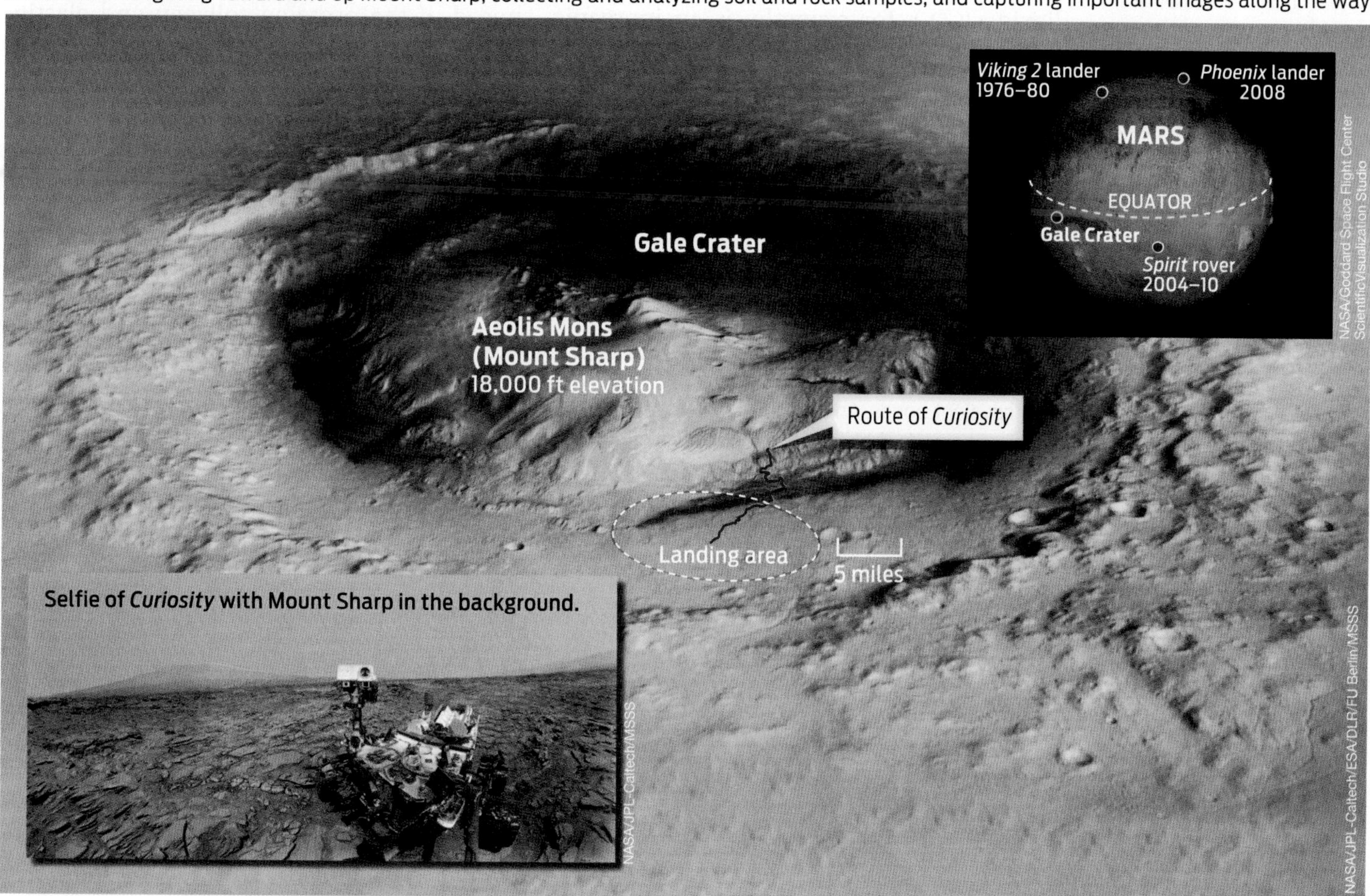

Selfie of *Curiosity* with Mount Sharp in the background.

According to Mahaffy, SAM is particularly curious about the carbon on Mars—does it appear linked to other carbon atoms in chains or rings, as it does in living things on Earth? One way that atoms link together is by sharing electrons. When two atoms share a pair of electrons, one from each atom, a strong attraction called a **covalent bond** forms between the atoms. Atoms linked by covalent bonds form **molecules.** Different atoms can form different numbers of covalent bonds, depending on the number of electrons in their outer shell. Carbon atoms have four outer-shell electrons with which to form covalent bonds, giving the element enormous versatility in forming molecules.

Living things on Earth are made up of **organic** molecules, which have a backbone of interconnected carbon atoms and at least one carbon attached to a hydrogen atom. Most organic molecules require living things to make them, which is why they are often telltale signs of life. An example of a simple organic molecule is glucose, a type of sugar. Its molecular formula (a shorthand for representing the composition of a molecule) is $C_6H_{12}O_6$. This means that each molecule of glucose has 6 carbon atoms, 12 hydrogen atoms, and 6 oxygen atoms. Glucose is a ring-shaped molecule, with the carbon atoms forming the backbone of the ring.

Nonliving things can also contain carbon, but this carbon is **inorganic:** inorganic

COVALENT BOND
A strong interaction resulting from the sharing of a pair of electrons between two atoms.

MOLECULE
Atoms linked by covalent bonds.

ORGANIC
Describes a molecule with a carbon-based backbone and at least one C–H bond.

INORGANIC
Describes a molecule that lacks a carbon-based backbone and C–H bonds.

CARBOHYDRATE
An organic molecule made up of one or more sugars.

PROTEIN
An organic molecule made up of linked amino acid subunits.

LIPIDS
Organic molecules that generally repel water.

molecules do not have a carbon-carbon backbone and a carbon-hydrogen bond. Carbon dioxide (CO_2), for example, is an inorganic molecule, one found in the atmospheres of both Mars and Earth **(INFOGRAPHIC 2.3)**.

Living things on Earth are made up of just four types of organic molecules: **carbohydrates, proteins, lipids,** and **nucleic acids.** Every molecule in the human body can be classified as one of these organic molecules. Your skin, for example, is composed of the proteins collagen and elastin, and the protein hemoglobin carries oxygen in your blood. The padding in your soft spots is composed of lipids called triglycerides, also known as fats. And in your liver and muscle cells, a carbohydrate called glycogen helps store energy. All of these organic

INFOGRAPHIC 2.3 Carbon Is a Key Component of Life's Molecules

Molecules are atoms linked by covalent bonds. An atom of the element carbon can form four covalent bonds with other atoms. This leads to a great deal of diversity in the shapes and compositions of biologically important molecules. Carbon is found in both organic and inorganic molecules.

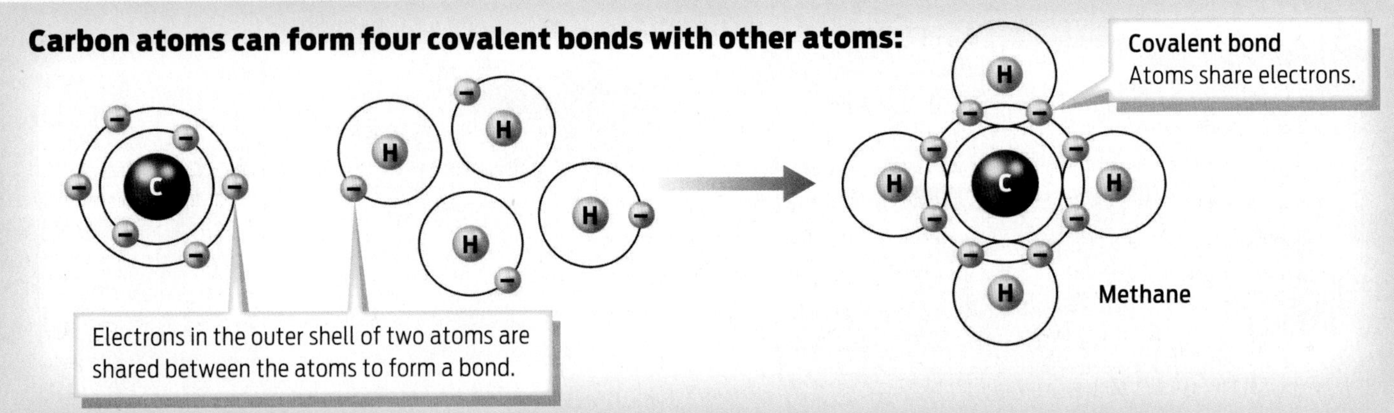
Carbon atoms can form four covalent bonds with other atoms:
Covalent bond
Atoms share electrons.
Methane
Electrons in the outer shell of two atoms are shared between the atoms to form a bond.

Organic molecule contain C–C and C–H bonds:

Organic molecules form when carbon atoms are covalently bound to other carbon and hydrogen atoms to form a diversity of chained, branching, and ring structures found in the molecules of life.

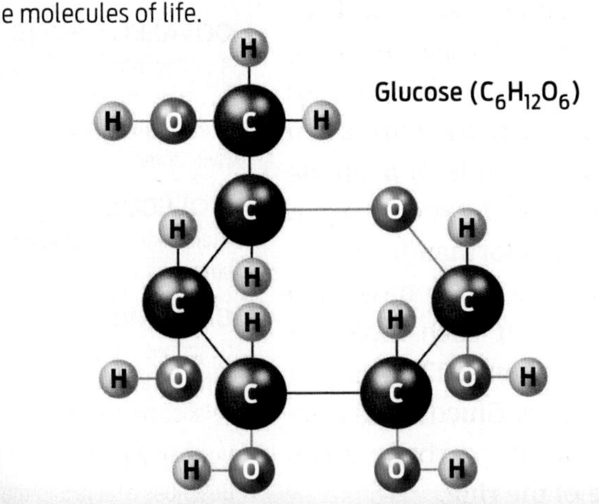
Glucose ($C_6H_{12}O_6$)

Inorganic molecules do not have C–C or C–H bonds:

Carbon atoms may also be found in inorganic molecules bound to non-carbon atoms. Inorganic molecules cannot form the large chained polymers found in organic molecules.

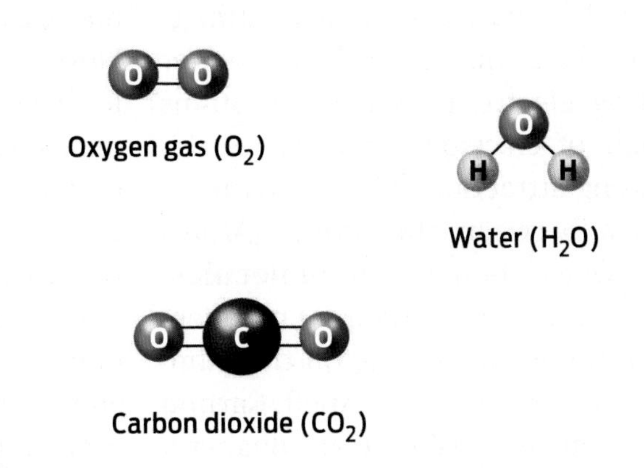
Oxygen gas (O_2)
Water (H_2O)
Carbon dioxide (CO_2)

? How many covalent bonds does every carbon in glucose have? How many bonds does the carbon of CO_2 have?

molecules have a backbone of interlinked carbon atoms.

Organic molecules can be quite large and are therefore considered **macromolecules.** Macromolecules share a similar organization in that they are composed of subunits called **monomers** linked together in a chain. When two or more monomers join, they form a **polymer.** Carbohydrates, for example, are polymers made up of linked monomers called **monosaccharides;** similarly, proteins are made up of subunits called **amino acids** that are bonded together; and nucleic acids are polymers composed of **nucleotides** that form long chains. Lipids are a more diverse group of molecules, and are not classified as polymers (see **UP CLOSE: MOLECULES OF LIFE**).

Since *Curiosity* landed in August 2012, it has spent much of its time exploring a region of Gale Crater right around the landing site, an area called Yellowknife Bay. By drilling into the sedimentary rocks there and analyzing the samples in SAM, the rover has found evidence of hydrogen, oxygen, carbon, nitrogen, sulfur, and phosphorus—all elements that are critical for life as we know it.

SAM has also detected organic molecules inside a rock it drilled into in May 2013. In the gases given off when this rock sample was baked, SAM sniffed a cornucopia of carbon-containing compounds—the first time that such organics were definitively detected on Mars. At this point, it's not possible to say whether the molecules were produced by living or nonliving means, but the discovery does show that Mars has many of the chemical building blocks necessary to make life.

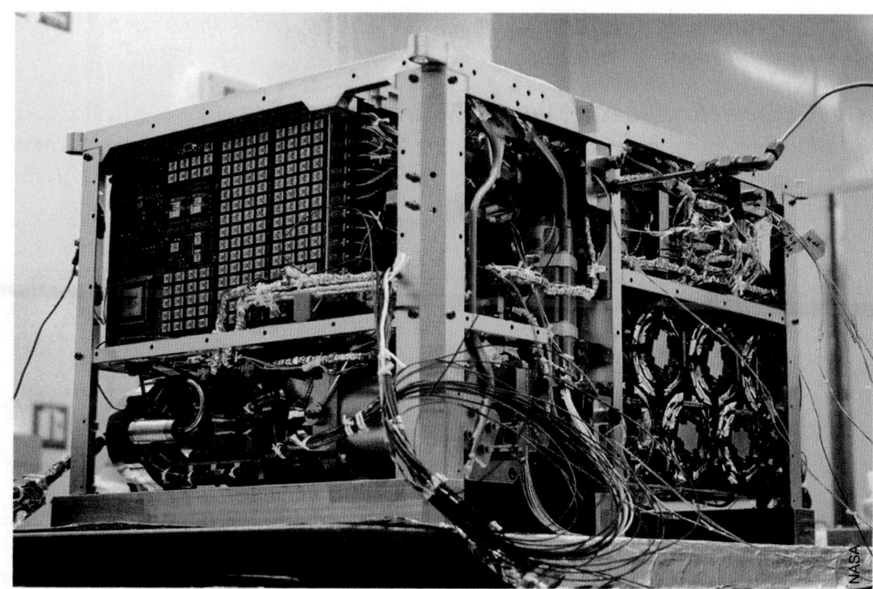

Meet SAM, *Curiosity's* organic molecule detector.

"My Favorite Earthlings"

When not helping to send rovers into space, McKay spends his time researching more terrestrial habitats—including ones nearly as foreboding as Mars, such as Chile's Atacama Desert and the Dry Valleys of Antarctica. Deserts are intriguing to him, he says, because they are good analogs, or approximations, of Mars. They allow scientists to ask questions about the Martian environment without actually traveling there.

McKay's office is full of rock souvenirs from deserts all over world. Rocks are home to some of his favorite creatures—cyanobacteria—which he calls "my favorite earthlings."

Cyanobacteria form a layer of green beneath desert rocks, where they live on light and moisture that is trapped there as in a greenhouse. Cyanobacteria represent one of the few organisms that can survive in these extreme environments. That they can survive is perhaps not surprising, since cyanobacteria are some of the most ancient organisms on Earth, having first evolved some 2.5 billion years ago, when the world was a much more hostile place. Not only that, but through photosynthesis (see Chapter 5) they filled the Earth's atmosphere with oxygen, making it habitable for the rest of life.

Not bad for an organism made up of only a single **cell.** Cells are the basic structural unit of life on Earth. They are like microscopic bricks making up the walls and ceilings of living things. Some organisms, like cyanobacteria, are only a single cell, while large organisms like humans contain trillions of them.

NUCLEIC ACIDS
Organic molecules made up of linked nucleotide subunits; DNA and RNA are examples of nucleic acids.

MACROMOLECULES
Very large organic molecules that make up living organisms; they include carbohydrates, proteins, and nucleic acids.

MONOMER
One chemical subunit of a polymer.

POLYMER
A molecule made up of individual subunits, called monomers, linked together in a chain.

MONOSACCHARIDE
The building block, or monomer, of a carbohydrate.

AMINO ACID
The building block, or monomer, of a protein.

NUCLEOTIDE
The building block, or monomer, of a nucleic acid.

CELL
The basic structural unit of living organisms.

2 UP CLOSE Molecules of Life

a. Carbohydrates Are Made of Monosaccharides

Carbohydrates are made up of repeating subunits of simple sugars known as monosaccharides. Carbohydrates act as energy-storing molecules in many organisms. Other carbohydrates provide structural support for cells.

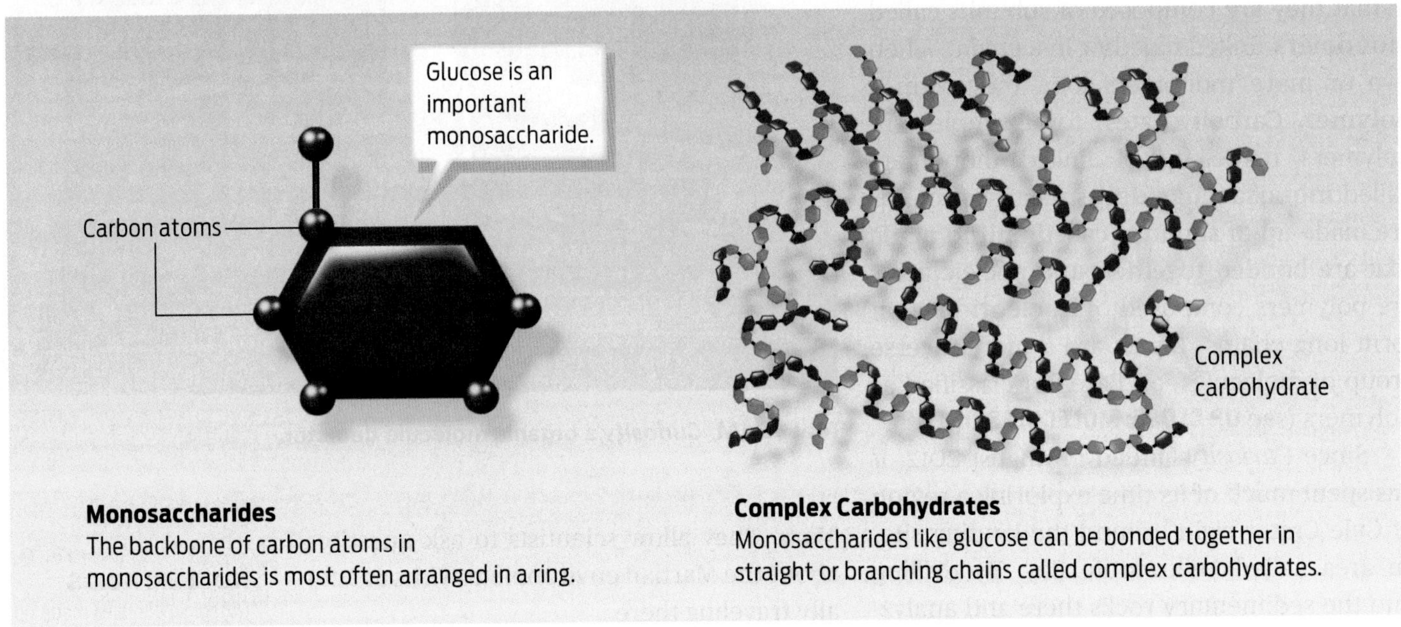

Monosaccharides
The backbone of carbon atoms in monosaccharides is most often arranged in a ring.

Complex Carbohydrates
Monosaccharides like glucose can be bonded together in straight or branching chains called complex carbohydrates.

b. Proteins Are Made of Amino Acids

Proteins are folded polymers of small repeating units called amino acids. Proteins carry out many functions in cells. They help speed up the rate of chemical reactions. They also move things through and around cells and even help entire cells move.

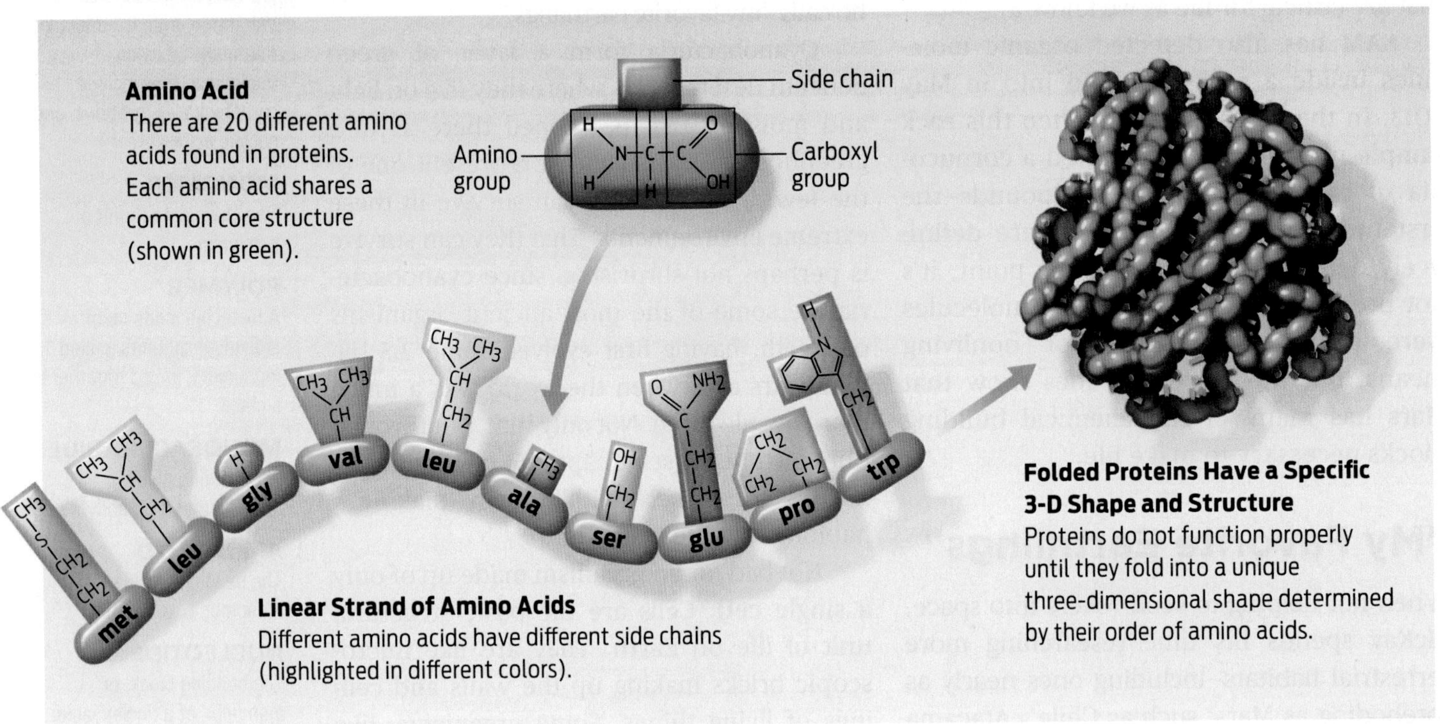

Amino Acid
There are 20 different amino acids found in proteins. Each amino acid shares a common core structure (shown in green).

Linear Strand of Amino Acids
Different amino acids have different side chains (highlighted in different colors).

Folded Proteins Have a Specific 3-D Shape and Structure
Proteins do not function properly until they fold into a unique three-dimensional shape determined by their order of amino acids.

c. Lipids Are Hydrophobic Molecules

There are different types of lipid, each with a distinct structure and function. Lipids are not polymers of repeating subunits like the other molecules of life, but they are all organic molecules. They are also all hydrophobic molecules, meaning they don't mix with water.

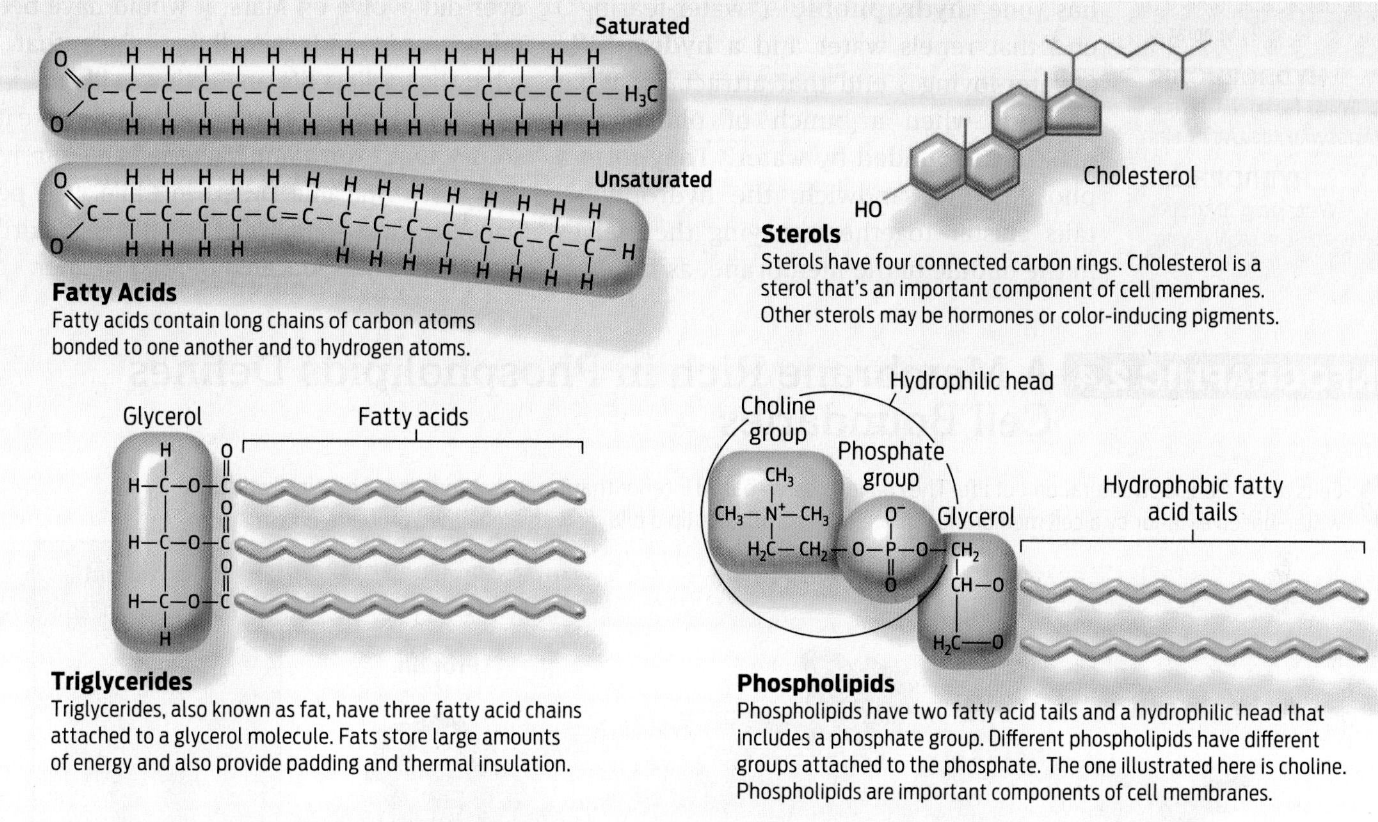

Fatty Acids
Fatty acids contain long chains of carbon atoms bonded to one another and to hydrogen atoms.

Sterols
Sterols have four connected carbon rings. Cholesterol is a sterol that's an important component of cell membranes. Other sterols may be hormones or color-inducing pigments.

Triglycerides
Triglycerides, also known as fat, have three fatty acid chains attached to a glycerol molecule. Fats store large amounts of energy and also provide padding and thermal insulation.

Phospholipids
Phospholipids have two fatty acid tails and a hydrophilic head that includes a phosphate group. Different phospholipids have different groups attached to the phosphate. The one illustrated here is choline. Phospholipids are important components of cell membranes.

d. Nucleic Acids Are Made of Nucleotides

Nucleic acids are polymers of repeating subunits known as nucleotides. There are two types of nucleic acid, DNA and RNA, each of which is made up of slightly different types of nucleotide. DNA and RNA are critical for the storage, transmission, and execution of genetic instructions.

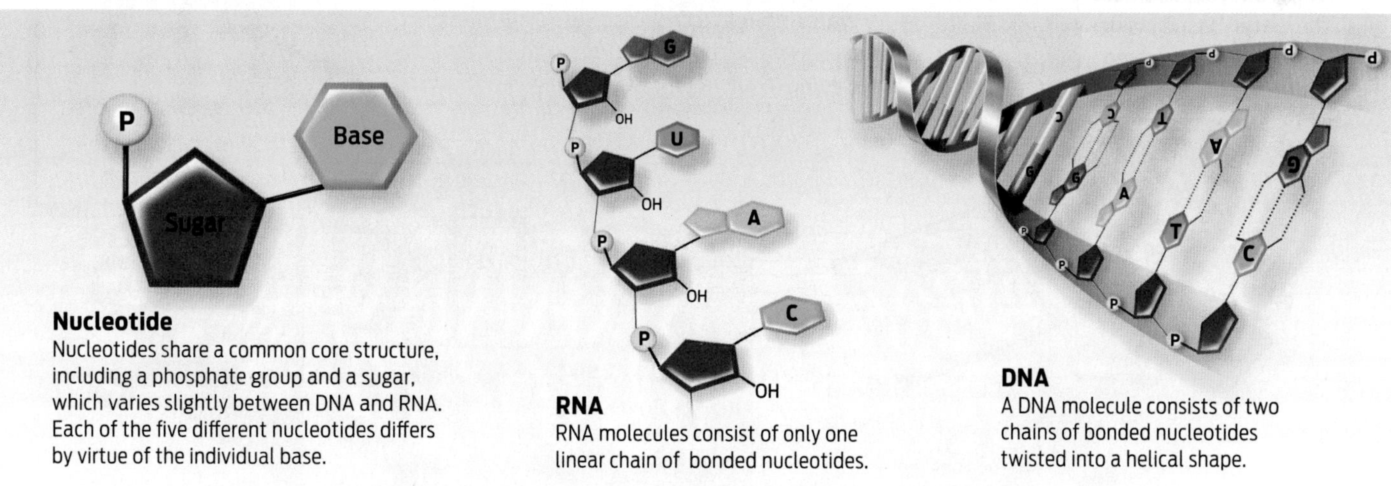

Nucleotide
Nucleotides share a common core structure, including a phosphate group and a sugar, which varies slightly between DNA and RNA. Each of the five different nucleotides differs by virtue of the individual base.

RNA
RNA molecules consist of only one linear chain of bonded nucleotides.

DNA
A DNA molecule consists of two chains of bonded nucleotides twisted into a helical shape.

CELL MEMBRANE A phospholipid bilayer with embedded proteins that forms the boundary of all cells.

PHOSPHOLIPID A type of lipid that forms the cell membrane.

HYDROPHOBIC "Water-fearing"; hydrophobic molecules will not dissolve in water.

HYDROPHILIC "Water-loving"; hydrophilic molecules dissolve easily in water.

All cells have the same basic structure: they are water-filled sacs bounded by a **cell membrane.** The cell membrane is a two-ply layer of **phospholipids** in which proteins are embedded. Each phospholipid has one **hydrophobic** ("water-fearing") end that repels water and a **hydrophilic** ("water-loving") end that attracts it. What happens when a bunch of phospholipids are surrounded by water? They form a phospholipid sandwich: the hydrophobic tails cluster together, burying themselves in the middle of the membrane, as far away from water as possible, while the hydrophilic heads face out, toward the water **(INFOGRAPHIC 2.4)**.

No one knows whether or not cells exist on Mars, but researchers suspect that if life ever did evolve on Mars, it would have been microscopic and unicellular, since that is what the earliest life on Earth was like.

McKay knows of only one desert where it's so dry that even cyanobacteria can't survive, and that's Atacama Desert in Chile. It is perhaps the most Mars-like of any place on Earth. In fact, as dry as Atacama is, Mars is drier.

INFOGRAPHIC 2.4 A Membrane Rich in Phospholipids Defines Cell Boundaries

Cells are the basic structural unit of life. They have a water-based interior that is separated from a chemically distinct water-based exterior by a cell membrane. The cell membrane is a lipid bilayer with embedded proteins.

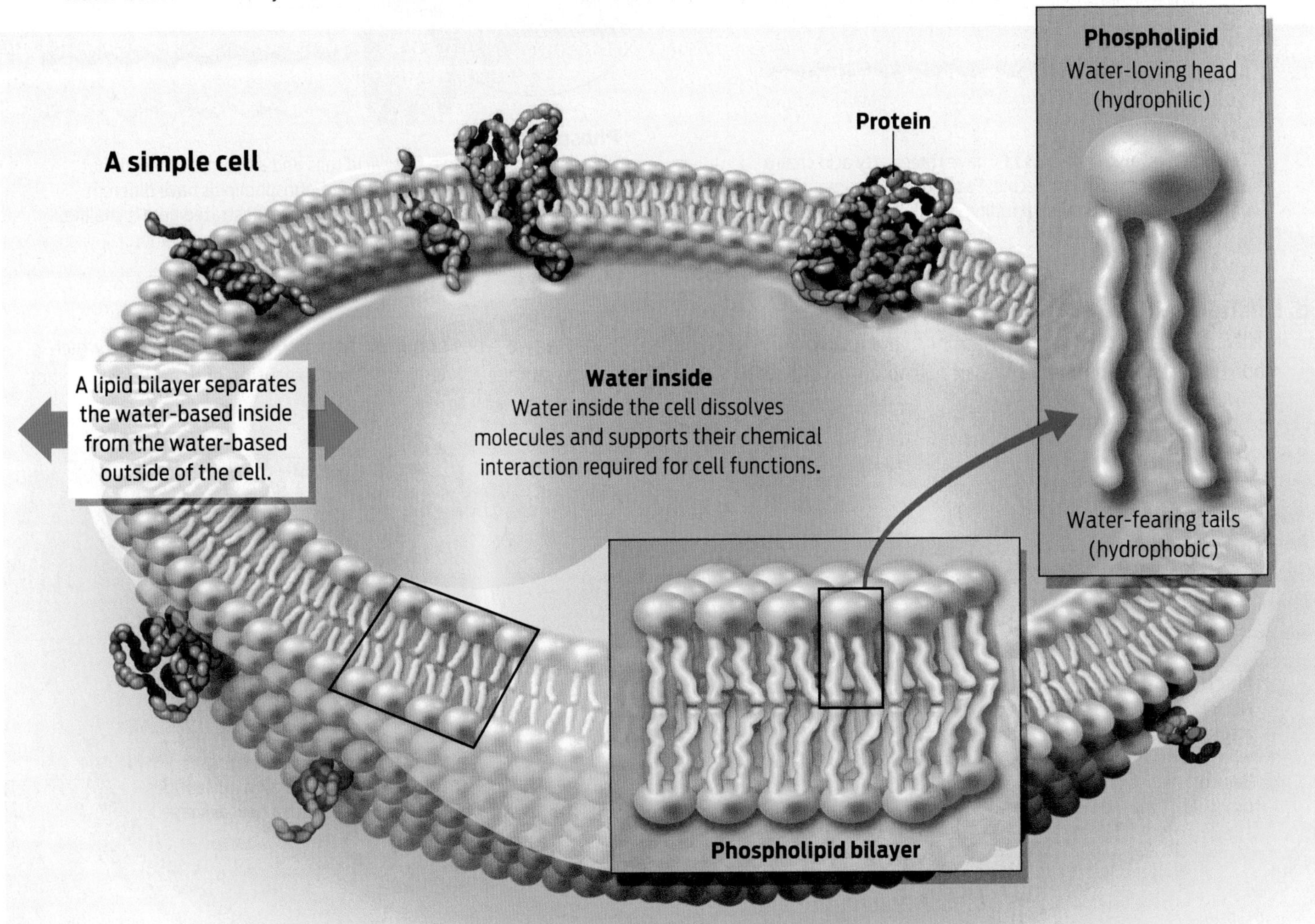
A simple cell
Protein
Phospholipid
Water-loving head (hydrophilic)
Water-fearing tails (hydrophobic)
A lipid bilayer separates the water-based inside from the water-based outside of the cell.
Water inside
Water inside the cell dissolves molecules and supports their chemical interaction required for cell functions.
Phospholipid bilayer

? What two molecules make up cell membranes?

Mars wasn't always that way. Scientists believe the planet was a lot moister in its earlier days–3 or 4 billion years ago, when it had a thicker atmosphere, and could hold this moisture in.

Follow the Water

In their search for extraterrestrial life, astrobiologists use a rule of thumb: "Follow the water." Water is viewed as a proxy for life because it is so crucial to life on Earth. Water makes up more than 75% of a cell's weight. All of life's chemical reactions take place in water, and many living things can survive only a few days without it.

A water molecule–H_2O–is shaped like Mickey Mouse's head (the oxygen atom is the face, and the two hydrogens are the ears). Many of water's life-conducive properties are a function of this simple shape. Water is a **polar molecule,** meaning that the electrons in the bonds between oxygen and the hydrogens are shared unequally; the oxygen atom has a stronger pull on electrons than the hydrogen atoms do, giving water a partial negative charge on one side and a partial positive charge on the other. When water molecules are near one another, the partial negative and partial positive charges of different water molecules attract one another, forming weak electrical interactions known as **hydrogen bonds** (**INFOGRAPHIC 2.5**).

While individual hydrogen bonds are relatively weak, the large number of hydrogen bonds means that they add up to a significant attractive force. All those hydrogen bonds make water "sticky," acting as a kind of glue holding water molecules together. The stickiness of water allows water molecules to cling to one another (demonstrating **cohesion**) or to a surface (demonstrating **adhesion**). You can see evidence of water's stickiness wherever you look: a drop of water clinging to a

POLAR MOLECULE
A molecule in which electrons are not shared equally between atoms, causing a partial negative charge at one end and a partial positive charge at the other. Water is a polar molecule.

HYDROGEN BOND
A weak electrical attraction between a partially positive hydrogen atom and an atom with a partial negative charge.

COHESION
The attraction between molecules (or other particles).

ADHESION
The attraction between molecules (or other particles) and a surface.

INFOGRAPHIC 2.5 Water Is Polar and Forms Hydrogen Bonds

Water is a polar molecule because electrons are not shared equally between the oxygen and the hydrogen atoms in each of its covalent bonds. Oxygen has a stronger pull on electrons, resulting in a slight negative charge on the oxygen and a slight positive charge on each hydrogen. When many water molecules are near one another, the partially positive hydrogen atoms of some molecules are attracted to the partially negative oxygen atoms of nearby water molecules. These weak electrical attractions are hydrogen bonds.

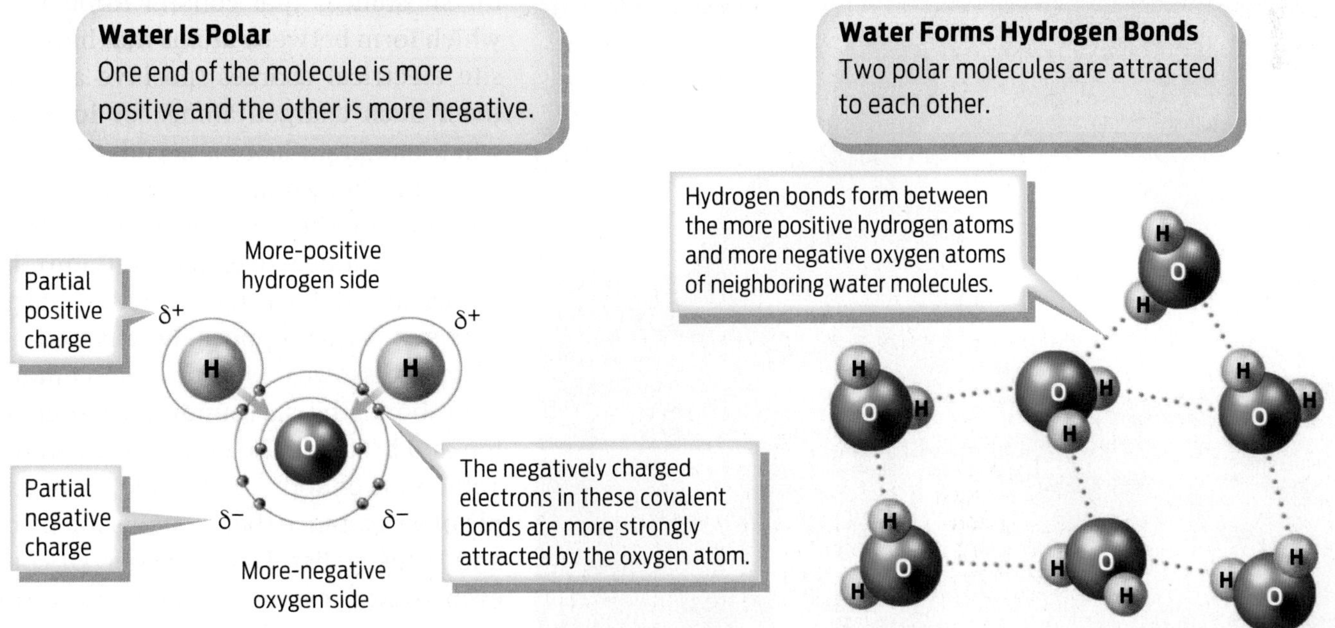

Water Is Polar
One end of the molecule is more positive and the other is more negative.
Water Forms Hydrogen Bonds
Two polar molecules are attracted to each other.
Partial positive charge
More-positive hydrogen side
δ+
δ+
H
H
O
Partial negative charge
δ−
δ−
More-negative oxygen side
The negatively charged electrons in these covalent bonds are more strongly attracted by the oxygen atom.
Hydrogen bonds form between the more positive hydrogen atoms and more negative oxygen atoms of neighboring water molecules.

? Look at the water molecule on the bottom right of the right panel. Draw another water molecule hydrogen bonded to this one.

SOLVENT
A substance in which other substances can dissolve. Water is a good solvent.

SOLUTE
A dissolved substance.

SOLUTION
The mixture of solute and solvent.

IONIC BOND
A strong electrical attraction between oppositely charged ions formed by the transfer of one or more electrons from one atom to another.

ION
An electrically charged atom, the charge resulting from the loss or gain of electrons.

leaf despite the downward pull of gravity, or an insect able to walk on the surface of a pond.

Compared to other molecules its size, water has a large liquid range–freezing at 0°C (32°F) and boiling at 100°C (212°F). That's because water molecules can absorb a lot of energy before they get hot and vaporize, or turn into a gas–again because of the many hydrogen bonds present. Water's liquid range can be extended even further: add salt and you can lower the freezing point to –46°C (–50°F); increase the pressure and you can bump up the boiling point to over 343°C (650°F). It's because there is so much salt in seawater that most oceans don't freeze in winter.

> "We don't see any other element that has the sort of flexibility and utility that carbon has."
>
> — Chris McKay

Unlike most substances on Earth, water has the unusual property of being less dense as a solid than as a liquid: ice floats. When water freezes, the water molecules form an ordered array, with individual water molecules spaced at an equal and fixed "arm's length" from one another. This increased separation between water molecules decreases the density of ice, allowing ice to float on water. And because ice floats, fish can live beneath frozen lakes in winter and not turn into ice cubes (**INFOGRAPHIC 2.6**).

Perhaps the most important life-conducive property of water is that it is a good **solvent:** its structure makes it capable of dissolving many substances. In fact, no other solvent can dissolve as many substances as water. Water transports all of life's dissolved molecules, or **solutes,** from place to place–whether through a cell, a body, or an ecosystem. Life, in essence, is a water-based **solution.** Many biological molecules, like proteins and DNA, have the specific shapes they do only because of the surrounding water with which they interact.

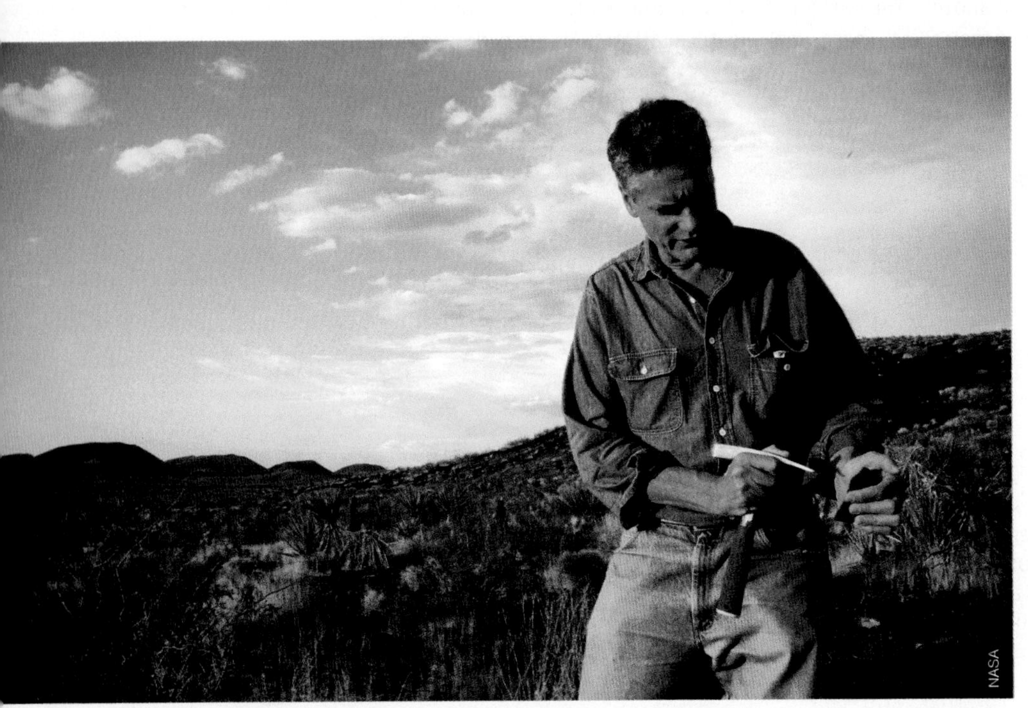

Astrobiologist Chris McKay studies desert analogs of Mars.

Water is an excellent solvent for other polar molecules with partial charges and for substances that contain **ionic bonds,** which form between atoms that have opposite electrical charges (positive and negative). Such charged atoms, or **ions,** form when one atom loses a negatively charged electron (becoming a positively charged ion) and another atom gains that electron (becoming a negatively charged ion). Ionic bonds are the strong attractions formed between these oppositely charged ions. Water dissolves substances containing ionic bonds by surrounding each charged ion and breaking the bonds between them. Table salt, or sodium chloride, is an example of a substance that contains ionic bonds and is easily dissolved by water. Of course, even water cannot dissolve everything; you have probably seen firsthand what happens when you try to mix fats, like

INFOGRAPHIC 2.6 Hydrogen Bonds Give Water Its Unique Properties

Because so many hydrogen bonds can form in a body of water (or ice), even these weak bonds can collectively give water its unique properties.

age fotostock/Superstock

Water Is Sticky
Hydrogen bonding allows water molecules to stick to one another (cohesion) and to other surfaces (adhesion), making them wet.

iStockphoto/Thinkstock

Water Can Absorb a Lot of Energy
It takes a lot of heat energy to disrupt the many hydrogen bonds between water molecules. That is why water is a liquid at a wide range of temperatures.

igorad1/Shutterstock

Ice Is Less Dense than Liquid Water
When water freezes, the bonds between the water molecules become rigid and expand, increasing the overall volume of the water. This makes ice less dense than liquid water. That is why ice floats.

? These properties are all the result of hydrogen bonding between water molecules. Draw two water molecules that are hydrogen bonded to each other.

butter or oils in salad dressing, with water (**INFOGRAPHIC 2.7**).

When astrobiologists speak about the importance of water for life, they make an important qualification: *liquid* water. Frozen water is found throughout the universe; there are abundant quantities on Mars and on other planets and moons in our solar system, for example. But only on Earth does water exist primarily in its liquid form at room temperature. "Liquid water is the key requirement in the search for life," says McKay. "The other worlds of the solar system have enough light, enough carbon, and enough of the other key elements for life. Water in the liquid form is rare."

Though liquid water is not present on the surface of Mars today, scientists suspect that liquid water–lots of it–may have once covered the planet. Clues to this ancient water can be seen all over the surface of the planet, which in many places is carved out like sections of the Grand Canyon. There are also salt deposits like those you can see when

INFOGRAPHIC 2.7 Water Is a Good Solvent

Because water molecules have partial charges, they can interact with charged ions and other hydrophilic molecules, allowing water to coat and then dissolve these solutes.

Ions Form Ionic Bonds

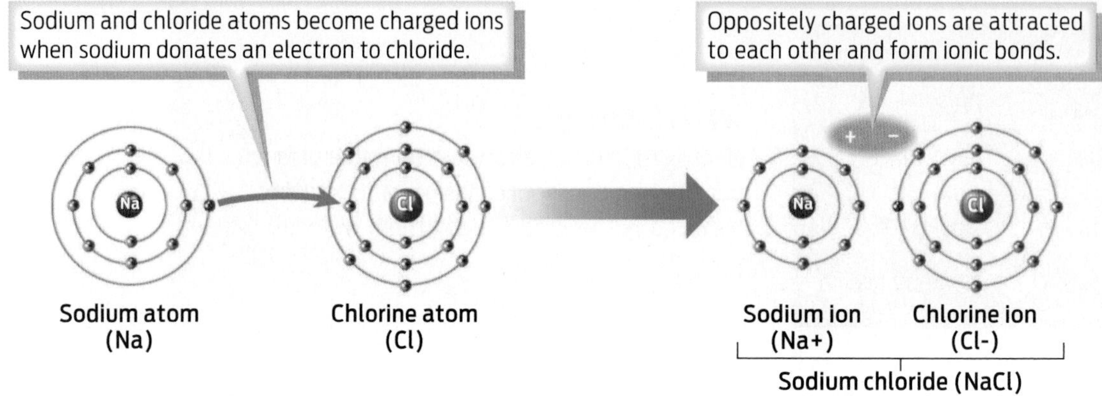

Charged Substances Dissolve in Water

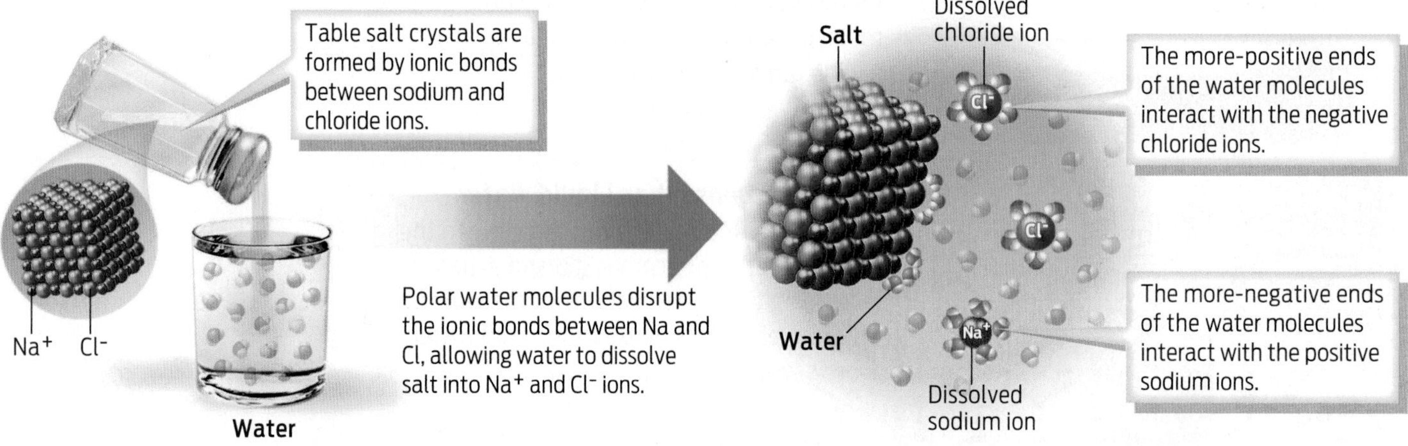

Table salt crystals are formed by ionic bonds between sodium and chloride ions.
Na+ Cl-
Water
Polar water molecules disrupt the ionic bonds between Na and Cl, allowing water to dissolve salt into Na+ and Cl- ions.
Salt
Dissolved chloride ion
The more-positive ends of the water molecules interact with the negative chloride ions.
Water
Dissolved sodium ion
The more-negative ends of the water molecules interact with the positive sodium ions.

? Why is a sodium ion positively charged? Why is a chloride ion negatively charged?

seawater evaporates, another telltale sign of liquid water's past on the surface of Mars. Gale Crater, where *Curiosity* landed, also has clay deposits and rock formations that suggest it was once the site of a large lake.

"From the *Curiosity* rover, we now know that Mars once was a planet very much like Earth, with warm, salty seas, with freshwater lakes, probably snow-capped peaks, and clouds and a water cycle," NASA scientist John Grunsfeld said at a September 2015 press conference.

Where all this water went, no one knows. For flowing water to have once existed on Mars, the planet's atmosphere must have been much thicker and warmer than it is today. Something happened to the atmosphere over time such that now the planet's atmosphere is thin and blisteringly cold. Exactly how this happened remains a mystery.

Some evidence suggests that liquid water may still lurk beneath the surface of the planet, and may even bubble to the surface periodically. Recent photographs taken by

NASA's *Mars Reconnaissance Orbiter*, which has been circling the planet since 2006, show what look to be rivulets of water flowing down rocky slopes. The flows occur when temperatures warm up and stop when it gets colder (**INFOGRAPHIC 2.8**).

If there is water on or within Mars, would it have the properties of Earth water, and could it therefore support life? Depending on what's dissolved in it, water can have a wide range of characteristics–from caustic drain cleaner to calming chamomile tea. The

INFOGRAPHIC 2.8 Evidence of Liquid Water on Mars

Visual and chemical data collected by *Curiosity* and the *Mars Reconnaissance Orbiter* provide evidence for liquid water flowing on Mars. Under a thicker atmosphere, water once flowed in rivers and gathered in lakes on the planet's surface. Now liquid surface water is limited to briny streaks that darken the soil in the warm season.

Water once flowed in ancient rivers and lakes

The Slope of Gale Crater
Smaller weathered sediments and clay deposits at the bottom of the crater are evidence of long-standing water in the ancient lakebed.
NASA/JPL-Caltech/MSSS

Liquid water once flowed on the surface of Mars, carving channels like this one that slope into the Gale Crater and transporting eroded rock into the basin.

Water flows today in salty streaks down the sides of steep craters

Minerals detected in these locations suggest that salty water flows when the temperature in these areas climbs above −10°F (−23°C).

? Why is the discovery of liquid water on Mars important?

pH
A measure of the concentration of H^+ in a solution.

ACID
A substance that increases the hydrogen ion concentration of solutions, making them more acidic.

BASE
A substance that reduces the hydrogen ion concentration of solutions, making them more basic.

different chemical properties of water-based solutions reflect their **pH,** the concentration of hydrogen ions (H^+) in a solution, denoted as a range from 0 to 14. In every water-based solution, water molecules (H_2O) split briefly into separate hydrogen (H^+) and hydroxide (OH^-) ions. In pure water, the number of separated H^+ ions is by definition exactly equal to the number of separated OH^- ions, and the pH is therefore 7, or neutral–the dead center of the 0 to 14 scale. Acidic solutions, or **acids,** have a higher concentration of hydrogen ions (H^+) and a pH closer to 0. When acids are added to water, they increase the concentration of hydrogen ions and make the solution more acidic. Basic solutions, or **bases,** on the other hand, have a lower concentration of H^+ ions and a pH closer to 14. Bases remove H^+ ions from a solution, thereby increasing the proportion of OH^- ions.

> "Our experience with life is limited to familiar Earth life and we know that all known life on Earth descends from a last universal common ancestor."
>
> — Carol Cleland

Strong acids and bases are highly reactive with other substances, which makes them destructive to the molecules in a cell. Also, many biochemical reactions take place only at a certain pH. Living things are thus extremely sensitive to changes in pH, and most function best when their pH stays within a specific range. The pH of human blood is about 7.4. If that pH were to fall even slightly, to 7, our biochemistry would malfunction and we would die. Previous missions to Mars have calculated the pH of Martian soil as 7.7–mild enough to grow asparagus, as one NASA chemist put it **(INFOGRAPHIC 2.9).**

Evidence for an ancient freshwater lake on Mars

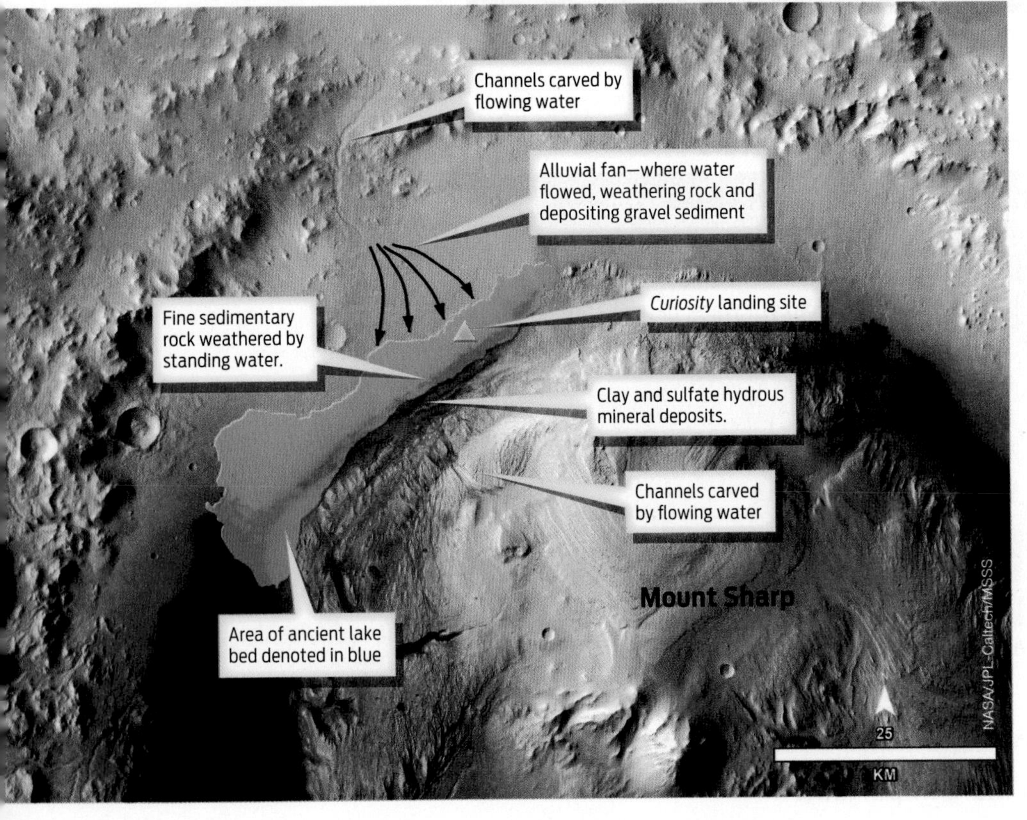

"Weird Life"

So far, NASA's search for life on Mars has stuck very closely to our understanding of life on Earth, where living things seem to share certain chemical and structural properties, like carbon-based molecules and cells. But how can we be sure that life on other planets will resemble life on Earth?

"The question 'What is life?' has taken on increasing scientific importance in recent years," says Carol Cleland, a philosopher at the University of Colorado and a member of NASA's Astrobiology Institute. It's become a subject for debate, she says, as molecular biologists attempt to create life from scratch in the lab, and as astrobiologists grapple with how alien life might differ from life on Earth.

"The problem," says Cleland, "is that our experience with life is limited to familiar Earth life and we know that all known life on Earth descends from a last universal common ancestor." Life's last universal common ancestor (or LUCA) lived approximately 3.5 billion years ago, and all present-day organisms on Earth are descended from this ancestor. But if life on other planets exists and descended from different ancestors, unique to each planet, who's to say that all life will look and behave just as it does on this planet?

More and more, scientists are accepting that they need to keep an open mind about what life on other planets might look like. In 2007, the National Academy of Sciences issued a "weird life" report suggesting that NASA should not be so narrowly focused on water and organic molecules in its search for life on other planets. True, water may be crucial to life on Earth, but that doesn't mean that other solvents–ammonia or methane, for example–could not support life elsewhere. The report also urged the space agency to avoid being "fixated on carbon," even though carbon forms the scaffold of life on Earth. Other elements, for example silicon, could hypothetically provide a functional scaffold for life on other planets.

Even on Earth, there appear to be a few exceptions, or boundary cases, that bend the "rules" of life as they are currently conceived. **Viruses,** for example, reproduce and pass their genetic information on to new viruses, but they are not made of cells. Instead, they consist of a protein shell that encloses genetic information. Viruses reproduce by infecting a host cell and hijacking its cellular machinery to make copies of itself. Other noncellular, self-reproducing entities include **prions,** infectious proteins that are responsible for mad cow disease and related illnesses. Whether or not viruses and prions are alive is hotly debated among scientists.

In 2010, researchers with the U. S. Geological Survey reported a very unusual finding that seemed to confirm that our definition of life was too narrow: they had discovered a bacterium in a lake full of arsenic that apparently could substitute arsenic for phosphorus in its DNA. In other words, this bacterium was an apparent exception to the "universal" chemical formula of life on Earth. And not only

INFOGRAPHIC 2.9 Solutions Have a Characteristic pH

The pH of a solution is a measure of the concentration of hydrogen ions (H^+) in it. Solutions with a low concentration of H^+ ions have a basic pH (greater than pH 7). Solutions with a high concentration of H^+ ions have an acidic pH (a pH of less than 7). Both acids and bases can be damaging because they are highly reactive with other substances. A neutral solution has a pH of 7.

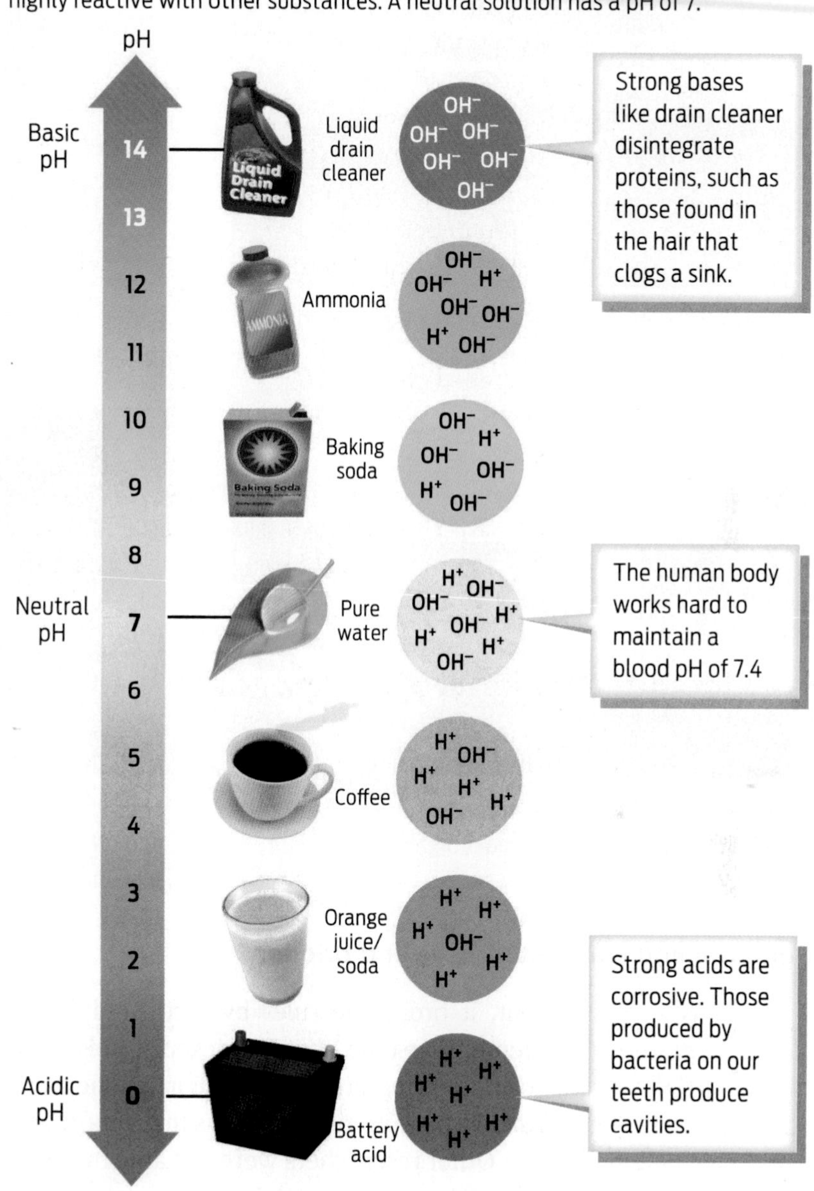

? What is the normal pH of human blood?

VIRUS
An infectious agent made up of a protein shell that encloses genetic information.

PRION
A protein-only infectious agent.

INFOGRAPHIC 2.10 On the Fringe

Many organisms defy our criteria for living organisms, and many infectious agents that are not technically alive have powerful impacts on living organisms.

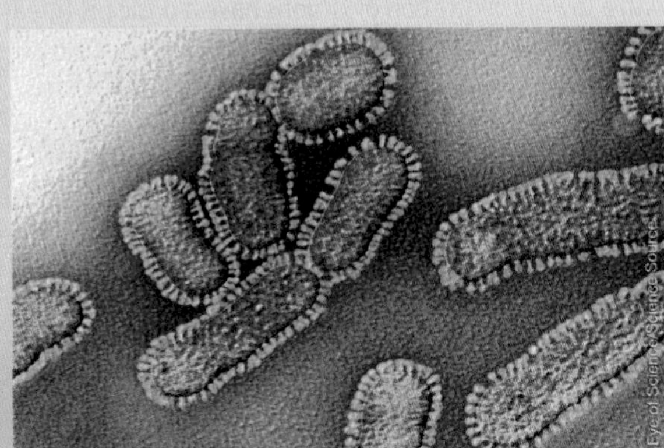

Viruses are not cellular. They infect other cells and use host cell machinery to replicate.

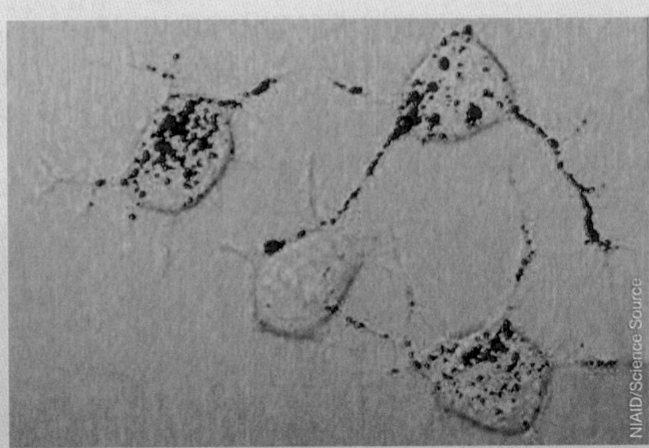

Prions are not cellular. They are infectious proteins that replicate in cells, causing disease.

These bacteria live in a high concentration of arsenic, a chemical that is toxic to most organisms.

These bacteria live in conditions of extreme heat and pressure that would kill most organisms.

? Compare and contrast viruses and prions.

that, it broke the rules by incorporating an element–arsenic–that is wickedly poisonous to most living creatures on Earth. The news made a big splash when it was first reported.

Other researchers were not able to reproduce the controversial findings, however, and scientists now think that this "exception" does not hold up: the bacteria do need phosphorus, and the universal formula of life stands unbroken.

Nevertheless, the fact that this bug can live happily in a lake full of poison does challenge our notions of what life looks like and where it can survive. Microscopic organisms have been found living just about anywhere on Earth, from radioactive waste and deep-sea vents to frozen Antarctic lakes submerged under miles of ice. Such extreme-loving organisms reveal that life is nothing if not adaptive **(INFOGRAPHIC 2.10)**.

Could similarly adaptive organisms have once inhabited Mars? Might they still? NASA's Chris McKay, for one, is cautiously optimistic: "I spend my time and energy in the search for evidence of life on Mars," he says. "Obviously, this is because I think there must have been life there and we have a good chance of finding evidence of it." ■

CHAPTER 2 SUMMARY

- On Earth, living organisms share a number of fundamental properties: they grow and reproduce, maintain homeostasis, sense and respond to their environment, and rely on energy to carry out their functions.
- All matter is composed of elements, of which there are 118 known in the universe. Each element has a unique atomic structure, with a particular number of protons, neutrons, and electrons.
- When atoms share pairs of electrons they form covalent bonds, building molecules.
- On Earth, living organisms are made up of organic molecules, which contain a backbone of the element carbon and carbon–hydrogen bonds.
- Four types of carbon-based organic molecules make up living things: carbohydrates, proteins, nucleic acids, and lipids.
- Living organisms on Earth are made of cells, which contain water and are bounded by a cell membrane; cells are the smallest unit of life.
- Water is a polar molecule, with the hydrogens carrying a partial positive charge and the oxygen carrying a partial negative charge.
- Because of its partial charges, a water molecule can form hydrogen bonds (weak attractions between these opposite partial charges) with other water molecules and interact with other charged molecules.
- Water has many properties that make it a crucial component of life on Earth: it is "sticky," it regulates heat well, it floats when frozen, and it is a good solvent.
- When atoms lose or gain electrons, they become ions. Oppositely charged ions can form ionic bonds—strong electrical attractions. Water is a good solvent of substances with ionic bonds.
- Substances that easily dissolve in water, like salt, are considered hydrophilic; substances that do not dissolve in water, like lipids, are hydrophobic.
- The concentration of H^+ ions in a solution determines its pH. Most chemical reactions in cells take place at a nearly neutral pH.
- If life is found on other planets, it may or may not use the chemical framework used by life on Earth.

MORE TO EXPLORE

- **NASA, *Mars Science Laboratory* (aka *Curiosity*) mission page: http://www.nasa.gov/mission_pages/msl**
- **Video: *Curiosity*'s Seven Minutes of Terror: http://www.jpl.nasa.gov/video/details.php?id=1090**
- **Follow *Curiosity* on Twitter @MarsCuriosity.**
- **Special issue the journal *Science* devoted to *Curiosity*: http://www.sciencemag.org/site/extra/curiosity/**
- **McKay, C. P. (2004) What is life—and how do we search for it in other worlds? *PLoS Biol* 2(9):1260–1263.**
- ***The Limits of Organic Life in Planetary Systems* (aka the "Weird Life" Report). (2007) Washington, DC: National Academies Press.**

CHAPTER 2 Test Your Knowledge

DRIVING QUESTION 1 How do we define life, and does this definition apply when we search for life on other planets?

By answering the questions below and studying Infographics 2.1, 2.3 and 2.10, you should be able to generate an answer for the broader Driving Question above.

KNOW IT

1 Which of the following is *not* a generally recognized characteristic of most (if not all) living organisms?

a. the ability to reproduce
b. the ability to maintain homeostasis
c. the ability to obtain energy directly from sunlight
d. the ability to sense and respond to the environment
e. the ability to grow

2 What is homeostasis? Why is it important to living organisms?

3 What does it mean to say that a macromolecule is a polymer? Give an example.

4 A collection of amino acids could be used to build a

- **a.** protein.
- **b.** complex carbohydrate.
- **c.** triglyceride.
- **d.** nucleic acid.
- **e.** cell.

USE IT

5 How would you assess whether or not a possibly living organism from another planet were truly alive?

6 Which of the characteristics of living organisms (if any) allow you to distinguish between living and formerly living (that is, dead) organisms? Explain your answer.

7 You are searching for life in a sample of dirt. If you had evidence that carbon dioxide was being consumed and converted to glucose, what could you conclude about the presence of a living organism in your sample? Explain your answer.

DRIVING QUESTION 2 How is matter organized into the molecules of living organisms?

By answering the questions below and studying Infographics 2.2, 2.3, and Up Close: Molecules of Life, you should be able to generate an answer for the broader Driving Question above.

KNOW IT

8 What subatomic particles are located in the nucleus of an atom?

- **a.** protons
- **b.** neutrons
- **c.** electrons
- **d.** protons, neutrons, and electrons
- **e.** protons and neutrons

9 When an atom loses an electron, what happens?

- **a.** It becomes positively charged.
- **b.** It becomes negatively charged.
- **c.** It becomes neutral.
- **d.** Nothing happens.
- **e.** Atoms cannot lose an electron because atoms have a defined number of electrons.

10 Glucose (a monosaccharide) has the molecular formula $C_6H_{12}O_6$. How many carbon atoms are in each glucose molecule?

USE IT

11 Consider the types of lipid.

- **a.** How does a sterol, such as cholesterol, differ from a triglyceride?
- **b.** Structurally, what do triglycerides and phospholipids have in common?

12 If a cell were unable to take up or make sugars, which class of molecule(s) would it be unable to make?

- **a.** carbohydrates
- **b.** proteins
- **c.** lipids
- **d.** nucleic acids
- **e.** all of the above
- **f.** a and d

DRIVING QUESTION 3 What is the basic structural unit of life, and what are its properties?

By answering the questions below and studying Infographics 2.4 and 2.10, you should be able to generate an answer for the broader Driving Question above.

KNOW IT

13 The basic building blocks of life are

- **a.** DNA molecules.
- **b.** cells.
- **c.** proteins.
- **d.** phospholipids.
- **e.** inorganic molecules.

14 The cell membrane is made of

- **a.** water.
- **b.** proteins.
- **c.** phospholipids.
- **d.** nucleotides.
- **e.** b and c

USE IT

15 What are the arguments for and against considering viruses living organisms?

16 Why do phospholipids form a bilayer in water-based solutions?

DRIVING QUESTION 4 Why is water so important for life and living organisms?

By answering the questions below and studying Infographics 2.4, 2.5, 2.6, 2.7, 2.8 and 2.9, you should be able to generate an answer for the broader Driving Question above.

KNOW IT

17 Is olive oil hydrophobic or hydrophilic? What about salt? Explain your answer.

18 The "stickiness" of water results from the ____________ bonding of water molecules.

a hydrogen
b. ionic
c. covalent
d. acidic
e. hydrophobic

19 Coffee or tea with sugar dissolved in it is an example of a water-based solution.

a. What is the solvent in such a beverage?
b. What is the solute in such a beverage?
c. Given that the sugar has dissolved in the beverage, are sugar molecules hydrophobic or hydrophilic?

20 As an acidic compound dissolves in water, the pH of the water

a. increases.
b. remains neutral.
c. decreases.
d. doesn't change.
e. becomes basic.

21 In a water molecule, the bond between the oxygen atom and a hydrogen atom is a(n) __________ bond.

a. covalent
b. hydrogen
c. ionic
d. hydrophobic
e. noncovalent

22 How do ionic bonds compare to hydrogen bonds? What are the similarities and differences?

USE IT

23 Why do olive oil and vinegar (a water-based solution) tend to separate in salad dressing? Will added salt dissolve in the oil or in the vinegar? Explain your answer.

24 Which of the following is/are most likely to dissolve in olive oil?

a. a polar molecule
b. a nonpolar molecule
c. a hydrophilic molecule
d. a and c
e. b and c

INTERPRETING DATA

apply YOUR KNOWLEDGE

25 Look at Infographic 2.9. For the substances drain cleaner, coffee, and soda, answer the following questions: Is the substance an acid or a base? What is the hydrogen ion concentration relative to a solution with a neutral pH?

MINI CASE

apply YOUR KNOWLEDGE

26 One approach to finding out if there is life on Mars is to bring Martian dirt samples to Earth for analysis. What are possible considerations for science and society if a Martian life form is released on Earth? Given that *Curiosity* has landed on Mars, what are the possible consequences if an Earth life form is released on Mars? What steps can mission control take to minimize these risks?

BRING IT HOME

apply YOUR KNOWLEDGE

27 Your tax dollars are being invested in projects such as the *Curiosity* rover project. Investigate the NASA website to learn more about NASA's rationale for the investment in this mission. Now draft a letter to your congressional representative that expresses your opinion about this expenditure of taxpayer dollars. If you agree, state specific reasons why you think this a good investment of your money. If you disagree, state your reasons, and describe at least two other scientific programs that you would prefer to see funded, providing a rationale for why these are more important.

3 Cell Structure and Function

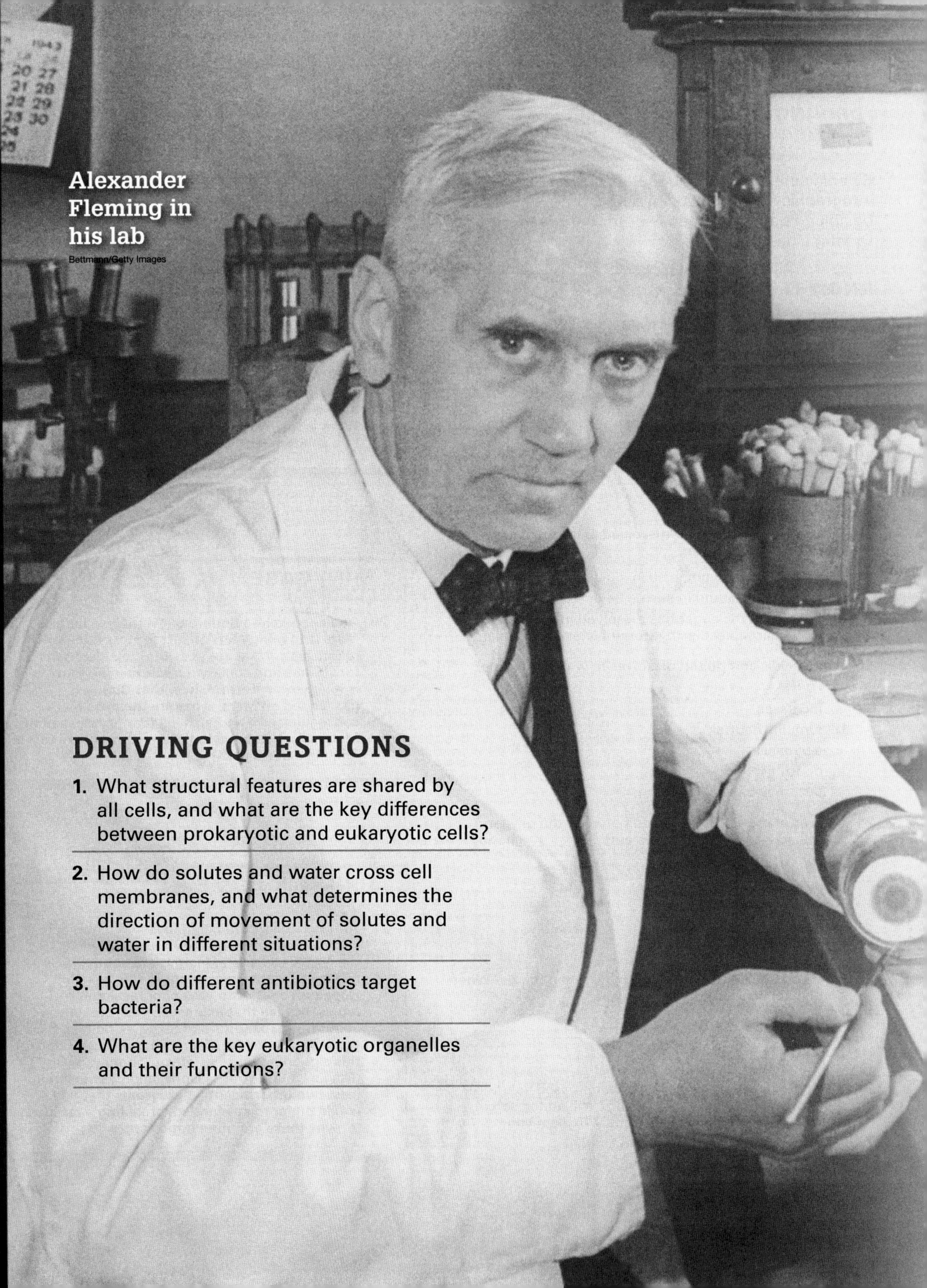

Alexander Fleming in his lab

Bettmann/Getty Images

DRIVING QUESTIONS

1. What structural features are shared by all cells, and what are the key differences between prokaryotic and eukaryotic cells?
2. How do solutes and water cross cell membranes, and what determines the direction of movement of solutes and water in different situations?
3. How do different antibiotics target bacteria?
4. What are the key eukaryotic organelles and their functions?

Wonder DRUG

How a chance discovery in a London laboratory revolutionized medicine

ON A SEPTEMBER MORNING IN 1928, biologist Alexander Fleming returned to his laboratory at St. Mary's Hospital in London after a short summer vacation. As usual, the place was a mess. Flasks were scattered everywhere, and his workbench was strewn with the petri plates on which he was growing bacteria. On this day, as Fleming sorted through the plates, he noticed that one was growing a patch of fluffy white mold. It had been contaminated, likely by a rogue mold spore that had drifted in from a neighboring laboratory.

Fleming was about to toss the plate in the sink when he noticed something unusual: wherever mold was growing, there was a zone around the mold where the bacteria did not seem to grow. Curious, he looked under a microscope and saw that the bacterial cells near the mold had lysed, or burst. Something in the mold was killing the bacteria.

Fleming scooped a bit of mold from the plate, grew it in a broth of nutrients, then tested the liquid in a battery of additional experiments. The results were clear and dramatic: even when diluted 800 times, this "mold juice"–Fleming's term–was a potent killer of many different kinds of bacteria. What's more, no other fungus that he tested–including one obtained from a pair of moldy old shoes–had this remarkable killing power. Fleming published his results in 1929 in the *British Journal of Experimental Pathology*. He named the antibacterial substance "penicillin," after the fungus that produces it, *Penicillium notatum*. This was the birth of the first **antibiotic** (**INFOGRAPHIC 3.1**).

INFOGRAPHIC 3.1 How Penicillin Was Discovered

A fortuitous observation by Fleming led to the discovery of the first antibiotic. He realized that the fungus on his culture plate was somehow inhibiting the reproduction of bacteria.

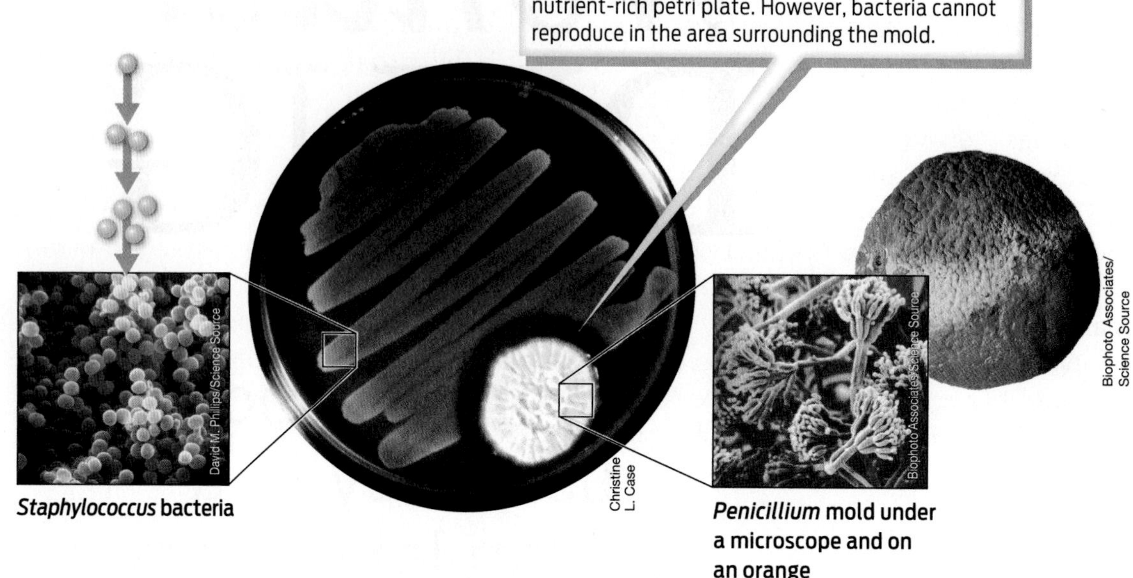

1. A single bacterial cell lands on the surface of a nutrient-rich plate.
2. Nutrients in the plate support the growth and division of the bacterial cells.
3. After many rounds of cell division, enough cells accumulate to be visualized as noticeable bacterial growth on the plate.
Both bacteria and mold can grow and divide on the nutrient-rich petri plate. However, bacteria cannot reproduce in the area surrounding the mold.
David M. Phillips/Science Source
Christine L. Case
Biophoto Associates/Science Source
Biophoto Associates/Science Source

Staphylococcus bacteria

Penicillium mold under a microscope and on an orange

? What would the plate look like if the fungus did not inhibit the growth of bacteria?

Fleming was not the first to notice the bacteria-killing property of *Penicillium*, but he was the first to study it scientifically and publish the results. In fact, Fleming had been looking for bacteria-killing substances for a number of years, ever since he had served as a medical officer in World War I and witnessed soldiers dying from bacteria-caused infections. He had already discovered one such antibacterial agent–the chemical lysozyme–which he detected in his own tears and nasal mucus. Nevertheless, his 1928 discovery was serendipitous: "Penicillin started as a chance observation," Fleming said many years later. "My only merit is that I did not neglect the observation and that I pursued the subject as a bacteriologist."

Most people have seen the *Penicillium* fungus growing on a piece of moldy bread or rotting fruit. It doesn't look very impressive, but the chemical it produces ushered in a whole new age of medicine. For the first time, doctors had a way to treat many illnesses caused by bacteria, such as pneumonia, syphilis, gonorrhea, and meningitis. Before penicillin, there was nothing much doctors could do for a patient with a serious bacterial infection. Now, they had a powerful weapon on their side. As physician and author Lewis Thomas recalled in his 1992 memoir, "We could hardly believe our eyes on seeing that bacteria could be killed off without at the same time killing the patient. It was not just amazement, it was a revolution."

ANTIBIOTIC
A chemical that can slow or stop the growth of bacteria; many antibiotics are produced by living organisms.

CELL THEORY
The concept that all living organisms are made of cells and that cells are formed by the division of existing cells.

Bug Bullet

What makes antibiotics special is not just their ability to kill bacteria. After all, bleach kills bacteria just fine. The important thing about antibiotics is that they exert their destructive effects on bacteria without (typically) harming the human or animal host. Although Fleming didn't know it at the time, penicillin and other antibiotics preferentially kill bacteria because they target what is unique about bacterial cells.

Cells are the basic building blocks of life. According to the **cell theory,** all living things are made of cells, and every new cell comes from the division of a pre-existing one. But not all cells are alike. Cells come in many shapes and sizes and perform a variety of different functions. Moreover, they fall into one of two fundamentally different categories: **prokaryotic** or **eukaryotic.** Prokaryotic cells are relatively small and lack internal membrane-bound compartments, called **organelles.** Eukaryotic cells, by contrast, are much larger and contain many different organelles.

Like all prokaryotic organisms, or prokaryotes, bacteria exist as single cells. They include the harmless *Escherichia coli* that live in our gut and the *Lactobacillus acidophilus* that make our yogurt, as well as the many bacteria that make us sick. Eukaryotes may be single-celled or multicellular. Plants, animals, protists, and fungi are all examples of eukaryotic organisms **(INFOGRAPHIC 3.2)**.

To understand why antibiotics affect bacterial and human cells differently, it helps to first understand what these cells have in common. All cells, both prokaryotic and eukaryotic, are surrounded by a **cell membrane** composed of phospholipids and proteins. This flexible yet sturdy structure forms a boundary between the external environment and the cell's watery interior; it separates inside from outside and literally defines the cell. The solution inside every cell, enclosed

PROKARYOTIC CELLS
Cells that lack internal membrane-bound organelles.

EUKARYOTIC CELLS
Cells that contain membrane-bound organelles, including a central nucleus.

ORGANELLES
The membrane-bound compartments of eukaryotic cells that carry out specific functions.

CELL MEMBRANE
A phospholipid bilayer with embedded proteins that forms the boundary of all cells.

INFOGRAPHIC 3.2 Cell Theory: All Living Things Are Made of Cells

All living organisms are composed of cells. These cells arise from the reproduction of existing cells. Different cells have different structures and functions.

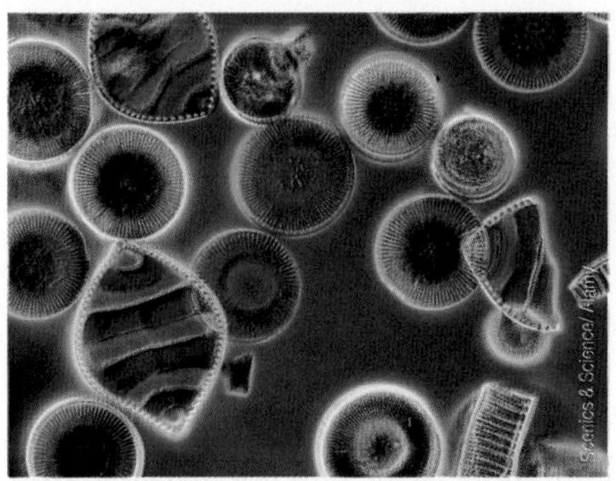

Diatoms (algae): single-celled eukaryotes

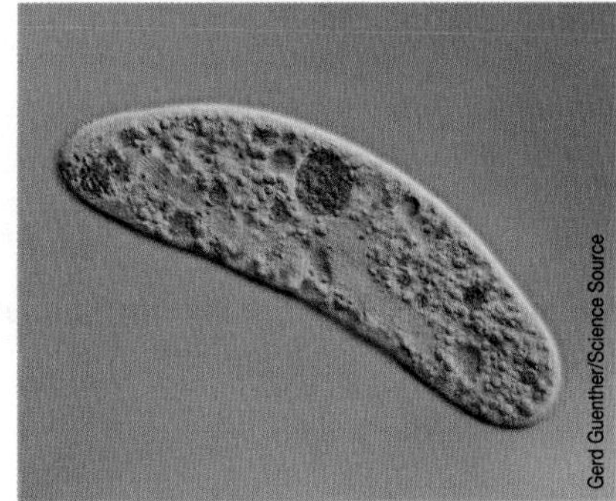

Amoeba (a protist): a single-celled eukaryote

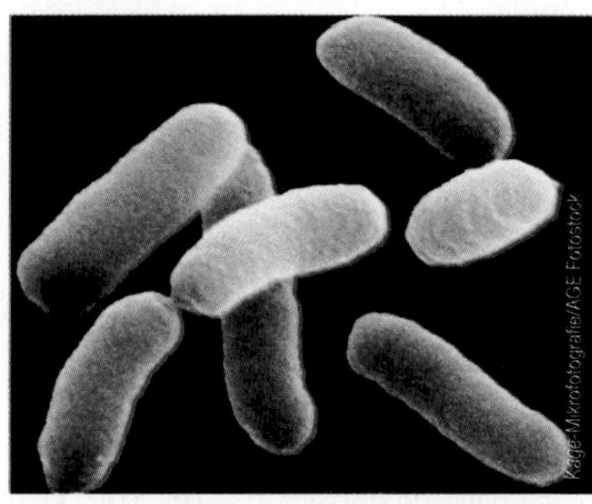

Bacteria: single-celled prokaryotes

Molds (fungi): single and multicellular eukaryotic cells

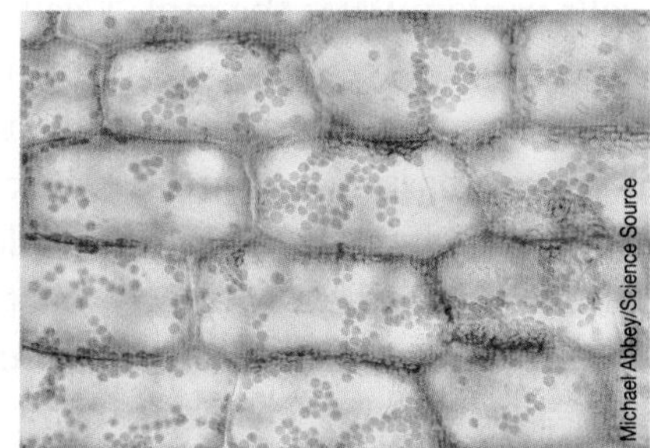

Elodea (an aquatic plant): a multicellular eukaryote

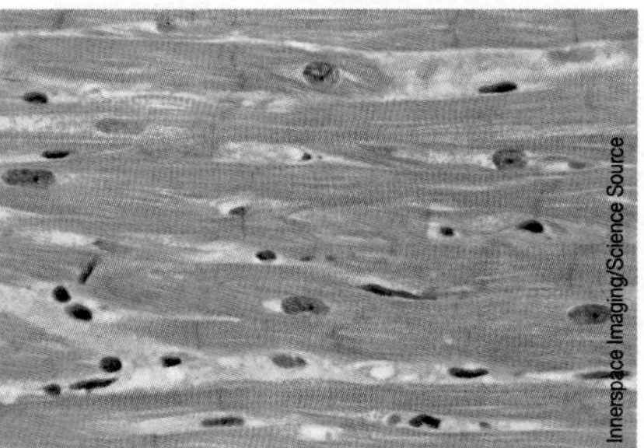

Humans (these are heart cells): multicellular eukaryotes

? Which of the non-human organisms shown have cells of the same structural type as human cells?

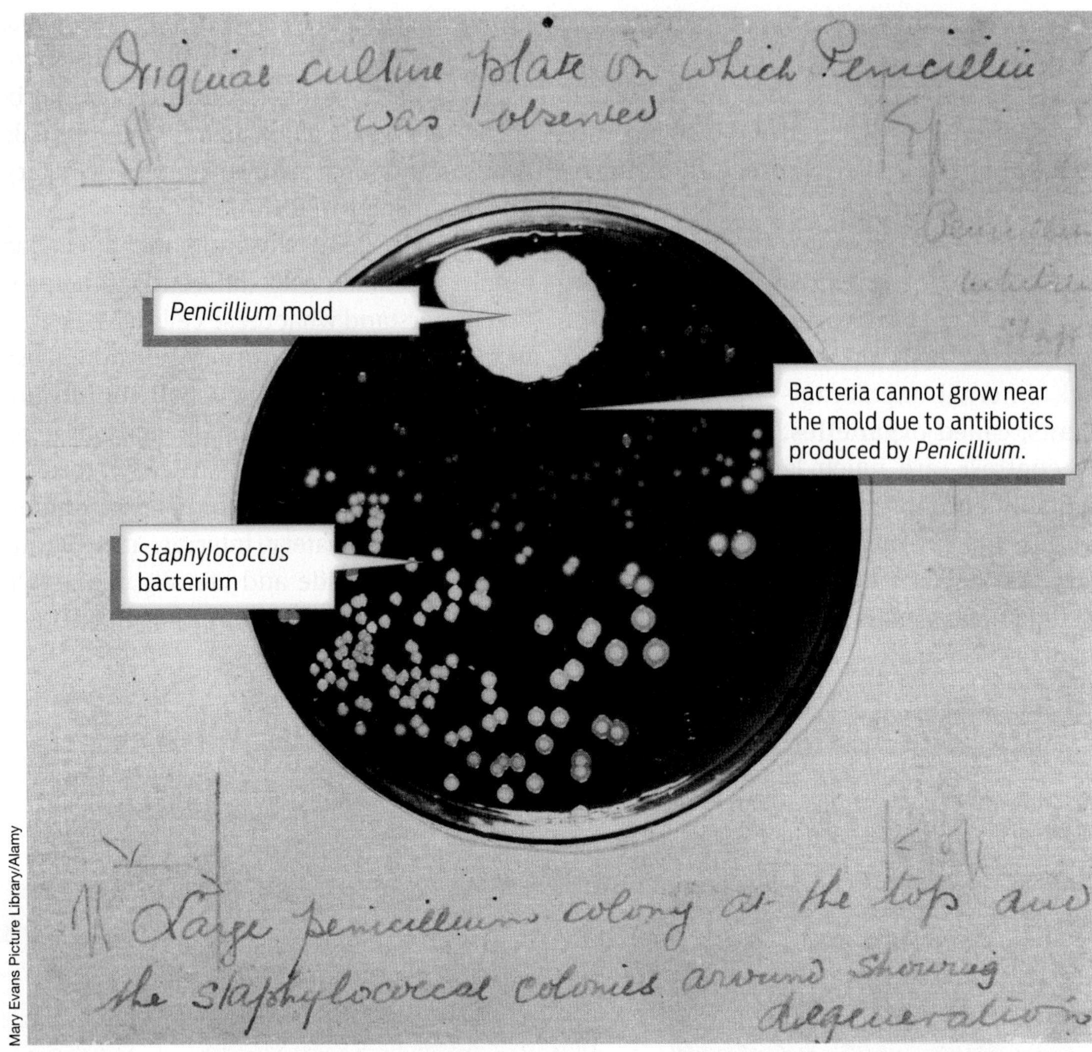
Penicillium mold
Bacteria cannot grow near the mold due to antibiotics produced by Penicillium.
Staphylococcus bacterium
Mary Evans Picture Library/Alamy

The original plate on which Alexander Fleming observed *Penicillium* mold inhibiting the growth of *Staphylococcus aureus* bacteria.

CYTOPLASM
The gelatinous, aqueous interior of all cells.

RIBOSOME
A complex of RNA and proteins that carries out protein synthesis in all cells.

NUCLEUS
The organelle in eukaryotic cells that contains the genetic material.

CELL WALL
A rigid structure present in some cells that encloses the cell membrane and helps the cell maintain its integrity.

OSMOSIS
The diffusion of water across a membrane from an area of lower solute concentration to an area of higher solute concentration.

by the cell membrane, is the **cytoplasm.** In addition, all cells have **ribosomes,** which synthesize the proteins that are crucial to cell function. And all cells contain DNA, the molecule of heredity.

Beyond these common features, however—cell membrane, cytoplasm, ribosomes, and DNA—the two cell types are structurally quite different. For one thing, the DNA in a prokaryotic cell floats freely within the cell's cytoplasm, while in a eukaryotic cell it is housed within a central "command center" called the **nucleus.** The nucleus is one of many organelles found within eukaryotic cells but not in their simpler prokaryotic cousins **(INFOGRAPHIC 3.3).**

Penicillin selectively kills bacteria because of an important difference between prokaryotic and eukaryotic cells. Most prokaryotic cells, including bacteria, are surrounded by a **cell wall,** while many eukaryotic cells, including those of humans and other animals, lack this structure. The rigid cell wall, which encloses the cell membrane, is what allows bacteria to survive in watery environments—for example, in the intestines, in blood, or in a pond.

Water has a tendency to move across cell membranes in order to balance the solute concentrations on each side of the membrane, a process called **osmosis.** Water will predictably move from the solution with the

INFOGRAPHIC 3.3 Features of Prokaryotic and Eukaryotic Cells

While all cells have a cell membrane, cytoplasm, ribosomes, and DNA, there are specific structural differences between prokaryotic and eukaryotic cells. Eukaryotic cells contain a variety of membrane-enclosed organelles, while prokaryotic cells do not. All prokaryotic cells have a cell wall surrounding the cell membrane, while many eukaryotic cells do not.

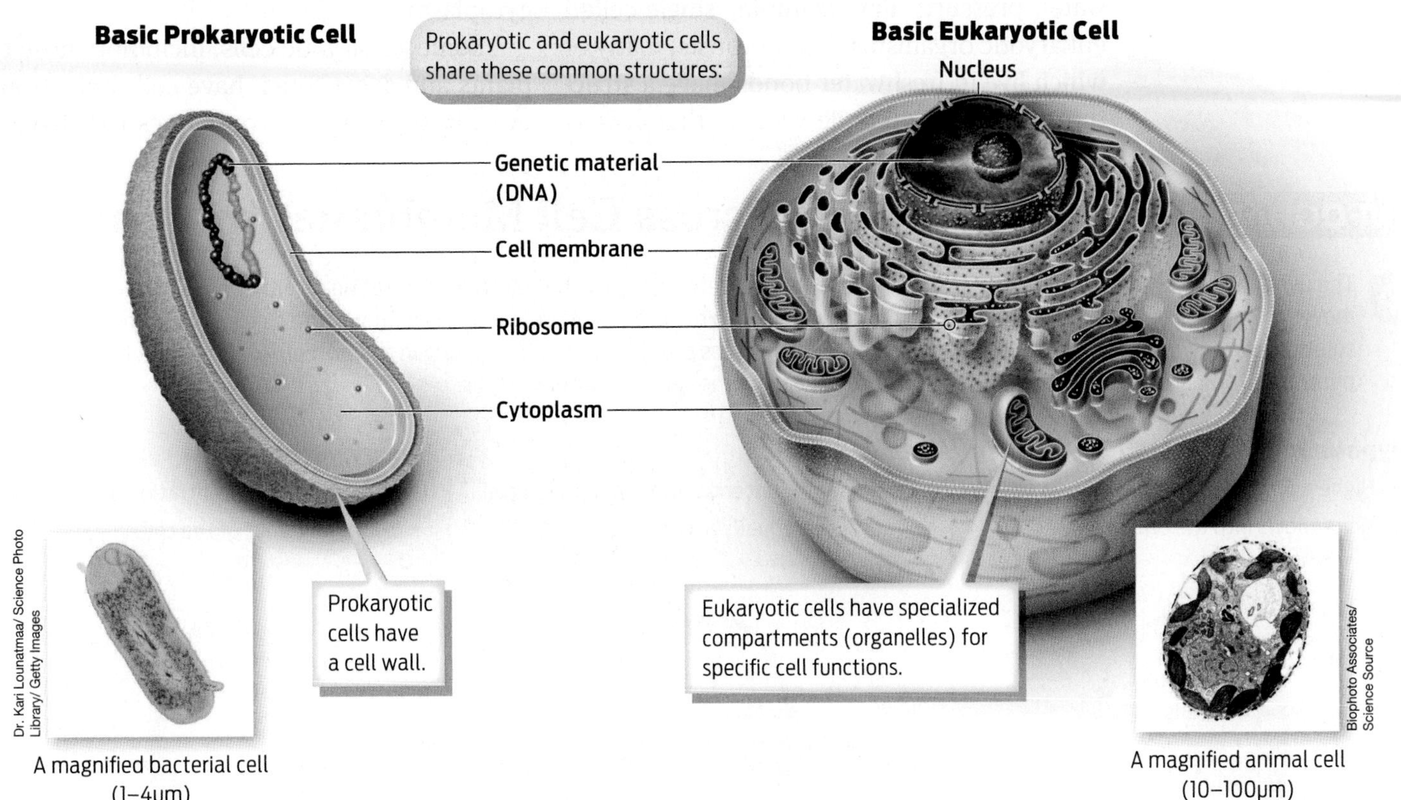

Basic Prokaryotic Cell
Prokaryotic and eukaryotic cells share these common structures:
Basic Eukaryotic Cell
Nucleus
Genetic material (DNA)
Cell membrane
Ribosome
Cytoplasm
Prokaryotic cells have a cell wall.
Eukaryotic cells have specialized compartments (organelles) for specific cell functions.

Dr. Kari Lounatmaa/ Science Photo Library/ Getty Images

A magnified bacterial cell (1–4µm)

Biophoto Associates/ Science Source

A magnified animal cell (10–100µm)

? Describe at least one difference between eukaryotic and prokaryotic cells.

lower solute concentration to the solution with the higher solute concentration. For example, cells placed in a lower-solute, or **hypotonic,** solution will tend to take up water and swell. On the other hand, cells placed in a higher-solute, or **hypertonic,** solution will tend to lose water and shrivel. In an **isotonic** solution, where the solute concentration is the same as the cell's cytoplasm, there is no net movement of water into or out of the cell. In all cases, water moves in a direction that will tend to even out the solute concentrations on each side of the membrane.

The environments that many bacteria find themselves in tend to be hypotonic. Water then enters the bacterial cells by osmosis, causing them to swell. This swelling would be fatal to bacteria were it not for the cell wall, which limits how much water can enter the cell. The rigid cell wall counteracts the pressure of the incoming water, preventing the cell from swelling to the bursting point.

What makes the bacterial cell wall rigid is the molecule **peptidoglycan,** a polymer made of sugars and amino acids that link to form a chainlike sheath around the cell. Different bacterial cell walls can have different structures, but all have peptidoglycan, which is found only in bacteria. And here's where penicillin comes in: by interfering with the synthesis of peptidoglycan, penicillin weakens the bacterial cell wall, which is then no

HYPOTONIC
Describes a solution surrounding a cell that has a lower concentration of solutes than the cell's cytoplasm.

HYPERTONIC
Describes a solution surrounding a cell that has a higher concentration of solutes than the cell's cytoplasm.

ISOTONIC
Describes a solution surrounding a cell that has the same solute concentration as the cell's cytoplasm.

PEPTIDOGLYCAN
The macromolecule found in all bacterial cell walls that gives the cell wall its rigidity.

longer able to withstand the pressure of the incoming water. Eventually, the bacterial cell bursts (**INFOGRAPHIC 3.4**).

Eukaryotic cells lacking cell walls have other ways of protecting themselves from water pressure. For example, single-celled eukaryotic organisms called protists, many of which live in freshwater ponds, have a structure called a contractile vacuole that acts as a water pump, continually pumping excess water out of the cell. In multicellular eukaryotic organisms, the environments surrounding cells (e.g., blood) are often maintained as an isotonic solution with respect to the cells' cytoplasm.

Some eukaryotic cells, including those of plants and some fungi, have cell walls. However, bacteria are the only ones that have a

INFOGRAPHIC 3.4 Water Flows Across Cell Membranes by Osmosis

The direction of water movement across the cell membrane is determined by the solute concentration on either side. Water always moves toward the side with the higher solute concentration. In a hypotonic solution, water will flow into the cell, and in a hypertonic solution, water will flow out of a cell. The bacterial cell wall helps protect the cell from lysing in a hypotonic environment. In the presence of some antibiotics the cell wall is disrupted, leaving the cell susceptible to lysis.

Hypotonic Solution
- Higher solute concentration inside cell
- Water flows into the cell.

Isotonic Solution
- Equal solute concentration in and out of cell
- Water flows equally in both directions.

Hypertonic Solution
- Higher solute concentration outside cell
- Water flows out of the cell.

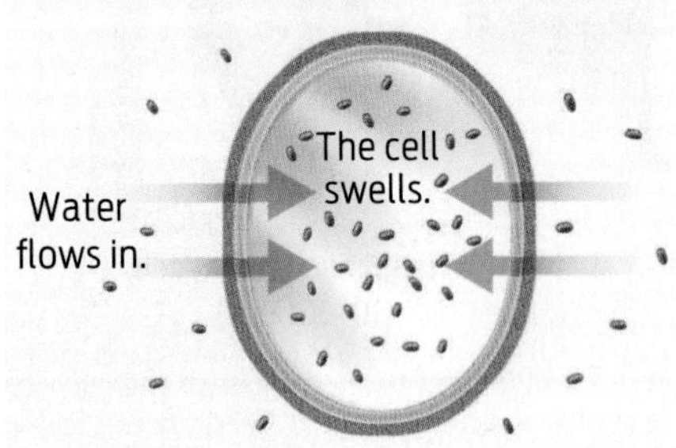

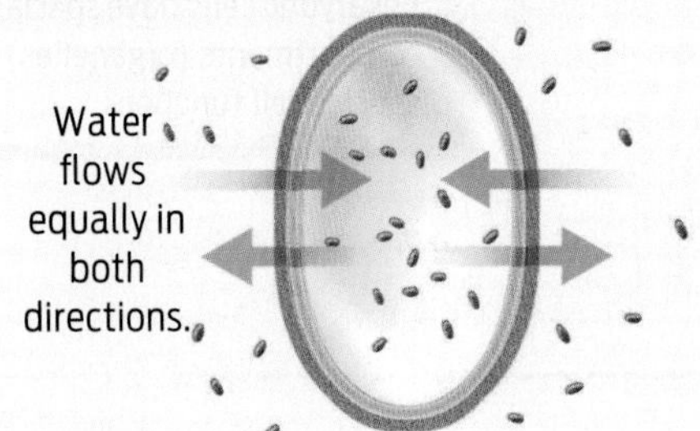

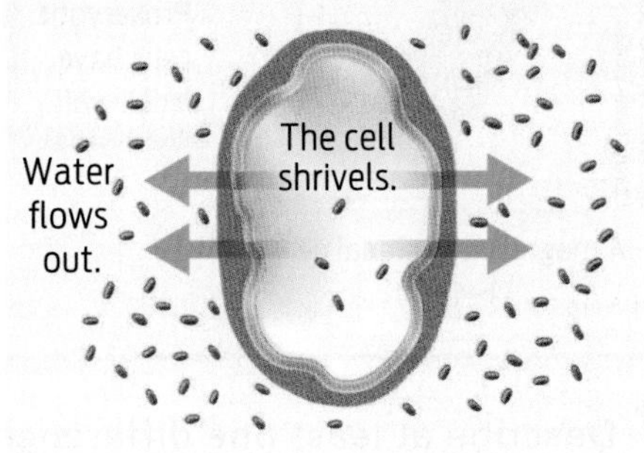

In the presence of penicillin:

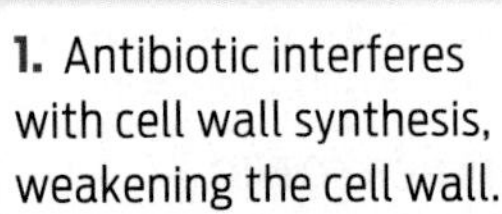

1. Antibiotic interferes with cell wall synthesis, weakening the cell wall.

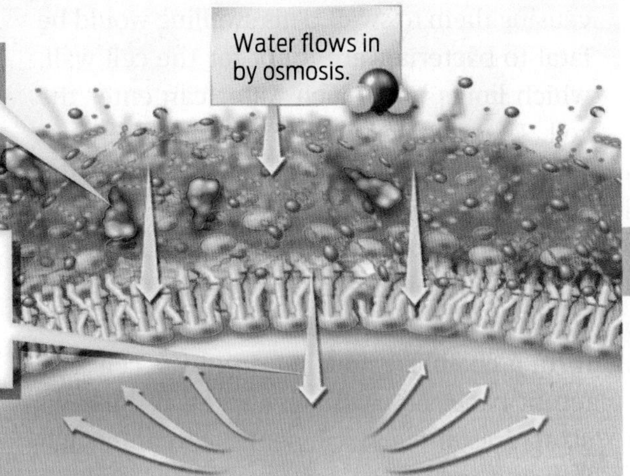

Water flows in by osmosis.
2. In a hypotonic solution, water flows into the cell, applying pressure on the cell wall.

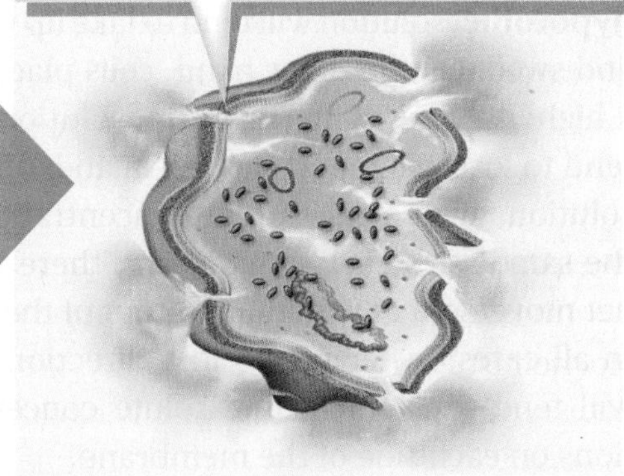

3. The weakened bacterial cell wall cannot resist the internal pressure of the water. The cell ruptures in this environment.

? Human cells do not have cell walls. What will happen to a human cell placed in a hypotonic environment?

cell wall made of peptidoglycan—which is why penicillin is such a selective bacteria killer.

Ironically, despite its remarkable killing powers, penicillin was not immediately recognized as a medical breakthrough when it was first discovered. In fact, Fleming didn't think his mold had much of a future in medicine. "I had not the slightest suspicion that I was at the beginning of something extraordinary," recalled Fleming years later.

At the time, the idea that an antiseptic agent could kill bacteria without also harming the patient was unheard of, so Fleming never considered that penicillin might be taken internally. Nor was he a chemist, so he lacked the expertise to isolate and purify the active ingredient from the mold. While he found that his mold juice made a "reasonably good" topical antiseptic, he noted that "the trouble of making it seemed not worthwhile," and largely gave up working on it.

Ten years would pass before anyone reconsidered Fleming's mold. By then, history had intervened and given new urgency to the search for antibacterial medicines.

From Fungus to Pharmaceutical

On September 1, 1939, Germany invaded Poland, plunging the world into war for the second time in a generation. With the horrors of World War I seared into memory, many feared the death toll that would result from the hostilities. Millions of soldiers and civilians had died in World War I, many not as a direct result of combat injuries, but instead from infected wounds. With few other antibacterial medicines available, penicillin suddenly became the focus of research during World War II.

In 1938, Ernst Chain, a German-Jewish biochemist, was working in the pathology department at Oxford University, having fled Germany for England in 1933 when the Nazis came to power. Both Chain and his supervisor, Howard Florey, were interested in the biochemistry of antibacterial substances. Chain stumbled across Fleming's 1929 paper on penicillin and set about trying to isolate and concentrate the active ingredient from the mold, which he succeeded in doing by 1940. Chain's breakthrough allowed Florey's group to begin testing the drug's clinical efficacy. They injected the purified chemical into bacteria-infected mice and found that the mice were quickly rid of their infection. Human trials followed next, in 1941, with the same remarkable result.

As encouraging as these results were, there was one nagging problem: it took up to 2,000 liters (more than 500 gallons) of mold growing in liquid culture to obtain enough pure penicillin to treat one person. The Oxford doctors used almost their whole supply of the drug treating their first patient, a policeman ravaged by a staphylococcal infection. The team stepped up their purification efforts—even culturing the mold in patients' bedpans and repurifying the drug from patients' urine—but there was no way they could keep up with demand.

The turning point came in 1941, when Oxford scientists approached the U.S. government and asked for help in growing penicillin on a large scale. The method they devised took advantage of something America had in abundance: corn. Using a by-product of large-scale corn processing as a culture medium in which to grow the fungus, the scientists were able to produce penicillin in much greater quantities.

At first, all the penicillin harvested from U.S. production plants came from Fleming's original strain of *Penicillium notatum*. But researchers continued to look for more potent strains to improve yields. In 1943,

Daily Herald Archive/Getty Images

Manufacturing penicillin in 1943: culture flasks are filled with the nutrient solution in which penicillin mold is grown.

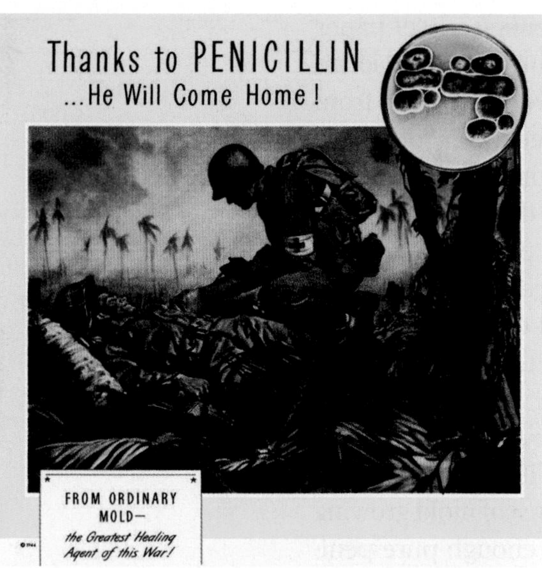

Penicillin and the war effort: feelings of wartime patriotism were enlisted to support production of the drug.

they got lucky when researcher Mary Hunt discovered one such strain growing on a ripe cantaloupe in a Peoria, Illinois, supermarket. This new strain, *Penicillium chrysogenum,* produced more than 200 times the amount of penicillin as the original strain. With it, production of the drug soared. By the time the Allies invaded France on June 6, 1944—D-day—they had enough penicillin to treat every soldier who needed it. By the following year, penicillin was widely available to the general public.

The optimism with which patients and doctors greeted the new bacteria-killer cannot be overstated. "Penicillin seemed to justify a carefree attitude to infection," says science historian Robert Bud, principal curator of the Science Museum in London. "In Western countries, for the first time in human history, most people felt that infectious disease was ceasing to be a threat, and sexually infectious disease had already been conquered. For many it seemed cure would be easier than prevention."

Yet, as effective as penicillin was, it was effective only against certain types of bacteria. Against others, it was powerless.

Stockpiling the Antibiotic Arsenal

As Fleming knew, most of the bacterial world falls into one of two categories, **Gram-positive** or **Gram-negative.** These names reflect the way bacterial cell walls trap a dye known as the Gram stain (after its discoverer, the Danish scientist Hans Christian Gram). Gram-positive bacteria retain the dye, while Gram-negative bacteria do not. Fleming found that while penicillin easily killed Gram-positive bacteria like *Staphylococcus*

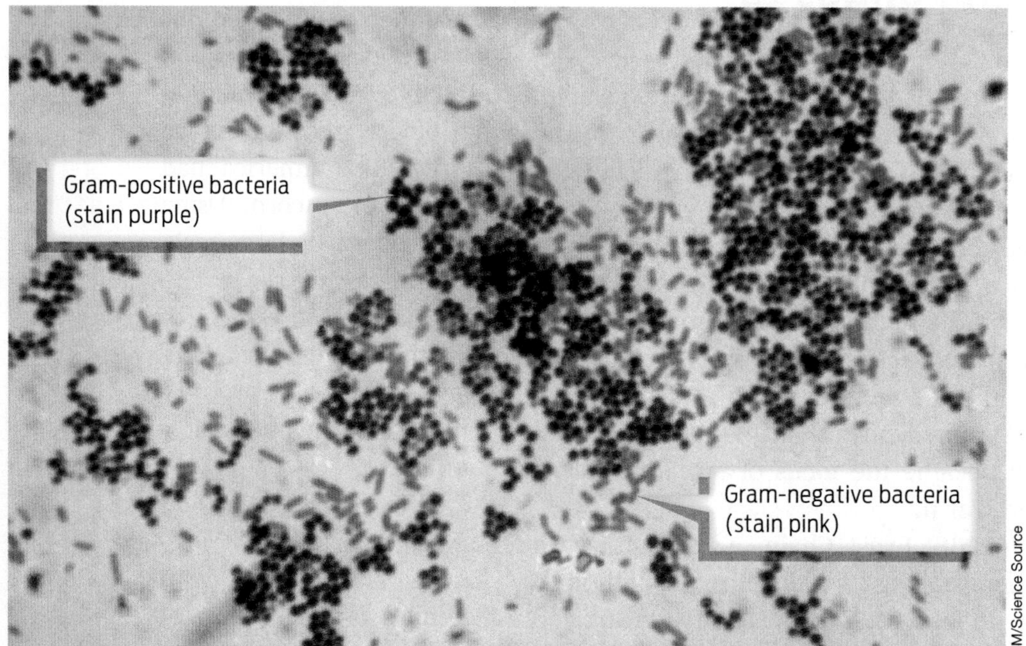

To identify bacteria, researchers treat them with a violet Gram stain. Gram-positive bacteria retain the dye well and so appear purple or blue under a microscope. Gram-negative bacteria do not retain the dye as well, but do retain a second dye, and so appear red or pink.

GRAM-POSITIVE
Describes bacteria with a cell wall that includes a thick layer of peptidoglycan that retains the Gram stain.

GRAM-NEGATIVE
Describes bacteria with a cell wall that includes a thin layer of peptidoglycan surrounded by an outer lipid membrane that does not retain the Gram stain.

and *Streptococcus*–the microbes that cause staph infections and strep throat, respectively–it had little effect on Gram-negative bacteria like *E. coli.* and *Salmonella*.

It turns out that the cell wall of Gram-negative bacteria includes an extra layer of lipids surrounding the peptidoglycan layer. This extra lipid layer prevents penicillin from reaching the peptidoglycan underneath.

The discovery that penicillin was effective mostly against Gram-positive bacteria led researchers in the 1940s to look for other antibiotics that could kill Gram-negative bacteria. The first such broad-spectrum antibiotic was streptomycin, discovered in 1943 by researchers at Rutgers University. In addition to killing Gram-negative bacteria, streptomycin was the first effective treatment for the deadly bacterial disease tuberculosis. The reason for its effectiveness? Streptomycin has a chemical structure that allows it to pass more easily through the outer lipid layer of the Gram-negative bacterial cell wall.

Once inside the cell, streptomycin works by interfering with protein synthesis by bacterial ribosomes. Ribosomes are the molecular machines that assemble a cell's proteins. While both eukaryotic and prokaryotic cells have ribosomes, their ribosomes are of different sizes and have different structures. Because streptomycin targets features specific to bacterial ribosomes, it doesn't harm the eukaryotic cells of the human who is taking it (**INFOGRAPHIC 3.5**).

INFOGRAPHIC 3.5 Some Antibiotics Inhibit Prokaryotic Ribosomes

Ribosomes are responsible for the synthesis of proteins in both prokaryotic and eukaryotic cells, but their structure is slightly different in the two types of cell. Antibiotics that interfere with prokaryotic ribosomes leave eukaryotic ribosomes unaffected.

Prokaryotic Ribosome

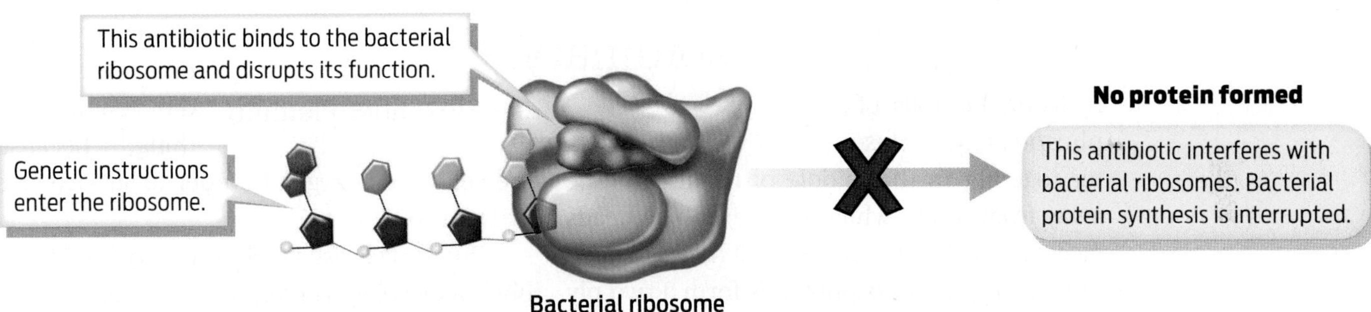

This antibiotic binds to the bacterial ribosome and disrupts its function.
Genetic instructions enter the ribosome.
Bacterial ribosome
No protein formed
This antibiotic interferes with bacterial ribosomes. Bacterial protein synthesis is interrupted.

Eukaryotic Ribosome

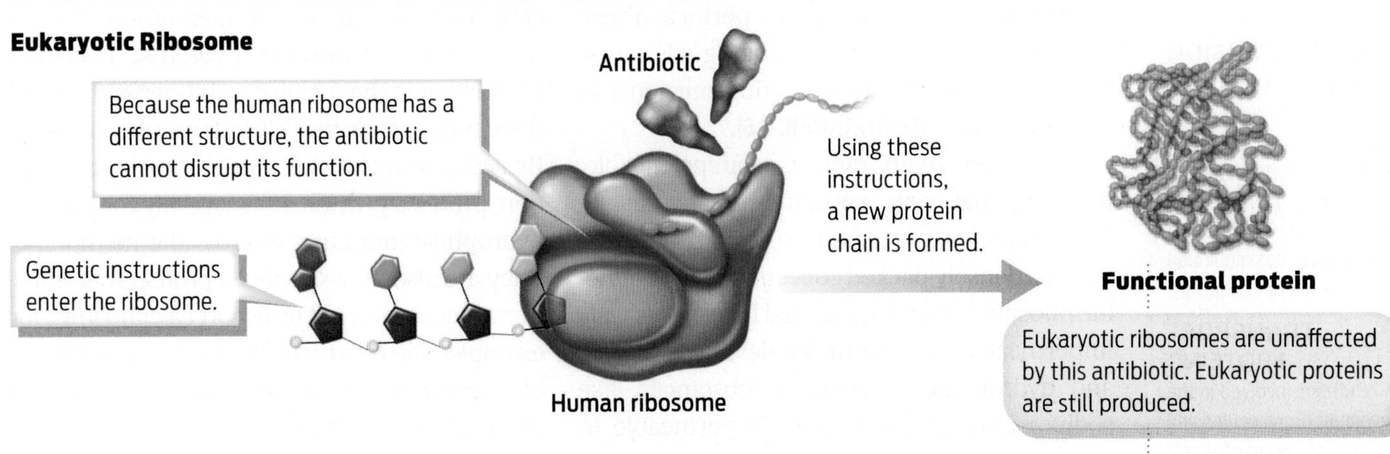

Antibiotic
Because the human ribosome has a different structure, the antibiotic cannot disrupt its function.
Genetic instructions enter the ribosome.
Using these instructions, a new protein chain is formed.
Human ribosome
Functional protein
Eukaryotic ribosomes are unaffected by this antibiotic. Eukaryotic proteins are still produced.

? If this antibiotic targets ribosomes, why can eukaryotic cells continue to synthesize proteins in its presence?

Other antibiotics work in different ways–by inhibiting a bacterium's ability to make a critical vitamin or to copy its DNA before dividing, for example. When this happens, the bacterium dies instead of reproducing.

Crossing Enemy Lines

For any drug to be effective, it has to reach its designated target. In the case of many antibiotics, that means getting inside the cell to do their work. But the cell membrane, which surrounds all cells, acts as a barrier to the free flow of substances into the cell. How do antibiotics get past this barrier?

To understand how substances move across the cell membrane, it helps to know a bit more about its structure. The cell membrane is a flexible yet durable structure composed of phospholipids and proteins. A phospholipid has two main parts: a hydrophilic "head" and a hydrophobic "tail." In the watery context of a cell, the hydrophobic tails of phospholipids cluster together in the middle of the membrane away from water, while the hydrophilic heads face outward, toward water. When arranged in this way, the phospholipids form a two-ply structure called a bilayer. Proteins sit nestled in the lipid bilayer, where they perform a variety of functions such as relaying signals across the membrane and transporting nutrients in and wastes out (**INFOGRAPHIC 3.6**).

> "I had not the slightest suspicion that I was at the beginning of something extraordinary."
>
> — Alexander Fleming

The cell membrane is semipermeable, meaning that only substances with certain characteristics can cross it easily without help. With its densely packed collection of hydrophobic phospholipid tails, the lipid bilayer is largely impermeable to large molecules, like glucose, and hydrophilic or charged substances, like sodium ions, and only weakly permeable to water. In fact, the only things that do cross the lipid bilayer easily and without assistance are small, uncharged molecules like oxygen (O_2) and carbon dioxide (CO_2) gas, which cross by **simple diffusion.** Diffusion is the natural tendency of dissolved substances to move from an area of higher concentration to one of lower concentration–think of food coloring dispersing in a glass of water. In simple diffusion, the substance moves directly through the phospholipids of the membrane from the side of the membrane with a higher concentration to the side with a lower concentration, thus balancing the concentrations on both sides. Because substances naturally move from an area of higher concentration to an area of lower concentration, no additional energy is required for this movement beyond what is stored in the concentration difference, or gradient, itself.

SIMPLE DIFFUSION The movement of small, uncharged solutes across a membrane from an area of higher concentration to an area of lower concentration without the aid of transport proteins; does not require an input of energy.

TRANSPORT PROTEINS Proteins involved in the movement of molecules and ions across the cell membrane.

Take oxygen, for example. The concentration of oxygen molecules, which are small and uncharged, is often higher outside the cell and lower inside it. This concentration gradient allows oxygen to diffuse easily into the cell–a good thing, because the cell needs oxygen in order to survive (see Chapter 6).

But the cell also needs some large or hydrophilic molecules in order to survive–one of them is glucose, the cell's energy source. To move glucose molecules across the membrane, the cell makes use of **transport proteins.** Transport proteins sit in the membrane bilayer with one of their ends outside the cell and the other inside it. By acting as a channel, carrier, or pump, transport proteins provide a passageway for large or hydrophilic molecules to cross the membrane. They are also very specific: a protein that transports glucose will not transport calcium ions, for example. The cells of the body contain hundreds of different transport proteins, each specific for a different type of cargo.

Transport proteins can move substances either with or against their concentration

INFOGRAPHIC 3.6 Membranes: All Cells Have Them

All cells are enclosed by a cell membrane made of phospholipids and proteins. The hydrophilic heads of the phospholipids face "out" (on either side of the membrane) and interact with water, while the hydrophobic tails cluster together inside the membrane away from water.

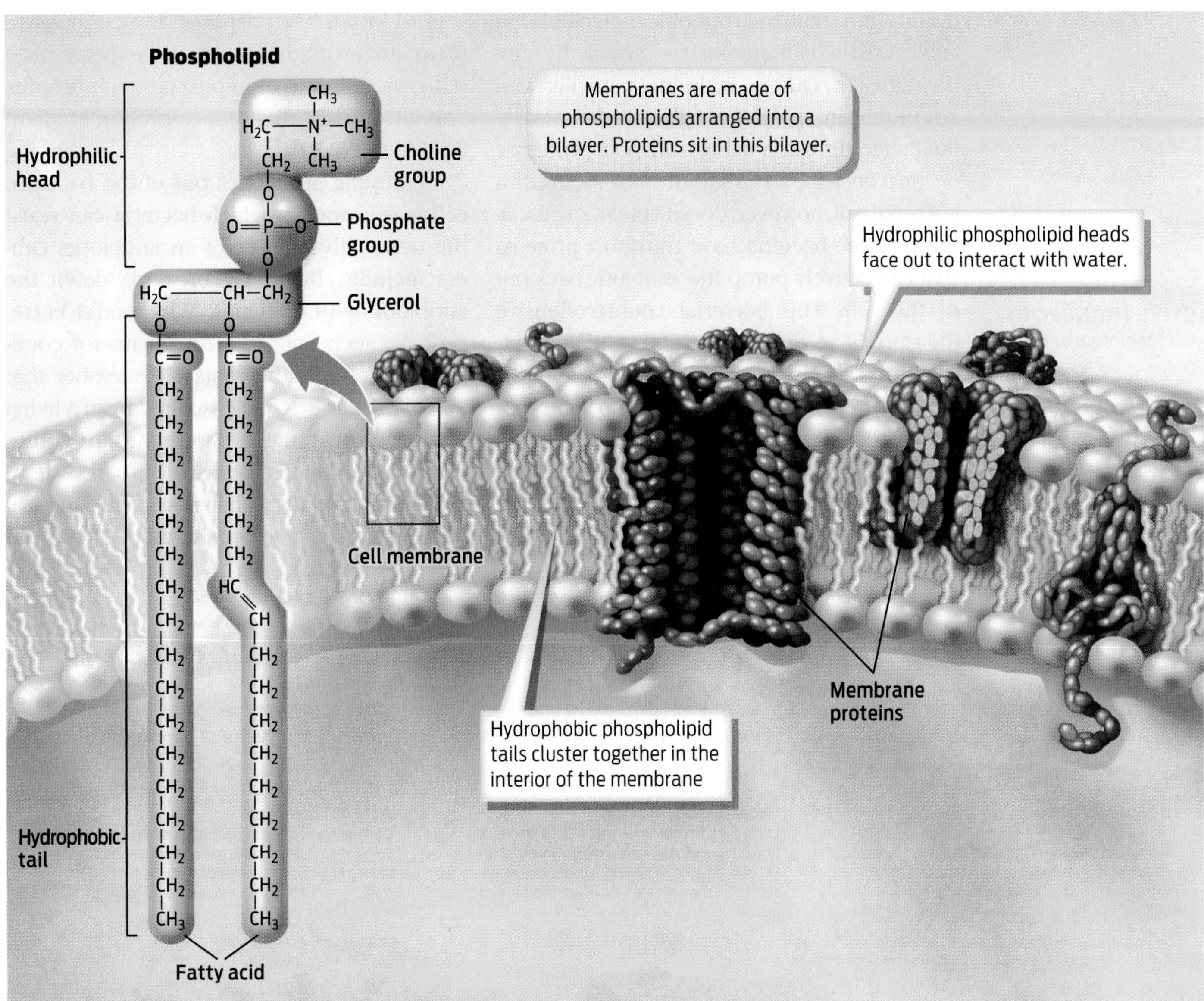

Phospholipid
Hydrophilic head
Choline group
Phosphate group
Glycerol
Hydrophobic tail
Fatty acid
Membranes are made of phospholipids arranged into a bilayer. Proteins sit in this bilayer.
Hydrophilic phospholipid heads face out to interact with water.
Cell membrane
Membrane proteins
Hydrophobic phospholipid tails cluster together in the interior of the membrane

? Which portion of a phospholipid is found in the middle (the interior) of a membrane?

gradient–either "downhill" or "uphill" across the membrane. When a substance moves "downhill" by a transport protein from an area of higher concentration to an area of lower concentration, the process is called **facilitated diffusion.** Like simple diffusion, facilitated diffusion requires no additional energy besides that stored naturally in the concentration gradient. For this reason, facilitated diffusion is also sometimes known as passive transport. Many substances enter the cell by facilitated diffusion. In addition to glucose, ions and water also move across the cell membrane by facilitated diffusion.

Osmosis, the movement of water across membranes discussed previously, relies on both simple diffusion and facilitated diffusion through transport proteins. In both cases, the

FACILITATED DIFFUSION
The process by which large, hydrophilic, or charged solutes move across a membrane from an area of higher concentration to an area of lower concentration with the help of transport proteins; does not require an input of energy.

movement is passive–i.e., it does not require an input of energy.

Antibiotics move across membranes in a number of ways. Some antibiotics, like tetracycline, are small hydrophobic molecules that can cross the cell membrane directly by simple diffusion. Others, including penicillin and streptomycin, pass through membranes by facilitated diffusion using transport proteins.

Just because an antibiotic makes it inside a bacterial cell, however, doesn't mean it will stay there. Some bacteria have transport proteins that can actively pump the antibiotic back out of the cell. This bacterial counteroffensive measure is an example of **active transport,** in which proteins pump a substance "uphill" from an area of lower concentration to an area of higher concentration. Unlike facilitated diffusion, active transport requires an input of chemical energy. Active transport keeps the antibiotic concentration in the bacterial cell low, but the cell must expend energy to keep pumping the antibiotic out against its concentration gradient.

ACTIVE TRANSPORT
The process by which solutes are pumped from an area of lower concentration to an area of higher concentration with the help of transport proteins; requires an input of energy.

Eukaryotic cells use active transport for several important purposes–for example, to maintain ion gradients across the membranes of nerve cells and to import certain nutrients "uphill" against their concentration gradient **(INFOGRAPHIC 3.7)**.

Pumping antibiotics out of the bacterial cell is one way in which bacteria can resist the destructive power of an antibiotic. Others include chemically breaking down the antibiotic with enzymes. Why would bacteria have such built-in mechanisms for counteracting or resisting drugs? Remember that penicillin was originally isolated from a living organism, a fungus. Streptomycin comes from microorganisms living in soil. Microorganisms like these have evolved chemical defenses as a way to protect themselves from

INFOGRAPHIC 3.7 Molecules Move Across the Cell Membrane

The cell membrane is semipermeable: only a few substances can cross it unassisted. Solutes move across the membranes depending on solute size and charge, the relative concentration on each side of the membrane, and the presence or absence of transport proteins.

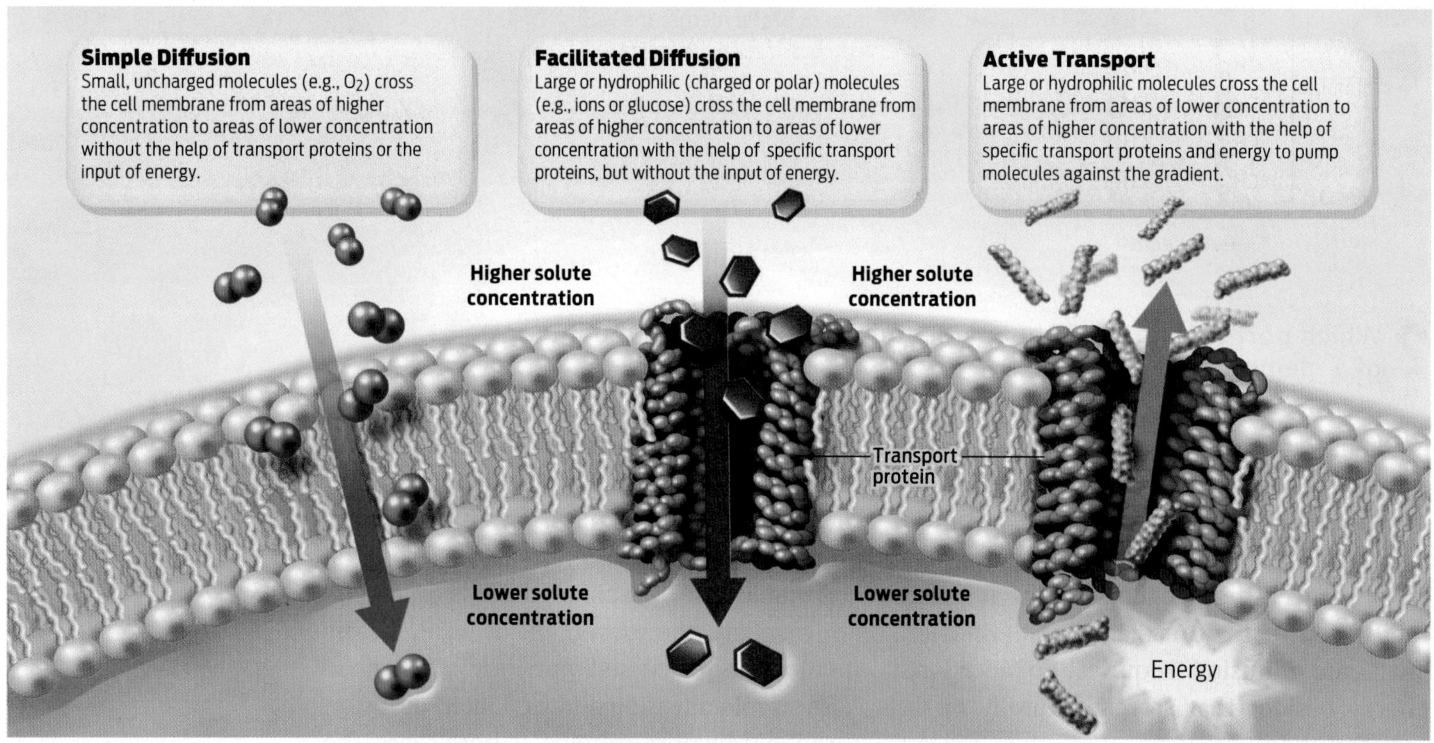

Simple Diffusion
Small, uncharged molecules (e.g., O_2) cross the cell membrane from areas of higher concentration to areas of lower concentration without the help of transport proteins or the input of energy.
Facilitated Diffusion
Large or hydrophilic (charged or polar) molecules (e.g., ions or glucose) cross the cell membrane from areas of higher concentration to areas of lower concentration with the help of specific transport proteins, but without the input of energy.
Active Transport
Large or hydrophilic molecules cross the cell membrane from areas of lower concentration to areas of higher concentration with the help of specific transport proteins and energy to pump molecules against the gradient.
Higher solute concentration
Higher solute concentration
Transport protein
Lower solute concentration
Lower solute concentration
Energy

? What is a major difference between diffusion and active transport?

other organisms. In turn, their combatants have evolved countermeasures that give them resistance. Humans thus find themselves embroiled in a battle originally waged solely between microorganisms.

Your Inner Bacterium

Antibiotics kill bacteria but leave humans unharmed because bacterial and human cells have different structures. Of all the ways that prokaryotic and eukaryotic cells differ, the most obvious is the complexity of eukaryotic cells compared to their smaller prokaryotic cousins. In particular, eukaryotic cells are characterized by the presence of multiple, distinct membrane-bound organelles (**INFOGRAPHIC 3.8**).

You can think of a eukaryotic cell as a miniature factory with an efficient division of labor. Each organelle is separated from the cell's cytoplasm by a membrane similar to the cell's outer membrane, and each performs a distinct function.

The nucleus is the defining organelle of eukaryotic cells (from the Greek *eu,* meaning "good" or "true," and *karyon,* meaning "nut" or "kernel"). It is surrounded by the **nuclear envelope,** a double membrane made of two lipid bilayers dotted by small openings called pores. The nucleus encloses the cell's DNA and acts as a kind of control center. Important reactions for interpreting the genetic instructions contained in DNA take place in the nucleus.

Other organelles in a eukaryotic cell perform other specialized tasks. **Mitochondria** are the cell's "power plants"–they use oxygen to extract energy from food and convert that energy into a useful form.

NUCLEAR ENVELOPE
The double membrane surrounding the nucleus of a eukaryotic cell.

MITOCHONDRIA (SINGULAR: MITOCHONDRION)
Membrane-bound organelles responsible for important energy-conversion reactions in eukaryotes.

INFOGRAPHIC 3.8 Eukaryotic Cells Have Organelles

Humans and other animals, as well as plants, fungi, and protists, are eukaryotes–they are made up of eukaryotic cells that contain internal organelles. All eukaryotic cells have a nucleus, endoplasmic reticulum, ribosomes, mitochondria, and Golgi apparatus. Some eukaryotic cells have additional structures, like the water vacuole, cell wall, and chloroplasts of a plant cell, that provide them with unique characteristics.

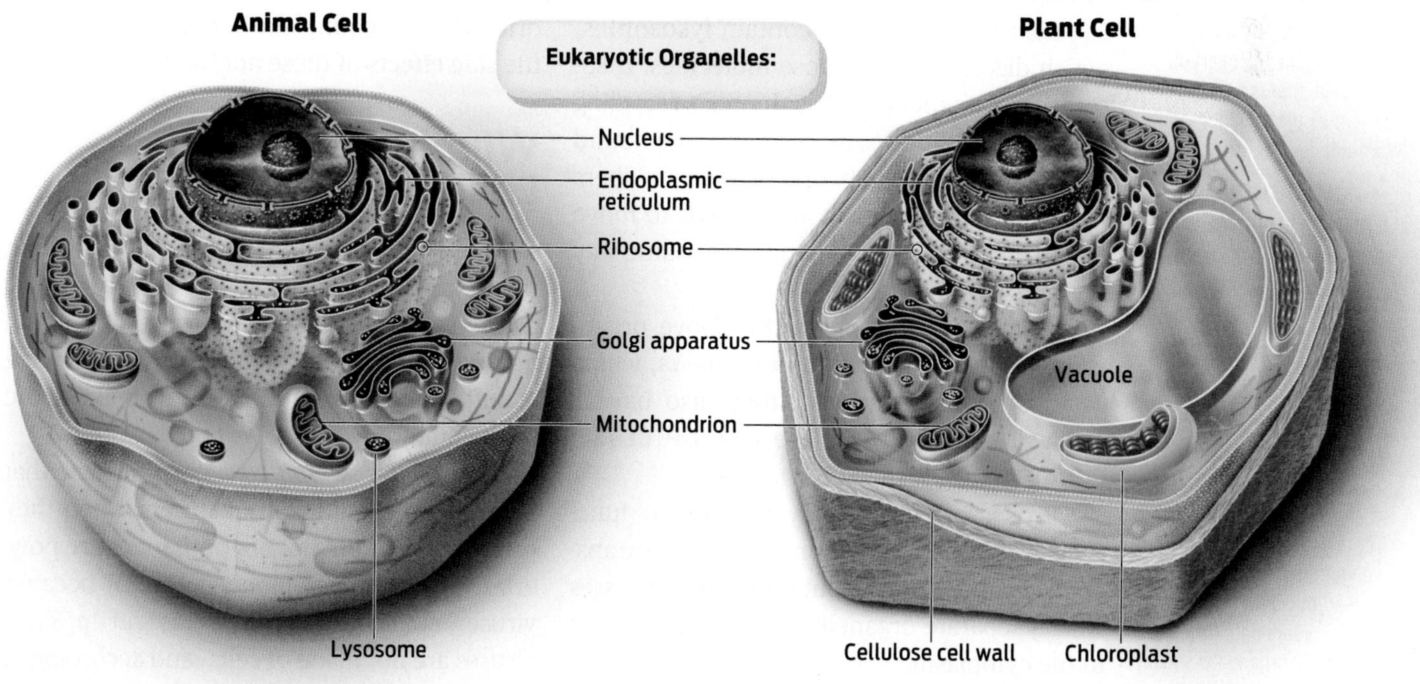

? Name and give the functions of at least three organelles present in both plant and animal cells.

All eukaryotes–including plants–have mitochondria. Humans who inherit or develop defects in their mitochondria usually die–an indication of just how important these organelles are.

The **endoplasmic reticulum (ER)** is a vast network of membranes that serves as a kind of assembly line for the manufacture of proteins and lipids. The "rough" ER is studded with ribosomes making proteins; the "smooth" ER makes lipids. Newly made proteins travel from the ER to the **Golgi apparatus,** an organelle that packs the protein "cargo" into vesicles and then ships them to specific destinations, such as the cell membrane, other organelles, and the bloodstream. The nucleus, ER, and Golgi apparatus thus work together to make and transport proteins to specific locations in and out of the cell.

> "Penicillin seemed to justify a carefree attitude to infection. In Western countries, for the first time in human history, most people felt that infectious disease was ceasing to be a threat."
>
> — Robert Bud

Eukaryotic cells also contain **lysosomes,** which digest and repurpose molecules. Lysosomes can be thought of as the cell's recycling centers. In addition to these membrane-bound structures, a vast network of protein fibers called the **cytoskeleton** allows cells to move and maintain their shape, much the same way the human skeleton does.

Finally, in addition to the above organelles, plant cells contain **chloroplasts,** which carry out photosynthesis; they also have a cell wall made of cellulose (see **UP CLOSE: EUKARYOTIC ORGANELLES**).

Prokaryotic cells carry out similar functions of energy conversion and protein transport, but they don't contain these processes within separate organelles; everything occurs in the cytoplasm.

How did eukaryotic cells develop their factory-like compartments? That question has long intrigued biologists. One fascinating hypothesis was proposed in the 1960s by biologist Lynn Margulis, who argued that eukaryotic organelles such as mitochondria and chloroplasts were once free-living prokaryotic cells that became incorporated–engulfed–by other free-living prokaryotic cells.

Although many considered this notion of **endosymbiosis** a crazy idea at first, a wealth of evidence now supports it (see **Milestone 1: Scientific Rebel**). Mitochondria and chloroplasts are about the same size and shape as bacteria. Both mitochondria and chloroplasts have circular strands of DNA, just like prokaryotic cells. They also contain ribosomes that are similar in structure to prokaryotic ribosomes–so similar, in fact, that some antibiotics that target prokaryotic ribosomes can affect the ribosomes in eukaryotic mitochondria, which accounts for both the toxicity and the side effects of these antibiotics.

ENDOPLASMIC RETICULUM
A network of membranes in eukaryotic cells where proteins and lipids are synthesized.

GOLGI APPARATUS
An organelle made up of stacked membrane-enclosed discs that packages proteins and prepares them for transport.

LYSOSOME
An organelle in eukaryotic cells that is filled with enzymes that can degrade worn-out cellular structures.

CYTOSKELETON
A network of protein fibers in eukaryotic cells that provides structure and facilitates cell movement.

CHLOROPLAST
An organelle in plant and algal cells that is the site of photosynthesis.

ENDOSYMBIOSIS
The scientific theory that free-living prokaryotic cells engulfed other free-living prokaryotic cells billions of years ago, forming eukaryotic organelles such as mitochondria and chloroplasts.

Winning the Battle, Losing the War

To those who first benefited from its healing powers, penicillin seemed like a truly wonder drug. A once-lethal bacterial infection could now be cleared in a matter of days with a course of antibiotic. Today, antibiotics are some of the most commonly prescribed medications.

Antibiotics are so common, in fact, that many people routinely take them when they catch a cold or the flu. But antibiotics are powerless against these illnesses. That's because viruses, not bacteria, cause colds and flu. Since viruses are not made of cells–and according to the cell theory are not even considered to be alive–they can't be killed with an antibiotic.

But that hasn't stopped people from trying. The Centers for Disease Control and Prevention (CDC) estimate that as many as 50% of the antibiotics prescribed for upper respiratory illnesses are unnecessary, since they are being used (ineffectively) to treat colds and other viral infections. What's more, many patients who are prescribed antibiotics for bacterial infections use them improperly. Taking only part of a prescribed dose, for example, can spare some harmful bacteria living in the body, and those bacteria that survive are often heartier and more resistant to the antibiotic than the ones that were killed. This overuse and misuse of antibiotics has led to an epidemic of antibiotic-resistance, which the CDC calls "one of the world's most pressing public health problems."

Fleming foresaw this very danger. In his own research, he found that whenever too little penicillin was used or when it was used for too short a time, populations of bacteria emerged that were resistant to the antibiotic. Fleming warned that improper use of penicillin among patients could lead to the emergence of virulent strains of bacteria that are resistant to the drug. He was right. In 1945, when penicillin was first introduced to the public, virtually all strains of *Staphylococcus aureus* were sensitive to it. Today, more than 90% of *S. aureus* strains are resistant to the antibiotic that once defeated it. (For more on antibiotic-resistant bacteria, see Chapter 13.)

Because of the alarming growth in antibiotic-resistant superbugs, drug companies and researchers are working to develop new antibiotics. One strategy they employ is to tweak the chemical structure of existing antibiotics just enough that a bacterium cannot disable it. Another approach is to look for antibiotics that target other bacterial weaknesses.

But all these efforts would be nothing without the man who gave a moldy petri dish a second glance nearly a century ago. That famous dish now sits in the British Museum in London. For his pioneering research, Alexander Fleming—along with Oxford researchers Howard Florey and Ernst Chain—won the Nobel Prize for Medicine or Physiology in 1945. ■

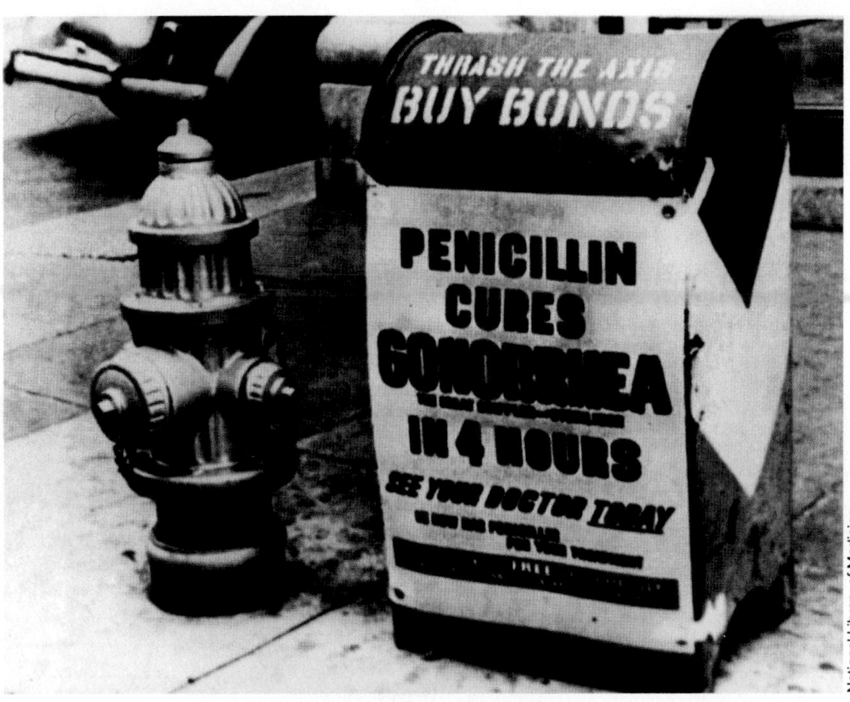

National Library of Medicine

Once perceived as a magic bullet, penicillin is now largely ineffective at treating many infections, including gonorrhea.

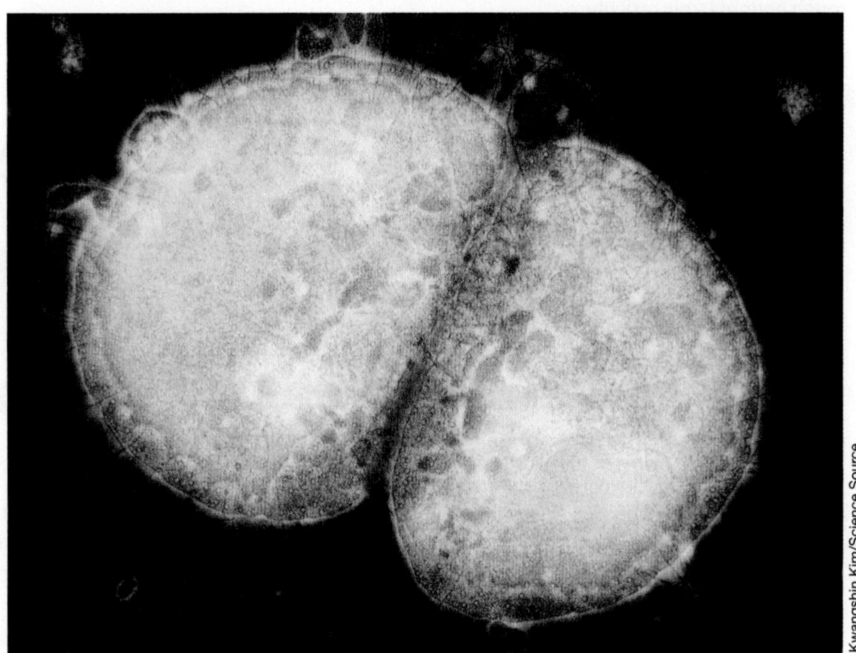

Kwangshin Kim/Science Source

Scanning electron micrograph of *Neisseria gonorrhoeae*, the bacterium that causes gonorrhea.

UP CLOSE Eukaryotic Organelles

Nucleus

The nucleus is the defining organelle of eukaryotic cells. The nucleus is separated from the cytoplasm by a double membrane (two phospholipid bilayers) known as the nuclear envelope. The nuclear envelope controls the passage of molecules between the nucleus and cytoplasm. The nucleus contains the DNA, the stored genetic instructions of each cell. In addition, important reactions for interpreting these genetic instructions occur in the nucleus.

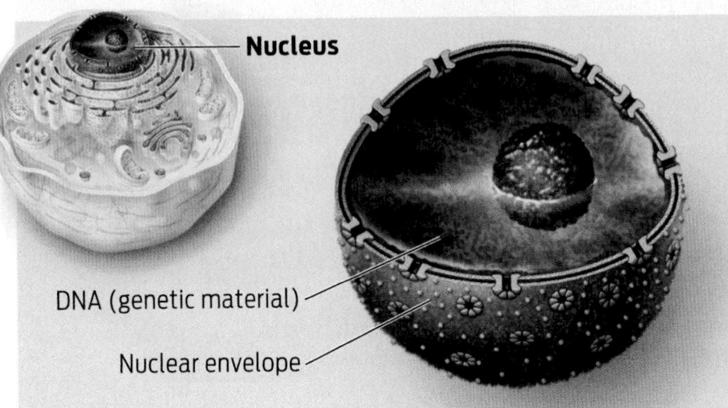

Endoplasmic Reticulum

The endoplasmic reticulum (ER) is an extensive, membranous intracellular "plumbing" system that is critical for the production of new proteins. The "rough" ER has a rough appearance because it is studded with ribosomes that are making proteins. The rough ER is contiguous with the "smooth" ER, the site of lipid production.

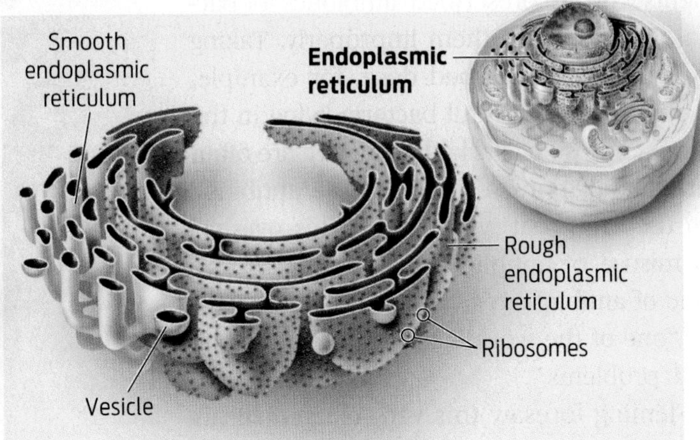

Golgi Apparatus

The Golgi apparatus is a series of flattened membrane compartments, whose purpose is to process and package proteins produced in the rough endoplasmic reticulum.

Golgi apparatus
1. Vesicles deliver proteins from the endoplasmic reticulum to the Golgi apparatus.
2. As the proteins make their way through the Golgi apparatus, they are processed to complete their structure and identify them for transport to specific locations in the cell.

The Nucleus, Endoplasmic Reticulum, and Golgi Apparatus Work Together to Produce and Transport Proteins

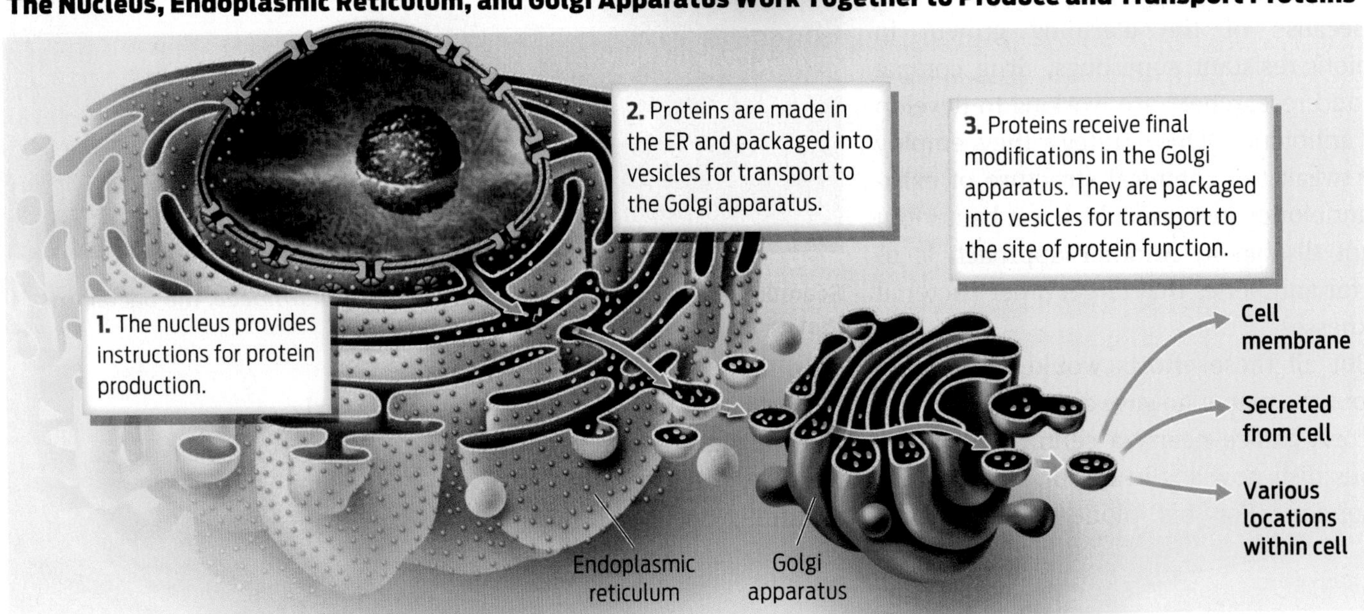

2. Proteins are made in the ER and packaged into vesicles for transport to the Golgi apparatus.
3. Proteins receive final modifications in the Golgi apparatus. They are packaged into vesicles for transport to the site of protein function.
1. The nucleus provides instructions for protein production.
Cell membrane
Secreted from cell
Various locations within cell
Endoplasmic reticulum
Golgi apparatus

Mitochondria

Mitochondria are found in almost all eukaryotes, including plants. Mitochondria have two membranes surrounding them. The inner one is highly folded. Mitochondria carry out critical steps in the extraction of energy from food, and the conversion of that "trapped" energy to a useful form. They are the cell's "power plants."

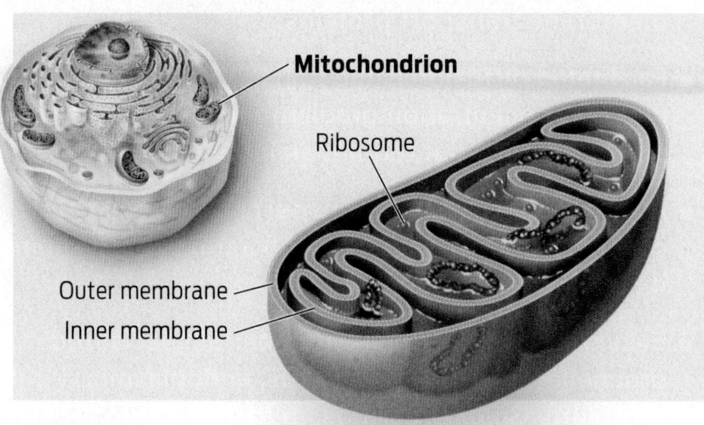

Chloroplast

Chloroplasts are organelles found in algae and in the green parts of plants. Chloroplasts have two membranes surrounding them, as well as an internal system of stacked membrane discs. Chloroplasts are the sites of photosynthesis, the reactions by which plants capture the energy of sunlight and convert it to a usable form.

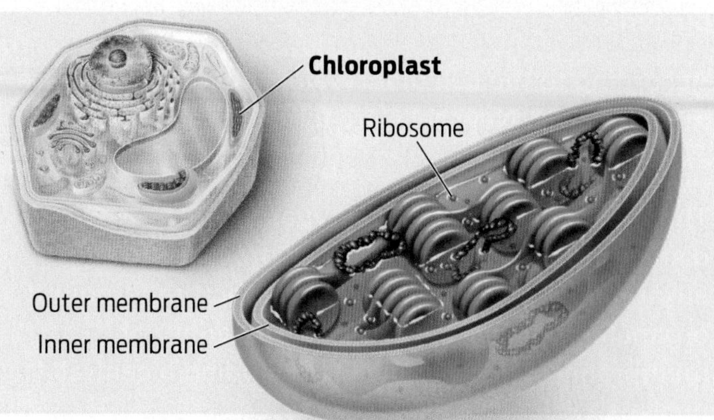

Lysosome

Lysosomes are the cell's digestive and recycling centers. They break down food and other molecules taken into the cell. The digested parts are then used in the cell for various function. Lysosomes also recycle worn-out organelles and other cellular components so their subunits can be reused for new cell structures.

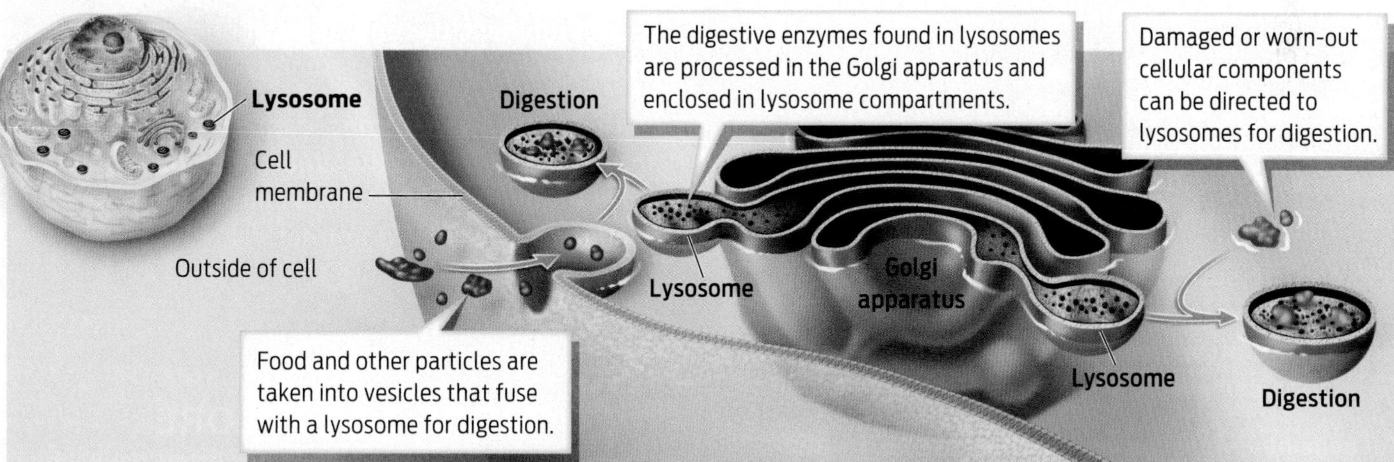

Cytoskeleton

The cytoskeleton is a meshwork of protein fibers that carry out a variety of functions, including cell support, cell movement, and movement of structures within cells. There are three main types of cytoskeletal fibers: microfilaments, intermediate filaments, and microtubules.

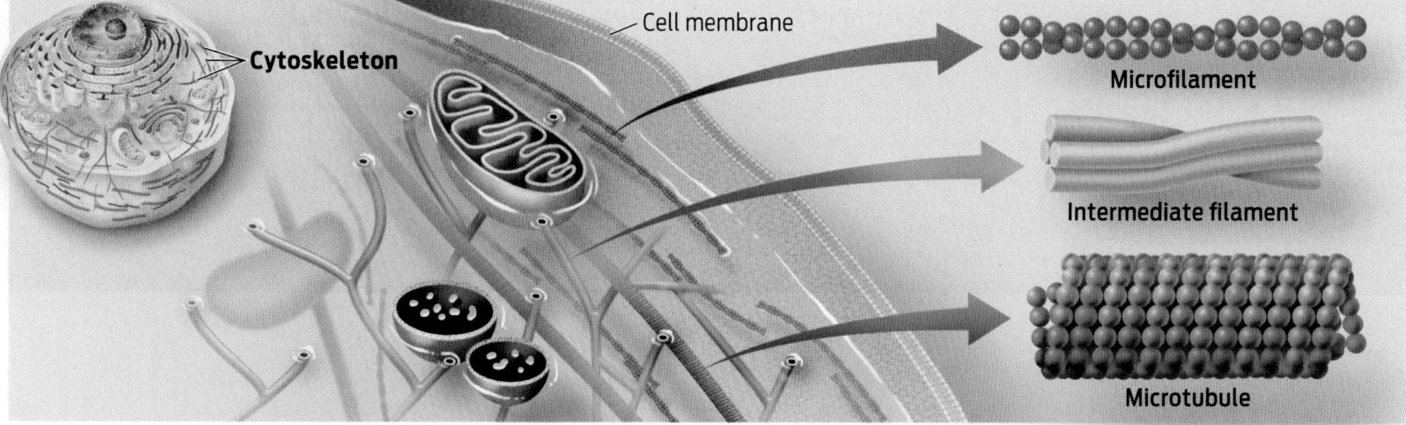

CHAPTER 3 SUMMARY

- Antibiotics are chemicals, originally produced by living organisms, that selectively target and kill bacteria.
- According to the cell theory, all living organisms are made of cells. New cells are formed when an existing cell divides.
- There are two fundamental types of cells, distinguished by their structure: prokaryotic and eukaryotic. Prokaryotic cells lack membrane-bound organelles; eukaryotic cells have a variety of membrane-bound organelles, including a central nucleus.
- All cells are enclosed by a cell membrane made up of phospholipids and proteins. Some cells also have a cell wall surrounding the cell membrane that imparts additional strength.
- Water crosses cell membranes by osmosis in order to balance the solutes on each side. In a hypotonic solution, water moves into the cell, causing it to swell; in a hypertonic solution, water leaves the cell, causing it to shrivel.
- Bacteria are surrounded by a cell wall containing peptidoglycan, a molecule not found in eukaryotes. Some antibiotics, like penicillin, work by preventing peptidoglycan synthesis, weakening the cell wall.
- All cells have ribosomes, complexes of RNA and proteins that synthesize new proteins. Some antibiotics, like streptomycin, work by interfering with prokaryotic ribosomes.
- The cell membrane is semipermeable: only substances with certain characteristics can cross it freely without help.
- Small hydrophobic molecules can cross cell membranes by simple diffusion, a process that does not require an input of energy.
- Large, hydrophilic, or charged molecules are transported across the membrane with the help of membrane transport proteins.
- Facilitated diffusion is the transport of molecules down a concentration gradient through a transport protein; it does not require an input of energy.
- Active transport is the transport of molecules up a concentration gradient through a transport protein; it requires an input of energy.
- Eukaryotic cells contain a number of specialized organelles, including a nucleus, endoplasmic reticulum, Golgi apparatus, mitochondria, lysosomes, and chloroplasts, each of which carries out a distinct function.
- Eukaryotic cells likely evolved as a result of endosymbiosis, the engulfing of one single-cell prokaryote by another.
- Increased and sometimes inappropriate use of antibiotics has led to the emergence of antibiotic-resistant bacteria. Infections caused by these bacteria are very hard to treat.

MORE TO EXPLORE

- **Video: Penicillin Acting on Bacteria: http://www.hhmi.org/biointeractive/penicillin-acting-bacteria**
- **The Centers for Disease Control and Prevention, Antibiotic Resistance: http://www.cdc.gov/drugresistance/**
- **Fleming, A. (1929) On the antibacterial action of cultures of a penicillium. *British Journal of Experimental Pathology* 10:226–236.**
- **Lax, R. (2005) *The Mold in Dr. Florey's Coat: The Story of the Penicillin Miracle.* New York: Owl Books.**
- **Bud, R. (2007) *Penicillin: Triumph and Tragedy.* Oxford: Oxford University Press.**

CHAPTER 3 Test Your Knowledge

DRIVING QUESTION 1 What structural features are shared by all cells, and what are the key differences between prokaryotic and eukaryotic cells?

By answering the questions below and studying Infographics 3.2, 3.3, 3.5, 3.6 and 3.8, you should be able to generate an answer for the broader Driving Question above.

KNOW IT

1 What does the cell theory state?

2 Which of the following statements best explains why bacteria are considered living organisms?

a. They can cause disease.
b. They are made up of biological macromolecules.
c. They move around.
d. They are made of cells.
e. They contain organelles.

3 What are the two main types of cells found in organisms?

4 Which of the following is/are *not* associated with human cells?

a. cell membrane
b. ribosomes
c. DNA
d. cell wall
e. All of the above are associated with human cells.

5 Bacteria are _____ cells, defined by the _____.

a. prokaryotic; presence of a cell wall
b. eukaryotic; presence of organelles
c. eukaryotic; absence of a cell wall
d. prokaryotic; absence of organelles
e. eukaryotic; absence of organelles

6 Which of the following is/are associated with eukaryotic cells but *not* with prokaryotic cells?

a. cell membrane
b. cell wall
c. DNA
d. ribosomes
e. nucleus

USE IT

7 According to the cell theory, all living organisms are made of cells. More specifically, what do all living organisms have in common? For example, do all living organisms carry genetic instructions? Do their cells all have a nucleus? What other features do they have in common?

8 In the soil of a forest, you find a single-celled organism with a cell wall—could this organism be an animal? Why or why not? Which of the following facts would convince you that the organism is a bacterium and not a plant?

a. The cell wall is made of cellulose.
b. The DNA is contained in a nucleus.
c. The cell wall is made of peptidoglycan.
d. a and b
e. b and c

DRIVING QUESTION 2 How do solutes and water cross cell membranes, and what determines the direction of movement of solutes and water in different situations?

By answering the questions below and studying Infographics 3.4 and 3.7, you should be able to generate an answer for the broader Driving Question above.

KNOW IT

9 The two major components of cell membranes are _____ and _____.

a. phospholipids; DNA
b. DNA; proteins
c. peptidoglycan; phospholipids
d. peptidoglycan; proteins
e. phospholipids; proteins

10 If a solute is moving through a phospholipid bilayer from an area of higher concentration to an area of lower concentration without the assistance of a protein, the manner of transport must be

a. active transport.
b. facilitated diffusion.
c. simple diffusion.
d. any of the above, depending on the solute.
e. Solutes cannot cross phospholipid bilayers.

11 Consider the movement of molecules across the cell membrane.

a. What do simple diffusion and facilitated diffusion have in common?

b. What do active transport and facilitated diffusion have in common?

12 Water is moving across a membrane from solution A into solution B. What can you infer?

a. Solution A must be pure water.

b. Solution A must have a lower solute concentration than Solution B.

c. Solution A must have a higher solute concentration than Solution B.

d. Solution A and Solution B must have the same concentration of solutes.

e. Solution B must be pure water.

USE IT

13 Why does facilitated diffusion require membrane transport proteins while simple diffusion does not?

14 Sugars are large, hydrophilic molecules that are important energy sources for cells. How can they enter cells from an environment with a very high concentration of sugar?

a. by simple diffusion

b. by osmosis

c. by facilitated diffusion

d. by active transport

e. by using ribosomes

15 Many foods—for example, bacon and salt cod—are preserved with high concentrations of salt. How can high concentrations of salt inhibit the growth of bacteria? (Think about the high solute concentration of the salty food relative to the solute concentration in the bacterial cells. What will happen to the bacterial cells under these conditions?)

apply YOUR KNOWLEDGE

MINI CASE

16 Marc, a first-year college student, starts out on a backpacking trip in southern New Mexico. It is September, so the daytime temperatures are quite high, and the desert air is very dry. He has a portable water filter to treat river and stream water that he finds on his planned route through the Gila wilderness. On the second day of his weeklong trip his water filter breaks. He is afraid of contracting giardiasis (a protozoal disease spread through water contaminated by animal feces), so he drinks only the small amount of water that he can boil on his camp stove at night. By the fifth night he is feeling weak and thirsty, and starts to hike out. He makes it to a local highway and collapses. A passing motorist calls 911 for an ambulance.

a. Given that Marc has sweat a lot, and that sweat causes the loss of more water than solutes, what has happened to the solute concentration of his blood as a result of his dehydration?

b. From the solute concentration of his blood, what is likely to be happening to those of his body cells that are in contact with his blood and related fluids (e.g., lymph and cerebral spinal fluid)?

c. The paramedics have available three saline solutions. One is a "normal" isotonic saline—0.9% NaCl. One is a hypertonic saline (3% NaCl). The last is a "half normal" saline (0.45% NaCl). Which one would you use to treat Marc? Why?

DRIVING QUESTION 3 How do different antibiotics target bacteria?

By answering the questions below and studying Infographics 3.1, 3.4 (bottom), and 3.5, you should be able to generate an answer for the broader Driving Question above.

KNOW IT

17 Penicillin interferes with the synthesis of _____.

a. bacterial cell membranes

b. peptidoglycan

c. the nuclear envelope

d. membrane proteins

e. ribosomes

18 Would phospholipids of the cell membrane be a good target for an antibiotic? Explain your answer.

USE IT

19 A bacterial infection is being treated with an antibiotic that stops bacterial reproduction by blocking DNA replication (bacterial cells cannot reproduce if they cannot replicate their DNA). The physician decides to add penicillin, which inhibits the production of new peptidoglycan. Would this use of penicillin be effective (i.e., will it have any additional impact on treating the infection)? Explain your answer.

20 If bacterial cells were placed in a nutrient-containing solution (one that supports their growth) that had the same solute concentration as the cytoplasm, and which also contained penicillin, would the cells burst? Explain your answer. What if the same experiment were repeated with lysozyme, an enzyme that degrades intact peptidoglycan? What if the two experiments were repeated in solutions that have lower solute concentrations than the cytoplasm, and did *not* contain growth-supporting nutrients?

21 Fungi are eukaryotic organisms. Scientists have found it more challenging to develop treatments for fungal infections (e.g., yeast infections, athlete's foot, and certain nail infections) than for bacterial infections. Why is this so?

apply YOUR KNOWLEDGE

INTERPRETING DATA

22 Bacteria can be characterized as sensitive, intermediately resistant, or fully resistant to different antibiotics. If a strain of bacteria is sensitive to an antibiotic, we can prescribe that antibiotic to treat an infection caused by that strain and have confidence that it will work. If the strain is fully resistant to an antibiotic, that antibiotic cannot treat that infection. In cases of intermediate resistance, it is better to try to find an antibiotic to which the strain is sensitive, as the infection may not respond to antibiotics to which it has intermediate resistance.

The table shows the concentrations of antibiotics that determine how a bacterial species will respond to those antibiotics. A sensitive strain will be killed by the concentration of antibiotic shown in the "sensitive" column. A strain with intermediate resistance will only be affected by concentrations in the range indicated in the "intermediately resistant" column. And a fully resistant strain requires concentrations shown in the "fully resistant" column in order to be killed.

Antibiotic	Effective Dose for Sensitive Bacterial Strains (μg/ml)	Effective Dose for Intermediately Resistant Bacterial Strains (μg/ml)	Effective Dose for Fully Resistant Bacterial Strains (μg/ml)
Oxacillin	≤2 μg/ml		≥4 μg/ml
Vancomycin		8–16 μg/ml	≥32 μg/ml
Erythromycin	≤0.5 μg/ml	1–4 μg/ml	≥8 μg/ml
Tetracycline	≤4 μg/ml		≥16 μg/ml
Levofloxacin	≤2 μg/ml	4 μg/ml	≥8 μg/ml

A hospital patient has a *Staphylococcus aureus* infection. As part of laboratory testing, the *S. aureus* from the patient was grown in different concentrations of various antibiotics. For oxacillin, the lowest concentration that inhibited the growth of the strain was 8 μg/ml; for vancomycin, 4 μg/ml; for erythromycin, 16 μg/ml; for tetracycline, 32 μg/ml; and for levofloxacin, 8 μg/ml. Which antibiotic should be used to treat the infection in this patient?

apply YOUR KNOWLEDGE

BRING IT HOME

23 Many patients attempt to pressure their physician to prescribe antibiotics for colds. Is this a good idea? Why or why not?

DRIVING QUESTION 4 What are the key eukaryotic organelles and their functions?

By answering the questions below and studying Infographic 3.8 and Up Close: Eukaryotic Organelles, you should be able to generate an answer for the broader Driving Question above.

KNOW IT

24 Briefly describe the structure and function of each of the following eukaryotic organelles:

a. mitochondrion
b. nucleus
c. endoplasmic reticulum
d. chloroplast

25 Which of the following is *not* a cytoskeletal fiber in eukaryotic cells?

a. macrotubules
b. intermediate filaments
c. microfilaments
d. microtubules
e. All of the above are cytoskeletal fibers.

26 Insulin is a protein hormone secreted by certain pancreatic cells into the bloodstream. Which of the following organelles is/are involved in the synthesis and secretion of insulin?

a. rough ER
b. Golgi apparatus
c. ribosomes
d. all of the above
e. a and c

USE IT

27 Some inherited syndromes, for example Tay-Sachs disease and MERRF (myoclonic epilepsy with ragged red fibers), interfere with the function of specific organelles. MERRF disrupts mitochondrial function. From what you know about mitochondria, why do you think the muscles and the nervous system are the predominant tissues affected in MERRF? (Think about the activity of these tissues compared to, say, skin.)

28 Which organelle would cause the most damage to cytoskeletal fibers in the cytoplasm if its contents were to leak into the cytoplasm?

a. smooth ER
b. nucleus
c. lysosome
d. Golgi apparatus
e. rough ER

29 Cystic fibrosis is an inherited condition that affects the lungs and digestive tract (see Chapter 11). In many people with cystic fibrosis, a membrane channel protein is found in the rough endoplasmic reticulum instead of the cell membrane. How could a cell membrane protein end up in the rough endoplasmic reticulum? (Hint: Look at the box about the cooperation of the nucleus, endoplasmic reticulum, and Golgi apparatus in Up Close: Eukaryotic Organelles.)

1

Milestones in Biology

SCIENTIFIC REBEL

Lynn Margulis and the theory of endosymbiosis

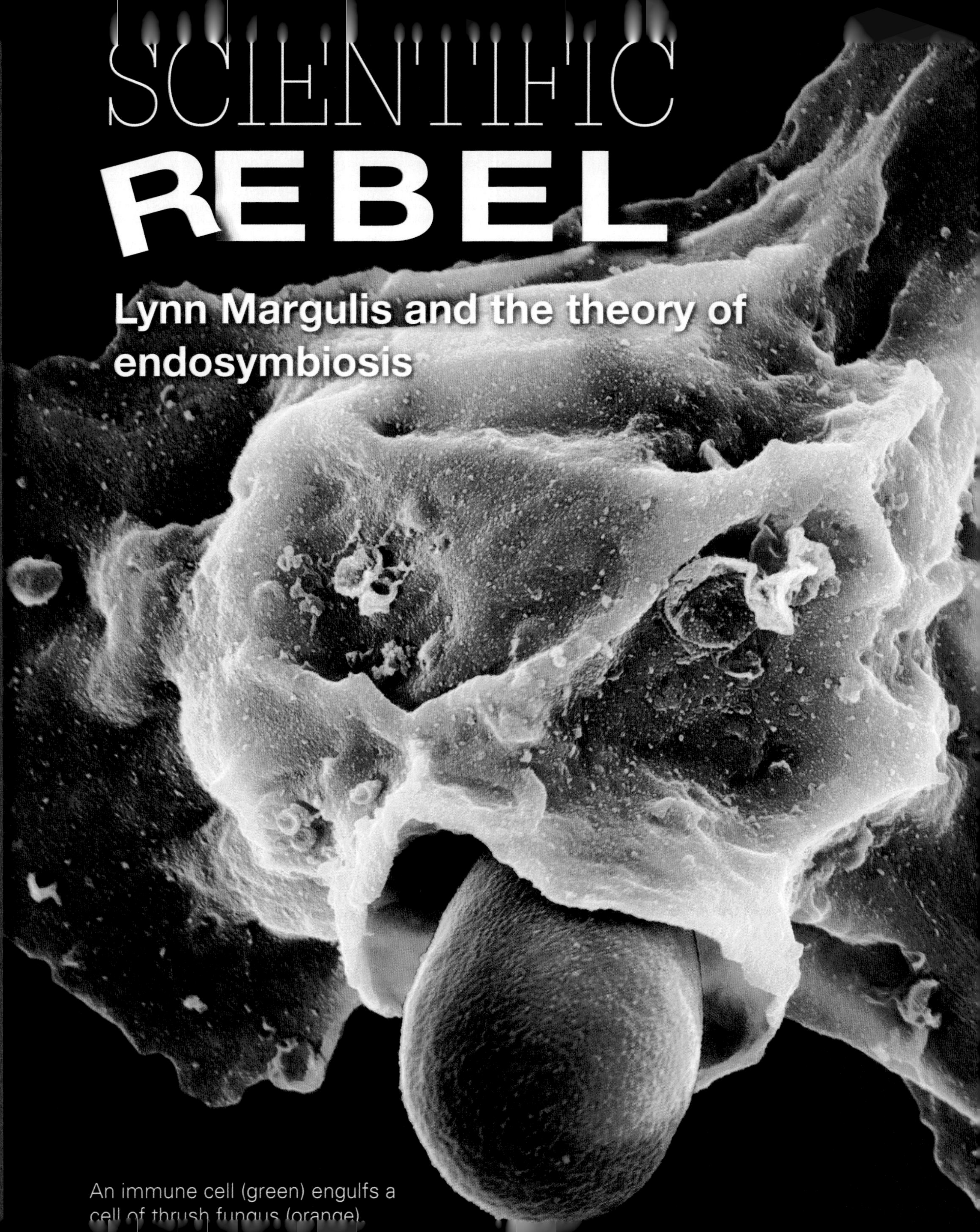

An immune cell (green) engulfs a cell of thrush fungus (orange).

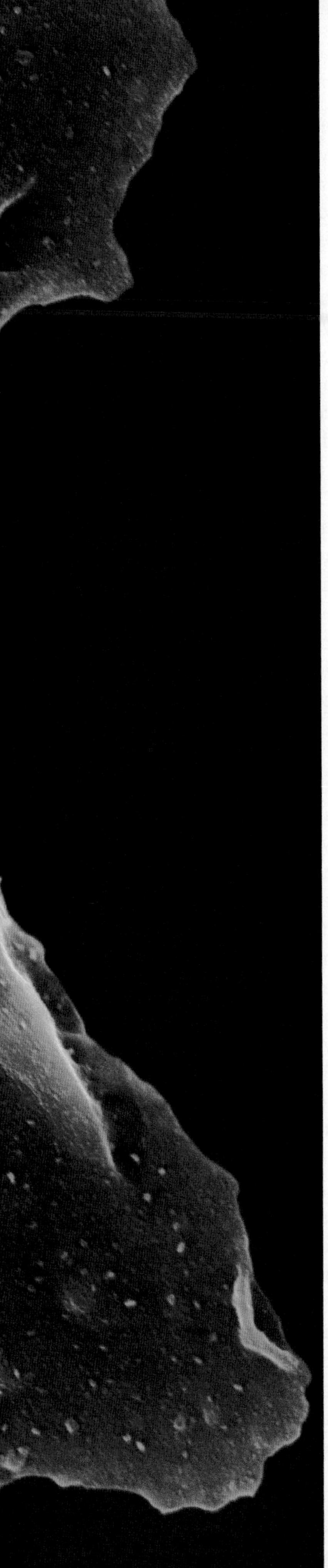

DRIVING QUESTIONS

1. What does endosymbiosis say about the origin of organelles such as mitochondria and chloroplasts?
2. What evidence supports the proposed origin of organelles such as mitochondria and chloroplasts?

> "Life on Earth is such a good story you can't afford to miss the beginning."
>
> —Lynn Margulis

LYNN MARGULIS NEVER MET A MICROBE she didn't like. From the time she first peered through a microscope as a teenager at the University of Chicago and witnessed a world of unicellular organisms swimming in a drop of pond water, she was hooked. Other biologists were impressed by the adaptations of plants and animals, but Margulis was smitten with microscopic life. Long before there were animals on the planet, she pointed out, there were microbes, and if it weren't for them, we humans wouldn't be here.

"Life on Earth is such a good story you can't afford to miss the beginning," Margulis said in an interview with the University of Wisconsin alumni magazine. "Do historians begin their study of civilization with the founding of Los Angeles? This is what studying natural history is like if we ignore the microcosm."

In 1966, when she was 28, Margulis used her knowledge of microbial life to propose a radical hypothesis about how cells had come to be. What distinguishes eukaryotic cells from prokaryotic cells, Margulis knew, was the presence in eukaryotic cells of internal membrane-bound organelles (see Chapter 3). Margulis proposed that eukaryotic cells, with their internal organelles, had formed when one prokaryotic cell engulfed–ate–another. Instead of being digested, the ingested cell survived and took up residence in its new host.

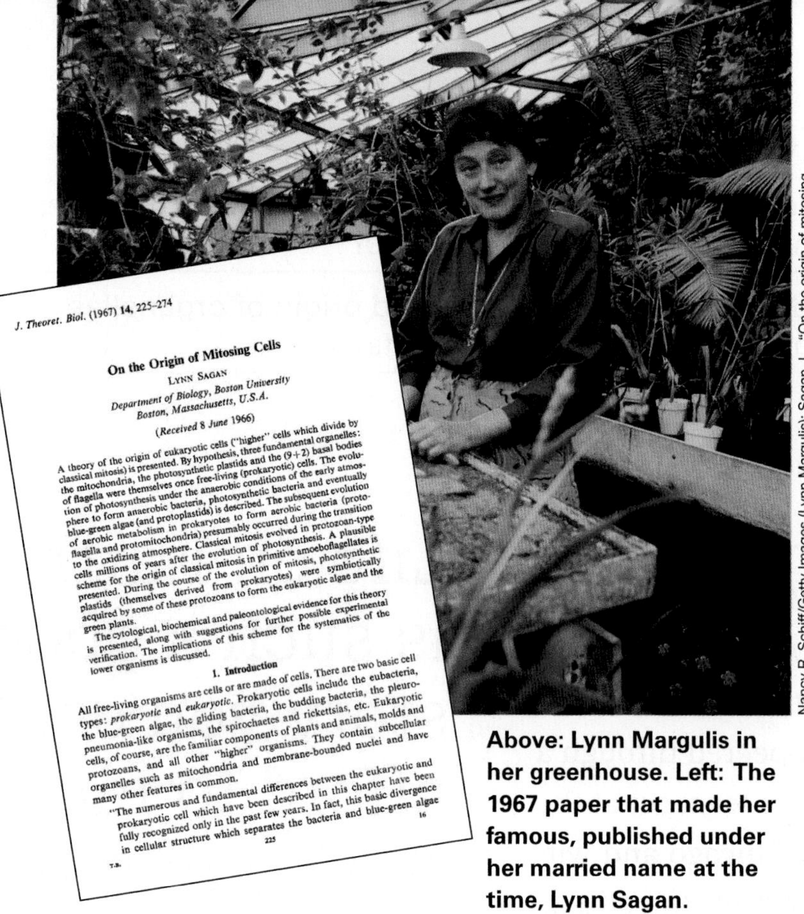

J. Theoret. Biol. (1967) **14**, 225–274

On the Origin of Mitosing Cells

LYNN SAGAN

Department of Biology, Boston University
Boston, Massachusetts, U.S.A.

(*Received 8 June 1966*)

A theory of the origin of eukaryotic cells ("higher" cells which divide by classical mitosis) is presented. By hypothesis, three fundamental organelles: the mitochondria, the photosynthetic plastids and the (9+2) basal bodies of flagella were themselves once free-living (prokaryotic) cells. The evolution of photosynthesis under the anaerobic conditions of the early atmosphere to form anaerobic bacteria, photosynthetic bacteria and eventually blue-green algae (and protoplastids) is described. The subsequent evolution of aerobic metabolism in prokaryotes to form aerobic bacteria (protoflagella and protomitochondria) presumably occurred during the transition to the oxidizing atmosphere. Classical mitosis evolved in protozoan-type cells millions of years after the evolution of photosynthesis. A plausible scheme for the origin of classical mitosis in primitive amoeboflagellates is presented. During the course of the evolution of mitosis, photosynthetic plastids (themselves derived from prokaryotes) were symbiotically acquired by some of these protozoans to form the eukaryotic algae and the green plants.

The cytological, biochemical and paleontological evidence for this theory is presented, along with suggestions for further possible experimental verification. The implications of this scheme for the systematics of the lower organisms is discussed.

1. Introduction

All free-living organisms are cells or are made of cells. There are two basic cell types: *prokaryotic* and *eukaryotic*. Prokaryotic cells include the eubacteria, the blue-green algae, the gliding bacteria, the budding bacteria, the pleuropneumonia-like organisms, the spirochaetes and rickettsias, etc. Eukaryotic cells, of course, are the familiar components of plants and animals, molds and protozoans, and all other "higher" organisms. They contain subcellular organelles such as mitochondria and membrane-bounded nuclei and have many other features in common.

"The numerous and fundamental differences between the eukaryotic and prokaryotic cell which have been described in this chapter have been fully recognized only in the past few years. In fact, this basic divergence in cellular structure which separates the bacteria and blue-green algae

16

225

T.B.

Above: Lynn Margulis in her greenhouse. Left: The 1967 paper that made her famous, published under her married name at the time, Lynn Sagan.

Eventually, the two cells formed a mutually beneficial relationship, and the result was a complex eukaryotic cell. Margulis called her idea endosymbiosis (from the Greek "endo" meaning "within," and "symbiosis" meaning "living together").

Many scientists dismissed the idea outright as a crackpot notion. The paper she wrote proposing the idea, which she titled "On the Origin of Mitosing Cells," was rejected by about 15 scientific journals before finally being accepted by the *Journal of Theoretical Biology* in 1967. But an interesting thing happened after it was published: people couldn't stop talking about it.

Microscopic Clues

In one sense, the idea of endosymbiosis was not entirely new. A few biologists in the late 19th century and early 20th century had suggested it after observing the remarkable resemblance between free-living bacteria and certain organelles of the eukaryotic cell. But in those early days it was not possible to test the idea, and so it was largely ignored. In 1925, Edmund Wilson, a prominent cell biologist, wrote, "To many, no doubt, such speculations may appear too fantastic for present mention in polite biological society." But he went on to suggest that those speculations might "someday call for serious consideration."

In 1960, while working on her master's degree in biology at the University of Wisconsin, Margulis first became intrigued by the notion of endosymbiosis. Her advisors, Walter Plaut and Hans Ris, had recently made the startling discovery that chloroplasts–the tiny green organelles inside plant cells that carry out photosynthesis–have their own DNA, the molecule of heredity. Most DNA in a eukaryotic cell is housed in the nucleus, where it serves as the genetic blueprint for life. What was DNA doing in a chloroplast? To Margulis, the discovery suggested that chloroplasts had once led a separate existence as independent, free-living cells, and thus needed DNA to reproduce, much as bacteria do.

She pursued this idea further as part of her Ph.D. work at the University of California at Berkeley. In particular, she studied the small unicellular eukaryote called euglena, which lives in water and contains numerous chloroplasts. She used radioactively labeled nucleotides, the building blocks of DNA, to show that the little circular squiggle inside a euglena chloroplast was indeed DNA, as her advisors had claimed. (Cells incorporate the nucleotides into DNA when they divide, and the radioactivity can be detected with photographic film.) This work lent additional support to the idea of endosymbiosis.

The more Margulis looked, the more evidence she found. Not only did chloroplasts have their own circular DNA, but they also had ribosomes–the structures that, in the cytoplasm of both prokaryotic and eukaryotic cells, synthesize proteins–and these ribosomes were essentially the same size as those in prokaryotic cells. Chloroplasts were also about the same size as bacteria, and to reproduce they divided much the same way bacteria divide. Further investigation revealed that mitochondria, the organelles that serve as the cell's "power plants," also contained circular DNA and prokaryotic-size ribosomes, and divided in a fashion similar to bacteria. This evidence suggested that mitochondria, too, had once been free-living cells (**INFOGRAPHIC M1.1**).

In her 1967 paper, Margulis gave credit to the biologists who had previously suggested endosymbiosis, but she did more than simply rehash old ideas. She brought together for the first time all the existing evidence from cell biology and biochemistry and wove it together into a coherent account. She also put the idea into evolutionary context and offered a rough timeline of when these events happened.

In Margulis's view, the first cells on Earth were prokaryotic and arose some 3.5 billion years ago. These early bacteria-like cells were anaerobic—they did not require oxygen to live (which makes sense, since the early Earth had no substantial oxygen). These cells evolved and diversified for a billion years until, about 2.5 billion years ago, some developed the capacity to harvest the energy of sunlight while also splitting water molecules and releasing oxygen gas as a by-product. As oxygen built up in the atmosphere as a result of this photosynthesis, it produced an

INFOGRAPHIC M1.1 Chloroplasts and Mitochondria Share Traits with Bacteria

Margulis observed that chloroplasts and mitochondria shared several traits with free-living bacteria.

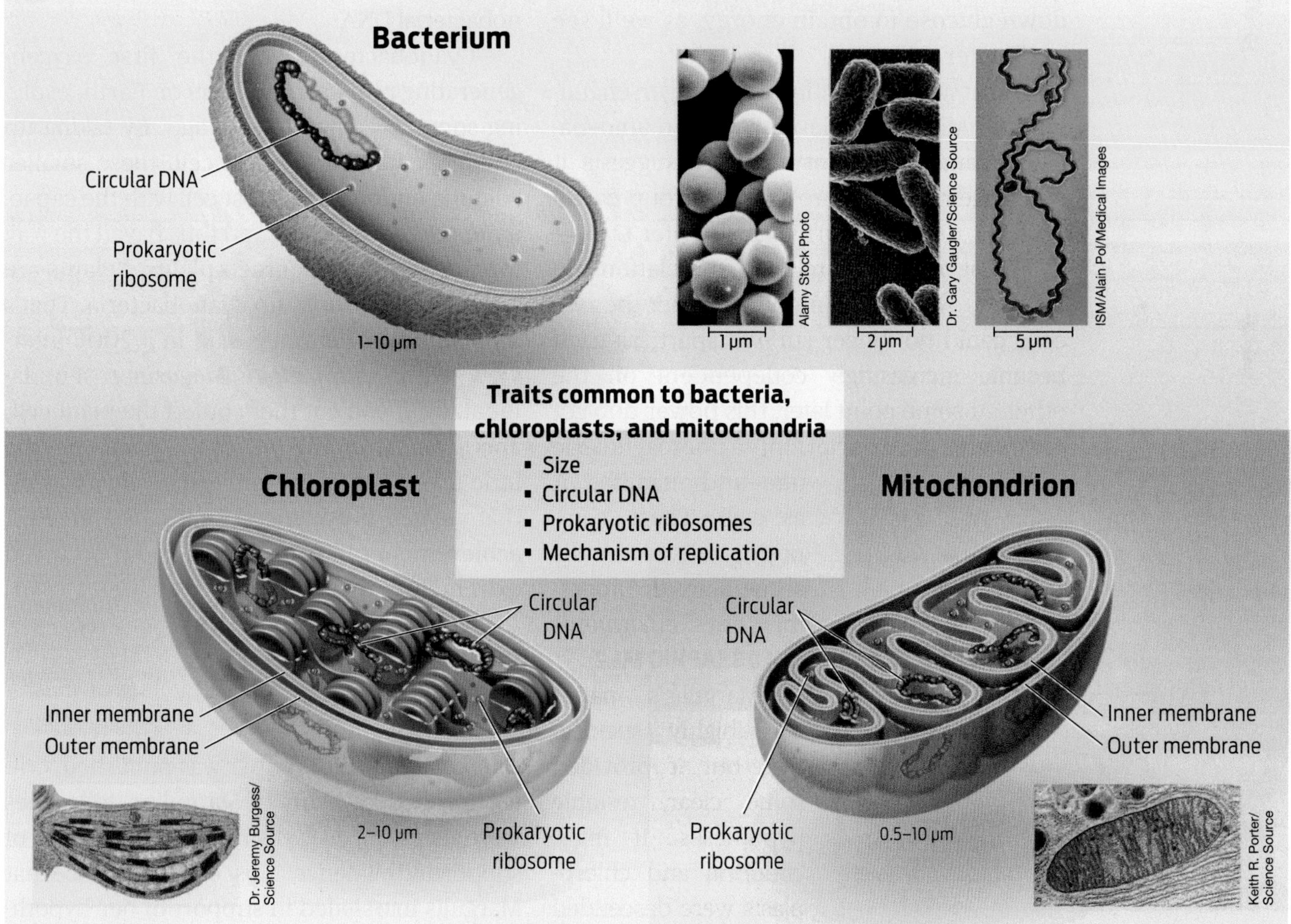

? What is the shape of the DNA molecule in bacteria, chloroplasts, and mitochondria?

environment that strongly favored bacteria that could use oxygen to live.

At that point, according to Margulis, one such oxygen-using, or aerobic, bacterium was engulfed by a larger anaerobic prokaryotic cell. The ingested bacterium was not destroyed; instead, the two cells formed a symbiotic, mutually beneficial relationship. The smaller aerobic bacterium enabled the larger anaerobic cell to use oxygen to obtain energy, and the larger anaerobic cell provided the smaller bacterium with a source of sugars. (These ideas bear a strong resemblance to the way that eukaryotic cells still, to this day, break down glucose to obtain energy, as we'll see in Chapter 6.)

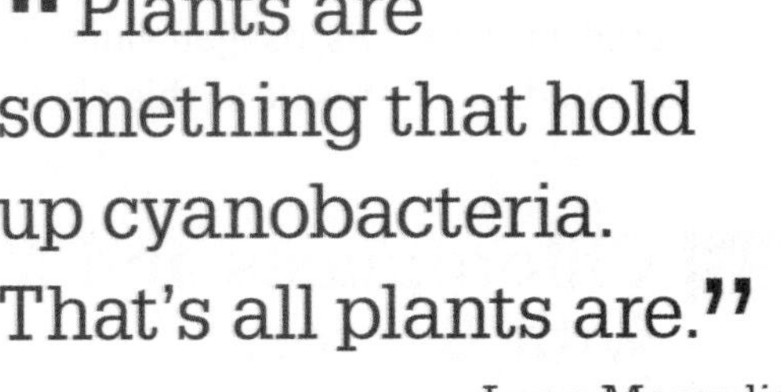

What was the identity of this larger, engulfing cell? Margulis believed it was an amoeba-like organism; current evidence suggests it was a member of a prokaryotic group called Archaea, discussed further in Chapter 17.

What began as a mutualistic relationship became, over time, an obligate one: the two cells could no longer survive apart, as each became increasingly codependent on the other. At some point later, this power duo got cozy with a third bacterium—a photosynthetic one—and engulfed it as well. These composite cells, Margulis argued, are the ancestors of eukaryotes (**INFOGRAPHIC M1.2**).

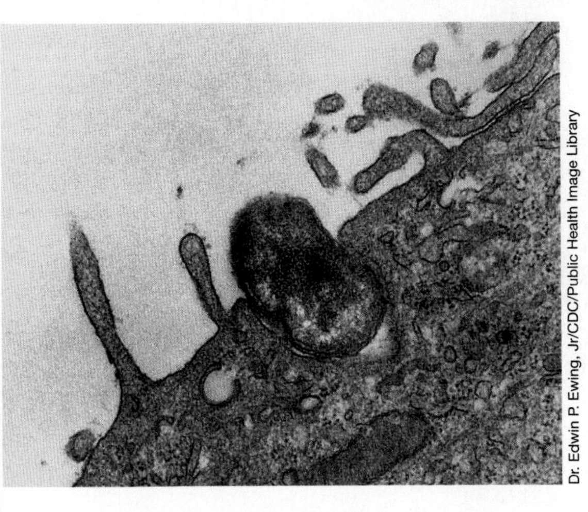

Dr. Edwin P. Ewing, Jr/CDC/Public Health Image Library

Electron micrograph of a *Rikettsia* bacterium burrowing inside a mouse cell.

Margulis's paper was highly speculative, but it provided some clear, testable hypotheses. If mitochondria and chloroplasts were descended from free-living bacteria, then it should be possible to determine from their DNA what those free-living bacteria were. At the time Margulis wrote her paper, chemical analysis of DNA was in its infancy, and she couldn't use this technique. But by the mid-1970s, analyzing DNA had become routine, and it was DNA evidence that clinched her case: by comparing the sequences of mitochondrial and chloroplast DNA to a wide range of bacterial DNA, researchers discovered that mitochondrial DNA closely resembled DNA from a small bacterium called *Rikettsia*—interestingly, a type of intracellular parasite that burrows inside other cells in order to live. Chloroplast DNA, they discovered, is essentially the same as cyanobacterial DNA.

Cyanobacteria were the first oxygen-generating photosynthesizers on Earth, evolving some 2.5 billion years ago. By taking up residence in a larger host cell, these smaller bacteria endowed the host cell with the capacity to photosynthesize and thus paved the way for the evolution of green plants. "Plants are something that hold up cyanobacteria. That's all plants are," Margulis said in a 2006 interview with *Astrobiology Magazine*. "Fundamentally, if you cut them out of the plant cell, and throw away the rest of the plant cell, the little green dot is the only thing that can do that oxygen production. That is the greatest achievement of life on Earth, and it occurred extremely early in the history of life."

From Heresy to Orthodoxy

As soon as her 1967 paper was published, criticism rolled in. Many of Margulis's colleagues were skeptical, even dismissive, citing a lack of supporting evidence. Most of the evidence that Margulis marshaled in support of her hypothesis was circumstantial rather than direct (the conclusive DNA evidence did not come until

INFOGRAPHIC M1.2 The First Eukaryotes Were Products of Endosymbiosis

All eukaryotic cells are characterized by the presence of a membrane-enclosed nucleus and other organelles. While the origin of the nucleus is not yet completely understood, there is good evidence that mitochondria and chloroplasts arose when ancient eukaryotic ancestors engulfed bacteria.

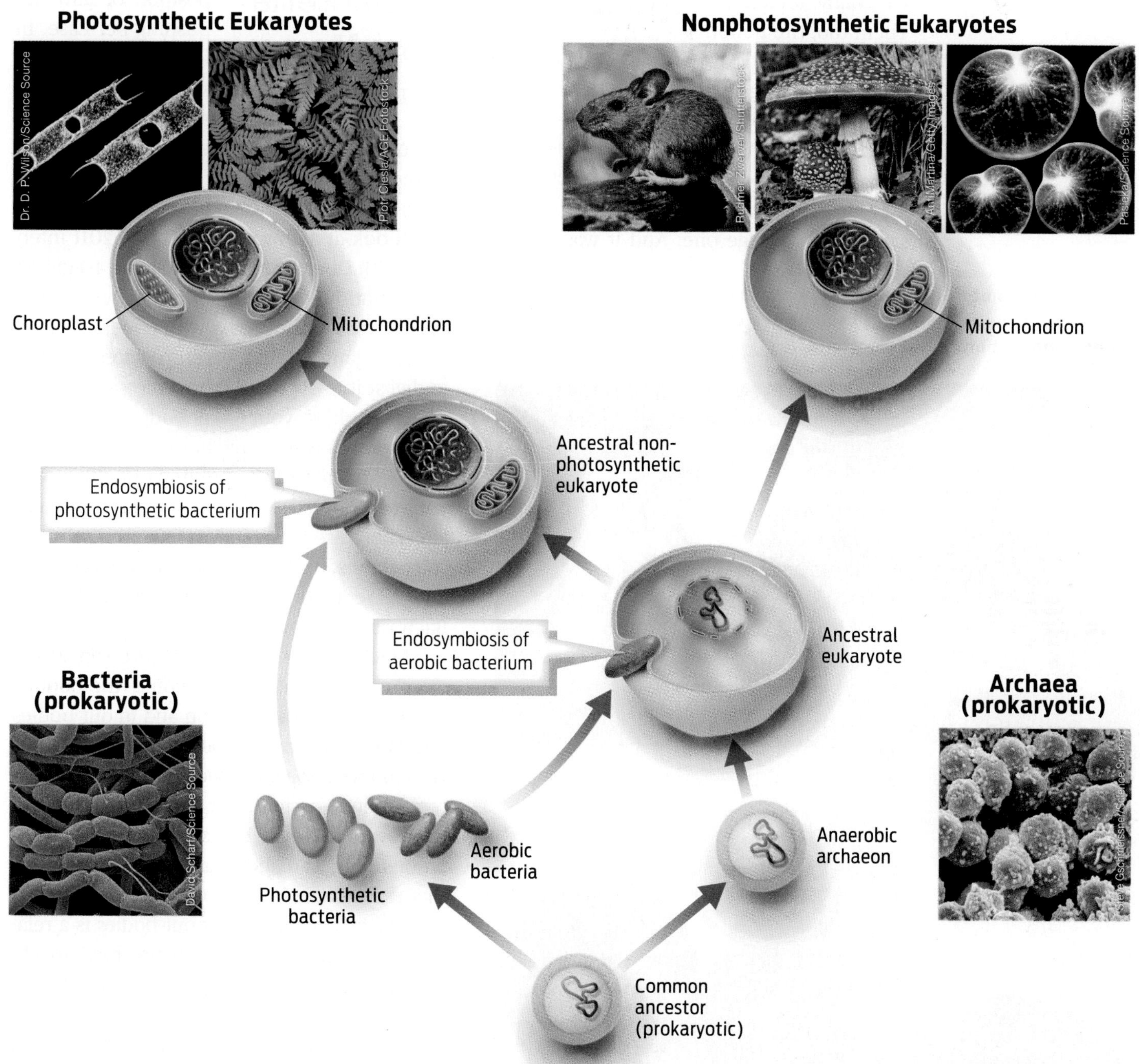

? Are mitochondria most closely related to archaea, photosynthetic bacteria, or non-photosynthetic bacteria?

1978). There was philosophical opposition, too. To many people, bacteria were "bad" because they were known to cause disease; the idea that we have bacteria to thank for our very existence was difficult for many to accept. Then, too, the notion of endosymbiosis seemed to go against evolutionary dogma, which held that evolution occurred in small steps as a result of an individualistic "struggle for existence." Endosymbiosis was not a small step, it was a huge one. And it wasn't about competition as much as cooperation. For these reasons, it just didn't sit well with many hard-nosed Darwinists.

> "We take for granted their influence. Bacteria are our ancestors."
>
> —Lynn Margulis

For her part, Margulis never shrank from her position or gave up pushing her case. In fact, the resistance she encountered seemed almost to embolden her, and she spent years uncovering many other examples of symbiotic relationships at work in nature.

"Look at a cow," she said in a 2011 interview with *Discover* magazine. "It is a 40-gallon fermentation tank on four legs. It cannot digest grass and needs a whole mess of symbiotic organisms in its overgrown esophagus to digest it."

Or look at your own body. "There are hundreds of ways your body wouldn't work without bacteria," she pointed out. "Between your toes is a jungle; under your arms is a jungle. There are bacteria in your mouth, lots of spirochetes, and other bacteria in your intestines. We take for granted their influence. Bacteria are our ancestors."

Much current research is focused on understanding our human microbiome—the population of bacteria that lives on and in our bodies and influences many aspects of our health. In addition to helping us digest food and shaping our immune system, our unique microbiome may even influence our susceptibility to conditions such as diabetes and obesity. The idea that these bacteria are not passive freeloaders but crucial constituents of our bodies is a relatively new way of looking at things—but it would hardly have come as a surprise to Margulis.

When Margulis died in 2011 at the age of 73, she was remembered as the person who most fundamentally changed our view of cells. These days, the idea that mitochondria and chloroplasts started as free-living organisms is accepted as fact by the scientific mainstream, and we now refer to the theory of endosymbiosis, acknowledging the abundant evidence and wide support it has.

Animals live symbiotically with a vast universe of bacteria, collectively known as the microbiome—from bacteria in a cow's digestive tract (top), to the bacteria on human skin (bottom).

Janice Carr/CDC/Public Health Image Library (inset); Peter Zijlstra/Shutterstock (cows).

DWaschnig/Shutterstock (sweating man); CNRI/Science Source (inset).

"The evolution of the eukaryotic cells was the single most important event in the history of the organic world," said Ernst Mayr, the grandfather of modern evolutionary studies. "Margulis's contribution to our understanding the symbiotic factors was of enormous importance." Richard Dawkins, a prominent British evolutionary biologist, described the theory of endosymbiosis as "one of the great achievements of twentieth-century evolutionary biology." Botanist Peter Raven said the idea caused "nothing less than a revolution" in our thinking about the cell.

Not all of Margulis's ideas have gained widespread acceptance. In particular, her claim that the whiplike tail of a sperm cell derives from a formerly free-living bacterium called a spirochete lacks convincing evidence and is not accepted as fact by the scientific establishment. But on one of the most important questions of modern biology, her intellectual daring paid off.

PAUL HOSEFROS/The New York Times/Redux Pictures

Margulis receiving the National Medal of Science from President Bill Clinton in 1999.

Asked by the *Discover* interviewer how she felt about being the source of so many controversial ideas over the years, Margulis responded in typical fashion: "I don't consider my ideas controversial. I consider them right." ■

MORE TO EXPLORE

- Sagan, L. (1967) On the origin of mitosing cells. *Journal of Theoretical Biology* 14: 225–274.
- Margulis, L., and Sagan, D. (1986) *Microcosmos: Four Billion Years of Evolution from Our Microbial Ancestors*. New York: Summit Books.
- Teresi, D. (2011) Interview with Lynn Margulis. *Discover*. http://discovermagazine.com/2011/apr/16
- Sapp, J. (2012) "Too Fantastic for Polite Society: A Brief History of Symbiosis Theory." In *Lynn Margulis: The Life and Legacy of a Scientific Rebel*, edited by Dorian Sagan. White River Junction, Vermont: Chelsea Green Publishing.

MILESTONES IN BIOLOGY **1** Test Your Knowledge

1 What is the function of mitochondria? Of chloroplasts?

2 What evidence did Margulis present to support her hypothesis that organelles had once been free-living prokaryotic organisms?

3 On the basis of DNA sequence analysis,

- **a.** which bacteria are likely the closest relatives of the chloroplast?
- **b.** which bacteria are likely the closest relatives of the mitochondria?

4 Which of the following is *not* a trait shared by chloroplasts and prokaryotic cells?

- **a.** the size of their ribosomes
- **b.** the shape of their DNA molecule
- **c.** the presence of a nucleus
- **d.** their mechanism of replication

5 From what you have read here about endosymbiosis:

- **a.** Could you live without your endosymbiotic organelles? Why or why not?
- **b.** Could you live if plants did not have their endosymbiotic organelles? Explain your answer.

4 Nutrition, Enzymes, Metabolism

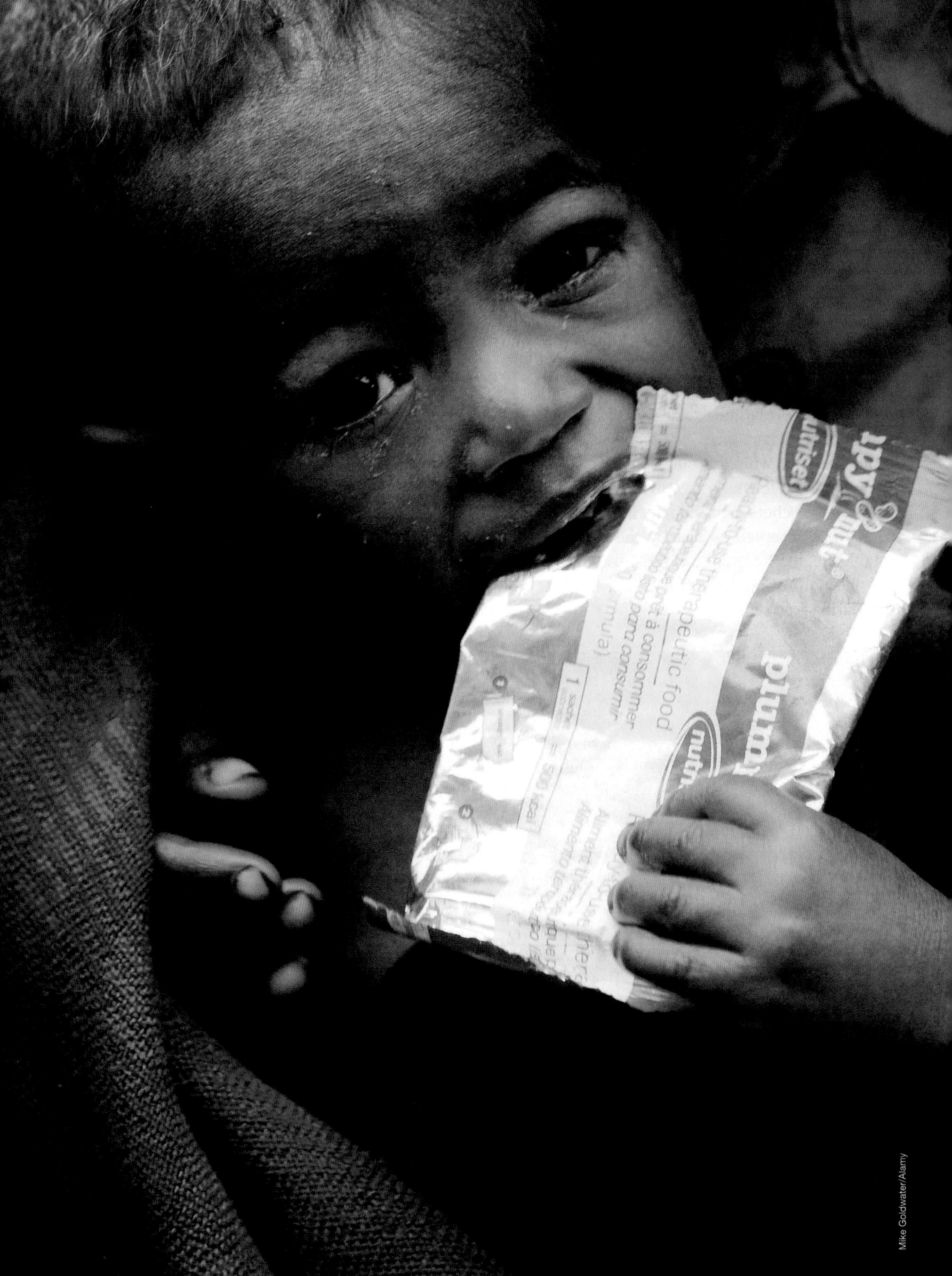

Mike Goldwater/Alamy

The PEANUT BUTTER Project

One doctor's crusade to end malnutrition in Africa, a spoonful at a time

DRIVING QUESTIONS

1. What are the macronutrients and micronutrients provided by food?
2. What are essential nutrients?
3. What are enzymes, how do they work, and how do they contribute to reactions of metabolism?
4. What are the consequences of a diet lacking sufficient nutrients?

AT A MEDICAL CONFERENCE IN 2003, during a panel on malnutrition, someone in the audience stood up and shouted at the presenter, "You're killing children!" Mark Manary was discussing a new method of treating malnutrition in children that he had used successfully in some of the poorest countries in Africa. But it flew in the face of guidelines endorsed by the World Health Organization (WHO). Now, leaders in the field were angry.

Manary, a pediatrician from St. Louis, Missouri, had firsthand experience with the standard WHO treatment. For much of the 1990s, he had devoted his life to it: in hospital wards in Malawi, a small, landlocked country in southeastern Africa, he faithfully administered fortified milk solutions to sickly children in accordance with WHO guidelines for treating malnutrition. But even with this treatment, children continued to die, and only a small percentage ever got better. "On our best days only 10% of the kids would die. But still only 25% would recover," he says.

Manary decided to find a better way. After searching around for potential alternatives, he settled on a surprising solution: peanut butter.

Peanut butter, he found, is well suited to the purpose. It's packed with nutrients, it doesn't require cooking, and–most important–it doesn't spoil easily and can be kept unrefrigerated in tropical climates for up to 3 months.

Manary began testing his peanut butter treatment in 2001. He gave his patients an ample supply of peanut butter and released them from the hospital. "Basically, what we did was empty the place out and send everybody home." Manary says.

The hospital staff was appalled by his brashness, but the gamble paid off. Within weeks, 95% of the children eating peanut butter had fully recovered.

Manary is now the director of Project Peanut Butter, a nonprofit organization using the American pantry staple to end malnutrition in Africa. Spin-offs of the organization have sprung up all around the globe, including in Haiti and South Asia.

Once criticized by the international aid community, Manary's method is now being hailed as a kind of miracle cure. Supporters of the approach have even likened it to the discovery of penicillin, a treatment with the potential to save millions of lives every year.

The Elephant in the Room

In 1994, Manary became a visiting faculty member at the University of Malawi School of Medicine. When he arrived, he asked the hospital director, "What's your biggest problem?" The director replied, "The malnutrition ward." So that's where Manary chose to work.

The ward was a dismal and discouraging place–a large room where 50 starving children lay close to death on crowded cots, their bodies little more than skin and bone. About a third of the children would die, despite the doctors' best efforts at rehabilitating them. This poor response rate was, Manary says, "the elephant in the room."

Percent undernourished worldwide, 2015

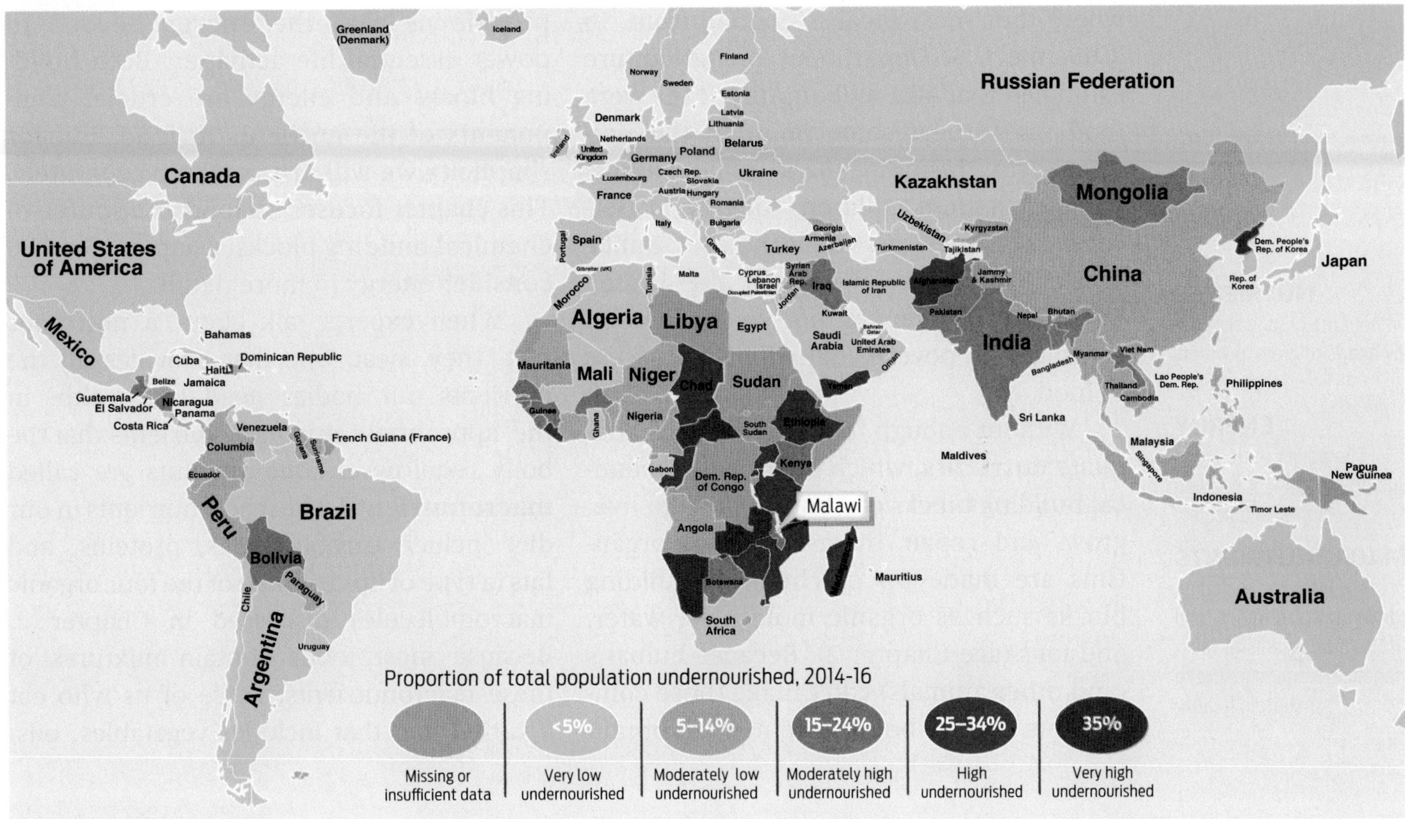

Data courtesy World Food Programme

Food is the most basic of human requirements for survival, but on average 1 in 8 people goes to bed hungry each night.

Severe acute **malnutrition** is the number one killer of children in the world. An estimated 3.5 million children die from malnutrition every year—more than the number who die from AIDS, tuberculosis, and malaria combined. Most of these deaths occur in sub-Saharan Africa, where grinding poverty is endemic, and food is scarce for large portions of the year.

In Malawi, the crisis is particularly acute. The World Food Programme of the United Nations estimates that 2.8 million people, or about 16% of the country's population, experienced food shortages at some point in 2016. Historically, more than half of all Malawian children are chronically malnourished, and nearly one in four die from consequences of this malnutrition.

Why is food so scarce? Malawi is one of the poorest countries in the world, with a population of mostly subsistence farmers. The primary agricultural crops are corn and soybeans. Farmers also grow commercial cash crops for export, including tobacco, sugar cane, coffee, and tea. But agriculture is not easy in Malawi. Rain comes only once a year, from December to March, when it might rain every day. New crops are planted during this time, but the harvest won't be available until March. Quite often, supplies from the previous year's harvest run out before the next one is in. Locals call this time of food insecurity "the hungry season."

Food insecurity is by no means restricted to developing nations like Malawi. Even in the

MALNUTRITION
The medical condition resulting from the lack of essential nutrients in the diet. Malnutrition is often, but not always, associated with starvation.

United States, many families do not know where their next meal is coming from. In 2015, the U.S. Department of Agriculture estimated that 48.1 million Americans were food insecure at some point in the previous year, including 32.8 million adults and 15.3 million children—roughly 14 percent of American households. Unlike in the developing world, in the United States the problem is not so much food shortages as poverty that prevents access to food.

NUTRIENTS
Components in food that the body needs to grow, develop, and repair itself.

ENERGY
The ability to do work, including the work of building complex molecules.

MACRONUTRIENTS
Nutrients, including carbohydrates, proteins, and fats, that organisms must ingest in large amounts to maintain health.

Without enough food, people lack adequate **nutrients,** which provide the chemical building blocks our bodies need to live, grow, and repair themselves. All organisms are made up of chemical building blocks such as organic molecules, water, and ions (see Chapter 2). Because humans (and other animals) can't make these components in their bodies, we need to obtain them from our diet. These nutrients also provide us with the **energy** needed to power essential life activities. Both building blocks and energy are crucial components of the nutrients in food, but for simplicity we will discuss them separately. This chapter focuses on food as a source of chemical building blocks; Chapters 5 and 6 consider energy in more detail.

When experts talk about a nutritious diet, they mean one that provides all the nutrients our bodies need for health, in the appropriate amounts. Nutrients that the body requires in large amounts are called **macronutrients.** The macronutrients in our diet include carbohydrates, proteins, and fats (a type of lipid)—three of the four organic macromolecules discussed in Chapter 2. Because most foods contain mixtures of these macronutrients, those of us who eat a varied diet that includes vegetables, oils,

Food insecurity in the US, 2014

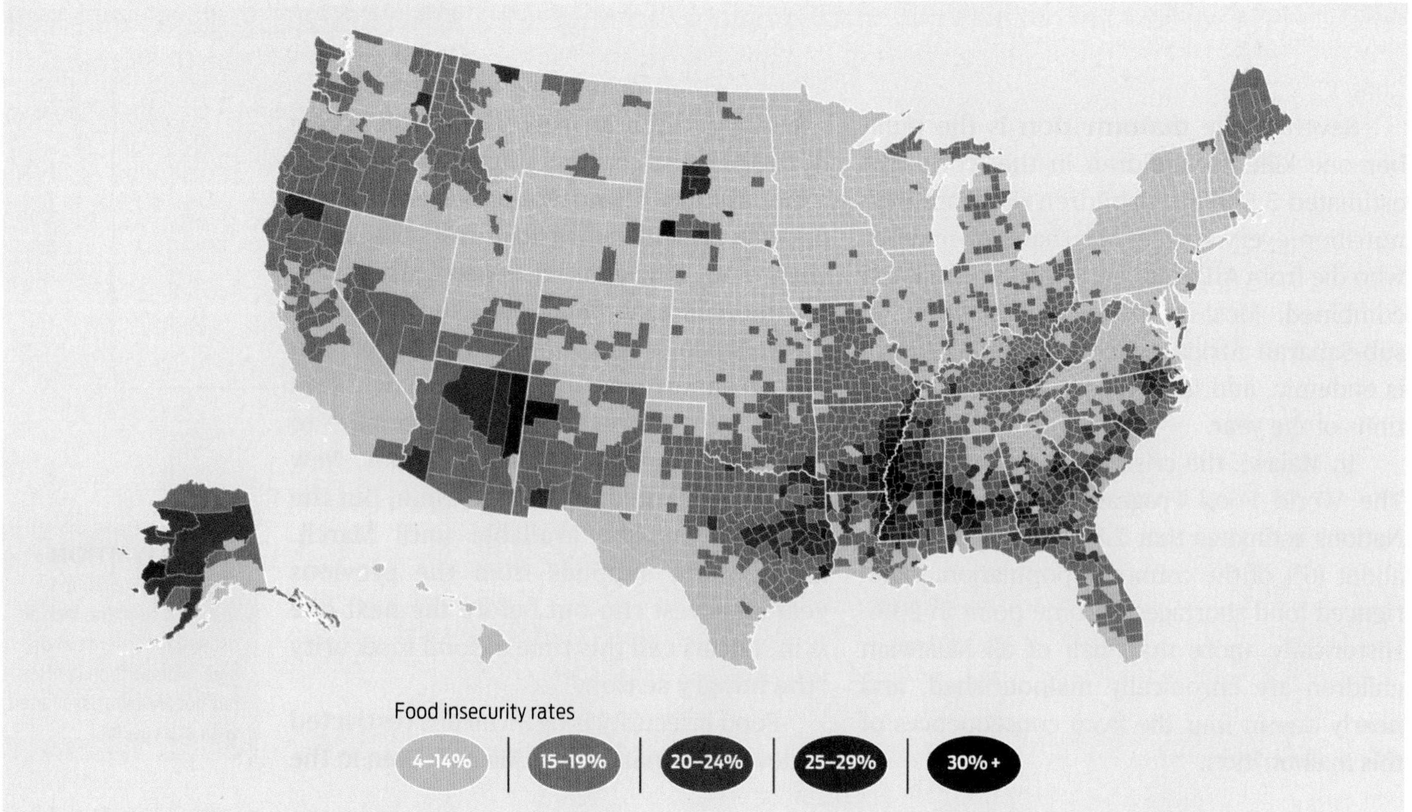

Data from C. Gundersen, A. Dewey, A. Crumbaugh, M. Kato & E. Englehard. Map the Meal Gap 2016. Food Insecurity and Child Food Insecurity Estimates at the County Level. Feeding America, 2016

Food insecurity means a lack of access to enough nutritionally adequate food for a healthy life. Food availability is unequally distributed across the United States.

grains, meat, and dairy products can easily obtain all the macronutrients our bodies need (**INFOGRAPHIC 4.1**).

Macronutrients from the diet cannot be used directly by our bodies, in part because they are too large to be absorbed into the bloodstream from our digestive tract. To be useful, macronutrients must first be broken down into smaller subunits by digestion (see Chapter 27). These subunits are small enough to be absorbed from the digestive tract, taken up by cells, and used to build the macromolecules our cells need. For example, dietary carbohydrates from bread and pasta are broken down into simple sugars, which our bodies use to build an energy-storing carbohydrate called glycogen in liver and muscle tissue. Proteins from a steak are broken down into amino acids, which can be taken up and used to build new proteins, like those making up our muscles. Fats from our diet are broken down into fatty acids and glycerol, which are used to assemble the phospholipids that make up cell membranes.

The food we eat also contains nucleic acids, the fourth macromolecule making up cells. Although not considered macronutrients

INFOGRAPHIC 4.1 Food Is A Source of Macronutrients

The most important dietary macronutrients are carbohydrates, proteins, and fats (a type of lipid). While most foods contain all of these, one or two macronutrients predominate in each food type. A well-balanced diet is one that includes a variety of foods to ensure that the body gets enough of each macronutrient to grow and remain healthy.

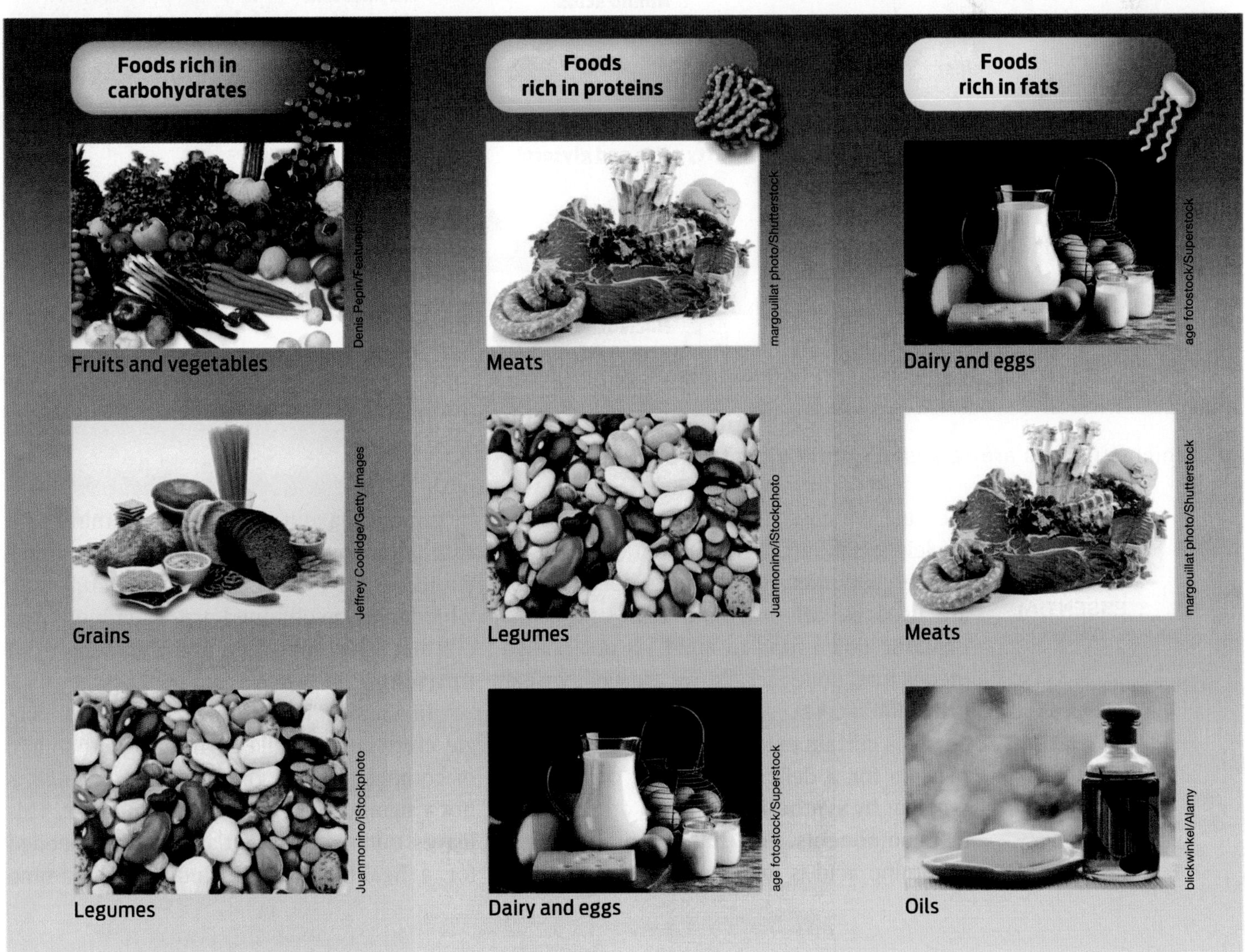

? Which macronutrient would not be well represented in a diet rich in rice and beans?

INFOGRAPHIC 4.2 Macronutrients Build and Maintain Cells

The four macromolecules that make up cells are carbohydrates, proteins, lipids (including fats), and nucleic acids. Cells synthesize these macromolecules from the subunits released by the digestion of macronutrients in food.

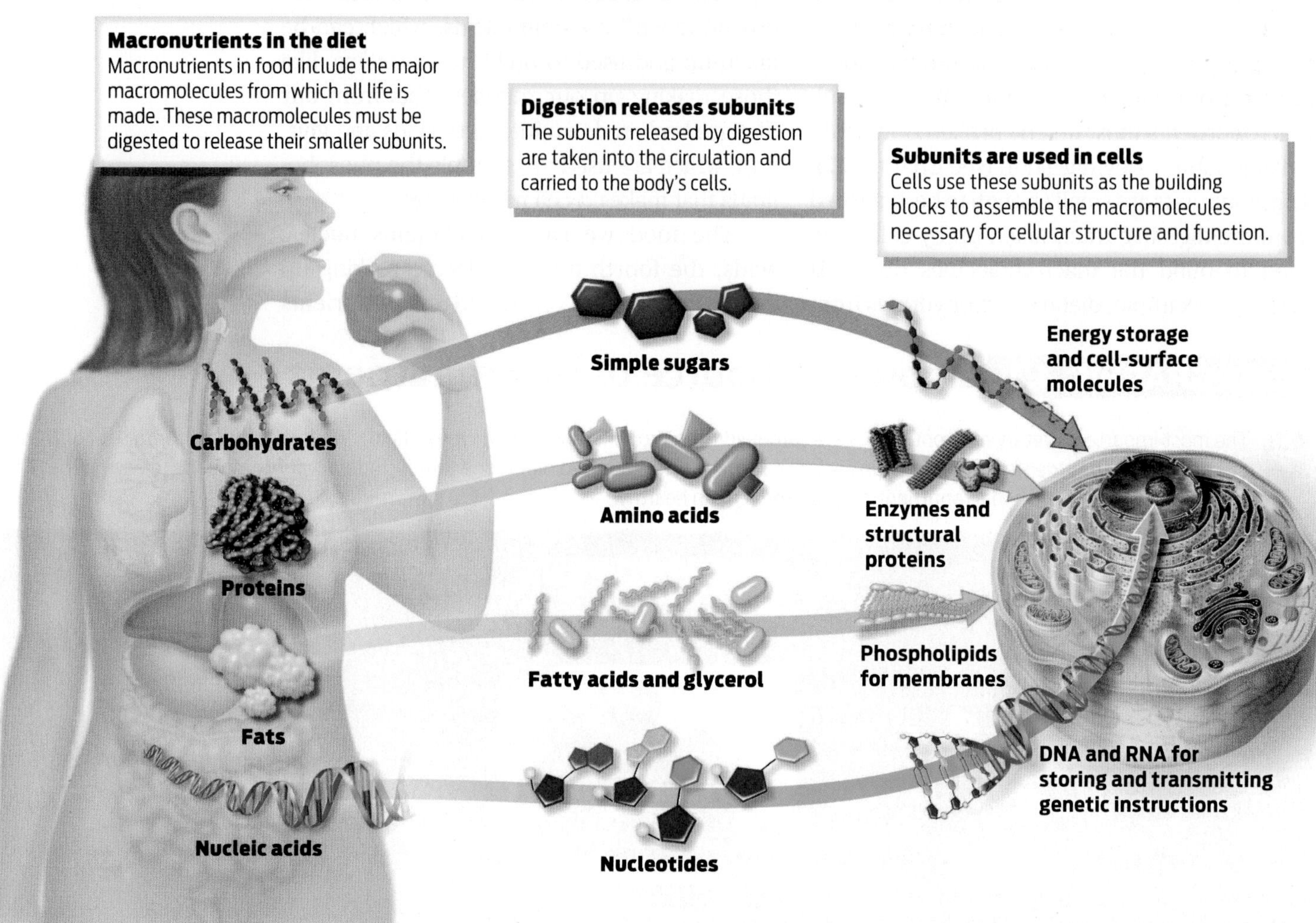

Macronutrients in the diet
Macronutrients in food include the major macromolecules from which all life is made. These macromolecules must be digested to release their smaller subunits.
Digestion releases subunits
The subunits released by digestion are taken into the circulation and carried to the body's cells.
Subunits are used in cells
Cells use these subunits as the building blocks to assemble the macromolecules necessary for cellular structure and function.
Carbohydrates
Simple sugars
Energy storage and cell-surface molecules
Proteins
Amino acids
Enzymes and structural proteins
Fats
Fatty acids and glycerol
Phospholipids for membranes
Nucleic acids
Nucleotides
DNA and RNA for storing and transmitting genetic instructions

? Which subunits are released by the digestion of proteins?

(because we need them in smaller amounts), nucleic acids are also broken down into smaller subunits called nucleotides, which are used by cells to build DNA and RNA. Breaking down food to build up our bodies means that, quite literally, we are what we eat **(INFOGRAPHIC 4.2)**.

ESSENTIAL NUTRIENTS
Nutrients that can't be made by the body, and so must be obtained from the diet.

STARCH
A complex plant carbohydrate made of linked chains of glucose molecules; a source of stored energy.

To a certain extent, our bodies can compensate for a deficiency in one or another nutrient by synthesizing it from other chemical components. For example, if a particular amino acid is in short supply, cells may be able to make it from another amino acid that is in excess. But there are some nutrients that our bodies can't manufacture and which must be obtained pre-assembled from our diet. These nutrients are called **essential nutrients.**

In Malawi, most families subsist on a single crop—corn. While corn is a good source of complex carbohydrates, like **starch,** it is not a significant source of protein or fat. This leaves out many essential nutrients needed for a healthful diet, including vital amino acids. From starting materials in food, adults can synthesize 11 of the 20 amino acids they need to make proteins. The other 9 must be obtained pre-assembled from our diet. Because our body can't manufacture them, these 9 amino acids are called **essential amino acids.** A few more amino acids are considered essential for infants and children. Animal products such as meat, eggs, fish, and dairy are the richest sources of essential amino acids, but in many places around the globe, these foods are luxuries people can't afford.

Once criticized by the international aid community, Manary's method is now being hailed as a kind of miracle cure.

In 1999, when Manary first looked around for an alternative treatment to combat severe hunger, he considered sending kids home with a bag of ingredients, such as corn flour and soy, to make traditional meals. He quickly realized, however, that this would not work. First, such foods would need to be cooked, and cooking is extra work for an already overburdened family. Second, severely malnourished children would need to eat many bowls of these foods per day in order to obtain enough nutrients to recover. Third, the food wouldn't keep.

Peanut butter, on the other hand–especially when it has been fortified with additional nutrients–is a highly effective way of delivering the most nutrients per spoonful of food. Technically, it's known as a ready-to-use-therapeutic food (RUTF), which means it is a complete source of nutrition. "If you eat RUTF you don't need to eat anything else," says Manary. "You're getting everything you need–period."

The RUTF that Manary uses is made up of four main ingredients: full-fat milk powder, sugar, vegetable oil, and peanut butter. Peanut butter is a useful treatment for malnutrition partly because it is energy-dense, which helps children put on weight fast. "Peanuts are the only plant I know of that has 50% fat, when you get rid of the water," says Manary. But the fattening quality is only part of peanut butter's appeal. Because it contains very little water and has a pasty consistency that keeps out air, peanut butter is naturally resistant to spoiling; bacteria can't grow without water. It also doesn't require cooking and is therefore ready to eat at any time. And it's full of protein, a crucial macronutrient for growing children. All these things make peanut butter a near-perfect therapeutic food, supplying children with the necessary fats, carbohydrates, proteins, and essential amino acids (from the milk powder) that they would otherwise lack.

Incidentally, many people in industrialized countries have peanut allergies. But in developing countries, such allergies are rare or nonexistent. That has more to do with the way we train our immune systems in our hyperscrubbed world, Manary explains, than with any intrinsic quality of peanuts.

Life in the Village

In 1999, having spent 5 years working in a hospital in Malawi, Manary took an unusual step for an American-trained doctor. He left the hospital and went to live in a rural village for 10 weeks. He wanted to get a first-hand view of the way people lived and the obstacles they faced. The thing that struck him the most was the sheer monotony of village life.

ESSENTIAL AMINO ACIDS
Amino acids that can't be made by the body, and so must be obtained pre-assembled from the diet.

"Every day is the same," he says. "It starts with going to the water pump, hauling water. Then it's finding firewood. Then it's cooking food. It's work all day, 365 days a year."

Instantly, he understood why childhood malnutrition is such a terrible trap for families. Families live perilously close to the edge of sufficiency, and any slight change in their circumstances can push them over. Even one extra mouth to feed can create food shortages. Moreover, for many rural populations, the nearest hospital is miles away. Taking a sick child to the hospital is a major disruption to family life, so most mothers do it only as a last resort, when the child is already quite sick. Lastly, telling mothers–as they routinely were told by hospital doctors–that they needed to feed their malnourished child seven times a day was not a workable solution. As Manary explains, "It'd be like me telling you 'Hey, you need to walk back and forth from New York to Connecticut every day.'" It just wasn't realistic.

> "If you eat RUTF you don't need to eat anything else. You're getting everything you need—period."
>
> —Mark Manary

Malnutrition-related problems usually start when children are 6 months old. Babies younger than 6 months are sustained easily by breast-feeding alone. Breast milk is a perfect food for children at this age, explains Manary. It is nutritionally appropriate for what their bodies can handle, and it provides protection against infections and disease. Since 2001, the World Health Organization has recommended that women breast-feed exclusively until a child is 6 months old (and they are encouraged to continue for 2 years and beyond, even after infants begin to eat other food). Virtually all women in Malawi breast-feed.

At 6 months, however, children's growing bodies require more nourishment than breast milk can provide, but no other food is as plentiful or accessible. Malnutrition due to chronic undereating sets in after a period of time, usually between ages 1 and 3.

Scientists sometimes refer to the first 1,000 days of life, from gestation to age 2, as the "golden interval." "Babies go through a huge amount of physical and cognitive development during this period of 1,000 days,"

In Malawi, villages are sometimes miles from the nearest health facility.

Dr. Mark Manary

Physician Mark Manary tends to a malnourished child.

says Mary Arimond, a nutrition expert at the University of California, Davis, who studies nutrition in the developing world. "If deficits occur in this window, they are difficult to reverse, especially in continued conditions of poverty."

Growth and development are essentially a series of chemical reactions requiring nutrients as starting materials. These reactions begin as soon as the digestion of food begins. Reactions that break down larger structures into smaller ones are **catabolic reactions.** Reactions that build new structures from smaller subunits are **anabolic reactions.** Together, all the chemical reactions occurring in the body constitute the body's **metabolism.**

To proceed normally, metabolic chemical reactions require the assistance of helper proteins called **enzymes.** Nearly every chemical reaction in the body requires enzymes in order for it to happen at a rate that is fast enough to keep up with the needs of a living organism. For example, digestive enzymes made by cells in the digestive tract help us digest food molecules into their constituent subunits. In the absence of such digestive enzymes, food molecules would remain stable and intact; they would not break down spontaneously at a rate that would suit our metabolic needs. When cells divide–for instance in bone marrow to generate new blood cells, or in skin to replace cells lost to death or injury–they rely on enzymes to copy the DNA in cells. Similarly, enzymes produced by cells in bone carry out anabolic reactions that contribute to the formation of new bone.

Enzymes work by speeding up, or catalyzing, chemical reactions–a process called **catalysis.** In order to accelerate a chemical reaction, an enzyme must bind to the molecules involved in the reaction. The molecules that enzymes bind to are called **substrates.** The part of the enzyme that binds to a substrate is called its **active site.** Each enzyme has an active site that fits only one particular substrate molecule or molecules, making each enzyme specific for the reaction that it catalyzes. Given the thousands of reactions occurring in organisms, organisms have to produce and rely on the activity of thousands of different enzymes, each catalyzing a specific reaction. The digestion of a cheeseburger, for example, requires a variety of different digestive enzymes, some that specifically digest protein, some that digest carbohydrates, and one that digests fat.

Enzymes speed up reactions by facilitating the breaking or formation of chemical bonds

CATABOLIC REACTION
Any chemical reaction that breaks down complex molecules into simpler molecules.

ANABOLIC REACTION
Any chemical reaction that combines simple molecules to build more complex molecules.

METABOLISM
All biochemical reactions occurring in an organism, including reactions that break down food molecules and reactions that build new cell structures.

ENZYME
A protein that speeds up the rate of a chemical reaction.

CATALYSIS
The process of speeding up the rate of a chemical reaction (e.g., by an enzyme).

SUBSTRATE
A molecule to which an enzyme binds and on which the enzyme acts.

ACTIVE SITE
The part of an enzyme that binds to a substrate.

in substrates. In the case of catabolic reactions, the binding of the enzyme to its substrate puts stress on a chemical bond, causing it to break. In an anabolic reaction, the enzyme brings substrate molecules together in close proximity, increasing the likelihood that a bond between them will form (**INFOGRAPHIC 4.3**).

In effect, enzymes catalyze reactions by lowering the amount of energy required to nudge a chemical reaction into motion. Enzymes

INFOGRAPHIC 4.3 Enzymes Facilitate Chemical Reactions

Cells require enzymes to break down and build up macromolecules. Enzymes are proteins that speed up chemical reactions.

Catabolic Reactions: Bonds are broken

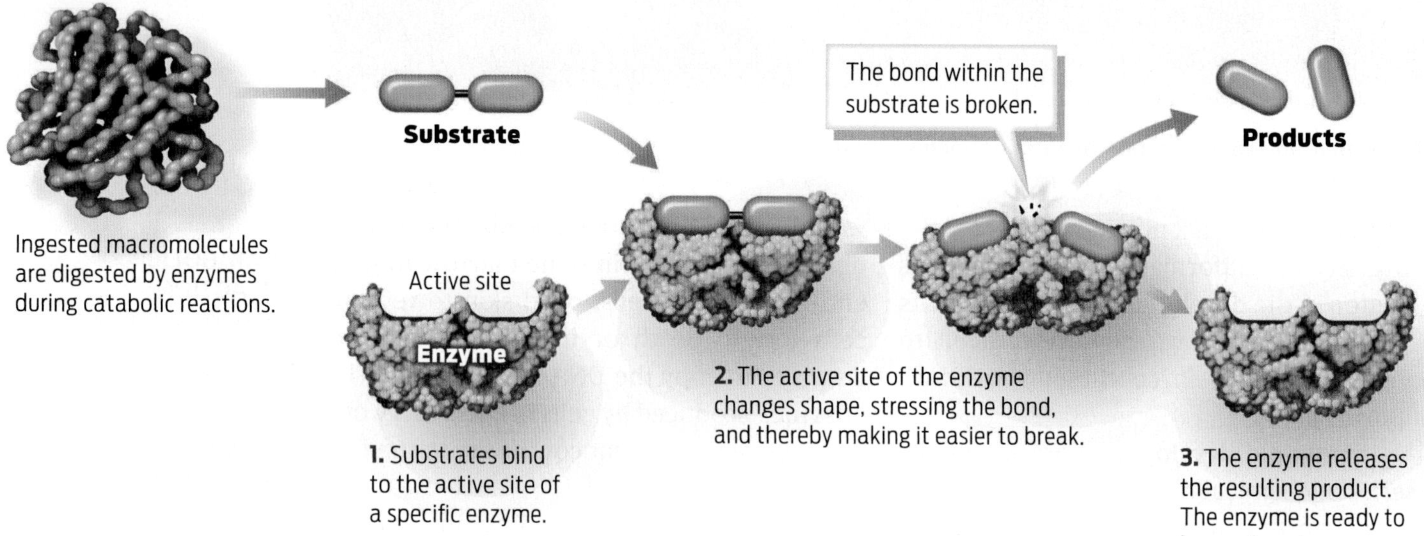

Substrate
The bond within the substrate is broken.
Products
Ingested macromolecules are digested by enzymes during catabolic reactions.
Active site
Enzyme
1. Substrates bind to the active site of a specific enzyme.
2. The active site of the enzyme changes shape, stressing the bond, and thereby making it easier to break.
3. The enzyme releases the resulting product. The enzyme is ready to be used again.

Anabolic Reactions: Bonds are created

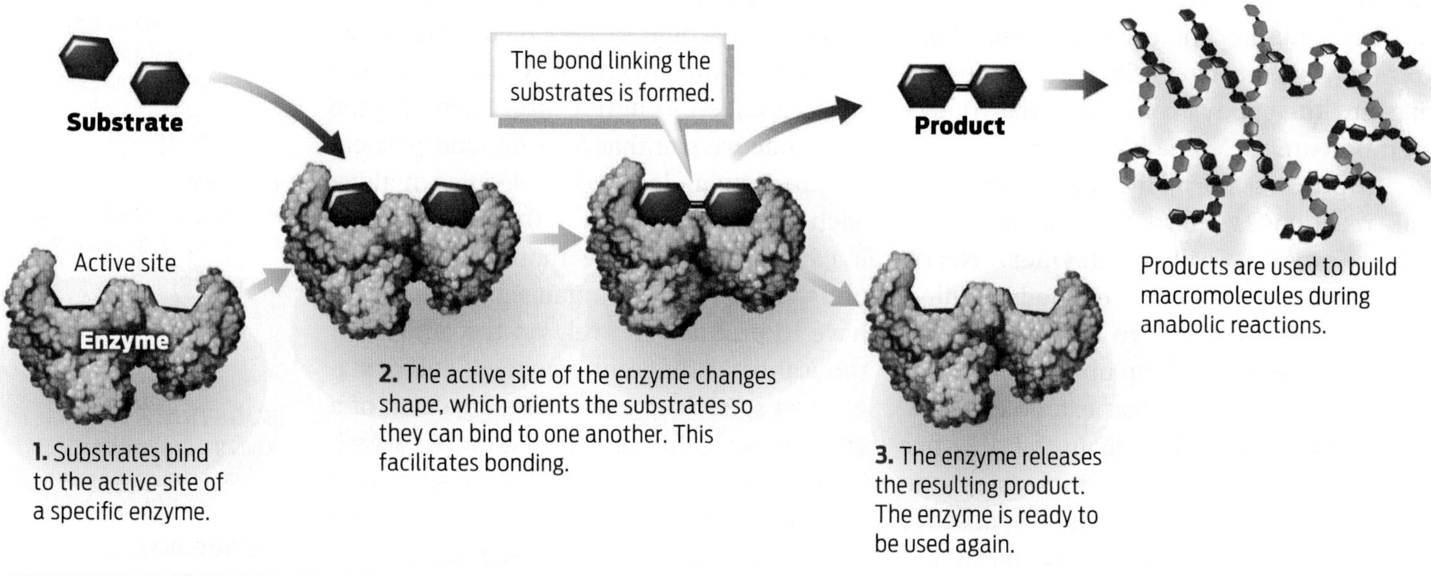

Substrate
The bond linking the substrates is formed.
Product
Active site
Enzyme
Products are used to build macromolecules during anabolic reactions.
1. Substrates bind to the active site of a specific enzyme.
2. The active site of the enzyme changes shape, which orients the substrates so they can bind to one another. This facilitates bonding.
3. The enzyme releases the resulting product. The enzyme is ready to be used again.

? Is the synthesis of proteins from amino acids anabolic or catabolic?

INFOGRAPHIC 4.4 Enzymes Catalyze Reactions by Lowering Activation Energy

The activation energy is the energy that must be put into a reaction in order to make it "go." Enzymes reduce the activation energy in both anabolic and catabolic reactions, making them occur more rapidly.

Reaction without enzyme:

Total energy

Activation energy

More Activation Energy required to initiate the reaction

Reactant

Once the activation energy has been achieved, the reaction proceeds without the input of additional energy.

Products

Progress of reaction

Aflo/AGE Fotostock

Reaction with enzyme:

Total energy

Activation energy

Less Activation Energy required to initiate the reaction

Reactant

Enzyme

Lowering the activation energy makes it easier for products to be produced.

Products

Progress of reaction

AiFangXiaNuo/Getty Images

? How does an enzyme affect the energy of the reactants, the energy of the products, and the activation energy?

substantially reduce this **activation energy,** allowing the reaction to occur more easily. You can think of activation energy as the amount of energy an athlete needs to surmount a hurdle; the steeper the hurdle, the more energy required. Enzyme-catalyzed reactions are leaping over a much lower bar, so less energy is required to clear it **(INFOGRAPHIC 4.4)**.

If a child doesn't get enough to eat, the enzymes that carry out the anabolic reactions necessary for growth do not have substrates upon which to act, so these anabolic reactions cannot occur. What's more, in a malnourished child, catabolic reactions will begin to break down muscle protein to obtain amino acids needed for other purposes. The results are the telltale signs of malnutrition: thin arms with skin wrinkling over wasted muscle; painful swelling of the legs and feet caused by a buildup of fluid; blond or rust-colored hair resulting from a deficiency of protein; and a distended, bloated stomach.

Because a healthful diet requires a balance of many different kinds of nutrient, it's not only the sheer quantity of food that is important, but also the specific type of food that matters. Some nutrients can be missing from a diet, even if a child is otherwise well fed.

Hidden Hunger

When Manary first arrived at the Malawi hospital in 1994, he made a few changes to the then-standard treatment regimen of fortified milk. One thing he did right away was to add extra potassium to the milk the children were drinking. Immediately, the fatality rates dropped—from 33% to 10%. Potassium is required for proper muscle contraction and

ACTIVATION ENERGY
The energy required for a chemical reaction to proceed. Enzymes accelerate reactions by reducing their activation energy.

nerve function. It is one of many **minerals** that we need to keep healthy.

Because most minerals are required only in small amounts, they are known as **micronutrients** (as opposed to macronutrients, which are needed in much larger quantities). **Vitamins** are another kind of micronutrient. Just because our bodies require only small amounts of micronutrients doesn't mean they aren't important. Micronutrient deficiency can have serious health consequences. Iron deficiency can impair the blood's ability to carry oxygen, causing anemia, for example, and lack of vitamin C causes a tissue-deteriorating disease called scurvy.

> "Babies go through a huge amount of physical and cognitive development during this period of 1,000 days."
>
> —Mary Arimond

In Malawi, as in many parts of the developing world, people often suffer deficiencies of vitamin A and zinc, which can cause impairment of vision and the immune system. Such micronutrient deficiencies are sometimes known as "hidden hunger," because the problem is not lack of food per se, but a lack of necessary micronutrients. Manary says the problem is widespread because most of the world's staple crops–foods like rice, corn, wheat, and cassava–do not contain adequate micronutrients.

Food producers routinely add to foods some micronutrients that are hard to obtain from natural sources. Iodine, for example, is added to table salt (in "iodized" salt) to prevent goiter, an abnormal thickening of the neck caused by an enlarged thyroid gland due to a lack of dietary iodine.

The peanut butter RUTF that Manary uses contains a mineral and vitamin mix that is 1.6% of the paste by weight. Although peanut butter naturally contains many micronutrients, the amounts malnourished children need are greater than what is normally found in most foods. So the peanut paste is deliberately enriched with micronutrients (**TABLE 4.1**).

Vitamins and minerals play numerous roles in the body. Some play structural roles–the mineral calcium, for example, is a primary component of bones and teeth. Others play functional roles, helping other molecules to act. Perhaps their most critical role is serving as **cofactors** that assist enzymes.

Cofactors are accessory or "helper" substances that enable enzymes to function. Cofactors include inorganic metals such as zinc, copper, and iron. Cofactors can also be organic molecules, in which case they are called **coenzymes**. Most vitamins are important coenzymes. Without cofactors and coenzymes that bind to enzymes and enable them to bind to substrates, cell metabolism would grind to a halt (**INFOGRAPHIC 4.5**).

By adding extra vitamins and minerals to the milk that malnourished children were receiving in the hospital, Manary found that he could get the death rate to go way down, but the recovery rate still wouldn't budge. That's when he knew it was time for a new approach.

Emptying the Wards

By the time Manary began seriously thinking about home-based therapy, he had been working in Malawi for about 5 years, witnessing the failure of the standard WHO treatment, and he had spent time in villages. In his head, he'd been tossing around the idea of something new and different. Out of the blue, he got an email from a doctor in France, André Briend, who was also thinking about home-based therapy as a treatment for malnutrition.

The two scientists corresponded for about a year by email, weighing the pros and cons of various foods that could be eaten at home yet still pack a nutrient wallop. After considering various options–they considered biscuits, pancakes, even Nutella–the two researchers

MINERAL
An inorganic chemical element required by organisms for normal growth, reproduction, and tissue maintenance; examples are calcium, iron, potassium, and zinc.

MICRONUTRIENTS
Nutrients, including vitamins and minerals, that organisms must ingest in small amounts to maintain health.

VITAMIN
An organic molecule required in small amounts for normal growth, reproduction, and tissue maintenance.

COFACTOR
An inorganic substance, such as a metal ion, required to activate an enzyme.

COENZYME
A small organic molecule, such as a vitamin, required to activate an enzyme.

INFOGRAPHIC 4.5 Vitamins and Minerals Have Essential Functions

Vitamins and minerals are micronutrients, nutrients that are essential for health but required in far smaller amounts than macronutrients.

Some Vitamins and Minerals Are Structural Elements

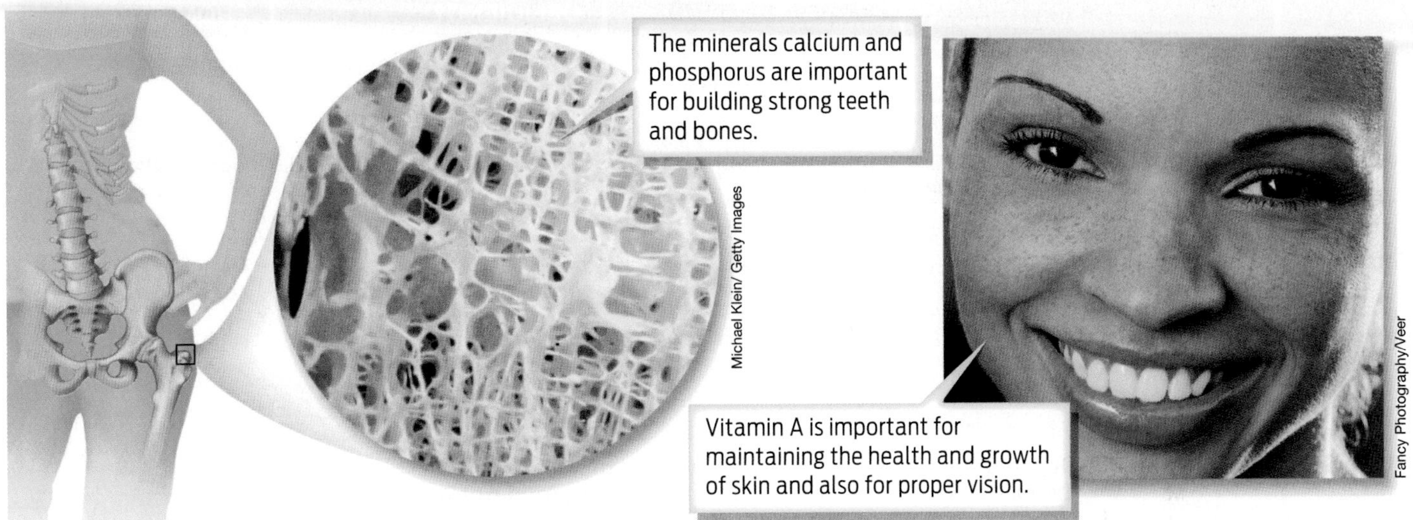

The minerals calcium and phosphorus are important for building strong teeth and bones.
Michael Klein/ Getty Images
Vitamin A is important for maintaining the health and growth of skin and also for proper vision.
Fancy Photography/Veer

Vitamins and Minerals Are Cofactors That Allow Enzymes to Function

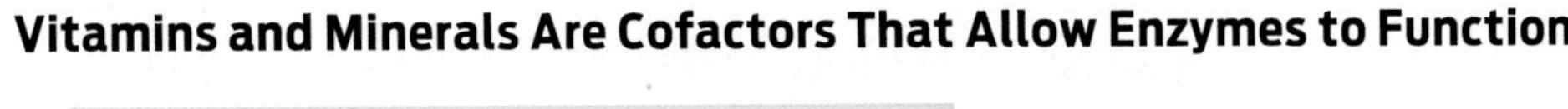

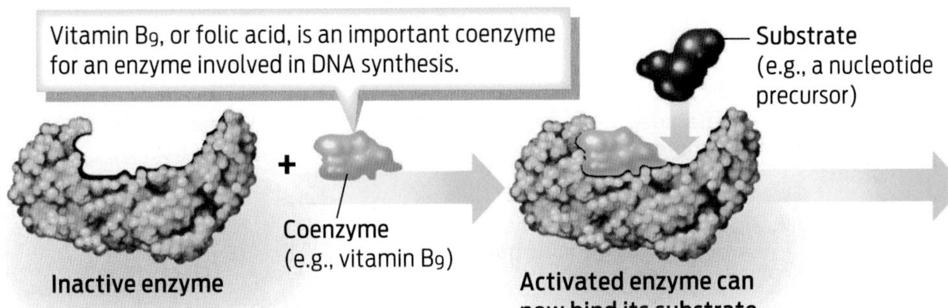

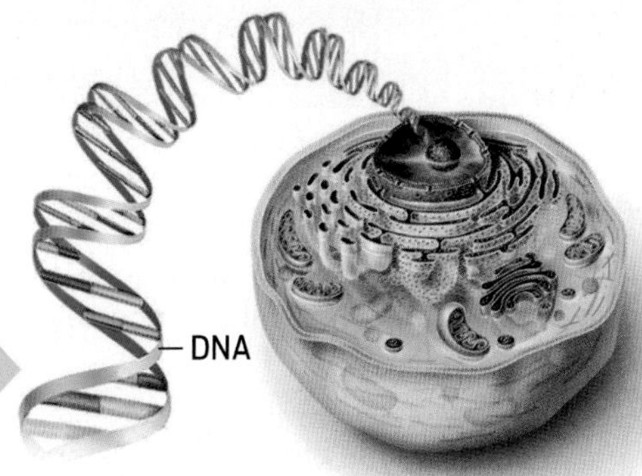

? Does the fact that vitamins and minerals are micronutrients make them less important than macronutrients?

eventually decided on peanut butter as the best choice. In 2001, they decided to test the idea.

Briend had some of the peanut butter RUTF made up in France, sealed in foil packets, and shipped to Malawi. The two scientists then carefully designed a scientific study to test the product. After a brief stabilization phase in the hospital (during which antibiotics were given, if necessary), the children were discharged and sent home on one of three different treatment regimens: (1) ample amounts of traditional food–corn flour and soy; (2) a small amount of peanut butter-based RUTF, to be used as a supplement to the normal diet at home; and (3) the full dose of peanut butter-based RUTF, with sufficient nutrients to meet the total nutritional needs of the children. The goal of the study was to see which treatment regimen was most effective.

Within a few months, the results were clear–and impressive: 95% of the children who received the full peanut butter-based RUTF recovered. Those who received traditional food or supplemental RUTF also did pretty well–about 75% of them recovered, but still the full peanut butter-based RUTF was better. And all the home-based treatments were significantly better than standard

TABLE 4.1 A Sample of Micronutrients in Your Diet

MINERALS: Inorganic elements not synthesized by the body.

MINERAL	FUNCTIONS INCLUDE	FOOD SOURCES	CONDITIONS RESULTING FROM DEFICIENCY	CONDITIONS RESULTING FROM EXCESS
Calcium	Bone and teeth formation, blood clotting, cell signaling, nerve function	Dairy products, green vegetables, legumes	Osteoporosis, stunted growth	Kidney stones
Iron	Component of hemoglobin in red blood cells; carries oxygen throughout the body	Green vegetables, beef, liver	Anemia, fatigue, dizziness, headaches, poor concentration	Constipation, risk of type 2 diabetes
Potassium	Electrolyte balance, muscle contraction, nerve function	Fruits, vegetables, meat	Muscle weakness, neurological disturbances	Muscle weakness, heart failure
Sodium	Electrolyte balance, muscle contraction, nerve function	Salt, bread, milk, meat	Muscle cramps, reduced appetite, neurological disturbances	High blood pressure

WATER-SOLUBLE VITAMINS: Organic molecules not synthesized by the body.
Excess vitamin is excreted in urine and so does not harm health.

VITAMIN	FUNCTIONS INCLUDE	FOOD SOURCES	CONDITIONS RESULTING FROM DEFICIENCY	CONDITIONS RESULTING FROM EXCESS
B_1 (thiamine)	Cofactor for enzymes involved in energy metabolism and nerve function	Leafy vegetables, whole grains, meat	Heart failure, depression	None
Folate	Cofactor for enzymes involved in DNA synthesis and cell production	Dark green vegetables, nuts, legumes, whole grains	Neural tube defects, anemia	None
B_{12}	Cofactor for enzymes involved in the breakdown of fatty acids and amino acids and nerve cell maintenance	Meat, milk, eggs	Anemia, neurological disturbances	None
C	Cofactor for enzymes involved in collagen synthesis; improves iron absorption and immunity	Citrus fruits	Scurvy, poor wound healing	None

FAT-SOLUBLE VITAMINS: Organic molecules not synthesized by the body (except vitamin D).
Excess vitamin is stored in fat cells and can harm health.

VITAMIN	FUNCTIONS INCLUDE	FOOD SOURCES	CONDITIONS RESULTING FROM DEFICIENCY	CONDITIONS RESULTING FROM EXCESS
A (retinol)	Component of eye pigment, supports skin, bone, and tooth growth, supports immunity and reproduction	Fruits and vegetables, liver, egg yolk	Skin problems, blindness	Headaches, intestinal pain, bone pain
D	Calcium absorption, bone growth	Fish, dairy products, eggs	Bone deformities	Kidney damage
E	Antioxidant, supports cell membrane integrity	Green leafy vegetables, legumes, nuts, whole grains	Neural tube defects, anemia, digestive-health problems	Fatigue, headaches, blurred vision, diarrhea
K	Supports synthesis of blood clotting factors	Green leafy vegetables, cabbage, liver	Abnormal blood clotting, bruising	Liver damage, anemia

hospital-based milk therapy, which historically had a 25%–40% recovery rate.

Manary couldn't quite believe how effective the treatment was. "I said, 'Damn, this stuff really works!'" **(INFOGRAPHIC 4.6)**.

One of the main reasons that peanut butter RUTF—which locals call chiponde, or "nutpaste"—is more effective than standard therapy is that it can be administered safely at home. When children are malnourished, their immune systems aren't functioning at optimal levels, which means that hospitals are often the worst place for them to be because of the risk of infection from other sick patients. Peanut butter RUTF is also something that children can eat on their own, without help from their parents. And, because they clearly like the taste of it (one doctor described it as tasting like the inside of a Reese's Peanut Butter Cup), they gobble it up.

As dramatic as Manary's initial results were, however, not everyone was convinced. In fact, most leaders in nutrition science were

INFOGRAPHIC 4.6 Peanut Butter–Based RUTF Saves More Children

Studies show that significantly more children recover when treated at home with peanut butter–based RUTF compared to a corn/soy flour diet or their regular diet supplemented with a small amount of the RUTF supplement.

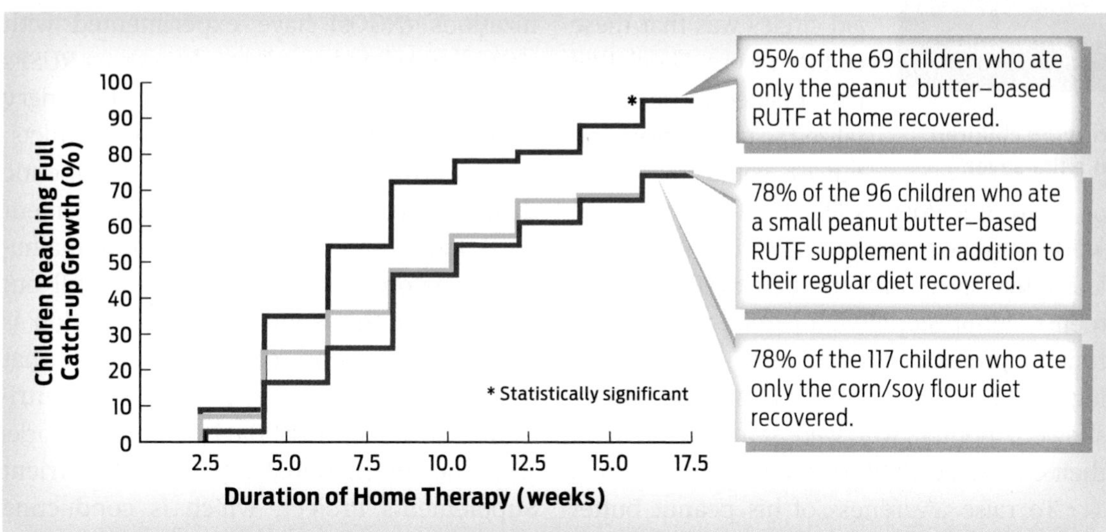

RUTF (Ready-to-Use Therapeutic Food) Recipe:

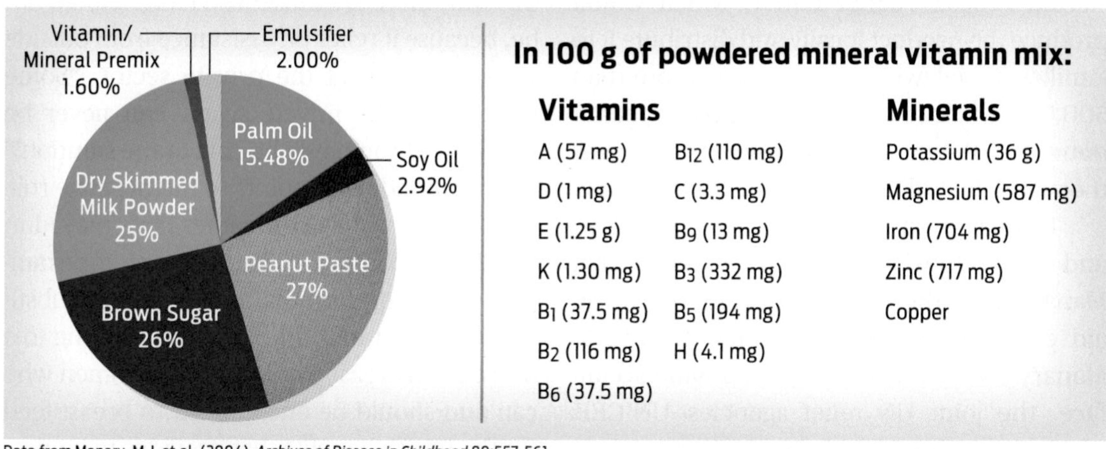

Vitamin/ Mineral Premix 1.60%
Emulsifier 2.00%
Palm Oil 15.48%
Soy Oil 2.92%
Dry Skimmed Milk Powder 25%
Peanut Paste 27%
Brown Sugar 26%
In 100 g of powdered mineral vitamin mix:
Vitamins
A (57 mg)
B12 (110 mg)
D (1 mg)
C (3.3 mg)
E (1.25 g)
B9 (13 mg)
K (1.30 mg)
B3 (332 mg)
B1 (37.5 mg)
B5 (194 mg)
B2 (116 mg)
H (4.1 mg)
B6 (37.5 mg)
Minerals
Potassium (36 g)
Magnesium (587 mg)
Iron (704 mg)
Zinc (717 mg)
Copper

Data from Manary, M.J. et.al. (2004) *Archives of Disease in Childhood* 80:557-561

? In this study, did the RUTF work better as a full dietary replacement or as a supplement to the normal diet?

Before and after: For many starving children, peanut butter RUTF has been a life-saver.

vehemently opposed to the approach, which is why Manary was heckled at the conference in 2003.

Manary wasn't the only person in the field to face a backlash. Other doctors working in this area faced similar opposition. "It was pretty nasty," says Steve Collins, a physician and early advocate of home-based treatment whose work was also criticized.

At that time, around 2003, the conventional wisdom in humanitarian aid circles was that these children were so sick that they needed to be hospitalized so that doctors and nurses could take care of them. But doctors like Manary, who were working in the trenches, knew differently. They knew that even with the best available treatments–administered with careful precision by trained professionals–children were recovering only 25%–40% of the time. And, as Manary showed, mothers who were sent home to treat their children with RUTF could do much better.

To raise awareness of his peanut butter approach and to start making the treatment available on a wider basis, in 2004 Manary started Project Peanut Butter, which would produce the product locally and distribute it to families in Malawi. So far, he says, more than 500,000 children have been helped by peanut butter RUTF–not only in Malawi, but all across the world, including in Haiti and Sierra Leone.

Thanks in part to Project Peanut Butter, and the extensive body of evidence that Manary and others have collected, the world aid community eventually came around to Manary's view: in 2007, in a dramatic about-face, the joint UN relief agencies–UNICEF, WHO, and the World Food Programme–issued an official statement saying that home-based therapy with peanut butter RUTF is the preferred way to treat acute malnutrition. Manary's approach was vindicated.

Preventing Hunger?

Once made in a broom closet in a Malawi hospital, peanut butter RUTF is now produced in giant industrial factories and used all over the world. The largest producer is the France-based company Nutriset, which sells its product (called Plumpy'Nut) to relief agencies like UNICEF and Doctors without Borders, which then distribute it free to countries in need.

The undeniable success of peanut butter RUTF in the treatment of malnutrition has inspired some aid groups to want to expand its reach into the realm of prevention. Several UN agencies and nongovernmental organizations (NGOs) have experimented with providing related products, known as RUSFs (ready-to-use supplemental foods), to hungry children around the world. These products, which provide less than the full therapeutic dose of nutrients, are meant to supplement local foods and thus help prevent malnutrition before it starts. The jury is still out on whether this approach to prevention is effective. "The state of the evidence is that there is not enough evidence," says nutritionist Arimond of UC-Davis, who also works with the International Lipid-Based Nutrient Supplements Project, which is conducting some of this research.

Even if malnutrition could be prevented this way, there are those who feel it should not be, because it relies on assistance from outside groups by way of the private sector. "Some people feel the private sector can never be trusted and can never be part of the solution," says Arimond, who points to the negative role that infant formula companies have played in developing countries over the years–for example, marketing formula as a healthful substitute for breast milk and thus contributing to a decrease in breast-feeding among women who can and should be encouraged to breast-feed for the health benefits that provides to babies.

For others, the concern is less whether the profit motive is involved, and more who

INFOGRAPHIC 4.7 A Balanced Diet

A balanced diet includes all the nutrients needed for full health.

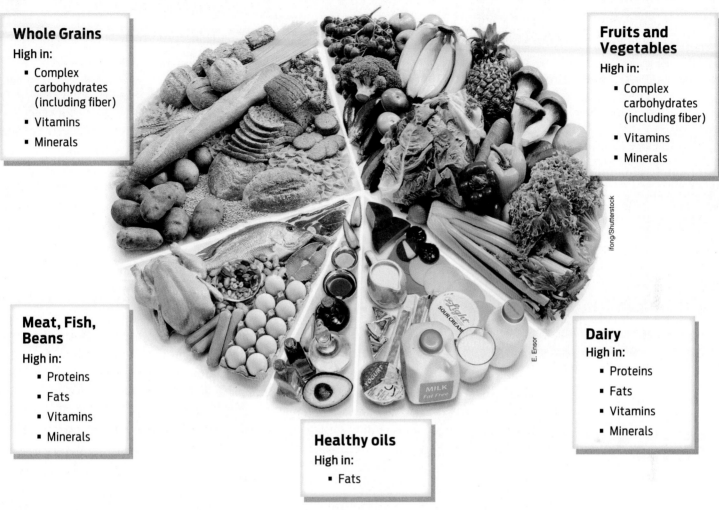

Whole Grains
High in:
- Complex carbohydrates (including fiber)
- Vitamins
- Minerals

Fruits and Vegetables
High in:
- Complex carbohydrates (including fiber)
- Vitamins
- Minerals

ifong/Shutterstock

Meat, Fish, Beans
High in:
- Proteins
- Fats
- Vitamins
- Minerals

E. Ensor

Dairy
High in:
- Proteins
- Fats
- Vitamins
- Minerals

Healthy oils
High in:
- Fats

? Some diet plans restrict or eliminate entire food groups. How easily will these diet plans fit into a balanced diet?

benefits from those profits. "If it's benefiting U.S. agriculture, or European agriculture, or European manufacturing and European transporters, it's not benefiting the right people," says Steve Collins, the physician advocate of home-based treatment quoted earlier. For that reason, nonprofits like Project Peanut Butter and the Haiti-based Meds & Foods for Kids have made a clear commitment to producing RUTF locally. By using local peanuts, and working with local farmers, these groups contribute directly to the health and growth of the local economy, and thus help to alleviate the conditions that spawn malnutrition in the first place (**INFOGRAPHIC 4.7**).

Manary doesn't disagree. He believes that the ultimate solution to malnutrition is prevention through improved agriculture. To that end, he is also working on a project sponsored by the Bill and Melinda Gates Foundation that aims to use genetic engineering to improve the nutritional quality of subsistence crop plants in Africa. Manary says he has high hopes for the project, and envisions a day when children in this region do not routinely die from hunger. "My professional goal is to fix malnutrition for kids in Africa," he says. ■

CHAPTER 4 SUMMARY

- Food is a source of nutrients. Nutrients provide the chemicals required to build and maintain cells and the energy cells need to function.
- Macronutrients are nutrients required in large amounts. Micronutrients are nutrients required in smaller amounts. Both are essential for good health.
- Macronutrients include carbohydrates, proteins, and lipids, including dietary fats; these are among the organic macromolecules that make up our cells.
- Digestion breaks down macromolecules into smaller subunits, which are then used by cells to build cell structures and carry out cell functions.
- Enzymes are proteins that accelerate the rate of chemical reactions. Nearly all reactions in the body require enzymes, including those reactions required for growth and development.
- Enzymes speed up reactions by binding specifically to substrates and reducing the activation energy necessary for a reaction to occur. Enzymes mediate both bond-breaking (catabolic) and bond-building (anabolic) reactions.
- Many enzymes require small "helper" chemicals called cofactors to function. Micronutrients such as minerals and vitamins, found abundantly in fruits and vegetables, are important cofactors (referred to as coenzymes in the case of vitamins).
- Eating a balanced diet that contains abundant quantities of fruits and vegetables is the best way to ensure proper nutrition.
- Malnutrition results when adequate macronutrients or micronutrients are lacking in the diet.
- Malnutrition is especially dangerous for children, whose bodies are, or should be, growing rapidly.

MORE TO EXPLORE

- **Project Peanut Butter: http://www.projectpeanutbutter.org**
- **UN World Food Programme, Malawi: https://www.wfp.org/countries/malawi**
- **U.S. Department of Agriculture, Food Security in the U.S.: http://www.ers.usda.gov/topics/food-nutrition-assistance/food-security-in-the-us.aspx**
- **Film: *A Place at the Table:* http://www.magpictures.com/aplaceatthetable/. Documents food scarcity in America.**
- **Manary, M. J., et al. (2004) Home based therapy for severe malnutrition with ready-to-use food. *Archives of Disease in Childhood* 89:557–561.**
- **Diamond, J. (1999) *Guns, Germs, and Steel: The Fates of Human Societies.* New York: W. W. Norton. Provides useful historical context for underdevelopment and lingering wealth disparities among nations.**

CHAPTER 4 Test Your Knowledge

DRIVING QUESTION 1 What are the macronutrients and micronutrients provided by food?

By answering the questions below and studying Infographics 4.1 and 4.2 and Table 4.1, you should be able to generate an answer for the broader Driving Question above.

KNOW IT

1 A macronutrient is

a. a nutrient with a large molecular weight.
b. a nutrient that is abundant in the diet.
c. a nutrient that is required in large amounts.
d. a nutrient that is stored in large amounts in the body.
e. a nutrient that the body makes in large quantities.

2 Which of the following is/are macronutrient(s)?

a. protein
b. iodine
c. vitamin C
d. fats
e. all of the above
f. a and d

3 A multivitamin supplement is a(n) __________ supplement.

a. macronutrient
b. micronutrient
c. mineral
d. enzyme
e. a and b

4 Which of the following foods is a rich source of protein?

a. lean meat, such as chicken breast
b. whole grains (e.g., whole wheat bread)
c. olive oil
d. leafy greens
e. berries (e.g., blueberries and raspberries)

USE IT

5 Explain the difference between macronutrients and micronutrients.

6 A typical multivitamin supplement contains vitamin A, vitamin C, vitamin D, vitamin E, vitamin K, vitamin B_1, vitamin B_2, vitamin B_6, biotin, calcium, iron, magnesium, zinc, selenium, copper, manganese, and chromium. Explain your answers to the following questions.

a. Are all of these vitamins? If there are ingredients that are not vitamins, what are they?
b. Are all of these micronutrients?

DRIVING QUESTION 2 What are essential nutrients?

By answering the questions below and studying Infographics 4.2 and 4.4 and Table 4.1, you should be able to generate an answer for the broader Driving Question above.

KNOW IT

7 What subunits are proteins broken down into during digestion?

a. fatty acids
b. amino acids
c. glycerol
d. nucleotides
e. simple sugars

8 Where (or how) do we obtain essential amino acids?

a. from carbohydrates in our diet
b. by synthesizing them from other amino acids
c. from oils in our diet
d. from bright orange fruits and vegetables
e. from protein in our diet

USE IT

9 Our bodies cannot synthesize vitamin C, but require it. Therefore, vitamin C is

a. an essential micronutrient.
b. an essential mineral.
c. an essential macronutrient.
d. a nonessential vitamin.
e. a nonessential amino acid.

10 Which component of peanut butter RUTF supplies essential amino acids?

a. milk powder
b. peanut butter
c. sugar
d. vegetable oil
e. powdered vitamins and minerals
f. a and c

11 Corn lacks the essential amino acids isoleucine and lysine. Beans lack the essential amino acids tryptophan and methionine. Soy contains all the essential amino acids.

a. Could someone survive on a diet with a corn-based protein alone? Why or why not?
b. Why do many traditional diets combine corn (e.g., in tortillas) with beans?
c. Why did one of the home-based feeding therapies in Malawi combine soy flour with corn flour?

DRIVING QUESTION 3 What are enzymes, how do they work, and how do they contribute to reactions of metabolism?

By answering the questions below and studying Infographics 4.3, 4.4, and 4.5, you should be able to generate an answer for the broader Driving Question above.

KNOW IT

12 The substrate of an enzyme is

a. an organic accessory molecule.
b. the molecule(s) released at the end of an enzyme-facilitated reaction.
c. the shape of the enzyme.
d. one of the amino acids that makes up the enzyme.
e. what the enzyme acts on.

13 Compare and contrast enzyme cofactors and coenzymes.

14 Enzymes speed up chemical reactions by

a. increasing the activation energy.
b. decreasing the activation energy.
c. breaking bonds.
d. forming bonds.
e. releasing energy.

15 How is folate (folic acid) best described?

a. as a substrate of an enzyme
b. as a nucleotide
c. as an organic cofactor (coenzyme)
d. as an enzyme
e. a and b

USE IT

16 If the shape of an enzyme's active site were to change, what would happen to the reaction that the enzyme usually speeds up?

17 Considering the function of folate (folic acid) given in Infographic 4.6, why would you say pregnant women (and women who could become pregnant) should ensure that they have adequate levels of folate in their diets?

DRIVING QUESTION 4 What are the consequences of a diet lacking sufficient nutrients?

By answering the questions below and studying Infographics 4.5 and 4.6 and Table 4.1, you should be able to generate an answer for the broader Driving Question above.

KNOW IT

18 When vitamins are consumed:

a. Why does excess vitamin E cause problems, but excess vitamin C does not?
b. If you were to take a supplement with a high amount of vitamin C, what would happen to all that vitamin C? Would it all be used? Would some of it be stored in your body?

19 What ingredient(s) in RUTF peanut paste specifically help bone growth? (Hint: Refer to Table 4.1.)

a. calcium
b. vitamin D
c. potassium
d. all of the above
e. a and b

apply YOUR KNOWLEDGE

INTERPRETING DATA

20 Infographic 4.6 shows the results of a study examining three different home-based therapies for malnourished children in Malawi.

a. From the data shown, how many of all the children in the study reached full catch-up growth?
b. What percentage of the children in the study does this number represent? How does this compare to previous recovery rates of 25%–40% for children who had received standard hospital therapy?

The children who received the RUTF were given enough of it to supply 730 kJ of energy per kg of body weight. This is sufficient energy to meet their needs.

c. A malnourished 2-year-old girl weighs a mere 6 kg (~13 pounds; an average 2-year-old American girl weighs approximately 28 pounds). If she had been in the RUTF group in the study, how many daily kJ would she have obtained from the RUTF?
d. If the same malnourished 2-year-old had been in the RUTF supplement group, she would have received 2,100 kJ per day from the supplement. What percentage of her daily energy needs would this represent? (Hint: Use your answer to part c.)
e. Children in the RUTF supplement group ate a traditional diet of corn/soy flour to make up the rest of their diet. Corn/soy flour contains 4 kJ per gram. How many grams of the traditional mix would this 2-year-old need to consume (on top of the RUTF supplement) to meet her daily needs?

MINI CASE — *apply* YOUR KNOWLEDGE

21 A college student returns home at the end of the school year. His mother is shocked by the large number of unhealed scrapes and sores on his knees and arms. She also notices that he has put on a few pounds. The student tells his mother that the scrapes are just left over from a skateboarding mishap a few weeks ago and that he guesses he could cut back on some of his snacks. A week after coming home, he goes to the dentist for his yearly checkup. The dentist is alarmed by his bleeding and swollen gums. When asked about his diet, the student notes that he and some of his friends challenged one another to see who could go the longest eating nothing but eggs, mac 'n' cheese, and toast with butter. He proudly announces that he had stayed on this diet for 6 months.

a. Could this student be suffering from malnutrition? Explain your answer.

b. What mineral(s) or vitamin(s) (or both) are you most concerned about, given the symptoms noted by the dentist?

c. What dietary recommendations would you make for this student?

BRING IT HOME — *apply* YOUR KNOWLEDGE

22 Is malnutrition a problem solely in developing countries? Do some investigative research on at least two food-aid programs, local, federal, or international. What criteria would you consider before deciding to donate money to a food-aid program? Explain your answer.

5 Energy and Photosynthesis

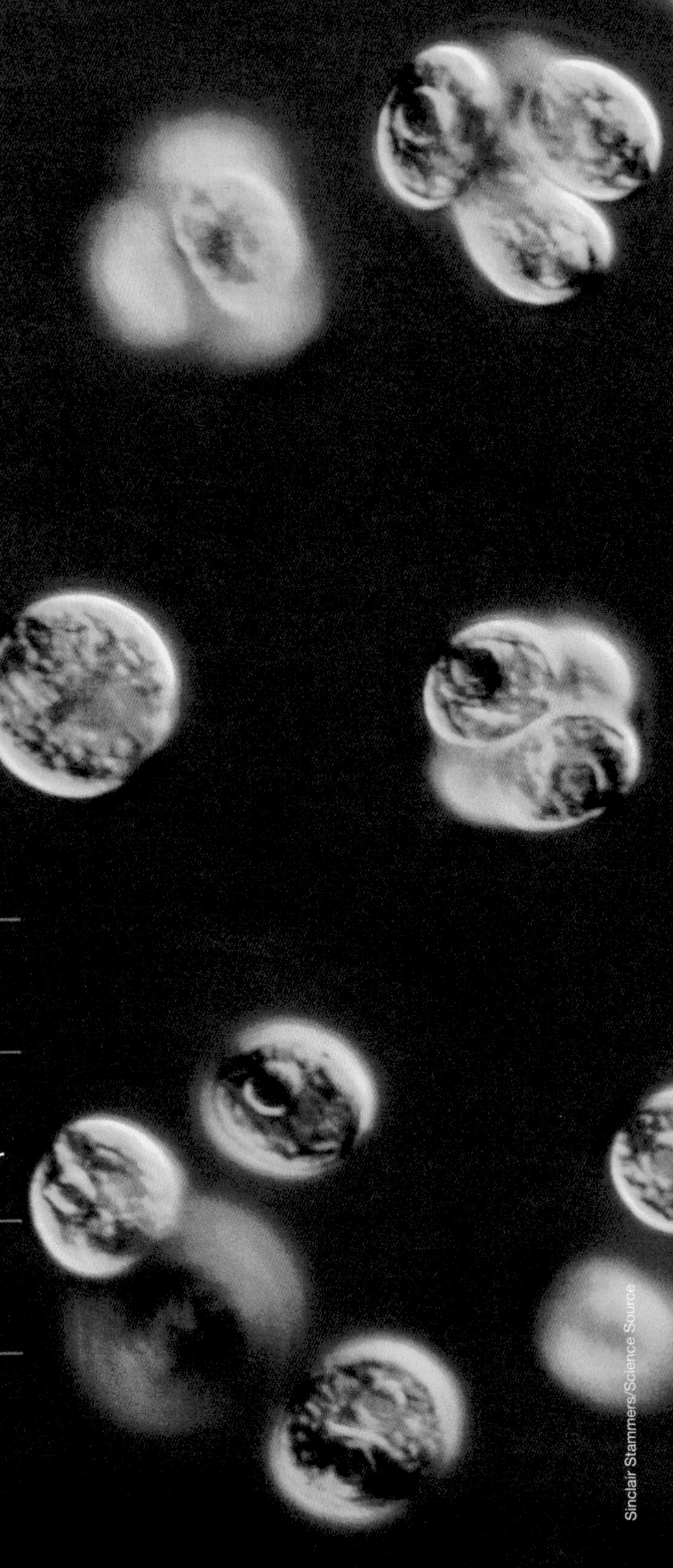

Sinclair Stammers/Science Source

DRIVING QUESTIONS

1. What are the photosynthetic organisms on the planet, and why are they so important?
2. What are the different types of energy, and what transformations of energy do organisms carry out?
3. How do plants and algae convert the energy in sunlight into energy-rich organic molecules? (And why can't other organisms, including humans, do this?)
4. How do algal biofuels compare to other fuels in terms of costs, benefits, and sustainability?

the FUTURE of *FUEL*?

Scientists seek to make algae the next alternative energy source

AS AN ENGINEER WORKING FOR the Navy Seals in 1978, Jim Sears took a nighttime scuba dive off the coast of Panama City, Florida, one of many he took to do underwater research. The dive started out routinely, but then, suddenly, glowing phosphorescent algae appeared as if out of nowhere. When Sears put his hands out in front of him, sparkling streamers of microbes trickled off his fingertips. "It was magical," he recalls.

Sears is an inventor with many and varied devices to his credit. In the 1970s and 1980s, he built an underwater speech descrambler and a portable mine detector, among other gadgets. Later, he moved on to more creative technologies, including a "hump-o-meter" that could tell farmers when their animals were in heat or mating.

But the seeds of his real claim to fame weren't sown until 2004, when Sears was working in agricultural electronics. That's when he turned his attention toward what he felt was the world's biggest problem: dwindling fossil fuel reserves. After he did some thinking and a little research, the tiny, glowing organisms that had wowed him during his nighttime dive more than two decades earlier came to mind. He realized suddenly that they might be able to help.

Algae are perhaps best known as the layer of green scum coating the surfaces of ponds and swimming pools, but they have other claims to fame as well. Like plants, algae have the impressive ability to capture the **energy** of sunlight and convert it into a form that other organisms can use. Even more remarkable, algae trap much of this energy in the form of oils ideally suited to making fuel. The oil that

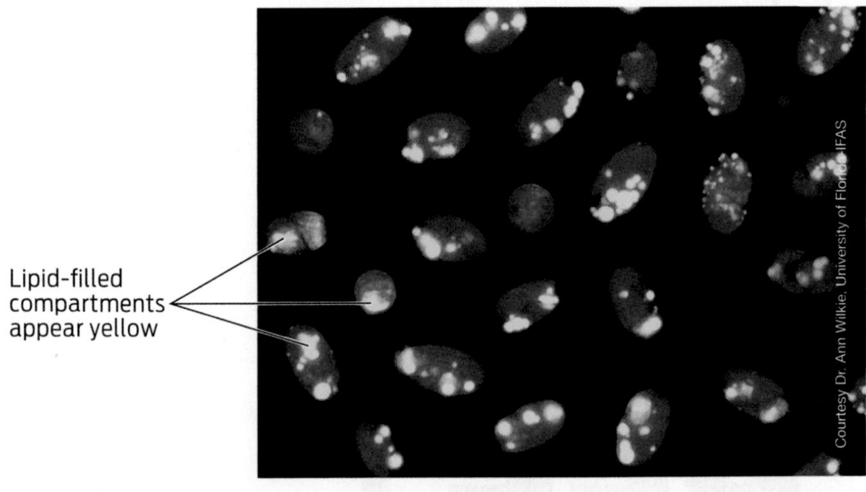

Some algae capture energy from sunlight and convert a portion of it into oil.

ENERGY
The capacity to do work. Cellular work includes processes such as building complex molecules and moving substances into and out of the cell.

BIOFUELS
Renewable fuels made from living organisms (e.g., plants and algae).

algae produce is very similar to common vegetable oil. It accumulates inside the microbes' cells, and once extracted, it can be processed to make biodiesel, gasoline, or jet fuel. "The more I looked into them, the more amazing they were," Sears says.

And that's good news, because America is desperate for new fuels. After all, Americans burn through about 375 million gallons of gasoline a day, enough to fill about 560 Olympic-size swimming pools. And despite the fact that our fuel demand will likely increase over the next 25 years, the sources of our precious gasoline–oil reserves buried deep underground–are finite, take millions of years to replenish, and largely lie outside U.S. borders.

Confronted with this looming crisis, scientists and politicians are increasingly turning toward alternative energy sources such as **biofuels**–renewable fuels made from living organisms. In an effort to reduce our dependence on oil, in 2007 President George W. Bush signed the Energy Independence and Security Act, which requires the United States to produce 36 billion gallons of biofuels fuels by 2022. Biofuels include fuels like biodiesel made from plant scraps, ethanol made from sugar cane, and natural gas harvested

Distribution of Recoverable Oil Reserves

The gasoline and diesel used to power cars and trucks begins as oil formed deep in the ground over millions of years. The United States depends heavily on oil recovered from other countries for its fuel supply.

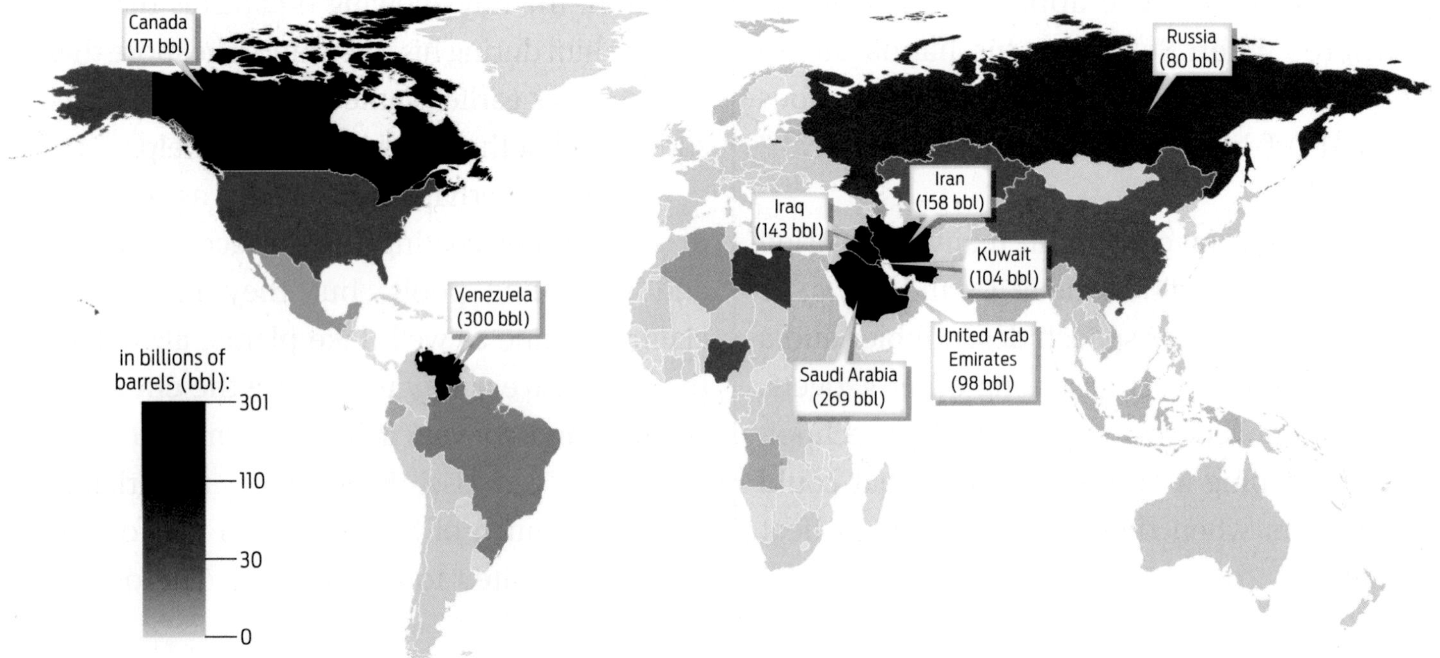

Data from the Central Intelligence Agency, *The World Factbook*, 2015

from municipal wastewater–as well as fuels made from algae.

Convinced of the promise of algae biofuels, in 2006, Sears founded Solix, one of the first biotechnology companies working to mass-produce biodiesel from algae. In 2009, the company began commercially producing its algae-based fuel, with the goal of making 3,000 gallons of biodiesel per acre of cultivated algae per year. Other companies have joined the race as well, including San Diego-based Sapphire Energy and Bay Area-based Solazyme. In September 2009, a modified Toyota Prius dubbed Algaeus drove 3,750 miles across the country powered in part by Sapphire's algae-based gasoline. In November 2011, United Airlines flew a passenger plane fueled by Solazyme's algae-based jet fuel. And in November 2012, California gas stations began a test project pumping Solazyme's algae-derived biodiesel. Algae, some say, are the fuel source of the future.

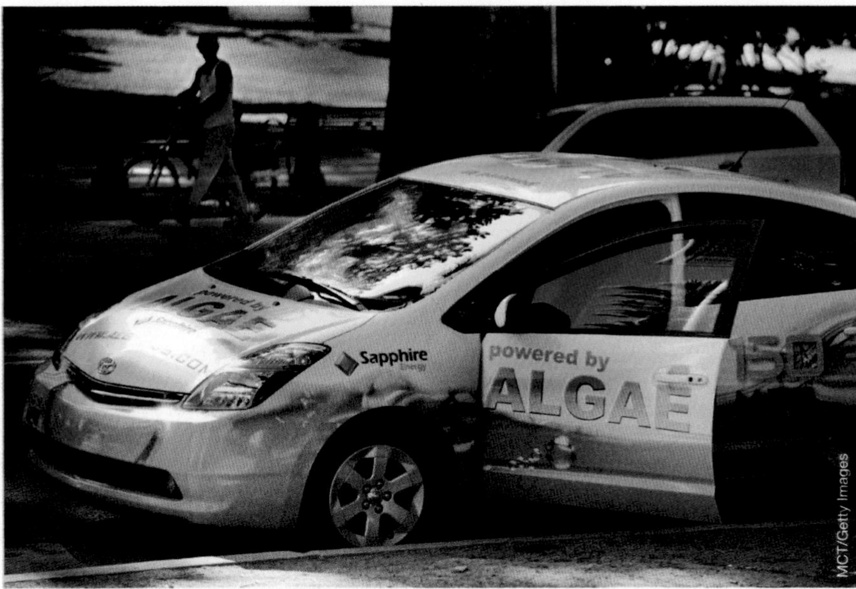

The Toyota Algaeus, powered by algae biofuel, gasoline, and batteries.

United Airlines Boeing 737-800 Eco-Skies airplane, powered in part by algae biofuel.

Pond Scum Power

To power nearly everything in our modern lives, we need energy. Americans get their energy from several sources, but most of it comes from **fossil fuels**–oil (petroleum), natural gas, and coal. The compressed remains of once-living organisms, fossil fuels are excellent sources of energy because they release so much of it when they are burned–to heat a home, for example, or power an engine. But there are downsides to these fuels, too. They are essentially nonrenewable, meaning they won't last forever. And extracting them from the earth can cause environmental damage, like oil spills and climate change. Those are two big reasons why renewable sources of energy are an attractive alternative to fossil fuels. Currently only about 10% of our energy comes from renewable sources like wind, solar, and biofuels, but that proportion is growing (**INFOGRAPHIC 5.1**).

Energy isn't only needed to power our technology. Living things need energy, too. Energy powers every activity we perform, from the more obvious ones like breathing, thinking, and running to less obvious ones like building the molecules that make up our bodies. Without a continuous source of energy, all life on Earth would grind to a halt, like a cell phone with a dying battery.

FOSSIL FUELS
Carbon-rich energy sources, such as petroleum, natural gas, and coal, which are formed from the compressed, fossilized remains of once-living organisms.

INFOGRAPHIC 5.1 U.S. Energy Consumption Relies on Non-renewable Resources

The United States leads the world in petroleum consumption, and is the second largest net consumer of fossil fuels behind China. Fossil fuels are considered non-renewable because they take millions of years to create by natural processes. As we continue to deplete fossil fuels, new energy sources are being developed that reduce our demand on petroleum and other fossil fuels.

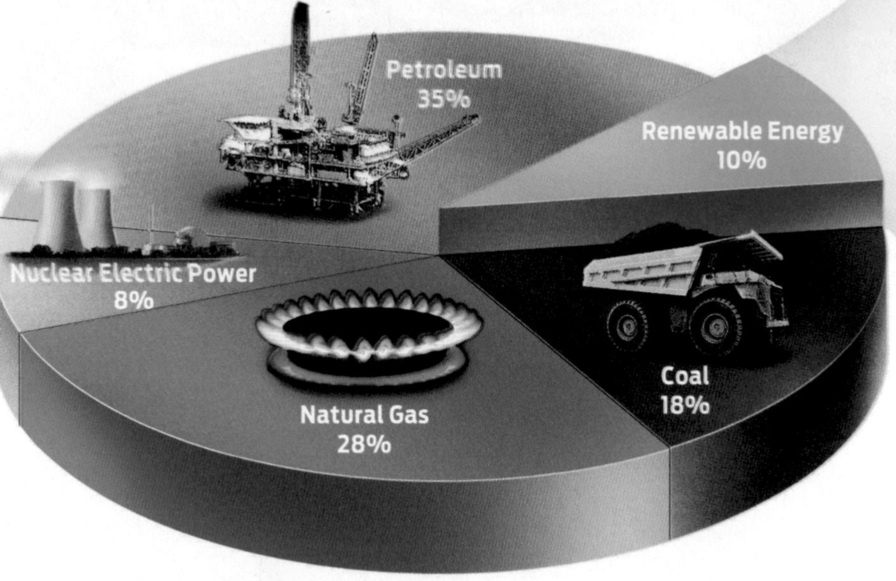

Data from *U.S. Energy Information Administration, Monthly Energy Review (2015)*, preliminary data

? Of the energy sources shown, which are fossil fuels?

CHEMICAL ENERGY
Potential energy stored in the bonds of biological molecules.

Humans and other animals obtain the energy they need by eating food. We've already seen that food is a source of organic molecules–the carbohydrates, fats, and proteins that our digestive systems break down into smaller subunits (see Chapter 4). The chemical bonds in these subunits represent a form of stored **chemical energy** that can be used to power cell functions.

Algae and plants, on the other hand, get their energy from the sun. They trap the energy of sunlight and store it in the form of molecules inside their cells. Algae are a promising source of biofuels because they are very good at trapping and storing energy–the oil they produce is rich in chemical energy. Even more impressive, all they need to make this oil is sunlight, carbon dioxide, and water (plus a few nutrients). Give them these basic ingredients, and algae grow rapidly–some strains double their volume in 12 hours–all the while accumulating gobs of oil inside their cells. In addition to this oil, which can be used to make biodiesel, algae also make carbohydrates (sugars), which can be converted into other biofuels like ethanol and butanol, and proteins, which can be used for a variety of purposes, such as feeding farm animals **(INFOGRAPHIC 5.2)**.

Sears wasn't the first to consider algae's fuel potential. In 1978–the same year Sears took his fateful night dive–the U.S. Department of Energy started its Aquatic Species Program, with the goal of exploring algae's fuel possibilities. Algae are a diverse group of hundreds of different species of aquatic,

INFOGRAPHIC 5.2 Algae Capture Energy in Their Molecules

Algae can use sunlight, carbon dioxide, water, and nutrients to produce a high volume of oil readily available to produce biofuel, in addition to carbohydrates and proteins that can be useful as additional energy sources.

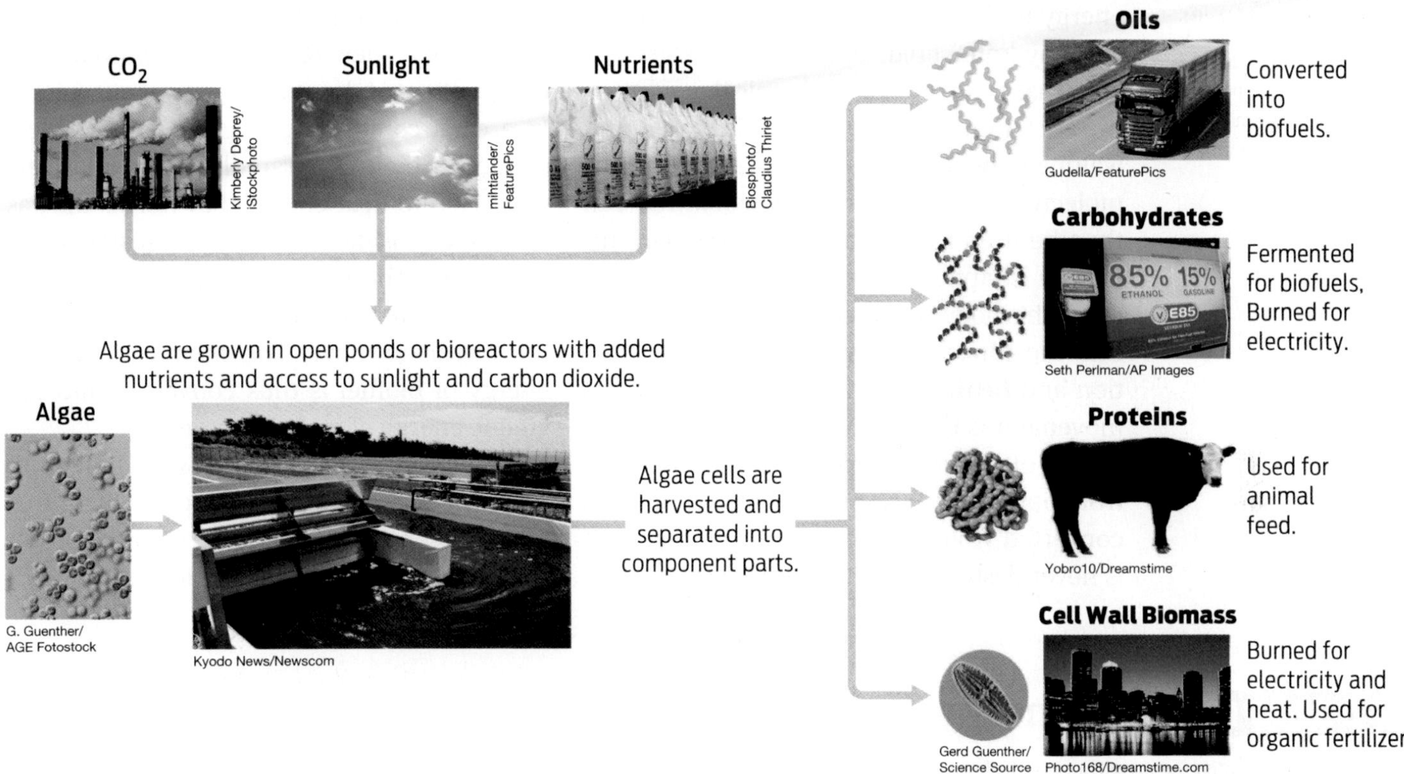

? Which algal component is used to produce biofuels? Which can be used to generate electricity?

plantlike organisms. There are large multicellular forms, like the seaweed that washes up on a beach after a storm, and microscopic single-celled forms, like those that turn ponds green. One major discovery of the Aquatic Species Program was that, for making biofuels, the algae that are most useful are the single-celled pond scum variety. These algae produce some of highest amounts of oil per cell—as much as 70% of the cell's dry weight.

The Aquatic Species Program was a direct response to the oil crisis of the 1970s, during which the cost of oil skyrocketed (quadrupling to $60 a barrel in today's dollars), supplies were rationed, and long lines formed at the pump. But when oil fell to $20 a barrel in 1996, the government abandoned the program, assuming that oil made from algae would always be too expensive. Now, with oil prices much more volatile, and the cost of alternatives coming down, biofuel from algae has once again become an attractive option.

Follow the Energy

With the topic of dwindling energy reserves frequently in the news, it's tempting to think that energy is something that we simply use up over time. But in fact, energy cannot be created or destroyed. When energy is used to power our cars—or our brain cells—that energy is not destroyed but rather changes form, a principle known as the **conservation of energy.** One of the laws of thermodynamics, this principle is key to understanding

CONSERVATION OF ENERGY
The principle that energy cannot be created or destroyed, but can be transformed from one form to another.

POTENTIAL ENERGY
Stored energy.

KINETIC ENERGY
The energy of motion or movement.

HEAT
The kinetic energy generated by random movements of molecules or atoms.

energy use, as applicable to people as it is to cars.

Consider a cyclist who eats a power bar before a ride. The bar contains chemical energy in the form of chemical bonds in the carbohydrate, lipid, and protein molecules that make up the bar. Chemical energy is **potential energy,** energy that is stored and waiting to be used. Digestion breaks up these molecules into smaller subunits, which can then be used as fuel for cells (discussed further in Chapter 6). As the cyclist begins to pedal, his body converts this potential energy into the **kinetic energy** of muscle contraction and **heat.** The kinetic energy of muscle movement is then converted into the kinetic energy of moving wheels. From start to finish, from power bar to spinning wheels, energy is converted from one form into another, but is never destroyed (**INFOGRAPHIC 5.3**).

Energy from biofuel has a similar life story. Oil from algae is rich in chemical energy: the lipids in the oil store energy in their bonds. When these lipids undergo chemical reactions–for example, when they are burned–they release large amounts of energy that can be used to power machines. In a car's combustion engine, the chemical energy in biofuel is rapidly and explosively converted to heat energy that warms the gas molecules inside a chamber, causing them to expand. The expansion of the heated gas molecules pushes against the pistons, causing the wheels to move. The chemical potential energy of biofuel is thus converted into the kinetic energy of car movement.

If energy is never destroyed, only converted, why do we need to keep filling our tanks? Essentially it's because energy doesn't stay in the car system (or in your body).

INFOGRAPHIC 5.3 Energy Is Conserved

Energy in the universe is neither created nor destroyed, but is converted from one form to another. Stored potential energy in food, for example, can be converted to kinetic energy as it powers a cyclist's movements.

Potential Energy

Food is a source of chemical potential energy, stored in the chemical bonds of food molecules (carbohydrates, proteins, and lipids).

Digestion breaks down food molecules into smaller subunits that are used as fuel by the body's cells, including muscle cells.

Kinetic Energy

Chemical energy is converted into kinetic heat energy that is lost from the body.

Chemical energy is converted into the kinetic of muscle movement.

Kinetic energy of muscle movement is converted into kinetic energy of wheel movement.

KENZO TRIBOUILLARD/Getty Images

The bike moves forward powered by energy converted from chemical to kinetic form.

? Into what forms of energy is the chemical energy in the food converted as the cyclist completes his training ride?

INFOGRAPHIC 5.4 Energy Conversion Is Not Efficient

As energy is converted from one form to another, only some of the available energy is fully converted to the next form. Some of it is converted into heat that escapes into the environment, and some energy is not converted at all.

Unconverted fuel energy is removed in exhaust.

Heat energy (kinetic) escapes

Heat energy (kinetic) escapes

Heat energy (kinetic) escapes

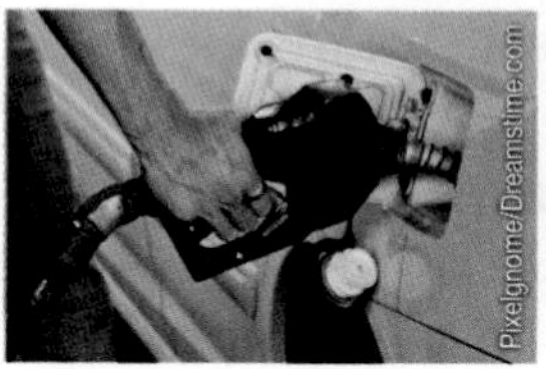

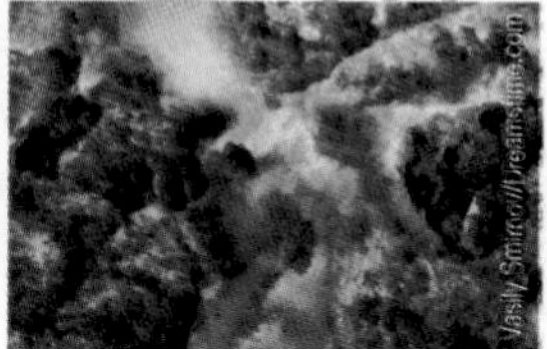

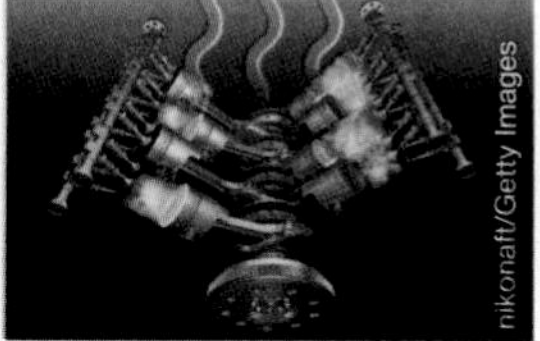

Fuel Is Chemical Energy: The chemical bonds of biofuel molecules store potential energy.

Fuel Combustion: Chemical energy is converted to heat (kinetic) energy.

Pistons Fire: Kinetic energy of heated air molecules is converted to kinetic energy of piston movement.

Tires Roll: Kinetic energy of pistons is converted to kinetic energy of tire movement.

Energy in Car Is Depleted: More chemical energy is required to keep tires rolling.

? When the vehicle is fueled, what form of energy is being added to the tank?

Energy flows from the fuel to the engine to the tires to the brakes and eventually leaves the car system as heat. What's more, the conversion of energy from one form to another isn't 100% efficient. With every energy conversion, a bit of energy is "lost" to the environment as heat. This is why car engines are warm after being driven (and why our bodies heat up when we exercise). Just how well your car converts the chemical energy of gas into the kinetic energy of car speed determines how many miles per gallon you get. If an engine doesn't combust efficiently, some of the fuel molecules will undergo chemical reactions and be converted to other molecules—like pollutants—rather than generating heat to power the pistons. If the pistons can't use the heat efficiently, the heat will leave the car without powering the wheels. If you drive a heavy SUV, more of the energy in gasoline will be required to reach a given speed compared to a lighter compact model. At each step of energy transformation, energy is lost from the car system and into the environment, and we're back to the fuel pump once more **(INFOGRAPHIC 5.4)**.

Solar-Powered Cells

To grow algae for biofuel, scientists are experimenting with several different production techniques. One involves growing algae in large open-air tanks—the closest to the natural process of algae growing on ponds, and the one that Sears's company Solix uses. Another method uses closed bioreactors in which algae are not exposed to outside air, carbon dioxide is pumped in, and the individual tanks can be stacked vertically to reduce the overall footprint of the operation and better prevent contamination. A third approach involves using algae to convert plant sugars

PHOTOSYNTHESIS The process by which plants and algae harness the energy of sunlight to make sugar from carbon dioxide and water.

into oil in large vats. In all cases, the goal is the same: convert as much energy from sunlight as possible into biofuel.

Plentiful and free, sunlight is a biofuel maker's dream. These qualities are the main reason why so much research has gone into developing technologies to capture solar energy. Yet figuring out how to tap and store this energy cheaply has proved challenging. So far, even with our best technology, humans can't beat what plants and algae do naturally–which is why many biofuels rely on these organisms as their starting point.

The secret to their success is **photosynthesis.** Through this process, plants, algae and a few other organisms capture the energy of sunlight and convert it into the chemical energy of sugar molecules. That may sound simple, but photosynthesis surely ranks as one of the most sophisticated and consequential achievements of evolution. By performing this energy conversion, photosynthesis makes possible nearly all life on Earth (**INFOGRAPHIC 5.5**).

There's another name for the energy-rich products of photosynthesis: food. Organisms

INFOGRAPHIC 5.5 Photosynthesis Converts Light Energy into Chemical Energy

Through the process of photosynthesis, plants, algae, and some bacteria are able to convert light energy from the sun into chemical energy stored in glucose.

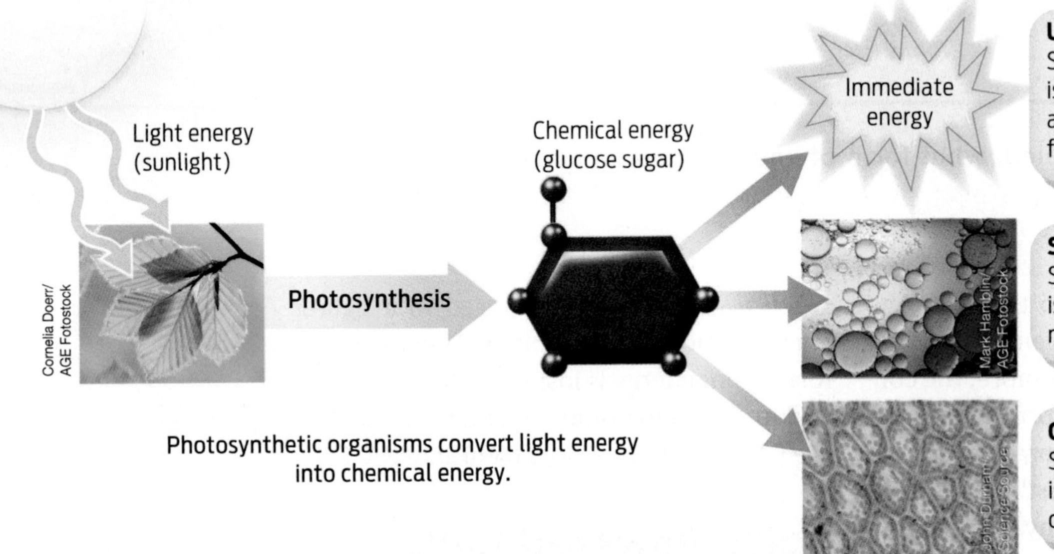

Usable Energy
Some of the chemical energy is converted into a form that is available to power cellular functions.

Stored Energy
Some of the chemical energy is stored as potential energy in molecules like oil.

Cell Structures
Some of the chemical energy is used as building blocks for cell structures.

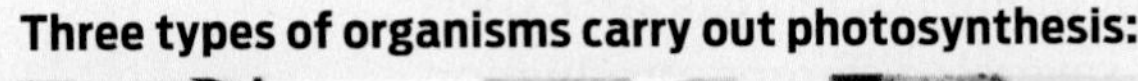

Three types of organisms carry out photosynthesis:

Plants

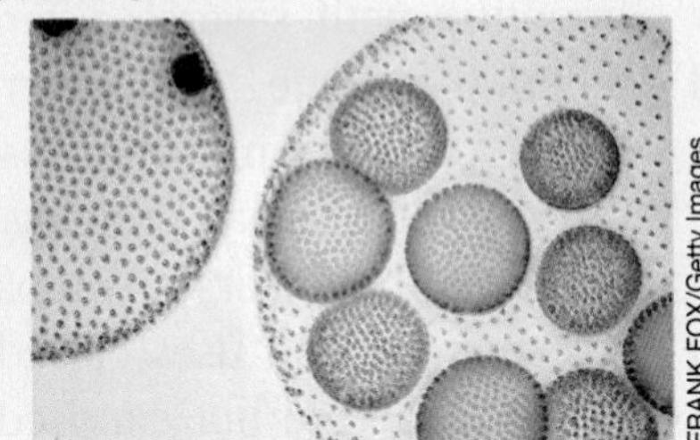

Algae

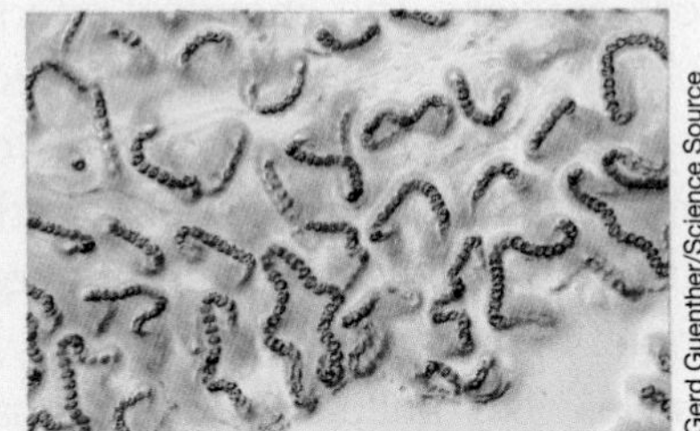

Some bacteria (e.g., cyanobacteria)

? How can the glucose produced in photosynthesis be used by the plant?

such as plants, algae, and certain bacteria that can make food from inorganic (nonliving) starting materials–from carbon dioxide, water, and sunlight, for example–are called **autotrophs.** (Their name literally means "self-feeders.") Autotrophs include not just crop plants like wheat, corn, and soybeans, but also all flowering plants, trees, and bushes. Oceans and lakes have an abundance of autotrophs, too, including much of the planet's algae and photosynthetic bacteria.

Organisms that can't make their own food and must consume organic molecules produced by other living organisms to obtain energy are called **heterotrophs** ("other-feeders"). This group includes all animals, fungi, and most bacteria. Photosynthesis, in other words, is what makes life possible for the rest of us.

Photosynthesis: How It Works

So how do plants and algae accomplish this unique feat of creating their own food? The process of photosynthesis can be divided into two steps: a "photo" step and a "synthesis" step. During the "photo" step, light energy is captured in chemical form. In the process, water is split and oxygen is released as a by-product. During the "synthesis" step, this chemical energy powers the formation of glucose molecules, using the carbon atoms of carbon dioxide. This synthesis phase does not directly require sunlight, but it does require the products of the "photo" reactions. The entire process of photosynthesis takes place inside an organelle called the **chloroplast,** which is present in the leaf cells of plants and in the cells of photosynthetic algae (**INFOGRAPHIC 5.6**).

The glucose made by photosynthesis is used by plants (or algae) in a variety of ways: it can be used for growth–to build new plant parts, like stems and fruit–or as an energy source to power cellular reactions.

While glucose is the major product of photosynthesis, other smaller sugars are produced during the "synthesis" reactions. Glucose and these other sugars provide the building materials for a variety of molecules in the cell–for example, amino acids to make proteins and lipids to make oils.

Like all photosynthesizers, algae take in carbon dioxide from the atmosphere and release oxygen. This is advantageous for life on Earth, since many creatures–including humans–use this oxygen to breathe. Nearly all the breathable oxygen that exists on our planet comes from photosynthesis. And, since humans and other organisms give off carbon dioxide (CO_2) as a waste product when we breathe out, we provide the raw material for photosynthesis to continue: a win-win situation for life on Earth.

Though carbon dioxide, like sunshine, is free and readily available to power photosynthesis, the amount required to make biofuel can exceed what is found in the surrounding air. That means that biofuel producers

AUTOTROPHS
Organisms such as plants, algae, and certain bacteria that can make their own food from inorganic starting materials (e.g., CO_2).

HETEROTROPHS
Organisms, such as humans and other animals, that obtain energy by consuming organic molecules that were produced by other organisms.

CHLOROPLAST
The organelle in plant and algae cells where photosynthesis occurs.

Mark Boster/Getty Images

Algae farms like this one in San Diego, California, circulate algae as they are grown for biofuel production.

INFOGRAPHIC 5.6 Photosynthesis Captures Sunlight and Carbon Dioxide to Make Food

Photosynthesis is the process by which plants, algae, and some other autotrophs use the energy of sunlight and carbon dioxide to make food. In plant and algae cells, photosynthesis occurs in an organelle called the chloroplast (found in cells that make up the green parts of the plant). Photosynthesis has two main steps, one that converts light energy into chemical energy and a second that uses chemical energy to build sugar molecules from carbon dioxide.

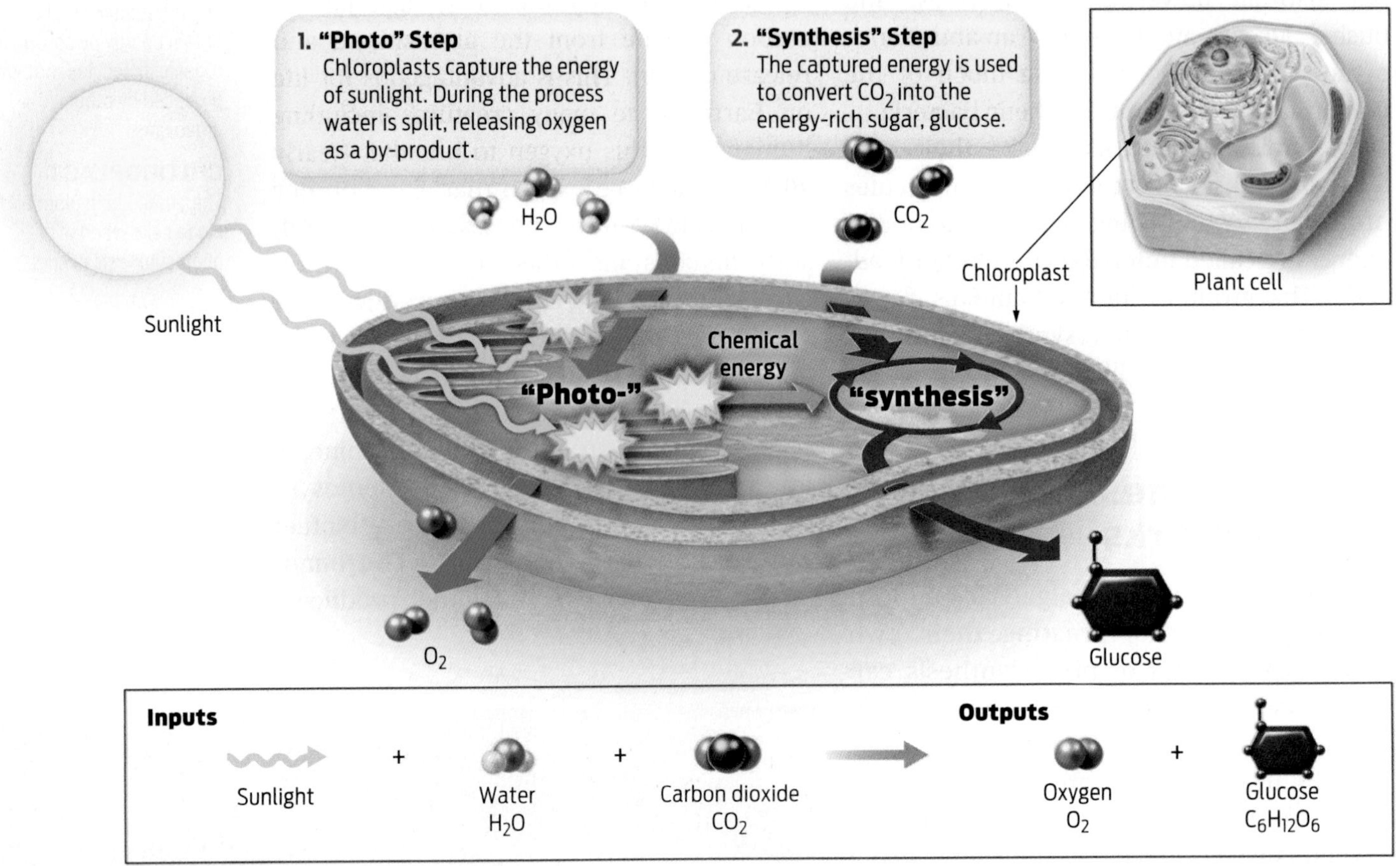

Inputs						Outputs		
Sunlight	+	Water H_2O	+	Carbon dioxide CO_2	→	Oxygen O_2	+	Glucose $C_6H_{12}O_6$

? What is the source of the carbon atoms in the glucose and of the oxygen atoms in the O_2 produced by photosynthesis?

sometimes have to supply another source of carbon dioxide, which can be costly. Sears's company, Solix, has set up its first biofuel production plant next to a beer manufacturer that produces carbon dioxide as a by-product of brewing. The company simply siphons off this carbon dioxide and feeds it to its algae, thus helping them grow. Other companies are looking to place their algae bioreactors next to carbon dioxide-emitting power plants—thus turning waste into profit.

There are advantages to photosynthesizers' sucking up so much carbon dioxide from the atmosphere. Carbon dioxide (CO_2) is not just the gas that plants and algae take in during photosynthesis, it is also the gas that is released by burning fossil fuels. If you think about it, this makes perfect sense: fossil fuels such as coal, petroleum, and natural gas are the compressed remains of once-living photosynthetic organisms that have formed over millions of years; burning these fuels releases their stored carbon dioxide, sending it back into the atmosphere. Carbon dioxide is a greenhouse gas that accumulates in the atmosphere and is partly responsible for the increasing temperatures around the globe and other signs of climate change (see Chapter 22).

In contrast, when plant and algae products are converted to fuel and burned, the CO_2 released is the same CO_2 that they took up by photosynthesis. This means that they are not contributing additional CO_2 to the atmosphere when they are burned (in other words, they are carbon neutral). By allowing more fossil fuels to stay in the ground, algae biofuels could help with the problem of climate change, too.

From Sun to Fuel

You may be wondering how something as intangible as sunlight can carry energy. If you've ever walked barefoot across a sandy beach on a hot summer day, you know that sunlight is a potent source of heat energy. You may also have a sense that certain colors absorb or reflect sunlight better than others—on a sunny day, wearing a reflective white shirt keeps you cooler than a black one that absorbs more of the sun's rays.

These properties of sunlight reflect the nature of **light energy,** a type of electromagnetic radiation (which also includes X-rays, microwaves, and radio waves). Light energy from the sun travels to Earth in waves. These waves of light are made up of discrete packets of energy called **photons.** Photons with different wavelengths contain different amounts of energy, and different objects on Earth absorb and reflect different wavelengths of light. Some of these wavelengths of light energy, when viewed by the human eye and interpreted by the human brain, appear to us as different colors. This is the visible light portion of the electromagnetic spectrum (**INFOGRAPHIC 5.7**).

When sunlight hits a green plant, for example, its leaves absorb red and blue wavelengths and reflect green wavelengths, which is why plants appear green to our eyes. The molecule within the plant cells that absorbs and reflects these wavelengths of light is the pigment **chlorophyll**—a crucial player in photosynthesis. It is chlorophyll that actually captures the energy of sunlight. During the "photo" reaction, chlorophyll molecules within chloroplasts absorb energy from the red and blue wavelengths of sunlight. In addition to chlorophyll, plants and algae contain other pigment molecules that absorb and reflect other wavelengths of light, giving them their distinctive colors. But chlorophyll is the main pigment involved in photosynthesis. When red and blue photons of sunlight hit chlorophyll, the electrons in its atoms become excited—boosted to a higher energy level. This is the initial step in the conversion

LIGHT ENERGY
A type of electromagnetic radiation that includes visible light.

PHOTONS
Packets of light energy, each with a specific wavelength and quantity of energy.

CHLOROPHYLL
The pigment present in the green parts of plants that absorbs photons of light energy during the "photo" reactions of photosynthesis.

Pete Atkinson/Getty Images

A spirulina cyanobacteria farm from the air.

INFOGRAPHIC 5.7 The Energy in Sunlight Travels in Waves

The sun emits electromagnetic radiation with a spectrum of wavelengths. The majority of the electromagnetic radiation emitted by the sun is ultraviolet (UV), visible, and infared (IR), each with a specific wavelength.

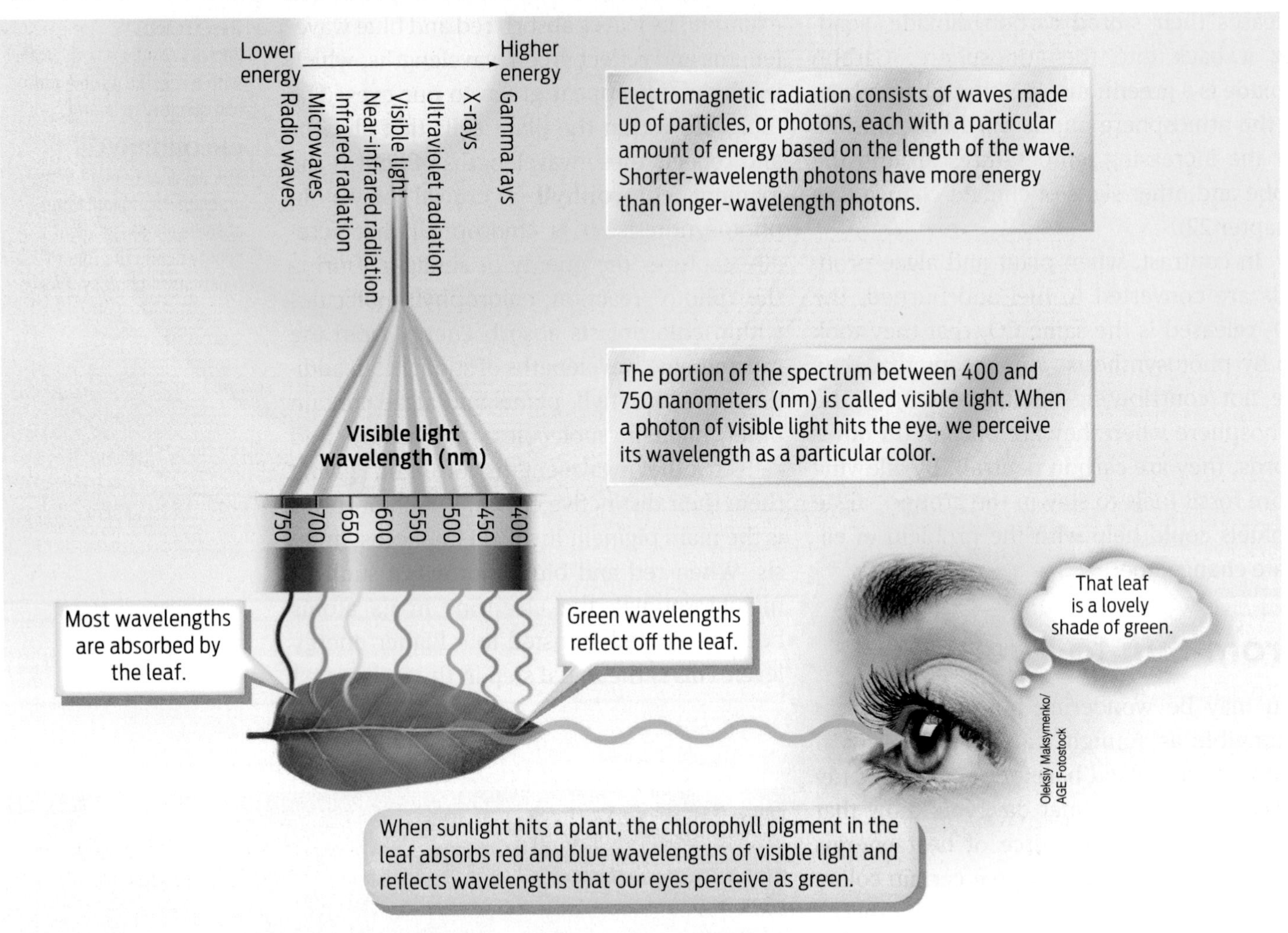

Lower energy
Higher energy
Gamma rays
X-rays
Ultraviolet radiation
Visible light
Near-infrared radiation
Infrared radiation
Microwaves
Radio waves
Electromagnetic radiation consists of waves made up of particles, or photons, each with a particular amount of energy based on the length of the wave. Shorter-wavelength photons have more energy than longer-wavelength photons.
Visible light wavelength (nm)
400
450
500
550
600
650
700
750
The portion of the spectrum between 400 and 750 nanometers (nm) is called visible light. When a photon of visible light hits the eye, we perceive its wavelength as a particular color.
Most wavelengths are absorbed by the leaf.
Green wavelengths reflect off the leaf.
That leaf is a lovely shade of green.
Oleksiy Maksymenko/ AGE Fotostock
When sunlight hits a plant, the chlorophyll pigment in the leaf absorbs red and blue wavelengths of visible light and reflects wavelengths that our eyes perceive as green.

? Which color of visible light has the most energetic photons?

ADENOSINE TRIPHOSPHATE (ATP) The molecule in cells that powers energy-requiring functions.

of light energy into chemical energy during photosynthesis.

Next, these electrons are put to use, powering the gears of the photosynthesis machinery. The excited electrons in chlorophyll are eventually knocked off the molecule completely. They go on to power the formation of an energy-carrying molecule called **adenosine triphosphate (ATP),** which is used in the "synthesis" part of photosynthesis as an energy source to make sugar. (We'll talk more about ATP in Chapter 6.) Meanwhile, the electrons lost by chlorophyll are replaced when water is split into oxygen (O_2) and hydrogen (and electrons). At the end of the photo reactions, the excited electrons from chlorophyll are captured by a molecule called NADPH. This molecule, an electron carrier, then brings the electrons to the "synthesis" reactions. Energy from these excited electrons, as well as from the breakdown of ATP, is then used in the synthesis reactions to incorporate carbon dioxide molecules into sugar **(INFOGRAPHIC 5.8).**

INFOGRAPHIC 5.8 Photosynthesis: A Closer Look

1. Light "Photo" Reactions
Chlorophyll pigments within internal chloroplast membranes absorb photons. Chlorophyll electrons (e^-) become excited and enter a series of reactions that generate energy-carrying molecules called ATP and electron-carrying molecules called NADPH, which are used in the synthesis reactions.
2. Carbon "Synthesis" Reactions
Energy from the breakdown of ATP and electrons from NADPH are used in the carbon reactions to fix carbon dioxide into organic sugar molecules, a form of stored chemical energy.
Water (H_2O)
Water is split during the light reactions. Split water molecules release electrons that replace electrons lost by excited chlorophyll molecules.
Carbon dioxide (CO_2)
CO_2 gas enters plant cells from the atmosphere. The carbon atoms are incorporated into organic sugar molecules.
H_2O
CO_2
Sunlight
Light reactions
Chemical energy from light reactions
Carbon reactions
Chloroplast
Internal chloroplast membranes containing chlorophyll
e^-
ATP
NADPH
O_2
Glucose
Oxygen (O_2)
This gas is a by-product of water splitting during the light reactions.
Glucose ($C_6H_{12}O_6$)
Glucose, the carbohydrate product of photosynthesis, contains the chemical energy converted from sunlight and the fixed carbon from atmospheric CO_2.

? Which molecules provide the energy to fix carbon dioxide in the carbon reactions?

Capturing Carbon

In the end, photosynthesis accomplishes two main things. First, it converts light energy from the sun into chemical energy that can be used to make food and fuel for plants and animals. Second, it captures inorganic carbon dioxide gas from the air and incorporates those carbon atoms into organic sugar molecules in a process called **carbon fixation.** What is being "fixed" is the inorganic carbon in CO_2, which is pretty much useless to non-photosynthetic organisms. We must rely on plants, algae and other photosynthesizers to "fix" carbon into organic sugars that our bodies can use.

Carbon fixation is Jim Sears's favorite topic these days. When he left Solix in 2007, Sears became president of A2BE Carbon Capture, a company based in Boulder, Colorado, that is looking for ways to reduce carbon dioxide levels in the atmosphere. Since plants, algae,

CARBON FIXATION
The conversion of inorganic carbon (e.g., CO_2) into organic forms (e.g., sugars like glucose, $C_6H_{12}O_6$).

and other photosynthetic organisms all help to temper the effects of climate change by pulling carbon dioxide out of the atmosphere and fixing it into organic sugars, scientists are looking for ways to enhance this natural process.

Believe it or not, according to Sears the healthy soil in your backyard is actually photosynthetic. Tiny organisms called cyanobacteria thrive in healthy soil, and they perform photosynthesis. Cyanobacteria absorb sunlight and capture carbon from the atmosphere. They then convert the carbon into forms that provide energy and nutrition to other microorganisms buried deep within the soil. One square meter of healthy, undisturbed soil can remove 30 g of atmospheric carbon per year, according to Sears. Thus, cyanobacteria are good not only for other microorganisms in the soil, but also for the entire planet.

Believe it or not, the healthy soil in your backyard is actually photosynthetic.

The problem is that approximately 2 billion out of Earth's 13 billion total hectares of landmass have been damaged by human activity—construction and fires are among the biggest culprits. According to Sears, it can take anywhere from 30 to 3,000 years for soil microorganisms to regenerate after being destroyed—and in the meantime, the damaged soil is unable to remove carbon dioxide from the atmosphere.

But Sears thinks he has a solution. His company takes small samples of microorganisms from healthy soil, grows them in a contained facility, and then transplants them to damaged soil, where they spread out and thrive. He estimates that if 1 billion hectares of land were restored in this way, one-seventh of the world's greenhouse gas problem would be solved because of the vast amounts of carbon dioxide that would be pulled out of the atmosphere by the photosynthetic cyanobacteria in the regenerated soil.

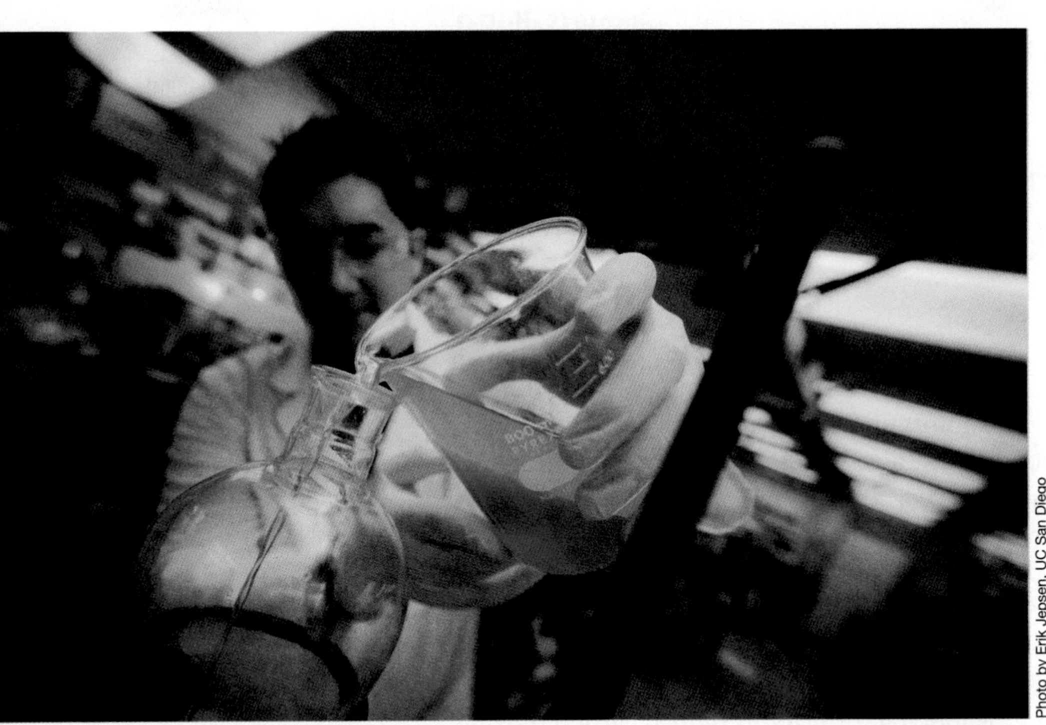

Photo by Erik Jepsen, UC San Diego

California Center for Algae Biotechnology, UC San Diego

UC–San Diego scientists have produced an algae-based sustainable surfboard.

Practical Hurdles

Despite considerable excitement, algae biofuels remain at the prototype stage. It's not currently possible to fill up your car's gas tank with algae biofuel (though a test at California gas stations in 2012 showed that customers were willing to buy it).

Other biofuels are available right now. Ethanol derived from corn is routinely added to gasoline in the United States, where it helps to offset carbon emissions from fossil fuels. Soybeans and rapeseed are used to make biodiesel, which is blended with regular diesel for use in diesel vehicles.

The trouble with both of these biofuels is that the plants used to make them are grown on land that might otherwise be used for food crops. This means they compete with food production and can raise the price of food. Algae don't present this problem. They can be grown on land that is unsuitable for growing food crops because they don't require fertile soil.

Compared to corn, soybeans, and rapeseed, algae have the potential to produce much more fuel for the amount of space they take up. By some estimates, more than three times the total area of cropland in the United States would need to be devoted to soybean cultivation to meet just 50% of U.S. transport fuel needs, but only 3% of total U.S. cropland would be required if algae were grown instead.

Algae do require significant amounts of water to grow, but they aren't as picky as crop plants and can grow in brackish (salty) water, which is not suitable for agriculture. Researchers with NASA are even exploring the use of wastewater from toilets as a source of nutrient-rich water for growing algae (**Table 5.1**).

Today, the biggest hurdle for algae biofuels is cost. Compared to gasoline, diesel, or even corn-based ethanol, the cost of algae biofuels is currently prohibitively expensive. This high cost is mostly a function of the energy-intensive technology needed to contain and protect the algae as it grows, to keep CO_2 continuously pumping into the pond or reactor, and to harvest the oil. With current strains of algae in use and current manufacturing processes, algae biofuel runs about $8

TABLE 5.1 How Green Are Biofuels?

Biofuels are attractive alternatives to fossil fuels, but not all biofuels are alike in terms of their environmental, energy, and land-use impacts. Compared to conventional crop plants, algae have some of the lowest environmental costs per similar quantity of oil produced.

CROP	GREENHOUSE GAS EMISSIONS*	WATER USE	FERTILIZER USE	PESTICIDE USE	ENERGY INPUTS**	LAND USE (M^2/KG BIODIESEL)	OIL YIELD (L/HA)	BIODIESEL PRODUCTIVITY (KG/HA)
Corn	81–85	High	High	High	High	66	172	152
Soybeans	49	High	Low	Medium	Low	18	18	562
Rapeseed, canola	37	High	Medium	Medium	Low	12	974	862
Microalgae (low–high oil content strains)	–183	Medium	Low	Low	High	0.1–0.2	59,000–137,000	52,000–121,000

* kg of CO_2 produced per megajoule of energy.
** Energy used during growing, harvesting, and refining of fuel.

Data from Groom, M. J., et al. (2008), *Conservation Biology* 2:602–609; Mata, T. M., et al. (2010), *Renewable and Sustainable Energy Reviews* 14:217–232; Christi, Y. (2007), *Biotech Advances* 25(3):294–306.

per gallon–far more expensive than the $3 gasoline we have in 2016.

To help bring costs down, researchers are experimenting with ways to make biofuel production less energy intensive and more efficient–through genetic engineering, for example. In 2011, ExxonMobil signed a deal with a San Diego-based company started by biotech entrepreneur Craig Venter to develop genetically modified algae strains with higher oil content and faster growth. In the interim, to remain economically viable, many algae companies have turned to producing "co-products" like cosmetics, foods–even surfboards–made from algae.

> "Algae truly are the foundations of our entire planet."
>
> — Jim Sears

Cost aside, the need for alternative fuels remains. We have quite a ways to go before we reach the goal set by the 2007 Energy Independence and Security Act of producing 36 billion gallons of renewable fuels by 2022. The interim goal for 2016 was 22.25 billion gallons, but the actual amount will likely be closer to 16 billion gallons.

The U.S. government seems to be betting on algae to help us make up the difference. In 2015, the Department of Energy awarded $18 million to six laboratories that are exploring ways to make algae biofuel cheaper, with a goal of reaching $3 per gallon by 2030 and bringing algae to the pump.

That seemingly simple organisms like algae are so vital to life on Earth is surprising enough. To think they may one day be an important alternative fuel source gives new meaning to the term "green energy." All this from single-celled organisms that have just one major claim to fame: they soak up sunlight and carbon dioxide to make sugar. "Algae truly are the foundations of our entire planet," Sears says. ■

CHAPTER 5 SUMMARY

- All living organisms require energy to live and grow. The ultimate source of energy on Earth is the sun.
- Energy is neither created nor destroyed, but is converted from one form into another; this principle is known as the conservation of energy.
- Kinetic energy is the energy of motion and includes heat energy and light energy. Potential energy is stored energy and includes chemical energy.
- Energy flows from the sun, is captured and transferred through living organisms, and then flows back into the environment as heat.
- Energy conversions are inefficient. With every conversion of energy, some energy is lost to the environment as heat.
- Photosynthesis is a series of chemical reactions that captures the energy of sunlight and converts it into chemical energy in the form of sugar. This energy is used by all living organisms to fuel cellular processes.
- Photosynthesis can be divided into two main parts: a "photo" part, during which the pigment chlorophyll captures light energy and water is split to produce high-energy electrons, and a "synthesis" part, during which captured energy and electrons are used to fix carbon dioxide into glucose.
- Photosynthetic organisms are known as autotrophs ("self-feeders"); they include plants, algae, and some bacteria. Animals do not photosynthesize; they are known as heterotrophs ("other-feeders").
- Photosynthetic algae convert glucose into energy-rich oils that can be used as fuel to power automobiles and aircraft. These biofuels show promise as alternatives to fossil fuels.

MORE TO EXPLORE

- **Energy 101: Algae-to-Fuels: http://energy.gov/eere/bioenergy/algal-biofuels**
- **PBS, NOVA, From Pond Scum to Power: http://www.pbs.org/wgbh/nova/tech/algae-biodiesel.html**
- **Can Algae Feed the World and Fuel the Planet? A Q&A with Craig Venter: http://www.scientificamerican.com/article/can-algae-feed-the-world-and-fuel-the-planet/**
- **Jones, C. S., & Mayfield, S. P. (2012). Algae biofuels: versatility for the future of bioenergy. *Current Opinion in Biotechnology*, 23(3), 346–351.**
- **Hunter-Cevera, J., et al. (2012). Sustainable development of algal biofuels. National Academy of Sciences Report.**

CHAPTER 5 Test Your Knowledge

DRIVING QUESTION 1 What are the photosynthetic organisms on the planet, and why are they so important?

By answering the questions below and studying Infographics 5.2 and 5.5, you should be able to generate an answer for the broader Driving Question above.

KNOW IT

1 **What do algae, cyanobacteria, and plants have in common?**

2 **Can animals directly use the energy of sunlight to make their own food (in their own bodies)? Briefly explain your answer.**

3 What organelle(s) would a nonphotosynthetic alga need to obtain in order to carry out photosynthesis?

a. mitochondria
b. nucleus
c. chloroplast
d. solar transformer
e. cell membrane

4 Why do many species of algae appear green?

5 Compare and contrast the ways photosynthetic algae and animals obtain and use energy.

USE IT

6 What would happen to humans and other animals if algae, cyanobacteria, and plants were wiped out? Would we only lose a food source (e.g., plants), or would there be other repercussions?

7 Why would a dark dust cloud that prevented sunlight from reaching Earth's surface be potentially devastating to animal life?

DRIVING QUESTION 2 What are the different types of energy, and what transformations of energy do organisms carry out?

By answering the questions below and studying Infographics 5.3 and 5.4, you should be able to generate an answer for the broader Driving Question above.

KNOW IT

8 The fuel energy you provide your car is best described as

a. kinetic.
b. chemical.
c. heat.
d. potential.
e. both chemical and potential.

9 The energy in a cereal bar is ___________ energy. The energy of a cyclist pedaling is ___________ energy.

a. light; chemical
b. potential; chemical
c. chemical; kinetic
d. potential; potential
e. kinetic; potential

10 Kinetic energy is best described as

a. stored energy.
b. light energy.
c. the energy of movement.
d. heat energy.
e. any of the above, depending on the situation

USE IT

11 If you wanted to get the most possible energy from photosynthetic algae, should you eat algae directly or feed algae to a cow and then eat a burger made from that cow? Explain your answer.

DRIVING QUESTION 3 How do plants and algae convert the energy in sunlight into energy-rich organic molecules? (And why can't other organisms, including humans, do this?)

By answering the questions below and studying Infographics 5.6, 5.7, and 5.8, you should be able to generate an answer for the broader Driving Question above.

KNOW IT

12 Which of the following photon wavelengths contains the greatest amount of energy?

a. violet
b. red
c. green
d. yellow
e. blue

13 Glucose is a product of photosynthesis. Where do the carbon atoms in glucose come from?

a. starch
b. cow manure
c. molecules in air
d. water
e. soil

14 Mark each of the following as an input (I) or an output (O) of photosynthesis.

Oxygen _____
Carbon dioxide_____
Photons _____
Glucose _____
Water _____

15 Photosynthetic algae are

a. eukaryotic autotrophs.
b. prokaryotic autotrophs.
c. eukaryotic heterotrophs.
d. prokaryotic heterotrophs.

USE IT

16 Global warming is linked to elevated atmospheric carbon dioxide levels. How might this affect photosynthesis? If global warming should cause ocean levels to rise, in turn causing forests to be immersed in water, how would photosynthesis be affected?

17 Why are energy-rich lipids from algae more useful as a fuel than energy-rich sugars and other carbohydrates produced by photosynthetic organisms like corn and wheat?

18 Outline (with words or a diagram) the process of photosynthesis. Include the following forms of energy and molecules: sunlight; carbon dioxide; glucose (stored chemical energy); water; ATP; heat.

DRIVING QUESTION 4 How do algal biofuels compare to other fuels in terms of cost, benefits, and sustainability?

By answering the questions below and studying Infographic 5.1 and Table 5.1, you should be able to generate an answer for the broader Driving Question above.

KNOW IT

19 Which of the following is/are necessary for biofuel production by photosynthetic algae?

a. sunlight
b. sugar
c. CO_2
d. soil
e. all of the above
f. a and b
g. a and c

20 Why are algae considered more valuable for biofuel than plants (such as corn)?

a. because their photosynthetic products are an oil
b. because they are cheaper to grow
c. because they do not require as much CO_2
d. because they do not require as much fertilizer
e. all of the above

USE IT

21 Many types of algae can divert the sugars they make by photosynthesis into lipids that can be used to make biodiesel. Biodiesel is a promising replacement for fossil fuels. Describe the energy conversions required to make algal lipids for biodiesel and explain why biodiesel might be a more promising fuel than fuel from lipids extracted from animals.

22 What do you think are some of the advantages and disadvantages of growing algae in enclosed tubes or bags compared to growing them in open vats? Make a table listing the advantages and disadvantages of each approach and explain your reasoning.

23 Many biofuels, such as corn-derived ethanol, require arable land (land that is suitable for agriculture) for their production. Discuss competing needs for arable land in the context of human requirements for food and fuel, and how algae may alleviate this tension.

apply YOUR KNOWLEDGE

MINI CASE

24 A CEO of a new algal biofuel company is trying to select the site for a production facility. There are three possible options:

The desert of southern New Mexico (sunny, hot, mild winters, nonarable land, remote)

Denver, Colorado (sunny, cold winters, urban area with CO_2 emissions from factories and cars)

Central Washington State (sunny, hot, a rich agricultural zone)

Discuss the pros and cons of each site and make a recommendation.

apply YOUR KNOWLEDGE

BRING IT HOME

25 The airline Virgin Atlantic has committed to using a "green" fuel produced by microbes that use carbon monoxide (CO) from industrial emissions (such as from steel factories) as its carbon and energy source. Through a fermentation process that occurs in a reactor chamber, the microbes convert CO into usable ethanol, a viable "green" fuel, which can be converted to jet fuel. This fuel is predicted to reduce CO_2 emissions by 60% relative to conventional jet fuel. Consider how this fuel compares and contrasts with algal biofuel and corn ethanol. If this "green" fuel venture is successful, would their use of "green" fuel influence your decision to choose Virgin Atlantic over another air carrier? Why?

apply YOUR KNOWLEDGE

INTERPRETING DATA

26 The United States currently uses approximately 19 million barrels of petroleum per day. Of this, approximately 8 million barrels per day is imported, and the rest is from U.S. sources (which includes biofuels). The table shows the production cost estimated for different petroleum sources. (The actual cost is driven by a variety of market and geopolitical factors, so we will use production cost as a substitute for actual cost.)

a. Using the data for cost per barrel from various sources, calculate the production cost to meet current U.S. daily use. Assume that approximately 40% of the imports are from Canada, approximately 49% are from OPEC and Persian Gulf countries with production costs similar to Saudi Arabia, and 11% are split evenly between Mexico and Venezuela.

b. Let's say that the United States replaces half of its current oil imports from OPEC and Persian Gulf countries with domestically produced algal biofuel. What will this do to the cost of production to meet our daily needs?

Petroleum Source	Production Cost per Barrel (Average)
United States	\$36.20
Canada	\$41.00
Saudi Arabia	\$9.90
Venezuela	\$23.50
Mexico	\$29.10
Algal biofuel	~\$318

Data from: https://www.eia.gov/tools/faqs/faq.php?id=727&t=6

c. How could algal biofuel companies work to decrease production costs of their product?

6 Dietary Energy and Cellular Respiration

DRIVING QUESTIONS

1. How is obesity defined, and what are some hypotheses proposed to explain current obesity rates?
2. How does the body use the energy in food?
3. How does aerobic respiration extract useful energy from food?
4. When does fermentation occur, and why can't a human survive strictly on fermentation?

The sitting disease

Understanding the causes and consequences of obesity

WHEN JAMES LEVINE WAS 11 YEARS OLD, he began a science experiment in his bedroom. His test subjects? Snails.

In glass-walled fish tanks he built himself, Levine collected pond snails from nearby Regents Park in his native London. He then methodically monitored and recorded their movements.

"My idea was that every snail has a built-in hard-wired style of movement," explains Levine. "One snail will do swirly-whirly-whirly, while another snail will always move in a straight line."

To test this hypothesis, every night, between 9:00 P.M. and 5:00 A.M., he'd wake up hourly, cued by an alarm clock, and mark on the glass where the snail had moved. At the end of the night, he'd trace the snail's journey.

"Not surprisingly, I was constantly asleep at school," Levine confesses.

By the time he finished his experiment, Levine had 270 snail tracings. What he discovered was that snails didn't quite sort the way he thought. But they did have stereotypical styles of movement.

"Joanna"–he named the snails–"always does ziggidy-zaggidy-ziggidy-zaggidy. And John always moves in a smooth way."

The snail experiment made a lasting impression on Levine, who eventually went on to medical and graduate school. Now, 35 years later, Levine has turned his number-crunching fixation to another slow-moving creature: the human couch potato.

Levine, a professor of medicine at the Mayo Clinic campus in Phoenix, Arizona, and at Arizona State University, is an expert on the physiology of weight gain and loss, with a special focus on obesity. Obesity–having an unhealthy amount of body fat–has been called America's number one health crisis. Obesity has been linked to a whole host of health problems, including heart disease, diabetes, stroke, Alzheimer's disease, hypertension, and even cancer. It is also a major killer: in the United States, only tobacco use causes more premature deaths.

Rates of obesity have skyrocketed over the past four decades, leading many to refer to an "obesity epidemic." As of 2017, more than one-third of U.S. adults are obese, and another third are overweight. Rates in children are not far behind **(INFOGRAPHIC 6.1)**.

INFOGRAPHIC 6.1 Body Mass Index and Increasing Obesity Rates

A body mass index (BMI) chart provides an indirect measure of body fat based on the ratio of body height to weight. Using the BMI to categorize the U.S. population, studies show that the percentage of U.S. adults who are obese or extremely obese has increased since the 1976–1980 reporting period.

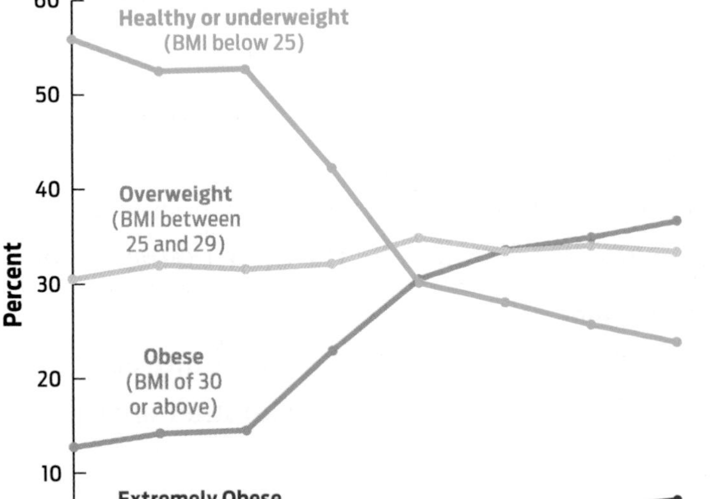

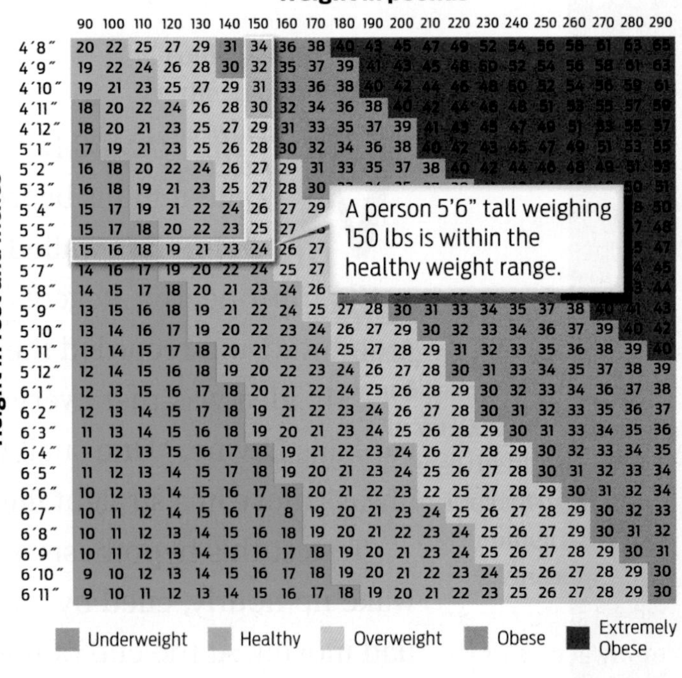

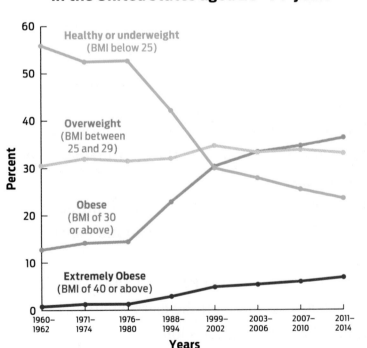

Data from CDC NCHS, Health, United States, 2015, table 58; Flegal, K.M. et al. (1998) *Intl. J. Obesity*, 22(1), 39–47.

? Is someone who is 5 feet, 10 inches tall and weighs 210 pounds considered underweight, of normal weight, overweight, or obese?

Definitions of healthy and unhealthy amounts of body fat are based on a tool called the **body mass index (BMI),** which provides an easy-to-digest estimate of body fat based on one's height and weight. People with a BMI between 19 and 24 are considered to be at a healthy weight; people with a BMI between 25 and 29 are considered **overweight;** people with a BMI of 30 and above are considered **obese;** and people with a BMI of 40 or above are considered extremely obese.

While pretty much everyone agrees that obesity is bad–for individuals and for society as a whole–scientists are divided about just what is causing our waistlines to expand at an ever-alarming pace. Is increased food intake–linked perhaps to larger portion sizes on menus and in supermarkets–mostly to blame? Or is decreased energy expenditure–a result of our increasingly sedentary lifestyle–the bigger player? No one can say for sure.

Scientists are also divided about how to best curtail this growing epidemic. But now, with the help of some clever experiments using sophisticated undergarments, Levine thinks he has found a solution to the puzzle.

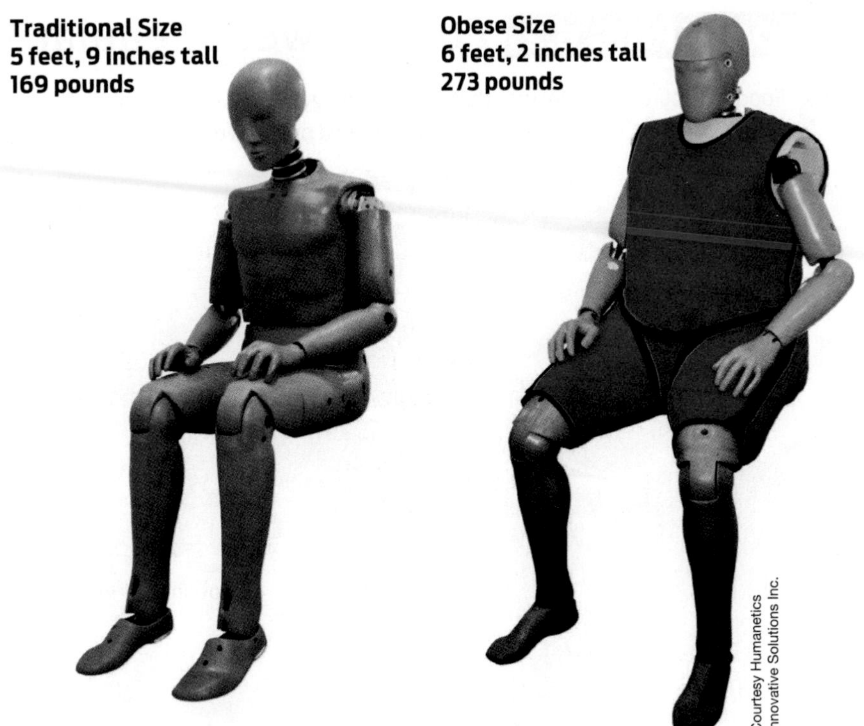

Humanetics, a crash test dummy developer, is increasing the size of some of its models to reflect our growing girth as Americans.

A Creeping Problem

To understand the obesity epidemic, says Levine, you have to realize that it didn't happen overnight. "Obesity doesn't occur over minutes and hours," says Levine. "Obesity occurs over years, decades, and generations."

What that means is that small changes in the way we live can seriously add up. Even something as simple as the amount of time we spend sitting or standing–if multiplied consistently across time–can profoundly affect how much energy we store or use.

Energy, defined as the capacity to do work, is what powers our biological lives. We obtain energy from the food we eat, and this energy fuels all of our activities–everything from thinking and digesting to sleeping and running. Obesity is fundamentally a problem of energy imbalance: taking in, over time, much more energy than we expend in our activities.

The energy in food is often measured in units called calories. A **calorie** (in lower case) is the amount of energy required to raise the temperature of 1 g of water by 1°C. On most food labels, the amount of energy stored is listed in kilocalories, which are also referred to as kcals or Calories (with a capital "C"). One **Calorie** is equal to 1,000 calories, which is equal to 1 kcal.

Different foods contain different amounts of energy, and therefore allow us to perform different amounts of work. Of all the organic molecules, fats are the most energy dense: each gram of fat stores approximately 9 Calories in its chemical bonds. Proteins and carbohydrates are about half as energy dense: they store about 4 Calories per gram. A 200-g serving of bacon contains many more

BODY MASS INDEX (BMI)
An estimate of body fat based on height and weight.

OVERWEIGHT
Having a BMI of 25 or more but less than 30.

OBESE
Having 20% more body fat than is recommended for one's height, as measured by a body mass index equal to or greater than 30.

calorie
A calorie (spelled with a lower-case "c") is the amount of energy required to raise the temperature of 1 gram of water by 1°C.

Calorie
A Calorie (spelled with a capital "C") is 1,000 calories or 1 kilocalorie (kcal). The Calorie is the common unit of energy used in food nutrition labels.

INFOGRAPHIC 6.2 Food Powers Cellular Work

Food is a source of molecules that provide energy to cells. The energy in food is measured in units called Calories. Fats provide more than twice the Calories than carbohydrates and proteins do, while nucleic acids are not a significant source of energy for a cell.

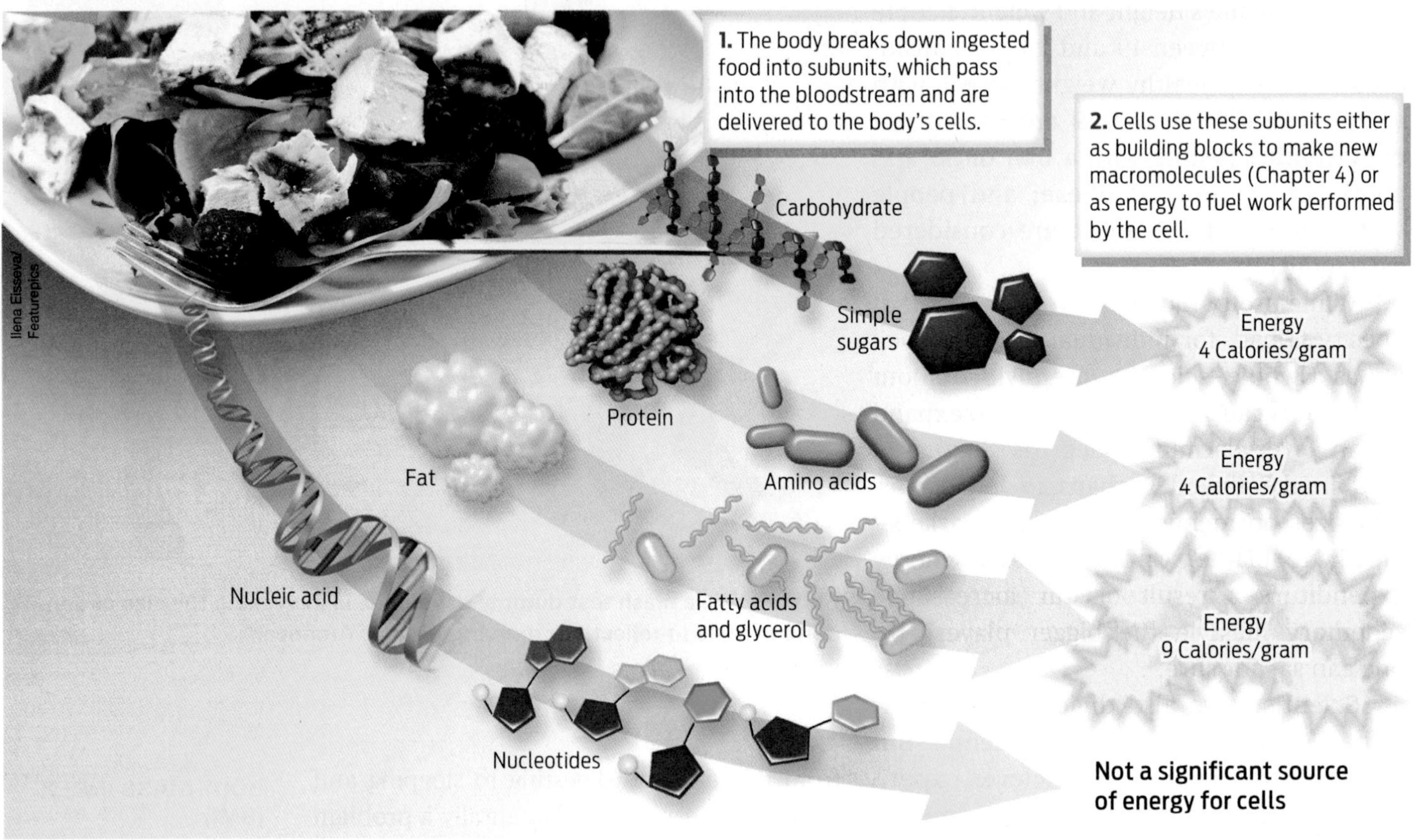

1. The body breaks down ingested food into subunits, which pass into the bloodstream and are delivered to the body's cells.
2. Cells use these subunits either as building blocks to make new macromolecules (Chapter 4) or as energy to fuel work performed by the cell.
Carbohydrate
Simple sugars
Energy 4 Calories/gram
Protein
Amino acids
Energy 4 Calories/gram
Fat
Fatty acids and glycerol
Energy 9 Calories/gram
Nucleic acid
Nucleotides
Not a significant source of energy for cells
Ilena Elsseva/ Featurepics

? Why does reducing the amount of fat in the diet have a greater impact on total Calories than reducing the amount of carbohydrates?

Calories than does a 200-g serving of asparagus (**INFOGRAPHIC 6.2**).

Likewise, different activities require different amounts of energy. Jogging, for example, burns about 500 Calories an hour, whereas sitting burns only about 5 Calories per hour. Deliberate exercise makes up a relatively small portion of our total daily energy expenditure. By far, the largest portion is made up of our basal metabolism–the many thousands of chemical reactions that keep our cells and organs functioning, and us alive.

How many Calories a person needs to eat in order to meet daily energy needs largely depends on gender, age, genetics, body type, and physical activity levels. A sedentary college-age average-size male, for example, would need to consume anywhere from 2,200 and 2,400 Calories per day to power his activities and maintain his weight, whereas a football player would need more than 3,200 Calories a day to power and maintain his (**INFOGRAPHIC 6.3**).

The only way to gain weight is by taking in more Calories than we expend through our activities. If we take more food energy into our bodies than we use to power cellular reactions and physical movement and generate heat, the excess is stored as fat. In other words, our waistlines obey the principle of conservation of energy: energy is neither created nor destroyed but merely converted from one form into another.

Most of us are surrounded by an abundance of easy-to-obtain, tasty foods. Food manufacturers also spend billions of dollars designing foods that will trigger our "bliss-point" and marketing these products to us.

So is overeating the cause of our obesity epidemic? Are we simply being goaded into eating more food than we used to, and paying the price in bigger waistbands? While that may be part of the explanation, Levine and his colleagues think it's not that simple.

A NEAT Cause of Weight Gain

We all have observed people who seem to eat anything thing they wish and never gain weight. Likewise, there are those who, it seems, merely have to look at food to put on pounds. It turns out there is scientific support for these subjective observations. Studies that have looked at weight gain in response to overfeeding have shown that people vary greatly in how much fat they accumulate. The biological mechanism that allows some individuals to resist weight gain more than others, however, has not been identified.

In 1999, Levine set out to solve the mystery. He and a team of researchers at the Mayo Clinic in Rochester, Minnesota recruited 16 nonobese adults (12 males and 4 females, ranging in age from 25 to 36 years). These individuals participated in a 10-week study to determine the effect of daily overfeeding on weight gain. For 2 weeks, the subjects were monitored to determine their daily caloric needs. Then, for the next 8 weeks, each participant was fed 1,000 Calories each day over the amount required to maintain his or her current weight. Meals were prepared and consumed in a research facility at the Mayo Clinic, and activity levels were strictly monitored through daily interviews and accelerometers worn by the study participants.

Because each person in the study was overeating to the same extent, you might expect that each person would gain the same

INFOGRAPHIC 6.3 Balancing Energy In with Energy Out

Different foods provide different numbers of Calories, and different activities expend different numbers of Calories. Ultimately, the balance between Calorie input and expenditure determines whether a person gains, maintains, or loses weight.

Food	Calories (kcal)
Medium apple (125 g)	65
1 hard-boiled egg	70
1 slice whole wheat bread	79
¼ cup cooked white rice	102
4 oz chicken breast	120
12 oz nonfat milk	120
1 slice thick-crust pizza	256
4 oz sirloin steak	280
Subway Turkey Breast Sandwich (6 in.)	282
1 Starbucks Grande Mocha Frappucino with whipped cream	360
Burger King Whopper	670

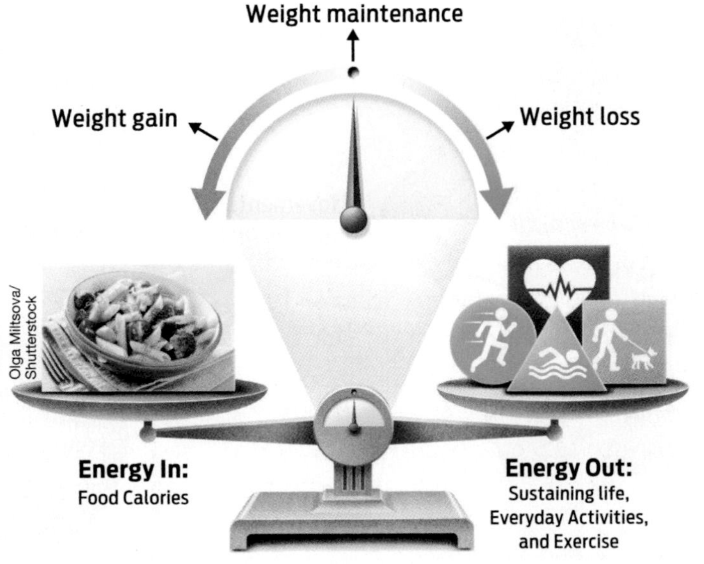

Activity	Calories (kcal)/hr
Sitting	5
Standing	15
Gum chewing	20
Walking (2 mph)	120
Stair climbing	200
Biking (moderate)	450+
Jogging (5 mph)	500+
Swimming (active)	500+
Hiking	500+
Cycling (stationary bike)	650

? How long would a person have to stand in order to burn off the Calories in a slice of whole wheat bread?

amount of fat. However, the researchers found that fat gain varied 10-fold among individuals in the study, ranging from a gain of about a pound to a gain of nearly 10 pounds. Where was all the extra energy provided by overfeeding going if not into the fat cells of study participants? One obvious possibility is different amounts of energy expenditure. In Levine's study, he found that test participants had markedly different levels of physical activity. Interestingly, intentional exercise was not the crucial difference in activity level.

Levine designed his study in such a way that intentional exercise was kept at a constant and minimum level across all test subjects. Therefore, the differences in physical activity were principally not related to exercise. Levine has a name for this type of activity: he calls it **NEAT,** short for "non-exercise activity thermogenesis."

> "People who are resistant to weight gain and can stay thin are people who can switch on their NEAT in response to overfeeding and never gain a pound."
>
> — James Levine

NEAT
Non-exercise activity thermogenesis, the amount of energy expended in everyday activities.

NEAT includes all the activities of daily living, such as household chores, yard work, shopping, going to a job, walking the dog, or playing a musical instrument. It also includes the energy expended to maintain posture, and it includes spontaneous movements such as fidgeting, pacing, or even chewing gum.

In Levine's study, energy expenditure via NEAT varied greatly among individuals, by as much as 700 Calories per day. More important, changes in NEAT were inversely correlated with fat gain: participants who increased their NEAT the most gained the lowest amount of fat. In other words, NEAT accounted for how these individuals resisted fat gain. "People who are resistant to weight gain and can stay thin are people who can switch on their NEAT in response to overfeeding and never gain a pound," Levine says **(INFOGRAPHIC 6.4)**.

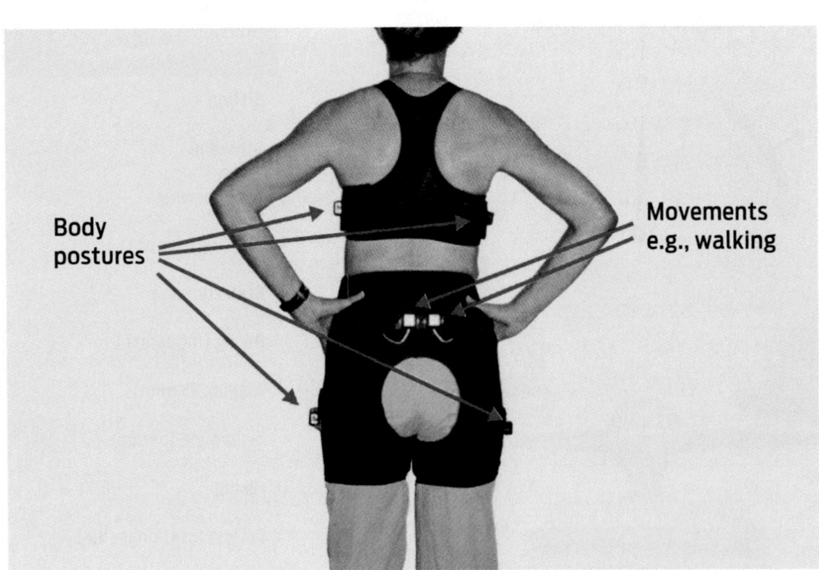

Dr. James Levine.

To study NEAT, Levine employs what he calls "magic underwear"—specially designed undergarments with sensors that detect movement.

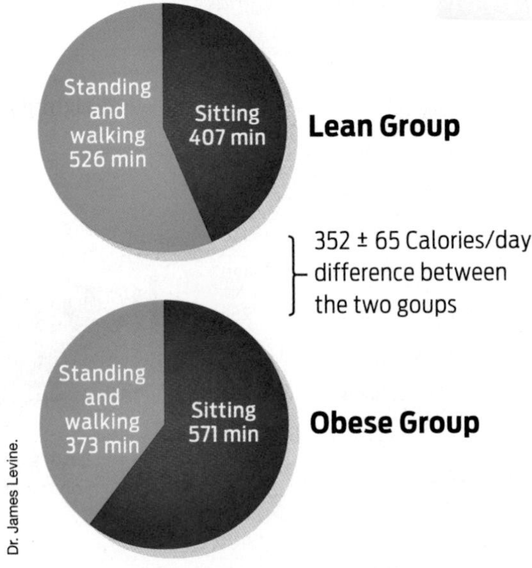

Data from Ravussin, E. (2005) *Science* Vol. 307: 530–531.

INFOGRAPHIC 6.4 NEAT Activities Influence Resistance to Fat Gain and Obesity

When fed 1,000 Calories per day more than needed to maintain weight, participants who increased their NEAT to a greater extent gained less fat. In another study, lean people had higher levels of NEAT (for example, higher levels of standing and moving around).

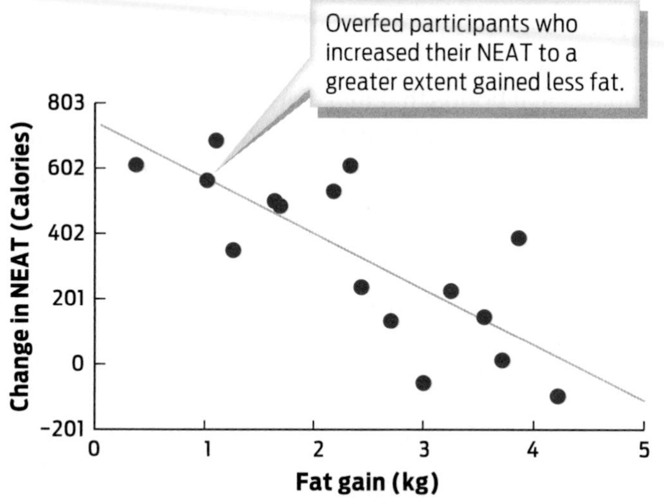

Data from Levine JA, Eberhardt NL, Jensen MD. *Science* 1999; 283: 212–4.

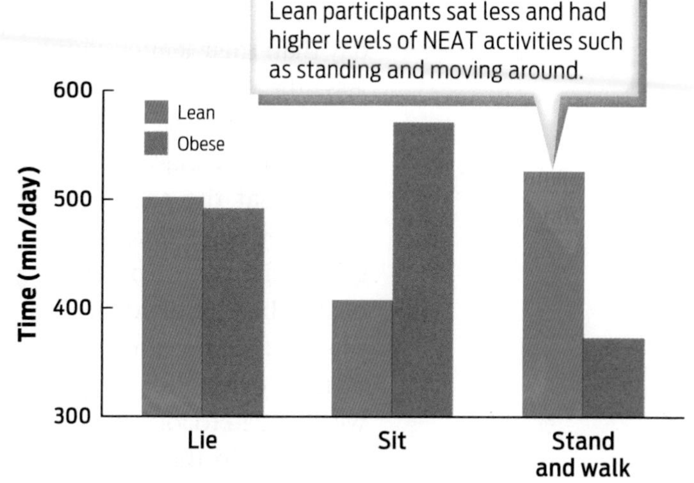

Data from Levine JA, Lanningham-Foster LM, McCrady SK, et al. *Science* 2005; 307: 584–6.

NEAT includes non-exercise activities such as walking the dog, shopping, dishwashing, social activities, and fidgeting.

connel/Shutterstock; CSP_monkeybusiness/AGE Fotostock; SolStock/Getty Images; Chris Whitehead/AGE Fotostock; Phanie/Science Source

? Consider someone who cannot handle a kickboxing class but who walks the dog and weeds the garden. What evidence suggests that these activities may help this person avoid fat gain?

Next, Levine wanted to know whether NEAT plays a role in obesity. Do lean and obese people differ in their levels of NEAT, for example? To get at this question, in 2005, Levine and his colleagues recruited 20 healthy volunteers who were self-proclaimed "couch potatoes." Ten participants (five females and five males) were lean (BMI ~23) and 10 participants (five females and five males) were mildly obese (BMI ~33). The volunteers agreed to have all their movements measured for 10 days. They were instructed to continue their normal daily activities and not to adopt new exercise regimens.

To measure NEAT, Levine and colleagues devised a novel way to track activity levels in test participants. They built a special kind of undergarment outfitted with electronic sensors that detect movement. The undergarment, which Levine calls "magic underwear," was built to allow people to wear it essentially all the time—even while going to the bathroom and having sex.

"We literally have snapshots as to how real people live their lives every half second of every day for days and days and days on end," Levine says.

Over the 10-day period, Levine's team collected 25 million data points on NEAT for each individual. From those measurements, they calculated how much time each person spent standing, walking, sitting, or sleeping per day, and how many calories each person expended as a result.

The bottom line? Obese individuals in the study sat on average 2.25 hours longer per day than their leaner counterparts. By sitting less, the lean people burned an additional 350 Calories a day. Moreover, this tendency seemed to be ingrained: previously lean individuals sat the same amount of time even after they were forced to gain weight, and previously obese individuals sat the same amount of time even after they were forced to lose it.

> "What's really cool about NEAT is that everyone can do it."
> — James Levine

What do the results of Levine's studies (both reported in the journal *Science*) mean for the obesity epidemic? According to Levine, they mean that the energy we expend in everyday activities–our NEAT–is far more important in controlling weight than anyone previously imagined. And, they suggest that a good way to combat obesity would be to get people up out of their chairs.

ATP: The Energy Currency

All physical activity–both NEAT and deliberate exercise–burns energy. Much in the same way that an automobile burns gasoline to power its pistons, the human body burns food molecules like sugars and fats to carry out its many chemical reactions. Both processes are a kind of controlled combustion, but there are important differences, too.

In order for the energy released from food molecules to be useful to the body, it has to be captured in a form that can participate in the cell's chemical reactions. That form is a molecule called **adenosine triphosphate (ATP).**

ADENOSINE TRIPHOSPHATE (ATP) The molecule that cells use to power energy-requiring functions.

AEROBIC RESPIRATION A series of reactions that occurs in the presence of oxygen and converts energy stored in food into ATP.

To make ATP, our bodies first break down food molecules into their smaller subunits by digestion: carbohydrates into sugars, fats into fatty acids and glycerol, proteins into amino acids. Once released from food, these subunits leave the small intestine and enter the bloodstream, which transports them to the body's cells. Inside the cells, enzymes break apart the bonds holding these subunits together. The energy stored in those bonds is then captured and transferred into the chemical bonds that make up ATP. When cells need energy, they break these bonds in ATP, releasing the stored energy, which can then participate in chemical reactions.

You can think of food as being like a bar of gold: it has a great deal of value, but if you brought that gold bar to your local convenience store, you wouldn't be able to buy even a cup of coffee with it. You would first have to convert your gold bar into bills and coins. ATP is the energetic equivalent of bills and coins; it's currency that your body can actually spend (**INFOGRAPHIC 6.5**).

ATP is spent anytime a muscle contracts or a neuron fires. When we go for a run, or even when we sit in a chair but tap our feet, our muscle cells break the bonds in ATP. The energy released allows muscle fibers to contract, powering movement. In our large brain, which runs principally on glucose, ATP is spent to move ions across cell membranes, enabling our neurons to fire. Of all the organs in the body, the brain is the largest consumer of energy. When we die, the rigor mortis that sets in occurs because ATP is required for muscle proteins to slide past one another; without it, they stay locked in place, and become stiff (and so do you). Ensuring a steady supply of ATP is therefore one of life's most critical activities.

Burning Food for Energy

The primary process that eukaryotic organisms–animals, plants, fungi, and protists–use to convert food energy into ATP is called aerobic cellular respiration, or just **aerobic respiration.** "Aerobic" means "in

INFOGRAPHIC 6.5 ATP: The Energy Currency of Cells

Just as a gold bar must be converted to currency in order to buy merchandise, the energy in food must be converted to ATP before it can be used by the cell.

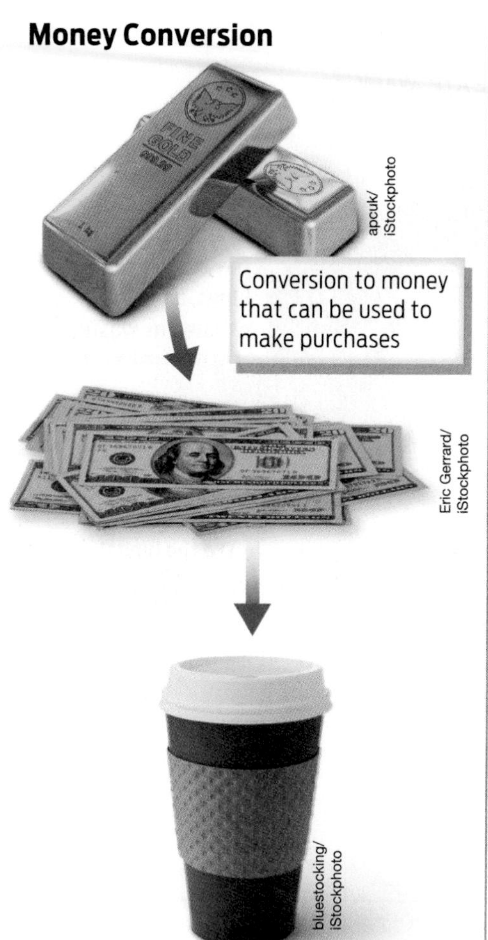

Money Conversion
Conversion to money that can be used to make purchases
apcuk/ iStockphoto
Eric Gerrard/ iStockphoto
bluestocking/ iStockphoto

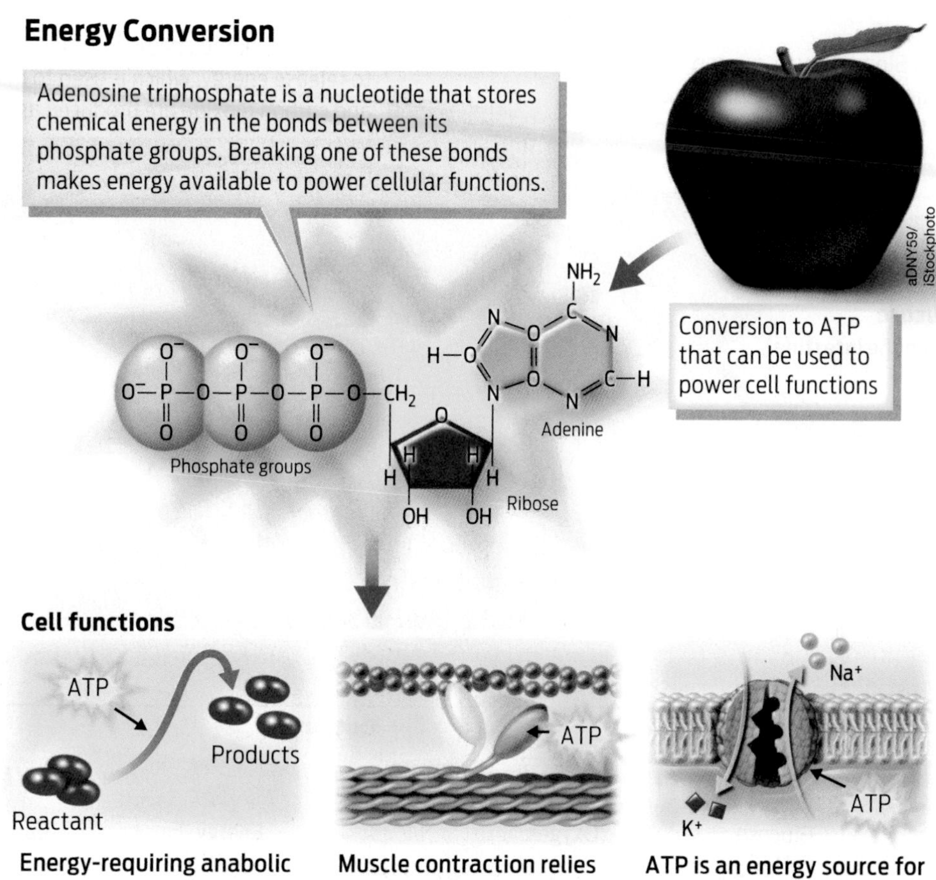

Energy Conversion
Adenosine triphosphate is a nucleotide that stores chemical energy in the bonds between its phosphate groups. Breaking one of these bonds makes energy available to power cellular functions.
Conversion to ATP that can be used to power cell functions
aDNY59/ iStockphoto
Phosphate groups
Adenine
Ribose
Cell functions
ATP
Products
Reactant
ATP
Na+
ATP
K+

Energy-requiring anabolic reactions can use ATP as their energy source.

Muscle contraction relies on releasing the energy stored in ATP.

ATP is an energy source for active transport of solutes/ions across membranes.

? What do muscle contraction and active transport have in common?

the presence of oxygen," and, as the term suggests, this process requires a continual source of oxygen. Sugars, fats, and amino acids from our diets can all be burned in aerobic respiration to make ATP. For simplicity, we focus here on glucose, which is the most common fuel source for all organisms, from bacteria to humans. During aerobic respiration of glucose, oxygen is consumed, energy is released and captured in the bonds of ATP, and carbon dioxide is given off as waste **(INFOGRAPHIC 6.6)**.

Aerobic respiration is a three-stage process that takes place in different parts of the cell. The first stage, **glycolysis,** takes place in the cell's cytoplasm. Glycolysis is a series of chemical reactions that splits glucose in half, into two smaller molecules of pyruvate. The pyruvate molecules then enter the cell's mitochondria.

During the second stage, the **citric acid cycle,** a series of reactions strips electrons from the bonds between the carbon and hydrogen atoms that were originally in glucose and are now in pyruvate. In the process, the pyruvate is broken down into smaller and smaller carbon-based molecules, and eventually exhaled as carbon dioxide from the lungs.

As the energy-rich bonds in glucose and pyruvate are broken, some of the energy released is used to make a small amount of ATP. The rest of the energy is stored in electrons released from the broken bonds. These electrons are picked up by a molecule

GLYCOLYSIS
A series of reactions that breaks down sugar into smaller units; glycolysis takes place in the cytoplasm and is the first stage of both aerobic respiration and fermentation.

CITRIC ACID CYCLE
A set of reactions that takes place in mitochondria and helps extract energy (in the form of high-energy electrons) from food; the second stage of aerobic respiration.

INFOGRAPHIC 6.6 Aerobic Respiration Transfers Food Energy to ATP

During aerobic respiration, our cells use the oxygen we inhale to help extract energy from food. Cells convert the energy stored in food molecules into the bonds of ATP, the cell's energy currency.

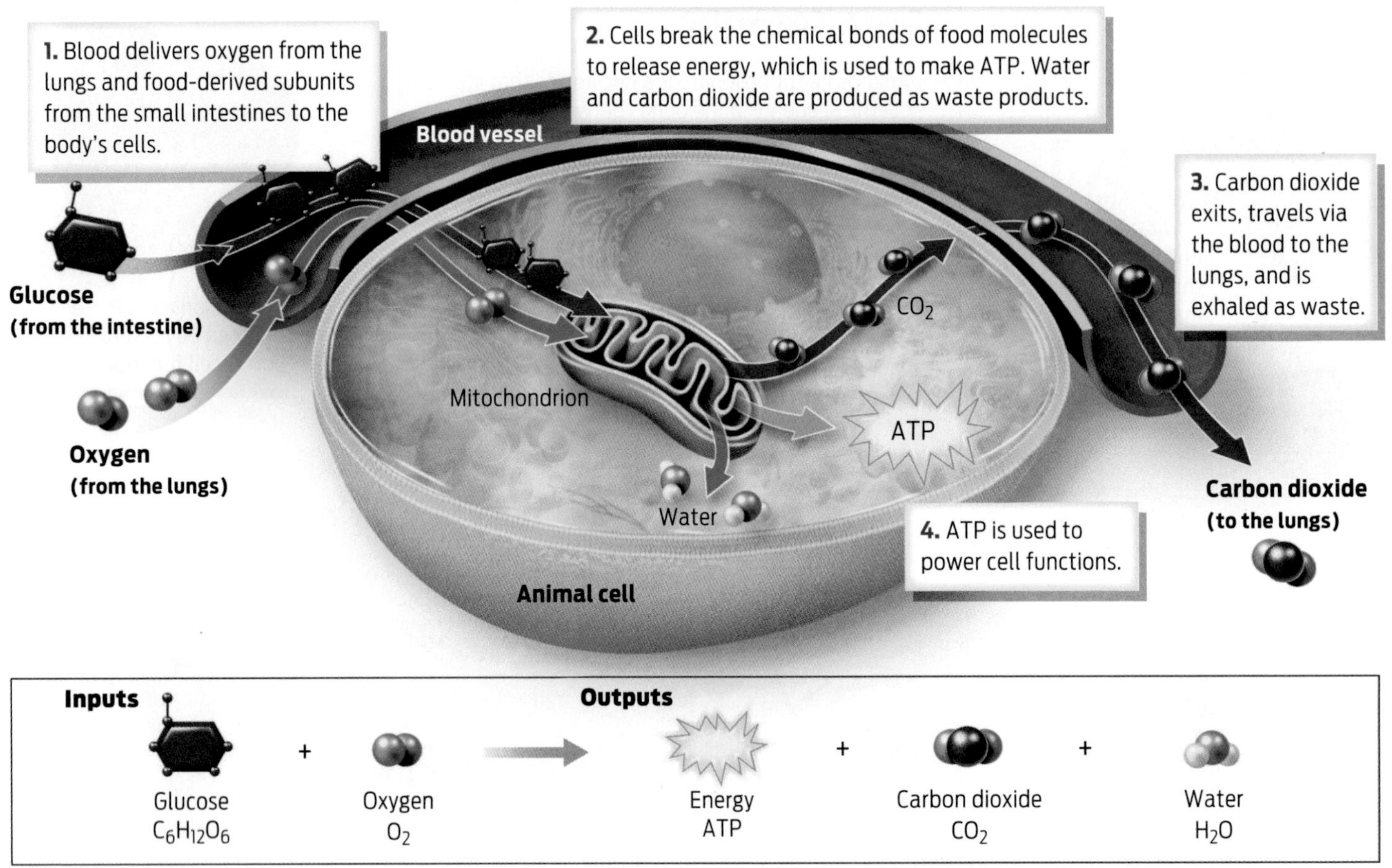
1. Blood delivers oxygen from the lungs and food-derived subunits from the small intestines to the body's cells.
2. Cells break the chemical bonds of food molecules to release energy, which is used to make ATP. Water and carbon dioxide are produced as waste products.
Blood vessel
3. Carbon dioxide exits, travels via the blood to the lungs, and is exhaled as waste.
Glucose
(from the intestine)
CO2
Mitochondrion
ATP
Oxygen
(from the lungs)
Water
Carbon dioxide
(to the lungs)
4. ATP is used to power cell functions.
Animal cell

Inputs			Outputs				
Glucose $C_6H_{12}O_6$	+	Oxygen O_2	→	Energy ATP	+	Carbon dioxide CO_2	+ Water H_2O

? What is the source of the glucose and the oxygen used in aerobic respiration?

NAD⁺
An electron carrier. NAD⁺ can accept electrons, becoming NADH in the process.

ELECTRON TRANSPORT CHAIN
The transfer of electrons that takes place in mitochondria and produces the bulk of ATP during aerobic respiration; the third stage of aerobic respiration.

called **NAD⁺.** When NAD⁺ picks up electrons, it becomes NADH (the electron-carrying form of the molecule). NADH then carries the electrons to the inner membrane of the mitochondria, where NADH gives them up (reverting to NAD⁺). The electrons then go through the third and last stage of aerobic respiration: the **electron transport chain.**

During electron transport, the energetic electrons are passed like hot potatoes down a chain of molecules in the inner mitochondrial membrane. As electrons pass down the chain, they release their stored energy, which is used to power reactions that form many molecules of ATP. Eventually the electrons are passed to oxygen molecules, which combine with hydrogen atoms to produce water **(INFOGRAPHIC 6.7)**.

When Oxygen Is Scarce

Aerobic respiration requires a steady supply of oxygen, which is transported to cells of the body in blood traveling from the lungs. Occasionally, when we perform very strenuous activities, the rate at which

INFOGRAPHIC 6.7 Aerobic Respiration: A Closer Look

Nearly all eukaryotic organisms carry out aerobic respiration. The three main stages of aerobic respiration occur in specific locations within the cell and yield distinct products.

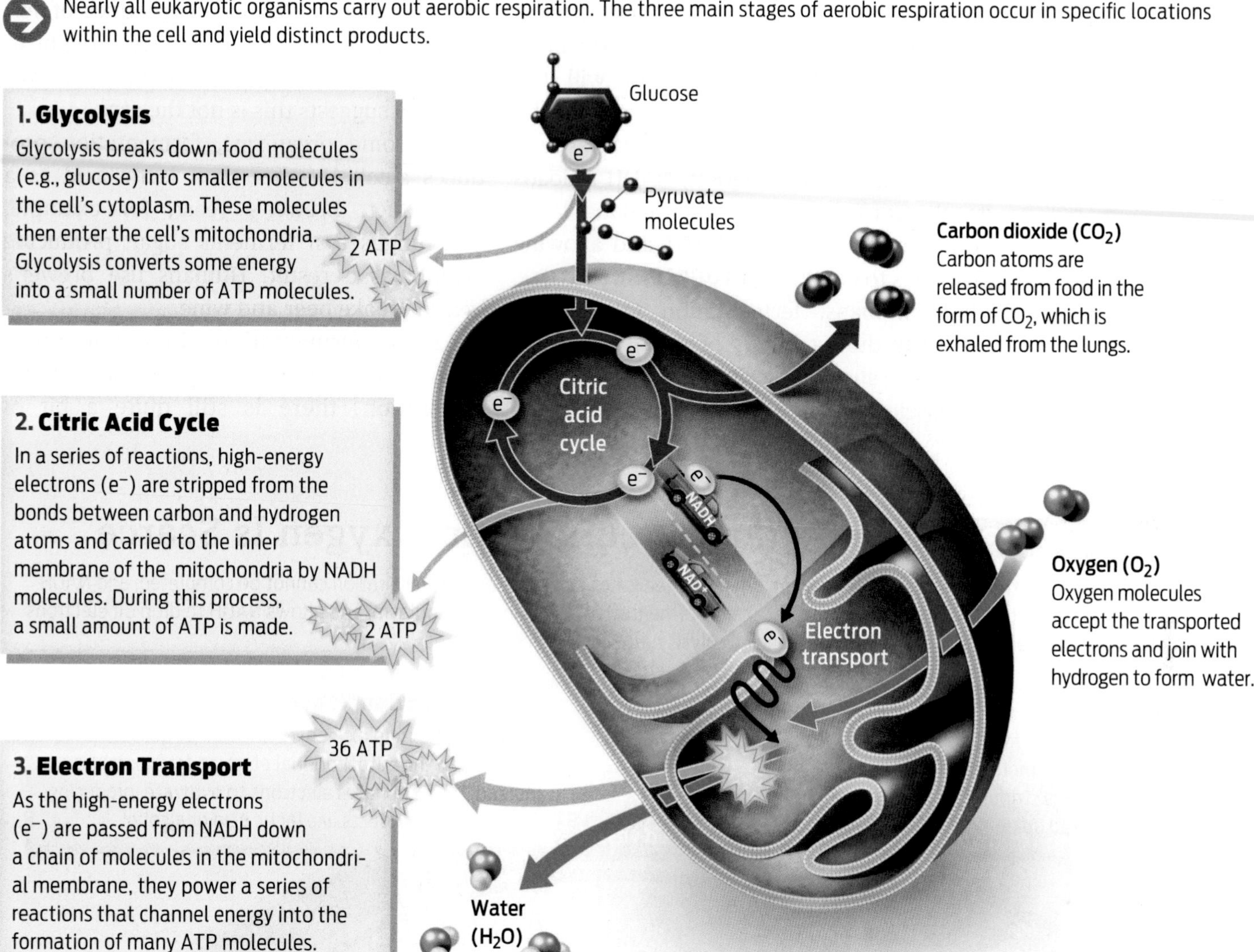

1. Glycolysis
Glycolysis breaks down food molecules (e.g., glucose) into smaller molecules in the cell's cytoplasm. These molecules then enter the cell's mitochondria. Glycolysis converts some energy into a small number of ATP molecules.
2 ATP
Glucose
e⁻
Pyruvate molecules
Carbon dioxide (CO_2)
Carbon atoms are released from food in the form of CO_2, which is exhaled from the lungs.
Citric acid cycle
2. Citric Acid Cycle
In a series of reactions, high-energy electrons (e⁻) are stripped from the bonds between carbon and hydrogen atoms and carried to the inner membrane of the mitochondria by NADH molecules. During this process, a small amount of ATP is made.
2 ATP
NADH
NAD^+
Oxygen (O_2)
Oxygen molecules accept the transported electrons and join with hydrogen to form water.
Electron transport
36 ATP
3. Electron Transport
As the high-energy electrons (e⁻) are passed from NADH down a chain of molecules in the mitochondrial membrane, they power a series of reactions that channel energy into the formation of many ATP molecules.
Water (H_2O)

? List the three main steps of aerobic respiration. Where in the cell does each step occur?

oxygen can be delivered to muscles is lower than the rate at which oxygen is consumed. Without oxygen to accept electrons, the electron transport chain stops, and aerobic respiration grinds to a halt. When this happens, another form of metabolism, **fermentation,** comes into play. This is an anaerobic process, which means that it occurs without oxygen. During fermentation, the products of glycolysis do not go through the citric acid cycle and the electron transport chain. Instead, they are shunted into a different set of reactions, which takes place in the cell's cytoplasm and produce lactic acid as a by-product.

Fermentation does not actually produce any more ATP beyond what is produced by glycolysis, so you might wonder: why do cells do it? In essence, it's a way to keep glycolysis running. As glucose is converted to pyruvate in glycolysis, glucose gives up electrons to NAD^+. In the absence of oxygen, NADH

FERMENTATION
A series of chemical reactions beginning with glycolysis and taking place in the absence of oxygen. Fermentation produces far less ATP than does aerobic respiration.

cannot unload its electrons to the electron transport chain, and therefore after a while there is no NAD^+ available to pick up electrons from glucose (since all the NAD will be in the NADH form). Soon glycolysis will stop making ATP unless there is some way for the cell to regenerate NAD^+. This is what fermentation does. In fermentation, NADH unloads its electrons onto pyruvate (making lactic acid), thus regenerating NAD^+ and allowing glycolysis to continue **(INFOGRAPHIC 6.8)**.

In humans, fermentation takes place primarily during bursts of energy-intensive activities, such as sprinting, when oxygen in muscles is scarce. Red blood cells–which carry lots of oxygen but lack mitochondria–also must rely on glycolysis and fermentation to produce ATP. (The product of fermentation, lactic acid, was once thought to be the cause of muscle cramping, but newer research suggests this is not the case.)

In some organisms, fermentation produces alcohol rather than lactic acid as a by-product. Brewer's yeast, for example, is a fungus that ferments sugar, producing alcohol as a result. Humans use brewer's yeast to make beer and wine.

Since fermentation does not break down glucose as completely as does aerobic respiration, there is still quite a bit of

INFOGRAPHIC 6.8 Fermentation Occurs When Oxygen Is Scarce

Glycolysis occurs whether or not oxygen is present. When oxygen is not available, aerobic respiration cannot be completed, and NADH cannot drop off electrons to the electron transport chain. Fermentation occurs in the cytoplasm, and allows NADH to drop off electrons to pyruvate. This regenerates NAD^+, which can keep glycolysis running.

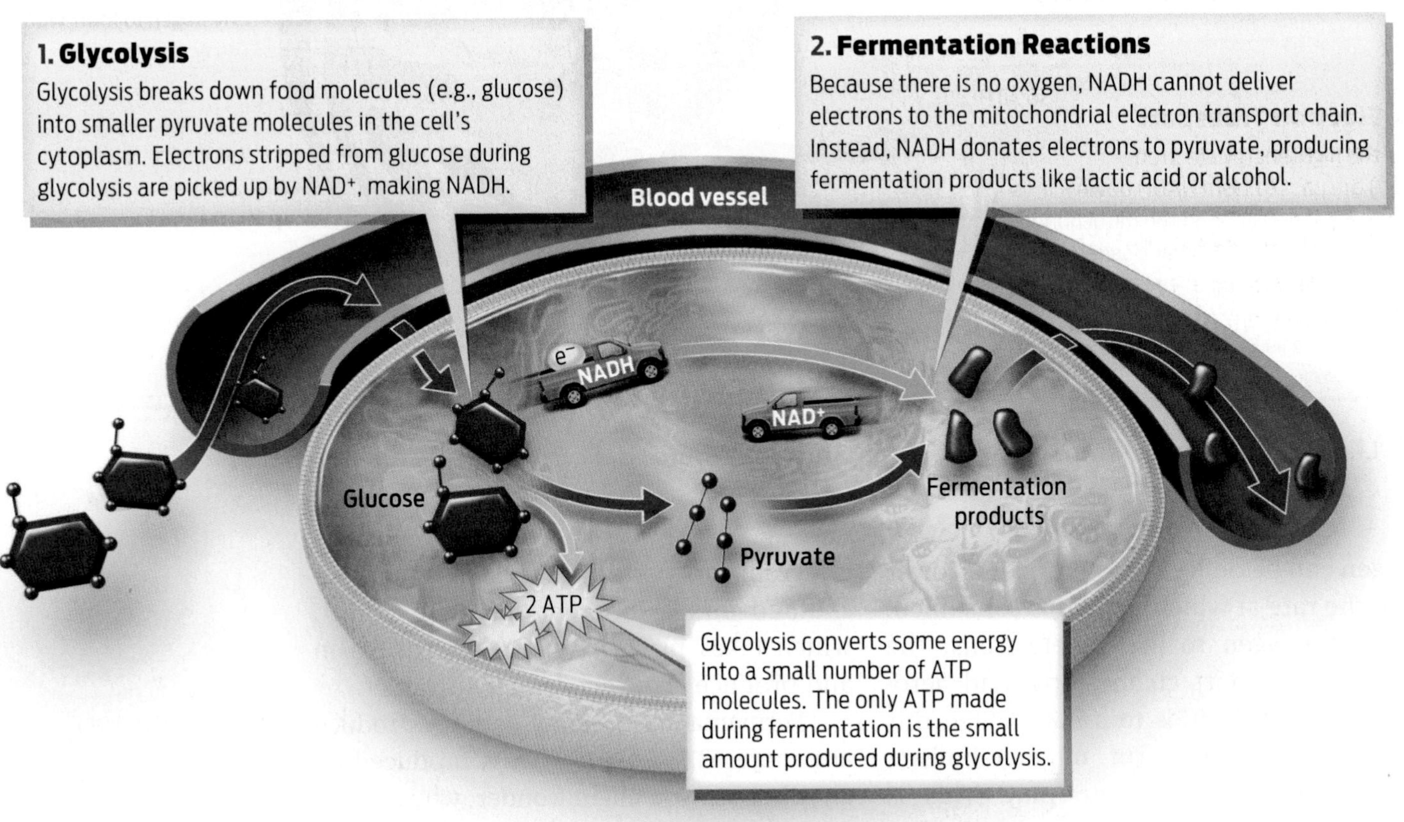

1. Glycolysis
Glycolysis breaks down food molecules (e.g., glucose) into smaller pyruvate molecules in the cell's cytoplasm. Electrons stripped from glucose during glycolysis are picked up by NAD+, making NADH.
2. Fermentation Reactions
Because there is no oxygen, NADH cannot deliver electrons to the mitochondrial electron transport chain. Instead, NADH donates electrons to pyruvate, producing fermentation products like lactic acid or alcohol.
Blood vessel
e⁻
NADH
NAD+
Glucose
Pyruvate
Fermentation products
2 ATP
Glycolysis converts some energy into a small number of ATP molecules. The only ATP made during fermentation is the small amount produced during glycolysis.

? During fermentation, what accepts electrons from NADH? What is the product?

carbohydrate energy left in such beverages as beer and wine, about 7 Calories per gram—which explains why most weight-loss diets eliminate alcohol.

Modern Times, Modern Problems

A hundred and fifty years ago, 90% of the world's population lived in agricultural regions. Much like our distant evolutionary ancestors, they walked to work, performed manual labor, and walked home at the end of the day. They prepared their own food and washed their clothes by hand.

Today, in developed countries, most people live in cities and work behind a computer. They sit during their drive to work, sit all day at work, sit to drive home, and sit in the evening watching television, surfing the Internet, or playing video games. They even sit as the washing machine washes, rinses, and spins their laundry.

"In a mere 150 years," says Levine, "*Homo sapiens* has become addicted to the chair."

In adopting this sedentary lifestyle, humans have decreased their NEAT by approximately 1,500 Calories per day, says Levine. At the same time, we have easier access to high-caloric foods and can pretty

Occupational Activity

Jobs today require less physical activity intensity (measured in metabolic equivalents: METS) and expend less energy than they did in 1960.

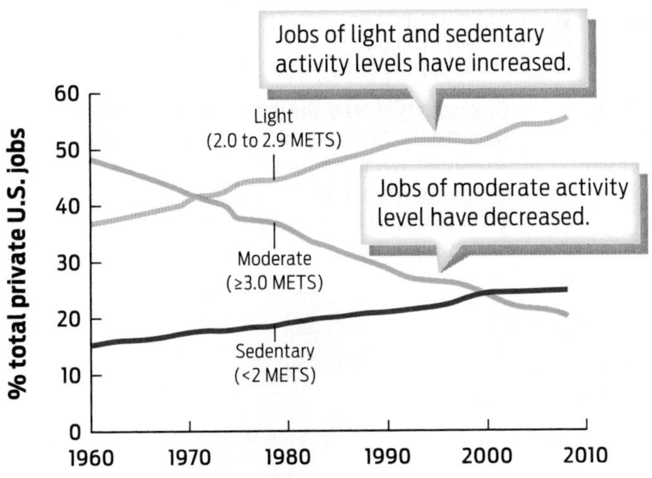

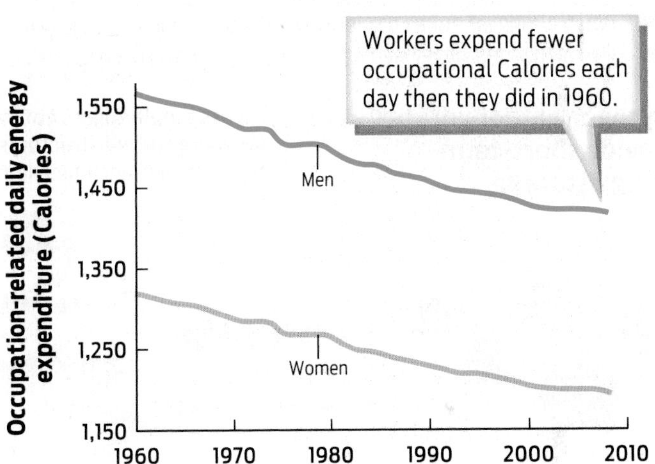

Data from Church T.S. et al. (2011) *PLoS One*. 6(5): 1–7.

Domestic Machines

Increased sales of domestic machines in the U.S. correlates with increases in obesity rates.

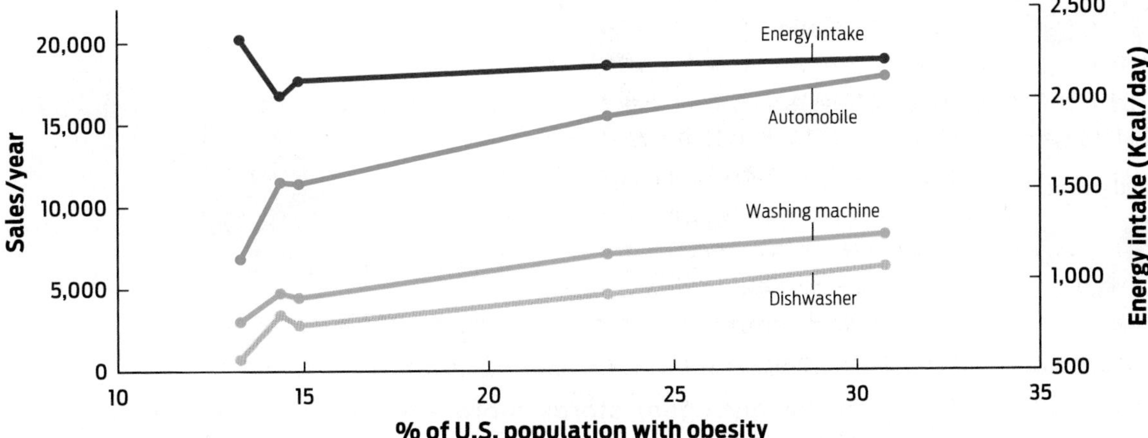

Data from Levine, J., et al. (2006) *Arteriosclerosis, Thrombosis, and Vascular Biology*. 26: 729–736.

GLYCOGEN
A complex animal carbohydrate, made up of linked chains of glucose molecules, that stores energy for short-term use.

TRIGLYCERIDES
A type of lipid found in fat cells that stores excess energy for long-term use.

much eat whenever we want. As a result, we are experiencing a population-wide positive energy balance: more Calories in than Calories out.

For the large majority of us, when we eat Calories beyond what our bodies expend, the extra energy is stored in one of two forms: as glycogen in muscle and liver cells, or as triglycerides (fat) in fat cells. **Glycogen,** a polymer of glucose, is the energy-storing carbohydrate found in animal cells. It's essentially a short-term storage system. When we require short bursts of energy–in a sprint, for example–the body breaks down glycogen into its component glucose molecules, and uses those in aerobic respiration to obtain usable energy (ATP). However, because a gram of glycogen stores only half as many Calories as a gram of fat (about 4 Calories per gram versus 9), our bodies would have to carry around twice as much glycogen to store the same amount of energy. So our bodies store most excess energy as **triglycerides** in fat cells, which actually allows us to carry around less weight overall. The body burns this fat only after it has already used up food molecules in the bloodstream and used up its stored glycogen **(INFOGRAPHIC 6.9)**.

INFOGRAPHIC 6.9 Glycogen and Fat Store Excess Calories

When we ingest more Calories than our bodies need, they are stored as glycogen molecules in muscle and liver cells. Once the body's glycogen stores have been replenished, any excess Calories are stored as triglyceride molecules in fat cells.

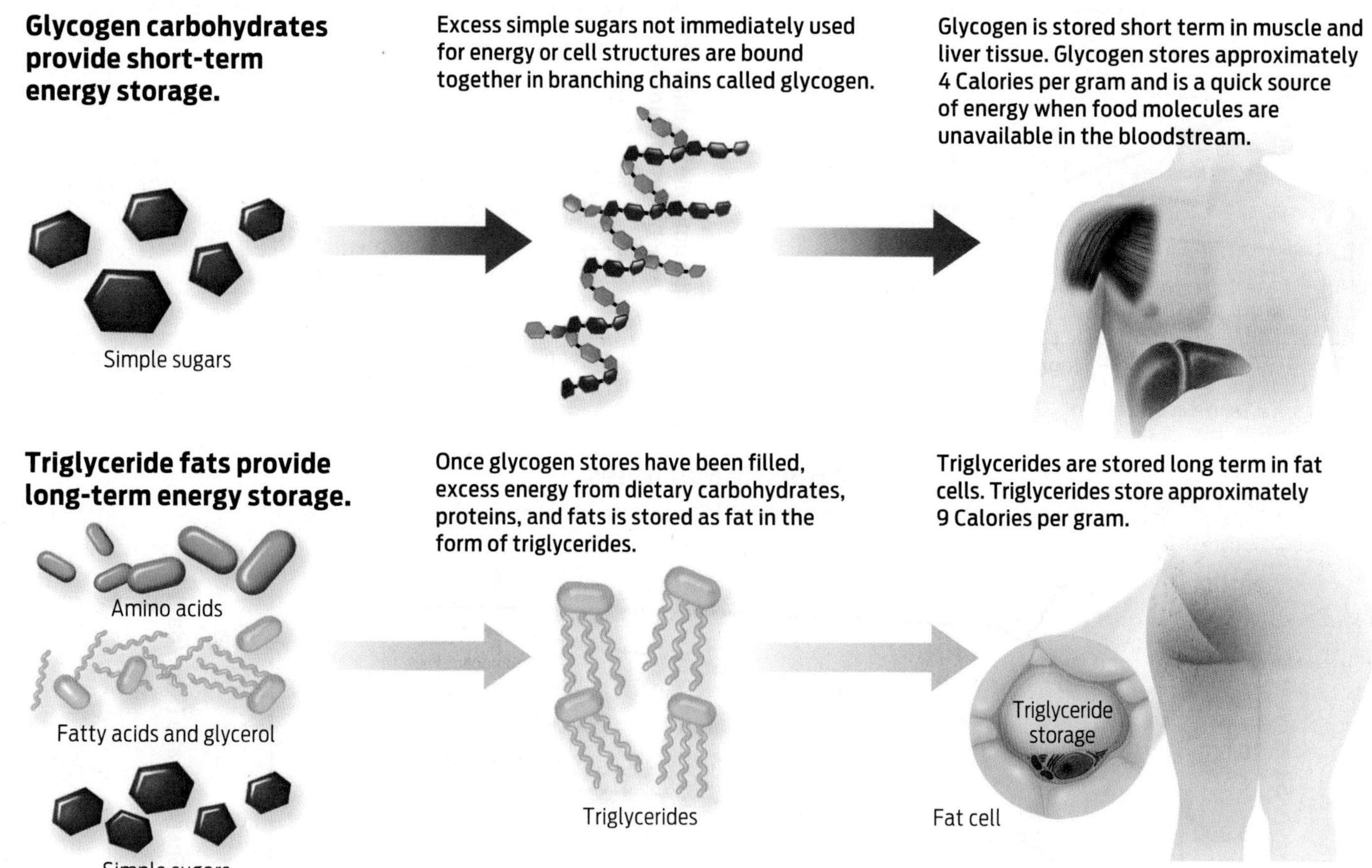

Glycogen carbohydrates provide short-term energy storage.
Excess simple sugars not immediately used for energy or cell structures are bound together in branching chains called glycogen.
Glycogen is stored short term in muscle and liver tissue. Glycogen stores approximately 4 Calories per gram and is a quick source of energy when food molecules are unavailable in the bloodstream.
Simple sugars
Triglyceride fats provide long-term energy storage.
Once glycogen stores have been filled, excess energy from dietary carbohydrates, proteins, and fats is stored as fat in the form of triglycerides.
Triglycerides are stored long term in fat cells. Triglycerides store approximately 9 Calories per gram.
Amino acids
Fatty acids and glycerol
Simple sugars
Triglycerides
Triglyceride storage
Fat cell

? Which molecule (a triglyceride or glycogen) stores more energy per gram? Where is this molecule stored?

The average adult American today weighs 26 pounds more than in the 1970s. Clearly, as a population we are taking in more energy than we use. Scientists generally agree that some combination of consuming more Calories and expending less energy is at work, but pinning down the exact contributions have been difficult.

"There is a huge debate about this," says Eric Ravussin, director of the Nutrition Obesity Research Center at the Pennington Biomedical Research Center at Louisiana State University. "Portion sizes have increased, the number of meals, the amount of snacking," he notes, "but also sitting more in front of computers, not walking as much to go to work." Separating out these influences can be challenging.

Many chroniclers of the obesity epidemic have noted that our collective weight has crept up in tandem with a change in what we are eating. Dietary guidelines issued by the U.S. government shifted in the late 1970s, for the first time recommending that Americans cut back on the amount of fat (particularly saturated fat) that they consume. The goal of this recommendation was to lower the risk of heart disease, which some studies had linked to consuming certain types of fats. While a good idea in principle, the problem is that people responded by replacing fats in their diet with carbohydrates–particularly starchy, refined carbohydrates like those in bread, pasta, and sugary drinks. Food companies even began creating and marketing fat-free cookies–still loaded with sugar–as a healthful alternative to eating fat.

> "You can train somebody who is a sitter to become a mover."
>
> — James Levine

While a Calorie is a Calorie regardless of whether it comes from fat or sugar (or protein), there is some evidence that eating lots of refined carbohydrates leaves us less satiated and more hungry than eating foods with fat. As a result, we consume more Calories overall. This, say some experts, could be a contributing factor to the obesity epidemic.

Regardless of what we eat, there is no escaping the laws of thermodynamics. All the food we eat–whether burger, broccoli, or bread–originally gets its energy from the sun, by way of photosynthesis (see Chapter 5). Plants capture the energy of sunlight and convert it into chemical energy stored in sugar. We then eat this sugar (or eat animals that have eaten this sugar), and that stored energy becomes available to us. It's because energy is never destroyed, only converted from one form into another, that an energy excess can find its way onto our belly, hips, and thighs **(INFOGRAPHIC 6.10)**.

By the same logic, the only way to lose weight is to shift our energy balance to the negative: expend more energy than we take in. While cutting back on food and increasing our deliberate exercise are two ways of doing this, many people find doing either to be difficult–especially over the long term. Levine's work suggests a powerful third way–through increasing our NEAT.

Researchers know that a person's baseline level of NEAT is controlled in part by chemicals in the brain. When researchers inject a hormone called orexin into the brains of lab rats, they find that the rats' NEAT quotient goes up dramatically. "It's like the rats discovered Starbucks," Levine says. They go from lolling about in their cages to darting about, rearing up on their hind legs, and grooming more avidly. Interestingly, in rats bred to be lean, orexin triggers higher levels of NEAT than in rats bred to be obese, suggesting that the brains of lean and obese rats have different thresholds of response to this hormone. (Individual snails, too, have a characteristic style of NEAT.)

While our baseline level of NEAT may have a genetic influence, that doesn't mean

CHRISTINA PAOLUCCI/AP Images

James Levine, working at his computer in his office at the Mayo Clinic, uses a specially designed elevated desk that includes a treadmill.

we can't change it. Through deliberate practice and conscious choice, Levine says, we can retrain our brains to become "NEATER." "You can train somebody who is a sitter to become a mover," he says. We can also change our environment in ways that encourage us to increase our NEAT, which over time will add up to big Calorie differences.

A Moratorium on the Chair

When we first caught up with Levine to discuss NEAT, we found him in transit, walking to his office in downtown Phoenix, Arizona. Levine walks *a lot*. He routinely conducts meetings, interviews, and many other tasks on the go. "Contrary to popular belief, I do sit from time to time," he says.

Asked just how bad sitting is for you, Levine rattles off a list of 16 associated health risks, including obesity, diabetes, hypertension, high cholesterol, cardiovascular disease, depression, swollen ankles, joint problems, back pain, depression, and cancer. Even one's creativity, he suggests, may be dulled from sitting too much.

That's why Levine has called for a "moratorium on the chair." He believes it is time to fundamentally redesign our environments so that higher NEAT is the norm. Toward that end, he is working with the mayor of Phoenix on initiatives to encourage more walking among commuters, and also works with businesses and school systems to make workplaces and classrooms more active—or, at Levine put it, "NEATER."

INFOGRAPHIC 6.10 Photosynthesis and Aerobic Respiration Form a Cycle

Photosynthesis and respiration form a continuous cycle, with the outputs of one process serving as the inputs of the other.

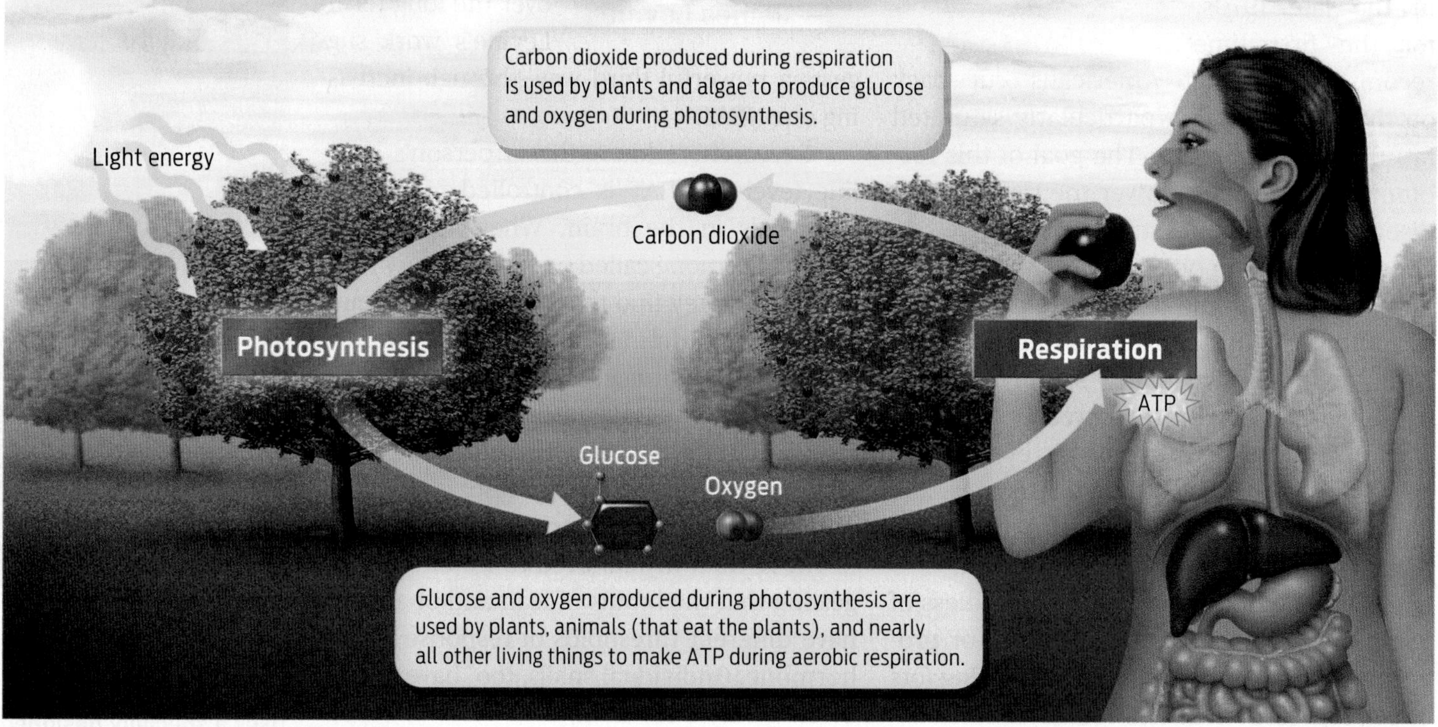

Carbon dioxide produced during respiration is used by plants and algae to produce glucose and oxygen during photosynthesis.
Light energy
Carbon dioxide
Photosynthesis
Respiration
ATP
Glucose
Oxygen
Glucose and oxygen produced during photosynthesis are used by plants, animals (that eat the plants), and nearly all other living things to make ATP during aerobic respiration.

? What is the role of carbon dioxide in photosynthesis and in cellular respiration?

INFOGRAPHIC 6.11 Small Changes in NEAT Behavior Add Up to Big Calorie Rewards

Small, consistent changes in lifestyle can tip the scales considerably toward more Calories burned per week. Over time this may be all that is necessary for some to maintain a healthy weight.

NEAT Calories Burned per Hour
(Calories burned above those burned at rest for a 150-pound person)

0–50 Calories	50–100 Calories	100–200 Calories
Sitting Activities: Driving, Desk work	**Standing Activities:** Cooking, Stretching, Ironing	**Walking Activities:** Dog walking, House cleaning, Gardening

Sedentary Activities	Calories burned in one week
Park by the door and take the elevator to and from your third floor office or apartment.	210
Relax with your feet up while eating during your 45-minute lunch break.	175
Spend an hour sitting while talking on the phone.	105
Drive to and from the convenience store for milk.	70
Spend an hour on your favorite social media sites.	70
Order pizza and sit while waiting for it.	53
Watch TV for 30 minutes after dinner.	53
Hire a maid.	20
Total Calories:	**756**

NEAT Activities	Calories burned in one week
Park several blocks away, walk and take the stairs to and from your third floor office or apartment.	840
Get out for a 30-minute walk, then sit down for a quick 15-minute lunch.	875
Use a headset to walk or pace while making phone calls.	910
Walk 15 minutes to and from the convenience store and carry the milk home.	350
Spend an hour playing cards or working a puzzle.	210
Move about the kitchen making dinner for 30 minutes.	490
Take a 30-minute walk after dinner.	700
Clean one room each day for 15 minutes.	210
Total Calories:	**4,585**

Data from Levine, J. and Yeager, S. *Move a Little, Lose a Lot.* Three Rivers Press, 2009 and *The Compendium of Physical Activities Tracking Guide*, 2011.

? What three NEAT activities will allow you to burn approximately 1,500 Calories in a week?

"What's really cool about NEAT is that everyone can do it," says Levine. "You can have somebody who's 150 pounds going for a 'walk and talk' meeting with someone who's 350 pounds. You can promote NEAT in all people without having to change their clothes." (**INFOGRAPHIC 6.11**).

The degree of positive energy balance that has produced our obesity epidemic has been termed our "energy gap." By some estimates, a lifestyle change that reduced energy intake or increased energy expenditure by 100 Calories per day would completely abolish the energy gap, and hence weight gain, for most of the population. Walking just one extra mile each day would increase one's energy expenditure by about 50–100 Calories per day, depending on body weight. Similarly, taking just a few fewer bites of food at each meal would reduce energy intake by 100 Calories per day. The problem, of course, is that our current environment strongly discourages both of these good habits.

Levine once referred to NEAT as the "crouching tiger, hidden dragon" of societal weight gain. What he meant was that the behaviors that lead to obesity may be both sneakier and more deadly than we ever imagined.

"Whoever thinks about the amount of time they spend sitting?" Levine asks. "No one does." ■

CHAPTER 6 SUMMARY

- The macronutrients in our food (proteins, carbohydrates, and fats) are sources of dietary energy.
- Fats are the most energy-rich organic molecules in our diet. Fats contain twice as many Calories per gram as carbohydrates and proteins.
- When we consume more Calories than we use, our bodies store the excess energy in the bonds of glycogen in muscle and liver cells, and of triglycerides in fat cells.
- Cells carry out chemical reactions that break down food to obtain usable energy in the form of ATP.
- In the presence of oxygen, aerobic respiration produces large amounts of ATP from the energy stored in food.
- Aerobic respiration occurs in three stages: (1) glycolysis, (2) the citric acid cycle, and (3) electron transport. The first stage occurs in the cytoplasm, the latter two in the mitochondria. Electron transport produces the bulk of ATP.
- In the absence of oxygen, fermentation follows glycolysis and produces lactic acid in animals (or, in some organisms, alcohol). Fermentation produces far less ATP than does aerobic respiration.
- Activity—both intentional exercise and NEAT—helps burn stored Calories. A combination of eating fewer Calories and being active will result in weight loss.
- During exercise, glycogen is used first. Stored fats are tapped only when glycogen stores have been depleted, as may occur during long periods of exercise.
- The ultimate source of energy in food is the sun. Photosynthesizers such as plants trap the energy of sunlight and convert it into the chemical energy of sugar. Animals then eat this sugar either directly or indirectly. Both plants and animals can use this sugar as an energy source for aerobic cellular respiration.
- Photosynthesis and respiration form a cycle: the carbon dioxide given off by animals, plants, and all organisms that perform aerobic respiration is used by photosynthesizers to make glucose and oxygen during photosynthesis.

MORE TO EXPLORE

- **NOVA's Secret Life of Scientists and Engineers (2014), James Levine: "I Came Alive as a Person": https://www.youtube.com/watch?v=fLgGf0BO4tw**
- **James Levine (2014) *Get Up!: Why Your Chair Is Killing You and What You Can Do About It.* New York: St. Martin's Griffin.**
- **David Ludwig (2016) *Always Hungry? Conquer Cravings, Retrain Your Fat Cells, and Lose Weight Permanently.* Grand Central Life & Style.**
- **Marion Nestle and Malden Nesheim (2012) *Why Calories Count: From Science to Politics.* University of California Press.**
- **Gary Taubes (2016) *The Case Against Sugar.* New York: Knopf.**
- **Frontline PBS (2004) *Diet Wars*: http://www.pbs.org/wgbh/pages/frontline/shows/diet/**

CHAPTER 6 Test Your Knowledge

DRIVING QUESTION 1 How is obesity defined, and what are some hypotheses about current obesity rates?

By answering the questions below and studying Infographics 6.1–6.4 and 6.11, you should be able to generate an answer for the broader Driving Question above.

KNOW IT

1 A 6'0" male weighs 230 pounds. Use Infographic 6.1 to determine his BMI. Would he be considered underweight, of normal weight, overweight, or obese?

2 If a person wants to lose weight, which of the following will contribute to the necessary Calorie imbalance?

- **a.** fidgeting more
- **b.** eating less
- **c.** exercising more
- **d.** all of the above
- **e.** b and c

USE IT

3 Consider the 6'0", 230-pound male from question 1. If you learned that he was an NFL quarterback, would you reconsider how to interpret his BMI? Explain your answer. (Hint: Muscle is denser than fat. A given volume of muscle will weigh more than the same volume of fat.)

4 Which snack will provide the highest number of Calories?

- **a.** 25 g sugar, 5 g protein, 0 g fat
- **b.** 30 g sugar, 0 g protein, 5 g fat
- **c.** 10 g sugar, 10 g protein, 10 g fat
- **d.** 0 g sugar, 15 g protein, 15 g fat
- **e.** 10 g sugar, 25 g protein, 0 g fat

5 If you wanted to balance Calories consumed in a snack of an apple and a hard-boiled egg, how long would you have to stand? How long would you have to walk (at 2 mph)?

6 Two friends are trying to lose weight. They adopt the same diet, but one briskly walks her dog for 30 minutes twice a day, while the other jogs slowly for 15 minutes a day. Predict the outcomes in 90 days. Will there be a substantial difference in weight loss? Why or why not?

apply YOUR KNOWLEDGE

INTERPRETING DATA

7 Look at Infographic 6.11. The data are shown for one week. If you adopted all the NEAT activities shown for an entire month (4 weeks), how many additional Calories would you burn compared to a month (4 weeks) of sedentary activities? Assuming that it takes ~3,500 Calories to lose 1 pound of fat, how many additional pounds of fat would you lose during a month of NEAT activities?

DRIVING QUESTION 2 How does the body use the energy in food?

By answering the questions below and studying Infographics 6.5 and 6.9, you should be able to generate an answer for the broader Driving Question above.

KNOW IT

8 Which type of organic molecule serves as long-term energy storage in humans?

- **a.** proteins
- **b.** starch
- **c.** nucleic acid
- **d.** fats (triglycerides)
- **e.** b and d

9 Which part of an ATP molecule is associated with its ability to store energy?

- **a.** the adenine
- **b.** the ribose
- **c.** the phosphate groups
- **d.** all of the above

10 If you exercise for an extended period of time, you will use energy first from _____, then from _____.

- **a.** fats; glycogen
- **b.** proteins; fats
- **c.** glycogen; proteins
- **d.** fats; proteins
- **e.** glycogen; fats

USE IT

11 Compare the weight of 1,000 Calories stored as glycogen with the weight of 1,000 Calories stored as fat.

MINI CASE

apply YOUR KNOWLEDGE

12 Consider a well-trained 130-pound female marathon runner. She has just loaded up on a carbohydrate meal and has the maximum amount of stored glycogen (6.8 g of glycogen per pound of body weight).

a. How many grams of glycogen is she storing?
b. How many Calories does she have stored as glycogen?
c. If this same number of Calories were stored as fat, how much would it weigh?
d. Suppose she decides to go for a run at a pace of 9 mph (she will be running 6.5-minute miles). Given her weight, she will burn 885 Calories per hour at this pace. How long will it take her to deplete her glycogen stores? How many miles can she run before her glycogen supplies run out? Will she be able to complete a 26.2-mile marathon?
e. Once her glycogen supplies run out, what has to happen if she wants to keep running?

DRIVING QUESTION 3 How does aerobic respiration extract useful energy from food?

By answering the questions below and studying Infographics 6.6, 6.7, and 6.10, you should be able to generate an answer for the broader Driving Question above.

KNOW IT

13 Which process is *not* correctly matched with its cellular location?

a. glycolysis—cytoplasm
b. citric acid cycle—mitochondria
c. glycolysis—mitochondria
d. electron transport—mitochondria
e. none of the above; they are all correctly matched

14 In the presence of oxygen we use _____ to fuel ATP production. What process do plants use to fuel ATP production from their stored sugars?

a. aerobic respiration; photosynthesis
b. aerobic respiration; aerobic respiration
c. fermentation; aerobic respiration
d. fermentation; photosynthesis
e. glycolysis; photosynthesis

15 Given 1 g of each of the following, which would yield the greatest amount of ATP by aerobic respiration?

a. fat
b. protein
c. carbohydrate
d. nucleic acid
e. alcohol

16 During aerobic respiration, what molecule has (and carries) electrons stripped from food molecules?

a. NAD^+
b. NADH
c. O_2
d. H_2O
e. pyruvate

17 During aerobic respiration, how does NADH give up electrons to regenerate NAD^+?

a. by giving electrons to O_2
b. by giving electrons to pyruvate
c. by giving electrons to glucose
d. by giving electrons to the electron transport chain
e. by giving electrons to another NAD^+

USE IT

18 Draw a carbon atom that is part of a CO_2 molecule such as you just exhaled. In a written description or a diagram, trace what happens to that carbon atom as it is absorbed by the leaf of a spinach plant and then what happens to the carbon atom when you eat that leaf in a salad.

19 If you ingest carbon in the form of sugar and use it to generate ATP via cellular respiration, how is that carbon released from your body?

a. as sugar
b. as fat
c. as CO_2
d. as protein
e. in urine

DRIVING QUESTION 4 When does fermentation occur, and why can't a human survive strictly by fermentation?

By answering the questions below and studying Infographic 6.8, you should be able to generate an answer for the broader Driving Question above.

KNOW IT

20 Compared to aerobic respiration, fermentation produces _____ ATP.

a. much more
b. the same amount of
c. a little less
d. much less
e. no

21 Which process is most directly prevented in the absence of adequate oxygen?

a. citric acid cycle
b. glycolysis
c. electron transport chain
d. a, b, and c
e. glycolysis and the citric acid cycle

22 During fermentation, how does NADH give up electrons to regenerate NAD^+?

a. by giving electrons to O_2
b. by giving electrons to pyruvate
c. by giving electrons to glucose
d. by giving electrons to the electron transport chain
e. by giving electrons to another NAD^+

23 Where in the cell does fermentation take place?

a. cytoplasm
b. mitochondria
c. nucleus
d. cytoplasm and mitochondria
e. Fermentation doesn't occur in cells, it occurs in the liquid portion of blood.

USE IT

24 Explain how the presence or absence of oxygen affects ATP production. (The terms *aerobic respiration* and *fermentation* should be in your answer.)

25 Consider fermentation.

a. How much ATP is generated during fermentation?
b. How does the amount of ATP generated by fermentation compare to aerobic respiration?
c. In humans, why can't fermentation sustain life? (Hint: Think of two reasons—one is related to the product of fermentation and what happens if it accumulates.)

apply YOUR KNOWLEDGE

BRING IT HOME

26 A 60-year-old CEO wants to lose some weight, but he has bad knees, so can't work out as he used to. He is skeptical about a weight-loss plan not based on exercise. What specific activities can you suggest to him, and what information can you provide to persuade him to give a NEAT plan a try?

7 DNA Structure and Replication

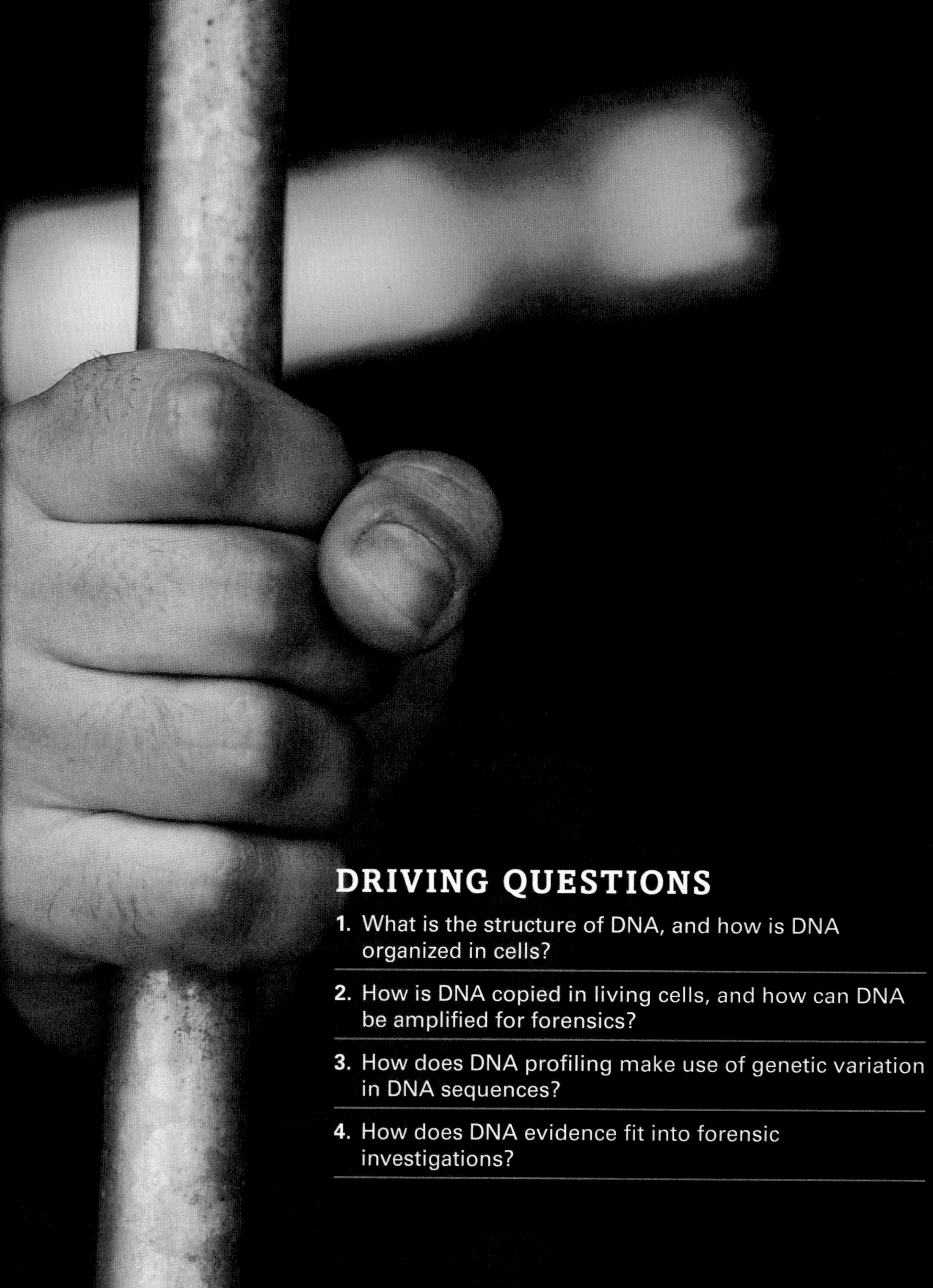

DRIVING QUESTIONS

1. What is the structure of DNA, and how is DNA organized in cells?
2. How is DNA copied in living cells, and how can DNA be amplified for forensics?
3. How does DNA profiling make use of genetic variation in DNA sequences?
4. How does DNA evidence fit into forensic investigations?

Biologically Unique

How DNA helped free an innocent man

Ken Kaminesky/Getty Images

ROY BROWN THOUGHT THE POLICE were just checking up on him when an officer knocked on his door one day in May 1991. Brown, a self-professed hard drinker who earned a living selling magazine subscriptions, had only a week before been released after serving an 8-month prison term. His crime: threatening to kill the director of the Cayuga County Department of Social Services in upstate New York. A caseworker had deemed Brown unfit to care for his 7-year-old daughter. Furious, Brown made a series of threatening phone calls to the director. But he had served his time. What could the officer want from him now?

Three days earlier, police had found the battered body of a woman lying in the grass about 300 feet from the farmhouse where she had lived. Someone had burned the place to the ground. The body was identified as that of Sabina Kulakowski, a social worker at the Cayuga County Department of Social Services. The crime was horrific. The murderer had beaten the 49-year-old Kulakowski, bitten her several times, dragged her outside, and then stabbed and strangled her. It was obvious that Kulakowski had struggled; her body was covered with defensive wounds.

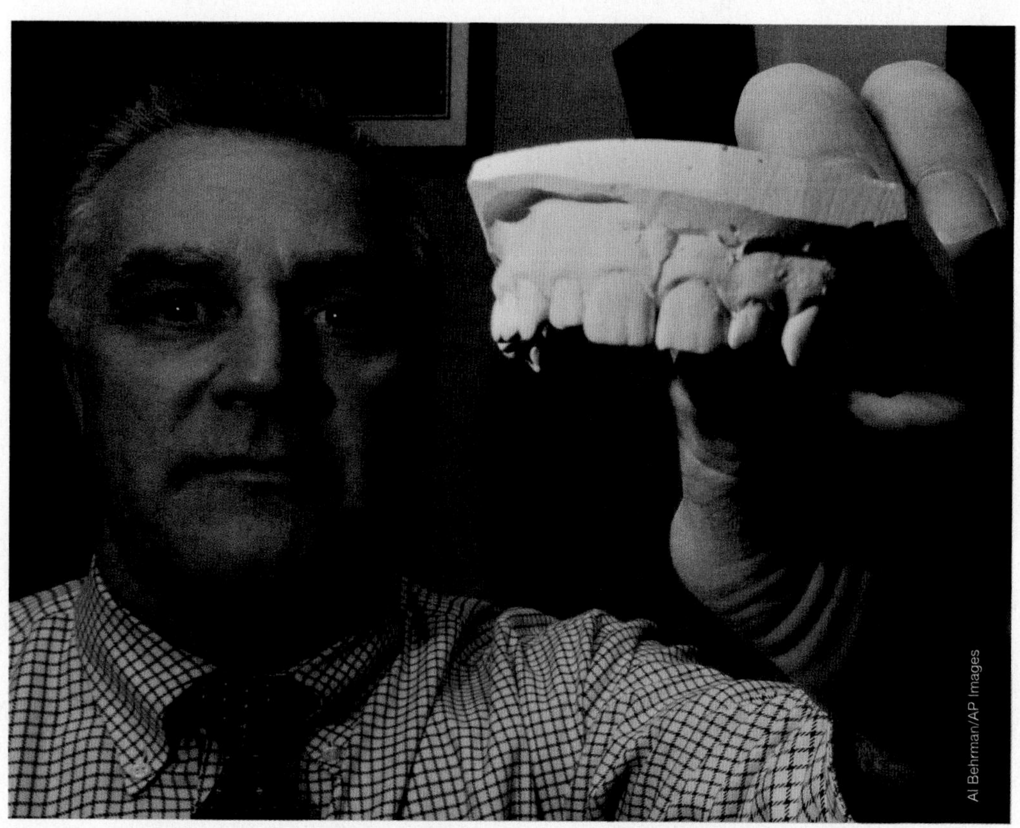

Al Behrman/AP Images

Dr. Frank Wright, a forensic dentist, uses tooth molds like this one (which is not Roy Brown's) for bite-mark analysis.

Although Kulakowski had not been involved in Brown's case, officers arrested Brown that day on suspicion of murder. Eight months later, a jury found Brown guilty of homicide and sentenced him to prison for 25 years to life. The prosecution argued that Brown's motive was revenge against the Department of Social Services. But what really nailed the case was testimony from an expert who stated that bite marks on the victim's body matched Brown's teeth.

Brown, however, maintained his innocence. "I never knew Ms. Kulakowski, and I had nothing to do with that woman's death . . . I am truly innocent," he told the court and onlookers after the verdict had been announced.

Even from prison, Brown never stopped trying to prove his innocence. He repeatedly petitioned, in vain, for a retrial.

Then something unexpected happened: Brown uncovered additional evidence that strongly suggested he was not the perpetrator. The evidence was so compelling, in fact, that in late 2004, after Brown had spent 12 years in prison, his lawyers decided to contact the Innocence Project, a nonprofit organization founded in 1992. Its mission: to use DNA evidence to free people wrongly convicted of crimes.

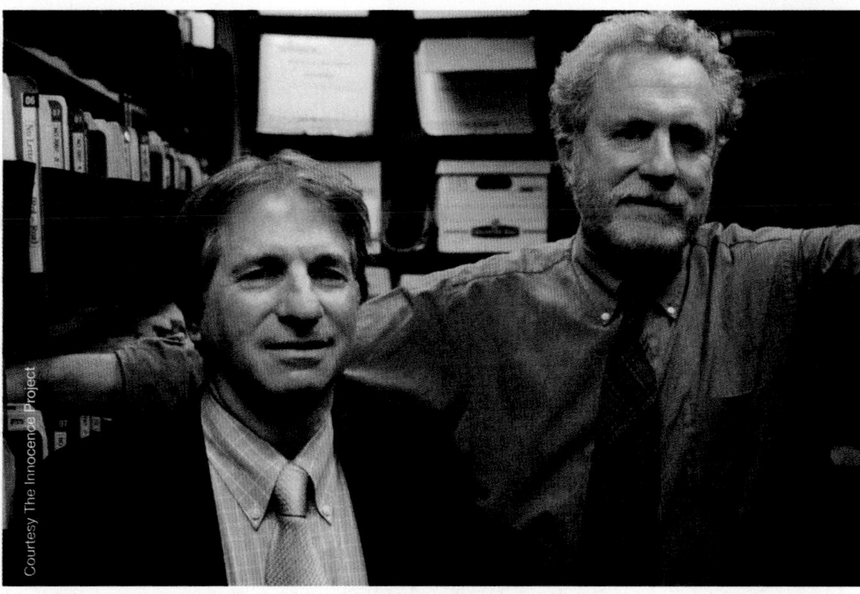

Courtesy The Innocence Project

Barry Scheck and Peter Neufeld, the lawyers who founded the Innocence Project.

DNA as Evidence

When the jury convicted Brown in 1992, DNA testing was still in its infancy, so DNA was rarely used as evidence in criminal cases. But over the next decade, DNA testing became a standard part of court cases, as science increasingly showed that it was an extremely accurate way to match crime scene evidence to perpetrators.

How can scientists use DNA to identify a person? The answer lies in the chemical makeup of this molecule, often referred to as the blueprint of life. **Deoxyribonucleic acid,** or **DNA,** stores biological information and serves as the instruction manual from which we are built. It is passed from parents to offspring during reproduction, and is the reason children resemble their biological relatives. All life forms—from bacteria to plants to humans—use DNA as the storehouse of their biological information and transmit it to their offspring. DNA is the hereditary molecule of life, establishing the uniqueness of each of us.

Where can you find DNA? In eukaryotic cells, DNA is found inside the nucleus, where it exists in the form of chromosomes. A **chromosome** is a single, large DNA molecule wound around proteins. Human cells have 23 pairs of chromosomes; we inherit one chromosome of each pair from our mother and the other from our father, for a total of 46 chromosomes. The 23rd chromosome pair consists of the sex chromosomes, X and Y, which determine a person's sex. Men have an X and a Y, and pass on one or the other during reproduction. Women have two Xs, and therefore can only pass on an X. Children who inherit a Y from dad and an X from mom are males, while children who inherit one X from each parent are females. Since nearly all cells contain DNA, scientists can collect evidence such as blood, skin, semen, saliva, or hair from a crime scene and extract DNA from it to identify a perpetrator (**INFOGRAPHIC 7.1**).

DNA testing has helped the Innocence Project free more than 300 people from prison since 1992, including 18 who served time on death row. The technology has not only given these people their lives back but has also thrown a spotlight on flaws in our criminal justice system. Why were people wrongly convicted and placed on death row? Innocence Project lawyers have found several culprits: dishonest witnesses, apathetic or overburdened lawyers, overzealous police officers, and mistakes in eyewitness identification. But fortunately, new developments such as those in DNA technology are helping to improve the system.

"DNA is only one example of how advances in science have made the criminal justice system more reliable," says Peter Neufeld, who, along with Barry Scheck, both from the Benjamin N. Cardozo School of Law in New York City, founded the Innocence Project. "But what we really hope to do now is use DNA as the gold standard of reliability to weed out junk science."

DEOXYRIBONUCLEIC ACID (DNA)
The molecule of heredity, common to all life forms, that is passed from parents to offspring.

CHROMOSOME
A single, large DNA molecule wrapped around proteins. Chromosomes are located in the nuclei of eukaryotic cells.

INFOGRAPHIC 7.1 What Is DNA and Where Is It Found?

Deoxyribonucleic acid, or DNA, is the hereditary molecule common to all living organisms. It is the instruction manual from which an organism is built.

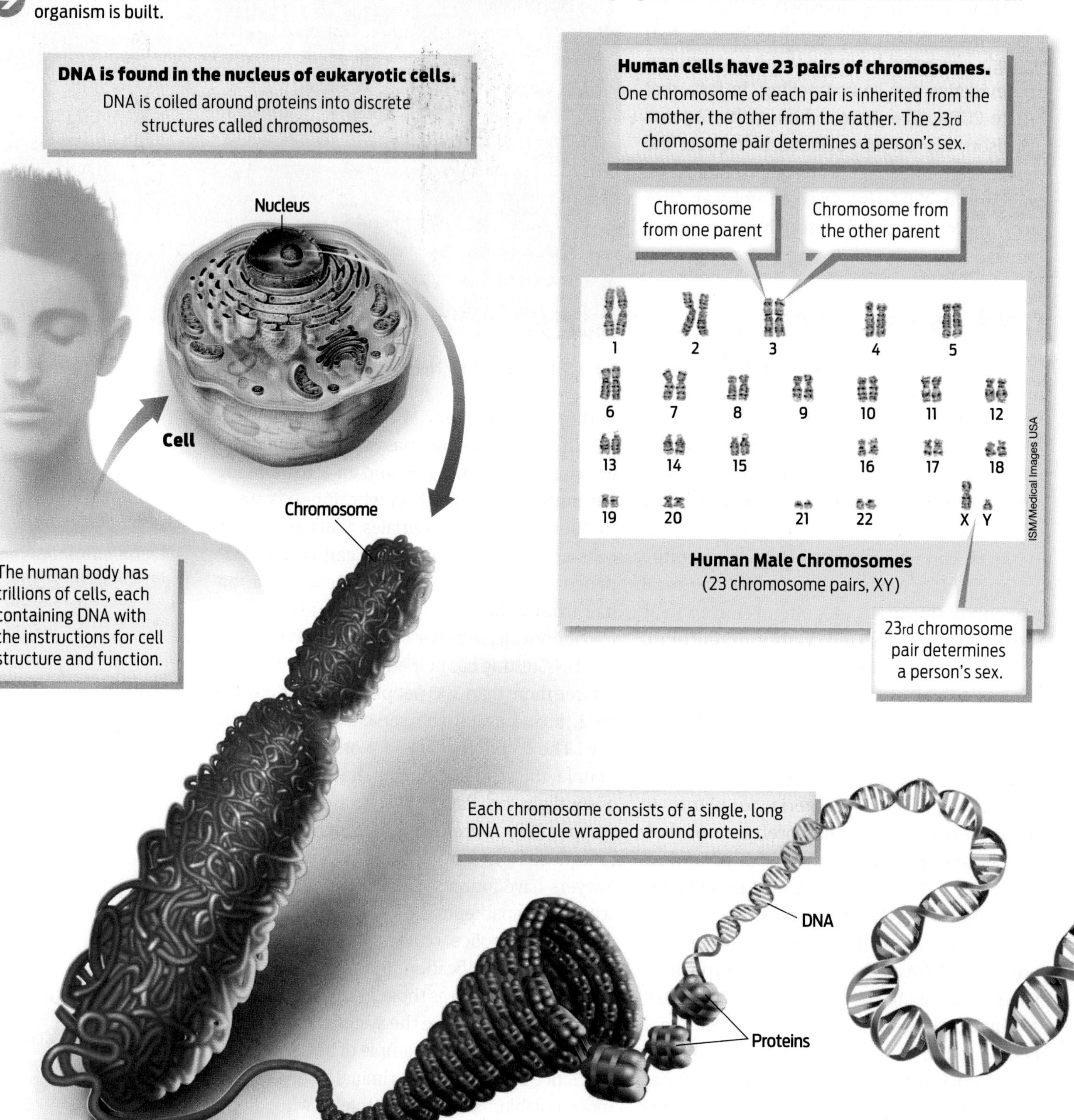

DNA is found in the nucleus of eukaryotic cells.
DNA is coiled around proteins into discrete structures called chromosomes.
Nucleus
Cell
Chromosome
The human body has trillions of cells, each containing DNA with the instructions for cell structure and function.
Human cells have 23 pairs of chromosomes.
One chromosome of each pair is inherited from the mother, the other from the father. The 23rd chromosome pair determines a person's sex.
Chromosome from one parent
Chromosome from the other parent
1
2
3
4
5
6
7
8
9
10
11
12
13
14
15
16
17
18
19
20
21
22
X
Y
ISM/Medical Images USA
Human Male Chromosomes
(23 chromosome pairs, XY)
23rd chromosome pair determines a person's sex.
Each chromosome consists of a single, long DNA molecule wrapped around proteins.
DNA
Proteins

? How many DNA molecules are in the nucleus of each human cell?

Unreliable Evidence

Indeed, it was "junk science" that convicted Roy Brown. The only physical evidence linking Brown to the case was his teeth. A dentist hired by the prosecution testified that the bite marks on Kulakowski's body matched Brown's teeth. But as the defense pointed out, the bite marks came from someone with six upper teeth–Brown had only four. The prosecution's witness argued that Brown could have twisted Kulakowski's skin while biting her and filled in the gaps–an argument that ultimately proved convincing to the jury.

Bite-mark analysis is a particularly troubling form of evidence. No widely accepted rules or standards govern its use, and no government or outside scientific commission has ever validated its claims. In fact, studies show error rates–the rate at which experts have falsely identified bite marks as belonging to a particular person–as high as 91%.

Hair analysis, another common type of evidence, can be equally unreliable. In dozens of cases, Innocence Project lawyers found that forensic scientists had testified that hairs from crime scenes matched the accused, explains Neufeld. But when scientists subsequently tested the DNA inside the follicle cells from those hairs, the DNA didn't match.

The problem is that hair analysis, performed under a microscope, can reveal only certain characteristics: it can distinguish whether hair is human or not, or show a person's ancestry (because of ethnic differences in hair texture); hair analysis can tell whether the hair has been dyed, cut in a certain way or pulled out, and where on the body it came from. Hair samples can exclude a suspect, but not positively identify one.

By contrast, each person's DNA is unique. To understand how DNA varies from person to person, consider its structure. A DNA molecule is made up of two strands of subunits linked together in long chains. As we learned in Chapter 2, each subunit–called a **nucleotide**–has three parts: a sugar, a phosphate group, and a base. In each DNA strand, the phosphate group of one nucleotide binds to the sugar of the next nucleotide to form a chain of interlinked nucleotides. The two strands of linked nucleotides pair up and twist around each other to form a spiral-shaped **double helix.** The sugars and phosphates form the outside "backbone" of the helix and the bases point toward its center, forming internal "rungs," like steps on a twisting ladder. The bases in one strand associate with bases from the other

NUCLEOTIDES
The building blocks of DNA. Each nucleotide consists of a sugar, a phosphate group, and a base. The sequence of nucleotides (As, Cs, Gs, Ts) along a DNA strand is unique to each person.

DOUBLE HELIX
The spiral structure formed by two strands of DNA nucleotides held together by hydrogen bonds.

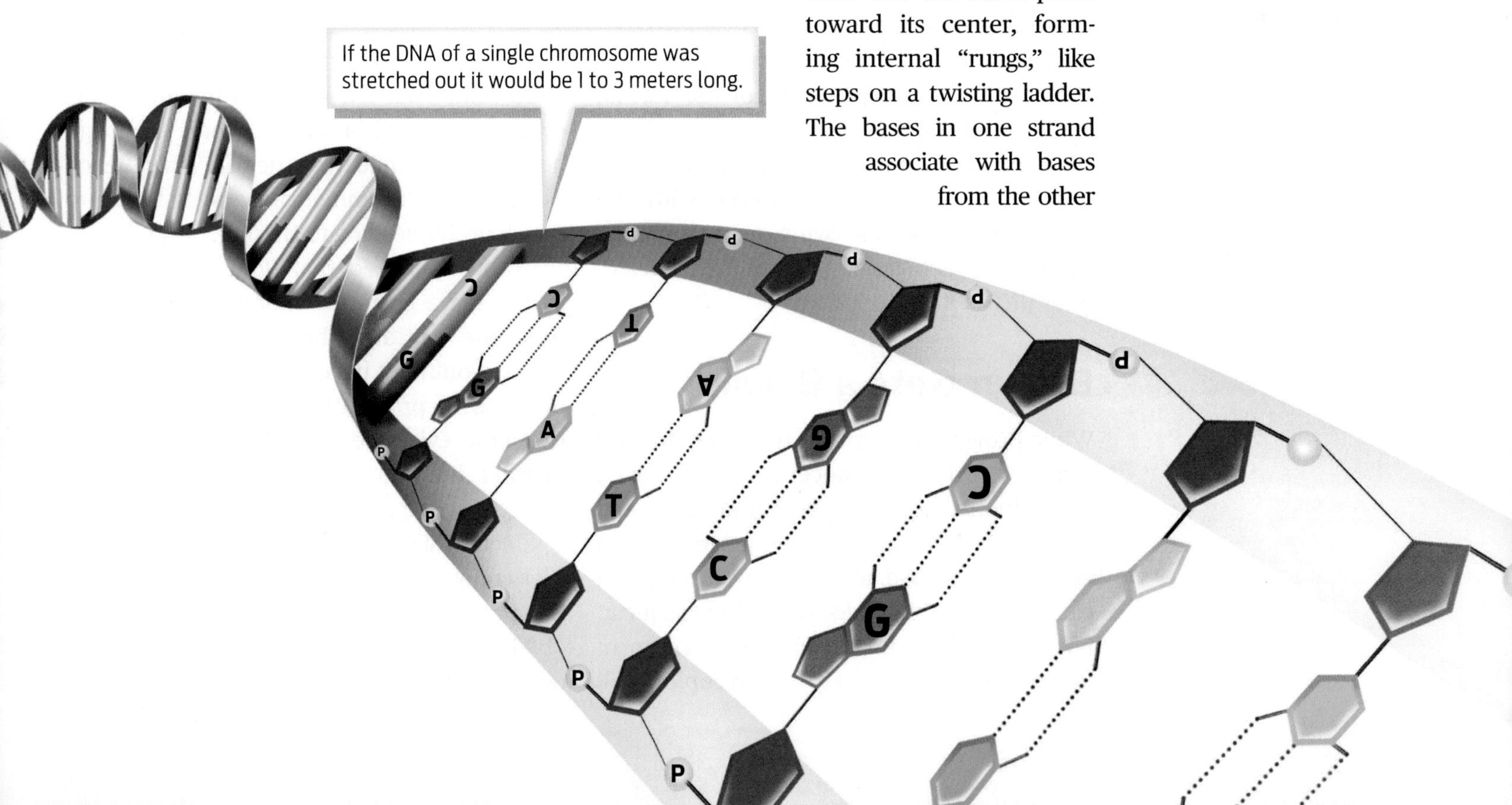

strand through hydrogen bonds, which hold the DNA double helix together.

The rungs of the DNA ladder, made up of the bases, are the most useful part of DNA for profiling. There are four different possible nucleotide bases: adenine (A), thymine (T), guanine (G), and cytosine (C). In DNA, these four nucleotide bases are repeated over and over, billions of times, in different orders along a DNA strand. The order of nucleotide bases in a strand is a key form of genetic information in cells–it provides the instructions for making proteins (see Chapter 8). Human DNA sequences are about 99.9% the same from to person to person (defining, in part, what makes us human, as opposed to some other species). In that last 0.1% are unique sequence variations that give us our individuality–having brown eyes versus blue, for example, or light skin versus dark.

To identify perpetrators, forensic scientists examine the specific sequence of nucleotide bases along one strand of a person's DNA–the precise order of As, Ts, Gs, and Cs–and compare that sequence with the DNA found in crime scene evidence. DNA evidence is convincing, because–with the exception of identical twins–no two people share the exact same order of DNA nucleotides (**INFOGRAPHIC 7.2**).

To identify perpetrators, forensic scientists examine the specific sequence of nucleotide bases along one strand of a person's DNA.

Brown Gets a Break

While Brown sat in prison, he never stopped trying to prove his innocence. For 11 years, he petitioned for an appeal and requested specifically that DNA tests be performed on evidence collected at the crime scene. On each occasion, the judge denied his request. Finally, in 2003, Brown filed a Freedom of Information Act request to obtain copies of all documents relating to his case. That's when he made a surprising discovery.

The additional evidence Brown obtained included four affidavits collected by the Cayuga County Sheriff's Department the day after the murder–documents that neither Brown nor his lawyers had ever seen. In the affidavits, four people described the suspicious behavior of another man: Barry Bench. Bench was the brother of Kulakowski's former boyfriend. The Bench family owned the farmhouse in which Kulakowski had been living.

The affidavits, which included sworn testimony from neighbors as well as from Bench's then girlfriend, Tamara Heisner, stated that on the day of Kulakowski's murder, Bench argued with Heisner, went to a local bar, and returned home between 1:30 and 1:45 A.M.–the same time the victim's neighbors alerted the fire department that the farmhouse was ablaze.

The statements further noted that Bench, who came home highly intoxicated, had left the bar at approximately 12:30 A.M. That left 60 to 75 minutes unaccounted for until he arrived home–although he lived only a mile from the bar. When Bench came home, he immediately went inside to "wash up," according to Heisner.

Brown realized that Bench would have had to drive by the farmhouse to get home from the bar and thought it strange that Bench would not have noticed the raging fire on his own property. While not conclusive, this new evidence was enough to prompt Brown's lawyers to contact the Innocence Project for help.

Meanwhile, Brown decided to write Bench a letter detailing what he had found and urging him to confess. He warned him of his intent to obtain a DNA test on evidence from the murder. "Judges can be fooled and juries make

INFOGRAPHIC 7.2 DNA Is Made of Two Strands of Nucleotides

DNA is a double-stranded molecule. Each of the two strands consists of a chain of subunits called nucleotides that are bonded together. There are four types of nucleotides: adenine (A), thymine (T), guanine (G), and cytosine (C). Nucleotides are the building blocks of DNA.

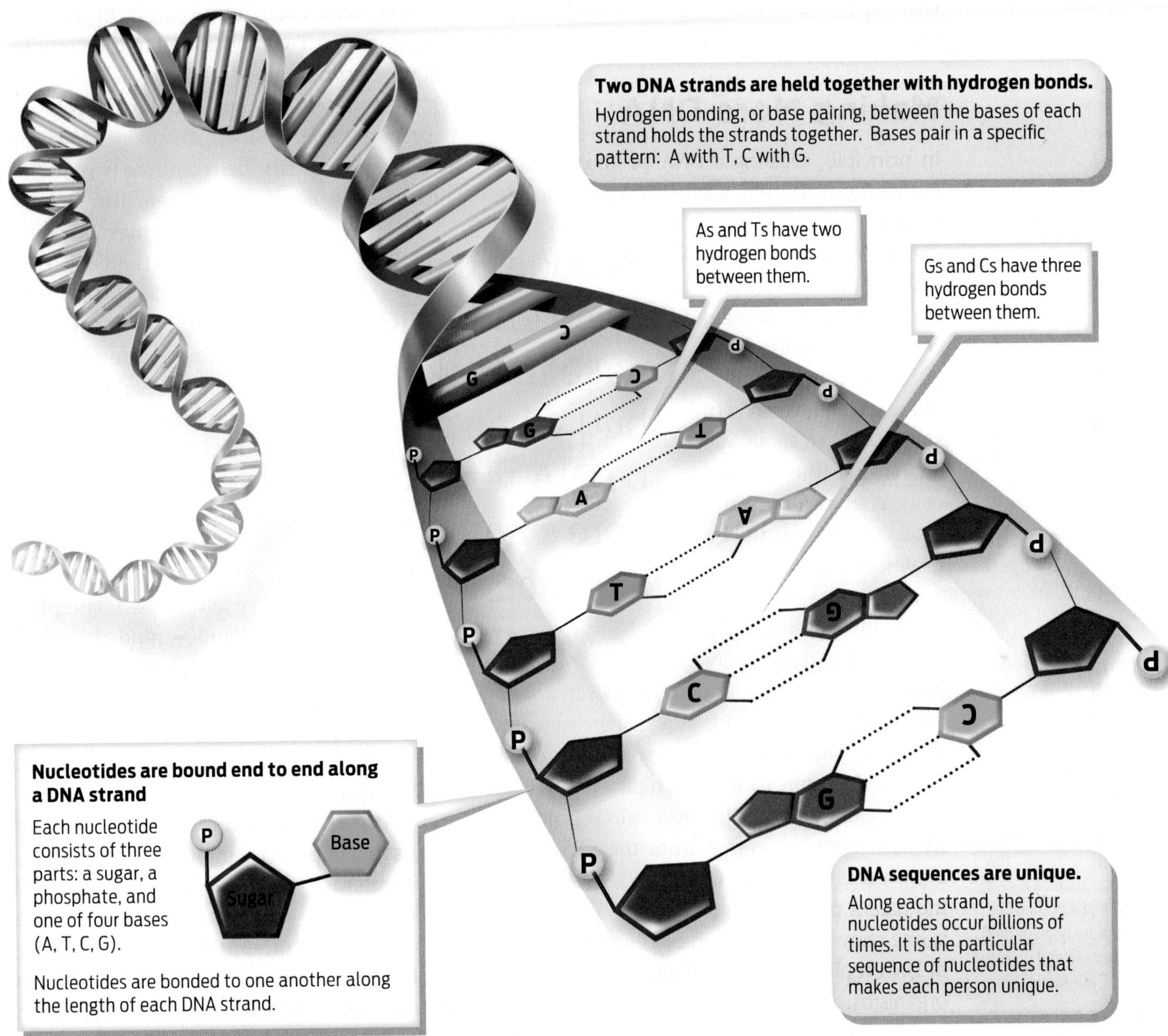

Two DNA strands are held together with hydrogen bonds.
Hydrogen bonding, or base pairing, between the bases of each strand holds the strands together. Bases pair in a specific pattern: A with T, C with G.
As and Ts have two hydrogen bonds between them.
Gs and Cs have three hydrogen bonds between them.
Nucleotides are bound end to end along a DNA strand
Each nucleotide consists of three parts: a sugar, a phosphate, and one of four bases (A, T, C, G).
P
Sugar
Base
Nucleotides are bonded to one another along the length of each DNA strand.
DNA sequences are unique.
Along each strand, the four nucleotides occur billions of times. It is the particular sequence of nucleotides that makes each person unique.

? One strand of a DNA molecule has the sequence ATGCTGA. What is the corresponding sequence on the second strand?

mistakes," he wrote, "[but] when it comes to DNA testing there's no mistakes. DNA is God's creation and God makes no mistakes."

Five days after Brown mailed his letter, Bench threw himself in front of an Amtrak train and was killed instantly.

Soon after, the Innocence Project team took on Brown's case and filed a motion to have Kulakowski's nightshirt tested for DNA at a New York State crime lab. The nightshirt had been found in some tall grass near the body. It was not only bloodstained, it was also stained with saliva. Since both saliva and blood contain cells that carry DNA, scientists could chemically extract the DNA from the cells to create a DNA profile of the perpetrator. A **DNA profile** is a readout of DNA sequences that is unique to a single individual—akin to a genetic fingerprint.

Making More DNA

In principle, creating a DNA profile is simple enough, but there is a huge practical hurdle: having enough crime scene DNA to analyze. Although all body fluids and materials contain cells that house our DNA, the amount left at crime scenes is often very small. Without some way to increase the amount of DNA in a saliva stain, for example, it would be useless as evidence.

> "Judges can be fooled and juries make mistakes, [but] when it comes to DNA testing there's no mistakes."
>
> —Roy Brown

Forensic scientists solve this problem by using a method to amplify (that is, increase) the amount of DNA in a sample. The method takes advantage of the way cells normally make more DNA—through a process called **DNA replication.** DNA replication occurs throughout our lives whenever cells reproduce. Because each cell comes from the division of a pre-existing cell (see Chapter 3), the DNA of the parent cell must be replicated so that there is one copy for each daughter cell. The process of DNA replication is essentially the same in all organisms.

To understand how DNA replication works, note that the two strands of nucleotides in a DNA helix do not pair up randomly, but in a consistent pattern: A pairs with T, and G pairs with C. These nucleotides pair preferentially because they are the right shape to form stable hydrogen bonds with each other. Because of this patterned pairing, the two strands are said to be **complementary,** meaning that they fit together like pieces of a puzzle.

During DNA replication, each strand of DNA serves as a template for the creation of a new complementary strand. The new strand will have bases complementary to the original strand, following base-pairing "rules." For example, wherever there is an A on the template strand, a T will be added to the complementary strand being formed.

The steps of replication happen in a precise order: First, an enzyme called **helicase** unwinds the helix, and the two strands "unzip" from each other. Then, the enzyme **DNA polymerase** builds a new strand of DNA along each unzipped strand. Free nucleotides floating inside the cell's nucleus are added to each new strand in a sequence that is complementary to the nucleotide sequence on the original template strand, A pairing with T and C with G. The end result is two complete double-stranded molecules of DNA. Because each replicated DNA molecule is made up of one original and one new strand, DNA replication is said to be **semiconservative** (**INFOGRAPHIC 7.3**).

DNA replication is a remarkably accurate process that happens at mind-boggling speeds, the polymerase enzyme adding about 1,000 nucleotides per second and rarely making a mistake (though the mistakes it does make are often consequential; see Chapter 10). On a human scale, that's like a car speeding down the highway at 300 miles per hour, weaving in and out of traffic without hitting any other cars.

Forensic scientists make use of this natural cellular process when they need to amplify the DNA in a crime scene sample. The method they use, the **polymerase chain reaction (PCR),** is similar to DNA replication in cells—but it takes place in a test tube.

Here's how it works: To a small sample of DNA, scientists add nucleotides, the DNA

DNA PROFILE
A visual representation of a person's unique DNA sequence.

DNA REPLICATION
The natural process by which cells make an identical copy of a DNA molecule.

COMPLEMENTARY
Fitting together; two strands of DNA are said to be complementary in that A in one strand always pairs with T in the other strand, and G always pairs with C.

HELICASE
An enzyme that unwinds and unzips the DNA double helix during DNA replication.

DNA POLYMERASE
An enzyme that "reads" the nucleotide sequence of a DNA strand and incorporates complementary nucleotides into a new strand during DNA replication.

SEMI-CONSERVATIVE
DNA replication is said to be semiconservative because each newly made DNA molecule has one original DNA strand and one new DNA strand.

POLYMERASE CHAIN REACTION (PCR)
A laboratory technique used to replicate, and thus amplify, a specific DNA segment.

INFOGRAPHIC 7.3 DNA Structure Provides a Mechanism for DNA Replication

When cells reproduce, they must first replicate their DNA so that each new cell contains a copy of the original DNA molecule. Complementary base pairing between DNA strands guides replication of two new strands.

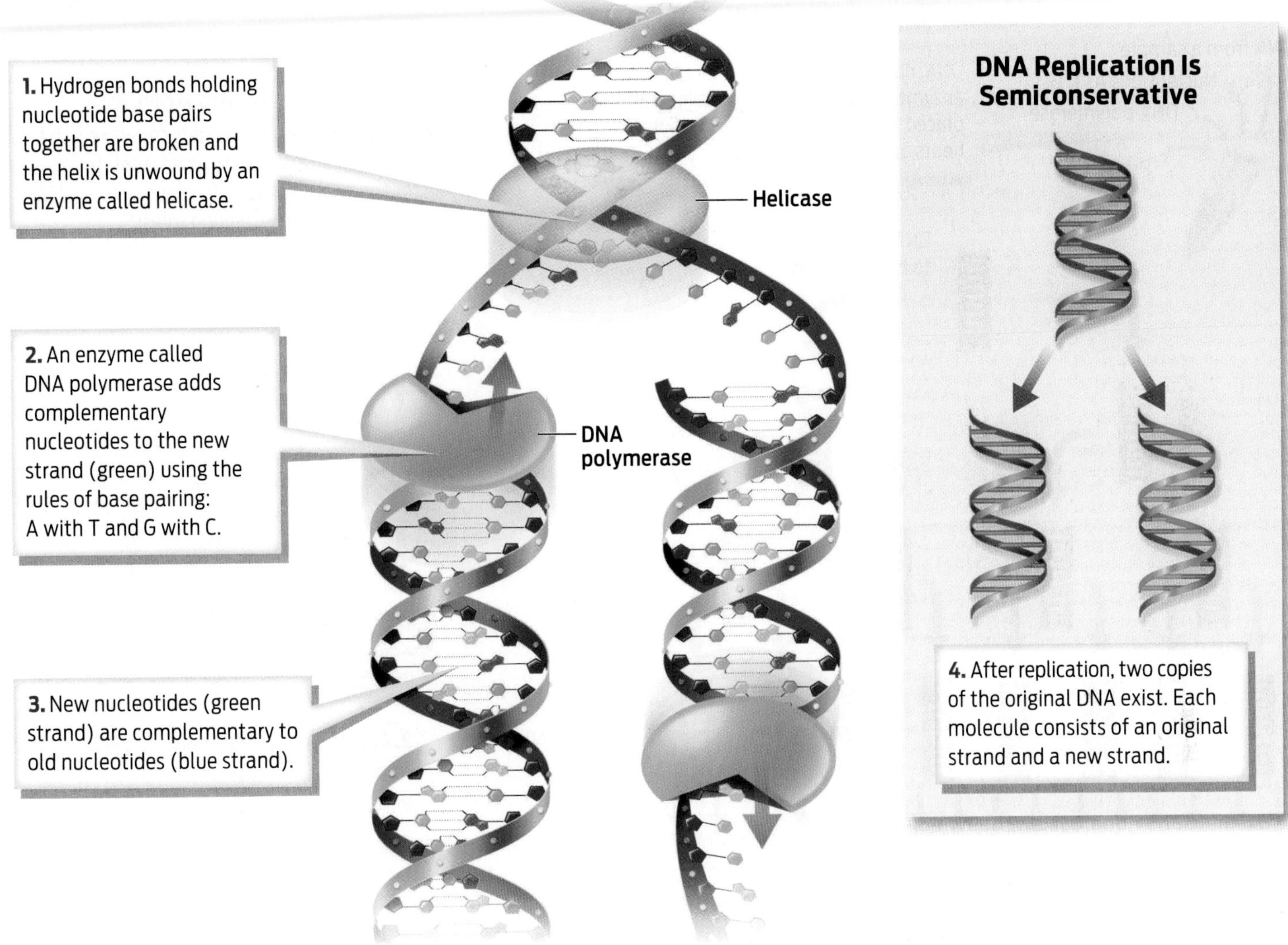

1. Hydrogen bonds holding nucleotide base pairs together are broken and the helix is unwound by an enzyme called helicase.
Helicase
2. An enzyme called DNA polymerase adds complementary nucleotides to the new strand (green) using the rules of base pairing: A with T and G with C.
DNA polymerase
3. New nucleotides (green strand) are complementary to old nucleotides (blue strand).
DNA Replication Is Semiconservative
4. After replication, two copies of the original DNA exist. Each molecule consists of an original strand and a new strand.

? Which enzyme adds complementary nucleotides during DNA replication?

polymerase enzyme, and primers—short segments of DNA that act as guideposts and flag the section to which DNA polymerase should bind to begin replication. The DNA is first heated to separate the strands, and then cooled to allow the primers to associate with the DNA and new nucleotides to be added by DNA polymerase. From a starting sample of just a few DNA molecules, PCR can make billions of copies of a specific region of the DNA in less than a few hours (**INFOGRAPHIC 7.4**).

DNA from the Crime Scene

In 2006, a New York State crime lab used PCR to amplify DNA from various items of evidence collected by Cayuga County law enforcement officials during their original investigation of Kulakowski's murder. The evidence included remnants of cotton swabs used to sample bite marks on the victim and the nightshirt stained with saliva and blood.

INFOGRAPHIC 7.4 The Polymerase Chain Reaction Amplifies Small Amounts of DNA

The polymerase chain reaction (PCR) is similar to naturally occurring DNA replication, except that it occurs in a test tube and only specific regions of the original DNA sequence are amplified. From a starting sample of just a few molecules of DNA, PCR can make billions of copies.

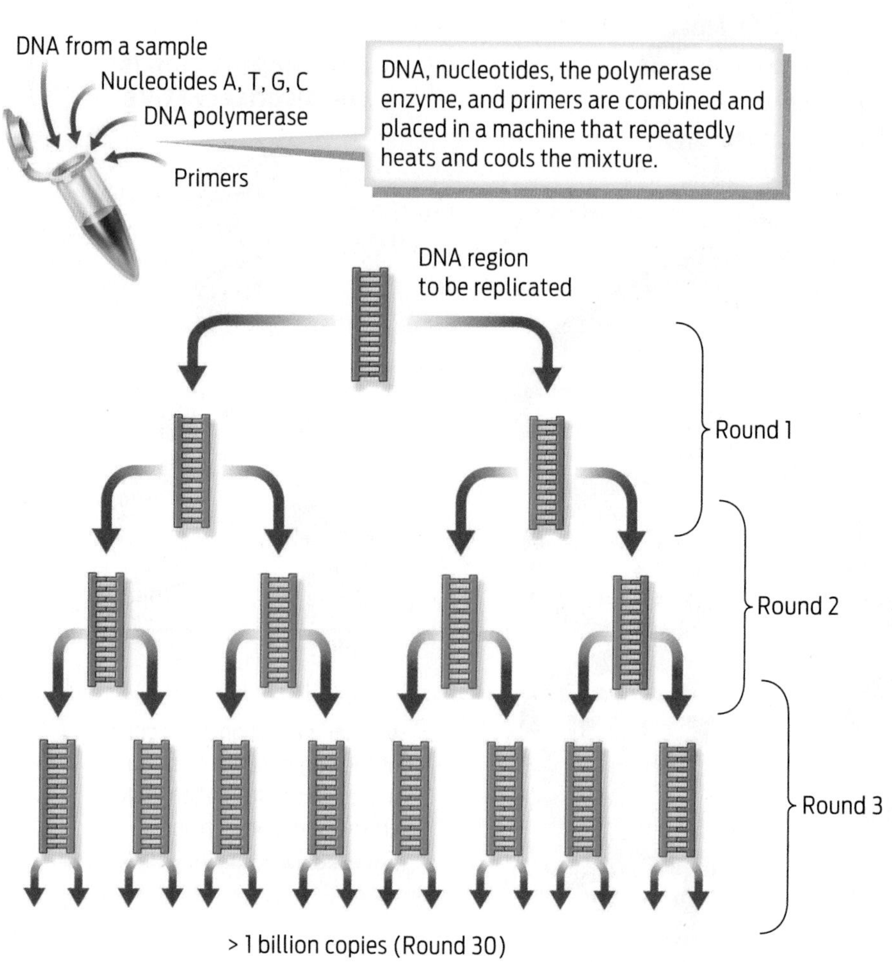

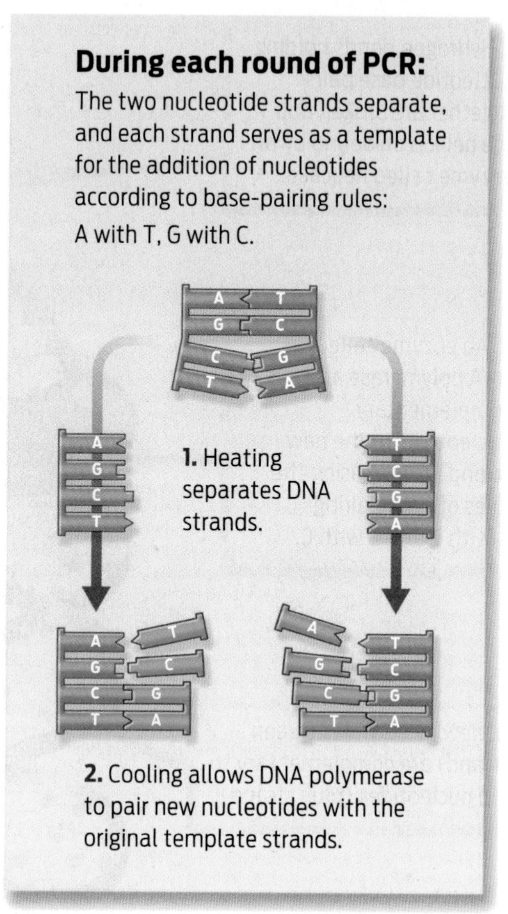

During each round of PCR:
The two nucleotide strands separate, and each strand serves as a template for the addition of nucleotides according to base-pairing rules:
A with T, G with C.
1. Heating separates DNA strands.
2. Cooling allows DNA polymerase to pair new nucleotides with the original template strands.

? Which enzyme adds complementary nucleotides during PCR?

To extract DNA from a forensic sample, scientists typically use chemicals to separate cells from other material, like fabric. Then, a device called a centrifuge, which spins samples at high speeds to separate out materials from a mixture, is used to further extract DNA from cells. DNA extraction is usually the most painstaking step of the process because it can be difficult to obtain sufficient cells in a forensic sample to yield enough DNA for PCR. Also, improperly stored samples can degrade too much to be useful. Samples can also become contaminated with foreign DNA from improper handling, which would render results useless.

In Brown's case, the laboratory's first report on the tested evidence was disappointing. Technicians hadn't been able to obtain any DNA from the bite-mark swab. The lab's second report was more conclusive: seven different pieces of the victim's nightshirt contained DNA. Moreover, the report went on to state that six of the pieces contained DNA from two different people, the victim and another person who was male.

DNA Profiling: How It Works

Once DNA from crime scene evidence is obtained, the next step is to analyze it. Human cells contain vast amounts of DNA–on the order of 3 billion nucleotide base pairs within the full stretch of the human **genome**, or one complete set of genetic instructions. Figuring out the sequence of every nucleotide in the genome would be extremely time consuming and expensive. So instead, forensic scientists use a shortcut–they employ PCR to amplify specific segments of DNA and analyze just these segments. These segments are known as short tandem repeats.

Short tandem repeats (STRs) are blocks of repeated DNA sequences found at points along our chromosomes. These sequences are noncoding, meaning they do not contain instructions for making proteins. They are a bit like nonsense words in our DNA: the sequence AGCT repeated over and over again, for example. While all of us have STRs in the same places along our chromosomes, the exact length of each STR varies from person to person. At a single STR site, one person may carry the AGCT sequence repeated six times while another person might carry the AGCT sequence repeated four times. Also, since we inherit two copies of every chromosome, every person has two copies of each STR, and they can be of two different lengths. Forensic scientists use these differences in STR lengths to distinguish between individuals **(INFOGRAPHIC 7.5)**.

GENOME
One complete set of genetic instructions encoded in the DNA of an organism.

SHORT TANDEM REPEATS (STRs)
Sections of a chromosome in which short DNA sequences are repeated.

INFOGRAPHIC 7.5 DNA Profiling Uses Short Tandem Repeats

No two people have the same exact nucleotide sequence. The specific regions of DNA that forensic scientists analyze are those that contain short tandem repeats (STRs). STRs are short stretches of repeated DNA sequences. People differ in the number of copies of an STR sequence found along their chromosomes.

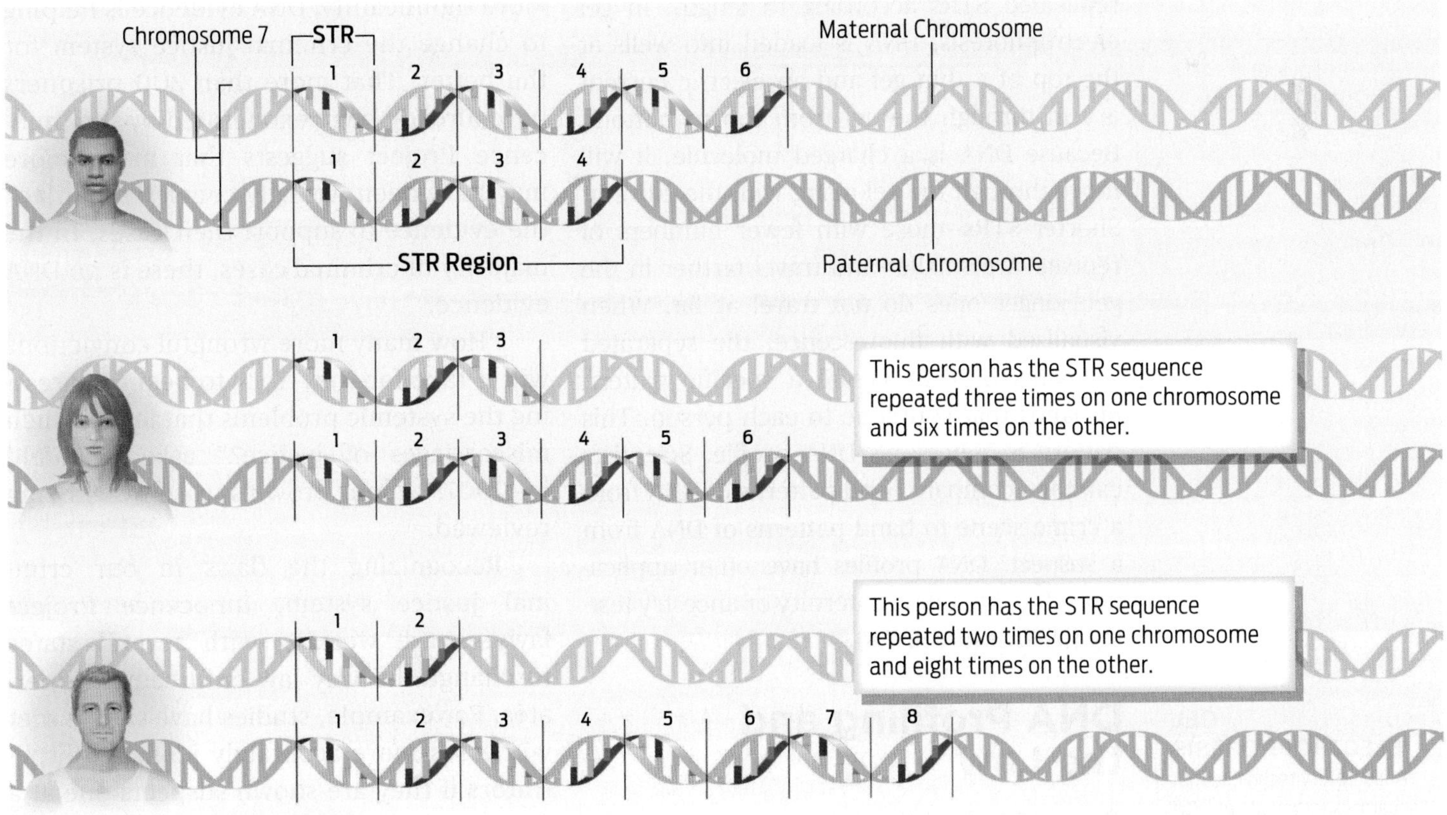

? How many STR copies did the top person inherit from his mother and father?

Tek Image/Science Source

Each person's DNA sequence is unique and can be used as a kind of molecular "fingerprint."

To create a DNA profile, scientists first use PCR to increase the amount of DNA at multiple STR regions. Then they use a method called **gel electrophoresis** to separate the replicated STRs according to length. In gel electrophoresis, DNA is loaded into wells at the top of a thin gel and an electric current is run through the gel from top to bottom. Because DNA is a charged molecule, it will move through the gel along with the current. Shorter STRs–those with fewer numbers of repeats–are smaller and travel farther in the gel; longer ones do not travel as far. When visualized with fluorescence, the separated segments of DNA create a specific pattern of bands that is unique to each person. This pattern is a person's DNA profile. Scientists can then compare band patterns of DNA from a crime scene to band patterns of DNA from a suspect. DNA profiles have other applications, too, including paternity or ancestry testing (**INFOGRAPHIC 7.6**).

GEL ELECTROPHORESIS
A laboratory technique that separates fragments of DNA by size.

DNA Profiling and the Law

Since 1994, the federal government has been collecting DNA profiles of offenders and storing this information in the Combined DNA Index System (CODIS), a computer database that contains more than 10 million profiles from criminals convicted of specific crimes in all 50 states. Each profile consists of a banding pattern that represents 15 specific STR regions scattered throughout our genomes. Forensic scientists typically describe the likelihood that any two unrelated people will have the same number of repeated sequences at all 15 regions as 1 in some number of quintillions (billions of billions) (**INFOGRAPHIC 7.7**).

So far, the database of DNA profiles has proved helpful in more than 190,000 cases. More significantly, DNA evidence is helping to change the criminal justice system for the better. That more than 300 prisoners have already been exonerated by the Innocence Project suggests that many more may have been wrongly convicted but lack the evidence to support their cases. In the majority of criminal cases, there is no DNA evidence.

"How many more wrongful convictions will it take for New York to begin addressing the systemic problems that lead to such miscarriages of justice?" asked Neufeld in 2007, when Brown's case was being reviewed.

Recognizing the flaws in our criminal justice system, Innocence Project lawyers are working with several states to change the way law enforcement operates. For example, studies have shown that witnesses more accurately identify perpetrators if they are shown suspects one at a time instead of in a group line-up. Project lawyers are also helping to force changes in the way interrogations are conducted, calling for them to be videotaped to reduce the possibility of forced confessions. In addition, they are lobbying for legislation to ensure that evidence from crime scenes is properly collected and maintained, since DNA evidence can be ruined or contaminated during collection. They also advocate that anyone convicted of a crime be able to gain access to DNA testing.

"The key is that DNA really gives us an opportunity to start making the other institutions in the system more scientific and reliable as well," says Neufeld.

Vindication

The DNA that the New York State crime lab extracted from the victim's nightshirt contained a mixture of DNA from the victim and from another person, who was male. Analysis showed that this male DNA, however, did not match Roy Brown's. DNA evidence excluded him as Kulakowski's murderer.

Additional testing eventually linked that DNA evidence to Barry Bench. After Bench's suicide, of course, he couldn't provide DNA directly. So lawyers pursued the next best option: a DNA sample voluntarily donated by Bench's biological daughter, Katherine Eckstadt. Because we all receive one set of chromosomes from our mother and one set from our father, half of Katherine Eckstadt's DNA would have come from her father, and therefore would show great similarity to his. The test yielded dramatic results—a 99.99% probability that the man who deposited his saliva on Sabina Kulakowski's nightshirt was Eckstadt's father, Barry Bench.

INFOGRAPHIC 7.6 Creating a DNA Profile

Cells left at crime scenes contain DNA that can be profiled. The DNA is isolated from the cells (evidence). STRs are amplified by PCR, and the profiles are compared to possible suspects.

1. Collect cells from crime scene evidence and extract DNA.

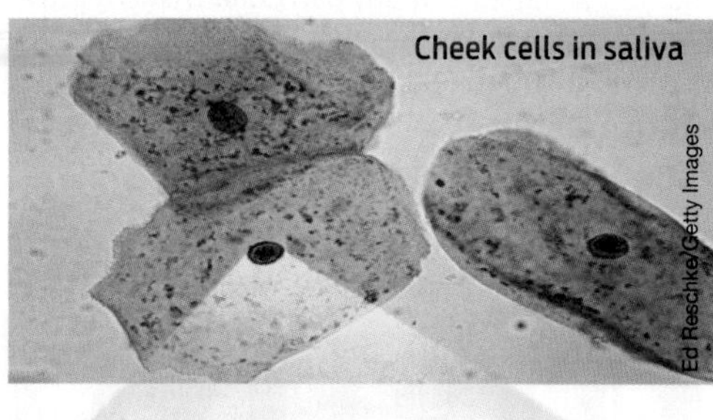

2. Amplify multiple STR regions by PCR.

3. Separate STRs by gel electrophoresis.

PCR products are loaded into chambers at the top of a gel. An electric current applied to the gel causes the charged DNA to migrate through it. Shorter fragments of DNA travel faster—and thus farther—than longer fragments in a given amount of time. All DNA fragments of the same length will travel together, forming a band of DNA that can be visualized on the gel.

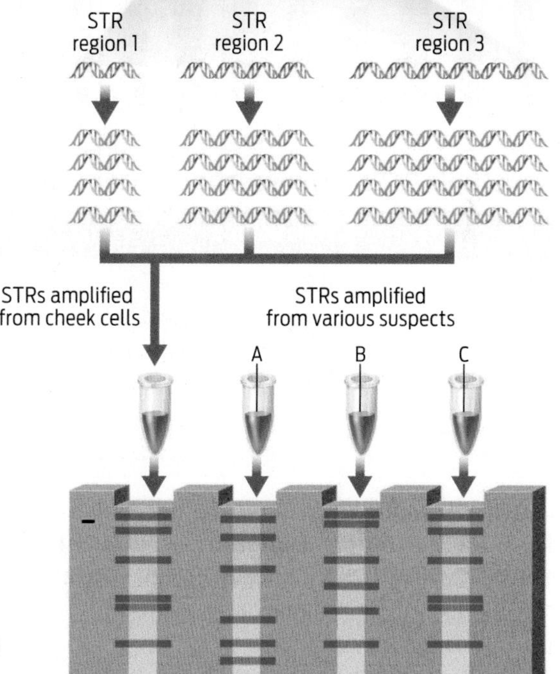

4. Compare STR banding patterns.

The gel shows the results of three different STR regions (green, red, and blue bands) analyzed from the DNA in a crime scene sample and in three suspects.

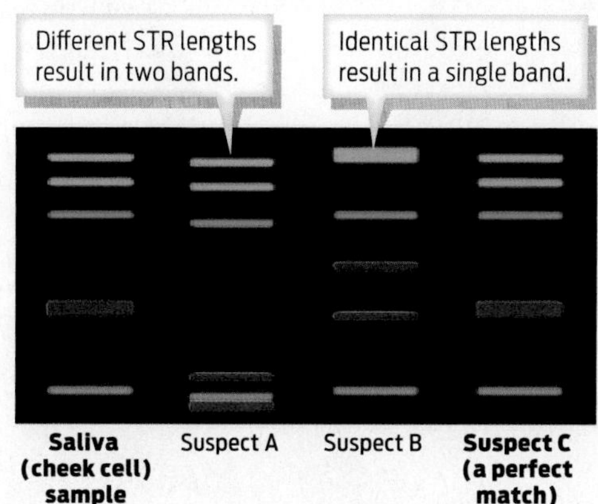

? In a paternity test, how many of the child's STR bands should match bands from the father?

INFOGRAPHIC 7.7 DNA Profiling Uses Multiple STRs to Identify Genetically Unique Individuals

To create a DNA profile, scientists analyze 15 different STRs (yellow boxes) scattered among our chromosomes. Sharing the same number of repeats at any particular STR is relatively common—typically 5% to 20% of people share the same pattern at any one STR site. But it is the combined pattern of STR repeats at multiple sites that is unique to a person; the more STRs tested, the more discriminating the test becomes.

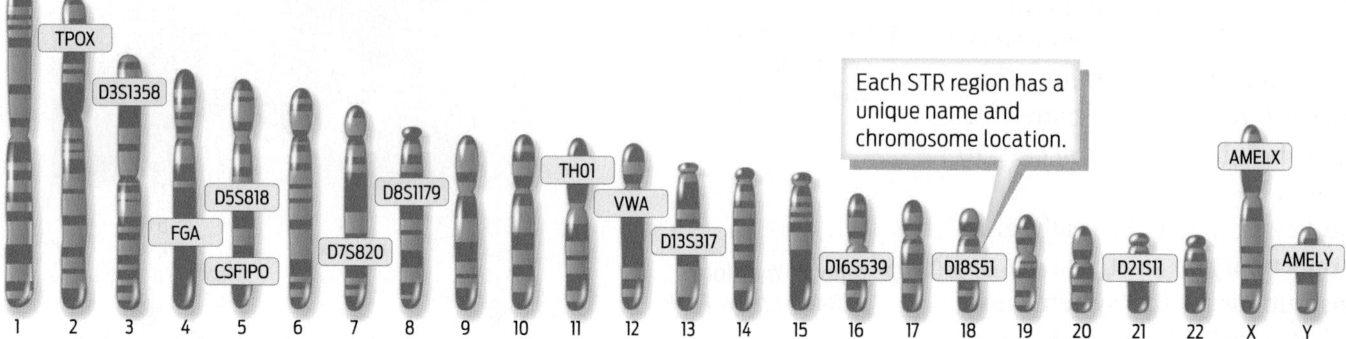

Using multiple STRs increases the chances of a profile's being unique to a single individual:

Analysis	Calculation	People sharing this profile
1 STR region	0.2	1 in 5
2 STR regions	0.2 x 0.2 = 0.04	1 in 25
5 STR regions	0.2 x 0.2 x 0.2 x 0.2 x 0.2 = 0.00032	1 in 3,125
15 STR regions	0.2 x 0.2 x 0.2 x 0.2 x 0.2 x 0.2 x 0.2 x 0.2 x 0.2 x 0.2 x 0.2 x 0.2 x 0.2 x 0.2 x 0.2 = 3.3×10^{-11}	1 in several quintillion

? If an STR analysis were carried out with three STRs, what proportion of people would be expected to share that profile (if the probability of matching at each STR is 0.2)?

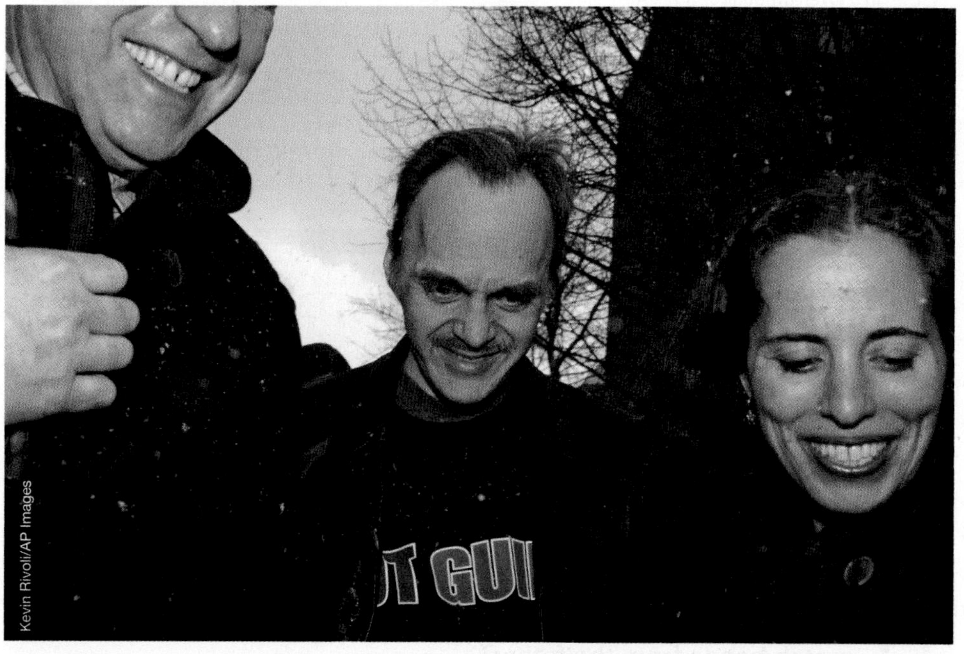

Roy Brown leaves the courthouse a free man in 2007 with his lawyers, Peter Neufeld and Nina Morrison.

To clinch the case, Cayuga County prosecutors eventually agreed to have Bench's body exhumed for DNA tests–which matched the DNA from the saliva stains.

"We've had a lot of crazy cases," says Nina Morrison, the Innocence Project attorney who handled Brown's case, "but this is really up there with the best of them . . . the client solving his own case . . . it's insane." After serving more than 25 years in prison for a crime he didn't commit, Brown was ultimately cleared of all charges and released from prison in 2007. ■

CHAPTER 7 SUMMARY

- DNA is the hereditary molecule of all living organisms. DNA contains instructions for building an organism.
- DNA sequences determine the genetic uniqueness and relatedness of individuals.
- The DNA in a eukaryotic cell is packaged into chromosomes located in the nucleus.
- Humans have 23 pairs of chromosomes in their cells—one chromosome of each pair inherited from the mother, the other from the father.
- DNA is a double-stranded molecule that forms a spiral structure known as a double helix.
- Each strand of DNA is made of nucleotides bonded together in a linear sequence.
- There are four distinct nucleotides in DNA: adenine (A), thymine (T), guanine (G), and cytosine (C).
- The two linear strands of a DNA molecule are bound together by hydrogen bonds between base pairs. Base pairing is complementary, A with T and G with C.
- Complementary pairing of DNA strands guides DNA replication, a fundamental part of cell reproduction.
- PCR enables scientists to vastly increase the number of copies of specific DNA sequences.
- Forensic scientists use noncoding DNA sequences known as STRs to create a DNA profile.
- STRs are blocks of repeated sequences of DNA. People differ in the number of times the sequences are repeated along their chromosomes.
- A DNA profile is more accurate and reliable than many other forms of evidence.

MORE TO EXPLORE

- **The Innocence Project: http://www.innocenceproject.org/**
- **National Human Genome Research Institute: http://www.genome.gov**
- **Scheck B., et al. (2003) *Actual Innocence: When Justice Goes Wrong and How to Make It Right.* New York: NAL Trade Paperbacks.**
- **NOVA, Forensics on Trial, October 17, 2012: http://www.pbs.org/wgbh/nova/tech/forensics-on-trial.html**
- **Cold Spring Harbor Laboratory: DNA from the Beginning: An animated primer of 75 experiments that made modern genetics. http://www.dnaftb.org**

CHAPTER 7 Test Your Knowledge

DRIVING QUESTION 1 What is the structure of DNA, and how is DNA organized in cells?

By answering the questions below and studying Infographics 7.1 and 7.2, you should be able to generate an answer for the broader Driving Question above.

KNOW IT

1 Which of the following is *not* a nucleotide found in DNA?

a. adenine (A)
b. thymine (T)
c. cytosine (C)
d. guanine (G)
e. uracil (U)

2 If the sequence of one strand of DNA is AGTCTAGC, what is the sequence of the complementary strand?

a. AGTCTAGC
b. CGATCTGA
c. TCAGATCG
d. GTCGACGC
e. GCTAGACT

3 In addition to the base, what are the other components of a nucleotide?

a. sugar and polymerase
b. phosphate group and sugar
c. phosphate group and polymerase
d. phosphate group and helix
e. helix and sugar

4 The __________ chromosomes in a typical human cell are found in the __________.

a. 46; cytoplasm
b. 23; nucleus
c. 24; cytoplasm
d. 46; nucleus
e. 22; nucleus

5 Each chromosome contains

a. DNA only.
b. proteins only.
c. DNA and proteins.
d. the same number of genes and STRs.
e. the entire genome of a cell.

USE IT

6 You can detect DNA that is specifically from the X chromosome in a DNA sample from a person. Can you definitively determine the sex of that person (male or female) from the presence of the X chromosome? Explain your answer.

7 Human red blood cells are enucleated (that is, they do not have nuclei). Is it possible to isolate DNA from red blood cells? Why or why not?

DRIVING QUESTION 2 How is DNA copied in living cells, and how can DNA be amplified for forensics?

By answering the questions below and studying Infographics 7.3 and 7.4, you should be able to generate an answer for the broader Driving Question above.

KNOW IT

8 DNA replication is said to be semiconservative because a newly replicated, double-stranded DNA molecule consists of

a. two old strands.
b. two new strands.
c. one old strand and one new strand.
d. two strands, each with a mixture of old and new DNA.
e. any of the above, depending on the cell type

9 Which of the following statements about PCR is true?

a. DNA polymerase is the enzyme that copies DNA in PCR.
b. Primers are not necessary for PCR.
c. PCR does not require nucleotides.
d. PCR does not generate a complementary DNA strand.
e. PCR can make only a few copies of a DNA molecule.

USE IT

10 Complete the statements below, and then number them to indicate the order of these two major steps necessary to copy a DNA sequence during PCR.

Step # __________ The enzyme __________ "reads" each template strand and adds complementary nucleotides to make a new strand.

Step # __________ The two original strands of the DNA molecule can be separated by __________.

11 Given this segment of a double-stranded DNA molecule, draw the two major steps involved in DNA replication:

ATCGGCTAGCTACGGCTATTTACGGCATAT
TAGCCGATCGATGCCGATAAATGCCGTATA

DRIVING QUESTION 3 How does DNA profiling make use of genetic variation in DNA sequences?

By answering the questions below and studying Infographics 7.5, 7.6, and 7.7, you should be able to generate an answer for the broader Driving Question above.

KNOW IT

12 Which STR will have migrated farthest through an electrophoresis gel?

a. GAAG repeated twice
b. GAAG repeated three times
c. AGCT repeated five times
d. GAAG repeated seven times
e. AGCT repeated seven times

13 An individual's STR may vary from the same STR of another individual by

a. the order of nucleotides.
b. the specific bases present.
c. the specific chromosomal location of the STR.
d. the number of times the sequence is repeated.
e. the number of coding regions.

14 Which of the following represents genetic variation between individuals?

a. whether or not G pairs with C or T
b. the presence of STRs in their genomes
c. the number of chromosomes in the nucleus
d. the sequence of nucleotides along the length of each chromosome
e. the number of chromosomes received from each parent

15 **A person has an STR with the sequence GACCT repeated six times on one chromosome and eight times on the other chromosome. The STR is amplified by PCR, and the PCR product is run on a gel. Which lane (A–D) in this gel shows the banding pattern you would expect to see? The marker lane (M) has DNA fragments starting at 10 nucleotides (at the bottom) and increasing in 10-nucleotide increments.**

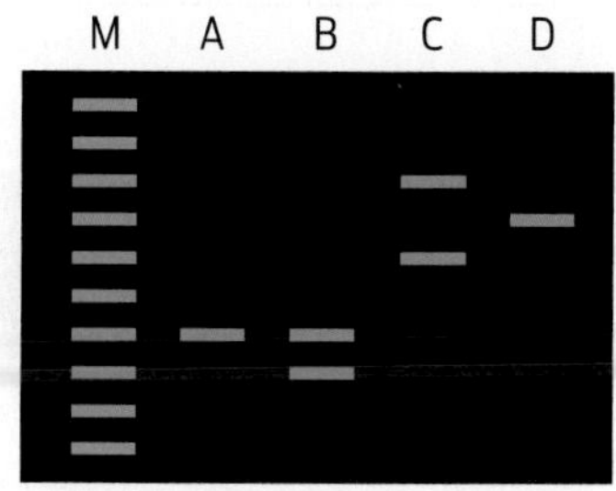

Gel for Question 15

USE IT

16 **A series of statements is presented below. Mark each statement as true (T) or false (F).**

a. ________ G pairs with T.

b. ________ Genetic information is passed on to the next generation in the form of DNA molecules.

c. ________ All DNA sequences encode information to produce proteins.

d. ________ Each person carries the same number of STR repeats on both maternal and paternal chromosomes.

e. ________ DNA evidence can be obtained from saliva left in a bite mark.

17 **Explain why the statements that you marked as true in Question 16 are in fact true.**

18 **Rewrite the statements that you marked as false in Question 16 to make them true.**

DRIVING QUESTION 4 How does DNA evidence fit into forensic investigations?

By answering the questions below and studying Infographics 7.1, 7.5, 7.6, and 7.7, you should be able to generate an answer for the broader Driving Question above.

KNOW IT

19 **This gel shows the DNA profile of STRs from four sources: blood from crime scene evidence (E), suspect A, suspect B, and the victim (V). An eyewitness identified suspect A as fleeing the apartment building where the crime occurred. Suspect B was picked up at a local convenience store after using bloodstained money.**

a. From the DNA profiles shown, can you draw any conclusions about where the crime scene DNA came from?

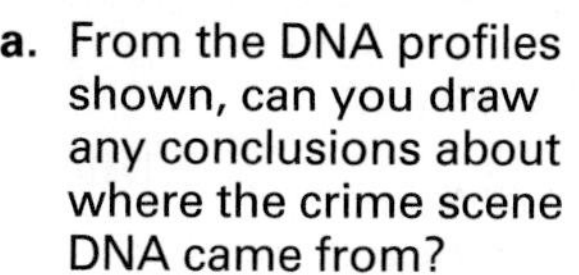

b. Can you draw any conclusions about relationships among the people profiled? Explain your reasoning.

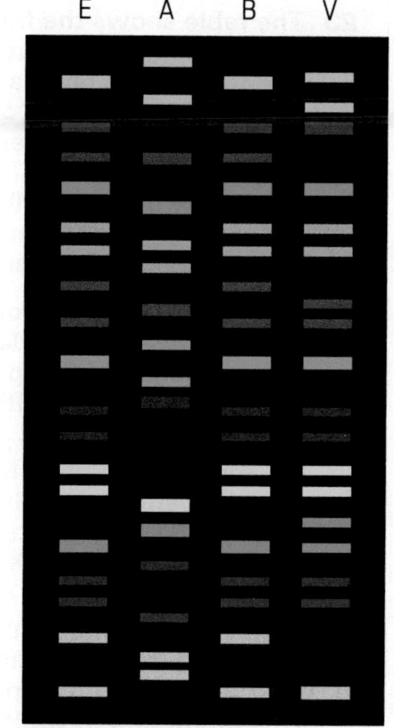

Gel for Question 19

USE IT

20 **Look at Infographic 7.7. From the STRs used in forensic investigations, which STRs on which chromosomes would be particularly useful in determining whether crime scene evidence was left by a female or a male?**

21 **Explain your response to Question 20, stating the number of STR copies you would expect to see if the perpetrator was female and if the perpetrator was male.**

22 **This gel shows a DNA profile using five STRs. The lane labeled W is a mother and the lane labeled C is her child. The lanes labeled M1 and M2 are two men, either of whom, according to the mother, could be the father of the child.**

a. Circle the STR bands that the child (C) inherited from its mother (W).

b. Use the DNA profiles to determine which man is the father of the child.

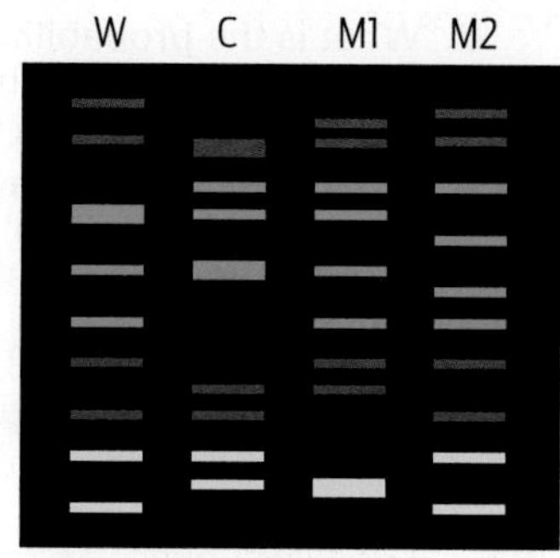

Gel for Question 22

INTERPRETING DATA

apply YOUR KNOWLEDGE

23 The table shows the frequencies for STR lengths (repeat number) in different U.S. populations. Since we inherit two copies of every chromosome, every person has two copies of each STR. You can determine the probability of a particular combination of STRs by multiplying the frequencies.

If the STR lengths on each chromosome are identical, the probability of this pattern can be determined by multiplying the frequency by itself.

For example, the probability of a Hispanic person having 14 D3S1358 repeats (0.079) on both chromosomes is 0.079 x 0.079 = 0.0062.

If the STR lengths on each chromosome are different, the probability of this pattern can be determined by multiplying the two frequencies together. Then, because there are two ways of inheriting this pattern (getting a 14 from dad and an 18 from mom or getting the 18 from dad and the 14 from mom), this number must be multiplied by 2.

For example, the probability of a Hispanic person having 14 D3S1358 repeats (0.079) *and* 18 D3S1358 repeats (0.125) is 0.079 X 0.125 X 2 = 0.0198.

CODIS* STR	No. of Repeats	Frequency (Caucasian)	Frequency (African American)	Frequency (Hispanic)
D3S1358	14	0.103	0.089	0.079
	15	0.262	0.186	0.293
	16	0.253	0.248	0.286
	17	0.215	0.242	0.204
	18	0.152	0.155	0.125
TH01	5	0.002	0.004	0
	6	0.232	0.124	0.214
	7	0.190	0.421	0.096
	8	0.084	0.194	0.096
	9	0.114	0.151	0.150
D18S51	10	0.008	0.006	0.004
	11	0.017	0.002	0.011
	12	0.127	0.078	0.118
	13	0.132	0.053	0.111
	14	0.137	0.072	0.139

Data from Butler et al. (2003) Allele frequencies for 15 autosomal STR loci on U.S. Caucasian, African American and Hispanic populations. *Journal of Forensic Sciences* 48(4): 908–911.

*CODIS stands for "Combined DNA Index System," a government database of DNA profiles from offenders, crime scenes, and missing persons.

a. What is the probability of a Caucasian American having a 16, 17 combination for D3S1358?

b. What is the probability of an African American having a 16, 17 combination for D3S1358?

c. What is the probability of a Hispanic American having a 16, 17 combination for D3S1358?

d. What is the probability of a Caucasian American having a 16, 17 combination for D3S1358, and a 5, 9 combination for TH01?

e. What is the probability of a Caucasian American having a 16, 17 combination for D3S1358, a 5, 9 combination for TH01 and an 11, 14 combination for D18S51?

f. Consider your answers to a, d, and e. Why does forensic analysis use many CODIS STRs (and not just one or two)?

MINI CASE

apply YOUR KNOWLEDGE

24 A female eyewitness has identified a Hispanic American male as the man who stole her car. The eyewitness stated that the man was bleeding profusely from a head wound. Her car was recovered, and male blood with a 16, 17 combination for D3S1358, a 5, 9 combination for TH01, and an 11, 14 combination for D18S51 was found on the driver's seat and steering wheel. Does this finding call the eyewitness evidence into question? Explain your answer.

BRING IT HOME

apply YOUR KNOWLEDGE

25 Scientists used DNA from Barry Bench's daughter to pinpoint Bench as a possible suspect, as his DNA was not on file anywhere. Similarly, in cases of disasters, DNA evidence is sometimes required to identify victims. If a victim doesn't have a DNA profile on file, identity must be reconstructed by comparing the victim's DNA profile to that of relatives. These situations illustrate that a DNA profile database has the potential to be useful in cases in which DNA-based identification is required. However, maintenance of such a database is controversial. What arguments can you make for and against banking DNA profiles in a database? If such a database existed, what restrictions would you place on it? Would you choose to register your DNA or your child's DNA in such a database?

2 Milestones in Biology

DRIVING QUESTIONS

1. Who were the major players in the discovery of the structure of DNA?
2. What pieces of scientific knowledge were assembled in order to elucidate the structure of DNA?

The Model Makers

Watson, Crick, and the structure of DNA

IN 1953, JAMES WATSON AND FRANCIS CRICK announced to a crowd in their favorite pub in Cambridge, England, that they had found the secret of life. They had every right to boast. In revealing the structure of DNA, they had solved one of the greatest mysteries of science—the chemical basis of inheritance. Their discovery would revolutionize biology and push forward the study of anthropology, evolution, and medicine.

But Watson and Crick's success wasn't merely the result of a marriage between two great minds. Science is a collective enterprise and progresses through the work of many people. Breakthroughs happen when scientists are well positioned to build upon foundations laid down by other scientists. And so it was with DNA: in addition to their own insight, Watson and Crick built on the discoveries of others. They also had luck on their side—they were in the right place at the right time.

Watson and Crick met in 1951 at Cambridge University, in England. Watson was an American scientist who had just accepted a research position at the university. Crick was then a Ph.D. student, studying protein structure with a technique called x-ray crystallography.

The men didn't appear obvious collaborators. Watson was a prodigy: twelve years younger than Crick, he had earned his Ph.D. at 22. Crick, by contrast, was a late bloomer: he was 38 years old by the time he had his Ph.D. What they did share was intellectual curiosity. Both had changed their research focus several times. By their own admission, both were more

INFOGRAPHIC M2.1 The DNA Puzzle

It was known that DNA was made of nucleotides that included a deoxyribose sugar, a phosphate group, and one of four nitrogenous bases. But no one had yet figured out how the nucleotides fit together to produce a DNA molecule.

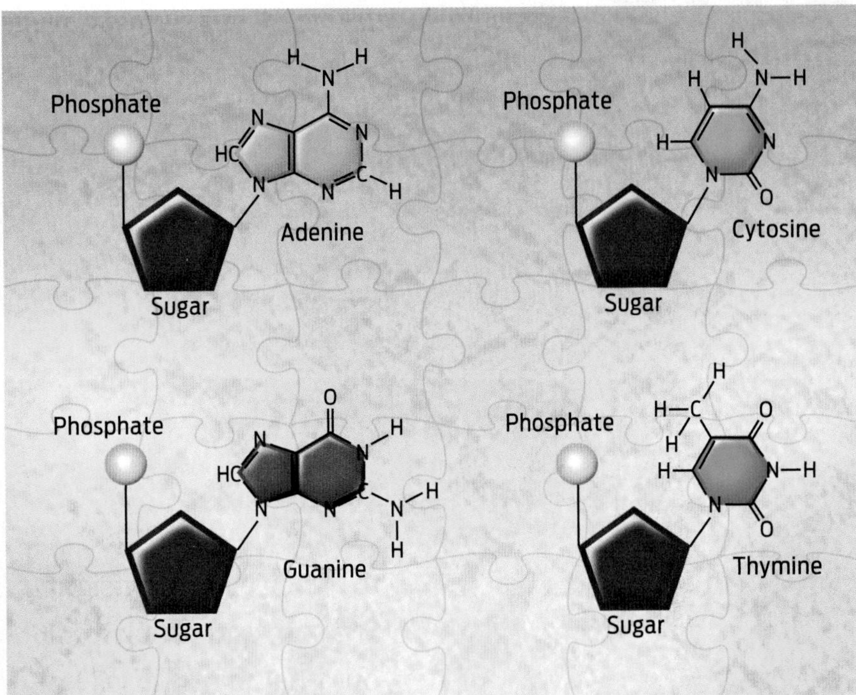

? What do all nucleotides have in common?

interested in solving current hot topics in science—like the structure of DNA—than pursuing the more obscure science that each had trained to do.

Although DNA was first observed in cell nuclei in the late 1860s, it took almost a century before scientists realized its importance. For a long time, the prevailing belief was that proteins carried the genetic information. But by the 1950s, scientists had pegged DNA as the more likely candidate. A key experiment carried out in 1952 convincingly demonstrated that DNA, not protein, was the genetic material. This set the stage for a race among scientists to understand the structure of DNA and how it carried genetic information.

When Crick and Watson came to the problem, they already knew, from the work of other scientists, that DNA was made up of nucleotides, each containing a sugar, a phosphate group, and one of four bases: adenine (A), thymine (T), cytosine (C), or guanine (G). But how did the elements fit together? To answer this question, they took inspiration from the chemist Linus Pauling. Pauling had been studying the structure of proteins and had built a molecular model showing that some proteins exist as a single-stranded, twisting helix. He backed up his model with lab experiments to prove his structure was correct. If an eminent scientist like Pauling could model a protein structure without first conducting laboratory experiments, Watson and Crick thought they might be able to do the same with DNA. And, since Pauling was one of several scientists now chasing after DNA's structure, they hoped to—as Watson put it—"imitate Linus Pauling and beat him at his own game" (**INFOGRAPHIC M2.1**).

Using wire and metal, Watson and Crick began building scale models of DNA on the basis of existing evidence about the chemical structure of nucleotides. They initially built a three-helix model with the phosphate groups on the inside and the bases radiating outward. But colleagues who analyzed the structure deemed it chemically unstable.

Then came a crucial finding. In 1951, Watson attended a lecture by a 31-year-old scientist named Rosalind Franklin. In her laboratory at King's College London, she had been making x-ray diffraction pictures of DNA. X-ray diffraction analyzes the way x-rays bounce off molecules in order to determine their chemical structure. Franklin had observed that increasing the humidity of DNA caused it to elongate. To Franklin, this suggested that water molecules were being attracted to the helix, coating it and causing it to stretch out. And, since water is a polar molecule (see Chapter 2), attracted to charged (hydrophilic) molecules, this further suggested that the charged, water-loving phosphate groups of DNA must therefore be on the *outside* of the helix—not, as others had suggested, on the inside.

Franklin's contribution didn't end there. She also discovered other important facts about the structure of DNA. Working with a graduate student, Raymond Gosling, she found that her x-ray diffractions confirmed that the elongated form of DNA had all the characteristics of a twisting helix (**INFOGRAPHIC M2.2**).

Maurice Wilkins, who was Franklin's peer and worked in the same laboratory, was also studying DNA structure at the time. In early 1953, Wilkins saw Franklin's best unpublished x-ray picture of DNA and showed it to Watson without Franklin's knowledge. "The instant I saw the picture my mouth fell open and my pulse began to race," Watson recalled in his memoir of the discovery, *The Double Helix*, published in 1968. The sneak preview "gave several of the vital helical parameters."

With that clue in hand, Watson and Crick then took a crucial conceptual step and suggested that the molecule was made of two chains of nucleotides. Each formed a helix, as Franklin had found, but because the DNA molecule was symmetrical–it looked the same when flipped upside down and backward–they realized that the two chains of nucleotides must be oriented in opposite directions.

To construct the model, Watson and Crick also built on a discovery made a few months earlier. In 1952, Erwin Chargaff had found that any given DNA sample, no matter the

INFOGRAPHIC M2.2 Rosalind Franklin and the Shape of DNA

Franklin's 1951 x-ray diffraction studies of DNA showed that the structure was likely helical, involving two strands that run in opposite directions, and that the phosphate groups were on the outside of the molecule.

Rosalind Franklin
July 25, 1920–April 16, 1958

X-ray diffraction images of DNA

X-ray diffraction involves shooting x-rays at a crystallized version of the molecule and recording on film how the x-rays scatter when they bounce off its surface. The images shown here are views from the end of a DNA molecule looking down its center.

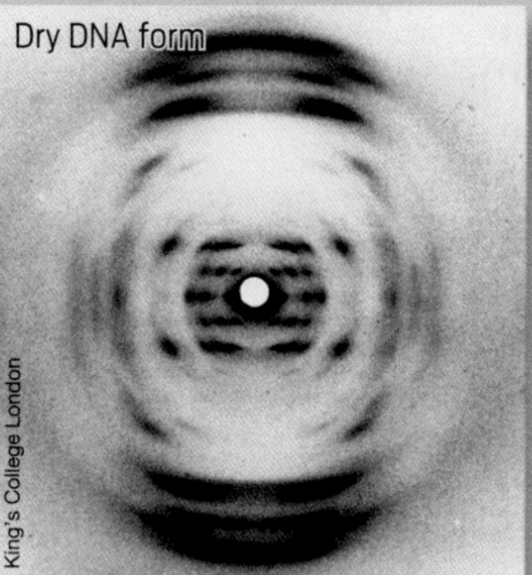

Humidity

Wet DNA form

King's College London

The signature "X" in the middle of this picture suggests a double-stranded helical structure.

The symmetry of the image suggests that the molecule is uniform in width.

DNA elongates in the presence of water, allowing the molecule to take on a regular helical shape, shown by the clarity of the image on the right. This evidence suggests that the water-loving phosphate groups are on the outside of the DNA molecule.

? Which part of a nucleotide (the sugar, phosphate, or base) can interact with water?

organism, always contained equal amounts of adenine and thymine, and equal amounts of guanine and cytosine. This information was critical. As the helix had a smooth shape and a uniform thickness, and as the bases had to point toward the inside of the helix, the different-size bases somehow had to fit together in a way that allowed for a consistent width of the helix. Following a tip about the structure of bases, Watson and Crick were able to use a model to show that A-T pairs and G-C pairs were exactly the same width, explaining the consistent helix shape, and also accounting for Chargaff's finding that the amounts of A and T were equal to each other, as were the amounts of G and C **(INFOGRAPHIC M2.3)**.

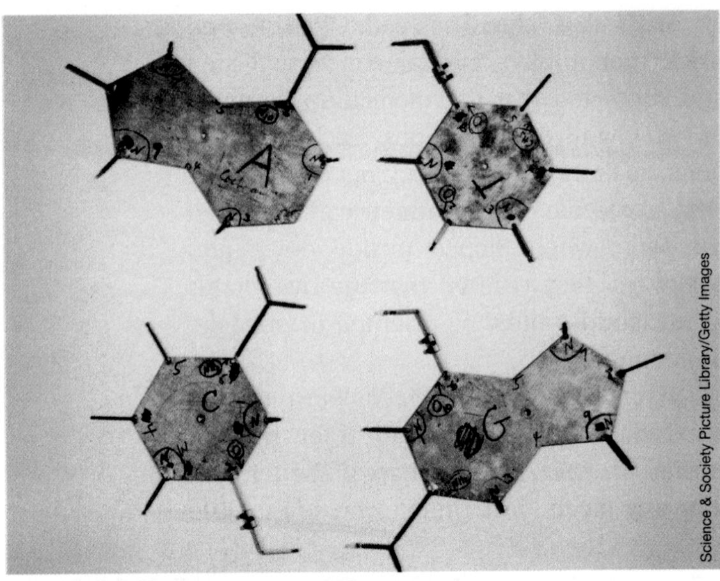

The aluminum bases Watson and Crick used in their DNA model.

INFOGRAPHIC M2.3 Erwin Chargaff's Work Provided a Clue to Base Pairing

Erwin Chargaff studied the nitrogenous bases of DNA. He found that no matter which DNA molecule he analyzed, it always contained equal amounts of adenine and thymine bases and equal amounts of cytosine and guanine bases. Additionally, the width of an A–T pair is the same as the width of a C–G pair. These data suggest that in a double-stranded DNA molecule, adenine must pair with thymine and cytosine must pair with guanine. This pairing is consistent with a uniform base-pair width down the length of the DNA molecule, as Franklin's data suggested.

Chargaff's Rule of Base Pairing

Within any DNA molecule:
% adenine = % thymine
% cytosine = % guanine

Erwin Chargaff
August 11, 1905–June 20, 2002

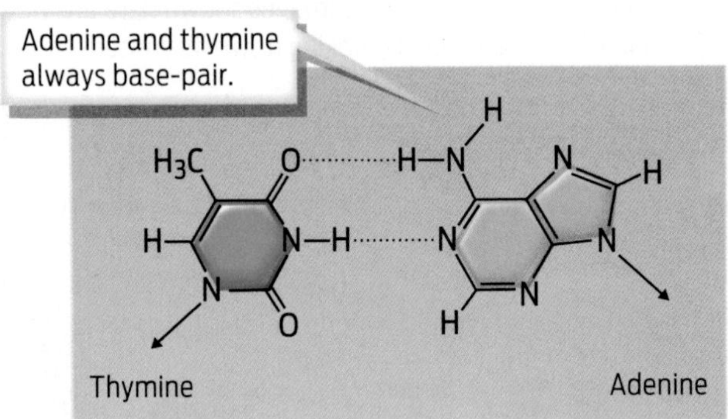

Adenine and thymine always base-pair.
Thymine
Adenine

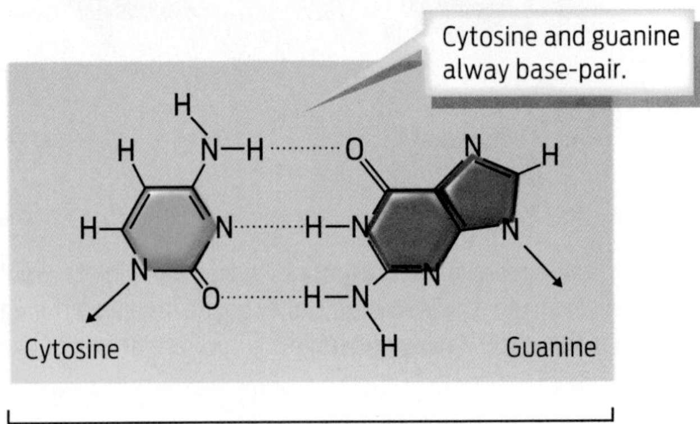

Cytosine and guanine alway base-pair.
Cytosine
Guanine

Uniform base-pair width

? If a DNA molecule is 20%G, what is the %C? What is the %A?

The final double-helix model so perfectly fit the experimental data that the scientific community almost immediately accepted it. Watson and Crick published their paper on the structure of DNA in April 1953 in the prominent journal *Nature*. In it, with considerable understatement, they wrote: "It has not escaped our notice that the specific pairing we have postulated immediately suggests a possible copying mechanism for the genetic material." Indeed, the model that Watson and Crick proposed solved at once both the structure of DNA and how this structure provides a mechanism for DNA replication: each strand of an unzipped helix serves as the template for the creation of a complementary strand, thus reproducing the original pattern (see Chapter 7). The mystery of heredity had finally been unlocked.

Alongside their paper, in the same issue, individual papers by Wilkins and Franklin discussed their respective x-ray diffraction results, which supported the Watson-Crick model. In 1962, Watson, Crick, and Wilkins shared the Nobel Prize in Physiology or Medicine (**INFOGRAPHIC M2.4**).

But what about Franklin? She had died of ovarian cancer in 1958, at the age of 37. By Nobel Foundation rules, she was ineligible for nomination since prizes are not awarded posthumously.

Controversy over whether Franklin has been adequately recognized continues.

INFOGRAPHIC M2.4 The Structure Is Finally Known: The DNA Double Helix

The structure of DNA that Watson and Crick proposed fit all the experimental evidence—the x-ray crystallography data, the base-composition evidence, and the placement of the phosphates on the exterior of the double helix.

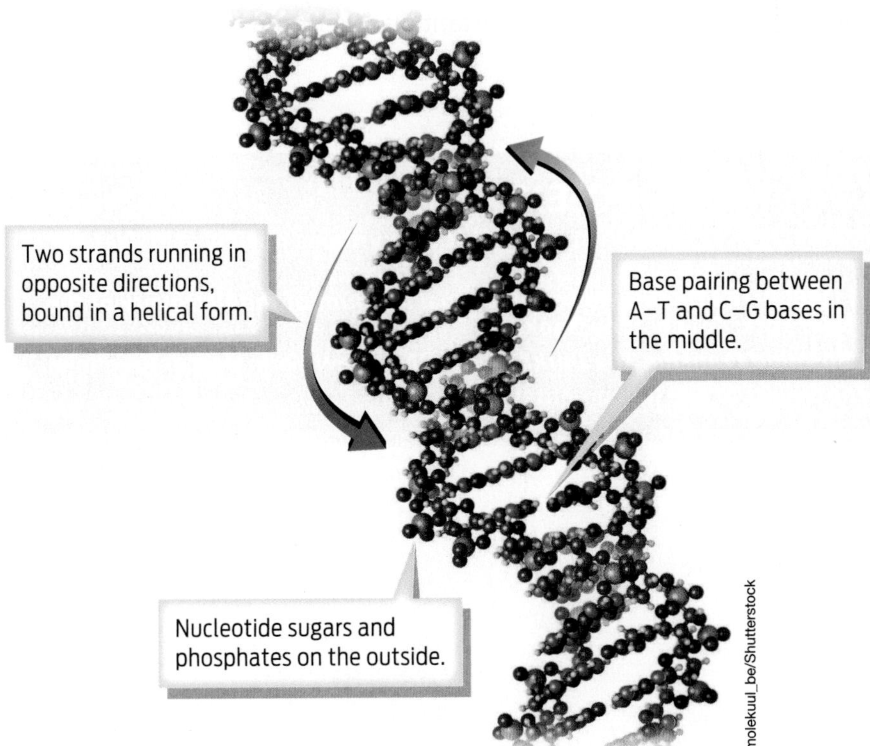

Two strands running in opposite directions, bound in a helical form.
Base pairing between A–T and C–G bases in the middle.
Nucleotide sugars and phosphates on the outside.
molekuul_be/Shutterstock

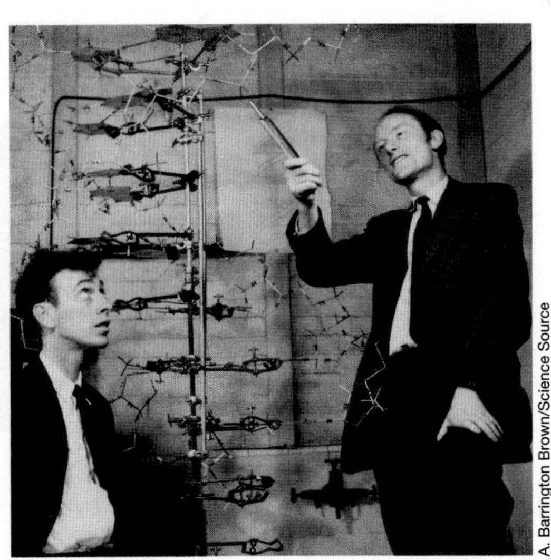

A. Barrington Brown/Science Source

? Describe the structure of a DNA molecule.

Although Watson and Crick acknowledged her input in their article in *Nature,* the extent to which her x-ray pictures helped them build their DNA model was revealed only much later in Watson's 1968 book, published 10 years after Franklin's death. For example, at the time *Nature* published the papers on DNA structure, Franklin's paper was perceived as merely supporting evidence. But it was her data that helped Watson and Crick clinch the structure. Some historians argue that sexist attitudes prevented her from receiving the acclaim she deserved before she died. At the time, female scientists in the biomedical sciences were few and were frequently confronted by negative attitudes from their male peers. "I'm afraid we always used to adopt–let's say, a patronizing attitude towards her," Crick publicly commented after Watson's book was published. He added that if Franklin had lived, "It would have been impossible to give the prize to Maurice [Wilkins] and not to her" because "she did the key experimental work."

> "It would have been impossible to give the prize to Maurice [Wilkins] and not to [Franklin because] she did the key experimental work."
>
> —Francis Crick

Although it is quite normal for colleagues to share data, some have even argued that Wilkins showed Franklin's critical x-ray diffraction photos to Watson out of jealousy or disdain. But despite controversy, Franklin's contribution to the discovery has never been completely ignored, and she is now recognized as having been a top-notch scientist: her notebooks show that without her thorough scientific research and original ideas, we would have had to wait much longer for what is still considered to be one of the most important discoveries in biology. ■

MORE TO EXPLORE

- Watson, J. D., and Crick, F. H. C. (1953) A structure for deoxyribose nucleic acid. *Nature* 171:737–738.
- Watson, J. D. (2001 [1968]) *The Double Helix: A Personal Account of the Discovery of the Structure of DNA.* New York: Touchstone Books.
- Maddox, B. (2003) *Rosalind Franklin: The Dark Lady of DNA.* New York: Harper Perennial.
- PBS (2003) *DNA: The Secret of Life* (documentary)

MILESTONES IN BIOLOGY 2 Test Your Knowledge

1 What technique did Rosalind Franklin use to examine the structure of DNA?

a. mass spectrometry
b. gel electrophoresis
c. x-ray diffraction
d. model building
e. all of the above

2 Which of the following statements about DNA structure is *not* true?

a. DNA is a double helix.
b. The phosphate groups are on the outside of the helix.
c. The two strands run in opposite orientations.
d. A pairs with A, T with T, C with C, and G with G.
e. The helix has a constant diameter along its length.

3 Describe the scientific contributions of Watson, Crick, Franklin, and Wilkins to the discovery of the structure of DNA.

4 How did differing amounts of water in the DNA crystals help explain the x-ray diffraction patterns?

5 Summarize the structure of a DNA double helix.

8 Genes to Proteins

DRIVING QUESTIONS

1. What determines the shape of a protein molecule?
2. What are the steps of gene expression, and where in the cell do they occur?
3. How can organisms be genetically modified to produce recombinant proteins?
4. What are some pros and cons of genetically modified organisms?

mycteria/Shutterstock

BULLET PROOF

Scientists hope to spin spider silk into the next indestructible superfiber

PETER PARKER'S SPIDEY SENSE is tingling from the latest news out of biotech: genetically engineered spider silk, produced from spider DNA but assembled inside an entirely different organism.

For more than two decades, scientists have sought ways to harness the unique properties of spider silk–a near-miracle fiber that is pound for pound stronger than steel but also lightweight and flexible. Yet harvesting silk from spiders is challenging.

"Spiders don't like neighbors," says David Kaplan, a professor of biomedical engineering at Tufts University. "They are territorial and cannibalistic."

This makes them much harder to work with than, say, silkworms–another source of silk fibers. "If I put a hundred silkworms together on a table with enough food, I can come back in 30 days and find a hundred cocoons," Kaplan explains. "If I put a hundred spiders together with enough food and come back 30 days later, there'd be only one spider."

The other approach–collecting spider silk from webs in the wild–is exceedingly tedious. It would take about 1 million spider webs to make a single spider silk garment. Which is why, recently, scientists have turned to putting spider DNA into other, more congenial organisms.

The potential applications of genetically engineered spider silk are vast. They include more-durable, environmentally friendly clothing; safer medical products such as biocompatible bone screws and sutures; and lightweight military vests that deflect bullets better than Kevlar. While some of these products are years away, others debuted in 2016.

> "On an equal weight basis, spider silk has a higher toughness than steel and Kevlar."
>
> — David Kaplan

PROTEIN
A macromolecule made up of repeating subunits called amino acids, which determine the shape and function of a protein. Proteins play many critical roles in living organisms.

AMINO ACIDS
The building blocks of proteins. There are 20 different amino acids.

Stronger than Steel

If you think about it, a spider's web is, in essence, a device for stopping a speeding projectile–such as a flying insect. It's no wonder, then, that the fibers making up the web are uncommonly strong.

"On an equal weight basis, spider silk has a higher toughness than steel and Kevlar," says Kaplan, whose lab is using genetically engineered spider silks to improve medical technologies. But it's also flexible and elastic, which allows it to absorb the energy of whatever hits it. A rope of spider silk the diameter of a pencil could theoretically stop a jet landing on an aircraft carrier, though no one has tried that yet.

The physical properties of spider silk reflect the structure of the **proteins** making it up. Recall from Chapter 2 that proteins are one of the four main macromolecules that make up cells, along with carbohydrates, nucleic acids, and lipids. Proteins are the cell's workhorse molecules. They perform myriad functions inside cells, and in turn help our bodies perform countless tasks–everything from contracting our muscles and sensing light to regulating blood sugar and fighting infections. Spiders use their silk proteins to build webs for trapping prey, sacs to protect eggs, and bungee cord-like draglines that catch them when they fall.

All proteins are made of the same building blocks, which are called **amino acids**.

Mechanical properties of spider silk

Spider silk exhibits a unique combination of strength and elasticity, enabling silk fibers to absorb a lot of energy before breaking (toughness).

Material	Strength (MPa) Spider silk is as strong as high-grade alloy steel and about half as strong as Kevlar.	Extensibility (%) Silk is extremely elastic, and some silks are able to stretch up to four times their relaxed length.	Toughness (MJ/m^3) The combination of strength and extensibility make silks tougher than many engineered polymers like Kevlar.
Spider dragline silk	880–1500	21%–27%	136–194
Silkworm silk	600	18%	70
Nylon	950	18%	80
Kevlar 49 fiber	3600	2.7%	50
High-tensile steel	1500	0.8%	6

Data from Anna Rising and Jan Johansson. 2015. *Nature Chem Biol*. 11:309–315. Lin Römer and Thomas Scheibel. 2008. *Prion* 2(4):154–161.

Spider silk characteristics and applications

Spider silk is tough, flexible, and elastic, making it well suited for military and industrial uses. It is also biocompatible, making it ideal for a variety of surgical applications, and better for the environment, too.

Bulletproof vests

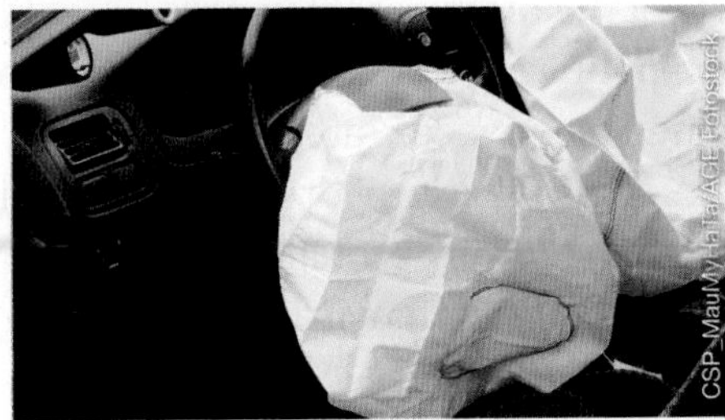

Strong, elastic airbags

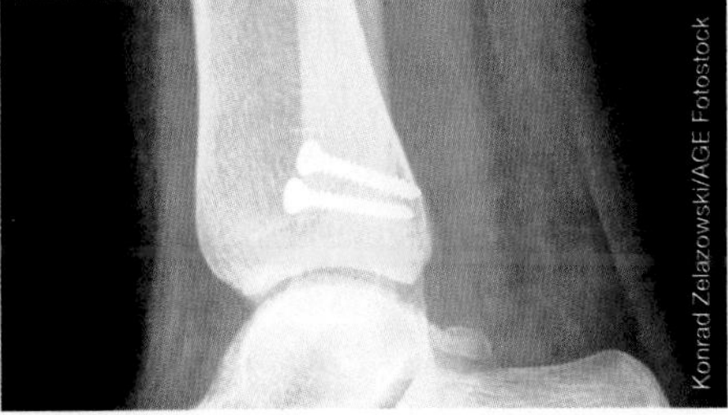

Biocompatible and biodegradable screws and sutures

Stronger skin grafts

Durable ecotextiles

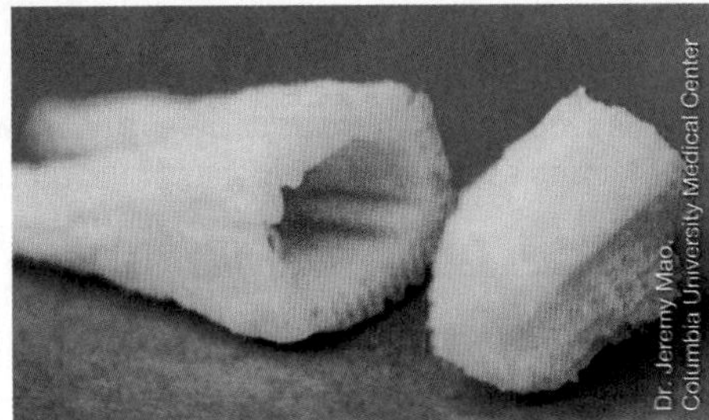

Scaffolds for growing tissues

There are 20 different amino acids in cells. All amino acids have the same basic core structure, but each of the 20 also has a unique chemical side chain that distinguishes it from all the others. To form proteins, amino acids link together in linear chains. Spider silk proteins, called spidroins, are chains of about 3,500 amino acids. This is longer than many animal proteins, which average around 400 amino acids in length. But proteins can be much longer or much shorter. The longest human protein, titin (involved in muscle contraction), is a single chain of 34,350 amino acids. Insulin, a protein that helps regulate blood sugar, has only 51.

In cells, the amino acid chain of a protein folds into a distinct three-dimensional shape, or conformation, that underlies a protein's function. Some proteins, like insulin, are made up of just one folded chain. Other proteins, such as the antibodies of our immune system or the hemoglobin that carries oxygen in our red blood cells, are made up of multiple folded amino acid chains bound together. A spider silk thread is made up of many spidroin proteins linked together into a fiber.

The particular sequence of amino acids in a chain determines how the chain will fold. Interactions between amino acid side chains, and between these side chains and the surrounding water, influence the precise folding pattern. Hydrophobic amino acid side chains tend to clump together, away from water, while hydrophilic amino acids face out toward water. The distinct three-dimensional shape that forms as a result is what ultimately determines how a protein functions.

According to Kaplan, the first thing you'd notice about the amino acid sequence of a spider silk protein is its repetitive nature. Most of the protein–roughly 90%–is composed of repeated sequences of relatively hydrophobic amino acids. It's these repetitive sequences that are responsible for the toughness of spider silks. Regions rich in the hydrophobic amino acid alanine pack closely together, away from water, and form flat, interlocking structures called beta sheets (a folded structure found in many proteins but at a much higher frequency in silk). These regions impart strength to the protein. Other regions, rich in the amino acid glycine, form more flexible parts of the protein, which confer elasticity. Nonrepetitive regions that flank these core repeated sequences are made of charged amino acids. These hydrophilic end regions keep the spidroin proteins dissolved in the silk gland's watery environment and prevent silk proteins from crystalizing spontaneously into fibers inside the spider, which would kill the animal **(INFOGRAPHIC 8.1)**.

Silk proteins lack the structural complexity you would see in an enzyme or an antibody, in which every region has a different function or activity. In that sense, Kaplan says, "They're very much like synthetic polymers." A polymer is a large molecule made up

INFOGRAPHIC 8.1 Amino Acid Sequence Determines Protein Shape and Function

The sequence of amino acids determines how a chain will fold into a three-dimensional shape and potentially interact with other chains to establish the final shape (and function) of that protein.

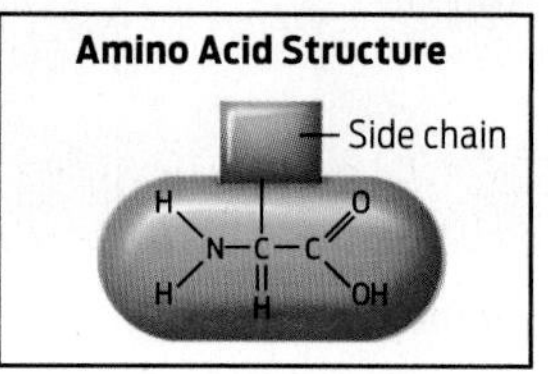

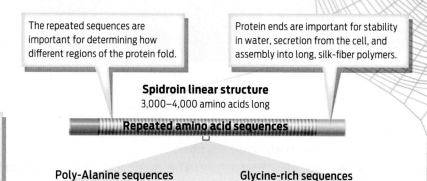

1. Linear Amino Acid Chain
Amino acids bind together in linear chains. In this linear form, a chain of amino acids does not yet have a specific function.

2. Three-Dimensional Protein Folding
Interactions among amino acid core structures, among amino acid side chains, and between side chains and water all direct three-dimensional folding. The overall shape of the protein, including placement of its side chains, determines its ultimate function.

3. Polymer Assembly
Within the protein, beta-sheet crystalline regions, which confer strength, alternate with more flexible regions, which confer flexibility. Overall, the protein is strong and flexible.

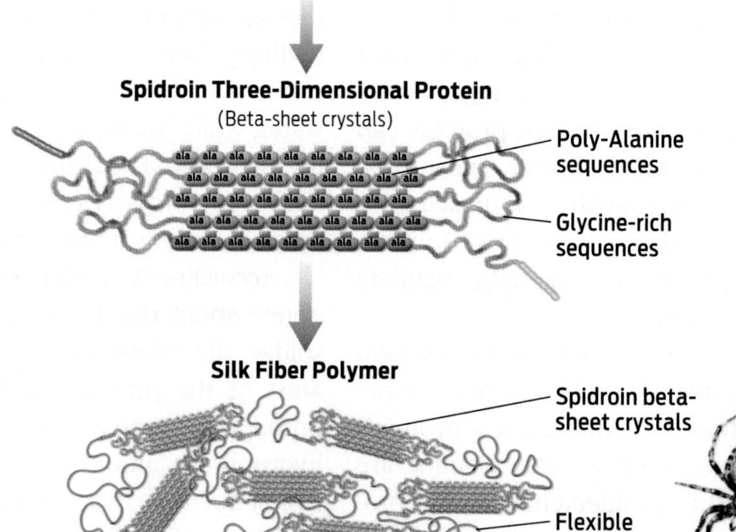

? Based on their side chains, are alanine and glycine hydrophobic or hydrophilic? How will this characteristic influence their tendency to interact with water?

of many repeating subunits. Natural polymers include spider silk, as well as other fiber-like proteins in cells, such as those making up the filaments of the cytoskeleton.

Where do spiders get the information to build such unique proteins? As with all organisms, the instructions to make proteins are encoded in the DNA, in genes. A **gene** is a sequence of DNA that provides instructions for making one or more proteins. These instructions come in the form of the particular DNA nucleotide sequence making up the gene. Genes are found along the length of chromosomes, with each specific chromosome carrying a unique set of genes.

The synthesis of a protein from the information encoded in a gene is called **gene expression.** When a cell makes the protein encoded by that gene, the gene is said to be "expressed" **(INFOGRAPHIC 8.2).**

Silk proteins are made in the spiders' silk glands. Each silk gland is connected to a microscopic "spigot" that protrudes from a larger structure called a spinneret on the rear end of the spider. Most spiders have three

GENE
A sequence of DNA that contains the information to make at least one protein.

GENE EXPRESSION
The process of using DNA instructions to make proteins.

INFOGRAPHIC 8.2 Chromosomes Include Gene Sequences That Code for Proteins

Chromosomes have many genes along their length. Each gene contains instructions to make at least one protein. Depending on the needs of the cell at any given time, each gene may be expressed (making protein) or silenced.

Chromosome
Other gene
Spidroin gene
Other gene
Gene
A section of DNA that contains a nucleotide sequence with the instructions to make at least one protein.
Protein
When a gene is "turned on," or expressed, the encoded protein is produced.
Spidroin gene expression
DNA instructions are used to produce protein.
Spidroin protein

? Why is spidroin protein being produced, but not the proteins encoded by the other genes?

A colored scanning electron micrograph (SEM) of silk fibers emerging from a spider's spinneret.

spinnerets, each with multiple spigots that extrude silk protein polymers.

Silk starts as a protein-rich liquid made by cells in the silk gland. As the liquid travels through the gland, water is gradually removed and the remaining solution containing the dissolved silk proteins is dowsed in an acid bath, which precipitates the silk proteins into a solid form. During "spinning," or thread formation, the silk proteins link up and pack together tightly, forming a silk protein polymer. This polymer is squeezed like toothpaste from the spigot as a fiber about 50 micrometers thick–roughly half the width of a human hair. Spiders can extrude silk fibers from multiple spigots simultaneously, lending additional strength to the webs and draglines they produce.

Back to Nature

Humans have used silk for thousands of years to make lightweight yet durable textiles. The silk in these textiles comes primarily from silkworms–the larvae of silk moths. During their development, silkworms produce protective cocoons made of silk, which is harvestable by humans.

In addition to gowns and neckties, silk makes good medical bandages and suturing thread because it is sturdy yet biocompatible. Until World War II, silk was also used to make military products like parachutes and flak vests. But the scarcity of the material during wartime opened the way for new materials to move into silk territory, beginning with the synthetic petroleum-based polymer called nylon.

Cheap, sturdy, and elastic, nylon is a nearly miracle fiber. DuPont introduced the fiber in 1939, and it is now found in everything from women's stockings ("nylons") to parachutes, tires, and toothbrushes. But for all its versatility, it has two very large drawbacks: (1) it's not biodegradable, which means it sticks around in the environment for a long time; (2) producing it requires nonrenewable resources like petroleum, the extraction of which contributes to climate change. The same is true of other synthetic fibers made from petroleum, such as Lycra, polyester, and acrylic.

"Manmade fabrics like nylon and polyester have transformed the fashion industry, for better and for worse," says Dan Widmaier, CEO of Bolt Threads, a San Francisco Bay Area company that is developing clothing made from genetically engineered spider silk. But, he says, "The use of hydrocarbon polymers in these textiles has created a lingering toxic problem for the environment." Although garments with synthetic fibers can be recycled, roughly 90% still end up in the waste stream. Spider silks, by contrast, are biodegradable.

A former graduate student in chemical biology at the University of California, San Francisco, Widmaier started Bolt Threads in 2009 with the help of three friends and fellow scientists. Their original goal was to make lightweight bulletproof vests that would appeal to the defense industry. But Widmaier–whose wife is a fashion designer at Old Navy–eventually decided to shift the company's focus to the consumer textile industry, which represents a much larger market and has a much greater impact on the environment.

Besides Bolt Threads, several other biotech companies are also hoping to snag a piece of the spider silk action, including Spiber Inc. in Japan and Kraig Biocraft Laboratories in Ann Arbor, Michigan.

Spider Silk Factories

Scientists have tried making genetically engineered spider silk in a variety of different organisms, including bacteria, insects, plants–even goats. (The goats can be engineered to produce the silk in their milk.) But each of these organisms has drawbacks. Silks are large proteins and can be tricky to produce in the prokaryotic cells of bacteria. Animals like goats reproduce slowly and take up a lot of space and resources. And it's hard to scale up from insects like silkworms.

Bolt Threads seeks to overcome these hurdles by relying on a different unicellular organism: yeast. Better known for its role in making bread and beer, yeasts have several attributes that make them good silk producers. For one, they are eukaryotic cells, which means their protein production machinery is much like that of a spider's (another eukaryote). But they are easier to house and cheaper to feed than more complex multicellular animals like silkworms and goats. When yeasts are fed a simple diet of sugar, they grow and divide plentifully. In the process, they synthesize new proteins, including–if they have been engineered to contain the silk gene–an abundance of silk. By conducting the entire process in large industrial vats, scientists can make literally tons of silk protein this way.

The process of making genetically engineered silk begins with isolating spider DNA from spider cells or synthesizing it from scratch. This bit of DNA is then inserted into a cell of a new organism, which is then coaxed to make the protein encoded by the spider gene. Organisms that have been genetically modified to contain genes from other species are called **transgenic** ("trans" means "across"–in this case, across species, from one to another).

The method that scientists use to make a transgenic organism relies on the fact that every gene has two parts: a regulatory sequence and a coding sequence. **Regulatory sequences** are like on-off switches for genes: they determine when, where, and how much protein is produced from a gene. **Coding sequences** determine the identity of a protein: they specify the order, or sequence, of amino acids **(INFOGRAPHIC 8.3)**.

By combining the regulatory sequence of one species with the coding sequence of another, scientists can coax an unrelated organism to make the desired protein. To make a transgenic yeast, for example, scientists first fuse the coding sequence of a spider silk gene to the regulatory sequence of a yeast gene. The combination is called a **recombinant gene,** since it mixes and matches segments of genes that weren't naturally found together. Next, using **genetic engineering** techniques, which manipulate DNA, scientists insert the recombinant gene into a piece of

TRANSGENIC
Refers to an organism that carries one or more genes from a different species.

REGULATORY SEQUENCE
The part of a gene that determines the timing, amount, and location of protein production.

CODING SEQUENCE
The part of a gene that specifies the amino acid sequence of a protein. Coding sequences determine the identity, shape, and function of proteins.

RECOMBINANT GENE
A genetically engineered gene that contains portions of genes not naturally found together.

GENETIC ENGINEERING
Altering or manipulating the DNA of organisms by modern laboratory techniques.

INFOGRAPHIC 8.3 The Two Parts of a Gene

Genes are organized into two parts. Regulatory sequences determine when a protein is made from a gene and in which cells, and how much protein a gene makes. Coding sequences determine the amino acid sequence of the encoded protein, which determines its shape and function.

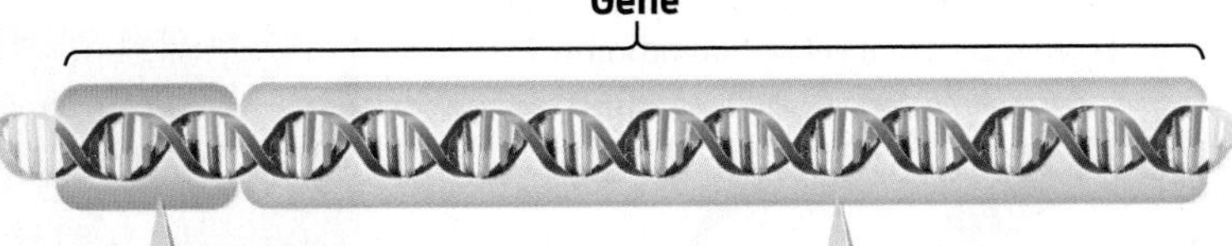

Regulatory Sequence
Controls the timing, location, and amount of gene expression

Coding Sequence
Determines the sequence of amino acids in the protein

? Why does a milk protein gene (expressed in mammary glands) have both a different regulatory sequence and a different coding sequence from the insulin gene (expressed in the pancreas)?

VECTOR
A DNA molecule used to deliver a recombinant gene to a host cell.

DNA that can carry the recombinant gene into the yeast cell, and ultimately into a yeast chromosome. The carrier DNA molecule is called a **vector.** The final step is gene expression, when the yeast protein machinery "reads" the instructions in the recombinant gene and synthesizes spider silk protein **(INFOGRAPHIC 8.4).**

As the yeast grow, they make abundant quantities of silk protein, which they secrete into the surrounding culture media. The scientists then harvest these proteins. Once the silk proteins are in hand, the next step in the manufacturing process is spinning them into fibers, using a process that mimics what happens naturally inside spiders. The wet silk

INFOGRAPHIC 8.4 Making a Transgenic Organism

Transgenic organisms contain genes from other organisms. In order for the foreign gene to be expressed in the new host, it needs to be modified. The modified recombinant gene contains a regulatory sequence from the host organism and the coding sequence from the gene of interest. The recombinant gene is then inserted into and expressed in the host.

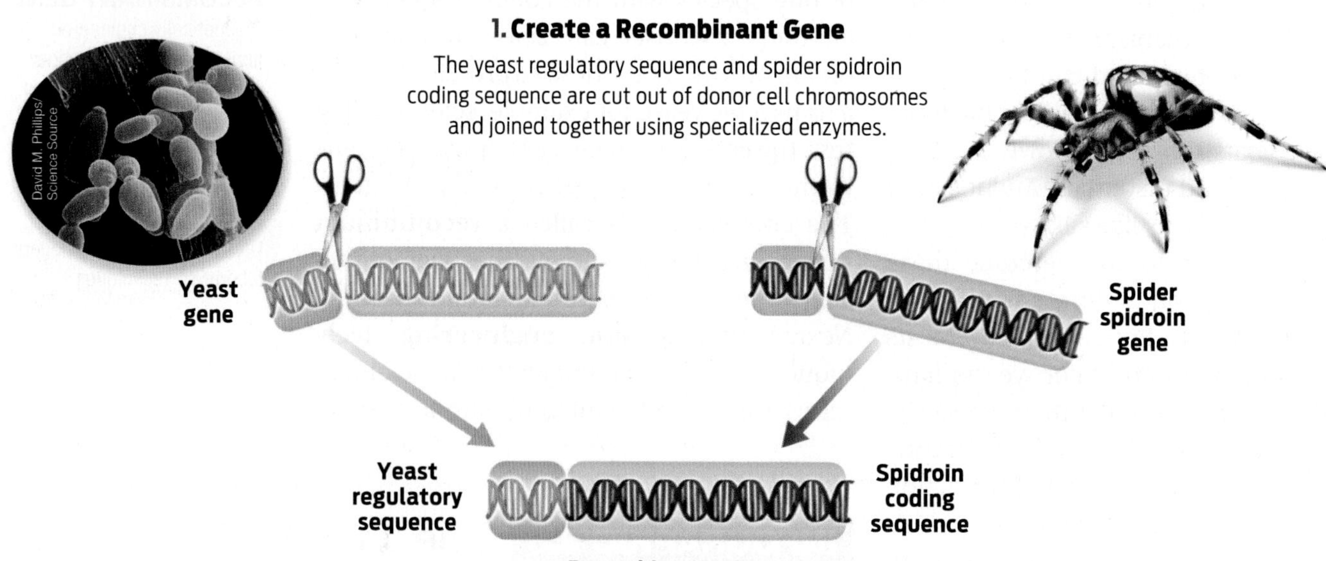

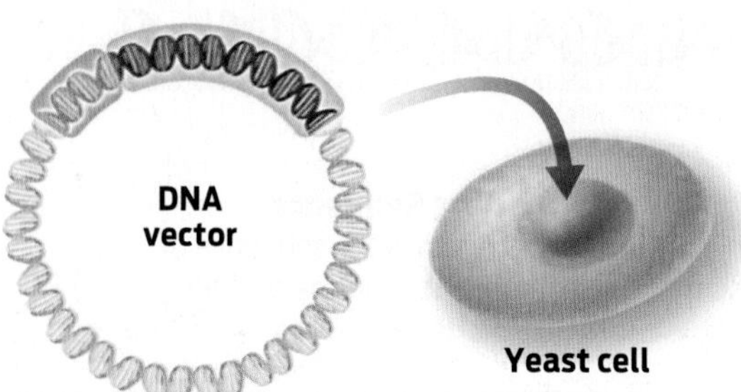

2. Insert the Recombinant Gene into a Yeast Chromosome
The recombinant gene is inserted into a small piece of DNA. This recombinant DNA is added to yeast nuclei where it becomes part of a yeast chromosome.
DNA vector
Yeast cell

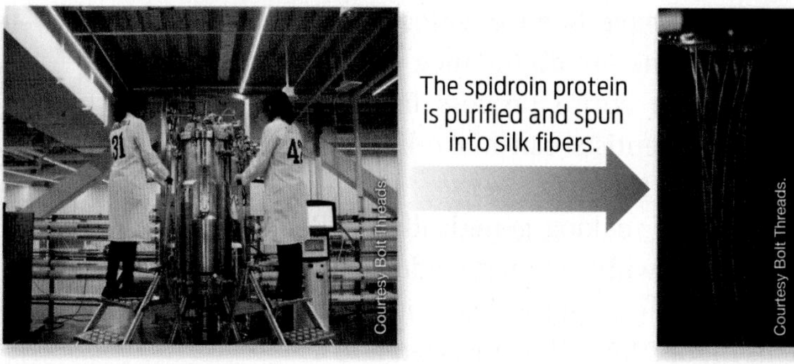

3. Spidroin Secretion and Harvest
Yeast cells are grown in a large tank under conditions that allow them to express and secrete spidroin protein.
Courtesy Bolt Threads.
The spidroin protein is purified and spun into silk fibers.
Courtesy Bolt Threads.

? Why did the scientists give the recombinant spidroin gene a yeast regulatory sequence?

protein solution produced by yeast is squeezed through the small holes of a metal container into a liquid acid bath that transforms the liquid proteins into solid fibers. The fibers are then collected and can be woven together to make fabrics.

Making Proteins, or How Genes Are Expressed

Whether produced inside a spider's abdomen or inside yeast grown in steel vats, all spider silk proteins are made from silk genes in essentially the same way.

In order to get from a gene to a protein, cells carry out two major steps: transcription and translation. **Transcription** is the process of using DNA to make a **messenger RNA (mRNA)** copy of the gene. **Translation** is the process of using this mRNA copy as a set of instructions to assemble amino acids into a protein **(INFOGRAPHIC 8.5)**.

Why two separate steps? As the names "transcription" and "translation" imply, the

TRANSCRIPTION
The first stage of gene expression, during which cells produce molecules of messenger RNA (mRNA) from the instructions encoded within genes in DNA.

MESSENGER RNA (mRNA)
The RNA copy of an original DNA sequence made during transcription.

INFOGRAPHIC 8.5 Gene Expression: An Overview

Gene expression is the process of converting the genetic information of DNA into the amino acid sequence of a protein. Gene expression has two main steps: transcription and translation.

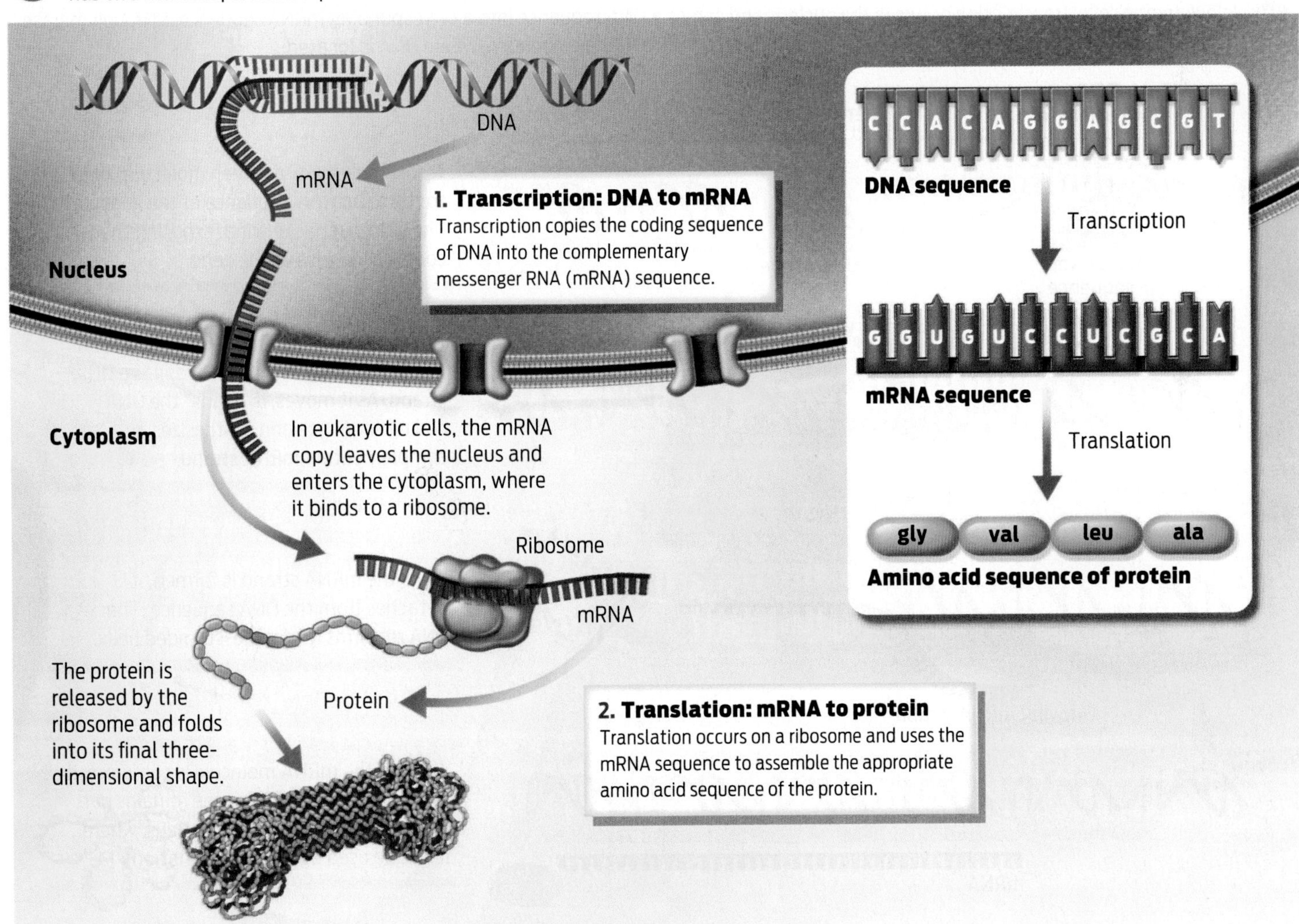

DNA
mRNA
1. Transcription: DNA to mRNA
Transcription copies the coding sequence of DNA into the complementary messenger RNA (mRNA) sequence.
Nucleus
Cytoplasm
In eukaryotic cells, the mRNA copy leaves the nucleus and enters the cytoplasm, where it binds to a ribosome.
Ribosome
mRNA
The protein is released by the ribosome and folds into its final three-dimensional shape.
Protein
2. Translation: mRNA to protein
Translation occurs on a ribosome and uses the mRNA sequence to assemble the appropriate amino acid sequence of the protein.
C C A C A G G A G C G T
DNA sequence
Transcription
G G U G U C C U C G C A
mRNA sequence
Translation
gly val leu ala
Amino acid sequence of protein

? What are the products of transcription and translation?

TRANSLATION
The second stage of gene expression, during which mRNA sequences are used to assemble the corresponding amino acids to make a protein.

RNA POLYMERASE
The enzyme that carries out transcription. RNA polymerase copies a strand of DNA into a complementary strand of mRNA.

process of gene expression is like copying a text and then converting it into another language. In this case, the text to be translated is a valuable, one-of-a-kind document: DNA. Just as you would be forbidden to borrow a rare manuscript from the library at school and would instead have to copy the text into your notebook or laptop, the cell cannot take DNA out of its "library"–the nucleus. It must first make a copy–the mRNA. The cell can then take this mRNA copy into the cytoplasm, where it is translated into a new language: protein.

Transcription begins in the nucleus of a cell when an enzyme called **RNA polymerase** binds to DNA at a gene's regulatory sequence, located just ahead of the coding sequence. At that site, RNA polymerase separates the two strands of the DNA double helix and begins moving along one DNA strand. As it moves, the RNA polymerase reads the DNA sequence and synthesizes a complementary mRNA strand according to the rules of base pairing. The same rules that govern DNA base pairing apply here, with one difference: RNA nucleotides are made with the base uracil (U) instead of thymine (T). So the complementary base pairs are C with G and A with U **(INFOGRAPHIC 8.6)**.

As its name implies, messenger RNA serves to relay information. Once the mRNA

INFOGRAPHIC 8.6 Transcription: A Closer Look

In eukaryotic cells, transcription occurs in the nucleus and copies a DNA sequence into a corresponding mRNA sequence. RNA polymerase is the key enzyme involved. In prokaryotic cells, transcription occurs in the cytoplasm, where DNA is located.

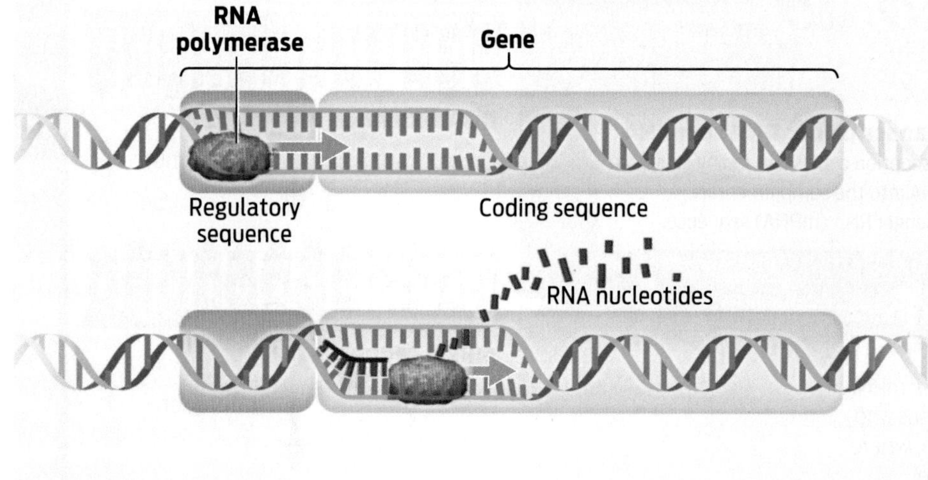

1. RNA polymerase (green molecule) binds to the regulatory sequence of the gene. The DNA strands unwind, exposing the coding sequence of the gene.

2. RNA polymerase moves along the DNA strand. As it moves, it "reads" the DNA coding sequence and synthesizes a complementary mRNA strand.

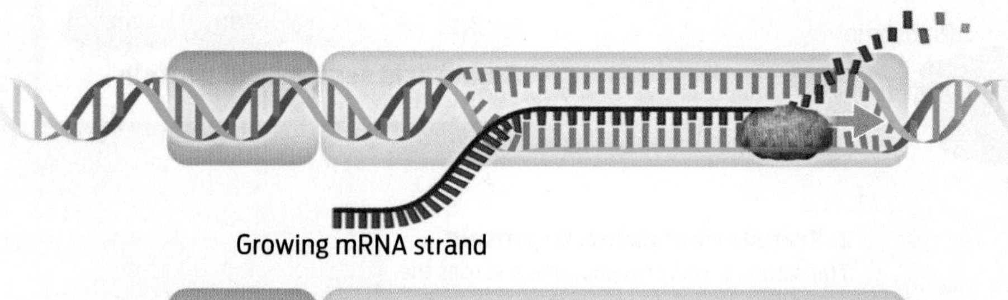

3. As the mRNA strand is formed, it detaches from the DNA sequence. The DNA re-forms its double-stranded helix.

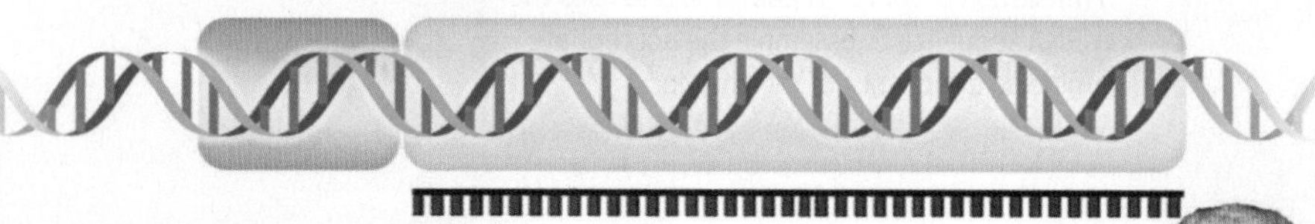

4. Once the mRNA molecule is complete, it leaves the nucleus. The gene remains part of the chromosome in the nucleus, where it can be used again in transcription.

? What enzyme is responsible for transcription, and what, specifically, does it do?

copy, or transcript, is made, it leaves the nucleus and attaches to a complex piece of cellular machinery in the cytoplasm called the ribosome. This is the start of translation.

During translation, the **ribosome** reads the mRNA transcript and translates it into a chain of amino acids. The sequence of nucleotides in the mRNA transcript specifies which amino acids should be joined together in the newly forming protein chain. Each of the 20 amino acids that occur in proteins is specified by a group of three mRNA nucleotides called a **codon** that functions like a word: for example, the codon GGU specifies the amino acid glycine.

The actual building blocks of proteins—amino acids—are physically delivered to the ribosome by another type of RNA, called **transfer RNA (tRNA).** Each tRNA molecule serves as a kind of adaptor, with one end binding to a specific amino acid and the other end binding to the mRNA codon for that particular amino acid. The part that binds mRNA is called the **anticodon** because it base-pairs in a complementary fashion with the mRNA codon. When the amino acid-toting tRNA finds its mRNA codon match, it releases the amino acid to the ribosome, which adds it to the growing protein chain (**INFOGRAPHIC 8.7**).

The vast majority of mRNA codons specify a specific amino acid, but there are a few with other functions. The "start codon," which in eukaryotes codes for methionine, is the first

RIBOSOME
The cellular machinery that assembles proteins during translation.

CODON
A sequence of three mRNA nucleotides that specifies a particular amino acid.

TRANSFER RNA (tRNA)
A type of RNA that transports amino acids to the ribosome during translation.

ANTICODON
The part of a tRNA molecule that binds to a complementary mRNA codon.

INFOGRAPHIC 8.7 Translation: A Closer Look

In the cytoplasm, the ribosome reads the mRNA sequence and translates it into a chain of amino acids to make a protein.

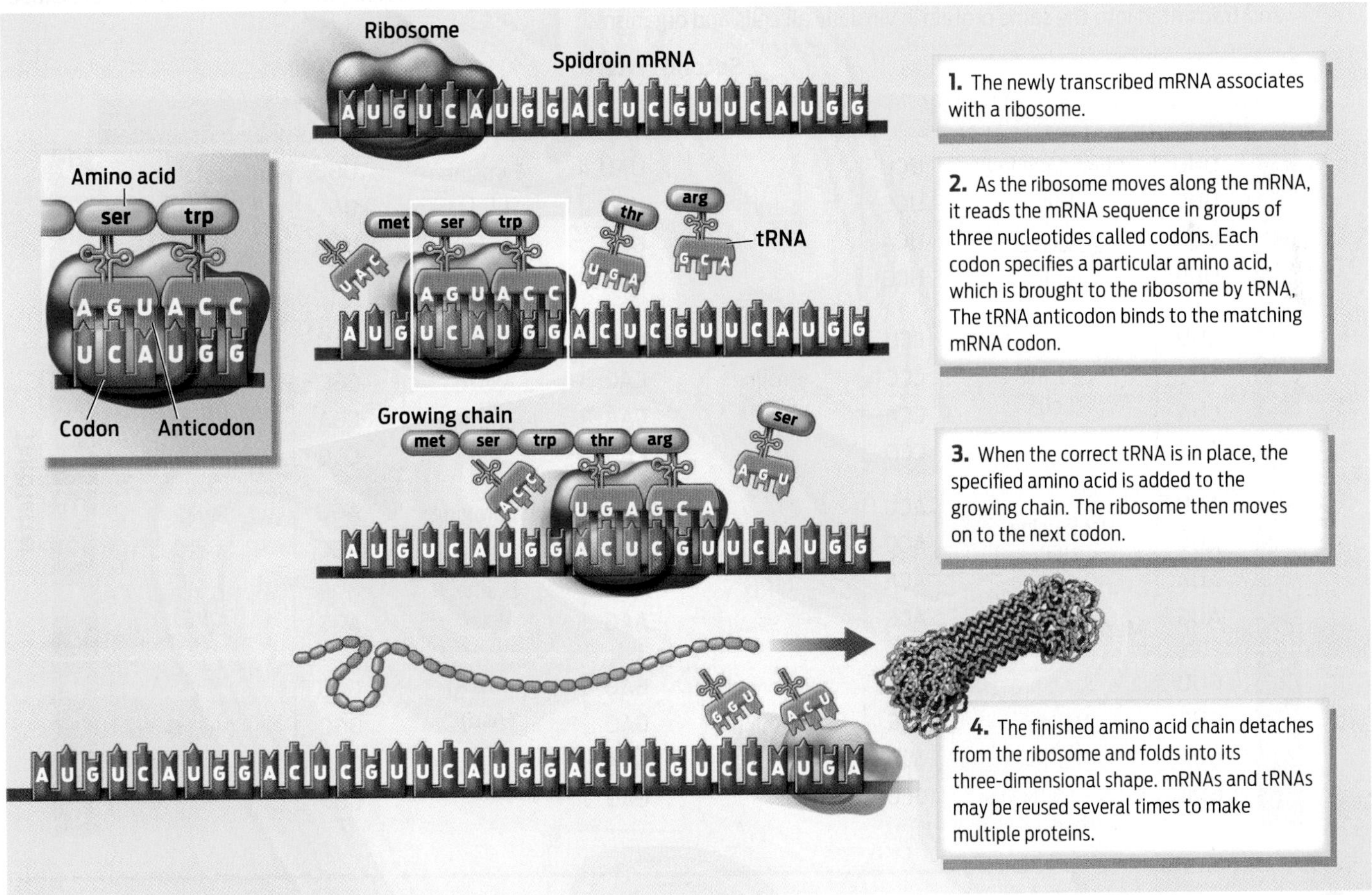

1. The newly transcribed mRNA associates with a ribosome.

2. As the ribosome moves along the mRNA, it reads the mRNA sequence in groups of three nucleotides called codons. Each codon specifies a particular amino acid, which is brought to the ribosome by tRNA. The tRNA anticodon binds to the matching mRNA codon.

3. When the correct tRNA is in place, the specified amino acid is added to the growing chain. The ribosome then moves on to the next codon.

4. The finished amino acid chain detaches from the ribosome and folds into its three-dimensional shape. mRNAs and tRNAs may be reused several times to make multiple proteins.

? What molecule reads the mRNA codons and brings amino acids to the growing protein? How does it do this?

GENETIC CODE
The set of rules relating particular mRNA codons to particular amino acids.

codon of a coding sequence; it tells the ribosome to start translating and begin adding amino acids. "Stop codons" (there are three) tell the ribosome to stop translating and not add any more amino acids to the growing chain.

Although the human genome encodes many thousands of different proteins, each one is pieced together from the starting set of just 20 amino acids. In the same way that the 26 letters in our alphabet can spell hundreds of thousands of words, the basic set of amino acids can make hundreds of thousands of proteins. The set of rules dictating which mRNA codons specify which amino acid is called the **genetic code.** Scientists have pieced together this code by systematically studying how changes to the letters of a codon alter the specified amino acid, so that we now know what amino acid each codon stands for. Each codon specifies one and only amino acid.

Two additional features of the genetic code stand out. (1) The code is redundant: multiple codons specify the same amino acid. In many cases, a codon will differ at the third nucleotide position without changing the amino acid that is specified. (Note that while the code is redundant it is not ambiguous–the same codon will not specify more than one amino acid.) (2) The genetic code is universal, which means that it is the same in all living organisms. It is because the code is universal that a yeast cell carrying a spider gene can express that gene and produce spider silk **(INFOGRAPHIC 8.8)**.

INFOGRAPHIC 8.8 The Genetic Code Is Universal

Codons are three-nucleotide sequences within chains of mRNA. Most codons specify a particular amino acid. One codon specifies where to start translation (start codon) and others specify where to end (stop codons). There is redundancy in the genetic code, as 64 possible codons code for only 20 different amino acids. Since the genetic code is universal, the same gene will be transcribed and translated into the same protein in virtually all cells and organisms.

Second letter

First letter	U		C		A		G		Third letter
U	UUU	Phenylalanine (Phe)	UCU	Serine (Ser)	UAU	Tyrosine (Tyr)	UGU	Cysteine (Cys)	U
	UUC		UCC		UAC		UGC		C
	UUA	Leucine (Leu)	UCA		UAA	Stop	UGA	Stop	A
	UUG		UCG		UAG	Stop	UGG	Tryptophan(Trp)	G
C	CUU	Leucine (Leu)	CCU	Proline (Pro)	CAU	Histidine (His)	CGU	Arginine (Arg)	U
	CUC		CCC		CAC		CGC		C
	CUA		CCA		CAA	Glutamine (Gln)	CGA		A
	CUG		CCG		CAG		CGG		G
A	AUU	Isoleucine (Ile)	ACU	Threonine (Thr)	AAU	Asparagine (Asn)	AGU	Serine (Ser)	U
	AUC		ACC		AAC		AGC		C
	AUA		ACA		AAA	Lysine (Lys)	AGA	Arginine (Arg)	A
	AUG	Start Methionine (Met)	ACG		AAG		AGG		G
G	GUU	Valine (Val)	GCU	Alanine (Ala)	GAU	Aspartic acid (Asp)	GGU	Glycine (Gly)	U
	GUC		GCC		GAC		GGC		C
	GUA		GCA		GAA	Glutamic acid (Glu)	GGA		A
	GUG		GCG		GAG		GGG		G

? What is the sequence of the start codon, and what amino acid does it specify?

By tweaking the specific sequence of nucleotides in the spider genes they introduce into yeast, scientists can produce spider proteins with unique amino acid sequences and therefore unique properties–perhaps sturdier than the native form, or stickier, or more elastic. With genetic engineering techniques, the possibilities to produce designer proteins are nearly endless.

Brave New World?

Bolt Threads announced in May 2016 that it was partnering with clothing manufacture Patagonia to develop a new line of eco-friendly clothing, which it plans to bring to market in the coming years. The Japan-based company Spiber, which has partnered with North Face, unveiled a winter parka in 2016, making it the first clothing made from genetically engineered spider silk to be sold in stores. (It currently costs $1,000).

The process of gene expression is like copying a text and then converting it into another language.

Yet these are just the tip of the iceberg when it comes to spider silk products in development. "Nature provides a great starting point and then once we're outside of the spider we can expand the set of materials and uses," says bioengineer Kaplan. His lab at Tufts University is focused on medical applications–like genetically engineered spider fibers that can substitute for bone and ligaments.

Other researchers have their eyes set on industrial uses, like a less bruising material for car airbags. Eventually, the goal is to make superstrength products that substitute spider silk for Kevlar, such as in bulletproof vests. Even superhuman tissues might not be out of the question. In 2013, Dutch researchers grew human skin cells together with spider silk proteins to make what they call "bulletproof skin," which can stop a bullet traveling at half the normal speed of a typical bullet.

Genetically engineered spider silks capture headlines, but other uses of genetic engineering are already common in daily life. Much of the corn we eat today is transgenic, as are the soybeans that we feed to farm animals. Transgenic organisms are examples of **genetically modified organisms (GMOs)**–organisms whose genomes have been altered through modern genetic engineering techniques, sometimes to contain new genes. Transgenic crops such as corn and soybeans usually contain genes for natural pesticides, which help the plants fight pests and reduce the amount of pesticide a farmer must use. Other varieties of GM crops contain herbicide-resistance genes, allowing farmers to spray herbicides on fields to kill weeds without at the same time killing the crops (see Chapter 24).

Genetic engineering also has important medical applications. The drug insulin, used for treating diabetes, is commonly produced inside a genetically engineered bacterium–one into which the (human) insulin gene has been inserted. Through **gene therapy**–replacing a defective human gene with a healthy one–scientists hope to one day be able to treat, cure, or even prevent several inherited genetic disorders, including cystic fibrosis, Huntington's disease, and hemophilia.

Despite the many actual and potential benefits of genetic engineering, the practice inspires debate among scientists, environmentalists, and the general public alike. Some groups object to humans' meddling with the biology of organisms that have evolved naturally because they are afraid that eating GMOs might have negative effects on health. Others worry about the consequences to our environment if, for example, pesticide genes were to spread in natural populations. And for many, the idea of tampering with human genes to build "better" people raises the specter of eugenics–the early twentieth century practice of trying to weed out the "unfit" from society.

GENETICALLY MODIFIED ORGANISM (GMO)
An organism whose genome has been altered through genetic engineering techniques, often to contain a gene from another species.

GENE THERAPY
A treatment that aims to cure, treat, or prevent human disease by replacing defective genes with functional ones.

Spiber, Inc.

Moon Parka™, a prototype by Spiber and The North Face Japan, is the first apparel product to integrate synthetic spider silk with existing industrial manufacturing technology, and is made using Spiber's Qmonos™ fiber.

Those who may find nothing ethically troubling about using gene therapy to treat human diseases might nonetheless find the prospect of using it to change physical appearance or psychological traits like sexual orientation much more problematic. Even in the case of treating disease, using gene therapy to modify human embryos or germ cells is more troubling to some than using it to treat adults with a medical condition. That's because an adult person can consent to the procedure, which will affect only his or her cells; changes to an embryo or germ cell will be inherited, affecting future generations who did not consent.

Disquieting or not, genetic engineering appears to be speeding ahead. In 2015, a group of researchers in China reported that they had used a new genetic engineering technique called CRISPR (see Chapter 24) to edit the genome of a developing human embryo. The group used the technique, which can very precisely change DNA sequences, to repair a faulty gene that causes a rare blood disorder. They were roundly criticized by scientists in the West, who have urged caution when applying such gene editing techniques to humans. Although the United States does not officially ban the editing of DNA in human embryos, the National Institutes of Health, which funds most biomedical research in this country, currently prohibits it.

Charged with both hope and fear, debates about the ethical use of genetic engineering are unlikely to go away any time soon. And, given the rapid progress being made in the related field of bioengineering, opportunities to remake our world—perhaps even ourselves—are sure to proliferate. But it will be at least a few years before Spiderman has any real competition. ■

CHAPTER 8 SUMMARY

- Proteins are folded chains of amino acids that perform many functions in cells, such as transmitting signals, catalyzing chemical reactions, and generating force for movement.
- The order and identity of amino acids in a protein chain determine the shape and function of the protein.
- Genes provide instructions to make proteins. The process of using the information in genes to make proteins is called gene expression.
- Every gene has two parts: a coding sequence and a regulatory sequence. The coding sequence determines the identity of a protein; the regulatory sequence determines where, when, and how much of the protein is produced.
- Gene expression occurs in two stages, transcription and translation, which take place in separate compartments in eukaryotic cells.
- Transcription is the first step of gene expression, copying the information stored in DNA into mRNA. Transcription occurs in the nucleus.
- Translation, the second step of gene expression, uses the information carried in mRNA to assemble a protein. Translation occurs in the cytoplasm.
- Proteins are assembled by ribosomes with the help of tRNA, which delivers amino acids to the ribosome.
- The genetic code is the set of rules by which mRNA sequences are translated into protein sequences; the code is redundant and universal—shared by all living organisms.
- Through genetic engineering, genes from one species of organism can be inserted into the genome of another species of organism to make a transgenic organism.
- Transgenic organisms have numerous uses in biotechnology and health.

MORE TO EXPLORE

- **Hayashi, C. (2010) TED Talk: The magnificence of spider silk: https://www.ted.com/talks/cheryl_hayashi_the_magnificence_of_spider_silk**
- **Tokareva, O., et al. (2013) Recombinant DNA production of spider silk proteins. *Microbial Biotechnology* 6(6), 651–663.**
- **Center for Genetics and Society: http://www.geneticsandsociety.org**
- ***Nature* Special Report (2015): CRISPR: The Good, the Bad and the Unknown. http://www.nature.com/news/crispr-1.17547**
- **Kevles, D. J. (1995) *In the Name of Eugenics: Genetics and the Uses of Human Heredity.* Cambridge: Harvard University Press.**

CHAPTER 8 Test Your Knowledge

DRIVING QUESTION 1 What determines the shape of a protein molecule?

By answering the questions below and studying Infographic 8.1, you should be able to generate an answer for the broader Driving Question above.

KNOW IT

1 A protein is made up of a chain of _____.

a. nucleotides
b. amino acids
c. lipids
d. fatty acids
e. simple sugars

2 What determines a protein's function?

a. the sequence of amino acids
b. the three-dimensional shape of the folded protein
c. the location of its gene on the chromosome
d. all of the above
e. a and b

USE IT

3 Spidroin proteins are in an unfolded state in the spider's silk gland before they are extruded through the spinneret. In their unfolded state, will they have the same properties as spider fibers in a web? Explain your answer.

4 If the repeated alanines in spidroin were changed to amino acids with hydrophilic side chains, would they still cluster together away from water? Explain your answer.

DRIVING QUESTION 2 What are the steps of gene expression, and where in the cell do they occur?

By answering the questions below and studying Infographics 8.2, 8.3, 8.5, 8.6, 8.7 and 8.8, you should be able to generate an answer for the broader Driving Question above.

KNOW IT

5 "A gene contains many chromosomes. Each chromosome encodes a protein." Is this statement accurate? If not, explain why not, and rewrite the statement to make it correct.

6 What is the final product of gene expression?

a. a DNA molecule
b. an RNA molecule
c. a protein
d. a ribosome
e. an amino acid

7 For each structure or enzyme listed, indicate by N (nucleus) or C (cytoplasm) where it acts in the process of gene expression in a eukaryotic cell.

_____ RNA polymerase
_____ Ribosome
_____ tRNA
_____ mRNA

8 What is encoded by a single codon?

a. a single protein
b. an RNA nucleotide
c. a DNA nucleotide
d. an amino acid
e. any of the above, depending on the organism

9 A gene has the sequence ATCGATTG. What is the sequence of the complementary RNA?

a. ATCGATTG
b. AGCTAAC
c. GTTAGCTA
d. UAGCUAAC
e. CAAUCGAU

USE IT

10 If a spider wasn't making the normal amount of its spidroin protein, would you suspect a problem in the regulatory or coding sequence of the spidroin gene? Explain your answer.

11 If you wanted to try to increase the amount of spidroin protein a spider produces, would you modify the regulatory sequence or the coding sequence? Explain your answer.

12 A change in DNA sequence can affect gene expression and protein function. What would be the impact of each of the following changes? How, specifically, would each change affect protein or mRNA structure, function, and levels?

a. a change that prevents RNA polymerase from binding to a gene's regulatory sequence
b. a change in the coding sequence that changes the amino acid sequence of the protein
c. a change in the regulatory sequence that allows transcription to occur at much higher levels
d. a combination of the changes in b and c

13 The insulin gene is normally expressed in specific cells in the pancreas, but not in a type of immune cell known as a B cell. On the other hand, B cells express large amounts of antibody proteins. What would you have to do get a B cell to express insulin? (Hint: Remember that all cells in an organism have the same set of chromosomes and associated genes.)

DRIVING QUESTION 3 How can organisms be genetically modified to produce recombinant proteins?

By answering the questions below and studying Infographic 8.4, you should be able to generate an answer for the broader Driving Question above.

KNOW IT

14 Why is recombinant protein production in yeast an efficient strategy?

a. because yeast can easily be grown in large quantities
b. because yeast can secrete large amounts of recombinant proteins into their growth medium
c. because yeast are multicellular, so have a variety of cell types for recombinant gene expression
d. all of the above
e. a and b

15 **What is the purpose of the vector in generating a transgenic organism?**

16 **Describe the recombinant gene that would be needed to create a transgenic spider that produces a yeast protein in its silk glands.**

USE IT

17 **Why is it important that the transgenic yeast expressing recombinant spidroin proteins secrete the protein into their culture (growth) medium? (Hint: What has to happen to spidroin to convert it into actual silk?)**

18 **Melanin is a pigment expressed in skin cells; melanin gives skin its color. If you wanted to express a different gene in skin cells, which part of the melanin gene would you use? Why? If you wanted to produce melanin in yeast cells, what part of the melanin gene would you use? Why?**

19 **Lysozyme is a protein secreted in tears and saliva in all mammals. Amylase is a protein secreted in mammalian saliva.**

- **a.** Describe the recombinant gene that you would assemble to express recombinant human lysozyme in the tears of goats.
- **b.** Describe the recombinant gene that you would assemble to express recombinant human amylase in goat saliva.

DRIVING QUESTION 4 What are some pros and cons of genetically modified organisms?

By answering the questions below and studying Infographic 8.4, you should be able to generate an answer for the broader Driving Question above.

KNOW IT

20 **Why is transgenic technology needed to produce large quantities of spider silk?**

21 **Why is spider silk such a valuable product?**

USE IT

22 **Type 1 diabetes results from a loss of insulin production from the pancreas. People with diabetes take recombinant human insulin expressed in bacteria.**

- **a.** Describe the gene construct necessary for expression of human insulin in bacteria.
- **b.** Describe the gene construct necessary to produce human insulin in goat's milk.
- **c.** If you were to attempt gene therapy (genetically modifying the human's genome so that insulin would be produced in the human's pancreas), would you need a recombinant form of the insulin gene? Explain your answer.

apply YOUR KNOWLEDGE

INTERPRETING DATA

23 **A biotechnology company has created a number of strains of transgenic yeast with a recombinant spidroin gene. Each strain is grown in 1 L of culture medium. The cells are separated from the culture medium. All of the spidroin protein present in the culture medium is isolated and quantified. Similarly, all the cells are lysed (broken open) and all the spidroin present within the cells in quantified. The results are shown in the table below.**

Strain	Spidroin protein isolated from cells (mg)	Spidroin protein present in culture medium (mg)
Nontransgenic yeast	0	0
Transgenic strain 1	80	0
Transgenic strain 2	40	40
Transgenic strain 3	0	80
Transgenic strain 4	60	20

- **a.** How much protein (total) is being produced by each strain in 1 L of culture?
- **b.** What are the differences in spidroin production in the different strains?
- **c.** Which strain should the company use to commercialize spidroin production? Explain your answer,

apply YOUR KNOWLEDGE

MINI CASE

24 **A physician is stationed in a military hospital in Iraq. She often has to deal with severe wounds caused by sniper shots. Infection is always a concern, and current bandages are not always flexible enough to permit movement of the affected area as the wound heals. Often, the treated soldiers need to return to duty before their stitches are ready to be removed.**

Given the scenario described, what case could a spider silk biotechnology company representative make to the army to support spider silk research?

apply YOUR KNOWLEDGE

BRING IT HOME

25 **A number of concerns have been expressed about GMOs. Search the Internet for reliable sources about a particular GMO that you have heard of or in which you are interested (e.g., Golden Rice or genetically modified salmon). List what you consider to be the pros and cons of at least two GMOs. Has what you have read in this chapter about other genetically modified organisms and the transgenic yeast changed your opinion about GMOs? What restrictions (if any) would you place on GMOs?**

3 Milestones in Biology

SEQUENCE SPRINT

Venter and Collins race to decode the human genome

DRIVING QUESTIONS

1. Why is knowing the sequence of the human genome important?
2. What were some of the challenges and surprises associated with sequencing the human genome?

Some simply deemed the task impossible, given the state of sequencing technology at the time.

IT STARTED WITH A VISIONARY IDEA: the entire sequence of human DNA spelled out, nucleotide by nucleotide, to be read like a book. This epic text would be an indispensable medical tool. Scientists could, for example, scan the genome for genes that confer susceptibility to disease, which might lead to better treatments. It would enable diagnostic tests that could help predict the risk of developing certain diseases. Fields other than medicine would benefit, too. Comparing the human genome to the genomes of other organisms, for example, would shed light on our own evolution. The possible benefits to science were endless.

But when an international group of scientists met in the early 1980s and first floated the idea of sequencing the human genome, they were met with skepticism. Some found the idea absurd, especially given its then-estimated $3 billion price tag. Others thought the project a waste of time, and that efforts should be placed elsewhere. Some simply deemed the task impossible, given the state of sequencing technology at the time.

Over the course of the decade, however, as the fields of molecular biology and genetics progressed, the idea gained both scientific and political support. In 1988, Congress funded both the National Institutes of Health (NIH) and the U.S. Department of Energy to explore the novel concept. By 1990, the collaborative effort to sequence the entire string of more than 3 billion As, Ts, Gs, and Cs in a human genome—the Human Genome Project (HGP)—was officially under way.

Map First, Sequence Later

When the project began, the NIH appointed James Watson, the co-discoverer of the structure of DNA, to head and coordinate it. U.S.-based labs led the mammoth undertaking, receiving valuable assistance from partner labs in the United Kingdom, France, Germany, China, and Japan.

Initially, the researchers set about constructing a map of genes, determining where among the 23 human chromosomes a particular gene sat. With these landmarks in place, it would be much easier to eventually sequence the stretches of DNA between them. This approach, dubbed "map first, sequence later," reflected the limited capacity of DNA sequencing technology at the time. In the early 1990s, a DNA sequencing lab could sequence only about 500 nucleotides a day, working around the clock—a pace much too slow to meet the 15-year deadline that project leaders set for themselves. Developing a map of the genome would save time later, while allowing sequencing technology to progress in the meantime (**INFOGRAPHIC M3.1**).

As the map of genes grew, and as genes were assigned to specific chromosomal locations, it became much easier for scientists to pinpoint the locations of genes they were studying. Since the information was uploaded to a public online database, any scientist could view the map to see where exactly his or her gene of interest was located and apply that information to further research. By 1997, a map of 16,000 genes had been assembled, including genes for Parkinson's disease, prostate cancer, breast cancer, and Huntington's disease. But criticism began to mount that sequencing had lagged behind: only about 3% of the genome had been actually sequenced at that point.

One particular scientist was especially frustrated with the slow pace of sequencing. Craig Venter was a surfer-turned-scientist who was working at NIH. He had earlier proposed an alternate sequencing approach that relied on machines and computers to identify genes. Using this method, his lab had already identified hundreds of potential gene sequences. But Watson, head of the genome project, dismissed the method as sloppy and inadequate. At one point, he even called Venter's work "brainless," and said that his operation "could be run by monkeys." Angered by his rejection and lured by a hefty salary, Venter left NIH in 1992 to found his own investment-backed institute, called The Institute for Genome Research (TIGR).

Meanwhile, also in 1992, Watson resigned as head of the HGP after a disagreement with

David Parker/Science Source

Large-scale DNA sequencing at the Sanger Center in 1999.

INFOGRAPHIC M3.1 DNA Sequencing: A Primer

One method of DNA sequencing is known as "chain termination," or Sanger sequencing. In this method, fluorescent nucleotides, each of the four bases a unique color, are incorporated into an in vitro DNA synthesis reaction. When one of these nucleotides is incorporated into a growing DNA strand, the strand terminates with that fluorescent nucleotide and color. The result is a mixture of strands, each of a different length, and each ending with a particular fluorescent nucleotide. By reading the nucleotide at the end of successively longer strands, the complete sequence can be determined.

1. In vitro DNA synthesis is used to make a complement to one strand of the molecule to be sequenced.

DNA polymerase

5'– T G G G C T A A C A A G C A A T C G A A G T C

3'– A C C C G A T T G T T C G T T A G C T T C A G T C A A A T T G T C T –5'

One strand of a DNA molecule of unknown sequence

2. Newly synthesized strands will terminate at different points, generating strands of different lengths. Each one ends with a fluorescently tagged nucleotide.

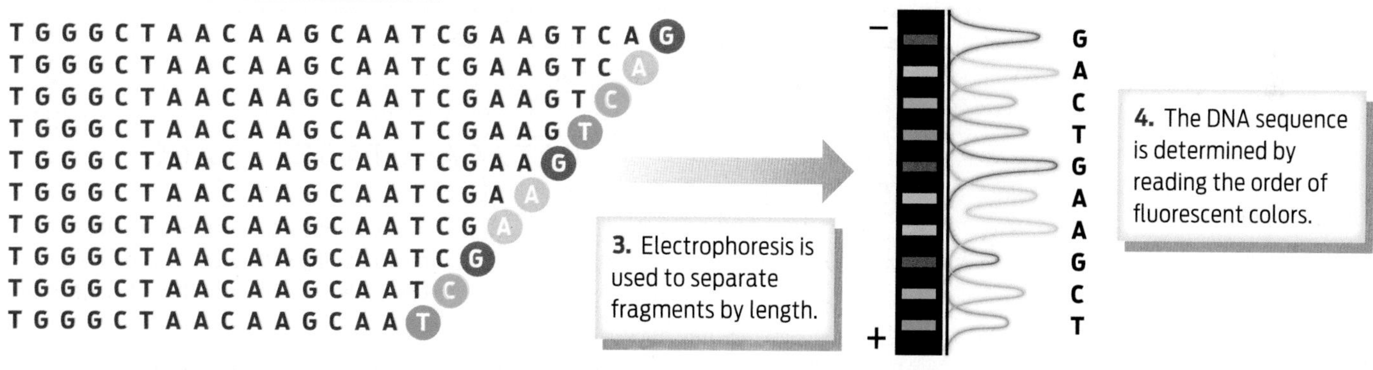

3. Electrophoresis is used to separate fragments by length.

4. The DNA sequence is determined by reading the order of fluorescent colors.

? Look at the sequence of fluorescent nucleotides generated so far, and look at the sequence of the original DNA molecule. What will be the nucleotide at the end of the next-longer strand?

the NIH director over the issue of patenting genes (Watson was opposed). In early 1993, NIH appointed a new head, the geneticist Francis Collins. The idea of speeding up the project by using Venter's shortcut didn't sit well with Collins, either, and the decision was made to stay the course.

Then, in 1995, Venter shocked the scientific community by successfully sequencing the entire genome of a bacterium in a short time, using a new sequencing method developed at his institute. This approach involved breaking the entire genome into many small fragments and sequencing them simultaneously–without first mapping them to specific chromosome locations, as Collins's group was doing. Venter argued that the same technique could be used on the human genome. But the human genome was much larger and contained many repetitive sequences–how would the multitude of randomly generated sequences be aligned properly? Critics claimed that the sequence data would be riddled with mistakes **(INFOGRAPHIC M3.2)**.

INFOGRAPHIC M3.2 Assembling the Sequence of the Human Genome

Human chromosomes are too long to sequence from end to end. The human genome was therefore cut into random large fragments, which were sequenced and reassembled in the correct order using computer algorithms to determine overlapping sequences.

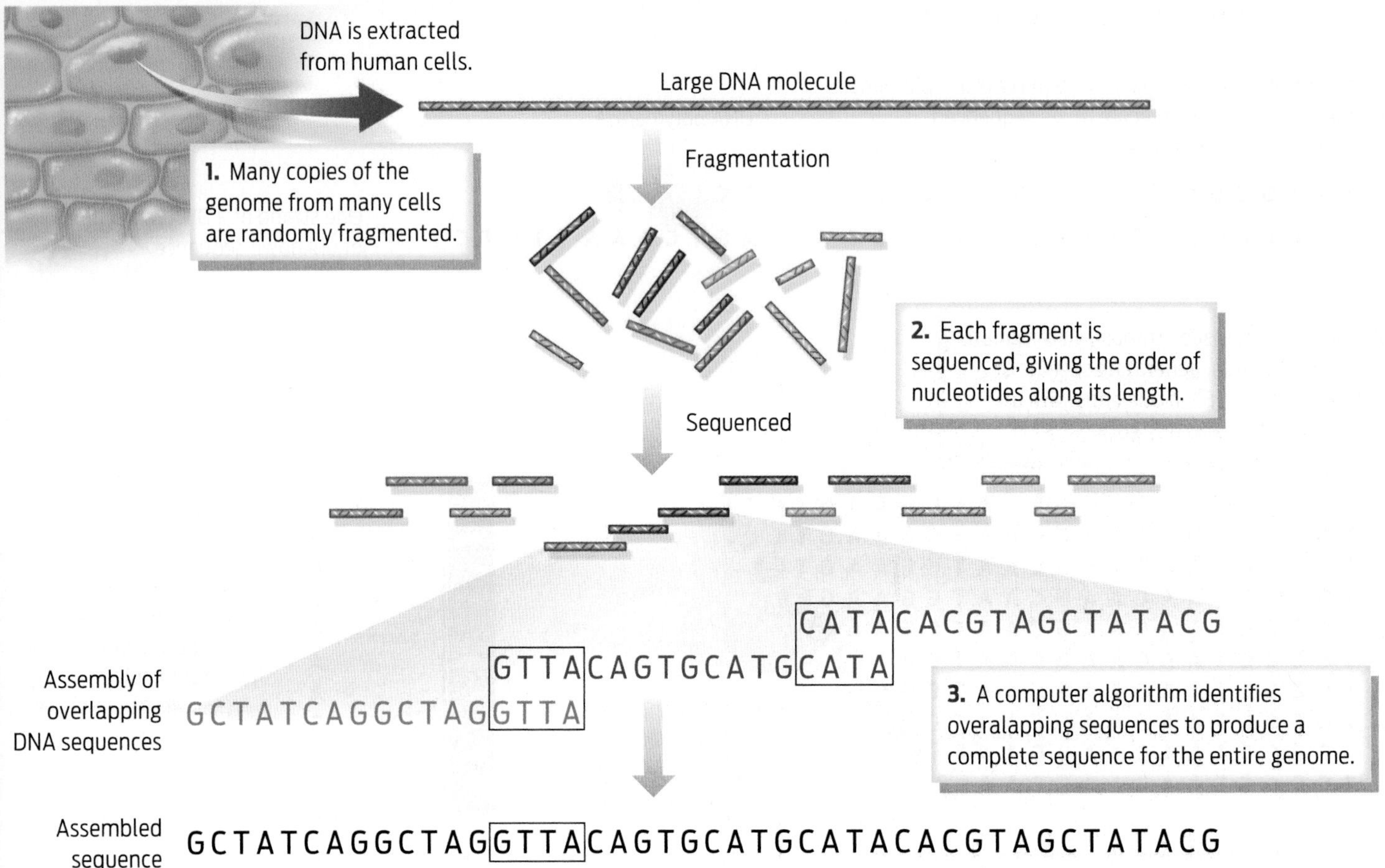

DNA is extracted from human cells.
Large DNA molecule
Fragmentation
1. Many copies of the genome from many cells are randomly fragmented.
2. Each fragment is sequenced, giving the order of nucleotides along its length.
Sequenced
CATACACGTAGCTATACG
GTTACAGTGCATGCATA
GCTATCAGGCTAGGTTA
Assembly of overlapping DNA sequences
3. A computer algorithm identifies overalapping sequences to produce a complete sequence for the entire genome.
Assembled sequence
GCTATCAGGCTAGGTTACAGTGCATGCATACACGTAGCTATACG

Why were more nucleotides sequenced across the fragments than are actually present in the assembled sequence?

Venter was undeterred. In 1998, he dropped another bomb: he had made a deal with the Perkin-Elmer Corporation, which was about to unveil a new automated sequencing machine. Together they would create a new company, called Celera Genomics (from the Latin word *celeritas*, meaning "speed"), that intended to sequence the entire human genome in just 3 years for a mere $300 million–a fraction of the cost of the publicly funded consortium.

Collins and other leaders of the public project were troubled. The U.S. Congress might favor Celera's approach and stop funding the public project altogether. Collins was also concerned that Celera would try to patent its sequence data, which would have restricted public access.

The race was on. Collins and his colleagues stepped up the pace. Venter wasn't the only one who had access to new automated sequencing machines and powerful computers that could process large amounts of data. Publicly financed scientists, too, had access to these and other new tools. Such technological advances dramatically cut the amount of time it took to sequence each nucleotide and simultaneously cut costs (**INFOGRAPHIC M3.3**).

INFOGRAPHIC M3.3 New Technology Cut DNA Sequencing Time and Cost

New "Next Generation" or "NextGen" sequencing technologies have dramatically reduced the cost to sequence a human genome since approximately 2008. Having the original reference human genome sequence has decreased the time to sequence an individual human genome, as sequences can be directly compared to the reference genome, eliminating the need for complex assembly.

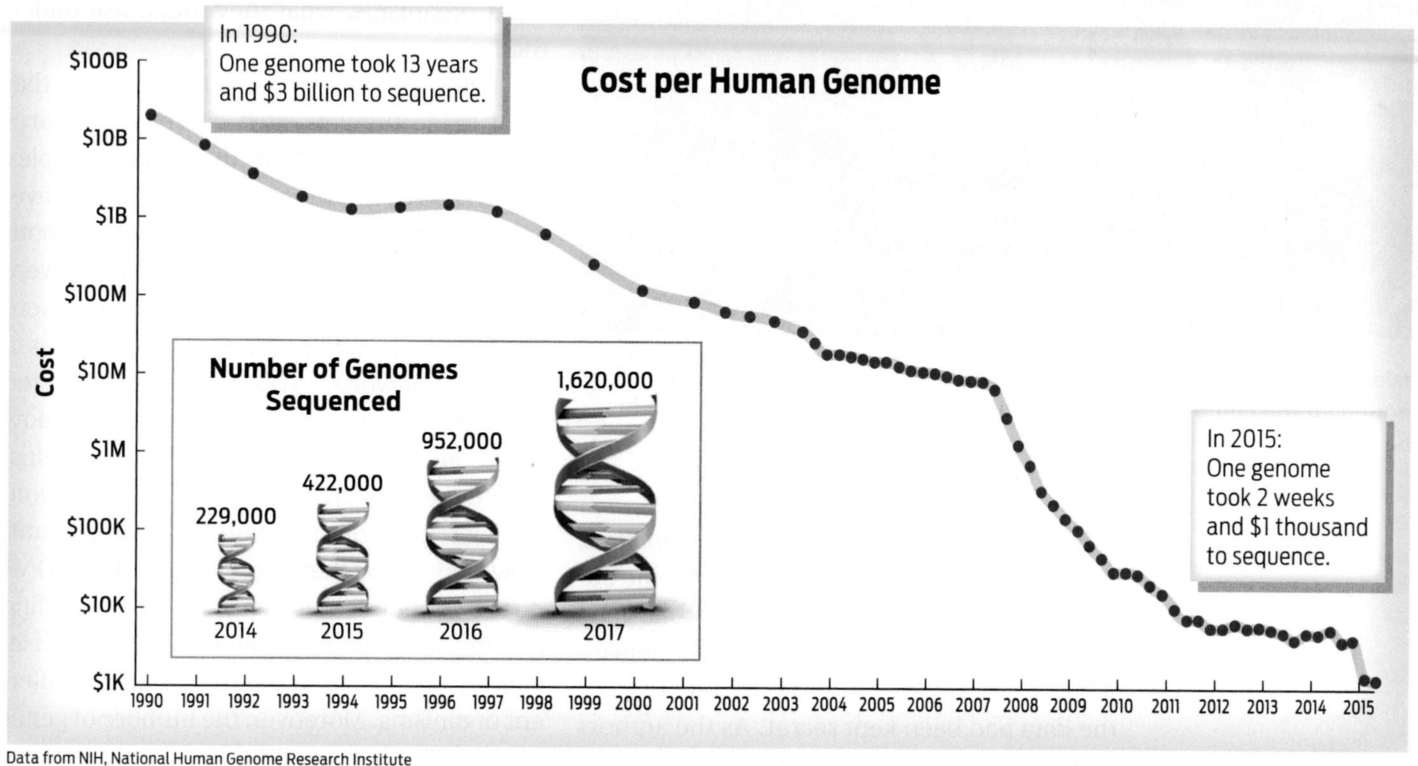

Data from NIH, National Human Genome Research Institute

? What was the cost to sequence a human genome in 2001, 2006, and 2015? Would you spend $1,000 to have your genome sequenced?

Photo Finish

About 6 months after Venter's announcement, Collins announced that the public consortium would complete its genome sequence by 2003—2 years ahead of schedule. The consortium also planned to produce a rough draft of the genome by 2001, which was about the same time that Venter planned to finish his draft. Collins justified his decision by stating that scientists were clamoring for the data even in rough form.

For a few years the contest between the privately funded Celera and the publicly funded HGP was bitter, each side criticizing the other's methods. The two sides eventually agreed to share the glory and appeared at a White House press briefing on June 26, 2000, together with U.S. President Bill Clinton and British Prime Minister Tony Blair, to announce the completion of a rough draft. In February 2001, both groups published their drafts of the human genome simultaneously in the journals *Nature* and *Science*.

Just whose genome was in fact sequenced? Geneticists working on the publicly funded project had collected blood samples from anonymous donors. The ultimate sequence is thus a composite pieced together from the genome sequences of several individuals. Celera's sequence data come mostly from the DNA of Venter himself.

Did one group "win" the race? Some say Venter won, because his sequence was further

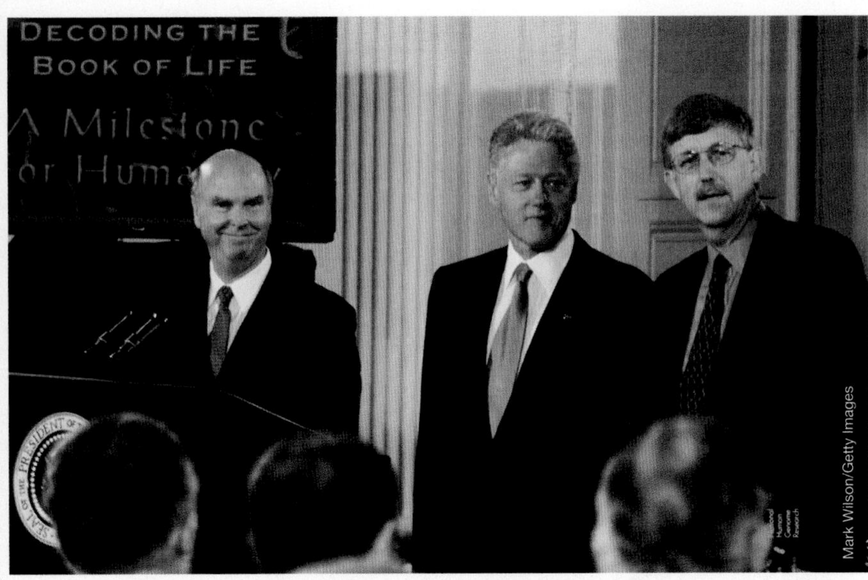

President Bill Clinton, with J. Craig Venter (left) and Francis Collins (right), announcing the completion of the human genome at the White House in June 2000.

along when the two groups presented their findings. But others were quick to point out that Venter's group had used the public gene map in order to assemble their sequence–which they wouldn't have been able to do if the data had been kept secret. As the authors of a 2002 scientific report analyzing the two approaches noted, "When speed truly matters, openness is the answer."

Ultimately, when the HGP was officially completed in 2003, 99% of the genome had been sequenced, with an accuracy of 1 mistake in 100,000 nucleotides. The achievement was hailed as one of the greatest scientific accomplishments of the 20th century. Some even consider it the greatest achievement ever in biology. Not only did it reveal new characteristics of the human genome, it also shed light on how we differ from other organisms.

Surprises and Applications

As the human genome was being sequenced, scientists also sequenced the genomes of several other organisms, including the fruit fly, mouse, yeast, and roundworm. Since the HGP was completed, they've continued to sequence the genomes of different organisms–17,479 so far–including the dog, Tasmanian devil, kangaroo, and broccoli. When scientists compared our genome to the genomes of these other organisms, what they found astounded them.

Before the HGP, scientists thought that what made humans such complex creatures was gene number–that the more complex the creature, the more genes it should have. They estimated that humans likely had about 100,000 genes. But data from the HGP eventually revealed that humans carry a mere 20,000 genes–about the same number as a lowly roundworm. This evidence suggested that gene number isn't as important as how those genes are expressed into proteins. Before the project, scientists didn't think noncoding regions of the genome were important. Now they know that about 98.5% of our DNA does not code for proteins. These noncoding regions actually regulate genes and consequently contribute to the complexity of different organisms. Moreover, the number of genes an organism has says nothing about the number of proteins that are produced from those genes. Through a variety of mechanisms that are still being studied, it is becoming clear that a relatively small number of genes can produce an enormous number of diverse proteins. For example, the human genome may encode 250,000-plus proteins from just 20,000 genes.

While the full potential of the HGP has yet to be tapped, it has already inspired many useful applications across many fields, from medicine to evolution. For example, the National Center for Genome Resources in Santa Fe, New Mexico, uses human genome sequences to improve diagnostics, control, and treatment of disease. They have designed a screening test that lets potential parents know if they are carriers for any of 448 debilitating or fatal diseases. This information can help parents understand their risk of having a child with one of these diseases.

Thanks to the HGP, researchers have been able to identify genes that are linked to diseases such as cancer and develop specific therapies that are targeted directly to these faulty genes. More than 70% of melanomas (a type of skin cancer) have a mutation in a gene called *BRAF*, for example. Drugs targeting the protein made from this faulty gene are now a mainstay of cancer treatment **(INFOGRAPHIC M3.4)**.

INFOGRAPHIC M3.4 The Human Genome: Outcomes and Applications

The human genome sequence has revealed basic information about our genetic makeup and facilitated the development of genetic tests for various diseases, as well as genetic-based tests for ancestry.

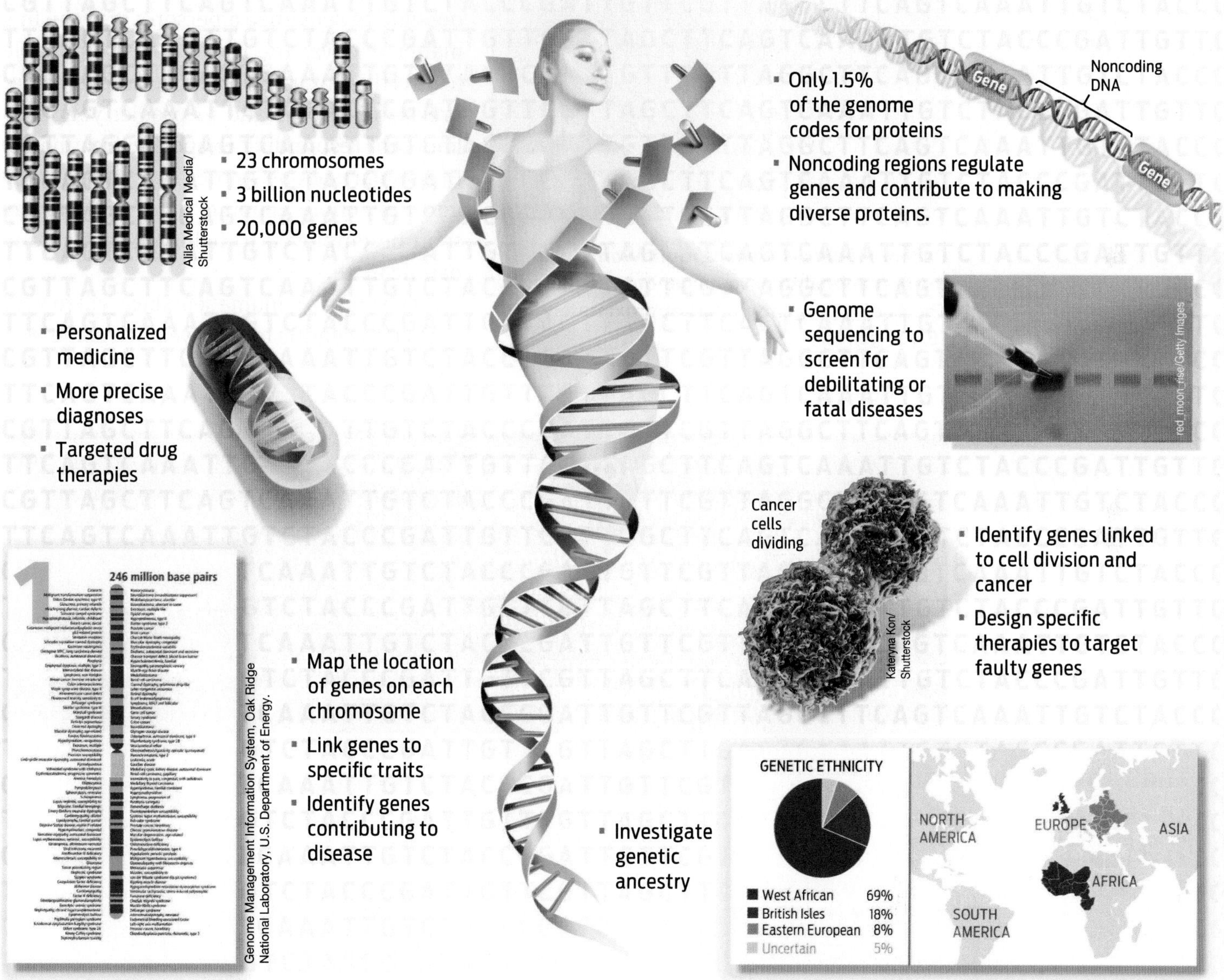

- 23 chromosomes
- 3 billion nucleotides
- 20,000 genes

Alila Medical Media/Shutterstock

- Only 1.5% of the genome codes for proteins
- Noncoding regions regulate genes and contribute to making diverse proteins.

Gene

Noncoding DNA

Gene

- Personalized medicine
- More precise diagnoses
- Targeted drug therapies

- Genome sequencing to screen for debilitating or fatal diseases

red_moon_rise/Getty Images

Cancer cells dividing

Kateryna Kon/Shutterstock

- Identify genes linked to cell division and cancer
- Design specific therapies to target faulty genes

1

246 million base pairs

Genome Management Information System, Oak Ridge National Laboratory, U.S. Department of Energy.

- Map the location of genes on each chromosome
- Link genes to specific traits
- Identify genes contributing to disease

- Investigate genetic ancestry

GENETIC ETHNICITY

West African	69%
British Isles	18%
Eastern European	8%
Uncertain	5%

NORTH AMERICA

EUROPE

ASIA

AFRICA

SOUTH AMERICA

? Which part of the genome, coding or noncoding, is most likely to be included in a genetic test for cystic fibrosis, which is caused by a faulty membrane protein?

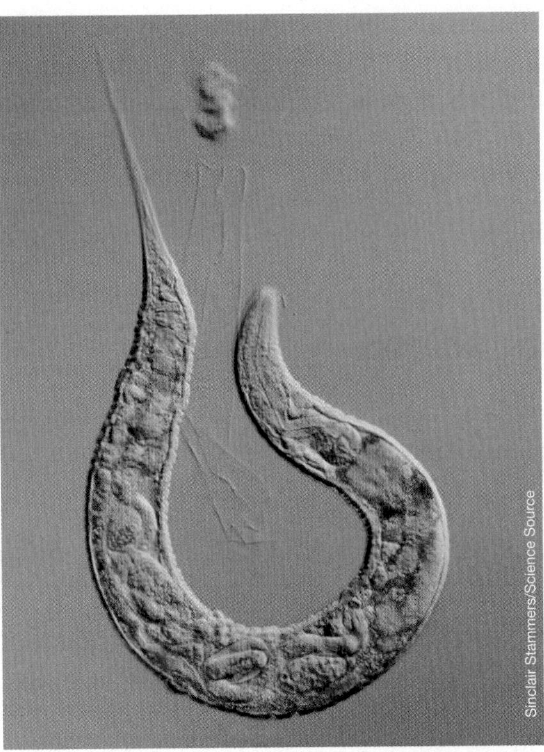

Caenorhabditis elegans, **a roundworm, was the first animal to have its genome sequenced. The *C. elegans* genome has approximately 19,000 genes.**

In addition to sophisticated medical tests ordered by physicians, anyone (anyone with $99 or $199) can submit a DNA sample to a commercial company, 23andme, for a variety of genetic analyses. The tests can reveal information about ancestry, carrier status for over 35 diseases, as well as inherited traits like male-pattern baldness and lactose intolerance.

Perhaps the most exciting application of the information is the prospect of "personalized medicine"–treatments based on your genome, tailored specifically to you. It is now possible to sequence a person's genome in a matter of days, and it may soon be possible to use this information to make ever more precise diagnoses and to predict your likelihood of responding to specific treatments. Even smart phone app creators are getting into the action: genomic data may one day soon be employed to help you find a compatible mate.

The human genome sequence will remain a crucial scientific tool for years to come. ■

MORE TO EXPLORE

- The Human Genome Project Archive: http://www.ornl.gov/sci/techresources/Human_Genome/home.shtml
- Mukherjee, S. (2016) *The Gene: An Intimate History*. New York: Scribner's.
- NOVA (video), *Cracking the Code of Life* (2001).
- Davies, K. (2001) *Cracking the Genome: Inside the Race to Unlock Human DNA*. New York: Free Press.
- Kevles, D., and L. Hood, eds. (2001) *The Code of Codes: Scientific and Social Issues in the Human Genome Project*. Cambridge: Harvard University Press.

MILESTONES IN BIOLOGY 3 Test Your Knowledge

1 If only 1.5% of the 3 billion nucleotides in our haploid genome encodes 20,000 genes, what would you predict about the average length (in nucleotides) of a human gene?

2 How many genes are encoded in the genomes of each of the following organisms? (Hint: You may need to do some online research.)

- **a.** humans
- **b.** baker's yeast (*Saccharomyces cerevisiae*)
- **c.** a nematode roundworm (*Caenorhabditis elegans*)
- **d.** the Norway spruce (*Picea abies*)
- **e.** a pufferfish (*Tetraodon nigroviridis*)

3 The first human genome sequence was a "draft" sequence. Draft sequences generally cover about 90% of the genome, with about 99.9% accuracy. For the human genome, how many nucleotides were included in the draft genome? Of these, how many were possibly inaccurate?

4 What were some surprises in the human genome?

- **a.** The number of nucleotides was much smaller than expected.
- **b.** The number of genes was much smaller than expected.
- **c.** A small new chromosome was discovered.
- **d.** The sequence is extremely similar between different people.
- **e.** all of the above

5 The National Center for Genome Resources has developed a test that allows couples to determine if they are carriers for mutations in genes that can cause devastating and lethal genetic diseases in their children. Would you want to take this test before having a child? What do you think you would do if you and your partner were found to be at risk for having a child with a lethal genetic disease?

9 Cell Cycle and Cell Differentiation

Wake Forest Institute for Regenerative Medicine

GROW YOUR OWN

Is regenerative medicine the solution to organ transplantation?

DRIVING QUESTIONS

1. How and why do cells divide?
2. How does one specialized cell type differ from another, and how do stem cells differentiate into these specialized cells?
3. What is regenerative medicine, and what are specific approaches to repairing or replacing organs?

BY THE TIME LUKE MASSELLA WAS 10 YEARS OLD, he'd already had 16 surgeries to repair problems stemming from a birth defect that affected his spine. One was proving hard to fix: his bladder was malfunctioning and causing urine to back up into his kidneys. These organs, in turn, were beginning to fail.

Luke faced a lifetime of spending hours every week hooked up to tubes and machines to undergo a treatment called dialysis that would help compensate for his failing kidneys but would mean his dreams of playing sports and living a normal life were over. There was one other option, however: an experimental surgery that might solve his bladder problems once and for all.

In 2001, Luke became one of the first people in the world to receive this experimental treatment, in which an engineered human bladder, one grown from his own cells, was implanted to replace his faulty one. "They take a piece of your bladder out. They grow it in a lab for 2 months into a new bladder that's your own. And they put it back in," Luke explained to reporters in 2012.

Thanks to the engineered bladder he received in 2001, Luke Massella went on to become captain of his high school wrestling team and attend college at the University of Connecticut. Today, he coaches high school wrestling and is pursuing a career in education.

The bladder-growing technique that saved Luke's life is the brainchild of Anthony Atala, director of the Wake Forest University Institute for Regenerative Medicine. In 2006, Atala announced that he and his colleagues had successfully transplanted engineered human bladders into several children and teenagers, surprising the medical community. Although scientists had for years been transplanting hearts, livers, and kidneys, the bladders were the first transplanted organs that were grown in the lab.

"It was very significant work," says William Wagner, deputy director of the McGowan Institute for Regenerative Medicine at the

University of Connecticut Photo

Luke Massella

University of Pittsburgh, of Atala's accomplishment. "He's overcome a huge number of challenges."

The potential applications of Atala's organ-growing technique are enormous. Each year, the demand for transplant organs such as hearts, livers, and kidneys vastly exceeds supply. In 2016, surgeons transplanted about 30,000 organs, according to the Organ Procurement and Transplantation Network. Meanwhile, there are about 120,000 people waiting for an organ transplant. And even when an organ does become available, the recipient's body may reject it because the donor tissue and recipient immune system are not compatible, leaving the patient sicker than before the transplant. Growing organs from a person's own cells not only sidesteps organ rejection, it also eliminates the need for donors.

The goal of growing organs outside the body dates back decades, but it's only in the past 20 years that scientists have made real progress toward bringing this idea to fruition. Central to this progress were advances in our understanding of cell division and cell differentiation. Increasingly, scientists are able to prod cells into dividing on cue. They're also deciphering the specific signals that tell a cell to develop into one cell type versus another–a muscle cell, say, rather than a skin cell.

Already, scientists are using this knowledge to construct new organs in the lab, which could one day be implanted into humans who need them. One particularly eye-catching method, called "bioprinting," uses computer graphics and cellular "ink" to manufacture organs from scratch. Need a new bladder? Just hit print.

Organ Transplantation in the United States

The number of organ transplants performed in the United States is not keeping up with the number of people in need.

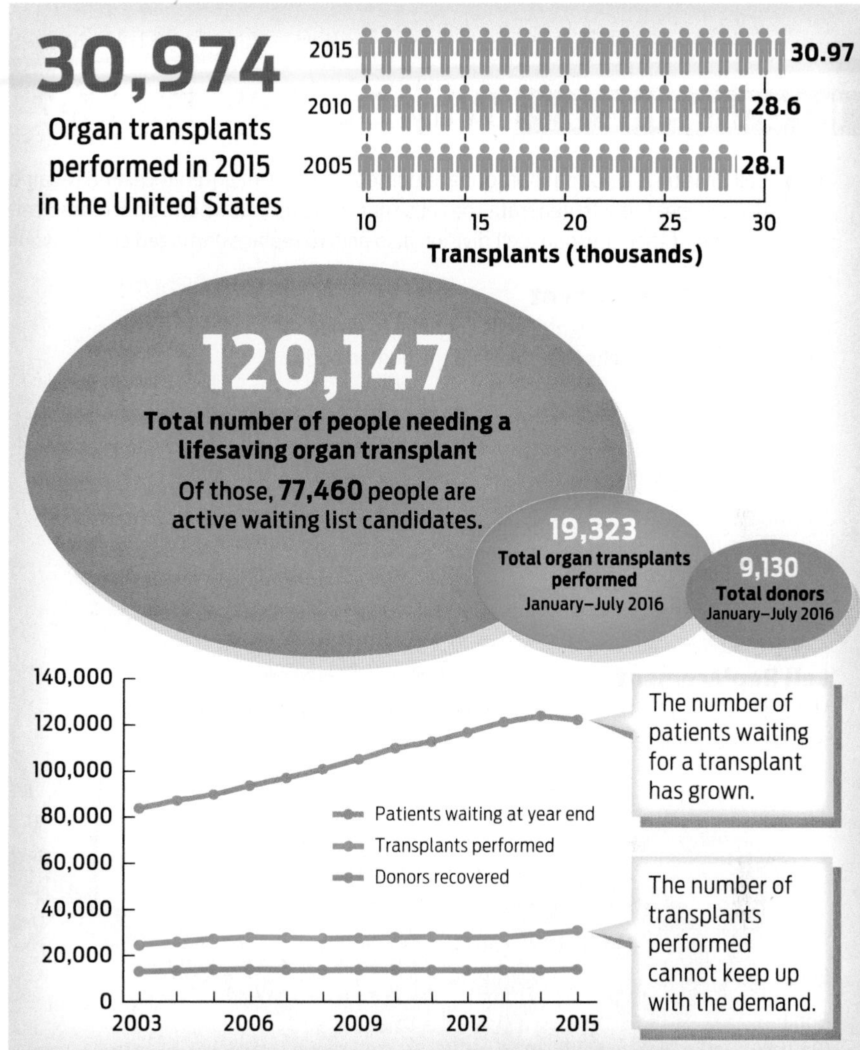

Data from Organ Procurement and Transplantation Network/HRSA (Data as of 8/16/2016)

The Body in Flux

We tend to think of our bodies as relatively fixed structures, formed of **tissues** made up of different cell types working together to carry out a particular function. But most of our tissues are in constant flux as cells divide periodically to replace cells that have reached the end of their life span. **Cell division** is a normal part of the development, growth, maintenance, and repair of the body. In fact, cell division in our bodies begins long before we are born. During embryonic development, a single fertilized egg cell divides, and its daughter cells divide again and again, eventually forming trillions of cells by the time a baby

TISSUE
An organized group of different cell types that work together to carry out a particular function.

CELL DIVISION
The process by which a cell reproduces itself; cell division is important for normal growth, development, maintenance, and repair of an organism.

is born. During childhood, cell division helps us grow larger. As we age, our tissues continually discard old cells and generate new ones in their place. And when we cut or injure ourselves, cells in the area divide to heal the wound **(INFOGRAPHIC 9.1)**.

INFOGRAPHIC 9.1 Why Do Cells Divide?

Cell division is a normal part of development, growth, maintenance, and repair of the body. A fertilized egg cell and its descendants divide repeatedly to generate the cells that make up a newborn. Specialized cells in tissues divide to replace cells that have reached the end of their lifespan. Cell division also acts to replace damaged cells in wounds.

Embryonic Development

A fertilized egg and its daughter cells continue to divide to create the trillions of cells that make up the human body.

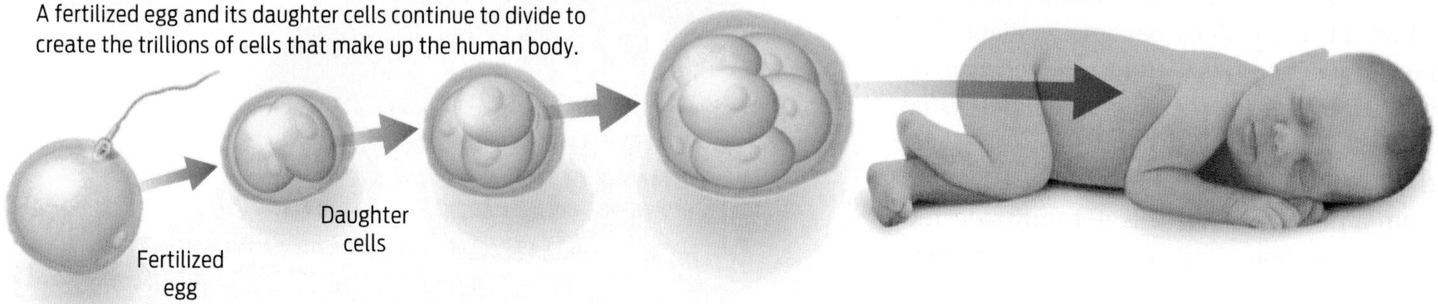

Cell Replacement

Most cells have a finite life span. Cell division within tissues regularly replaces the dying cells and maintains healthy tissues.

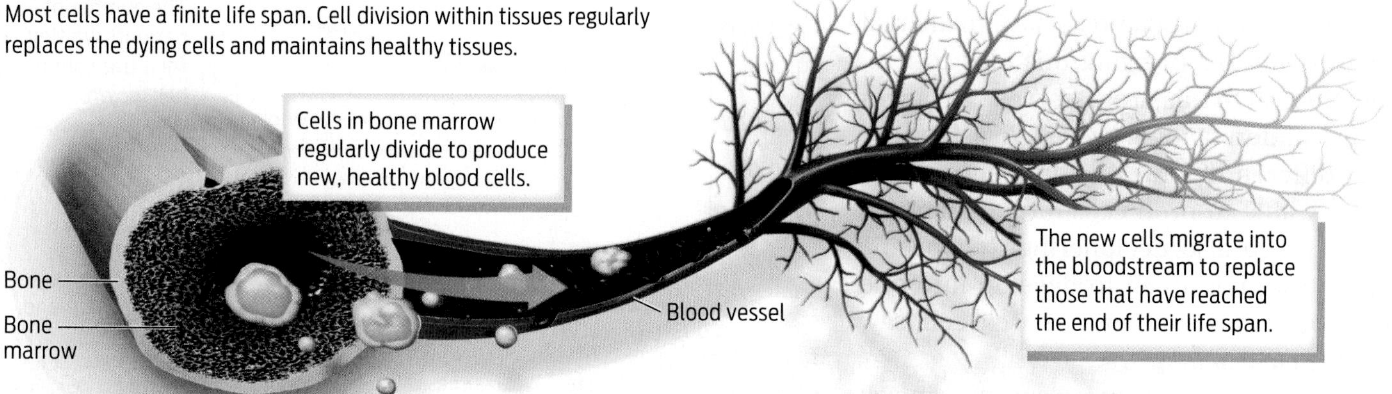

Cells in bone marrow regularly divide to produce new, healthy blood cells.
Bone
Bone marrow
Blood vessel
The new cells migrate into the bloodstream to replace those that have reached the end of their life span.

Repairing Damaged Tissue

Injury triggers cell division to replace damaged cells.

Cells divide to replace cells lost in the injury process.

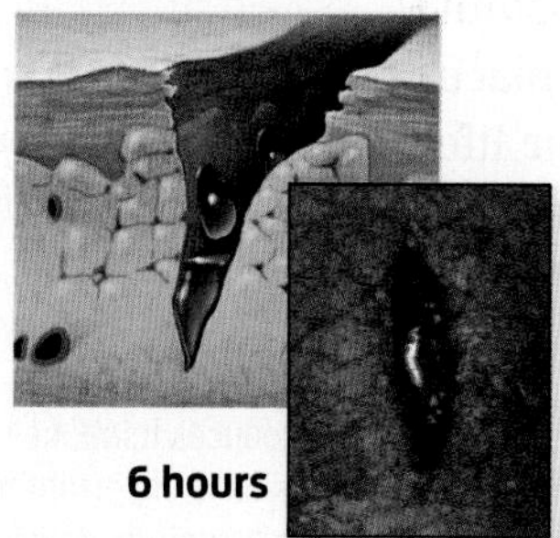

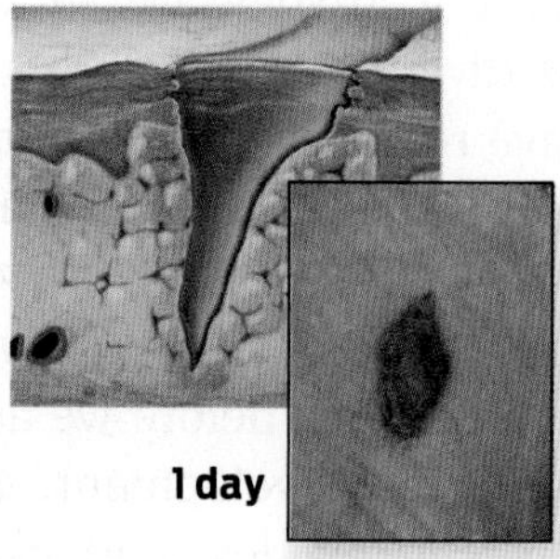

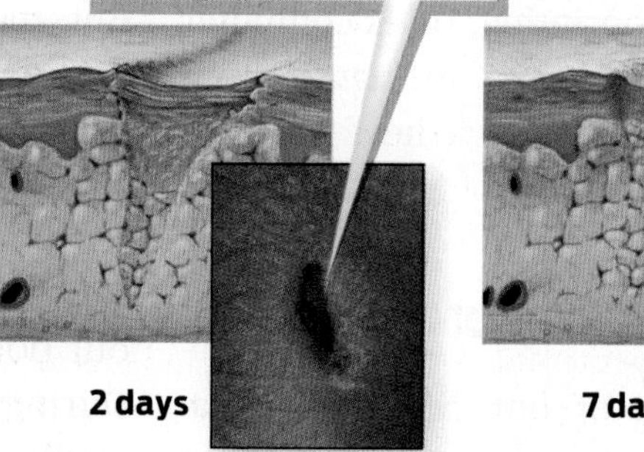

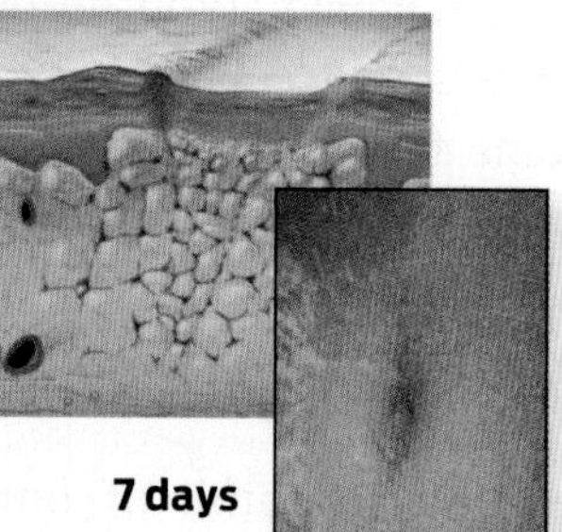

Wound repair photos republished with permission of Elsevier, from Targeting Connexin43 Expression Accelerates the Rate of Wound Repair. (2003). Qiu, C., et al. Current Biology, 13(19), 1697-1703; permission conveyed through Copyright Clearance Center, Inc.

? When considering cell replacement and repairing damage, which of these processes is more likely to involve "scheduled" cell division at a particular rate?

If cell division can repair a wound on our finger, why can't it repair a damaged body part or organ? Actually, in some cases it can. Children who lose the tip of a finger can sometimes regrow it, fingernail and all. Broken bones can heal by growing new bone. And certain organs, such as the liver, can regrow to essentially full size after surgery to remove a damaged part or to obtain tissue for a liver transplant in another patient. But these are the exceptions: most human organs, including the bladder, cannot repair damage on their own.

A surgeon by training, Atala sought to help his patients whose bladders were not functioning normally because of cancer, injury, or—as with Luke—a birth defect. For more than a century, doctors have repaired defective or injured bladders by stitching in pieces of the person's intestine or bowel. But because the intestine and bowel are built for different functions from the bladder's, this treatment is not ideal—it can lead to problems down the line, including leaks, repeat surgeries, and even cancer. Growing a new bladder to replace the damaged organ, Atala reasoned, would be much better.

> "They take a piece of your bladder out. They grow it in a lab for 2 months into a new bladder that's your own. And they put it back in."
>
> — Luke Massella

But that is easier said than done. When Atala began his research in the 1990s, very little was known about how to keep bladder cells alive in a dish outside the body, let alone grow enough of them to make a bladder. There was also the matter of getting cells to assume the round, three-dimensional shape of a bladder. That required developing a biodegradable scaffold on which bladder cells would grow and thrive, even when scaffold and bladder cells are implanted into the body. It took Atala more than a decade to understand what makes bladder cells tick. Much of that time was spent trying to understand bladder cell division.

To produce new cells by cell division, cells pass through a series of stages collectively known as the **cell cycle.** During the cell cycle, one cell becomes two. A cell doesn't simply split in half, however. If it did, each resulting cell would be smaller than the original, and with each division, each cell would lose half its contents. So before a cell divides, it first makes a copy of its contents so that each new cell has the same amount of organelles, DNA, and cytoplasm as the original cell. This preparatory stage of the cell cycle, called **interphase,** has separate subphases: G_1 phase, when the cell grows larger and begins to produce more cytoplasm and organelles; synthesis phase (S), when DNA is replicated and therefore chromosomes are duplicated forming linked **sister chromatids**; and G_2 phase, during which the cell prepares to enter the division phases. In a cell that takes approximately 24 hours to divide, interphase takes about 22 hours to complete.

Once the cell duplicates its contents, it enters the division phases of the cell cycle: **mitosis,** when the sister chromatids of each replicated chromosome are evenly divided between the two daughter cells; and **cytokinesis,** when the two daughter cells physically separate. Together, mitosis and cytokinesis take about 2 hours to complete. Once the cell is finished dividing, it may enter a resting phase of the cell cycle, called G_0, during which the cell goes about business as usual **(INFOGRAPHIC 9.2)**.

Though it takes up only a fraction of the total cell cycle, mitosis is in some ways the star of the whole performance. It's mitosis that ensures that each new daughter cell has the

CELL CYCLE
The ordered sequence of stages through which a cell progresses in order to divide; the stages include preparatory phases (G_1, S, G_2) and division phases (mitosis and cytokinesis).

INTERPHASE
The stage of the cell cycle in which dividing cells spend most of their time, preparing for cell division. There are three distinct subphases: G_1, S, and G_2.

SISTER CHROMATIDS
The two identical DNA molecules that result from the replication of a chromosome during S phase.

MITOSIS
The segregation and separation of replicated chromosomes during cell division.

CYTOKINESIS
The physical division of a cell into two daughter cells.

INFOGRAPHIC 9.2 The Cell Cycle: How Cells Reproduce

The purpose of the cell cycle is to replicate cells, creating two new daughter cells that are genetically identical to the original parent cell. The cell cycle consists of preparatory phases collectively known as interphase, as well as the division phases, mitosis, and cytokinesis.

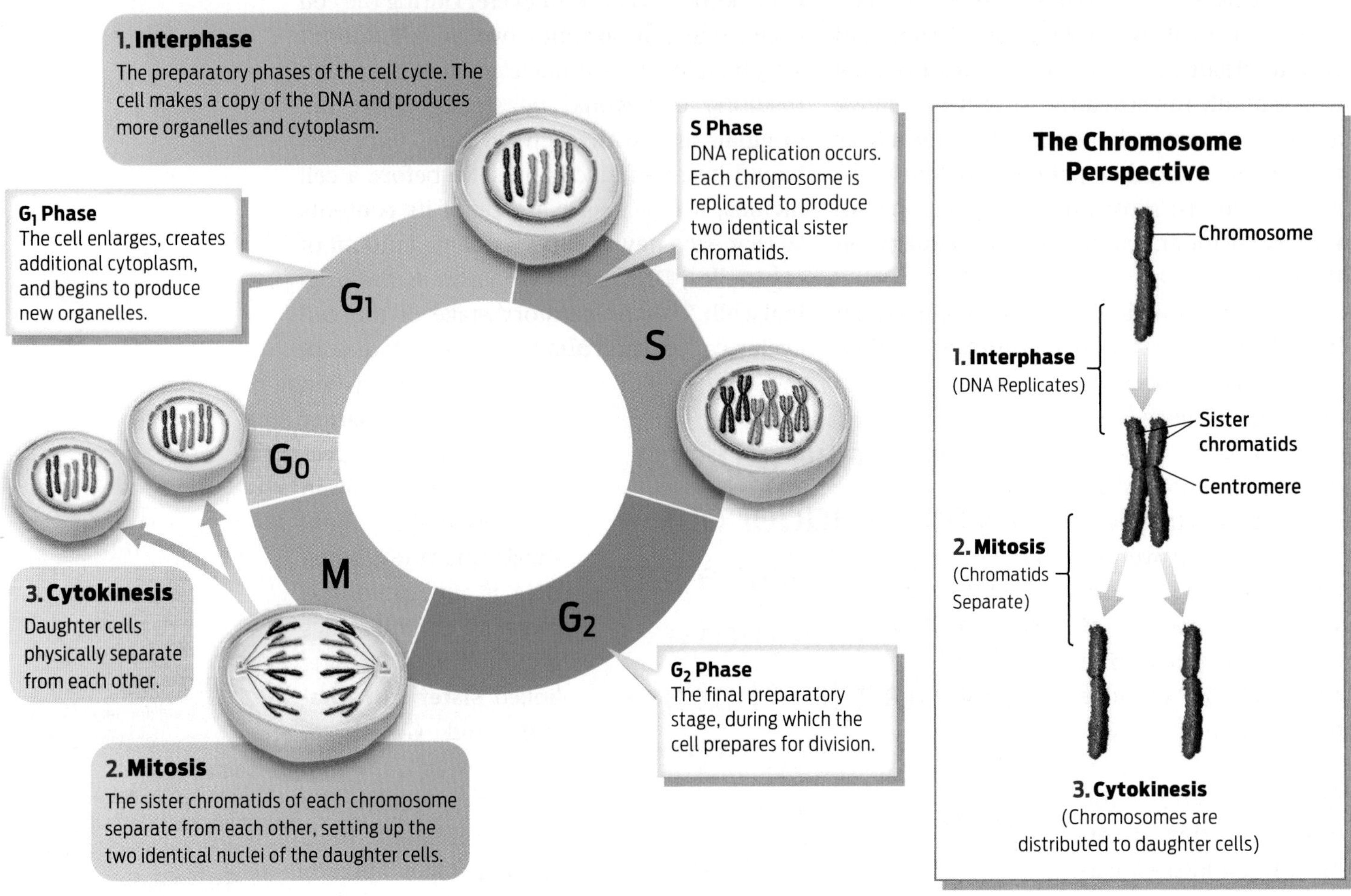
1. Interphase
The preparatory phases of the cell cycle. The cell makes a copy of the DNA and produces more organelles and cytoplasm.
S Phase
DNA replication occurs. Each chromosome is replicated to produce two identical sister chromatids.
G1 Phase
The cell enlarges, creates additional cytoplasm, and begins to produce new organelles.
G1
S
G0
M
G2
3. Cytokinesis
Daughter cells physically separate from each other.
G2 Phase
The final preparatory stage, during which the cell prepares for division.
2. Mitosis
The sister chromatids of each chromosome separate from each other, setting up the two identical nuclei of the daughter cells.
The Chromosome Perspective
Chromosome
1. Interphase
(DNA Replicates)
Sister chromatids
Centromere
2. Mitosis
(Chromatids Separate)
3. Cytokinesis
(Chromosomes are distributed to daughter cells)

? What is the general function of interphase, and what stages of the cell cycle are part of interphase?

CENTROMERE
The specialized region of a chromosome where the sister chromatids are joined; it is critical for proper alignment and separation of sister chromatids during mitosis.

proper number of chromosomes–no more, no less than, in humans, a complete set of 46. Each chromosome contains a unique set of genes that instruct cells how to function, and so getting the distribution right is critical. If any chromosome is left out of a daughter cell, all the genetic information contained in that chromosome will be missing, which could spell disaster for the cell.

You can think of mitosis as a carefully choreographed dance of chromosomes. At the beginning of the dance, the replicated chromosomes are present in the nucleus as loosely gathered threads. To prevent the chromosomes from becoming entangled during the dance, they condense into tightly wrapped, rod-shaped structures that can move about the cell more easily. If you peered at the cell through a microscope at this point, you'd see that the two identical sister chromatids of a replicated chromosome are connected at a region called the **centromere.** As mitosis progresses, these linked sister chromatids move across the stage and line up along the midline of the cell, like dancers in formation.

The chromosomes do not get there by chance. Rather, they are pushed and pulled into this position by protein fibers of a structure called the **mitotic spindle**. The mitotic spindle is formed when microtubules of the cell's cytoskeleton rearrange into this structure as the nuclear membrane dissolves. Microtubule spindle fibers extend from each end of the cell and snag the centromere region of the chromosomes, with one spindle fiber attaching to the centromere of each sister chromatid. As spindle fibers lengthen and shorten, the attached chromosomes are moved around the cell, much like marionettes on strings. Once aligned at the midpoint of the cell, the two sister chromatids of each replicated chromosome separate from each other and are pulled to opposite sides of the cell by the spindle fibers, which contract to reel the chromatids in. Each chromatid will eventually form one of two genetically identical chromosomes—one for each daughter cell (see **UP CLOSE: PHASES OF MITOSIS**).

Replaceable You

Normally, cells divide only when they are signaled to do so by molecules called growth factors. This is generally a good thing, since if cells divided without receiving the proper cues, there would be so many cells that a body could not maintain its structure. (And in fact, this is what happens in cancer, which results from uncontrolled cell division; see Chapter 10.) But this need for division signals becomes a liability when growing organs, since scientists have to figure out what those signals are. For some cells—like nerve cells, liver cells, and pancreas cells—researchers are still looking for the right signals, and so these cells remain very difficult to grow outside the body. But for bladder cells, Atala successfully figured out the signaling recipe, and by 1999 he was ready to begin engineering organs for his patients.

For each patient, Atala cut a piece of tissue smaller than a postage stamp from inside the bladder and extracted two types of cell—muscle cells and epithelial cells. He then mixed these cells with growth factors that promoted cell division, producing a stockpile of millions of cells. Next, he layered the two types of cells onto a biodegradable scaffold, which he had sculpted to resemble a human bladder. Atala likens this process to building a layer cake; each layer of cells is prepared one by one. Finally, he placed the scaffold in an incubator to simulate conditions inside the human body. The cells went through several more cell

MITOTIC SPINDLE
The microtubule-based structure that separates sister chromatids during mitosis.

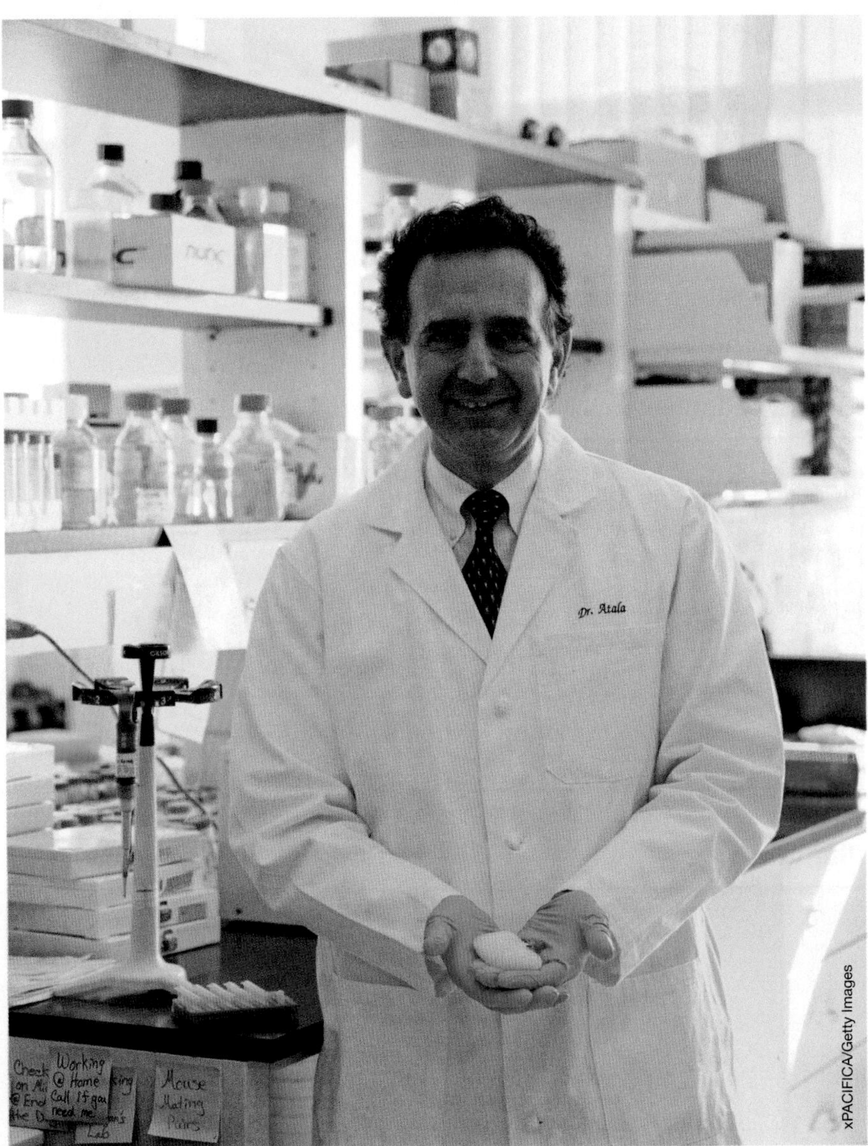

xPACIFICA/Getty Images

Anthony Atala, seen here in his lab at Wake Forest University Institute for Regenerative Medicine, holds a kidney scaffold made by a 3-D printer.

UP CLOSE Phases of Mitosis

Interphase

- Each chromosome replicates in S phase, resulting in two sister chromatids connected at the centromere.
- Chromosomes are loosely gathered in the nucleus.

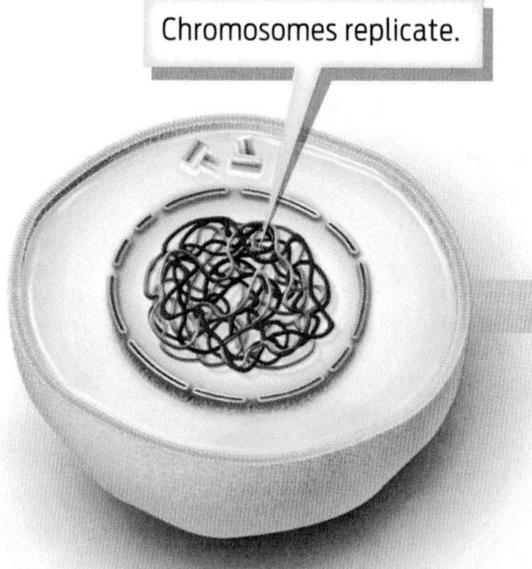
Chromosomes replicate.

Mitosis

Prophase

- Replicated chromosomes begin to coil up.
- The nuclear membrane begins to disassemble.
- Microtubule fibers begin to form the mitotic spindle.

Chromosomes coil.

Metaphase

- Microtubule spindle fibers from opposite ends of the cell attach to the sister chromatids of each chromosome.
- Replicated chromosomes become aligned along the middle of the cell.

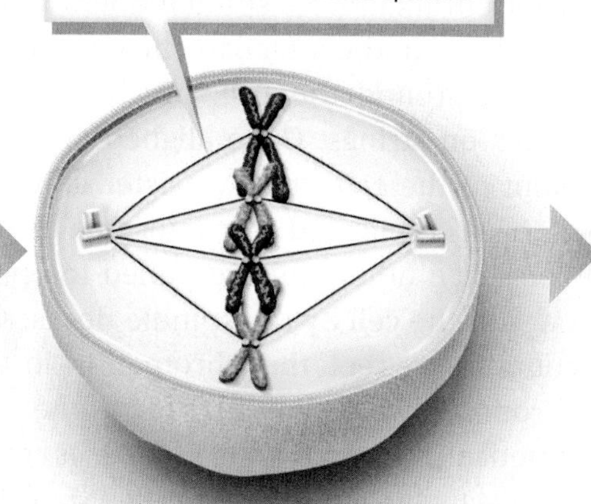
Microtubules form mitotic spindle.

Animal Cells:

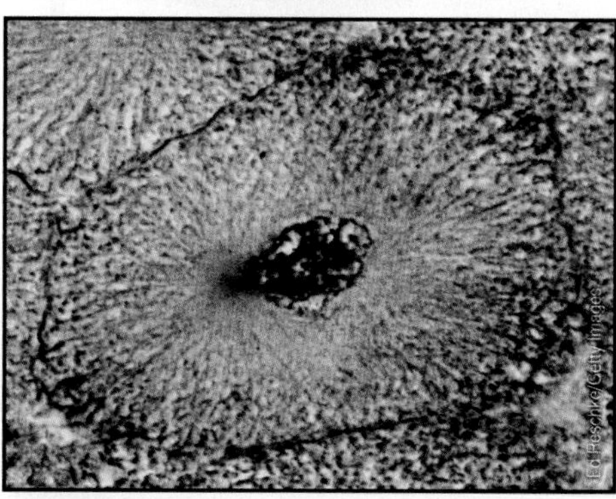
Ed Reschke/Getty Images

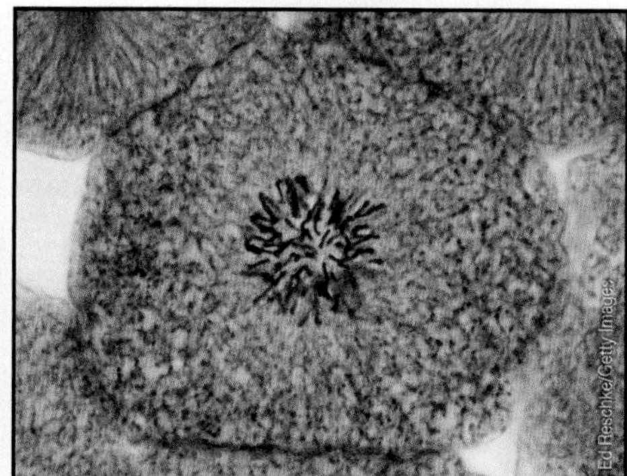
Ed Reschke/Getty Images

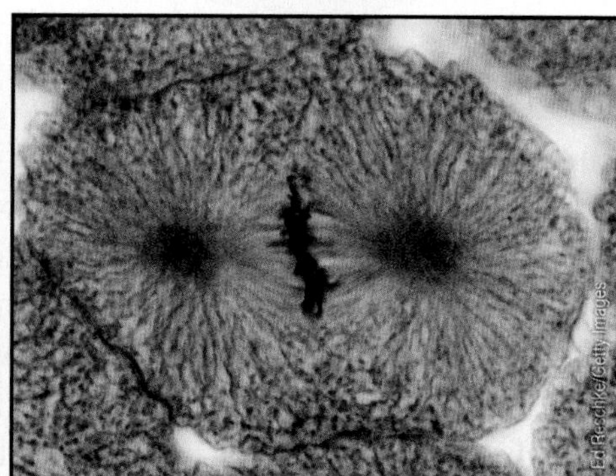
Ed Reschke/Getty Images

Plant Cells:

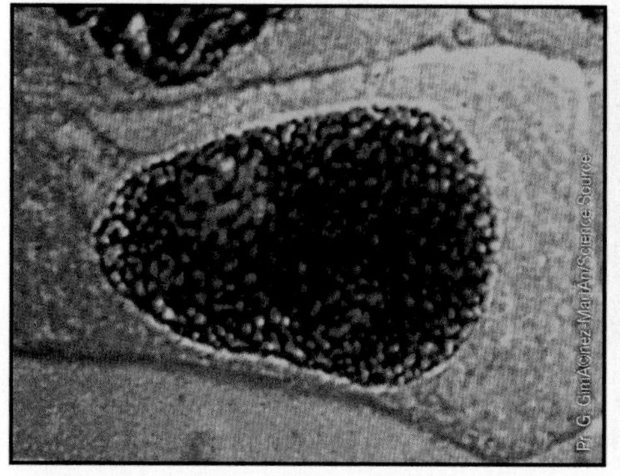
Pr. G. Gimenez-Martin/Science Source

Pr. G. Gimenez-Martin/Science Source

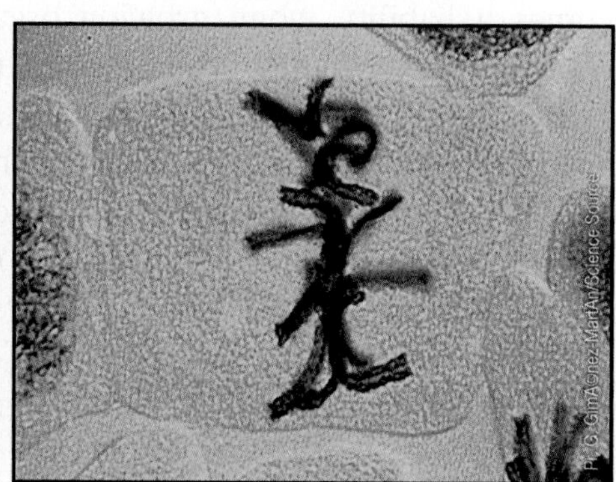
Pr. G. Gimenez-Martin/Science Source

Anaphase

- Microtubules shorten, pulling the sister chromatids to opposite ends of the cell.

Telophase

- Identical sets of chromosomes reach each pole.
- Microtubule spindle fibers disassemble.
- Nuclear membrane forms around each set of chromosomes, forming the daughter cell nuclei.

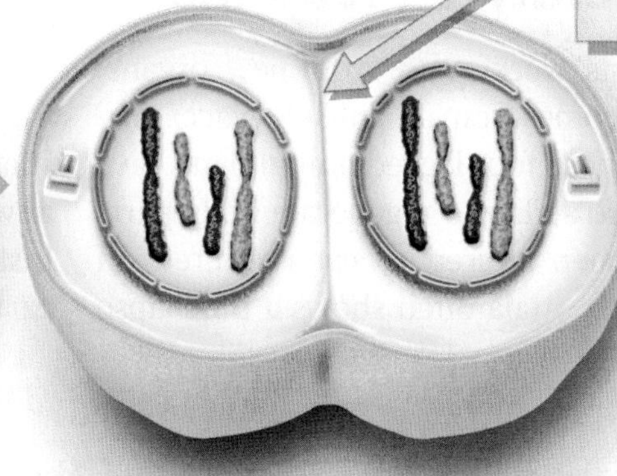

Cytokinesis
▪ Cytoplasm divides.
▪ Two nuclei become separated into daughter cells.

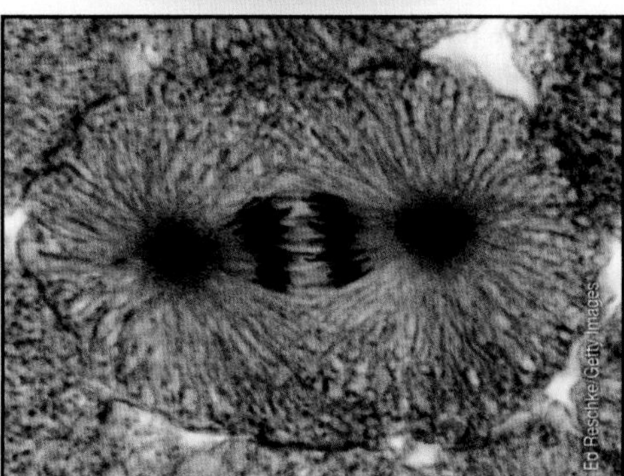

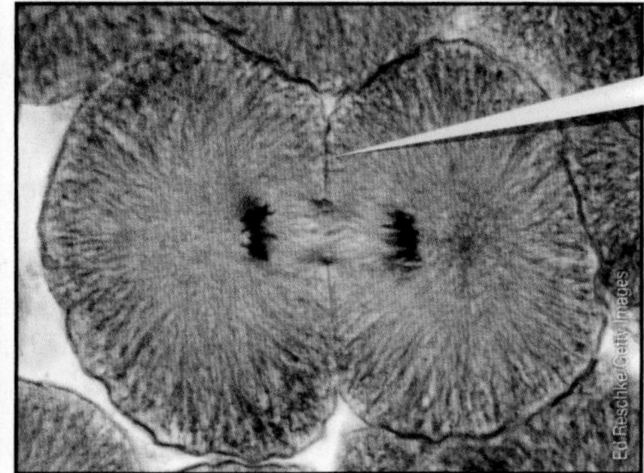

During cytokinesis in animal cells the cell membrane pinches in to separate the daughter cells.

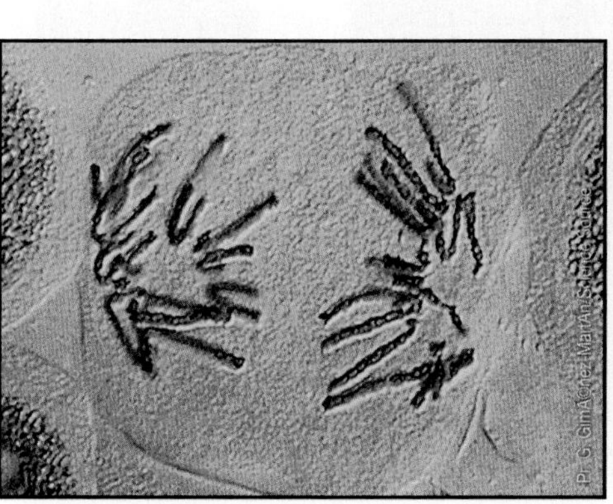

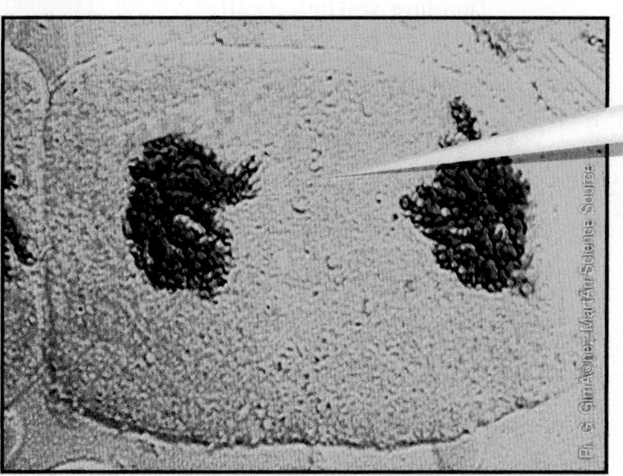

During cytokinesis in plant cells, a new cell wall is synthesized between the daughter cells.

divisions and established connections to one another. Two months later, surgeons reconstructed the patient's bladder using the new bladder tissue **(INFOGRAPHIC 9.3)**.

The treatment, so far, seems remarkably successful. Not only does the technique improve bladder function, it also avoids the complications that result from using intestine or bowel for bladder repair. While the approach is not yet approved by the FDA, it is being tested in clinical trials, and its developers hope that a green light from regulators will be forthcoming soon.

Regenerative Medicine

For scientists working to grow new organs, inspiration comes from an unlikely place: salamanders. These four-legged amphibians have a remarkable ability to regenerate body parts that have been injured or even severed. In presentations, Atala often shows a time-lapse video of a salamander growing back an entire front limb, complete with webbed digits.

If a salamander can regenerate tissues, Atala asks, why can't a person? In fact, he says, they can. "The human body is constantly regenerating," he explains. The whole superficial layer of our skin turns over every couple of weeks, while the cells lining our intestines turn over about every week. Even something as seemingly solid as bone is completely replaced about every 10 years. "Basically you don't have any bone cells left over that were there 10 years prior," Atala says.

Scientists have known for decades that some cells, like skin and blood cells, divide continually to repair and replace these tissues. But only within the past 15 years have they discovered that many, if not most, tissues in the human body are refurbished with new cells. In 2005, biologists at the Karolinska Institute in Sweden used a type of radiometric carbon dating (see Chapter 16) to determine

INFOGRAPHIC 9.3 Engineering a Human Organ for Transplant

Cells are isolated from a patient's tissue. These cells are coaxed to divide and populate a scaffold that resembles the organ. This engineered organ is then transplanted into the patient, providing a matched tissue that will not be rejected by the immune system.

1. Diseased organ A bladder is made of two main tissue layers consisting of specialized cell types.

2. Tissue is removed A healthy portion of the patient's bladder with the two main tissues is removed.

3. Cells divide in culture Cells from each tissue type are stimulated to divide, which produces large numbers of cells.

4. Cells form tissues on scaffold Cells are grown on a biodegradable scaffold, where they form the tissues of a functional bladder.

5. Engineered tissue is transplanted Surgeons replace the diseased bladder tissues with the newly grown tissues to restore normal bladder function.

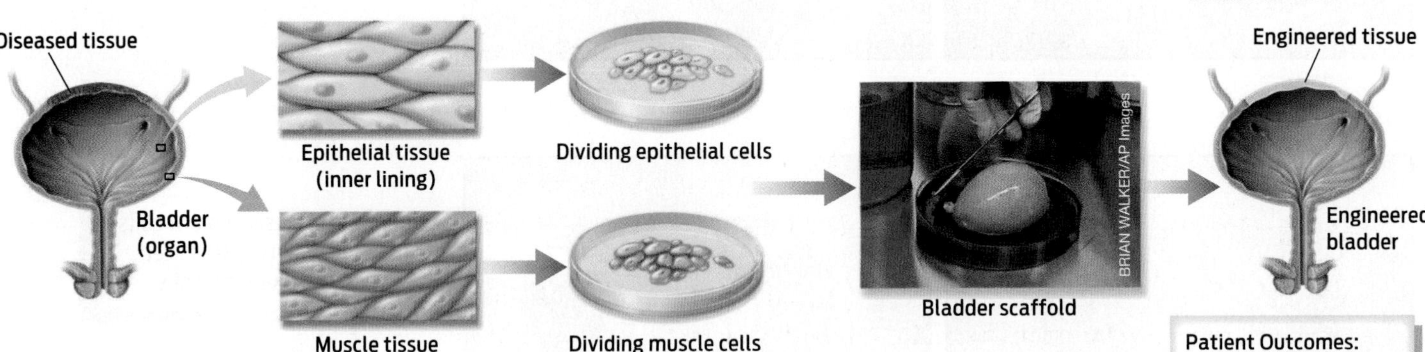

Patient Outcomes:
1. Improved urinary function
2. No immune rejection

? What is the advantage of engineering an organ over transplanting an organ into a patient in need?

TABLE 9.1 How Old Are You?

Tissues are as old as the cells making them up. Cells in tissues are replaced when they wear out or reach the end of their life span. Scientists have dated the age of tissues in the human body and have found that some are much younger than you may think.

TISSUE TYPE	TURNOVER RATE
Epidermis (skin surface)	2 to 3 weeks
Red blood cells	120 days
Liver	300 to 500 days
Bones	About 10 years
Gut (except lining)	15.9 years
Rib muscle	15.1 years
Lens of the eye	Never replaced
Neurons of the cerebral cortex	Never replaced

Data from Spalding, K. L., et al. (2005) Retrospective birth dating of cells in humans. *Cell*, 122(1):133–143.

the age of different tissues in the human body. They discovered that many tissues in the body are much younger than commonly thought because the cells within them are continually replaced with new cells **(Table 9.1)**.

Where do these new cells come from? Many tissues in the body contain what are called **stem cells**–immature cells that can divide and give rise to cells of different types. When a stem cell divides, one of the daughter cells remains an immature stem cell, while the other one "grows up," developing into a more specialized cell. In this way, stem cells contribute to tissue repair but retain the capacity for further cell division in the future.

The first stem cells to be discovered were those that produce all the various cells making up blood (these are called hematopoietic stem cells). Scientists have since discovered that many other tissues contain stem cells as well. Stem cells have been found in the brain, blood vessels, muscle, skin, teeth, heart, gut, liver, ovaries, and testes, for example. Stem cells also make up part of a developing embryo, where they give rise to a wide range of cell types. Stem cells found in adult tissues are sometimes called adult stem cells to distinguish them from the embryonic stem cells found in an embryo.

When scientists discovered that many tissues had their own pool of stem cells, the search began for ways to use these stem cells therapeutically–to engineer new organs, for example. This field of research has been called regenerative medicine. "We have not lost the ability to regenerate," Atala says. "The question is how can we harness that potential again?"

How Cells Grow Up

To repair tissue damaged by injury or disease, stem cells must do more than simply divide repeatedly. The new cells must also go through a process of specialization to develop into the specific cell types appropriate to the tissue in need of repair. Remember that during embryonic development a single cell becomes millions as the embryo grows. These dividing cells eventually become specialized as muscle cells, kidney cells, heart cells, and more than 200 other cell types by the time we are born.

STEM CELLS
Immature cells that can divide and differentiate into specialized cell types.

CELL DIFFERENTIATION
The process by which a cell becomes specialized to carry out a specific role by turning specific genes "on" and "off" and making different suites of proteins.

This process, in which a stem cell develops into a more specialized cell type, is called **cell differentiation.** Stem cells differentiate into specialized cells by turning some genes "on" and others "off." Through this unique pattern of gene expression (see Chapter 8), different cell types make different suites of proteins that are specific to those cell types. So while every cell in our body carries the exact same DNA, it is a cell's pattern of gene expression–and therefore the proteins produced from those genes–that defines it as one cell type or another.

Take, for example, two cell types with very distinct characteristics: muscle cells and B cells of the immune system. Muscle cells make proteins like actin and myosin that allow them to contract and make our bodies move. B cells produce antibody proteins that protect the body from the viruses and bacteria that cause infections. Both muscle cells and B cells contain exactly the same DNA, and therefore contain the same genes. But only a subset of those genes is turned on in each cell type. Stem cells with "muscle genes" turned on become muscles, while stem cells with "B cell genes" turned on become B cells. As a result of this differential gene expression, each cell type produces a unique set of proteins that distinguish one cell type from another **(INFOGRAPHIC 9.4)**.

INFOGRAPHIC 9.4 Specialized Cells Express Different Genes

Every cell in your body has the same genes, or genome. What distinguishes one cell type from another is the pattern of gene expression and, consequently, the proteins each cell makes. A muscle cell makes a set of proteins that help muscles contract, whereas B cells express high levels of antibody proteins, which help the body fight infections.

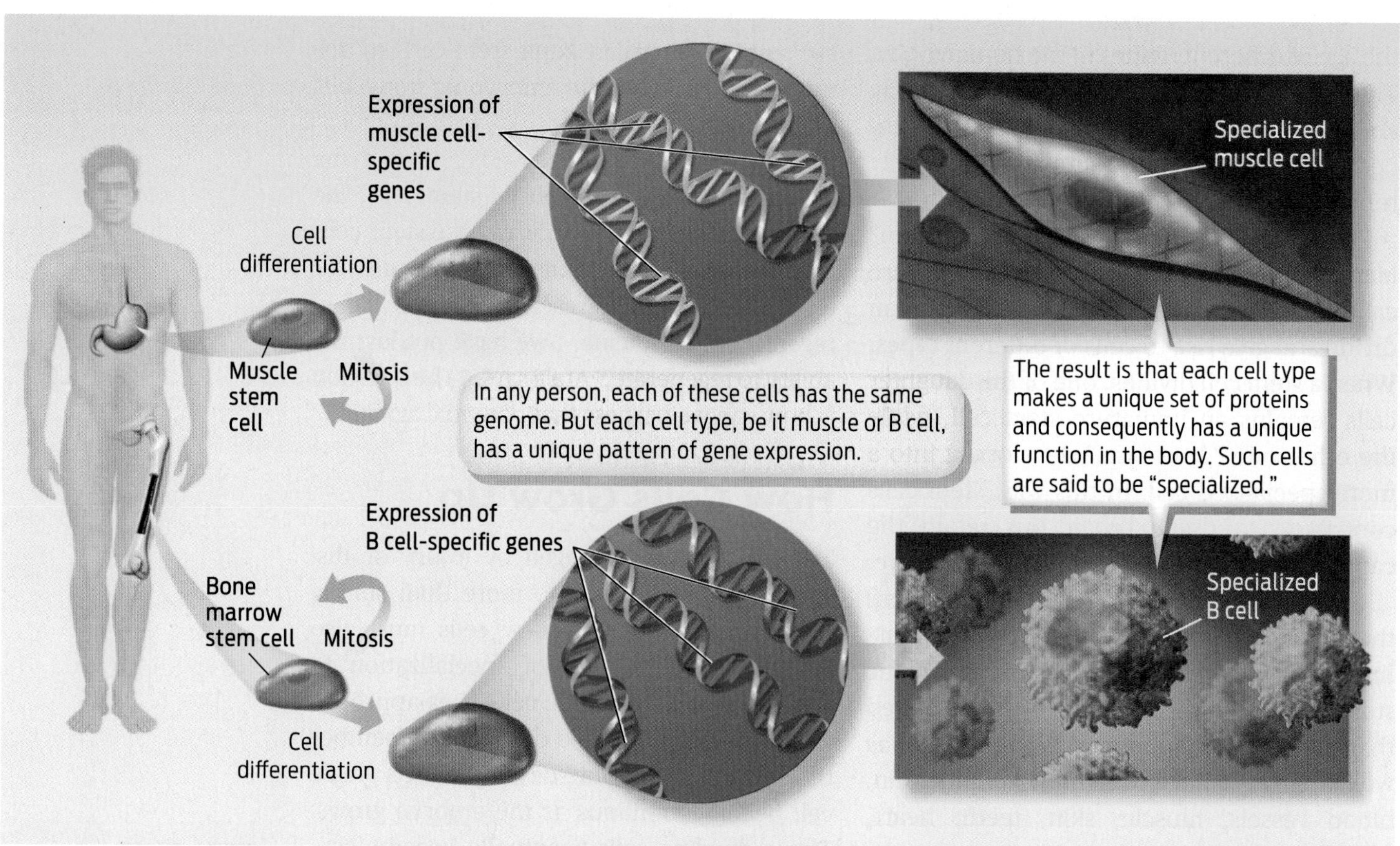
Expression of muscle cell-specific genes
Cell differentiation
Muscle stem cell
Mitosis
Specialized muscle cell
In any person, each of these cells has the same genome. But each cell type, be it muscle or B cell, has a unique pattern of gene expression.
The result is that each cell type makes a unique set of proteins and consequently has a unique function in the body. Such cells are said to be "specialized."
Expression of B cell-specific genes
Bone marrow stem cell
Mitosis
Cell differentiation
Specialized B cell

? Would you predict the pattern of gene expression of a heart cell to be more similar to that of a muscle cell in your arm or to a B cell?

With regenerative medicine, scientists hope to coax cells to both divide and differentiate on cue—not only to produce enough cells to build an organ, but also to have them be the right sort of cells, capable of carrying out the unique functions of a particular tissue.

Scientists are exploring several different approaches to regenerative medicine, although not all of them are equally far along. One approach uses drugs to stimulate stem cells in the body to grow and differentiate into the tissues that need repairing. In some ways, this approach represents the ultimate goal of regenerative medicine, but its use is still a long way off, mostly because scientists still have a lot to learn about stem cells.

"Even the dumbest stem cell is smarter than the smartest neuroscientist," says Evan Snyder, program director at the Burnham Institute for Medical Research in La Jolla, California. "The cells are making stuff that we might not be able to identify for centuries."

Another approach, somewhat further along, is to remove stem cells from the body, stimulate them to reproduce and differentiate, and then re-implant the cells into a patient with a damaged tissue or organ. This approach has been used in patients with damaged heart tissue following heart attacks and in those with macular degeneration of the eye—two cases where it has shown promise.

The third approach is one that Atala has used to create new bladders, urethras, ears, and other organs: seeding cells on a biodegradable scaffold outside the body, growing a new organ, then implanting it in a patient. He has also begun using cadaver-derived organs as the scaffold. These organs are washed of their cells, leaving only the protein (mostly collagen) matrix that imparts structure and shape to the organ. This matrix then serves as a scaffold for new cells from the patient's own body to grow on and in **(INFOGRAPHIC 9.5)**.

Building Organs in 3-D

In Atala's lab at Wake Forest, approximately 300 researchers are working to engineer more than 30 different body parts. They have had the most success so far with simple structures, such as bladders, urethras, and ears, which don't have a lot of blood vessels or nerve connections. These organs are being used now to treat patients who need them. When it comes to building more complex solid organs, such as a heart or a kidney, they still have a ways to go. But the stakes could not be higher: about 90% of people on the transplant list are waiting for a kidney.

A major hurdle in building more-complex organs is the intricacy of the design. A bladder is essentially a balloon. But solid organs are much more complicated. A kidney, for example, consists of millions of tiny filters called nephrons that clean blood. And each nephron consists of a delicate array of tubes and vessels that cannot feasibly be re-created by hand. One way that scientists are attempting to solve this problem of structural complexity is by using computers to design the delicate structures on-screen and then print the results.

Until recently, most bioprinters were simply modified ink jet printers with cartridges filled with cells instead of ink. The printers that now do this work are specifically designed and

Hyun-Wook Kang, Wake Forest University Institute of Regenerative Medicine

A bioprinter filled with cellular "ink" printing a human kidney.

INFOGRAPHIC 9.5 Regenerative Medicine: Four Approaches

Regenerative medicine techniques coax stem cells to divide and differentiate into specific types of cells for use in repairing damaged tissue. Stem cells can be stimulated within the tissues of the body, or stimulated outside body before implantation.

Therapeutic drugs stimulate stem cells within the body to repair damaged tissue.

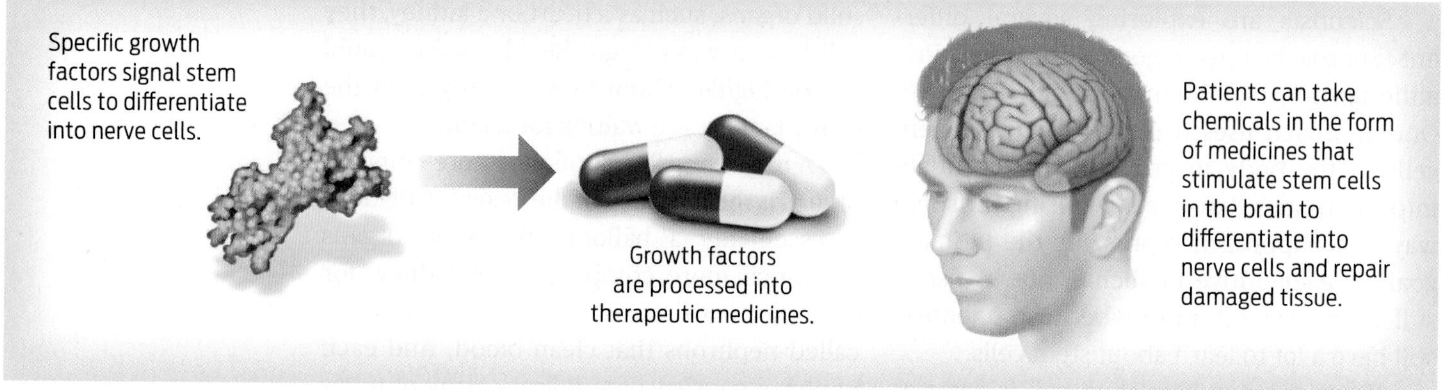

Stem cells are stimulated outside the body and then injected to repair damaged tissue.

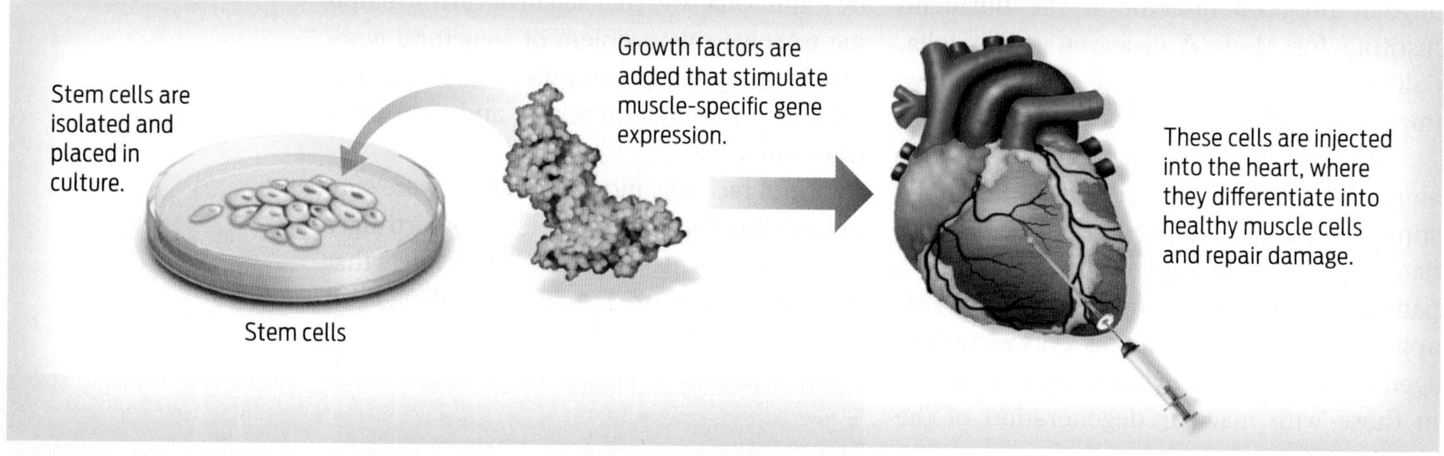

Stem cells are isolated and placed in culture.
Stem cells
Growth factors are added that stimulate muscle-specific gene expression.
These cells are injected into the heart, where they differentiate into healthy muscle cells and repair damage.

built for the purpose of making organs. They have different wells for different types of cells and the synthetic polymers that make up the biodegradable scaffold. Atala and his group at Wake Forest have used bioprinters to build miniature kidneys that reproduce the internal complexity of the organ. He printed one of the organs before a live audience during a TED Talk in 2011. These experimental prototypes aren't ready to be used in people, but they do show what is possible with the technology.

A second difficulty with engineering more-complex solid organs like livers, hearts, and kidneys is ensuring that all the cells in the organ are linked to an adequate blood supply. In normal tissues, an intricate web of capillaries supplies blood throughout the tissue (see Chapter 27). Without this blood supply, an engineered organ cannot be more than a centimeter or so thick—otherwise, the cells in the middle of the structure will die from lack of oxygen and nutrients. For that reason, Atala's group is experimenting with printed organs designed to incorporate tiny channels that pervade the organ. The hope is that these

Biodegradable scaffolds are used to grow tissues and organs outside the body, which are then transplanted to replace damaged tissue.

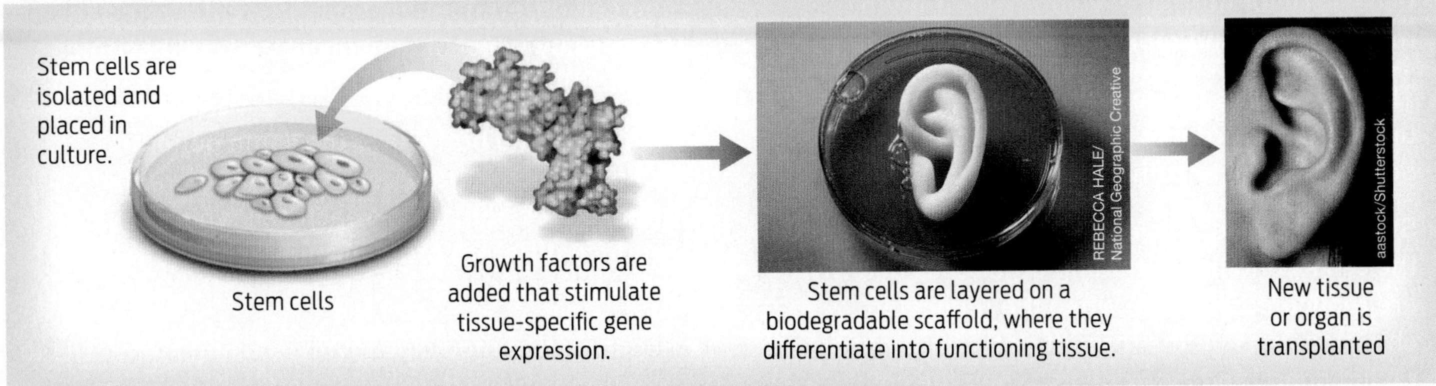

Cadaver organs are washed of their cells and reseeded with stem cells from the patient before they are transplanted to replace diseased organs.

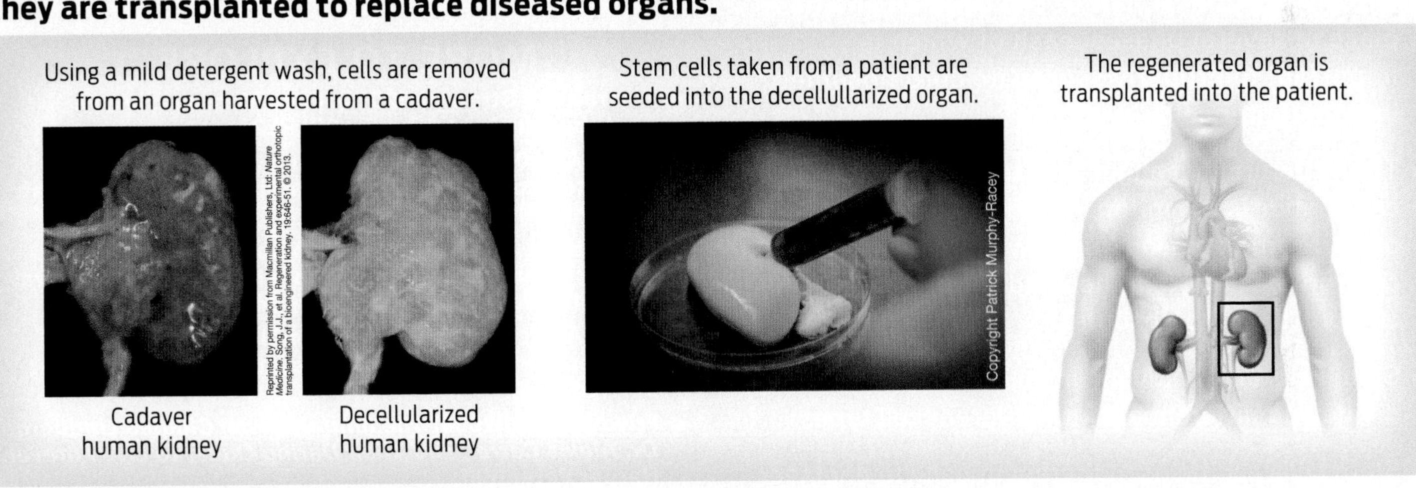

Using a mild detergent wash, cells are removed from an organ harvested from a cadaver.
Reprinted by permission from Macmillan Publishers, Ltd: Nature Medicine. Song, J.J., et al. Regeneration and experimental orthotopic transplantation of a bioengineered kidney, 19:646-51, © 2013.
Cadaver human kidney
Decellularized human kidney
Stem cells taken from a patient are seeded into the decellullarized organ.
Copyright Patrick Murphy-Racey
The regenerated organ is transplanted into the patient.

? Which approaches rely on the use of chemicals to stimulate cells to divide or differentiate, and which do not?

channels can be seeded with cells to recreate a functioning network of capillaries. "It's a tough challenge, but I think it's doable," Atala says.

Cadavers could help, too. Cadaver-derived scaffolds have the advantage of retaining the body's natural design of channels and other spaces. Atala's group has used the approach to regrow rabbit penises in the lab that can be successfully transplanted and used by the bunny recipients. This research, which is being supported by the U.S. military, may one day help soldiers whose pelvic regions were harmed by explosive devices.

Atala looks forward to the day, in the not too distant future, when patients will be able to order just about any organ they need, designed to work perfectly in their own body. No longer will patients die while waiting for an organ donor. The goal is straightforward, he says: "With the field of regenerative medicine, the hope is to be able to have tissues and organs available for patients that wouldn't otherwise have them." ■

CHAPTER 9 SUMMARY

- Cell division is a fundamental feature of life, necessary for normal development, growth, maintenance, and repair of the body.
- The cell cycle is the sequence of steps that a cell undergoes in order to divide. During the cell cycle, one parent cell becomes two daughter cells.
- Stages of the cell cycle include interphase, when the cell's contents, including organelles and DNA, are replicated; mitosis, when the replicated chromosomes (in the form of sister chromatids) are pulled apart; and cytokinesis, when the cell physically divides into two daughter cells.
- Mitosis is a critical part of cell division and takes place in several phases, each of which is important to properly segregate chromosomes into daughter cells.
- The structure that aligns and segregates sister chromatids is the mitotic spindle; it is made up of microtubule-based fibers, which attach to centromeres and lengthen and shorten to push and pull the chromosomes.
- Stem cells are relatively unspecialized cells that can divide and differentiate into different cell types.
- Stem cells can be used therapeutically to engineer or regenerate tissues and organs.
- Making new tissues requires both cell division and cell differentiation. Cell differentiation is the process by which an unspecialized cell becomes a specialized cell with a unique function.
- All cells in the body have the same genome but express different genes. Such differential gene expression causes each cell type to produce different proteins and to have different functions.
- Regenerative medicine involves using stem cells to repair or regrow damaged tissues and organs.

MORE TO EXPLORE

- **Portraits of Strength (2014): Luke Massella: www.youtube.com/watch?v=5-q9BLRL2ns**
- **Atala, A. (2011) TED Talk: Printing a human kidney: www.ted.com/talks/anthony_atala_printing_a_human_kidney.html**
- **Kang, H. W., et al. (2016) A 3D bioprinting system to produce human-scale tissue constructs with structural integrity. *Nature Biotechnology* 34(3): 312–319.**
- **Atala, A., et al. (2006) Tissue-engineered autologous bladders for patients needing cystoplasty. *The Lancet* 367:1241–46.**
- **Friedman, D. (2008) *The Immortalists: Charles Lindbergh, Dr. Alexis Carrel, and Their Daring Quest to Live Forever.* Harper Perennial.**

CHAPTER 9 Test Your Knowledge

DRIVING QUESTION 1 How and why do cells divide?

By answering the questions below and studying Infographics 9.1 and 9.2 and Up Close: Phases of Mitosis, you should be able to generate an answer for the broader Driving Question above.

KNOW IT

1 What process is critical for embryonic development, wound healing, and replacement of blood cells? (Hint: All these processes require new cells.)

2 In the cell cycle, DNA is replicated during

a. mitosis.
b. G_1.
c. S.
d. G_2.
e. cytokinesis.

3 Following mitosis and cytokinesis, daughter cells

a. are genetically unique.
b. are genetically identical to each other.
c. are genetically identical to the parent cell.
d. contain half of the parent cell's chromosomes.
e. b and c

4 During which stage of the cell cycle do sister chromatids separate from each other?

5 During which stage of the cell cycle are sister chromatids initially produced?

USE IT

6 A chemical is added to cells growing in a culture dish. This chemical blocks the completion of DNA replication. In what stage of the cell cycle will the cells be stuck?

7 Many drugs interfere with cell division. Why shouldn't pregnant women take these drugs?

8 What would be the result if a cell completed interphase and mitosis but failed to complete cytokinesis—how many cells would there be at that point, and how many chromosomes would those cells have in comparison to the parent cell?

DRIVING QUESTION 2 How does one specialized cell type differ from another, and how do stem cells differentiate into these specialized cells?

By answering the questions below and by studying Infographic 9.4, you should be able to generate an answer to the broader Driving Question above.

KNOW IT

9 You scraped your knee when you tripped over a curb playing Pokemon Go. As your injury heals, what cell type is likely to be dividing to replace the cells lost when you injured yourself?

10 In order to coax a stem cell to differentiate into a specific type of specialized cell, what has to happen?

a. It must acquire new genes
b. it must eliminate unnecessary genes
c. it must be express a specific subset of genes
d. a and b

11 Relative to one of your liver cells, one of your skin cells

a. has the same genome (that is, the same genetic material).
b. has the same function.
c. has a different pattern of gene expression.
d. a and c
e. b and c

12 Which of the following is/are differences between stem cells and mature, specialized cells?

a. most mature specialized cells do not divide
b. stem cells can divide indefinitely
c. stem cells have more genes in their genomes than mature specialized cells
d. all of the above
e. a and b

13 Is the genome of stem cells larger than that of specialized cells?

a. yes, because stem cells need the genes found in every cell type, whereas specialized cells need only a subset of all the genes
b. yes, because stem cells express more genes than do specialized cells
c. no, because all cells in a person have the identical set of genes in their genome
d. no; stem cells have a smaller genome, because they are equivalent to sperm and eggs (which are haploid—they have half the chromosome number) in that they can potentially create an entire individual
e. no; stem cells have a smaller genome because they express only a subset of genes

USE IT

14 Specific proteins expressed by specific cell types can be used as markers to both identify and isolate specific cell types from a population of cells. The table below provides a list of markers specifically associated with different cell types.

Marker Protein	Cell Type
Collagen type II	Cartilage cells
MAP2	Neurons
Myosin heavy chain	Cardiac muscle
Telomerase	Stem cells

A researcher has isolated stem cells and is coaxing them to differentiate into cardiac muscle cells to treat a patient with damaged heart muscle due to a heart attack. After the treatment, three populations of cells have been identified:

- Population A expresses MAP2
- Population B expresses Collagen type II
- Population C expresses Myosin heavy chain

Which population should be injected into the patient? Explain your answer.

15 The surface of the very upper layer of your skin consists of specialized cells known as keratinocytes. You shed cells from this layer of skin every day. If you sustain a burn and lose skin tissue, could your shed keratinocytes be used to generate replacement tissue in a dish? Why or why not?

DRIVING QUESTION 3 What is regenerative medicine, and what are specific approaches to repairing or replacing organs?

By answering the questions below and by studying Infographics 9.3 and 9.5, you should be able to generate an answer to the broader Driving Question above.

KNOW IT

16 What are the pros and cons of receiving an organ transplant versus growing a replacement organ from one's own cells?

17 Nerve cells (neurons) are highly specialized and generally don't divide. How does this characteristic help explain the severity of spinal cord injuries?

18 Cancer chemotherapy can damage the epithelial cells that line the digestive tract, leading to nausea and vomiting. These symptoms generally resolve after chemotherapy. What cell type is likely responsible for the regeneration of the lining of the digestive tract?

a. smooth muscle cells
b. epithelial cells
c. stem cells
d. neurons

apply YOUR KNOWLEDGE

BRING IT HOME

19 There are specific types of stem cells in umbilical cord blood from newborns. These stem cells have the capacity to differentiate into any type of blood cell and can replace bone marrow cells. Some parents choose to "bank," or preserve, cord blood containing these stem cells, in case their child might benefit from a matched transplant in the future. One company offers this banking service starting at $1,750 for the first year. If you were a new parent, would you consider banking your baby's cord blood? Discuss the factors contributing to your decision.

apply YOUR KNOWLEDGE

MINI CASE

20 A patient's airway (the trachea) has been severely damaged. The trachea consists of a hollow tube made of cartilage, chondrocytes (cells that help maintain the cartilage by producing the proteins that make up cartilage), and a lining made up of specialized epithelial cells. Design a regenerative medicine approach to help this patient. Include the components that you will use (and their source), the steps you will follow, and a justification for your approach.

apply YOUR KNOWLEDGE

INTERPRETING DATA

21 A great deal of research is being carried out to design scaffolds for replacement organs. One group has been studying "decellularized" liver scaffolds. These researchers obtain human livers and treat them to remove all the cells, leaving just a scaffold. When these scaffolds are placed in mice, there is no evidence of inflammation or immune rejection, suggesting that these decellularized scaffolds are not rejected by the immune system. The researchers then wanted to determine whether or not these decellularized scaffolds will support the growth of liver cells. They seeded small sections of the scaffolds with human liver cancer cells, incubated for 1, 2, and 3 weeks, and counted the number of living cells at each time.

Incubation Time	No. of Cells
7 days	50
14 days	120
21 days	140

Data from Mazza et al. (2015) Scientific Reports 5:13079 | DOI: 10.1038/srep13079

a. Graph these data.

b. Do the decellularized scaffolds appear to support the growth of human cells?

c. What are the implications of this experiment for regenerative medicine?

10 Mutations and Cancer

FIGHTING FATE

When cancer runs in the family, ordinary measures are not enough

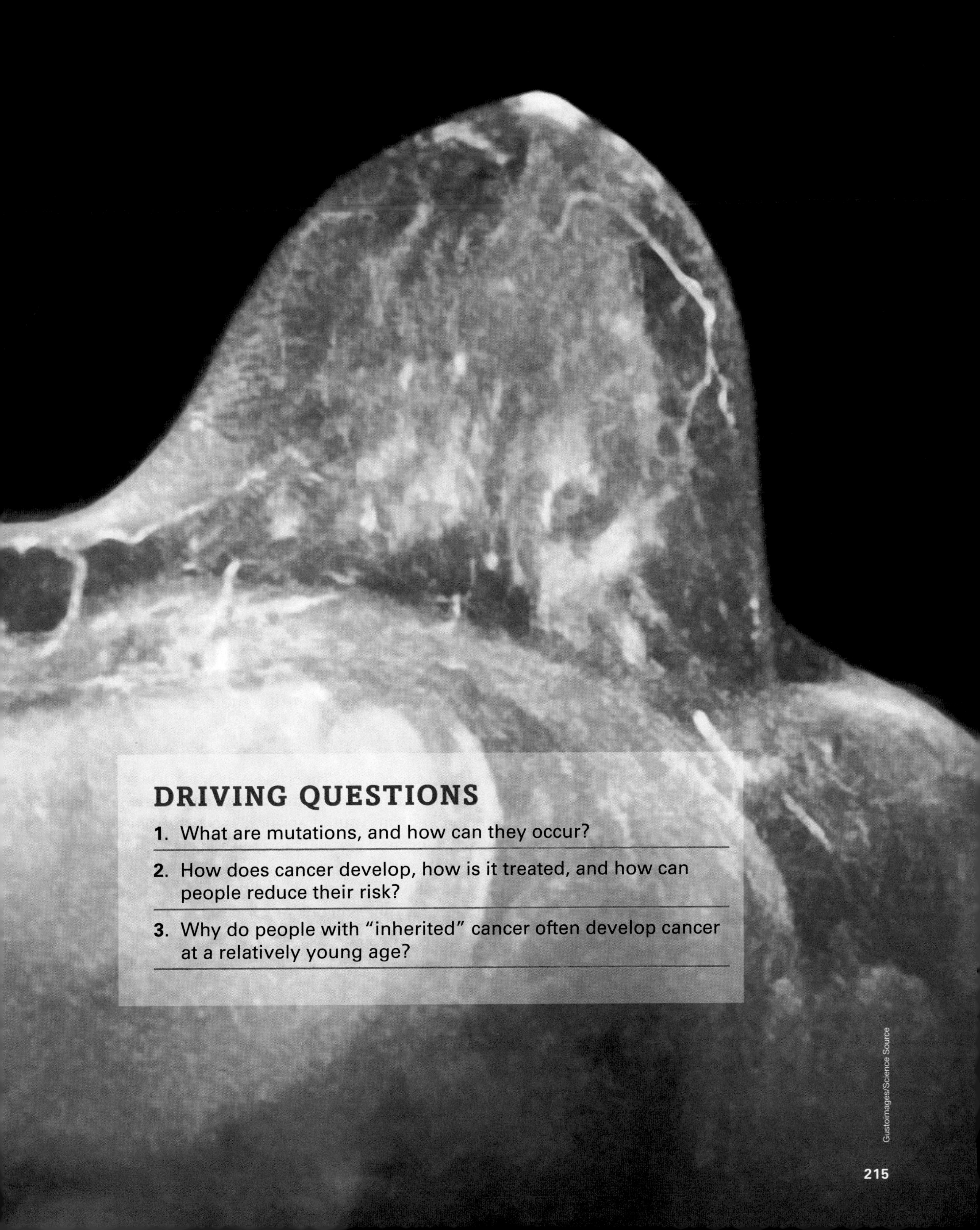

DRIVING QUESTIONS

1. What are mutations, and how can they occur?
2. How does cancer develop, how is it treated, and how can people reduce their risk?
3. Why do people with "inherited" cancer often develop cancer at a relatively young age?

Gustoimages/Science Source

Lorene Ahern of Twinsburg, Ohio, wasn't totally surprised when she was diagnosed with breast cancer. "Half of me was expecting it all my life and part of me was saying, 'No, this won't happen to me,'" says the 47-year-old mother of two. She knew that her risk of cancer might be higher than average—her mother had died of cancer at 49. But until the day she was diagnosed, Ahern, who took good care of herself and had a healthy lifestyle, had never fully believed she would develop cancer.

There was more bad news in store for Ahern. About a year after she received the diagnosis of breast cancer, Ahern had DNA extracted from her blood and tested for mutations in two genes—*BRCA1*, located on chromosome 17, and *BRCA2*, located on chromosome 13 (*BRCA* stands for "*br*east *ca*ncer"). **Mutations** are changes in the nucleotide sequence of DNA. Women who are born with mutations in either of the two genes Ahern had tested have a much higher risk of developing breast and ovarian cancers; men with these mutations are at higher risk for breast and prostate cancers. The test was positive: Ahern had a mutation in one of her copies of the *BRCA1* gene, which meant that she was at high risk for other cancers as well. Moreover, she could have passed on this mutation to her two children, putting them at risk.

MUTATION
A change in the nucleotide sequence of DNA.

Aside from nonmelanoma skin cancer, breast cancer is the most common cancer to affect women worldwide. More than 250,000 women in the United States are diagnosed with breast cancer each year, according to the National Cancer Institute. For most women, the lifetime risk of developing breast cancer is about 12%, or 1 in every 8 women. For women with mutations in *BRCA1* or *BRCA2*, however, the risk is much higher: on average, a 45% to 65% lifetime risk of developing breast cancer and a 10% to 40% risk of developing ovarian cancer. In some families with particular *BRCA1* mutations, the risk of getting breast cancer can run as high as 80% **(INFOGRAPHIC 10.1)**.

The good news is that studies have shown that diet and lifestyle changes can dramatically cut a woman's risk of getting cancer—just quitting smoking cuts the risk by 30%. The bad news is that prevention is not that simple for women with inherited predispositions to breast cancer—for this group, diet and lifestyle changes make less of a difference. That's because, as scientists have learned, a genetic mistake stacks the deck against them, creating biological conditions conducive to cancer development. What's more, even after receiving treatment for cancer, women with *BRCA* mutations are at risk of developing a new cancer. But these women do have options that can drastically reduce their risk

INFOGRAPHIC 10.1 Mutations in the *BRCA* Genes Increase the Risk of Cancer

People born with certain mutations in the genes *BRCA1* or *BRCA2* have a higher risk of developing cancer than people with only normal versions of the *BRCA* genes. A genetic test can ascertain whether a person carries any of the *BRCA* high-risk mutations.

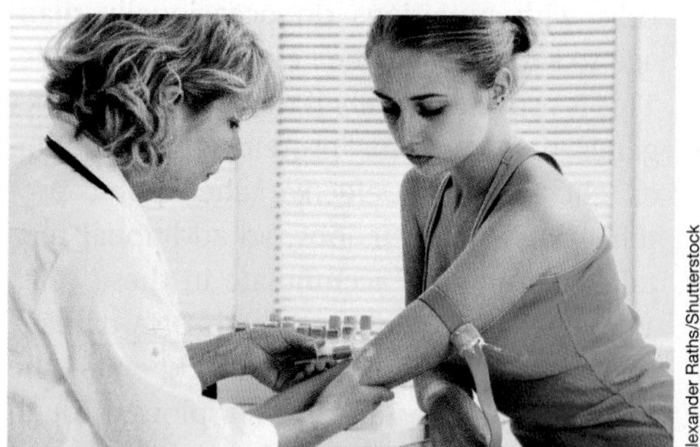

Genetic testing is performed by extracting DNA from the nuclei of white blood cells and looking for mutations in the genes of interest.

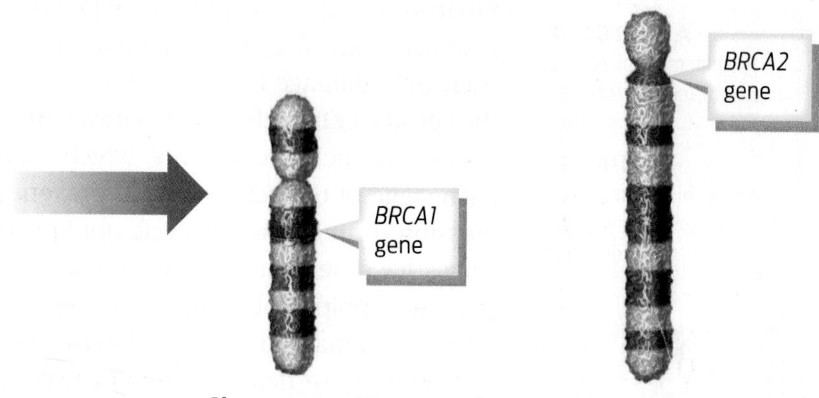

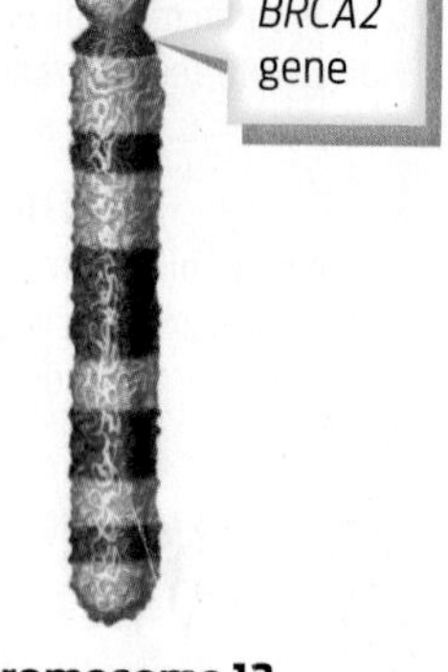

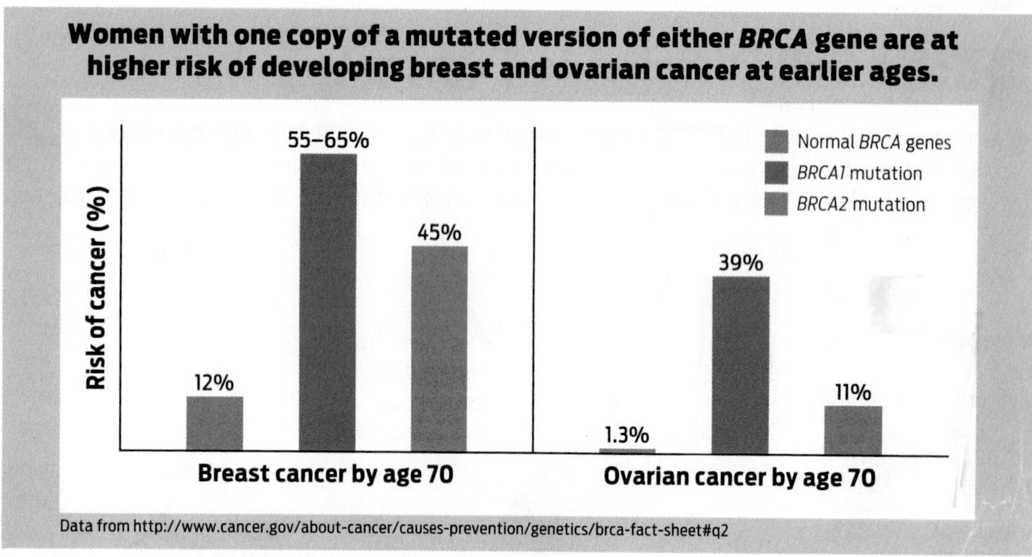

? Which *BRCA* gene causes the greater increase in cancer risk when it is mutated?

of getting cancer and options for treating it if it occurs.

What Is Cancer?

Cancer is a disease of unregulated cell division. As we saw in Chapter 9, cell division is an essential part of the growth and repair of tissues. Normally, cells divide only in response to appropriate growth signals. When cells no longer need to divide—for example, when a wound has healed or worn-out tissues have been replaced—these growth signals are turned off and cells take a break. Cancer cells, on the other hand, divide even in the absence of growth signals and without stopping. Cancer is cell division run amok.

What causes cells to "go rogue"? Cancer results when cells accumulate DNA sequence changes—mutations—that interfere with the

CANCER
A disease of unregulated cell division: cells divide inappropriately and accumulate, in some instances forming a tumor.

CELL CYCLE CHECKPOINT
A cellular mechanism that ensures that each stage of the cell cycle is completed accurately.

APOPTOSIS
Programmed cell death; often referred to as cellular suicide.

TUMOR
A mass of cells resulting from uncontrolled cell division.

orderly steps of the cell cycle. Mutations can happen in a variety of ways. Every time a cell replicates its DNA, for example, there is a small chance that it will make a mistake–insert the wrong nucleotide, for instance. Environmental insults, like smoking or a bad sunburn, can also damage DNA. Normally, such DNA damage is caught and repaired by the cell at a **cell cycle checkpoint.** Cells have a series of such checkpoints, which monitor each stage of the cell cycle and prevent progression through the cell cycle until previous stages have been successfully completed. At one checkpoint, for example, proteins scan DNA for damage or incorrect base pairing. If problems are detected, one of two things happens: either the cell ramps up DNA repair mechanisms, giving itself time to fix DNA mistakes, or, in the case of severe and irreparable damage, the checkpoints direct a cell to commit suicide in a process called **apoptosis** (**INFOGRAPHIC 10.2**).

Even with these checkpoint mechanisms, cells with mutations do occasionally manage to slip through the cell cycle and divide. This is especially likely if the mutations affect genes that code for proteins that function as checkpoints. Defective checkpoint proteins don't do their jobs, so additional mistakes continue to accumulate in these cells. When cells accumulate enough DNA damage to interfere with multiple checkpoints, the result is cancer. Cancer cells plow through the cell cycle unimpeded, divide uncontrollably, and in many cases eventually form a mass of cells called a **tumor**.

INFOGRAPHIC 10.2 Cell Division Is Tightly Regulated

Normal cells have mechanisms to ensure that cell division is carried out accurately and only when necessary. Cell cycle checkpoints regulate a cell's progress through the cell cycle. Checkpoints prevent a cell from progressing to the next stage until it accurately finishes the current stage. Regulated cell division ensures that adequate cell number and healthy tissue structure are maintained in the body.

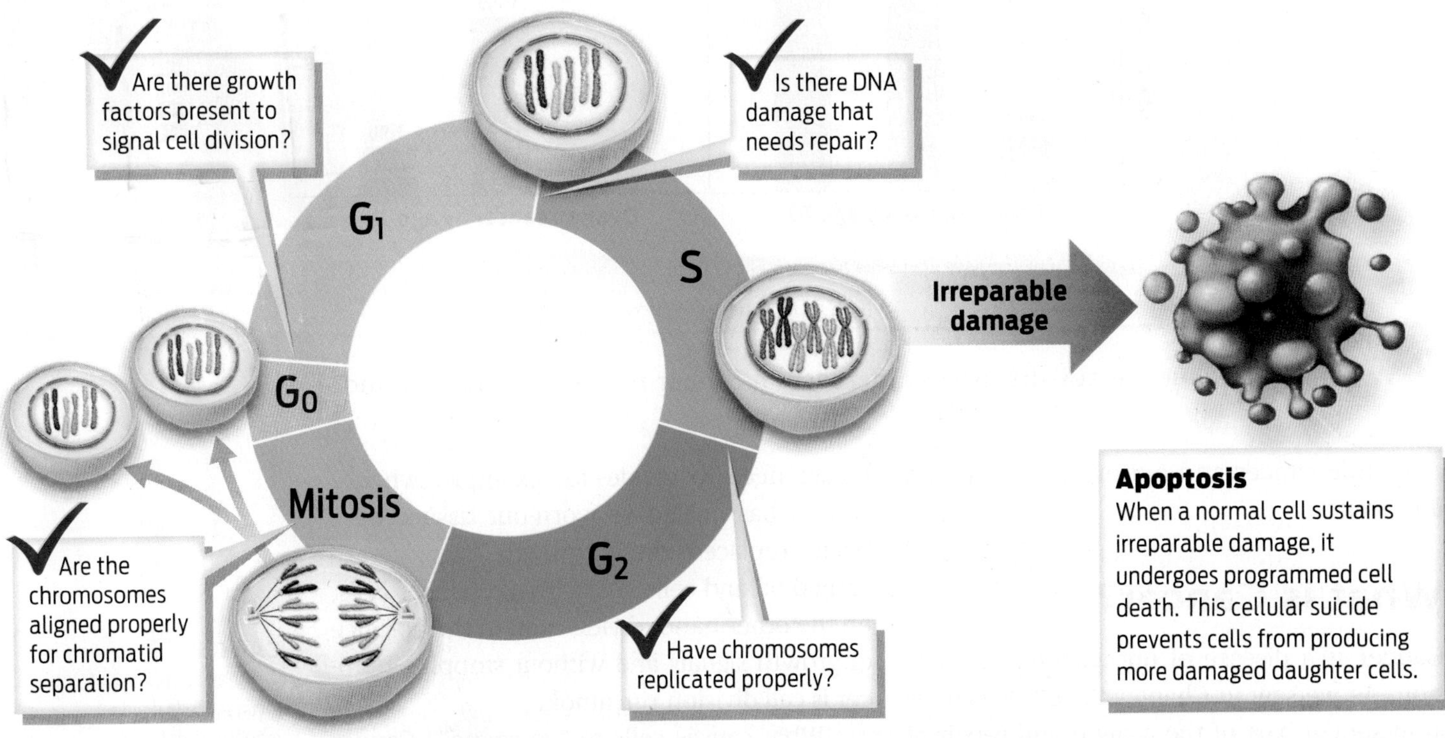

? Why is it important to detect and repair DNA damage (including mutations) before S phase?

According to the American Cancer Society, one in two males and one in three females will develop cancer in their lifetimes. The two most common non-skin cancers in males are prostate and lung; in females they are breast and lung. Cancer is the second leading cause of death among both men and women in the United States, right after heart disease. Cancer kills by crowding out normal cells and invading organs, causing them to fail. Cancer cells also secrete a variety of chemicals that wreak havoc on the body's biochemistry.

Women who, like Ahern, have a genetic predisposition to getting cancer are often said to have "hereditary" or "inherited" cancer. This does not mean that their cancer was passed from parent to child, the way that eye or hair color is. It means that they have inherited a genetic mutation, from one or both parents, that makes the development of cancer much more likely. In other words, the cancer itself is not inherited, but the risk of getting it is.

Doctors have known for decades that breast cancer runs in certain families: a woman's chance of getting the disease is greatly increased if she has a sister or mother who has it. But it wasn't until the 1990s that researchers homed in on the reason: mutations in a particular region of chromosome 17 are unusually common in these families. By the middle of the decade, scientists had isolated the specific mutated gene, which they called *BRCA1*, and determined its function. *BRCA2* was not far behind.

In their normal form, the BRCA1 and BRCA2 proteins help to repair DNA damage, specifically a type of damage called a double-strand break. When these proteins are mutated, cells can no longer properly repair this type of damage. As a result, additional genetic mistakes–additional mutations–accumulate in these cells, eventually leading to cancer.

Scientists estimate that inherited mutations like those in *BRCA* genes account for about 5%-10% of all cancers. The other 90%-95% of cancers are caused by non-inherited (that is, acquired) mutations that occur during a person's lifetime.

Inherited vs. Acquired Cancer Mutations

The vast majority of cancers are caused by mutations that occur during a person's lifetime—acquired mutations. A small minority of cancers run in families, due to inheritance of mutations in cancer-associated genes.

Breast Cancers

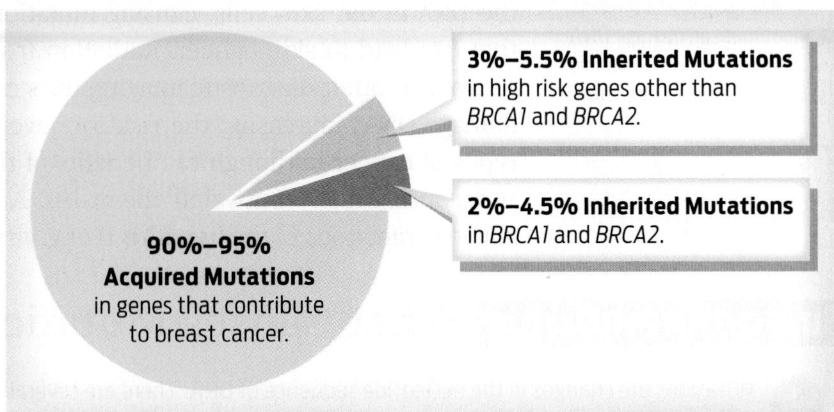

All Cancers

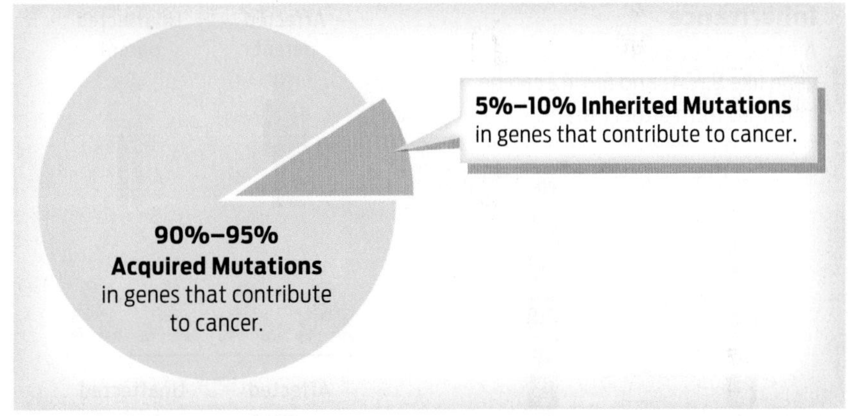

Causes and Consequences of Mutation

Though checkpoint mechanisms ensure that the vast majority of DNA replication mistakes are fixed, on average 1 uncorrected mistake occurs for every 10 billion DNA base pairs that are replicated. That may not seem like a lot, but when you consider that each human cell has 6 billion total base pairs, and that cells are replicated trillions of times in our lifetimes, the number of mutations that can occur in any given cell quickly adds up.

Moreover, our DNA is continually being bombarded by environmental insults that can

CARCINOGEN
Any substance that causes cancer. Most carcinogens are mutagens.

MUTAGEN
Any chemical or physical agent that can damage DNA by changing its nucleotide sequence.

add to our mutation burden. Environmental insults from chemicals, ultraviolet light, radiation, and infectious agents like viruses can damage DNA, causing mutations. Exposure to ultraviolet light, for example, damages the DNA in our skin cells, causing mutations that can lead to skin cancer. Radiation from X-rays and other diagnostic imaging tests can damage DNA, increasing the risk for several types of cancer (although the benefits of the procedure usually outweigh the risks). And chronic infections of the hepatitis B or C virus can damage liver cells, causing mutations that can lead to liver cancer.

Substances that are known to cause cancer, or increase its risk of forming, are called **carcinogens.** Most carcinogens are also **mutagens**–physical or chemical agents that cause mutations in DNA. Not all mutagens originate outside the body. For example, some of the reactions that occur in the mitochondria during cellular respiration (see Chapter 6) produce DNA-damaging molecules called free radicals **(INFOGRAPHIC 10.3)**.

INFOGRAPHIC 10.3 What Causes Mutations?

Mutations are changes in the nucleotide sequence of DNA. There are several ways that a person can end up with a mutation: it may have been inherited; it may have occurred randomly during DNA replication; or it may have been the result of an environmental insult.

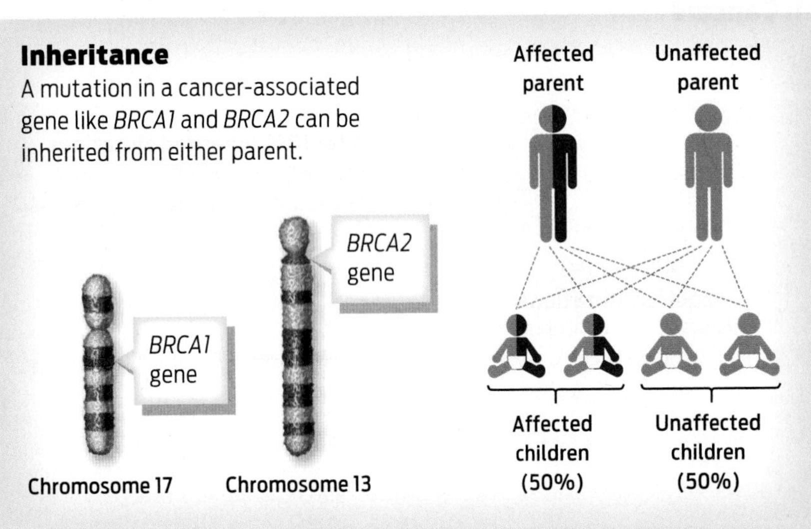

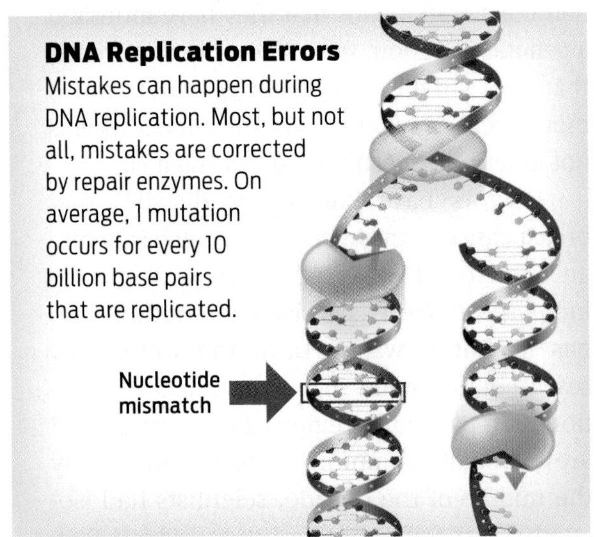

DNA Replication Errors
Mistakes can happen during DNA replication. Most, but not all, mistakes are corrected by repair enzymes. On average, 1 mutation occurs for every 10 billion base pairs that are replicated.
Nucleotide mismatch

Mutagens

Many components of the environment, our food, and even our cells can cause mutations.

Radiation

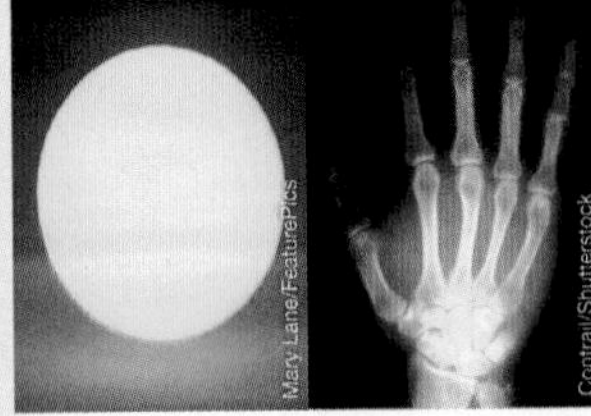

UV radiation X-rays

Chemicals

Pollution and pesticides Smoking

Alcohol

Char (blackened bits) on meats cooked at high temperatures

Infectious Agents

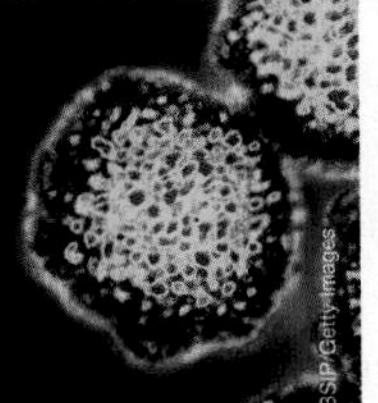

Some viruses, like hepatitis C

Cellular Reactions

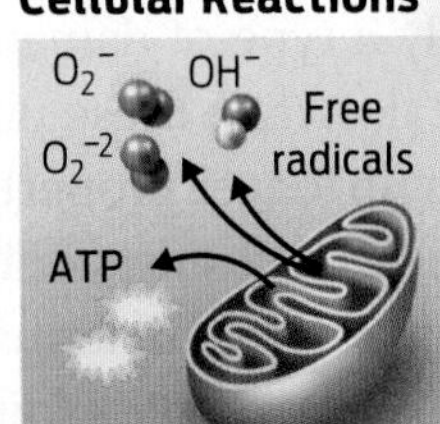

Cellular processes produce mutagenic free radicals

? What three things could you do to reduce your risk of developing a mutation in *BRCA1*?

TABLE 10.1 Types of Mutations and Their Effects

TYPE OF MUTATION	EXAMPLE	EFFECT ON PROTEIN FUNCTION
Original DNA sequence: No mutation	DNA HAS ALL YOU CAN ASK FOR	
Silent point mutation: Change one nucleotide to another but do not change the amino acid sequence of the protein.	DNA HAS ALL YOO CAN ASK FOR	The amino acid does not change, so the protein functions normally.
Missense point mutation: Change one nucleotide to another, resulting in a different amino acid sequence of the protein.	DNA HAS ALL LOU CAN ASK FOR	Slight amino acid alteration changes the shape and function of the protein.
Nonsense point mutation: Change the DNA sequence in a way that results in an early stop codon.	DNA HAS ALL YOU	The protein is too short, which makes it not functional.
Insertion mutation (frameshift): Insert one or more extra nucleotides into the DNA sequence. Insertions often alter the reading frame of the gene such that every codon after the insertion is altered.	DNA HAS ALL YYO UCA NAS KFO R	All amino acids after the insertion are altered. The protein made is not functional.
Deletion mutation (frameshift): Remove one or more nucleotides from the DNA sequence such that the reading frame is altered. Every codon after the deletion is altered.	DNA HAS ALY OUC ANA SKF OR ↑	All amino acids after the deletion are altered. The protein is not functional.
Inversion mutation: A group of DNA nucleotides are turned over to be read in reverse order. This changes the amino acid sequence in that location.	DNA HAS ALL YOC UAN ASK FOR	Slight amino acid alteration changes the shape and function of the protein.
Translocation mutation: The movement of segments of DNA from one chromosome to another can fuse portions of different genes together	DNA HAS ALL YOU CAN EAT ⇅ THE DOG AND CAT ASK FOR	The addition of readable codons adds amino acids to the protein, which can result in a protein with a novel function.

Mutations can take several different forms. A mutation can be a point mutation, a change in just one DNA nucleotide. Depending on where that point mutation occurs, it may change the amino acid sequence of the protein, or it may not. Point mutations that change the amino acid sequence are called missense mutations; those that do not change the sequence are called silent mutations. In other cases, one or more DNA nucleotides may be inserted or deleted from genes, changing the reading frame of that gene—that is, changing the way the nucleotides are divided into the triplet letters of DNA codons.

Whole blocks of DNA can be rearranged as a result of mutation. These kinds of rearrangements can include a segment of DNA that "flips" within its normal chromosomal location (this is called an inversion), or segments of DNA that trade places between different chromosomes (translocations). Large inversions and translocations can fuse portions of different genes together, creating new proteins with novel activity (**TABLE 10.1**).

Ultimately, the impact of a mutation on the function of the encoded protein depends on where in the gene the mutation occurs and on the effect of this change on the amino acid sequence of the protein. Some mutations have no effect on the amino acid sequence and therefore do not alter the function of the protein. In many cases where the sequence is changed, the new sequence alters the shape of the protein in a way that makes it nonfunctional. In other cases, the mutation changes the shape of the protein in a way that makes it overly active. If the proteins encoded by these genes play roles in the cell cycle, the mutations and corresponding changes in protein activities may eventually lead to cancer. The most common mutations in *BRCA* genes are deletions that hobble the function of the encoded proteins, making them ineffective at DNA repair **(INFOGRAPHIC 10.4)**.

Mutations that occur in body cells, such as those making up the breast or colon, are called acquired (or sporadic) mutations. These mutations are passed from cell to cell through mitosis, but not from parent to offspring, and so they are not inherited. Occasionally, mutations will occur in a germ cell—an egg or a sperm cell. These mutations can be passed on to offspring through sexual reproduction (see Chapter 11). Mutations inherited via germ cells are present in every cell of the adult person's body. And if that person has children, their children can inherit those mutations, too. This is how *BRCA* mutations were introduced into the population.

INFOGRAPHIC 10.4 Mutations in Genes Can Alter Amino Acid Sequence and Therefore Protein Function

Mutations alter the nucleotide sequence of DNA. If a mutation changes the coding region of a gene, the corresponding protein may be dysfunctional. When the protein in question helps regulate the cell cycle, cancer may result.

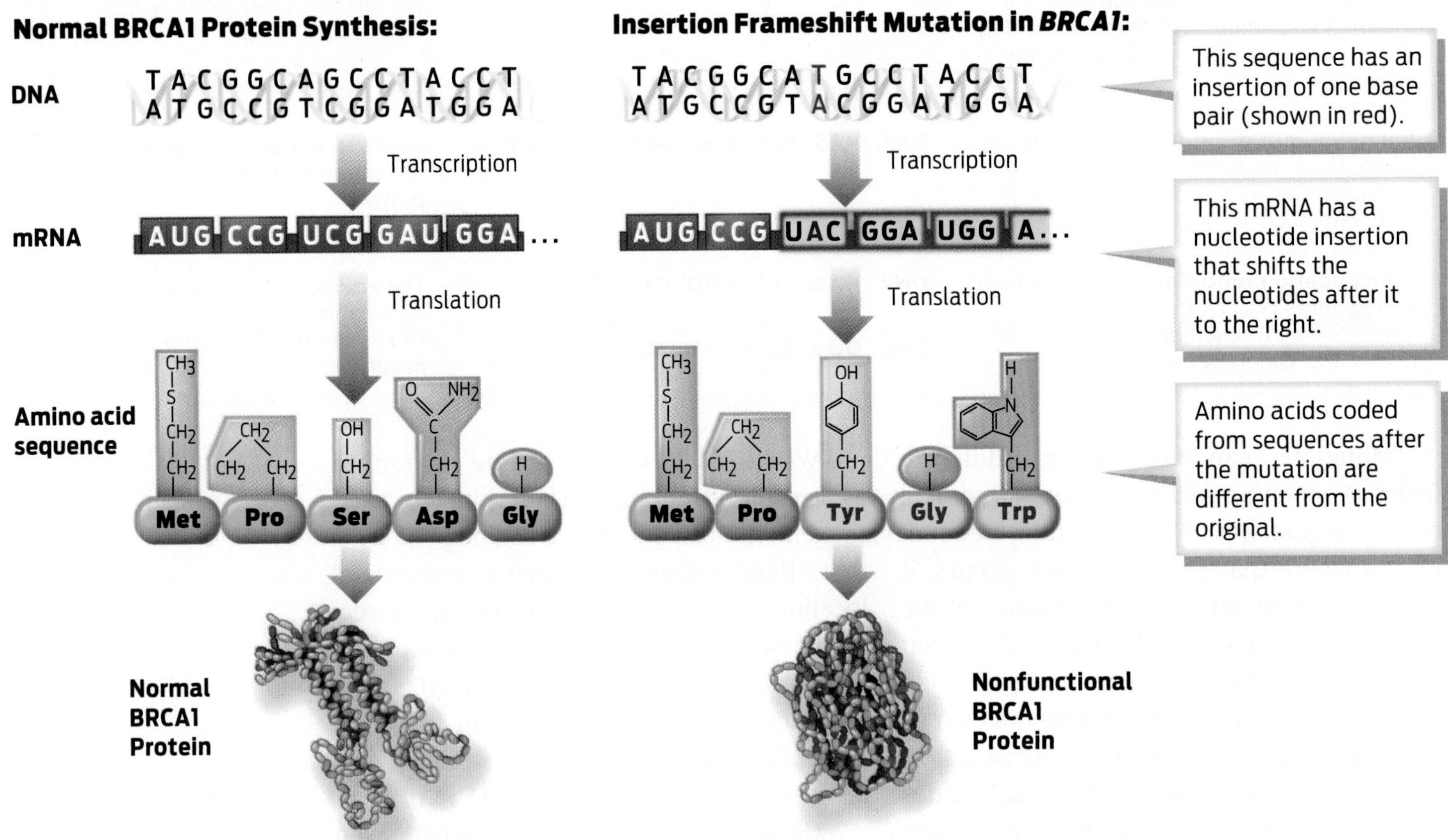

? Would a missense mutation have changed all the codons that follow it?

Inherited mutations aren't always bad; they do more than cause cancer and other genetic diseases. In fact, inherited mutations are the source of all the interesting variations we see from individual to individual. Whether you have blue eyes or brown eyes, red or black hair, are musically inclined or athletic–all these traits reflect the influence of different variations of genes, each originally the result of mutation. Different versions of the same gene, produced through mutation, are called **alleles.** Alleles are the basis of the genetic inheritance that makes each of us unique (see Chapter 11). People with blue eyes, for example, have alleles for eye color that are different from the ones people with brown eyes have.

Environmental insults, like smoking or a bad sunburn, can also damage DNA.

What's more, inherited mutations are the source of the genetic variation on which evolution acts. Without mutation, there would be no evolution at all, and the great diversity of living creatures we see around us would not have been possible. When it comes to surviving and reproducing, mutations can be harmful, neutral, or beneficial. Harmful and beneficial mutations will affect the fitness of an individual and may influence his or her ability to survive and reproduce, leading to evolution by natural selection (see Chapter 13).

Ethnic Groups and Genetic Disease

Though any family may carry mutated *BRCA* genes, certain ethnic groups are more likely than others to carry them. Ahern's background is Ashkenazi Jewish, meaning that her ancestors were Jews of German and Eastern European descent. Ahern's father was born in Germany, immigrating to the United States in 1939; her maternal grandfather was born in Russia.

A number of historical factors have made the Ashkenazi Jewish population more susceptible than others to genetic diseases. First, they descend from a small group of people that separated from a larger group. And second, members of the population usually marry within the community. Consequently, Ashkenazi Jews are an example of an ethnic group that has a more uniform genetic background than the general population, and is more likely to pass on alleles of genes associated with certain genetic diseases to future generations.

Scientists have discovered more than 1,000 genetic diseases in the general population, but most of them are rare. In Ashkenazi Jews, however, the prevalence of some of these diseases is increased 100-fold or more. Tay-Sachs disease, Gaucher disease, and Bloom syndrome are genetic diseases that all occur more frequently in this ethnic group than in the general population; approximately 1 in 25 Ashkenazi Jews carries disease alleles for at least one of these disorders **(TABLE 10.2)**.

Ashkenazi Jews are not the only ethnic group to have a higher incidence of certain genetic diseases than occurs in the general population. For example, people from Mediterranean, African, and Asian countries have higher rates of thalassemias–blood disorders that cause anemia. Sickle-cell anemia, another type of hereditary anemia, is more common among people of African descent. And the Amish have higher than typical numbers of people with polydactyly–having more than the typical number of fingers or toes–in this case, the result of Ellis-van Creveld syndrome, which has a genetic cause.

Ashkenazi Jews are also more likely than the general population to carry mutations in *BRCA1* and *BRCA2*. Some studies have found that approximately 3% of Ashkenazi women carry a mutated *BRCA1* gene, compared to only 0.2% of other women.

In Ahern's case, an inherited mutation in one of the copies of her *BRCA1* gene predisposed her to cancer. One mutated copy of

ALLELE
Alternative versions of the same gene that have different nucleotide sequences.

TABLE 10.2 Incidence of Hereditary Diseases in Different Populations

HEREDITARY DISEASE	CARRIER RATE IN IDENTIFIED POPULATIONS	CARRIER RATE IN GENERAL POPULATION
Ashkenazi Jewish Population		
Bloom syndrome	1 in 134 (0.7%)	Rare/unknown
BRCA1 and *BRCA2* mutations	1 in 40 (2.5%)	1 in 500 (0.2%)
Canavan's disease	1 in 55 (1.8%)	Rare/unknown
Cystic fibrosis	1 in 24 (4.2%)	1 in 38 (2.6%)
Familial dysautonomia	1 in 31 (3.2%)	Rare/unknown
Familial hyperinsulinism	1 in 68 (1.5%)	Rare/unknown
Gaucher disease, type 1	1 in 15 (6.7%)	1 in 100 (1.0%)
Tay-Sachs disease	1 in 27 (3.7%)	1 in 250 (0.4%)
African/African American Populations		
Sickle-cell anemia	1 in 14 (7.1%)	1 in 63 (1.6%)
Mediterranean Population		
Beta-thalassemia	1 in 6 (18.0%)	1 in 67 (1.5%)
Old Order Amish Population		
Ellis-van Creveld syndrome	1 in 8 (12.5%)	1 in 126 (0.8%)

Data from Jewish Genetic Diseases Consortium, Centers for Disease Control, Lahiry, P. (2008) *The Open Hematology Journal* 2(1), and D'Asdia et al. (2013) *Europ. J. Med. Genet.* 56: 80–87, 2013.

PROTO-ONCOGENE
A gene that codes for a protein that helps cells divide normally.

TUMOR SUPPRESSOR GENE
A gene that codes for a protein that monitors and checks cell cycle progression. When these genes mutate, tumor suppressor proteins lose normal function.

ONCOGENE
A mutated and overactive form of a proto-oncogene. Oncogenes drive cells to divide continually.

the *BRCA1* gene, inherited from one parent, is enough to disrupt DNA repair and significantly increase the risk of getting cancer.

How Cancer Develops

Inheriting certain *BRCA* alleles dramatically increases the risk of cancer, but it doesn't mean that the disease will necessarily develop. In most cases, hereditary cancer occurs only when additional, nonhereditary (that is, acquired) mutations in a cell accumulate. Similarly, the accumulation of harmful acquired mutations in a cell can lead to cancer, even in someone with no genetic predisposition. This is known as the multi-hit model of cancer, in which each "hit" is a mutation, and multiple hits are needed to cause the disease.

Mutations that cause cancer typically occur in two types of genes that influence the cell cycle: **proto-oncogenes** and **tumor suppressor genes.** Normal proto-oncogenes activate cell division, but only in response to appropriate signals. When proto-oncogenes are mutated, they can become permanently turned on, stimulating cells to divide all the time. In this state they are called **oncogenes**–literally, "genes that cause cancer." In other words, oncogenes are proto-oncogenes that have mutated to become permanently activated. *Her2*, a gene frequently mutated in certain types of breast cancer, is an example of a proto-oncogene.

Tumor suppressor genes make proteins that normally pause cell division, repair damaged DNA, or initiate apoptosis. Tumor suppressor genes cause cancer when they are inactivated by mutation. More than 50% of all cancers have a mutation in the *p53* tumor suppressor gene, for example. Without a properly functioning p53 protein, cells cannot commit suicide in response to DNA damage, and mutations accumulate in cells as a result. The BRCA proteins are also tumor suppressors.

"You can think of the suppressors as brakes and the oncogenes as the accelerators," says Thomas Sellers, a cancer biologist at the H. Lee Moffitt Cancer Center and Research Institute in Tampa, Florida. "They are sort of the yin and yang of each other." If a proto-oncogene is mutated, it's as if the accelerator is stuck down and cell division keeps going; if a tumor suppressor gene is mutated, it's as if the brakes don't work and the cell division cannot be stopped **(INFOGRAPHIC 10.5)**.

INFOGRAPHIC 10.5 Mutations in Two Types of Cell Cycle Genes Cause Most Types of Cancer

Checkpoint proteins regulate progression through the cell cycle. Proto-oncogenes push the cell to divide when the appropriate signals are present. Activating mutations in these genes push the cell to divide even in the absence of signals to divide. Tumor suppressors act to detect problems (e.g., DNA damage) and pause the cell cycle, or cause it to carry out apoptosis. Mutated tumor suppressors allow the cell to continue to divide, even if there is DNA damage.

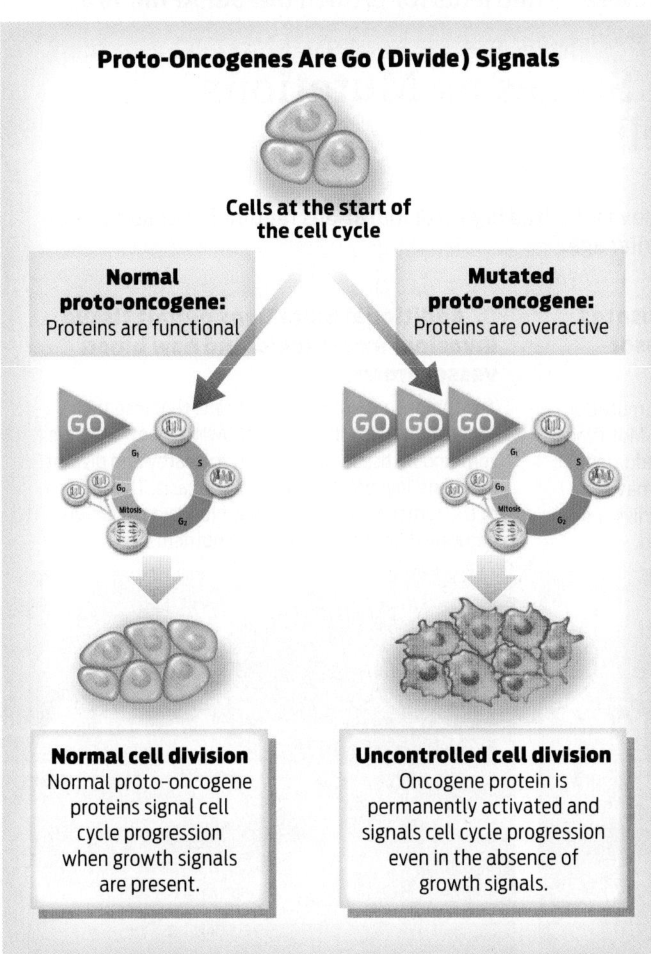
Proto-Oncogenes Are Go (Divide) Signals
Cells at the start of the cell cycle
Normal proto-oncogene: Proteins are functional
Mutated proto-oncogene: Proteins are overactive
GO
GO GO GO
Normal cell division
Normal proto-oncogene proteins signal cell cycle progression when growth signals are present.
Uncontrolled cell division
Oncogene protein is permanently activated and signals cell cycle progression even in the absence of growth signals.

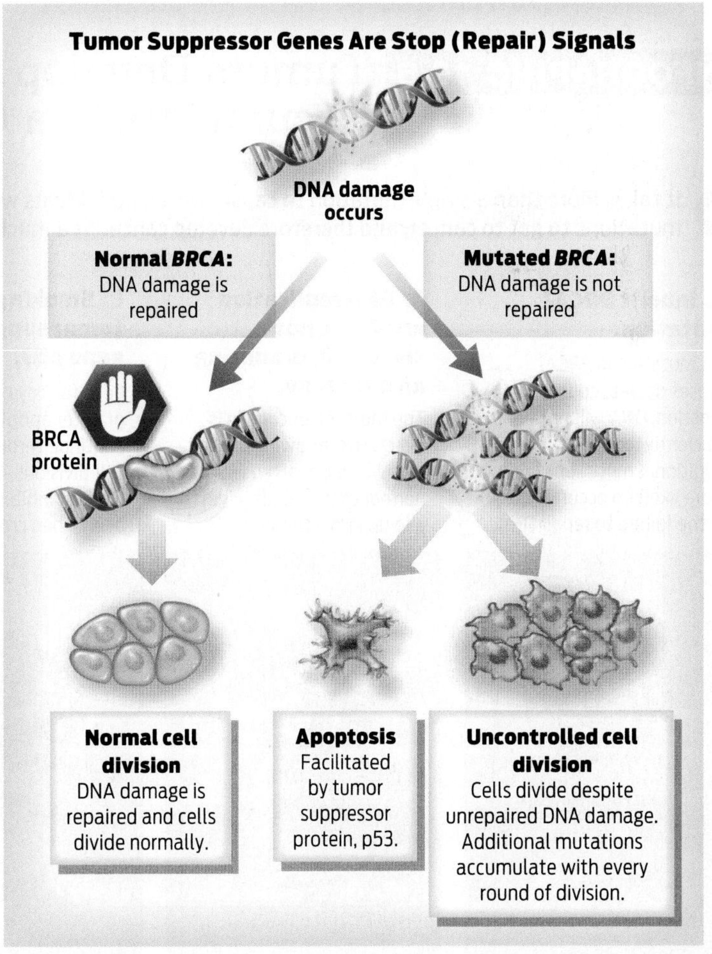
Tumor Suppressor Genes Are Stop (Repair) Signals
DNA damage occurs
Normal BRCA: DNA damage is repaired
Mutated BRCA: DNA damage is not repaired
BRCA protein
Normal cell division
DNA damage is repaired and cells divide normally.
Apoptosis
Facilitated by tumor suppressor protein, p53.
Uncontrolled cell division
Cells divide despite unrepaired DNA damage. Additional mutations accumulate with every round of division.

? A cell is dividing even though its DNA is damaged. What type of gene is most likely mutated in this cell?

BENIGN TUMOR A noncancerous tumor whose cells will not spread throughout the body.

MALIGNANT TUMOR A cancerous tumor whose cells can spread throughout the body.

METASTASIS The spread of cancer cells from one location in the body to another.

CONTACT INHIBITION A characteristic of normal cells that prevents them from dividing once they have filled a space and are in contact with their neighbors.

ANCHORAGE DEPENDENCE The need for normal cells to be in physical contact with another layer of cells or a surface.

It usually takes more than a single mutation in a cell to cause cancer. In most cases, a cell will become cancerous only after it has acquired multiple mutations–multiple "hits"–in several genes that regulate the cell cycle or repair DNA damage. The collection of mutated genes can include a combination of tumor suppressor genes that have lost their function and proto-oncogenes that have been activated to oncogenes. This is one reason cancer affects people more as they age: as cells accumulate mutations over time through exposure to carcinogens and repeated rounds of cell division, the chances increase that a cell will accumulate enough mutations to become cancerous.

After one or two hits, a **benign tumor** may form. Cells in a benign tumor divide more frequently than do cells in normal tissues, but they do not spread to other areas and so are usually much less dangerous. After several more mutations, a **malignant tumor** may result. Malignant tumors have the capacity to **metastasize,** their cells invading other tissues and spreading to other parts of the body.

Cancer cells have other distinctive properties that set them apart and contribute to malignancy. Normal cells stop dividing once they come into contact with neighboring cells, a property called **contact inhibition.** As genetic mutations accumulate, cancerous cells usually lose this property. The result is a pile of cells growing on top of each other. Normal cells also usually require connections to the tissue to which they belong; this is called **anchorage dependence.** Cancer cells typically lose this constraint as well, and so are enabled to detach and spread. Finally, cancer cells promote the growth of new blood vessels, or **angiogenesis,** to acquire oxygen and nutrients for growth **(INFOGRAPHIC 10.6).**

INFOGRAPHIC 10.6 Tumors Develop in Stages as Mutations Accumulate in a Cell

It takes more than a single mutation to cause cancer. Individuals who have inherited high-risk mutations require fewer additional mutations to get to cancer, and therefore develop cancer at a much earlier age.

A. Inherit *BRCA1* mutation.

If a mutation in *BRCA1* causes decreased BRCA function, DNA will not be efficiently repaired. Additional mutations are more likely to occur because of the failure to repair DNA.

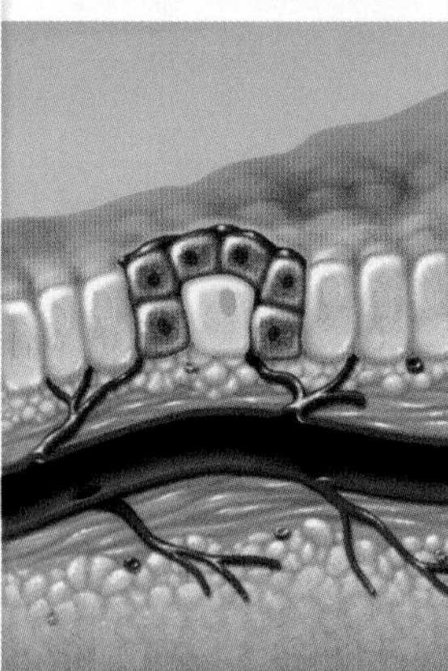

B. DNA replication errors are not corrected, producing an oncogene.

The mutated oncogene is overactive, allowing the cell to divide more often without normal checks. Cells begin to pile up into a tumor.

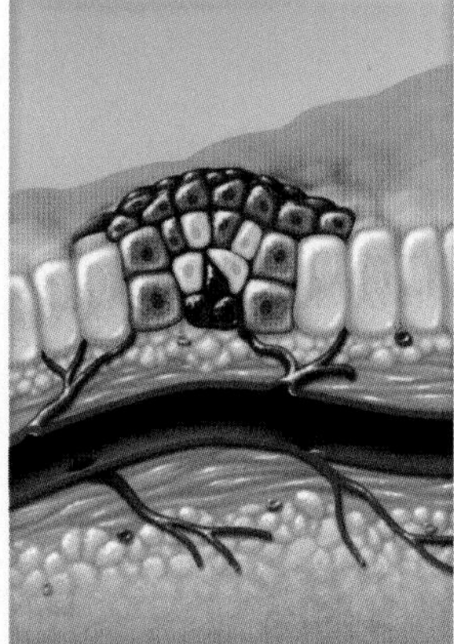

C. Smoking mutates tumor suppressor gene *p53*.

Cells fail to correct mutations or enter apoptosis. Mutations continue through additional cell divisions. Cells have lost contact inhibition, dividing even when crowded.

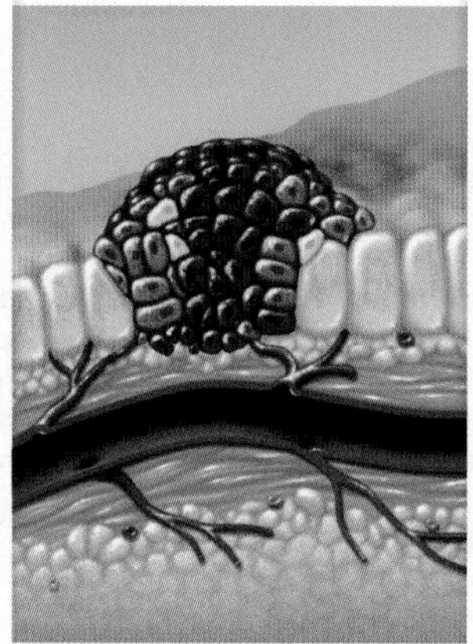

D. Additional mutations permit tissue invasion, metastasis, and new blood vessel growth.

The tumor contains malignant cells that lose their connections with the tissue, allowing them to invade surrounding tissues and spread to nearby and distant locations, in a process called metastasis. The cells of the tumor also promote new blood vessel growth, allowing delivery of oxygen and nutrients.

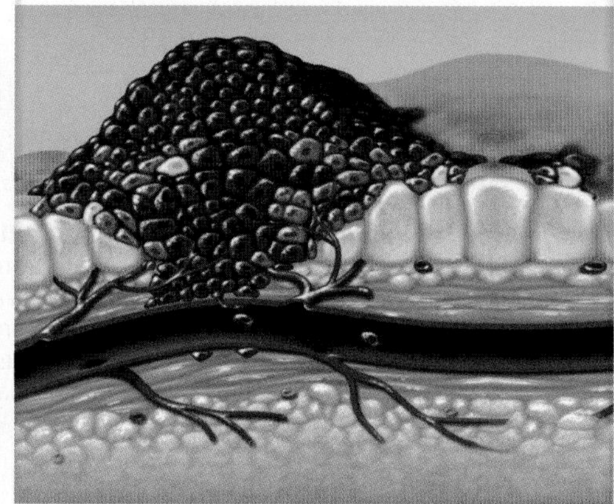

? If a person with an inherited *BRCA1* mutation develops breast cancer, is the *BRCA1* mutation likely to be the only mutation in the malignant cells?

A person who inherits a *BRCA* mutation starts life with at least one cancer-predisposing mutation, so he or she requires fewer additional mutations to develop cancer. Because it disrupts DNA repair, a single mutated *BRCA* allele is enough to increase the chances that additional mutations will occur in a cell; DNA damage will go unrepaired—or be repaired incorrectly—leading to a change in the nucleotide sequence of DNA. Daughter cells will accumulate mutations at a faster rate, which is why hereditary breast cancer often strikes women who are in their 30s and 40s—much younger than women who have no inherited predisposition to cancer. Typically, by the time a woman with an inherited predisposition develops breast cancer, her other (normal) copy of the *BRCA* gene will have been mutated as well.

Since the *BRCA1* and *BRCA2* genes are expressed in many cell types in addition to breast tissue, mutations in either gene raise the risk of cancers in other organs, too. Scientists have linked mutations in both genes to a higher than average risk of prostate, colon, and pancreatic cancers, among others, in both men and women. But the breasts and ovaries are at especially high risk of developing cancers because they respond to the hormone estrogen, which causes cells in these organs to divide more often; more cell divisions means more chances for mutations to occur.

Since the *BRCA* genes were discovered, nearly a dozen of other genes that predispose to cancers have been discovered, including genes that predispose to skin, breast, colon, ovarian, and brain cancers. Genetic testing for these mutations is now routinely available at major medical centers.

Treating Cancer

For many types of cancer, the first line of treatment is surgery to remove the tumor. In the case of breast cancer, that can be complete removal of the breast, called a mastectomy, or removal of just the tumor plus a small amount of surrounding tissue, called a lumpectomy. Surgery is generally effective for solid tumors that are diagnosed early, but not for cancers that have spread to other parts of the body, or for cancers that do not produce tumors, like leukemia. In these cases, the best option is usually **chemotherapy**—using toxic chemicals to kill the cancer cells in the body. Most chemotherapy drugs work by interfering with one or more steps of the cell cycle in dividing cells.

Doctors may also treat a tumor with radiation. In **radiation therapy,** beams of ionizing (high-energy) radiation are focused on a tumor. The radiation damages DNA in dividing cells. Cancer cells are less able to repair this DNA damage than are normal cells, and so they die.

The downside of both chemotherapy and radiation is that they can cause severe side effects. That's because neither therapy is very specific—both treatments damage all dividing cells in their path, including healthy ones. Chemotherapy routinely kills healthy cells lining the intestinal tract, cells in hair follicles, and cells in the bone marrow—all of them cells that normally divide frequently. The side effects of these unintended cell deaths can include vomiting, bruising, hair loss, and susceptibility to infection.

In recent years, new forms of cancer treatment have become more common, including targeted therapy and immunotherapy. **Targeted therapies** are designed to kill cancer cells specifically, often by combating a specific defect in the cancer cell, such as the mutated protein made from an oncogene or tumor suppressor. For example, drugs called PARP inhibitors are an effective treatment for *BRCA*-mutated cancers, including breast, ovarian, and prostate cancers. PARP inhibitors block the activity of an enzyme that contributes to DNA repair. Normal cells with functioning BRCA proteins can repair DNA damage in the presence of PARP inhibitors, but *BRCA*-mutant cells cannot and consequently die. Women with HER2-positive breast cancer commonly receive the targeted drug Herceptin (trastuzumab), which blocks the action of the protein made

ANGIOGENESIS
The growth of new blood vessels.

CHEMOTHERAPY
Treatment by toxic chemicals that kill cancer by interfering with cell division.

RADIATION THERAPY
The use of ionizing (high-energy) radiation to treat cancer.

TARGETED THERAPY
A cancer therapy that is specific for cancer cells and not harmful to normal cells.

INFOGRAPHIC 10.7 Cancer Treatments

There are many strategies to treat cancer. Surgery removes a tumor or at-risk tissue; chemotherapy and radiation therapy kill both normal and cancer cells. Newer treatment approaches specifically target cancer cells or improve the ability of the immune system to recognize and attack cancer cells.

Conventional Treatments

Conventional therapies are generally nonspecific, affecting both normal and cancer cells.

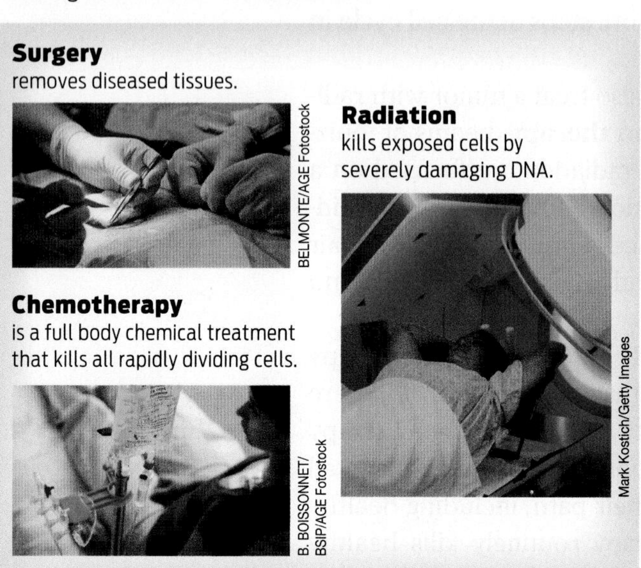

Surgery
removes diseased tissues.
BELMONTE/AGE Fotostock
Radiation
kills exposed cells by severely damaging DNA.
Chemotherapy
is a full body chemical treatment that kills all rapidly dividing cells.
B. BOISSONNET/ BSIP/AGE Fotostock
Mark Kostich/Getty Images

Targeted Therapy

Targeted therapies kill cancer cells specifically by exploiting weaknesses caused by oncogenes or mutated tumor suppressor genes.

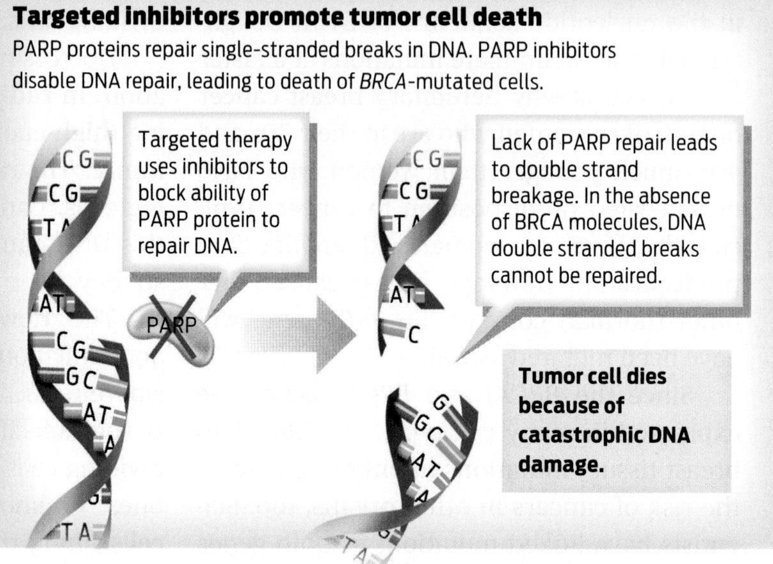

Targeted inhibitors promote tumor cell death
PARP proteins repair single-stranded breaks in DNA. PARP inhibitors disable DNA repair, leading to death of *BRCA*-mutated cells.
Targeted therapy uses inhibitors to block ability of PARP protein to repair DNA.
PARP
Lack of PARP repair leads to double strand breakage. In the absence of BRCA molecules, DNA double stranded breaks cannot be repaired.
Tumor cell dies because of catastrophic DNA damage.

? What happens when an immunotherapeutic antibody blocks PD-L1 on cancer cells?

from this oncogene. Targeted therapies exist for many other cancers, too, including Gleevec (imatinib mesylate) for certain leukemias and Zelboraf (vemurafenib) for BRAF-mutated melanoma.

Immunotherapies are drugs or treatments that stimulate our own immune system to find and fight cancer in the body. For some types of cancer–including melanoma, lung, and kidney cancer–this approach has proven to be very effective. There are several forms of immunotherapy. In one common approach, drugs are used to release a natural brake on immune cells that keeps the immune system in check. Once this brake is released, the immune cells can attack the cancer. Not everyone responds to these drugs, but those who do can sometimes experience dramatic results. In 2016, former U.S. President Jimmy Carter received an immunotherapy drug called Keytruda (pembrolizumab), which blocks a protein called PD-1, which acts as a brake on immune cells. He experienced a complete remission of his metastatic melanoma (**INFOGRAPHIC 10.7**).

When it came time for treatment, Ahern chose a lumpectomy to remove her breast tumor, followed by chemotherapy and radiation. But she also did something less conventional: she had her ovaries removed. The ovaries are a major source of estrogen in the female body. Removing this source of estrogen cuts down on cell division in breast tissue, and therefore reduces the likelihood that mutations in breast cells will occur. As estrogen is also produced in small amounts

IMMUNOTHERAPY
A cancer therapy that uses the immune system to recognize and destroy cancer cells.

Immunotherapy

Some cancers can evade the immune system because they produce a molecule on their surface that signals immune cells to leave the cancer cell alone. Immunotherapy uses antibodies to block these cancer molecules so the cell can be recognized and killed by the immune system.

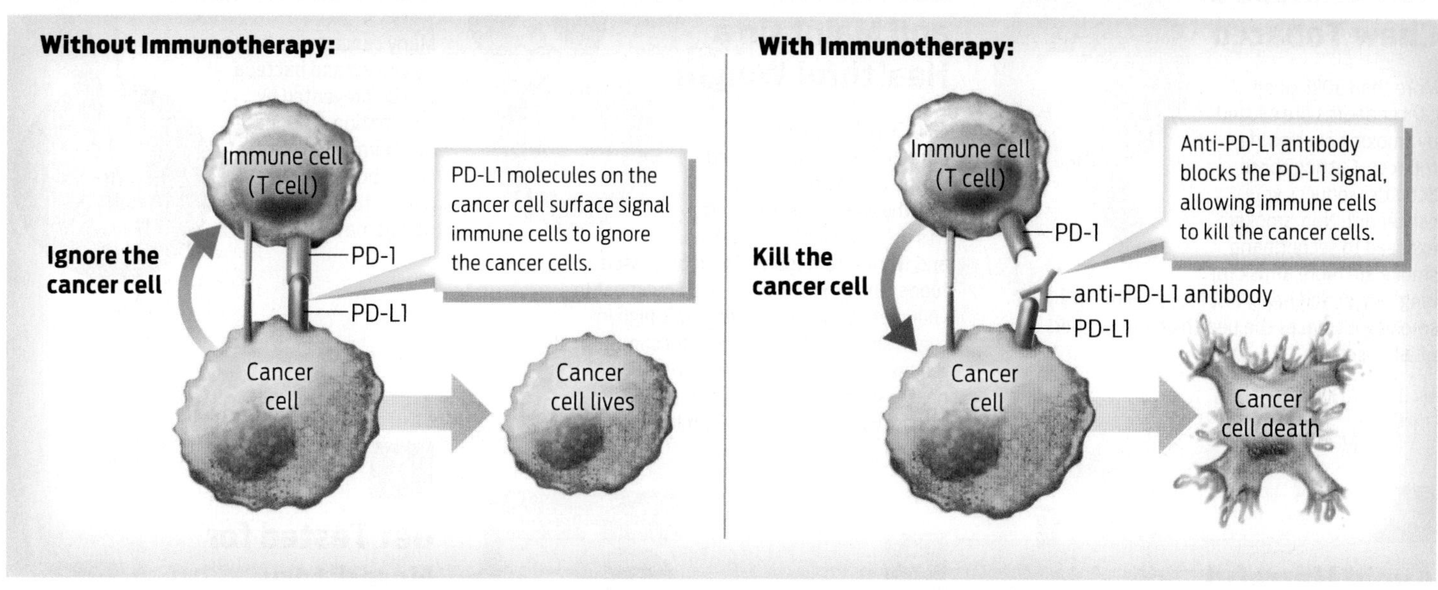

from other organs, Ahern also takes an estrogen-blocking drug called tamoxifen. All of these treatments are designed to reduce the chance that the cancer will recur in her body.

An Ounce of Prevention

If inherited mutations account for only a small fraction of cancers (about 5%–10% of the total), that means that most cancers are the result of mutations we acquire during our life. Some of these we have control over.

According to the American Association for Cancer Research, more than half of all cancer deaths each year are related to preventable causes. At the top of the list is tobacco use, followed closely by obesity and cancer-causing infections. For that reason, your best bet for avoiding cancer is to quit smoking, maintain a healthy weight, and get vaccinated. The human papillomavirus (HPV) vaccine, for example, prevents the infections that cause a large portion of genital cancers and head and neck cancers–provided people get the vaccine when they are children, before they are exposed.

Screening is also very important to detect cancers early. As tumors in the breast are hard to detect by a breast self-exam alone, the best primary screening method is a **mammogram,** an X-ray of the breast. Doctors recommend that women without a family history of breast cancer begin getting mammograms at age 45 or 50. For women with such a history or a known *BRCA* mutation, earlier mammograms are recommended **(INFOGRAPHIC 10.8)**.

MAMMOGRAM
An X-ray of the breast.

INFOGRAPHIC 10.8 Reducing Your Risk of Cancer

By being aware of risk factors for cancer, people can minimize their exposure and therefore reduce their risk of developing cancer. Other preventative strategies include screening and genetic testing for people with a family history of cancer.

Don't Smoke or Chew Tobacco

More than 30% of all cancer deaths are caused by smoking or chewing tobacco. Cutting these activities reduces your cancer risk. Non-smokers exposed to secondhand smoke are more at risk for lung cancer and heart disease, so eliminating smoking increases the health of all members of a household.

Eat Healthy and Maintain a Healthful Weight

4%–7% of new cancer cases in the U.S. are related to obesity. Maintaining a healthy weight can decrease this risk. Eat fruits, vegetables, and whole grain foods. Avoid processed foods, and limit the amount of red meat in your diet. Avoid processed foods high in sugar, salt, and fat, and limit consumption of red meat. Avoid alcohol, as 3.5% of all cancer deaths in the U.S. are alcohol related. Be physically active at least 30 minutes each day.

Get Vaccinated

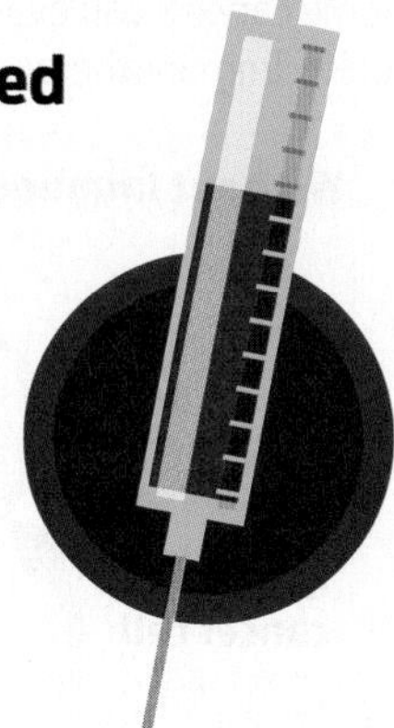

Many cancers caused by viruses and bacteria can be prevented by vaccination. Vaccinate newborns for hepatitis B and boys and girls age 11–12 for human papillomavirus (HPV).

4

Avoid Harmful Environments

Environmental pollutants are responsible for 4%–19% of all cancers. Avoid pollutants in the workplace by following safety procedures. Minimize radiation exposure in your house by monitoring radon levels. By age sixty-five, 40–50 percent of Americans will have developed skin cancer. Wear protective clothing, and sunscreen daily, even when the sun is not shining brightly.

Screen Regularly for Cancer

Screening for cancer or conditions that could lead to cancer increases the odds of successful treatment. Regular Pap tests screen for cervical cancer, and mammograms for women over 45 check for breast cancer. Colon cancer screening is recommended after age 50. Check with your doctor for screening programs near you.

Get Tested for Hereditary Cancers

Genetic testing for high-risk cancer genes is a good idea if cancer runs in your family. Testing allows you to plan ahead for prevention and treatment options. Those who've inherited high-risk cancer gene alleles may consider surgery to remove the high-risk tissue, as in the removal of breast and ovarian tissue for those carrying mutated *BRCA* genes.

What is the recommended screening method for breast cancer?

Proactive Measures

With genetic testing for *BRCA* mutations becoming increasingly common, more and more women find themselves facing a difficult choice: wait to see if they develop cancer or take proactive action. Actor Angelina Jolie made headlines in 2013 when she elected to undergo a double mastectomy as a preventative measure. Jolie had tested positive for a *BRCA1* mutation, and her mother, grandmother, and aunt had all died of either breast or ovarian cancer. In 2015, Jolie went further and had her ovaries removed as well.

A decade ago, such operations would have seemed drastic and unwarranted. But

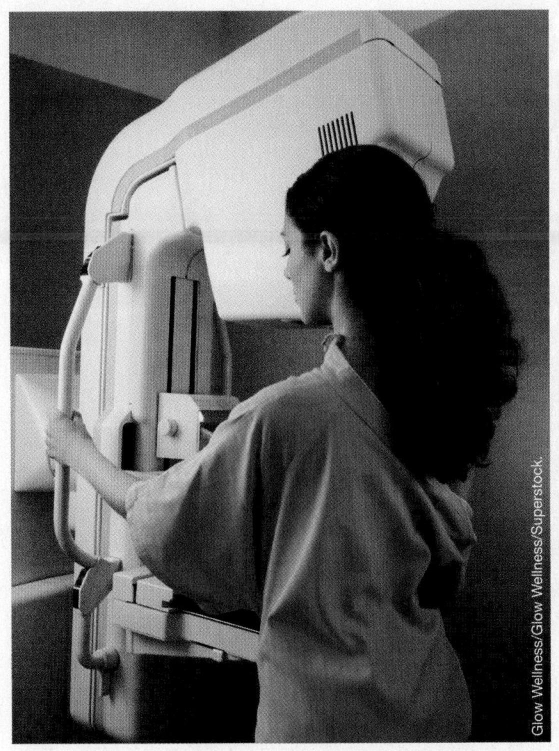

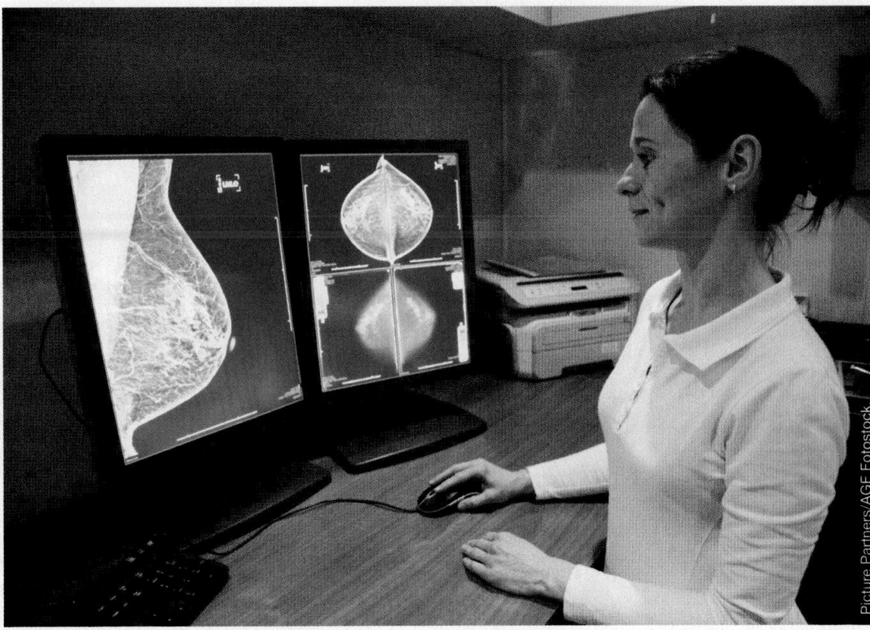

Mammograms are made by X-rays that image the breasts. Pathologists analyze the images, looking for abnormalities that could indicate cancer.

over the years, the value of prophylactic surgery to prevent future cancers has been validated by science. "The difference is that now we have empirical data," Sellers comments.

Studies in women with mutated *BRCA* genes who have not yet developed cancer show that ovary removal cuts the risk of breast cancer by 50% and the risk of ovarian cancer by 90%. Double mastectomy cuts the risk of breast cancer by 95%. While these surgeries also lower the risk of cancer recurring in women, like Ahern, who have already had cancer, the numbers aren't quite as high because it's sometimes difficult to assess whether a cancer has already metastasized to other parts of the body.

Nonetheless, these aren't easy decisions for a woman to make, and they can take their physical and psychological toll. Though Ahern says she feels "pretty good" right now, there was a time when she was visiting online breast cancer discussion groups every evening after work and all weekend long. They not only helped her cope emotionally but also helped to inform her about her disease and her treatment options. Patient-oriented support groups can be a valuable source of information for people with cancer, in addition to what health professionals provide. However, this may not be the best route to support for everyone.

According to Sue Friedman, executive director of Facing Our Risk of Cancer Empowered (FORCE), there is still a lot of misinformation out there about hereditary breast cancer. People still assume that diet and lifestyle changes will cut the risk of cancer in people with hereditary cancer. "Those factors may help, but not enough in our community," says Friedman, who has had cancer herself. Women also have options regarding prophylactic surgery–when to have it and how much is necessary–that aren't always effectively communicated by health care professionals.

Doctors admit that surgery isn't the most palatable treatment. "Surgery cuts your risk substantially, but it's still pretty traumatic," says Sellers. "It would be nice to say we've got a medication you can take and you'll have the same effect. But we just don't have that kind of treatment right now." ■

CHAPTER 10 SUMMARY

- Cancer is uncontrolled cell division caused by mutations—changes in the nucleotide sequence of DNA.
- Cancer cells have lost the ability to regulate cell division and reproduce uncontrollably, in most cases eventually forming a tumor.
- Cell cycle checkpoints ensure accurate progression through the cell cycle; repair mechanisms at each checkpoint can fix mistakes that occur, such as improper base pairing or DNA damage.
- In the absence of proper checkpoint function, cells may fail to properly repair DNA mistakes, leading to mutations that are passed on to daughter cells.
- Mutations occur spontaneously during DNA replication. They can also be caused by environmental triggers such as tobacco, UV radiation, chemicals, and viruses, and by chemicals naturally produced by the body.
- Mutations in two types of genes, proto-oncogenes and tumor suppressors, cause most cancers.
- Multiple mutations must occur in the same cell for the cell to become cancerous.
- People with "hereditary" cancer inherit predispositions to the disease in the form of specific genetic mutations. These mutations are present in all body cells and can serve as the first genetic "hit."
- Women with *BRCA* mutations have a much higher risk of developing cancer, and at an earlier age, than women without these mutations.
- Alternate forms of the same gene, with different nucleotide sequences, are called alleles.
- Certain alleles are more common in ethnic groups that were at some point small in number and genetically isolated, their members often marrying within their own group.
- Cancer is often treated with a combination of surgery, chemotherapy, and radiation. New and promising cancer treatments include targeted therapy and immunotherapy.
- Women who test positive for *BRCA* mutations may consider prophylactic removal of the ovaries or breasts to prevent cancer developing in these organs.

MORE TO EXPLORE

- **NIH, Genetics Home Reference: https://ghr.nlm.nih.gov/condition/breast-cancer**
- **FORCE: http://www.facingourrisk.org/**
- **Sedic, M., and Kuperwasser, C. (2016) BRCA1-hapoinsufficiency: Unraveling the molecular and cellular basis for tissue-specific cancer. *Cell Cycle* 15.5:621–627.**
- **King, M.-C. (2014). "The race" to clone *BRCA1. Science* 343(6178), 1462–1465.**
- **Skloot, R. (2010) *The Immortal Life of Henrietta Lacks.* New York: Random House.**
- ***Cancer: The Emperor of All Maladies* (2015), a PBS documentary: http://www.pbs.org/kenburns/cancer-emperor-of-all-maladies/home/**

CHAPTER 10 Test Your Knowledge

DRIVING QUESTION 1 What are mutations, and how can they occur?

By answering the questions below and studying Infographics 10.3 and 10.4 and Table 10.2, you should be able to generate an answer for the broader Driving Question above.

KNOW IT

1 A mutation causes a substitution of one amino acid for another in the encoded protein. What type of mutation is this?

a. silent
b. nonsense
c. missense
d. insertional frameshift
e. deletional frameshift

2 Which of the following is a known mutagen?

a. cigarette smoke
b. sunlight
c. charred meat cooked at high temperatures
d. X-rays
e. all of the above

3 Why does wearing sunscreen reduce cancer risk?

a. Sunscreen can repair damaged DNA.
b. Sunscreen can activate checkpoints in skin cells.
c. Sunscreen can reduce the chance of mutations caused by exposure to UV radiation present in sunlight.
d. It doesn't; sunscreen causes mutation and actually increases cancer risk.
e. Sunscreen can prevent cells with mutations from being destroyed.

4 Look at the mutagens illustrated in Infographic 10.3. Of these, which are most easily avoidable, and which are not avoidable?

USE IT

5 The mutation illustrated in Infographic 10.4 inserted an A in the third codon of the mRNA shown. Use the genetic code (Infographic 8.8, p. 178) to match each mutation below (all are mutations of the normal mRNA sequence shown in Infographic 10.4) with both its effect on the protein and the type of mutation. For each mutation, put a check mark next to the corresponding effect and type of mutation.

Mutation	Effect of Mutation	Type of Mutation
Substitution of an A for a C in the third codon	_____ Protein will have an incorrect amino acid in its sequence. _____ No impact on the protein _____ Protein will be shorter than normal.	_____ Nonsense _____ Missense _____ Silent
Substitution of a C for the U in the fourth codon	_____ Protein will have an incorrect amino acid in its sequence. _____ No impact on the protein _____ Protein will be shorter than normal.	_____ Nonsense _____ Missense _____ Silent
Substitution of an A for the first C in the second codon	_____ Protein will have an incorrect amino acid in its sequence. _____ No impact on the protein _____ Protein will be shorter than normal.	_____ Nonsense _____ Missense _____ Silent

6 Are all mutations bad? Explain your answer.

DRIVING QUESTION 2 How does cancer develop, how is it treated, and how can people reduce their risk?

By answering the questions below and studying Infographics 10.2, 10.5, 10.6, 10.7, and 10.8, you should be able to generate an answer for the broader Driving Question above.

KNOW IT

7 In an otherwise normal cell, what happens if one mistake is made during DNA replication?

a. Nothing; mistakes just happen.
b. A cell cycle checkpoint detects the error and pauses the cell cycle so the error can be corrected.
c. The cell will begin to divide out of control, forming a malignant tumor.
d. A checkpoint will force the cell to carry out apoptosis, a form of cellular suicide.
e. The mutation will be inherited by the individual's offspring.

8 What are some differences and some similarities between tumor suppressor genes and oncogenes?

9 Which of the following can cause cancer to develop and progress?

a. a proto-oncogene
b. an oncogene
c. a tumor suppressor gene
d. a mutated tumor suppressor gene
e. b and d
f. b and c

10 Which form of breast cancer treatment is the least specific for the cancer cells?

a. chemotherapy
b. targeted therapy
c. immunotherapy
d. lumpectomy

USE IT

11 What would you say to a niece if she asked you how she could reduce her risk of breast cancer? (Assume there is no family history of breast cancer.) How might each of your suggestions reduce her risk?

12 Why is age a risk factor for cancer?

DRIVING QUESTION 3 Why do people with "inherited cancer" often develop cancer at a relatively young age?

By answering the questions below and studying Infographics 10.1, 10.3, 10.6, and 10.7, you should be able to generate an answer for the broader Driving Question above.

KNOW IT

13 A woman with a *BRCA1* mutation

a. will definitely develop breast cancer.
b. is at increased risk of developing breast cancer.
c. must have inherited it from her mother because of the link to breast cancer.
d. will also have a mutation in *BRCA2.*
e. b and c

14 What is the role of *BRCA1* in normal cells?

15 Which of the following family histories most strongly suggests a risk of inherited breast cancer due to *BRCA1* mutations?

a. many female relatives who were diagnosed with breast cancer in their 70s
b. many relatives with skin cancer
c. many relatives diagnosed with skin cancer at an early age
d. many female relatives diagnosed with breast cancer at an early age
e. many female relatives with both early breast cancer and ovarian cancer

USE IT

16 A 28-year-old male graduate student was born with an inherited predisposition to colon cancer due to a mutation in a DNA repair gene called *MLH1.* He has recently been diagnosed with colon cancer. At the cellular and genetic level, was he born with colon cancer? Was he born with a predisposition to colon cancer? At birth, were cells in his colon genetically identical to cells in his liver? Now that he has colon cancer, are his cancer cells genetically identical to his normal colon cells? Explain your answers.

17 Which of the following women would be most likely to benefit from genetic testing for breast cancer?

a. a 25-year-old woman whose mother, aunt, and grandmother had breast cancer
b. a healthy 75-year-old woman with no family history of breast cancer
c. a 40-year-old woman who has a cousin with breast cancer
d. a 55-year-old woman whose older sister was just diagnosed with breast cancer
e. All women can benefit from genetic testing for breast cancer.

18 People like Lorene Ahern have inherited a mutated version of *BRCA1.* Why does this mutation pose a problem? Why are these people at high risk of developing breast cancer when they still have a functional *BRCA1* allele? Describe how the protein encoded by normal *BRCA1* compares to that encoded by mutant alleles of *BRCA1.*

apply YOUR KNOWLEDGE

MINI CASE

19 Nellie has a family history similar to Lorene Ahern's. Nellie's mother died at an early age from breast cancer, as did her maternal aunt (her mother's sister). Nellie is not yet 35 but has started having annual mammograms. She has also been tested for *BRCA1* and *BRCA2* mutations. She has a *BRCA2* mutation and is considering prophylactic surgery. Her younger sister, Anne, doesn't want to know the results of Nellie's genetic testing because if Nellie has a *BRCA2* mutation, then there is a chance that Anne could have inherited the same mutation from their mother. Does Nellie or Nellie's doctor have an obligation to tell Anne about the test results? What about Nellie's older brother? Should he be told? There are personal and medical benefits and risks to consider here.

INTERPRETING DATA

apply YOUR KNOWLEDGE

20 José is a 32-year-old landscaper living in Phoenix, Arizona. He and his 64-year-old father, Ray, were both diagnosed with metastatic melanoma within 2 months of one another. Both had their tumors biopsied to look for potential targets for targeted therapy. The *BRAF* proto-oncogene from each of their tumors was sequenced. Their cancer cells were analyzed for expression of PD-L1 (see Infographic 10.7). The data are shown in the table below.

	BRAF gene sequence (corresponding to amino acids 598–602; note that the DNA groups shown below correspond to mRNA codons)	PD-L1 expression
José	CGA TGT CTC TTT AGA	No
Ray	CGA TGT CAC TTT AGA	Yes

a. Transcribe and translate the *BRAF* gene sequences from José's and Ray's tumors. What amino acid is at position 600 in each?

b. Zelboraf is a drug that stops division of metastatic melanomas by blocking the activity of a mutant (oncogenic) BRAF protein that has a glutamic acid at position 600 (the proto-oncogene has a valine at position 600). Keytruda is an antibody drug that blocks the interaction between PD-L1 on cancer cells and its binding partner (PD-1) on immune cells (see IG 10.7). Given their individual tumors, what treatment(s) are available for José and for Ray? Consider both targeted and traditional therapies and justify your answer.

c. From the information presented, do you think this is more likely to be a case of inherited cancer, or two cancers that just happened to occur in these two family members? Explain your answer, and consider other risk factors that may be involved for José and Ray.

BRING IT HOME

apply YOUR KNOWLEDGE

21 If you wanted to change your lifestyle to reduce your risk of developing cancer, what specific steps could you take with respect to each of the following? Be as specific as you can. Take your age and gender into consideration as you consider each factor.

a. alcohol consumption

b. sun exposure

c. tobacco use

d. exposure to pesticides

e. meat preparation (cooking method)

11

Simple Inheritance and Meiosis

DRIVING QUESTIONS

1. How does the organization of chromosomes, genes, and their alleles contribute to human traits?
2. How does meiosis produce genetically diverse gametes?
3. Why do different traits have different patterns of inheritance?
4. What are some practical applications of understanding the genetic basis of human disease?

CATCHING BREATH

One woman's mission to outrun a genetic disease

EMILY SCHALLER NEVER THOUGHT she'd be running marathons one day. From the time she was a teenager, the Detroit, Michigan, native has struggled just to take a deep breath without coughing. She spent her youth in and out of hospitals, getting treatment for the recurrent lung infections that plague people with her condition.

Emily has cystic fibrosis (CF), a genetic disorder she inherited from her parents. CF has many symptoms, and the most dangerous is mucus that clogs airways in the lungs and makes it difficult to breathe. People with CF also have trouble digesting food–mucus blocks the passageways through which digestive enzymes travel to the intestines–and so they must swallow enzymes before each meal to ensure they get enough nutrients.

Emily's struggle with the condition reached a tipping point in 2007. Her health had declined to the point that she qualified for social security and disability insurance at age 25. "I thought to myself, 'I'm sick of being sick. I have to do something.'"

So she started running–slowly and painfully at first, but eventually with more ease. "It took me 3 to 4 months just to run 2 miles–coughing and spitting every block," she says. Eventually, through running, Emily managed to raise her lung function from 50% to 75%. "Running totally changed my life," she says.

ALLELE
Alternative versions of the same gene that have different nucleotide sequences.

DIPLOID
Having two copies of every chromosome.

HOMOLOGOUS CHROMOSOMES
A pair of chromosomes that both contain the same genes. In a diploid cell, one chromosome in the pair is inherited from the mother, the other from the father.

Emily's successes haven't come from sheer determination alone. She's also benefited greatly from progress in medical science. Through research, scientists have come to understand CF better. They now know what genes are affected and are learning why some people with CF have more severe forms of the disease. This research is leading to new drugs that specifically target the defect that occurs in CF and make it possible for someone with the disease to live and breathe more easily.

A Faulty Gene

Every year, approximately 2,500 babies in the United States are born with CF, making it the most common life-threatening genetic disease in this country. The current life expectancy for someone with CF is about 40 years of age. When Emily's CF was diagnosed in 1983, it was 18.

It wasn't until 1989 that scientists understood what caused the disease. That year, a team of scientists led by Lap Chee Tsui at Toronto's Hospital for Sick Children and Francis Collins, then at the University of Michigan, homed in on the genetic basis for the condition: a mutation in a specific gene that sits on chromosome 7.

Cystic Fibrosis Life Expectancy

As a result of medical research and the development of effective treatments, the life expectancy of people living with cystic fibrosis has increased by about 10 years in the last two decades, and doubled since Emily's diagnosis in 1983.

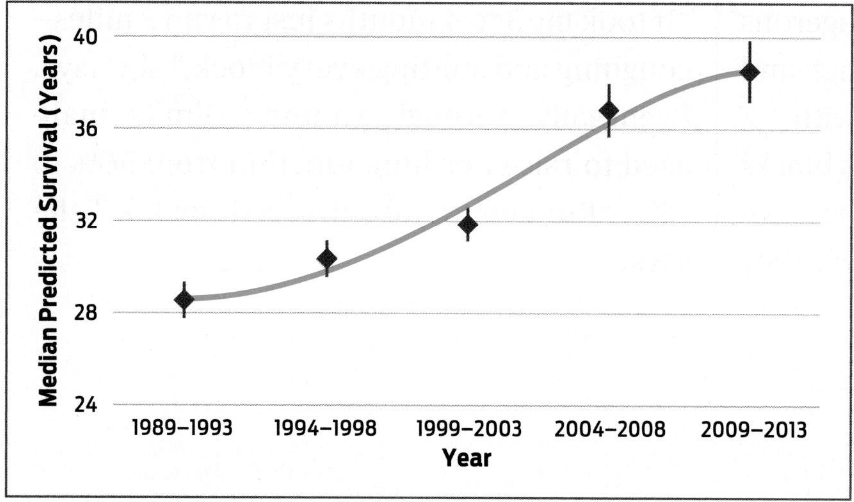

Data from Cystic Fibrosis Foundation

As we saw in Chapter 10, mutations are changes in the nucleotide sequence of DNA. They result when mistakes occur during the process of DNA replication, or when environmental insults cause DNA damage that isn't repaired.

Mutations create altered versions of genes called **alleles**. Because genes code for proteins, a change in the sequence of a gene can change the shape of the encoded protein, and therefore its function.

You can think of alleles as being like words with different spellings. Some of these variant spellings are harmless, and variants retain the same meaning (for example, color, colour; theater, theatre). Others change the meaning of the word entirely (here, hear; read, red). Still other changes to a word's spelling can render it nonsensical (green, grken; song, sxng). A change in one letter–one nucleotide–is sometimes enough to alter the function of a protein, or make it nonfunctional.

Tsui and Collins (who would later head the Human Genome Project at the National Institutes of Health) found that the CF mutation occurs in a gene that provides the instructions to make a protein that shuttles chloride ions into and out of cells. They called the gene *CFTR*, for *cystic fibrosis transmembrane conductance regulator*.

The 1989 discovery was a milestone. Now that they knew the gene responsible, scientists could study how mutations in the gene make people sick. A variety of mutations in the *CFTR* gene can cause CF. The most common is a deletion of three nucleotides within the *CFTR* gene. People who carry this mutation are missing a key amino acid in the CFTR protein. As a result, the CFTR protein is faulty and cannot do its job **(INFOGRAPHIC 11.1)**.

Luck of the Draw

When Emily's mother, Debbie, learned that her daughter had CF, she was shocked. She and her husband, Lowell, were both healthy, and they already had two healthy sons. How did their daughter develop a genetic disease that neither Debbie nor her husband had?

The answer has to do with how genes are inherited–how they are passed down from generation to generation. Genes, which provide instructions for making proteins, are the units of inheritance, physically transmitted from parents to children. The particular alleles of genes you received from your parents are the reason you resemble your mother and father, and possibly also an uncle or a grandparent. But not every child of a couple receives exactly the same set of parental alleles, and so children can and do differ from their parents and from one another.

Consider Emily's parents. Like all humans, they are **diploid** (from the Greek for "double"), meaning that each of their body cells carries two copies of each chromosome. Of the 46 total chromosomes in our cells, 23 come from our mother and 23 come from our father. These 23 individual chromosome pairs, one from mom, one from dad, are called **homologous chromosomes.**

Because our chromosomes come in pairs, so too do our genes. Humans have two copies of nearly every gene in our body cells. These two gene copies can either be the same, or slightly different, in the specific nucleotide sequences making them up. In other words, each chromosome may carry a different allele of the same gene.

Because we have two alleles for a given gene, a nonfunctional allele doesn't always spell disaster. In the case of the *CFTR* gene, for example, a person can have one defective allele and remain healthy if his or her other *CFTR* allele is normal and produces a functional CFTR protein. This is why Emily's parents, Debbie and Lowell, are healthy: they each have one normal *CFTR* allele that makes up for the defective copy (**INFOGRAPHIC 11.2**).

But as Debbie and Lowell illustrate, it's not always possible to know what genes a person has just from his or her outward appearance. For that reason, geneticists make a distinction between a person's observable or measurable

INFOGRAPHIC 11.1 CF Is Caused by Mutations in the *CFTR* Gene

Cystic fibrosis (CF) is caused by a variety of mutations in the cystic fibrosis transmembrane regulator (*CFTR*) gene that sits on chromosome 7. One such mutation consists of a deletion of three consecutive nucleotides, which creates a CF-associated mutant allele. Consequently, the protein expressed from this allele is missing an amino acid, which renders the protein nonfunctional.

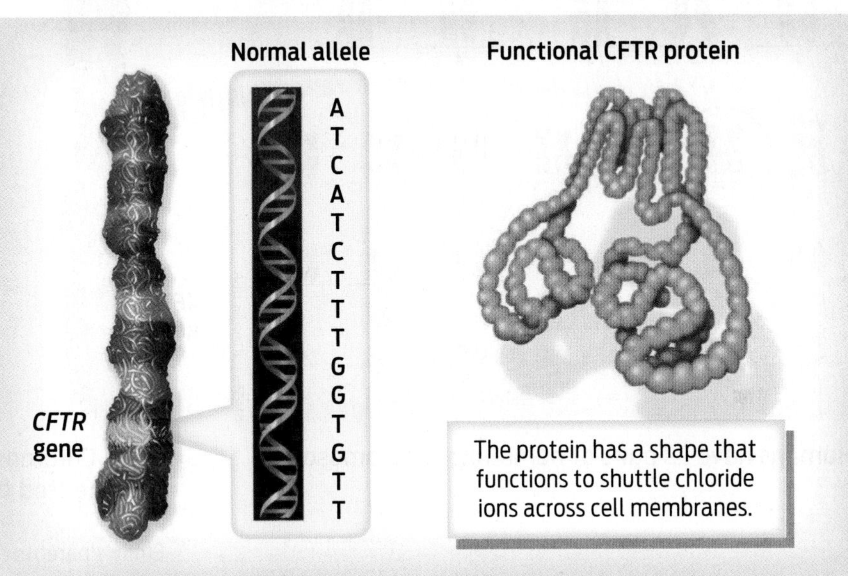
Normal allele
Functional CFTR protein
A T C A T C T T T G G T G T T
CFTR gene
The protein has a shape that functions to shuttle chloride ions across cell membranes.

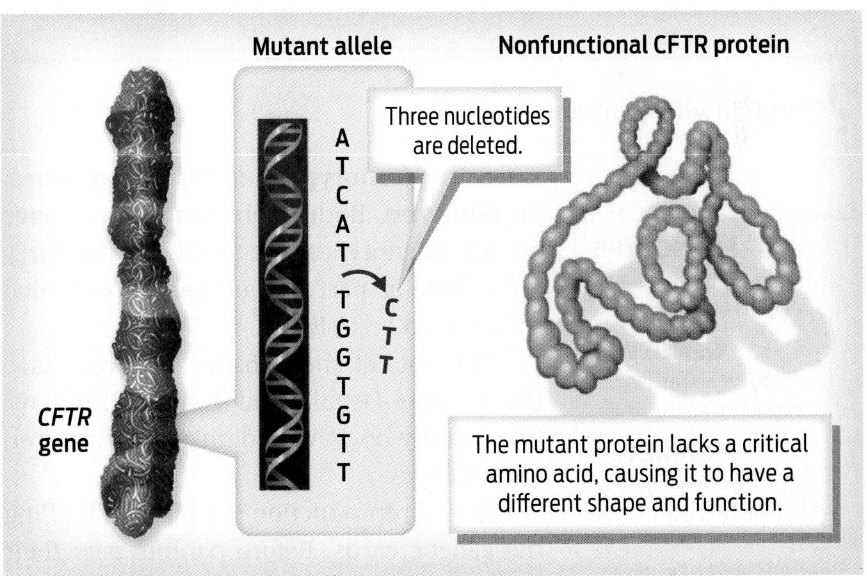
Mutant allele
Nonfunctional CFTR protein
Three nucleotides are deleted.
A T C A T T G G T G T T
C T T
CFTR gene
The mutant protein lacks a critical amino acid, causing it to have a different shape and function.

? How many nucleotides differ between the mutant and normal alleles of the *CFTR* gene illustrated? How many amino acids differ between the proteins encoded by the two alleles?

INFOGRAPHIC 11.2 In Humans, Genes Come in Pairs

Humans are diploid organisms, meaning they have two complete sets of chromosomes—one set inherited from each parent. Because chromosomes come in pairs, so too do the genes located on them. The two gene copies, known as alleles, can either be the same or different. In the case of CF, having at least one normal allele is sufficient to remain healthy.

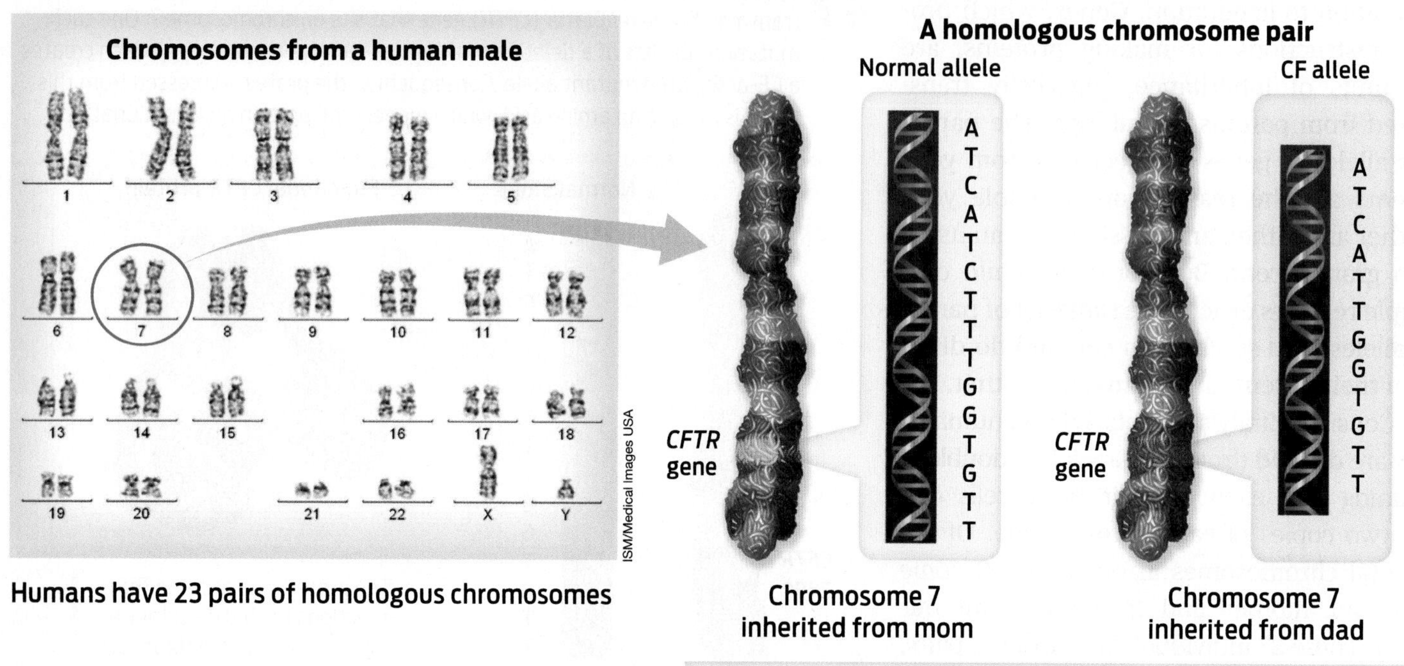

Emily's parents remain healthy because each has a normal allele, which produces enough of the normal protein for cells to function properly.

? Explain why humans are diploid.

PHENOTYPE
The visible or measurable features of an individual.

GENOTYPE
The particular genetic makeup of an individual.

GAMETES
Specialized reproductive cells that carry one copy of each chromosome (that is, they are haploid). Sperm are male gametes; eggs are female gametes.

HAPLOID
Having only one copy of every chromosome.

MEIOSIS
A type of cell division that generates genetically unique haploid gametes.

traits, or **phenotype,** and his or her genes, or **genotype.** Both Debbie and Lowell have normal phenotypes, but they both also carry a CF allele as part of their genotype, which they passed on to Emily.

But not all the Schaller children have the disease—Debbie and Lowell also have two healthy boys. Why didn't these children inherit CF?

Sexual reproduction is a bit like shuffling the genetic cards. Before parents pass their genes to their offspring, the alleles of genes on homologous chromosomes are first separated from one another and divvied up differently, such that not every child is dealt the same combination of alleles.

To reproduce sexually, organisms must first create sex cells called **gametes.** In humans, these are the egg and sperm cells. Unlike the rest of the body's cells, which are diploid, gametes carry only one copy of each chromosome, which makes them **haploid.**

Haploid gametes are created from diploid cells through a type of cell division called **meiosis.** In humans, meiosis halves the total number of chromosomes in a cell from 46 total (23 pairs) to 23 individual chromosomes. When a haploid sperm (with 23 chromosomes) fertilizes a haploid egg (with 23 chromosomes), the result is a diploid **zygote** (with 46 chromosomes). This zygote will then divide repeatedly by mitosis to become an **embryo,** which will eventually grow into a human child **(INFOGRAPHIC 11.3).**

It's useful to explore how meiosis differs from mitosis, the type of cell division that reproduces body cells, which we encountered in Chapter 9. In mitosis, replicated chromosomes (consisting of a pair of sister chromatids) line up along the midline of the

INFOGRAPHIC 11.3 Gametes Pass Genetic Information to the Next Generation

To reproduce sexually, diploid organisms produce specialized sex cells called gametes, which are haploid—they carry only one copy of each chromosome. When a sperm fertilizes an egg the resulting diploid zygote divides by mitotic cell division, eventually generating enough cells to form a baby. The baby is diploid.

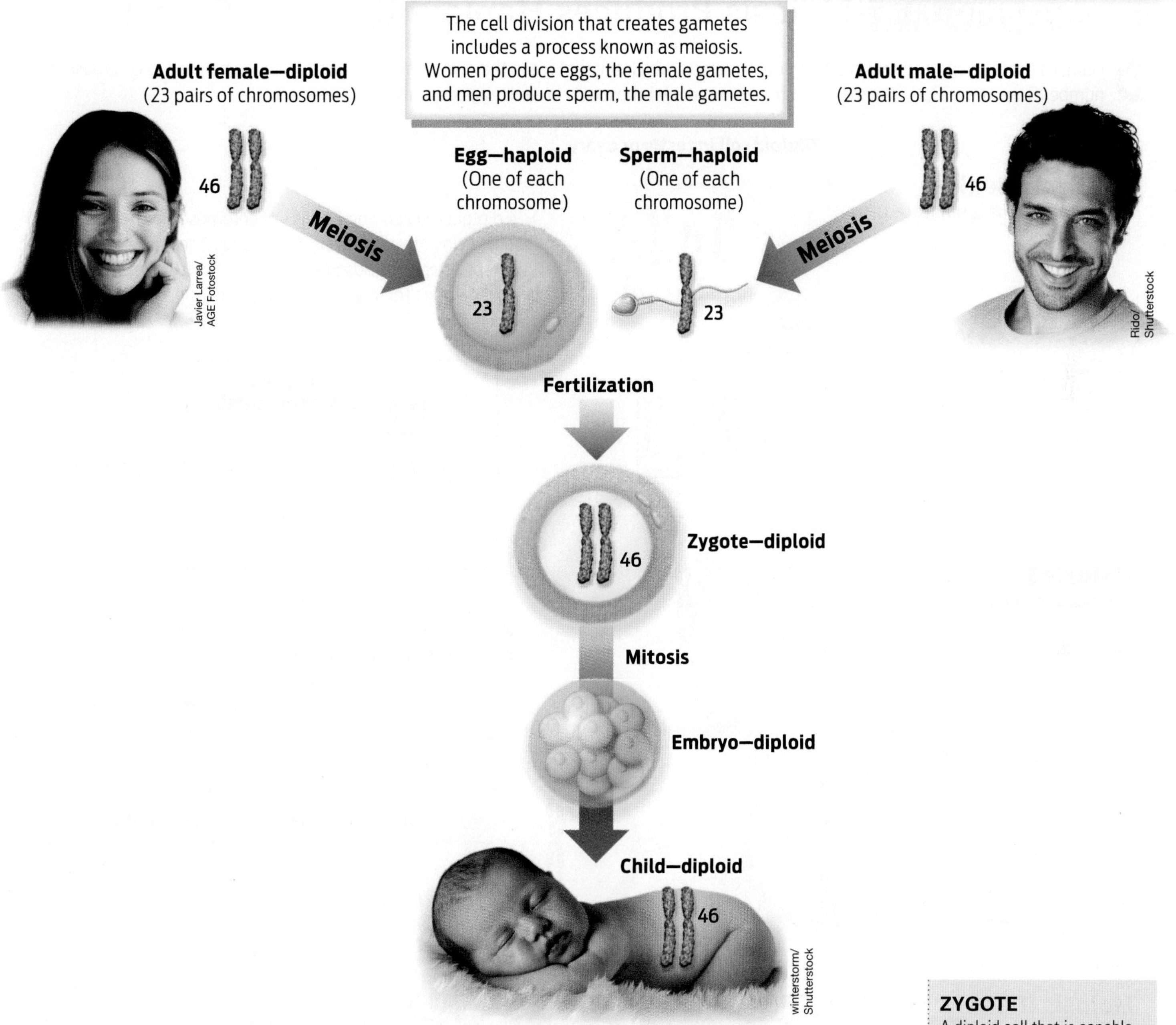

? Which of the following is/are haploid: a skin cell; a zygote; a sperm?

cell. The two sister chromatids of each chromosome then separate, becoming independent chromosomes. Each daughter cell ends up with an identical (diploid) complement of 46 chromosomes. In many ways, meiosis is similar to mitosis, except that in meiosis there are two separate divisions, producing four daughter cells rather than two, and

ZYGOTE
A diploid cell that is capable of developing into an adult organism. The zygote is formed when a haploid egg is fertilized by a haploid sperm.

EMBRYO
An early stage of development reached when a zygote undergoes cell division to form a multicellular structure.

giving each cell only one copy of each chromosome. The first division separates homologous chromosomes from each other, while the second division, like mitosis, separates sister chromatids. At the end of meiosis, one (diploid) cell has divided into four (haploid) cells, which develop into egg or sperm (**INFOGRAPHIC 11.4**).

INFOGRAPHIC 11.4 Meiosis Produces Haploid Eggs and Sperm

Humans produce egg and sperm cells through meiosis, which takes place in the ovaries and the testes. Meiosis halves the chromosome number from 46 to 23, by placing one complete set of chromosomes in each egg or sperm cell.

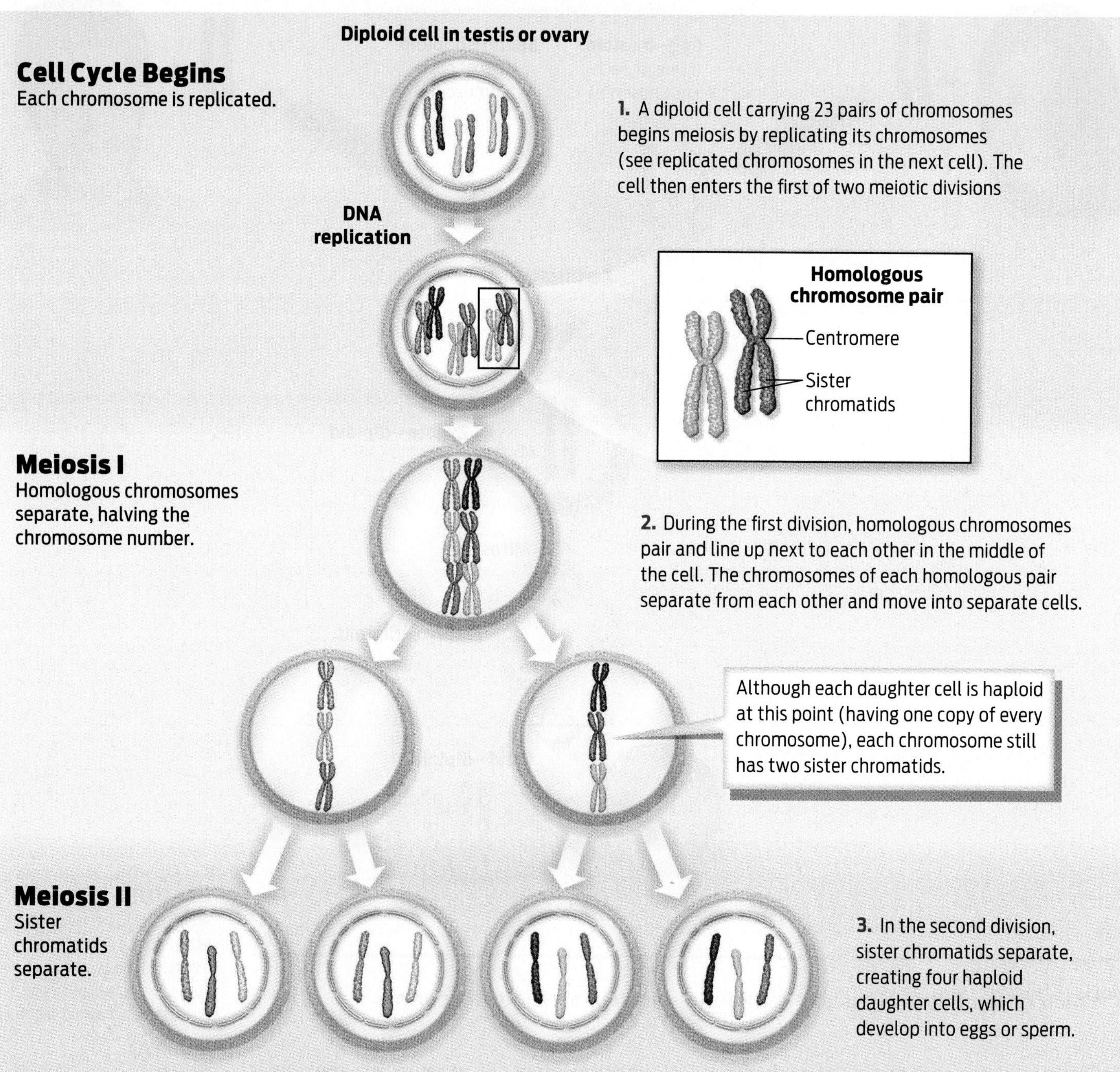

Diploid cell in testis or ovary
Cell Cycle Begins
Each chromosome is replicated.
1. A diploid cell carrying 23 pairs of chromosomes begins meiosis by replicating its chromosomes (see replicated chromosomes in the next cell). The cell then enters the first of two meiotic divisions
DNA replication
Homologous chromosome pair
Centromere
Sister chromatids
Meiosis I
Homologous chromosomes separate, halving the chromosome number.
2. During the first division, homologous chromosomes pair and line up next to each other in the middle of the cell. The chromosomes of each homologous pair separate from each other and move into separate cells.
Although each daughter cell is haploid at this point (having one copy of every chromosome), each chromosome still has two sister chromatids.
Meiosis II
Sister chromatids separate.
3. In the second division, sister chromatids separate, creating four haploid daughter cells, which develop into eggs or sperm.

? Which meiotic division reduces the chromosome number from diploid to haploid?

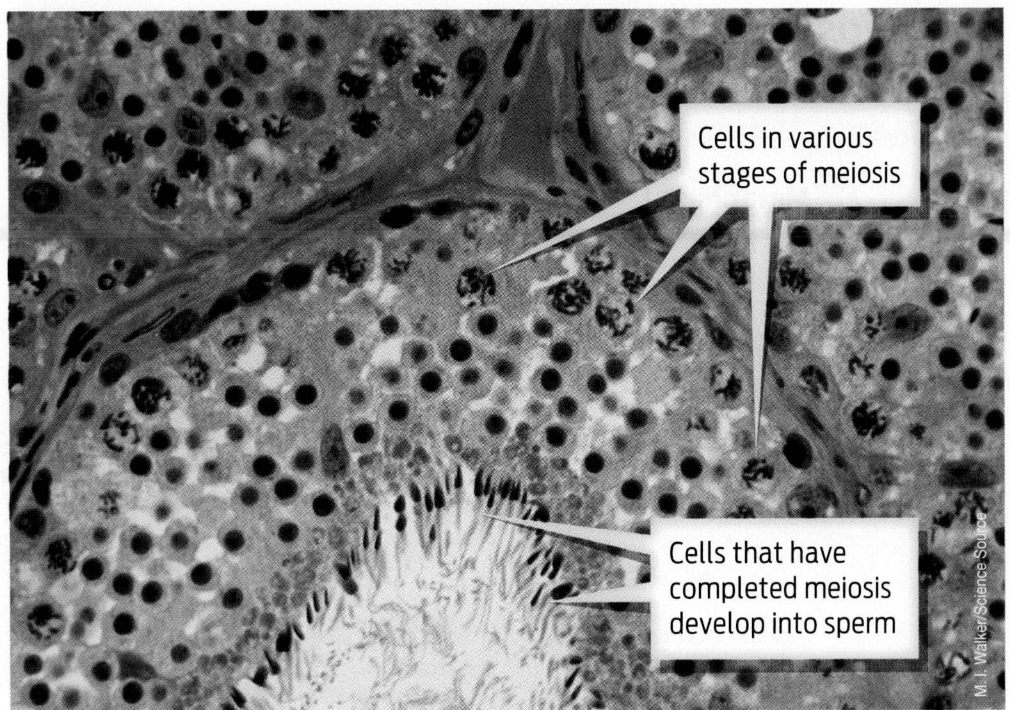

Cross section of a testis, the site of meiosis in males.

Like mitosis, meiosis is a finely orchestrated process that can be divided into defined stages based on the movement of chromosomes. As in mitosis, replicated chromosomes condense early in the process. The nuclear envelope dissolves and a spindle apparatus forms. The movement of chromosomes is choreographed by spindle fibers that attach to the centromeres of chromosomes and push and pull them to the proper location in preparation for the two separate divisions. The first division, which separates homologous chromosomes from one another, is termed meiosis I; the second division, which separates sister chromatids, is called meiosis II (see **UP CLOSE: PHASES OF MEIOSIS**).

Importantly, no two sex cells produced by the same parent are identical, and that is because of two major events that occur during meiosis. The first event is **recombination,** in which homologous maternal and paternal chromosomes pair up next to each other in the cell and physically exchange segments of DNA before they separate during meiosis I. As a result of recombination, which is also called crossing over, maternal chromosomes contain segments (and therefore alleles) from paternal chromosomes. Likewise, paternal chromosomes contain segments (and alleles) from maternal chromosomes.

The second event is **independent assortment,** in which the two chromosomes of each homologous pair distribute into daughter cells randomly with respect to all other chromosome pairs. As a consequence of independent assortment, the alleles of different genes are distributed independently of one another, not as a package. This means, for example, that what alleles of hair color you inherit have no bearing on what alleles for height you inherit. During meiosis I, homologous maternal and paternal chromosomes line up along the midline of the dividing cell. Each chromosome pair can line up in two different ways: sometimes the maternal chromosome is on the left, sometimes the paternal chromosome is. All the chromosomes on the left are pulled into one daughter cell, while all the chromosomes on the right are pulled into the other daughter cell. Because the identity of the chromosomes on the left and right vary, the exact combination of maternal and

RECOMBINATION
An event in meiosis during which maternal and paternal chromosomes pair and physically exchange DNA segments.

INDEPENDENT ASSORTMENT
The principle that alleles of different genes are distributed independently of one another during meiosis.

UP CLOSE Phases of Meiosis

Interphase

- Each chromosome replicates in S phase, resulting in two sister chromatids, connected at the centromere.
- Chromosomes are loosely gathered in the nucleus.

Meiosis I

Prophase I

- Replicated chromosomes begin to coil up.
- The nuclear membrane begins to disassemble.
- Microtubule fibers begin to form the meiotic spindle.

Metaphase I

- Microtubule spindle fibers from opposite ends of the cell attach to each of the two homologous chromosomes.
- Paired homologous chromosomes become aligned along the middle of the cell.

Anaphase I

- Microtubules shorten, pulling the homologous chromosomes to opposite ends of the cell.

Telophase I

- Homologous chromosomes reach each pole.
- Microtubule spindle fibers disassemble.
- Nuclear membrane forms around each set of chromosomes, forming the haploid daughter cell nuclei.

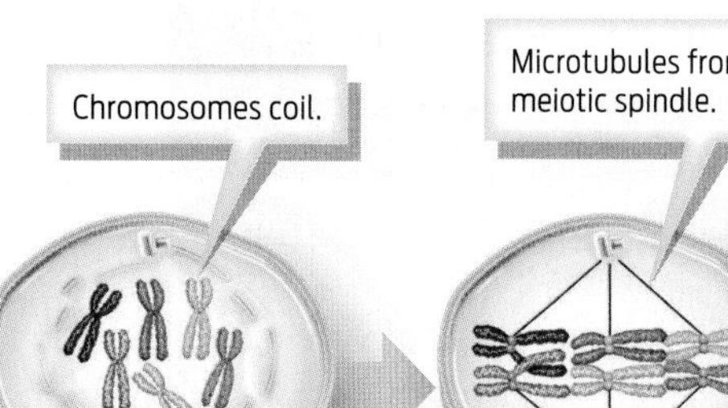

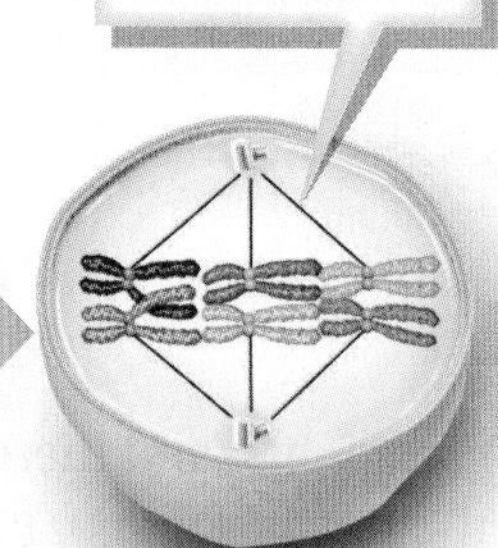

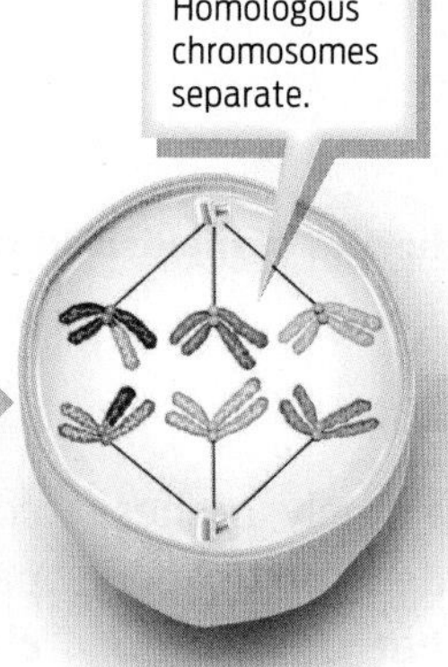

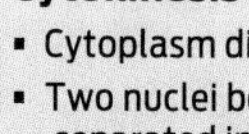

Cytokinesis
- Cytoplasm divides.
- Two nuclei become separated into daughter cells.

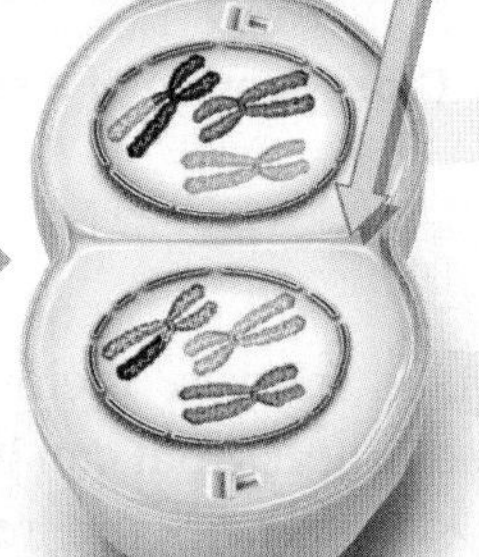

Plant Pollen Cells:

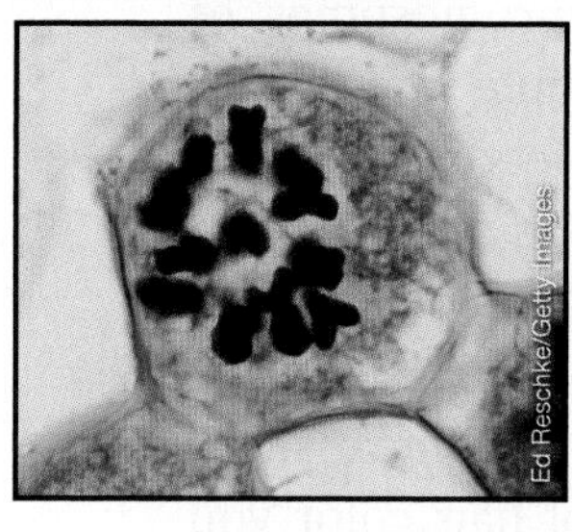

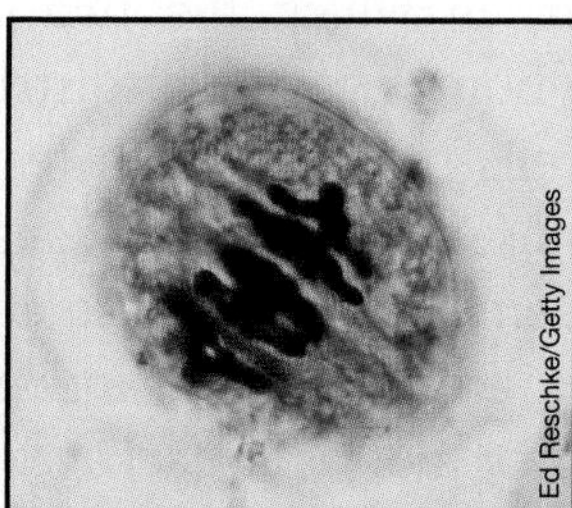

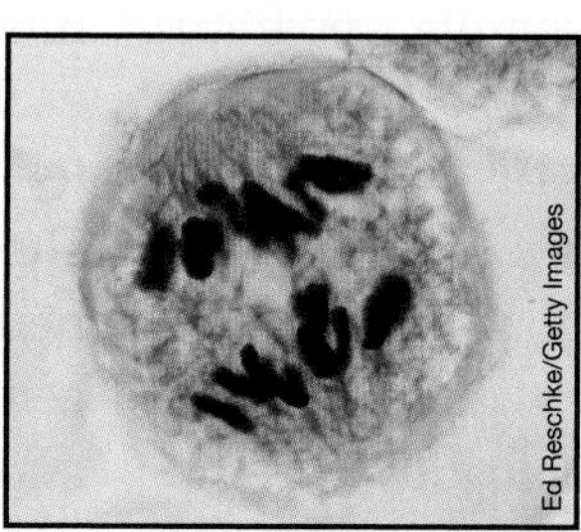

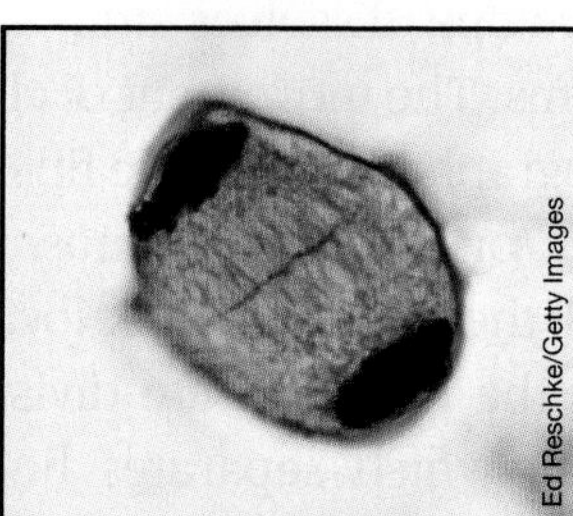

paternal chromosomes that each sperm or egg inherits is different every time meiosis occurs.

Because of recombination and independent assortment, no two gametes produced by an individual are exactly alike. Each gamete will contain one copy of every gene, but the particular combination of alleles in each gamete will be unique (**INFOGRAPHIC 11.5**).

Meiosis is the reason that not everyone in the Schaller family has CF. Some of the Schaller parents' gametes will carry the CF allele and others will not. If by chance a sperm that carries a CF allele fertilizes an egg

Meiosis II

Prophase II
- The nuclear membrane disassembles.
- Microtubule fibers form the meiotic spindle.

Metaphase II
- Microtubule spindle fibers from opposite ends of the cell attach to the sister chromatids of each chromosome.
- Replicated chromosomes become aligned along the middle of the cell.

Anaphase II
- Microtubules shorten, pulling the sister chromatids to opposite ends of the cell.

Telophase II
- Identical sets of chromosomes reach each pole.
- Microtubule spindle fibers disassemble.
- Nuclear membrane forms around each set of chromosomes, forming the haploid daughter cell nuclei.

Daughter Cells
- Cells develop into eggs or sperm.
- Cells are haploid (n).

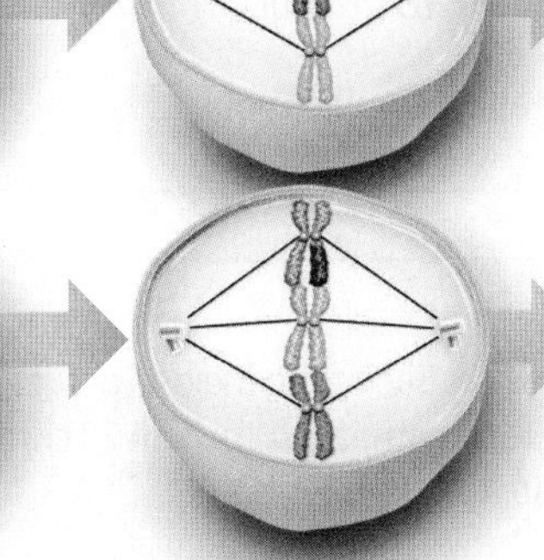
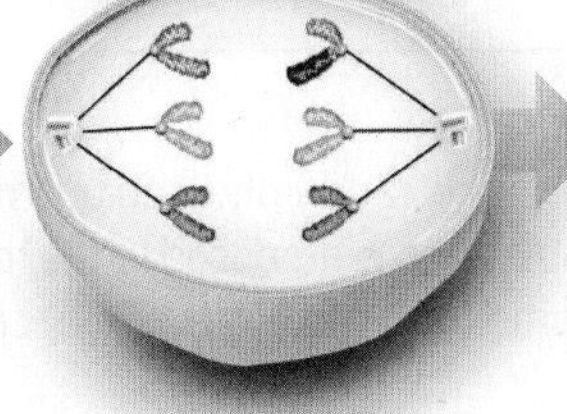
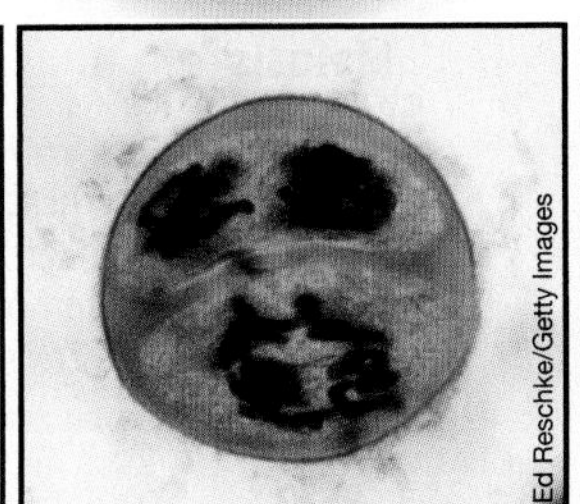
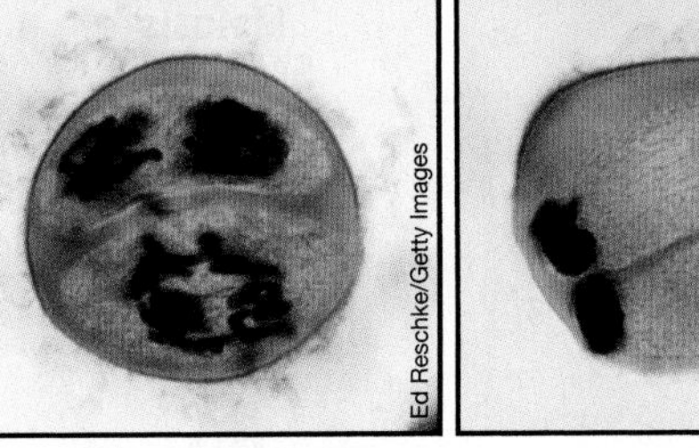

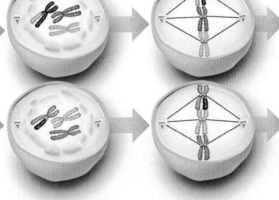

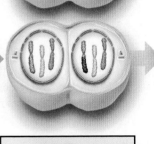

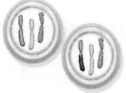

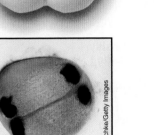

that also carries a CF allele, the resulting child will have CF.

After Emily's cystic fibrosis was diagnosed, both Debbie and Lowell learned that their parents had relatives who had died at a very young age. At the time, the cause of death was thought to be a respiratory illness such as pneumonia. But these relatives most likely had CF, Debbie now thinks; doctors at the time simply did not have the tools to diagnose the disease.

The Schallers now knew that the disease ran in both sides of the family. But they could still not help Emily. "They told us she would

INFOGRAPHIC 11.5 Meiosis Produces Genetically Diverse Eggs and Sperm

Meiosis produces haploid gametes that are genetically unique. Each egg and sperm has its own distinct combination of alleles. The two events that create this diversity are recombination and independent assortment.

Recombination:

Before separating at meiosis I, the maternal and paternal chromosomes line up next to each other and physically exchange segments of DNA. Consequently, maternal chromosomes contain segments (and therefore, alleles) from paternal chromosomes and vice versa.

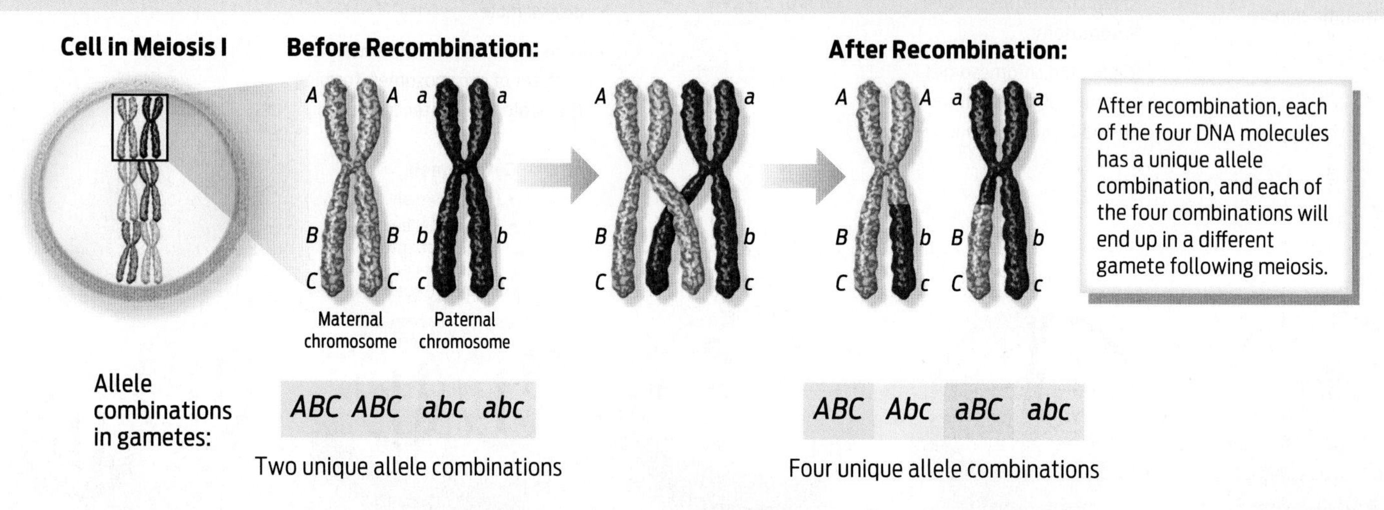
Cell in Meiosis I
Before Recombination:
After Recombination:
A A a a
B B b b
C C c c
Maternal chromosome
Paternal chromosome
After recombination, each of the four DNA molecules has a unique allele combination, and each of the four combinations will end up in a different gamete following meiosis.
Allele combinations in gametes:
ABC ABC abc abc
Two unique allele combinations
ABC Abc aBC abc
Four unique allele combinations

Independent Assortment:

Homologous chromosome pairs separate according to how they have randomly lined up in the cell. Each time meiosis occurs, the chromosome pairs line up differently, and thus a different chromosome combination is produced in the resulting gametes. When all 23 chromosome pairs are considered, there are more than 8 million unique chromosome (and therefore, allele) combinations possible.

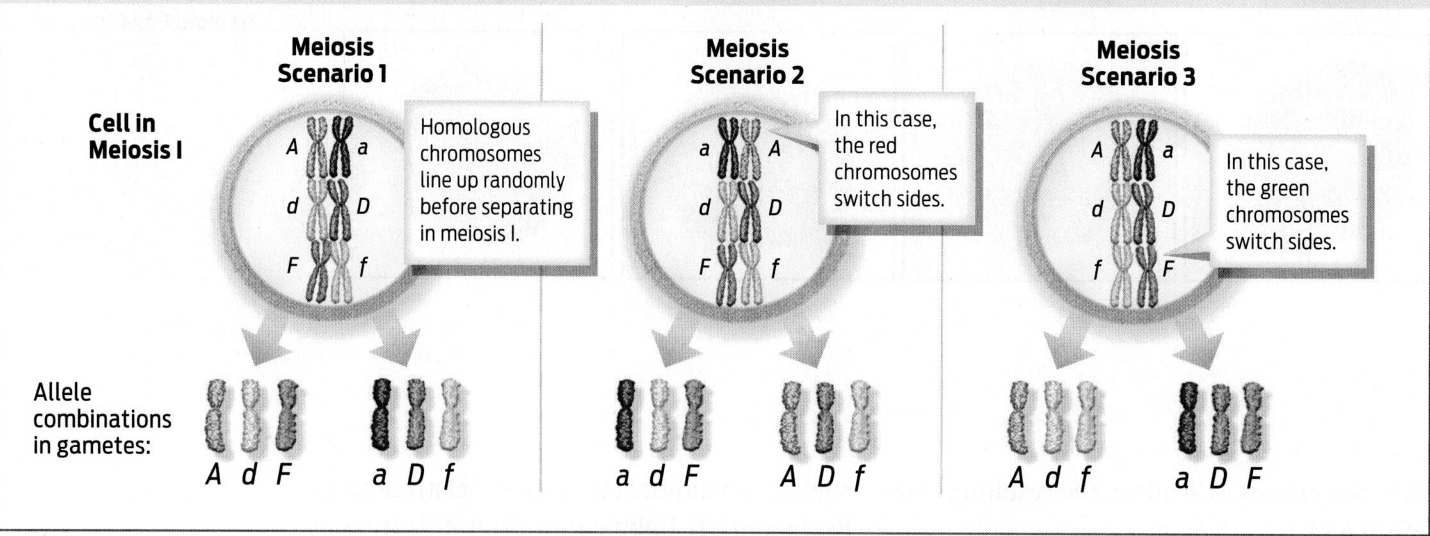
Cell in Meiosis I
Meiosis Scenario 1
Homologous chromosomes line up randomly before separating in meiosis I.
Meiosis Scenario 2
In this case, the red chromosomes switch sides.
Meiosis Scenario 3
In this case, the green chromosomes switch sides.
Allele combinations in gametes:
A d F
a D f
a d F
A D f
A d f
a D F

? Which process (recombination or independent assortment) mixes up combinations of alleles of genes on a single chromosome? Explain your answer.

only live to be about 12 years old," Debbie recalls, adding, "We just put ourselves in the hands of medical professionals."

Living with the Disease

Growing up, Emily was scarcely aware of her own disability. The visits to doctors and periodic stays in the hospital were just a part of life. All her teachers and friends knew that she had CF. "My family and friends were all so supportive," she says. In high school she played volleyball, basketball, and soccer, and participated in many walkathons to raise money for CF research.

It wasn't until the end of high school that she started to be impeded by the disease. "My lung function started to decline, so I wasn't able to keep up with the other kids," Emily says. She began to be hospitalized more often.

In healthy people, the CFTR protein acts as a channel through a cell's membrane that allows chloride ions to move in and out of the cell, keeping the cell's chemistry in balance. But in people with CF, the channel is distorted or absent altogether, and the mechanism goes awry. The result is that mucus—a slippery substance that lubricates and protects the linings of the airways, digestive system, reproductive system, and other tissues—becomes abnormally thick and sticky.

This abnormal mucus blocks ducts throughout the body, including in the pancreas, where it causes digestive problems. (The name "cystic fibrosis" refers to the characteristic fibrous lumps or cysts that form within the pancreas of those with the condition.) The most problematic symptom, however, is that thick mucus builds up in the lungs. People with CF have trouble breathing, and the mucus provides fertile ground for bacteria and other organisms to grow. Over time, repeated infections permanently damage the lungs. As a result, people with CF may slowly lose their ability to breathe, eventually dying of suffocation (**INFOGRAPHIC 11.6**).

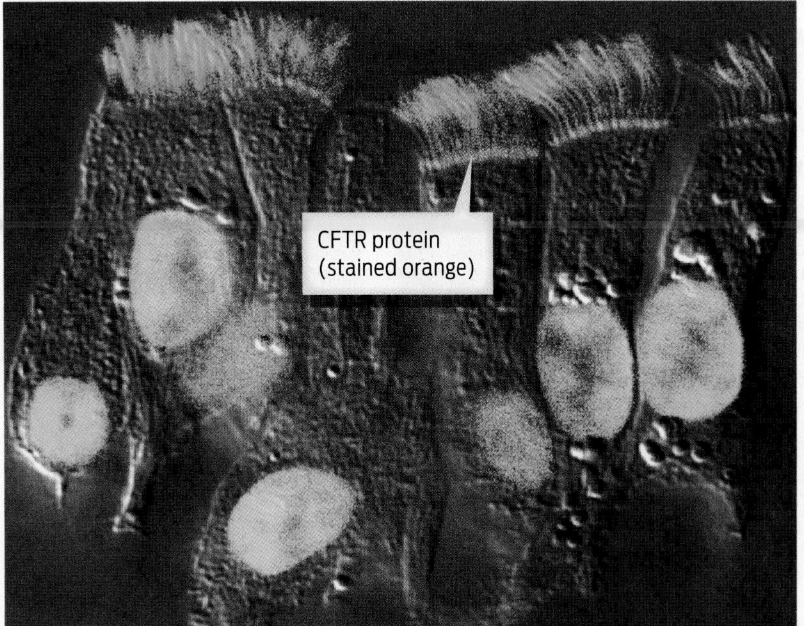

CFTR protein expression in human lung cells.

Cystic fibrosis patients like Emily wear vibrating vests to loosen the mucus in their lungs while inhaling a saltwater mist to thin the mucus.

To avoid lung damage, every morning Emily wears an inflatable vest that vibrates to loosen mucus in her lungs. For 30 minutes she inhales a saltwater mist and another medication to thin her mucus, which she then coughs out periodically. To that regime she adds two

INFOGRAPHIC 11.6 The CFTR Protein and Cystic Fibrosis

The CFTR membrane protein facilitates the movement of chloride ions. When the CFTR protein is not working, chloride ions are trapped in the cell. The result is a thick, sticky mucus on the surface of cells that traps bacteria, leading to infections.

Normal Lung

In cells lining the lungs, the CFTR membrane protein allows passage of chloride ions. When ions can leave the cell, water is able to flow across the membrane freely, keeping the mucus on the outside of the cell thin and slippery. Bacteria and particles trapped in mucus are easily cleared from the lungs.

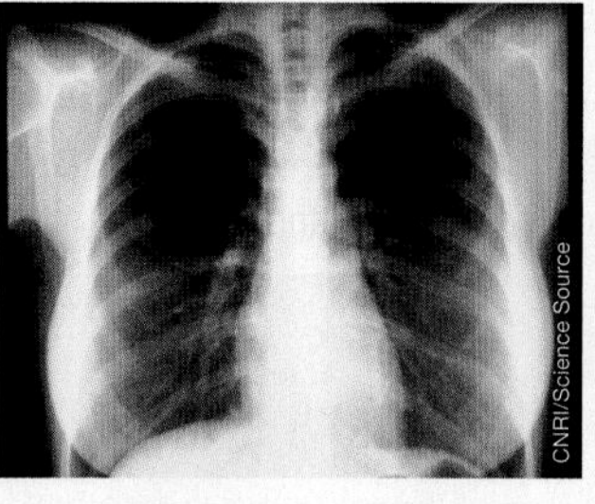
CNRI/Science Source

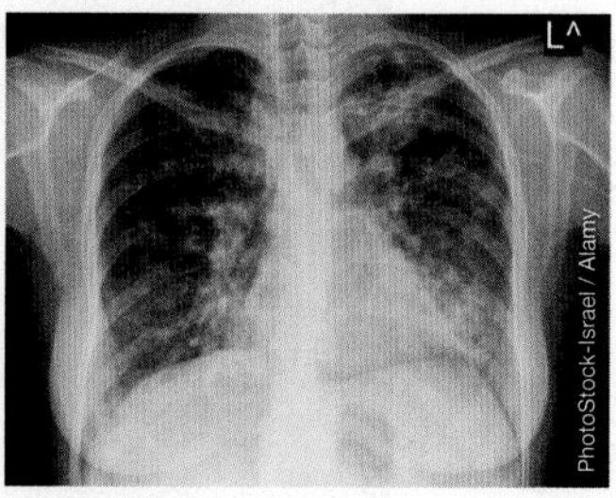
L^
PhotoStock-Israel / Alamy

Cystic Fibrosis Lung

The CFTR protein is not working in these cells, disrupting chloride ion flow. These cells retain more water, causing the mucus outside the cell to become thicker and hard to clear out of the lungs. Bacteria trapped in this mucus remain in the lungs and cause infection.

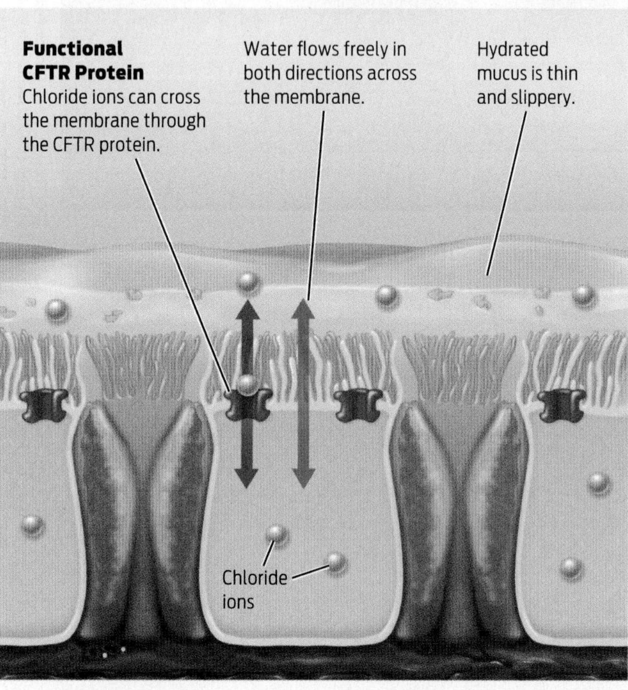
Functional CFTR Protein
Chloride ions can cross the membrane through the CFTR protein.
Water flows freely in both directions across the membrane.
Hydrated mucus is thin and slippery.
Chloride ions

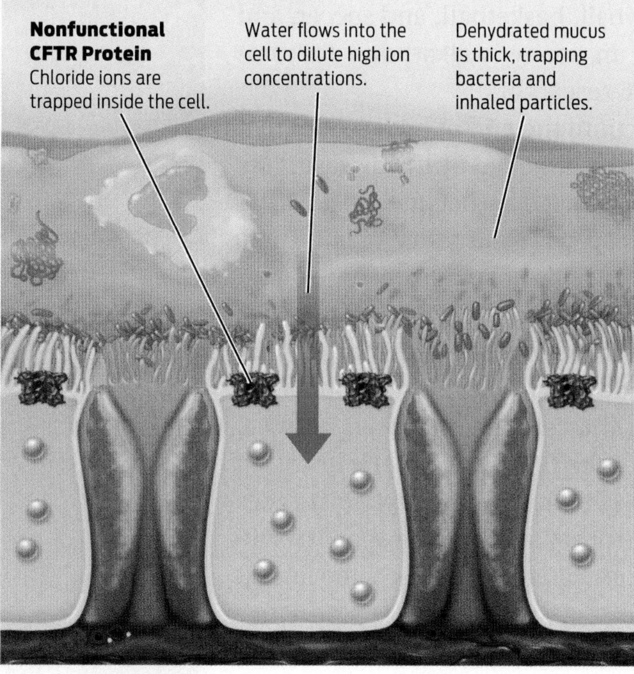
Nonfunctional CFTR Protein
Chloride ions are trapped inside the cell.
Water flows into the cell to dilute high ion concentrations.
Dehydrated mucus is thick, trapping bacteria and inhaled particles.

? Explain the relationship between the CFTR protein, chloride ions, water movement, and how thin or sticky the mucus is.

other medications three times a week to kill infections and to keep her lungs from becoming inflamed. And she runs every morning, which helps keep her lung capacity up.

RECESSIVE ALLELE
An allele that reveals itself in the phenotype only if a masking dominant allele is not present.

DOMINANT ALLELE
An allele that can mask the presence of a recessive allele.

HETEROZYGOUS
Having two different alleles for a given gene.

Emily hasn't ruled out having a family of her own one day. Even though she has CF, her children will not necessarily have the disease.

Why not? Remember that since Emily has CF, her parents, Lowell and Debbie, both must carry a CF allele. The CF allele produces a nonfunctioning CFTR protein, but Debbie and Lowell are healthy because they each have a normal *CFTR* allele that encodes a normally functioning CFTR protein. The effect of the CF allele in each of them is thus hidden. When one allele of a gene masks the effect of the other, the hidden allele is described as a **recessive allele** (designated by a lower-case letter, e.g., *a*). The other allele, which masks the effect of the recessive allele, is known as the **dominant allele** (designated by a capital letter, e.g., *A*). The inheritance of traits controlled by single genes with alleles that are either dominant or recessive is sometimes called Mendelian, after Gregor Mendel, the scientist who discovered this pattern (see **Milestone 4: Mendel's Garden**).

Geneticists call Lowell's and Debbie's genotypes, each with two distinct alleles,

heterozygous. Their two healthy sons are either heterozygous like their parents, or they have two normal alleles–that is, their genotype is **homozygous** for the normal allele. A genotype made up of two dominant alleles is known as homozygous dominant. Emily's genotype is homozygous recessive: she inherited one recessive CF allele from each parent, which is why she has the disease.

What were the chances that Debbie and Lowell would have a child with CF? To figure out the likelihood that parents will have a child with a particular trait, it helps to plot the possibilities on a **Punnett square** (named after the geneticist Reginald Punnett, who devised it). A Punnett square matches the possible parental gametes and shows the likelihood that particular parental alleles will combine. As heterozygous individuals, Debbie and Lowell each have a 50% chance of passing on their CF allele to a child, which means they have a 25% chance of having a child with CF and a 75% chance of having a healthy child. The chance that a child will be a heterozygous **carrier**–that is, that the child will carry the recessive allele for CF but will not have the disease because the allele's effect is masked by the dominant allele–is 50% **(INFOGRAPHIC 11.7)**.

HOMOZYGOUS
Having two identical alleles for a given gene.

PUNNETT SQUARE
A diagram used to determine probabilities of offspring having particular genotypes, given the genotypes of the parents.

CARRIER
An individual who is heterozygous for a recessive allele and can therefore pass it on to offspring without showing any of its effects.

INFOGRAPHIC 11.7 How Recessive Traits Are Inherited

Cystic fibrosis is a recessive trait, which means that the disease phenotype is caused by inheriting two recessive alleles, as in Emily's case. Emily's parents do not have CF because, even though they each carry one recessive CF allele, they also each possess one dominant normal CFTR allele. In other words, they are both heterozygous carriers of the recessive CF allele. To calculate the probability that Debbie and Lowell will have a child with CF, we can create all possible alleles in their gametes and then join all possible combinations of these sperm and egg in a Punnett square.

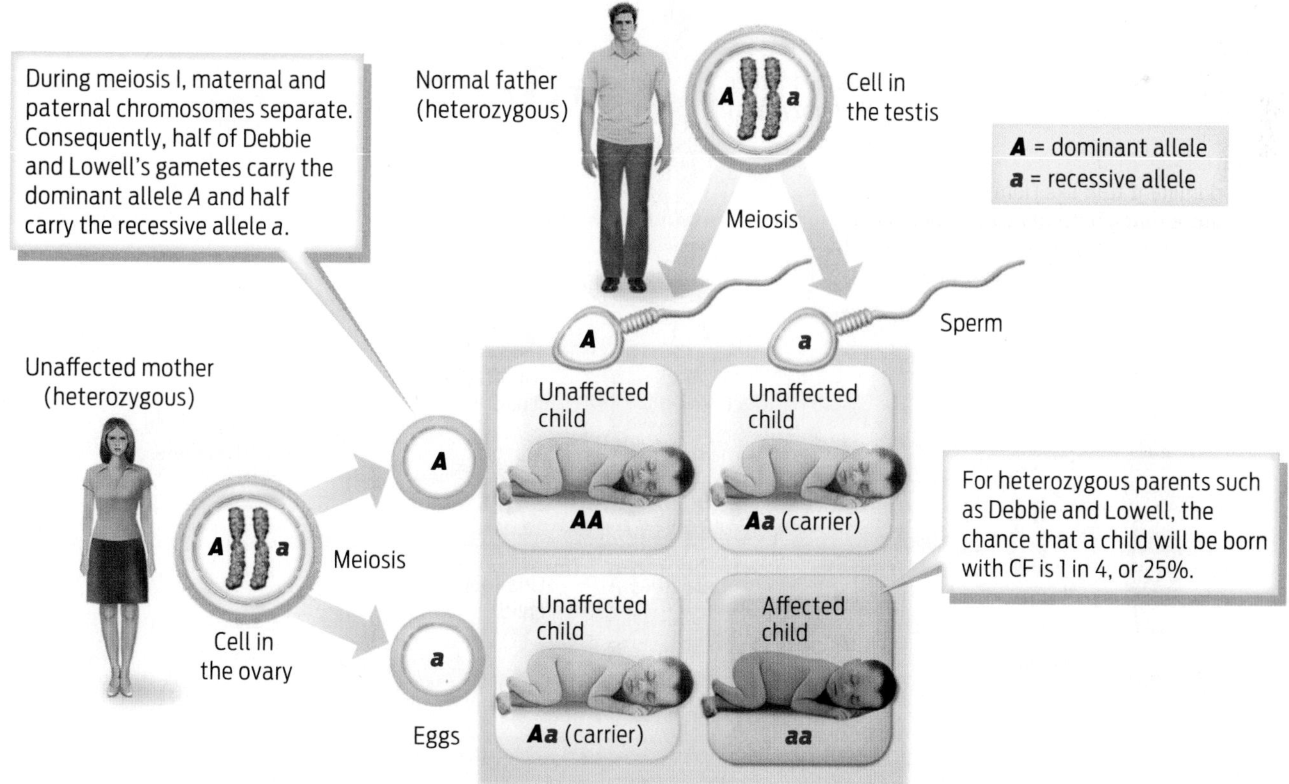
During meiosis I, maternal and paternal chromosomes separate. Consequently, half of Debbie and Lowell's gametes carry the dominant allele *A* and half carry the recessive allele *a*.
Normal father (heterozygous)
A
a
Cell in the testis
A = dominant allele
a = recessive allele
Meiosis
Sperm
A
a
Unaffected mother (heterozygous)
Unaffected child
AA
Unaffected child
Aa (carrier)
For heterozygous parents such as Debbie and Lowell, the chance that a child will be born with CF is 1 in 4, or 25%.
Meiosis
Cell in the ovary
Eggs
Unaffected child
Aa (carrier)
Affected child
aa

? What is the probability that these two parents will have a child who is a carrier for cystic fibrosis?

Just as Emily's genotype is different from her parents' genotype, Emily's children will have different genotypes from her own. Whether or not her children develop CF depends on the father's genotype. Since Emily is homozygous, she can contribute only recessive CF alleles to her children. If Emily were to have children with a man who had two normal alleles, for example, none of her children would have the disease–they would all have a heterozygous genotype but a normal phenotype. But as carriers, they could pass on the recessive CF allele to their children. If Emily had children with a man who was heterozygous for the CF gene, then her children would have a 1 in 2, or 50%, chance of having CF.

Not all recessive alleles cause disease. Many physical traits are the result of inheriting two recessive alleles of a gene. For example, people with blue eyes or red hair have inherited recessive alleles that prevent the deposition of dark pigment. And not all genetic diseases are caused by recessive alleles: some, such as the neurodegenerative disorder Huntington disease, are determined by dominant alleles. Diseases caused by dominant alleles have a higher probability of showing up in the next generation because it takes only one disease allele to cause the trait **(INFOGRAPHIC 11.8)**.

In all cases, anyone with a genetic disease is at risk for passing it on to his or her

INFOGRAPHIC 11.8 How Dominant Traits Are Inherited

Some genetic conditions are caused by dominant alleles. Examples are Huntington disease, a degenerative neurological disease, and polydactyly, having more than five fingers or toes per limb. Many common traits such as dark eyes and dimples are also determined by dominant alleles. In these cases, having one copy of the dominant allele is sufficient to display the trait.

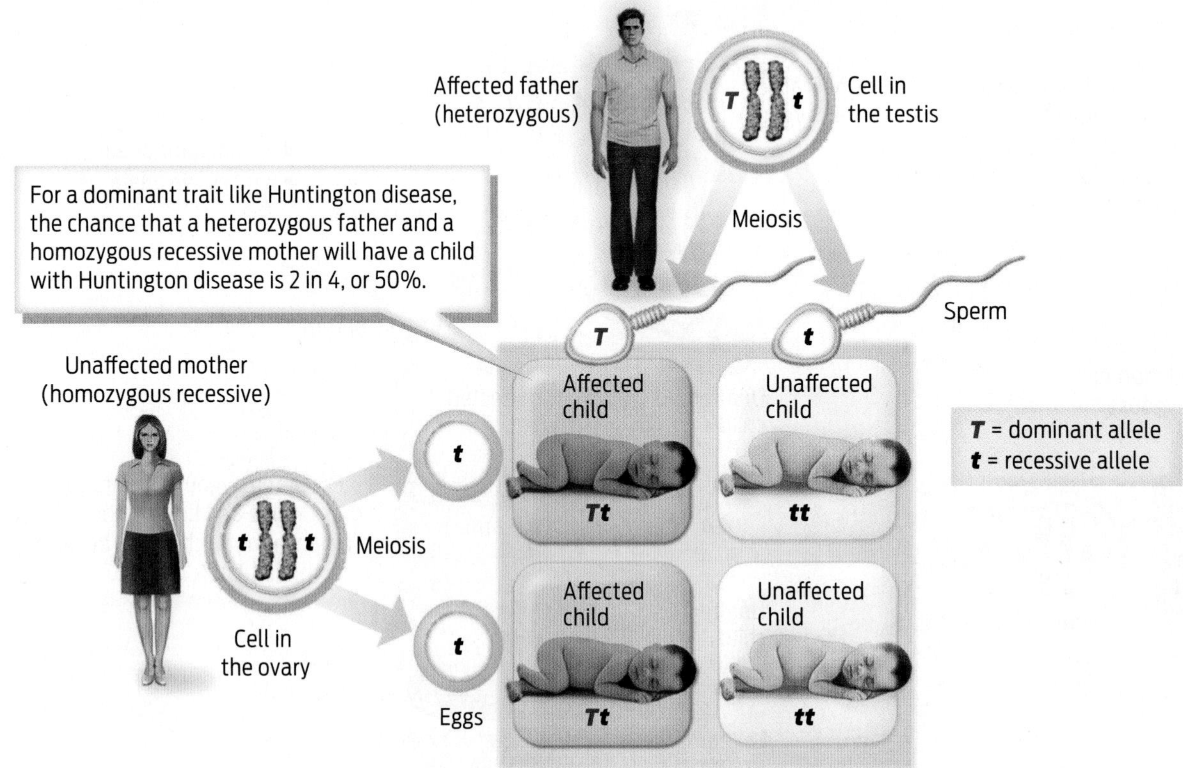

? What is the probability of these parents having a child with two copies of the dominant (*T*) allele?

TABLE 11.1 Inherited Genetic Conditions in Humans

RECESSIVE TRAIT	PHENOTYPE
Albinism	Lack of pigment in skin, hair, and eyes
Cystic fibrosis	Excess mucus in lungs, digestive tract, and liver; increased susceptibility to infections
Sickle-cell disease	Sickled red blood cells; damage to tissues
Tay-sachs disease	Lipid accumulation in brain cells; mental deficiency, blindness, and death in childhood
DOMINANT TRAIT	**PHENOTYPE**
Huntington disease	Mental deterioration and uncontrollable movements; onset at middle age
Freckles	Pigmented spots on skin, particularly on face and arms
Polydactyly	More than five digits on hands or feet
Dimples	Indentations in the skin of the cheeks
Chin cleft	Indentation in chin

children. The risk merely varies, depending on whether the disease-related alleles are dominant or recessive, and on the genotype of the partner (**TABLE 11.1**).

Couples who carry disease alleles needn't feel that having children is a roll of the dice, however. There are ways to ensure that their children won't develop the diseases they could otherwise inherit. Many couples in this situation use a technology called pre-implantation genetic diagnosis to detect and select embryos that do not carry defective alleles. Through in vitro fertilization, a man's sperm can fertilize a woman's eggs outside the body (see Chapter 31). The genes of each resulting embryo are then examined for specific alleles, and then only embryos that don't contain defective alleles are implanted into the mother. Hundreds of thousands of babies have been born by this technique. Some couples, however, may choose not to undergo assisted reproduction because of religious or other personal reasons.

For children who do inherit a genetic disorder, there may be ways to treat the symptoms or reduce their severity. In the case of CF, new treatments are coming online that could help Emily, and her children and grandchildren, too. Furthest along are a class of medications that, when inhaled, can restore the balance of ions inside affected cells in the lungs. Scientists are presently testing more than 20 experimental drugs in humans, and a few are already on the market.

New Research in the Pipeline

Through basic research, scientists continue to learn more about cystic fibrosis. Over the past 25 years, scientists have discovered more than 1,000 different alleles of the *CFTR* gene. About one in 25 people in the general population are carriers of a CF mutation. The most common is Δ*F508*. About 70% of people with cystic fibrosis are homozygous for this mutation, and around 90% have at least one Δ*F508* allele. People with this particular CF allele have more severe disease because it causes the CFTR protein to be absent from the cell membrane altogether.

Researchers have long puzzled over why the severity of CF often varies significantly between any two people with the disease. Even those who are both homozygous for the *△F508* allele may vary in how their disease progresses. Researchers long thought that perhaps environmental factors such as diet, social relationships, and exercise were responsible.

But in recent years, scientists have learned that there is more to the story. Other genes on different chromosomes contribute to the severity of CF symptoms. The genes so far discovered predominantly influence the immune system, which helps the body fight off infections.

For example, scientists have found that one allele of a gene called *TGFβ1*, located on chromosome 19, is associated with more-severe lung disease in people with CF. This gene influences the immune response to infection. Scientists suspect that people with this *TGFβ1* allele mount a more vigorous response to infections than those with other *TGFβ1* alleles. Such a heightened immune response can cause lung tissue to scar. So if a person with CF also inherited this specific allele of *TGFβ1*, his or her lungs are more likely to scar in response to infections. The impact of such "modifier genes" on the CF phenotype makes it more complicated to predict how disabling any particular person's CF disease will be—but it is not impossible.

Parents who are heterozygous carriers of a CF allele, for example, have a 1 in 4, or 25%, chance of having a child who has CF. If these two parents are also heterozygous for *TGFβ1*, then their chance of having a child who is homozygous recessive for *TGFβ1* is also 1 in 4 (25%). The chance of two independent events occurring together is calculated by multiplying the two independent chances together. So the probability of being homozygous recessive for both *CFTR* and *TGFβ1* is $1/4 \times 1/4$, or 1 in 16. This probability can also be calculated by using a Punnett square **(INFOGRAPHIC 11.9)**.

Understanding how these modifier genes contribute to the disease may point the way to even more therapies. Drugs that reduce inflammation by targeting the *TGFβ1* protein, for example, may help reduce scarring in the lungs.

Emily recently had her genotype tested. While she doesn't carry *△F508*, she does have another allele associated with severe disease, *G551D*. This allele is much rarer than *△F508*—only about 4% of those with CF have it. People with this mutation have a CFTR channel that doesn't open properly. The good news for people with this mutation is that a new FDA-approved drug, called Kalydeco (its generic name is ivacaftor), is remarkably effective at restoring function to their malfunctioning CFTR channel. Unlike other drugs, which merely treat the symptoms of the disease, this one addresses the underlying cause: the malfunctioning protein.

> "I just took the first deep breath I've ever taken in my life."
>
> — Emily Schaller

Emily started taking the drug in 2010. By day 4, she knew it was making a difference. "I just took the first deep breath I've ever taken in my life," she recalls telling her brother as they were walking down the street. She didn't cough, and there wasn't any mucus in her lungs.

Kalydeco is not a cure for CF—the genetic mutation is still present in the cells of someone with CF—but for those with the *G551D* allele, it helps immensely. Emily nicknamed the drug "blue lightning" because of how quickly the small blue pill helped her breathing.

Keenly aware of how medical progress has extended her life, Emily conducts her own share of fund raising and education. In 2007, she started the Rock CF Foundation, a nonprofit devoted to improving the lives of people with cystic fibrosis and raising

INFOGRAPHIC 11.9 Tracking the Inheritance of Two Genes

People with CF differ in the severity of their disease. Some of this variability is influenced by alleles of other genes that sit on other chromosomes. One such gene, *TGFβ1*, is located on chromosome 19. We can use a Punnett square to follow the inheritance of two genes, as in the example below. In this example, the *CFTR* gene is represented by the symbol A and the *TGFβ1* gene by the symbol D.

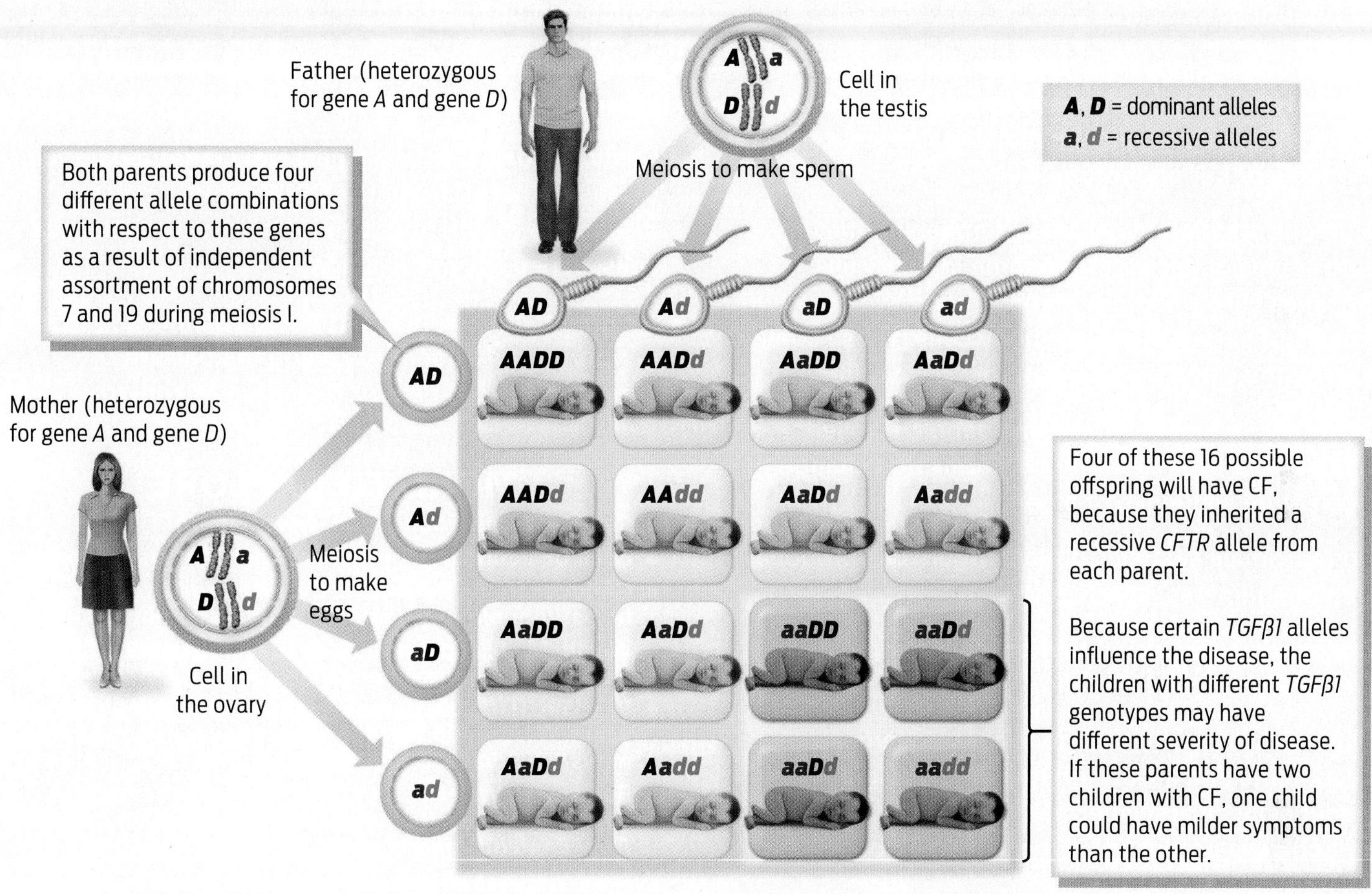

Father (heterozygous for gene *A* and gene *D*)
Cell in the testis
Meiosis to make sperm
A, **D** = dominant alleles
a, **d** = recessive alleles
Both parents produce four different allele combinations with respect to these genes as a result of independent assortment of chromosomes 7 and 19 during meiosis I.
Mother (heterozygous for gene *A* and gene *D*)
Cell in the ovary
Meiosis to make eggs
AD
Ad
aD
ad
AADD
AADd
AaDD
AaDd
AAdd
Aadd
aaDD
aaDd
aadd
Four of these 16 possible offspring will have CF, because they inherited a recessive *CFTR* allele from each parent.
Because certain *TGFβ1* alleles influence the disease, the children with different *TGFβ1* genotypes may have different severity of disease. If these parents have two children with CF, one child could have milder symptoms than the other.

? What is the probability of having a child with two dominant *CFTR* and two dominant *TGFß1* alleles?

awareness of the disease. The name of the foundation comes from Emily's rock 'n' roll roots. In the early 2000s, she played drums in an all-female rock band called Hellen. "I favor the old stuff, AC-DC, Led Zeppelin, the Ramones," she says.

Emily's since retired her drumsticks but continues to work actively with the foundation, which sponsors an annual half marathon and also provides sneakers to kids with CF who want to take up running. Through her foundation, Emily hopes to lessen the load for others living with the disease, the way hers has been lightened.

In 2016, she gave a talk in Baltimore, Maryland, called, "Running Down a Dream: Emily's Mission to Rock CF." She told the assembled audience about her lifelong struggle to beat the odds against CF, her passion for running as a way to stay healthy, and the optimism she has for the future.

"I know I'm going to be old and see my nieces and nephews grow up and it's pretty damn cool," she said. ■

CHAPTER 11 SUMMARY

- Genes, which code for proteins, are the units of inheritance, physically passed down from parents to offspring.
- An organism's physical traits constitute its phenotype; its genes constitute its genotype. You can't always determine an individual's genotype from his or her phenotype.
- Humans are diploid organisms, meaning they have two copies of each chromosome in their cells. Because chromosomes come in pairs, we have two copies of each gene in our body cells. These copies can be the same or different from each other.
- Different versions of the same gene are called alleles. Alleles arise from mutations that change the nucleotide sequence of a gene.
- To reproduce sexually, diploid organisms produce haploid gametes, sperm and egg.
- Meiosis is cell division that produces haploid gametes from diploid cells.
- During meiosis, homologous chromosomes recombine and assort independently to generate genetically diverse sperm and eggs. No two sperm or egg cells produced by the same person will be exactly alike.
- Alleles may be dominant or recessive. Dominant alleles can mask the effects of recessive alleles, which can be hidden.
- Many traits result from carrying two recessive alleles; others result from carrying one dominant allele.
- Cystic fibrosis (CF) is a recessively inherited genetic disease. Alterations in the CFTR protein cause disease by interfering with ion and water balance in cells, especially in the lungs.
- A Punnett square can help predict a child's genotype and phenotype when the pattern of inheritance, dominant or recessive, is known.

MORE TO EXPLORE

- **Cystic Fibrosis Foundation: www.cff.org/**
- **Rock CF Foundation: http://letsrockcf.blogspot.com/**
- **Genetics Home Reference, Cystic Fibrosis: https://ghr.nlm.nih.gov/condition/cystic-fibrosis**
- **Welsh, M. J., and Smith, A. E. (December 1995) Cystic fibrosis. *Scientific American*.**
- **Pearson, H. (2009) Human genetics: One gene, twenty years. *Nature* 460:164–169. http://www.nature.com/news/2009/080709/full/ 460164a.html**
- **Ramsey, B. W., et al. (2011) A CFTR potentiator in patients with cystic fibrosis and the *G551D* mutation. *New England Journal of Medicine* 365:1663–1672.**

CHAPTER 11 Test Your Knowledge

DRIVING QUESTION 1 How does the organization of chromosomes, genes, and their alleles contribute to human traits?

By answering the questions below and studying Infographics 11.1, 11.2, 11.3, and 11.6, you should be able to generate an answer to the broader Driving Question above.

KNOW IT

1 **How do the two alleles of the *CFTR* gene in a lung cell differ?**

a. They were inherited from different parents.
b. One is on chromosome 7 and one is on chromosome 3.
c. Only one is expressed.
d. all of the above
e. There is no difference because they are both the same gene.

2 Consider a liver cell.

a. How many chromosomes are present?
b. How many alleles of each gene are present?

3 Consider a gamete.

a. How many chromosomes are present?
b. How many alleles of each gene are present?

USE IT

4 A diploid cell of baker's yeast has 32 chromosomes.

a. How many pairs of homologous chromosomes are present in a diploid cell of baker's yeast?
b. How many chromosomes are present in each of its haploid gametes?

5 In diploid organisms, having two homologues of each chromosome can be beneficial if one allele of a gene encodes a nonfunctional protein. Can haploid organisms survive having nonfunctional alleles? Explain your answer.

6 From which parent did Emily inherit cystic fibrosis? Explain your answer.

DRIVING QUESTION 2 How does meiosis produce genetically diverse gametes?

By answering the questions below and studying Infographics 11.3, 11.4, 11.5, and Up Close: Phases of Meiosis, you should be able to generate an answer to the broader Driving Question above.

KNOW IT

7 A human female has _____ chromosomes in each skin cell and ______ chromosomes in each egg.

a. 46; 46
b. 23; 46
c. 46; 23
d. 23; 23
e. 92; 46

8 A woman is heterozygous for the CF-associated gene (the alleles are represented here by the letters *A* and *a*). Assuming that meiosis occurs normally, which of the following represent eggs that she can produce?

a. *A*
b. *a*
c. *Aa*
d. *AA*
e. *aa*
f. *A* or *a*
g. *A*, *a*, or *Aa*

9 Draw a maternal version of chromosome 7 in one color and a paternal version of chromosome 7 in another color. Maintaining this color distinction, now draw a possible version of chromosome 7 that could end up in a gamete following meiotic division.

USE IT

10 An alien has 82 total chromosomes in each of its body cells. The chromosomes are paired, making 41 pairs. If the alien's gametes are produced by meiosis, what are the number and arrangement (paired or not) of chromosomes in one of its gametes? Give the reason for your answer.

11 Describe at least two major differences between mitosis (discussed in Chapter 9) and meiosis.

12 If meiosis were to fail and a cell skipped meiosis I, so that meiosis II was the only meiotic division, how would you describe the resulting gametes?

DRIVING QUESTION 3 Why do different traits have different patterns of inheritance?

By answering the questions below and studying Infographics 11.7, 11.8, and 11.9, you should be able to generate an answer to the broader Driving Question above.

KNOW IT

13 What is the genotype of a person with CF?

a. homozygous dominant
b. homozygous recessive
c. heterozygous
d. any of the above
e. none of the above

14 Strictly on the basis of the following *CFTR* genotypes, what do you predict the phenotype of each person to be?

a. heterozygous
b. homozygous dominant
c. homozygous recessive

15 How many copies of the CF-associated allele does a person with CF have in one of his or her lung cells? How does this compare to someone who is a carrier for CF? How does it compare to someone who is homozygous dominant for the gene *CFTR*?

16 Women can inherit alleles of a gene called *BRCA1* that puts them at higher risk for breast cancer. The alleles associated with elevated cancer risk are dominant. Of the genotypes listed below, which carries the lowest genetic risk of developing breast cancer?

a. *BB*
b. *Bb*
c. *bb*
d. *BB* and *Bb* carry less risk than *bb*.
e. All carry equal risk.

USE IT

17 A person has a heterozygous genotype for a gene associated with a particular inherited disease. However, this person does not have the disease phenotype. Does this disease have a dominant or a recessive inheritance pattern?

18 Assume that Emily (who has CF) decides to have children with a man who does not have CF and who has no family history of CF.

a. What combination of gametes can each of them produce?

b. Place these gametes on a Punnett square and fill in the results of the cross.

c. On the basis of the Punnett square results, what is the probability that they will have a child with CF?

d. On the basis of the Punnett square results, what is the probability that they will have a child who is a carrier for CF?

apply YOUR KNOWLEDGE

MINI CASE

19 You are a genetic counselor. A 21-year-old college student has scheduled an appointment because his 47-year-old mother has Huntington disease, and he is worried about developing this disease himself. You ask about other family members. The student's maternal grandmother (his mother's mother) does not have Huntington disease. The student's father is 62 years old and does not have Huntington disease. From this information, you draw a Punnett square to determine the probability that the student will develop Huntington disease.

a. What could you tell the student about his risk?

b. The student has a half sister. She is 19 years old and has the same mother but a different father. Her father is 45 and does not have Huntington disease. However, the father's mother died of Huntington disease. How does the half sister's risk compare to her brother's (the student's) risk? Could you give her a definitive answer about her risk? Why or why not?

20 From the discussion in this chapter, why might one person with a homozygous recessive *CFTR* genotype have a somewhat different phenotype from another person who also has a homozygous recessive *CFTR* genotype?

21 Phenylketonuria is considered to be an inborn error of metabolism. It is a recessive genetic condition in which the enzyme that breaks down the amino acid phenylalanine is defective or missing. Testing of all newborns allows this condition to be detected at birth. A special diet that severely minimizes phenylalanine (e.g., avoiding diet sodas and most usual sources of protein) can treat the condition. If two carriers of both cystic fibrosis and phenylketonuria were to have a child, what is the probability that the child will have

a. both cystic fibrosis and phenylketonuria?

b. cystic fibrosis and be a carrier for phenylketonuria?

c. neither condition?

d. neither condition and not be a carrier for either?

DRIVING QUESTION 4 What are practical applications of understanding the genetic basis of human disease?

By answering the questions below and studying Infographics 11.6, 11.7, 11.8, and 11.9, you should be able to generate an answer to the broader Driving Question above.

KNOW IT

22 Can cystic fibrosis be diagnosed prenatally by examining the fetal chromosomes (as shown in the inset in Infographic 11.2)? Why or why not?

INTERPRETING DATA

apply YOUR KNOWLEDGE

23 Ivacaftor (trade name Kalydeco) is a drug designed to enhance the activity of the CFTR protein encoded by the *G551D* allele. The protein encoded by this allele is present in the cell membrane, but it is not very active. The drug has been shown in laboratory studies to enhance the activity of this protein. To determine if ivacaftor has an impact on disease symptoms in people with cystic fibrosis, the drug was tested in a randomized, double-blind clinical trial.

a. From the information provided here and in the graph below, describe the experimental design of the trial. Include who the participants were, how participants were assigned to the experimental and the control group, and what treatment(s) were given to participants in each group. Why was it important that this was a double-blind trial?

b. Before treatment started, baseline measurements of lung function were obtained. One of these measurements is the amount of air that a participant could forcibly exhale from the lungs in 1 s (expressed as % of total lung volume). This measurement is known as the predicted FEV. The drug was given to the experimental group every 12 hours; the predicted FEV was measured, and expressed as a change from the baseline measurement. The results are shown in the graph. The number of subjects (*N*) is shown for each group at various time points during the study.

- How soon did the experimental group experience an improvement in lung function as measured by predicted FEV?
- How long was the improvement sustained?
- By the end of the study, what was the absolute improvement of the experimental group relative to baseline?

c. The *ΔF508* allele causes the CFTR protein to be absent from the cell membrane. Is ivacaftor likely to be a viable treatment for patients whose CF is caused by this allele? Why or why not?

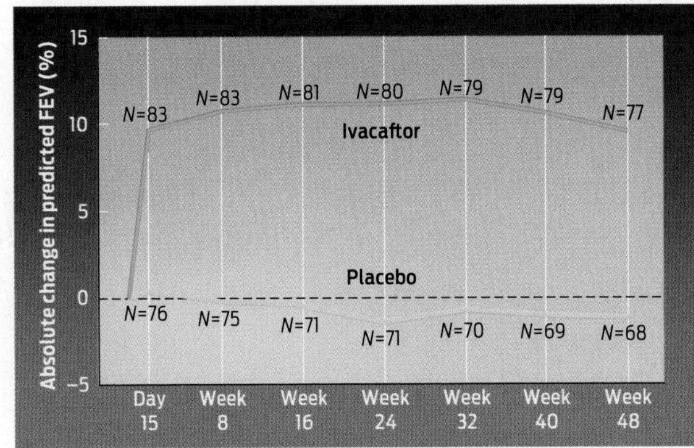

Data from Ramsey, B. W., et al. (2011) A CFTR potentiator in patients with cystic fibrosis and the G551D mutation. Figure 1A. *New England Journal of Medicine* 365:1663–1672.

BRING IT HOME

apply YOUR KNOWLEDGE

24 Emily took a genetic test to determine which CF alleles she inherited. The results revealed she has a *G551D* allele, making her a candidate for ivacaftor. Since taking this drug, Emily's breathing and lung function has improved. In this case, the genetic test opened up a treatment option for a patient. In other cases (e.g., cystic fibrosis patients with the *ΔF508* allele) treatments are limited, and for some genetic diseases, such as Huntington disease, treatment is limited and there is no cure. If you were faced with the decision to take a genetic test, especially for a disease for which there is no cure, would you take the test? Why or why not?

4 Milestones in Biology

The garden outside the Augustinian Abbey in Brno, where Gregor Mendel performed his experiments.

DRIVING QUESTIONS

1. How was Mendel able to recognize the transmission of alleles before the discovery of DNA?
2. What do Mendel's two laws state about how offspring inherit alleles from their parents?

BIOPHOTO ASSOCIATES/Getty Images

Mendel's *Garden*

An Austrian monk lays the foundation for modern genetics

GREGOR JOHANN MENDEL was an unlikely father of genetics. He was a melancholy Austrian monk who by all accounts suffered from debilitating test-taking anxiety, failing his teaching exam twice. Mendel nevertheless collected the first research suggesting that each parent passes discrete "elements" to each child that determine specific traits. These elements remain intact and can be passed on indefinitely to future generations without being diluted. Although he couldn't say at the time what these elements were, Mendel had in fact discovered what came to be called genes. We now know that genes come in pairs, and that they exist in multiple discrete forms called alleles (see Chapter 11).

Many scientists of Mendel's day–the mid-19th century–believed that parental traits were blended together in offspring, like mixing paint colors. For example, a tall mother and short father would have children of medium height, who would then pass on that trait–medium height–to their own children. Other scientists clung to an old idea that a sperm or egg contained a miniature adult waiting to be born. But through a series of simple yet elegant experiments conducted in a monastery garden, Mendel provided a new explanation for heredity, decades before the word "genetics" was coined (**INFOGRAPHIC M4.1**).

INFOGRAPHIC M4.1 Ideas of Inheritance Before Mendel

Preformationist ideas, popular for centuries before Mendel, held that the next generation of life already existed fully formed in miniature inside the egg or sperm. More common in Mendel's time was the idea that substances from the mother and father blend together during conception to produce traits of the offspring.

Preformationist Ideas

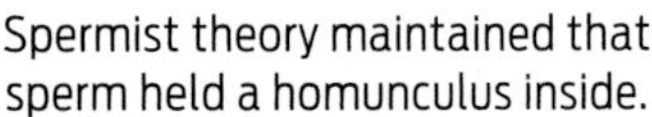

Spermist theory maintained that sperm held a homunculus inside.

Ovist theory proposed that the preformed human was inside the egg.

Blending Ideas

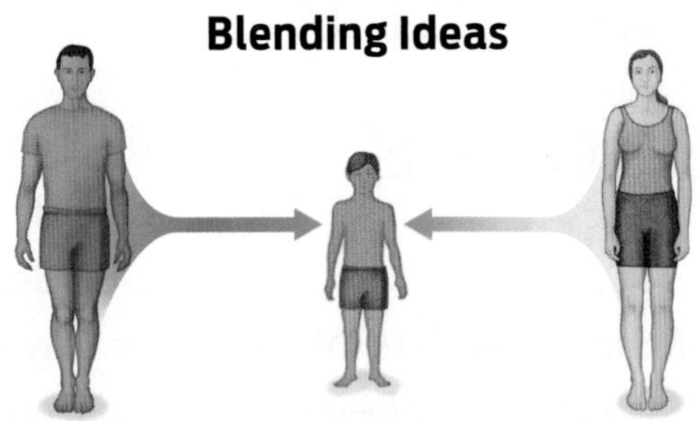

Fluid or particles from the father and mother blend together to make the traits of the offspring.

? What did the spermist and ovist theories have in common?

In 1843, Mendel became a monk at the Augustinian Abbey of St. Thomas in Brünn (now Brno, in the Czech Republic). He studied theology and was ordained a priest in 1847. When he failed his teaching exam (the Augustinians were a teaching order), the abbot at St. Thomas sent Mendel to the University of Vienna to brush up. For 2 years he studied math, physics, zoology, and botany, but once again he flunked the test. This depressing result encouraged him to turn from teaching to research.

Mendel returned to the monastery in 1853, and a year later began researching a topic that had sparked his interest in school: hybrids, the offspring of two different breeds or varieties. Mendel hoped to explain what he and many others had observed about hybrids—that physical traits (size, color, etc.) can sometimes skip a generation, disappearing in one generation only to show up again in the next. For example, hybrid crop plants would often produce offspring that looked much like themselves. But occasionally, they produced offspring that resembled earlier generations.

Mendel began his research using mice, which he kept in cages in his two-room flat. He intended to breed gray mice with albino (white) mice to see what color fur the offspring would have. He didn't get very far down this path because, as Robin Henig wrote in her 2001 book *The Monk in the Garden*, the local bishop found "toying with the reproduction of animals simply too vulgar an undertaking for a priest." So Mendel decided to work instead with pea plants, which proved a better model organism anyway. The plants grew quickly, he could better control their environment and breeding, and they smelled better too.

Mendel began by choosing specific traits that he could see and study, among them seed texture, seed color, pod shape, pod color, flower color, and stem length. Each of the traits he chose to study existed in two forms. For example, seed texture was either round or wrinkled; seed color was either green or yellow. He started his breeding experiments with plants that "breed true"—that is, produce offspring that carry the same traits as the parents, generation after generation. Only then could he study what happened to particular traits when purebred plants of one variety were mated with purebred, or true-breeding, plants of another variety.

Pea plants can self-pollinate, which means that the pea flower contains both male and female sexual organs and a single plant can fertilize itself to produce offspring. To produce true-breeding plants, Mendel covered pea flowers with a small bag so that he could control fertilization, manually fertilizing plants with their own pollen and preventing pollen from another plant from entering. Once he had established true-breeding plants, he could then cross-pollinate two different plants. What would happen if he crossed a true-breeding green-seeded plant with a true-breeding plant that produced yellow seeds? Or a purple-flowered plant with a white-flowered plant? For each cross, Mendel painstakingly pollinated individual flowers from the two plants by hand. He also prevented self-pollination by removing the male reproductive parts from the plants to be fertilized.

Mendel noticed that when he crossed a true-breeding white-flowering plant with a true-breeding purple-flowering plant, the first generation of offspring (what we now call the F_1 generation) all had purple flowers. That the flowers were true purple rather than pale purple suggested that parental traits were not blended, as earlier hypotheses of inheritance would have predicted. But the trait for white flowers did not disappear completely, either. When Mendel randomly selected two F_1 purple-flowering plants to breed, he found that on average 1 out of every 4 plants of the second generation of offspring (the F_2 generation) had white flowers. Mendel reasoned that a hidden white element must be present in the purple F_1 plants. So each F_1 plant must have two such elements, one representing purple (the trait that appeared) and the other representing white (the hidden trait) **(INFOGRAPHIC M4.2)**.

INFOGRAPHIC M4.2 Mendel's Experiments

When Mendel cross-pollinated true-breeding white flowering plants with true-breeding purple flowering plants, all the F_1 offspring had purple flowers. When these F_1 offspring were crossed, some of F_2 had purple flowers, and some had white flowers, in a predictable ratio. This allowed Mendel to recognize that there are alternative "elements" for the flower color trait. These elements remain intact over generations and do not blend with one another.

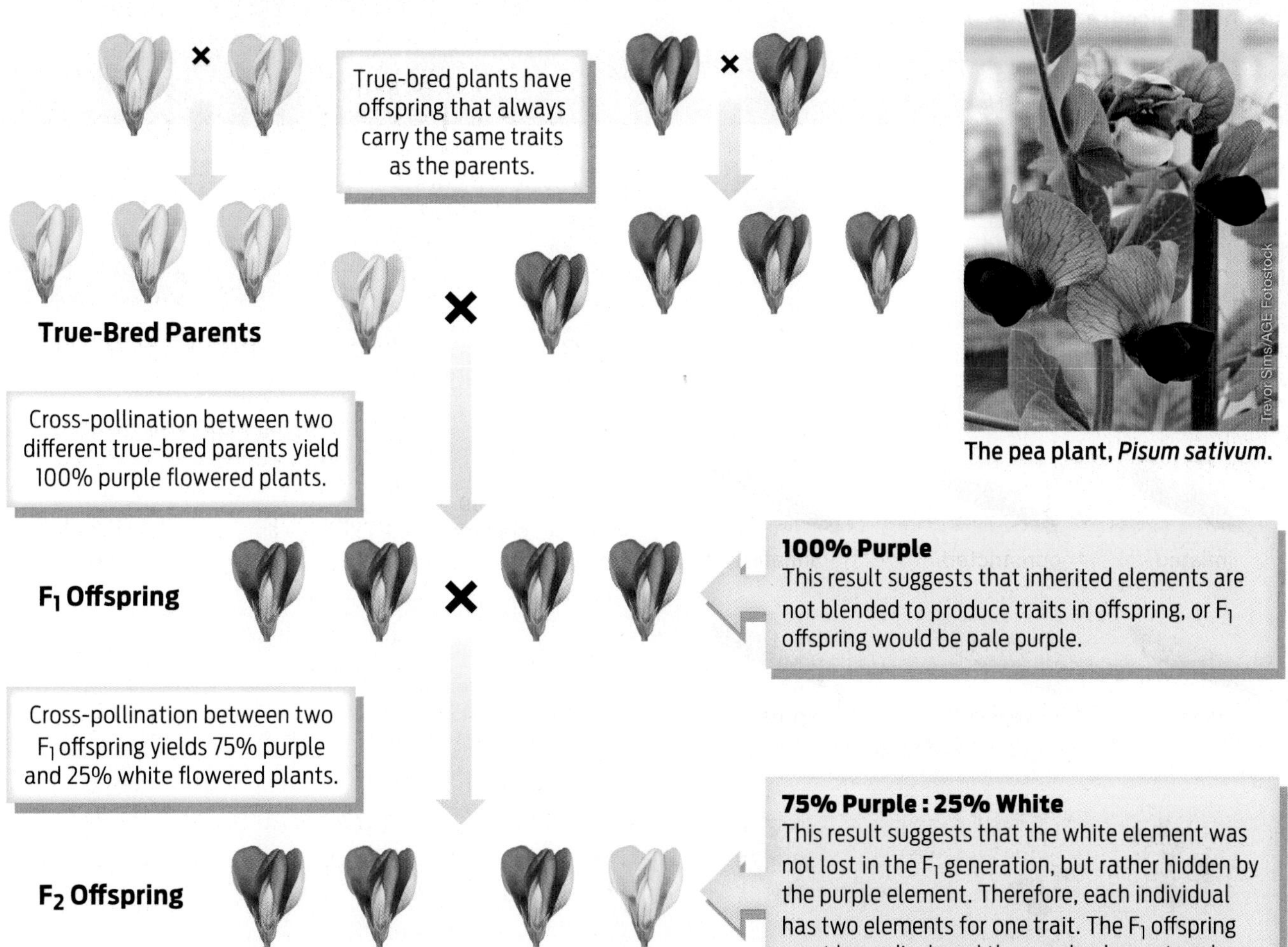

True-bred plants have offspring that always carry the same traits as the parents.
True-Bred Parents
Cross-pollination between two different true-bred parents yield 100% purple flowered plants.
F₁ Offspring
100% Purple
This result suggests that inherited elements are not blended to produce traits in offspring, or F₁ offspring would be pale purple.
Cross-pollination between two F₁ offspring yields 75% purple and 25% white flowered plants.
F₂ Offspring
75% Purple : 25% White
This result suggests that the white element was not lost in the F₁ generation, but rather hidden by the purple element. Therefore, each individual has two elements for one trait. The F₁ offspring must have displayed the purple element and carried the hidden white element.
Trevor Sims/AGE Fotostock

The pea plant, *Pisum sativum*.

? Which flower color (purple or white) is dominant?

If these results sound familiar, there's a good reason for that: they reflect dominant and recessive patterns of inheritance, which we discussed in Chapter 11. Purple flower color is dominant over white, which is recessive. Mendel was the first to gather evidence showing that traits could be inherited in a dominant or recessive fashion, and was in fact the one who coined these terms. While earlier scientists had noticed that traits could disappear in one generation and reappear in later generations, Mendel was the first to offer a coherent explanation of *why* they did.

Over 7 years, Mendel cultivated and tested more than 28,000 pea plants, and in the process discovered the basic principles of inheritance. He published his results in 1866 (**TABLE M4.1**).

Today we know that Mendel's "elements" are alleles of genes, and that genes are located on chromosomes. The principles he discovered have been formalized into two laws. The first of these is Mendel's law of segregation, which states that for any diploid organism, the two alleles of each gene segregate separately into gametes. That is, every gamete receives only one of the two alleles and the

TABLE M4.1 Mendel's Experimental Data

Mendel was meticulous in his data collection. He cultivated, crossed, and analyzed thousands of offspring to ensure accurate ratio calculations.

TRAIT	P_1 CROSS	F_1 GENERATION	F_2 GENERATION	ACTUAL RATIO	PROBABILITY RATIO
Seed texture	round X wrinkled	round	5,474 round : 1,850 wrinkled	2.96 : 1	3 : 1
Seed color	yellow X green	yellow	6,022 yellow : 2,001 green	3.01 : 1	3 : 1
Pod shape	inflated constricted	inflated	882 inflated : 299 constricted	2.95 : 1	3 : 1
Pod color	green yellow	green	428 green : 152 yellow	2.82 : 1	3 : 1
Plant height	tall short	tall	787 tall : 277 short	2.84 : 1	3 : 1

INFOGRAPHIC M4.3 Mendel's Law of Segregation

Mendel's experiments enabled him to formulate the law of segregation. This law has held up over time, although today we know that Mendel's "elements" are alleles of genes. The law of segregation states that when an organism produces gametes, the two alleles for any given gene separate so that each gamete receives only one allele. When gametes come together in fertilization, the resulting offspring will have traits that reflect the particular combination of inherited alleles.

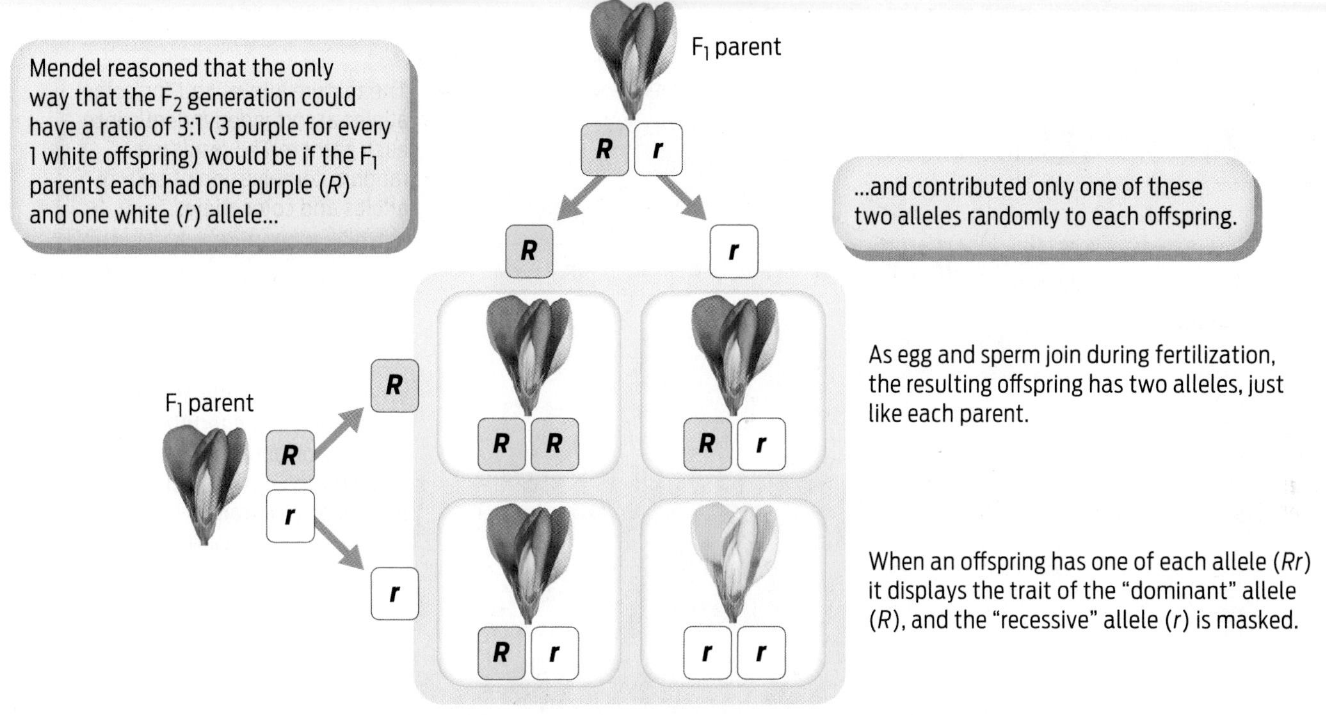

F_1 parent
Mendel reasoned that the only way that the F_2 generation could have a ratio of 3:1 (3 purple for every 1 white offspring) would be if the F_1 parents each had one purple (*R*) and one white (*r*) allele...
R *r*
...and contributed only one of these two alleles randomly to each offspring.
R
r
F_1 parent
R *R*
R *r*
As egg and sperm join during fertilization, the resulting offspring has two alleles, just like each parent.
R *r*
r *r*
When an offspring has one of each allele (*Rr*) it displays the trait of the "dominant" allele (*R*), and the "recessive" allele (*r*) is masked.

? How many alleles for any given gene are segregated into a single gamete?

specific allele that any one gamete receives is random. The physical basis of segregation is the separation of maternal and paternal chromosomes during meiosis (**INFOGRAPHIC M4.3**).

The second law, the law of independent assortment, states that the two alleles of any given gene segregate independently from any two alleles of another gene. Because of independent assortment, offspring can display any combination of the different traits, rather than inheriting the traits together. We now know this holds true only for genes that are located on different chromosomes, or far enough away from each other on a single chromosome to readily recombine. It was mere happenstance that Mendel chose traits for which the genes are

Gregor Mendel at work in his laboratory.

INFOGRAPHIC M4.4 Mendel's Law of Independent Assortment

Mendel went on to study how multiple traits are inherited—for example, seed color and seed texture. Tracing two traits at a time helped him formulate the law of independent assortment. This law states that the two alleles of any given gene will segregate independently from the alleles of other genes when passed on to gametes. Consequently, each gamete may acquire any possible combination of alleles, and therefore traits.

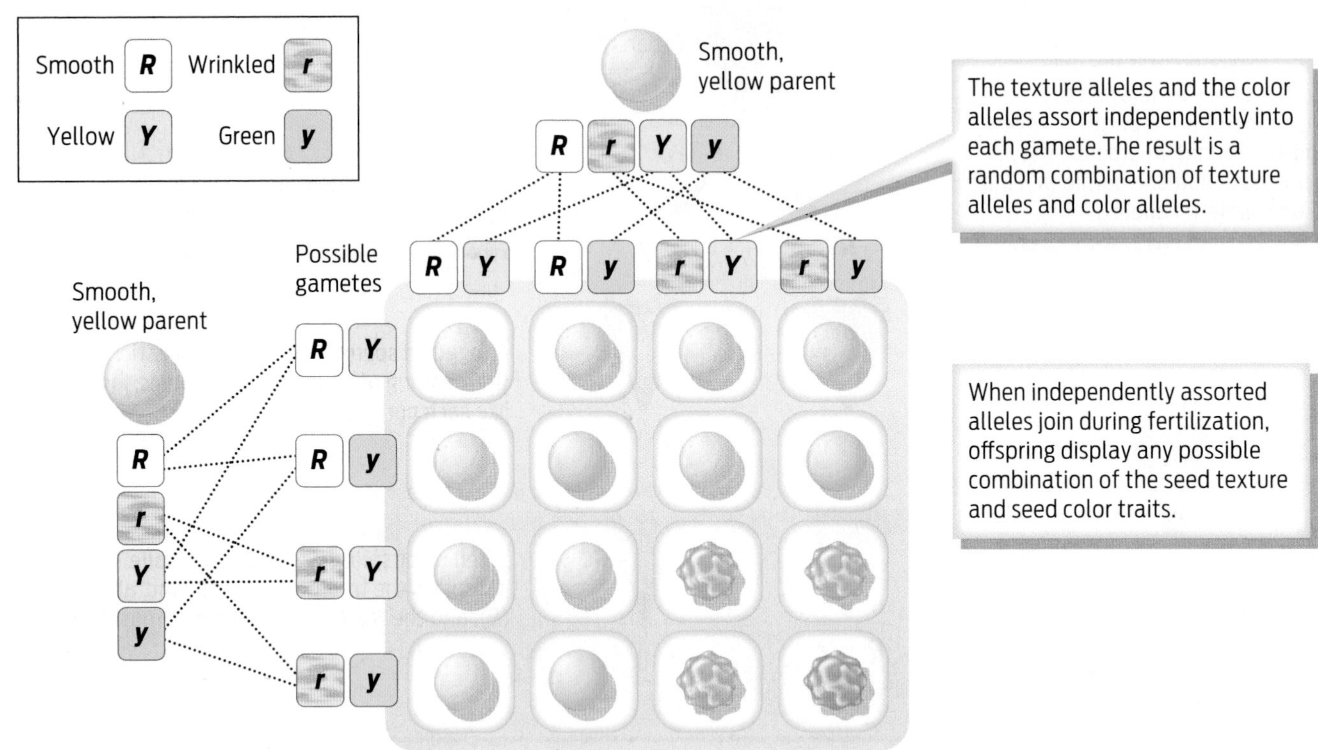

? Why are there not two alleles of the *R* gene or two alleles of the *Y* gene in any gamete?

located on different chromosomes and so assort independently (**INFOGRAPHIC M4.4**).

Despite Mendel's groundbreaking research, no one realized the significance of his results at the time—not even Charles Darwin, whose *Origin of Species* was published in 1859. In 1868, Mendel was elected abbot of St. Thomas and largely shifted his focus from science to monastic life and the administration of the abbey. Although Mendel's research was cited by other scientists, he didn't receive much notice until three botanists who were also studying how traits are inherited in plants rediscovered his work over 30 years later, in 1900. While preparing to publish their ideas about inheritance, they looked through the research literature and found that Mendel had beaten them to the punch. Mendel was finally recognized as the researcher who had solved a crucial mystery of inheritance many years before. ■

MORE TO EXPLORE

- Abbott, S., and Fairbanks, D. J. (2016) Experiments on Plant Hybrids by Gregor Mendel. *Genetics*. Vol. 204, no. 2:407–422: www.genetics.org/content/204/2/407
- Henig, R. M. (2001) *The Monk in the Garden: The Lost and Found Genius of Gregor Mendel, the Father of Genetics*. New York: Mariner Books.
- DNA Learning Center, Cold Spring Harbor Laboratory, Gregor Mendel and Pea Plants: www.dnalc.org/view/16002-gregor-mendel-and-pea-plants.html
- Mukherjee, S. (2016) *The Gene: An Intimate History*. New York: Scribner.

MILESTONES IN BIOLOGY 4 Test Your Knowledge

1 You have a tall pea plant.

a. What are its possible genotypes?

b. What crosses could you do to determine whether or not it is true breeding? For each cross, give the expected phenotypes of the offspring.

2 In crossing pea plants:

a. If a true-breeding tall pea plant with purple flowers is crossed with another true-breeding tall pea plant with purple flowers, what will the offspring pea plants look like?

b. Will the offspring be true breeding? Explain your answer.

3 Half of the gametes of a heterozygous parent will carry the dominant allele, and half will carry the recessive allele. Which of Mendel's laws explains this?

4 Consider Mendel's laws.

a. Which of Mendel's laws would be violated if the offspring of two heterozygous tall, purple-flowered pea plants (denoted as *TtPp*) were only tall, purple-flowered plants or short, white-flowered plants?

b. What would such a violation suggest about the two genes?

12

Complex Inheritance

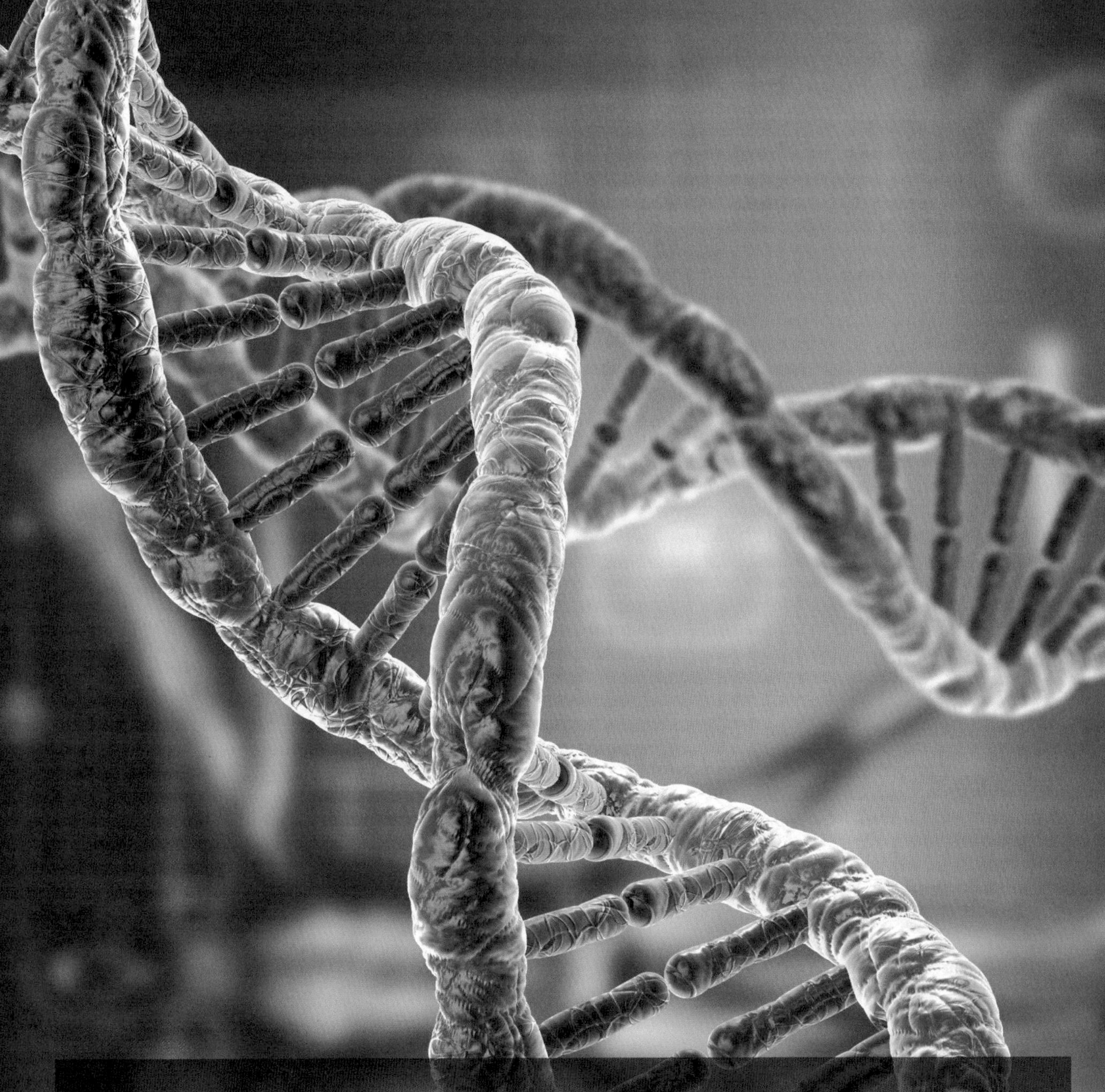

DRIVING QUESTIONS

1. How do chromosomes determine sex, and how does sex influence the inheritance of certain traits?
2. Some traits are not inherited in simple dominant or recessive inheritance patterns. What are some complex inheritance patterns?
3. How does nondisjunction lead to numerical abnormalities of chromosomes, and what are the consequences of these abnormalities?

Genetics Q&A

Complexities of human genetics, from sex to depression

In Chapter 11, we discussed the basic principles of inheritance, including how gametes are formed and how simple dominant and recessive traits are passed from parents to offspring. In this chapter, we examine more-complex cases of inheritance, including examples of sex-linked traits, multi-gene traits, and the role of the environment in shaping phenotypes.

SEX DETERMINATION

Q: What makes a man?

A: A botched circumcision in 1966 on a little boy named Bruce Reimer in time became a landmark example of how biology shapes gender identity—the subjective experience of oneself as male or female. Doctors at the hospital where Reimer was circumcised used an experimental procedure that involved burning off the foreskin. The procedure went awry, and Bruce's penis was singed nearly completely off. On the advice of John Money, a well-known psychologist at Johns Hopkins University in Baltimore, Maryland, who had written much about the importance of environment in determining a person's gender identity, Bruce's parents decided to have their little boy surgically reassigned as a girl and to rear him as "Brenda." Bruce's testes were removed and he was given hormone treatments to promote female development.

But Brenda never behaved like a typical girl. She didn't like playing with girls' toys, didn't enjoy wearing dresses, and often got into fistfights at school. By the time Brenda reached puberty, her behavior had become so troublesome that her father broke down and told her what had happened to her. Brenda felt relieved rather than angry: "All of a sudden everything clicked. For the first time things made sense and I understood who and what I was."

Brenda eventually had reconstructive surgery to remove her breasts and create a penis, took hormones, and changed her name to David. David told his story in a book published in 2000, *As Nature Made Him: The Boy Who Was Raised as a Girl*, by John Colapinto. By that time, it had become increasingly clear that gender identity is significantly influenced by biology. Studies had shown that prenatal exposure to fetal sex hormones such as testosterone not only determine whether a fetus will develop female or male genitalia, but also act on a developing baby's brain, influencing behavior.

Reuters Pictures.

Brenda Reimer in elementary school; David Reimer as an adult.

Sex hormones are produced by sex organs called **gonads**—ovaries in females, testes in males. Sex hormones include **androgens** (from *andros*, Greek for "man"), such as testosterone, and **estrogens** (from the Greek *oistros*, meaning "frenzy" or "passion") such as estradiol. Both males and females produce androgens and estrogens, but in most cases males produce higher levels of androgens and females produce higher levels of estrogens. In a developing fetus, these hormones shape the development of both internal and external sexual anatomy.

Whether a fetus develops male or female gonads depends on the set of chromosomes it receives from its parents. Humans have 23 pairs of chromosomes; 22 pairs are **autosomes** and 1 pair are the **sex chromosomes,** X and Y. Females have two X chromosomes, while males have one X and one Y. Sons inherit their Y chromosome from their father and their X chromosome from their mother. Daughters inherit one X chromosome each from their mother and father. The Y chromosome contains a gene called *SRY* that signals testes to develop. (*SRY* stands for "*s*ex-determining *r*egion on the *Y* chromosome.") The testes, in turn, produce masculinizing hormones such as testosterone that lead to the development of a male body. In the absence of a Y chromosome, a fetus will develop into a female. Thus, fathers determine the sex of a baby by providing either an X or a Y sex chromosome in the sperm that fertilizes a mother's egg (**INFOGRAPHIC 12.1**).

GONADS
Sex organs: ovaries in females, testes in males.

ANDROGENS
A class of sex hormones, including testosterone, that are present in higher levels in men and cause male-associated traits like deep voice, growth of facial hair, and defined musculature.

ESTROGENS
A class of sex hormones, including estradiol, that are higher in women than in men and support female sexual development and function.

AUTOSOMES
Paired chromosomes present in both males and females; all chromosomes except the X and Y chromosomes.

SEX CHROMOSOMES
Paired chromosomes that differ between males and females. Females have XX, and males have XY.

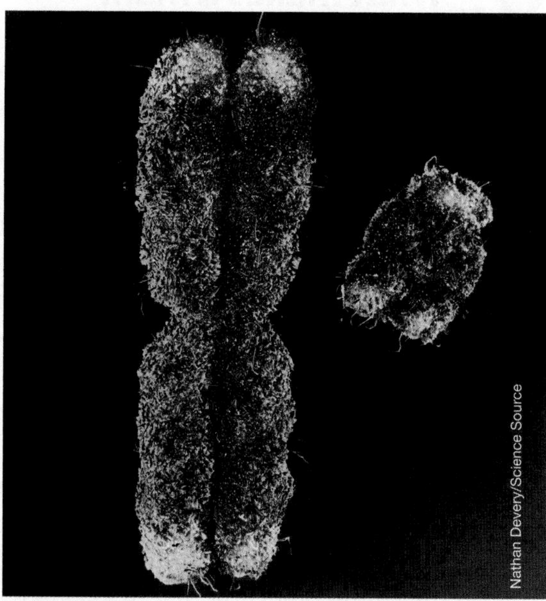

Nathan Devery/Science Source

X and Y chromosomes. The *SRY* gene on the Y chromosome signals the development of testes. In the absence of the *SRY* gene on the Y chromosome, the embryo will develop as a female.

INFOGRAPHIC 12.1 X and Y Chromosomes Determine Human Sex

Human chromosomes can be isolated from cells, photographed under a microscope, and then arranged in their pairs for examination. As shown below, males and females differ by virtue of a pair of sex chromosomes. Females have two X chromosomes and males have a single X and a single Y chromosome. Every person must have at least one X chromosome, but it's the presence of a gene on the Y chromosome that initiates male development.

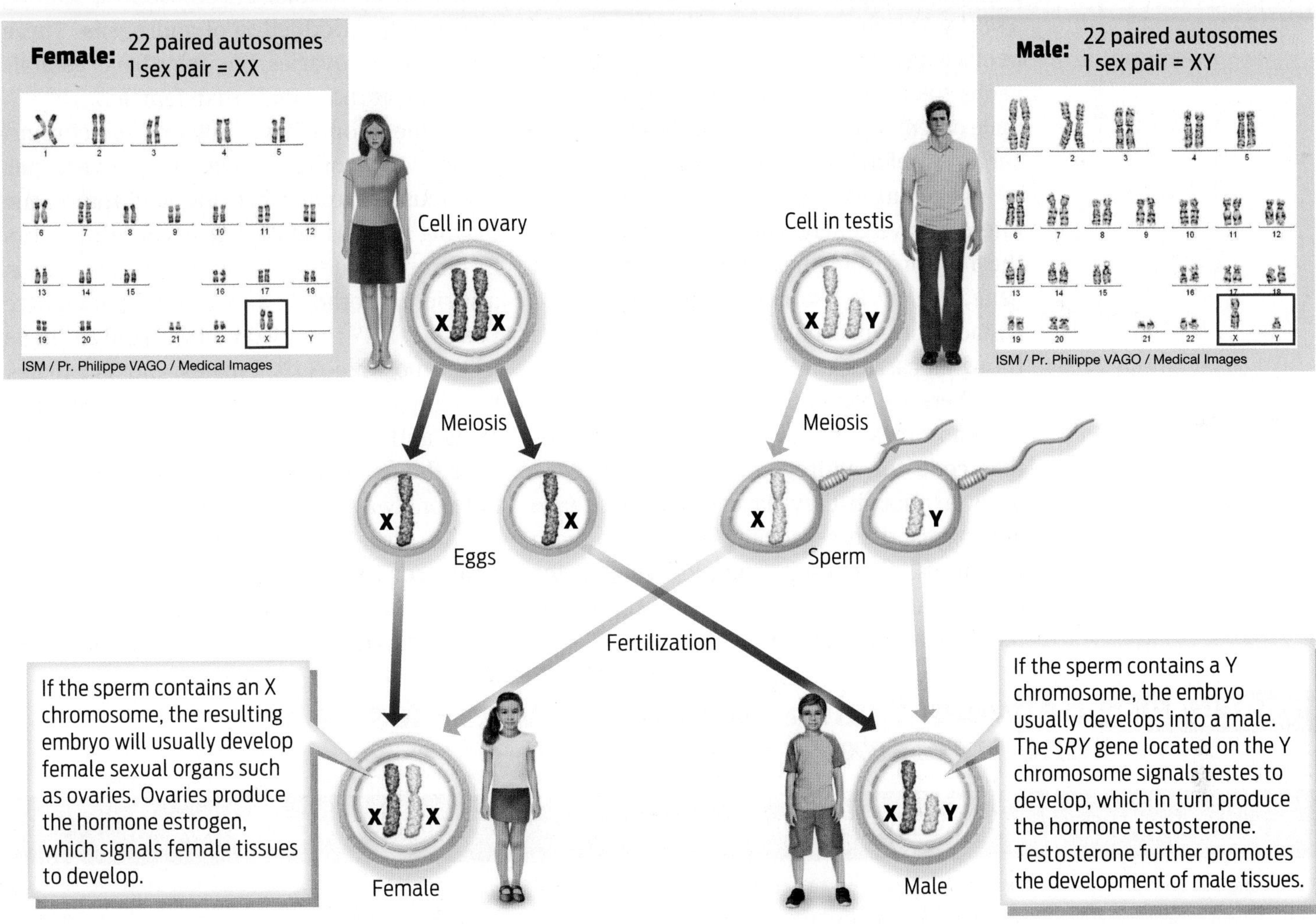

Female: 22 paired autosomes
1 sex pair = XX
ISM / Pr. Philippe VAGO / Medical Images
Male: 22 paired autosomes
1 sex pair = XY
ISM / Pr. Philippe VAGO / Medical Images
Cell in ovary
Cell in testis
Meiosis
Meiosis
Eggs
Sperm
Fertilization
If the sperm contains an X chromosome, the resulting embryo will usually develop female sexual organs such as ovaries. Ovaries produce the hormone estrogen, which signals female tissues to develop.
If the sperm contains a Y chromosome, the embryo usually develops into a male. The SRY gene located on the Y chromosome signals testes to develop, which in turn produce the hormone testosterone. Testosterone further promotes the development of male tissues.
Female
Male

? What would you predict would be the biological sex of a person born with two X chromosomes and a Y chromosome?

Although most individuals have internal and external genitalia that are either clearly male or clearly female, there are exceptions. Each year about 1 in every 1,500 babies born in the United States falls into an intermediate sex category termed "intersex." An intersex person is someone whose external genitalia do not match his or her internal sex organs or genetic sex—for example, a person with an XX chromosome pair who has internal ovaries but external genitalia that appear male. Also included in this category are babies born with ambiguous genitalia—for example, a penis that is very small or a clitoris that is exceptionally long.

Debate over the experience of David Reimer and other similar cases, as well as research showing how strongly biology influences gender identity, has changed the care of intersex babies. In the past, doctors would typically operate on intersex babies immediately after birth to "fix" the genital anomaly. They might shorten a large clitoris in genetic females, for example. In genetic males with a very small penis (termed a micropenis), they might remove the penis and testes and assign the baby a female gender. Sometimes this would be done even without consulting with parents. Today, such babies are often assigned a sex by parents and doctors only after a period of observation to assess behavior patterns. Surgeons then perform surgery to create either male or female genitalia. Or, parents may forgo surgery, preferring that their child remain as is.

What causes intersex? Some instances of intersex have genetic causes. For example, if the Y chromosome has a mutation in the *SRY* gene, a baby is likely to have undeveloped gonads with external female genitalia, even though it carries an XY chromosome pair.

There are also cases of people with XY sex chromosomes who have mutations in genes that code for androgen receptors on cells. So even though they have testes that make androgens, their cells fail to respond to these chemicals. As a result, complete male external genitalia do not develop, and these people appear to be female.

Similarly, there are people with XX sex chromosomes who have male genitalia. In some cases, this is caused by a condition called congenital adrenal hyperplasia. These individuals have one or more mutations in genes on autosomal chromosomes, leading to excessive production of androgens. These people have ovaries but may have genitals that appear more male than female.

Some people have only one sex chromosome, while others have three. Every person must have at least one X chromosome (having none is fatal), but because of errors in chromosome segregation during meiosis, a variety of other X and Y combinations are possible: XXY men, women with only a single X chromosome, XXX females, and XYY males. In many of these cases, a person's physical traits and genitalia reveal that they do not have the usual makeup of sex chromosomes, but not always (**TABLE 12.1**).

As these varieties of intersex demonstrate, sex isn't always so clear cut. There are genetic, hormonal, anatomical, and psychological aspects to sex, and these elements

TABLE 12.1 Between Male and Female: Varieties of Sex and Intersex

SEX CATEGORY	CHROMOSOMES	GONADS	GENITALIA	OTHER CHARACTERISTICS
Female	XX	Ovaries	Female	
Male	XY	Testes	Male	
Female pseudo-hermaphroditism	XX	Ovaries	Male	Infertile
Male pseudo-hermaphroditism	XY	Testes	Female or ambiguous	Infertile
True gonadal intersex	XX and/or XY	Ovaries and testes	Male, female, or ambiguous	Infertile; historically called true hermaphrodites
Triple X syndrome	XXX	Ovaries	Female	Fertile; taller than average; learning disabilities
Klinefelter syndrome	XXY	Testes	Male	Infertile; enlarged breast tissue
47, XYY syndrome	XYY	Testes	Male	Fertile; taller than average; elevated risk of learning and emotional disabilities
Turner syndrome	X	Ovaries	Female	Infertile; broad chest; webbed neck

don't always align. Defining what counts as "masculine" and "feminine" is even more complicated. For example, some men have characteristics that we typically identify as female, such as a high voice and sparse body hair, yet they are genetically and anatomically male. And many women have what are considered to be more masculine features, such as angular faces and more muscle as compared to body fat. Yet, they are genetically and anatomically female. In fact, there are very few physical or mental characteristics that are entirely male or entirely female. In addition, some people—for example, transgender individuals—may mentally identify with one sex even though their genitalia and chromosomal makeup classify them as the other.

For biologists, the story of David Reimer—known in the medical literature as the "John/Joan case"—seriously undermined the idea that children are born psychosexually "neutral." Contrary to what some researchers had claimed, it was not possible to mold someone into whichever gender identity you wanted through surgery and child rearing—at least not without damaging psychological fallout.

> "For the first time things made sense and I understood who and what I was."
>
> — David Reimer

David Reimer eventually married a woman and adopted her three children. Though he could not have children of his own (since his testes had been removed), he took the responsibilities of fatherhood seriously. "From what I've been taught by my father," he told Colapinto, "what makes you a man is: You treat your wife well. You put a roof over your family's head. You're a good father. . . . That, to me, is a man."

David said he told his story so that others would be spared the nightmarish experience he went through. That experience may have led to the depression that cost him his life. David killed himself in 2004 at the age of 38.

SEX-LINKED INHERITANCE

Q: Why do some genetic conditions affect sons more often than daughters?

A: Some 10 million American men—about 7% of the male population—either cannot distinguish red from green, or they perceive these hues differently from the way other people do. Such red-green color blindness affects only 0.4% of women. Similarly, 1 in 5,000 boys worldwide is born with hemophilia, a blood-clotting disorder, yet hemophilia rarely afflicts girls.

Why these disparities? These conditions are caused by genes found on the X chromosome. When a gene is located on either of the sex chromosomes, daughters and sons don't share the same probability of inheriting it.

Take the neuromuscular condition Duchenne muscular dystrophy (DMD), for example. DMD is a disease in which muscles slowly degenerate, leading to paralysis. About 1 in 2,400 boys worldwide is born with the condition each year. Most affected boys are in wheelchairs by the time they are teenagers, and they rarely live past 30.

Why does DMD primarily affect males? Recall that a female has two X chromosomes. For a recessive trait like DMD, a normal allele on one X chromosome masks the effect of a recessive DMD allele on the other X chromosome. A male, on the other hand, has a single X chromosome, and therefore will show the effects of any recessive alleles on that X chromosome. Statistically, it is much more likely for a male to inherit one rare DMD allele than it is for a female to inherit two.

Because females can carry an allele for DMD without having the disease, they may

X-LINKED TRAIT
A phenotype determined by an allele on an X chromosome.

PEDIGREE
A visual representation of the occurrence of phenotypes across generations.

not even know they are carriers who can pass it on to their sons. Males with a DMD allele can pass it to daughters, but not to sons. Traits such as DMD, hemophilia, and red-green color blindness that are inherited on X chromosomes are called **X-linked traits** (**INFOGRAPHIC 12.2**).

Sex-linked traits–those influenced by genes on either the X or Y chromosome–have distinct patterns of inheritance compared with traits influenced by genes on the autosomes. In the latter case, males and females inherit the trait with equal probability. In the former case, males and females are affected differently.

Determining what pattern of inheritance a particular trait has is easy to do with plants or laboratory animals, where you can perform a carefully controlled genetic cross–as Mendel did with his peas (see **Milestone 4: Mendel's Garden**). With humans, it's not possible to conduct such crosses. Instead, scientists rely on a tool called a **pedigree** to determine a pattern of inheritance in humans. With the help of a pedigree that extends back several generations, scientists can determine whether a trait is inherited on autosomes or sex chromosomes.

One of the most famous pedigrees of all time is that of Queen Victoria of England and her descendants, several of whom were afflicted with the blood-clotting disorder hemophilia. People with hemophilia cannot clot blood in response to an injury. Before there were medicines for the condition, people sometimes bled to death as a result of a relatively minor cut.

INFOGRAPHIC 12.2 X-linked Traits Are Inherited on X Chromosomes

Duchenne muscular dystrophy (DMD) is an example of an X-linked trait. Recessive mutations of the *dystrophin* gene on the X chromosome cause the disease. DMD primarily affects males because they inherit only one copy of the X chromosome (from their mothers). Therefore, the single DMD allele they inherit determines their phenotype. Since females have two X chromosomes, they may carry the DMD allele but have a healthy phenotype.

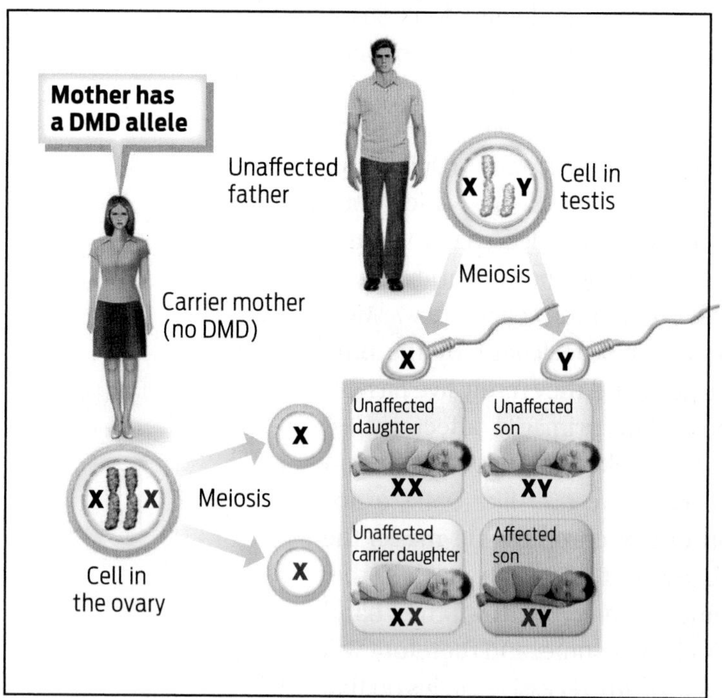

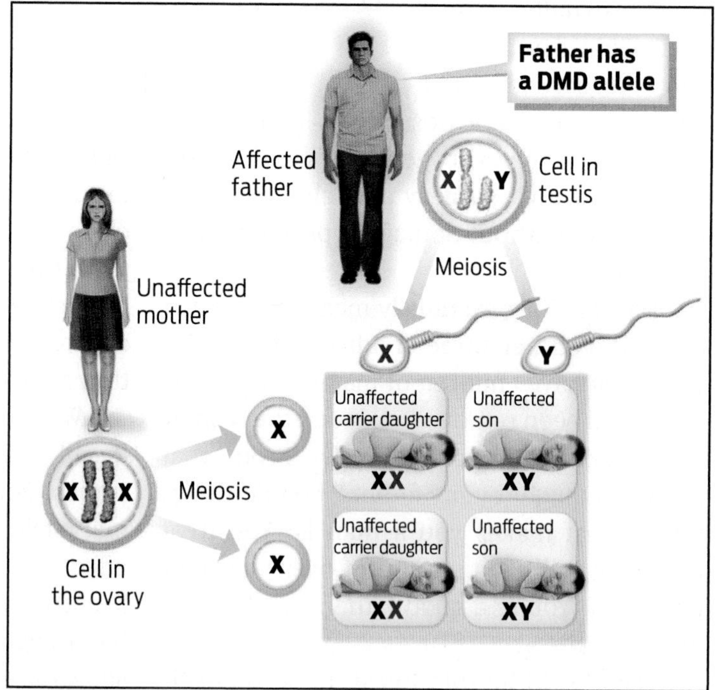

? Red–green color blindness is most commonly the result of inheriting recessive alleles of specific genes on the X chromosome. A woman and her husband both have the same type of red–green color blindness. Predict the frequencies of sons and daughters with red–green color blindness that this couple could have.

Victoria and her husband, Prince Albert, had nine children, five girls and four boys. One of those boys, their son Leopold, had hemophilia. He died at age 30. Their daughter Alice had several children, including one son who had hemophilia. His sister, Alexandra, did not have hemophilia, but her son, Alexei, did. By the early 20th century, 10 of Victoria's descendants had hemophilia—all of them male. This pattern of a trait appearing more frequently in males, but inherited via mothers, is typical of rare, recessive X-linked traits (**INFOGRAPHIC 12.3**).

Y-CHROMOSOME ANALYSIS

Q: Did Thomas Jefferson father children with a slave?

A: Thomas Jefferson was the third president of the United States, the principal architect of the Declaration of Independence, and founder of the University of Virginia. He was also a slaveholder. Historians have long debated the meaning of this and other seeming contradictions in the

INFOGRAPHIC 12.3 A Pedigree Analysis Can Help Determine a Sex-linked Pattern of Inheritance

This diagram, known as a pedigree, shows how an X-linked recessive trait (in this case hemophilia) passes through generations. Female carriers of the allele do not express the hemophilia trait, but their sons who inherit the allele have hemophilia.

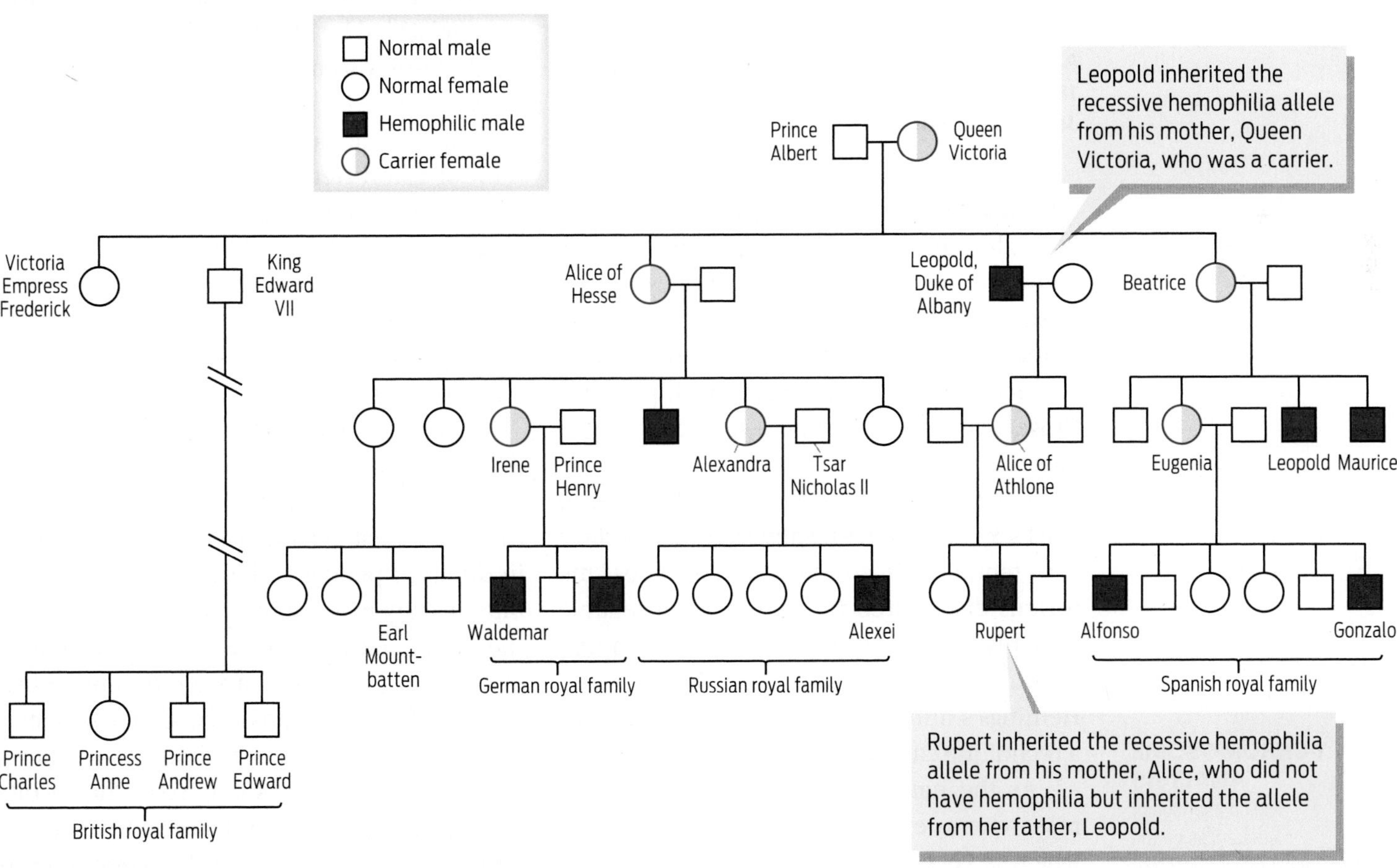

? Imagine that Irene married Maurice instead of Prince Henry. What would you predict about their sons and daughters?

A slave named Lucy, born at Monticello in 1811. Slaves at Monticello often cared for the Jefferson children.

founding father's life and politics. For example, although Jefferson's writings clearly show that he did not support the institution of slavery, he owned at least 200 slaves. He made disparaging comments about slaves, yet maintained close relationships with those living in his house—sometimes very close.

Jefferson was rumored to have fathered at least six children with Sally Hemings, a slave who tended to his family. For decades, historians discredited the rumor as unreliable oral history. In 1998, scientists attempted to find out by testing the DNA of both Hemings's and Jefferson's descendants with a technique called **Y-chromosome analysis.**

Y-CHROMOSOME ANALYSIS
The comparison of sequences on the Y chromosomes of different individuals to examine paternity and paternal ancestry.

As the name implies, in this technique researchers examine the Y chromosome, which is very small and contains few genes. Sons inherit their Y chromosome from their fathers. These Y chromosomes are passed through generations, from fathers to sons, largely unchanged. That's because Y chromosomes have no homologous partner chromosome with which to pair and exchange DNA during meiosis (see Chapter 11). In other words, the Y chromosome rarely undergoes genetic recombination. Consequently, the Y chromosome that a son inherits from his father is almost identical to the Y chromosome that his father inherited from his father. Comparing DNA sequences on Y chromosomes, therefore, can reliably establish paternity (**INFOGRAPHIC 12.4**).

During Jefferson's own time, people commented on the resemblance of Hemings's children to the president. But later historians either explained the resemblance away or proposed other explanations—for example, that one of Jefferson's nephews had fathered her children.

To set the record straight, in 1998 a team of geneticists led by Eugene A. Foster compared the Y chromosomes of three groups of men: descendants of Thomas Jefferson's paternal uncle Field Jefferson; one male descendant of Eston Hemings, Sally Hemings's son; and descendants of Jefferson's sister's sons (Jefferson's nephews). Since Jefferson's only surviving child from his wife was a daughter, he did not have any direct male descendants, which is why scientists tested descendants of Jefferson's uncle.

The team analyzed 11 short tandem repeats (STRs) on the Y chromosome. (Recall from Chapter 7 that STRs are short regions of DNA that are repeated a different number of times in different people and are used in DNA profiling.) They made some startling discoveries: the results clearly showed that descendants of Thomas's nephews have a different Y chromosome from the man descended from Eston Hemings, thus ruling out Thomas's nephews as the father of Sally's children. The results also clearly showed that Eston's descendant—John Weeks Jefferson—has the same Y chromosome as the descendants of Field Jefferson. Consequently, Thomas Jefferson *could* have fathered Eston Hemings. The study does not

INFOGRAPHIC 12.4 Y Chromosomes Pass Largely Unchanged from Fathers to Sons

Y-chromosome analysis for paternity testing relies upon the fact that the Y chromosome does not undergo recombination during meiosis and so passes unchanged from a father to his sons.

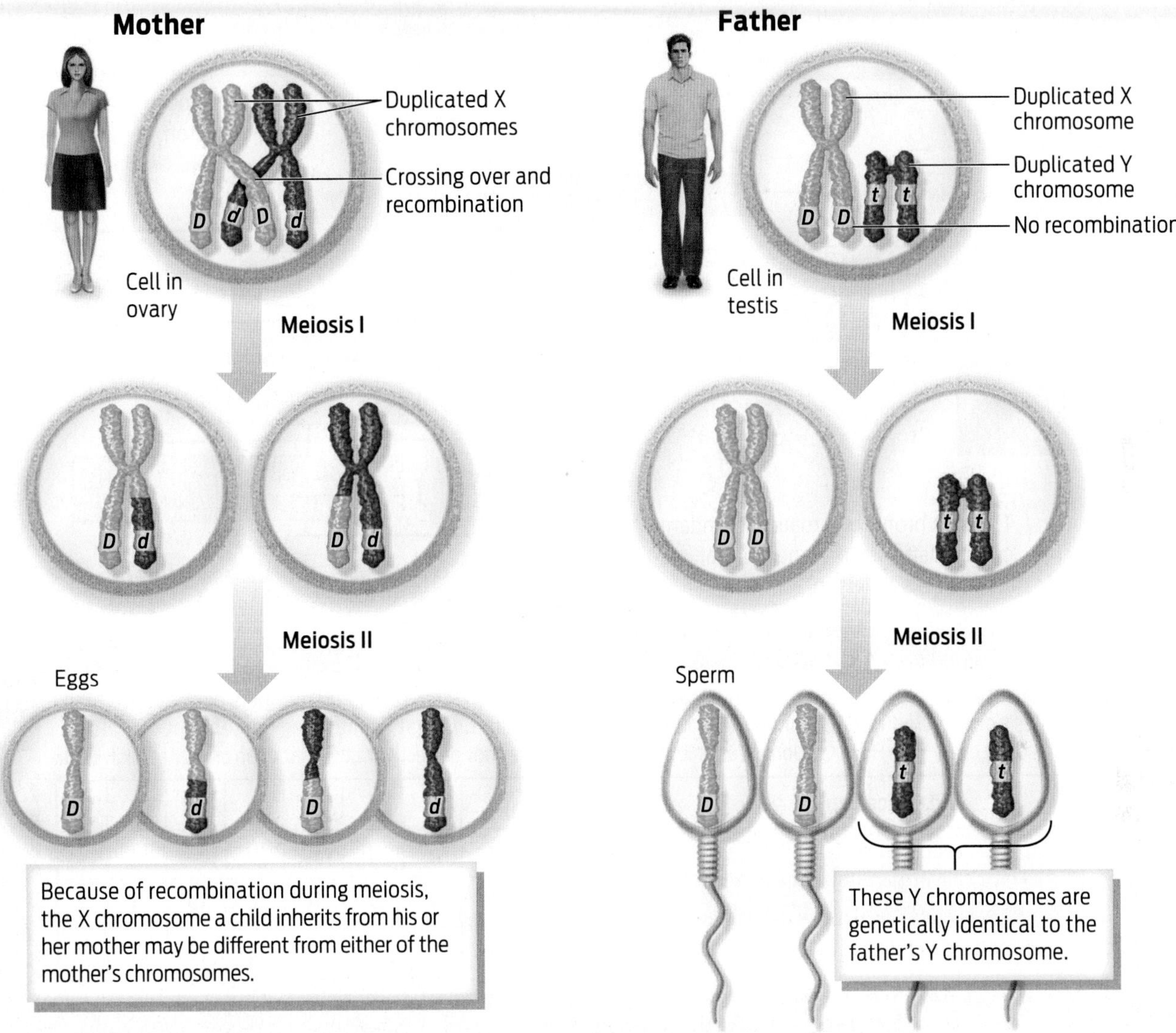

? A male has a particular dominant allele *F* on his Y chromosome. Which of his children (sons and/or daughters) will inherit this allele?

prove that he is the father, but it does show that the father was definitely a male Jefferson **(INFOGRAPHIC 12.5)**.

Some historians have argued that Thomas's younger brother Randolph Jefferson could have fathered Eston. (Or, indeed, that any of the other seven Jefferson males who periodically visited Monticello, where Thomas and Sally lived, could be the father.) But other experts have argued that historical evidence—for example, records of the president's travels—place him rather than Randolph under the same roof as Sally at the time of her conceptions. A 2000 report by the Thomas Jefferson Foundation concluded that the preponderance of historical and biological evidence points to a "strong likelihood" of a sexual relationship between Thomas and Sally.

For the descendants of Eston Hemings, the DNA study was powerful vindication, even if

INFOGRAPHIC 12.5 DNA Links Sally Hemings's Son to Jefferson

Scientists compared DNA sequences on the Y chromosome of Sally Hemings's and Thomas Jefferson's grandfather's descendants. The DNA sequences match at the 11 different STR locations analyzed.

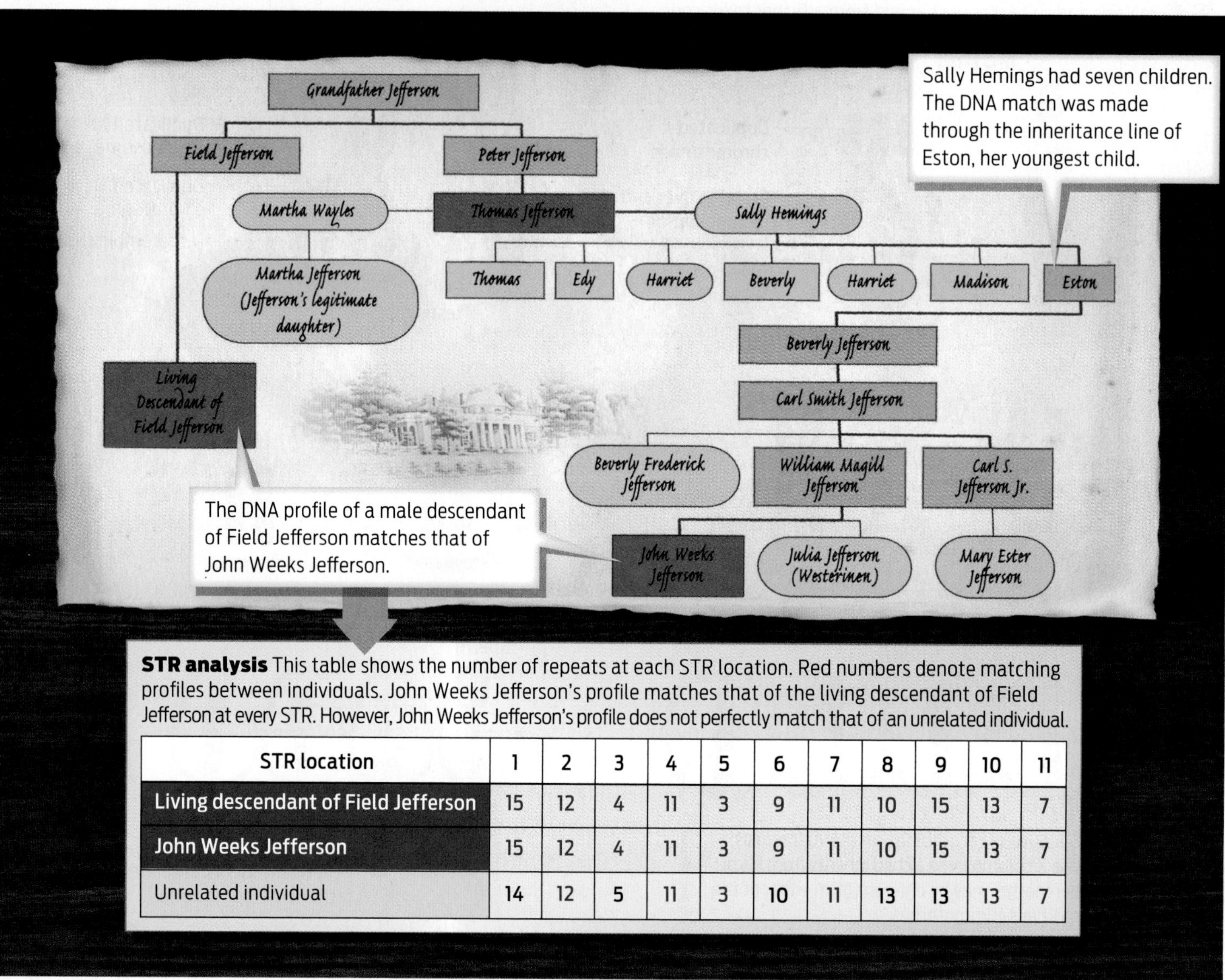

STR analysis This table shows the number of repeats at each STR location. Red numbers denote matching profiles between individuals. John Weeks Jefferson's profile matches that of the living descendant of Field Jefferson at every STR. However, John Weeks Jefferson's profile does not perfectly match that of an unrelated individual.

STR location	1	2	3	4	5	6	7	8	9	10	11
Living descendant of Field Jefferson	15	12	4	11	3	9	11	10	15	13	7
John Weeks Jefferson	15	12	4	11	3	9	11	10	15	13	7
Unrelated individual	14	12	5	11	3	10	11	13	13	13	7

? Of the proposed children of Thomas and Sally, who is predicted to have the same Y chromosome as Thomas? If Harriet had a son, would he share his Y chromosome with his grandfather Thomas?

debate continues. They had long argued that they were descended from Thomas Jefferson, but without hard evidence, most historians disregarded their claims. "I feel wonderful about it," Julia Jefferson Westerinen, a Staten Island artist and Eston's great-great-granddaughter, told the *New York Times* when the study results were published, "I feel honored."

As for the relationship between Jefferson and Sally Hemings, historians continue to debate whether it was consensual or forced. "I was a history major," said Jefferson Westerinen, "And we learned not to say, 'I feel this, I think that,' without knowing the facts. They had a relationship of 38 years. I would like to think they were in love, but how would I know?"

We'll likely never know the truth. Neither Thomas Jefferson nor Sally Hemings left any written evidence of their relationship.

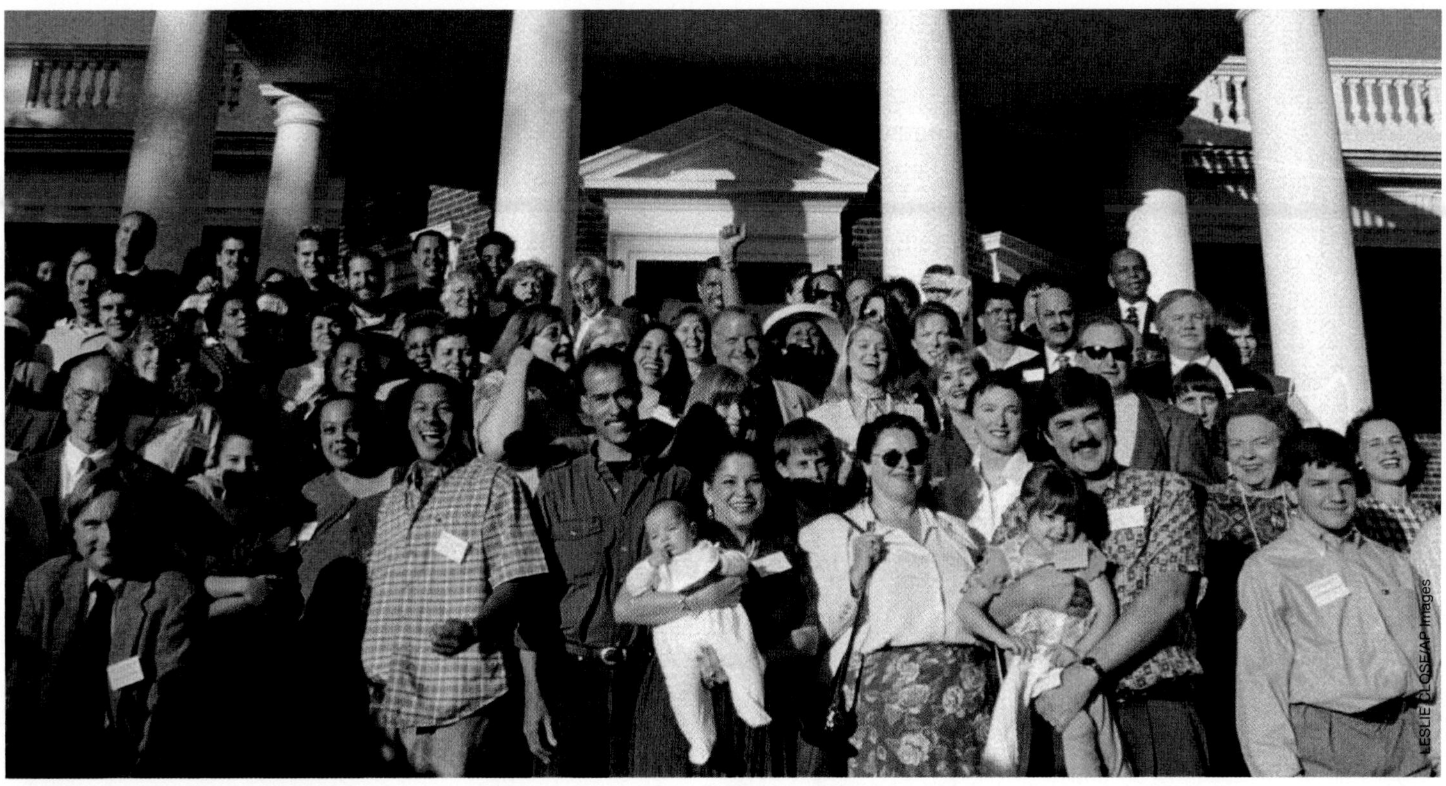

Are these people descendants of Thomas Jefferson and Sally Hemings? The group gathered for a photo at Monticello in 1999.

INCOMPLETE DOMINANCE

Q: Why do some traits appear to blend in offspring?

A: We saw in Chapter 11 and Milestone 4: Mendel's Garden that many traits are inherited in a dominant or a recessive pattern. Emily Schaller has cystic fibrosis because she inherited two copies of the mutated *CFTR* allele, which is recessive. Her parents, however, are healthy because they each have one normal copy of the *CFTR* allele, which is dominant. And as Mendel discovered, pea plants with purple flowers crossed with pea plants with white flowers produce offspring with purple flowers (at least in the first generation) because the purple allele is dominant to the white allele. One of Mendel's greatest contributions was to show that alleles are discrete entities that don't blend away like mixed paint colors.

Yet many traits we see around us do appear to blend in the offspring. In snapdragon plants, for example, red-flowering plants mated with white-flowering plants produce plants with pink flowers. The children of a mother with curly hair and a father with straight hair are likely to have wavy hair.

When heterozygotes have a phenotype that is intermediate between the two homozygotes, the inheritance pattern is called **incomplete dominance.** Incomplete dominance is shown when one allele does not completely mask the presence of the other allele, and there is a measurable effect on the phenotype of having one versus two copies of a dominant allele. (This is in contrast to complete dominance, in which one allele masks the other and homozygotes and heterozygotes have indistinguishable phenotypes.)

An example of incomplete dominance in humans is familial hypercholesterolemia (FH). This dangerous condition results from inherited mutations in the gene encoding

INCOMPLETE DOMINANCE
A form of inheritance in which heterozygotes have a phenotype that is intermediate between the two homozygotes.

the low-density lipoprotein (LDL) receptor, which helps remove harmful cholesterol from the blood (see Chapter 28). Heterozygotes have elevated levels of blood cholesterol and a tendency to suffer heart disease. Homozygotes for the mutant allele have much higher levels of cholesterol and a worse outcome; without treatment, they may die from heart attack or stroke in their teenage years (**INFOGRAPHIC 12.6**).

Why do some traits behave as dominant, others as recessive, and still others as incompletely dominant? It helps to think about what is happening on the molecular level. Many genes encode proteins that serve as enzymes, which catalyze biochemical reactions in the body (see Chapter 4). For most enzymes, only a very small quantity of the working molecule is needed to do the job. Someone who is a heterozygote, with one working and one

INFOGRAPHIC 12.6 The Inherited Trait of High Cholesterol Exhibits Incomplete Dominance

In incomplete dominance, heterozygotes display a phenotype intermediate between that of either homozygous genotype. In familial hypercholesterolemia (FH), blood levels of LDL cholesterol are determined by the genotype of the LDL receptor gene. The LDL receptor normally helps clear LDL cholesterol from the blood. People who are homozygous for the mutant allele have dangerously high levels of LDL cholesterol, heterozygotes have moderately elevated levels of LDL cholesterol, and people who are homozygous for the normal allele have normal cholesterol levels.

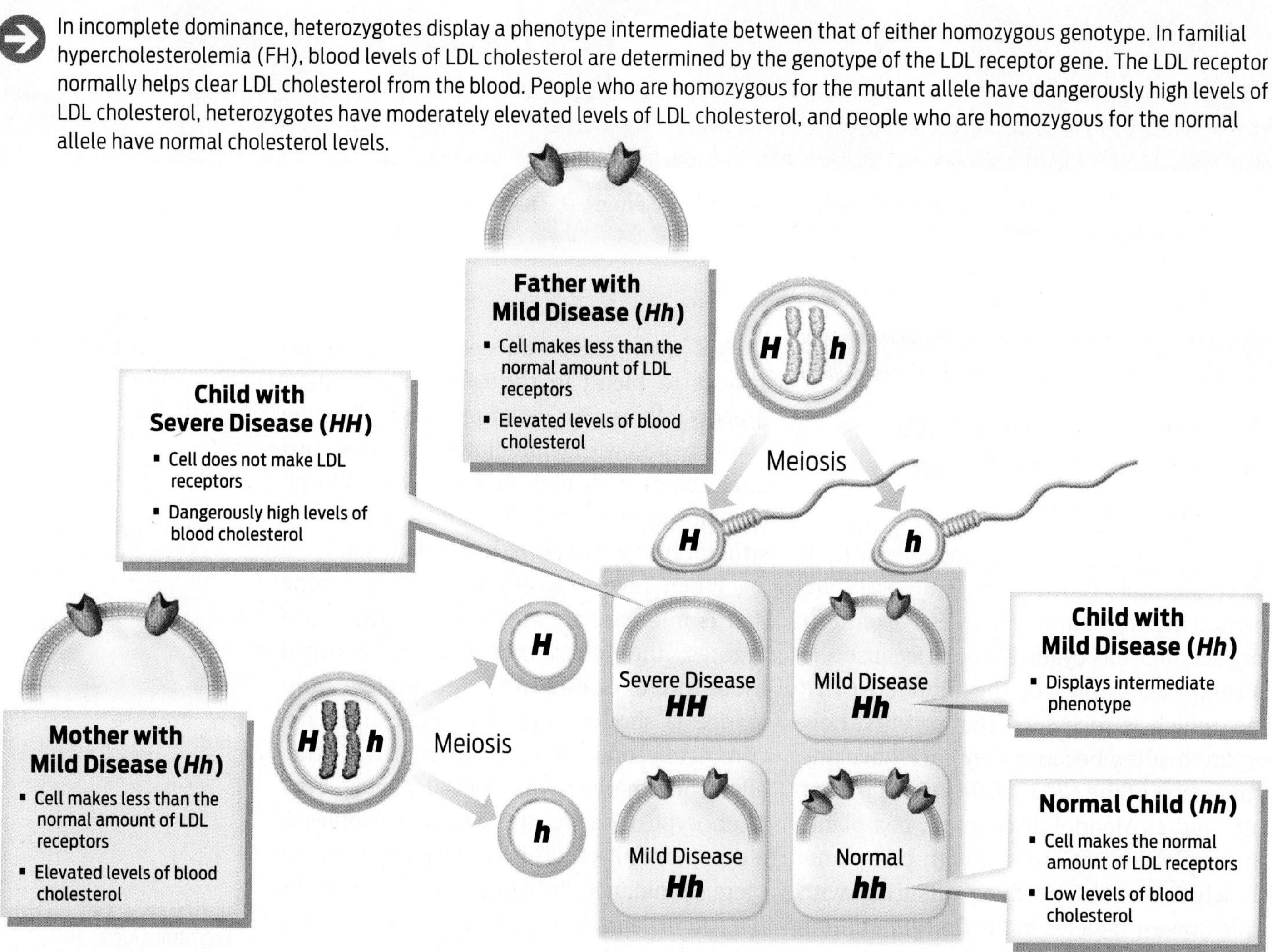

? A woman has the severe form of FH, and her husband does not have FH at all. What proportion of their children will have severe FH, mild FH, and/or no FH?

dysfunctional copy of a particular enzyme, will likely still produce plenty of the normal enzyme to catalyze the reaction. So this mutant allele would behave as recessive, and you'd need two mutant versions to get sick. Phenylketonuria, also called PKU, is a good example. PKU is a recessive inherited condition that results from inheriting two defective copies of the enzyme that metabolizes (that is, breaks down) the amino acid phenylalanine. People with two defective copies of the gene are quite sick, but people with one working copy are just fine.

For processing cholesterol, however, being a heterozygote and having only half the number of normal LDL receptors is not sufficient to remain healthy. So this mutant allele behaves as incompletely dominant. (The state of heterozygotes experiencing a deficit as a result of having only half the normal amount of protein is also sometimes called haploinsufficiency.)

For the neurodegenerative condition called Huntington disease, having one mutant version of the protein is enough to cause the illness. That's because the mutant version acts as a "poison" that interferes with the action of the normal protein. This allele is therefore dominant.

Though concepts of dominance and recessive are very useful in explaining inheritance patterns, it's helpful to recall that these terms were coined before scientists had any notion of how genes encoded proteins, and what proteins did in the body. One gene product doesn't really "mask" another. Both are present, but sometimes we only see the consequences of one, or of one more than the other.

CODOMINANCE

Q: Who can be a universal blood donor?

A: When someone needs a blood transfusion, the donated blood cannot come from just anyone. The transfused blood must be compatible in ways that are determined by genetics. The two most important genetic attributes are ABO blood type and Rhesus (Rh) factor, both of which must be compatible between donor and recipient. Mixing incompatible blood leads to an immune reaction against the genetically mismatched cells, causing blood cells to clump together and ultimately form life-threatening blood clots.

Your ABO blood type indicates the presence or absence of specific markers on the surface of your red blood cells. For example, if you have type A blood, your cells display A markers, whereas if you are type B, your cells have B markers. If you have type O blood, then you lack both A and B markers. And if you have type AB blood, then you have both A and B markers.

Which blood type you have is determined by a single gene, *ABO*, found on chromosome 9. There are three alleles of the *ABO* gene: *A*, *B*, and *O*. The protein, an enzyme, made from the *A* allele deposits A markers on the red blood cell surface; the enzyme made from the *B* allele adds B markers; the enzyme made

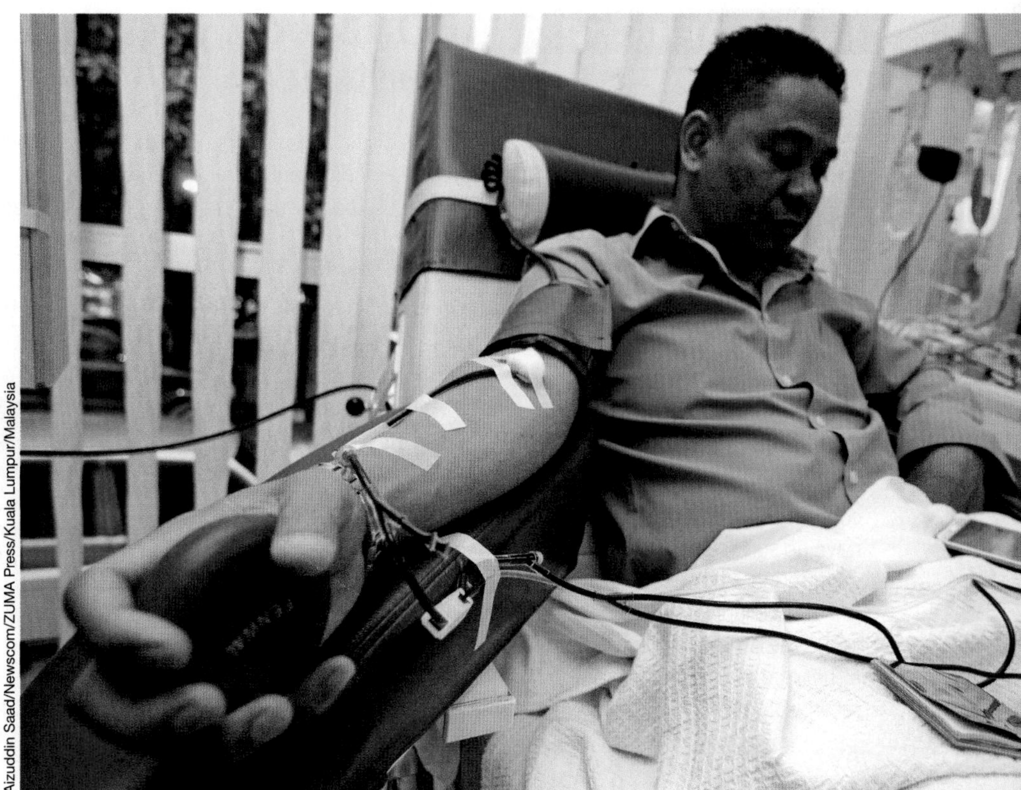

Aizuddin Saad/Newscom/ZUMA Press/Kuala Lumpur/Malaysia

Blood from donors can be used for transfusions but must be compatible with the blood type and Rh factor of the recipient.

CODOMINANCE
A form of inheritance in which the effects of both alleles are displayed in the phenotype of a heterozygote.

from the *O* allele is nonfunctional and deposits neither A nor B markers on the cell surface. Since we inherit one allele from each parent, the possible combinations of the three alleles are *OO, AO, BO, AB, AA,* and *BB*.

Blood type AB is an example of **codominance**, in which the contributions of both the maternal allele and the paternal allele in a heterozygous individual are displayed in the phenotype. Unlike incomplete dominance, in which heterozygotes have an intermediate phenotype, codominant alleles share the limelight: heterozygotes express both traits.

Blood type alleles *A* and *B* are codominant, while *O* is recessive to both *A* and *B*. Consequently, if you have blood type A, your genotype will be either *AA* homozygous or *AO* heterozygous. The same goes for blood type B: you will have a genotype of either *BB* or *BO*. People with type AB blood have an *AB* genotype and express both A and B markers on their cells **(INFOGRAPHIC 12.7)**.

Another important blood type marker is Rh factor, a protein found on the surface of red blood cells. The Rh factor is encoded by a gene found on chromosome 1. Your Rh status, (+) or (-), indicates the presence or absence of the Rh factor protein on the surface of your red blood cells. The inheritance pattern for Rh factor involves dominant and recessive alleles: positive Rh factor alleles (*Rh+*) are dominant over negative Rh factor alleles (*Rh-*), which do not produce the protein. So if a person carries one positive and one negative allele, the positive allele will dominate and that person will have an Rh-positive phenotype.

Type O Rh-negative donors are known as universal donors because their blood can be transfused to people of any other blood type. Red blood cells from these donors lack

INFOGRAPHIC 12.7 ABO Blood Type Demonstrates Codominant Inheritance

In codominant inheritance, heterozygotes display the effects of both alleles in their phenotype. ABO blood type in humans is an example. Alleles for blood type code for the presence of different surface markers on red blood cells. A person with type AB blood, for example, displays both A and B markers, while type O blood displays neither A nor B markers. A person's ABO blood type must be considered when he or she gives or receives blood.

Blood Transfusions
The ability to donate or receive blood is based on immune rejection. If two people have the same surface markers, then their blood will be compatible. People with type O blood have no surface markers that provoke an immune response in a recipient (so O is the universal donor). People with type AB blood will not recognize either marker as foreign, so can receive blood from any donor.

	Type A markers	Type B markers	Type A and B markers	Neither A nor B markers
Red blood cell type				
Genotype	*AA* or *AO*	*BB* or *BO*	*AB*	*OO*
Can donate to	Type A or AB recipient	Type B or AB recipient	Type AB recipient	Type A, B, AB, or O recipient
Can receive from	Type A or O donor	Type B or O donor	Type A, B, AB, or O donor	Type O donor

? Which blood type exemplifies codominance? What are the possible blood types of the parents of someone who has the codominant blood type?

the surface markers that the immune system recognizes as foreign and so will not trigger an immune response in a recipient. Because any person can receive O Rh-negative blood, O-negative donors are always in demand. Blood banks can fall short of O-negative blood during such disasters as earthquakes or hurricanes in which many people are hurt and require blood. People with type AB Rh-positive blood can receive any type of blood and are known as universal recipients (**INFOGRAPHIC 12.8**).

INFOGRAPHIC 12.8 A Mismatched Blood Transfusion Causes Immune Rejection

If donor and recipient are not compatible in ABO blood type and Rh factor, a recipient can have a life-threatening immune reaction to donated blood. A person with type B blood, for example, cannot donate blood to a person with type A blood.

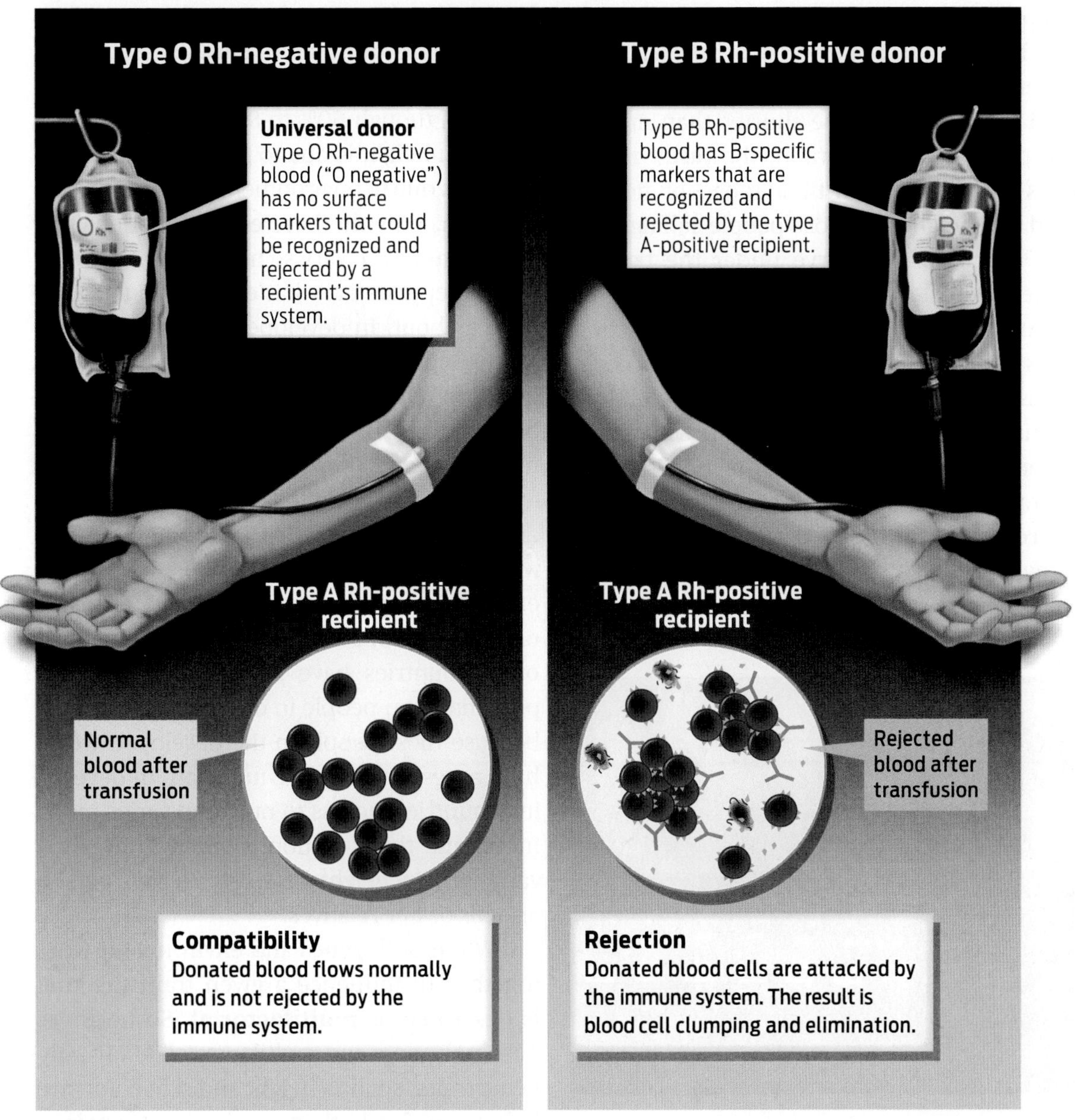

Type O Rh-negative donor
Universal donor
Type O Rh-negative blood ("O negative") has no surface markers that could be recognized and rejected by a recipient's immune system.
Type A Rh-positive recipient
Normal blood after transfusion
Compatibility
Donated blood flows normally and is not rejected by the immune system.
Type B Rh-positive donor
Type B Rh-positive blood has B-specific markers that are recognized and rejected by the type A-positive recipient.
Type A Rh-positive recipient
Rejected blood after transfusion
Rejection
Donated blood cells are attacked by the immune system. The result is blood cell clumping and elimination.

? Who can safely receive the blood of someone with AB positive blood?

CONTINUOUS VARIATION Variation in a population showing an unbroken range of phenotypes rather than discrete categories.

POLYGENIC TRAIT A trait whose phenotype is determined by the interaction among alleles of more than one gene.

MULTIFACTORIAL INHERITANCE An interaction between genes and the environment that contributes to a phenotype or trait.

POLYGENIC INHERITANCE

Q: How much of human height is genetic?

A: The short answer is, a lot. Geneticists estimate that height is 60% to 80% genetic. This means that genes account for 60% to 80% of the difference in height you see from person to person. But there isn't one single gene that determines height, there are many (more than 400 at current count). This is why we see such a range of heights among us.

In the United States, most people fall somewhere between 5 feet and 6 feet, 2 inches tall, and women tend to be shorter than men. If we plot height on a graph, the result resembles a bell curve, with a range of heights and the heights of most people falling near the middle of the curve under the bell. In other words, height is a trait that shows **continuous variation** in the population. This is in contrast to the discrete, or discontinuous, traits we've encountered, in which individuals have one of only two or three possible phenotypes for a given trait—Mendel's round or wrinkled peas (see **Milestone 4: Mendel's Garden**), or ABO blood type, for example. Why does height show an unbroken range of phenotypes rather than discrete categories like tall or short? The main reason is that human height is a **polygenic trait**—one that is influenced by more than one gene. When multiple genes act together, their effects add up to produce a range of phenotypes. Another example of a polygenic trait is skin color.

The fact that height is largely genetic means that, all other things being equal, two tall parents are very likely to have a child who is also tall; the same goes for two short parents. If you were to plot the height of students in your class against the average of their parents' height, most tall children would come from tall parents, and most short children would come from short parents. But genes aren't the whole story.

Even though 60% to 80% of the variation we see in height is due to genes, another 20% to 40% is due to environmental factors such as nutrition. Why do these estimates of environmental influence vary so much? The answer is that it depends on what environment you're talking about. In developed countries, where most people have access to adequate nutrition, height is more than 80% inherited, or heritable. This means that when scientists compare the height of a person to his or her relatives, they find that height varies less than 20% among close relatives. In developing countries, where many people are still malnourished, environment plays a larger role. Another way of looking at this is that more people in developed countries have reached their genetic potential than people in developing countries because most people in the developed world have access to adequate nutrition. In developing countries, access to nutrition varies much more, and this variation is reflected in larger variations in height between a given person and his or her relatives.

When both genes and environment work together to influence a given trait, the trait is described as **multifactorial.** So height is both polygenic and multifactorial. Many polygenic traits, such as height and skin color, are influenced by both genes and the environment and thus may be considered multifactorial (**INFOGRAPHIC 12.9**).

Height is an example of a trait that shows continuous variation in any given population. In humans it is determined by as many as 697 alleles of 423 genes.

INFOGRAPHIC 12.9 Human Height Is Both Polygenic and Multifactorial

Multiple genes as well as environmental factors such as diet, nutrition, and overall health act together to determine how tall we become.

Polygenic: multiple genes working together
Many genes act together to determine one's height. The combination of alleles a person inherits (*aabbcc*, *AabbCc*, etc.) predicts a distinct height phenotype. Within a population, the result of these polygenic interactions is a range of height phenotypes, from very short to very tall.

A Polygenic Punnet Square
A mating between two people of medium height (where three genes control height: *AaBbCc* × *AaBbCc*) produces seven distinct phenotypes, determined by the number of dominant alleles inherited.

	abc	*abC*	*aBc*	*Abc*	*aBC*	*AbC*	*ABc*	*ABC*
abc	*aabbcc*	*aabbCc*	*aaBbcc*	*Aabbcc*	*aaBbCc*	*AabbCc*	*AaBbcc*	*AaBbCc*
abC	*aabbCc*	*aabbCC*	*aaBbCc*	*AabbCc*	*aaBbCC*	*AabbCC*	*AaBbCc*	*AaBbCC*
aBc	*aaBbcc*	*aaBbCc*	*aaBBcc*	*AaBbcc*	*aaBBCc*	*AaBbCc*	*AaBBcc*	*AaBBCc*
Abc	*Aabbcc*	*AabbCc*	*AaBbcc*	*AAbbcc*	*AaBbCc*	*AAbbCc*	*AABbcc*	*AABbCc*
aBC	*aaBbCc*	*aaBbCC*	*aaBBCc*	*AaBbCc*	*aaBBCC*	*AaBbCC*	*AaBBCc*	*AaBBCC*
AbC	*AabbCc*	*AabbCC*	*AabbCc*	*AAbbCc*	*AaBbCC*	*AAbbCC*	*AABbCc*	*AABbCC*
ABc	*AaBbcc*	*AaBbCc*	*AaBBcc*	*AABbcc*	*AaBBCc*	*AABbCc*	*AABBcc*	*AABBCc*
ABC	*AaBbCc*	*AaBbCC*	*AaBBCc*	*AABbCc*	*AaBBCC*	*AABbCC*	*AABBCc*	*AABBCC*

Multifactorial: genes and environment working together
Human populations show a continuous range of heights (red line), rather than a genetically predicted number of phenotypes (blue bars). Environmental influences blur the genetic boundaries, resulting in a seamless continuity across the phenotype range.

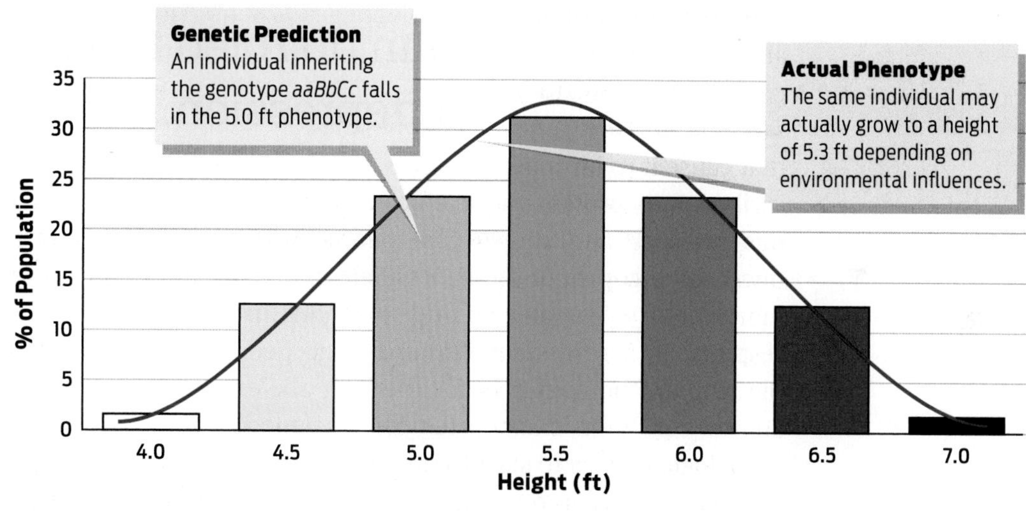
Genetic Prediction
An individual inheriting the genotype *aaBbCc* falls in the 5.0 ft phenotype.
Actual Phenotype
The same individual may actually grow to a height of 5.3 ft depending on environmental influences.
% of Population
35
30
25
20
15
10
5
0
4.0
4.5
5.0
5.5
6.0
6.5
7.0
Height (ft)

? A person is 5 feet, 6 inches (5.5 feet) tall. Consider at least two ways that this person could reach this height (factor in the number of inherited dominant alleles and nutritional status).

MULTIFACTORIAL INHERITANCE

Q: Are some people genetically predisposed to depression?

A: Depression is a serious mental health condition that affects more than 15 million adults in the United States each year—nearly 7% of the population. Stressful life events involving loss of a loved one or a job can influence the onset and course of depression. But not everyone who experiences stressful life events becomes clinically depressed.

Researchers Terrie Moffitt and Avshalom Caspi, a husband-and-wife team of psychologists at King's College London, wanted to understand why that is. They speculated that perhaps a person's genetics shapes their response to difficult life events, and that some people are genetically more likely to be wounded by these events.

In 2003, they conducted an epidemiological study looking at whether different alleles of a gene encoding a protein called the serotonin transporter might be responsible for this sensitivity. Serotonin is a chemical in the brain that affects mood. Common drugs used to treat depression called selective serotonin reuptake inhibitors (SSRIs) work by targeting the action of this protein.

There are two common alleles of the serotonin transporter gene, which is located on chromosome 17: a short version and a long version. The long version contains about 44 extra base pairs in the regulatory sequence of the gene that cause it to be transcribed into protein more efficiently. Previous research had shown that people with one or two copies of the short allele exhibit more intense fear and anxiety reactions in response to fearful stimuli compared to people who are homozygous for the long allele. So the serotonin transporter gene seemed like a logical place to start to understand individual difference in responses to stressful life events.

> The combination of hard knocks and short alleles more than doubled the risk of depression.

To conduct their analysis, Moffitt and Caspi turned to a long-term health study of almost 900 New Zealanders. They identified these participants' serotonin transporter alleles and interviewed them about traumatic experiences in early adulthood—experiences such as a major breakup, a death in the family, or serious injury—to see if these difficulties brought out an underlying genetic tendency toward depression.

The results were striking: clinical depression was diagnosed in 43% of participants who had two copies of the short allele and who had experienced four or more tumultuous events. By contrast, only 17% of participants who had two copies of the long allele and who had endured four or more stressful events had become depressed. Participants with two short alleles who experienced no stressful events fared pretty well, too—they were no more likely to become depressed than those with two long alleles. Clearly, it was the combination of hard knocks and short alleles that more than doubled the risk of depression **(INFOGRAPHIC 12.10)**.

Caspi and Moffitt's study lent strong support to the idea that mental illnesses like depression and other complex diseases cannot be explained by genetic or environmental factors alone, but often arise from an interaction between the two. In other words, mental illnesses exhibit multifactorial inheritance.

Multifactorial inheritance is a common pattern of inheritance. Diseases such as asthma, diabetes, and heart disease are all caused by a combination of several genes and their interaction with the environment.

NONDISJUNCTION

Q: Why does the risk of having a baby with Down syndrome go up as a woman ages?

A: At age 25, a woman's risk of having a baby with Down syndrome is 1 in 1,250 births (that is, 1 out of every 1,250 babies born to women who are 25 is likely to have Down syndrome). At age 40 her risk skyrockets to 1 in 100 births.

As women age, the risk of giving birth to a baby with any chromosomal abnormality

INFOGRAPHIC 12.10 Depression Can Result from a Combination of Genotype and Life Experiences

In 2003, Terrie Moffitt and Avshalom Caspi showed that a specific allele of the serotonin transporter gene—a gene that influences levels of the signaling molecule serotonin in the brain—in combination with stressful life events can cause depression. The gene comes in long and short versions.

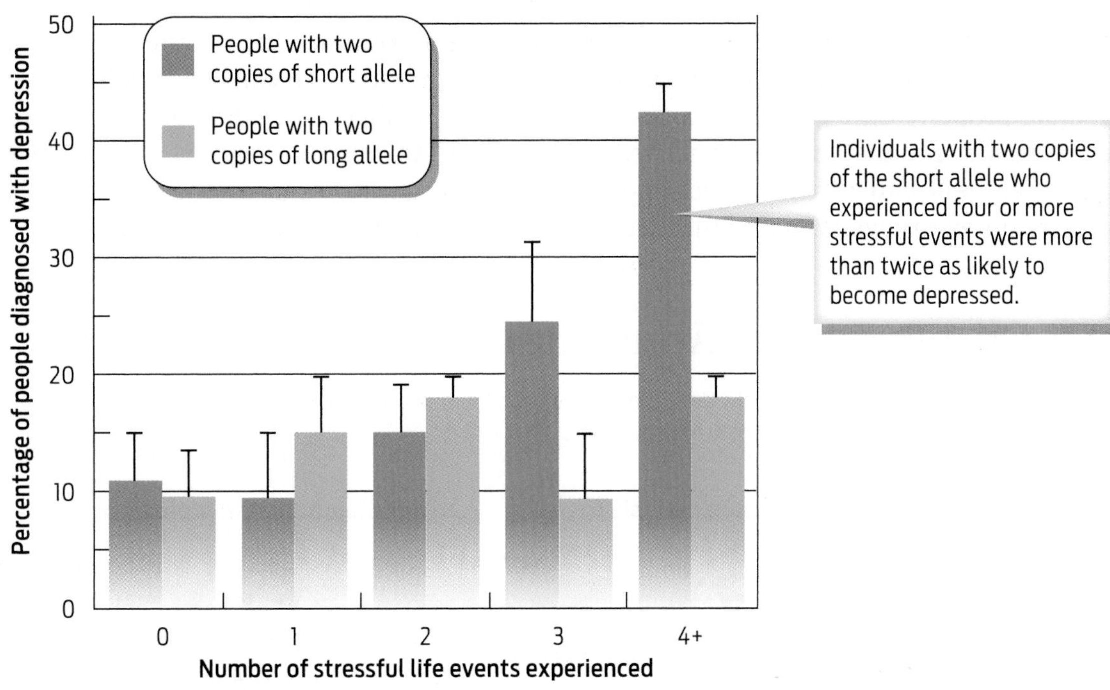

Data from Caspi, A., et al. (2003) *Science* 301(5631):386–389.

? Do these results support a polygenic or a multifactorial inheritance model for depression?

increases. That's because as a woman ages, so do her eggs. All the eggs that a woman will ever have were formed before she was born, during fetal development, and they have been aging like the rest of the cells in her body. Until puberty, a woman's developing eggs are paused in the middle of meiosis (at meiosis I); they haven't yet completed their cell division. During a menstrual cycle, one of these cells resumes meiosis and is ovulated. In older women, as eggs complete meiosis and are ovulated, they are more likely to experience an error in chromosome segregation, leading to an abnormality of chromosome number in the egg.

Markus Moellenberg/Getty Images

Most Down syndrome children have learning disabilities that range from mild to moderate.

ANEUPLOIDY
An abnormal number of one or more chromosomes (either extra or missing copies).

NONDISJUNCTION
The failure of chromosomes to separate accurately during cell division; nondisjunction in meiosis leads to aneuploid gametes.

TRISOMY 21
Having an extra copy of chromosome 21; also known as Down syndrome.

A numerical chromosomal abnormality means that a developing fetus carries a chromosome number that differs from the usual 46. The most common numerical abnormalities in humans are called **aneuploidies,** deviations from the normal number of chromosomes in which single chromosomes are either duplicated or deleted. Most aneuploidies arise during meiosis, as the parents' sex cells are being formed. If chromosomes do not separate properly during meiosis, an occurrence called **nondisjunction,** the resulting gamete will either lack a chromosome or carry an extra one. When that gamete is fertilized by a normal gamete, the resulting zygote can have an abnormal number of chromosomes. In most cases, the abnormality is so severe the zygote spontaneously aborts (**INFOGRAPHIC 12.11**).

There are, however, cases in which the abnormality is not life threatening but does cause severe disability. The most common of these is **trisomy 21,** another name for Down syndrome. Trisomy 21 results when an embryo inherits an extra copy of chromosome 21. Anyone can conceive a child with the abnormality, but older women are at much higher risk.

INFOGRAPHIC 12.11 Chromosomal Abnormalities: Aneuploidy

Birth defects can arise when chromosomes fail to separate normally during meiosis, a phenomenon called nondisjunction. The resulting gametes carry an abnormal number of chromosomes, a condition called aneuploidy.

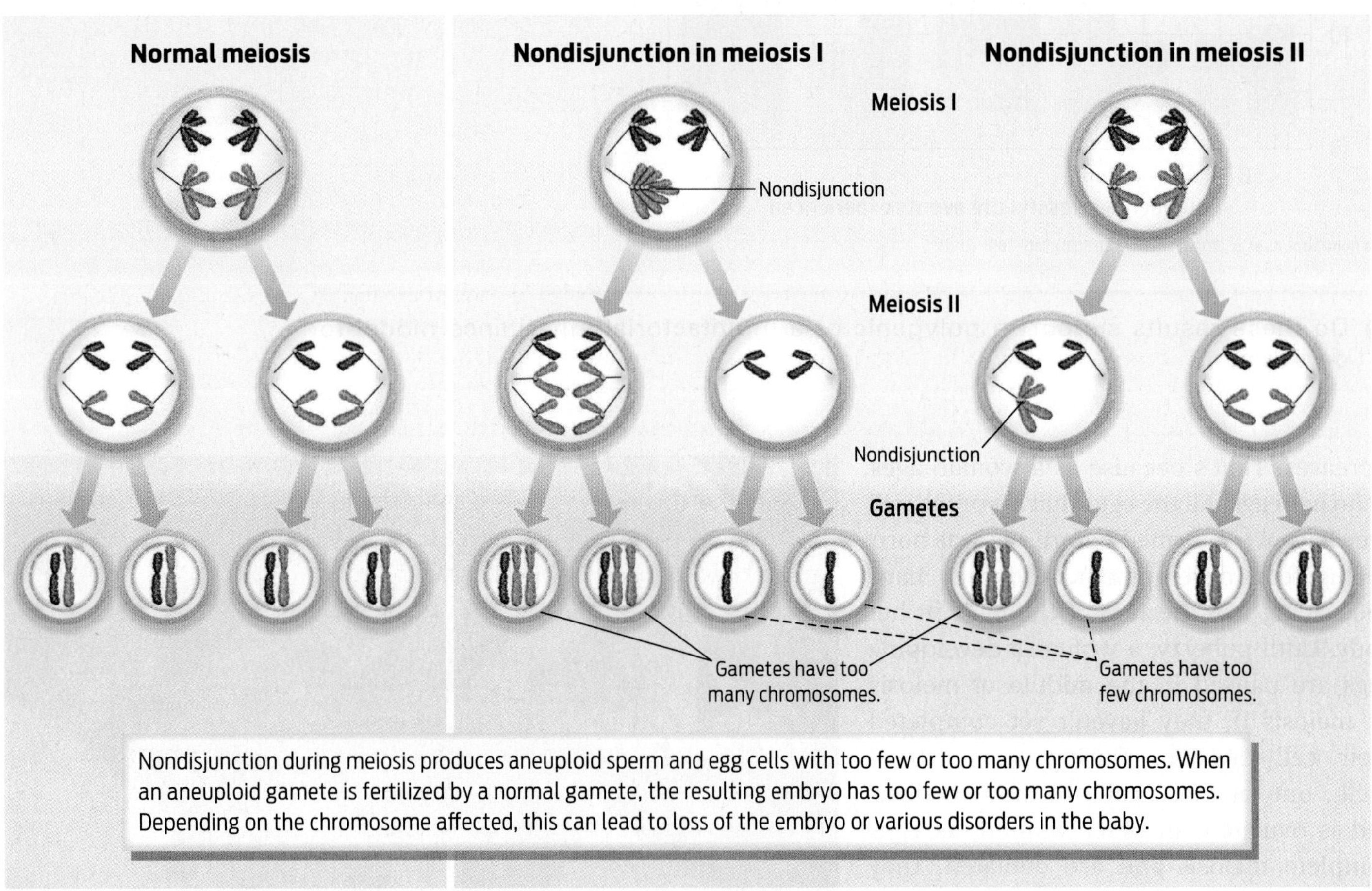

Nondisjunction during meiosis produces aneuploid sperm and egg cells with too few or too many chromosomes. When an aneuploid gamete is fertilized by a normal gamete, the resulting embryo has too few or too many chromosomes. Depending on the chromosome affected, this can lead to loss of the embryo or various disorders in the baby.

? An egg experienced nondisjunction during meiosis I. It has 22 autosomes and no sex chromosome. If this egg is fertilized by a sperm containing 22 autosomes and a Y chromosome, how many autosomes, sex chromosomes, and total chromosomes will the zygote have? Which of the egg, sperm, and zygote can be described as aneuploid?

INFOGRAPHIC 12.12 Amniocentesis Provides a Fetal Karyotype

Doctors perform a procedure called amniocentesis to obtain fetal cells and diagnose chromosomal abnormalities such as Down syndrome. A karyotype analysis is done on the fetal cells to look for chromosomal abnormalities, in particular missing or extra chromosomes.

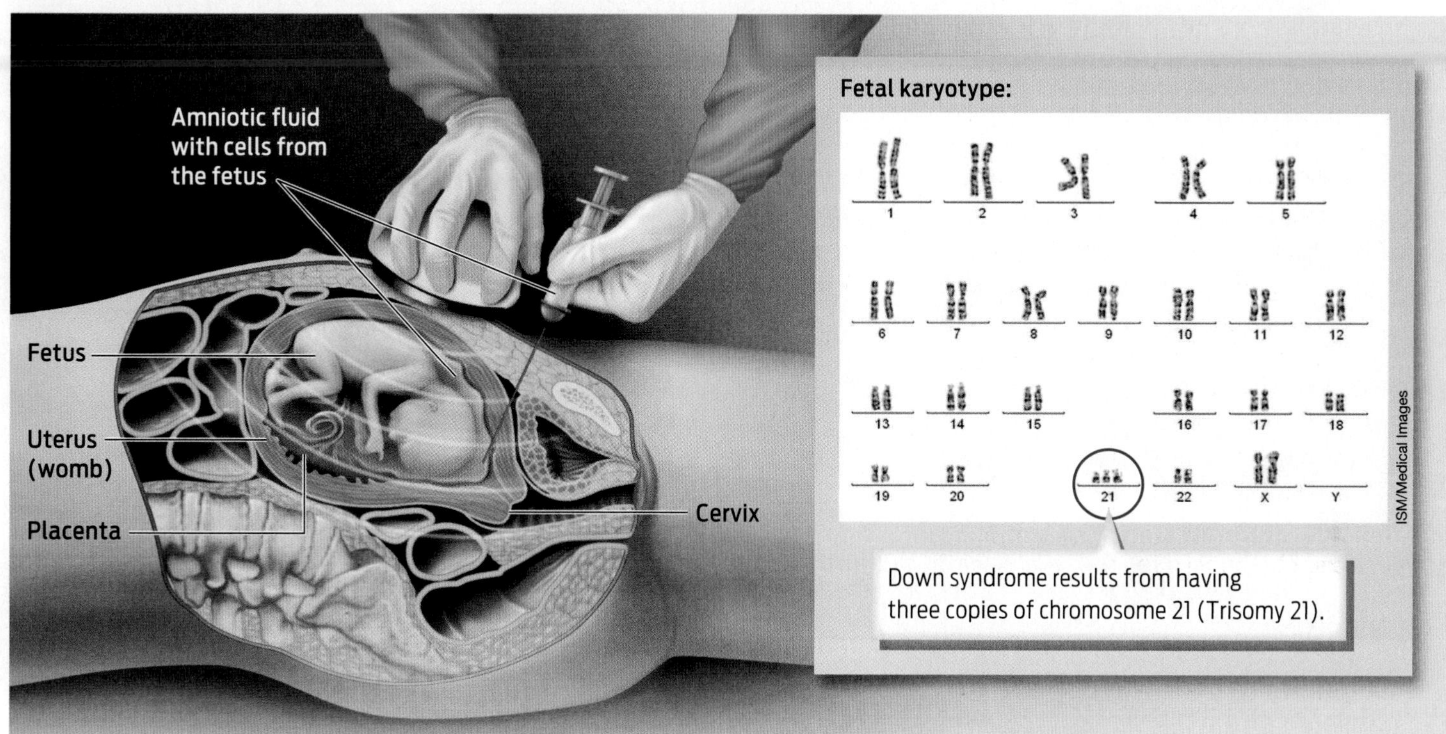

? What would you observe in the karyotype of a fetus with Turner syndrome? (Hint: Refer to Table 12.1.)

Most Down syndrome children have learning disabilities that range from mild to moderate, but some have profound mental disability. They are also at higher risk for other diseases and typically don't live beyond 50 years of age.

A variety of screening tests are used to detect chromosomal abnormalities early in pregnancy. Some are based on the levels of specific proteins in the mother's blood, some are based on analyzing fetal DNA found in the mother's circulation, and some are based on collecting fetal cells and examining their chromosomes. This last analysis relies on **amniocentesis**, a procedure in which a long, thin, hollow needle is inserted through a woman's abdominal wall and into her uterus. Through the needle, the equivalent of 2 to 4 teaspoons of amniotic fluid, which surrounds the growing fetus, is removed. This fluid contains fetal cells that contain the fetus's DNA. From that fluid, technicians analyze the fetal **karyotype**–that is, the chromosomal makeup in its cells (**INFOGRAPHIC 12.12**).

The reasons to undergo amniocentesis vary from couple to couple. But if a test comes back positive, couples have options: they can begin to plan for a disabled child, or make the decision not to carry the child to term.

Although scientists have linked some of the most obvious birth defects to the age of a woman's eggs, recent research also shows that a man's age affects the quality of his sperm. Men who father children after age 45 are more likely to have children with cognitive disorders such as autism, for example. Male fertility declines over time, too, although much more gradually than does female fertility. Research shows that the older the man, the more likely he is to produce sperm with genetic defects. ■

AMNIOCENTESIS
A procedure that removes fluid surrounding the fetus to obtain and analyze the chromosomal makeup of fetal cells.

KARYOTYPE
The chromosomal makeup of cells. Karyotype analysis can be used to detect chromosomal disorders prenatally.

CHAPTER 12 SUMMARY

- Humans have 23 pairs of chromosomes. One of these pairs is the sex chromosomes: XX in females and XY in males. It is the presence of the Y chromosome that determines maleness, and therefore it is fathers who determine the sex of a baby.
- Sex as a body phenotype is determined by genetics working in combination with hormones; variations in either genes or hormones can lead to intersex or cases of ambiguous genitalia.
- Disorders and other traits inherited on X chromosomes are called X-linked traits, and are more common in males than in females.
- Because the Y chromosome in a male does not have a homologous partner, it does not recombine during meiosis. The Y chromosome a son inherits from his father is essentially identical to the Y chromosome his father inherited from his father (the grandfather), a fact that can be used to establish paternity.
- Familial hypercholesterolemia is a trait that exhibits incomplete dominance, a form of inheritance in which heterozygotes have a phenotype intermediate between the two homozygotes.
- ABO blood type is an example of a codominant trait—both maternal and paternal alleles contribute equally and separately to the phenotype.
- Many traits are polygenic—that is, they are influenced by the additive effects of multiple genes. Polygenic traits often show a continuous, bell-shaped distribution in the population. Human height is an example.
- In many cases, a person's phenotype is determined by both genes and environmental influences; this type of inheritance is described as multifactorial. Depression and cardiovascular disease are examples of multifactorial illnesses.
- Some genetic disorders result from having a chromosome number that differs from the usual 46. Down syndrome, or trisomy 21, is caused by an extra copy of chromosome 21.
- Aneuploidy, having one or more extra or missing chromosomes, is the result of nondisjunction—when chromosomes fail to separate properly during meiosis, generating aneuploid gametes.

MORE TO EXPLORE

- **Colapinto, J. (2000) *As Nature Made Him: The Boy Who Was Raised as a Girl*. New York: HarperCollins.**
- **Intersex Society of North America (ISNA): http://www.isna.org**
- **Foster, E. A., et al. (1998) Jefferson fathered slave's last child. *Nature* 396:27–28.**
- **Allen, E. G., et al. (2009) Maternal age and risk for trisomy 21 assessed by the origin of chromosome nondisjunction: a report from the Atlanta and National Down Syndrome Projects. *Human Genetics* 125(1):41–52.**
- **Cowan, Ruth Schwartz (2008) *Heredity and Hope: The Case for Genetic Screening*. Cambridge: Harvard University Press.**

CHAPTER 12 Test Your Knowledge

DRIVING QUESTION 1 How do chromosomes determine sex, and how does sex influence the inheritance of certain traits?

By answering the questions below and studying Infographics 12.1–12.5 and Table 12.1, you should be able to generate an answer for the broader Driving Question above.

KNOW IT

1 Which of the following most influences the development of a female fetus?

a. the presence of any two sex chromosomes
b. the presence of two X chromosomes
c. the absence of a Y chromosome
d. the presence of a Y chromosome
e. either b or c

2 Why are more males than females affected by X-linked recessive genetic diseases?

3 If a man has an X-linked recessive disease, can his sons inherit that disease from him? Why or why not?

USE IT

4 Which of the following couples could have a boy with Duchenne muscular dystrophy (DMD)?

a. a male with Duchenne muscular dystrophy and a homozygous dominant female
b. a male without Duchenne muscular dystrophy and a homozygous dominant female
c. a male without Duchenne muscular dystrophy and a carrier female
d. a and c
e. none of the above

5 Predict the sex of a baby with each of the following chromosomal makeups. Use your answer to check your answer to Question 1.

a. XX
b. XXY
c. XY
d. X

6 Consider your brother and your son.

a. If you are female, will your brother and your son have essentially identical Y chromosomes? Explain your answer.
b. If you are male, will your brother and your son have essentially identical Y chromosomes? Explain your answer.

7 A wife is heterozygous for Duchenne muscular dystrophy alleles, and her husband does not have DMD. Neither has any other notable medical history. What percentage of their sons, and what percentage of their daughters, will have

a. Duchenne muscular dystrophy (which is determined by a recessive allele on the X chromosome)?
b. an X-linked dominant form of rickets (a bone disease)?

DRIVING QUESTION 2 Some traits are not inherited as simple dominant or recessive inheritance patterns. What are some complex inheritance patterns?

By answering the questions below and studying Infographics 12.6–12.10, you should be able to generate an answer for the broader Driving Question above.

KNOW IT

8 What aspects of height make it a polygenic trait?

9 Which of the following inheritance patterns includes an environmental contribution?

a. polygenic
b. X-linked recessive
c. X-linked dominant
d. multifactorial
e. none of the above

10 What is the difference between polygenic inheritance and multifactorial inheritance?

11 How does incomplete dominance differ from codominance?

12 If you are blood type A-positive, to whom can you safely donate blood? Who can safely donate blood to you? List all possible recipients and donors and explain your answer.

USE IT

13 If two women have identical alleles of the suspected more than 400 height-associated genes, why might one of those women be 5 feet, 5 inches tall and the other 5 feet, 8 inches tall?

14 Look at Infographic 12.10. How do the data given support the hypothesis that both genes and the environment influence at least some cases of clinical depression?

15 Look at Infographic 12.10. At approximately how many stressful experiences does the homozygous short-allele genotype begin to influence the depression phenotype?

16 From what you have read in this chapter, how can you account for two people with the same genotype for a predisposing disease allele having different phenotypes?

apply YOUR KNOWLEDGE

MINI CASE

17 A serious car crash on a freeway has resulted in multiple injuries causing substantial blood loss in three members of a family—a mother, a father and their 2-year-old daughter. The local blood bank will be challenged to supply blood, as their supplies of all blood types were drained after the roof of a shopping plaza collapsed the week before and many transfusions were required.

- **a.** The EMTs must give blood immediately to all three members of the family. What blood type should they use (consider both ABO blood type and Rh factor)? Explain your answer.
- **b.** Both parents carry a blood donor card. The mother is O-negative and the father is A-positive. From this information, what (if any) additional blood types (beyond your answer to part a) can be given to either parent? Explain your answer.
- **c.** Does knowing the parents' blood types give you enough information about the daughter's possible blood type to use a different blood type for her transfusion? Why or why not? (Hint: Consider possible blood types for the daughter and the implications of, for example, using A-negative donor blood. Could you guarantee that this would be safe?)

DRIVING QUESTION 3 How does nondisjunction lead to numerical abnormalities of chromosomes, and what are the consequences of these abnormalities?

By answering the questions below and studying Infographics 12.11 and 12.12, you should be able to generate an answer for the broader Driving Question above.

KNOW IT

18 What is the normal chromosome number for each of the following?

- **a.** a human egg
- **b.** a human sperm
- **c.** a human zygote

19 When looking at a karyotype, for example to diagnose trisomy 21 in a fetus, is it possible to use that analysis also to tell if the fetus has inherited a cystic fibrosis allele from a carrier mother?

USE IT

20 Which of the following can result in trisomy 21?

- **a.** an egg with 23 chromosomes fertilized by a sperm with 23 chromosomes
- **b.** an egg with 22 chromosomes fertilized by a sperm with 23 chromosomes
- **c.** an egg with 24 chromosomes, two of which are chromosome 21, fertilized by a sperm with 23 chromosomes
- **d.** an egg with 23 chromosomes fertilized by a sperm with 24 chromosomes, two of which are chromosome 21

21 From what you have read in this chapter, which of the possibilities in Question 20 is most likely? Explain your answer.

apply YOUR KNOWLEDGE

BRING IT HOME

22 What factors would lead you to consider prenatal genetic testing? In your opinion, what is the value of having this information? Use the table below to help organize your thoughts, then make a conclusion about the use of prenatal genetic testing.

Factor	Value
Consider known risks (e.g., family history, mother's age).	*Consider how the information may be used.*

apply YOUR KNOWLEDGE

INTERPRETING DATA

23 The graph at right shows the average ("mean") age of women who had children with trisomy 21 ("Cases"), of those who did not ("Controls"), and the average age of all women giving birth in the population studied. The data are presented for 15 years.

a. In general, how does the age (at time of birth) of women giving birth to a baby with Down syndrome compare to the age of women giving birth to a baby without Down syndrome?

b. During which year was the average age of the cases closest to the average age of the controls? How close were the average ages in this year?

c. During which year was the average age of the cases the most different from the average age of the controls? How different were the average ages in this year?

d. Using the data points for each year over this 15-year period, calculate the overall average age of women having babies with Down syndrome and the overall average age of women having babies who do not have Down syndrome.

Maternal Age and Trisomy 21, 1989–2004

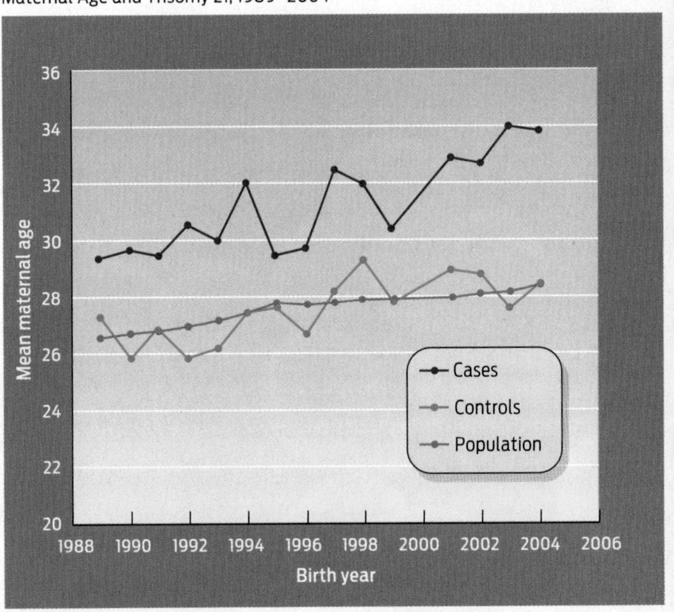

Data—Graves Allen, E., et al. (2009) Maternal age and risk for trisomy 21 assessed by the origin of chromosome nondisjunction a report from the Atlanta and National Down Syndrome Projects. Human Genetics 125(1):41–52.

13 Natural Selection and Adaptation

David M. Phillips/Science Source

Bugs THAT RESIST Drugs

Drug-resistant bacteria are on the rise. Can we stop them?

In January 2008, sixth-grader Carlos Don, an active footballer and skateboarder, boarded a bus headed for a class trip, happy and healthy. A month later he was dead.

In April 2006, 17-year-old Rebecca Lohsen was a model student at her high school; she was on the honor roll and was a member of the swim team. Four months later Rebecca was dead.

In December 2003, Ricky Lannetti was a college senior, a star football player and all-around athlete. A few weeks later Ricky was dead.

DRIVING QUESTIONS

1. What is staph, and can a person have it in the absence of an infection?
2. How do bacteria resist the effects of antibiotics?
3. How do populations evolve, and what is the role of evolution in antibiotic resistance?

The list of sudden deaths like these goes on and on. But surprisingly, these young people weren't killed in accidents or by violence; they were all killed by an infectious bacterium known as methicillin-resistant *Staphylococcus aureus* (MRSA), which has become widespread in recent years and which is difficult to treat with many existing antibiotics.

MRSA (pronounced "mer-sa") sickened some 75,000 people in the United States in 2012 and killed nearly 10,000. Formerly, outbreaks of MRSA were confined mainly to hospitals, where people are often very ill and have compromised immune systems. But since the mid 1990s, a growing number of healthy people are becoming infected outside hospitals. Daycare facilities, military bases, and schools nationwide have been reporting outbreaks, and young, healthy people are getting sick.

For more than 70 years, we've successfully combated potentially deadly bacterial infections with **antibiotics**–those "wonder drugs" that Alexander Fleming first discovered in a moldy petri plate nearly a century ago (see Chapter 3). But many of our most trusted antibiotics are no longer effective at killing the bacteria they once defeated. Over time, these bacteria have changed genetically–evolved–to become resistant. As a result, scientists are now seeing bacterial infections that don't respond to any known antibiotics, leading many to fear the day when we run out of treatment options altogether.

Scientists are now seeing bacterial infections that don't respond to any known antibiotics, leading many to fear the day when we run out of treatment options altogether.

"This is a major public health imperative," says Robert Daum, professor of microbiology at the University of Chicago, who studies antibiotic resistance. "We need a plan of attack now."

ANTIBIOTIC
A chemical that can slow or stop the growth of bacteria; many antibiotics are produced by living organisms.

Staph the Microbe

Bacteria are everywhere: in the air, in food, on toothbrushes and computer keyboards–even on and inside you. Your body hosts a wide variety of bacteria, ranging from helpful ones like *Lactobacilli* and *Bifidobacteria* that live in the gut and aid digestion to harmful ones like *Salmonella* that cause food poisoning. In fact, there are more bacteria living on and in you than there are human cells making up your body. The helpful bacteria on your skin produce acids that make your skin inhospitable to other, less helpful bacteria. At any given moment, a fair number of us harbor potentially dangerous bacteria on our skin. Most of the time, this isn't a problem. But in certain circumstances, these bacteria can outgrow your helpful skin bacteria and cause serious infections **(INFOGRAPHIC 13.1)**.

MRSA infection is caused by the bacterium *Staphylococcus aureas*–often called simply "staph." Although all strains of staph can cause infection, the medical community is especially concerned about *S. aureus* strains that have developed resistance to the antibiotic drugs that once effectively killed them. "MRSA" is actually a misnomer because the antibiotic methicillin is no longer used to treat staph infections. Drug-resistant strains of staph are usually resistant to several different types of antibiotics, including penicillins and cephalosporins.

Surprisingly, staph bacteria cause no harm to most of the people who carry them. About one-third of the U.S. population carries staph on their skin or in their nose, and about 2% of the population carries MRSA, according to the Centers for Disease Control and Prevention (CDC). If you carry staph of any strain

INFOGRAPHIC 13.1 The Human Body Hosts a Huge Number of Bacteria

A huge diversity of microbes is found in and on the human body. These resident bacteria are essential to human health. They help block pathogens from establishing an infection, and many produce helpful compounds such as vitamins. In some cases, potential pathogens can also take up residence and cause disease if they can outgrow the other resident bacteria.

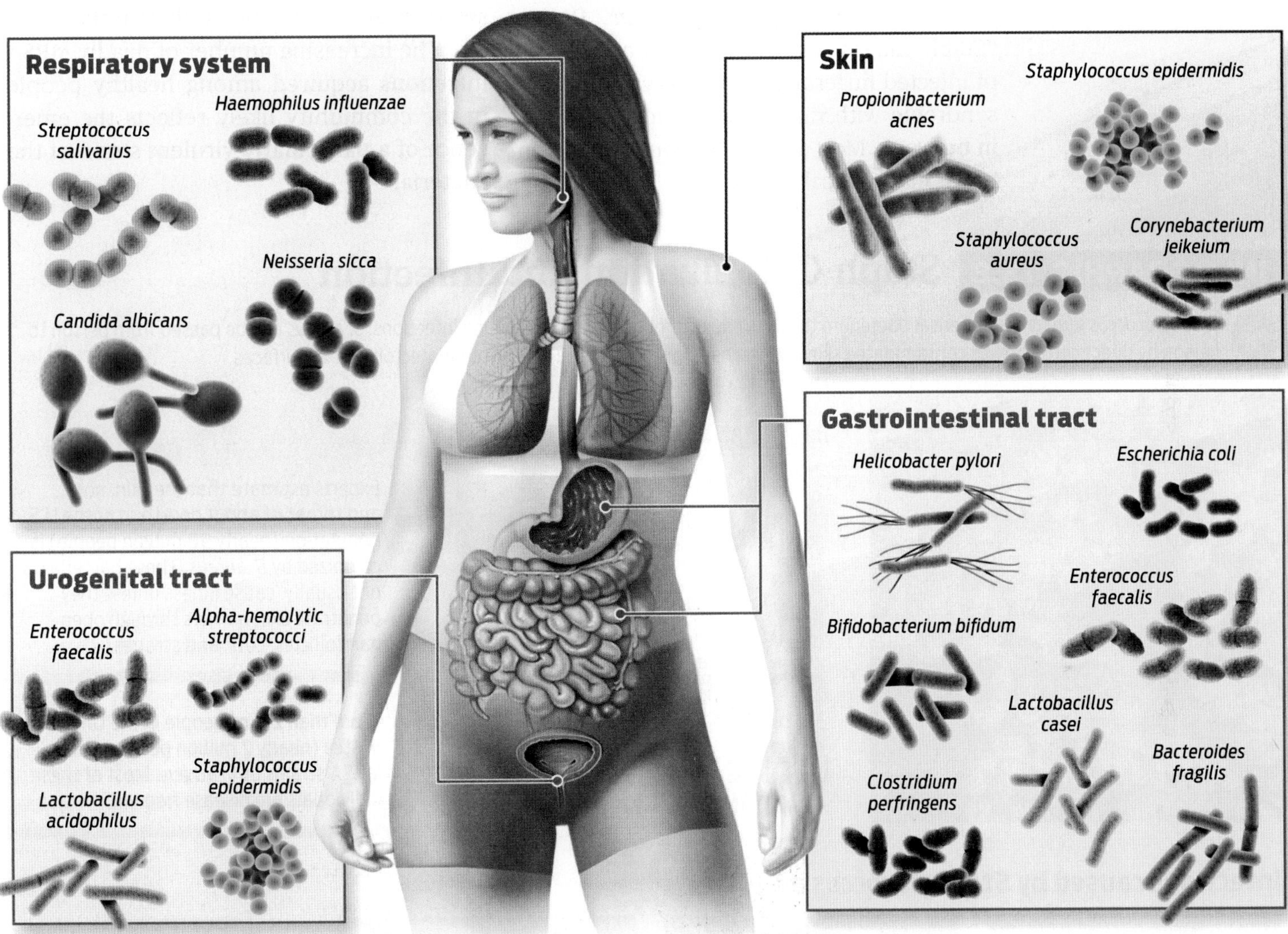

? Of the body sites highlighted, where is staph most likely to be found?

but aren't sick, you are "colonized" but not infected. Staph spreads from person to person through skin-to-skin contact or through shared contaminated items such as towels and bars of soap.

Most healthy people can be colonized with any staph strain, including MRSA, and not become ill—for the most part our skin and our immune systems protect us. But infections can occur if staph bacteria come into contact with a wound or otherwise enter the body. For example, athletes who have cuts and scrapes may acquire a staph infection in locker rooms or during contact sports. Staph infection usually causes only minor skin eruptions such as boils or pustules that can resemble spider bites. But if staph manages to enter the bloodstream or travel deeper into the body–becoming invasive–then more serious complications can occur (**INFOGRAPHIC 13.2**).

"Every one of us has probably had a staph infection at some point," explains Daum. "Staph ranges from the commonest cause of infected fingernails all the way to a severe syndrome with rapid death, and everything in between. Most staph infections don't even result in a medical encounter."

Staph bacteria can cause such a range of symptoms in part because there are many different strains of staph. Each strain differs from all others in its genetic makeup. MRSA, for example, is composed of a number of unique strains of drug-resistant staph, and some cause more serious disease than others. The increasing number of deadly MRSA infections acquired among healthy people in the community likely reflects the emergence of a particularly virulent strain of the bacteria.

INFOGRAPHIC 13.2 Staph Colonization and Infection

Staphylococcus aureus is a spherical bacterium that can cause pimples, boils, and wound infections. *S. aureus* can be passed from person to person by direct contact with contaminated skin or by transfer of the bacteria via contaminated objects or surfaces.

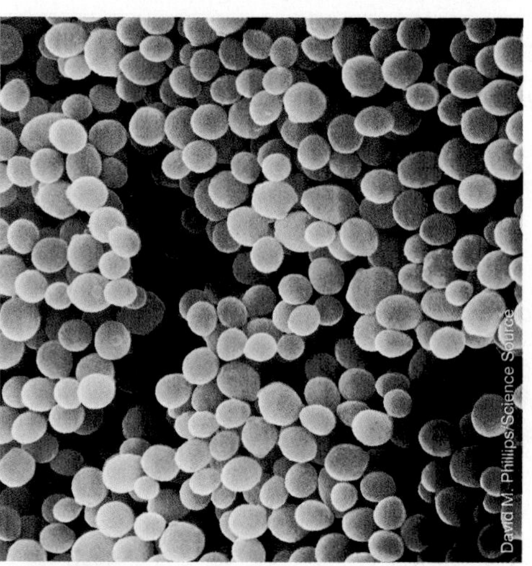
David M. Phillips/Science Source

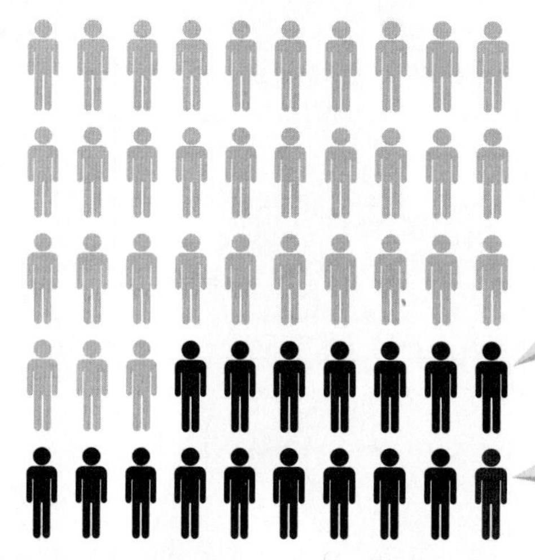

Experts estimate that the skin, nose, and throat of about one-third of the U.S. population (nearly 90 million people) is colonized by *S. aureus*. These bacteria do not usually cause illness unless they penetrate skin barriers through open hair follicles, cuts, and scrapes.

More than 2% of people in the United States (nearly 2 million people) carry MRSA strains of *S. aureus*. Most of these individuals are disease free.

Infections caused by *Staphylococcus aureus*:

Lauren Shear/Science Source

S. aureus can colonize normal skin surfaces without causing infection.

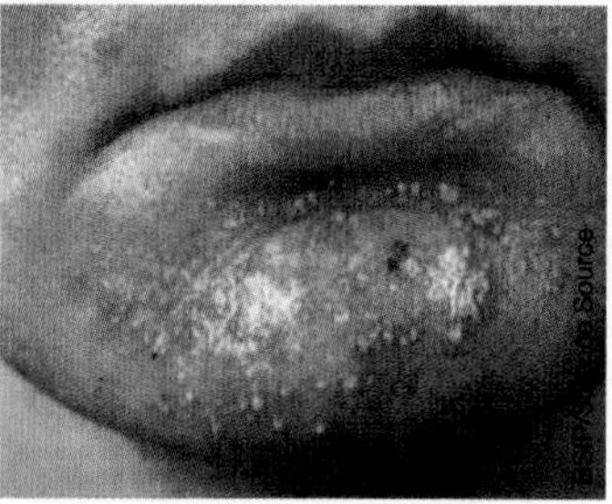

S. aureus can cause mild skin infections like pimples and boils.

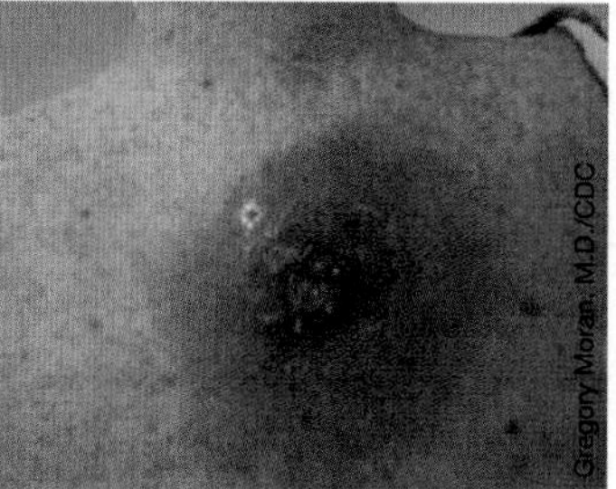
Gregory Moran, M.D./CDC

Methicillin-resistant *S. aureus* can cause more severe illness that is difficult to treat with existing medications.

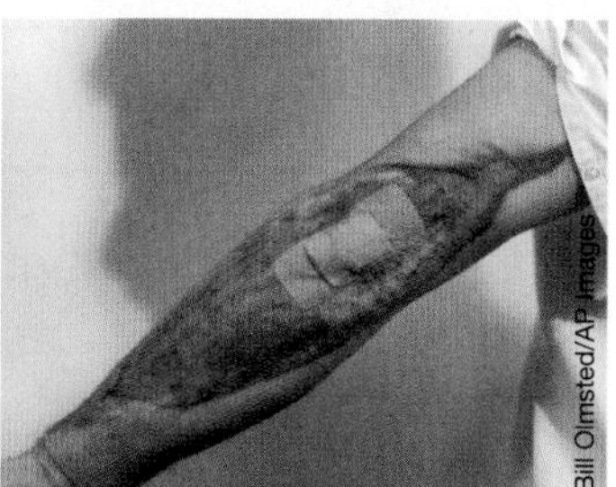
Bill Olmsted/AP Images

Some extremely virulent strains of MRSA can eat through skin and soft tissue.

? A college wrestler has developed a serious (invasive) MRSA infection. List three possible sources of the MRSA bacteria causing his infection.

Ricky Lannetti, for example, was a healthy 21-year-old football player at Lycoming College in Williamsport, Pennsylvania. "He was strong as an ox and he ran like a deer," says his mother, Theresa Drew. In early December 2003, Ricky came down with the flu. He wasn't recovering, and on the morning of December 6, Drew drove her son to Williamsport Hospital. By the time he was admitted, his blood pressure was dangerously low and his body temperature was erratic. As each hour passed, his condition worsened. His lungs began to fail. Doctors tried five different antibiotics, all without success. When his heart began to weaken, doctors prepared him to be flown to the cardiac center at a bigger hospital in Philadelphia. But it was too late; Ricky died that night.

It was only after an autopsy was performed that doctors discovered what had killed him: MRSA had infected Ricky's bloodstream and attacked his organs. Although doctors couldn't be sure how Ricky contracted MRSA, they suspect that it entered his body through a pimple on his buttocks. From there, it spread to his internal organs.

"Doctors tried every antibiotic imaginable, including vancomycin," says his father, Rick Lannetti. But the treatment was too late. Ricky's immune system was already weak because of the flu. When he contracted MRSA, his body was unable to fight back as well as it otherwise would have. "In the end," his father says, "MRSA had broken every one of his organs beyond repair."

Athletes have an elevated risk of skin infections

Staphylococcus aureus and other harmful bacteria can transfer from person to person by direct or indirect contact. For this reason athletes are at special risk of infection. To raise awareness, the CDC teamed up with the National Collegiate Athlete Association to create a series of educational posters.

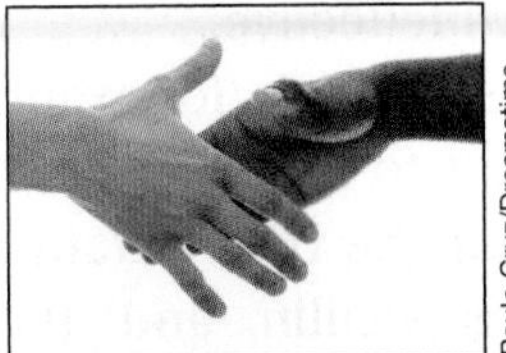

Paulo Cruz/Dreamstime

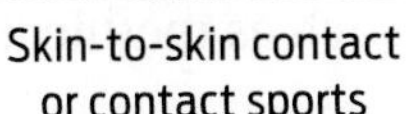

Skin-to-skin contact or contact sports

DNY59/iStockphoto; Offscreen/Dreamstime.com

Used soap, towels, and equipment

Genevieve Astrelli/iStockphoto

Contaminated surfaces

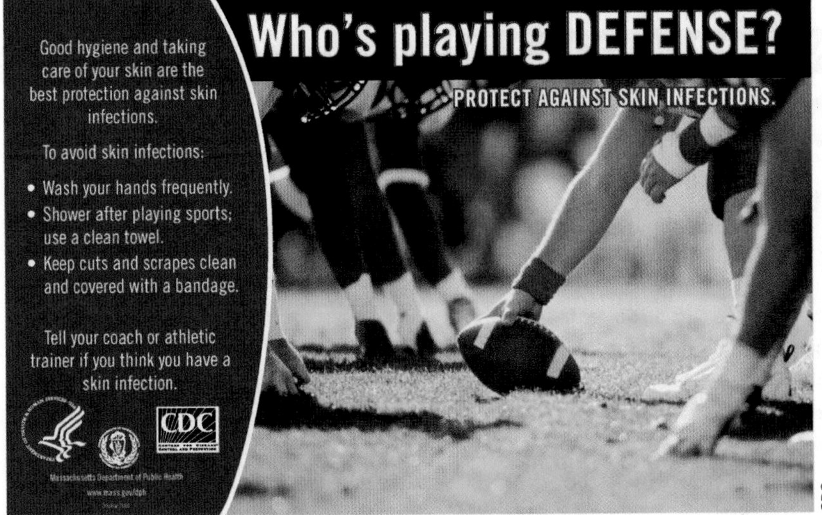

Who's playing DEFENSE?
PROTECT AGAINST SKIN INFECTIONS.
Good hygiene and taking care of your skin are the best protection against skin infections.
To avoid skin infections:
• Wash your hands frequently.
• Shower after playing sports; use a clean towel.
• Keep cuts and scrapes clean and covered with a bandage.
Tell your coach or athletic trainer if you think you have a skin infection.
CDC

CDC

Lycoming College Sports Information; Courtesy Theresa Drew

Ricky Lannetti as a healthy 21-year-old football player. Right: Ricky with his mother, Theresa Drew.

Acquiring Resistance

Bacterial infections were a common cause of death before the 1940s, when antibiotics first became widely available. Since then, we've relied on antibiotics to treat most common bacterial infections, including ones that might otherwise have been killers. But almost immediately after antibiotics were introduced, bacteria that could survive antibiotics—drug-resistant bacteria—began to emerge. Within the last decade, drug-resistant bacterial strains have become much more common. Most people infected with drug-resistant bacterial strains are still treatable, but they have fewer treatment options. And sometimes—as in Ricky Lannetti's case—existing drugs are completely ineffective.

Drug-resistant strains of staph, for example, are typically resistant to an entire class of antibiotic drugs called the beta-lactams. Beta-lactams include penicillin and the cephalosporin antibiotics, such as methicillin and cephalexin. Beta-lactams are the most commonly prescribed class of antibiotics, used to treat ear infections, bronchitis, and urinary tract infections, among others. They work by interfering with a bacterium's ability to synthesize cell walls (**INFOGRAPHIC 13.3**).

Vancomycin, a non-beta-lactam drug, is the antibiotic of choice when a serious MRSA infection is confirmed. But even vancomycin isn't always effective; there are now staph strains resistant to vancomycin, too.

INFOGRAPHIC 13.3 How Beta-Lactam Antibiotics Work

Antibiotics are grouped into classes, one of which is called the beta-lactams. Beta-lactam antibiotics interfere with a bacterium's ability to synthesize cell walls.

In the absence of antibiotic:

The cell wall is a strong, rigid structure.

Water flowing into the cell exerts pressure on the cell wall.

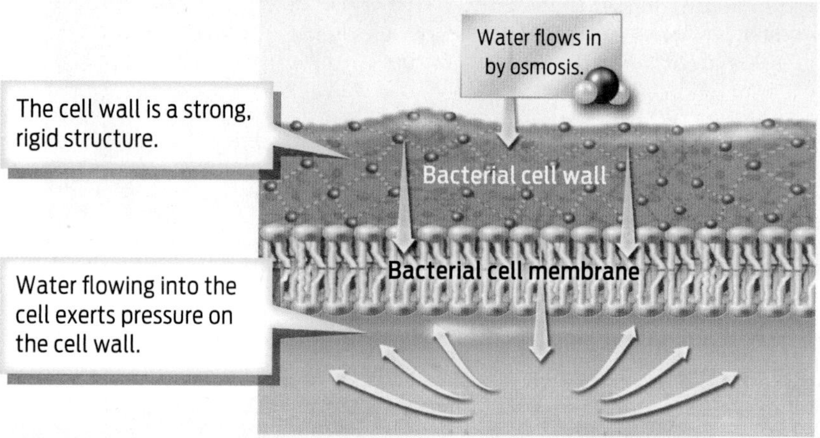
Water flows in by osmosis.
Bacterial cell wall
Bacterial cell membrane

The strong cell wall remains intact.

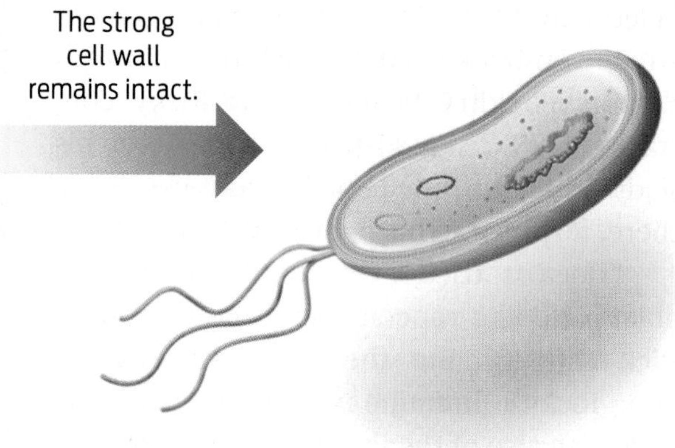

In the presence of antibiotic:

Beta-lactam antibiotics interfere with proper synthesis of the cell wall, causing it to be weak.

Water flowing into the cell exerts pressure on the cell wall.

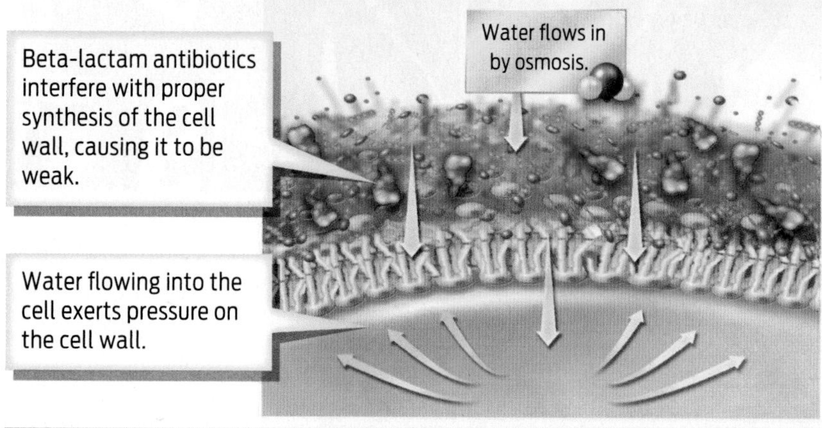
Water flows in by osmosis.

The weakened cell wall breaks apart.

Cell ruptures

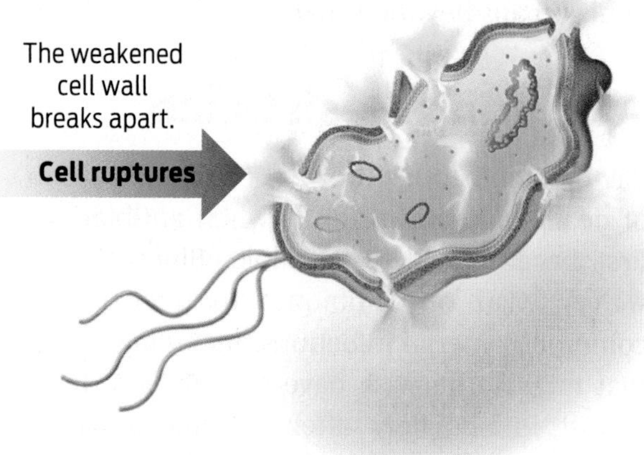

? If cells were grown in an isotonic solution that included a beta-lactam antibiotic, would they still rupture? Explain your answer.

Ricky Lannetti did not respond to vancomycin. Nor did Rebecca Lohsen, the 17-year-old high school swimmer. She was diagnosed with MRSA 2 days after she was admitted to the hospital for pneumonia. Antibiotics were ineffective in controlling the MRSA that attacked her lungs and then her heart. Twelve-year-old Carlos Don, who skateboarded and played football, also did not respond to vancomycin; he, too, died.

What's behind this rise in drug-resistant bugs? Like all organisms, bacteria can acquire mutations when they replicate their DNA during cell reproduction (see Chapter 10). These random mutations create new alleles, including ones that may confer antibiotic resistance.

Bacteria reproduce asexually by binary fission. Unlike sexual reproduction, in which gametes from two parents fuse, asexual reproduction does not require a partner. In **binary fission,** a single parental cell simply replicates its single chromosome, grows, and then splits into two daughter cells, each with a copy of the parental DNA (**INFOGRAPHIC 13.4**).

Each time DNA is replicated during binary fission, there is a chance that mutations will occur. When they do, new alleles are produced, which will then be carried into the daughter cells. Because bacteria reproduce much more rapidly than other organisms—a generation of bacteria can double in as little as 20 minutes—they accumulate mutations at a relatively high rate. An entire population of bacteria that is genetically different from the original cell can arise very quickly. A **population** is a group of individuals of the same species (in this case *Staphylococcus aureus*) living together in the same geographic area. That area could be a prairie, a pond, or a person's nose. New mutations can spread throughout a population of bacteria in a person's body over the course of an illness.

Mutations aren't the only way that populations of bacteria can acquire genetic variation. Bacteria can acquire new alleles, and even new genes, through a mechanism called **gene transfer,** in which pieces of DNA pass from one type of bacteria to another (**INFOGRAPHIC 13.5**).

Staph bacteria became resistant to drugs either by mutations in their own genes or by picking up resistance genes from other drug-resistant bacteria. The genetic changes ultimately altered staph proteins in ways that helped them dodge antibiotic drugs. The altered or acquired genes may code for proteins that can disable antibiotics. Or, they may code for proteins with altered shapes to which antibiotics can no longer bind. Some bacteria produce enzymes called beta-lactamases that chew up beta-lactam

BINARY FISSION
A type of asexual reproduction in which one parental cell divides into two.

POPULATION
A group of organisms of the same species living together in the same geographic area.

GENE TRANSFER
The process by which bacteria can exchange segments of DNA between them.

INFOGRAPHIC 13.4 How Bacteria Reproduce

Bacteria reproduce through binary fission. Binary fission is a form of asexual reproduction in which a single parent cell replicates its contents and then divides into two daughter cells. Note that each daughter cell inherits all its DNA from the single parent cell.

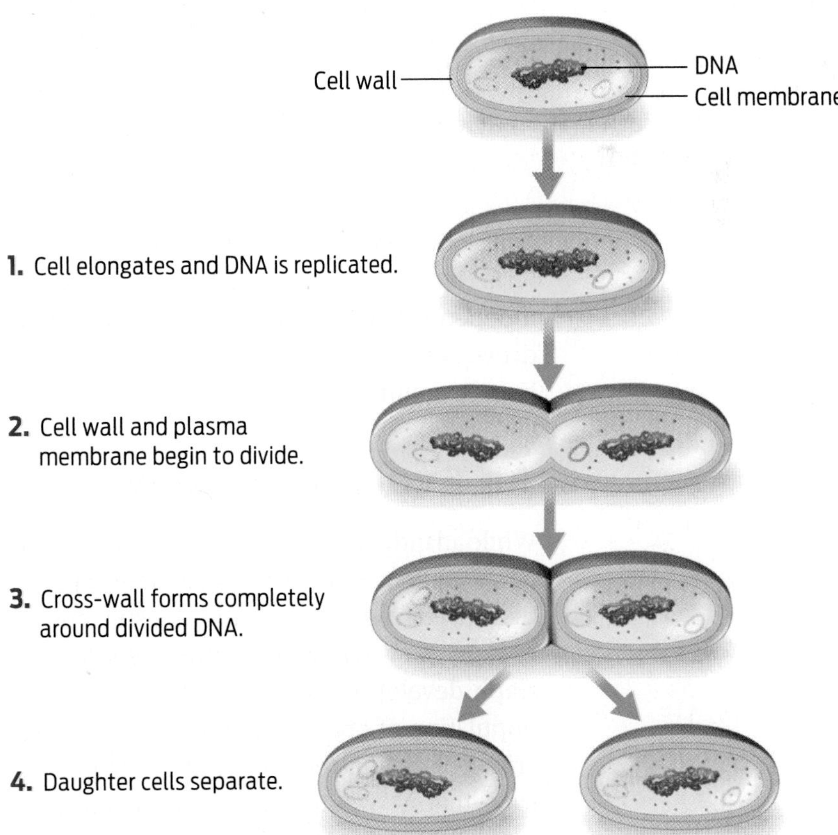
Cell wall
DNA
Cell membrane
1. Cell elongates and DNA is replicated.
2. Cell wall and plasma membrane begin to divide.
3. Cross-wall forms completely around divided DNA.
4. Daughter cells separate.

? How do the daughter cells produced by binary fission compare to each other and to the parent cell?

INFOGRAPHIC 13.5 How Bacterial Populations Acquire Genetic Variation

Asexually reproducing bacterial populations become genetically diverse by accumulating mutations and by picking up genes from other bacteria of the same or different species.

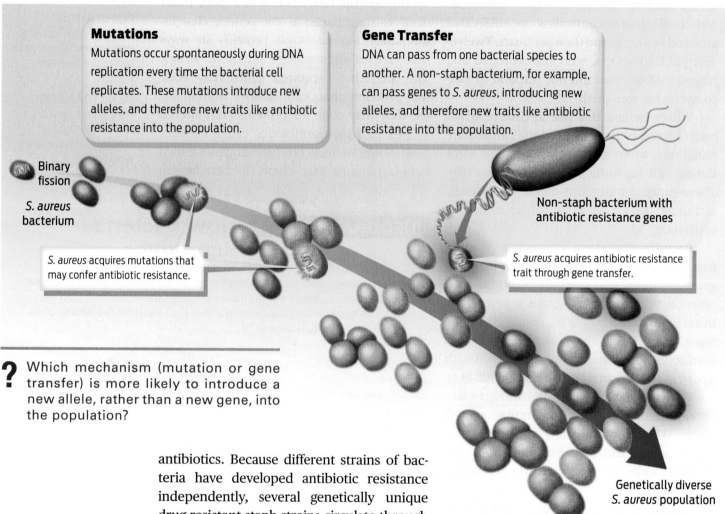

? Which mechanism (mutation or gene transfer) is more likely to introduce a new allele, rather than a new gene, into the population?

antibiotics. Because different strains of bacteria have developed antibiotic resistance independently, several genetically unique drug-resistant staph strains circulate through human communities at the same time.

An Evolving Enemy

While an individual bacterium—or any individual organism—can undergo genetic changes that give it new traits, this doesn't entirely explain how bacterial populations such as staph develop resistance to drugs. An entire population of organisms with a new trait can arise only when the environment favors that trait—that is, when the trait is advantageous to the organisms carrying it.

When a population's environment favors one trait over others, the frequency of alleles that code for that trait in the population changes over time. Take the trait for drug resistance, for example. As a result of random mutations and gene transfers happening all the time in any bacterial population, a genetically diverse population of bacteria will contain some individuals possessing alleles that confer resistance. In an environment free of antibiotics, any individual bacterium will have about the same chance of surviving and reproducing as any other, whether or not it carries a drug-resistance allele. In other words, the ability to resist antibiotics will confer neither an advantage nor a disadvantage, since there are no antibiotics around. But in the presence of an antibiotic, bacteria with an allele for drug resistance may survive, whereas other bacteria will die. The surviving bacteria reproduce and pass their alleles for drug resistance on to future generations. Consequently, the frequency of the resistance trait increases. This is how a population evolves. **Evolution** is defined as a change in the frequency of alleles in the population over time.

An organism's ability to survive and reproduce in a particular environment is called its **fitness.** The higher an organism's fitness, the more likely that alleles carried by that organism will be passed on to future generations and increase in frequency. In an environment in which antibiotics are abundant, drug-resistant bacteria are fitter than nonresistant bacteria (**INFOGRAPHIC 13.6**).

Currently, in the United States, antibiotic use is widespread. In 2013, doctors prescribed 268.6 million courses of antibiotics—equivalent to 849 prescriptions for every 1,000 people in the country. The abundance of antibiotics in the environment has created the perfect breeding ground for antibiotic-resistant staph.

In a different environment, however, one in which antibiotics are less common, these same resistant bacteria will not necessarily have an edge over other bacteria. In other words, fitness is always relative to the environment; organisms can be fit in one environment and not in another.

EVOLUTION
Change in allele frequencies in a population over time.

FITNESS
The relative ability of an organism to survive and reproduce in a particular environment.

Selecting for Superbugs

Ultimately, the interplay between an organism's traits—its phenotype—and its environment is what determines what traits will predominate in a population. When the environment favors the survival and reproduction of individuals with certain traits, those

INFOGRAPHIC 13.6 An Organism's Fitness Depends on Its Environment

The term "fitness" describes the relative ability of an organism to reproduce in a particular environment. Fitness is determined by the interaction between phenotype and environment. Antibiotic-resistant bacteria, for example, have high fitness in the presence of antibiotics.

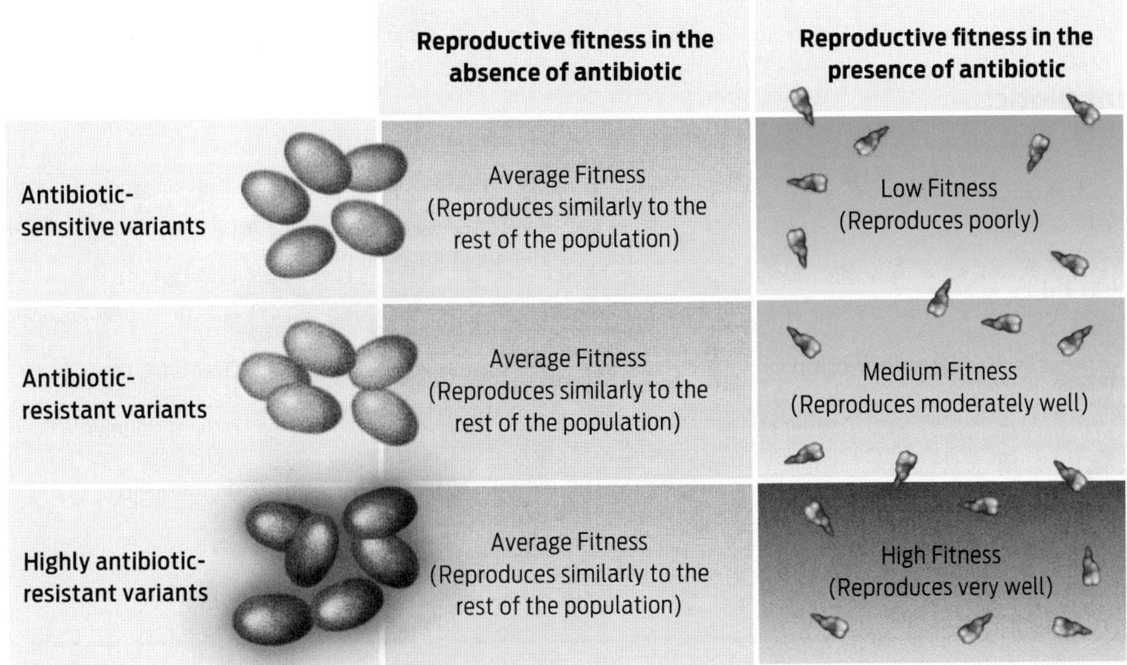

Reproductive fitness in the absence of antibiotic
Reproductive fitness in the presence of antibiotic
Antibiotic-sensitive variants
Average Fitness (Reproduces similarly to the rest of the population)
Low Fitness (Reproduces poorly)
Antibiotic-resistant variants
Average Fitness (Reproduces similarly to the rest of the population)
Medium Fitness (Reproduces moderately well)
Highly antibiotic-resistant variants
Average Fitness (Reproduces similarly to the rest of the population)
High Fitness (Reproduces very well)

? In the absence of antibiotics, is there a difference between the fitness of antibiotic-sensitive and antibiotic-resistant variants?

NATURAL SELECTION
Differential survival and reproduction of individuals in response to environmental pressure that leads to change in allele frequencies in a population over time.

traits become more common in the population. The differential survival and reproduction of individuals within a population in response to environmental pressure is called **natural selection.** Much in the way plant and animal breeders have for centuries practiced artificial selection to produce individuals with desired traits, the environment also, in a sense, selects individuals with certain traits.

When natural selection acts on a population over time, advantageous traits become more common, and the population becomes better suited to its environment. In other words, evolution by natural selection leads to **adaptation.** This is what we see with antibiotic-resistant bacteria: the population has become better suited, or adapted, to an environment in which antibiotics are

INFOGRAPHIC 13.7 Evolution by Natural Selection

In a typical population of organisms, individuals will vary genetically. When the environment favors some genetic variants over others, those variants will have higher fitness—they will survive better and reproduce more successfully. Over generations, the frequency of alleles that confer higher fitness will increase, while those that confer lower fitness will decrease. This nonrandom change in allele frequencies over generations is called evolution by natural selection.

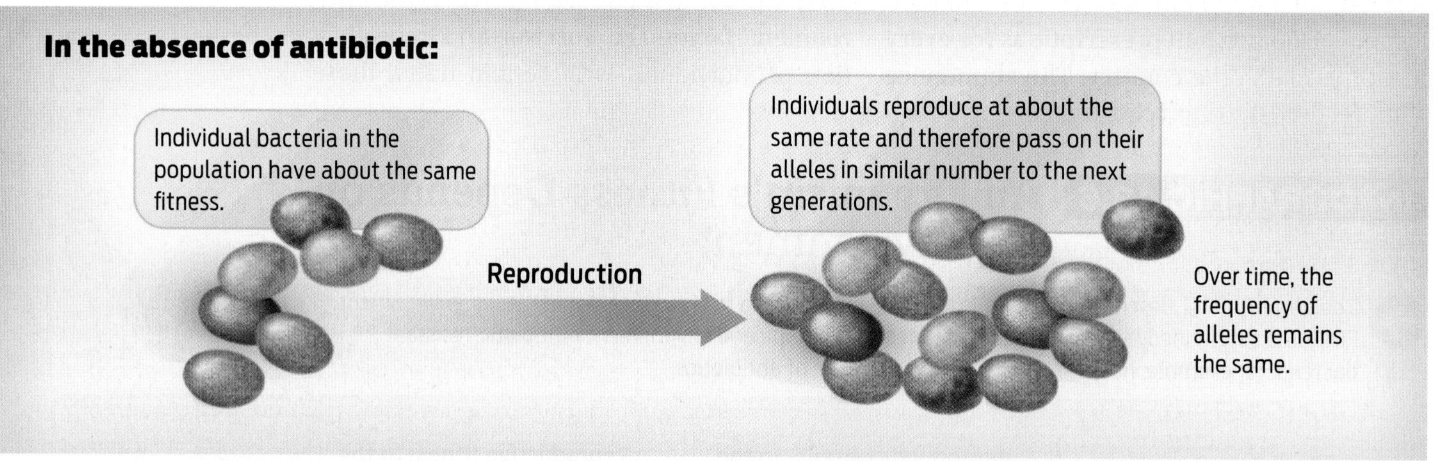

In the absence of antibiotic:
Individual bacteria in the population have about the same fitness.
Reproduction
Individuals reproduce at about the same rate and therefore pass on their alleles in similar number to the next generations.
Over time, the frequency of alleles remains the same.

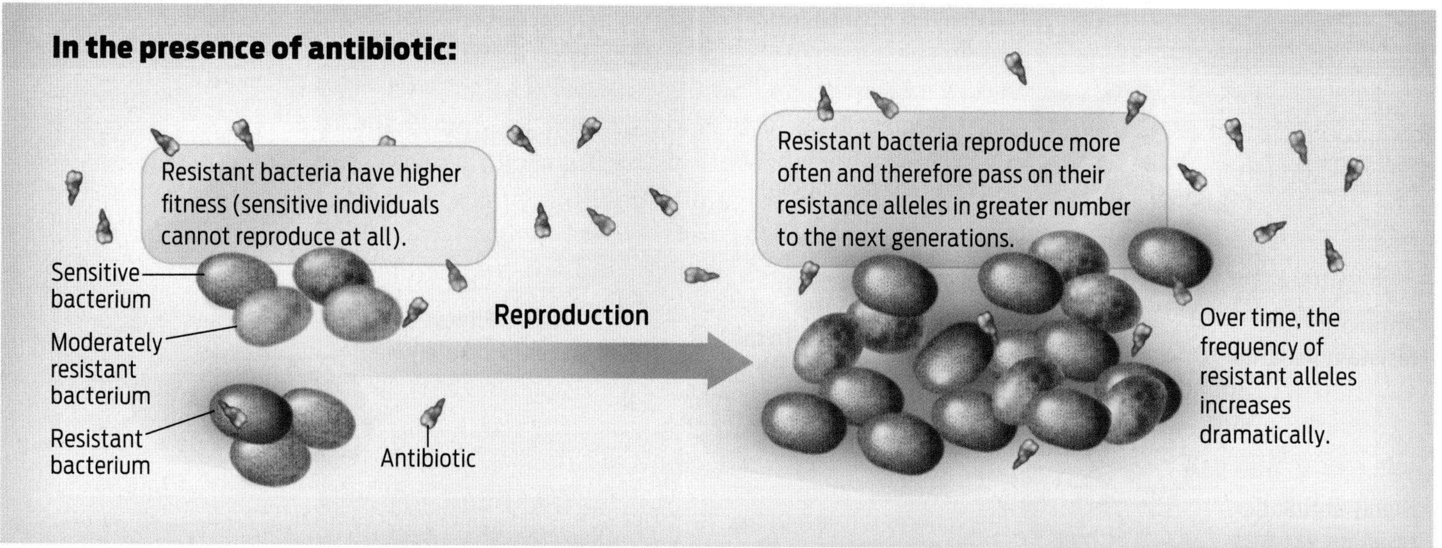

In the presence of antibiotic:
Resistant bacteria have higher fitness (sensitive individuals cannot reproduce at all).
Sensitive bacterium
Moderately resistant bacterium
Resistant bacterium
Antibiotic
Reproduction
Resistant bacteria reproduce more often and therefore pass on their resistance alleles in greater number to the next generations.
Over time, the frequency of resistant alleles increases dramatically.

? Some blue mussels can develop thicker shells because of a specific allele that controls shell thickness. Blue mussels with this allele are more likely to survive attacks by crabs. If crabs are introduced into a genetically diverse population of mussels, what will happen to the mussel population over time?

abundant. Charles Darwin was one of the first people to figure out how natural selection works and to study its results (see **Milestone 5: Adventures in Evolution**).

Note that evolution by natural selection occurs in populations, not individuals. Individual organisms do not experience a change in allele frequencies over time. Therefore, individual organisms do not evolve (**INFOGRAPHIC 13.7**).

Natural selection doesn't always affect populations in the same way. By studying how natural selection has shaped populations, scientists have identified three major patterns of natural selection. When the environment favors an extreme phenotype, causing the population to shift in one direction over time, **directional selection** has occurred. The emergence of antibiotic-resistant bacterial populations is a good example of directional selection. Another is fur color in rock pocket mice living in the American Southwest. Ancestral populations of rock pocket mice had light fur that blended in with the region's sandy soil. After a series of volcanic eruptions, black lava rocks were deposited throughout the area, and mice with darker fur have become more common–presumably because they are better able to evade predators.

When the environment favors the middle of the phenotypic spectrum, and extremes are selected against, we call this **stabilizing selection.** Human birth weight is an example. Both very small and very large babies are less fit than medium-size babies, who are hearty but still small enough to fit through the birth canal. Finally, when the environment favors the ends of a phenotypic spectrum, **diversifying selection** occurs. An example is finches in environments where only large seeds and small seeds are available; birds with medium-size beaks are not as successful at cracking either type of seed, and so big-beaked birds and small-beaked birds have the advantage (**INFOGRAPHIC 13.8**).

The particular pattern of natural selection a population follows depends on the interaction of phenotypes with the environment. So, for example, in the absence of antibiotics, populations of staph bacteria might have followed stabilizing or diversifying selection in response to some other pressure. Instead, the presence of antibiotics in the environment resulted in directional selection leading to the MRSA that killed Carlos Don, Rebecca Lohsen, and Ricky Lannetti.

MRSA in the Community

Drug-resistant staph strains first emerged in hospitals during the early 1960s, partly as a result of selection pressure from antibiotics. Since then, hospitals have remained hot spots for staph infections. The combination of heavy antibiotic use, lots of sick patients, and close quarters makes hospitals a fertile environment for the emergence of resistant bugs.

In response, many hospitals have implemented measures to reduce infections. For example, studies have shown that simply requiring all health care workers to wash their hands before handling each patient can dramatically reduce the number of infections. "Washing hands well and often is absolutely critical," says Ruth Lynfield, state epidemiologist and medical director of the Minnesota Health Department.

More alarming than MRSA infections in hospitals are MRSA infections in the community at large. Though previously almost unheard of, in the mid-1990s the number of community-acquired infections in the United States began to spike, says Daum, the University of Chicago microbiology professor. In a study that he published in 1998, Daum found that the prevalence of community-acquired MRSA infections requiring hospitalization increased from 10 cases per 100,000 hospital admissions in 1988-1990 to 259 cases per 100,000 admissions in 1993-1995–a more than 25-fold jump.

What happened in those years? New strains of *Staphylococcus aureus* emerged and flourished in community environments where antibiotic use is rampant. Daum and his

ADAPTATION
The process by which populations become better suited to their environment as a result of natural selection.

DIRECTIONAL SELECTION
A type of natural selection in which organisms with phenotypes at one end of a spectrum are favored by the environment.

STABILIZING SELECTION
A type of natural selection in which organisms near the middle of the phenotypic range of variation are favored by the environment.

DIVERSIFYING SELECTION
A type of natural selection in which organisms with phenotypes at both extremes of the phenotypic range are favored by the environment.

INFOGRAPHIC 13.8 Patterns of Natural Selection

There are three major patterns of natural selection, each resulting in a shift in the phenotype of the population. In all three cases, the environment is responsible for the specific pattern.

Directional selection
Occurs when a phenotype at one extreme is favored by the environment.

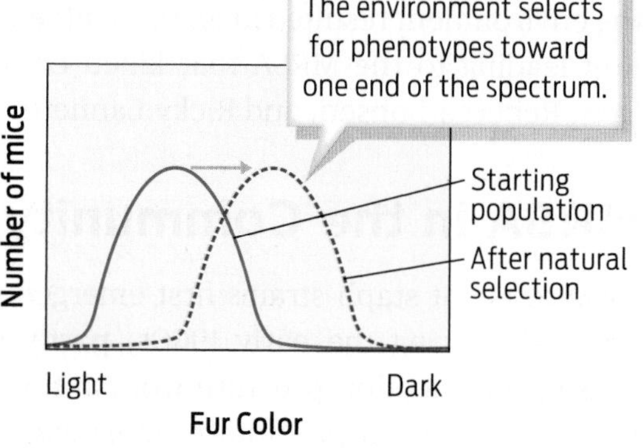

Example: Rock pocket mice fur color
In an environment covered with deposits of dark lava rocks, pocket mice with dark fur evade predators better than those with light fur. Dark mice become more frequent.

Hoekstra Lab, Harvard University

Stabilizing selection
Occurs when phenotypes at each end of the spectrum are less suited to the environment than individuals in the middle of the phenotypic range.

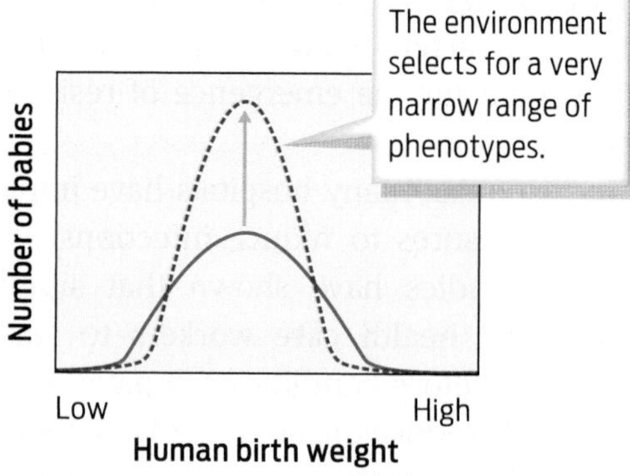

Example: Human birth weight
Human babies with very low birth weights do not survive as well as larger babies. Very large human babies are not easily delivered through the birth canal. Therefore, midrange babies are favored.

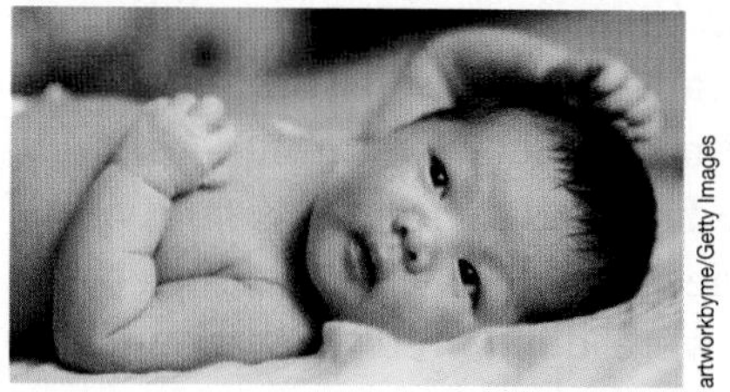
artworkbyme/Getty Images

Diversifying selection
Occurs when the extremes of the phenotypic range are better suited to the environment than individuals in the middle of the phenotypic range.

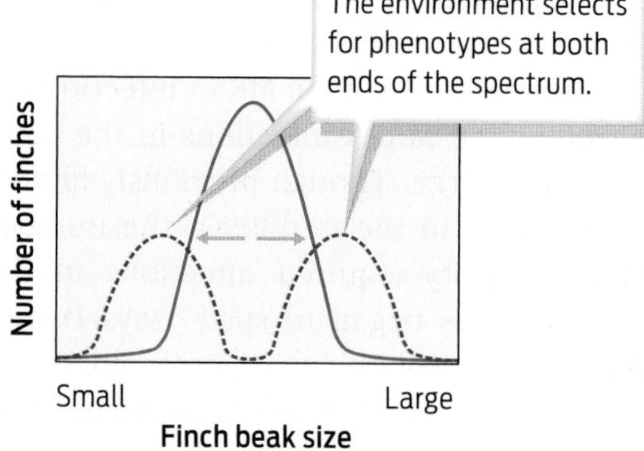

Example: Finch beak size
Finches living in an environment where only large, hard seeds and small, soft seeds are available are more successful if they have either large or small beak sizes. Medium beaks are not as successful at cracking either type of seed.

Alphotographic/Getty Images

McPHOTO/AGE Fotostock

? An aquatic environment has light rocks and deep shadows. What pattern of selection will allow oysters to hide from predators in this environment?

colleagues recently showed that a strain called USA300 is more virulent than other MRSA strains. "It appears to be juiced up," he says. Many USA300 genes are expressed at high levels. One gene in particular that controls expression of a number of staph toxins, which can damage cells and tissues, is turned on all the time.

The result is that people infected with these strains have more-severe disease, including necrotizing fasciitis, in which the bacteria literally eat through skin and soft tissues. These "superbugs" can also kill more quickly. Symptoms can appear so suddenly that, according to Daum, "You could be healthy at 1:00 in the afternoon and be dead by 1:00 in the morning."

It was likely USA300 that killed Carlos Don, Rebecca Lohsen, and Ricky Lannetti. "I've been an infectious disease guy for over 20 years now and we didn't talk about staph necrotizing pneumonia like we do now," says Daum.

Even more troubling, staph is continuing to evolve. There is evidence that when strains that are prevalent in the community mix with strains that are prevalent in hospitals, the risk increases that a staph strain even more virulent than either will emerge.

> "We really have to be careful about how we use antibiotics because antibiotic use is the biggest driver of antibiotic resistance."
> — Ruth Lynfield

Stopping Superbugs

Staph bacteria aren't the only ones that have grown resistant to antibiotics. It is getting harder to treat patients with severe *Salmonella* food poisoning caused by drug-resistant strains of this bug. *Neisseria gonorrhoeae*, the bacterium that causes gonorrhea, has become resistant to another important group of antibiotics, the fluoroquinolones. And there are now forms of pneumonia caused by strains of *Klebsiella* that are resistant to every available antibiotic.

Because the very use of antibiotics drives bacterial populations to evolve resistance, antibiotic resistance is inevitable. But humans have hastened the emergence of drug-resistant strains of bacteria by the haphazard use and overuse of antibiotics. From the moment antibiotics were first introduced, physicians began prescribing them for colds, coughs, and earaches, most of which are caused by viruses, rather than bacteria, and so aren't killed by antibiotics anyway.

Doctors aren't the only culprits. Approximately 80% of the antibiotics sold in the United States are used for livestock. Farmers give antibiotics in low doses to poultry, swine, and cattle to promote growth and to help prevent infections in crowded, unsanitary conditions. This practice can cause food-borne pathogens such as *Salmonella* or *Campylobacter* to develop antibiotic resistance. Eating improperly prepared foods from these animals can result in dangerous infections in humans, which then must be treated with even more antibiotics. What's more, undigested antibiotics in animal manure can leach into the environment through groundwater or when manure is used as fertilizer. In this environment drug-resistant bacteria are more fit and will thus become more prevalent over time through natural selection.

"We really have to be careful about how we use antibiotics because antibiotic use is the biggest driver of antibiotic resistance," says Lynfield, medical director of the Minnesota Health Department.

Clearly, developing stronger antibiotics isn't the only or the best solution to the problem of resistance because bacteria will ultimately adapt to those, too. Perhaps the best way to control resistance, say experts, is to change practices that enable resistant strains to thrive. It is critical, for example, that when an antibiotic is prescribed it is taken precisely as directed, for the full course of treatment, no matter how much better the patient may be feeling. If bacteria

Preventing and Treating Infection by Antibiotic-Resistant Bacteria

Calllo Images - Farming SA/Getty Images

Reduce antibiotics in livestock feed. Excessive antibiotics in the environment create continuous selective pressure for all bacteria.

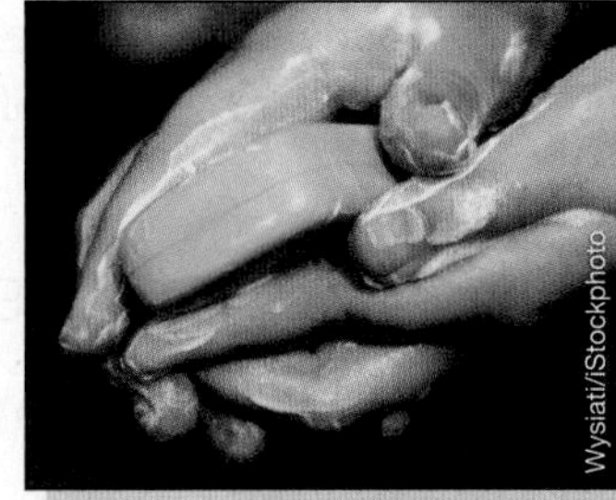
Wysiati/iStockphoto

Wash hands frequently. Both regular and antibacterial soap, as well as hand sanitizer, are effective at preventing the spread of bacterial infection.

walik/Getty Images

Keep locker rooms and sports equipment clean. Protect athletes from contact with contaminated surfaces.

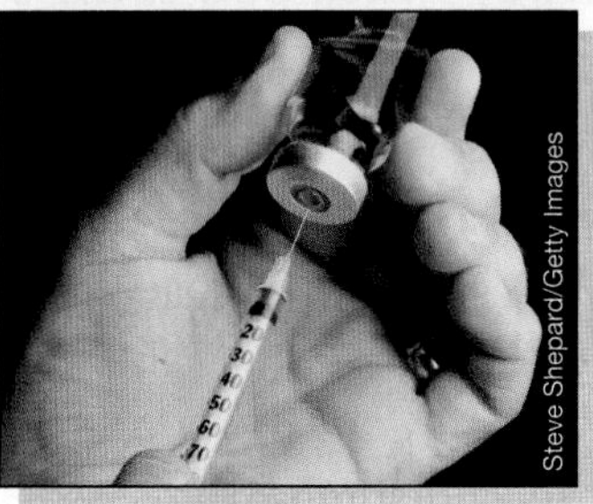
Steve Shepard/Getty Images

Research new vaccines. Vaccines prevent resistant bacterial strains from making people ill.

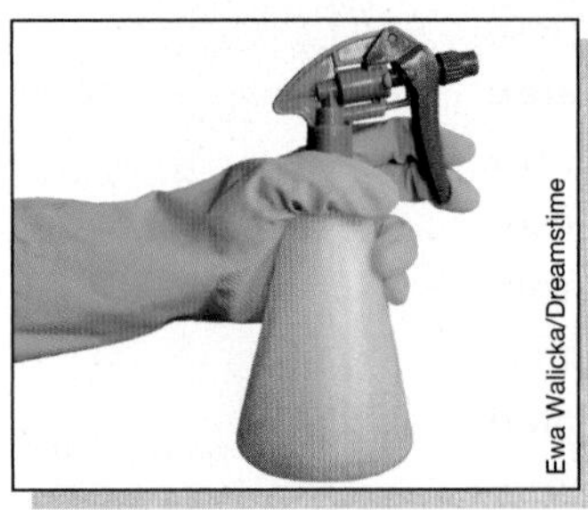
Ewa Walicka/Dreamstime

Disinfect common surfaces. Disinfection reduces the transmission of infection by contact with contaminated surfaces, especially in facilities that serve a lot of people.

Christine Schuhbeck/AGE Fotostock

Do not take antibiotics for viral infections. Viruses are not killed by antibiotics. Overuse of antibiotics causes resistant bacterial strains to become widespread.

are exposed to antibiotics at low levels or for short durations, the entire population may not be eradicated. The remaining bacteria may be resistant to the antibiotic and proliferate. And anyone taking antibiotics exposes all the bacteria in his or her body to the antibiotics, possibly enabling other drug-resistant bacteria to emerge. These drug-resistant bacteria might then be transmitted to other people.

At the community level, the more antibiotics are used, the more resistance will emerge. So doctors are heavily discouraged from prescribing antibiotics unnecessarily. And efforts are being made to crack down on the practice of feeding livestock low levels of antibiotics.

Of course, these measures won't fight resistant strains that are already circulating. But there are ways to reduce and perhaps prevent infections in this case, too. Careful hygiene is one important tool. A vaccine would be another way to prevent staph infections. A vaccine against *Streptococcus pneumoniae* introduced in 2000 caused the rate of infection—and especially the rate of drug-resistant infections—to drop dramatically. And not only did the rate of infection drop in vaccinated children, but other age groups benefited as well because the bacteria were not being transmitted as frequently. As another example of the impact that vaccines have on infections, Daum points to the bacterium *Haemophilus influenzae*, which frequently caused pneumonia, meningitis, and other serious diseases in children. Today, children are vaccinated against it. "When I was an intern we used to see 60 to 80 *Haemophilus* infections a month," he says. "Today we see none, it's gone. And MRSA needs to be gone, too." ■

CHAPTER 13 SUMMARY

- Populations are groups of individuals of the same species living together in the same geographic area.
- Populations of bacteria exist nearly everywhere, including on and in our bodies; most are harmless or even beneficial, but some can cause disease.
- Within any population, there is genetic variation among individuals. This variation exists in the form of different alleles that arise randomly by mutation.
- Bacterial populations can acquire genetic variation by both mutation and gene transfer—the direct exchange of DNA between bacterial cells.
- Genetic variation in a population gives rise to corresponding phenotypic variation in the population.
- Individuals with different phenotypes will have differing ability to survive and reproduce in a population; that is, they will have different fitness.
- The differential survival and reproduction of individuals in a population over time in response to environmental pressure is called natural selection.
- Natural selection is one cause of evolution, which is defined as a change in the allele frequency of a population over time.
- Individuals with higher fitness in a given environment reproduce and pass on their alleles more frequently than do individuals with lower fitness, resulting in evolution by natural selection.
- Over time, natural selection leads to adaptation: advantageous traits become more common in the population, which as a result becomes more suited, or adapted, to its environment.
- Natural selection can take several forms that shift phenotypes in a population in distinct ways: directional selection, diversifying selection, or stabilizing selection.
- Antibiotic-resistant populations of bacteria emerge by directional selection in the presence of antibiotics.

MORE TO EXPLORE

- **The Evolution of Bacteria on a "Mega-Plate" Petri Dish (Harvard Medical School, 2016): www.youtube.com/watch?v=plVk4NVIUh8**
- **Centers for Disease Control and Prevention, Antibiotic Resistance: www.cdc.gov/drugresistance/**
- **Herold B. C., et al. (1998) Community-acquired methicillin-resistant *Staphylococcus aureus* in children with no identified predisposing risk. *JAMA* 279(8): 593–598.**
- **McKenna, M. (2011) *Superbug: The Fatal Menace of MRSA*. New York: Free Press.**
- **Rock Pocket Mice animation: http://learn.genetics.utah.edu/content/selection/comparative/**

CHAPTER 13 Test Your Knowledge

DRIVING QUESTION 1 What is staph, and can a person have it in the absence of an infection?

By answering the questions below and studying Infographics 13.1 and 13.2, you should be able to generate an answer for the broader Driving Question above.

KNOW IT

1 Can *S. aureus* be present in or on a person who has no evidence of an infection?

a. no; S. *aureus* is associated only with infections
b. yes, but only non-MRSA strains are present in the absence of an infection
c. yes, but only for very short periods of time (between touching a contaminated surface and washing the hands)
d. yes; *S. aureus* is a common skin bacterium
e. yes; *S. aureus* is a common bacterium found in the bloodstream

2 The term "MRSA" as it is used today refers to

- **a.** *S. aureus* bacteria that are resistant to many antibiotics.
- **b.** a collection of skin and other infections caused by a type of bacteria.
- **c.** *S. aureus* bacteria that are found only in humans with certain types of skin infections.
- **d.** *S. aureus* bacteria that are normal residents of human skin in the vast majority of the human population.
- **e.** all bacteria that are resistant to antibiotics.

3 What is the difference between an *S. aureus* colonization and an *S. aureus* infection?

4 MRSA is most likely to be problematic if found

- **a.** on the surface of the skin.
- **b.** in nasal passages.
- **c.** in the bloodstream.
- **d.** on the fingernails.
- **e.** The presence of MRSA in any of those locations indicates a serious infection.

USE IT

5 An athlete has a nasty skin infection caused by MRSA. How might this infection have been contracted?

6 For the patient in Question 5, which general kinds of antibiotics would you choose (or avoid) in treating the infection? What other measures would you recommend to prevent spread of MRSA to the athlete's teammates and family? Explain your answer.

DRIVING QUESTION 2 How do bacteria resist the effects of antibiotics?

By answering the questions below and studying Infographics 13.3 and 13.5, you should be able to generate an answer for the broader Driving Question above.

KNOW IT

7 In the presence of penicillin:

- **a.** What happens to a penicillin-sensitive strain of *S. aureus*?
- b. What happens to a penicillin-resistant strain of *S. aureus*?

8 How do beta-lactam antibiotics kill sensitive bacteria?

- **a.** by attracting water into cells
- **b.** by destabilizing the cell membrane
- **c.** by preventing DNA replication during bacterial reproduction
- **d.** by destabilizing the cell wall
- **e.** all of the above, depending on the specific strain of bacteria

USE IT

9 Why do beta-lactam antibiotics affect sensitive bacterial cells but not eukaryotic cells? (You may want to review cell structure, discussed in Chapter 3, to answer this question.)

10 A sensitive *S. aureus* bacterium acquires a new gene that allows it to resist the effects of beta-lactam antibiotics (that is, the bacterium is now resistant). What might the protein encoded by that gene do?

- **a.** synthesize beta-lactam antibiotics
- **b.** digest beta-lactam antibiotics
- **c.** produce a toxin
- **d.** enhance colonization of human skin
- **e.** enhance entry into the bloodstream

DRIVING QUESTION 3 How do populations evolve, and what is the role of evolution in antibiotic resistance?

By answering the questions below and studying Infographics 13.4, 13.5, 13.6, 13.7, and 13.8, you should be able to generate an answer for the broader Driving Question above.

KNOW IT

11 What are the two major mechanisms by which bacterial populations generate genetic diversity?

- **a.** mutation and meiosis
- **b.** binary fission and evolution by natural selection
- **c.** gene transfer and mutation
- **d.** mutation and binary fission
- **e.** gene transfer and replication

12 What is the environmental pressure in the case of antibiotic resistance?

- **a.** the growth rate of the bacteria
- **b.** how strong or weak the bacterial cell walls are
- **c.** the relative fitness of different bacteria
- **d.** the presence or absence of antibiotics in the environment
- **e.** the temperature of the environment

13 What is the evolutionary meaning of the term "fitness"?

14 The evolution of antibiotic resistance is an example of

- **a.** directional selection.
- **b.** diversifying selection.
- **c.** stabilizing selection.
- **d.** random selection.
- **e.** steady selection.

15 In humans, very-large-birth-weight babies and very tiny babies do not survive as well as midrange babies. What kind of selection is acting on human birth weight?

- **a.** directional selection
- **b.** diversifying selection
- **c.** stabilizing selection
- **d.** random selection
- **e.** steady selection

USE IT

16 Binary fission is asexual. What does this mean? How could two daughter cells end up with different genomes at the end of one round of binary fission?

17 In what sense do bacteria "evolve faster" than other species?

apply YOUR KNOWLEDGE

INTERPRETING DATA

18 A single *S. aureus* cell gets into a wound on your foot. *S. aureus* divides by binary fission approximately once every 30 minutes.

a. Thirty minutes after the initial infection, how many *S. aureus* cells will be present?

b. In 1 hour, how many *S. aureus* cells will be present?

c. In 12 hours, how many *S. aureus* cells will be present? (Hint: The general formula is $2^{\text{number of generations}}$; you need to figure out how many generations occurred in 12 hours.)

d. Mutations occur at a rate of 1 per 10^{10} base pairs per generation. *S. aureus* has 2.8×10^6 base pairs in its genome. Therefore, approximately 0.0028 mutations will occur per cell in the population. At the end of 12 hours, how many mutations will be present in the population of *S. aureus* in the wound in your foot? What are the implications of this genetic diversity in the context of treating a possible infection?

19 If we take the fittest bacterium from one environment—one in which the antibiotic amoxicillin is abundant, for example—and place it in an environment in which a different antibiotic is abundant, will it retain its high degree of fitness?

a. yes; fitness is fitness, regardless of the environment

b. yes; once a bacterium is resistant to one antibiotic it is resistant to all antibiotics

c. not necessarily; fitness depends on the ability of an organism to survive and reproduce, and it may not do this as well in a different environment

d. no; what is fit in one environment will never be fit in another environment

It was on a short-cut through the hospital kitchens that Albert was first approached by a member of the Antibiotic Resistance.

20 If a single bacterial cell that is sensitive to an antibiotic—for example, vancomycin—is placed in a growth medium that contains vancomycin, it will die. Now consider another single bacterial cell, also sensitive to vancomycin, that is allowed to divide for many generations to become a larger population. If this population is placed into vancomycin-containing growth medium, some bacteria will grow. Why do you see growth in this case, but not with the transferred single cell?

21 Imagine that a genetically diverse population of garden snails occupies your backyard, in which the vegetation is many shades of green with some brown patches of dry grass.

a. If birds like to eat snails, but they can see only the snails that stand out from their background and don't blend in, what do you think the population of snails in your backyard will look like after a period of time? Explain your answer.

b. Suppose you move the population of snails to a new environment, one with patches of dark brown pebbles and patches of yellow ground cover. Will individual snails mutate to change their color immediately? As the population evolves and adapts to the new environment, what do you predict will happen to the phenotypes in your population of snails after several generations in this new environment? How did this occur? Include the terms *gametes, mutation, fitness, phenotype,* and *environmental selective pressure* in your answer.

apply YOUR KNOWLEDGE

MINI CASE

22 Your friend has had a virus-caused cold for 3 days and is still so stuffy and hoarse that he is hard to understand. He seems to be telling you that his doctor called in a prescription for an antibiotic for him to pick up at his pharmacy. You hope that you misunderstood him, but you realize that you heard him perfectly well.

a. Will the antibiotic help your friend's cold?

b. What are the risks to your friend if he takes the antibiotic? (Think about what might happen if he should develop a wound infection in the future.)

c. Your friend is a wrestler. What are the risks to his teammates or competitors if he takes the antibiotic?

apply YOUR KNOWLEDGE

BRING IT HOME

23 Your roommate has been prescribed an antibiotic for bacterial pneumonia. She is feeling better and stops taking her antibiotic before finishing the prescribed dose, telling you that she will save the remainder to take the next time she becomes sick. What can you tell your roommate to convince her that this is not a good plan?

5

Milestones in Biology

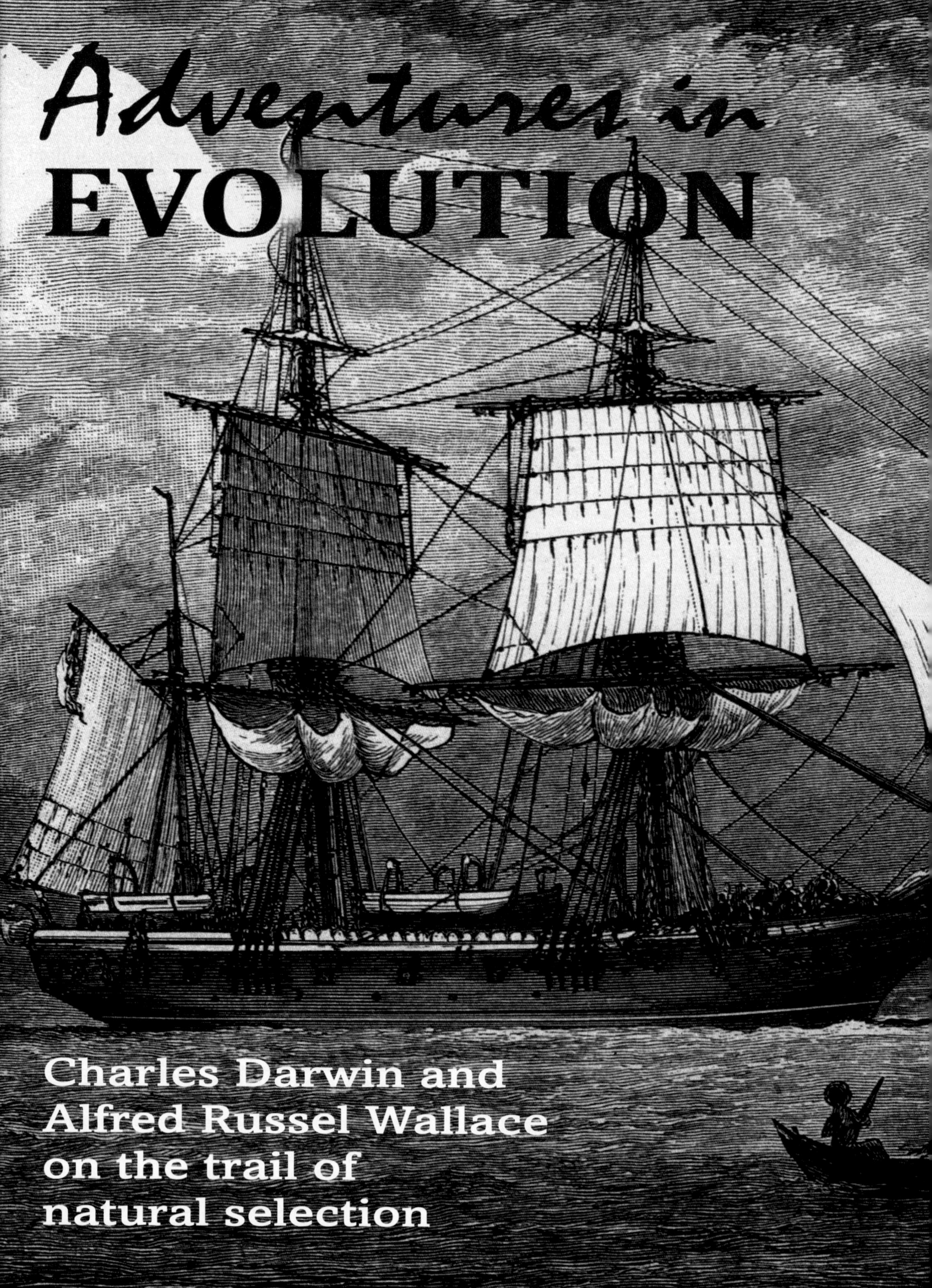

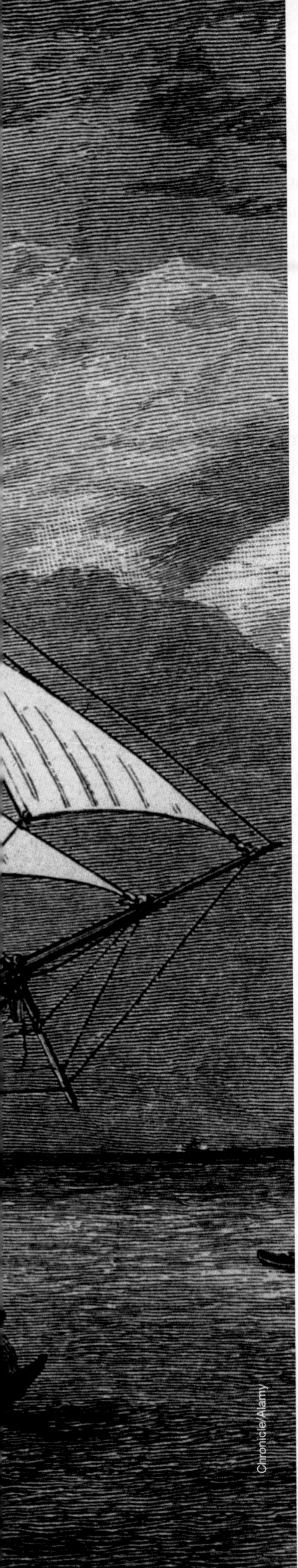

DRIVING QUESTIONS

1. What observations did Darwin make about nature that helped shape his thinking about evolution?
2. What works by other scientists shaped Darwin's thoughts about evolution?

> "The voyage of the *Beagle* has been by far the most important event in my life and has determined my whole career."
>
> — Charles Darwin

During his last term at Cambridge, Charles Darwin faced a dilemma: what should he do with himself after graduation? He'd considered becoming a physician, like his father. But the sight of blood made him queasy and he hated rote memorization. He changed his focus to theology, intending to become a clergyman, but his real passion was bug collecting. Only that wasn't going to pay the bills.

Then a professor told him about the internship of a lifetime: a 5-year, around-the-world trip as a naturalist aboard a British surveying ship, the HMS *Beagle*. The ship's captain, Robert FitzRoy, wanted a travel companion who would also collect specimens along the way. Unsure what he wanted to do with his life but eager to see the world, the 22-year-old Darwin jumped at the chance. He later said of the trip, "The voyage of the *Beagle* has been by far the most important event in my life and has determined my whole career."

Yet he almost didn't go. Darwin's father, Robert Darwin, thought his son should buckle down and prepare to enter the clergy. A trip around the world seemed to him a useless distraction—a "wild scheme," he called it—and he refused at first to let his son go. But eventually, at the cajoling of his family, he relented. Charles packed his bags, said goodbye to his girlfriend, Emma, and set sail for South America. It was December 1831.

The passage aboard the 90-foot vessel was frequently harrowing, and Darwin suffered debilitating bouts of seasickness, but his journey aboard the *Beagle* set in motion one of the great revolutions in science. What he saw on that trip planted the seeds of ideas that have completely changed the way we view the world and our place in it. As the evolutionary biologist Stephen Jay Gould put it, "The world has been different ever since Darwin."

> "The world has been different ever since Darwin."
>
> — Stephen Jay Gould

Journey to an Idea

Though Darwin is the most famous figure associated with the theory of evolution, he did not invent the idea. Nor was he alone among his contemporaries in studying it. In fact, the notion that species change gradually over time had been around for generations. To be sure, most people in the 1830s—Darwin included—still assumed that species were fixed and unchanging, created perfectly by God. But evidence to the contrary had been accumulating for some time. Explorers and naturalists were traveling to faraway lands and finding unusual plants and animals they had never seen before. Fossils were being uncovered, providing evidence that some species no longer seen on Earth had lived in the past. And anatomists were noting uncanny physical resemblances between different species, including chimpanzees and humans. Evolution was in the air when Darwin began thinking about it.

However, the ideas that people in Darwin's time had proposed to explain *how* species changed were flawed. One common misconception was Lamarckism, named after the French naturalist Jean-Baptiste Lamarck, who suggested that species could change through the inheritance of acquired characteristics. In the Lamarckian view, giraffes, for example, developed their long necks by continually stretching them to feed on tall trees. Once it acquired its long neck, a giraffe could then pass that advantageous trait on to its offspring. This idea of the inheritance of acquired characteristics, while incorrect, was a popular one in Darwin's time—one that even Darwin himself found it hard to fully shake off in his writings. (Mendel's work on pea plants, which helped establish our modern view of inheritance, was not rediscovered until 1900; see **Milestone 4: Mendel's Garden.**) **(INFOGRAPHIC M5.1)**

Though it would be years before Darwin proposed his own theory of evolution, his trip around the world provided him with an indispensable foundation. He kept a diary of his adventures, which was later published as *The Voyage of the Beagle* (1839).

While at sea, Darwin had plenty of time to read and think about the ideas then being discussed in scientific circles. He read, for instance, the work of Charles Lyell, whose *Principles of Geology* (1830-1833) argued that Earth was much older than the 6,000 years popularly accepted at the time (a figure based on a literal reading of the Bible), and that its geology had been shaped entirely by incremental forces operating over a vast expanse of time. Valleys, for example, were formed by the slow grinding forces of wind and water, not by catastrophic floods; mountains were pushed up gradually by the action of volcanoes and earthquakes. Lyell's view of incremental change producing dramatic results over great spans of time left an indelible impression on Darwin.

INFOGRAPHIC M5.1 Lamarckism: An Early Idea about Evolution

Lamarck believed that traits acquired in one's lifetime could be passed on to offspring. We now know this to be wrong.

Jean-Baptiste Lamarck
August 1, 1744–December 28, 1829
Bettmann/Getty Images
Reproduction
If a giraffe strained its neck to eat leaves from higher branches, then over time it would develop a longer neck.
It would then pass this acquired trait on to its offspring, and they would be born with long necks.

? In the Lamarck model, if you lift a lot of weights and develop a very muscular physique, would you expect your children to be equally "buff"?

Not long into his trip, he saw something that seemed to confirm Lyell's view. While docked at the Cape Verde islands, 300 miles off the west coast of Africa, he noticed a white layer of compressed seashells and corals embedded in a rock face 30 feet above sea level. This layer had clearly been formed in the sea, but had somehow been lifted up out of the water. Later on his trip, Darwin witnessed how this might happen. After an earthquake devastated the town of Concepción, Chile, he noticed that the coastline had been pushed up several feet, as shown by the rim of mollusks and barnacles that hovered out of the water around the bay. Lyell was right.

With such thoughts of an ancient, slowly changing Earth on his mind, Darwin studied the plants, animals, and geology at each stop on his trip, collecting fossils and specimens of local flora and fauna wherever he went (**INFOGRAPHIC M5.2**).

While exploring the shore of Argentina in August 1833, Darwin unearthed a particularly prized find: the fossilized remains of several large mammals embedded in a sea cliff, including one that looked like a giant armadillo and another that resembled a giant sloth. These animals had clearly lived long ago and were now extinct, since such oversize mammals no longer roamed the plains of Argentina. And yet the ancient creatures bore a striking resemblance to the much smaller armadillos and sloths that were indigenous to the area. Why should animals separated by such vast epochs of time share such similar anatomical structures? To Darwin, this suggested that these creatures were ancestrally related to one another, and also that the species had changed gradually over time. Darwin was beginning to question the conventional wisdom of his day.

In 1835, the young naturalist stepped ashore on the Galápagos Islands, off the coast of Ecuador. On this archipelago, or island chain, Darwin observed and collected many creatures, among them a variety of small birds. Months later, while studying the specimens back in England, he learned that they were all species of finch. Each species was distinguishable by a different size and shape of beak, but they all bore a family resemblance. Moreover, the different beak shapes–small and pointy, large and sturdy, for example–related to the different type of seeds and other food sources available on the particular island where the birds had been found. He later wrote in *The Voyage of the Beagle*, "One might really fancy that, from an original paucity of birds in this archipelago, one species had been taken and modified for different ends." An ancestral ground-dwelling, seed-eating finch, for example, may have arrived from South America and

INFOGRAPHIC M5.2 Darwin's Voyage on the Beagle

Darwin's 5-year-long trip around the globe aboard the HMS *Beagle* laid the foundation for his thinking about evolution. The plants, animals, fossils, and geological forces he encountered along the way led him to question the conventional wisdom of his day.

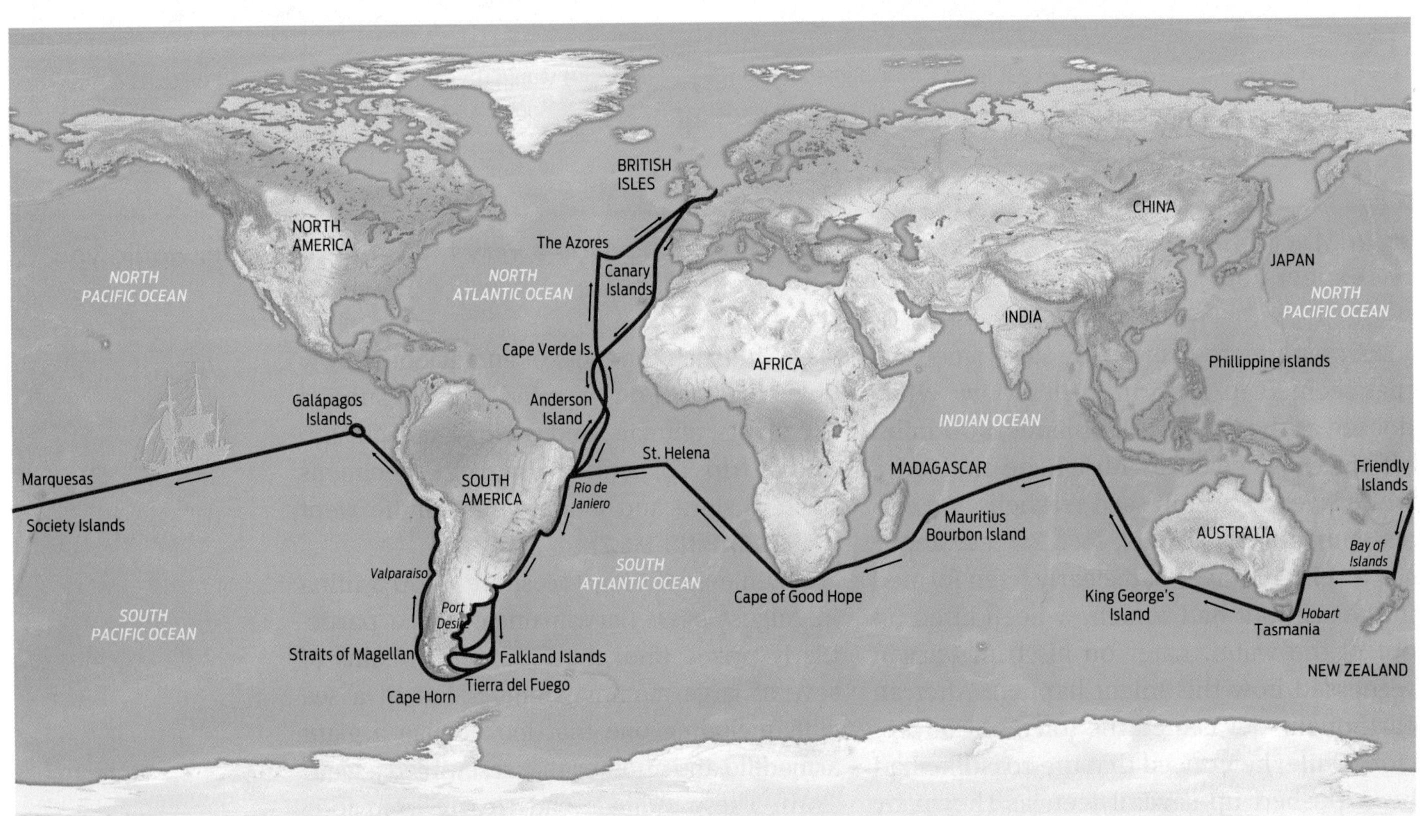

? Which parts of the globe were not part of Darwin's voyage?

colonized one island in the chain. Over time, as its descendants traveled to other islands, the birds developed new habits–including living in trees and eating different types of seeds and insects. Their beaks reflected these new habits.

This notion of one species giving rise over time to new species Darwin came to call "descent with modification," which he represented in his notebooks by a diagram that looked like a branching tree. (Darwin himself didn't like using the term "evolution" because he thought it gave a mistaken idea of progress or direction toward a goal, but "evolution" is the term that stuck.)

After 5 years traveling the globe, Darwin returned home to England in October 1836, but his intellectual journey was only just beginning. He began to think more seriously about how species might change over time. A key insight came to him in September 1838 while reading the work of the political economist Thomas Malthus, whose pessimistic book *An Essay on the Principle of Population* (1798) described how hunger, starvation, and disease would ultimately limit human population growth. The same must be true of plant and animal species, Darwin realized. If every individual in a population reproduced, even in a slowly reproducing population such as elephants, the world would be completely overrun with elephants in not that many generations. Since Earth is not overrun with elephants, factors must be limiting their population growth. Such limitations, Darwin reasoned, would lead to competition for resources that would put weaker individuals at a disadvantage. "It at once struck me," Darwin later wrote in his autobiography "that under these circumstances favourable variations would tend to be preserved, and unfavourable ones to be destroyed. The result of this would be the formation of new species."

> "It at once struck me that under these circumstances favourable variations would tend to be preserved, and unfavourable ones to be destroyed."
>
> — Charles Darwin

These favorable variations needn't be very pronounced, he realized. All that was needed was for certain individuals to have a slight edge over others in the competition to survive and reproduce. These helpful variations would then be inherited by offspring and become more common in the population. In effect, the environment was "selecting" for favorable traits, much as plant and animal breeders selected and perpetuated desirable varietals–a plant with especially large fruit, for instance, or the many breeds of dogs we see today.

This idea of "natural selection" (see Chapter 13) was Darwin's original contribution to the theory of evolution. Others had speculated at length about species change, but Darwin was the first to provide a clear mechanism of evolution. The philosopher of science Daniel Dennett has called natural selection "the single best idea anyone has ever had."

Among other things, natural selection provided a powerful explanation for the apparent design seen in nature. Before Darwin, the conventional view was that provided by William Paley, whose book *Natural Theology* (1802) Darwin had read in college. In that work, Paley made his famous "argument from design." That is, many creatures are so finely constructed and well adapted to their environment that their existence implies the existence of a designer–much in the way that the existence of a watch implies the existence of a watchmaker. To many

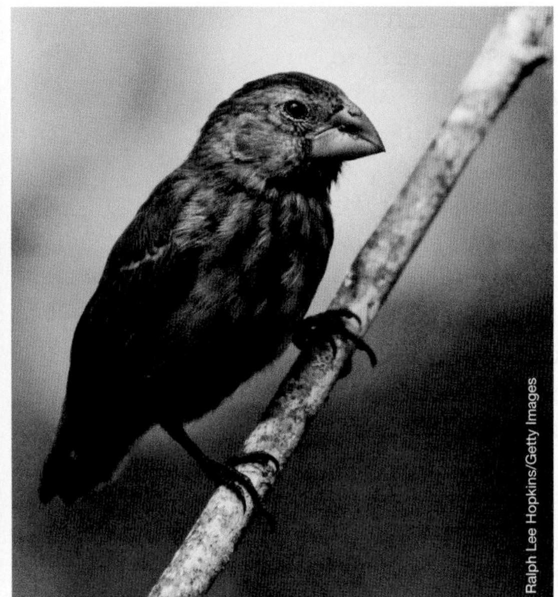

Ralph Lee Hopkins/Getty Images

The Large Ground Finch (*Geospiza magnirostris*), one of several finch species that Darwin studied, has a sturdy bill for cracking large seeds.

Danita Delimont/Getty Images

Barnacles are invertebrate animals that live attached to surfaces such as rocks near ocean shores. These plentiful and diverse creatures helped Darwin understand the importance of variation in nature.

people in the 19th century, that designer was God. With the idea of natural selection, Darwin did away with the need for a watchmaker. Natural selection could accomplish the same thing without any input from a designer.

By 1844, Darwin had developed his ideas into a 200-page manuscript that he hoped would be the definitive word on the subject. He did not rush his ideas about natural selection into print, however. He knew that his ideas would be controversial, contradicting as they did strongly held beliefs about God and the special creation of all animals, including humans. Other scientists with evolutionary ideas were causing quite a stir in England and being openly ridiculed. Even sharing his theory of evolution by natural selection with trusted colleagues, Darwin said, was "like confessing a murder." To withstand challenges, he knew he would need more detailed evidence.

And so, at age 37, Darwin began to investigate closely one large group of animals: barnacles, the small invertebrates that cling to ships or marine life. Darwin spent 8 years, from 1846 to 1854, carefully cataloguing the barnacles' tiny features, comparing and contrasting them with those of other known invertebrates. It was tedious work, leading Darwin to write, "I hate a Barnacle as no man ever did before." Yet the work proved valuable, for it reinforced his idea that a great deal of variation exists in nature–barnacles are nothing if not diverse–and it provided ample evidence of descent with modification since the creatures clearly shared adaptations with other invertebrates (**INFOGRAPHIC M5.3**).

In the summer of 1858, Darwin was hard at work on his "species" book when he received a letter from a young naturalist with whom he had a casual acquaintance, a collector named Alfred Russel Wallace who made a living selling rare butterflies and birds to other collectors and museums. The envelope was postmarked from an island in Indonesia. Inside was a 20-page manuscript describing the author's bold new idea about how species change over time, which he wanted Darwin to read and have published. Darwin had been scooped.

INFOGRAPHIC M5.3 The Evolution of Darwin's Thought

Darwin was influenced by the work of others, which informed the way he interpreted his own research and collections.

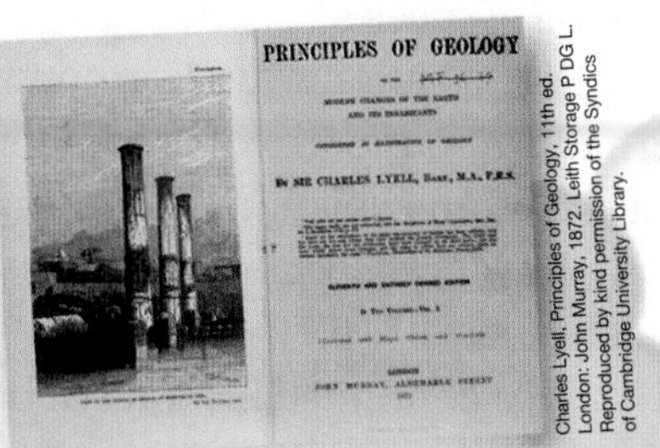

Charles Lyell, Principles of Geology, 11th ed. London: John Murray, 1872. Leith Storage P DG L. Reproduced by kind permission of the Syndics of Cambridge University Library.

Lyell's work (1833)

- Earth's geology is formed by slow-moving forces.
- Earth is much older than thought at the time.

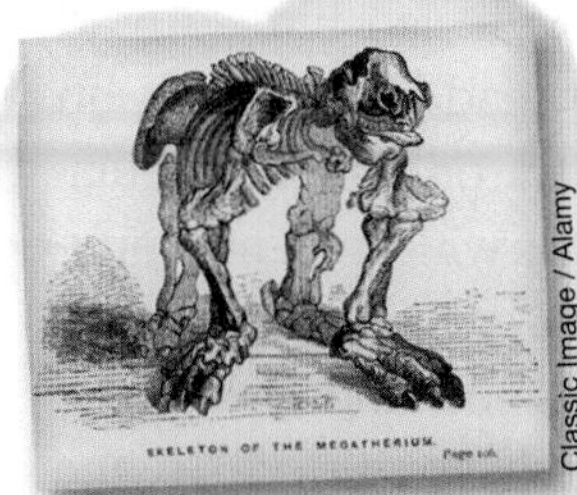

Classic Image / Alamy

Fossil of giant land sloth (1833)

- Buried in a sediment layer below a deposit of shells.
- The large creature was clearly extinct, but it resembled smaller, living South American animals. Could they be related?

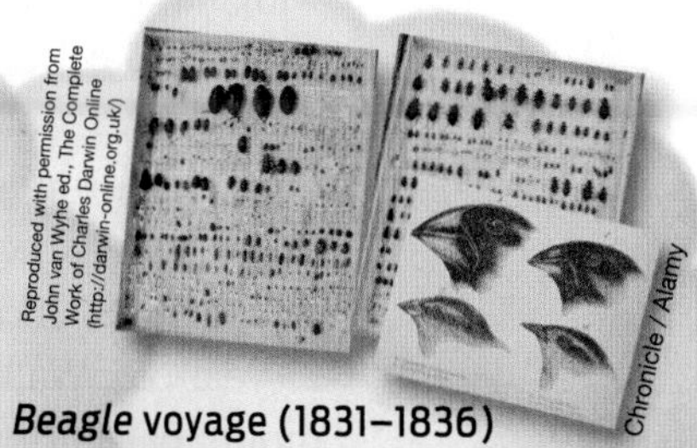

Reproduced with permission from John van Wyhe ed., The Complete Work of Charles Darwin Online (http://darwin-online.org.uk/)

Chronicle / Alamy

***Beagle* voyage (1831–1836)**

- Darwin collected plants, animals, and fossils from across the globe.
- Darwin observed similarities and differences and attempted to explain these characteristics.

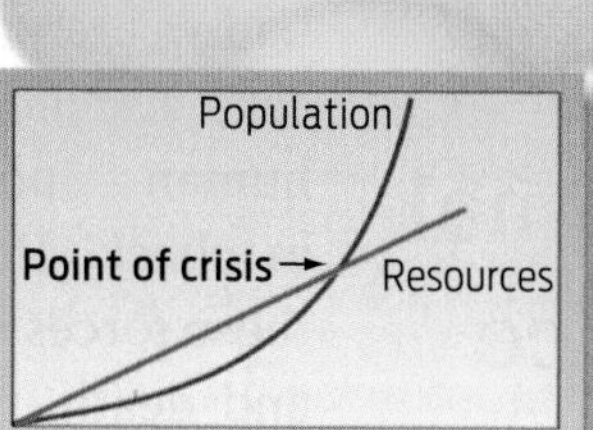

Malthus's work (1798)

- Populations are limited by a number of factors, including food, water, and disease.
- Some individuals die and some survive.

Reproduced with permission from John van Wyhe ed., The Complete Work of Charles Darwin Online (http://darwin-online.org.uk/)

Barnacle research (1846–1854)

- Darwin observed the many differences among barnacles, reinforcing for him the idea that much variation exists in nature.

Bettmann/Getty Images

Charles Darwin (February 12, 1809–April 19, 1882)

? How many years of work influenced Darwin's thoughts on evolution?

In Darwin's Shadow

Although we often credit Darwin with the discovery of natural selection, he was not alone in charting this intellectual territory. Another British naturalist was also hot on the trail. Like Darwin, Wallace was fascinated by natural history and had a thirst for adventure. In other ways, though, the two men couldn't have been more different. Darwin came from a wealthy family and had received a prestigious Cambridge education. He was greeted as a minor celebrity when he returned from his trip around the world and was accepted into the scientific establishment. Wallace, on the other hand, was a man of more humble origins, for whom nothing in life had come easily.

The eighth of nine children, Wallace could not afford a university education. He attended night school and supported himself as a builder and railroad surveyor. His budding fascination with natural history, though, led him to read widely. Like Darwin, he read Lyell's work on geology and Malthus's work on human population. He also devoured Darwin's recently published travel account, *The Voyage of the Beagle*.

In 1848, having scrimped and saved, the 25-year-old Wallace set sail for Brazil, to the mouth of the Amazon River. There he hoped to earn his reputation as a respectable scientist by understanding the origin of species. Exploring the rain forest of the Amazon, Wallace was struck by the distribution of distinct yet similar-looking (what he called "closely allied") species, which were often separated by a geographic barrier such as a canyon or river. For example, he noted that different species of sloth monkey were found on different banks of the Amazon River. Over the course of his 4-year trip, Wallace scoured the Amazon and collected thousands of specimens.

> How might these forces of disease and death, multiplied over time, influence animal populations, Wallace wondered?

Wallace was on his way home to London with his specimens in 1852 when disaster struck: his ship caught fire and sank. Wallace survived, but he lost everything–his notes, sketches, journals, and all his specimens. In spite of this catastrophe, Wallace was undeterred. Less than 2 years later, he was off on another collecting expedition, this time to the Malay archipelago (what is now Singapore, Malaysia, and Indonesia).

Wallace's first paper, "On the Law Which Has Regulated the Introduction of New Species," was published in September 1855. Based on his island work, it focused on the similar geographical distribution of "closely allied" species. For example, "the Galápagos Islands," he wrote "contain little groups of plants and animals peculiar to themselves, but most nearly allied to those of South America." From these observations, Wallace deduced this law, as he called it: "Every species has come into existence coincident both in space and time with a pre-existing closely allied species."

Wallace's article was groundbreaking, foreshadowing Darwin in a number of ways, but it lacked an explanation–a mechanism–of exactly how one species might have evolved from another.

Wallace continued his travels, but in early 1858 disaster struck again: he contracted malaria. Confined to bed, he let his mind wander. He thought about what Malthus had written about disease and how it kept human populations in check. How might these forces of disease and death, multiplied over time, influence animal populations, he wondered? Then came the flash of insight–like "friction upon the specially-prepared match," he recalled. In every generation, weaker individuals will die while those with the fittest variations will survive and reproduce; as a result species will change and adapt to their surroundings, eventually forming new species. Wallace had worked out the mechanism for evolution that was missing from his earlier work. He quickly wrote down his idea and sent it to the one naturalist he thought might be able to appreciate it. This was the 20-page manuscript that arrived on Darwin's doorstep on June 18, 1858 (**INFOGRAPHIC M5.4**).

Darwin was stunned. For 20 years he had been working diligently on the same idea and now it seemed someone else might get credit for it. "All my originality will be smashed," he wailed to his friend Lyell. Recognizing Darwin's predicament, Lyell and other colleagues devised a plan that would clearly establish Darwin's intellectual precedence. They would arrange to have papers by both men presented at a meeting of the

Linnaean Society in London. The meeting took place on July 1, 1858. The papers were dutifully read, but there was no discussion or fanfare. In fact, neither of the authors was even present. Wallace was still traveling in Malaysia, and Darwin was mourning the recent death of his young son and too distraught to attend.

The scientific meeting secured Darwin's reputation, but still he was unsettled. Wallace's communication had lit a fire under his feet. He needed to finish his book. That work, *On the Origin of Species by Means of Natural Selection*, was published in November 1859. It would become one of the most famous books of all time, going through six editions by 1872.

INFOGRAPHIC M5.4 The Evolution of Wallace's Thought

Like Darwin, Wallace was influenced by the writings and work of others, which shaped his interpretations of his own observations and research.

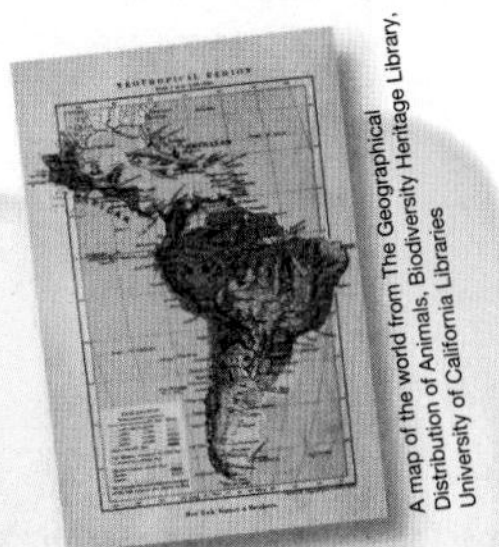

A map of the world from The Geographical Distribution of Animals, Biodiversity Heritage Library, University of California Libraries

Amazon trip (1848–1852)

- Wallace observed that related (or "closely allied") species occupied neighboring geographic areas.
- He noted the role of physical barriers (such as the Amazon River) in separating related species from one another.

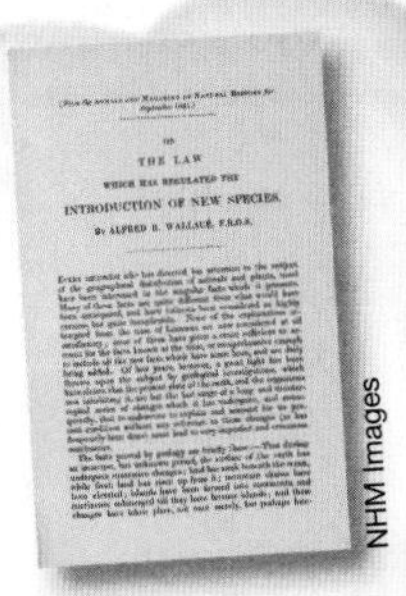

THE LAW WHICH HAS REGULATED THE INTRODUCTION OF NEW SPECIES.

NHM Images

First publication (1855)

- Wallace's publication dealt with the physical distribution of species.
- He introduced the idea that new species are temporally and spatially connected to a related species.

Disease and famine (1858)

- While suffering from malaria, Wallace pondered the role of disease and famine in keeping human populations in check.
- He wondered how these factors could apply to the evolution of animal species.

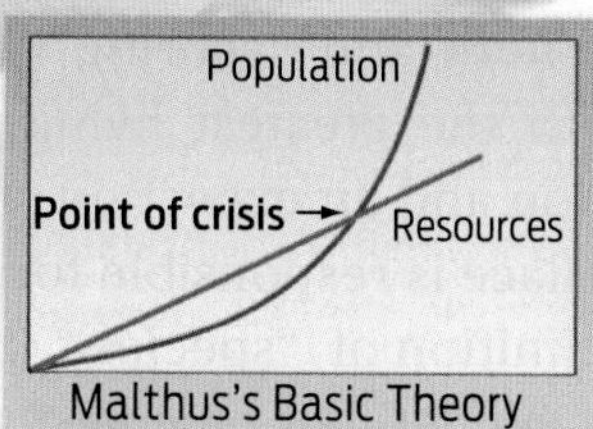

Malthus's work (1798)

- Populations are limited by a number of factors, including food, water, and disease.
- Some individuals die and some survive.

London Stereoscopic Company/Getty Images

Alfred Russel Wallace (January 8, 1823–November 7, 1913)

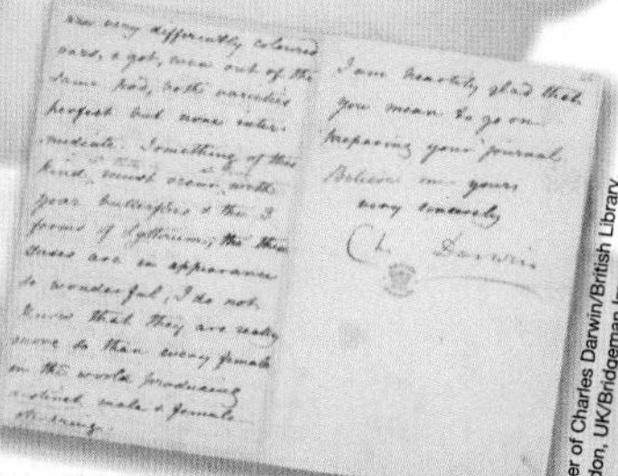

Letter of Charles Darwin/British Library, London, UK/Bridgeman Images

Wallace–Darwin correspondence (185?–1858)

- Wallace began corresponding with Darwin several years before sending Darwin his completed manuscript.
- The two men were clearly developing very similar ideas about the nature of evolutionary change.

? Both Darwin and Wallace made long trips that influenced their thinking. When and where did each travel?

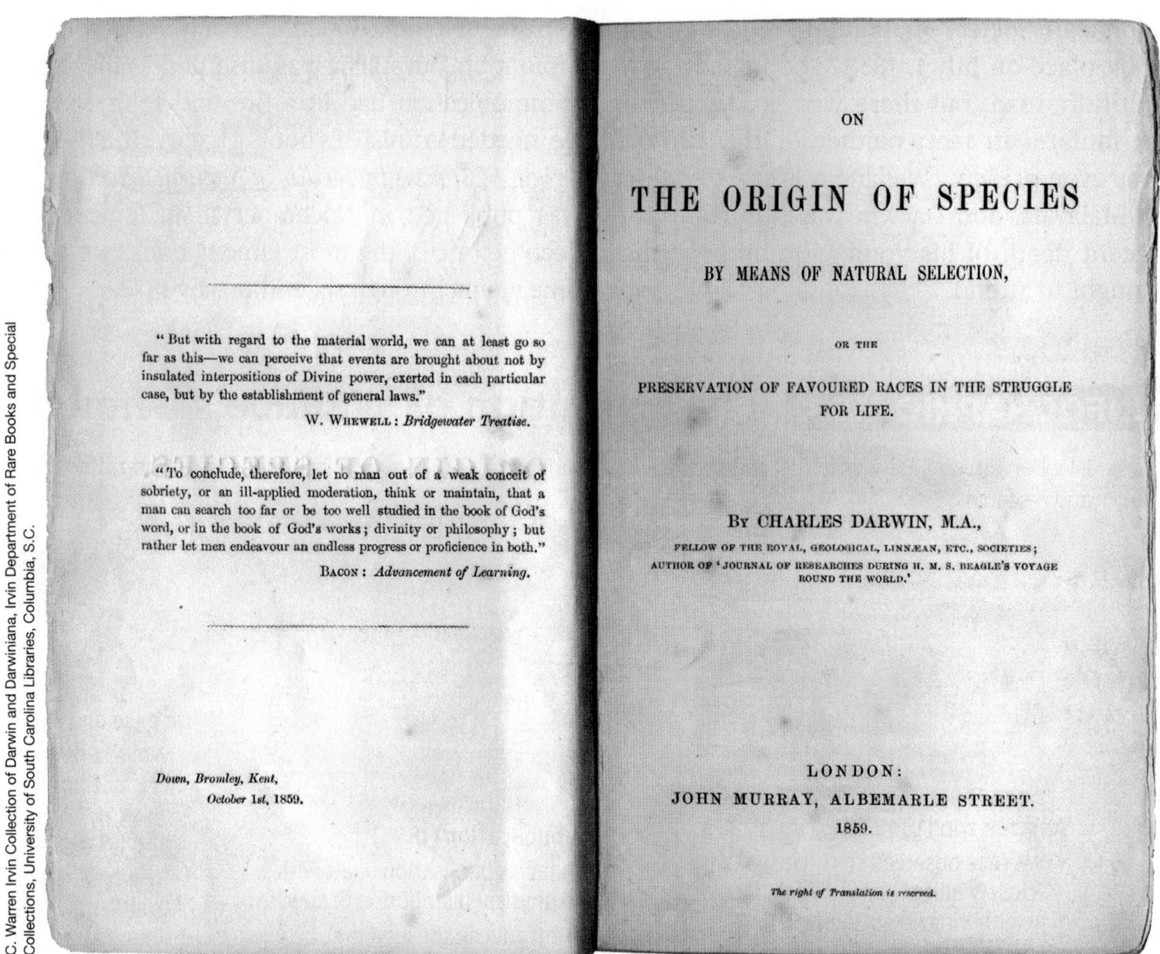

"But with regard to the material world, we can at least go so far as this—we can perceive that events are brought about not by insulated interpositions of Divine power, exerted in each particular case, but by the establishment of general laws."

W. Whewell: *Bridgewater Treatise.*

"To conclude, therefore, let no man out of a weak conceit of sobriety, or an ill-applied moderation, think or maintain, that a man can search too far or be too well studied in the book of God's word, or in the book of God's works; divinity or philosophy; but rather let men endeavour an endless progress or proficience in both."

Bacon: *Advancement of Learning.*

Down, Bromley, Kent,
October 1st, 1859.

ON

THE ORIGIN OF SPECIES

BY MEANS OF NATURAL SELECTION,

OR THE

PRESERVATION OF FAVOURED RACES IN THE STRUGGLE FOR LIFE.

By CHARLES DARWIN, M.A.,

FELLOW OF THE ROYAL, GEOLOGICAL, LINNÆAN, ETC., SOCIETIES;
AUTHOR OF 'JOURNAL OF RESEARCHES DURING H. M. S. BEAGLE'S VOYAGE ROUND THE WORLD.'

LONDON:
JOHN MURRAY, ALBEMARLE STREET.
1859.

The right of Translation is reserved.

Although it may seem that Wallace was cheated of his rightful recognition as a discoverer of evolution by natural selection, he was never bitter. On the contrary, he was delighted when he heard about his copublication with Darwin. He fully accepted that Darwin had formulated a more complete theory of natural selection before he did, and there is no trace of resentment in his later writings. In fact, Wallace titled his major work *Darwinism*, in recognition of the other man's intellectual influence.

After the presentation of 1858, Wallace stayed in the Malay archipelago for 4 more years, systematically recording its fauna and flora and securing his reputation as both the greatest living authority on the region and an expert on speciation. In fact, Wallace is responsible for our modern-day definition of "species." In work on butterflies, he defined "species" as groups of individuals capable of interbreeding with other members of the group but not with individuals from outside the group. This idea–known today as the biological species concept (see Chapter 14)–remains one of the most important in evolutionary theory. ■

MORE TO EXPLORE

- Darwin, C. (1989 [1839]) *The Voyage of the Beagle: Charles Darwin's Journal of Researches.* New York: Penguin Classics.
- A Trip Around the World, American Museum of Natural History: www.amnh.org/exhibitions/darwin/a-trip-around-the-world/
- Browne, J. (2008) *Darwin's Origin of Species.* New York: Grove Press.
- Wallace, A. R. (2016 [1869]) *The Malay Archipelago: The Land of the Orang-Utan and the Bird of Paradise.* Oxford, England: John Beaufoy Publishing.
- Secord, J. (2003) *Victorian Sensation: The Extraordinary Publication, Reception, and Secret Authorship of Vestiges of the Natural History of Creation.* Chicago: University of Chicago Press.

MILESTONES IN BIOLOGY 5 Test Your Knowledge

1 **What did the discovery of a fossil sloth in a sea cliff on the coast of Argentina suggest to Darwin?**

2 **Why was Thomas Malthus's book critical to Darwin's thinking about descent with modification?**

3 **How did Wallace use Thomas Malthus's book to inform his ideas about species?**

4 **What did the field experiences Darwin and Wallace had in observing the natural world at first hand, in addition to communication with other scientists and careful consideration, add to their understanding of evolution that perhaps reading and thinking alone couldn't provide?**

14 Nonadaptive Evolution and Speciation

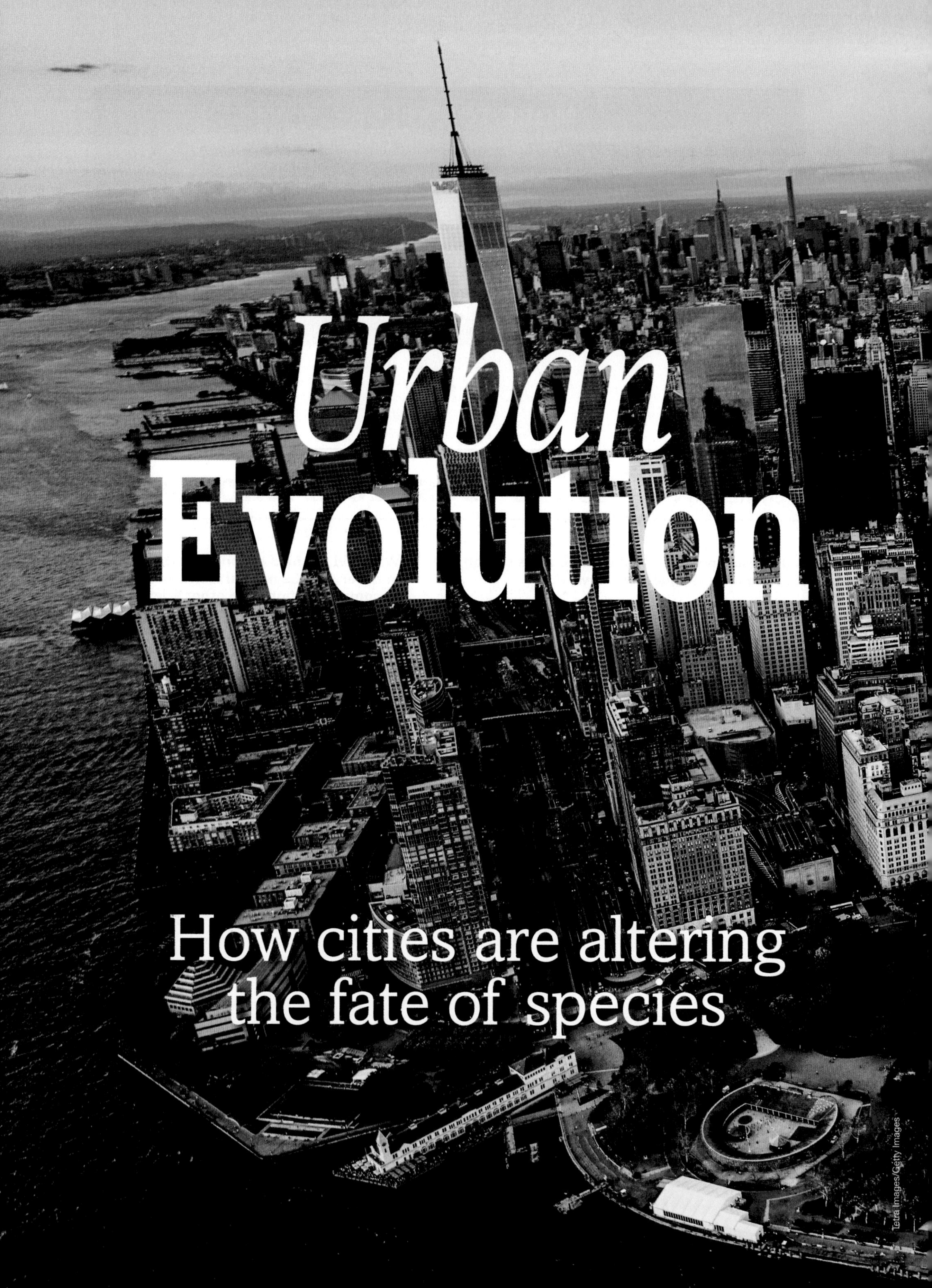

Urban Evolution

How cities are altering the fate of species

Tetra Images/Getty Images

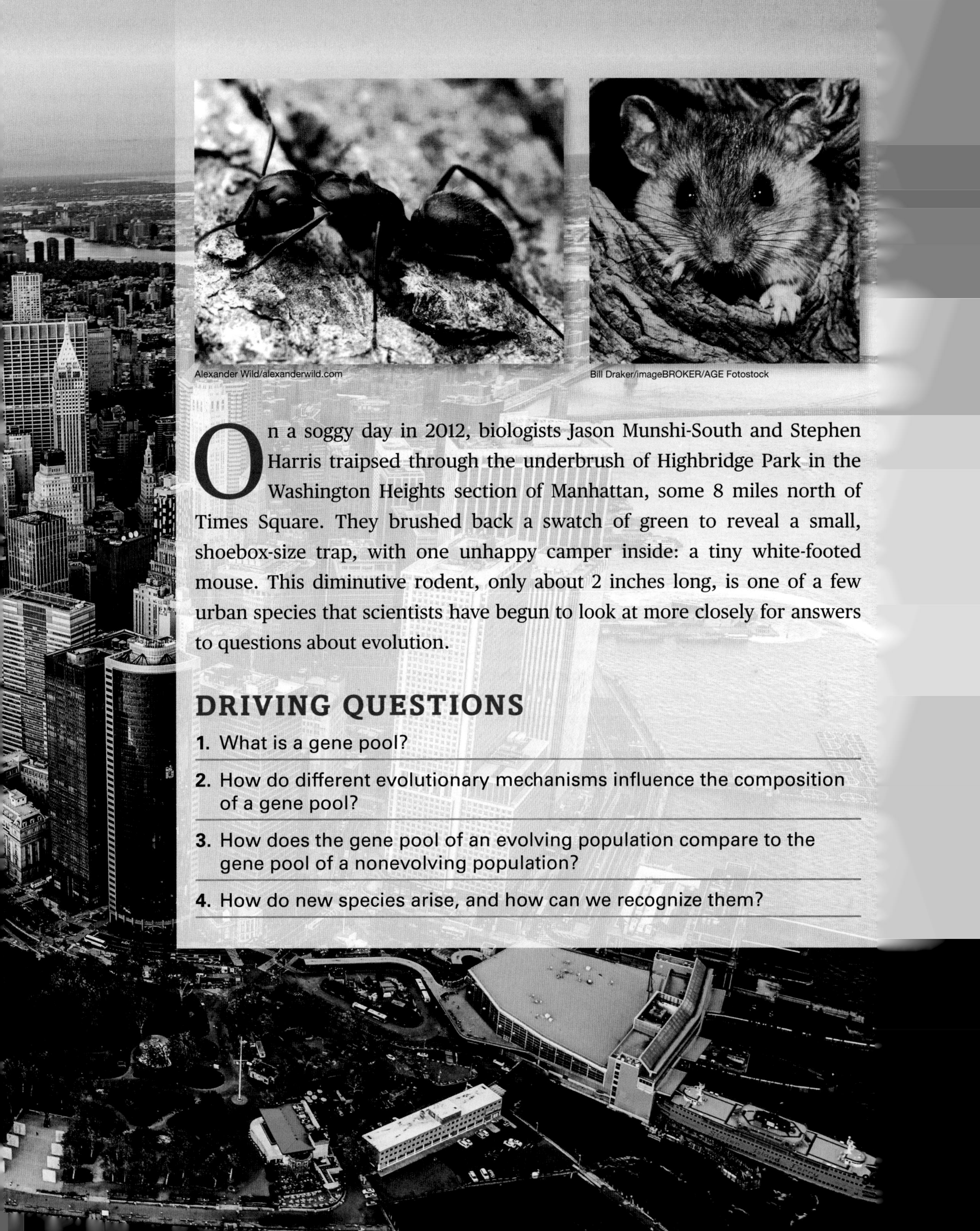

Alexander Wild/alexanderwild.com

Bill Draker/imageBROKER/AGE Fotostock

On a soggy day in 2012, biologists Jason Munshi-South and Stephen Harris traipsed through the underbrush of Highbridge Park in the Washington Heights section of Manhattan, some 8 miles north of Times Square. They brushed back a swatch of green to reveal a small, shoebox-size trap, with one unhappy camper inside: a tiny white-footed mouse. This diminutive rodent, only about 2 inches long, is one of a few urban species that scientists have begun to look at more closely for answers to questions about evolution.

DRIVING QUESTIONS

1. What is a gene pool?
2. How do different evolutionary mechanisms influence the composition of a gene pool?
3. How does the gene pool of an evolving population compare to the gene pool of a nonevolving population?
4. How do new species arise, and how can we recognize them?

Studying evolution in Manhattan—arguably the most unnatural place on the planet—might seem an odd choice. But Munshi-South and Harris belong to a new breed of biologist, one fascinated by the nature right under our noses. From rodents in parks to ants on median strips to the cockroaches and bedbugs inside buildings—even the bacteria brewing in our belly buttons—no location is too mundane for this scientific crew.

What's the advantage of staying local? Besides being quite convenient—Munshi-South works at Fordham University just north of Manhattan—cities like New York offer some unique opportunities for the evolutionary biologist.

"There isn't really anywhere on the planet that isn't impacted by human activity," says Munshi-South. "If we want to understand how species are actually evolving now, we need to understand the inputs from human activity."

And nowhere is that input more evident than in the Big Apple. With a population of 8 million people packed into just 300 square miles, New York City is the largest and most densely populated city in the United States. All those restless urbanites have left a profound mark on the landscape, changing it in both dramatic and subtle ways. Just look at Manhattan, New York City's most densely populated borough: once an island of thick forest, Manhattan is now a sliver of concrete interspersed with oases of green. The skyscrapers, the bridges, the subway system—not to mention the bars and coffee shops—have all transformed a once wild place beyond recognition. In the process, says Munshi-South, the city's wildlife has been subject to "a grand evolutionary experiment."

Biologist Jason Munshi-South laying a trap for white-footed mice.

Sex and the City

To study evolution, Munshi-South has traveled to some pretty far-flung locations: Southeast Asia, to study how the mating behavior of small mammals is affected by logging operations, and Africa, to research how the migration patterns and stress levels of forest elephants are affected by the petroleum operations there.

Mice in Manhattan might seem a far cry from that earlier research, but it's really not, he says. "Urbanism is basically one of these large-scale dramatic transformations of landscapes," he says. "And maybe one of the most complete transformations from a seminatural state to a use that's dominated by human activity." The common thread to this work, he says, is understanding how animals cope with a rapidly changing environment.

Urbanization isn't the only human-caused environmental change that animals face, of course—climate change is another big one—but it's one that has taken on pressing urgency in recent years.

"Already 50% of people live in cities," says Munshi-South. "It's going to be 60% in 15 years, and it's just going to keep going up and up." In fact, demographers predict that the human population will reach 9 billion in 2050, and that by then 70% of us—roughly 6 billion people—will live in cities. All those urban dwellers represent a significant evolutionary force to be reckoned with. So it's important to understand how our fellow animals are adapting—or not adapting—to city life.

If you're an evolutionary biologist and you want to understand how a group of organisms is coping with environmental changes, you need to know something about its underlying genetics—and not just the genetics of individuals, but the genetics of the population as a whole. For that, you need the tools of population genetics. **Population genetics** allows scientists to understand the nature of evolutionary change as it is reflected in the genes of a population. Essentially, it's a way to take stock of who's reproducing, who isn't, and the consequences for the population as a whole.

From a population genetics perspective, each distinct population of organisms—whether mice in Manhattan or elephants in Africa—has its own particular collection of alleles, which together constitute its **gene pool.** Within the gene pool, each allele is present in a certain proportion, or **allele frequency,** relative to the total number of alleles for that gene in the population. For example, if a particular allele for a gene is present 50 times out of a total of 1,000 alleles, its allele frequency is 0.05. Over time, several forces can change the frequency of alleles—that is, how common they are in the population. When the frequency of alleles changes over time, a population evolves. Recall from Chapter 13 that this is the definition of evolution (**INFOGRAPHIC 14.1**).

Evolutionary changes in a gene pool can have lasting consequences for a population. They can, for example, result in the population becoming more adapted to its environment—think of the antibiotic-resistant bacteria we met in Chapter 13. The evolutionary mechanism that results in adaptation is natural selection.

POPULATION GENETICS
The study of the genetic makeup of populations and how the genetic composition of a population changes.

GENE POOL
The total collection of alleles in a population.

ALLELE FREQUENCY
The relative proportion of an allele in a population.

INFOGRAPHIC 14.1 Population Genetics

Population geneticists study the gene pools of populations. If a gene pool changes (that is, if the allele frequencies have changed) over the course of generations, then evolution has occurred.

? The starting population (on the left) undergoes several generations of reproduction. In the resulting population, the frequency of the blue allele is 0.2, the frequency of the red allele is 0.4, and the frequency of the yellow allele is 0.4. Did evolution occur in this case? Why or why not?

NONADAPTIVE EVOLUTION
Any change in allele frequency that does not by itself lead a population to become more adapted to its environment; the mechanisms of nonadaptive evolution are mutation, genetic drift, and gene flow.

But natural selection isn't the only mechanism of evolution. Mutation, which introduces new alleles into a population (see Chapter 10), is another important evolutionary mechanism. Because it is rare, mutation by itself does not dramatically change allele frequencies. But that doesn't mean it isn't important—after all, mutation is the source of variation in a population, upon which natural selection acts. However, mutation is a fundamentally random process that does not by itself lead to a population becoming more adapted to its environment. In other words, mutation is a type of **nonadaptive evolution.** The other types of nonadaptive evolution are genetic drift and gene flow.

Nonadaptive evolution isn't necessarily "bad," or maladaptive. If mutations didn't introduce variation into a population, there would be no evolution at all. And many nonadaptive changes in allele frequency can be considered "neutral"—neither "good" nor "bad." But nonadaptive evolution can greatly influence the fate of a species, and so researchers are keen to study it.

Changing by Chance: Genetic Drift

Peromyscus leucopus—the white-footed mouse—is one of the oldest residents of Manhattan, long predating the arrival of the first colonists in the 1600s. This rodent squeaks out a living in the green spaces of the city—in any park that has canopy cover. There are more than a dozen distinct populations of white-footed mice living across the city.

When Munshi-South initially had the idea of studying mice in New York City, he thought there probably wouldn't be many genetic differences among the different populations living there; New York is a fairly young city, and evolutionary change generally happens slowly. But the results of his study clearly showed that was not the case.

"It turned out that actually the populations are fairly distinct in the parks," says Munshi-South. And these differences have come about over a relatively short period of evolutionary time—a few hundred years at the most.

Over three centuries, as New York City has gone from a rich tapestry of forest to a sparse patchwork of fragmented green spaces, populations of white-footed mice have become trapped in little islands of green, cut off from their distant cousins in the rest of the city. What this means is that each population of mice has its own distinct gene pool, and evolution is occurring differently in each local population.

To peer into these gene pools, Munshi-South and his colleagues analyzed mice from 15 different populations around New York City. Using mousetraps baited with birdseed, they caught a total of 312 mice from these populations. After catching each mouse, researchers cut off the tip of its tail and put the tissue sample in ethanol; DNA from the tail tissue could later be extracted and sequenced at specific chromosome locations (see Chapter 7). The mice, which were not seriously harmed by this procedure, were then released back into the wild.

When DNA sequences at 18 different chromosome locations from all 312 mice were assessed, the results showed distinct clustering of alleles, with mice within one population tending to share more alleles with one another than with mice from other populations. In fact, says Munshi-South, you can accurately predict where a mouse is from just by looking at its DNA (**INFOGRAPHIC 14.2**).

How did these genetic differences among populations come about? One possibility is that each population of mice evolved by natural selection as a result of local differences in the environment—perhaps each different green space has different predators or food sources, for example—which selected for individuals with different alleles. Given how close these green spaces are to one another, however—in some cases, less than a mile apart—and given also how similar the environments

INFOGRAPHIC 14.2 Gene Pools of New York City Mouse Populations

Researchers collected DNA from the tails of 312 mice at 15 locations in New York City. In order to determine how related the mouse populations were, researchers analyzed the DNA, assigning mice carrying similar genotypes the same color. When sorted by location, the data show that mice within a population shared more alleles with one another than they did with mice from other populations.

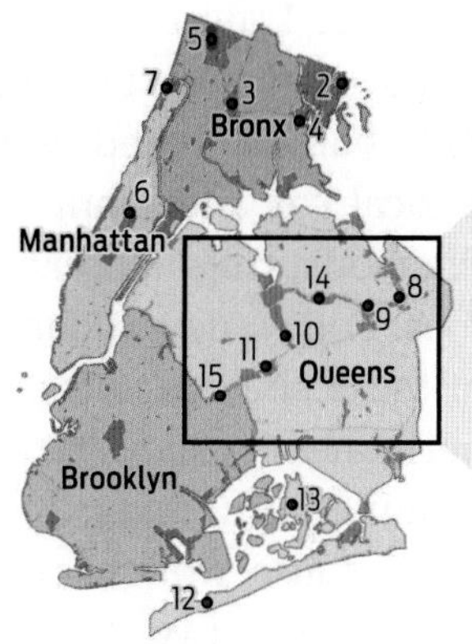

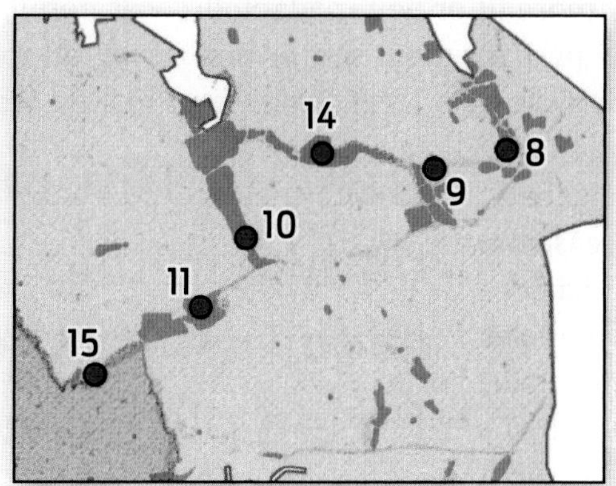

Habitat fragmentation has led to separated mouse populations in New York City.

The red dots show Munshi-South study sites. Shades of green represent park areas. Urbanization has fragmented green spaces into separate patches. Some patches are connected by "greenways" and some are not.

Mouse populations are genetically distinct from one another.

Each block represents a group of mice sampled from a particular location (population). The alleles of each sampled mouse are represented by a vertical bar within the block. When genetic sequences of these alleles match those of another mouse within the population they are given the same color. When sequences don't match, they are shown in different colors.

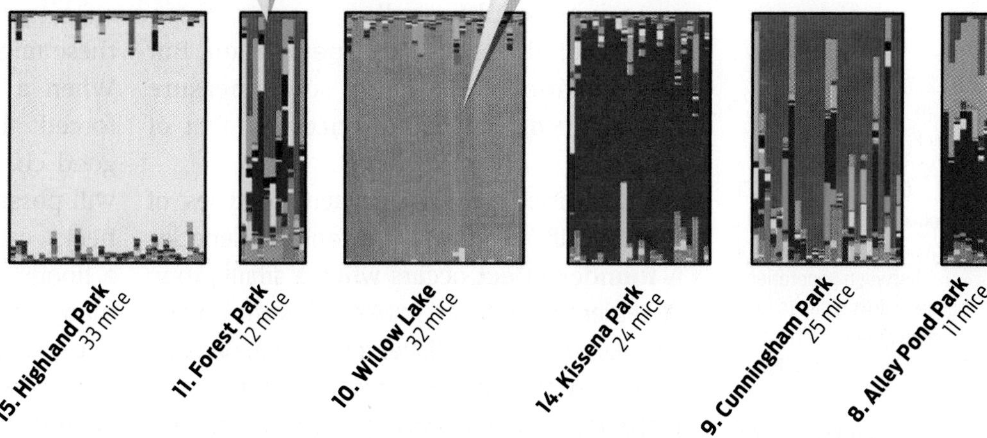

Overall, when you look at the blocks (populations), you can see that most mice in a given population have alleles (colors) that are similar to one another and distinct from other populations. For example, Highland Park mice have predominantly "yellow" alleles, while Willow Lake mice have predominantly "gray" alleles. Populations connnected by green corridors to other populations may have alleles that are common to the populations to which they are connected.

? From the allelic data, which populations are the most likely to be isolated (and not "trading" mice with other populations)?

are, this explanation isn't the most likely one. More likely, says Munshi-South, the cause is **genetic drift.**

"Genetic drift is a bit like rolling the evolutionary dice. By simple chance, some individuals survive and reproduce, and others do not." Those that pass on their genes aren't necessarily more fit or better adapted; they're just lucky–perhaps their nest or burrow wasn't swept away in a flash flood, for example.

Over time, genetic drift tends to decrease the genetic diversity of a population, as some alleles are lost completely and others sweep to 100% frequency. Genetic drift will have more dramatic effects in smaller populations than in larger ones. In a population with few individuals, any single individual that does not reproduce could spell the loss of alleles from the population. But all populations experience some measure of genetic drift, since chance is a fact of life.

Genetic drift is a bit like rolling the evolutionary dice. By simple chance, some individuals survive and reproduce, and others do not.

Biologists refer to two general types of genetic drift: founder effects and bottlenecks. A **founder effect** occurs when a small group of settlers ("founders") splits off from a main population and establishes a new one. Because a founding population is by definition small, there is a good chance that the particular alleles it carries will not be fully representative of the population it left. Thus, founder effects tend to reduce the genetic diversity of the new population.

If the founder population happens to contain a rare allele, then this allele may become much more common in the new population. Polydactyly (having extra fingers or toes) is an unusually common trait in the Amish population of eastern Pennsylvania, the result of Ellis-van Creveld syndrome, a recessive inherited condition. The Amish are a rural, isolated population stemming from a small number of German immigrants who moved to the area in the 18th century. The allele for Ellis-van Creveld syndrome arrived with a single couple who immigrated to the area in 1744 and has since spread throughout the population.

Polydactyly is also common in certain cat populations. The writer Ernest Hemingway adopted a six-toed cat when he lived in Key West, Florida. That cat mated with a local cat and many of their offspring had extra toes. Today, descendants of these six-toed cats are permanent residents of Hemingway's estate.

Because white-footed mice are native to the New York region, it is unlikely that they have experienced founder effects; the populations that are here have all been here for a very long time. More applicable to these mice is what's known as a **bottleneck.** When a population is cut down sharply–forced through a "bottleneck"–there's a good chance that the remaining population will possess a less-diverse gene pool. Bottlenecks can occur from natural causes–say, a flood that sweeps through the city, killing many individuals–or from human interference, such as the clearing of a forest. Either way, a population that is forced through a genetic bottleneck usually contains a fraction of the original diversity in the population (**INFOGRAPHIC 14.3**).

As an example of a genetic bottleneck, consider the cheetah (*Acinonyx jubatus*), the fastest land animal. Cheetahs almost became extinct 10,000 years ago when harsh conditions of the last ice age claimed the lives of many large vertebrates on several continents. Ultimately, a few cheetahs survived and reproduced, but the more than 12,000 individuals alive today are now so genetically similar that skin grafts between unrelated individuals do not cause immune rejection;

GENETIC DRIFT
Random changes in the allele frequencies of a population between generations; genetic drift tends to have more dramatic effects in smaller populations than in larger ones.

FOUNDER EFFECT
A type of genetic drift in which a small number of individuals leaves one population and establishes a new population, resulting in lower genetic diversity than in the original population.

BOTTLENECK EFFECT
A type of genetic drift that occurs when a population is suddenly reduced to a small number of individuals, and as a result alleles are lost from the population.

INFOGRAPHIC 14.3 Genetic Drift Reduces Genetic Diversity

Allele frequencies can change from one generation to the next purely as a result of chance: this is genetic drift. Drift has more dramatic effects in smaller populations than in larger ones. There are two types of genetic drift, the founder effect and the bottleneck effect.

Founder Effect

The founder effect is a type of genetic drift that occurs when a small group of "founders" leaves a population and establishes a new one. If, by chance, alleles from the original population are absent from the founders, they will also be absent from the new population. This will result in the founding population being less diverse than the original.

A large, diverse original population

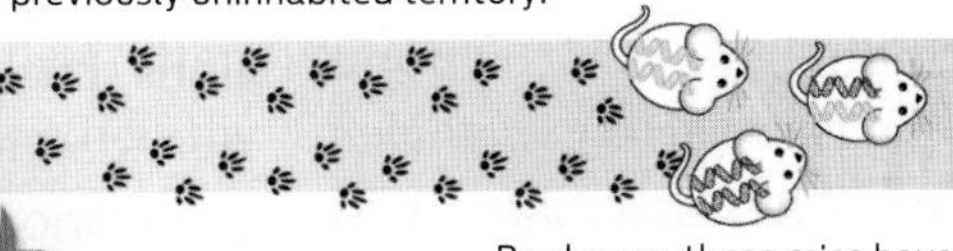

A few founders migrate to previously uninhabited territory.

By chance, these mice have a reduced frequency for some alleles compared to the original population.

A less-diverse founding population (after several generations)

Allele Frequencies before Drift

6/30 = 20% = 0.20

12/30 = 40% = 0.40

12/30 = 40% = 0.40

Blue alleles are not represented in the founding population.

Allele Frequencies after Drift

10/22 = 45% = 0.45

12/22 = 55% = 0.55

Bottleneck Effect

Genetic bottlenecks occur when a population loses a large proportion of its members. If the original population is large, the reduced population is likely to retain the same alleles present in the original population. But in a small starting population, bottlenecks are more consequential: the loss of individuals is more likely to result in the loss of alleles from the population.

A more-diverse original population

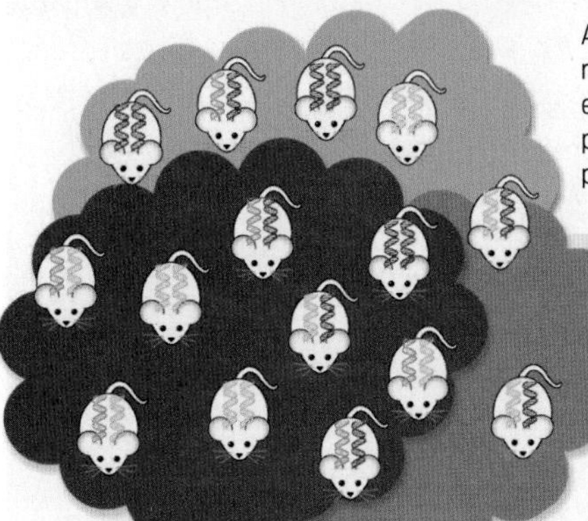

A bottleneck event, like rapid habitat loss, eliminates a large percentage of the population.

By chance, the surviving mice have a reduced frequency for some alleles compared to the original population.

A less-diverse bottleneck population (after several generations)

Allele Frequencies before Drift

6/30 = 20% = 0.20

12/30 = 40% = 0.40

12/30 = 40% = 0.40

Red alleles are lost from the gene pool.

Allele Frequencies after Drift

6/18 = 33% = 0.33

12/18 = 67% = 0.67

? Compare and contrast the bottleneck effect and the founder effect.

nearly all genetic diversity has been eliminated from the population.

Why does genetic diversity matter? You can think of a gene pool as being like a population's portfolio of financial assets. Having a diverse array of investments is a better strategy for long-term success than having all your money tied up in one kind of stock–especially if that stock loses value in changed economic times.

Say a population of mice suddenly finds itself in a more crowded and polluted environment than it did before. If the population carries with it a rich variety of alleles, some of these alleles (ones associated with stronger immune systems, for example) may help that population survive and reproduce in the altered conditions. Individuals with these alleles will be more fit in this environment, and the population will adapt by natural selection. With less diversity in the population, the opportunity for adaptation will be more limited, and the population may shrink. Preserving genetic diversity is thus a prime concern of conservation biologists interested in protecting natural populations from extinction, especially as humans encroach on more and more wild habitat.

The Daily Commute: Gene Flow

Once a population has lost genetic diversity because of genetic drift, there are only two ways that genetic diversity can be reintroduced: (1) by mutation, which as we saw in Chapter 10 continually introduces new alleles into the population, and (2) by **gene flow,** in which alleles move between populations as individuals leave and enter populations and breed with members of other populations. Like genetic drift, gene flow is a type of nonadaptive evolution that does not lead to a population becoming more adapted to its environment. Unlike genetic drift, gene flow tends to increase the genetic diversity of a population, not decrease it (**INFOGRAPHIC 14.4**).

Munshi-South and his colleagues found that different New York City mouse populations had different levels of gene flow, meaning that some populations exchanged alleles with other populations more than did others. They wanted to understand why, so they built a statistical model and tested whether it could explain the data they collected at their study sites. Their idea was this: gene flow should be possible between populations where there is a corridor of tree canopy connecting them.

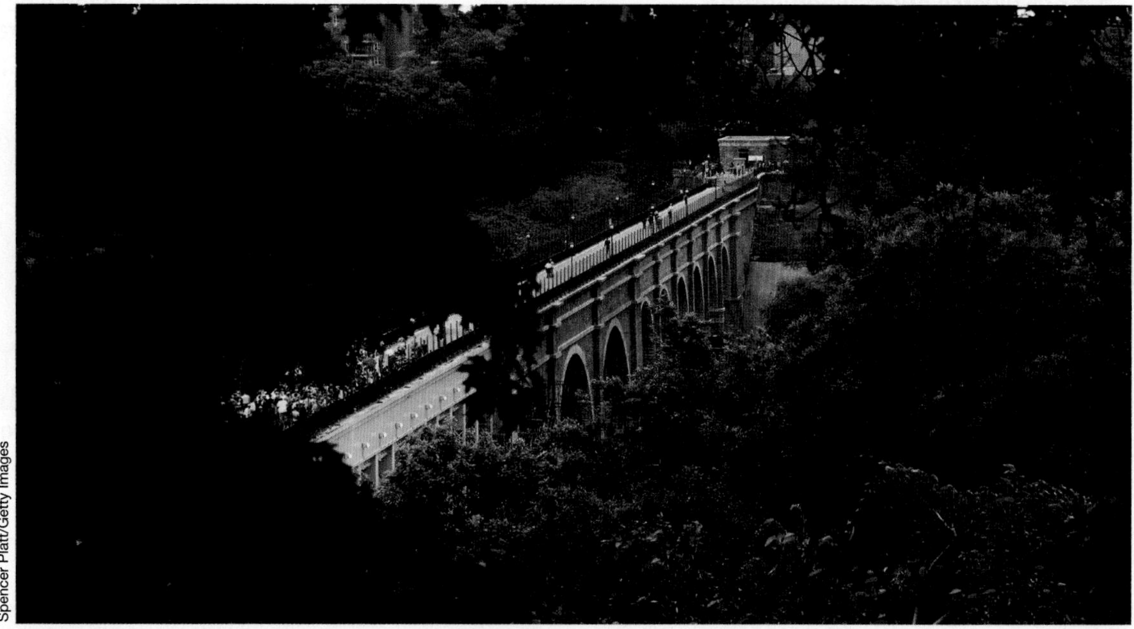

Spencer Platt/Getty Images

Man-made barriers, like New York City's Highbridge, interfere with gene flow between populations.

INFOGRAPHIC 14.4 Gene Flow Increases Genetic Diversity in Populations

Migration and interbreeding of individuals move alleles between populations. Populations that can interbreed with other populations have higher allele diversity than isolated populations.

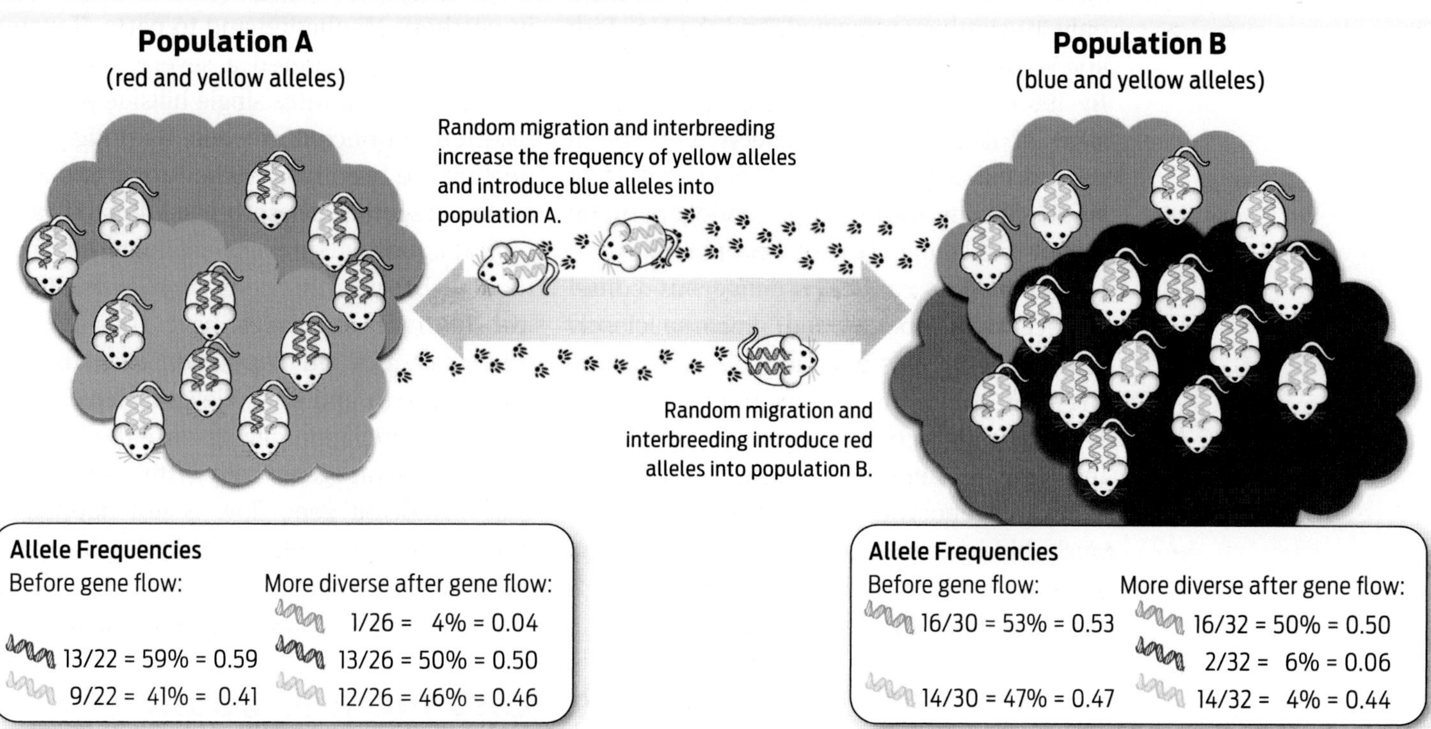

? Which new allele was introduced into population B by gene flow?

So they divided the city into categories based on the percentage of "green" landscape (tree cover) or "gray" landscape (concrete), and then tested how closely their actual data matched the hypothetical results. It was a simple idea, but it worked surprisingly well. "That pretty much explained the variation we saw in gene flow and drift between the populations," he says. In fact, the "green corridor" model explained the amount of gene flow even more than the absolute distance between the populations.

Putting it all together, the picture looks like this: urbanization has led to habitat fragmentation that has isolated and bottlenecked dozens of mouse populations, leaving each with a distinct gene pool. Populations that were completely isolated (for example, sites 10 and 15 in Infographic 14.2) have continued to diverge by genetic drift, while populations connected to other populations by a green corridor (for example, sites 8 and 9) have remained more similar to one another because of gene flow between them. Furthermore, mutation randomly adds new alleles to these populations, contributing to diversity between them. "So it's basically this interplay between genetic drift, gene flow, and mutation," says Munshi-South, that is influencing the level of genetic diversity in the populations.

One reason gene flow is important is that small, isolated populations can be damaged by lack of genetic diversity. Take the Florida panther (*Puma concolor*), for example. In the past, Florida panthers mated with puma populations from neighboring states where their ranges overlapped. This interbreeding–breeding among different populations of the same species–fostered an exchange of alleles that

GENE FLOW
The movement of alleles from one population to another, which may increase the genetic diversity of a population.

continually enriched the local populations' genetic diversity. By the mid-20th century, however, hunting and development had squeezed the Florida panther population into an isolated region at the state's southernmost tip. By 1967, only 30 panthers remained, and the U.S. Fish and Wildlife Service listed them as endangered. By 1980, the panthers showed unmistakable signs of ill health–birth defects, low sperm count, missing testes, and bent tails–that resulted from **inbreeding,** mating between closely related members of a population.

Inbreeding can have dangerous consequences for a population. Because closely related individuals are more likely to share the same alleles, the chance of two recessive harmful alleles coming together during mating is high. When that happens, homozygous recessive genotypes are created, and previously hidden recessive alleles start to affect phenotypes in negative ways. This effect is called **inbreeding depression.**

To counteract this dangerous trend, in 1995 the U.S. Fish and Wildlife Service brought in eight female pumas from Texas to mate with Florida's male panthers and thereby introduce genetic diversity. The program was successful: the hybrid kittens–30 in all–showed no signs of inbreeding depression. By 2007, more than 100 healthy panthers were roaming the swamps and grasslands of Florida, and their numbers continue to increase.

For the time being, says Munshi-South, mice in Manhattan seem to be doing just fine in maintaining adequate genetic diversity. While the populations are clearly distinct from one another, each has within it a fair amount of genetic diversity. This is probably because each population is still quite large and so the drift that is occurring has not dramatically reduced the number of alleles in each population. That, combined with mutation and occasional episodes of gene flow between some of the populations, has allowed these populations to maintain significant genetic diversity.

INBREEDING
Mating between closely related individuals. Inbreeding does not change the allele frequency within a population, but it does increase the proportion of homozygous individuals to heterozygotes.

INBREEDING DEPRESSION
The negative reproductive consequences for a population associated with having a high frequency of homozygous individuals possessing harmful recessive alleles.

The same cannot be said of other species that Munshi-South and his colleagues are studying in New York City. The northern dusky salamander (*Desmognathus fuscus*), which makes its home in freshwater streams seeping out of the ground, was extremely common throughout much of the city just 60 years ago. Today, it clings to a single hillside in northern Manhattan and to parts of Staten Island, having suffered a severe bottleneck. Not only that, but this single hillside has been bisected not once, but twice, by bridges connecting the boroughs of the Bronx and Manhattan. Together, these bridges host 14 lanes of traffic, which prevent salamanders from moving between sections of hillside. These human-made structures have thus divided this small, isolated population even further. Munshi-South and his colleagues have found that the populations living on either side of these bridges are genetically distinct–evidence that significant gene flow is not happening between them. Whether or not the dusky salamanders will be able to retain their tenuous hold along this hillside, only time will tell.

City Mouse, Country Mouse

Fifteen miles north of Manhattan, on farms and apple orchards tucked away in the New York countryside, white-footed mice lead slower-paced lives, free from the stresses of urban living.

Have city mice and country mice evolved differently as a result of living in different environments? To answer this question, Munshi-South and a graduate student, Stephen Harris, are collecting DNA samples from country mice in order to compare the gene pools of city and country mice. "The sort of low-hanging fruit that we hope to find are the genes that are consistently different between urban and rural populations," says Munshi-South.

They've only just begun this work, but already they are finding some interesting results. The allele frequencies for a number of genes do seem to differ consistently between urban and rural populations. These alleles have to do with traits like metabolizing different food resources and breaking down toxic chemicals. In urban environments, mice have increased access to high-fat human food waste products. They also must contend with higher levels of industrial toxins in the water and soil. City mice have higher frequencies of alleles that allow them to handle these changed circumstances. This makes perfect sense, Munshi-South says: the city mice and the country mice have experienced different selective pressures, and these selective pressures have led to differences in the gene pools of these two large populations **(INFOGRAPHIC 14.5)**.

While city mice have experienced selection for urban traits, this does not mean they are all the same. Many genetic differences still exist among the local populations of urbanized mice. That's because mutation, drift, and gene flow continue to occur even as natural selection

INFOGRAPHIC 14.5 Environment Influences Allele Diversity in Different Populations

Country mice and city mice differ most in genes associated with immunity, toxin detoxification, and nutrient sources, with urban populations having less allelic diversity in these genes. The reduction in allelic diversity in these genes in city mice may be reflecting directional selective pressures in the urban environment, such as exposure to pollution and high-fat foods, and competition for mates.

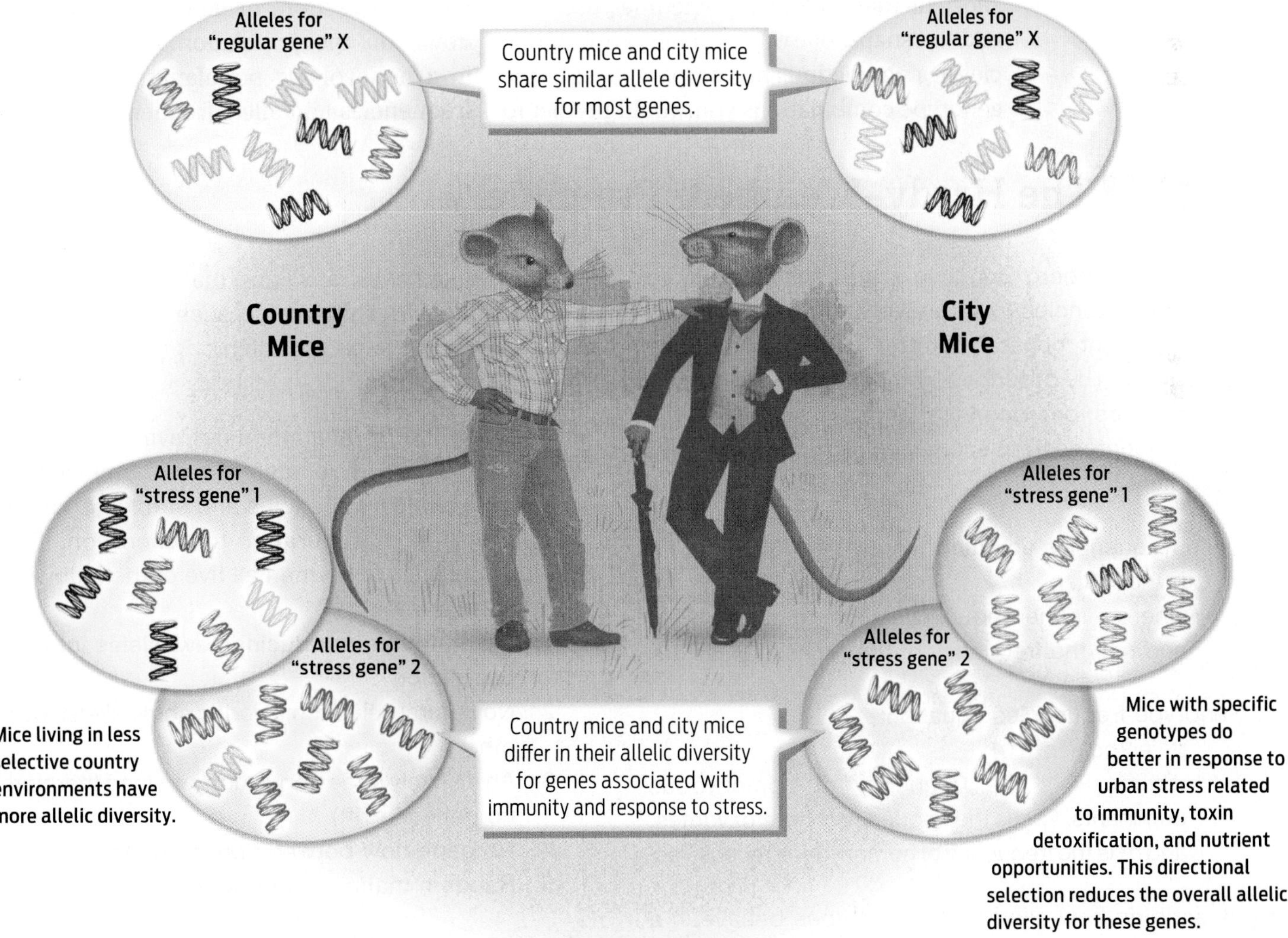

Alleles for "regular gene" X
Country mice and city mice share similar allele diversity for most genes.
Alleles for "regular gene" X
Country Mice
City Mice
Alleles for "stress gene" 1
Alleles for "stress gene" 2
Alleles for "stress gene" 1
Alleles for "stress gene" 2
Mice living in less selective country environments have more allelic diversity.
Country mice and city mice differ in their allelic diversity for genes associated with immunity and response to stress.
Mice with specific genotypes do better in response to urban stress related to immunity, toxin detoxification, and nutrient opportunities. This directional selection reduces the overall allelic diversity for these genes.

? Two populations of mice live in a similar warm climate with plentiful rainfall. One population is near a large farm and one is in undisturbed woods. Which of these genes would you predict to vary the most between the two populations: a gene involved in water balance, a gene involved in detoxification of chemicals (like pesticides), or a gene involved in temperature regulation?

HARDY–WEINBERG PRINCIPLE
The principle that, in a nonevolving population, both allele and genotype frequencies remain constant from one generation to the next.

is happening. In fact, for most natural populations, all four mechanisms of evolution–natural selection, mutation, gene flow, and genetic drift–are continually operating at the same time, fostering unique evolutionary outcomes **(TABLE 14.1)**.

How can scientists detect whether evolution is occurring in a population? The most direct way would be to determine the allele frequencies in a population, wait a few generations, and then determine the frequencies again. If the allele frequencies have changed, then the population has evolved (that's the definition of evolution).

But waiting for generations to go by isn't necessarily the most convenient approach. A shortcut that biologists sometimes use is to take a "snapshot" of the genotype frequencies in a population and compare these to the genotype combinations you would predict to find in a nonevolving population. If these differ, then you know the population is evolving.

How do genotypes behave in a nonevolving population? A common misconception is that dominant traits will tend to increase automatically in a population and recessives will be eliminated–thus removing variation from the population. Two mathematicians, G. H. Hardy and Wilhelm Weinberg, independently showed in the early 20th century that this was not the case. Through simple mathematical proofs, they showed that genotype frequencies (and therefore allele frequencies) in a population will remain constant, generation after generation, provided no evolutionary forces–such as genetic drift, natural selection, or gene flow–are bearing down on the population. In other words, in a nonevolving population, the genotype frequencies and allele frequencies will be in

UP CLOSE The Hardy–Weinberg Principle

The Hardy–Weinberg principle states that allele and genotype frequencies in a nonevolving population will remain constant generation after generation. These frequencies can be described mathematically. For a gene with two alleles, one dominant and one recessive, the allele and genotype frequencies can be calculated using two simple equations:

1. Allele frequencies equation:
   $p + q = 1$
   where p is the frequency of the dominant allele and q is the frequency of the recessive allele.

2. Genotype frequencies equation:
   $p^2 + 2pq + q^2 = 1$
   where p^2 is the frequency of homozygous dominants, $2pq$ is the frequency of heterozygotes, and q^2 is the frequency of homozygous recessives.

By describing how alleles and genotypes behave in a nonevolving population, the Hardy–Weinberg principle provides biologists with a baseline from which to evaluate whether or not evolution is occurring in a population. For example, if biologists find that the *actual* genotype frequencies in a population differ significantly from the *expected* values, then they know that some evolutionary force is acting to shift them—such as genetic drift, selection, gene flow, or mutation.

By definition, a population is not evolving (and is therefore in Hardy–Weinberg equilibrium) when it has stable allele frequencies and stable genotype frequencies from generation to generation. This can be achieved only when all five of the following conditions are met:

1. No mutation introducing new alleles into the population
2. No natural selection favoring some alleles over others
3. An infinitely large population size (and therefore no genetic drift)
4. No gene flow between populations
5. Random mating of individuals

Example:

The Hardy–Weinberg principle is also useful for calculating allele frequencies in a population when you know the frequency of at least one genotype. For example, let's

TABLE 14.1 Adaptive and Nonadaptive Mechanisms of Evolution

MECHANISM OF EVOLUTION	HOW ALLELE FREQUENCIES CHANGE	ADAPTIVE OR NONADAPTIVE?	HOW GENETIC DIVERSITY IS AFFECTED
Natural selection	Individuals with favorable alleles reproduce preferentially, increasing the frequency of these alleles.	Adaptive	Decreases—unfavorable alleles are eliminated from the population
Mutation	New alleles are created randomly.	Nonadaptive	Increases—new alleles are introduced into the population
Genetic drift (Founder effect and bottleneck effect)	Allele frequencies change due to chance events.	Nonadaptive	Decreases—alleles are eliminated from the population
Gene flow	Alleles move from one population to another.	Nonadaptive	Increases—new alleles are added to the population

equilibrium, and variation will be maintained. Moreover, genotype frequencies in a nonevolving population can be predicted from the allele frequencies. This mathematical observation has come to be known as the **Hardy-Weinberg principle** and is a cornerstone of population genetics (see **UP CLOSE: THE HARDY–WEINBERG PRINCIPLE**).

say you have a population of mice with two different alleles for a fur-color gene, *B* and *b*, where *B* is dominant and *b* is recessive. Let's further assume that there are two possible phenotypes associated with these genotypes, brown (*BB* or *Bb*) and white (*bb*).

Now let's say we find that white mice make up 9% of the mouse population. What, then, are the frequencies of the *B* and *b* allele in this population? We can use the Hardy–Weinberg equations to find out. Because white mice are *bb*, we know that the frequency of homozygote recessives, q^2, is 9%, or .09 (9/100). Taking the square root of this number, we get *q*, or the frequency of the *b* allele, which in this case is 0.3. Because there are only two alleles for fur color in the populations, their frequencies ($p + q$) must add up to 100%, or 1. Therefore, the frequency of the *B* allele is 1 − 0.3 = 0.7 or 70%.

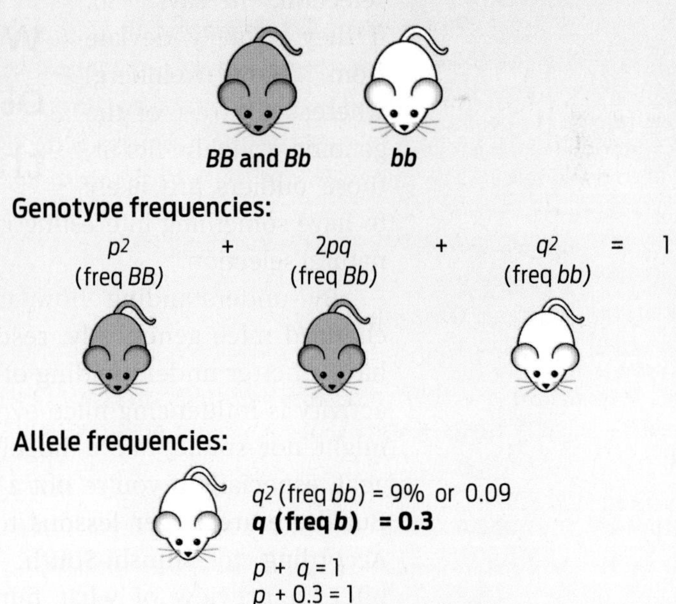

BB and Bb
bb
Genotype frequencies:
p^2 (freq BB) + $2pq$ (freq Bb) + q^2 (freq bb) = 1
Allele frequencies:
q^2 (freq bb) = 9% or 0.09
q (freq b) = 0.3
$p + q = 1$
$p + 0.3 = 1$
p (freq B) = 1 − 0.3 **= 0.7**

Now that we know the frequency of the two alleles, we know *p* and *q*, and we can calculate the genotype frequencies we would expect if this particular gene is in Hardy–Weinberg equilibrium in this population. For example, we would predict the frequency of heterozygotes to be $2pq$ (0.42). If we then sample mice in the population, analyze their genotypes, and find the frequency of heterozygotes to be 0.6, we would know that this gene is not in Hardy–Weinberg equilibrium in this population, and some evolutionary force is therefore acting to change the allele or genotype frequencies.

The Hardy-Weinberg principle can help researchers figure out, say, if genetic drift or natural selection is operating in a given population. Let's say biologists obtain samples of DNA from a random sampling of mice in a population and they look at the frequencies of genotypes at 10 different regions of DNA. Nine of those regions have genotype frequencies predicted by the Hardy-Weinberg principle, but one does not–it is far from Hardy-Weinberg equilibrium. Researchers then know that something interesting is happening at that one DNA location–some force of evolution is acting. In fact, this is how Munshi-South and his colleagues identified the candidate genes to compare between city and country mice.

"You can use certain deviations from Hardy-Weinberg equilibrium to find parts of the genome that are under selection," he says. "So, if they strongly deviate from Hardy-Weinberg, whereas the rest of the genome roughly fits it, those outliers are likely to have something interesting going on, like natural selection."

Manhattan offers a preview of what human activity will do to many other species in the coming years.

By understanding how city life has changed mice genetically, researchers will have a better understanding of how human activity is influencing mice evolution. That might not sound like a hugely important goal, especially if you're not a fan of mice. But there are larger lessons to take away. According to Munshi-South, "Manhattan offers a preview of what human activity will do to many other species in the coming years."

"Even just global warming alone is going to drive a lot of these processes in the future," he says. Indeed, some urban animals are already feeling the heat.

Biodiversity on Broadway

Just down the road from where Munshi-South works, researchers at Columbia University decided to look at what is perhaps the most urban of all green spaces: median strips. Their idea was to view these median strips as "islands" of wilderness within the city and to explore the diversity of ant species there. Were any species new to Manhattan, for example, or previously unidentified? And were native ant species able to compete successfully with introduced species?

The researchers collected 6,619 individual ants from 44 sites along three different avenues in New York–Broadway, Park Avenue, and the West Side Highway. Amid these crawling masses, they identified 13 different species of ant, including both native and introduced species. The most common ant species, found on nearly all medians, was an introduced species known as the pavement ant (*Tetramorium caespitum*), which hails originally from Europe, but ants from as far away as Japan were found. Somehow, despite the close quarters, all these different species coexist alongside one another while maintaining their distinct lifestyles. As the researchers noted in their study, "Manhattan is, if not quite a melting pot of ant species, at least a mixing bowl."

Rob Dunn, an entomologist at North Carolina State University, was one of the investigators in this ant study. From studying many urban ant populations, he has found evidence that that the ant species that seem to do best in urban settings are ones that hail from warmer climates, like the Southwest. Because of all the heat-absorbing concrete, cities on average tend to be warmer than rural and suburban areas. If cities are any indication, global climate change is going to alter the ranges and diversity of ant species in an area, with unpredictable consequences. It was while studying ants in New York City that Dunn first had the idea for a larger project: a nationwide study of ants, using citizens who are not professional scientists as research assistants to collect specimens. Now in its fifth year, the project is called, appropriately, School of Ants.

Anyone can enroll. All you need are index cards, ziplock bags, and a pecan sandy cookie–the bait of choice for seasoned ant collectors. Once the ants are collected, they're put in the freezer for a night and then shipped off to a lab at the North Carolina State University, where they will be studied and catalogued.

The aim of School of Ants is to create a map of ant species across the country, especially in urban areas, where ants have been little studied. "Basically, we're mapping what's actually out there," says Andrea Lucky, an entomologist at the University of Florida who, along with Dunn, now runs the project. "But instead of going off to the rain forest we're looking in people's backyards."

Some Ant Species in New York City

Locations of the 44 street medians included in this study: 1, Broadway; 2, West Side Highway; 3, Park Avenue.

Pecarevic et al (2010), Biodiversity on Broadway - Enigmatic Diversity of the Societies of Ants (Formicidae) on the Streets of New York City. PLoS ONE 5(10): e13222. doi:10.1371/journal.pone.0013222

Alexander Wild/alexanderwild.com

The pavement ant, *Tetramorium caespitum*
Nonnative species; found in 93.2% of medians

Alexander Wild/alexanderwild.com

The eastern black carpenter ant, *Camponotus pennsylvanicus*
Native species; found in 15.9% of medians

Photo by April Nobile / From www.antweb.org. Accessed 7 August 2013

The yellow-footed ant, *Nylanderian flavipes*
Nonnative species; found in 52.3% of medians

Alexander Wild/alexanderwild.com

The thief ant, *Solenopsis molesta*
Native species; found in 63.6% of medians

Alexander Wild/alexanderwild.com

The cornfield ant, *Lasius neoniger*
Native species; found in 11.4% of medians

BIOLOGICAL SPECIES CONCEPT
The definition of a species as a population whose members can interbreed to produce fertile offspring.

REPRODUCTIVE ISOLATION
Mechanisms that prevent mating (and therefore gene flow) between members of different species.

The most exciting part of the project, says Lucky, is discovering new ant species–as one School of Ant researcher did recently in Durham, North Carolina. The new species, which has extremely tiny eyes, behaves as a social parasite, taking over the worker ants of other colonies and making them into "slaves." The unusual species is currently in the process of being studied and named.

How did they know the ant species was new? Basically, says Lucky, you have to know a lot about what is already known. You look in books, in museum collections, and in research papers to see if your ant has been previously described. Then you compare your specimen to the existing ones. "You end up looking at a lot of ants," Lucky says.

Mostly, you're considering physical appearance–shape of the head, number of spines and segments of antennae, even how furry the creature is–but DNA evidence is important for recognizing more subtle differences. The idea is that these physical and genetic differences reflect adaptations to different environments that in turn prevent the species from interbreeding. In essence, that is what defines a species.

The term species comes from the Latin word for "kind" or "appearance." According to the **biological species concept**–the formal definition that biologists use–a species is a population of individuals whose members can interbreed and produce fertile offspring.

Members of different species cannot mate and produce fertile offspring with each other because their populations are reproductively isolated. Such **reproductive isolation** can be caused by a number of factors. For example, the two species may have different mating times, locations, or mating rituals–so, like ships passing in the night, they may never meet. This is true of many ant species, for example, which breed at different times of year. Or, two species may be able to mate–as zebras and horses can–but the hybrid offspring they produce is infertile **(INFOGRAPHIC 14.6)**.

Neil Emmerson/Robert Harding World Imagery

New York City medians are small islands of vegetation surrounded by concrete.

Lauren Nichols/Your Wild Life

Biologist Amy Savage, from Rutgers University, sampling ants on the Broadway median in New York City.

INFOGRAPHIC 14.6 Species Are Reproductively Isolated

Species are reproductively isolated in a variety of ways. These include isolation on the basis of habitat, behavior, genetics, and anatomy as well as timing and outcomes of reproduction.

Ecological Isolation—species live in different environments. The Arctic fox and the desert fox live in such different places, they never encounter one another.

Temporal Isolation—species have different mating or fertility timeframes. The leopard frog mates in early spring and the bullfrog mates in early summer.

Behavioral Isolation—species display different mating activities. The prairie chicken is not attracted to the mating display of the ring-necked pheasant.

Mechanical Isolation—species have incompatible mating organs. Plants pollinated by the hummingbird do not receive pollen from plants pollinated by the black bee.

Gametic Isolation—species have incompatible gametes. The egg and sperm from a dog and a cat cannot unite to form a zygote.

Hybrid Inviability—species' gametes unite but viable offspring cannot form. The goat and sheep can mate, but the zygote formed does not survive.

Hybrid Infertility—offspring are viable but cannot reproduce. Zebras and horses are different species because their hybrid offspring, zebroids, cannot produce offspring of their own.

? Which forms of reproductive isolation could not be overcome by in vitro fertilization?

Reproductive isolation explains why species remain separate–as do the variety of ant species that share a median strip–but how did the species form in the first place? New species form when a strong barrier to gene flow occurs between populations. That barrier could be physical–like a road or river that divides a forest in two–or climatic, like the different temperatures that occur at different elevations on a mountainside. Once this barrier forms, the separated gene pools will evolve independently by the mechanisms we have already encountered: mutation, genetic drift, and natural selection. Eventually, if enough genetic changes accumulate between populations of the same species to make them reproductively isolated, the two populations may diverge into separate species, a process called **speciation.**

Speciation is happening all the time in nature, but it can be hard to see because it occurs so slowly. It generally takes many thousands of years for species to diverge. We see the results of speciation whenever we look at the diversity of nature–there are more than 12,000 known ant species, for example–but observing speciation as it happens is much harder.

The classic, and still the best, example of speciation comes from Darwin himself: the finches he observed on the Galápagos Islands, near Ecuador, while traveling aboard the *Beagle* (see **Milestone 5: Adventures in Evolution**). The original finch species came from the mainland of Ecuador. Descendants of this mainland population then island-hopped through the archipelago, creating founder populations on each island that then adapted locally, by natural selection, to the unique environments they encountered. From one ancestral species of finch have emerged 13 different species of finch, each with a distinctive beak size and shape adapted to a unique environment (**INFOGRAPHIC 14.7**).

SPECIATION
The genetic divergence of populations, leading over time to reproductive isolation and the formation of new species.

In some respects, ants on median strips resemble Darwin's finches: they are geographically separated and do not typically interbreed. Given enough time, ants of one species located in different median strips might very well form different species, much like finches on different islands. The same goes for New York City's mouse populations: they, too, might one day form new species, provided the populations remain isolated and continue to diverge genetically. But the timescale involved makes it hard to predict. It took thousands of years for Darwin's finches to evolve into different species; the median strips and parks themselves may be long gone before that happens for ants and mice.

Empire State of Mind

If cities are the new evolutionary laboratory, then researchers Munshi-South, Dunn, and Lucky are like modern-day versions of Charles Darwin–mapping uncharted biological territory, albeit in places found right under our noses.

"I think a lot people are fairly shocked to find out that everything in North America isn't already named and catalogued and placed in a drawer somewhere," says Lucky. "There are multiple levels of discovery just waiting to happen." And you don't have to be a professional scientist to contribute to the process of discovery.

With more than half of all humans now living in urban environments, it makes sense for biologists to consider the effect that these many urban dwellers may have on populations of other organisms that share our living quarters. As Munshi-South points out, cities are active contributors to the evolutionary process, shaping the wildlife around us in profound ways. Along with climate change, urbanism is one of the main ways in which humans are directly altering the face of the planet and thereby shaping the fate of species. Scientists are only just

INFOGRAPHIC 14.7 Speciation: How One Species Can Become Many

New species may form when a strong barrier to gene flow occurs between populations. The first species of finch originated on the mainland of South America and spread to the Galápagos archipelago, a series of islands off the coast of Ecuador. As the original finch species moved from island to island, individuals were physically separated from one another into new populations that then evolved in isolation from one another. Ultimately, the separated populations evolved such that they could not interbreed, therefore becoming distinct finch species. At least 13 finch species have diverged from the original South American species.

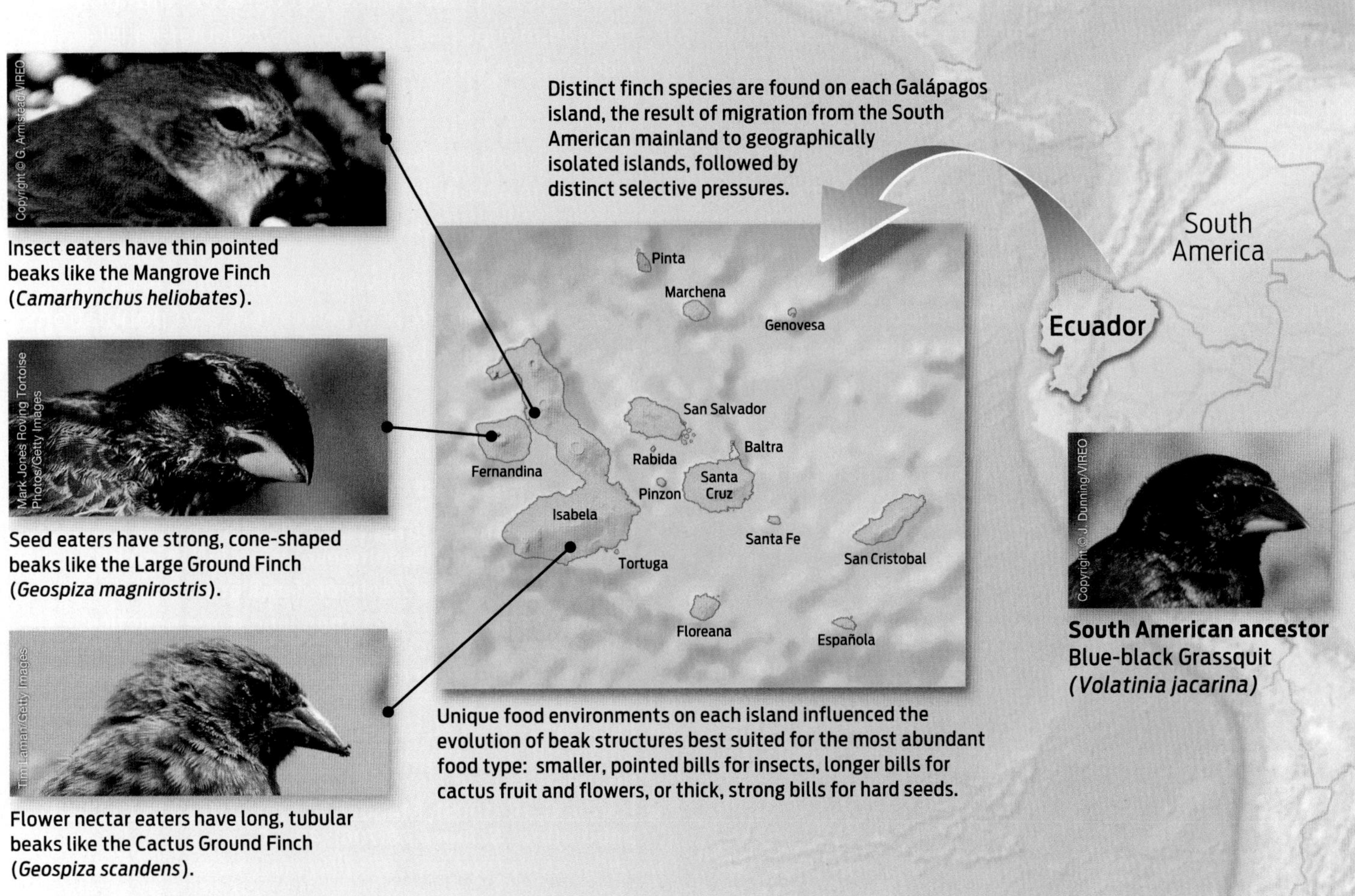

Copyright © G. Armistead/VIREO
Insect eaters have thin pointed beaks like the Mangrove Finch (*Camarhynchus heliobates*).
Mark Jones Roving Tortoise Photos/Getty Images
Seed eaters have strong, cone-shaped beaks like the Large Ground Finch (*Geospiza magnirostris*).
Tim Laman/Getty Images
Flower nectar eaters have long, tubular beaks like the Cactus Ground Finch (*Geospiza scandens*).
Distinct finch species are found on each Galápagos island, the result of migration from the South American mainland to geographically isolated islands, followed by distinct selective pressures.
Pinta
Marchena
Genovesa
San Salvador
Baltra
Rabida
Fernandina
Santa Cruz
Pinzon
Isabela
Santa Fe
Tortuga
San Cristobal
Floreana
Española
South America
Ecuador
Copyright © J. Dunning/VIREO
South American ancestor
Blue-black Grassquit
(*Volatinia jacarina*)
Unique food environments on each island influenced the evolution of beak structures best suited for the most abundant food type: smaller, pointed bills for insects, longer bills for cactus fruit and flowers, or thick, strong bills for hard seeds.

? The different species of finches have different beak shapes. Do these differences establish that the species are distinct?

beginning to understand how cities drive evolutionary change, but they will have ample opportunity for continued research. The trend of urbanism shows no signs of stopping.

Those curious about the future can catch a glimpse of what lies in store for us by studying those places that are already intensely urbanized, like New York City. In a way, say biologists, we are all New Yorkers now. ■

CHAPTER 14 SUMMARY

- From a genetic perspective, a population is identified by the particular collection of alleles in its gene pool.
- Genetic diversity, as reflected by the number of different alleles in a population's gene pool, is important for the continued survival of populations, especially in the face of changing environments.
- Evolution is a change in allele frequencies in a population over time. Evolution can be adaptive or nonadaptive. Natural selection is an adaptive form of evolution. Mutation, genetic drift, and gene flow are nonadaptive forms of evolution.
- The founder effect is a type of genetic drift in which a small number of individuals establishes a new population in a new location, with reduced genetic diversity as a likely result.
- The bottleneck effect is a type of genetic drift that occurs when the size of a population is reduced, often by a natural disaster, and the genetic diversity of the remaining population is reduced.
- Inbreeding of closely related individuals may occur in small, isolated populations, posing a threat to the health of a species.
- Gene flow is the movement of alleles between different populations of the same species, often resulting in increased genetic diversity of a population.
- Genetic diversity can be assessed by using DNA sequences to determine allele frequency.
- The Hardy–Weinberg principle describes the frequency of alleles and genotypes in a nonevolving population. It can be used to detect evolutionary change in a population, and to calculate allele frequencies when at least one genotype frequency is known.
- According to the biological species concept, a species is a population of individuals that can interbreed to produce fertile offspring.
- Speciation can occur when gene pools are separated, gene flow is restricted, and populations diverge genetically over time.

MORE TO EXPLORE

- **Munshi-South, J. (2012) TED Talk: Evolution in a big city: http://ed.ted.com/lessons/evolution-in-a-big-city**
- **School of Ants: http://schoolofants.org/**
- **Munshi-South, J. (2012) Urban landscape genetics: canopy cover predicts gene flow between white-footed mouse (*Peromyscus leucopus*) populations in New York City. *Molecular Ecology*. 21:1360–1378.**
- **Pećarević, M., et al. (2010) Biodiversity on Broadway—Enigmatic diversity of the societies of ants (Formicidae) on the Streets of New York City. *PLoS ONE* 5(10): e13222.**
- **Hardy, G. H. (1908). Mendelian proportions in a mixed population. *Science* 28(706): 49–50.**

CHAPTER 14 Test Your Knowledge

DRIVING QUESTION 1 What is a gene pool?

By answering the questions below and studying Infographics 14.1 and 14.2, you should be able to generate an answer for the broader Driving Question above.

KNOW IT

1 **Genetic diversity is measured in terms of allele frequencies (the relative proportions of specific alleles in a gene pool). A population of 3,200 mice has 4,200 dominant *G* alleles and 2,200 recessive *g* alleles. What is the frequency of *g* alleles in the population?**

2 The allele frequencies for a particular gene are given for three populations. From this information, which population is most likely to be an isolated population? Note that this gene has seven possible alleles: *A1, A2, A3, a1, a2, a3, and a4.*

Population A: 30% *A1*, 30% *a1*, and 40% *a4*
Population B: 25% *A2*, 25% *A3*, 25% *a2*, 25% *a3*
Population C: 20% *A2*, 20% *A3*, 30% *a2*, 30% *a3*

USE IT

3 A small population of 26 individuals has five alleles, *A* through *E*, for a particular gene. The *E* allele is represented only in one, homozygous, individual:

Five individuals are *D/A* heterozygotes.
Five individuals are *A/A* homozygotes.
Five individuals are *A/B* heterozygotes.
Five individuals are *C/D* heterozygotes.
Five individuals are *C/C* homozygotes.
One individual is an *E/E* homozygote.

a. Calculate the allele frequency of each of the five alleles.
b. Five *A/E* heterozygotes migrate into the population. Now what are the allele frequencies of each of the five alleles in the population?

4 Some populations, for example cheetahs, have gene pools with very few different alleles. What approach(es) could be taken to try and introduce new alleles into these kinds of populations?

5 The global human population continues to grow, and more people than ever are living in crowded cities. Given this situation, what selective pressures might the human population be currently facing or be expected to face in the near future?

DRIVING QUESTION 2 How do different evolutionary mechanisms influence the composition of a gene pool?

By answering the questions below and studying Infographics 14.3, 14.4, and 14.5 and Table 14.1, you should be able to generate an answer for the broader Driving Question above.

KNOW IT

6 Which of the following statements apply to the founder effect?

a. It is adaptive.
b. It decreases genetic diversity.
c. It is a type of genetic drift.
d. all of the above
e. b and c

7 Which of the following are examples of genetic drift?

a. founder effect
b. bottleneck effect
c. inbreeding
d. a and b
e. a, b, and c

8 A bottleneck is best described as

a. an expansion of a population from a small group of founders.
b. a small number of individuals leaving a population.
c. a reduction in the size of an original population followed by an expansion in size as the surviving members reproduce.
d. the mixing and mingling of alleles by mating between members of different populations.
e. an example of natural selection.

9 A population of ants on a median strip has 12 different alleles, *A* through *L*, of a particular gene. A drunk driver plows across the median strip, destroying most of the median strip and 90% of the ants. The surviving ants are all homozygous for allele *H*.

a. What is the impact of this event on the frequency of alleles *A* through *L*?
b. What type of event is this?

USE IT

10 From their gene pool and population size, which of the four populations in the accompanying table would you be most concerned about from a conservation perspective? Why would you be concerned?

Population	No. of Individuals	No. of Alleles, Gene 1	No. of Alleles, Gene 2	No. of Alleles, Gene 3
1	50	1	7	5
2	1,000	1	5	7
3	50	3	2	2
4	1,000	1	1	2

11 In humans, founder effects may occur when a small group of founders immigrates to a new country, for example to establish a religious community. In this situation, why might the allele frequencies in succeeding generations remain similar to those of the founding population rather than gradually becoming more similar to the allele frequencies of the population of the country to which they immigrated?

12 Why is genetic drift considered a form of evolution? How does it differ from evolution by natural selection?

apply YOUR KNOWLEDGE

INTERPRETING DATA

13 The figure below shows a bar plot of moles from different parks in New York City. As in Infographic 14.2, each vertical bar represents genotypes from 18 genomic locations in one animal. The bars are color coded, with similar genotypes represented by the same color. From the data presented in the figure:

a. Are these three populations genetically isolated from one another? Explain your answer. What factors could explain their isolation, or lack thereof?

b. Is one of the populations experiencing gene flow with another population? If so, which one, and how do you know?

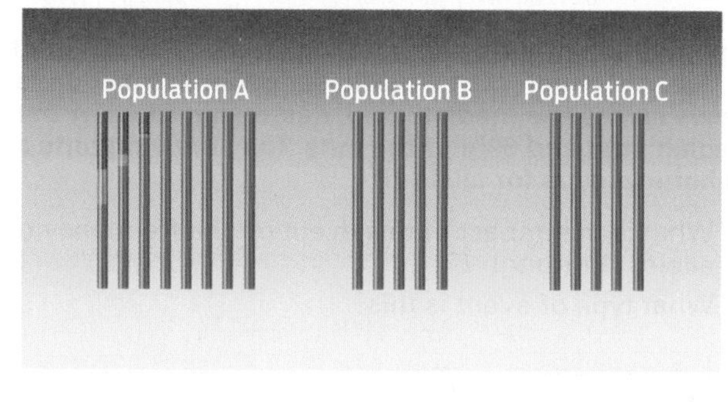

DRIVING QUESTION 3 How does the gene pool of an evolving population compare to the gene pool of a nonevolving population?

By answering the questions below and studying Infographic 14.5 and Up Close: The Hardy–Weinberg Principle, you should be able to generate an answer for the broader Driving Question above.

KNOW IT

14 Which of the following statements is/are true about a nonevolving population?

a. Allele frequencies do not change over generations.

b. Genotype frequencies do not change over time.

c. Organisms with the highest fitness are reproducing more frequently.

d. all of the above

e. a and b

15 A starting population of bacteria has two alleles of the *TUB* gene, *T* and *t*. The frequency of *T* is 0.8 and the frequency of *t* is 0.2. The local environment undergoes an elevated temperature for many generations of bacterial reproduction. After 50 generations of reproduction at the elevated temperature, the frequency of *T* is 0.4 and the frequency of *t* is 0.6. Has evolution occurred? Explain your answer.

16 Why is inbreeding detrimental to a population?

USE IT

17 The Hardy–Weinberg principle has important applications in public health. For example, it can be used to estimate the frequency of carriers (heterozygotes) of rare recessive diseases in a population. Phenylketonuria (PKU) is a rare, recessive genetic condition that affects approximately 1 in 15,000 babies born in the United States. (You may have noticed on products that contain aspartame the statement "Phenylketonurics: contains phenylalanine," a warning for people with PKU that they should avoid consuming that product.) For the purposes of this question, assume that the population is in Hardy–Weinberg equilibrium for this gene.

a. If PKU is an autosomal recessive condition, what is the genotype of people with PKU?

b. Express their genotype in terms of Hardy–Weinberg (that is, would they be *pp*, *pq*, or *qq*?)

c. Given the frequency of PKU in the population, what is the allele frequency of the recessive allele?

d. What is the frequency of the dominant allele?

e. Calculate the expected frequency of carriers in the U.S. population.

DRIVING QUESTION 4 How do new species arise, and how can we recognize them?

By answering the questions below and studying Infographics 14.6 and 14.7, you should be able to generate an answer for the broader Driving Question above.

KNOW IT

18 The biological species concept defines a species

a. on the basis of similar physical appearance.

b. on the basis of close genetic relationships.

c. on the basis of similar levels of genetic diversity.

d. on the basis of the ability to mate and produce fertile offspring.

e. on the basis of recognizing one another's mating behaviors.

19 How does geographic isolation contribute to speciation?

USE IT

20 Two populations of rodents have been physically separated by a large lake for many generations. The shore on one side of the lake is drier and has very different vegetation from that on the other side. The lake is drained by humans to irrigate crops, and now the rodent populations are reunited. How could you assess if they are still members of the same species?

21 If geographically dispersed groups of a given species all converge at a common location during breeding season, then return to their home sites to bear and rear their young, what might happen to the gene pools of the different groups over time?

MINI CASE

apply YOUR KNOWLEDGE

22 More than 50% of the global human population now lives in urban areas, and it is predicted that 70% will live in urban areas by 2050. Researchers have hypothesized that the emotional health of urbanites is influenced positively by interaction with nature. Given this information, and what you have read in this chapter, write a compelling paragraph on the need to conserve urban species and approaches to such conservation based on population genetics.

BRING IT HOME

apply YOUR KNOWLEDGE

23 The School of Ants is a citizen-scientist project to document the distribution and diversity of ants across the United States. Similarly, the Audubon Christmas Bird Count is a conservation-related project carried out by volunteer citizen-scientists. The Audubon Society uses the data collected by these volunteers to evaluate the health of bird populations and make informed decisions about conservation. There are many other citizen-scientist projects (for example, Project Squirrel and the Gravestone Project). Carry out an Internet search to find a citizen-scientist project that you find interesting. Now take the next step: enroll, collect some data, and contribute to science.

15 Evidence for Evolution

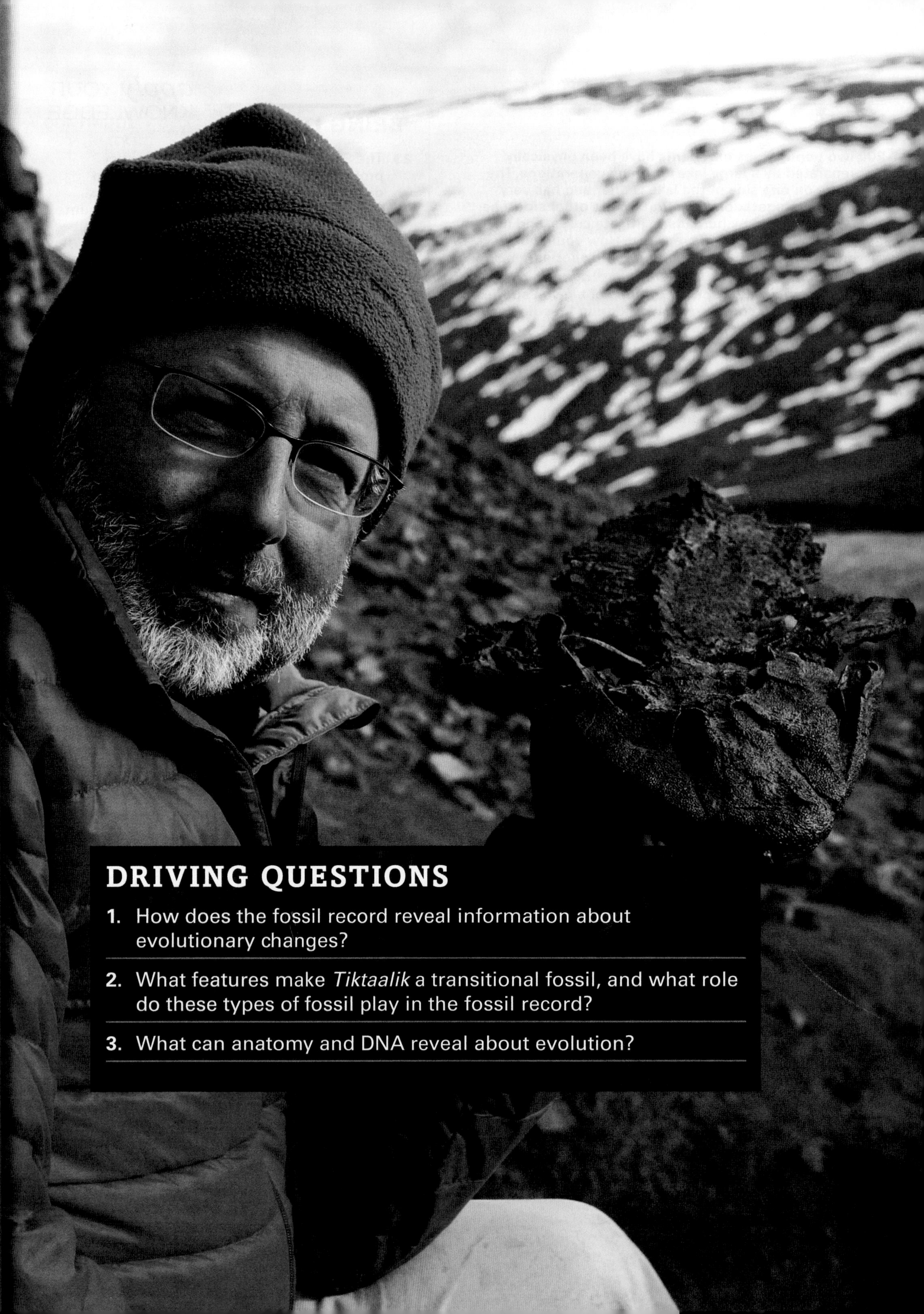

DRIVING QUESTIONS

1. How does the fossil record reveal information about evolutionary changes?
2. What features make *Tiktaalik* a transitional fossil, and what role do these types of fossil play in the fossil record?
3. What can anatomy and DNA reveal about evolution?

A Fish with Fingers?

A transitional fossil fills a gap in our knowledge of evolution

FOR 5 YEARS, BIOLOGISTS NEIL SHUBIN AND TED DAESCHLER spent their summers trekking through one of the most desolate regions on Earth. They were fossil hunting on remote Ellesmere Island, in the Canadian Arctic, about 600 miles from the north pole. Even in summer, Ellesmere is a forbidding place: a windswept, frozen desert where sparse vegetation grows no more than a few inches tall, where sleet and snow fall in the middle of July, and where the sun never sets. Only a handful of wild animals survive here, but those that do make for dangerous working conditions: hungry polar bears and charging herds of muskoxen are known hazards of working in the Arctic, says Daeschler, who carried a shotgun for protection.

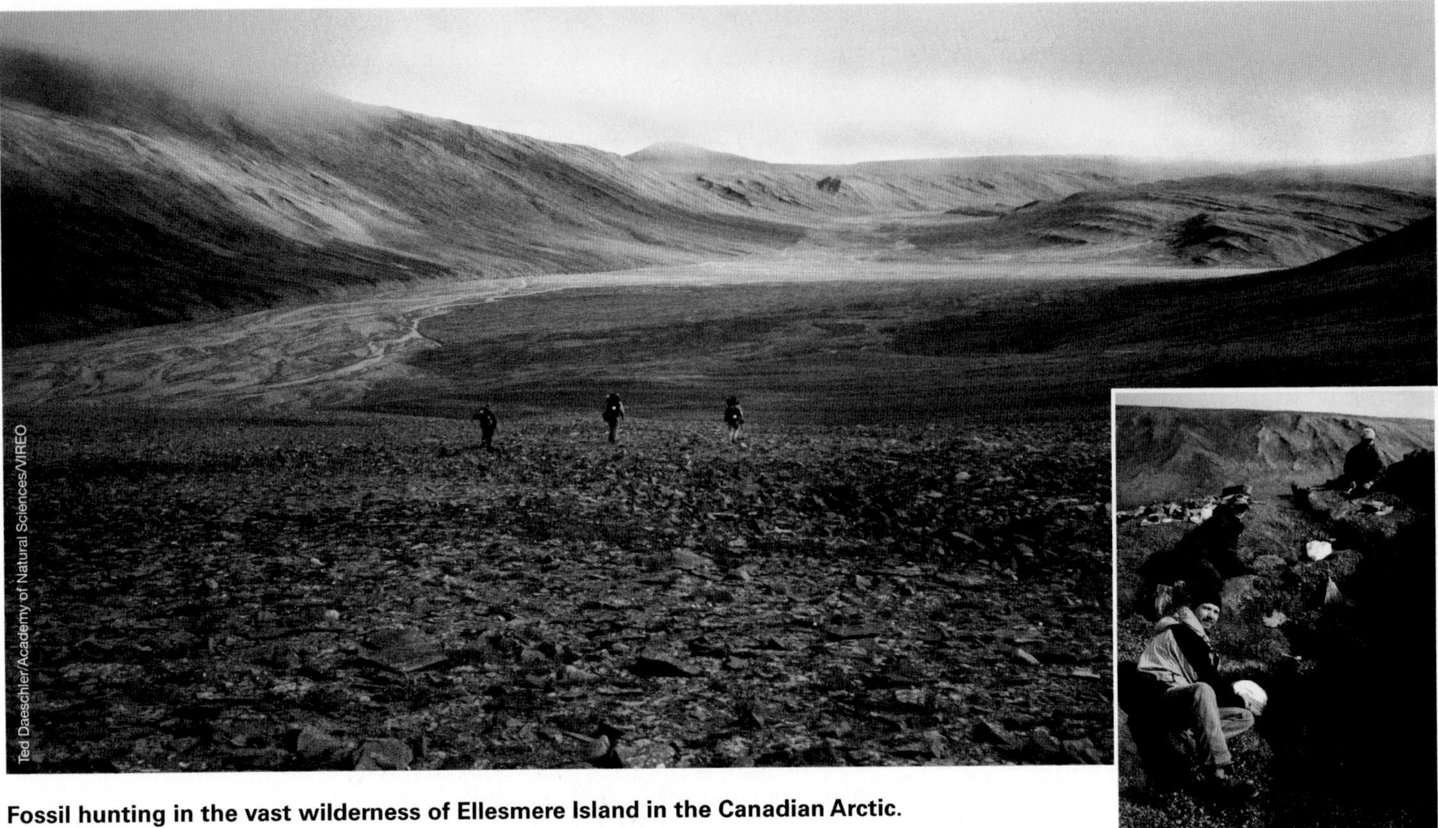

Fossil hunting in the vast wilderness of Ellesmere Island in the Canadian Arctic.

When not looking over their shoulders, the researchers drilled, chiseled, and hammered their way through rocks looking for fossils. Not just any rocks and fossils, but ones dating from 375 million years ago, when animals were taking their first tentative steps on land. For three summers, they scoured the site of what was once an active streambed but found little of interest. Then, in 2004, the team made a tantalizing discovery: the snout of a curious-looking creature protruding from a slab of pink rock. Further excavation revealed the well-preserved remains of several flat-headed animals between 4 and 9 feet long. In some ways, the animals resembled giant fish—they had fins and scales. But they also had traits that resembled those of land-dwelling amphibians—notably, a neck, wrists, and fingerlike bones. The researchers named the new species *Tiktaalik roseae*; *tiktaalik* (pronounced tic-TAH-lick) is a native word meaning "large freshwater fish." This ancient hybrid animal no longer exists, but it represents a critical phase in the evolution of four-legged, land-dwelling **vertebrates**—including humans.

VERTEBRATE
An animal with a bony or cartilaginous backbone.

Tiktaalik "splits the difference between something we think of as a fish and something we think of as a limbed animal," says Daeschler, a curator of vertebrate zoology at the Academy of Natural Sciences in Philadelphia. "In that sense, it is a wonderful transitional fossil between two major groups of vertebrates."

Today, of course, four-legged animals roam far and wide over land. But 400 million years ago it was a different story. Life was mostly aquatic then, restricted to oceans and freshwater streams. How life made the jump from water to land is a question that has long intrigued evolutionary biologists. In fact, scientists have been searching for evidence of this milestone ever since Charles Darwin first proposed that all life on the planet is related by a tree of common descent. According to Shubin, a professor of biology at the University of Chicago and the Field Museum of Natural History, *Tiktaalik* is the most compelling example yet of an animal that lived at the cusp of this important transition. Not only does it fill a gap in our knowledge, the discovery also provides persuasive evidence in support of Darwin's theory.

Reading the Fossil Record

The theory of evolution—what Darwin called **descent with modification**—draws two main conclusions about life on Earth: that all living things are related, and that the different species we see today have emerged over time as a result of natural selection operating over millions of years. Many lines of evidence support this theory (remember that in science a "theory" is an idea supported by a tremendous amount of evidence and which has never been disproved; see Chapter 1). One of the most compelling lines of evidence for evolution comes from **fossils,** the preserved remains or impressions of once-living organisms. Fossils are like snapshots of past life, capturing what life was like at particular moments in time.

Fossils are formed in a number of ways: an animal or plant may be frozen in ice, trapped in amber (hardened tree sap), or buried in a thick layer of mud. The entombed organism is thereby protected from being eaten by scavengers or rapidly decomposed by bacteria. Over time, if conditions are right—for example, if the mud encasing the specimen remains undisturbed long enough for hardening to occur—the organism's shape is preserved. Not all organisms are equally likely to form fossils, however: animals with bones or shells are more likely to be preserved than animals without such hard parts (think earthworms or jellyfish) that decay quickly. And conditions permitting fossilization are rare: the organism has to be in just the right place at just the right time **(INFOGRAPHIC 15.1).**

Because not all organisms are preserved, the **fossil record** is not a complete record of past life. Nevertheless, the existing fossil record is remarkably rich and offers a revealing window into the past. **Paleontologists,** scientists who study ancient life, have uncovered hundreds of thousands of fossils throughout the world, from many evolutionary time periods. When fossils are arranged in order of age, they provide a tangible history of life on Earth. The fossil record also allows biologists to test certain tenets of Darwin's theory.

For example, if all organisms have descended from a single common ancestor that lived billions of years ago, as the theory of evolution concludes they did, then we would expect the fossil record to show an ordered succession of evolutionary stages as organisms evolved and diversified. And, indeed, that is exactly what we see: prokaryotes appear before eukaryotes, single-celled organisms before multicellular ones, water-dwelling organisms before land-dwelling ones, fish before amphibians, reptiles before birds, and so on.

Moreover, we would expect to see changes over time within a family of organisms, and we do. One exceptionally well studied example is horses. Comparisons of modern-day horse bones with fossils of horse ancestors

DESCENT WITH MODIFICATION
Darwin's term for evolution, combining the ideas that all living things are related and that organisms have changed over time.

FOSSILS
The preserved remains or impressions of once-living organisms.

FOSSIL RECORD
An assemblage of fossils arranged in order of age, providing evidence of changes in species over time.

PALEONTOLOGIST
A scientist who studies ancient life by examining the fossil record.

Ted Daeschler/Academy of Natural Sciences/VIREO

A *Tiktaalik roseae* fossil.

INFOGRAPHIC 15.1 Fossils Form Only in Certain Circumstances

Not every organism that dies forms a fossil. Organisms are more likely to fossilize if they have bony skeletons or hard shells. In addition, the organism must be preserved quickly and kept undisturbed while mineralization or mud hardening occurs. Therefore, the fossil record is not a complete record of past life, but it has supplied an impressive body of evidence for evolution.

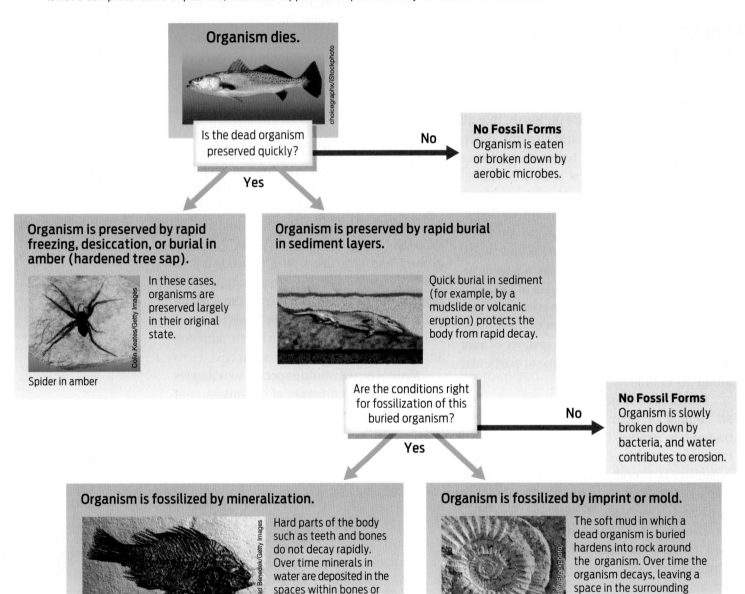

Organism dies.

choicegraphx/iStockphoto

Is the dead organism preserved quickly?

No →

No Fossil Forms
Organism is eaten or broken down by aerobic microbes.

Yes ↓

Organism is preserved by rapid freezing, desiccation, or burial in amber (hardened tree sap).

Colin Keates/Getty Images

Spider in amber

In these cases, organisms are preserved largely in their original state.

Organism is preserved by rapid burial in sediment layers.

Quick burial in sediment (for example, by a mudslide or volcanic eruption) protects the body from rapid decay.

Are the conditions right for fossilization of this buried organism?

No →

No Fossil Forms
Organism is slowly broken down by bacteria, and water contributes to erosion.

Yes ↓

Organism is fossilized by mineralization.

Arpad Benedek/Getty Images

Hard parts of the body such as teeth and bones do not decay rapidly. Over time minerals in water are deposited in the spaces within bones or replace the bone as it breaks down. The result is a mineralized fossil.

Organism is fossilized by imprint or mold.

Ralf Hettler/iStockphoto

The soft mud in which a dead organism is buried hardens into rock around the organism. Over time the organism decays, leaving a space in the surrounding rock. The space has the same shape as the exterior of the organism.

? Why are flies in amber more common than fossilized flies?

reveal how, in the course of evolution, horses have lost most of their toes. The fossils show a series of changes in bone structure over time, with the most recent fossils being the most similar to modern organisms, and the more ancient fossils being the most different. There are several branches and lineages of horse ancestors, including many that died out, but the fossils all clearly share a family resemblance (**INFOGRAPHIC 15.2**).

Descent with modification also predicts that the fossil record should contain evidence

INFOGRAPHIC 15.2 Fossils Reveal Changes in Species over Time

The fossil record of horses supports the theory of descent with modification. Forelimb fossils are similar to one another, but show changes over time from the earliest horse ancestors to modern-day horses as species diverged from a common ancestor. In the fossil record we can observe over time a reduction in toe number, as the central toe became dominant, allowing horses to move more rapidly in new prairielike environments.

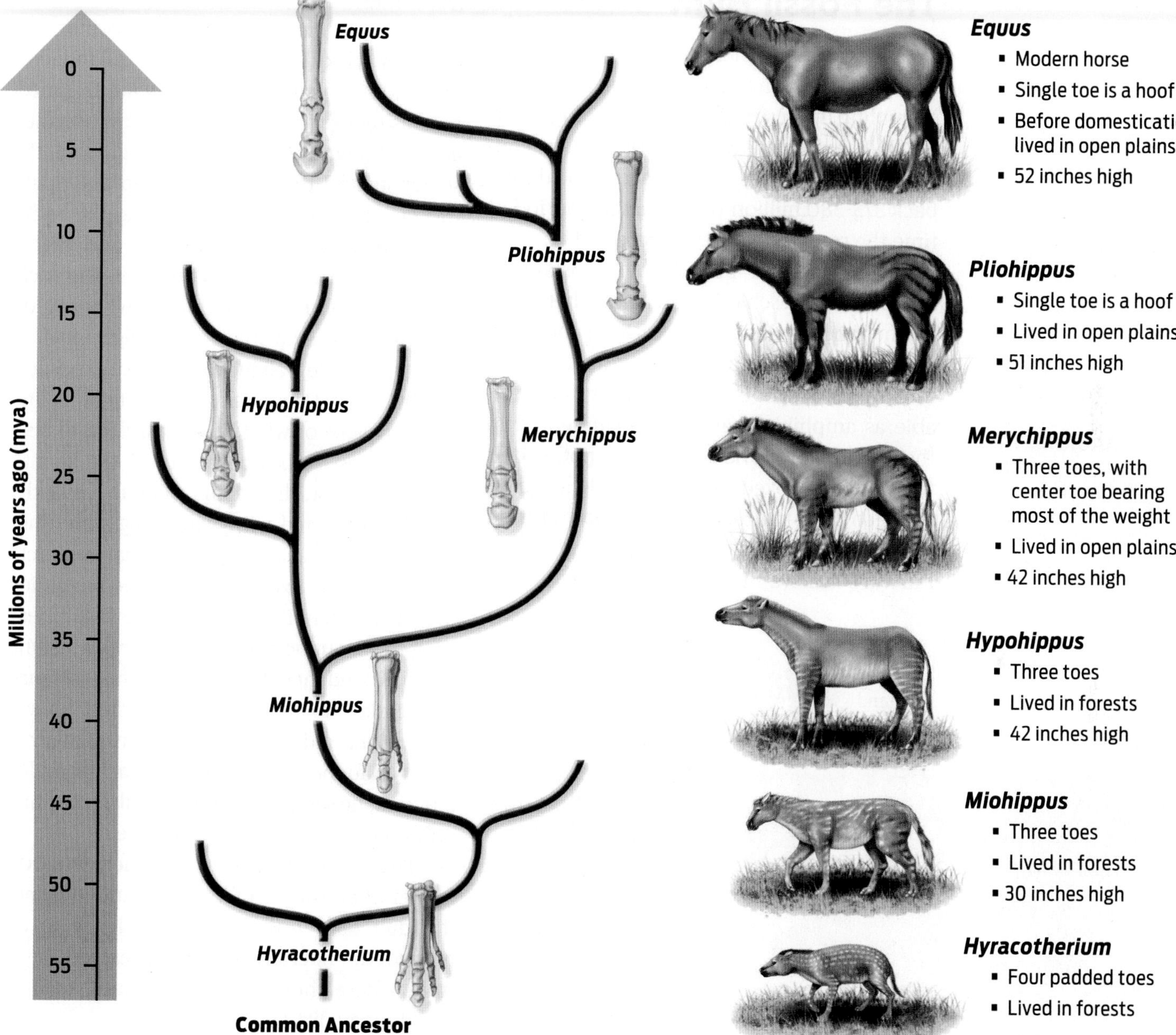

? What changes can be observed between three-toed *Miohippus* and three-toed *Merychippus*?

of intermediate organisms—those with a mixture of "old" and "new" traits. Darwin acknowledged in *The Origin of Species* that the fossil record of his day did not provide many examples of such intermediate organisms—a state of affairs he described as "probably the gravest and most obvious of all the many objections which may be urged against my views." Yet Darwin knew that if his hypothesis were correct, then intermediate fossils would eventually be found. And indeed many have been. Scientists have discovered animals with mixtures of reptile and bird characteristics and animals with mixtures of reptile and mammal characteristics. But the transition between fish and amphibians has so far remained more obscure.

The Fossil Hunt

Shubin and Daeschler began their hunt for fossils in the Canadian Arctic in 1999 after stumbling upon a map in an old geology textbook. The map showed that the region contained large swaths of exposed rock dating back 375-380 million years–just the period of time the researchers were interested in.

Why was this period so important to Shubin and Daeschler? They knew that there are no land-dwelling vertebrates in the fossil record before 385 million years ago. By 365 million years ago, organisms easily recognizable as amphibians are well documented in the fossil record. The scientists hypothesized that if they looked at rocks sandwiched in between these two time periods–those around 375 million years old–they might find one of Darwin's elusive "intermediates." Moreover, Ellesmere Island is one of only three places on Earth where rocks of this time period are exposed. To Shubin and Daeschler's knowledge, no other paleontologists had explored the area, which meant it was a potential fossil gold mine.

RADIOMETRIC DATING
The use of radioactive isotopes as a measure for determining the age of a rock or fossil.

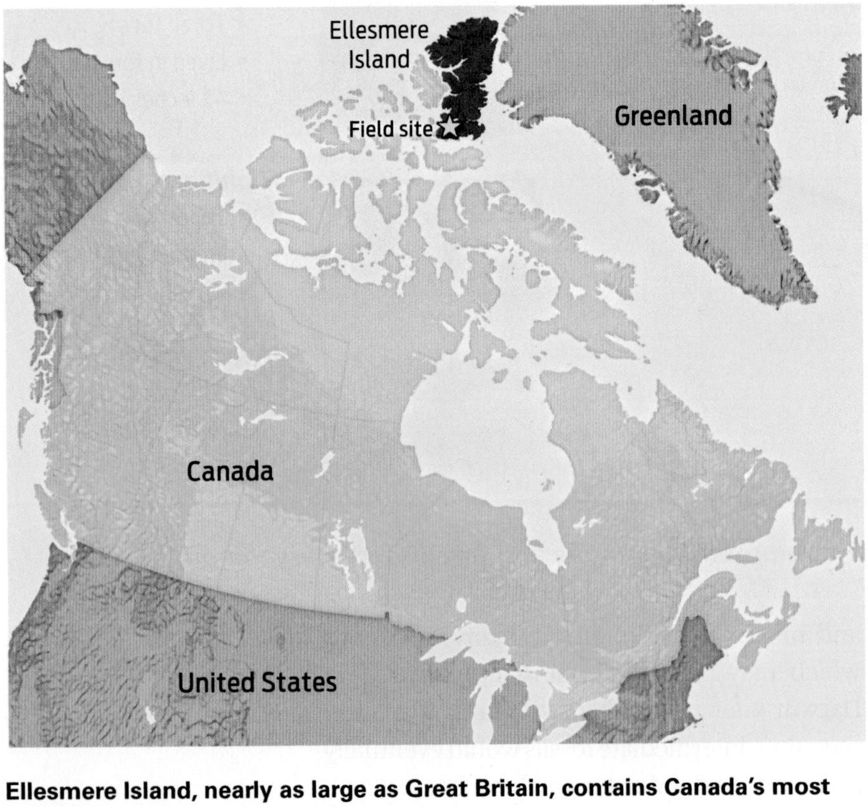

Ellesmere Island, nearly as large as Great Britain, contains Canada's most northern point.

Knowing exactly where to look for fossils was tricky, since Ellesmere Island covers 75,000 square miles. To locate the most promising dig site, the scientists first studied aerial photographs. Once on the ground, the scientists and their team split up and spent the first two seasons just walking the rocky exposures, prospecting for bits and pieces of fossils that had eroded out from the rock. When they found something interesting on the surface, they would start to dig.

It was while walking these rocky exposures in 2002 that Daeschler and his team found the first piece of what would turn out to be a *Tiktaalik* fossil–"basically part of the snout," he says. At first, they didn't think much of the find, but collected it anyway along with other fossil pieces. Back in Philadelphia, researchers cleaned the fossil, removing the remaining rock. Even then, Daeschler says, it wasn't clear what the snout belonged to. Not until a visiting graduate student remarked on the resemblance of the skull to one from the earliest known amphibians did the researchers realize what they had found. If ever there was a "lightbulb" moment, he says, this was it. But, alas, they had only one small piece of the creature.

The team returned to Ellesmere in 2004 for another round of hunting and digging. It didn't take long for their patience to be rewarded: "Literally inches," Daeschler says, from where they'd been excavating before, they hit pay dirt.

The fossils they found looked like the elusive intermediate creature the team had been hunting for. But how could they be sure it was the right age? Logically, fossils are at least as old as the rocks that encase them, so if you know the age of the rocks, then you know the age of the fossils, too. Some types of rocks can be dated directly by **radiometric dating,** in which the proportion of certain radioactive isotopes in rock crystals serves as a geologic clock (isotopes and radiometric dating are described further in Chapter 16). Fossils found in or near these layers can be dated quite precisely. If fossils are found in rock layers that cannot be directly dated by radiometric dating, they can be dated indirectly by their position with respect to rocks or fossils of known age that are either deeper or shallower, a technique called **relative dating.** Generally speaking, the deeper the fossils, the older they are. Using a combination of both methods, scientists have determined that the rocks where *Tiktaalik* was found are 375 million years old, which means *Tiktaalik* is that old as well **(INFOGRAPHIC 15.3)**.

RELATIVE DATING
Determining the age of a fossil from its position relative to layers of rock or fossils of known age.

INFOGRAPHIC 15.3 How Fossils Are Dated

When an organism dies and is preserved, its remains may in time become fossilized and buried under layers of sediment, which accumulate on top of even older layers of sediment. When a fossilized organism is uncovered, its age can be determined by two main forms of dating: radiometric and relative.

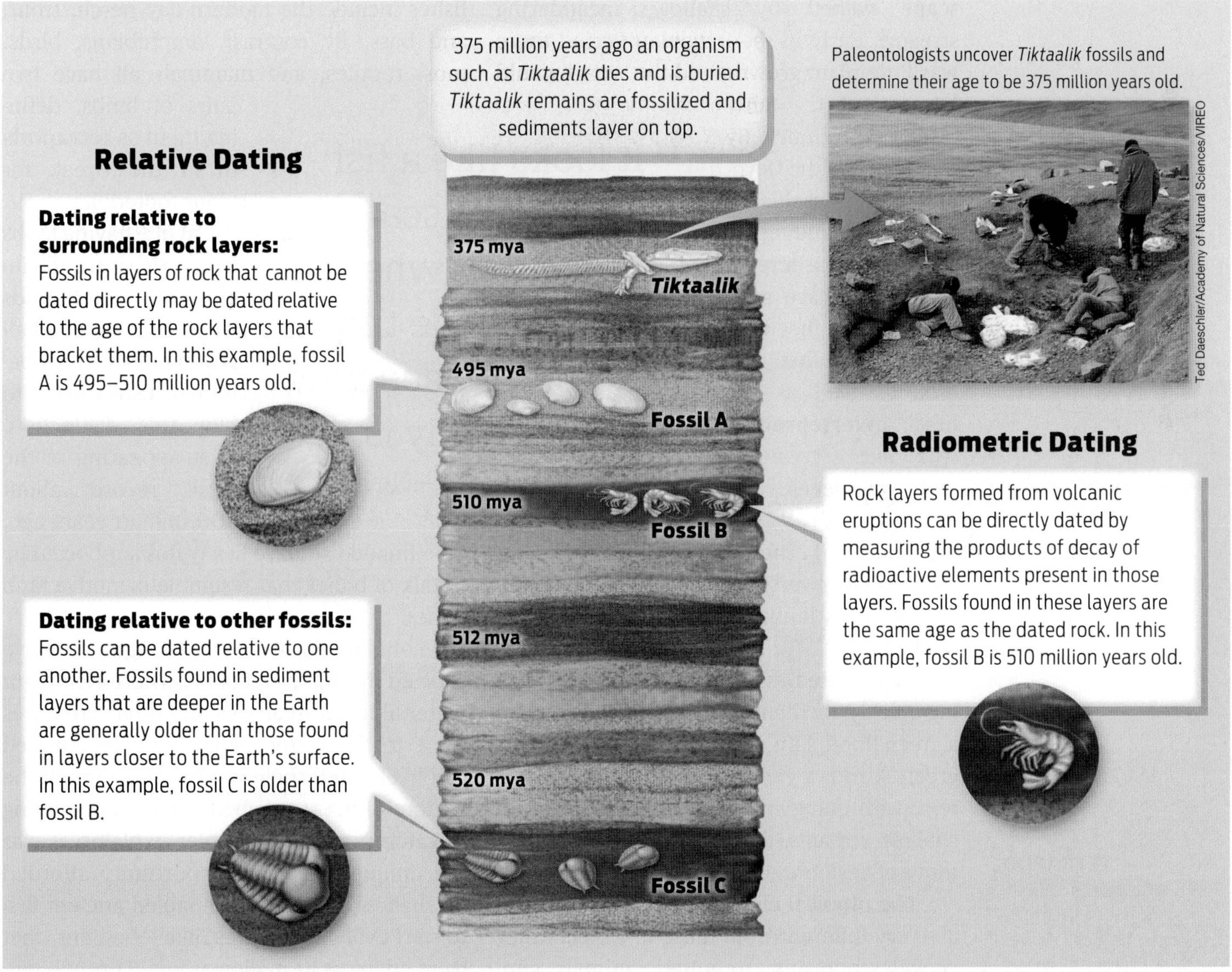

375 million years ago an organism such as *Tiktaalik* dies and is buried. *Tiktaalik* remains are fossilized and sediments layer on top.

Paleontologists uncover *Tiktaalik* fossils and determine their age to be 375 million years old.

Ted Daeschler/Academy of Natural Sciences/VIREO

Relative Dating

Dating relative to surrounding rock layers:
Fossils in layers of rock that cannot be dated directly may be dated relative to the age of the rock layers that bracket them. In this example, fossil A is 495–510 million years old.

375 mya
Tiktaalik
495 mya
Fossil A
510 mya
Fossil B
512 mya
520 mya
Fossil C

Dating relative to other fossils:
Fossils can be dated relative to one another. Fossils found in sediment layers that are deeper in the Earth are generally older than those found in layers closer to the Earth's surface. In this example, fossil C is older than fossil B.

Radiometric Dating

Rock layers formed from volcanic eruptions can be directly dated by measuring the products of decay of radioactive elements present in those layers. Fossils found in these layers are the same age as the dated rock. In this example, fossil B is 510 million years old.

? What is the approximate age of a fossil found in the blue-gray layer immediately below fossil B?

Setting the Stage for Life on Land

The geologic time period that Shubin and Daeschler are interested in is known as the Devonian–roughly 400-350 million years ago. Great transformations were occurring during the Devonian: jawed fishes, sharks, land plants, and insects all diversified in this period. Because sea levels were high worldwide, and much of the land lay under water, the Devonian Period has been called the Age of Fishes.

Back then, what is now the Canadian Arctic had a warm, wet climate and a landscape veined by shallow, meandering streams. Early in the Devonian Period there was little plant growth, and the world would have looked fairly brown and empty. By the middle of the Devonian, says Daeschler, if you were standing on the bank of a stream you would have seen some of the first land plants, the first forests, as well as the first **invertebrates**–spiderlike creatures and millipedes, for example–crawling on land. Still, there would have been no land-dwelling vertebrates at this time: nothing with bony limbs, nothing with a backbone or skull.

> "It looks like a fish in that it has scales and fins, but when you look inside the skeleton you see how special it really is."
>
> —Neil Shubin

By the late Devonian, things were changing quickly. By then, says Daeschler, "you had a green floodplain, a green world." It was this green world–a rich and productive ecosystem, with energy-rich leaf litter flowing into shallow streams–that set the stage for the move of vertebrates onto land.

The physical challenges of living on land are very different from those in water. Water provides buoyancy to aquatic animals, supporting their bodies and helping to keep them afloat. By contrast, animals that walk on land have to cope with gravity. Air doesn't support land animals, so their bodies need a sturdier structure. Animals on land can also dry out, which is dangerous for them because cells need water to function. And, of course, taking in oxygen is different on land and in water.

Of the many features that distinguish land animals from fish, biologists have singled out one as a key evolutionary milestone: limbs. Fish do not have limbs, in the sense of jointed, bony appendages with fingers and toes. Instead, they have webbed fins. In most fishes, the fin bones are thin and fan out away from each other. These so-called ray-finned fishes include the modern-day perch, trout, and bass. By contrast, amphibians, birds, most reptiles, and mammals all have two pairs of limbs, defining them as **tetrapods** (from the Greek for "four-footed").

While having limbs is a key feature distinguishing tetrapods from fish, one small group of fish–the lobe-finned fish–seems to blur this distinction. First appearing in the fossil record about 400 million years ago, lobe-finned fish have fleshy fins supported by a stalk of bones that resemble primitive limb bones.

Lobe-finned fish are thought to have evolved in shallow streams, where rich plant material lured small fish and other creatures close to the water's edge. The lobe-finned fish likely used their sturdy fins to touch the bottom of the streambed while maneuvering to catch prey. As Daeschler explains, it was the unique ecological opportunity afforded by shallow streams that enabled ancient fish to start evolving features, like lobed fins, that were adaptive in shallow water. Through natural selection, these traits would have become more common in the fish population. But

INVERTEBRATE
An animal without a backbone.

TETRAPOD
A vertebrate animal with four true limbs, that is, jointed, bony appendages with digits. Mammals, amphibians, birds, and reptiles are tetrapods.

A Comparison of Different Fins

Tiktaalik's forelimbs were technically fins. However, they more closely resembled lobe fins than ray fins.

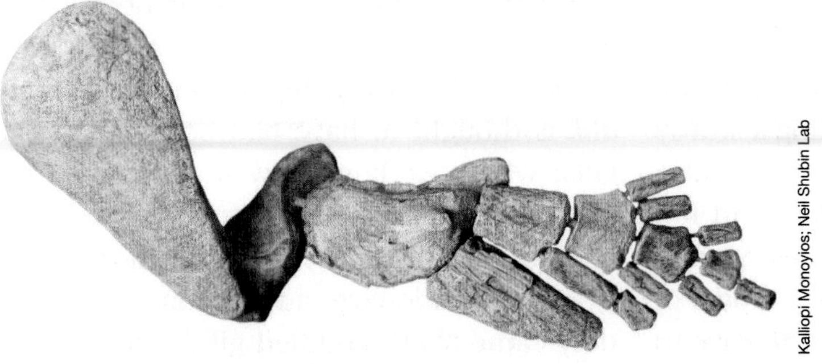

Kalliopi Monoyios; Neil Shubin Lab

A physical model of *Tiktaalik's* forelimb fin, made from replicas of fossil bones fused with flexible wire.

Tiktaalik Forelimb
Tiktaalik forelimbs had thick, weight-bearing bones covered in muscle and skin. The bones included a series of flexible wrist bones and digits ending in a small number of bony rays.

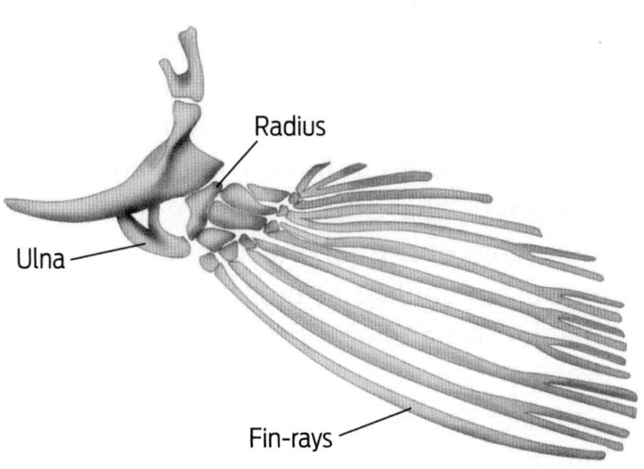

Humerus
Ulna
Radius
Fin-rays

Ray Fin
Ray-finned fishes have fins that consist of webs of skin supported by a row of thin bony rays that branch from a series of bones connected to the body of the fish.

Lobe Fin
Lobe-finned fishes have sturdy fins that consist of rows of thick, fleshy bones with short bony rays at the tips. These rows of bones all connect to a central bone, the humerus, which attaches to the body of the fish. The bony skeleton of the lobe fin is covered with muscle and skin.

lobe-finned fish were still very far from being true tetrapods. *Tiktaalik* is a step closer: "It looks like a fish in that it has scales and fins," Shubin told reporters in 2006 after the discovery, "but when you look inside the skeleton you see how special it really is."

The Fish That Did Push-Ups

Shubin and Daeschler were lucky: the fossils they found were so well preserved that they were able to study *Tiktaalik*'s skeletal anatomy in detail, even seeing how the bones interacted and where muscles attached. From these fossil bones, they determined that *Tiktaalik* was a predatory fish with sharp teeth, scales, and fins. In addition to these fishy attributes, it had a flat skull reminiscent of a crocodile head, as well as a flexible neck. To Shubin and Daeschler, the neck was one of the most surprising finds. Having a flexible neck meant that, unlike a fish, *Tiktaalik* could swivel its head independently of its body. This feature may have enabled it to catch a glimpse of predators sneaking up on it from behind, or to snap its jaws sideways like a crocodile. *Tiktaalik* also had the full-fledged ribs of a modern land animal, sturdy enough to support the animal's trunk out of water even against the force of gravity.

But it is *Tiktaalik*'s fins that have justly made it famous. While possessing many features of a lobe-finned fish, including a sturdy stalk of limblike bones, *Tiktaalik* appears also to have had a jointed elbow, wrist, and fingerlike bones. From the fossil pieces, Shubin and Daeschler were able to create a model of how the bones would have moved relative to one another, and they have visualized these movements on screen. The model shows that the bones and joints were strong enough to support the body and worked like those of the earliest known tetrapods–the early amphibians. "This animal was able to hold its fin below its body, bend the fin out toward what we think of as a wrist, and bend the elbow," explains Daeschler. In other words, it was a fish that could do a push-up.

With this hybrid anatomy, *Tiktaalik* was not galloping on land, of course. It probably lived most of the time in water, but Shubin and Daeschler suspect that *Tiktaalik* may have used its supportive fins to pull itself out of the water for brief periods. "This is a fish that can live in the shallows and even make short excursions onto land," Shubin said. The ability to crawl onto land would certainly have been a useful trait in the Devonian, when open water was a brutal fish-eat-fish world, whereas land was a predator-free paradise, full of nourishing bugs.

***Tiktaalik* likely used its sturdy forelimb fins to pull itself out of the water for short excursions on land.**

Like other fish living at the time, *Tiktaalik* is thought to have had both lungs and gills, which explains how it could breathe out of water for these short excursions. People sometimes assume that lungs were a late evolutionary adaptation, and that they came from modified gills, which modern-day fish use to breathe in water. But in fact, lungs–air-filled organs used for respiration–evolved very early in evolutionary history, more than 375 million years ago. They existed in ancient fish. We know this because fish fossils containing lung cavities and calcified bones surrounding them have been discovered in former mudbanks. Most modern fish have retained their gills but lost their lungs over time (the lungs have evolved into a balloonlike structure called a swim bladder, which helps fish float). Some modern fish known as lungfish, however, have retained this ancestral trait. Lungfish are lobe-finned fish closely related to the lobe-finned fish from which *Tiktaalik* is believed to have descended.

There was, of course, no forethought involved in the process of limb evolution or the emergence of other suitable land traits. Fish did not develop limbs for the purpose of walking on land. Rather, limbs first evolved in shallow water, where they proved adaptive and were thus retained in the descendants of the organisms who first developed them. Then, when there was an opportunity to take advantage of a tantalizing new habitat–land–the amphibious creatures already had the skeletal "toolkit."

For all its amphibian-like adaptations, *Tiktaalik* is still considered a fish because its limbs lack the true jointed fingers and toes that characterize tetrapod limbs (in other words, they're still fins). But it's by far the most tetrapod-like of all the ancient fishes discovered to date. Scientists have jokingly referred to it as a "fishapod" **(INFOGRAPHIC 15.4)**.

INFOGRAPHIC 15.4 *Tiktaalik*, an Intermediate Fossilized Organism

Tiktaalik possesses adaptations of both fish and tetrapods. The fish-like traits would have been useful when in water, while the tetrapod-like traits would have been useful when *Tiktaalik* ventured onto land. For all its amphibian-like characteristics, *Tiktaalik* is still technically a fish because its limbs lack true jointed fingers and toes, a defining feature of tetrapods.

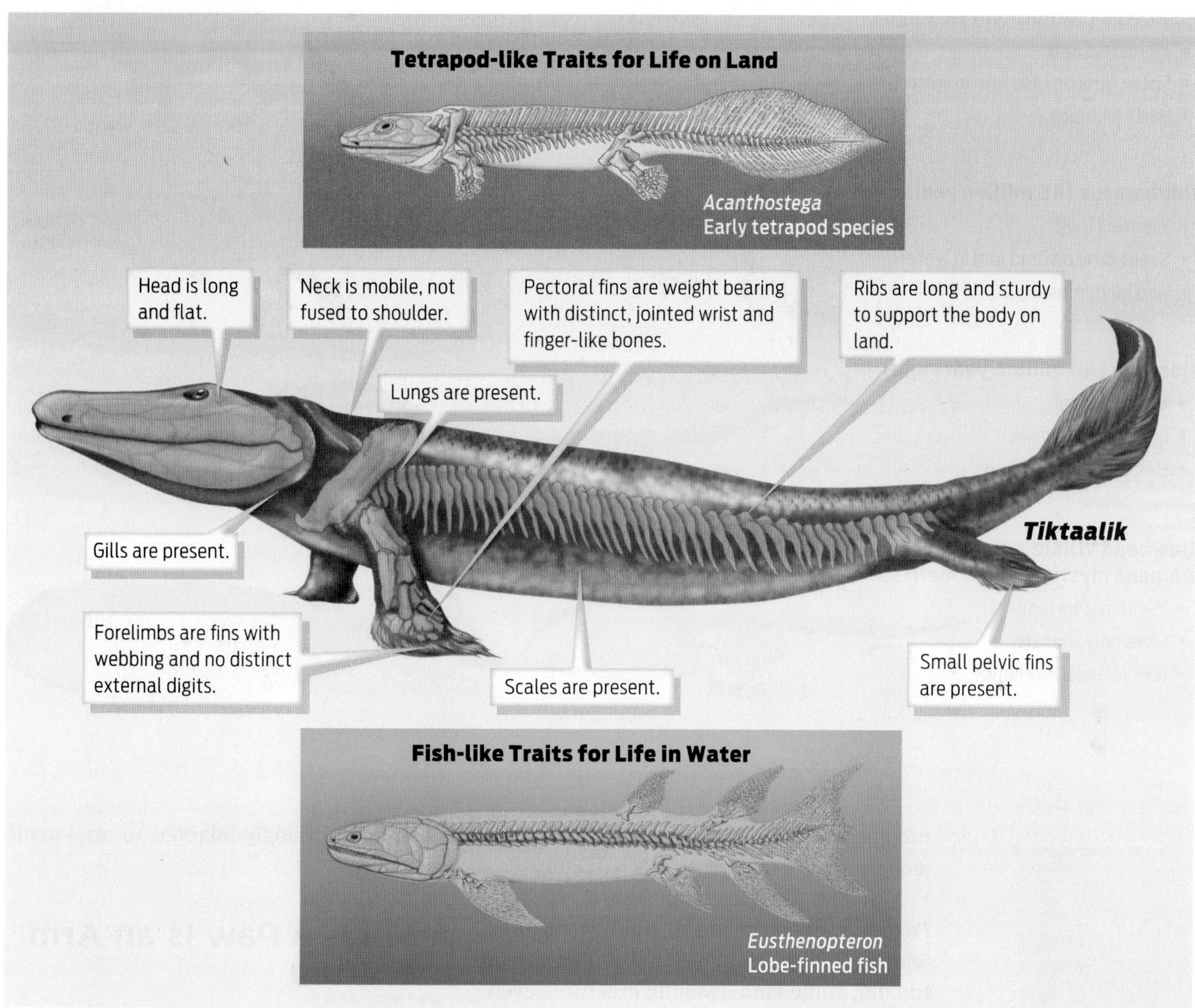

Tetrapod-like Traits for Life on Land
Acanthostega
Early tetrapod species
Head is long and flat.
Neck is mobile, not fused to shoulder.
Pectoral fins are weight bearing with distinct, jointed wrist and finger-like bones.
Ribs are long and sturdy to support the body on land.
Lungs are present.
Tiktaalik
Gills are present.
Forelimbs are fins with webbing and no distinct external digits.
Scales are present.
Small pelvic fins are present.
Fish-like Traits for Life in Water
Eusthenopteron
Lobe-finned fish

? Why would weight-bearing forelimbs be an advantage for *Tiktaalik*?

And that's what makes *Tiktaalik* such an important find: it occupies a midpoint between fish and tetrapods. "It very much fits in that gray area between things we typically call fish and things we typically call limbed animals," says Daeschler. Such intermediate, or transitional, fossils document important steps in the evolution of life on Earth. They help biologists understand how groups of organisms evolved, through natural selection, from one form into another. And they confirm that Darwin's theory of descent with

Another Evolutionary Transition: Land-dwelling mammals to ocean-dwelling whales

Ancestors of whales were terrestrial mammals. A very early whale, *Pakicetus*, was also terrestrial. Transitional fossils illustrate adaptations for an increasingly aquatic existence, including specializations of the ear and neck and almost complete loss of the hindlimbs.

***Pakicetus* (52 million years ago)**

- 1.8 meters long
- Spent time on land and in water
- Had hindlimbs

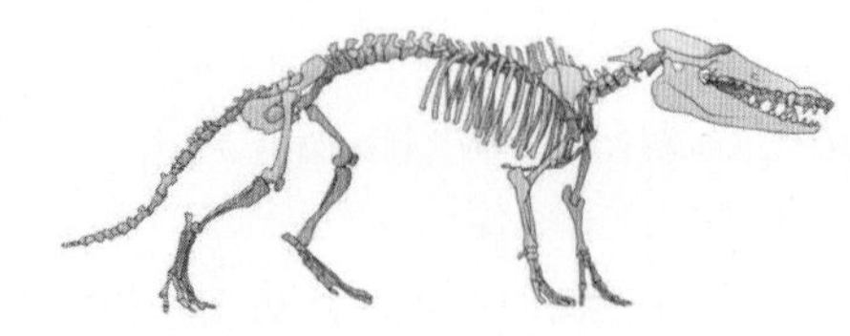

***Rodhocetus* (46 million years ago)**

- 3 meters long
- Spent time on land and in water
- Had hindlimbs

***Dorudon* (40 million years ago)**

- 6 meters long
- Lived only in water
- Had reduced hindlimbs

Bowhead Whale
***Balaena mysticetus* (modern species)**

- 15–18 meters long
- Lives only in water
- Has vestigial hindlimbs

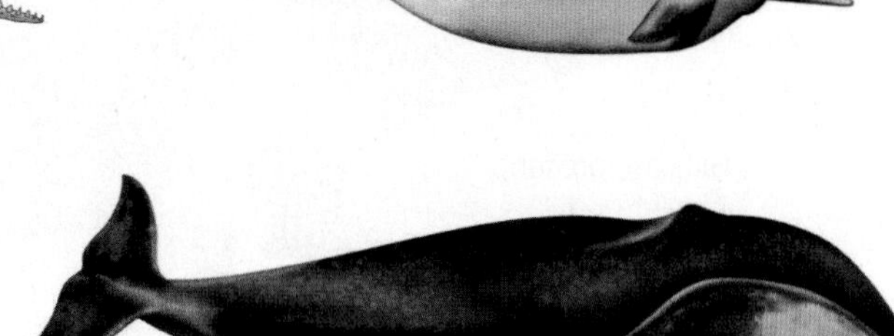

modification—which predicts such intermediate forms—is correct.

Another famous transitional fossil is *Pakicetus*, an early whale. Unlike tetrapods, which evolved first in water and then spread to land, some land-dwelling creatures eventually made their way back to the sea, adapting to an aquatic life once more. That group includes cetaceans—whales, porpoises, and dolphins. The ancestor of cetaceans was a wolf-size land-dwelling mammal that lived 50 million years ago. Although this animal, *Pakicetus*, had the body of a land animal, including four legs and paws, its head had the long skull shape reminiscent of a whale's. Fossils dating from the period since the time that *Pakicetus* lived show how whales became increasingly adapted to an aquatic existence.

A Fin Is a Paw Is an Arm Is a Wing

In *The Origin of Species*, Darwin asked, "What can be more curious than that the hand of a man, formed for grasping, that of a mole for digging, the leg of the horse, the paddle of the porpoise, and the wing of the bat, should all be constructed on the same pattern, and should include similar bones, in the same relative positions?" To Darwin, this uncanny similarity was evidence that all these organisms were related—that they share a common ancestor in the ancient past.

The fact that all tetrapods share the same forelimb bones, arranged in the same order, is an example of **homology**—a similarity due to common ancestry. Before Darwin, comparative anatomists had identified many such similarities in anatomy; what they lacked was a satisfactory explanation for why such similarity should exist. Darwin provided that explanation: homologous structures are ones that are similar because they are inherited from the same ancestor—in this case, an amphibious creature like *Tiktaalik*. Why is this significant? Think of it this way: every time you bend your wrist back and forth—to swipe a paint brush or hold a cell phone to your ear, for example—you are using structures that first evolved 375 million years ago in fish. As Shubin points out, "This is not just some archaic, weird branch of evolution; this is *our* branch of evolution" **(INFOGRAPHIC 15.5)**.

HOMOLOGY
Anatomical, genetic, or developmental similarity among organisms due to common ancestry.

INFOGRAPHIC 15.5 Forelimb Homology in Fish and Tetrapods

The number, order, and underlying structure of the forelimb bones are similar in all the groups illustrated below. The differences in the relative width, length, and strength of each bone contribute to the specialized function of each forelimb. This anatomical shape and function reflect evolutionary adaptations to different environments.

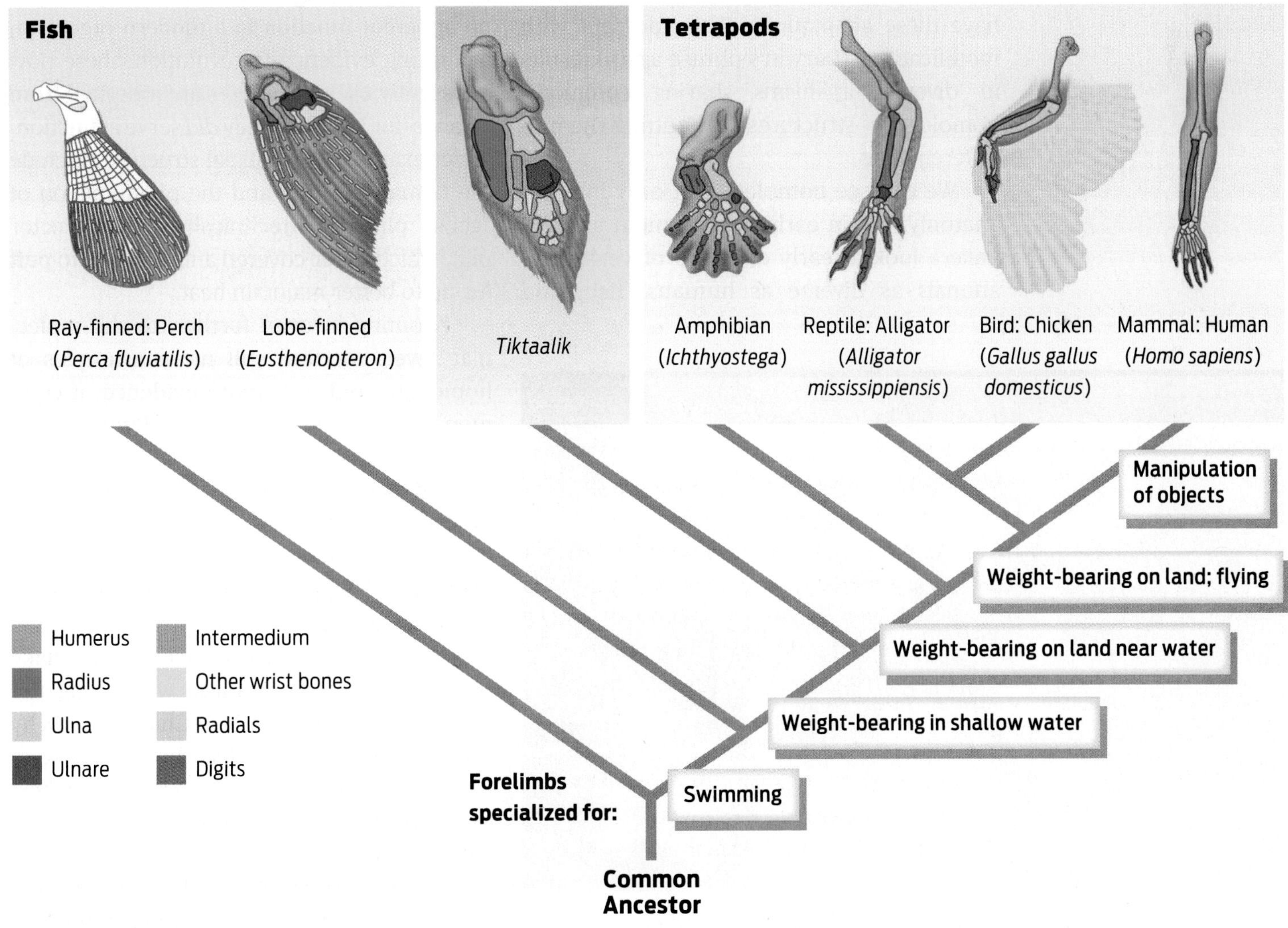

? Which bone is closest to the body in all the weight-bearing forelimbs?

VESTIGIAL STRUCTURE
A structure inherited from an ancestor that no longer serves a clear function in the organism that possesses it.

If they have the same bones, why then do a human arm and a bird wing look so different? Remember that during the process of cell division, mutations are continually being introduced into the DNA of genes, and that when these mutations occur in sperm or egg cells, they are inherited. Such mutations can produce subtle changes in the proteins encoded by those genes–proteins involved in constructing the bones that make up an arm or a wing, for example. Changes in bone proteins can result in slightly altered bones, for instance making them longer or thinner. When these modified bones are helpful to an organism's survival and reproduction, the advantageous traits are passed on to the next generation, and populations emerge that have these adaptations. This "descent with modification" (Darwin's phrase again) results in diverse organisms sharing common–homologous–structures and putting them to different uses.

We can see homology not only in adult anatomy, but in early development as well. Take a look at early embryos of vertebrate animals as diverse as humans, fish, and chickens and you'll see that they all look remarkably similar. Why should the embryonic stage of a human resemble the embryonic stage of a fish when the adults of each species look so different? Similar embryological structures are further evidence that all vertebrates have a common ancestor (**INFOGRAPHIC 15.6**).

Development helps us solve other evolutionary conundrums as well, such as why reptiles like snakes don't have limbs like other tetrapods. In fact, snake embryos *do* possess the beginnings of limbs, but these limb buds remain rudimentary and do not develop into full-fledged limbs (although you can still see stubby hindlimbs in some species of snake today). Such **vestigial structures,** which serve no apparent function in a modern organism, are strong evidence for evolution: these now apparently useless features are inherited from an ancestor in whom they *did* serve a function. Other examples of vestigial structures include the human tailbone and the phenomenon of "goose pimples" (technically called erector pili), which in fur-covered animals help to puff fur up to better maintain heat.

Zooming in even further, to the molecular level, we find still more examples of homology–and thus more evidence of common ancestry. Scientists have known since the 1950s that DNA is the molecule of heredity, and that it is shared by all living organisms on Earth. Every molecule of DNA–whether from fish, maple tree, bacterium, or human–is made of the same four nucleotides (A, C, T, and G), and all organisms use the information encoded by those nucleotides to make proteins in the same basic way, using the universal genetic code (see Chapter 8). Why should all living things use the same system of decoding genetic information? The best explanation is that this system was the one used by the ancient ancestor of all living organisms, passed on to all of its descendants, and preserved throughout billions of years of evolution.

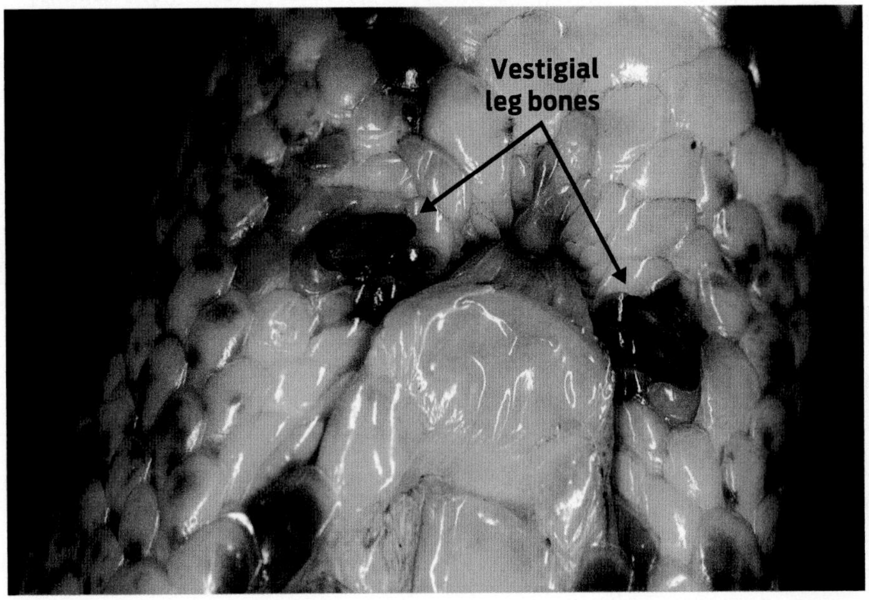

Vestigial leg bones

Simon D. Pollard/Science Source

Some snakes have remnants of pelvis and leg bones left over from their four-legged ancestors.

INFOGRAPHIC 15.6 Vertebrate Animals Share a Similar Pattern of Early Development

We can identify homologous structures by tracing their embryological development. Some of our middle ear bones, for example, are homologous with the jaw bones of reptiles and bones supporting gills in fish. We know this because all of these structures develop from the pharyngeal pouches that appear in all vertebrate embryos early in development. Similarly, all vertebrates have a post-anal tail (the extension of the spinal cord past the anus) during development. These developmental homologies are strong evidence that all vertebrate animals are related by common ancestry. Genetic changes over time have introduced modifications in later stages that give rise to distinct species with vast physical differences.

Early Embryos

Early-stage embryos of related organisms share common structures.

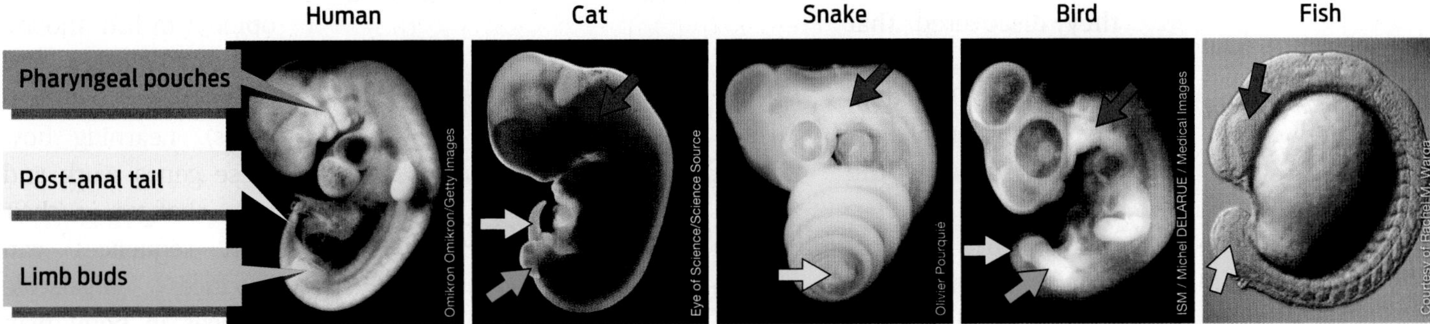

Adult Organisms

Later in development, these structures take on species-specific shape and function.

? Of the organisms shown, which have a post-anal tail as adults?

DNA and Descent

While all living organisms share DNA and the genetic code, no two species share the exact same sequence of DNA nucleotides. That's because (as described in Chapter 10) errors in DNA replication and other mutations are continually introducing variation into DNA sequences (and the proteins they encode). Over time, neutral and advantageous mutations will tend to be preserved, while harmful mutations will tend to be selected against and eliminated. In addition, much of our DNA consists of long stretches of non-coding sequences with no known function. Because mutations in these regions have no effect on an organism, they accumulate over time. As mutations are passed on to descendants, the number of sequence differences between the ancestor and its descendants grows–slowly in the case of sequences coding for critical proteins whose structures are well adapted to their functions, and more rapidly in the case of noncoding DNA, which is not involved in making proteins. Closely related species will therefore have more similar DNA sequences than species that are more distantly related.

For example, when scientists looked at one specific region of DNA–the cystic fibrosis transmembrane regulator (or *CFTR*) region–they discovered that human DNA in this region is 99% identical to chimpanzee DNA. The fact that the DNA of these two species is nearly identical reflects the fact that humans and chimps share a common ancestor that lived relatively recently–just 5-7 million years ago. By contrast, human DNA is 85% identical to the DNA of a mouse at this same region, which makes sense given that humans and mice share a common ancestor that lived between 60 and 100 million years ago. Less sequence identity would be seen between a human and a toad, whose common ancestor–a lobed-finned fish–lived roughly 375 million years ago. The more distantly related two species are, the more sequence differences in DNA sequences, and the less homology, you will see. DNA serves as a kind of molecular clock: each additional sequence difference is like a tick of the clock, showing the amount of time that has elapsed since the two species shared a common ancestor (**INFOGRAPHIC 15.7**).

> *Tiktaalik* "splits the difference between something we think of as a fish and something we think of as a limbed animal."
>
> —Ted Daeschler

When combined with evidence from the fossil record, anatomy, development (and biogeography, a fourth line of evidence, discussed in Chapter 16), molecular data become a powerful tool for understanding evolution. As we'll see in Chapter 16, DNA evidence is often a more reliable clue to common ancestry than is physical appearance, and can serve as a check on conclusions derived from the fossil record or anatomy. As well, DNA is deepening our knowledge of how limbs are constructed at the molecular level. Scientists working in Shubin's lab have shown that the same genes orchestrate limb development in animals as diverse as chickens, mice, and humans. In 2016, they discovered that these same genes also coordinate fin development in fish–though with a different end result (fins rather than limbs). Learning how these genes work and how changes in their DNA sequences can produce large-scale changes in body plan or limb structure is a hot area of biology right now, informally known as "evo-devo" (short for "evolutionary developmental biology").

Filling in the Gaps

Asked what he thinks is most interesting about the discovery of *Tiktaalik,* Daeschler homes in on what he says is a popular misconception about the fossil record–that it's "spotty" and "chaotic." But that's simply not true, he says. Despite the fact that it does not record *all* past life, the fossil record is still "very good"–so good, in fact, that you can use it to make and test predictions. You can, for example, look at the fossil record of fish and tetrapods and–suspecting on the basis of anatomy that the two groups are related–hypothesize that an intermediate-looking animal must have existed at some point. Then you can go look for it. Daeschler refers to this process as "filling in the gaps," and it's exactly what he and Shubin did with *Tiktaalik*. They knew, from the existing fossil

INFOGRAPHIC 15.7 Related Organisms Share DNA Sequences

Related organisms share DNA sequences inherited from a common ancestor. Over time, the sequence in each species acquires independent mutations. The more time that has passed, the greater the number of sequence differences that will be present. Thus, the percentage of nucleotides that differ between two species gives an indication of the evolutionary distance between them.

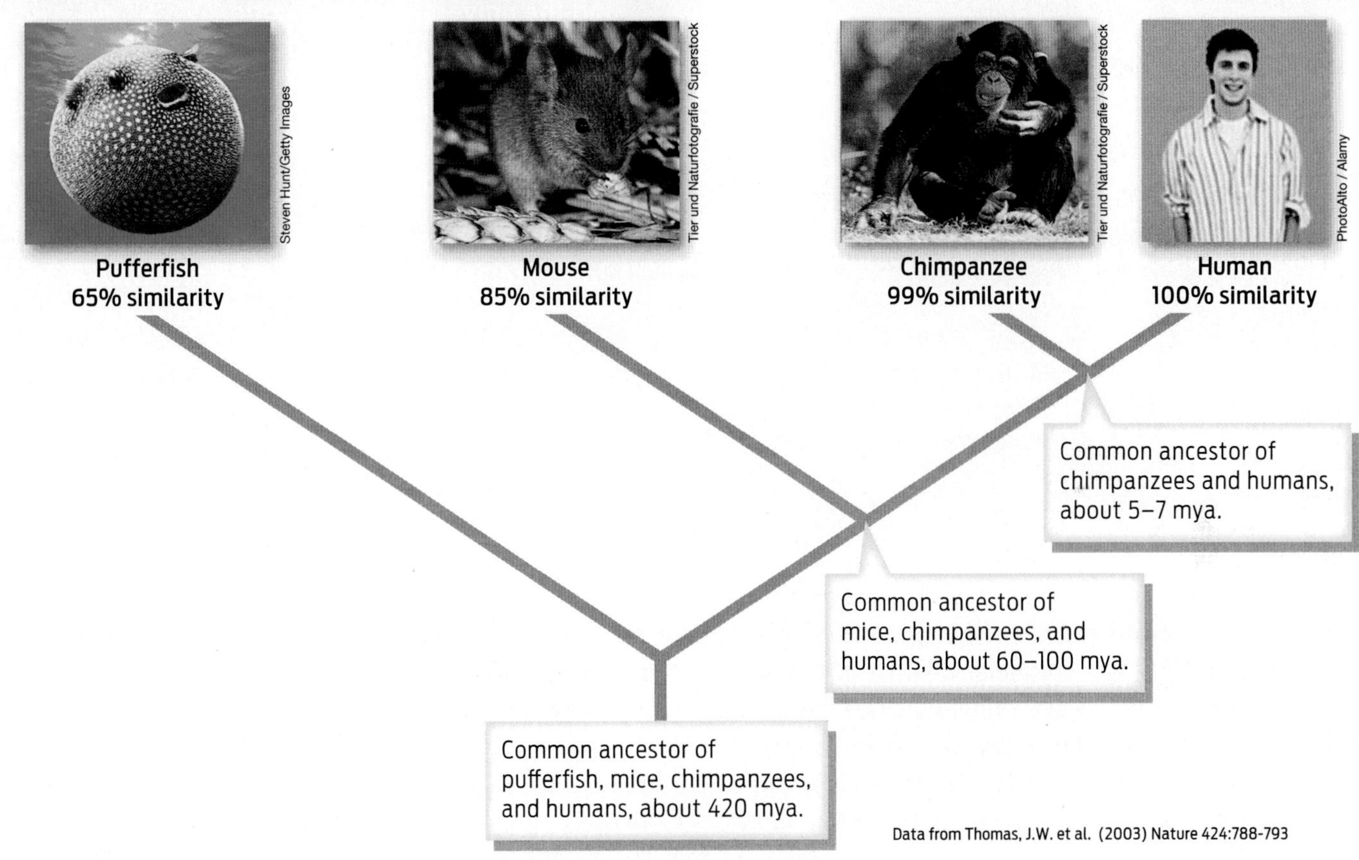

Data from Thomas, J.W. et al. (2003) Nature 424:788-793

? What is the percent similarity between the close relative and the distant relative shown in the top panel?

record, when such a creature was likely to have existed, so then it was just a question of where to look for it.

For Shubin and Daeschler, *Tiktaalik* is exciting because it shows that our understanding of evolution is correct: "It confirms that we have a very good understanding of the framework of the history of life," says Daeschler. "We predicted something like *Tiktaalik,* and sure enough, with a little time and effort, we found it." ■

CHAPTER 15 SUMMARY

- The theory of evolution—what Darwin called "descent with modification"—draws two main conclusions about life: that all living things are related, sharing a common ancestor in the distant past; and that the species we see today are the result of natural selection operating over millions of years.
- The theory of evolution is supported by a wealth of evidence, including fossil, anatomical, and DNA evidence.
- Fossils are the preserved remains or impressions of once-living organisms that provide a record of past life on Earth. Not all organisms are equally likely to form fossils.
- Fossils can be dated directly or indirectly: on the basis of the age of the rocks they are found in, or on their position relative to rocks or fossils of known ages.
- When fossils are dated and placed in sequence, they show how life on Earth has changed over time.
- As predicted by descent with modification, the fossil record shows the same overall pattern for all lines of descent: younger fossils are more similar to modern organisms than are older fossils.
- Descent with modification also predicts the existence of intermediate organisms, such as *Tiktaalik*, that possess mixtures of "old" and "new" traits. *Tiktaalik* has features of both fish and tetrapods (four-limbed vertebrates).
- An organism's anatomy reflects adaptation to its ecological environment. Changed ecological circumstances provide opportunities for new adaptations to evolve by natural selection.
- Homology—the anatomical, developmental, or genetic similarities shared among groups of related organisms—is strong evidence that those groups descend from a common ancestor.
- Homology can be seen in the common bone structure of the forelimbs of tetrapods, the similar embryonic development of all vertebrate animals, and the universal genetic code.
- Many genes, including those controlling limb development, are shared among distantly related species, an example of molecular homology owing to common ancestry.
- DNA can be used as a molecular clock: more-closely related species show greater DNA sequence homology than do more-distantly related species.

MORE TO EXPLORE

- *Tiktaalik roseae* home page: http://tiktaalik.uchicago.edu
- PBS, *Your Inner Fish* (2014): http://www.pbs.org/your-inner-fish/home/
- Shubin, N. (2008) *Your Inner Fish: A Journey into the 3.5-Billion-Year History of the Human Body*. New York: Random House.
- Nakamura, T., et al. (2016). Digits and fin rays share common developmental histories. *Nature* 537:225–228.
- Daeschler, E. B., et al. (2006) A Devonian tetrapod-like fish and the evolution of the tetrapod body plan. *Nature* 440:757–763.
- PBS, NOVA (2009) What Darwin Never Knew, based on Carroll, S. (2006) *Endless Forms Most Beautiful: The New Science of Evo Devo*. New York: Norton.

CHAPTER 15 Test Your Knowledge

DRIVING QUESTION 1 How does the fossil record reveal information about evolutionary changes?

By answering the questions below and studying Infographics 15.1, 15.2, and 15.3, you should be able to generate an answer for the broader Driving Question above.

KNOW IT

1 Which of the following is most likely to leave a fossil that represents most of the organism?

a. a jellyfish
b. a worm
c. a wolf
d. an octopus (an organism that lacks a skeleton)
e. All of the above are equally likely to leave a fossil.

2 Generally speaking, if you are looking at layers of rock, at what level would you expect to find the newest—that is, the youngest—fossils?

3 You are examining a column of soil that contains vertebrate fossils from deeper to shallower layers. Would you expect a fossil with four limbs with digits to occur higher or lower in the soil column relative to a "standard" fish? Explain your answer.

4 What can the fossil shown below tell us about the structure and lifestyle of the organism that left it? Describe your observations.

Mantonature/istockphoto

USE IT

5 You have molecular evidence that leads you to hypothesize that a particular group of soft-bodied sea cucumbers evolved at a certain time. You have found a fossil bed with many hard-shelled mollusks dating from the critical time, but no fossil evidence to support your hypothesis about the sea cucumbers. Does this cause you to reject your hypothesis? Why or why not?

6 A specific type of oyster is found in North American fossil beds dated from 100 million years ago. If similar oyster fossils are found in European rock, in layers along with a novel type of barnacle fossil, what can be concluded about the age of the barnacles? Explain your answer.

apply YOUR KNOWLEDGE

BRING IT HOME

7 Do an Internet search to find out about fossils discovered in your home state. Determine what kinds of organisms they represent, how old they are, and where in your state you would need to go in order to have a chance of finding fossils in the field.

DRIVING QUESTION 2 What features make *Tiktaalik* a transitional fossil, and what role do these types of fossils play in the fossil record?

By answering the questions below and studying Infographics 15.4 and 15.5, you should be able to generate an answer for the broader Driving Question above.

KNOW IT

8 Which of the following features of *Tiktaalik* is not shared with other bony fishes?

a. scales
b. teeth
c. a mobile neck
d. fins
e. none of the above

9 ***Tiktaalik*** **fossils have both fishlike and tetrapod-like characteristics. Which characteristics are related to supporting the body out of the water?**

USE IT

10 ***Tiktaalik*** **fossils are described as "intermediate" or "transitional" fossils. What does this mean? Why are transitional organisms significant in the history of life?**

11 ***Tiktaalik*** **has been called a "fishapod"—part fish, part tetrapod. Speculate on the fossil appearance of its first true tetrapod descendant—what features would distinguish it from** ***Titkaalik*****? How old would you expect those fossils to be relative to** ***Titkaalik*****?**

12 If some fish evolved modifications that allowed them to be successful on land, why didn't fish just disappear? In other words, why are there still plenty of fish in the sea if the land presented so many favorable opportunities?

DRIVING QUESTION 3 What can anatomy and DNA reveal about evolution?

By answering the questions below and studying Infographics 15.5, 15.6, and 15.7, you should be able to generate an answer for the broader Driving Question above.

KNOW IT

13 Compare and contrast the structure and function of an eagle wing with the structure and function of a human arm.

14 Vertebrate embryos have structures called pharyngeal pouches. What do these structures develop into in an adult human? In an adult bony fish?

15 You have three sequences of a given gene from three different organisms. How could you determine how closely the three organisms are related?

USE IT

16 What is the evolutionary explanation for the fact that both human hands and otter paws have five digits?

17 Could you use the presence of a tail to distinguish a human embryo from a chicken embryo? Why or why not?

18 If, in humans, the DNA sequence TTTCTAGGAATA encodes the amino acid sequence phenylalanine–leucine–glycine–isoleucine, what amino acid sequence will that same DNA sequence specify in bacteria?

19 Gene ***X*** **is present in yeast and in sea urchins. Both produce protein X, but the yeast protein is slightly different from the sea urchin protein. What explains this difference? How might you use this information to judge whether humans are closer evolutionarily to yeast or to sea urchins?**

apply YOUR KNOWLEDGE

INTERPRETING DATA

20 **The gene responsible for hairlessness in Mexican hairless dogs is called corneodesmosin (*CDSN*). This gene is present in other organisms. Look at the sequence of a portion of the *CDSN* gene from pairs of different species, given below. For each pair, determine the number of differences. From the variations in this sequence, which organism appears to be most closely related to humans? Which organism appears to be least closely related to humans?**

Species	Sequence
Homo sapiens (human)	ACTCCGGCCCCTACATCCCCAGCTCCCA
Canis lupus familiaris (dog)	ATTCTGGCTCCTACATTTCCAGCTCCCA
Homo sapiens (human)	ACTCCGGCCCCTACATCCCCAGCTCCCA
Pan troglodytes (chimpanzee)	ACTCCGGCCCCTACATCCCCAGCTCCCA
Homo sapiens (human)	ACTCCGGCCCCTACATCCCCAGCTCCCA
Sus scrofa (pig)	AGTCTGGCTCCTACATCTCCAGCTCCCA
Homo sapiens (human)	ACTCCGGCCCCTACATCCCCAGCTCCCA
Macaca mulatta (rhesus monkey)	ACTCTGGCCCCTACATCCCCAGCTCCCA

apply YOUR KNOWLEDGE

MINI CASE

21 **Fossils allow us to understand the evolution of many lineages of plants and animals. They therefore represent a valuable scientific resource. What if *Tiktaalik* (or an equally important transitional fossil) had been found by amateur fossil hunters and sold to a private collector? Do you think there should be any regulation of fossil hunting to prevent the loss of valuable scientific information from the public domain?**

16 Life on Earth

Ron Watts/Getty Images

DRIVING QUESTIONS

1. What do we know about the history of life on Earth, and how do we know it?
2. What factors help to explain the distribution of species on Earth, and how many species are there?
3. What are the major groups of organisms, and how are organisms placed in groups?

Q&A Evolution

From moon rocks to DNA, clues to the history of life on Earth

The history of life on Earth is a grand pageant, spanning eons, with a dizzying cast of characters—most of whom have long since exited the stage. Cataloguing all this diversity and understanding how it arose is a daunting challenge. To help make sense of this imposing sweep of history, scientists rely on a range of methods and disciplines, from geology and biochemistry to genetics and biogeography.

GEOLOGY

Q: How old is Earth, and how do we know?

A: When the *Apollo 11* astronauts Neil Armstrong and Buzz Aldrin returned to Earth from their historic 1969 moon walk, they carried with them a cargo of lunar rock chipped from the moon's surface. Embedded within these hunks of shimmering anorthosite lay clues to the earliest history of our solar system, including the planet we call home.

According to the most widely accepted explanation for the formation of our solar system, the planetary objects in our solar system are the result of a single event: the collapse of a swirling solar nebula, which formed both the sun and the planets out of cosmic dust. Since all the planets were formed at roughly the same time, we can date the age of the solar system by dating any planetary object within it. Of the many moon rocks obtained over the course of the six *Apollo* missions, the oldest have been calculated to be some 4.4 to 4.5 billion years old, which means that Earth is at least that old as well.

Why go to the moon to date Earth? With the exception of a few meteorite battle scars, the moon's surface has remained largely intact over the course of its existence. By comparison, Earth is a swirling ball of molten lava that continuously churns and digests its rocky outer crust. Because of this perpetual changing, it is difficult to find original, undisturbed rocks from Earth's earliest period. The oldest known rocks on Earth's surface, the Acasta Gneiss in a remote region of northern Canada, are 4.0 billion years old. Other ancient Earth rocks, from Western Australia, contain single mineral crystals that are 4.4 billion years old.

While these values do not establish an absolute age of Earth, they do provide a lower limit: Earth is at least as old as the materials that now make it up. From these Earth minerals and moon rocks, as well as material from meteorites

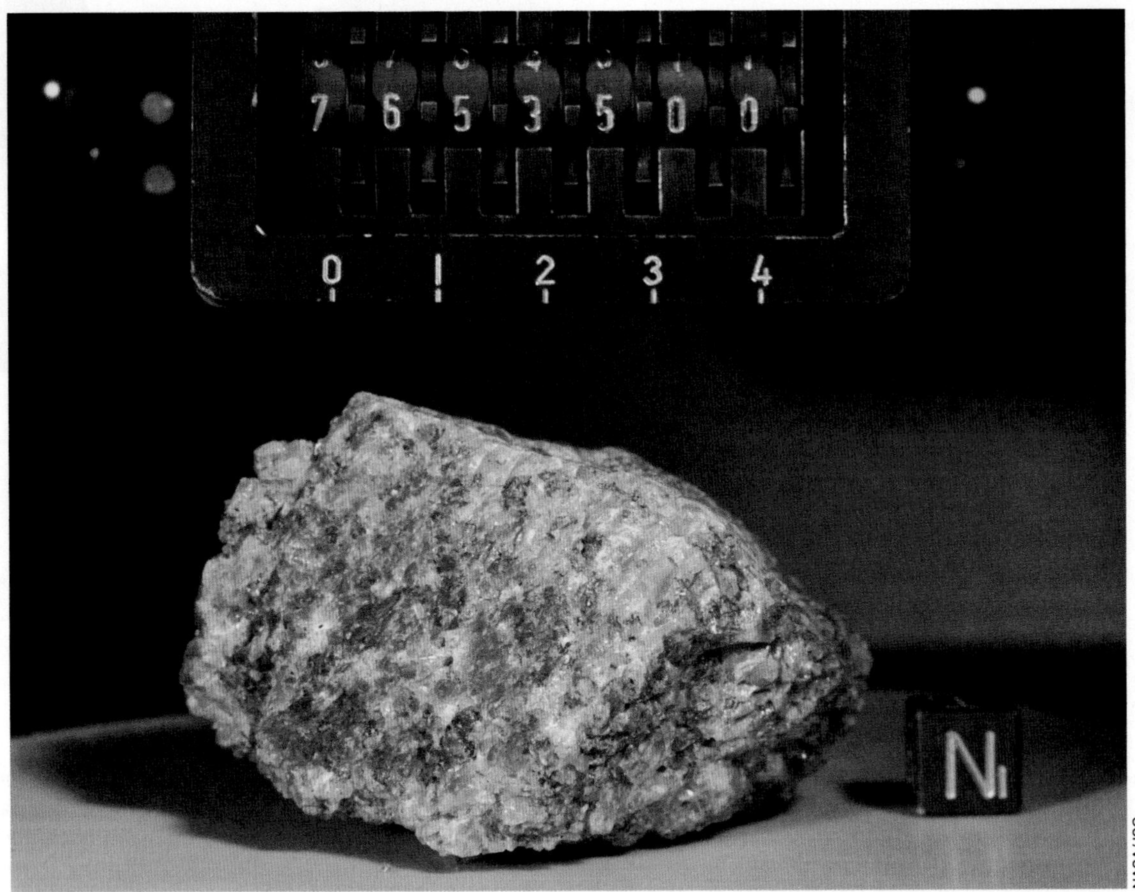

The Genesis Rock, a sample of lunar crust from about the time the moon was formed, was retrieved by *Apollo 15* astronauts James Irwin and David Scott in 1971. The scale at the top is in centimeters; the cube is 1 cm on a side.

RADIOMETRIC DATING
The use of radioactive isotopes as a measure for determining the age of a rock or fossil.

RADIOACTIVE ISOTOPE
An unstable form of an element that decays into another element by radiation, that is, by emitting energetic particles.

HALF-LIFE
The time it takes for one-half of a sample of a radioactive isotope to decay.

SEDIMENTARY ROCK
Rock formed from the compression of layers of particles eroded from other rocks.

that have fallen to Earth, scientists estimate that the age of Earth—and of the solar system more generally—is 4.54 billion years, give or take a few million years.

How are such rocks, extraterrestrial or earthly, dated? The most important method is **radiometric dating,** in which the amount of radioactivity present in a rock is used as a geologic clock. When rocks form, the minerals in them contain a certain amount of **radioactive isotopes**—unstable atoms of elements such as uranium, potassium, and rubidium—that decay into other atoms.

Radioactive isotopes decay by releasing high-energy particles from the nucleus, a change that causes one element literally to transform into another. For example, an atom of the radioactive isotope uranium-238 (the number designates the mass number—protons plus neutrons—of this particular isotope of uranium) eventually decays into a stable atom of lead-206. The time it takes for half the isotope in a sample to break down is called its **half-life.**

Different radioactive isotopes decay at different rates. Uranium-238 has a half-life of 4.5 billion years, whereas potassium-40 has a half-life of 1.3 billion years. The half-life of carbon-14 (useful for dating once-living, organic remains) is relatively short: it decays to nitrogen-14 in just 5,730 years. Because the isotopes decay at a known and constant rate, they can be used to determine the age of the materials in which they're found (**INFOGRAPHIC 16.1**).

As wind and water washed over rocks throughout Earth's history, they stripped off, or eroded, particles and carried them to other places. Sometimes the deposited particles were compressed over many years into new rock layers by water or by additional particles. Such rock, called **sedimentary rock,** can be seen in the distinctive striations, or stripes, marking successive layers of sandstone and limestone found in former riverbanks like those

INFOGRAPHIC 16.1 Unstable Elements Undergo Radioactive Decay

Radioactive isotopes are unstable versions of elements that undergo a process of radioactive decay, in which they emit energy and are converted to another element.

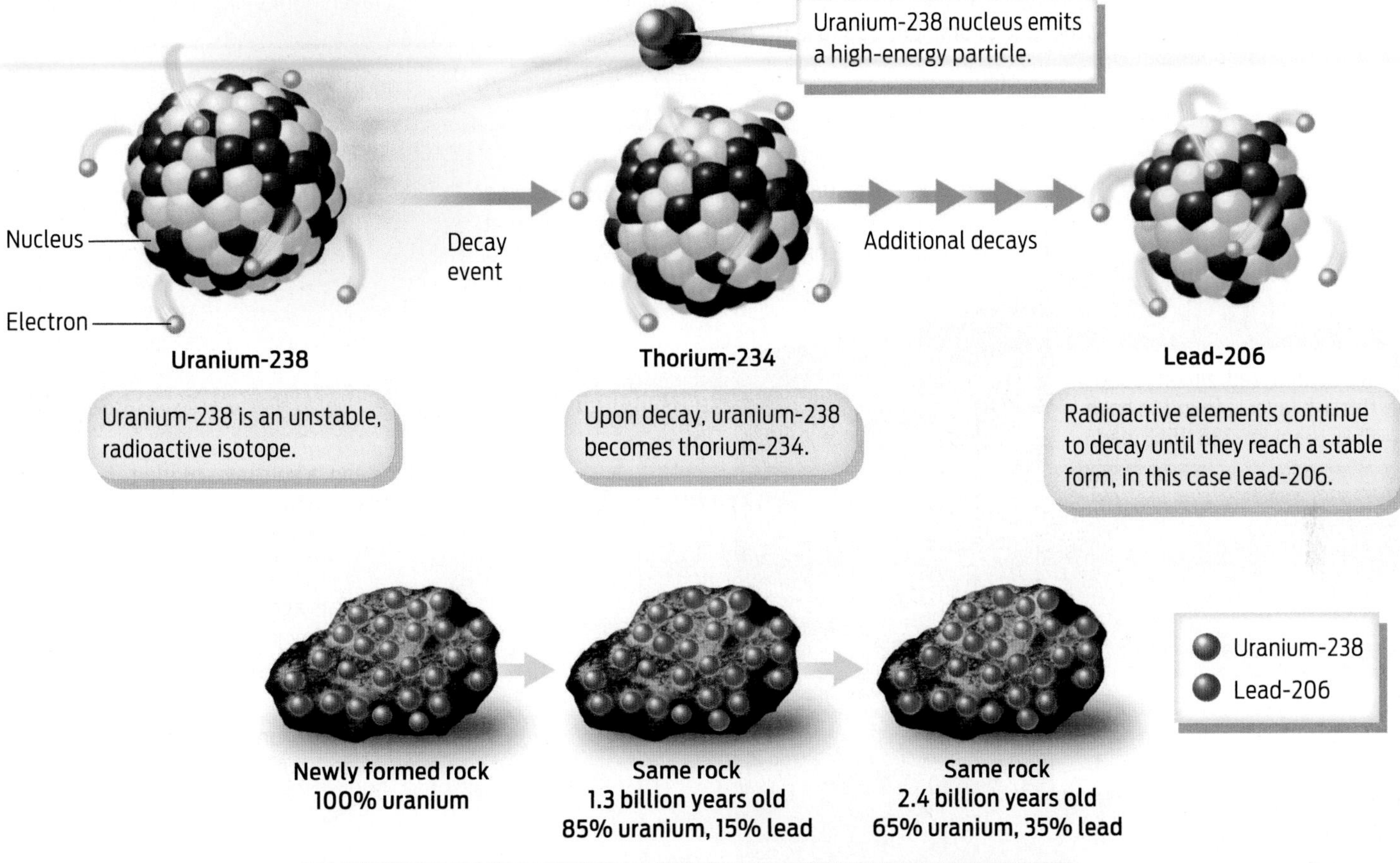
Uranium-238 nucleus emits a high-energy particle.
Nucleus
Electron
Decay event
Additional decays
Uranium-238
Thorium-234
Lead-206
Uranium-238 is an unstable, radioactive isotope.
Upon decay, uranium-238 becomes thorium-234.
Radioactive elements continue to decay until they reach a stable form, in this case lead-206.
Uranium-238
Lead-206
Newly formed rock
100% uranium
Same rock
1.3 billion years old
85% uranium, 15% lead
Same rock
2.4 billion years old
65% uranium, 35% lead
The half-life of uranium-238 is 4.5 billion years, meaning that it takes that long for half the amount of uranium in a sample to decay to lead-206. Over time, rock containing uranium-238 will have less radioactive uranium and more lead-206.

? Define the half-life of a radioactive element.

surrounding the Grand Canyon in Arizona. Most fossils are found in sedimentary rocks.

Rocks can also form suddenly as erupting volcanoes spew lava and ash over an area. When this molten debris cools and hardens, it forms **igneous rock** ("igneous" is from the Latin word for "fire").

Radiometric dating is typically performed on igneous rocks, with the uranium-lead method used to date the oldest igneous rocks. Here's how it works: When igneous rocks form, crystals of the mineral zircon are produced within the rock. Zircon crystals have a highly ordered structure that incorporates uranium but excludes lead; when these crystals first form, there is no lead present and the radioactive clock is set to zero. Over time, the uranium decays at a constant rate into lead-206. By measuring the ratio of uranium to lead in these crystals, scientists can calculate the age of the rock (**INFOGRAPHIC 16.2**).

Sedimentary rocks cannot be dated by radiometric methods because they are made up of particles from rocks of various ages. But when fossils are found in sedimentary rock between layers of igneous rock, the dating of the igneous layers gives an accurate estimate of the age range of the fossils sandwiched in between.

IGNEOUS ROCK
Rock formed from the cooling and hardening of molten lava.

INFOGRAPHIC 16.2 Radioactive Decay Is Used to Date Some Rock Types

Some rock types, like those produced during volcanic eruptions, contain radioactive minerals like zircon that can be used to determine the age of the rock. Because the isotope uranium-238 decays to lead at a constant rate, the age of rock layers containing these minerals can be calculated by measuring the ratio of uranium-238 to lead-206 in the mineral sample.

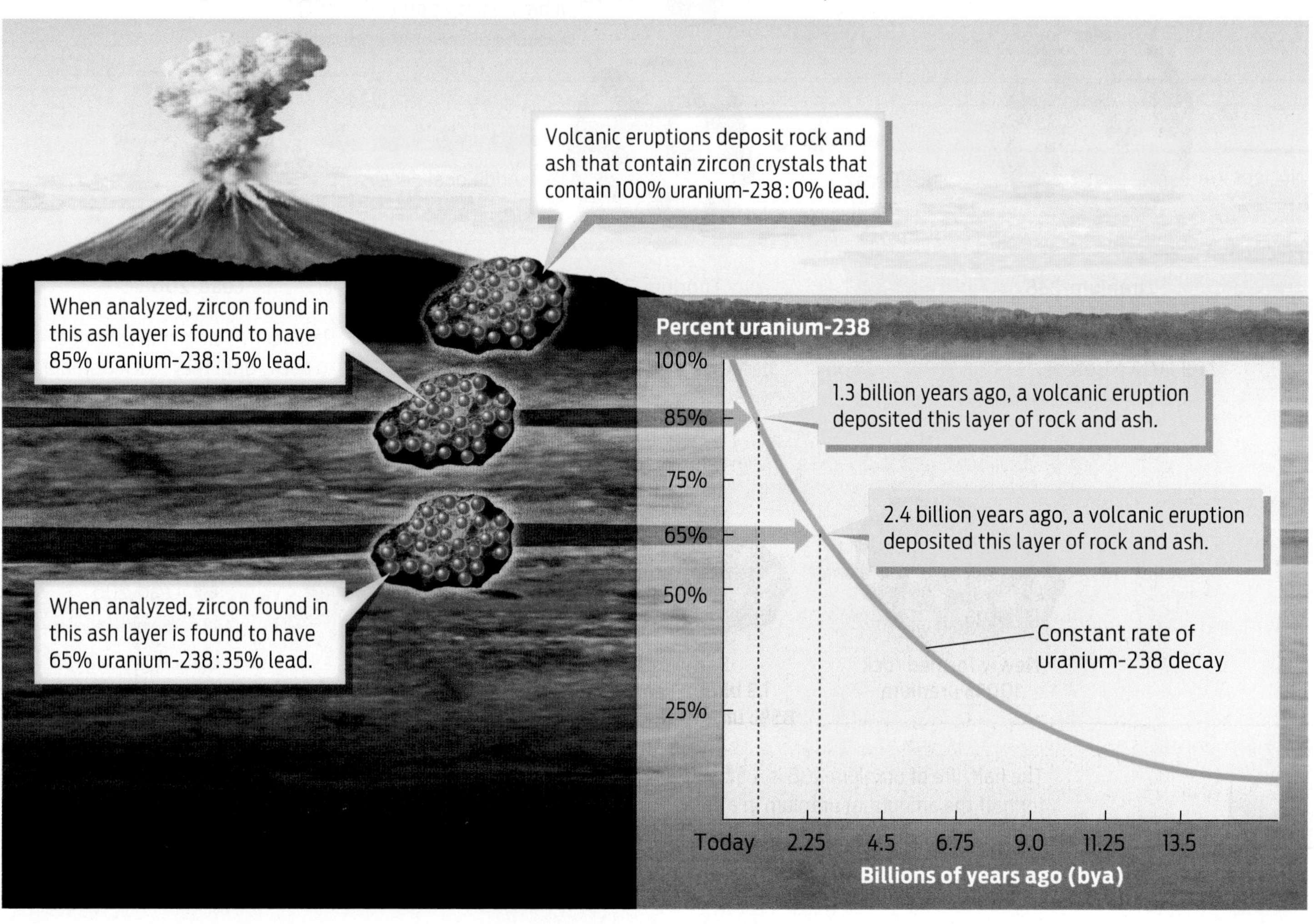

Volcanic eruptions deposit rock and ash that contain zircon crystals that contain 100% uranium-238:0% lead.
When analyzed, zircon found in this ash layer is found to have 85% uranium-238:15% lead.
When analyzed, zircon found in this ash layer is found to have 65% uranium-238:35% lead.
Percent uranium-238
100%
85%
75%
65%
50%
25%
1.3 billion years ago, a volcanic eruption deposited this layer of rock and ash.
2.4 billion years ago, a volcanic eruption deposited this layer of rock and ash.
Constant rate of uranium-238 decay
Today
2.25
4.5
6.75
9.0
11.25
13.5
Billions of years ago (bya)

? As rocks age, what happens to their amounts of uranium-238 and lead-206?

Dating rocks by radioactive isotopes is quite precise and can be confirmed by cross-checking with other radioactive isotopes. For example, scientists used three different radioactive isotopes to date minerals taken from layers of rock in Saskatchewan, Canada: the potassium-argon method yielded an age of 72.5 million years; the uranium-lead method yielded an age of 72.4 million years; and the rubidium-strontium method yielded an age of 72.54 million years.

By radioactive dating, scientists have established that Earth is indeed quite old—old enough for evolution to have been acting for billions of years. They have also been able to establish precisely the dates of key events in the timeline of life on Earth as represented in the fossil record.

BIOCHEMISTRY

Q: When and how did life begin?

A: At some point in Earth's distant past, life did not exist. Then, at a later point, it did. Where did this life come from? How did it start? The transition from nonliving to living occurred more than 3 billion years ago, a length of time so vast that reconstructing events from that period may seem hopeless. But we can still formulate hypotheses and investigate them scientifically.

Scientists have offered a number of hypotheses to explain how life began on Earth, including the idea that it arrived here fully formed on an asteroid or meteorite from outer space (an idea called panspermia). Others hypothesize that life emerged in stages over time, as inorganic chemicals combined into successively more complex molecules, including the ones making up living things. A landmark experiment lending support to this latter hypothesis was performed by University of Chicago chemist Harold Urey and his 23-year-old graduate student Stanley Miller in 1953 **(INFOGRAPHIC 16.3)**.

INFOGRAPHIC 16.3 Could Life Have Evolved in the Primordial Soup?

Urey and Miller replicated the chemical environment of the early Earth in a flask, then simulated lightning with electrical sparks. When the contents of the flask were analyzed, organic molecules, including amino acids, were present, formed from reactions between the inorganic precursors. The results suggest one way that life may have started.

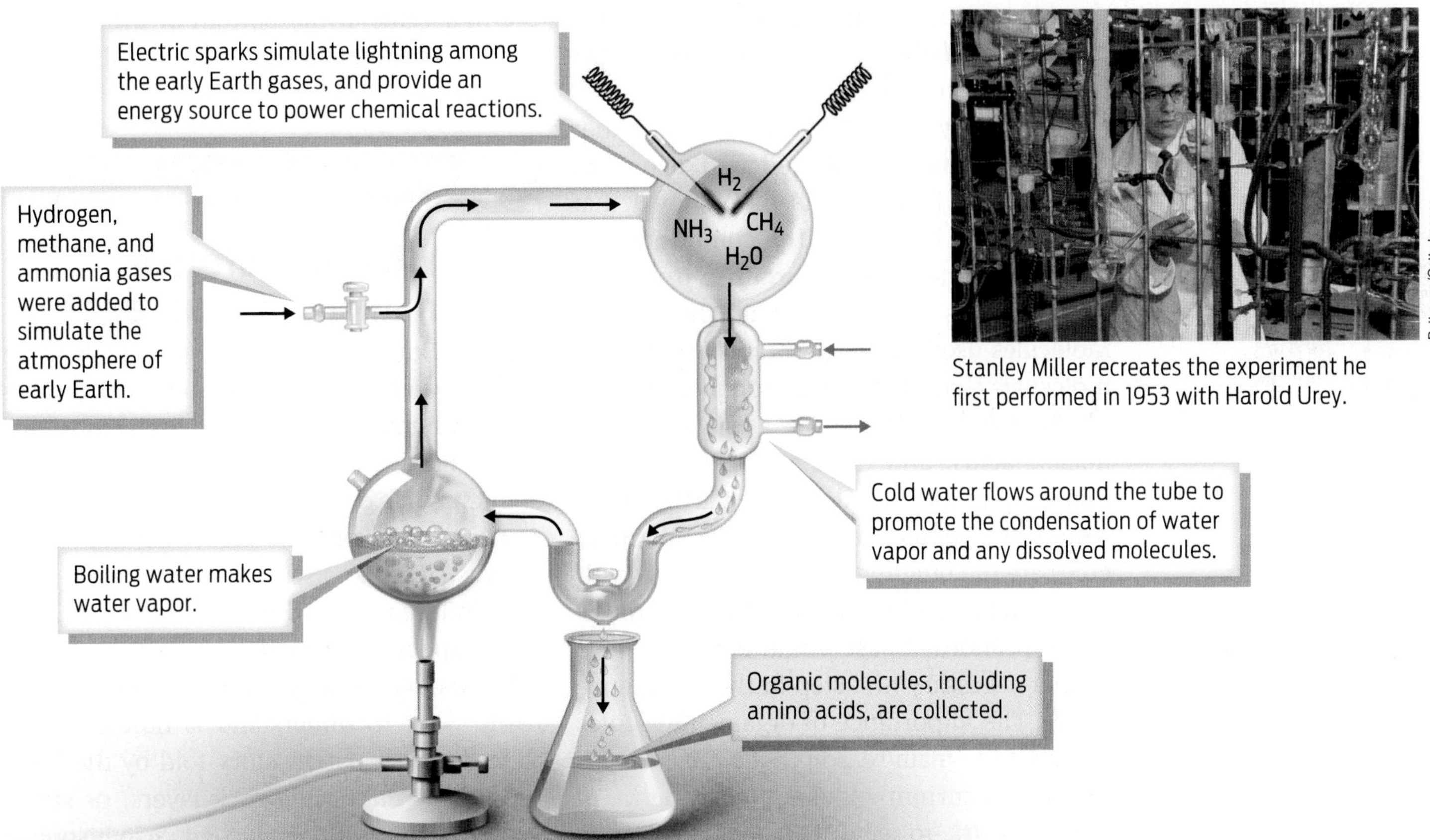

Stanley Miller recreates the experiment he first performed in 1953 with Harold Urey.

? Why was this experiment important to developing and testing hypotheses about the origin of life on Earth?

Urey and Miller hypothesized that they could synthesize organic molecules–the building blocks of life–by replicating the chemical environment of the early Earth. They combined the gases hydrogen (H_2), methane (CH_4), ammonia (NH_3), and water vapor (H_2O) in a flask filled with warm water–their best estimate of what the "primordial soup" was like. They then replicated lightning by discharging sparks into the chamber with an electrode.

The history of life on Earth is a grand pageant, spanning eons, with a dizzying cast of characters.

As the gases condensed and rained back into the flask, a host of new molecules–among them amino acids, the building blocks of proteins–formed from the basic ingredients. This landmark experiment showed for the first time that, given the right starting conditions, organic molecules could form from the inorganic materials believed to be present in the primordial soup.

Since Urey and Miller's experiment, other researchers have confirmed and extended their results. Experiments have shown that it is possible, by varying the composition of the starting materials, to produce from inorganic precursors essentially all the organic molecules used by living organisms. Organic molecules that have been produced include all 20 amino acids, as well as sugars, lipids, nucleic acids, and even ATP–the molecule that powers almost all life on Earth. These results are significant because organic molecules are the building blocks of life, and without them life could not exist. Their synthesis in the primordial past would therefore have been a necessary first step in the emergence of life. The importance of organic molecules for life is the main reason that NASA's *Curiosity* rover is currently looking for evidence of them on Mars–to see if Mars could have once supported or harbored life (see Chapter 2).

Although organic molecules are a prerequisite for life, they are not themselves alive. Life requires, among other things, the ability to grow and reproduce. Today, of course, cells carry out these life-sustaining functions. How then did living cells come about? Recall from Chapter 2 that cell boundaries are defined by a lipid membrane. Researchers hypothesize that at a certain point the lipid molecules in the primordial soup formed bubbles, a reasonable conclusion since lipids are hydrophobic and naturally form bubbles in water. The other organic molecules were incorporated into these bubbles and, over the course of millions of years, these membrane-bound bubbles filled with organic molecules eventually became cells, capable of reproducing. While these ideas are highly speculative, research on microorganisms living today in such unlikely places as hydrothermal vents at the bottom of the ocean is giving us concrete insights into how life might have begun (see Chapter 17).

PALEONTOLOGY

Q: What was life like millions of years ago?

A: Humans weren't around millions of years ago, so we have no cave paintings or other human records to help us picture what life on Earth was like. Most of what we know about past life on Earth comes from fossils–the preserved remains of once-living organisms, such as *Tiktaalik*, discussed in Chapter 15.

While each fossil find is a treasure, any single specimen reveals only a tiny slice of evolutionary history. What paleontologists really want to understand is how each fossil fits into the larger story told by the fossil record. By dating the rock layers, or strata, near where fossils are buried, scientists can determine when different organisms lived on Earth. Combined with geological evidence, the fossil record has enabled scientists to construct a timeline of life on Earth, organized by defining eras and periods (**INFOGRAPHIC 16.4**).

INFOGRAPHIC 16.4 Geologic Timeline of Life on Earth

The fossil record reveals changes in organisms over time, including the appearance and diversification of multicellular organisms as well as mass extinctions, the rapid loss of a large proportion of the organisms present at a given time.

Era	Time Period		Prominent Life Events
Precambrian	Archean	3,500 mya	First evidence of prokaryotic cells.
		2,500 mya	Prokaryotes dominate the oceans.
	Proterozoic	545 mya	First eukaryotes appear in water. Soft-bodied invertebrates develop in water.
Paleozoic Era	Cambrian	495 mya	Expansion of ocean animal diversity.
	Ordovician	439 mya	First coral reef and primitive fish appear. Ocean plants diversify. First fungi appear in water.
			Mass extinction
	Silurian	408 mya	Seedless plants, primitive insects and soft-bodied animals appear on land.
	Devonian	354 mya	Fish species diversify. First insects and seed-bearing plants appear on land.
			Mass extinction
	Carboniferous	290 mya	Amphibians appear and begin to diversify.
	Permian	251 mya	Coral species abundant in oceans. Reptiles appear on land.
			Mass extinction
Mesozoic Era	Triassic	206 mya	Ocean life diversifies in recovery from Permian extinction. Dinosaurs and mammals appear on land.
			Mass extinction
	Jurassic	144 mya	First flowering plants and bird species appear. Large dinosaurs are plant-eaters.
	Cretaceous	65 mya	Dinosaurs diversify. Cone-bearing and flowering plants dominate many habitats.
			Mass extinction
Cenozoic Era	Tertiary	1.8 mya	Mammals, birds and flowering plants diversify. Grasses appear. First primates and early humans appear.
	Quaternary	Today	Many large mammals become extinct. Appearance of modern species present today.

Life in Water

Life on Land

? During which period were prokaryotes the dominant organisms, and were they dominant on land or in water?

EXTINCTION
The elimination of all individuals in a species; extinction may occur over time or in a sudden mass die-off.

ADAPTIVE RADIATION
The spreading and diversification of organisms that occur when the organisms colonize a new habitat.

The timeline shows that during the 4.6 billion years that Earth has been around, its geography and climate have gone through dramatic changes. For the first few hundred million years or so Earth was a fiery inferno, coursing with seas of molten lava and bombarded by meteorites. Not until things simmered down and the surface cooled a bit, around 3.8 billion years ago, could it support life.

The oldest known fossils date from some 3.7 billion years ago, when Earth's climate was very different from what it is today, notably because of the absence of substantial amounts of oxygen (O_2). In this oxygenless world, the only organisms that could thrive were unicellular prokaryotes that did not rely on oxygen for their metabolic reactions. With the emergence and proliferation of unicellular photosynthetic organisms, between 3.0 and 2.5 billion years ago, oxygen began to accumulate in the atmosphere, opening the door for more-complex eukaryotic organisms to evolve.

The first multicellular, eukaryotic organisms to make use of this oxygen for cellular respiration were green algae, which appeared 1.2 billion years ago. Soft-bodied aquatic animals followed, about 600 million years ago, but it is only since about 545 million years ago, during the Cambrian Period, that we see fossil evidence of a truly diverse animal world. During the Cambrian explosion, as this event is known, ocean life swelled with a mind-boggling array of strange-looking creatures, including *Opabinia*, an organism with five eyes and a snout resembling a vacuum-cleaner hose, discovered in fossils from this period.

The first organisms to colonize land were primitive plants, appearing roughly 450 million years ago. By 350 million years ago, forests of seedless plants covered the globe.

Then, 250 million years ago, life was drastically cut down: roughly 95% of living species were extinguished in a mass die-off known as the Permian **extinction.** Scientists do not know what caused the Permian extinction, but some hypothesize that massive volcanic activity filled the atmosphere with heat-trapping gases that led to a rapidly changing climate. The Permian extinction wasn't bad for all organisms, though; some flourished as space and resources opened up for the survivors, who spread and diversified. This phenomenon is known as **adaptive radiation.** Among these surviving organisms were reptiles, who thrived in the hot, dry climate of the following Triassic Period. The most famous group of reptiles, the dinosaurs, dominated the land for nearly

Corey Ford/Stocktrek Images/ Science Source

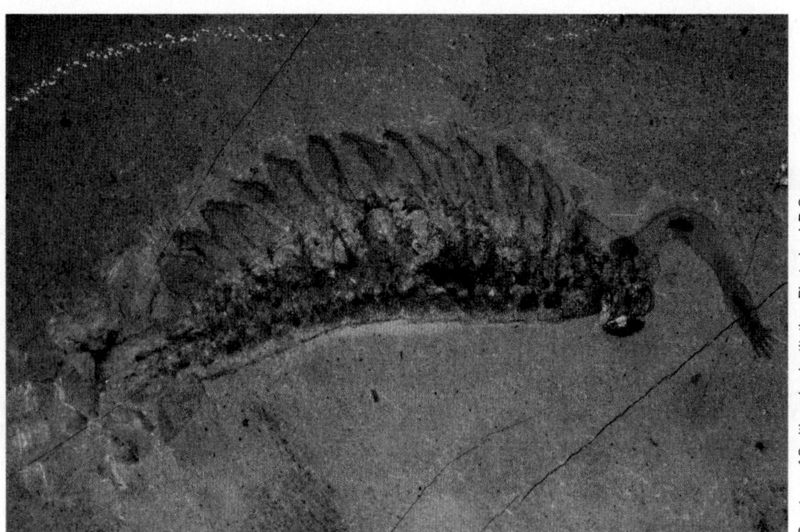
Courtesy of Smithsonian Institution. Photo by J-B Caron.

The Cambrian explosion was characterized by a huge diversity of organisms, including this five-eyed *Opabinia*.

200 million years, thanks to a combination of drought-resistant skin and fast-moving legs, until they died out in another **mass extinction** at the end of the Cretaceous Period, 65 million years ago.

The reason for the extinction of the dinosaurs was a mystery for many years. Evidence now suggests that what killed off the dinosaurs (and 60% of the other species living at the time) was a massive 6-mile-wide asteroid that plowed into Earth with almost unimaginable force, sending a thick layer of soot and ash into the atmosphere and blocking out the sun for months. A crater 110 miles wide in Mexico's Yucatán peninsula, near the town of Chicxulub, is the accepted impact site.

Evidence now suggests that what killed off the dinosaurs (and 60% of the other species living at the time) was a massive 6-mile-wide asteroid that plowed into Earth with almost unimaginable force.

With the extinction of the dinosaurs, it was mammals' chance to spread and diversify on land and thus give rise to many of the species of organisms we see on the planet today. This pattern of sudden change–extinctions followed by adaptive radiations–is seen in the fossil record multiple times. It is an example of **punctuated equilibrium,** in which most evolutionary change occurs in sudden bursts related to environmental change rather than taking place gradually.

BIOGEOGRAPHY

Q: Why are there no penguins at the north pole, and no polar bears at the south pole?

A: As places to live, the north pole and the south pole are pretty similar: cold, lots of ice, plenty of fish to eat. Yet polar bears are found only in the north and penguins are found only in the south. Why? If environmental conditions were the only factors, you might expect that polar bears and penguins would do equally well in either habitat and should therefore coexist in the same location (as they sometimes do in TV commercials). But the distribution of organisms around the world reflects the details of their evolutionary history–their connection in time and space to the species they are related to and descended from. In fact, this was one of the main pieces of evidence that Darwin and Wallace used to support their theory of evolution by natural selection (see **Milestone 5: Adventures in Evolution**). Their round-the-world voyages were among the first examples of **biogeography,** the study of the natural geographic distribution of species. Biogeography seeks to explain why particular organisms are found in some areas and not others.

Penguins, for example, make their home in the southern hemisphere, especially in the coastal regions of Antarctica. According to fossil evidence, penguins first appeared about 65 million years ago near what is now southern New Zealand, when that region was much closer to Antarctica (see next section). They are believed to have evolved from bird ancestors that could fly. Over time, penguins lost their ability to fly but gained the ability to swim, their wings essentially evolving into flippers. Penguins that migrated to Antarctica evolved insulation and other traits that helped them thrive in this cold, watery environment.

Polar bears, by contrast, live only in the Arctic. From fossil and DNA evidence,

MASS EXTINCTION
An extinction of between 50% and 90% of all living species that occurs relatively rapidly.

PUNCTUATED EQUILIBRIUM
Periodic bursts of species change as a result of sudden environmental change.

BIOGEOGRAPHY
The study of the distribution of organisms in geographical space.

scientists conclude that polar bears evolved from brown bears roughly 150,000 years ago in Siberia, when that region became isolated by glaciers. Like penguins, polar bears have adapted to life in a cold, icy environment.

But though their habitats are similar climatically, and might conceivably support the coexistence of both species, there's no easy way for them to travel from one pole to the other: the equator marks a warm tropical barrier for the cold-adapted species living on either side of it (**INFOGRAPHIC 16.5**).

BIOGEOGRAPHY

Q: Why are marsupials found in Australia and the Americas but nowhere else?

A: While traveling around the world, Darwin and Wallace were both struck by the presence on different landmasses of animals and plants that seemed to bear a family resemblance. Islands, for example, seemed

INFOGRAPHIC 16.5 Biogeography: Where Species Live Depends on Where Their Ancestors Evolved

Polar bears live at the north pole and penguins live at the south pole because that is where their respective ancestors evolved. Their adaptations for cold climates keep them from crossing the tropical equator.

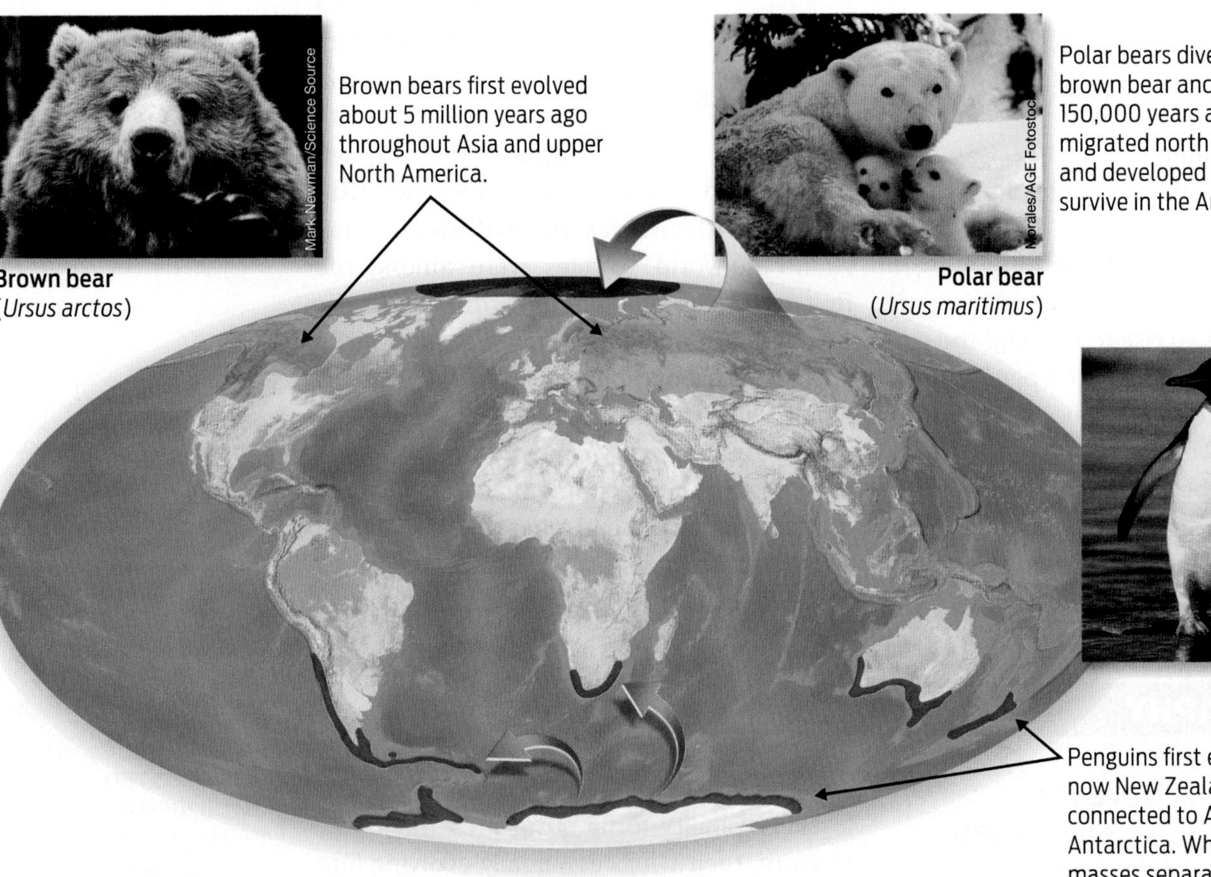

Mark Newman/Science Source
Brown bears first evolved about 5 million years ago throughout Asia and upper North America.
Brown bear
(*Ursus arctos*)
Polar bears diverged from their brown bear ancestors about 150,000 years ago, as they migrated north on ice sheets and developed adaptations to survive in the Arctic.
Polar bear
(*Ursus maritimus*)
Corbis/AGE Fotostock
Gentoo penguin
(*Pygoscelis papua*)
Penguins first evolved near what is now New Zealand, when it was still connected to Australia and Antarctica. When the southern land masses separated, penguins were able to migrate to South America and Africa, but not to the Arctic.

? If brown bears evolved in what is now New Zealand (instead of in Asia and upper North America), would polar bears and penguins coexist? If so, where?

to have species different from, but related to, those on the mainland. This distribution of species seemed to support the theory of evolution by natural selection that each scientist had been pondering. But some facts about species distribution remained curious in Darwin and Wallace's time. For example, why were Australia and South America the two places on the globe with a diversity of marsupials? (Marsupials are mammals that give birth to incompletely developed offspring, which they typically carry to term in a pouch on the outside of the mother's belly.) These continents are separated from each other by great distances, so it was unrealistic to suggest that a marsupial ancestor had swum between these regions.

It wasn't until the 20th century that scientists solved this conundrum. In 1915, Alfred Wegener, a German geologist, proposed that the distribution of plants and animal species on far-flung continents reflected **continental drift,** the movement of continents over time. Wegener proposed that at one time, the different continents we see today were all joined together into one large landmass, called Pangaea, that broke apart slowly over time. His ideas were mostly conjectural until the 1960s, when scientists learned about what came to be called **plate tectonics** and discovered a mechanism for how the continents move.

Convergent evolution shows that sometimes a good solution to a common problem can evolve more than once.

Continental drift helps to explain the distribution of marsupials we see today. The most recent common ancestor of marsupials and placental mammals (those, like humans, that bear young that have developed in a womb) appears to have lived about 120 million years ago in what is now China, when this region was part of a large landmass called Laurasia. Marsupials gradually diverged from the ancestor of placental mammals and began their march across Laurasia. About 65 million years ago, marsupials spread to what is now South America when it became connected to Laurasia through a land bridge.

Fossil evidence suggests that the most recent common ancestor of living marsupials lived in South America. Marsupial fossils dating from 40 million years ago have been discovered there. The oldest marsupial fossils in Australia are 30 million years old, suggesting they arrived later. But how? The theory of continental drift proposes that Australia, South America, and Antarctica were all once part of one large landmass called Gondwana, which began to break up about 180 million years ago. When marsupials arrived in South America, Australia was still connected to South America by what is now Antarctica (which was then ice free). This suggested to scientists that marsupials may have crossed over Antarctica to get to Australia. Evidence in support of this hypothesis comes from marsupial fossils discovered on Seymour Island off the coast of Antarctica that date from 40-35 million years ago **(INFOGRAPHIC 16.6).**

Tectonic plates continue to move, occasionally causing dramatic events such as earthquakes and volcanic eruptions. GPS measurements, used to track the direction and velocity of plate movement, show that some are moving at about 2 to 5 cm per year—about as fast as your fingernails grow.

CONVERGENT EVOLUTION

Q: Are creatures that look alike always closely related?

A: Polar bears share many traits with their brown bear cousins—both species are recognizable as bears, despite obvious differences in color. The fact that polar bears

CONTINENTAL DRIFT
The movement of the continents relative to one another over time.

PLATE TECTONICS
The movement of Earth's upper mantle and crust, which influences the geographical distribution of landmasses and organisms.

INFOGRAPHIC 16.6 Continental Drift Influences the Distribution of Species on Earth

Earth's landmasses are on plates that float on Earth's upper mantle and crust. Over geologic time, these plates have moved with respect to one another, affecting the ability of organisms to move among the landmasses. The geographic distribution of organisms on Earth today reflects continental drift.

180 mya

Laurasia

Gondwana

180 million years ago there were two supercontinents on Earth, Laurasia and Gondwana.

120 mya

About 120 million years ago marsupial and placental mammals diverged from a common mammal ancestor in China and migrated throughout Laurasia, which was not connected to Gondwana.

Australia koala
(*Phascolarctos cinereus*)

65 mya

About 65 million years ago, the part of Gondwana we now call South America became connected with North America, allowing migration of marsupials into this region.

50 mya

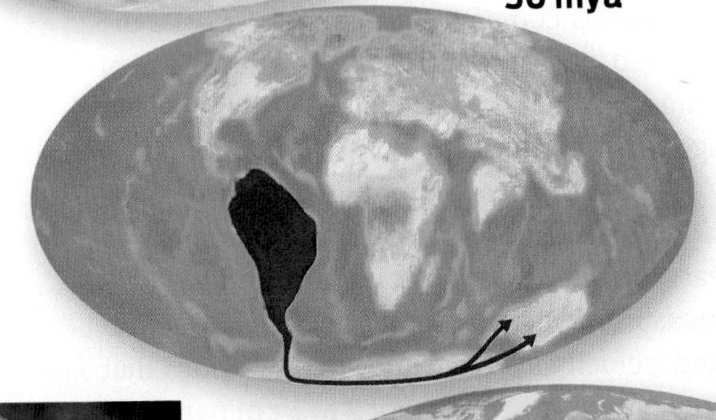

About 50 million years ago, marsupials migrated from South America through Antarctica and into Australia until about 35 million years ago when these pieces of Gondwana were no longer connected. The marsupials in Laurasia went through a mass extinction during this time and were almost completely eliminated from the northern land regions.

Today

Today, marsupials are found diversified only in South America and Australia. The disconnection of Antarctica from South America and Australia halted migration between these continents. As Antarctica migrated south its colder temperatures resulted in marsupial extinction there. Marsupial species are most diverse in Australia, where they lack competition from placental mammals that did not migrate to the continent before it broke away from Antarctica.

South American opossum
(*Didelphis marsupialis*)

? When did marsupials first migrate into the southern hemisphere?

resemble brown bears is suggestive evidence that the two species share a recent common ancestor. But common ancestry is not the only reason that two species might appear similar. Even species that are not closely related may share similar adaptations as a result of independent episodes of natural selection, a phenomenon called **convergent evolution.**

Take flying squirrels and sugar gliders. Flying squirrels are placental mammals that evolved in North America. Sugar gliders are marsupial mammals that evolved in Australia. Flying squirrels and sugar gliders are not closely related. However, they have both independently evolved gliding wings to leap from tree to tree in the forested environments in which they live. Convergent evolution shows that sometimes a good solution to a common problem can evolve more than once (**INFOGRAPHIC 16.7**).

DIVERSITY AND TAXONOMY

Q: How many species are there on Earth, and how do scientists keep track of them?

A: Current estimates of the total number of species on Earth range anywhere from 5 to 30 million, of which 1.5 million or so have been formally described. Many of these species are found in diversity hot spots like rain forests, and new species are constantly being discovered, on the order of 17,000 new species a year. Some noteworthy recent additions include the hog-nosed rat (*Hyorhinomys stuempkei*), a cartwheeling spider (*Cebrennus rechenbergi*), and the tiniest known snail (*Acmella nana*), which is smaller than the head of a matchstick (**INFOGRAPHIC 16.8**).

CONVERGENT EVOLUTION
The process by which organisms that are not closely related evolve similar adaptations as a result of independent episodes of natural selection.

INFOGRAPHIC 16.7 Convergent Evolution: Similar Solutions to Common Problems

Species that are not closely related can have very similar appearances because they independently evolved adaptations to similar environments and challenges.

Flying squirrels are placental mammals that evolved in North America.

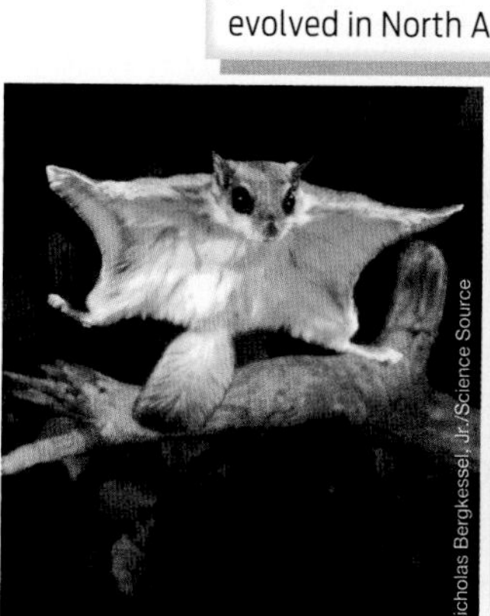

Southern flying squirrel
(*Glaucomys volans*)

Sugar gliders are marsupial mammals that evolved in Australia.

Sugar glider
(*Petarus breviceps*)

Flying squirrels and sugar gliders are not closely related. However, they have both evolved gliding wings to leap from tree to tree in the forested environments in which they live.

? What are some key differences between the placental flying squirrel and the marsupial sugar glider?

INFOGRAPHIC 16.8 How Many Species Are There?

The numbers of species in each group below represents only those that have been formally characterized and classified. The true number of species is likely to be much higher (estimated numbers are shown in parentheses).

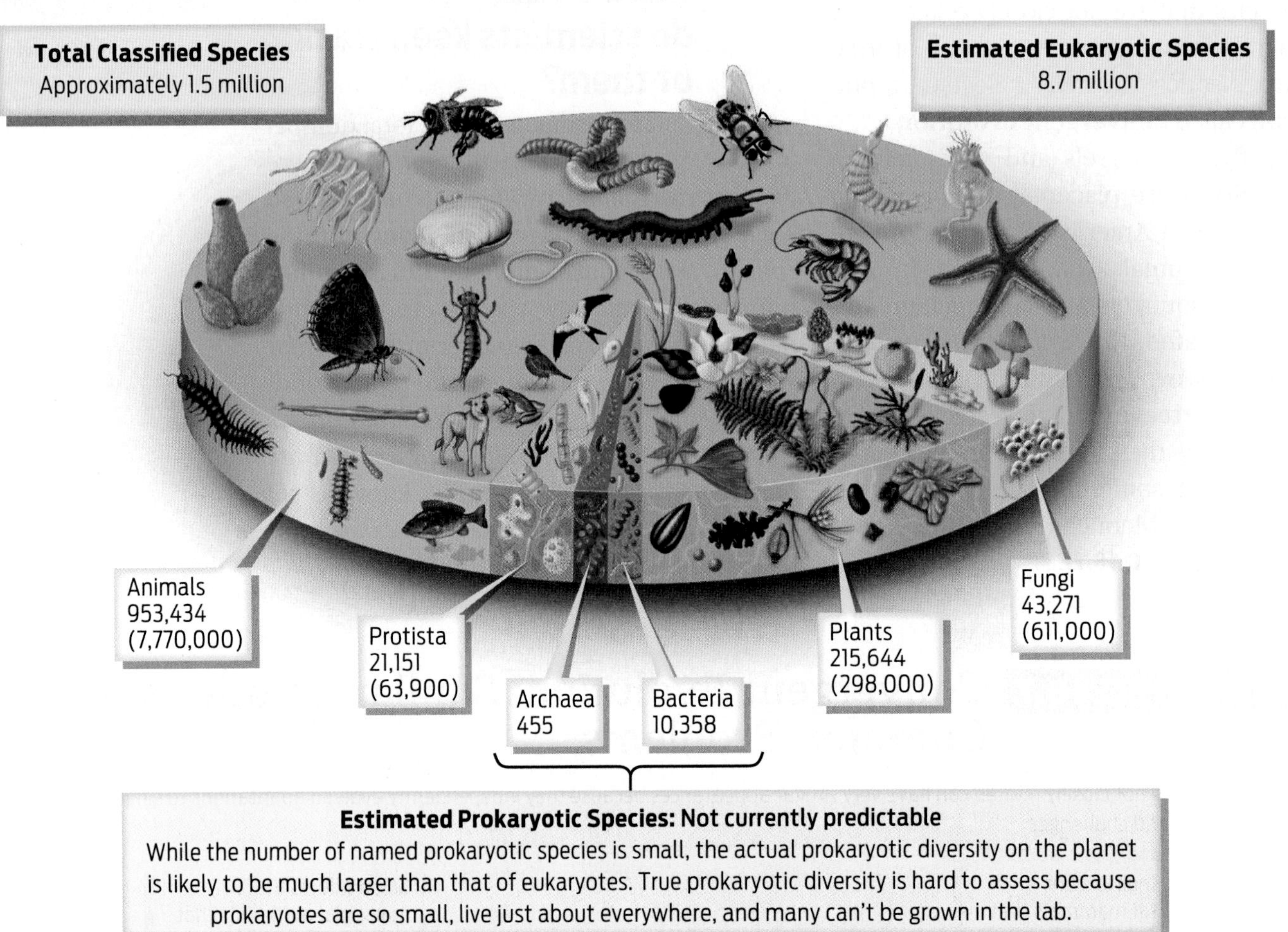

Estimated Prokaryotic Species: Not currently predictable
While the number of named prokaryotic species is small, the actual prokaryotic diversity on the planet is likely to be much larger than that of eukaryotes. True prokaryotic diversity is hard to assess because prokaryotes are so small, live just about everywhere, and many can't be grown in the lab.

? Which eukaryotic group has the greatest proportion of estimated species already described?

With so many species out there, how do scientists keep track of them all? The organizing principle by which scientists systematically identify, name, and classify organisms is called **taxonomy.** (Taxonomy is part of the broader study of systematics, the study of biological diversity of life on Earth.)

TAXONOMY
The identification, naming, and classification of organisms on the basis of shared traits.

Taxonomy is an attempt to impose a human sense of order on this vast array of species, categorizing them on the basis of features they have in common, such as whether their cells are eukaryotic or prokaryotic, whether or not they photosynthesize, whether or not they have four legs and fur.

To sort organisms, taxonomists use a system of nine progressively narrower categories: domain, supergroup, kingdom, phylum, class, order, family, genus, species. As you move down the list, from domain to species, the categories are increasingly exclusive, until finally only one specific type of organism is included. The genus and species names provide a useful scientific identifier for every living organism.

Because that scientific name is in Latin, it can be easily recognized in many languages.

Take humans, for example. Humans are animals, members of the domain Eukarya, the supergroup Opisthokont, and the kingdom Animalia. Within the animal kingdom, they belong to the phylum Chordata, a group that includes the **vertebrates** (animals with a rigid backbone). Further, humans are **mammals,** members of the class Mammalia; they share with all members of this class mammary glands and a body that is covered with hair. Humans belong to the Primate order, which also includes monkeys, apes, and lemurs. And humans are members of the Hominidae family, and so are closely related to their fellow hominids: chimpanzees, gorillas, and orangutans. Our scientific name–made up of our genus and species names–is *Homo sapiens* ("wise human") **(INFOGRAPHIC 16.9)**.

VERTEBRATES
Animals with a rigid backbone.

MAMMALS
Members of the class Mammalia; all members of this class have mammary glands and a body covered with hair.

INFOGRAPHIC 16.9 Classification of Species

Scientists classify organisms into groups that are increasingly exclusive. The broadest category is the domain. Closely related organisms are grouped based on morphological, nutritional, and genetic characteristics. There are far fewer organisms in an order than in a phylum.

Taxonomic group	Human group	Number of living species
Domain	Eukarya	~4-10 million
Supergroup	Opisthokonta	> 1 million
Kingdom	Animalia	> 1 million
Phylum	Chordata	~50,000
Class	Mammalia	~5,000
Order	Primate	~250
Family	Hominidae	7
Genus	Homo	1
Species	Homo sapiens	1

? Which of the following groups has the greatest number of living species: phylum, class, family, kingdom, or supergroup?

Classification would seem to be a simple matter–just observe, measure, and sort. But deciding which category an organism belongs in can sometimes be tricky, as the example of convergent evolution has shown. Sometimes, to properly classify organisms, scientists have to look a little deeper.

READING PHYLOGENETIC TREES

Q: Is a crocodile more closely related to a bird or to a lizard?

A: The fact that all land vertebrates have four limbs and the same forelimb bones indicates that they all share a common ancestor (see Chapter 15). But how precisely are they related? In other words, who's more closely related to whom? Biologists want not only to categorize organisms, but also to have those categories reflect **phylogeny,** the actual evolutionary history of the organisms. This evolutionary history is represented visually by a diagram called a **phylogenetic tree,** which is similar in some respects to a family tree.

Phylogenetic trees can be drawn in a number of ways, but most have certain features in common. At the base, or root, is the common ancestor shared by all organisms on the tree. Over time, and with different selective pressures, different groups of organisms diverged from that common ancestor and from one another, leading to separate branches on the tree. The points on the tree at which these branching points occur are called nodes. A node represents the common ancestor shared by all organisms on the branches above that node. At the very tips of the branches we find the most recent organisms in that lineage, including living organisms and organisms that became extinct. We can thus establish relationships between living organisms (at the tips of the branches) on the basis of the ancestors they share. The more recently two groups share a common ancestor, the more closely they are related (**INFOGRAPHIC 16.10**).

PHYLOGENY
The evolutionary history of a group of organisms.

PHYLOGENETIC TREE
A branching diagram of relationships showing common ancestry.

A phylogenetic tree is a visual representation of the best hypothesis we currently have for how species are related, based on a shared evolutionary history. The evidence for a phylogenetic tree comes from many sources, including the fossil record, physical traits, and shared DNA sequences. For many years, biologists relied solely on observable physical or behavioral features to construct phylogenetic, or evolutionary, trees. But with the genetic revolution, it's become common to include DNA evidence. Typically, researchers compare sequence differences in a gene that is found in all living organisms, such as the gene encoding a particular component of the ribosome, the cell's protein synthesis machinery. All cells have ribosomes, which is why this gene is useful for establishing phylogeny.

Sometimes the new genetic information yields surprises. Modern genetic evidence shows, for example, that crocodiles are more closely related to birds than they are to lizards, appearances notwithstanding. Genetics, you might say, is shaking the evolutionary tree.

CLASSIFICATION AND PHYLOGENY

Q: How many branches are on the tree of life?

A: Since each living species sits on its own branch of a phylogenetic tree, the complete tree of life has as many branches as there are species in the world. Today's species are like the thinnest twigs in the upper branches of an enormous oak tree. Closer to the bottom of the tree, nearer to the ancient trunk, however, we find significant forks. Just how many forks there are at the bottom of the tree is a question that has been debated for decades.

INFOGRAPHIC 16.10 How to Read a Phylogenetic Tree

Evolutionary history, or phylogeny, is represented visually by a phylogenetic tree. Trees have a common structure, with a root, nodes, and branches. To determine evolutionary relationships among living or extinct organisms, consider the most recent common ancestors.

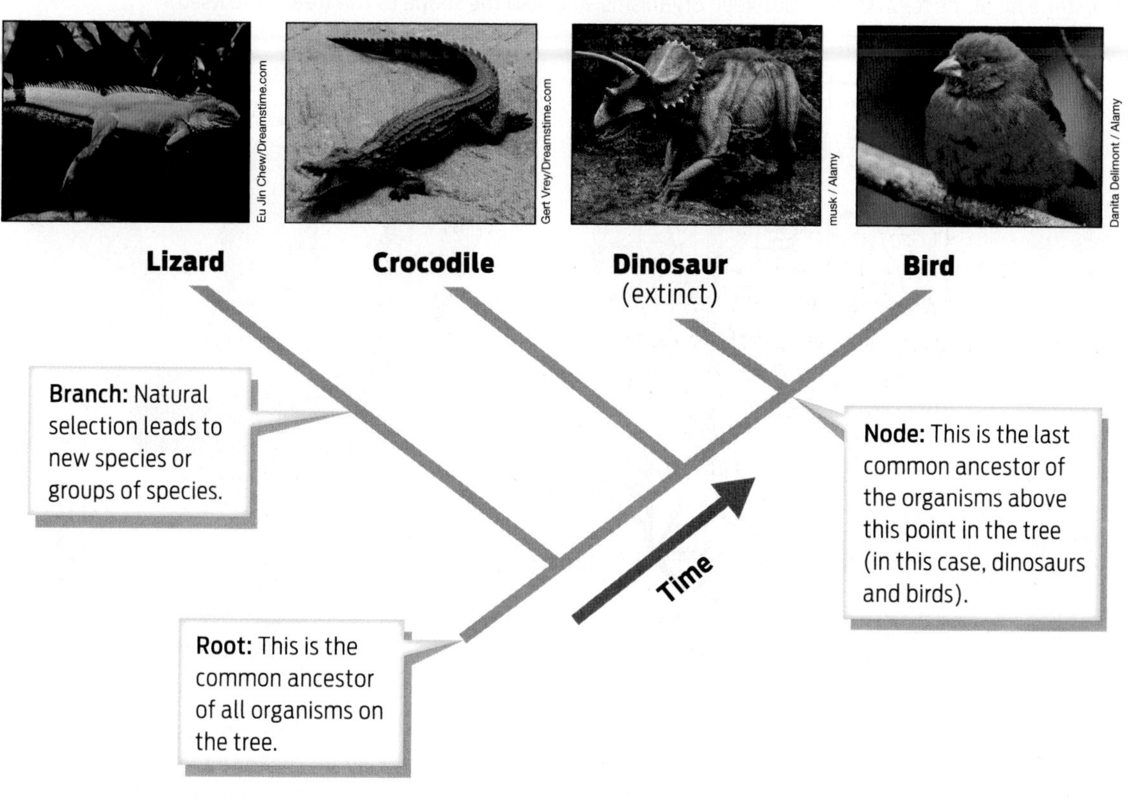
Eu Jin Chew/Dreamstime.com
Gert Vrey/Dreamstime.com
musk / Alamy
Danita Delimont / Alamy
Lizard
Crocodile
Dinosaur
(extinct)
Bird
Branch: Natural selection leads to new species or groups of species.
Node: This is the last common ancestor of the organisms above this point in the tree (in this case, dinosaurs and birds).
Time
Root: This is the common ancestor of all organisms on the tree.

? Is a crocodile more closely related to a bird or to a lizard?

Before the 18th century, biologists divided living things into just two main categories: animals and plants. This classification was based on whether an organism moved around and ate or did not move around and eat. By the mid-19th century, the microscope had revealed a whole new world of organisms, and so a third branch was added to life's tree. The German biologist and artist Ernst Haeckel, in 1866, was the first to draw a phylogenetic tree that included these microscopic organisms, which he termed Protista, or protists.

By the 1960s, taxonomists realized that even three branches did not fully capture the diversity of life. Many organisms–such as fungi–didn't fit neatly into any of these groups, and so another classification scheme was proposed. This one grouped all living organisms into five large kingdoms on the basis of how they obtained their food (by eating, photosynthesizing, or decomposing) and whether they had eukaryotic or prokaryotic cells. The five kingdoms were Animalia, Plantae, Fungi, Protista, and Monera. Protista included mostly single-celled eukaryotic organisms (such as the amoeba), and Monera included all prokaryotic organisms (such as bacteria).

Yet even this revised classification scheme eventually had to be overhauled as more information became available. In the 1970s, genetic studies by Carl Woese revealed that, on the basis of genetic relatedness, not all prokaryotes could be lumped together. Likewise, protists were too genetically diverse to

INFOGRAPHIC 16.11 The Evolution of the Tree of Life

Phylogenetic trees represent our best understanding of the relationships between living organisms. Early trees were based on obvious physical similarities between organisms. More detailed structural and functional analyses generated new groups. The use of gene sequences revealed that organisms previously thought to be closely related were not. Gene and even entire genome sequences are continuing to shape our understanding of the relationships between organisms, and also the shape of the tree of life itself.

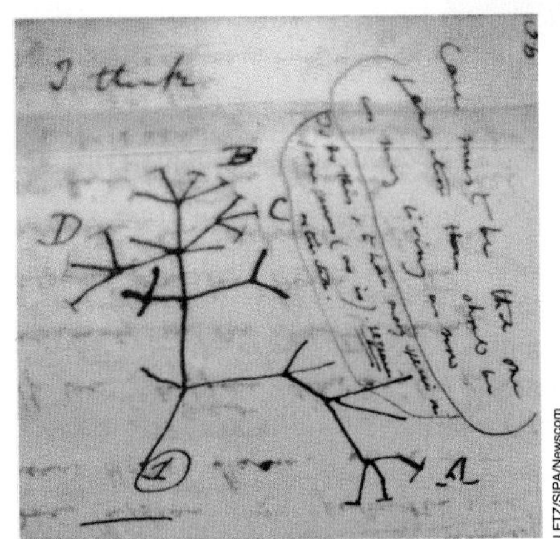

Darwin's first drawings show how his idea of natural selection implied relatedness between species.

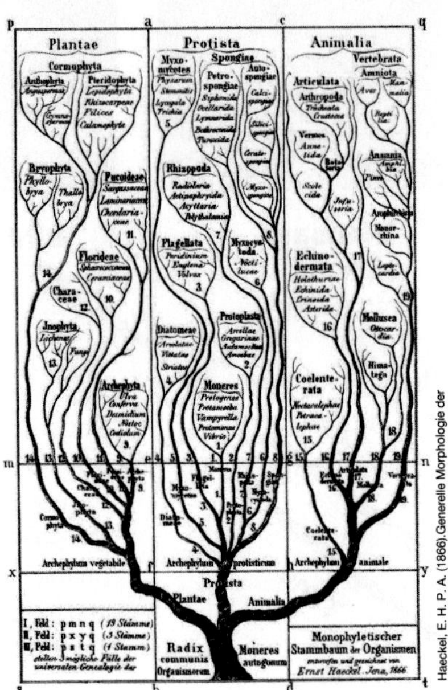

Ernst Haeckel's 1866 tree for the relationships of all living things included three major groups based on similar physical characteristics.

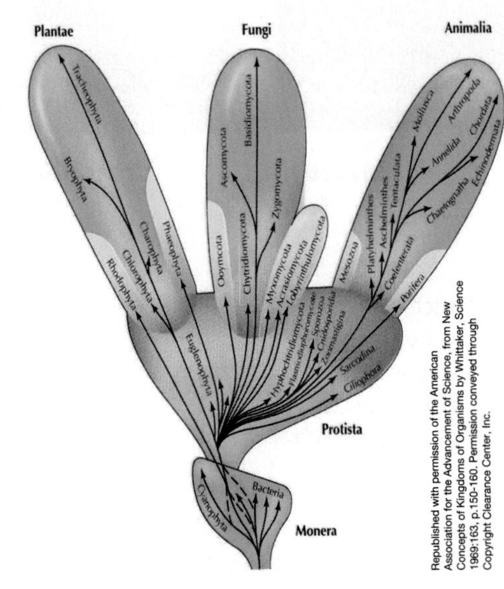

Whitaker's 1969 Five Kingdom Classification System placed prokaryotic organisms in a single group (based on their prokaryotic cell structure). Nutritional differences were used to place plants, fungi, and animals into different kingdoms.

Timeline of Phylogenetic Trees

DOMAIN
The highest (most inclusive) category in the modern system of classification. There are three domains: Bacteria, Archaea, and Eukarya.

be put in one category. Consequently, scientists now group organisms into one of three large **domains**–Bacteria, Archaea, and Eukarya–which represent three fundamental branch points in the trunk of the evolutionary tree. The original kingdom Monera, containing all prokaryotes, is now divided into two domains, Archaea and Bacteria, based on degree of genetic relatedness. Within the domain Eukarya, Animalia, Plantae, and Fungi remain recognized kingdoms, but the protists (members of the former kingdom Protista) are dispersed across the eukaryotic domain of life on the basis of genetic evidence (**INFOGRAPHIC 16.11**).

Since about 2012, researchers have recognized a new scheme for categorizing eukaryotic life. The supergroup, a new classification, is meant to help clarify where protists (former members of the now-defunct

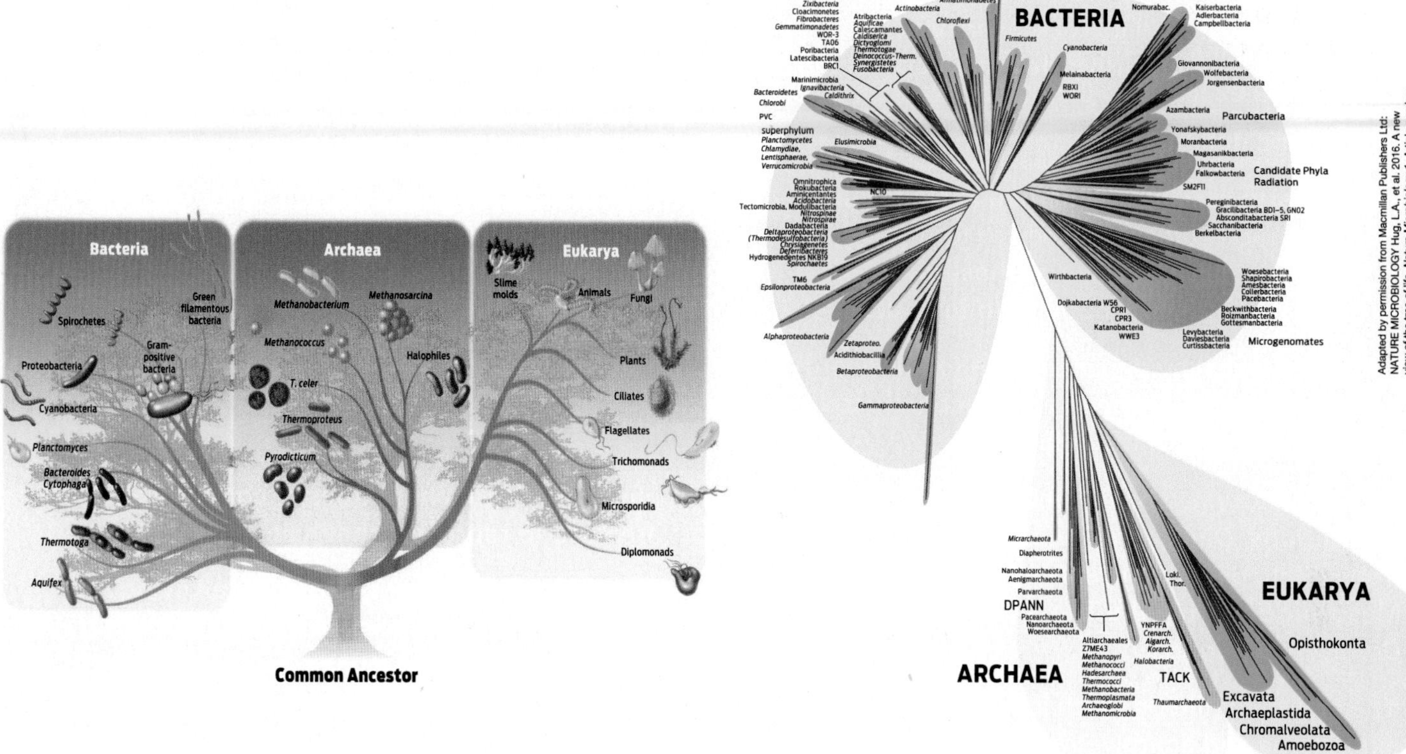

Woese's 1977 classification system split life into three domains, two of which are prokaryotic, based on analyzing DNA sequences.

A 2016 tree of life used whole genome sequences to confirm major eukaryotic supergroups and provide evidence for a wealth of previously uncharacterized bacterial diversity.

? Which two of the three domains are most closely related to each other?

kingdom Protista) should be placed on the eukaryotic tree. As mentioned before, grouping all single-celled eukaryotes together does not make sense from a genetic standpoint, since they are not all closely related. The new classification scheme attempts to classify protists by genetic similarity to other groups on the tree and, as a result, has produced some unexpected pairings. The supergroup Opisthokont, for example, groups certain single-celled parasites with members of the animal and fungi kingdoms. By comparing DNA sequences of genes not previously used to assemble phylogenetic trees, scientists have discovered that the members of this large supergroup are more closely related to one another than to other supergroups of the eukaryotic tree. The tree of life is truly a continual work in progress. ■

CHAPTER 16 SUMMARY

- The age of Earth and its rock layers can be determined by measuring the amount of radioactive isotopes present in rocks, a method known as radiometric dating.
- Life on Earth may have emerged in stages, as inorganic molecules combined to form organic ones in the primordial soup, and as these were incorporated into lipid bubbles to form cells.
- From geological evidence and the fossil record, paleontologists have been able to construct a geologic timeline of life on Earth.
- Earth's history can be divided into important eras and periods. Dinosaurs, for example, lived primarily from 250 to 65 million years ago, during the Mesozoic Era, from the Triassic through the Cretaceous periods.
- The history of life on Earth is marked by repeated extinctions and adaptive radiations, a phenomenon of intermittent rather than steady change known as punctuated equilibrium.
- Ancient movement of Earth's major landmasses affected the eventual distribution of species around the globe, the study of which is known as biogeography.
- Convergent evolution is the evolution of similar adaptations in response to similar environmental challenges in groups of organisms that are not closely related.
- Life is astoundingly diverse. Current estimates of the total number of species on Earth range anywhere from 5 to 30 million, of which 1.5 million have been formally described.
- Biologists sort organisms into a series of nested categories based on shared anatomical and genetic features: domain, supergroup, kingdom, phylum, class, order, family, genus, species.
- The scientific name of an organism is given by its genus and species names (the scientific name of humans is *Homo sapiens*).
- Both physical evidence and genetic evidence are used to understand evolutionary history, or phylogeny. Branching trees of common ancestry are used to represent that history visually.
- On the basis of genetic evidence, all living organisms can be classified into one of three domains: Bacteria, Archaea, or Eukarya.

MORE TO EXPLORE

- **Witze, Alexandra (2016) Claims of Earth's oldest fossils tantalize researchers. *Nature*, Aug 31.**
- **Gould, S. J. (1989) *Wonderful Life: Burgess Shale and the Nature of History*. New York: W. W. Norton.**
- **Mora, C., et al. (2011) How many species are there on Earth and in the ocean? *PLoS Biol* 9(8):e1001127.**
- **Woese, C.R., et al. (1990) Towards a natural system of organisms: proposal for the domains Archaea, Bacteria, and Eucarya. *Proceedings of the National Academy of Sciences* 87:4576–4579.**
- **Milius, Susan (2015) The tree of life gets a makeover: Schoolroom kingdoms are taking a backseat to life's supergroups. *ScienceNews*, 188, 22.**

CHAPTER 16 Test Your Knowledge

DRIVING QUESTION 1 What do we know about the history of life on Earth, and how do we know it?

By answering the questions below and studying Infographics 16.1–16.4, you should be able to generate an answer for the broader Driving Question above.

KNOW IT

1 **What do uranium-238, carbon-14, and potassium-40 have in common?**

2 **To date what you suspect to be the very earliest life on Earth, which isotope would you use: uranium-238, carbon-14, or potassium-40? Explain your answer.**

3 **Place the following evolutionary milestones in order from earliest (1) to most recent (7), providing approximate dates to support your answer.**

_____The first multicellular eukaryotes
_____The first prokaryotes
_____The Permian extinction
_____The Cambrian explosion
_____The first animals
_____The extinction of dinosaurs
_____ An increase in oxygen in the atmosphere

USE IT

4 **Consider a rock formed at about the same time as Earth was formed.**

a. How old is this rock?
b. How much of the original uranium-238 is likely to be left today in that rock?

5 **Diverse animal fossils are found dating from the Cambrian Period but not earlier. Why might these organisms have made their first appearance in the fossil record only then, even though their ancestors may have been living, and evolving, for a long time before the Cambrian? (Think about what kinds of new structures might have evolved during the Cambrian Period that would have allowed these organisms to leave fossils.)**

6 **You are a paleontologist with a particular interest in early microbes. A microbiologist has brought you a fossilized cell that appears to be prokaryotic. They claim that they used potassium-40 dating to date the fossil to 3.9 billion years ago. If the analysis is accurate, what percent of potassium-40 should remain in the specimen? If you used uranium-238 dating to verify, what proportion of uranium-238 should remain in the specimen?**

MINI CASE

apply YOUR KNOWLEDGE

7 **Along the banks of a river, some sedimentary rock strata have been revealed by erosion. By radiometric dating, the layer above these strata is determined to be about 290 million years old, and the layer beneath has been dated to about 354 million years ago. A paleontologist starts to uncover fossils in the sedimentary rock strata. The fossils are clearly land-dwelling vertebrates. Are they more likely to be reptiles or amphibians? Explain your answer.**

INTERPRETING DATA

apply YOUR KNOWLEDGE

8 **You have carried out radiometric analysis on four igneous rocks uncovered at several sites you are exploring. From the % lead you determine in each case, what is the approximate age of the rock?**

Rock A: 75% lead _____
Rock B: 50% lead _____
Rock C: 30% lead _____
Rock D: 10% lead _____

DRIVING QUESTION 2 What factors help to explain the distribution of species on Earth, and how many species are there?

By answering the questions below and studying Infographics 16.5–6.8, you should be able to generate an answer for the broader Driving Question above.

KNOW IT

9 **If two organisms strongly resemble each other in their physical traits, can you necessarily conclude that they are closely related? Explain your answer.**

10 **What did the arrangement of landmasses on Earth look like between 135 and 65 million years ago? What happened to these landmasses, and how does this change help explain the distribution of organisms found on the planet?**

USE IT

11 A cactus called ocotillo (*Fouquieria splendens*), which grows in New Mexico, looks very much like *Alluaudia procera,* a species of plant that grows in the deserts of Madagascar. These two plant species are not closely related—they are in different orders in the kingdom Plantae. Why then do they look so alike?

12 If penguins had evolved in upper North America and not New Zealand, what would you predict about the current distribution of polar bears and penguins?

13 Both bats and insects fly, but bat wings have bones and insect wings do not. Would you consider bat and insect wings to be a result of convergent evolution, or of homology—evolution based on inheritance of similar structures from a common ancestor? Explain your answer.

DRIVING QUESTION 3 What are the major groups of organisms, and how are organisms placed in groups?

By answering the questions below and studying Infographics 16.9–16.11, you should be able to generate an answer for the broader Driving Question above.

KNOW IT

14 Which of the following is *not* a domain of life?

a. Animalia
b. Eukarya
c. Bacteria
d. Archaea
e. Plantae
f. Neither a nor e is a domain of life.

15 Put the following terms in order from most inclusive (1) to least inclusive (6).

_____ Domain
_____ Species
_____ Supergroup
_____ Kingdom
_____ Genus
_____ Phylum

16 A phylogenetic tree represents

a. a grouping of organisms on the basis of shared structural features.
b. a grouping of organisms on the basis of cell type.
c. a grouping of organisms on the basis of complexity.
d. a grouping of organisms on the basis of evolutionary history.
e. a grouping of organisms on the basis of where they are found.

USE IT

17 Why was the classification of the kingdom Monera split into two domains? What are these two domains?

18 On the tree below, which number represents the most recent common ancestor of humans and corn?

a. 1
b. 2
c. 3
d. 4
e. Humans and corn do not share any ancestors.

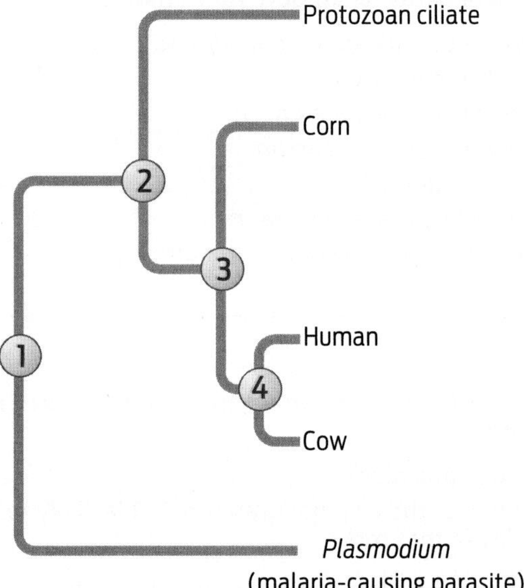

Bill Abbott/CartoonStock

BRING IT HOME

apply YOUR KNOWLEDGE

19 Carry out some online research on the fossils found in your home state. What groups of organisms are represented in the statewide fossil record? What is the oldest fossil found in your state? What do the fossils in your state suggest about the pattern(s) of evolution in your state?

17 Prokaryotic Diversity

LOST CITY

Probing life's origins at the bottom of the sea

Gretchen Früh-Green's heart never beat so fast. Late on a December evening in 2000, she was hunkered down in the control room of the research ship *Atlantis*, monitoring live video streaming up to the ship from a submersible camera swimming half a mile below. Strange white shapes began to appear in the blackness of her screen. As the camera panned, an underwater landscape of ghostly white towers suddenly came into focus. The huge limestone structures, which resembled the stalagmites found in caves, loomed above an otherwise empty seafloor. Früh-Green, a geologist with the Institute for Mineralogy and Petrology in Zurich, Switzerland, knew immediately that she was looking at something special and raced to tell her colleagues. "It was really quite exciting," she said. "It was late at night and kind of woke us all up."

DRIVING QUESTIONS

1. What are the prokaryotic domains of life?
2. What are the features of bacteria and of archaea?
3. What are the challenges faced by organisms living at Lost City, and how do they face them?

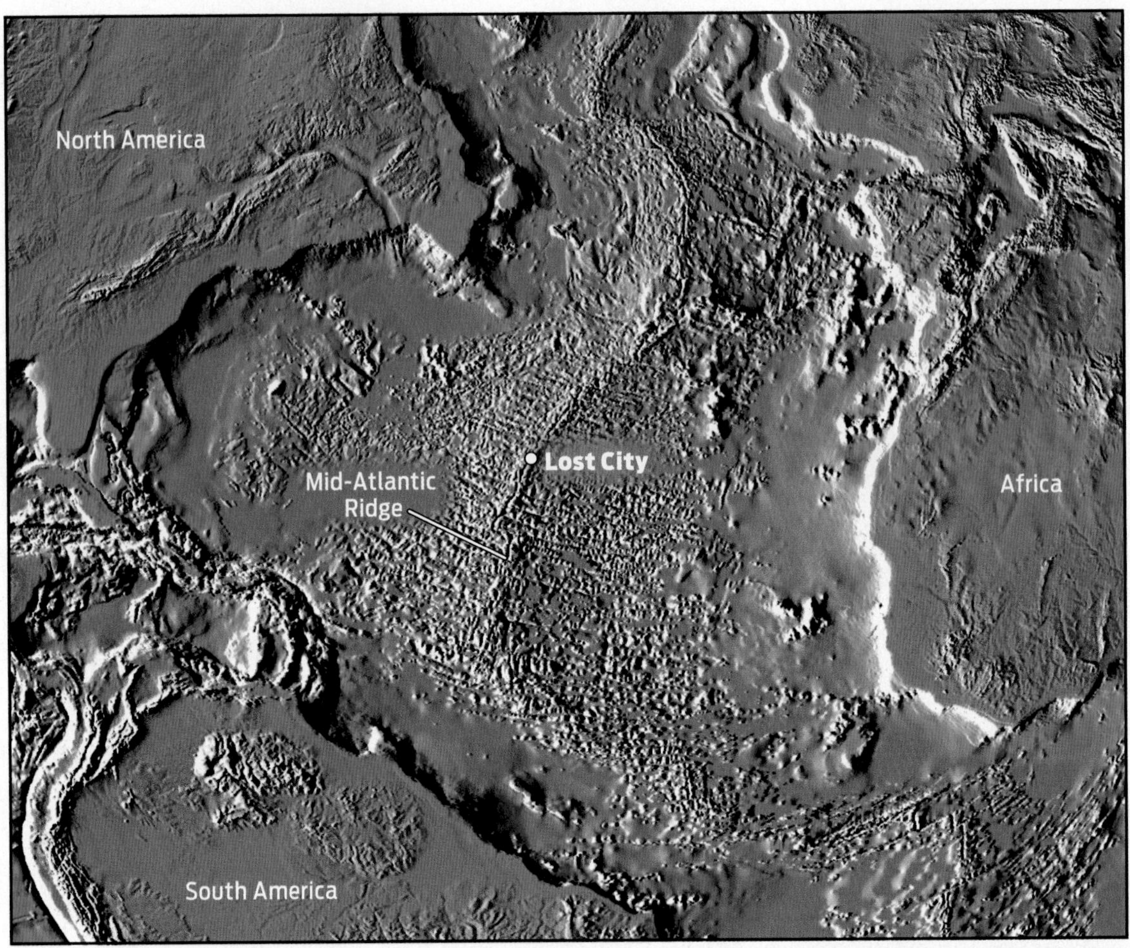

Lost City sits 2,600 feet under the sea on a mountain in the middle of the Atlantic Ocean.

The next day, the mission's chief scientist, Deborah Kelley, and two members of the team climbed into a submersible craft that dived 2,600 feet down to the site, which sits on an undersea mountain in the middle of the Atlantic Ocean. There they found a dense "cityscape" of rocky spires, stretching across roughly six football fields of dark seafloor. It was an uncharted undersea world no one had known existed. Kelley named it "Lost City," in homage to the research vessel that carried them. The tallest tower, measuring 200 feet, she dubbed "Poseidon," after the Greek god of the sea.

Subsequent research has shown that each tower is a type of underwater chimney, or spring, known as a hydrothermal vent. As rocks beneath Earth's crust come in contact with seawater, they react chemically, giving off a steady stream of heat and combustible gas that seep out of each vent. The resulting fluid is highly caustic, with a pH of 9 to 11—similar to that of drain cleaner—and temperatures as hot as 100°C (212°F).

Though deep-sea hydrothermal vents had been discovered before, the ones at Lost City were unique in their chemical and geological makeup. Nothing like them had ever been seen. "Rarely does something like this come along that drives home how much we still have to learn about our own planet," said Kelley, an oceanographer at the University of Washington in Seattle, who now leads the international team of scientists who are exploring Lost City's mysteries.

An extreme and seemingly inhospitable environment, the towers of Lost City are nonetheless home to a surprising number of life forms. The most prevalent living creatures are dense layers of unicellular microbes that coat the towers, inside and out (**INFOGRAPHIC 17.1**).

Microbial life exists pretty much everywhere on Earth—in frozen glaciers, in radioactive dirt, in the intestines of animals—yet the microbes at Lost City have earned the fascination of scientists. "They're living in boiling toilet bowl cleaner full of flammable gas," says Bill Brazelton, a professor of biology at the University of Utah, who studies the microbes.

The extreme conditions of this environment resemble what researchers think the ancient Earth was like some 3.8 billion years ago, when life was getting started. Scientists have long wondered how life managed to emerge and flourish in such inhospitable conditions. At Lost City, they are finding some tantalizing clues.

INFOGRAPHIC 17.1 Hydrothermal Vents of Lost City Provide an Extreme Environment for Life

The hydrothermal vents of Lost City are huge rock chimneys. Scalding fluid with an extremely basic pH flows out of the tops of these chimneys. The towering size and range of environments within and around a single Lost City spire mean it can host a variety of unique microbial communities.

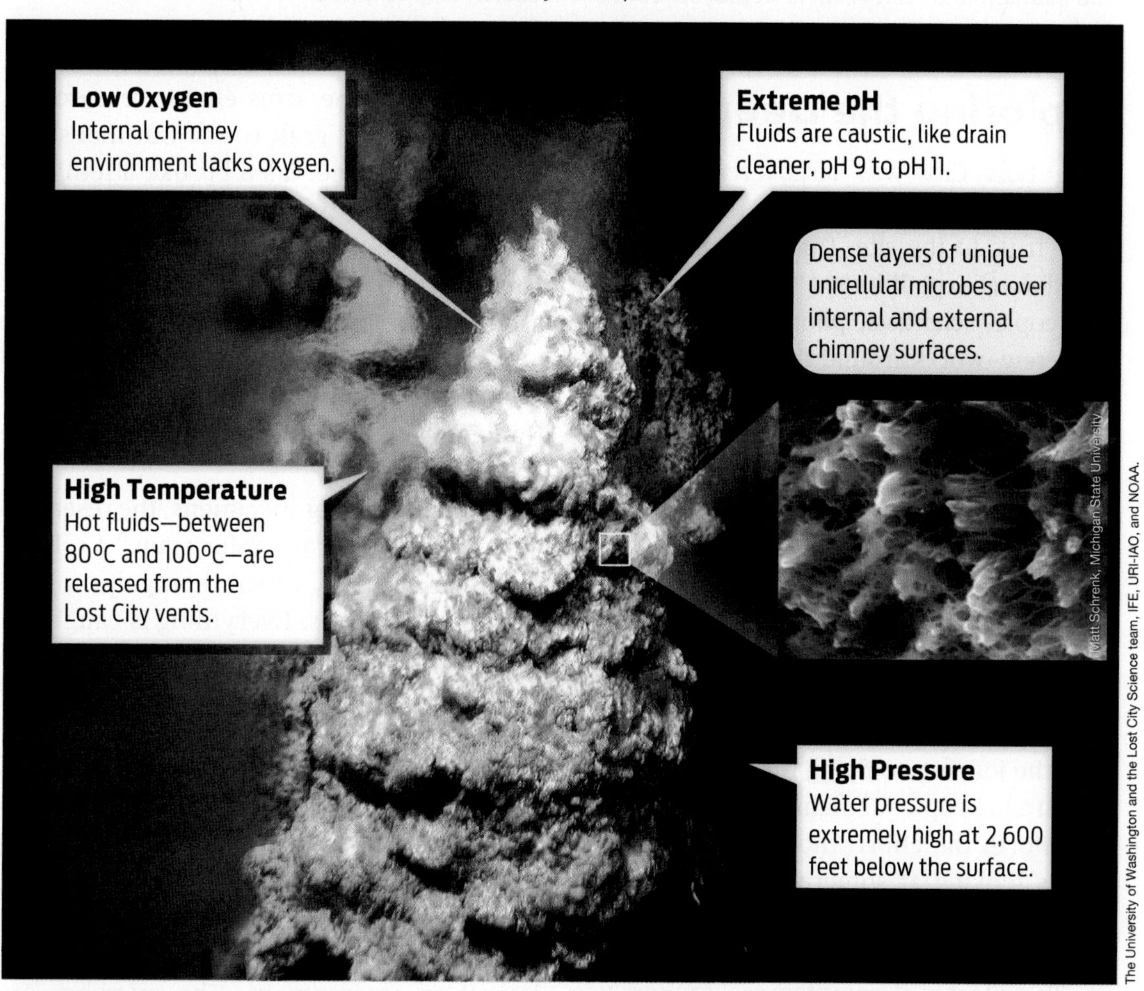
Low Oxygen
Internal chimney environment lacks oxygen.
Extreme pH
Fluids are caustic, like drain cleaner, pH 9 to pH 11.
Dense layers of unique unicellular microbes cover internal and external chimney surfaces.
High Temperature
Hot fluids—between 80°C and 100°C—are released from the Lost City vents.
High Pressure
Water pressure is extremely high at 2,600 feet below the surface.
Matt Schrenk, Michigan State University

? What four features make Lost City an extreme environment (at least from our perspective)?

Amy Nevala © Woods Hole Oceanographic Institution.

The submersible craft *Alvin* is deployed from the research ship *Atlantis*.

Exploring the Deep

Since Lost City was discovered in 2000, researchers have organized three exploratory trips to the site. During each of these month-long expeditions, a team of more than 50 people–scientists, pilots, engineers, and ship crew–works around the clock to orchestrate dives and obtain specimens for research. A squad of sophisticated robotic assistants aids the effort.

Researchers dive to the site in *Alvin*, a tiny three-person submersible craft. Each trip to the murky depths takes about 30 minutes and is a risky descent into dark, uncharted waters. This isn't flat seafloor, after all; the Lost City towers are like tall buildings, some as high as 18 stories. Team members compare the journey to flying through New York City in a helicopter at night with no lights. It's well worth the effort, though: "All the time you're looking at something that nobody's ever looked at before, and that's really cool," Brazelton says.

To collect rock samples, researchers use a pair of remotely operated mechanical vehicles, *Jason* and *Hercules*, each equipped with robotic arms. The arms enable researchers to reach out and grab rocks. But that's easier said than done: it's a bit like attempting to grab a toy with a shaky mechanical claw at the arcade, only underwater and with strong ocean currents whipping the claw around. The chalky limestone prizes can also be quite brittle, crumbling if squeezed too tightly. Often the researchers grab something only to have it fall down between the spires.

To collect living specimens, the researchers use what they call a "slurp gun." A robotic arm aims the gun, which then gulps a sample from the spires. Everything is caught on camera by *Hercules*. Eight hours later, the vehicles return to *Atlantis* with their cargo of samples.

Because many of the Lost City microbes cannot tolerate oxygen, the biologists have to be careful not to expose them to air. Using a special airtight bag with built-in gloves, the researchers transfer samples of microbes into test tubes without introducing oxygen into their environment. The microbes are then put in warm incubators and coaxed to grow (**INFOGRAPHIC 17.2**).

INFOGRAPHIC 17.2 Investigating Life in Lost City

Samples are obtained from fluids and rocks at Lost City. The samples are returned to the lab for analysis and identification.

Collecting Samples

Various types of life are collected from the surface of the Lost City spires and surrounding fluid. Collection is painstakingly completed with the use of robotic claws and suction devices that move samples into collection boxes for transport to the surface.

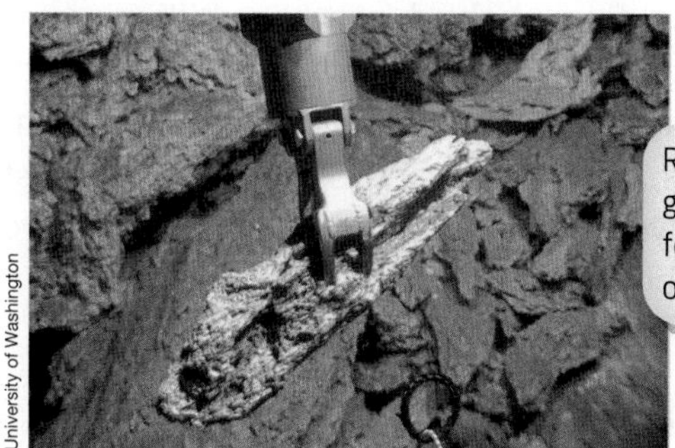

University of Washington

A "slurp gun" sucks up samples containing smaller organisms.

The University of Washington, IFE, URIIAO, Lost City Science Team and NOAA.

Processing Microbes

Once Lost City microbe samples arrive in the laboratory on board *Atlantis*, they are processed for growth and identification, or for DNA sequencing to identify organisms that can't be cultured.

A scientist processes microbial samples in an anaerobic glove bag that eliminates oxygen that may be harmful to Lost City microbes.

University of Washington, School of Oceanography

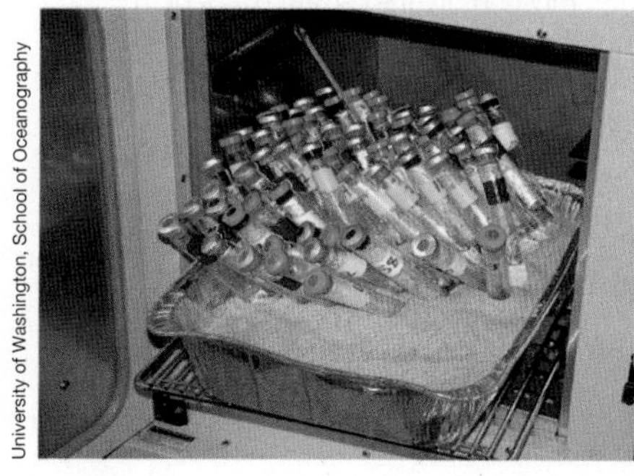

University of Washington, School of Oceanography

? When the researchers attempt to culture the organisms from Lost City, what incubation temperature will most likely support the microbes?

What Are They?

The microbial life in Lost City poses many riddles for the scientists trying to understand this mysterious deep-sea world: What are these creatures? How are they related to known organisms? What adaptations allow them to survive?

Lost City houses a community of life forms ranging from mats of microbes to translucent 1-cm-long animals and the larger fish that eat them. But most of the living things at Lost City are **prokaryotes**—microscopic organisms with cells that lack internal membrane-bound organelles. The DNA in a prokaryotic cell floats freely in the cytoplasm—rather than being housed in a nucleus, as in eukaryotes—and consists of a single, circular chromosome. Some prokaryotes contain additional, small circular DNA molecules called plasmids that may confer an advantage in certain environments. Prokaryotic organisms are tiny, on the order of 1-10 microns, which is about 1/10 the thickness of a human hair. The vast majority are unicellular, but some are able to form multicellular colonies that can function as a single unit. We've already encountered one type of prokaryote—bacteria—in earlier chapters **(INFOGRAPHIC 17.3)**.

PROKARYOTE
A (typically) unicellular organism whose cell lacks internal membrane-bound organelles and whose DNA is not contained within a nucleus.

What prokaryotes lack in size they make up for in numbers. Prokaryotes occupy virtually every niche on the planet, and most scientists agree that we have barely scratched the surface in cataloguing their numbers and diversity. There are more prokaryotic organisms in a handful of dirt than there are plants and animals in a rain forest. More prokaryotic organisms live on and in you right now than there are human cells in your body. At Lost City, up to 1 billion prokaryotic organisms inhabit each gram of chimney rock, forming a mucuslike biofilm several centimeters thick. It looks like the chimneys got sneezed on, says Brazelton.

As their numbers testify, prokaryotes are extraordinarily successful at making a living. From fossil evidence, we know that prokaryotes were the first colonizers of our planet, and for nearly 2 billion years its only life form. Having first evolved nearly 4 billion years ago, prokaryotic organisms have had plenty of time to adapt to a wide range of environments, including many that would kill most eukaryotes. In fact, prokaryotes can thrive just about anywhere (**INFOGRAPHIC 17.4**).

INFOGRAPHIC 17.3 Prokaryotic Cells Are Small and Lack Organelles

Prokaryotic cells are much smaller than eukaryotic cells and do not have the same internal organization. Prokaryotic cells lack organelles, instead carrying out all cellular functions in one central space or in association with the cell membrane. The DNA, including a single circular chromosome and smaller plasmids, is not found in a nucleus but instead floats freely in the cytoplasm.

Prokaryotic Organisms

- Typically single celled
- No organelles
- DNA floats freely in cytoplasm
- Unique cell walls

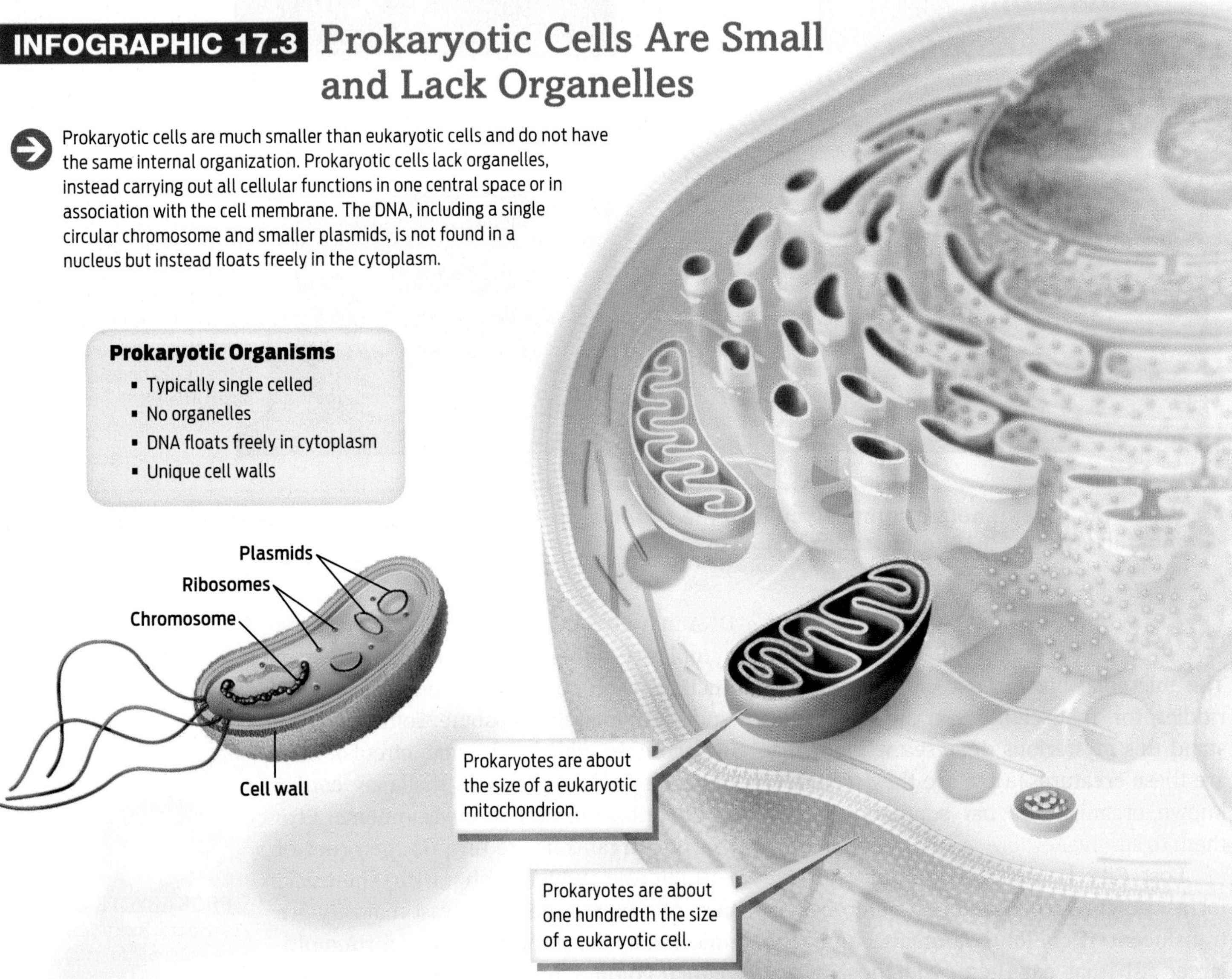

Plasmids
Ribosomes
Chromosome
Cell wall
Prokaryotes are about the size of a eukaryotic mitochondrion.
Prokaryotes are about one hundredth the size of a eukaryotic cell.

? List at least three features that distinguish prokaryotes from eukaryotes.

INFOGRAPHIC 17.4 Prokaryotes Are Abundant and Adapted to a Wide Variety of Environments

Even in seemingly inhospitable environments, there can be large numbers and many different types of prokaryotic microorganism.

Lost City Microbes

1 billion filamentous microbes per gram in the outer carbonate crust.

Methane-metabolizing microbes in fluids rich in volatile gases.

High-temperature, high-pH fluids

Desert Varnish

Dark biofilms from dozens of iron- and manganese-oxidizing bacterial species form on the surface of heat baked boulders.

Bleeding Glaciers

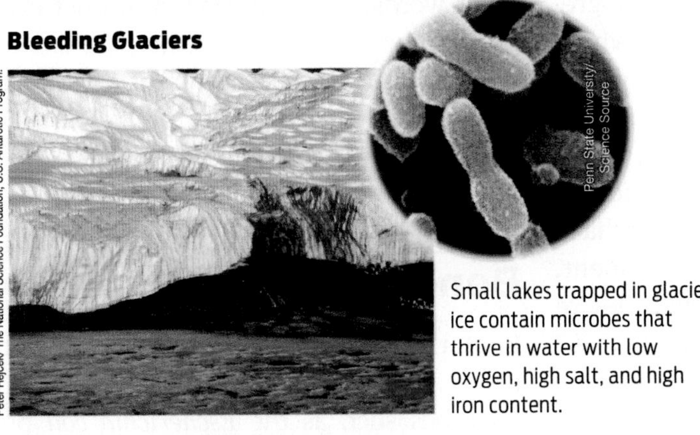

Small lakes trapped in glacier ice contain microbes that thrive in water with low oxygen, high salt, and high iron content.

Living Soils

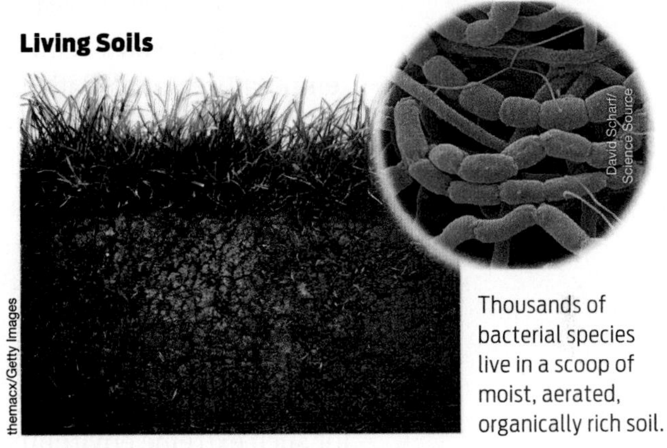

Thousands of bacterial species live in a scoop of moist, aerated, organically rich soil.

? If 1 g of Lost City crust has 1 billion microbes, how many microbes would there be in a pound of Lost City crust?

How do the prokaryotic organisms found at Lost City compare to those living elsewhere? When biologists want to identify a prokaryote, they can't always rely on physical appearance, since many prokaryotic organisms look similar under a microscope. Nor can every prokaryotic organism necessarily be grown in the laboratory. Many of the unusual prokaryotes at Lost City were almost impossible to culture in the lab, making it hard to study their features and behavior. Instead, biologists generally rely on DNA to identify prokaryotic organisms. The number of DNA sequence similarities between the new species and known ones establishes their degree of relatedness. Finding a unique DNA sequence in a sample means the researchers have discovered a new organism.

By looking at DNA sequences, researchers have discovered several new species of prokaryotic organisms living at Lost City. Two of these species fall into a group of prokaryotes known as archaea. Archaea aren't the only prokaryotes present at Lost City—the site is rich in bacterial populations, too—but it's the archaea that are most interesting to

Left: Gretchen Früh-Green examining rock samples collected from Lost City. Right: Deborah Kelley and a colleague monitoring a video feed from a robotic camera.

researchers. That's because the archaea are doing things that even bacteria can't do.

At Lost City, bacterial populations congregate on the outsides of the active vents, where temperatures are relatively mild and where oxygen is present in the seawater. Archaea, by contrast, are found inside the vents, where temperatures are hottest and where there is no oxygen. So far, just two species of archaea have been detected in this environment. "The conditions are so extreme that they're the only thing that has been able to survive," Brazelton says. Because of their preference for such extreme environments, archaea have been nicknamed "extremophiles."

As intriguing as these extreme-loving organisms are, however, they weren't even recognized as a distinct evolutionary group until quite recently. For many years all prokaryotes were lumped together into one large group–the kingdom Monera–a classification based largely on their cell structure. Then, in the late 1970s, Carl Woese and his colleagues at the University of Illinois made the surprising discovery that not all prokaryotic organisms are genetically similar enough to be classified as a single group. His work established an entirely new branch of prokaryotic organisms, now known as the archaea (see Chapter 16). While most archaea don't look that different from bacteria under the microscope–both are unicellular prokaryotes–genetically they are as different from bacteria as humans are. In other words, they represent a distinct evolutionary domain of life. Together, the domains Bacteria and Archaea represent a very large slice of the total diversity of life on Earth **(INFOGRAPHIC 17.5)**.

BACTERIA
One of the two domains of prokaryotic life; the other is Archaea.

Bounteous and (Mostly) Beneficial: Bacteria

Bacteria are the invisible workhorses of the planet. They include commonly encountered microbes such as the *Escherichia coli* present in the human gut and the *Staphylococcus aureus* colonizing our skin, as well as not so commonly encountered ones like the fluorescence-emitting bacteria living in the head of a sea squid. While all bacteria are prokaryotic, and most possess a cell wall, their genetic diversity translates into a wide variety of differences in nutrition, metabolism, shape, and lifestyle. Bacteria live just about everywhere on the planet, from briny lakes submerged under miles of ice, to hazardous waste disposal sites, and everything in between.

Like all organisms, bacteria can be categorized by what they eat. Some bacteria are autotrophs (literally, "self-feeders"): they are able to make their own food directly, using material from the nonliving environment,

INFOGRAPHIC 17.5 Bacteria and Archaea, Life's Prokaryotic Domains

All living organisms fall into one of the three domains of life. Within each domain, there are subgroups of organisms, grouped together based on their evolutionary relationships. Two of the domains of life, Bacteria and Archaea, have organisms with prokaryotic cells, but each has a distinct evolutionary history. Archaea are more closely related to eukaryotes than to bacteria.

? What is the basis for the classification of bacteria and archaea into two separate domains, despite their common prokaryotic structure?

from carbon dioxide to rocks. Others are heterotrophs (literally, "other feeders"): they must rely on other living organisms to provide them with food.

One of the largest and most important groups of autotrophic bacteria is the cyanobacteria, which are found in oceans and freshwater, as well as on exposed rocks and soil–virtually everywhere sunlight can reach. Cyanobacteria use the energy of sunlight to carry out photosynthesis in a manner similar to plants, taking in CO_2 and generating much of the oxygen that other organisms, including humans, rely on. Cyanobacteria are thought to be the oldest photosynthetic organisms on Earth, dating back roughly 2.5 billion years and playing a pivotal role in making the atmosphere breathable for the rest of us.

Not all autotrophic bacteria rely on sunlight for energy. Some, including those living at Lost City, can obtain energy directly from geological sources such as inorganic gases pouring out of hydrothermal vents–making them among the few organisms on Earth that do not rely on the sun's energy to survive.

Heterotrophic bacteria include those that obtain food by consuming material from living or dead organisms. Such heterotrophic bacteria play an important role in decomposition, allowing carbon and other elements–which would otherwise be trapped in dead organisms, sewage, or landfills–to be recycled. They are also useful in bioremediation projects. For example, some types of bacteria metabolize droplets of oil, much the way humans digest butter, so they help clean up oil spills naturally.

Bacteria break down their food molecules through a variety of metabolic pathways, some of which require oxygen, some of which do not. For example, many bacteria employ the anaerobic process of fermentation (see Chapter 6) to get energy from food. The products of fermentation can be valuable (and tasty) to humans. You may have seen "L. bulgaricus" listed as an ingredient of yogurt; live *Lactobacillus bulgaricus* bacteria are present and at work in the yogurt, fermenting sugars into lactic acid, which helps the milk solidify into yogurt and which gives yogurt its tangy taste. Other bacteria use oxygen to break down organic molecules, like the aerobic bacteria that feast on an oil spill.

In addition to nutritional and metabolic differences, bacteria display a variety of structural adaptations that suit their various lifestyles. They come in different shapes: spherical (in which case they are known as cocci), rod-shaped (bacilli), and spiral (spirochetes). Many bacteria are equipped with **flagella,** tiny whiplike structures that project from the cell and help it move. For example, the bacterium *Helicobacter pylori*, the most common cause of stomach ulcers, uses its flagella to propel itself through the gastric mucus of the stomach. **Pili** are shorter, hairlike appendages that enable bacteria to adhere to a surface. *Neisseria gonorrhoeae*, the bacterium that causes the sexually transmitted disease (STD) gonorrhea, uses its pili to remain attached to the lining of the urinary tract. Without pili, the bacteria would be flushed out by the flow of urine.

Other bacteria are surrounded by a **capsule,** a sticky outer layer that helps the cell adhere to surfaces and to avoid the defenses of the host. *Streptococcus mutans*, for example, produces a capsule that allows it to adhere to teeth, where it forms the plaque that can lead to cavities.

Highly resourceful, bacteria have a diverse range of living arrangements with other creatures. Many live in close association, or **symbiosis,** with other organisms–often to the benefit of one or both partners. Naturally occurring lactobacilli in the female vaginal tract, for example, obtain nourishment by fermenting naturally occurring sugars to lactic acid. The resulting acidity of the vaginal tract suppresses the growth of yeast, preventing yeast infections. Antibiotics taken for a bacterial infection are likely to kill the resident lactobacilli as well as the invaders, and a yeast infection is often an unhappy side effect.

> It was an uncharted undersea world no one had known existed. Kelley named it "Lost City."

Another example of beneficial bacterial symbiosis is *Vibrio fischeri,* a bioluminescent bacterium that lives and feeds inside the light organs of certain species of squid. The glow-in-the-dark *Vibrio* produces light beneath the squid and helps obscure the shadow that the squid might cast on a moonlit night, making it less noticeable to its prey as it hunts.

Unfortunately, not all bacteria are beneficial to the host. While the vast majority of bacteria do not cause human disease, some do. Organisms that cause disease are called **pathogens.** Many pathogenic bacteria cause disease by producing toxins that harm their hosts. For example, certain strains of the bacterium *Escherichia coli* secrete a potent toxin that causes bloody diarrhea and even sometimes kidney failure and death in its host. Keeping food refrigerated helps prevent food poisoning by slowing the growth of the bacteria and, therefore, production of its toxin.

FLAGELLA (SINGULAR: FLAGELLUM)
Whiplike appendages extending from some cells, used in movement of the cell.

PILI (SINGULAR: PILUS)
Short, hairlike appendages extending from the surface of some bacteria that enable them to adhere to surfaces.

CAPSULE
A sticky coating surrounding some bacterial cells that adheres to surfaces.

SYMBIOSIS
A relationship in which two different organisms live together, often interdependently.

PATHOGEN
A disease-causing agent or organism.

Not all pathogens produce toxins. Some cause disease by living and reproducing in the body and interfering with its normal processes–an example is the bacterium *Treponema pallidum,* which causes syphilis, a sexually transmitted disease.

Sometimes the line between harmless and harmful bacteria can be blurred. Organisms that can, but don't always, cause disease are known as opportunistic pathogens. For example, most of us have *Staphylococcus aureus* on our skin at many times during our lives. Most of the time, *S. aureus* does not cause any harm, but if it penetrates the skin–through a wound, for example–it can cause a serious infection and even death, as discussed in Chapter 13 **(INFOGRAPHIC 17.6)**.

For all their impressive diversity and abundance, bacteria are far from the totality of prokaryotic life. Moreover, their lifestyles and adaptations may seem tame next to those of the other domain of prokaryotic life: the Archaea.

INFOGRAPHIC 17.6 Exploring Bacterial Diversity

Bacteria live in every imaginable place on Earth and have a diverse array of lifestyles. Their unique structural and metabolic adaptations have enabled them to become a dominant force of life on the planet.

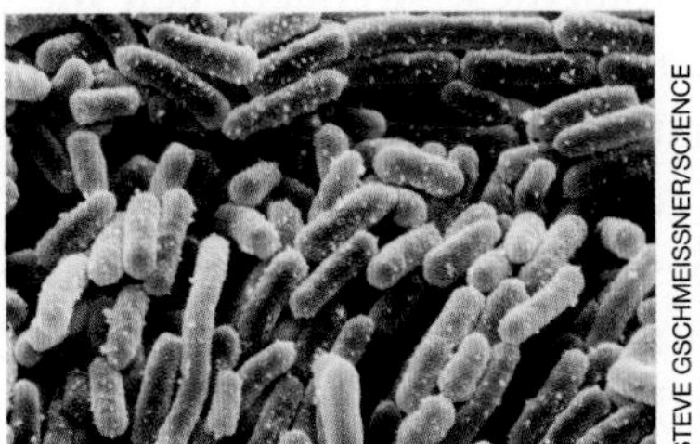

STEVE GSCHMEISSNER/SCIENCE PHOTO LIBRARY/Getty Images

Rod-shaped
Lactobacillus

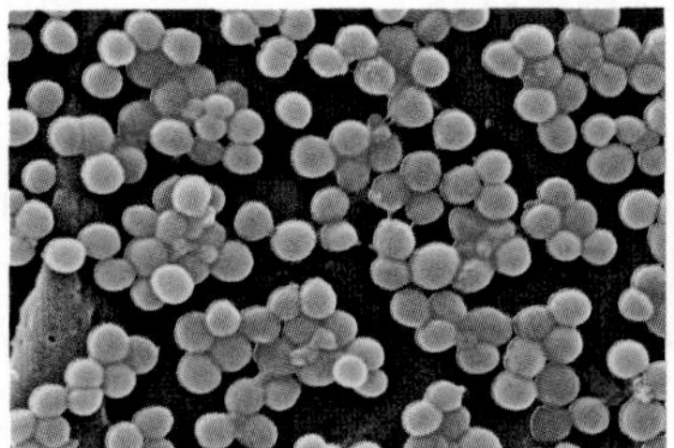

Janice Carr/Getty Images

Sphere-shaped
Staphylococcus

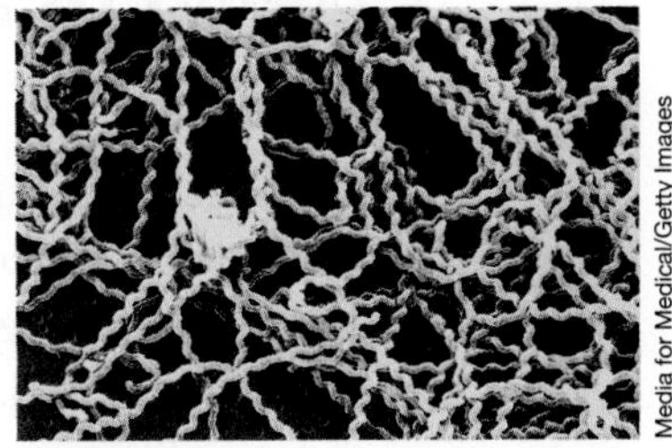

Media for Medical/Getty Images

Spiral-shaped
Leptospira

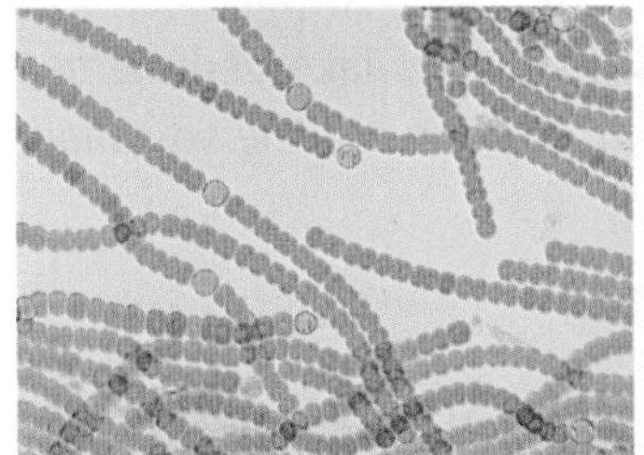

Carolina Biological Supply Company/Medical Images

Photosynthetic
Anabaena cyanobacteria

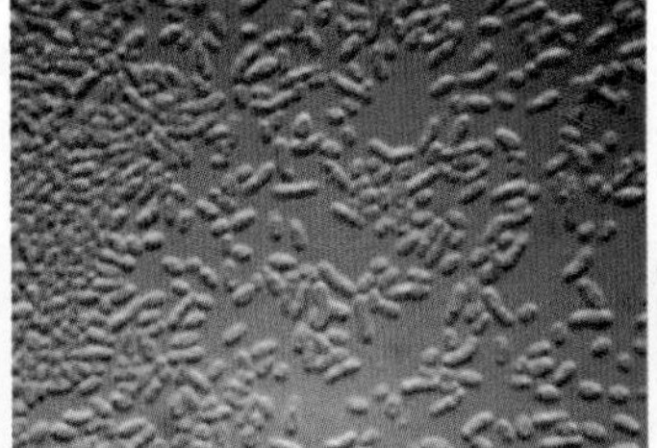

Michael Abbey/Science Source

Eats Oil
Ochrobactrum anthropi

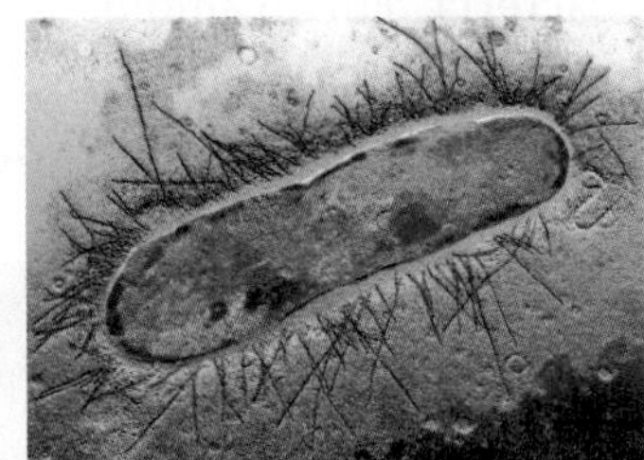

BSIP/Getty Images

Pathogenic
Escherichia coli

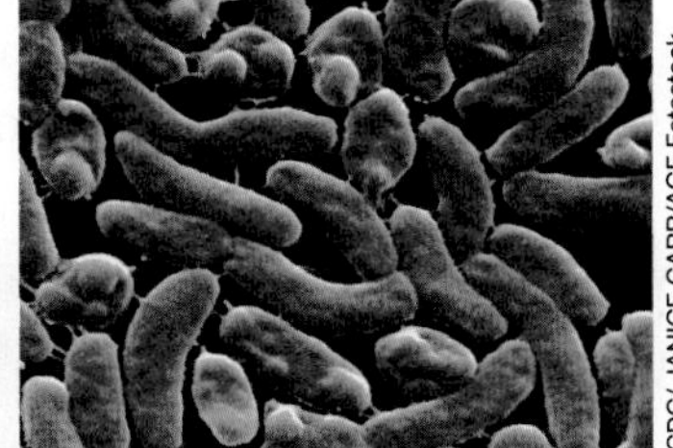

CDC/ JANICE CARR/AGE Fotostock

Thrives in a high salt environment
Halomonas elongata

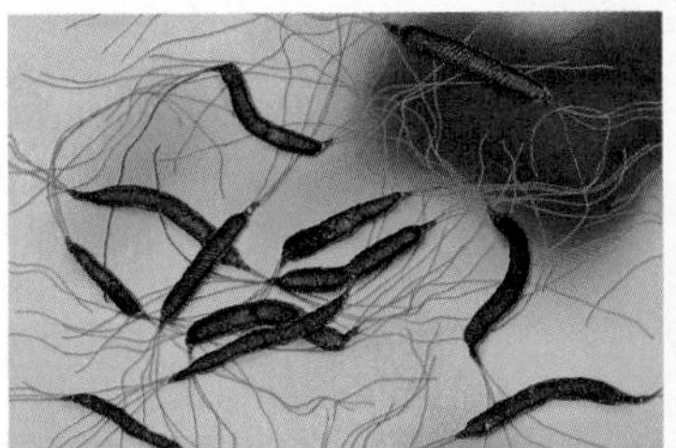

BSIP/UIG/Getty Images

Flagella
Helicobacter pylori

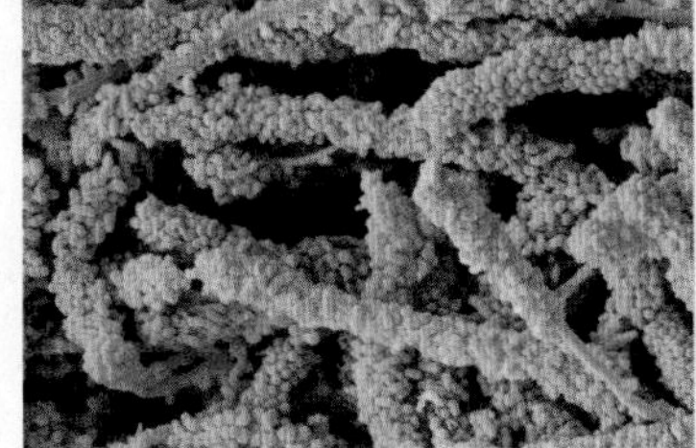

STEVE GSCHMEISSNER/SPL/Getty Images

Capsule
Streptococcus mutans

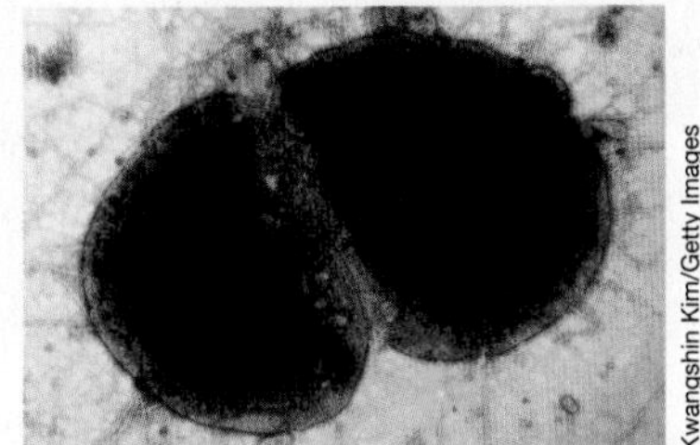

Kwangshin Kim/Getty Images

Pili
Neisseria gonorrhoeae

? Flagella, capsules, and pili can all be found on pathogenic bacteria. Pick one of these structures, and describe how it might contribute to the development of disease.

Symbiosis

Fluorescent bacteria reside in a light organ in the squid. The illumination from the bacteria helps the squid avoid casting a shadow on a moonlit night, making it less noticeable to prey.

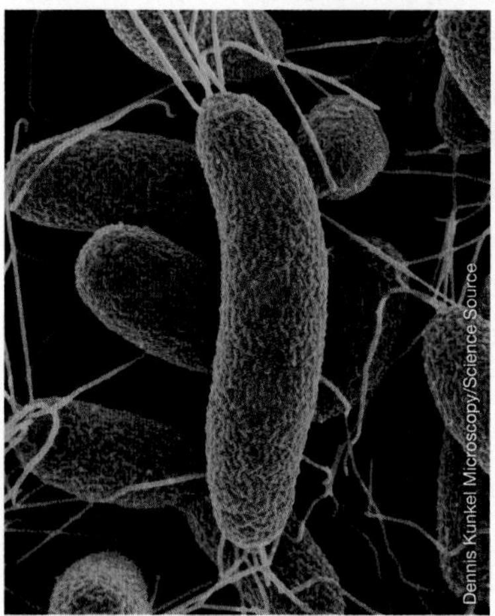

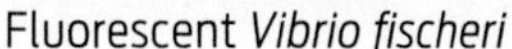
Fluorescent *Vibrio fischeri*

Euprymna scolopes

Going to Extremes: Archaea

Archaea are similar to bacteria in that archaea are single-celled organisms that lack a nucleus, but genetically they are as different from bacteria as humans are. All those genetic differences add up to a number of unique features that distinguish archaea from bacteria. For example, while bacteria have cell walls made of the molecule peptidoglycan, archaea have cell walls made of other molecules. Archaeal cell membranes can also have a different chemical composition that makes them less likely to be pulled apart, and archaeal DNA is protected by enzymes that resist heat. Archaea also have some unique forms of metabolism, which makes them of interest to researchers probing the origins of life.

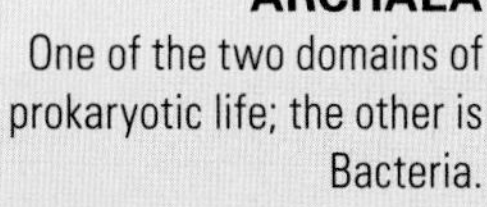
ARCHAEA
One of the two domains of prokaryotic life; the other is Bacteria.

Though archaea are found in many run-of-the-mill habitats such as rice paddies, forest soils, ocean waters, and lake sediments, the most well-known species are the so-called

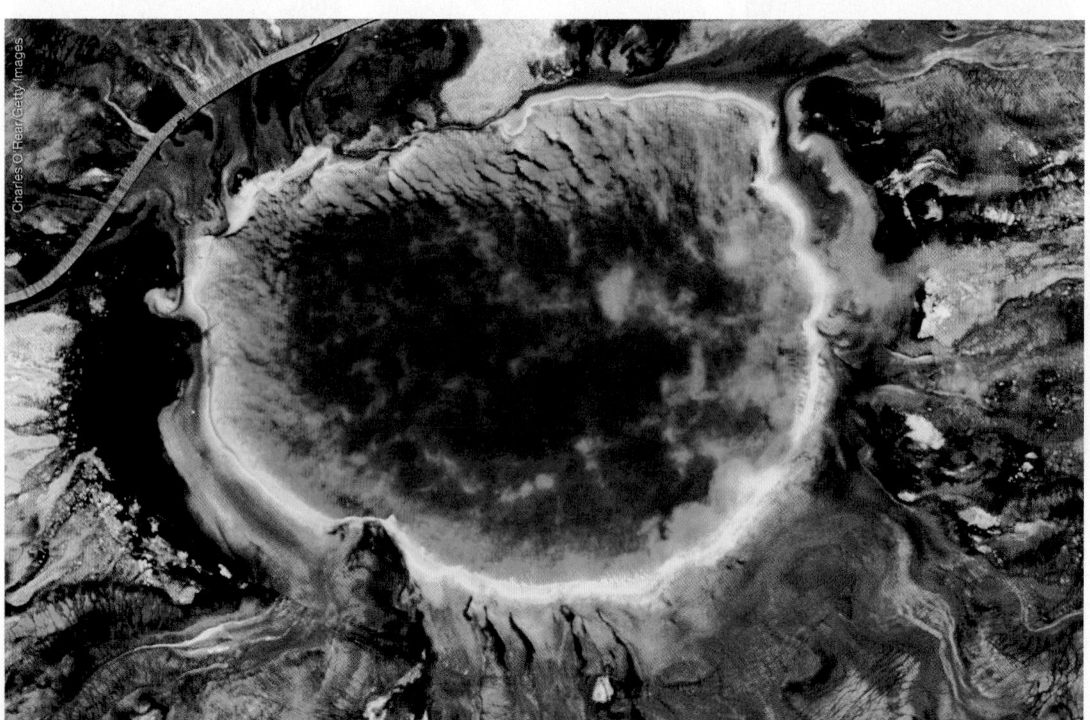

Yellowstone's Grand Prismatic Hot Spring

Heat-tolerant bacteria and archaea live in this 80°C–100°C, low pH and high sulfur aquatic environment that bubbles up from below Earth's crust. The rainbow colors that ring this hot spring come from light reflecting off the pigments made by different microbe species.

extremophiles. Many of these extreme-loving archaea, including those living in Lost City, are hyperthermophiles–organisms that can survive only at extremely high temperatures. Many hyperthermophilic archaea are anaerobic and rely on sulfur instead of oxygen in metabolism. Sulfur-rich hot springs like those in Yellowstone National Park are home to these archaea.

Other archaea are methanogens, which consume carbon dioxide and hydrogen and produce methane as a by-product. Because this gaseous meal is completely inorganic, these archaea are considered autotrophs. A methanogen that can survive in an even more extreme environment than Lost City is *Methanopyrus kandleri*, which lives in a type of hydrothermal vent known as a black smoker, where the temperature can be 121°C–thought to be the upper temperature limit of life. Methanogens occupy more mundane environments as well, including the digestive systems of methane-belching cows (and flatulence-emitting humans).

Some archaea are halophiles, or "salt lovers," and prefer a home saturated in salt, which would shrivel most other living things. Their presence is detectable by the colorful pigments they produce–bright reds, yellows, and purples–as seen in salt ponds in San Francisco Bay (**INFOGRAPHIC 17.7**).

Archaea's affinity for such extreme environments is suggestive of their evolutionarily ancient roots. The early Earth was a lot warmer than it is now, and it's long been a question how living things could withstand the heat that then prevailed. If the archaea at

INFOGRAPHIC 17.7 Exploring Archaeal Diversity

Archaea are sometimes known as "extremophiles." They live in diverse environments, often surviving and thriving in very harsh conditions.

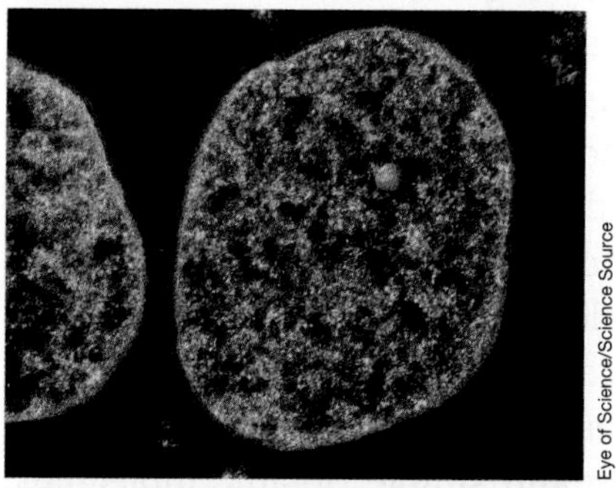
Eye of Science/Science Source

Sulfolobus archaea have flagella. They are considered extremophiles because they can grow at 70°C and at pH 2.0.

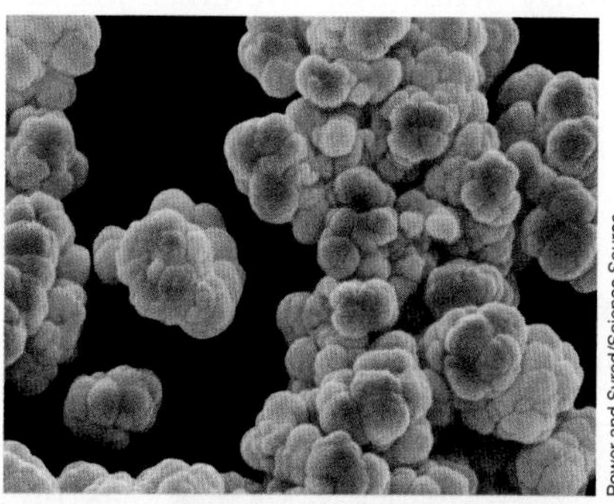
Power and Syred/Science Source

Methanosarcina have unusual cell walls and membranes, and produce methane.

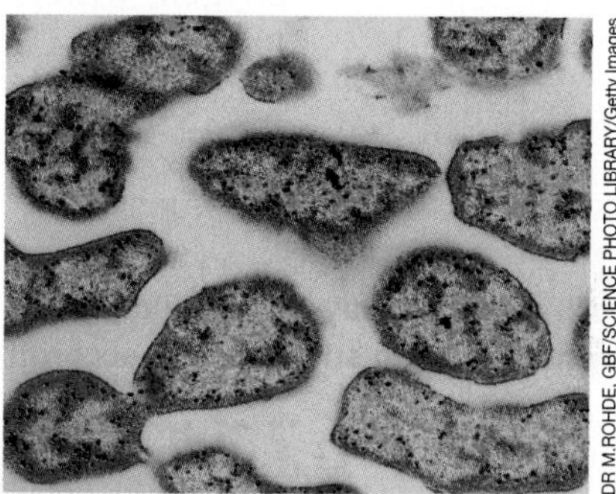
DR M.ROHDE, GBF/SCIENCE PHOTO LIBRARY/Getty Images

Methanococcoides grows in cold temperatures as low as -2.5°C.

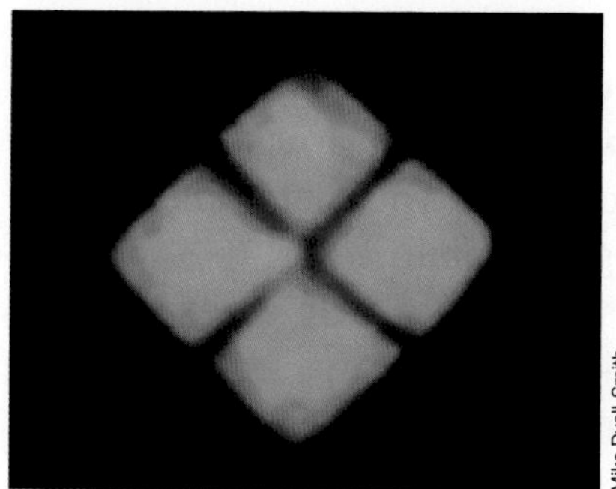
Mike Dyall-Smith

Haloquadratum walsbyi thrives in water temperatures well above boiling, at least 120°C.

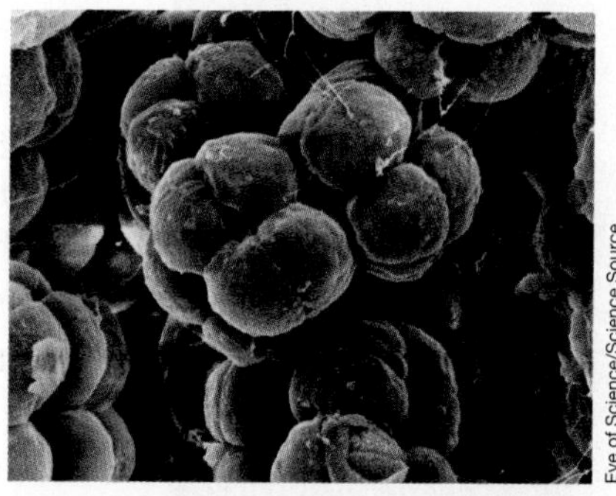
Eye of Science/Science Source

Halobacteriales archaea not only tolerate but require a high salt environment to live.

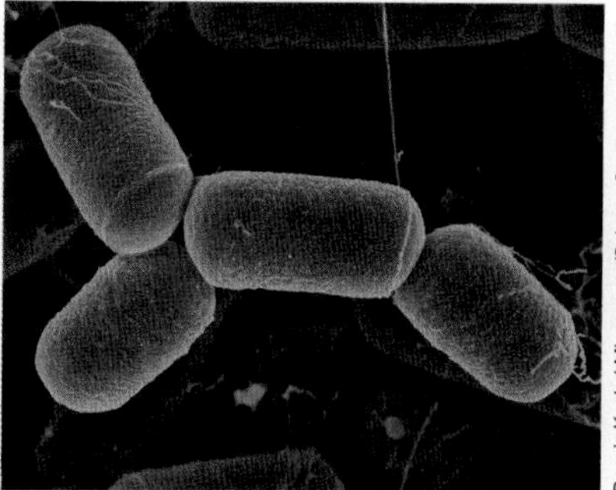
Dennis Kunkel Microscopy/Science Source

Methanobrevibacter smithii aids digestion in the human intestine.

? What are three extreme conditions in which archaea can survive?

Lost City are any indication, they would have done just fine.

In addition, some researchers suspect that archaeal metabolism may be among the most ancient forms of metabolism on Earth. Scientists who study the origin of life believe that the first organisms were likely autotrophs that obtained their carbon from carbon dioxide, using hydrogen as an energy source and emitting methane as a by-product–in other words, doing exactly what the archaea at Lost City do today.

So far, archaea are the only organisms known to perform this type of metabolism. What's more, they do it without the help of oxygen. That's a handy trait to have when you live in an environment lacking oxygen–which is exactly what the ancient Earth was.

> "[Lost City] is a good example of what we really don't know and what there is to still discover on the seafloor."
>
> — Gretchen Früh-Green

Coated with a mucusy layer of biofilm, the rocky chimneys at Lost City "look like they've been sneezed on," according to biologist Bill Brazelton.

Kelley, D.S., University of Washington and IFE, URI-IAO, UW, Lost City science party, and NOAA.

Lost City: Where Life Began?

In its early days, more than 4 billion years ago, Earth was mostly warm ocean, and its atmosphere lacked significant oxygen. Photosynthetic organisms did not exist, so there were no living things to harness the energy of sunlight to make organic molecules. Such conditions pose a conundrum for those who study the origin of life. With no photosynthetic organisms to make organic molecules, where did the energy and building blocks necessary to assemble living things come from? Lost City provides an important clue.

The hydrothermal vents at Lost City are driven by a geochemical process called serpentinization, which occurs any time a particular type of rock from Earth's mantle comes in contact with seawater. The rocks are exposed as a result of faulting and uplift of the seafloor. The reaction generates a lot of heat, releases hydrogen gas, and–when this hydrogen reacts with carbon from rocks or seawater–produces hydrocarbons such as methane and simple organic molecules. All this happens completely abiotically–that is, without the participation of living things. Such an environment would have been an ideal place for life to begin, says Brazelton. "You have energy and organic compounds and liquid water all in a warm spot, so that's a great place where you might imagine life could have got started."

Lost City may also help to explain the chicken-and-egg problem posed by the origin of life: organic molecules are needed to build living things, but living things are generally the source of organic molecules. So which came first, the chicken or the egg? "I think a big clue to the chicken-and-egg problem is that you don't need life to make organic compounds," says Brazelton. "We are studying environments right now where the organic compounds are literally pouring out of these chimneys, and they're being made without the help of life."

Scientists have also been struck by a further observation about the Lost City world: the chemical reactions taking place in sea water as a result of serpentinization are very similar to those occurring inside the archaea. Both use hydrogen gas as an energy source to convert carbon dioxide to methane. This chemical similarity raises the intriguing possibility that the first cells adopted, as a form of metabolism, a chemical reaction that was occurring in the surrounding vents: life imitating rocks (**INFOGRAPHIC 17.8**).

Many microbiologists would agree that deep-sea vents like Lost City likely represent some of the oldest habitats for microbial life on

INFOGRAPHIC 17.8 Energy from the Earth Fuels Life at Lost City

Many prokaryotes thrive in seemingly inhospitable environments at Lost City. Abiotic reactions at Lost City generate energy-rich and carbon-containing molecules that can sustain the organisms in this environment. These reactions may be similar to the ones that supported life on the early Earth.

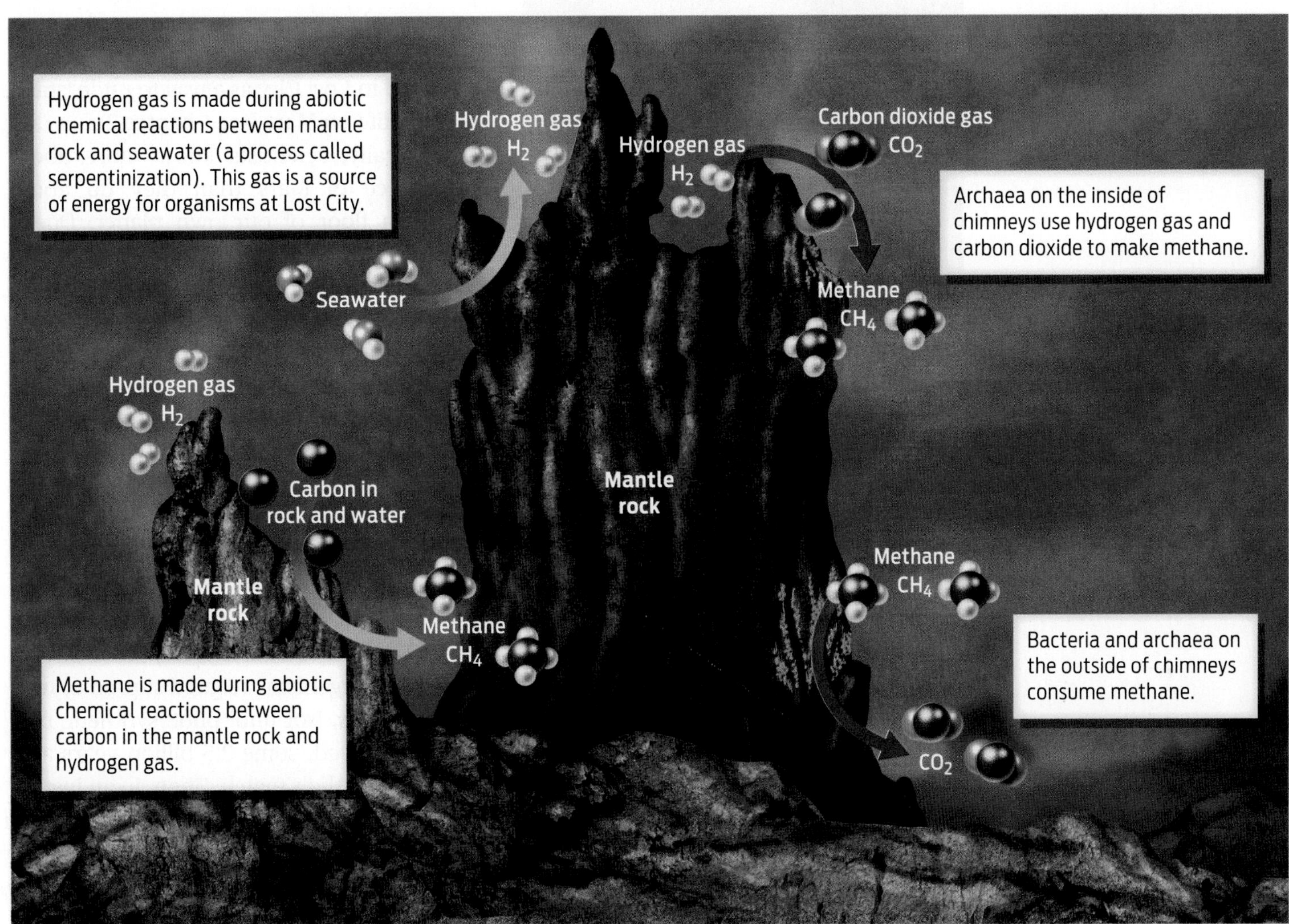

? Find one abiotic and one biotic source of methane in this diagram. Why is methane production important at Lost City?

At Lost City, single carbonate chimneys that grow out of cliff faces can reach nearly 10 stories in height.

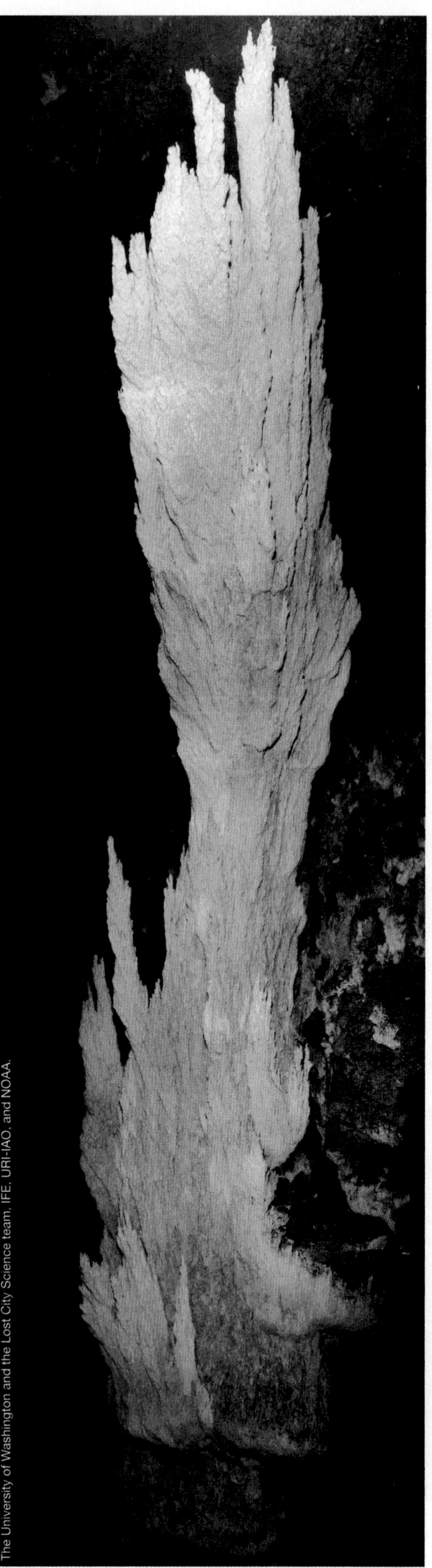
The University of Washington and the Lost City Science team, IFE, URI-IAO, and NOAA.

Earth. Radiometric dating of the rock layers indicates that Lost City's vents have been pumping strong for at least 100,000 years, and likely for much longer: Lost City sits on a layer of Earth's crust that is at least 1.5 million years old. Back then, rocks from the interior of Earth–from the mantle–were much closer to the surface than they are now, which means their reaction with seawater would have been more common. A journey to Lost City is thus like a journey back in time, to Earth's primordial past.

Lost City may even provide a clue to life beyond Earth. The rocks involved in serpentinization are quite common in the solar system. They are all over the surface of Mars, for example, and researchers suspect that the chemical reaction might be occurring right now beneath the surface of Mars, where recent evidence suggests that methane is being produced. NASA is therefore extremely interested in Lost City as a way to understand potential life on Mars (see Chapter 2).

Ironically, scientists know more about the surface of Mars than they do about the ocean floor of our own planet. Ocean covers 70% of Earth's surface, yet much of it remains unexplored. If Lost City is any indication, many scientific treasures await the patient explorer. "[Lost City] is a good example of what we really don't know and what there is to still discover on the seafloor," says geologist Früh-Green, who was lucky enough to get the first look at the underwater world.

If life on Earth did begin at hydrothermal vents like those at Lost City, then it could mean that these extreme-loving prokaryotes are the descendants of the most ancient form of life on Earth. For nearly 2 billion years, these earliest prokaryotic organisms reigned supreme, with no challengers. Not until photosynthetic prokaryotes evolved, some 2.5 billion years ago, did they meet their match. Then, in an instant (geologically speaking), life on our planet underwent a radical and unprecedented change: 2 billion years ago, one of these early prokaryotes engulfed another and the two cells began a symbiotic relationship. That was the birth of the first eukaryote. ■

CHAPTER 17 SUMMARY

- Prokaryotes are (typically) unicellular organisms that lack internal organelles and whose DNA is not contained in a nucleus.
- Prokaryotes are found in virtually every environment on Earth, even those with seemingly inhospitable conditions, such as hydrothermal vents on the ocean floor.
- Genetic analysis has led to the categorization of life into three domains: Bacteria, Archaea, and Eukarya. Each domain of life has a distinct evolutionary history.
- Both bacteria and archaea have prokaryotic cells, but they otherwise differ in their genetics, biochemistry, and lifestyles.
- Bacteria are a diverse group of prokaryotic organisms with many unique adaptations such as flagella and capsules that allow them to live and thrive in many environments.
- Some bacteria are disease-causing pathogens, but most are harmless and many even beneficial. Cyanobacteria, for example, are responsible for much of the photosynthesis that supports life on Earth.
- Often known as "extremophiles," archaea live in some of the most inhospitable conditions on Earth, such as hydrothermal vents. Many archaea flourish in less extreme environments as well.
- The harsh conditions of Lost City may resemble the conditions of the early Earth. The prokaryotic inhabitants of Lost City may be metabolically similar to the earliest known life.
- The energy that fuels life in Lost City comes from a geological source, rather than from sunlight, making Lost City one of the few communities on Earth that is not powered by photosynthesis.

MORE TO EXPLORE

- **Lost City Home: www.lostcity.washington.edu/**
- **Brazelton, W. J., et al. (2011) Physiological differentiation within a single-species biofilm fueled by serpentinization. *mBio* 2(4):e00127–11.**
- **Kelley, D. S. (2005) From the mantle to microbes: the Lost City hydrothermal field. *Oceanography* 18(3):32–45.**
- **Martin, W., et al. *(2008)*. Hydrothermal vents and the origin of life. Nature Reviews Microbiology 6:805–814.**
- **Yong, E. (2016) *I Contain Multitudes: The Microbes Within Us and a Grander View of Life*. New York: Ecco.**

CHAPTER 17 Test Your Knowledge

DRIVING QUESTION 1 What are the prokaryotic domains of life?

By answering the questions below and studying Infographics 17.3, 17.4, and 17.5, you should be able to generate an answer for the broader Driving Question above.

KNOW IT

1 Organisms are placed into one or another of the three domains of life on the basis of

a. cell type.
b. physical appearance.
c. evolutionary history as assessed by genetic relatedness.
d. ability to cause disease.
e. degree of sophistication, that is, how evolutionarily advanced they are.

2 Describe the major difference(s) between prokaryotic and eukaryotic organisms.

3 The absence of membrane-bound organelles in a cell tells you that the cell must be

a. from a member of the domain Bacteria.
b. from a member of the domain Archaea.
c. from a member of the domain Eukarya.
d. either a or b
e. either b or c

USE IT

4 Why were bacteria and archaea originally grouped together?

5 When first discovered, archaea were called "archaebacteria." Why do you suppose this was? What are the strengths and weaknesses of this earlier term?

DRIVING QUESTION 2 What are the features of bacteria and of archaea?

By answering the questions below and studying Infographics 17.4, 17.6, and 17.7, you should be able to generate an answer for the broader Driving Question above.

KNOW IT

6 The term *prokaryotic* refers to

a. a type of cell structure.
b. a domain of life.
c. a group with a shared evolutionary history.
d. a type of bacterium.
e. a type of archaea.

7 If you were looking for a bacterium, where would expect to find one?

a. on your skin
b. in soil
c. in the ocean
d. associated with plants
e. any of the above

8 When examining archaea, which of the following is *not* a trait that you could expect to find?

a. ability to grow at high temperatures
b. a nucleus
c. ability to grow at extremely acidic pH
d. a cell wall
e. ability to survive in a high-salt environment

9 If you are unable to culture archaea from an environmental sample, is it safe to conclude that there are no archaea present? Why or why not?

USE IT

10 Can you use cell structure to classify a cell as either bacterial or archaeal? Explain your answer.

11 Many prokaryotic organisms can carry out photosynthesis. How is this beneficial to humans?

12 Halophilic archaea are able to prevent osmotic water loss from their cells, even in high-salt environments. What is one mechanism by which they could prevent water loss and thrive in high-salt environments?

13 If *Neisseria gonorrhoeae* had no pili, would it still be a successful pathogen? Explain your answer.

DRIVING QUESTION 3 What are the challenges faced by organisms living at Lost City, and how do they face them?

By answering the questions below and studying Infographics 17.1, 17.2, and 17.8, you should be able to generate an answer for the broader Driving Question above.

KNOW IT

14 List the features that make Lost City a particularly harsh environment. For each feature, give a brief explanation of why it is inhospitable for many organisms.

15 If you were a prokaryotic organism and wanted to be successful at Lost City, what energy source must you be able to use?

a. sunlight
b. oxygen
c. hydrogen gas
d. electricity
e. None of the above is available at Lost City.

USE IT

16 What is the significance of methane at Lost City? (Think about both the origin of life and the sustenance of early life.)

17 If methane were not produced abiotically at Lost City, what would be the implications for early life?

18 Would you expect to find photosynthetic organisms at Lost City? Explain your answer.

19 Do you think that the scientists studying Lost City should be concerned about introducing microbial contaminants from their submersibles onto the towers of Lost City? How probable is this, given the conditions at Lost City and on the surface? If such an event could happen, what would be the implications?

INTERPRETING DATA

apply YOUR KNOWLEDGE

20 **Some of the chimneys at Lost City are actively venting. These chimneys have hot (80°–100°C) interiors that lack oxygen and have a pH range of 9–11. The exterior surfaces of the active chimneys are cooler (~7°C), contain oxygen, and have a pH of ~8.**

The inactive chimneys at Lost City are no longer venting hot fluids. Compared to the actively venting chimneys, their interiors are much cooler (7°–20°C), lack oxygen, and have a pH of 8–10. The exteriors of the inactive chimneys are very similar to the exteriors of the active chimneys.

From the properties of the organisms given in the table below, complete the table to indicate where in Lost City each of these organisms is most likely to be found.

Organism	Aerobic or Anaerobic?	Optimum pH	Optimum Temperature (°C)	Domain: Bacteria or Archaea?	Location
A	Anaerobic	9	15	Archaea	
B	Aerobic	8	6	Bacteria	
C	Aerobic	8.5	8	Bacteria	
D	Anaerobic	11	90	Archaea	
E	Aerobic	7.5	7	Bacteria	
F	Anaerobic	11	85	Archaea	

MINI CASE

apply YOUR KNOWLEDGE

21 **In many ways, the discovery of Lost City is about the discovery of life, as is the voyage of *Curiosity* to Mars (see Chapter 2).**

a. What features do these two environments, Lost City and Mars, share?

b. How do the challenges faced by *Curiosity* compare to those faced by *Jason*?

c. From the way microbes survive at Lost City, what properties might you expect Martian organisms to have?

BRING IT HOME

apply YOUR KNOWLEDGE

22 **Do you think that, instead of spending time studying microbes from extreme and remote environments, scientists should be studying microbes that are more apparently relevant to humans, such as ones that cause disease? In what ways might understanding the organisms at Lost City be useful to humans?**

18 Eukaryotic Diversity

AFP/Getty Images

Can **Rubber** Save the Rain Forest?

A small state in Brazil aims to find out

DRIVING QUESTIONS

1. What are eukaryotic organisms, and what factors influence their diversity?
2. How are plants defined, and what adaptations have enabled their success?
3. How are fungi defined, and what adaptations have enabled their success?
4. How are animals defined, and what adaptations have enabled their success?
5. What are protists, and why are they hard to classify?

During the 2016 summer Olympics, held in Rio de Janeiro, the Brazilian government distributed some 9 million condoms to athletes, locals, and visitors. Amid rising concerns over spread of the Zika virus and ongoing concerns about HIV/AIDS, the effort was a reasonable public health measure aimed at reducing sexually transmitted infections. It was also advertising—for Brazil's locally made, all natural, rubber-based condom.

A factory in the small Brazilian state of Acre manufactures the condoms from rubber harvested from the rubber tree (*Hevea brasiliensis*). The rubber tree is native to the Amazon rain forest, of which the largest part—about 60%—is located within Brazil's national borders.

Tapping trees for rubber is an ancient tradition in the Amazon and is still practiced the way it always has been, by rubber tappers skilled in the craft. The condom manufacturer, Natex, pays about 700 rubber-tapping families for the rubber that they collect. Natex is the world's only manufacturer of condoms made from rubber harvested from wild trees.

Acre's condom factory is just one of several efforts by the local government to create a market for sustainably derived forest products. The goal is to provide stable income for local people who would otherwise turn for subsistence to felling timber or raising cattle, both of which contribute to deforestation.

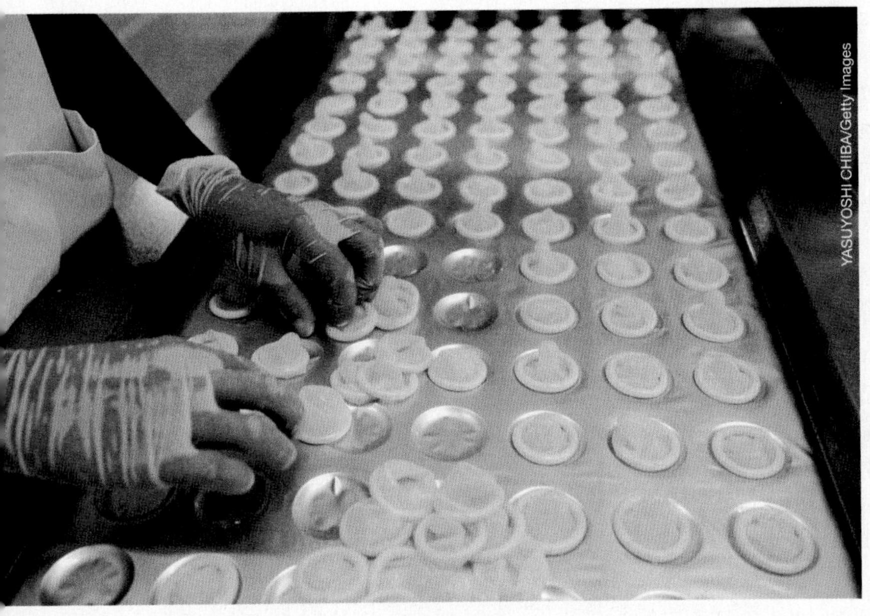

The Natex factory produces condoms made of rubber from native rubber trees.

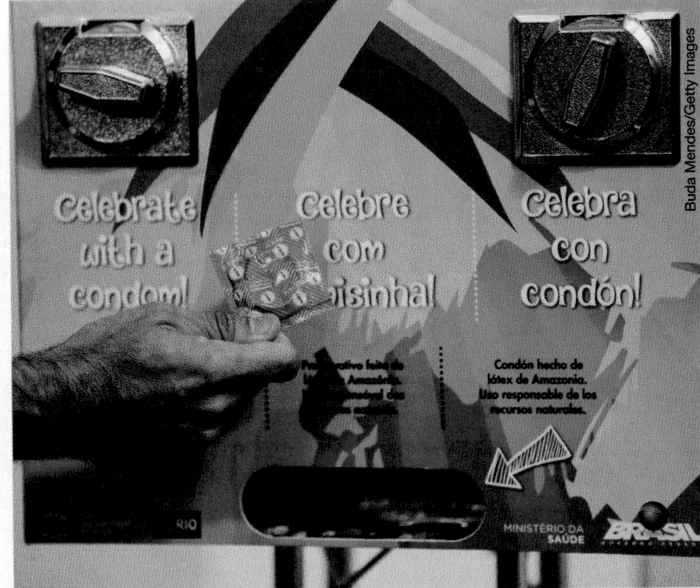

A condom dispenser at the 2017 Olympic Village.

Deforestation Is Highest in the Amazon Rain Forest

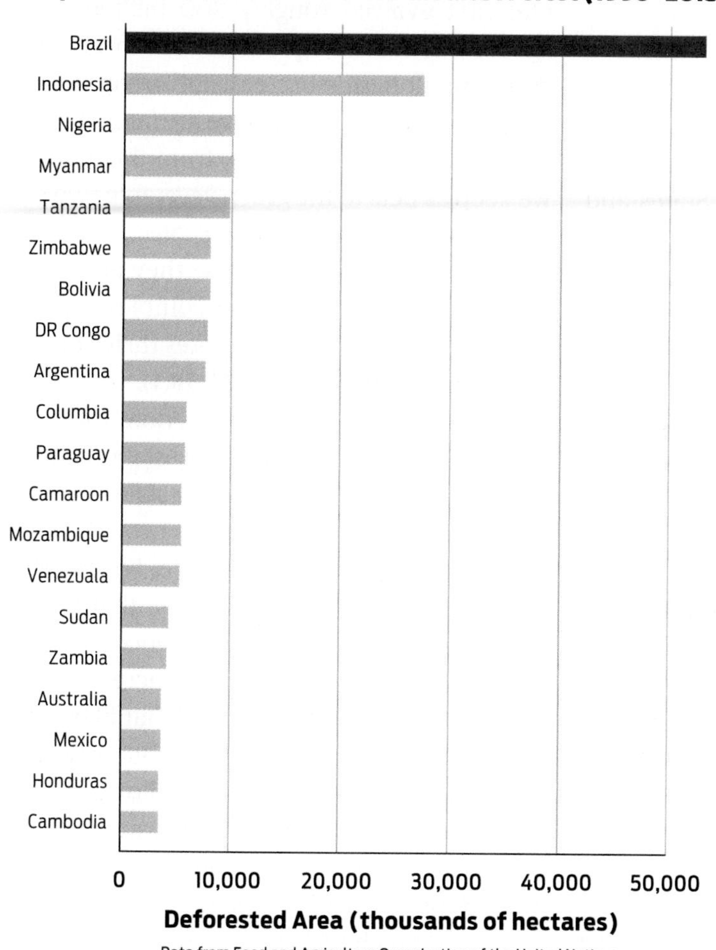

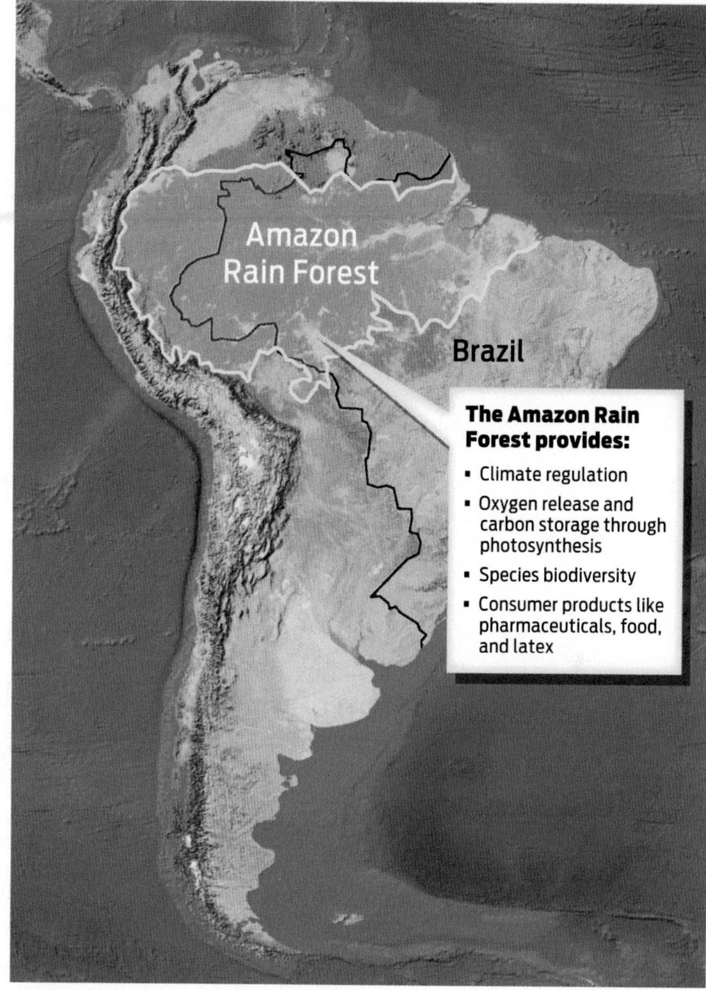

"Acre is kind of a laboratory for innovative forest policy," says Amy Duchelle, a senior scientist with the Center for International Forestry Research (CIFOR) in Bogor, Indonesia, who spent several years conducting research in the region. "The condom factory is a strategy to allow local people to profit from the rain forest without destroying it."

Through these efforts, Acre aims to halt deforestation at 18% of the state's surface area, which would make it the most protected region in Brazil. And environmentalists hope it will be a model for other states and countries to emulate.

That's crucial, because rain forests around the world are being cut down at an alarming rate. In the Amazon alone, nearly 20% of the forest has already been destroyed. Scientists warn that if deforestation continues at the current pace, the Amazon could be gone by the turn of the next century.

If that happens, Earth's **biodiversity** will suffer immeasurably. The Amazon is home to a staggering number of species, including many found there and nowhere else. Losing them from the Amazon means extinction from the planet. What's more, destruction of the Amazon would have far-reaching consequences for living things that don't live in the rain forest but rely on it for goods and services–tangible ones like pharmaceuticals, foods, and latex, and less tangible ones like climate regulation.

Can rubber make a difference? Environmentalists have high hopes that it can. If the past is any guide, the fate of the rubber tree

BIODIVERSITY
The number of different species and their relative abundances in a specific region or on the planet as a whole.

may be inextricably linked to the fate of the region as a whole.

Too Big to Fail

At 2.7 million square miles, the Amazon is the largest tropical rain forest on Earth. It's roughly the size of the continental United States and covers about 40% of the South American continent. The Amazon is sometimes referred to as the lungs of the planet, since the plants that grow here give off breathable oxygen as part of photosynthesis. But "lungs" is a bit of a stretch: compared to photosynthetic algae that live in the oceans, the Amazon is actually not a huge net producer of oxygen.

A more useful way to think about the rain forest is as the planet's air conditioner, helping to keep it from overheating. An enormous amount of water flows through the forest as it travels from snow in the Andes Mountains, down through the Amazon River, and out to the Atlantic Ocean. All that water plays a significant role in shaping local and regional climate, with the Amazon creating as much as 80% of its own rainfall.

"Acre is kind of a laboratory for innovative forest policy."

— Amy Duchelle

The plants that grow here also serve as a large reservoir of stored carbon–more than 100 billion metric tons. By capturing carbon into their organic molecules by photosynthesis, plants keep this carbon out of the atmosphere, reducing the atmospheric levels of carbon dioxide, a greenhouse gas (see Chapter 22).

It's partly because of its vast size and warm, wet climate that the Amazon ranks as the most biodiverse region on Earth: more numbers and types of organisms live here than anywhere else. In addition to about 16,000 species of tree, the Amazon is home to many thousands of other plant, animal, fungi, and protist species–the group of organisms we collectively call **eukaryotes.** These are organisms that all have a similar type of cell–one with a membrane-enclosed nucleus and organelles–owing to a linked evolutionary history. Eukaryotes first evolved roughly 800 million years ago and now make up one of life's three main domains, alongside bacteria and archaea.

EUKARYOTE
Any organism of the domain Eukarya; eukaryotic cells are characterized by the presence of a membrane-enclosed nucleus and organelles.

Most of the organisms we are familiar with in daily life are eukaryotes, from the animals we keep as pets to the green vegetables we put on our dinner plates. Eukaryotes like these are easy to recognize because they are multicellular and grow large and complex bodies, but many unicellular eukaryotes thrive on the planet as well **(INFOGRAPHIC 18.1)**.

Despite its richness, the rain forest is under serious threat from deforestation. Much of this deforestation is carried out to provide land for agriculture. For most of human history, that meant small subsistence farmers clearing a few acres to grow food for their families. But by the late 20th century, most agriculture-related deforestation was aimed at opening up wide tracts of land for large-scale operations, primarily cattle ranching.

Cattle ranchers have a natural incentive to want to cut down the forest: they need pasture to graze their cattle and tree-free areas to grow soybeans and other feed crops. But their needs put them in direct conflict with rubber tappers, who require standing forest to ply their trade.

The conflict between ranchers and rubber tappers reached fever pitch in 1988, when cattle rancher murdered Chico Mendes, a rubber tapper and union organizer.

"That really set off an international, national, and local outcry about human rights, and the rights of traditional people who were dependent on the forest for their livelihoods," says Duchelle, of CIFOR.

In reaction, Brazil moved to create protected forest areas called extractive reserves, where people can legally tap rubber, harvest Brazil nuts and other non-timber forest products, and even collect timber to a certain extent, provided it is done sustainably. The

INFOGRAPHIC 18.1 Tree of Life: The Eukaryotes

Eukaryotic organisms—members of the domain Eukarya—share a common ancestor, as well as structural features including cells with a membrane-bound nucleus and organelles. There is a tremendous amount of diversity represented among eukaryotes, from single-celled protists to multicellular plants, animals, and fungi. Using genetic evidence, scientists have constructed a phylogenetic tree of the eukaryotes, showing how these diverse organisms relate to one another.

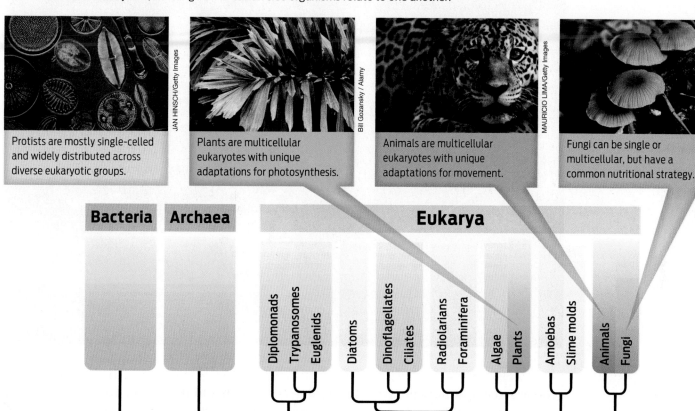

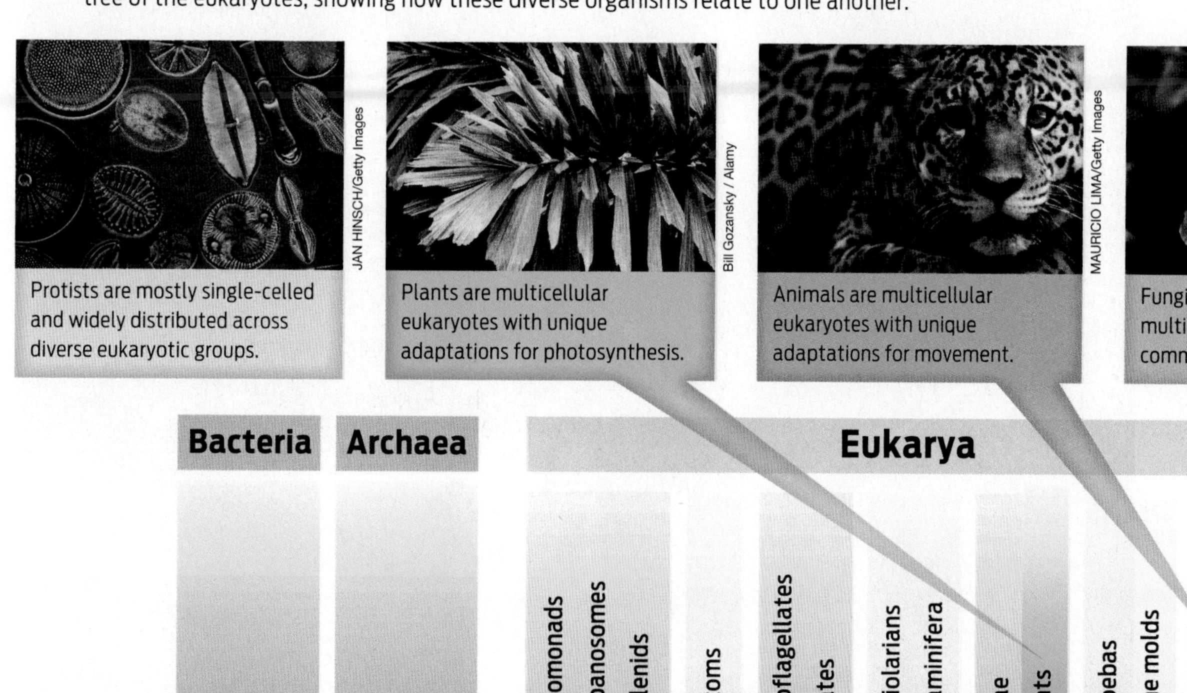

JAN HINSCH/Getty Images
Protists are mostly single-celled and widely distributed across diverse eukaryotic groups.
Bill Gozansky / Alamy
Plants are multicellular eukaryotes with unique adaptations for photosynthesis.
MAURICIO LIMA/Getty Images
Animals are multicellular eukaryotes with unique adaptations for movement.
Mark Green / Alamy
Fungi can be single or multicellular, but have a common nutritional strategy.
Bacteria
Archaea
Eukarya
Diplomonads
Trypanosomes
Euglenids
Diatoms
Dinoflagellates
Ciliates
Radiolarians
Foraminifera
Algae
Plants
Amoebas
Slime molds
Animals
Fungi
Common Ancestor

? Are fungi more closely related to animals or to plants?

first extractive reserve was created in Acre, where Mendes was from, but they've now spread across the Brazilian Amazon.

Natex, the condom manufacturer, is located in the town of Xapuri, across the river from the Chico Mendes Extractive Reserve, where a lot of the rubber is harvested.

"The goal with the condom factory is to both increase the price paid for rubber, and make sure that the value added benefits the collectors themselves, as opposed to going to middle men farther down the supply chain," Duchelle says.

White Gold

Natural rubber is a near-miracle substance. It is durable, waterproof, and, once treated or vulcanized, can maintain elasticity in heat or cold. About 40% of the world's rubber market is for natural rubber harvested from trees, while the rest is for synthetic versions. But for

Natural Rubber

Rubber tappers in the Amazon rain forest harvest latex from the native rubber tree (*Hevea brasiliensis*) by making diagonal cuts in the bark. Natural rubber is the first choice for products that can't afford to fail, like airplane tires, surgical gloves, and condoms.

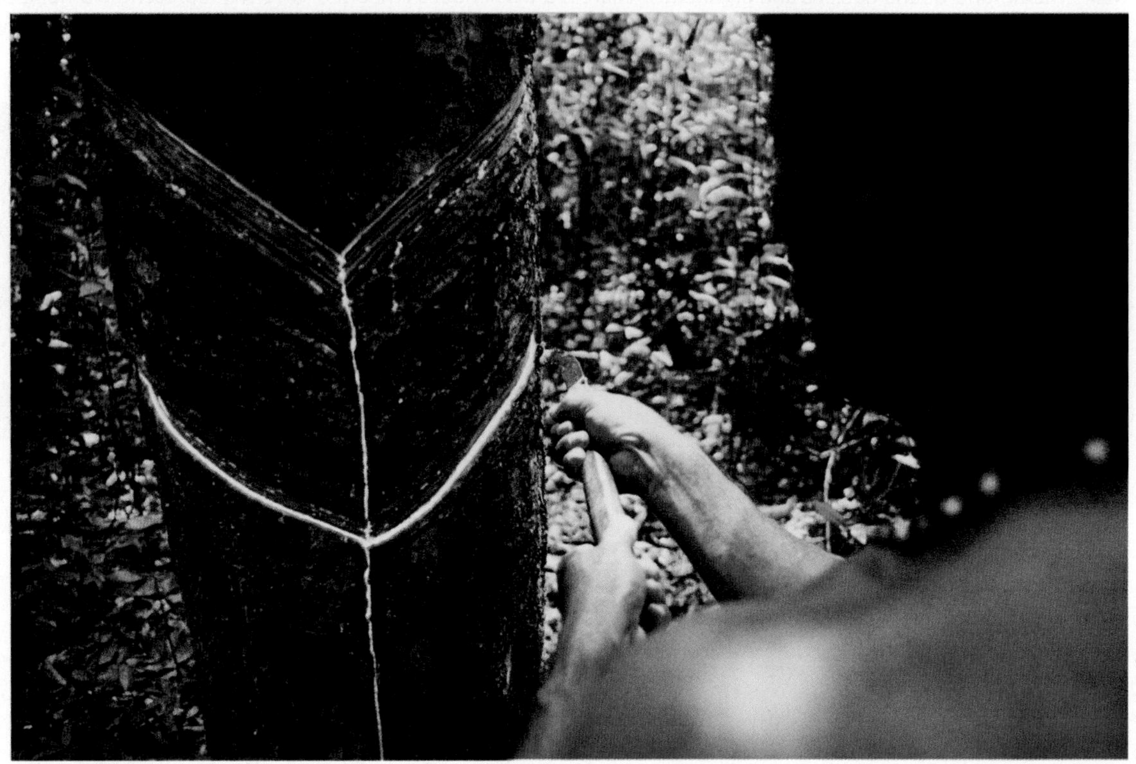

Marcelo Horn/Getty Images

products that truly must not fail–like airplane tires, surgical gloves, and condoms–natural rubber is the go-to choice.

Demand for rubber began to skyrocket in the late 19th century, with the rise of the automobile industry. Rubber was needed for hoses, belts, and–especially–tires. Many American and European "rubber barons" who sold rubber during this time became very wealthy.

But the rubber barons necessarily relied on the skill and know-how of rubber tappers who could identify rubber trees and coax rubber from them. Tapping rubber trees involves making a diagonal cut in the bark in line with the path of the spiraling vessels in the tree that carry the latex–a milky, white substance that coagulates in the air. Tappers then wind the congealing latex around a stick, forming heavy balls that they can sell to wholesalers and manufacturers.

Latex is obviously valuable to humans, but that's not why rubber trees make it. Latex is valuable to them, too. To understand just how valuable, it helps to know more about the rubber tree's evolution and its relationship to other eukaryotes.

The rubber tree is, of course, a **plant,** one of the main groups of multicellular eukaryotes. Every organism needs ways to obtain nourishment, reproduce, and defend itself, and plants are no exception. The particular ways that plants have solved these fundamental life challenges is what defines them as plants and distinguishes them from other eukaryotes–animals and fungi, for example.

Plants' signature talent is being able to make their own food. Through photosynthesis (see Chapter 5), plants capture the energy of sunlight and use it to produce sugar. This sugar nourishes not only the plant but also the many nonphotosynthetic creatures on Earth that depend on plants for food–including humans.

Plants are largely immobile–stuck in one place–and so they must obtain the starting

PLANT
A multicellular eukaryote that has cell walls, carries out photosynthesis, and is adapted to living on land.

materials for photosynthesis right where they are. Those materials include carbon dioxide that they take in through their leaves, and water and a few other nutrients that they obtain from the soil.

Most plants reproduce sexually, but unlike animals they can't go in search of mates. Instead, they rely on water, wind, or animals to bring male and female gametes together. Often this means producing an overabundance of sex cells in the hope that at least one will find a partner.

Since they are immobile, plants cannot run away from predators that seek to eat them, and so many have evolved other solutions to avoid becoming easy targets: thick, protective bark, for example, and defensive chemicals called **secondary metabolites;** some of these taste bitter and help to ward off leaf-munching insects and other herbivores. Latex is full of such secondary metabolites, and that is the main reason the rubber tree makes it **(INFOGRAPHIC 18.2)**.

SECONDARY METABOLITES
Chemicals produced by plants that are not directly involved in growth or reproduction but that help protect the plant by their impacts on other organisms.

INFOGRAPHIC 18.2 Plants: Getting Ahead while Standing Still

Plants are multicellular eukaryotes that carry out photosynthesis and are adapted to living on land. Because they are not mobile, they employ unique mechanisms of defense and reproduction to help them succeed despite being stuck in one place.

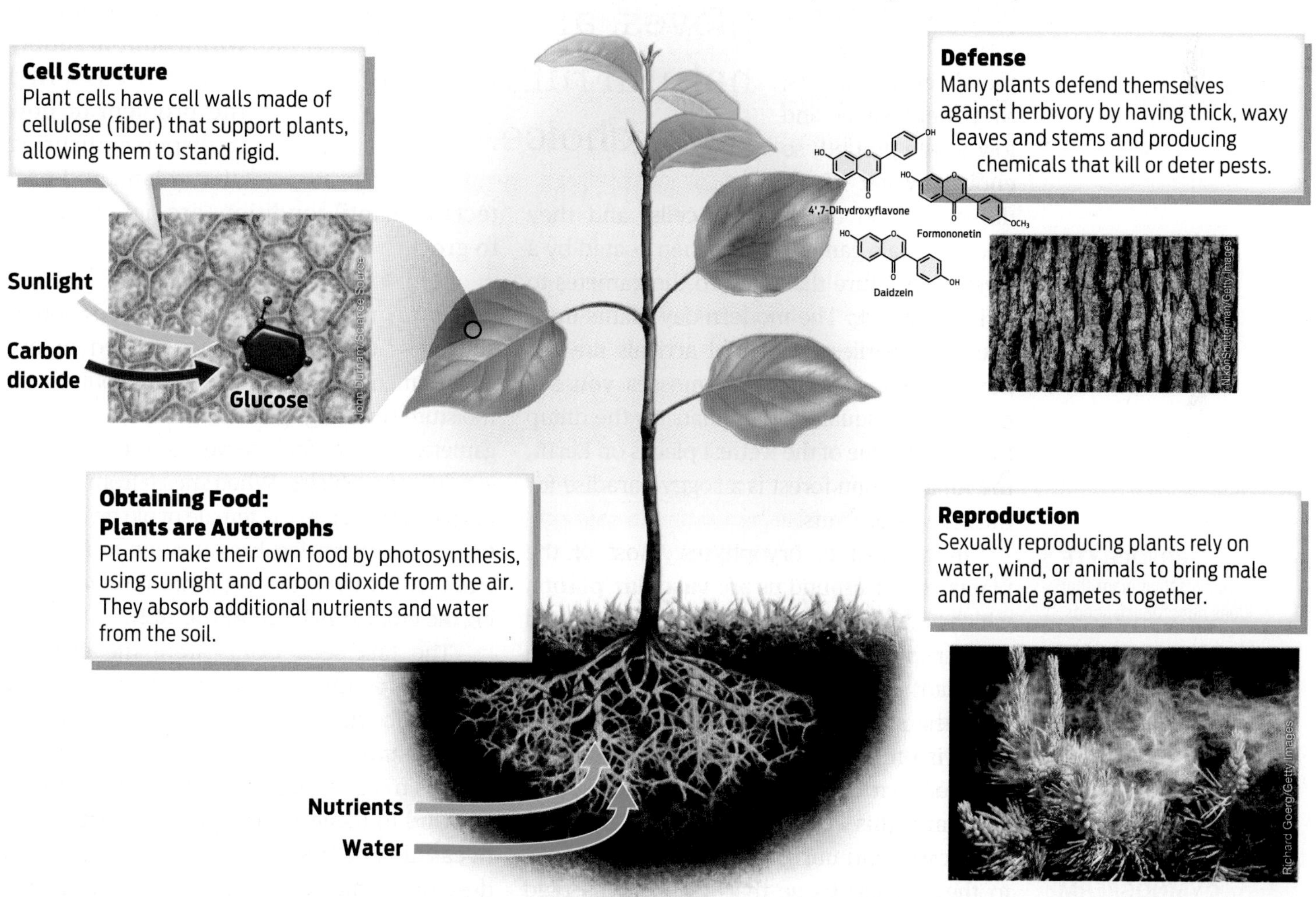

Cell Structure
Plant cells have cell walls made of cellulose (fiber) that support plants, allowing them to stand rigid.
Sunlight
Carbon dioxide
Glucose
John Durham/Science Source
Defense
Many plants defend themselves against herbivory by having thick, waxy leaves and stems and producing chemicals that kill or deter pests.
4',7-Dihydroxyflavone
Formononetin
Daidzein
©Nikon/Shutterman/Getty Images
Obtaining Food: Plants are Autotrophs
Plants make their own food by photosynthesis, using sunlight and carbon dioxide from the air. They absorb additional nutrients and water from the soil.
Nutrients
Water
Reproduction
Sexually reproducing plants rely on water, wind, or animals to bring male and female gametes together.
Richard Goerg/Getty Images

? List three characteristics of plants.

Though plants today cover every land habitat, they weren't always so well equipped to survive on land. Plants like those found in the Amazon first evolved from water-dwelling green algae about 450 million years ago, when life on Earth was confined primarily to seas. The water in which plant ancestors floated served several purposes: it prevented their cells from drying out, and it allowed male and female gametes to drift easily to each other. Before plants could diversify on land, they needed to evolve the means to cope with drier surroundings.

The earliest plants to make the transition from water to land naturally relied on some of same solutions as their ancestors. They lived in moist environments; they were small and low to the ground, so enough water could diffuse in and out of their cells; and they released male gametes only when coated by a layer of moisture that allowed the gametes to swim to an egg. The modern-day plants most like these earliest terrestrial arrivals are the **bryophytes,** including the mosses you see growing like squat, spongy mats on the damp forest floor. One of the wettest places on Earth, the Amazon rain forest is a soggy paradise for these ancient plants.

In contrast to bryophytes, most of the plants we see around us are **vascular plants,** which have specialized tubular tissues adapted for transporting water and nutrients through the plant body. These plants take up water from the soil through roots; thus as long as water is present in the soil, they can survive where the ground surface is dry. The first true vascular plants were **ferns.** Unlike bryophytes, ferns can stand upright and grow tall, thanks to the vascular tissue that keeps stems rigid and the plant well supplied with water. But ferns still must release their male gametes into a layer of moisture, so they are typically found growing in moist environments. At one time, ferns ruled the plant world, spreading their massive fronds across the entire landscape in the Carboniferous Period, 360 to 300 million years ago. But their reign was short lived. Soon another kind of plant evolved to challenge the ferns' dominance: those with seeds.

Seed plants first emerged about 360 million years ago, during the late Devonian Period. A seed, which envelops a plant's embryo, is an ideal package for withstanding harsh conditions and transporting the embryo to a location where it can grow into a new plant. The seed provides a protective shell for the plant embryo and comes complete with a built-in supply of food. Seeds solved the problem of how to disperse offspring on dry land and protect them until conditions are right for them to grow.

For products that truly must not fail—like airplane tires, surgical gloves, and condoms—natural rubber is the go-to choice.

Along with this innovative embryo-protection method, seed plants evolved another key innovation necessary to succeed in dry environments: pollen. These tiny but tough moisture-retaining packets contain the male gametes–sperm–and serve as durable gamete-delivery vehicles. When sperm from the pollen meets an egg in the female ovule on another plant, the sperm fertilizes the egg to form an embryo. The ovule develops into a seed containing the embryo (**INFOGRAPHIC 18.3**).

The first seed plants were the **gymnosperms,** a diverse group of plants that includes conifer, or cone-bearing, trees such as pine and spruce as well as foul-smelling ginkgo trees. Gymnosperms rely primarily on wind to disperse pollen. Pollen grains are released from male cones on the tree; seeds develop on the underside of scales on female cones and are released as the masses of "propellers" you've no doubt encountered on the ground in fall. Gymnosperm trees have

BRYOPHYTE
A nonvascular plant that does not produce seeds.

VASCULAR PLANT
A plant with tissues that transport water and nutrients through the plant body.

FERNS
The first true vascular plants; ferns do not produce seeds.

GYMNOSPERM
A seed-bearing plant with exposed seeds typically held in cones.

INFOGRAPHIC 18.3 Advantageous Adaptations: Vascular Tissue, Pollen, and Seeds

In order for plants to succeed and diversify on land, they needed to evolve adaptations for withstanding dry environments. Two key evolutionary milestones were the evolution of vascular tissue and of pollen and seeds.

Nonvascular

Plants like mosses are restricted to moist environments because they do not have vascular tissue. They absorb water and nutrients into their cells like a sponge, directly from their immediate environment.

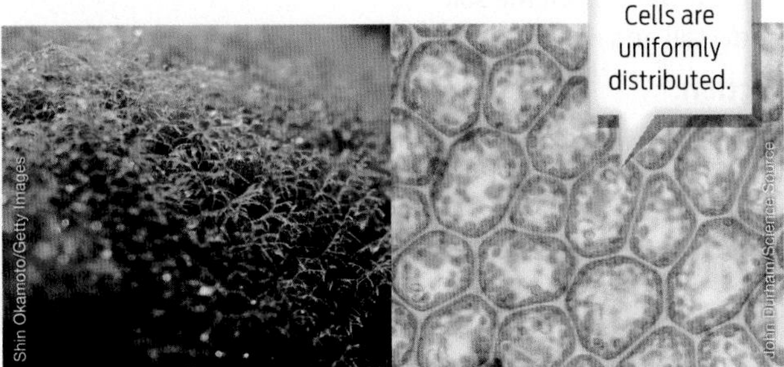

Vascular

Plants like rubber trees use vascular tissue to move water and nutrients. These structures, apparent as veins in leaves and vessels in the stem, allow plants to take up water and nutrients via roots that penetrate deep into soil, and to transport water, nutrients, and food throughout their bodies.

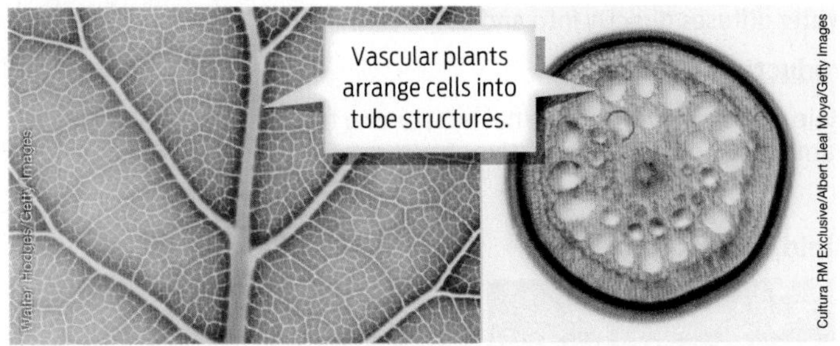

Cross section of a vascular stem

Sperm Survive Only in Water

Some plants have "swimming" sperm that must move through water to reach the female reproductive structures.

Pollen and Seeds Resist Drying

The male gametes of some plants are contained in pollen, which resists drying and can be transported in air. These plants also have seeds to protect the embryonic plant from drying.

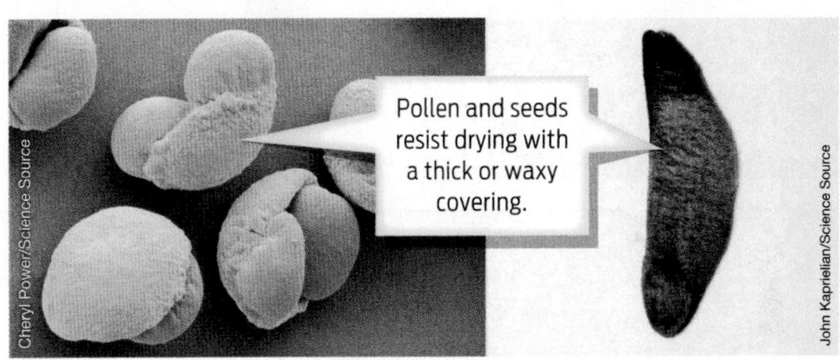

? Give two reasons why mosses are short plants that live near water.

colonized vast regions of the cold and dry northern regions of the planet.

Yet some environments posed a challenge to wind pollination. There is little wind in the understory of the Amazon rain forest, and the dense vegetation would prevent pollen from spreading far through the air. This is where the fourth major group of plants, the **angiosperms,** is believed to have evolved. Angiosperms have flowers, and the significance of flowers is that they make possible pollination by animals. Bees, insects, bats, and birds visit the flowers in search of pollen or nectar to eat, and the pollen sticks to their bodies. Some of the pollen they pick up is then dropped off at the next flower they visit. Because each type of pollinator visits only certain types of plants, they are like mailmen who carry a package to a specific address–in this way even a rare plant can receive pollen.

Seed plants were so successful that they quickly came to dominate forests by the time dinosaurs appeared in the Mesozoic Era, roughly 200 million years ago. Today, more than 90% of all living plants are seed

ANGIOSPERM
A seed-bearing flowering plant with seeds typically contained within a fruit.

INFOGRAPHIC 18.4 Evolution of Plant Diversity

All plants have evolved from an ancient algal ancestor that lived in water. Different groups of plants have developed different specializations that allow them to be successful on land.

Bryophytes
No Seeds, No Vascular Tissue

Key Features:

- Small and low to the ground
- Moist environments
- Water diffuses directly into and out of cells.

Reproduction:

- Male gametes swim in external moisture to fertilize female gametes in female reproductive structures.

Thalloid Liverwort

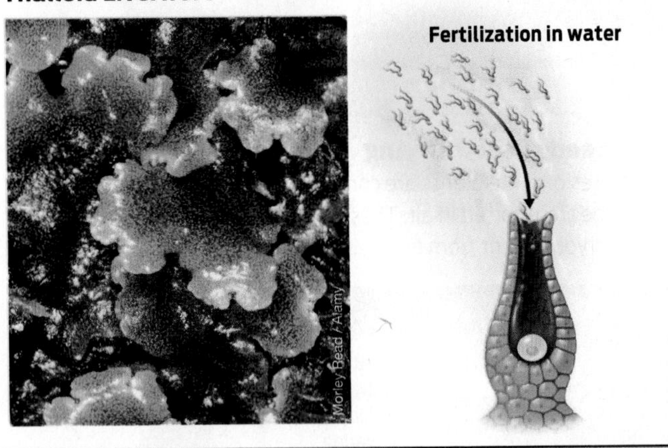

Ferns
Vascular, No Seeds

Key Features:

- Specialized tissues to transport nutrients and water
- Roots draw water from the soil
- Can grow tall and rigid

Reproduction:

- Male gametes swim in external moisture to fertilize female gametes in female reproductive structures.

Shuttlecock Fern

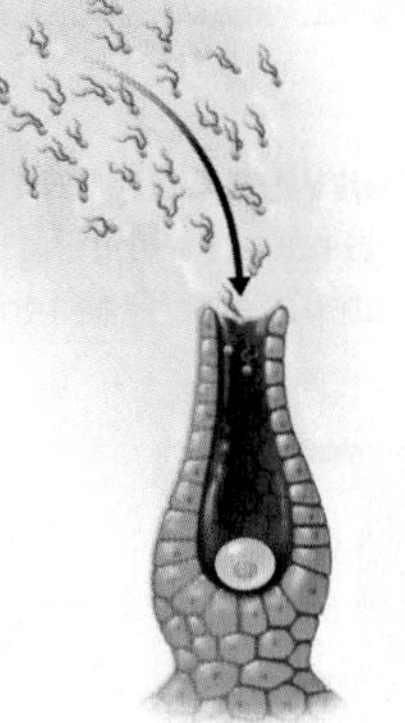

Vascular Tissue

Common Ancestor
450 million years ago

? Which evolved first: flowers, seeds, or vascular tissue?

plants. Gymnosperms have exposed seeds, as in a pinecone. ("Gymnos" is Greek for "naked," so the name literally means "naked seeds.") In angiosperms, seeds are contained in a fruit ("angio" is from the Greek for "vessel" or "container"). The rubber tree is an angiosperm whose fruits look like green, three-lobed heads of garlic, each containing three brown seeds inside **(INFOGRAPHIC 18.4)**.

Gymnosperms
Vascular, Seed-Producing in Cones

Key Features:

- Specialized tissues to transport nutrients and water
- Roots draw water from the soil
- Can grow tall and rigid

Reproduction:

- Male gametes in pollen, distributed to female structures by wind
- Produce seeds that develop in cones

Brazilian Pine

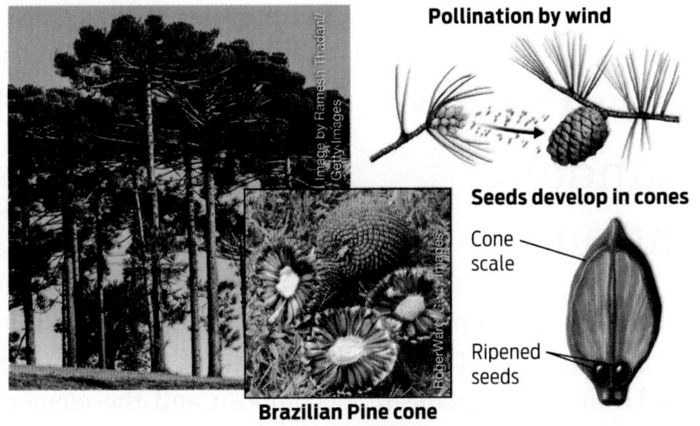

Brazilian Pine cone

Angiosperms
Vascular, Flowering, Seed-Producing in Fruit

Key Features:

- Specialized tissues to transport nutrients and water
- Roots draw water from the soil
- Can grow tall and rigid

Reproduction:

- Male gametes in pollen, distributed to female structures by pollinators attracted to flowers
- Produce seeds that develop in fruit

Amazon Rubber Tree

Amazon Rubber Tree fruit

Flowers and Fruits

Pollen and Seeds

Although native peoples had been extracting latex from rubber trees for many generations, it was only in 1839 that rubber caught the attention of industrialists. That was the year Charles Goodyear discovered the process of vulcanization, a chemical treatment that makes latex more durable, and launched the rubber boom in the Amazon.

The boom was short lived. In 1876, a British botanist spending time in Brazil absconded with 70,000 rubber plant seeds, which the British government used to establish rubber tree plantations in its eastern colonies. By 1910, there were 50 million rubber trees growing in Asia, and Brazil's rubber market had collapsed. A decade later, more than 90% of the world's rubber came from plantations in Asia.

"This was the first really well-known instance of biopiracy and Brazilians are still pissed off about it," says William Laurance, a conservation biologist at James Cook University in Australia, who lived and worked in Brazil for 20 years studying the rain forest.

The rubber industry in Brazil got a brief boost during World War II, when supplies of Asian rubber were blocked by Japan. Millions of Brazilian farmers moved into the forest to tap rubber to supply the United States with the material to support its war effort. But when the war ended, the market for Brazilian rubber collapsed again, leaving many of the local tappers struggling to make a living; there was just no way they could compete.

The Problem of Blight

In Brazil, rubber trees are spread throughout the forest at relatively low density (about two trees per hectare). In contrast, rubber trees in Asia are planted and harvested on plantations, where densities—and output—are much higher.

There are fungi everywhere in the Amazon, growing on, in, and under the abundant vegetation.

Every attempt to establish rubber tree plantations in the Amazon has ended in failure. Most famously, the car maker Henry Ford tried to establish a plantation and a town to accompany it (called, appropriately, Fordlandia), but he neglected to consult any botanists. Within a few years, all the trees—some 2 million of them—were dead from a disease called leaf blight.

Leaf blight is caused by a fungus called *Microcyclus ulei* that is native to the Amazon. This fungus spreads easily among rubber trees that grow close to one another. It kills by feasting on and destroying the trees' leaves, so the tree cannot feed itself through photosynthesis. The presence of this fungus in the Amazon means that closely planted rubber trees are a disaster waiting to happen.

Fungi are a second large group of eukaryotes. Unlike plants, fungi cannot photosynthesize and so they must obtain nutrients by feeding off other organisms. All fungi, including the one that causes leaf blight, obtain nutrients in a characteristic way: by secreting digestive enzymes onto their food source. These enzymes digest complex molecules into individual subunits such as amino acids and simple sugars. The fungi then absorb theses subunits into their cells. Because they digest their food externally, fungi have no need for a stomach or intestines.

Multicellular fungi have a body composed of threadlike structures called **hyphae.** Each individual hypha is a chain of many cells, capable of absorbing nutrients. Hyphae interweave to form a spreading mass called a **mycelium.** These structures can vary in size and location: the mold on a slice of bread has a small mycelium that's in plain view. By contrast, the mushrooms you see on the forest floor are merely one aboveground part of what can be a huge, underground mycelium.

Because they feast on dead organisms, fungi are **decomposers,** organisms that break down organic molecules trapped in dead organisms. Their activity is necessary for the recycling of nutrients in the forest. Without fungi, dead trees and animal carcasses would pile up in the forest and smother everything in it. Thanks to the action of fungi, however, the organisms decompose and the elements they contained will nourish many organisms throughout the environment.

Many fungi can reproduce both sexually and asexually, while others are strictly sexual. In all cases, fungi produce **spores**—tiny, one-celled reproductive units that resist drying out and disperse from the parent. In many multicellular fungi, the structure that releases spores is called a **fruiting body,** and is what we typically think of as a "mushroom," with a recognizable stalk and cap.

Like plants, fungi are immobile—a fact that influences their approaches to defense. To avoid being eaten by predators, many fungi live underground or produce poisonous toxins that dissuade would-be attackers, who may come to recognize their distinctive shapes, colors, and odors **(INFOGRAPHIC 18.5)**.

Like other eukaryotes, fungi got their start in the water. An early adaptation, achieved about 750 million years ago, was

FUNGUS
A unicellular or multicellular eukaryotic organism that obtains nutrients by secreting digestive enzymes onto organic matter and absorbing the digested product.

HYPHA (plural: HYPHAE)
A long, threadlike structure through which fungi absorb nutrients.

MYCELIUM
A spreading mass of interwoven hyphae that forms the often subterranean body of multicellular fungi.

DECOMPOSER
An organism such as a fungus or bacterium that digests and uses the organic molecules in dead organisms as sources of nutrients and energy.

SPORES (FUNGAL)
Fungal cells that are resistant to drying out and can be dispersed to new locations as part of sexual or asexual reproduction.

FRUITING BODY
A fungal structure that is specialized for the release of spores.

INFOGRAPHIC 18.5 Fungi: The Decomposers

Fungi are unicellular or multicellular eukaryotes that obtain nutrients by feeding on other organisms by external digestion. Their approaches to nutrition, reproduction, and defense reflect the fact that they are generally stuck in one place. Because they feed on dead organisms, fungi are important decomposers, returning nutrients to the soil as they break down organic molecules.

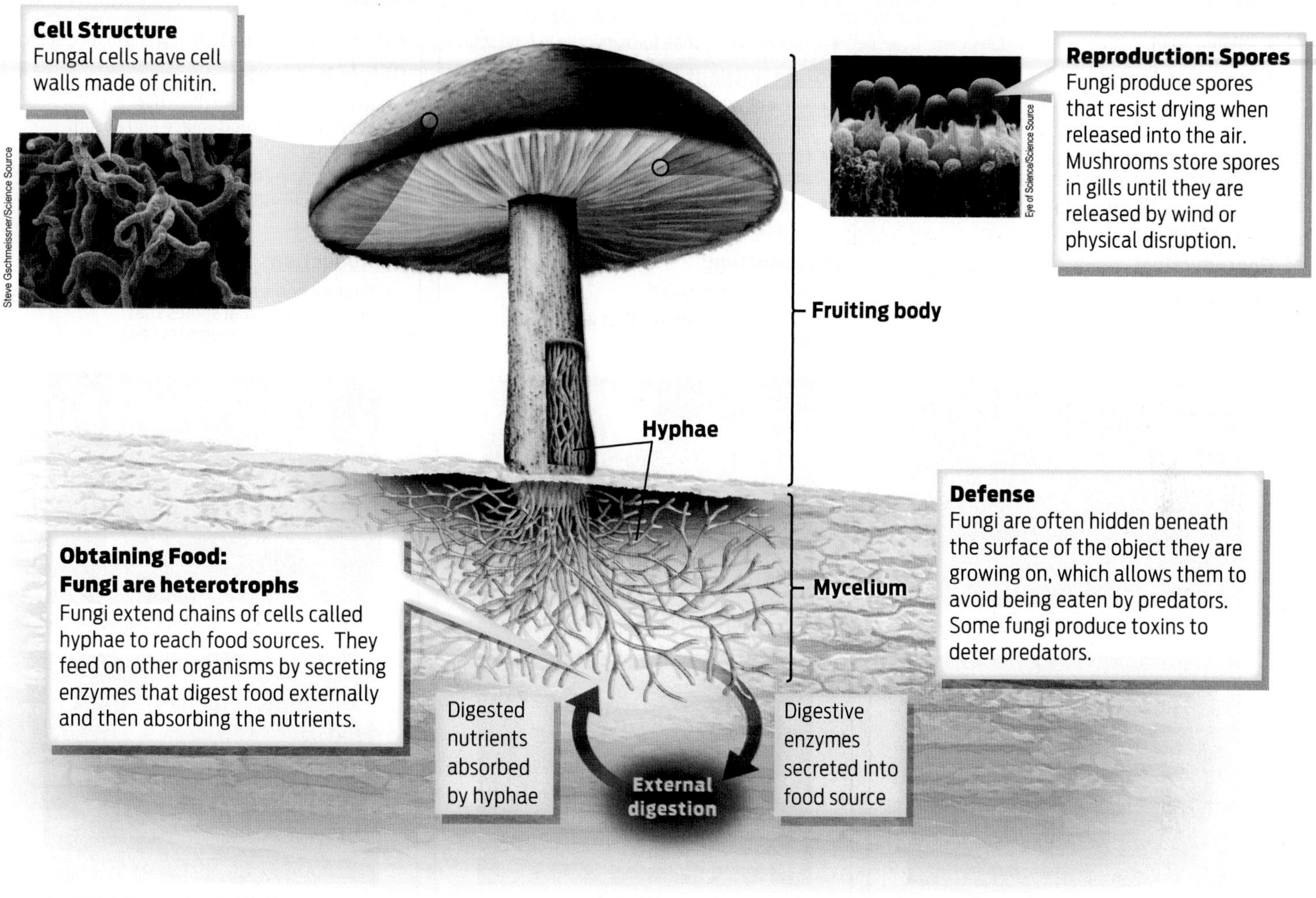

Cell Structure
Fungal cells have cell walls made of chitin.

Steve Gschmeissner/Science Source

Reproduction: Spores
Fungi produce spores that resist drying when released into the air. Mushrooms store spores in gills until they are released by wind or physical disruption.

Eye of Science/Science Source

Fruiting body

Hyphae

Mycelium

Defense
Fungi are often hidden beneath the surface of the object they are growing on, which allows them to avoid being eaten by predators. Some fungi produce toxins to deter predators.

Obtaining Food: Fungi are heterotrophs
Fungi extend chains of cells called hyphae to reach food sources. They feed on other organisms by secreting enzymes that digest food externally and then absorbing the nutrients.

Digested nutrients absorbed by hyphae

External digestion

Digestive enzymes secreted into food source

? List three characteristics of fungi.

the development of a cell wall containing the molecule chitin, which imparts strength and thus protection against the water pressure from osmosis. (Plant cells also have a cell wall, but it is made out of cellulose.)

Today, fungi come in a diverse array of forms. There are unicellular varieties collectively known as yeasts. Many species of yeast are useful to humans, including the ones that ferment sugar to alcohol to make wine and beer, and those that help make bread rise. Multicellular varieties of fungus include molds like the one that causes leaf blight and mushrooms such as you see growing on a tree trunk or sprouting from the ground.

There are fungi everywhere in the Amazon, growing on, in, and under the abundant vegetation. Some soil-dwelling species form a close relationship with the roots of many tree species. Their slender hyphae grow into microscopic spaces in the soil where the tree's roots can't fit, greatly enhancing a

INFOGRAPHIC 18.6 Fungal Diversity

Fungi are a diverse group of organisms with a variety of reproductive and feeding strategies. Fungi may be referred to as yeasts, molds, or mushrooms, depending on how they grow and reproduce.

Yeast

- Single-celled

Obtain Nutrients:

- Individual cells release digestive enzymes onto food source and absorb nutrients
- Some species contribute to tasty food and beverages when they ferment sugars
- Some species live symbiotically with plants or animals

Reproduction:

- Sexual and asexual

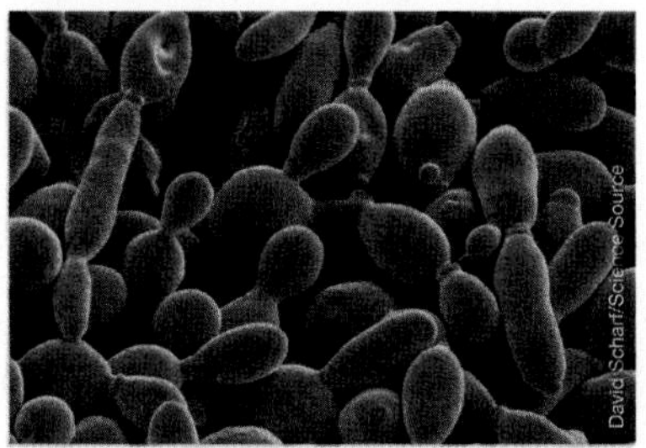

Saccharomyces cerevisiae yeast cells

Yeast on the surface of grapes

Molds

- Multicellular hyphae form mycelia in food source

Obtain Nutrients:

- Hyphal cells release digestive enzymes onto food source and absorb nutrients
- Many species decay leaf litter and dead organisms
- Some species associate with plant roots, supplying and receiving nutrients

Reproduction:

- Sexual and asexual
- Molds release spores that germinate to form new mycelia

Cordyceps fungus decomposing a moth

South American leaf blight (*Microcyclus ulei*)

Mushrooms

- Multicellular hyphae form mycelia in food source

Obtain Nutrients:

- Hyphal cells release digestive enzymes onto food source and absorb nutrients
- Many species decay leaf litter and dead organisms

Reproduction:

- Fruiting bodies emerge from mycelia and release sexual spores that germinate to form new mycelia

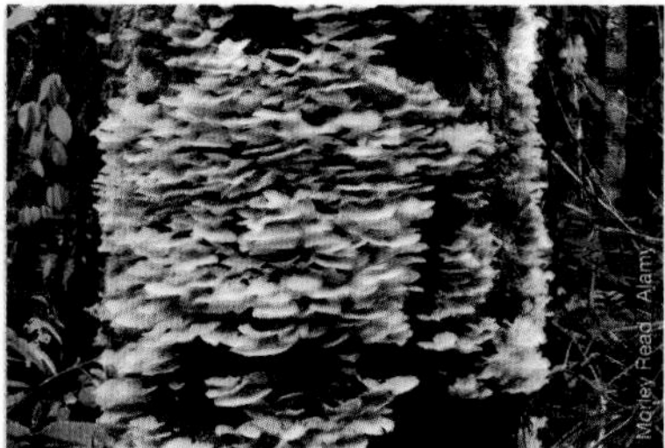

Bracket fungus on a tree

Mushroom hyphae on a dead log

? Of the three types of fungi described, which are multicellular?

root's ability to absorb water and nutrients. In turn, trees supply nutrients to the fungi (**INFOGRAPHIC 18.6**).

Perhaps as a result of having evolved together with the leaf blight fungus, rubber trees have developed seedpods that explode violently, sending seeds up to 120 feet away. Dispersing seeds this way is an adaptation that keeps the rubber trees from growing too close to one another in the forest, and therefore prevents leaf blight spores from spreading easily between them.

Leaf blight, like the rubber trees themselves, is native to the Amazon. The fungus does not currently exist in Asian countries, but some scientists think that the arrival of blight is inevitable. If or when that does happen, it could mean a devastating blow to the worldwide rubber market.

Diversity Engine

Intense battle, like that between rubber trees and the leaf blight fungus, is not unique to

INFOGRAPHIC 18.7 Pests Drive Rain Forest Diversity

Host-specific pests—including fungi, insects, and microbes—keep new plants of the same species from growing near the parent plant. This allows other plant species to grow nearby, maintaining a diversity of species in the rain forest.

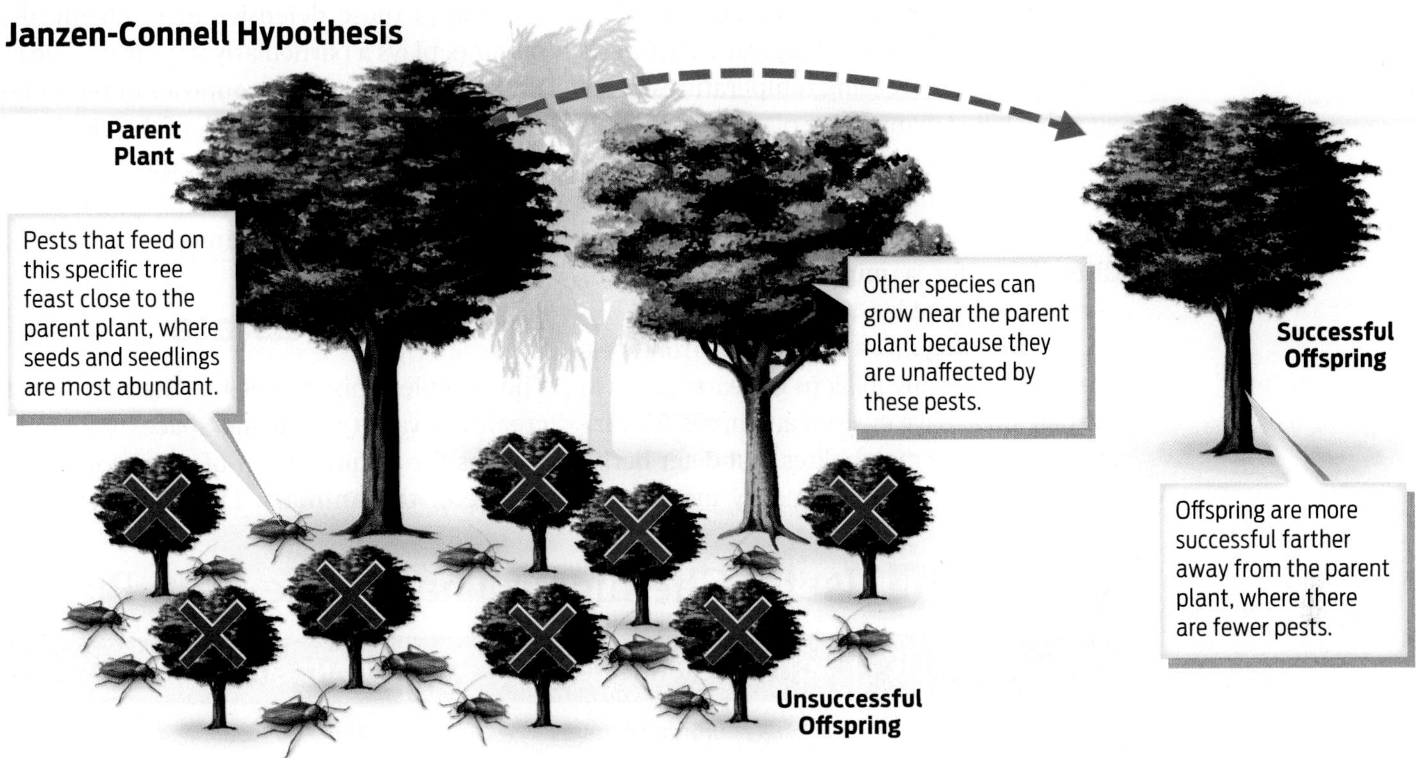
Janzen-Connell Hypothesis
Parent Plant
Pests that feed on this specific tree feast close to the parent plant, where seeds and seedlings are most abundant.
Other species can grow near the parent plant because they are unaffected by these pests.
Successful Offspring
Offspring are more successful farther away from the parent plant, where there are fewer pests.
Unsuccessful Offspring

? Why can plants of different species grow in close proximity in the rain forest, but plants of the same species cannot?

these two species of eukaryote. Fierce struggle over resources is common in the Amazon. Scientists believe that it's through these life-or-death clashes that the immense diversity of eukaryotic life here evolved in the first place and continues to be maintained.

If you were to compare the diversity of tree species in the Amazon to that of a typical northern forest, one thing would quickly stand out: northern (that is, temperate) forests are a lot more monotonous. They typically contain an abundance of one dominant type of tree—as for example a typical northern pine forest. By contrast, rain forests support thousands of species of tree in the same area. By one estimate, an area of rain forest of about 3 acres contains more than 650 species of tree—more than are found in the continental United States and Canada combined.

How this immense diversity is maintained over time is a question that has long intrigued scientists. In the early 1970s, two biologists, Daniel Janzen and Joseph Connell, independently came up with an answer. They proposed that rain forest tree diversity is maintained through the influence of pests—fungi, insects, and microbes—that feed on a specific tree species. These host-specific pests surround a tree of their preferred species and then feed on and kill any seedlings of that same tree species that happen to start to grow near the parent tree. This preferential feeding keeps any one tree species from taking over an area and helps to maintain tree diversity in the forest. This idea has become known as the Janzen-Connell hypothesis and is, according to Laurance, "one of the most prevailing and well-tested ideas in the tropics." It applies not just to rubber trees, he says, but to essentially all plants that live there (**INFOGRAPHIC 18.7**).

Northern forests have pests, too, of course, but the cold winters help to keep

ANIMAL
A eukaryotic multicellular organism that can move and that obtains nutrients by ingesting other organisms.

their populations in check. This is one reason temperate forests are generally less diverse than tropical ones (although no less important to the creatures that live there).

As well, the Amazon is much older than these northern forests. Tropical rain forests were spared the freezing temperatures of ice ages that killed many of the plants and animals in what are now the temperate regions of the planet. As a result, the plants and animals in tropical rain forests have been evolving and diversifying for a much longer continuous period of time—at least 140 million years.

Because they have faced such strong pest pressure over many millions of years, plants in the rain forest have evolved an impressive variety of secondary metabolites that deter herbivores. Some of these secondary metabolites are quite familiar to us: they include caffeine, nicotine, and cocaine. Many of our pharmaceuticals come from these plant secondary metabolites **(TABLE 18.1)**.

One of these defensive plant chemicals, quinine, plays a particularly important role in Brazil and surrounding countries of the Amazon basin. Found in the bark of the cinchona tree, which is native to Peru, quinine is one of the main treatments for malaria, a disease that is common in this region.

Prime Real Estate

The complex mix of trees in the rain forest creates a variety of living spaces and food sources for a third group of eukaryotes that live here: the **animals.** Like fungi, animals

TABLE 18.1 Useful Chemicals That Come from Plants

DRUG/CHEMICAL	ACTION/CLINICAL USE	PLANT SOURCE
Atropine	Blocks neurotransmitters	*Atropa belladonna*
Bromelain	Anti-inflammatory, proteolytic	*Ananas comosus*
Caffeine	Central nervous system stimulant	*Camellia sinensis*
Codeine	Analgesic, relieves cough	*Papaver somniferum*
Cyclosporine	Anti-transplant rejection drug	*Tolypocladium inflatum fungus*
Digitalin	Cardiac stimulant	*Digitalis purpurea*
Ephedrine	Antihistamine; stimulation of sympathetic nervous system	*Ephedra sinica*
Gossypol	Male contraceptive	*Gossypium species*
L-Dopa	Antitremor; treats Parkinson's disease	Mucuna species
Menthol	Dilates capillaries; treats muscle pain	Mentha species
Morphine	Analgesic; inhibits pain	*Papaver somniferum*
Novocaine	Local anesthetic	Coca plant (South America)
Quinine	Antimalarial, antifever	Cinchona tree
Salicin	Analgesic (aspirin)	*Salix alba (Willow bark tree)*
Sennosides	Laxative	Cassia species
Stevioside	Sweetner	*Stevia rebaudiana*
Taxol	Antitumor agent	*Taxus brevifolia*

INFOGRAPHIC 18.8 Animals: The Power of Movement

Animals are multicellular eukaryotes that obtain nutrients by eating other organisms. Animals have evolved the capacity for independent movement, which greatly aids the search for food and mates, as well as defense.

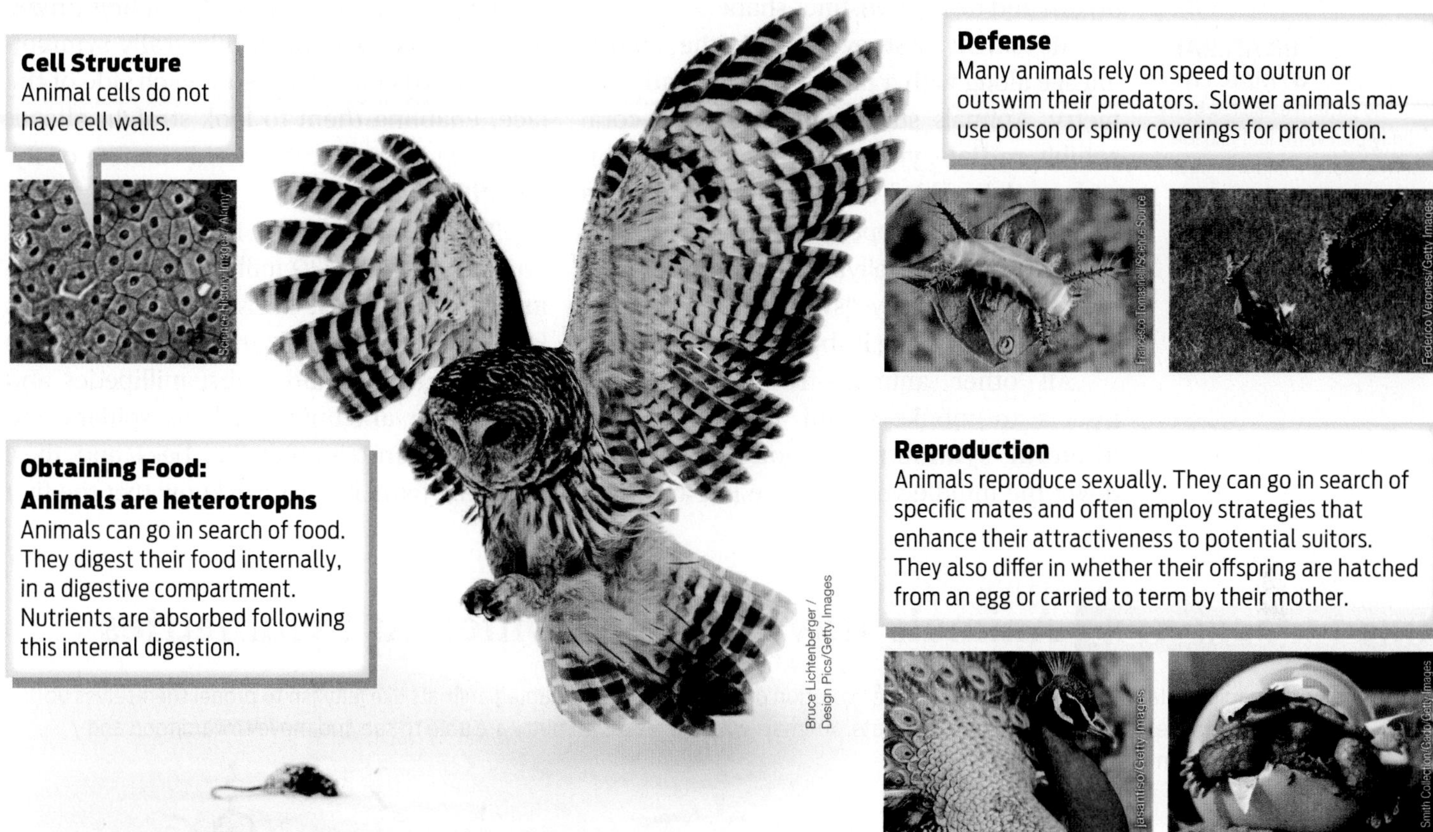
Cell Structure
Animal cells do not have cell walls.
Defense
Many animals rely on speed to outrun or outswim their predators. Slower animals may use poison or spiny coverings for protection.
Obtaining Food: Animals are heterotrophs
Animals can go in search of food. They digest their food internally, in a digestive compartment. Nutrients are absorbed following this internal digestion.
Reproduction
Animals reproduce sexually. They can go in search of specific mates and often employ strategies that enhance their attractiveness to potential suitors. They also differ in whether their offspring are hatched from an egg or carried to term by their mother.
Science History Images/Alamy
Francesco Tomasinelli/Science Source
Federico Veronesi/Getty Images
Bruce Lichtenberger / Design Pics/Getty Images
Jasantiso/Getty Images
Smith Collection/Gado/Getty Images

? List three ways in which animals benefit by being able to move.

are eukaryotes that cannot photosynthesize and so must obtain nutrients by feeding on other organisms. Unlike fungi, animals digest their food internally through a digestive system. Many animals are herbivores, eating only plants; others are carnivores, eating other animals. Animals reproduce sexually.

Perhaps what most distinguishes animals from plants and fungi is their nearly limitless capacity for movement. Whether it's snakes that slither quietly to sneak up on prey, mosquitoes that home in on blood, birds that soar down from the sky to snatch rodents in their talons, or salmon that swim upstream to congregate and breed, movement is a tremendously useful adaptation for solving life's fundamental problems: finding food, seeking mates, and avoiding predators.

Because animals can go in search of food, they don't have to make it for themselves (like plants) or grow until they find something to eat (like fungi). They don't have to rely on water, wind, or pollinators to disperse their gametes, hoping for the best; they can instead seek out and choose a particular mate. And instead of poisons to deter potential predators, many animals rely on speed to outrun them, although some of the slower ones, like slugs, also resort to poisons and warning coloration.

Although not every animal can move so effortlessly, the evolution of movement was a key evolutionary milestone that permitted animals to spread and diversify around the globe (**INFOGRAPHIC 18.8**).

The first animals were ocean-dwelling creatures, most likely resembling sponges.

RADIAL SYMMETRY
The pattern exhibited by a body plan that is circular, with no defined left and right sides.

BILATERAL SYMMETRY
The pattern exhibited by a body plan with right and left halves that are mirror images of each other.

ARTHROPOD
An invertebrate having a segmented body, a hard exoskeleton, and jointed appendages.

Sponges are stationary, so they cannot go in search of food. Instead, they feed by filtering nutrients out of seawater as it passes through their body. Sponges lack defined tissues or organs and have no distinct shape.

Movement first evolved in the animal lineage along with a body plan that had symmetry. Animals such as jellyfish and coral exhibit **radial symmetry:** they're circular, with no defined left and right sides. This is a useful shape for propelling oneself vertically through water, as jellyfish do. It also allows the animal to easily "swallow" its prey by essentially wrapping its body around it.

All other animals–from worms and insects to monkeys and humans–exhibit **bilateral symmetry:** if you draw a line down the middle you produce left and right halves that are mirror images of each other. Bilateral symmetry has become the most common body form in the animal kingdom because it is a useful adaptation for seeking out food, stalking prey, and avoiding predators. For instance, most bilaterally symmetrical animals have an eye on each side of the face, enabling them to look straight ahead, as well as paired limbs to aid running or flying **(INFOGRAPHIC 18.9)**.

The first animals to take tentative steps on land, roughly 420 million years ago, were invertebrates, most likely **arthropods.** This large group of animals includes crustaceans such as lobsters and crabs; millipedes and centipedes; arachnids such as spiders and scorpions; and insects like bees and flies. From the fossil record, we know that the first

INFOGRAPHIC 18.9 Animals Have Evolved Different Symmetries

The earliest animals lacked distinct symmetry. The evolution of radial symmetry enabled animals like jellyfish to propel themselves up and down in the water to find food and new habitats. Animals with bilateral symmetry are able to see and move toward food and mates, and away from danger.

Sponge

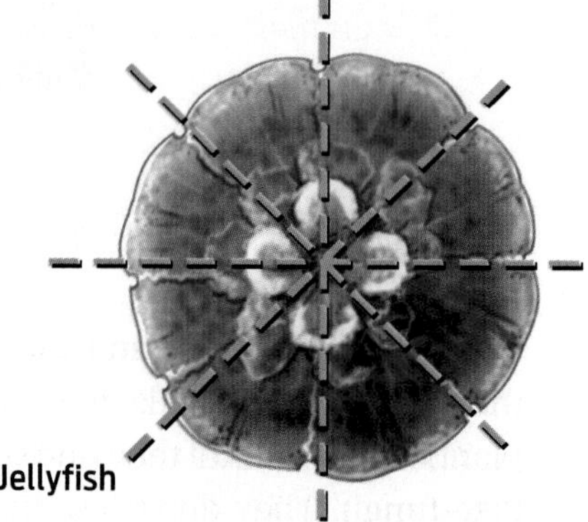
Jellyfish

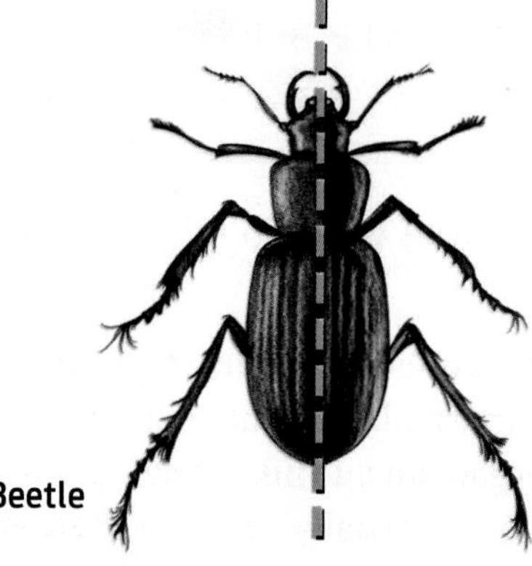
Beetle

Asymmetrical
Asymmetrical animals have no defined shape.

Animals without symmetry are not mobile as adults and rely on whatever floats by for food. They often have physical protective mechanisms.

Radial Symmetry
Animals with radial symmetry are symmetrical in all directions, with no distinct right or left side.

Animals with radial symmetry do not move in their adult life (corals and sea anemones) or depend on water currents and short-distance propulsion for mobility (jellyfish).

Bilateral Symmetry
Animals with bilateral symmetry have mirror-image left and right sides.

This type of symmetry enables purposeful, directed movement to find food and mates and to flee from danger.

? For each of the following animals, identify the symmetry of its body plan: sand dollar; parrot; jellyfish; earthworm; octopus.

arthropods to be successful on land resembled millipedes.

These early land animals faced many of the same problems that plants faced as they left the oceans: how to keep from drying out, how to support themselves without the natural buoyancy of water, and how to reproduce in ways that didn't rely on water to transport gametes.

A key adaptation that facilitated animals' move onto land was the evolution of an **exoskeleton.** This nonbony skeleton forms on the outside of the body and helps to support the animal against the pull of gravity. In most land arthropods, the exoskeleton is also waxy and waterproof, and so helps to retain water in dry environments. Much like the internal **endoskeleton** of humans and other vertebrates, the exoskeleton of arthropods provides attachment sites for muscles, permitting rapid movement.

Arthropods are, by far, the most successful animals on the planet, at least according to their numbers. There are an estimated 2-4 million species of arthropods, although only 855,000 have been officially described. The number of individual arthropods on the planet is estimated to be more than 10^{18}–that's 1 with 18 zeros after it. The vast majority of all arthropods are **insects**–arthropods with three pairs of jointed legs and a body with three distinct sections.

Though arthropods colonized land first, and still dominate it today, many other invertebrates have joined the arthropods on land, attracted to plentiful food sources and hiding places. Sliding quietly amid leaf litter are many specimens of **mollusk,** including slugs and snails. By digesting dead plant material, mollusks help recycle nutrients. And with their calcium-rich shells, snails provide this valuable mineral to the creatures that feast on them, such as rodents and birds. Some humans find mollusks a tasty treat as well: if you have enjoyed clams, oysters, or squid, then you have eaten some aquatic varieties of mollusk.

Lift any rock in the rain forest and you're likely to find many **annelids,** or segmented worms. Annelids such as earthworms perform a critical ecological service by creating passageways in the soil as they move. The passageways allow air and water to enter the soil, making it fertile for other life. By eating and digesting leaf and other plant litter, earthworms also make nutrients available for plants.

Last to arrive on land were our own closest animal relatives, the vertebrates. These are organisms that have a bony or cartilaginous backbone. Fish, birds, reptiles, amphibians, and mammals (including humans) are all vertebrates. Compared to the invertebrates, vertebrates make up a tiny fraction of the total animal diversity on the planet **(INFOGRAPHIC 18.10)**.

As impressive as animal diversity in the rain forest is, none of it would be possible without the plants that provide them with food and shelter. In fact, according to Laurance, there is a clear link between animal and plant diversity. "If you've got very high tree diversity, and then also a lot of these trees are growing very tall, you create a lot of different places for animals to live," he says. "So the animal diversity may be responding to the very high plant diversity."

The animals in the forest aren't simply freeloaders in this situation, however; they provide a valuable service to the many flowering plants in the forest. Insects, for example, play a crucial role in pollination of flowering plants. And larger animals help to disperse plants' seeds. Animals consume seeds when they eat a plant's fruit, then poop them out at a distance far enough away that they help the plant overcome the threat of pests. In this way, animals help contribute to forest diversity as well.

What may seem surprising, given all the life supported here, is that the soils in the Amazon are actually quite poor in nutrients. The incessant rains that fall here–about 9 feet every year–wash out nutrients from soil into the many smaller rivers that pour into the Amazon. As a result, essentially all the available nutrients in this ecosystem are trapped in the existing vegetation. As soon as a tree dies or a leaf falls, this organic material is decomposed by the abundant populations of bacteria and fungi and the nutrients are

EXOSKELETON
An external skeleton; in arthropods the exoskeleton is made up of proteins and chitin.

ENDOSKELETON
An internal body skeleton, typically made of cartilage or bone.

INSECT
An arthropod with three pairs of jointed legs and a body with three segments

MOLLUSK
An invertebrate with a soft, unsegmented body enclosed in a hard shell

ANNELID
An invertebrate with a soft, segmented body; annelids are commonly referred to as worms.

INFOGRAPHIC 18.10 Evolution of Animal Diversity

Over their 800-million-year history animals have adapted to many aquatic and terrestrial environments on Earth, enabled in part by their diverse body plans and varied adaptations for movement.

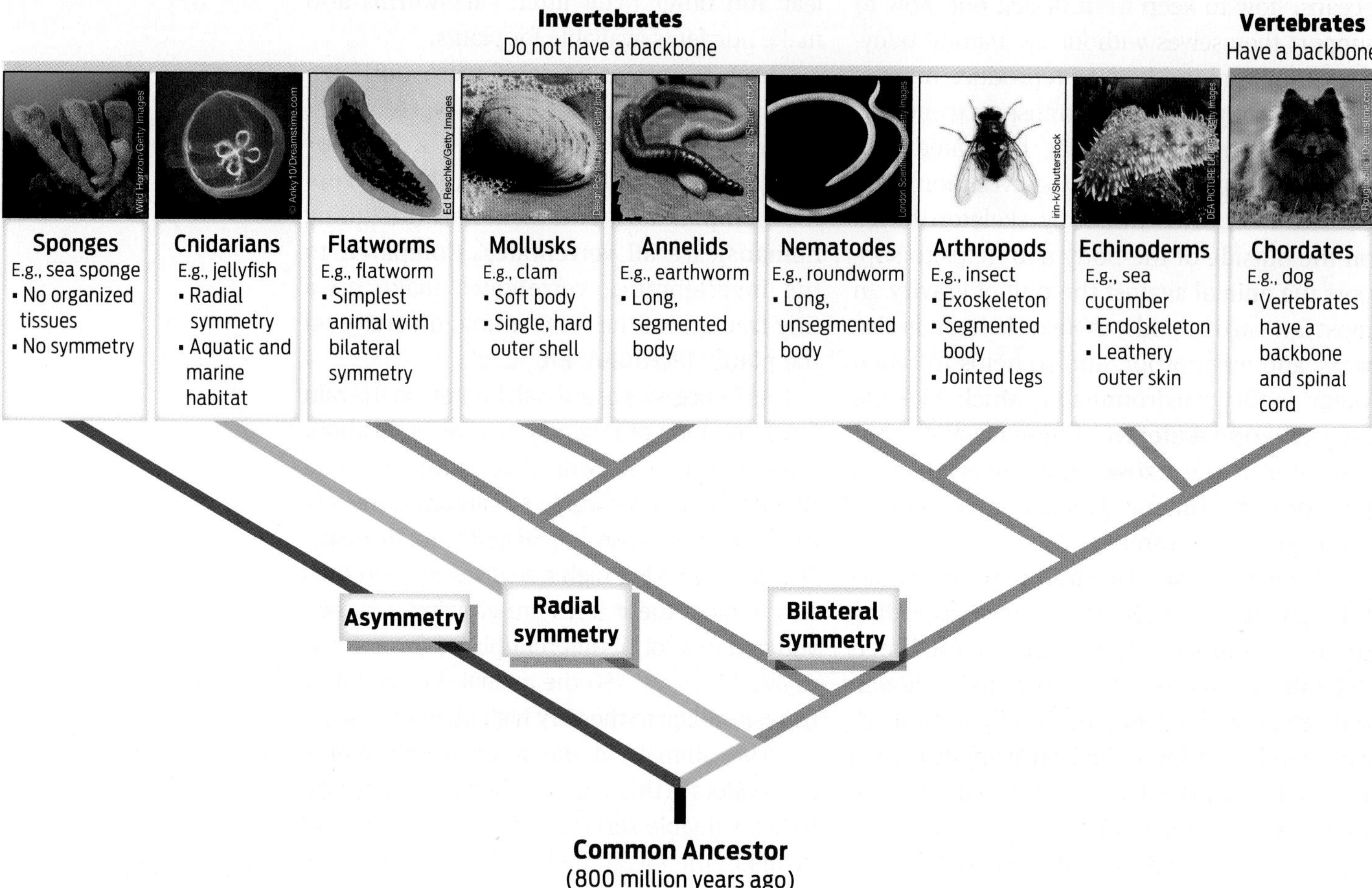

? List the three different animal body symmetries. For each type, name an animal group with that symmetry.

recycled, taken up into plant bodies as new plants grow. The rain forest is perhaps the most efficient ecosystem on Earth, but for that reason it's also one of the most fragile; once parts of it are cut down, they are not easily replaced.

Deforestation Casualties

Deforestation in the Amazon happens in a predictable pattern. It usually starts with an illegal road. Loggers penetrate deeper into the forest and fell the most valuable hardwood trees, like mahogany and ipe, and sell the timber to sawmills. This wood eventually finds its way to U.S. and European furniture and floor wholesalers, who may or may not be aware of its source.

Loggers then set fires, reducing the remaining trees to stubby ash skeletons. Next come the bulldozers, which push what's left into heaps and tear up the remaining roots from the soil. The land is now ready to be planted with soybeans or used as cattle pasture.

Burning trees deposits nutrients in the soil, which then support one or two seasons of crop growth. But because the underlying soil is naturally so nutrient-poor, these recently cleared lands quickly lose their fertility, leading to yet more slashing and burning.

Deforestation is threatening many of the eukaryotic species that live in the rain forest. According to a 2015 study by Laurance and his colleagues, roughly half of the 14,000 species of tree in the Amazon qualify as globally threatened under International Union for Conservation of Nature (IUCN) Red List criteria.

Among the animal species that could be put at risk by "business as usual" deforestation in the Amazon are many that live there and nowhere else, including the tree ocelot, hoary-throated spinetail bird, white-cheeked spider monkey, Rio Branco antbird, Brazilian tapir, yellow-headed poison dart frog, giant otter, and red-faced Uakari monkey.

Deforestation can have unexpected consequences. One such surprise is the 10-fold increase in the incidence of malaria infections that has occurred in the Amazon since 1970. Malaria is a disease caused by a parasite that is transmitted to humans and other animals by mosquitoes. Deforestation creates a perfect breeding ground for the mosquitoes that transmit the disease.

When humans slash and burn a tract of land, they often leave a few tall trees standing. These lone survivors provide partial shade that the mosquitoes find appealing. Burning of trees also increases the pH of soil, and mosquitoes like to breed in standing water that is slightly alkaline.

Though transmitted by mosquitoes, malaria is actually caused by a tiny, single-celled eukaryotic organism called *Plasmodium*. These tiny parasites live inside the mosquito and are transferred to other animals through bites.

Plasmodium is an example of a **protist,** a eukaryote that cannot be categorized as either plant, animal, or fungus. Protists are amazingly diverse, and therefore hard to classify. Most are unicellular, but there are also multicellular varieties, such as some types of **algae.** Multicellular algae float in water and they photosynthesize like plants, but unlike plants they lack specialized adaptations for living on land, such as vascular tissue. Other protists are similar to animals in that they eat other organisms, but since they are unicellular they are not technically animals. Some protist species have long slender bodies resembling fungi, but they are no more related to fungi than animals are. In fact, genetic evidence

PROTIST
A eukaryote that cannot be classified as a plant, animal, or fungus; usually unicellular.

ALGAE
A diverse collection of aquatic, photosynthetic organisms, including both unicellular and multicellular species.

Kitchin and Hurst/Getty Images

Yellow-headed poison dart frogs (*Dendrobates leucomelas*) make their home among the plants of the Amazon rain forest.

shows that protists do not form a cohesive evolutionary group; some may be as distinct from one another as plants are from animals (see **Milestone 6: Shaking the Tree**).

Despite their diversity, protists do share some common traits. They are all susceptible to drying out, so they are typically found in wet environments: lakes, oceans, ponds, moist soils, and living hosts **(INFOGRAPHIC 18.11)**.

Though protists are extremely resilient and unlikely to suffer in the same ways as other eukaryotes by destruction of rain forest habitats, it's clear that their impact on humans could shift dramatically as a result. This is especially true if, as is expected, climate change shifts the range of the insects that carry them, bringing the insects into regions where formerly they could not survive the winter.

A Tipping Point

Deforestation in the Amazon reached a tipping point in 2004, when some 11,000 square miles of forest was destroyed—an area about the size of Belgium. Since that time, the rate of deforestation has dropped dramatically—by about 80%.

What changed? Laurance credits several factors, including the high-profile murder of a Catholic nun, Dorothy Stang, who had been an outspoken critic of illegal logging and land theft. With the attention to her murder came a crackdown by the government on illegal logging and deforestation. Brazil also began using a satellite system to monitor deforestation from space.

Boycotts by U.S. companies have also likely played a role. In 2006, the nonprofit

INFOGRAPHIC 18.11 The Challenge of Classifying Protists

Protists are a diverse group of organisms that are difficult to classify. They share features with animals, plants, and fungi, but are not classified as any one of these. Nor do they have a single unifying characteristic that places them within a single evolutionary group. Protists are currently found in multiple groups within the domain Eukarya.

Animal-like, but not animals

Plant-like, but not plants

Fungus-like, but not fungi

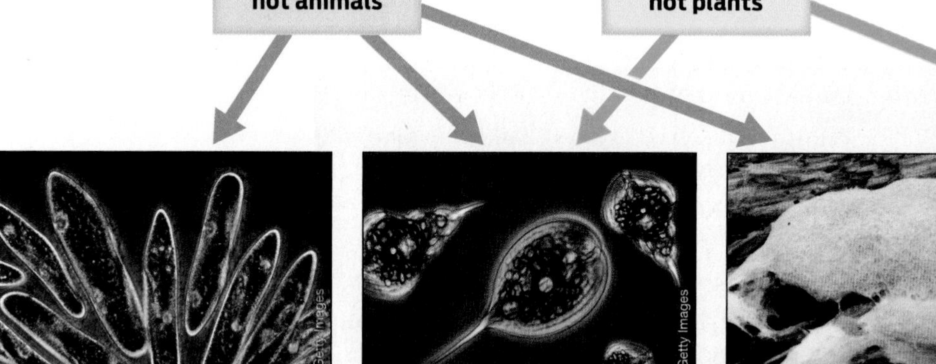

River and lake water is teeming with unicellular protists that feed on other organisms (for example, algae), as heterotrophic animals do.

Aquatic environments contain a diversity of unicellular organisms that, like plants, are photosynthetic. Some of these can move, a feature shared with animals.

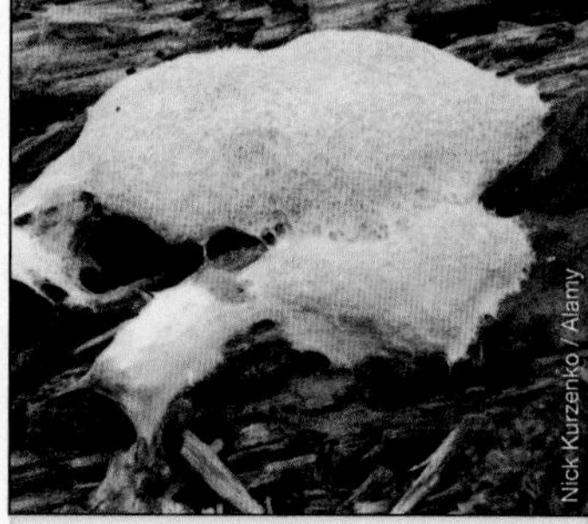

Dog vomit slime mold is a decomposer, like fungi, but may also eat other organisms as food, similar to heterotrophic animals.

Algae float in lakes and rivers, performing photosynthesis, like plants. Many algae form filamentous strands of cells, similar to many fungi.

? Are all protists closely related to one another? Explain your answer.

organization Greenpeace released a report showing that forests were being destroyed to grow soybeans to feed cattle, which were sold to companies like McDonald's and Wal-Mart. In response to international outrage, these companies agreed to stop buying meat from cattle fed soybeans grown on illegally deforested lands.

But deforestation has not stopped in Brazil. In 2014–a relatively "good" year, deforestation-wise–an area the size of Olympic National Park in Washington State was wiped out. And as of 2017, there are signs that deforestation may be ramping up again. That's why new ideas and new approaches are badly needed.

Brazilian Amazon Deforestation (1988–2016)

The area of forest removed (deforestation) increased between 1991 and 2004. Since then, annual deforestation declined steadily until 2015 and 2016, which have had higher losses than the previous years.

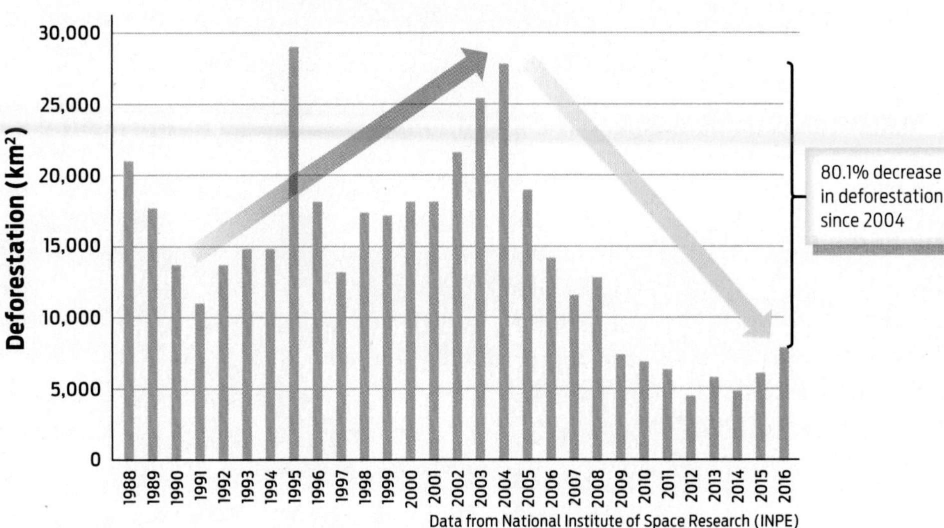

Innovation in Acre

In 2010, the government in Acre passed a law called the State System of Incentives for Environmental Services, with the goal of making standing forest more valuable by linking it to markets for the services it provides–rubber, Brazil nuts, fish, even carbon storage. The incentives reward small farmers, indigenous communities, and even cattle ranchers and loggers with cash payments for protecting forest resources.

As part of the law, Acre established the first state-based REDD+ program. REDD+ stands for "reducing emissions from deforestation and forest degradation, and enhancing carbon stocks." This is a program established by the United Nations that aims to reduce greenhouse gas emissions by encouraging tropical countries to better manage their forests. Through this program, governments can be eligible for payments after demonstrated results of keeping carbon in the trees. Germany has already invested $17.5 million into Acre, and California is currently considering whether to buy forest-based carbon credits from the state as part of its own cap-and-trade program.

How did Acre become such a leader in this area? "It really all started with Chico Mendes and the rubber tappers movement," says Duchelle, of the Center for International Forestry Research. Acre's self-proclaimed "Forest Government," which has held leadership in the state government for nearly 20 years, has worked on the premise that economic development need not be incompatible with forest sustainability.

In many ways, Acre is unique: its strong history of forest conservation movements makes it easier for sustainable practices to take hold here. But even much larger states in Brazil, with very different challenges–more soybean farmers, more cattle ranchers–have learned from Acre. Mato Grosso, a large state in central Brazil passed a REDD+ law in 2013 that was very much influenced by what Acre has done.

"It's a special environment," Duchelle says. "Again and again, Acre has pioneered innovative models for forest-based development." ■

CHAPTER 18 SUMMARY

- Rain forests like the Amazon are home to a great variety of eukaryotic organisms—plants, animals, fungi, and protists.
- Eukaryotic organisms all share a common evolutionary ancestor that lived roughly 800 million years ago. Eukaryotic cells have a membrane-bound nucleus, which contains DNA, and other organelles.
- Plants are multicellular eukaryotes that carry out photosynthesis and are adapted to living on land. Plant cells have a cell wall made of cellulose. They reproduce sexually.
- Some plants produce distasteful chemicals called secondary metabolites as a defense against herbivores; latex contains many such secondary metabolites.
- Plants can be divided into groups, such as the bryophytes, ferns, gymnosperms, and angiosperms, on the basis of their terrestrial adaptations.
- Two key evolutionary milestones in plants were the evolution of vascular tissues and reproduction using pollen and seeds.
- Fungi are decomposers, acquiring their nutrition by breaking down dead organic matter and absorbing the results. There are unicellular and multicellular fungi. Fungal cells have a cell wall made of chitin. They reproduce sexually and asexually.
- Animals are multicellular eukaryotic heterotrophs that obtain nutrients by ingestion. They reproduce sexually. Movement is a key evolutionary adaptation that has influenced the success of animals.
- Most animals are invertebrates (that is, they lack a backbone). The most abundant invertebrates by far are arthropods, especially insects.
- Protists are a diverse group of mostly unicellular eukaryotic organisms that do not cluster on a single branch of the evolutionary tree. They include photosynthetic plantlike algae, animal-like parasites, and many other organisms.
- Biodiversity in tropical rain forests reflects a continuous battle between plants and their animal, fungal, and microbial predators.
- Deforestation of rain forests threatens biodiversity on the planet and affects climate change.
- Subnational governments across the tropics are pioneering innovative solutions to conserve tropical forests and promote sustainable development.

MORE TO EXPLORE

- **NPR (November 12, 2015) "The Rain Forest Was Here": http://www.npr.org/series/455762230/the-rain-forest-was-here**
- **Duchelle, A.E., et al. (2014) Acre's State System of Incentives for Environmental Services (SISA), Brazil. In: Sills, E.O., et al., eds. *REDD+ on the ground: A case book of subnational initiatives across the globe*, pp. 68–85. CIFOR, Bogor, Indonesia**
- **Ter Steege, H., et al. (2015) Estimating the global conservation status of more than 15,000 Amazonian tree species. *Science Advances* 1, no. 10.**
- **Bagchi, R., et al. (2014) Pathogens and insect herbivores drive rainforest plant diversity and composition. *Nature* 506:85–88.**
- **Grandin, G. (2009) *Fordlandia: The Rise and Fall of Henry Ford's Forgotten Jungle City*. New York: Picador.**

CHAPTER 18 Test Your Knowledge

DRIVING QUESTION 1 What are eukaryotic organisms, and what factors influence their diversity?

By answering the questions below and studying Infographics 18.1 and 18.7, you should be able to generate an answer for the broader Driving Question above.

KNOW IT

1 Which of the following is *not* a characteristic found in at least some organisms in the domain Eukarya?

a. cell walls
b. photosynthesis
c. multicellularity
d. cells lacking a nucleus
e. unicellularity

2 What are the defining features of eukaryotes, members of the domain Eukarya?

3 What do a rubber tree and a human latex collector have in common?

USE IT

4 How do you think the diversity of eukaryotic organisms in each of the following areas would compare to the diversity in the Amazon rain forest? What factors might influence eukaryotic diversity in these areas?

a. Lake Michigan
b. the Sonoran Desert in Arizona
c. a high alpine meadow in the Colorado Rockies

5 Which of the following would you predict to have the greatest impact on the number of different eukaryotic species in the Amazon rain forest? Explain your answer, and consider the impact of each on all eukaryotes.

a. application of a herbicide
b. application of an insecticide
c. application of a fungicide

apply YOUR KNOWLEDGE

INTERPRETING DATA

6 A group of researchers designed an experiment to test the Janzen–Connell hypothesis, which they carried out in a rain forest in Belize. They began by analyzing seeds falling to the ground in a particular area, counting the number of seeds dropped by different species of plants. If all the seeds germinate, the seedlings present should match the diversity of species in the seeds. Then they analyzed seedlings that germinated in three small (1 m^2) areas in the same forest location: one that was sprayed weekly with water (control), one that was sprayed with an insecticide, and one that was sprayed with a fungicide. They identified and counted seedlings to determine diversity (number of species) and abundance (number of seedlings) (* indicates a significant difference from control).

	Control	Insecticide	Fungicide
Diversity (no. of species)	3.7	3.4	3.2*
Abundance (no. of seedlings)	30	58*	35

Data from Bagchi, R., et al. (2014) Pathogens and insect herbivores drive rainforest plant diversity and composition. Nature 506:85–88 doi:10.1038/nature12911

a. Plot these data on two graphs (one for species number and one for seedling number).
b. What can you conclude about the impact of insects on the number of seeds that germinate into seedlings? What can you conclude about the impact of fungi on the number of seeds that germinate into seedlings?
c. Do insects or fungi have a greater impact on maintaining plant diversity (defined as number of species) in the rain forest?

DRIVING QUESTION 2 How are plants defined, and what adaptations have enabled their success?

By answering the questions below and studying Infographics 18.2, 18.3, 18.4, and 18.7, you should be able to generate an answer for the broader Driving Question above.

KNOW IT

7 Which group of plants was the first to live on land? Why do we find these plants only in particular environments (after all, if they were the first, shouldn't they have spread everywhere by now)?

8 A major difference between a fern and a moss is

- **a.** the presence of seeds.
- **b.** the presence of flowers.
- **c.** the presence of cones.
- **d.** the presence of a vascular system.
- **e.** the ability to carry out photosynthesis.

USE IT

9 What is an advantage of having seeds? (Think about spreading to new locations and whether or not reproduction relies on water.)

10 What type of seed plant is likely to rely on hungry animals to spread its seeds? Explain your answer.

11 How did the evolution of vascular systems in plants change the landscape?

DRIVING QUESTION 3 How are fungi defined, and what adaptations have enabled their success?

By answering the questions below and studying Infographics 18.5 and 18.6, you should be able to generate an answer for the broader Driving Question above.

KNOW IT

12 Consider the eating habits of fungi.

- **a.** Can fungi carry out photosynthesis?
- **b.** Can fungi ingest their food?
- **c.** How do fungi obtain their nutrients and energy?

13 Which of the following meals include a fungus?

- **a.** a bread and blue cheese platter with fruit
- **b.** mushroom risotto
- **c.** a and b
- **d.** a fruit salad
- **e.** yogurt

USE IT

14 A very early classification scheme placed the fungi together with the plants. Why do you think fungi were grouped with plants? What features distinguish them from plants?

DRIVING QUESTION 4 How are animals defined, and what adaptations have enabled their success?

By answering the questions below and studying Infographics 18.8, 18.9, and18.10, you should be able to generate an answer for the broader Driving Question above.

KNOW IT

15 A sand dollar gets its name from its body shape—it resembles a large coin. What type of body symmetry does a sand dollar have?

- **a.** bilateral
- **b.** radial
- **c.** none (sand dollars are amorphous)
- **d.** hyphae
- **e.** mycelium

16 What do mosquitos, snails, and earthworms have in common?

- **a.** They are all insects.
- **b.** They are all mollusks.
- **c.** They are all arthropods.
- **d.** They are all invertebrates.
- **e.** They all have an exoskeleton.

17 Which of the following defense/predator avoidance features or strategies can be found in arthropods?

- **a.** an exoskeleton.
- **b.** venom
- **c.** flight
- **d.** secondary metabolites
- **e.** all of the above
- **f.** a, b, and c

18 Which of the following is *not* a feature of all animals?

- **a.** defined body symmetry
- **b.** internal digestion
- **c.** motility
- **d.** cells that lack cell walls
- **e.** multicellularity

USE IT

19 Many characteristics are used to classify animals. Why do we need to use so many different characteristics? Consider the following five animals: cockroach, earthworm, honeybees, parrot and slug; and the following three characteristics: ability to fly, two-legged, exoskeleton

a. Which of the five animals could be grouped by each characteristic?

b. Would this grouping reflect their real taxonomic relationships?

c. By what feature(s) would you put earthworms and slugs together in their own group? What about parrots and honeybees? Are the organisms in each of these pairs in the same taxonomic group?

DRIVING QUESTION 5 What are protists, and why are they hard to classify?

By answering the questions below and studying Infographic 18.11, you should be able to generate an answer for the broader Driving Question above.

KNOW IT

20 What do all members of the informal group known as protists have in common?

a. nothing

b. They are all eukaryotic.

c. They all carry out photosynthesis.

d. They are all human parasites.

e. They are all decomposers.

USE IT

21 Why do scientists no longer consider protists a separate kingdom? How are scientists placing protists into their new taxonomic "homes"?

22 Many protists have an organelle called the contractile vacuole that pumps out water that enters the cell by osmosis. Why is this a useful adaptation for a protist? What might happen to a protist if its contractile vacuole stopped working? (Think about where many protists live, and what happens to bacteria whose cell walls are disrupted by antibiotics.)

apply YOUR KNOWLEDGE

MINI CASE

23 Biopiracy has been described as the use of the knowledge of native peoples about nature and natural products by others for profit without obtaining permission of the indigenous people, and without sharing profits. Given the tremendous biodiversity in the Amazon and medicines that have already been developed from natural sources in the area, there are likely to be life-saving and lucrative medicines still to be discovered by bioprospecting rain forest resources. (Bioprospecting is the search for natural resources that provide valuable products.) The Convention on Biological Diversity attempts to address many of the concerns arising from biopiracy and bioprospecting.

a. Do some research on the Convention on Biological Diversity and identify the issues that would be particularly relevant for bioprospecting in the Amazon.

b. Draft a plan setting out ways in which a drug company could fairly bioprospect in the Amazon.

c. How do you think that the various stakeholders would respond to your plan?

apply YOUR KNOWLEDGE

BRING IT HOME

24 During his presidency, Barack Obama established over 30 new national monuments, protecting more than 550 million acres of land and water. Do some research to learn about one of these new national monuments.

a. Where is it?

b. What kind of eukaryotic biodiversity is present?

c. Are there any threats to the eukaryotic biodiversity present? Has the protected status of this national monument reduced any of these threats?

Shaking the TREE

A revised view of eukaryotic diversity may be the key to tackling deadly diseases

DRIVING QUESTIONS

1. How has genetic evidence transformed the classification of protists?
2. Why is it important to classify protists accurately?
3. What are current challenges in preventing and treating malaria?

As the war in Vietnam raged, soldiers on both sides of the conflict faced an unrelenting enemy: malaria. This parasite-caused illness–transmitted through the bite of a mosquito–produces devastating fever, headache, chills, and vomiting in victims. If not treated within 24 hours, the condition can be fatal.

Malaria had traditionally been treated with drugs such as quinine, derived from the bark of the cinchona tree, and its synthetic derivative chloroquine. But increasingly, these drugs were unable to stem the tide of infection; the parasites had begun to evolve resistance.

Worried that a malaria-weakened army would be unable to fight off U.S.-backed military forces, the prime minister of North Vietnam, Ho Chi Minh, turned to Communist China for help. Recognizing their common interest in defeating a shared set of enemies, China's leader, Mao Zedong, launched a secret mission to find a malaria cure.

Project 523, as it was called–it was launched on May 23, 1967–enlisted hundreds of Chinese scientists and traditional Chinese healers. They were charged with screening thousands of known plant compounds for antimalarial effects and scouring traditional sources of Chinese medicine for leads on promising new medicines.

Taking their cue from an ancient medical text, the scientists homed in on one particular herb, called qinghao (known in the West as *Artemisia annua* or sweet wormword). Qinghao had been used for centuries in treating "intermittent fevers," a description that aptly describes a symptom of malaria. The scientists tested the herb against malaria-infected mice but the results proved inconsistent.

During Project 523, Chinese scientists discovered artemisinin, a compound found in the herb *Artemisia annua* that effectively kills the malaria parasite *Plasmodium falciparum*.

Then, one scientist, Youyou Tu, got a flash of inspiration from another ancient text, which recommended soaking the herb in water and then drinking the liquid. Tu, a chemist, realized that the typical extraction process of boiling the herbs was likely destroying the chemical. She changed the extraction process and the results were dramatic: nearly 100% of the mice were cured by the herbal remedy.

Tu's efforts would eventually result in the drug artemisinin, one of today's main pharmaceutical weapons against malaria. The discovery of artemisinin was truly a milestone in medicine, helping to save the lives of millions of people worldwide each year. For her work, Tu shared the Nobel Prize for Physiology or Medicine in 2015.

But increasingly, scientists are seeing worrisome signs that malaria parasites are developing resistance to artemisinin, too. If that happens, millions of people around the world could be at risk for fatal malaria infections. The frightening prospect has scientists racing to find out more about these

Youyou Tu at work and with the 2015 Nobel Prize in Physiology or Medicine she received for her research on the development of artemisinin for the treatment of malaria.

tiny parasites and what makes them unique, in order to develop more effective weapons directed specifically against them.

What Are They?

The malaria parasite is a single eukaryotic cell. It belongs to a genus of organisms called *Plasmodium*. The most common malaria-causing species are *P. falciparum* and *P. vivax*, with *P. falciparum* causing the more severe form of the illness.

The parasite lives inside several different species of mosquitoes of the genus *Anopheles*. It is transmitted to humans through bites from female mosquitoes in search of a blood meal to feed their developing eggs. Once inside a human host, the malaria parasites first infect liver cells and then red blood cells. The parasites divide rapidly, eventually bursting out of these cells. The death of red blood cells and release of their contents are what produce the main symptoms of the disease. Eventually, the parasites are transmitted back to a mosquito as it feeds on the blood of an infected person, and the cycle continues **(INFOGRAPHIC M6.1)**.

In 2015, about 95 countries around the world, mostly in tropical regions of South America, Africa, and Asia, had ongoing

INFOGRAPHIC M6.1 Malaria Parasite Life Cycle

Malaria is caused by organisms in the genus *Plasmodium*. *Plasmodium* has a complex life cycle, involving a mosquito and a human host. In order to complete its life cycle and reproduce, various stages of *Plasmodium* mature and replicate in human liver and red blood cells, as well as in the gut of mosquitoes. The stages present in human blood cause the symptoms of malaria.

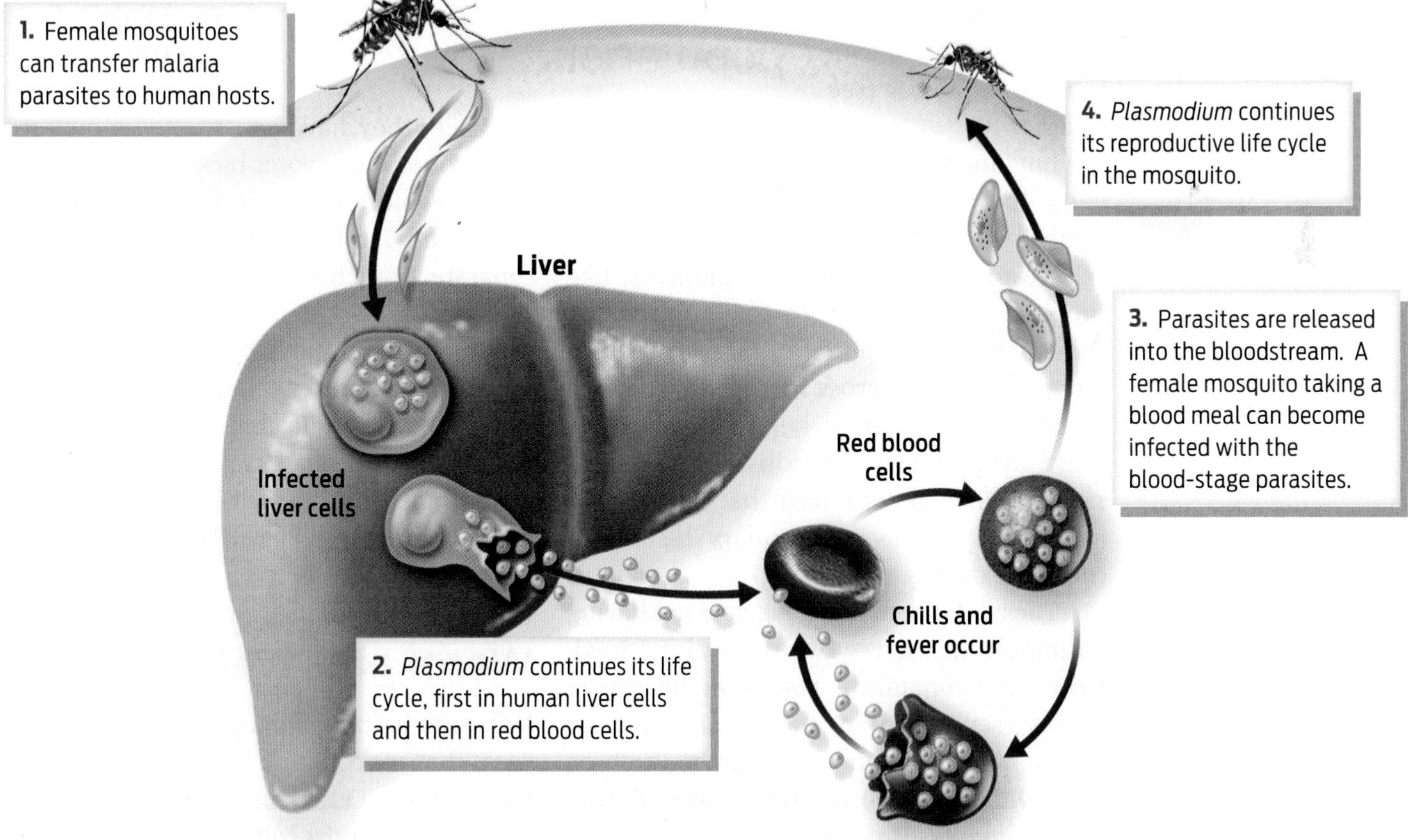

1. Female mosquitoes can transfer malaria parasites to human hosts.
Liver
Infected liver cells
2. *Plasmodium* continues its life cycle, first in human liver cells and then in red blood cells.
Red blood cells
Chills and fever occur
3. Parasites are released into the bloodstream. A female mosquito taking a blood meal can become infected with the blood-stage parasites.
4. *Plasmodium* continues its reproductive life cycle in the mosquito.

? Does the mosquito just carry the parasites between human hosts? Explain your answer.

malaria epidemics. About 3.2 billion people are at risk for the disease–nearly half the world's population. The malaria parasite infects roughly a quarter of a billion people every year and kills 1 million. The stakes of properly classifying and understanding this tiny parasite could not be higher.

Plasmodium is an example of a protist, a member of a diverse group of mostly single-celled eukaryotic organisms that we encountered in Chapter 18. Single-celled protists have long proved challenging for scientists to study and classify. Because of their small size, it's often hard to make useful discriminations on appearance alone. They are clearly eukaryotic; but equally clearly they lack the defining features of plants, fungi, or animals. For that reason, protists were historically lumped together into a single large group, the kingdom Protista–a kind of evolutionary grab bag (see Chapter 16).

About 3.2 billion people are at risk for the disease—nearly half the world's population.

But increasingly, scientists are realizing that protists are a highly diverse bunch–with many members more closely related to other eukaryotic organisms, like fungi or animals, than to one another. By sequencing multiple genes from representative organisms across the eukaryotic tree of life, scientists have begun to decipher these evolutionary relationships and construct a phylogenetic tree that more accurately reflects the evolutionary history of these organisms.

The results are surprising, and are having an impact on the way we think about medicine. The organism *Pneumocystis carinii* causes a severe form of pneumonia in people with a weakened immune system (such as those with HIV/AIDS). Scientists once believed that *Pneumocystis carinii* was an animal-like protist, but they now know, from genetic evidence, that it is a fungus. This revised classification has allowed scientists to develop more effective treatments for the infection–in this case, antifungal medications.

Similarly, the single-celled organism *Phytophthora infestans*, the cause of potato blight, was long considered to be a type of fungus. This organism was responsible for about 1 million deaths from starvation during the potato famine in Ireland between 1845 and 1851. Newer molecular data have revealed it to be a protist belonging to a group of organisms called Stramenopila, which are not closely related to fungi. Its revised classification explains why *Phytophthora* resists fungicides.

To make sense of these new data points, biologists have done away with the traditional eukaryotic tree that was divided into four kingdoms (with plants, animals, fungi, and protists each as a separate kingdom). In its place, they have constructed a new tree that groups eukaryotes based on their actual shared evolutionary history. Within this new tree, plants, animals, and fungi still exist as separate branches, but each of these four branches is part of a larger group in which they are surrounded by other organisms–many of them single-celled protists–that current science indicates are their genetic relatives. These new taxonomic divisions are called supergroups (see Chapter 16). *Plasmodium* belongs to the supergroup Rhizaria (**INFOGRAPHIC M6.2**).

Warm and Wet

Malaria is just one of many parasite-caused diseases that are common in regions with warm, wet climates. Others include Chagas disease, caused by a protist called *Trypanosoma cruzi* that is transmitted by blood-sucking assassin

INFOGRAPHIC M6.2 Protists Are Highly Diverse Members of the Domain Eukarya

Protists are typically single-celled eukaryotic organisms that lack distinguishing features of fungi, animals, and land plants. For this reason, they were historically lumped together in the kingdom Protista. An increasing amount of DNA sequence information has shown that protists are very diverse, sharing closer relationships with other eukaryotic supergroups than with one another.

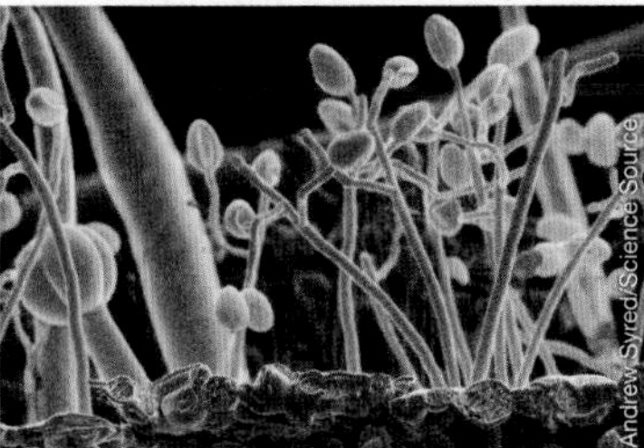

Phytophthora infestans was classified as a fungus until reclassified as a member of the Stramenopile supergroup.

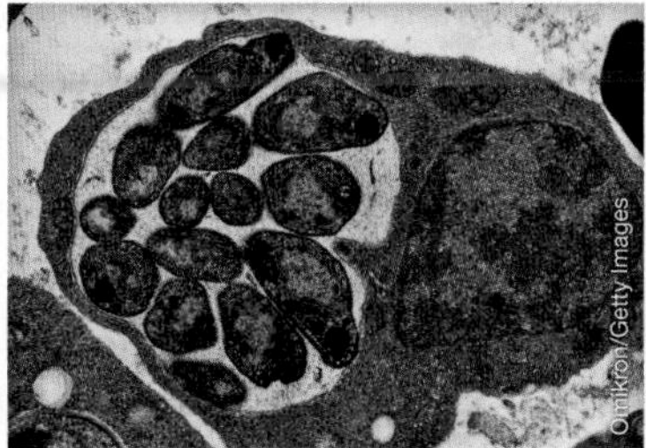

Plasmodium falciparum is a member of the Rhizaria supergroup, a diverse group joined together by DNA sequence similarities.

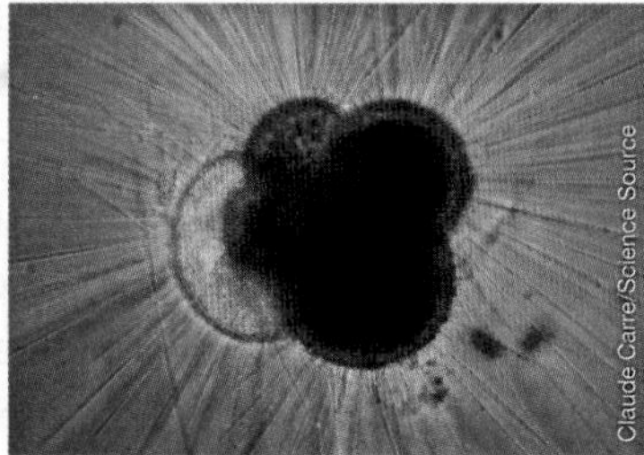

Photosynthetic *Globigerina bulloides* is a member of the supergroup Archaeplastida, which also includes land plants.

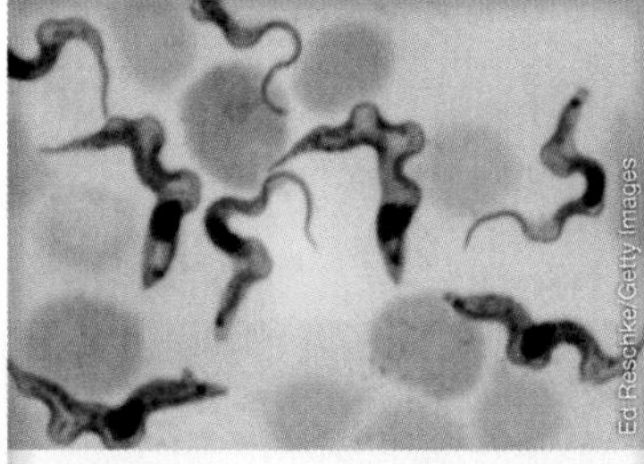

Trypanosoma brucei, the protist that causes African sleeping sickness, is classified in the supergroup Excavata. This group is named for a feeding structure that appears to be asymmetrically "excavated" from one side of the organism.

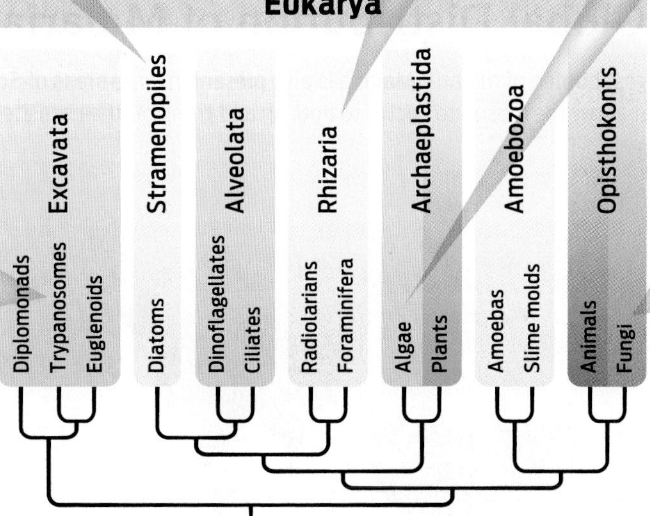

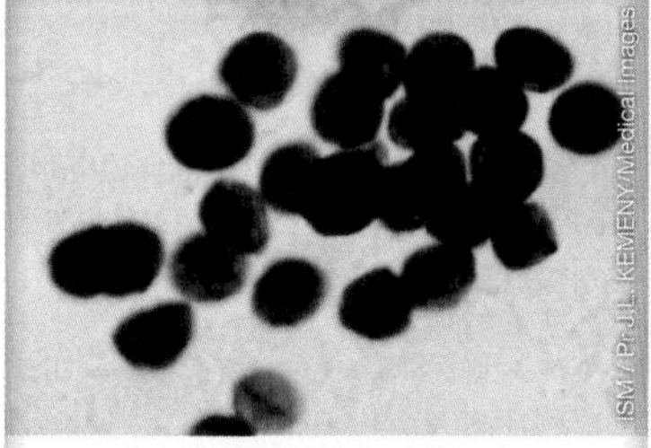

Pneumocystis carinii was classified as a protist until reclassified as a fungus in the Opisthokont supergroup that also includes animals.

? Which is more closely related to humans, *Phytophthora infestans* or *Pneumocystis carinii?*

bugs, and African sleeping sickness, caused by a protist called *Trypanosoma brucei* that is transmitted by the tsetse fly.

What makes tropical regions uniquely susceptible? Most likely, it's because the insect vectors that carry the parasites are able to grow unchecked in the warm year-round temperatures. And the heavy rains provide lots of moist spots in which the larvae of these various insects can develop. *Anopheles* mosquitoes, for example, lay their eggs in pools of water. The eggs eventually hatch the larvae, which develop into adult mosquitoes. Many tropical regions are also relatively poor and often lack adequate public health protections, including mosquito control measures.

Sub-Saharan Africa shoulders the highest burden of malaria worldwide: In 2015,

88% of malaria cases and 90% of malaria deaths occurred in sub-Saharan Africa (**INFOGRAPHIC M6.3**).

With the development of insecticides like DDT in the 1940s (see **Milestone 7: Progress or Poison**), there was hope that malaria could be eradicated completely by eliminating the mosquitoes that transmit the parasite. The World Health Organization (WHO) launched such an effort in 1955. As a result, malaria infections fell sharply in some countries, such as India and Sri Lanka. But the effort stalled when it became clear that mosquitoes were evolving resistance to the insecticides, including DDT. Some regions, including sub-Saharan Arica, were never part of the eradication program in the first place.

WHO currently endorses the use of DDT, sprayed indoors, as a way to control mosquitoes. But with resistance to the pesticide a large and growing problem, efforts that rely only on insecticides are likely to fail.

Another complicating factor is climate change, which may shift the ranges of the disease-carrying insects. Regions of the world not currently plagued by malaria may find themselves facing it should they become warmer and wetter.

Not all protists cause disease. Many are harmless and some are even beneficial components of our ecosystems. Close to 50% of all photosynthesis is carried out by photosynthetic protists called green algae, which thereby help to make the air breathable for

INFOGRAPHIC M6.3 Global Distribution of Malaria

Sub-Saharan Africa bears the largest burden of malaria. Malaria is also present in large areas of South America and southeast Asia. These areas all have climates that allow the mosquito vector to flourish and transmit the parasites.

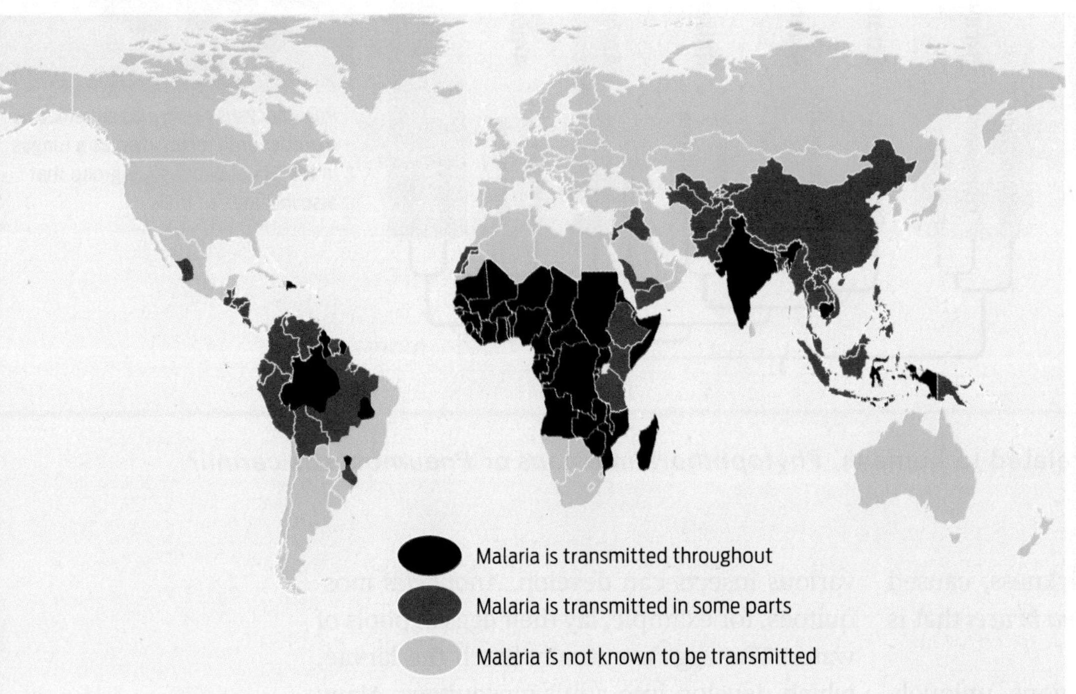

Data from the CDC.

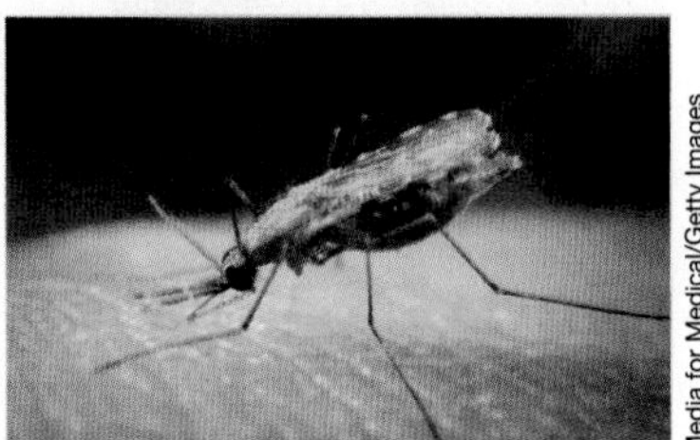

Media for Medical/Getty Images

The *Anopheles* mosquito transmits the *Plasmodium* parasite while feeding.

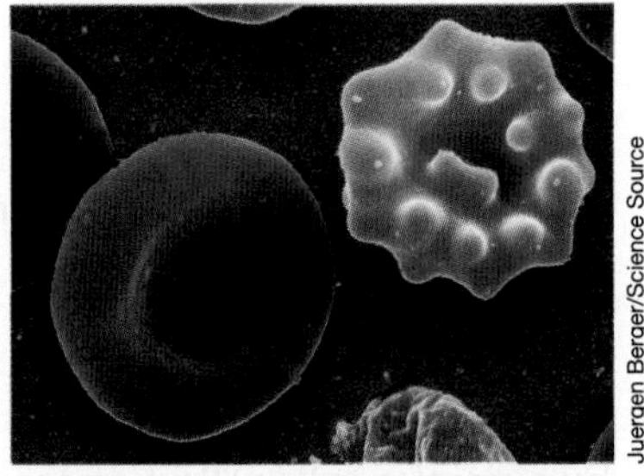

Juergen Berger/Science Source

The *Plasmodium* parasite reproduces inside infected red blood cells, giving them this lumpy appearance.

? The northern part of Australia has a warm and wet (tropical) climate. Why does the malaria parasite not appear to be transmitted there?

animals. Other protists produce valuable compounds—for example, red algae produce carrageenan, a polysaccharide that is used as a thickening agent for ice cream and yogurt.

Revised Beginnings

Protists have a unique claim to fame among eukaryotic organisms: they were the first eukaryotic organisms on the planet, and thrived here long before multicellular plants, animals, and fungi came along. Without protists, the other eukaryotes on the planet—including us—would not be here.

Scientists used to think that certain existing protists represented a kind of living relic—the closest thing to what ancient eukaryotes must have been like. *Giardia lamblia*, for example, is a unicellular parasite that lives in the intestinal tract of animals such as beavers; unsuspecting campers who drink untreated pond water can contract giardiasis (also called beaver fever). *Giardia* appears to lack mitochondria—the eukaryotic cell's power plants—and their absence suggested to some researchers that a *Giardia*-like organism may have been the primordial organism that, 2 billion years ago, engulfed a free-living bacterium to form the first mitochondrion through endosymbiosis (see **Milestone 1: Scientific Rebel**).

More recent evidence indicates that *Giardia* do have remnants of mitochondria and mitochondrial genes; they've just shrunk over evolutionary time, leaving truncated versions.

This kind of swapping and rearranging of organelles among early protists has left some remarkable legacies among today's eukaryotic organisms. And some of these unique attributes are proving useful when it comes to treating parasite-caused infections in humans.

The malaria parasite, for example, bears within its cell the remnants of a chloroplast that it borrowed from a photosynthesizing protist many millions of years ago. This lingering acquisition, called an apicoplast, still performs a crucial function in the parasite, even though it no longer photosynthesizes—and indeed, does not need to since the parasite lives in the dark inside animal hosts. That dependence makes it vulnerable to certain weed killers that target chloroplast enzymes. Scientists think that these or similar drugs might be the next potent weapons against the ever-evolving malaria parasite. Clinical trials of some of these drugs are under way (**INFOGRAPHIC M6.4**).

Battling Resistance

To help lower the risk of drug resistance, malaria is usually treated with a combination of drugs—typically, artemisinin and a quinine derivative such as piperaquine. Thanks to such combination therapy, WHO estimates that the incidence of malaria has fallen by 37% globally, and mortality rates have decreased by 60%. That's about 6 million deaths that have been avoided since 2001.

In 2015, 88% of malaria cases and 90% of malaria deaths occurred in sub-Saharan Africa.

But there are worrisome signs that malaria is once again becoming resistant to our most effective drugs. In parts of Cambodia and Thailand, *P. falciparum* has developed resistance to both artemisinin and the drugs with which it's combined. If these resistant strains spread to other regions, particularly sub-Saharan Africa, the consequences could be devastating. For that reason, WHO has

INFOGRAPHIC M6.4 Unique Protist Traits Can Be Pharmaceutical Targets

Many protists have the "signatures" of ancient endosymbioses. These include nuclear genes transferred from various symbionts, and remnants of chloroplasts. While these chloroplast remnants do not function in photosynthesis in *Plasmodium*, they do provide targets for drugs that target chloroplast enzymes.

Ancient endosymbiosis

A photosynthetic protist was engulfed by a nonphotosynthetic protist millions of years ago. Some of the genes from the engulfed cell and its chloroplast were transferred to the new host nucleus.

Modern protist with apicoplast

Over time, what remains of the chloroplast is an apicoplast that has limited DNA but provides a few essential functions for the protist.

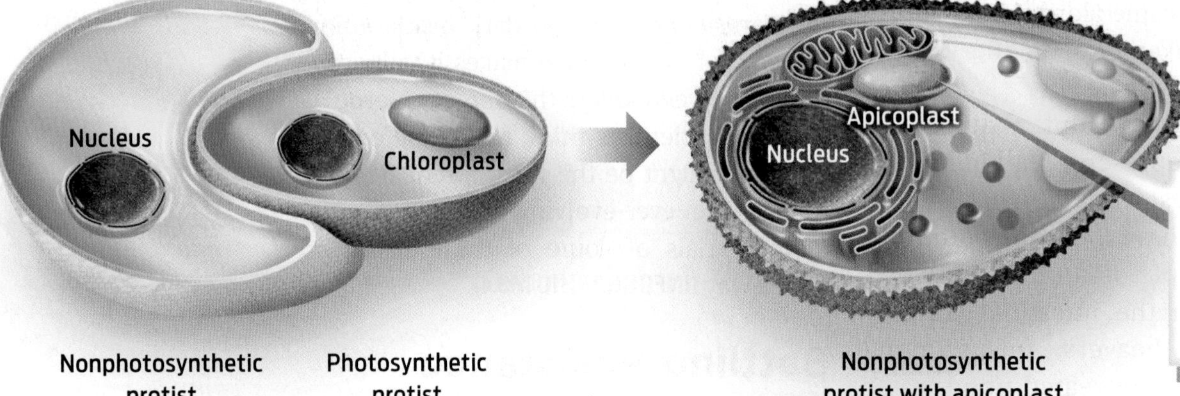

? Would you predict the nonphotosynthetic protist host (far left) to be susceptible to the antimalarial herbicide before it engulfed the photosynthetic endosymbiont? Why or why not?

endorsed a plan to eradicate *P. falciparum* in this region of Asia by 2030.

But its success may ultimately hinge on a better understanding of how these tiny protists evolved and where they fit in the overall eukaryotic tree of life. As the Chinese military general Sun Tzu once said, "the first rule of war is to know your enemy." ■

MORE TO EXPLORE

- Tu, Y. (2015) Nobel Prize lecture, "Discovery of Artemisinin: A Gift from Traditional Chinese Medicine to the World."
- Milius, S. (2015) The tree of life gets a makeover: Schoolroom kingdoms are taking a backseat to life's supergroups. *ScienceNews*. July 29.
- Yong, E. (2014) How malaria defeats our drugs. *Mosaic Science*.
- MacRae, J. I., et al. (2012) The Apicoplast: A Key Target to Cure Malaria. *Current Pharmaceutical Design* 18(24):3490–504.

MILESTONES IN BIOLOGY 6 Test Your Knowledge

1 About mosquitoes:

a. How is *Plasmodium falciparum* dependent on the *Anopheles* mosquito?

b. What is the evolutionary relationship between *Anopheles* mosquitoes and *Plasmodium falciparum*? (Hint: Consider their current classifications.)

2 Historically, *Phytophthora infestans* was classified as a fungus and *Pneumocystis carinii* and *Trypansoma brucei* were classified as protists. Review and explain the basis for their current classifications.

3 What are the practical implications of the reclassification of *Pneumocystis carinii*? Be specific in your answer.

4 Insecticides and drugs are part of antimalarial efforts. For each of the substances listed below, state whether it is directed against the mosquito or the parasite, and describe any current challenges to its use.

a. artemisinin

b. DDT

c. quinine

5 Why is climate change an important factor to consider in the context of malaria?

19 Human Evolution

Courtesy of American Anthropological Association

SKIN DEEP

Science redefines the meaning of racial categories

DRIVING QUESTIONS

1. What contributes to human skin color, and why is there so much variation in skin color among different populations?
2. Where did the earliest humans evolve, and how do we know?
3. What can genetics and the fossil record tell us about human evolution?

WHEN BARACK OBAMA WAS ELECTED FOR HIS FIRST TERM IN 2008, he was hailed as America's first black president. When Tiger Woods won the Masters Golf Tournament in 1997, he was lauded as the first black man to win. When Halle Berry won an Oscar in 2001 for best actress, she was commended as the first black woman to win in that category.

Why was skin color so remarkable? A 250-year history of slavery and racial discrimination in the United States has left a bitter legacy. More than 150 years after slavery was legally abolished in the United States, black Americans are still underrepresented in positions of power and prestige. Although the reasons for this underrepresentation are complex, the recognition of the achievements of Obama, Woods, and Berry signaled a major change: barriers to social advancement were beginning to come down.

Yet, to shoehorn any of these three people into a simple racial category is misleading: Barack Obama was born to a white mother and a black African father; Tiger Woods's background includes African, Chinese, Dutch, and Thai forebears; Halle Berry was born to a white mother and an African-American father. So what does the term "black" really mean?

Historically, racial categories like "white," "black," or "colored" were employed by one group to maintain power over another and to justify forms of oppression, including slavery. In the United States, racial categories were reinforced by laws like the "one drop" rule adopted by several states in the 1910s and 1920s, which held that any American with even "one drop" of African blood was to be considered black. People then continued to use these categories and their connotations to justify racial discrimination and, in some places, racial segregation.

Though social and political attitudes have changed, people continue to invoke racial categories like "black" or "white" for various reasons, including simple physical description. From a biological perspective, however, it is increasingly clear that racial categories have little meaning. Biologically distinct human races do not exist. Groups of people can and do share similar physical characteristics, such as skin color and other features, but these superficial differences obscure how fundamentally similar human beings are.

In fact, genetic studies comparing regions of the human genome from person to person show that each person's DNA is 99.9% identical to that of any other unrelated person. Of the 3 billion or so nucleotide base pairs making up the human genome, about 2,997,000,000 are exactly the same between any two individuals. It is only in the remaining 0.1% of our DNA that we differ, and it is this tiny fraction that accounts for the diversity of traits we see from person to person (**INFOGRAPHIC 19.1**).

INFOGRAPHIC 19.1 Humans Are Genetically Similar

Humans are 99.9% similar at the level of our DNA sequence. This similarity between humans includes all of the traits that define us as human. We are overwhelmingly more similar to one another than we are different and cannot be categorized into racial groups based on genetic analysis.

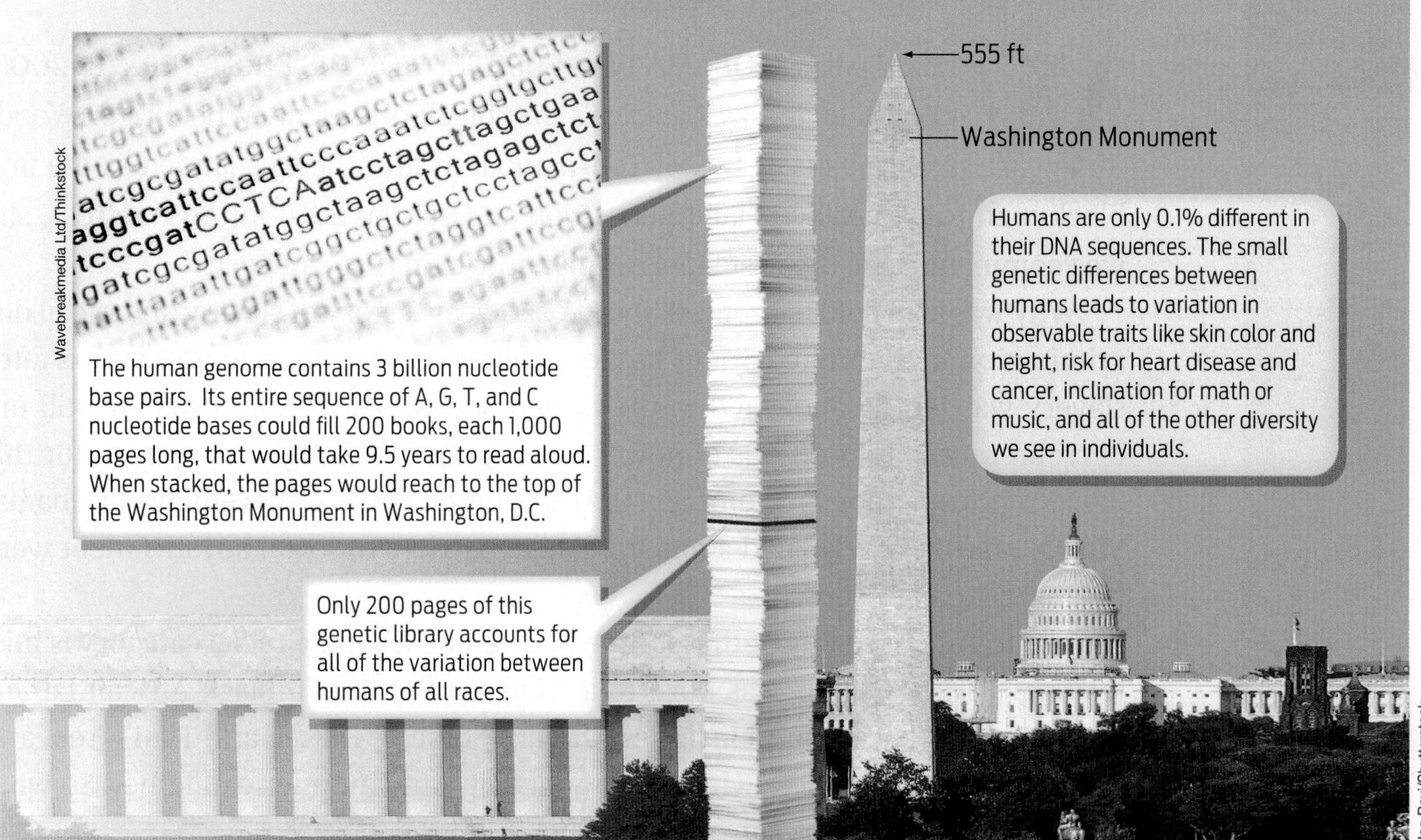

555 ft
Washington Monument
Wavebreakmedia Ltd/Thinkstock
The human genome contains 3 billion nucleotide base pairs. Its entire sequence of A, G, T, and C nucleotide bases could fill 200 books, each 1,000 pages long, that would take 9.5 years to read aloud. When stacked, the pages would reach to the top of the Washington Monument in Washington, D.C.
Humans are only 0.1% different in their DNA sequences. The small genetic differences between humans leads to variation in observable traits like skin color and height, risk for heart disease and cancer, inclination for math or music, and all of the other diversity we see in individuals.
Only 200 pages of this genetic library accounts for all of the variation between humans of all races.
Joe Ravi/Shutterstock

? Given variation between humans, how many nucleotide differences would you predict to find in a 1,000-nucleotide sequence in two unrelated people?

What's more, these differences have all come about very recently–in the blink of an eye, evolutionarily speaking. The human species itself–*Homo sapiens*–has only been around for about 200,000 years. In that time, humans have evolved many traits that helped them survive the environments they encountered as they migrated around the globe. People with African ancestry, for example, carry with high frequency an allele that helps them resist malaria, a disease endemic to Africa. An allele common among Northern Europeans enables them to digest milk better than other populations do, an indication that at some point in history, dairy products provided an important source of nutrition to this population. Tibetans have a high frequency of an allele that helps their red blood cells compensate for the low oxygen level in their high-altitude environment.

Skin color is another example of a trait that differs among human groups. For many years, scientists had wondered why different skin colors evolved. But it wasn't until the work of Nina Jablonski, an anthropologist at Pennsylvania State University, that they found a convincing answer.

MELANIN
A pigment produced by a specific type of skin cell that gives skin its color.

The Evolution of Skin Color

More than a decade ago, Jablonski and her husband, George Chaplin, a geographer, set out to understand why human populations evolved varying skin tones. They knew that skin tone largely reflects the amount of **melanin**, a pigment present in the skin. People naturally produce different levels of melanin. More melanin yields darker skin, less melanin yields lighter skin. Skin also responds to sunlight by producing more melanin and temporarily becoming darker **(INFOGRAPHIC 19.2)**.

INFOGRAPHIC 19.2 Melanin Influences Skin Color

Melanocytes are a type of cell located in the epidermis, the outermos layer of skin. Melanocytes make the pigment melanin and deposit it into other cells in the skin. A person's skin color depends largely on the amount and type of melanin that his or her skin melanocytes produce. Sunlight can also temporarily increase the amount of melanin in a person's skin.

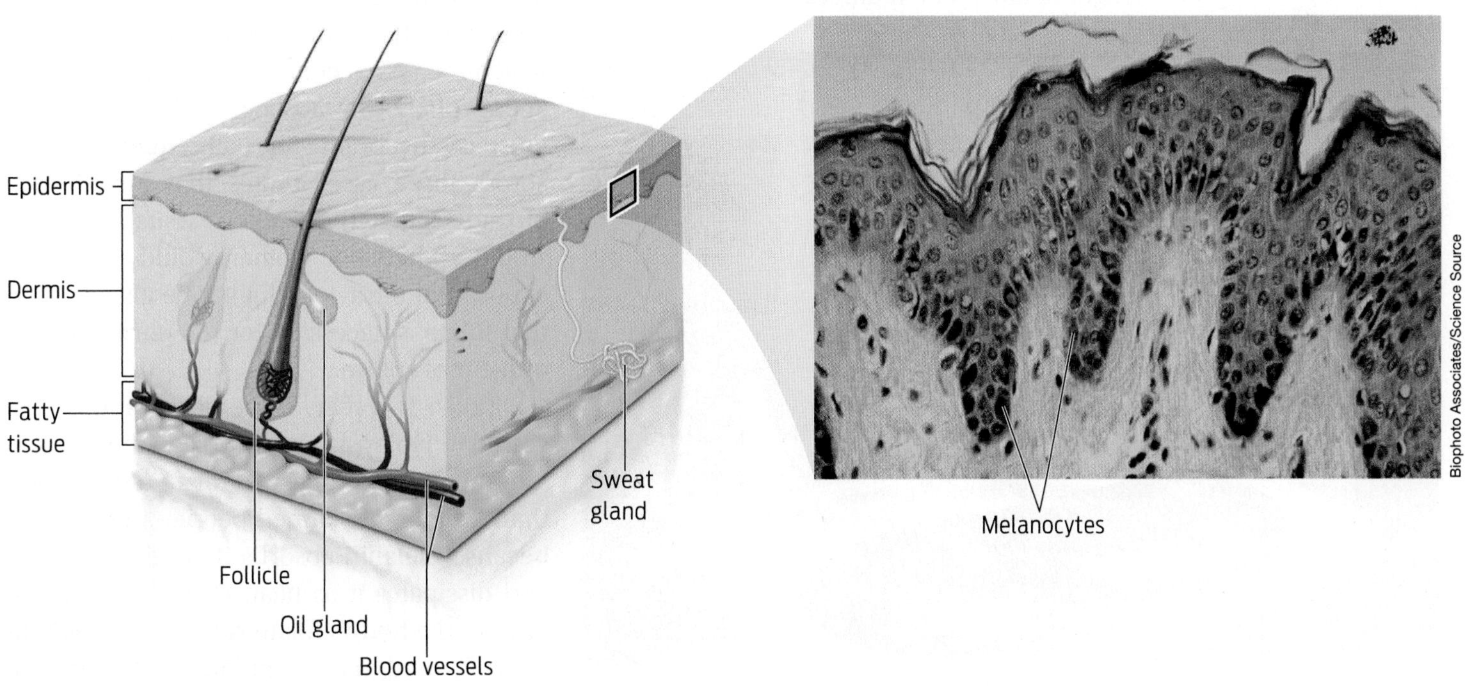

? For sunlight to increase melanin production in skin, what cell type must be active following exposure to sunlight?

Mark Wilson/Getty Images

In certain environments, is there an evolutionary advantage to light or to dark skin?

Jablonski and Chaplin also knew that, in general, skin tone correlates with geography: people from regions closer to Earth's poles tend to be lighter-skinned and those from areas closer to the equator tend to have darker skin. Jablonski wanted to understand this, so she searched the scientific literature. Might there be an evolutionary advantage to having light or dark skin in different environments?

FOLATE
A B vitamin also known as folic acid; folate is an essential nutrient, necessary for basic cellular processes such as DNA replication and cell division.

Nina Jablonski

George Chaplin and Nina Jablonski examine a map of predicted human skin color based on UV light intensity.

She found her first clue in a 1978 study showing that, in light-skinned people, an hour of intense sunlight can halve the level of an important vitamin called **folate**. Folate, also known as folic acid, is an essential nutrient, necessary for basic cellular processes like DNA replication and cell division.

Then, at a seminar, Jablonski learned that low folate levels can cause severe birth defects such as spina bifida, a condition in which the spinal column does not close, and anencephaly, the absence at birth of all or most of the brain. She subsequently came across three case studies that linked such birth defects to the mothers' visits to tanning studios, where the women would have been exposed to ultraviolet light. She also learned that folate is necessary for the normal development of sperm.

Taken together, these observations suggested to Jablonski that people with light skin are more vulnerable to folate destruction than are darker-skinned people—presumably because melanin absorbs damaging UV light and dissipates it as heat. Could the need to protect the body's folate stores from UV light have favored the evolution of darker skin shades? The supporting evidence was compelling. But then what was the advantage of having light skin at all, as many populations today do?

Jablonski considered a hypothesis first proposed in the 1960s by biochemist W. Farnsworth Loomis, who suggested that **vitamin D** might play a role in the evolution of skin color. Unlike folate, which is destroyed by excess sunlight, the production of vitamin D requires ultraviolet light. Vitamin D is crucial for good health: it helps the body absorb calcium and deposit it in bones. During pregnancy women need extra vitamin D to nourish the growing embryo. And because vitamin D is so important for healthy bone growth, too little might cause bone distortion, and a distorted pelvis would make it difficult for a woman to bear children **(INFOGRAPHIC 19.3)**.

VITAMIN D
A fat-soluble vitamin required to maintain a healthy immune system and to build healthy bones and teeth. The human body produces vitamin D when skin is exposed to UV light.

INFOGRAPHIC 19.3 Folate and Vitamin D Are Necessary for Reproductive Health

Folate, also known as folic acid, is especially critical during periods of rapid cell division, including embryonic and fetal development. Vitamin D is important for the absorption of calcium and phosphate from the small intestine, and also for bone mineralization. Folate is destroyed by UV, while vitamin D production is stimulated by UV. Skin color plays a role in folate protection and vitamin D production, with the balance between these depending on the amount of UV in the environment.

Low-UV-Light Environment Favors Light Skin

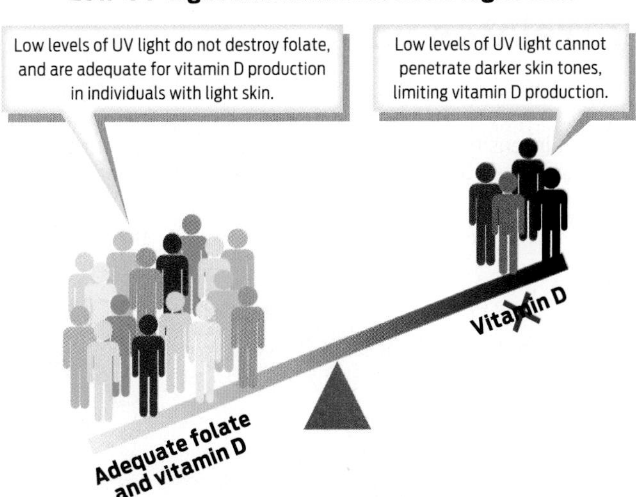

Low levels of UV light do not destroy folate, and are adequate for vitamin D production in individuals with light skin.
Low levels of UV light cannot penetrate darker skin tones, limiting vitamin D production.
Adequate folate and vitamin D
Vitamin D

High-UV-Light Environment Favors Dark Skin

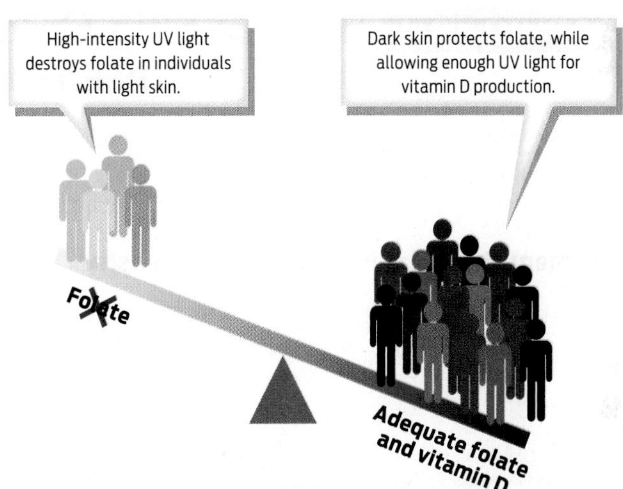

Vitamin D deficiency:

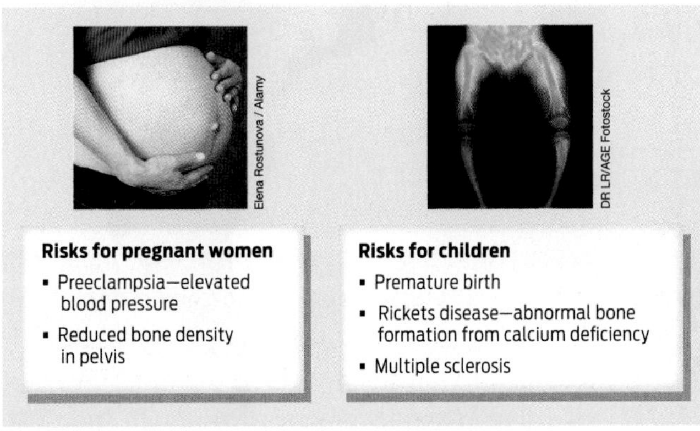

Elena Rostunova / Alamy
DR LR/AGE Fotostock
Risks for pregnant women
• Preeclampsia—elevated blood pressure
• Reduced bone density in pelvis
Risks for children
• Premature birth
• Rickets disease—abnormal bone formation from calcium deficiency
• Multiple sclerosis

Folate deficiency:

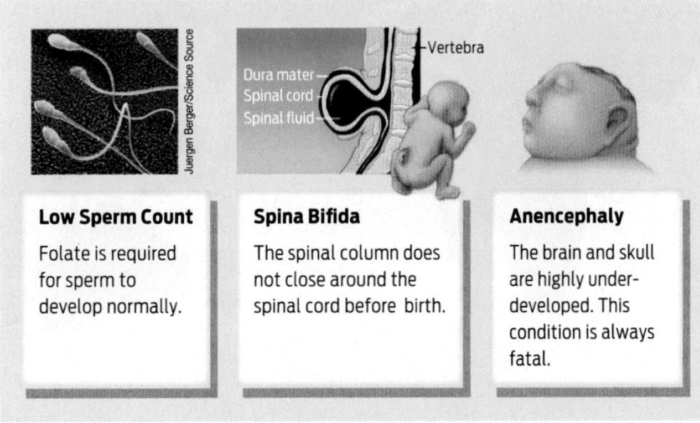

Juergen Berger/Science Source
Vertebra
Dura mater
Spinal cord
Spinal fluid
Low Sperm Count
Folate is required for sperm to develop normally.
Spina Bifida
The spinal column does not close around the spinal cord before birth.
Anencephaly
The brain and skull are highly under-developed. This condition is always fatal.

? Does dark skin confer more of an advantage in a low- or a high-UV environment? Explain your answer.

Building on this previous work, in 2000 Jablonski and Chaplin compared data on skin color in indigenous populations from more than 50 countries to levels of global ultraviolet light as measured by NASA satellites. They found a clear correlation: the weaker the ultraviolet light, the fairer the skin—a compelling suggestion that both dark and light skin are linked to levels of global sunlight. They published their results in the *Journal of Human Evolution*.

The researchers now had a complete hypothesis: light skin provided an evolutionary advantage in sun-poor parts of the world because it helped the body produce vitamin D, while dark skin was favored in sunny regions because it helped protect the body's folate stores. The body's need to balance levels of these two important nutrients given varied levels of UV light explains why there is so much variation in skin tone around the globe **(INFOGRAPHIC 19.4)**.

Since the publication of Jablonski and Chaplin's work, many other scientists have tested this hypothesis, and it is now the most widely accepted explanation for the evolution of human skin color. As Jablonski points out, "It synthesizes the available information on the biology of skin from anatomy, physiology, genetics, and epidemiology, and has not been contradicted by any subsequent data."

Out of Africa

If light and dark skin tones developed over time in tandem with human migrations, then that raises the question of what the earliest human

INFOGRAPHIC 19.4 Human Skin Color Correlates with UV Light Intensity

Nina Jablonski and George Chaplin used NASA satellite measurements of UV light intensity to predict the amount of skin pigment that would best block harmful UV rays yet still enable the body to produce sufficient vitamin D in populations around the globe. Their predictions closely match actual skin color variations around the world.

Predicted pigmentation of skin based on UV intensity

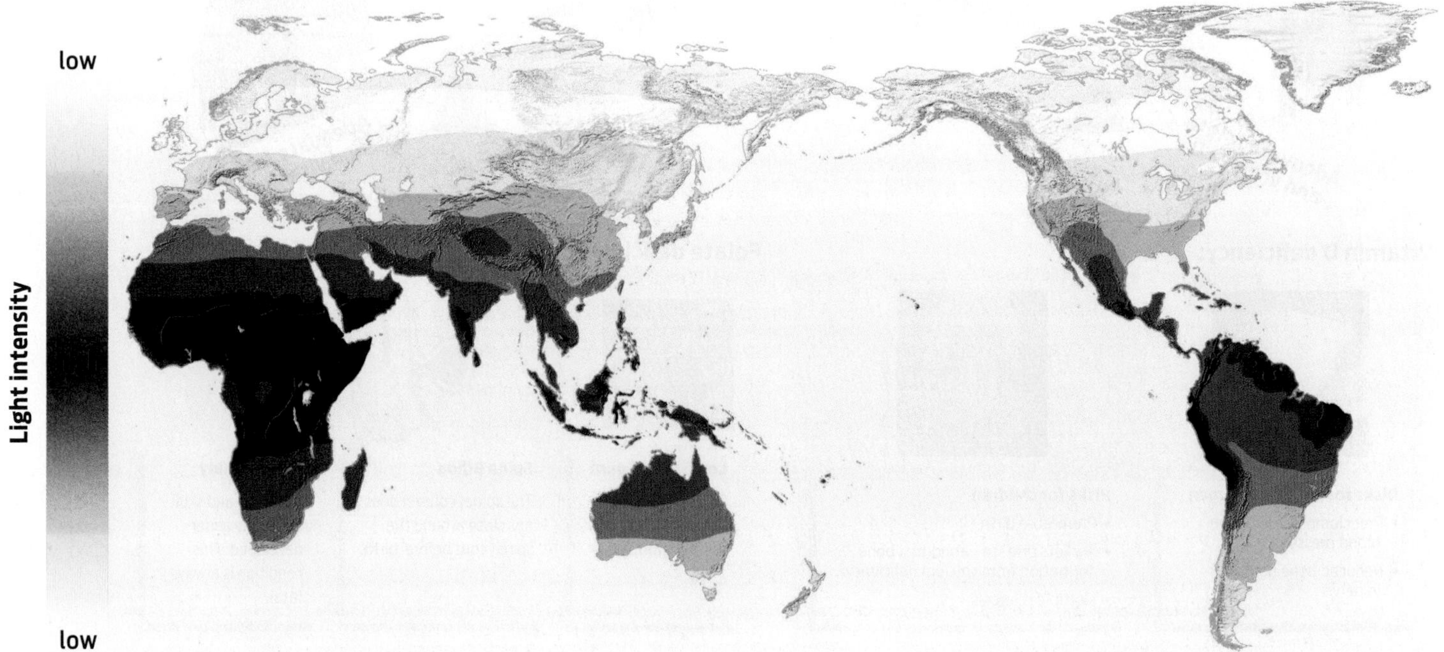

Data from Chaplin, G. (2004). *J.Phys. Anthro.* 125:292-302.
Map: Emmanuelle Bournay UNEP/GRID-Arendal

? In low-UV environments, what is the predominant pressure selecting for light skin?

skin tone was, and where humans began their journey. This is another way of asking: where did modern humans first evolve?

There are two main hypotheses concerning human origins. Both place the human birthplace in Africa, but they differ in how recently that birth occurred. According to one hypothesis (the multiregional hypothesis), ancient humans began migrating from Africa about 2 million years ago, journeying subsequently to Europe and Asia. Over the next 2 million years, these groups continued to evolve in these regions, as a single species, eventually becoming modern humans.

The second hypothesis (the out-of-Africa hypothesis) is that anatomically modern humans first appeared about 200,000 years ago, in Africa, and spread around the world from there—displacing earlier members of our genus, which became extinct as a result. Deciding between these two hypotheses remained controversial until the 1980s, when genetic evidence started to make the picture clear.

In 1987, a team of geneticists led by Allan Wilson of the University of California at Berkeley used **mitochondrial DNA (mtDNA)**—genetic material we inherit solely from our mothers—to construct an evolutionary tree of humanity. Mitochondrial DNA is DNA located in the mitochondria of our cells. Unlike nuclear DNA, which is inherited from both parents and which undergoes recombination during meiosis, mtDNA passes from mothers to offspring essentially unchanged. That's because sperm do not contribute their mitochondria to the newly formed zygote (**INFOGRAPHIC 19.5**).

Like nuclear DNA, mtDNA mutates at a fairly regular rate, but much faster. A mother with a mutation in her mtDNA will pass it to all her children, and her daughters will pass it to their children in turn. Because these mutations pass down without being combined and rearranged with paternal mitochondrial DNA, mtDNA is a powerful tool by which to track human ancestry back through thousands of generations.

Wilson and his colleagues Rebecca Cann and Mark Stoneking collected mtDNA from 147 contemporary individuals from Africa, Asia, Australia, Europe, and New Guinea. On the basis of the mtDNA sequence differences—that is, the number of individual nucleotide differences from one person to the next—researchers were able to determine how closely or distantly related the individuals were: individuals sharing nearly identical sequences are more closely related than individuals with many sequence differences between them. As mitochondrial DNA mutates at a constant rate, the researchers could also tell, based on the number of differences, how long it had been since groups diverged from one another. Using this information, the researchers created an evolutionary tree, grouping first individuals that shared the most similar mtDNA sequences, and then grouping these groups with respect to one another. A computer program helped them choose the statistically most probable tree.

> If every person on the planet were to construct a family tree that listed every female ancestor for thousands of generations back in time, they would all eventually converge at a single common female ancestor.

The researchers found that branches of the tree from all five geographical areas could be traced back to a single female ancestor who lived in eastern Africa some 200,000 to 150,000 years ago. In other words, if every person on the planet were to construct a family tree that listed every female ancestor for thousands of generations back in time, they would all eventually converge at a single common female ancestor.

MITOCHONDRIAL DNA (mtDNA)
The DNA within mitochondria; it is inherited solely from the mother.

INFOGRAPHIC 19.5 Mitochondrial DNA Is Inherited from Mothers

When egg and sperm fuse during fertilization, sperm contribute only nuclear DNA to the newly formed zygote. The egg provides all other organelles, including mitochondria. Consequently, only mothers contribute mitochondrial DNA (mtDNA) to their children.

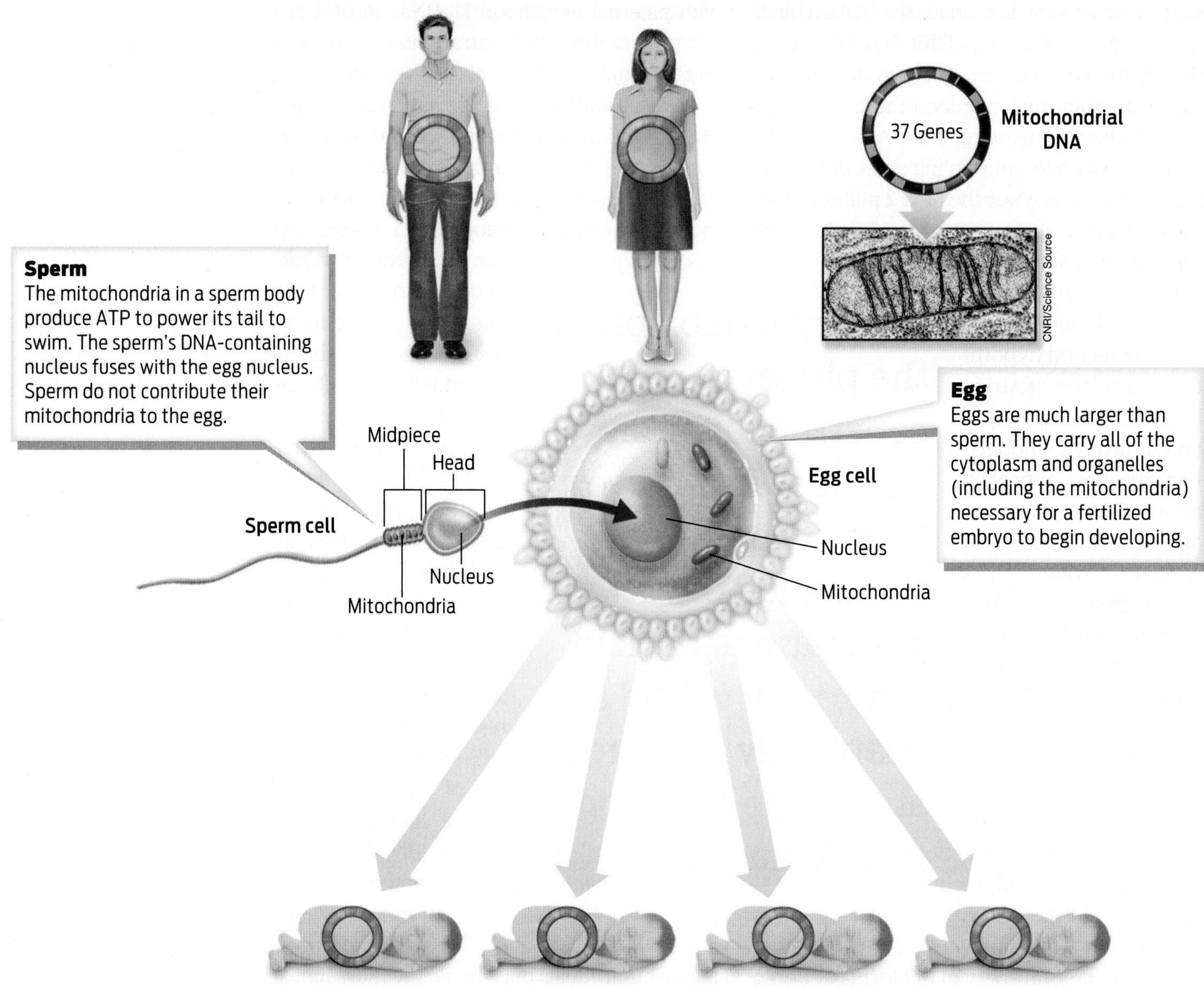

37 Genes
Mitochondrial DNA
CNRI/Science Source
Sperm
The mitochondria in a sperm body produce ATP to power its tail to swim. The sperm's DNA-containing nucleus fuses with the egg nucleus. Sperm do not contribute their mitochondria to the egg.
Midpiece
Head
Sperm cell
Nucleus
Mitochondria
Egg cell
Nucleus
Mitochondria
Egg
Eggs are much larger than sperm. They carry all of the cytoplasm and organelles (including the mitochondria) necessary for a fertilized embryo to begin developing.

? If biologists used mtDNA to trace the evolutionary history of a group of people, would they be following mothers, fathers, or both?

As a newspaper reporter put it, the researchers had found "Mitochondrial Eve."

This was clear evidence in support of a recent African origin for our species, and it made a big splash in popular culture. (*Newsweek* featured an African Adam and Eve on its cover.) But the finding also led to confusion. While catchy, the nickname "Eve" is misleading in a number of respects. This single female common ancestor wasn't the only female living at the time; she was merely one female in a population of many ancient humans. Nor was she the only female to leave descendants. But Eve's mitochondrial DNA is the only mitochondrial DNA that modern humans still carry today. How can this be? The reason is simple: while other females living at the time also had descendants, the lines of these descendants either died off, perhaps in a bottleneck event (see Chapter 14), or left only sons (who do not pass on their mtDNA to offspring). When that happened, their mitochondrial DNA lines became extinct. All humans today can trace their mitochondrial DNA back to Mitochondrial Eve (**INFOGRAPHIC 19.6**).

Interestingly, the tree Wilson generated had two major branches: one that leads to individuals now living in all five of the regions studied–Asia, Australia, Europe, New Guinea, and Africa–and one that leads only to modern-day Africans. The mtDNA of people on the exclusively African branch had acquired twice as many mutations as the mtDNA of people on the rest of the tree. The most likely interpretation of these data, the scientists reasoned, was that the African mtDNA had had more time to accumulate mutations, and was consequently older, evolutionarily speaking. This would mean that humans originated in Africa, where they likely formed several ancestral populations. After some period of time, one group of Africans left the continent, and their

INFOGRAPHIC 19.6 Modern Human Populations Share a Common Female Ancestor

Many women of Eve's generation left descendants, but only Eve's mitochondrial DNA survives among humans today.

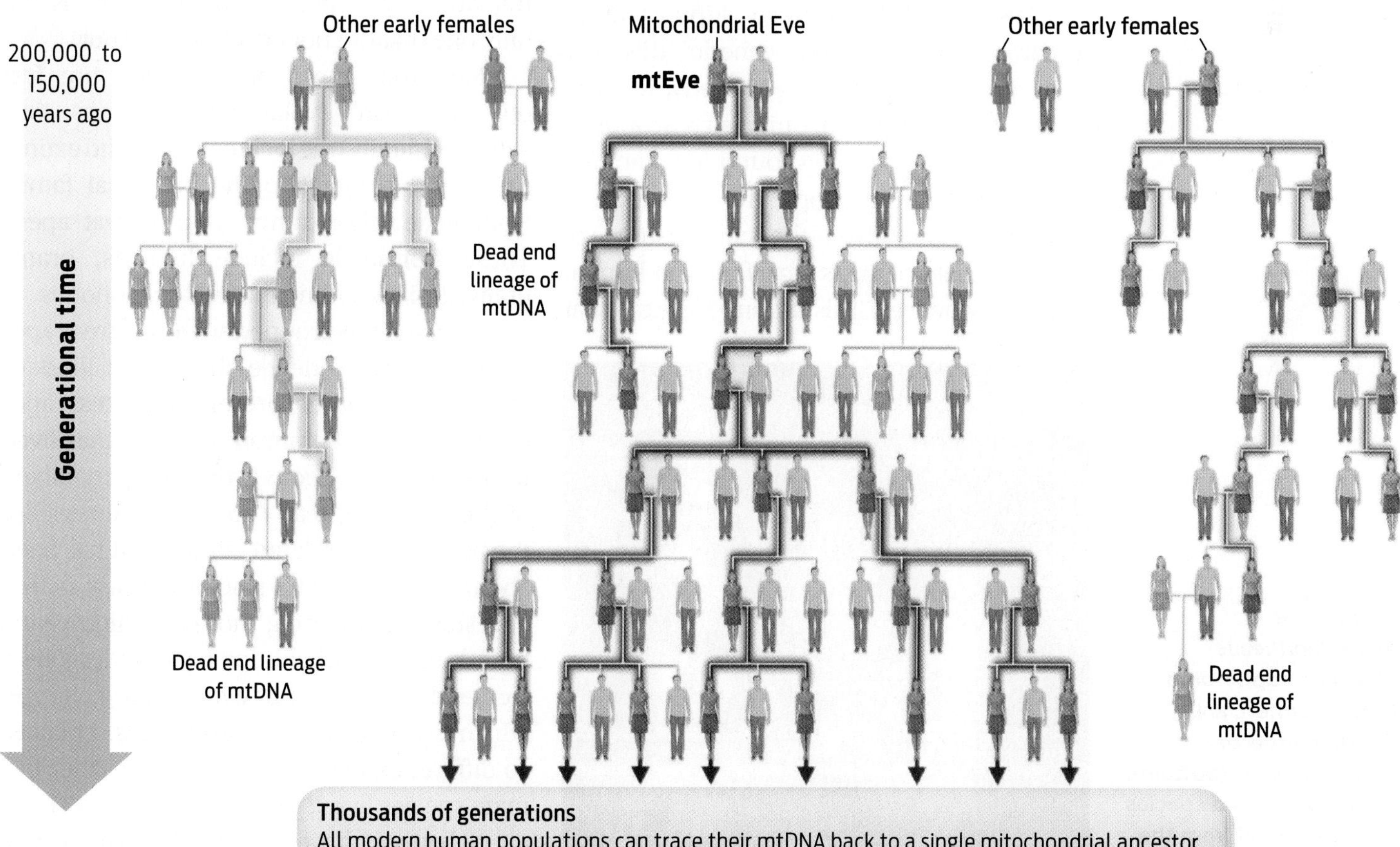

? If a male with the yellow mtDNA as shown here married a female with the purple mtDNA shown, which mtDNA would be passed on to their offspring?

HOMINIDS
Any living or extinct member of the family Hominidae, the great apes—humans, orangutans, gorillas, chimpanzees, and bonobos.

descendants continued to migrate to other continents, eventually becoming the ancestors of modern-day Asians, Australians, and Europeans. Evidence suggests this migration began around 70,000 years ago.

Since Wilson's study, additional evidence continues to back the out-of-Africa hypothesis. Fossils discovered in Ethiopia in 2003 and 2005 represent the oldest known fossils of modern humans, and they are 160,000 and 195,000 years old, respectively. Both sets of remains date precisely from the time Wilson and his colleagues think that Mitochondrial Eve lived in eastern Africa. The fossil discoveries provide evidence that anatomically modern humans were living in that region around the same time that Mitochondrial Eve lived, and provide further evidence that modern humans originated in Africa.

This hypothesis is also supported by research that sampled genetic diversity from nuclear DNA. In a 2008 study, Richard Myers, of the Stanford University School of Medicine, and colleagues found less and less genetic variation in people the farther away from Africa they lived—the same pattern of variation that scientists have found in human mtDNA sequences. This finding suggests that as each small group of people broke away to explore a new region, it carried only some of the parent population's alleles. Consequently, genetic diversity decreased in tandem with the distance people traveled away from Africa—a classic example of the founder effect described in Chapter 14 (**INFOGRAPHIC 19.7**).

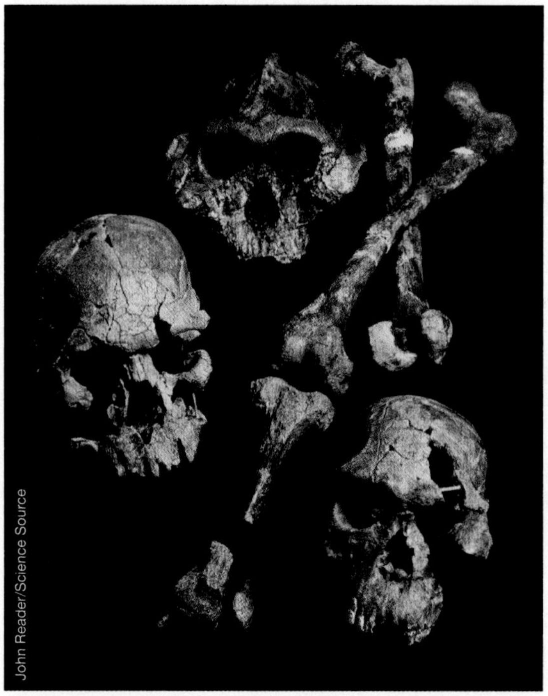

John Reader/Science Source

Fossil skulls of *Australopithecus africanus* (top), *Homo habilis* (center), and *Australopithecus robustus boisei* (bottom). All three hominid species were extracted from the same excavation site in East Turkana, Kenya, indicating that these species all lived at the same time, around 1.5 million years ago.

Becoming Human

Mitochondrial Eve shows that *Homo sapiens* is a relatively young species, having first evolved in Africa between 200,000 and 150,000 years ago. But Eve's family—our family—was just one of many other human families that existed at the time, in Africa and in other places around the world. The term "human" refers to our genus name—*Homo*. There were many other members of the genus *Homo*—many other ancient humans—that once roamed the globe. But our single species—*Homo sapiens*—is the only one that did not eventually die out.

Moreover, these other *Homo*s are just our immediate evolutionary relatives. Our extended family tree includes living and extinct primates belonging to the biological family Hominidae (also known as the great apes). These **hominids** include humans, orangutans, gorillas, chimpanzees, and bonobos.

Humans are grouped with the great apes because fossil evidence shows that modern humans and other present-day great apes evolved from a common ancestor that lived 13 million years ago. Of the living members of this group, humans and chimpanzees are the most closely related, although it has been about 7 million years since their last shared ancestor lived. During those 7 million years, both humans and chimps have undergone a tremendous amount of evolutionary change, which is why living humans look and behave so differently from chimps—or any other primate species living today.

Scientists haven't yet discovered fossil remains of the last common ancestor of chimps and humans. However, in October 2009 the first analyses of fossil remains of a 4.4-million-year-old hominid, *Ardipithecus ramidus*, nicknamed Ardi, were published.

INFOGRAPHIC 19.7 Out of Africa: Human Migration

Genetic evidence suggests that the earliest modern humans originated and evolved for thousands of years in Africa before a group of them migrated to the other continents. Because African populations are older, they have more genetic diversity. As migrating descendants only took a fraction of the alleles present in Africa, there is less genetic variation in other regions.

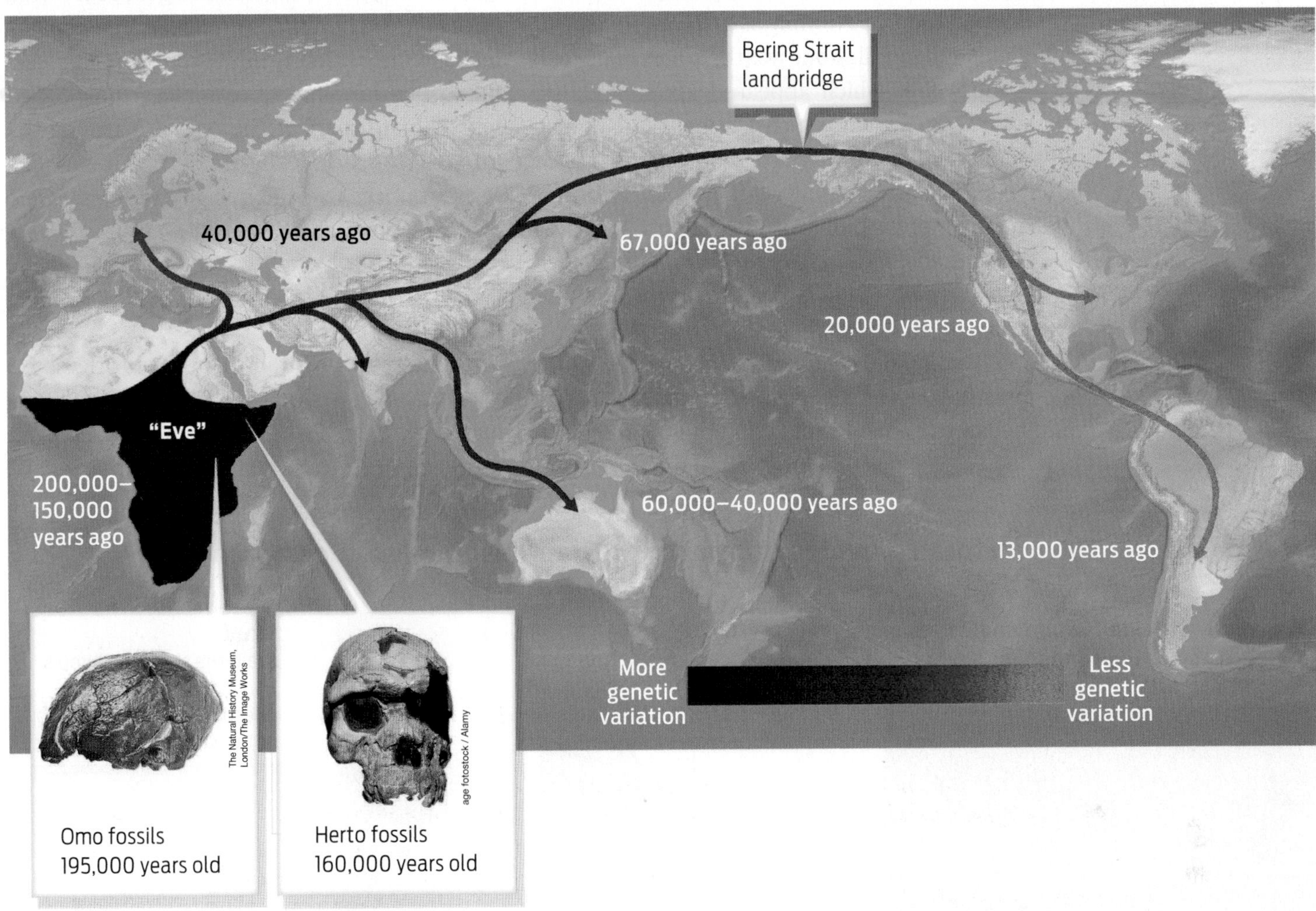

Omo fossils
195,000 years old

Herto fossils
160,000 years old

? If humans had originated and evolved in Australia and then, after thousands of years, migrated to Africa, what would you expect to be the level of genetic diversity in Australian and African populations?

Ardi's remains are among the oldest hominid fossils so far discovered and, as such, give tantalizing clues to early human origins.

Among the defining characteristics of *Homo sapiens* are the ability to walk upright and the possession of a big brain. An upright gait meant the hands were free to make and use tools, and a big brain enabled *H. sapiens* to develop complex language. Which came first, a big brain or walking upright? Ardi provides the answer: Ardi had a small brain, suggesting that she could not use complex language. But Ardi's bones clearly show that she could walk upright without dragging her knuckles, while still able to maneuver on all fours in trees. The ability to walk upright therefore evolved first and was a major step toward becoming human.

The fossil record after Ardi also reveals some of the major milestones in human evolution. For example, artifacts found at various archeological sites indicate that simple tool use began approximately 2.6 million years ago, most likely when our hominid ancestors began eating meat from large animals. The first hominid tool-users were members of the genus *Australopithecus*. This genus walked upright and appears to have lived on the ground, rather than in trees, as evidenced by the lack of an opposable big toe, which had helped the early hominids grip branches.

Another milestone was the ability to use and control fire, which appeared about 800,000 years ago. Evidence such as ashes of plant matter and fragments of burned bone found inside South African caves show that *Homo erectus,* a member of our human genus, was likely the first species able to control fire. Using fire enabled *Homo erectus* to cook meat, to stay warm, and probably to fight off predators.

Finally, at some point between 800,000 and 200,000 years ago, hominid brain size began to expand rapidly. Studies of fossilized organisms from ocean sediments show that this was also a time of climate instability. The temperature dramatically shifted from high to low and back several times during this period. Scientists hypothesize that a larger brain would have enabled better communication and problem solving, which would have been very useful to our hominid ancestors as they coped with climate change. This was also around the time that anatomically modern humans like Mitochondrial Eve and our own species, *Homo sapiens*, appeared **(INFOGRAPHIC 19.8)**.

Selection for Skin Color

That anatomically modern humans originated in Africa suggests that the first humans likely had dark skin. Dark skin probably evolved

The Ardi fossil provides clues about the earliest human ancestors.

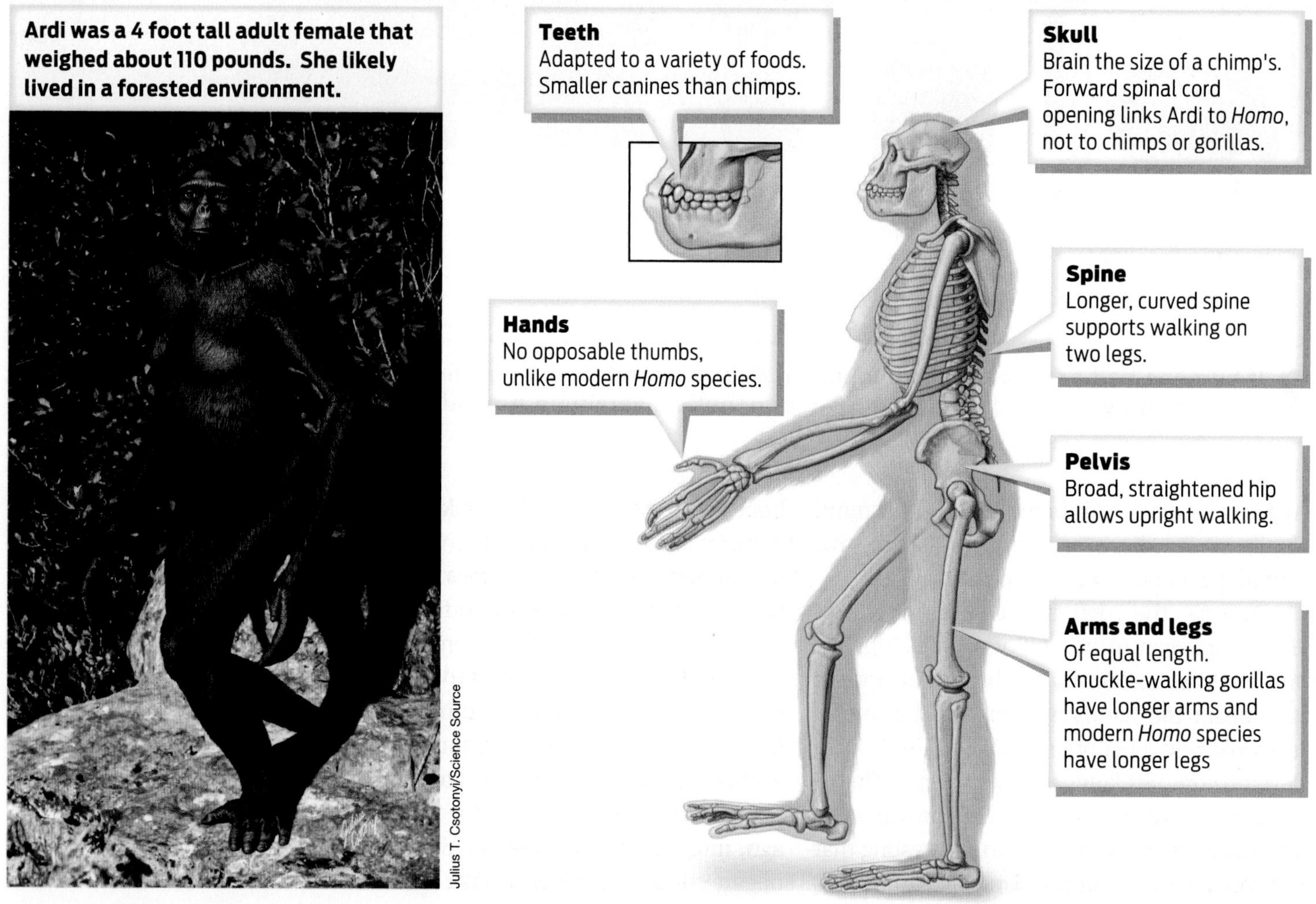

Ardi was a 4 foot tall adult female that weighed about 110 pounds. She likely lived in a forested environment.
Julius T. Csotonyi/Science Source
Teeth
Adapted to a variety of foods. Smaller canines than chimps.
Skull
Brain the size of a chimp's. Forward spinal cord opening links Ardi to Homo, not to chimps or gorillas.
Spine
Longer, curved spine supports walking on two legs.
Hands
No opposable thumbs, unlike modern Homo species.
Pelvis
Broad, straightened hip allows upright walking.
Arms and legs
Of equal length. Knuckle-walking gorillas have longer arms and modern Homo species have longer legs

INFOGRAPHIC 19.8 Humans: An Extended Family Tree

Homo sapiens is the only surviving lineage in the evolutionary history of humans. In other words, several hominids have existed or coexisted as related but distinct species in the past. The physical traits of modern-day humans, such as upright walking and big brains, reflect adaptations to evolutionary pressures.

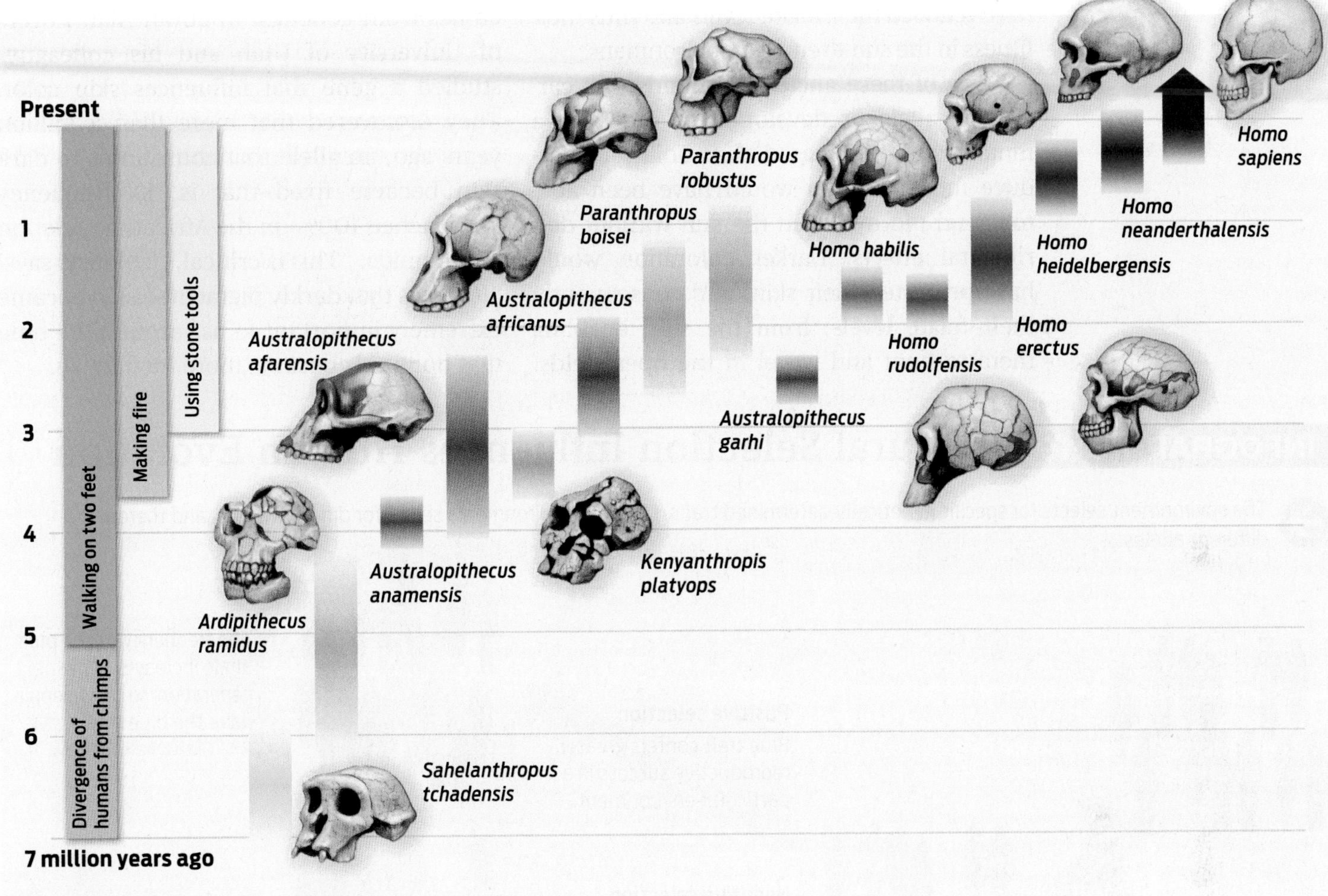

? Which human species were living at the same time as *Homo habilis*?

early in our history, in tandem with a reduction in body hair. Before dark skin evolved, our earliest ancestors would have had light skin, just as chimpanzees do today.

Fossil and genetic evidence suggests that about 2 million years ago hominids became "bipedal striders, long-distance walkers, and possibly even runners," according to Jablonski. To sustain such activities, hominids needed an effective cooling system, a feature they could have developed only by losing excessive body hair and gaining more sweat glands. In contrast, hairy chimpanzees, our closest living animal relatives, can sustain only short bouts of activity without getting overheated. "It's like sweating in a wool blanket," Jablonski explains. "After that blanket gets saturated, you can't lose very much heat."

Eventually, some factor–food scarcity, perhaps–forced ancient hominids out of the forests and into the open savannahs to hunt for food. Hominids with less hair and more sweat glands were likely better hunters because they could sustain long bouts of activity without getting overheated. Like modern-day chimpanzees, these hominids likely had fair skin under their hair. Without hair to protect their light skin, they were exposed to the intense African sun. And, according to Jablonski's hypothesis, exposure to the sun would have reduced their folate levels and thus their fitness in the sun-drenched environment.

Any of these ancient hominids that carried an allele or developed or inherited a mutation that increased their ability to produce more melanin would have been able to spend more time in the sun without detrimental effects. Darker coloration would have protected their skin, and consequently their folate levels, from the sun, enabling them to hunt and travel in the open fields. As a consequence, alleles for dark skin would have increased in frequency in the population (**INFOGRAPHIC 19.9**).

Evidence to support this hypothesis about the evolution of dark skin color comes from genetics. In 2004, Alan Rogers of University of Utah and his colleagues studied a gene that influences skin color. They discovered that more than a million years ago, an allele that contributes to dark skin became fixed—that is, its frequency approached 100%—in the African population of hominids. "This is critical," Jablonski says. "It shows that darkly pigmented skin became extremely important to us" around the time that hominids became more humanlike.

INFOGRAPHIC 19.9 Natural Selection Influences Human Evolution

The environment selects for specific genetically determined traits. Different environments select for different traits, and therefore different alleles.

1. Mutation introduces a new allele into a population.

2. Depending on the environment, the allele may or may not undergo selection.
Positive selection
Blue trait confers greater reproductive success in a particular environment.
Negative selection
Blue trait confers less reproductive success in a particular environment.
Neutral (no) selection
Blue trait does not influence reproductive success in a particular environment.

The frequency of the blue allele increases in the next generation, so more people have the blue trait.

The frequency of the blue allele decreases in the next generation, so fewer people have the blue trait.

The frequency of the blue allele does not change in the next generation, so the proportion of people with the blue trait remains the same.

? What kind of selection is predicted for alleles that contribute to dark skin in a low-UV environment?

The allele for darker skin was such an advantage in terms of survival and reproduction that hominids with darker skin left more offspring than their lighter-skinned relatives. Hominids born with light skin weren't able to survive and reproduce in great enough numbers for the trait to persist in the population. The allele for darker skin eventually increased in the population until it essentially reached 100%.

Populations that migrated north, away from the African sun, faced different environmental circumstances. Folate was not as easily destroyed in this lower-UV-light environment. But the high levels of melanin present in dark skin became a disadvantage; they prevented bodies from producing enough vitamin D. In this low-UV-light environment, fair skin allowed the body to soak up more ultraviolet light and produce essential vitamin D. Fair-skinned people thus were more fit and left more descendants than did dark-skinned people. Consequently, the frequency of light skin in northern climates increased with each generation.

If the earliest humans migrating from Africa to northern regions all had dark skin, where then did alleles for light skin come from? There are two possibilities. One is that the alleles were introduced as new mutations into the population either shortly before or shortly after humans migrated out of Africa and were subsequently maintained by natural selection. The other is that the alleles were still present in the African population, albeit at very low frequencies, and then subsequently became more common under selection in a new environment.

Both explanations may be true. Genetic studies show that the frequency of alleles for light skin increased and swept through populations as they migrated north—most likely more than once. Alleles of at least three different genes have been found to confer light skin in people from northwestern Europe and light skin in people from eastern Asia, suggesting that mutations for light skin arose independently and spread through those two populations separately. These mutations were genetically minor, yet phenotypically significant: single nucleotide changes in two of these genes have been shown to account for a large proportion of the difference in skin color between light- and dark-skinned people—a mere two nucleotides among the 3 billion total in our DNA. Skin color really is only skin deep **(INFOGRAPHIC 19.10)**.

An Evolving Explanation

There are other hypotheses for the evolution of skin tone. But of all of them, the folate-vitamin D hypothesis has the most supporting evidence and is consequently "the most reasonable," says Mark Stoneking of the Max Planck Institute for Evolutionary Anthropology in Leipzig, Germany, who collaborated on the Mitochondrial Eve study. In fact, Stoneking says, "Skin tone is one of the best examples of human evolution."

> "Skin tone is one of the best examples of human evolution."
> – Mark Stoneking

What's perhaps most remarkable is how quickly different skin tones arose—in the span of only a few thousand years. This is a clear example of how the phenotypic differences we see between human groups are the result of evolution acting on what amounts to very small genetic differences. The particular skin tone alleles that predominate in a particular population are ones that originally provided a selective advantage to its members by producing more or less melanin. Dark-skin alleles were favored in high-UV environments, while light-skin alleles were favored in low-UV environments. Skin color is thus an indicator of the geographic origin of our ancestors, but not much else.

Throughout human history, the lines between what we have come to call races have been fluid. Genetic studies show that

INFOGRAPHIC 19.10 The Evolution of Skin Color

Human skin color is an example of a trait that has undergone natural selection. Varying levels of UV light have selected for a range of skin tones around the globe. In each case, the amount of melanin represents a compromise between the need to protect folate and the need to make vitamin D.

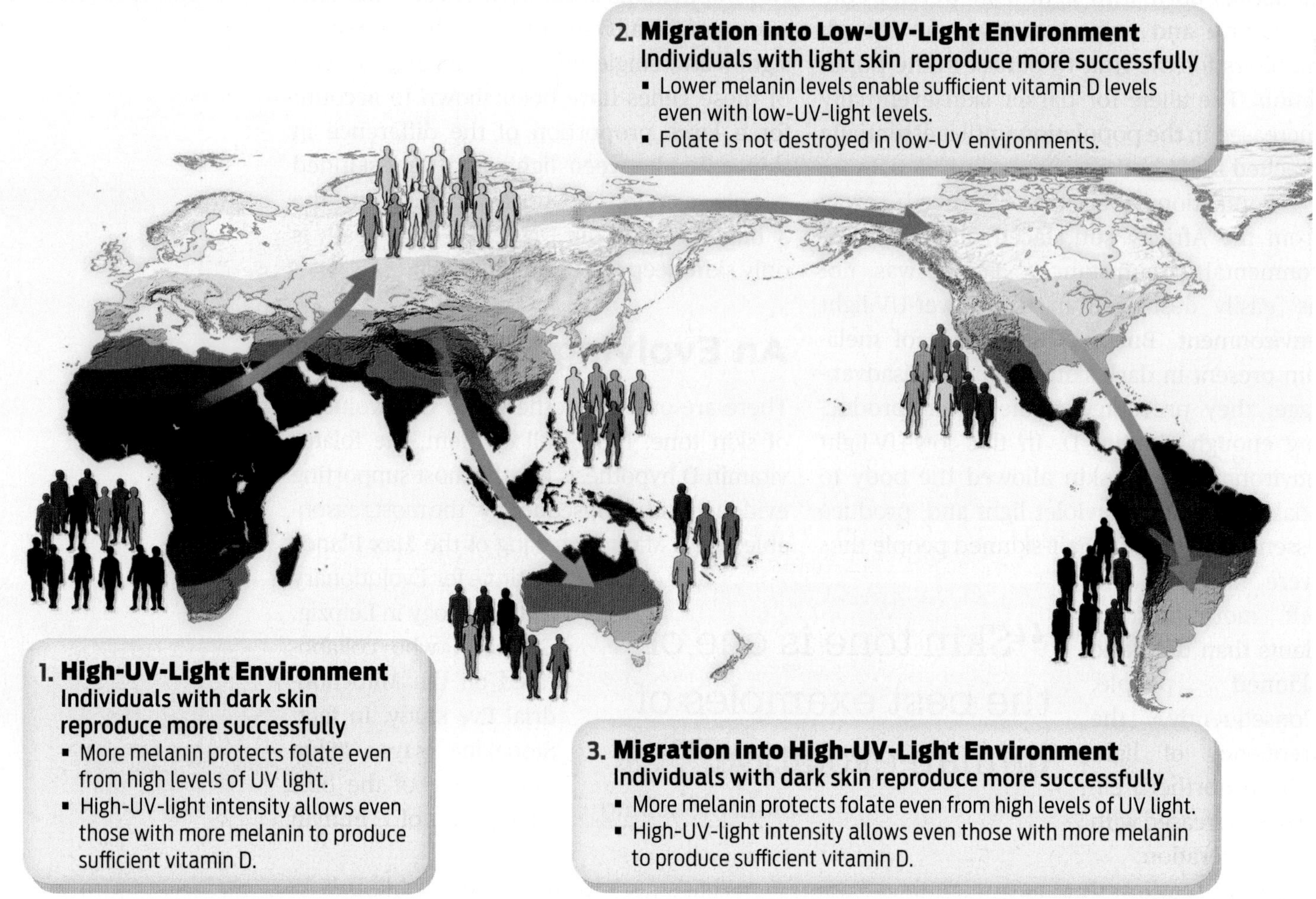

2. Migration into Low-UV-Light Environment
Individuals with light skin reproduce more successfully

- Lower melanin levels enable sufficient vitamin D levels even with low-UV-light levels.
- Folate is not destroyed in low-UV environments.

1. High-UV-Light Environment
Individuals with dark skin reproduce more successfully

- More melanin protects folate even from high levels of UV light.
- High-UV-light intensity allows even those with more melanin to produce sufficient vitamin D.

3. Migration into High-UV-Light Environment
Individuals with dark skin reproduce more successfully

- More melanin protects folate even from high levels of UV light.
- High-UV-light intensity allows even those with more melanin to produce sufficient vitamin D.

Data from Chaplin, G. (2004). *J.Phys. Anthro.* 125:292-302.
Map: Emmanuelle Bournay UNEP/GRID-Arendal

? Why is light skin a disadvantage in high-UV environments?

hardly any population is "pure" in the way that many have thought. As people moved around the globe, they settled and often conceived children with people they met along the way, introducing new alleles into the local gene pool. The particular environment people encountered favored some traits over others, and that is why populations that live in similar environments share similar features.

Though people tend to create racial groupings based on obvious physical characteristics, such features can be shared with other groups, says Jablonski. Not all Africans have equally dark skin and not all Europeans are light skinned, for example. And as humans travel more, settle in different areas, and intermarry, Jablonski says, "racial categories will get messier and messier." Perhaps in time the concept of race itself will disappear. ■

CHAPTER 19 SUMMARY

- Humans are 99.9% genetically identical to one another regardless of geographic origin. Biologically distinct human races do not exist.
- All humans are members of a single biological species, *Homo sapiens*, which evolved relatively recently—just 200,000 years ago.
- Physical features shared by people within populations reflect adaptations to specific environments.
- Alleles can be harmful, beneficial, or neutral in their effect on survival and reproduction.
- Skin color most likely evolved in response to environmental UV levels, an example of evolution by natural selection. Alleles for darker skin conferred an advantage in sunnier environments, while alleles for lighter skin conferred an advantage in regions that receive weak sunlight.
- Skin color represents an evolutionary trade-off between the need for vitamin D, which requires adequate sunlight for its production, and the need for folate, which is destroyed by too much sunlight.
- Fossil evidence shows that humans and apes descended from a common ancestor and that walking upright preceded development of a big brain. There were many species that could walk upright before *Homo sapiens* appeared.
- Fossil and DNA evidence shows that anatomically modern humans first emerged in Africa, approximately 200,000 years ago, and subsequently spread to other continents, beginning about 70,000 years ago.
- All modern-day humans can trace a portion of their genetic ancestry back to a single woman, Mitochondrial Eve, who lived 200,000 to 150,000 years ago in Africa.
- Humans evolved from apelike primate ancestors who likely had light skin. Darker skin emerged in tandem with loss of body hair as our hominid ancestors ventured into the hot savannah, while lighter skin emerged as humans migrated farther north.

MORE TO EXPLORE

- **Jablonski, N. G., and Chaplin, G. (2010) Human skin pigmentation as an adaptation to UV radiation. *Proceedings of the National Academy of Sciences* 107:8962–8968.**
- **Cann, R. L., and Wilson, A. C. (2003) The recent African genesis of humans. *Scientific American* 13(2):54–61.**
- **White, T. D., et al. (2009) *Ardipithecus ramidus* and the paleobiology of early hominids. *Science* 326(5949):64, 75–86.**
- **Stringer, C., and McKie, R. (1996) *African Exodus: The Origins of Modern Humanity*. New York: Henry Holt.**
- **Stoneking, M. (2008) Human origins: The molecular perspective. *EMBO reports* vol 9.**

CHAPTER 19 Test Your Knowledge

DRIVING QUESTION 1 What contributes to human skin color, and why is there so much variation in skin color between different populations?

By answering the questions below and studying Infographics 19.2, 19.3, 19.4, 19.9, and 19.10, you should be able to generate an answer for the broader Driving Question above.

KNOW IT

1 The ancestors of modern humans evolved in a high-UV environment. What does this suggest about their skin color?

2 In the course of human evolution, which of the following environmental factors likely influenced whether a population had mostly light-skinned individuals or mostly dark-skinned individuals?

- **a.** average annual temperature
- **b.** average annual rainfall
- **c.** levels of UV light
- **d.** the vitamin D content of the typical diet
- **e.** mitochondrial DNA inheritance

3 Jablonski and Chaplin hypothesized that darker skin is advantageous in _________ UV environments because darker skin __________.

- **a.** high-; reduces vitamin D production
- **b.** high-; protects folate from degradation
- **c.** high-; increases the rate of folate synthesis
- **d.** low-; allows more vitamin D to be produced
- **e.** low-; allows more folate to be produced

USE IT

4 If vitamin D production did not require exposure to UV, predict the skin color you might find in populations living at the equator; in populations living in Greenland. Explain your answers.

5 Which of the following would help darker-skinned people who live in low-UV environments remain healthy?

- **a.** folate supplementation
- **b.** sunscreen
- **c.** increased production of melanin
- **d.** vitamin D supplementation
- **e.** calcium supplementation

6 What can you infer about the skin-color alleles and the geographic origins of the ancestors of a light-skinned person and of a dark-skinned person?

7 Our closest primate relatives, chimpanzees, have light-colored skin yet live in tropical (high-UV) environments. How would the Jablonski–Chaplin hypothesis explain this observation?

- **a.** Chimpanzees don't need folate for successful reproduction.
- **b.** Chimpanzees are not susceptible to skin cancer.
- **c.** The hair of chimpanzees protects their light skin from UV light.
- **d.** Chimpanzees require much higher levels of vitamin D than humans do.
- **e.** In chimpanzees a light-colored pigment offers UV protection.

8 Vitiligo is a disease in which melanocytes are destroyed, with resulting loss of pigmentation.

- **a.** If a dark-skinned person develops vitiligo and therefore lighter-colored skin, will his or her race change?
- **b.** Given your answer to part a, can you think of one or more factors that might have led people to classify (or misclassify) themselves or others as members of one race or another?

DRIVING QUESTION 2 Where did the earliest humans evolve, and how do we know?

By answering the questions below and studying Infographics 19.5, 19.6, 19.7, and 19.8, you should be able to generate an answer for the broader Driving Question above.

KNOW IT

9 Who is mtEve, and when did she live?

10 The following three types of DNA can be used to trace evolutionary history. Match each DNA type with the best description of how it is transmitted in a population.

Type of DNA	Description
____ mtDNA	a. inherited by all children only from their mother
____ nuclear DNA	b. inherited by all children only from their father
____ Y-chromosome DNA	c. inherited only by sons from their father
	d. inherited by all children from both their mother and their father
	e. inherited only by daughters from their mother

11 According to the out-of-Africa hypothesis of human origins and migration, which group of people should show the highest level of genetic diversity?

- **a.** Africans
- **b.** Europeans
- **c.** Asians
- **d.** South Americans
- **e.** Australians

USE IT

12 Rank the levels of genetic diversity you would expect to find within the five populations listed in Question 11 from highest to lowest. Justify your ranking.

13 If there were many human females living ~200,000 years ago, why do we find that the mitochondrial DNA in all living humans is all related to a single woman from that time?

What kind of evidence could you look for to test your explanation? (Hint: Think about all the human fossils that have been uncovered, and consider that it is possible to extract DNA from fossils.)

apply YOUR KNOWLEDGE

MINI CASE

14 A mother with medium skin tone gives birth to a baby with darker skin than she has. Her lighter-skinned husband accuses her of infidelity.

a. How reasonable is this, given the genetics of skin color? (Hint: The genetics of skin color are similar to the genetics of height, shown in Infographic 12.9.)

b. Could an mtDNA analysis be used in a paternity test? Why or why not?

DRIVING QUESTION 3 What can genetics and the fossil record tell us about human evolution?

By answering the questions below and studying Infographics 19.1, 19.7, and 19.8, you should be able to generate an answer for the broader Driving Question above.

KNOW IT

15 What percentage of DNA sequences do all humans share?

a. 0%
b. 25%
c. 50%
d. 75%
e. >99%

16 Of the following traits that are associated with being human, which evolved most recently?

a. upright walking
b. the ability to control fire
c. social communication
d. tool use
e. a big brain

17 Place the following ancestors in order of earliest (1) to most recent (5).

_____ *Homo sapiens*
_____ Last common ancestor of chimpanzees and humans
_____ *Australopithecus*
_____ *Ardipithecus ramidus*
_____ *Homo erectus*

USE IT

18 Where would the last common ancestor of gorillas and humans fit into the ordering in your answer to Question 17? Explain.

19 Why would individual australopithicines who could make and use tools have had a selective advantage (that is, higher fitness) over individuals who could not make or use tools?

20 Ardi was partially arboreal (that is, her species could live in trees). The ability to move around in trees was facilitated by an opposable big toe that would help grip branches. Once ancient hominids moved permanently to a grounded lifestyle, would there have been any selective pressure to maintain an opposable big toe? Explain your answer.

21 Members of the genus *Australopithecus* walked upright, and their fossilized footprints show no evidence of an opposable big toe.

a. What foot structure and lifestyle might have been selected for if early hominid evolution occurred in a forested environment? In a grasslands environment? Would you predict any differences because of the selective pressures in each environment? Why or why not?

b. What other traits would you expect to be favored in a forested environment? In open grasslands?

apply YOUR KNOWLEDGE

INTERPRETING DATA

22 An extensive study of a hominid fossil dating from approximately 2 million years ago was published in 2013. For each of the features described below, consider whether they are closer to an ancestral state or to modern humans. On the basis of the features described, where would you place this fossil on the lineage between the chimpanzee–human ancestor and modern humans—what genus is it likely a member of?

- The shoulder structure and very long arms suggest the ability to climb and perhaps hang or swing.
- The spine and other skeletal features suggest an upright stance.
- There is no opposable toe.
- The heel is very narrow and pointed (not flat and wide, which would better accommodate body weight when walking).
- The skeleton suggests that in walking on two feet the feet would roll inward with each step.
- The skull is very small.
- The chest is not cylindrical but wider at the base and narrow at the shoulders, much like a triangle.

apply YOUR KNOWLEDGE

BRING IT HOME

23 The U.S. Census Bureau provides information on classifying race (http://www.census.gov/topics/population/race/about.html/). What races does the U.S. Census Bureau recognize? What about people of mixed race? What about people who identify themselves as Hispanic? How easy is it for you to identify yourself with respect to race given the racial categories on the U.S. Census?

DRIVING QUESTIONS

1. What is ecology, and what do ecologists study?
2. What are the different types of population growth?
3. What factors influence population growth and population size?

On the Tracks of **Wolves and Moose**

Ecologists learn big lessons from a small island

John Vucetich spent a freezing February day in 2010 trudging through knee-high snow on an island in Lake Superior. He was tracking a young gray wolf he called Romeo. The tracks led to a site in the forest where cracked branches and crimson-stained snow were evidence that a violent struggle had taken place hours earlier. Later, thawing in his cabin beside a wood-burning stove, Vucetich wrote in his field journal:

> *Teeth, hooves, blood, bruises, adrenaline, exhaustion. Romeo killed a moose. Very likely, this is the first moose he'd ever killed. He'd seen his parents, the alpha pair of Chippewa Harbor Pack, do it many times. . . . He'd wounded moose a couple of times this winter, but never killed one.*

ECOLOGY
The study of the interactions among organisms and between organisms and their nonliving environment.

POPULATION
A group of organisms of the same species living and interacting in a particular area.

COMMUNITY
Interacting populations of different species in a defined habitat.

For nearly 20 years, Vucetich has been shadowing wolves like Romeo and his kin on Isle Royale, a remote island about 15 miles off the Canadian shore in the northwest corner of Lake Superior. A 200-square-mile slice of roadless wilderness that is accessible only by boat and seaplane, Isle Royale may seem an unlikely place for a scientific laboratory, but that's exactly what it is for Vucetich and his colleagues. Every summer, and for a few weeks every winter, they investigate the island's packs of gray wolves (*Canis lupis*) and the herd of moose (*Alces alces*) that are their prey.

Begun in 1958, the Isle Royale wolf and moose study is the longest-running predator-prey study in the world. For six decades, researchers have studied how these two island inhabitants interact and coexist in this wild place. They are motivated by a simple yet increasingly pressing goal: to observe and understand the dynamic fluctuations of Isle Royale's wolves and moose in the hope that such knowledge will inspire a new, flourishing relationship with nature.

Islands have long provided biologists with important lessons about nature–think of Darwin and Wallace and their island adventures (see **Milestone 5: Adventures in Evolution**). But no island has ever produced such a mountain of data for researchers interested in **ecology**–the study of the interactions among organisms and between organisms and their nonliving environment–as Isle Royale. At a moment when wild populations are threatened all over the world, the lessons of Isle Royale couldn't have come at a better time.

In Nature's Laboratory

A number of features make Isle Royale an ideal place in which to study ecology. Because the island is uninhabited by humans and is protected as a national park, scientists can study moose-wolf interactions in a nearly natural environment, undisturbed by settlement, hunting, or logging.

Isle Royale is also an ideal distance from shore–close enough to the mainland for moose and wolves to have reached it, but far enough away that other animals do not easily migrate to it. Because there are no other predators or prey on the island, the only things eating moose are wolves, and moose are just about the only thing the Isle Royale wolves eat. These simplified conditions allow scientists to get a good look at the two residents' behavior and ecological impact.

Another feature that makes Isle Royale good for research is its size. The island is not so big as to have an unmanageably large population of moose, and not so small as to be unsupportive of a wolf population. "It's a little bit of the Goldilocks thing," says Vucetich, a professor in the School of Forest Resources and Environmental Science at Michigan Technological University. "Isle Royale is not too big and it's not too small and it's not too close and not too far. It's just the right size to have a population of wolves and moose that we can study."

Vucetich began studying wolves as a student at Michigan Tech in the early 1990s. In 2001, he became coleader of the Isle Royale study, working alongside his former teacher and mentor, Rolf Peterson. It's challenging work at times, but Vucetich says he may have been destined for this career path: "*Vuk*"–the root of his last name–"is the Croatian word for wolf."

Ecologists study organisms at a number of levels. They can look at an individual organism, such as a single moose or wolf, studying how it fares in its surroundings. They may also look at a group of individuals of the same species living in the same place–a herd of moose, or a pack of wolves, for example–watching what happens to this **population** over time. Two or more interacting populations of different species constitute a **community.** Isle Royale, for example, is home to a community of wolves, moose, and the plants the moose feed on.

Finally, ecologists may want to understand the functioning of an entire **ecosystem,** all the living organisms in an area and the nonliving components of the environment–such as temperature, rain, and sunlight–with which they interact. When moose eat trees, for example, they reduce the available habitat for other animals, such as birds. However, the heat of summer can reduce the ability of moose to feed, which in turn improves tree growth **(INFOGRAPHIC 20.1)**.

Vucetich was initially drawn to ecology as a way to experience the outdoors, his first love. Only later did he realize he was actually

ECOSYSTEM
All the living organisms in an area and the nonliving components of the environment with which they interact.

INFOGRAPHIC 20.1 Ecology of Isle Royale

Ecologists study organisms at a number of levels, each of which involves different types of interactions.

U.S. Fish & Wildlife Service

Individual

A single organism of a particular species

- One wolf

John Vucetich

Population

A group of individuals of the same species living and interacting in the same region

- A pack of wolves whose individuals live and reproduce with one another.

Accent Alaska.com / Alamy

Community

Interacting populations of different species

- Wolves prey on moose
- Ticks infest moose
- Moose feed on trees

sherwoodimagery/iStockphoto

Ecosystem

Species interacting with other species and the environment

- Moose eat trees. Less vegetation alters the landscape for other species.
- Hot summers reduce food for moose, affecting their winter survival

? Which level of ecology involves the interactions of organisms with their environment?

Small planes are used to conduct aerial surveys to determine the numbers of wolves and moose on Isle Royale.

quite good at something ecology has a lot of: math. "As a high school student, I didn't like math at all," he says. But when he saw that math allowed him to spend more time outdoors he became "interested and inspired to learn a great deal about math."

Vucetich is a population ecologist, and population ecology is all about numbers. On Isle Royale, the main numbers the researchers are interested in year after year are the numbers of wolves and moose. "In any given season there are more or less of those species and we want to understand why," says Vucetich. Answering the "why" involves a lot of time, patience, and—of course—counting.

Much of the counting is done from the air. Sitting one in front of the other inside a tiny two-seat plane, pilot and observer circle the island scanning for evidence of wolves and moose. Wolves are relatively easy to find and count, especially in the snow: "You follow the wolf tracks until you find the wolves," says Vucetich. The other thing that makes counting wolves easy is that they live in packs: if you find one wolf, you've generally found the others. And since there are usually no more than a dozen or so wolves on the island at any time, it's possible to count every one.

DISPERSION PATTERN
The way organisms are distributed in geographic space, which depends on resources and interactions with other members of the population.

It's a different story with moose. There can be more than a thousand moose on the island—too many to count all at once. Besides, moose are relatively solitary creatures, and their brown coloring makes them harder to spot against the backdrop of dark evergreen trees. When moose are feeding in the forest—which is much of the time—counting them is, according to Vucetich, "like trying to count fleas on a dog from across the room." It's simply not possible to count them all.

Instead, the team uses a shortcut: they count all the moose in a series of defined plots of land representing about 20% of the island, average the number of moose per plot, and then extrapolate to the rest of the island. But even this shortcut requires many careful hours of study in the plane, straining to see the moose through the trees. To help himself concentrate, Vucetich recites a sort of mantra: "Think moose, think moose, look for the moose." **(INFOGRAPHIC 20.2)**

The somewhat random distribution of individually roaming moose represents one type of **dispersion pattern** found in nature. Dispersion patterns generally reflect the way resources are distributed in an area, as well as characteristic interactions among other members of the population. Moose are found wherever there are edible trees. Being solitary

INFOGRAPHIC 20.2 Distribution Patterns Influence Population Sampling Methods

Wolves and moose have different lifestyles and distribution patterns on Isle Royale. Determining the size of each population requires a distinct counting strategy.

Individual wolves cluster together in packs, making them easier to spot, track, and count.

2013 Wolf Territories and Kill Locations

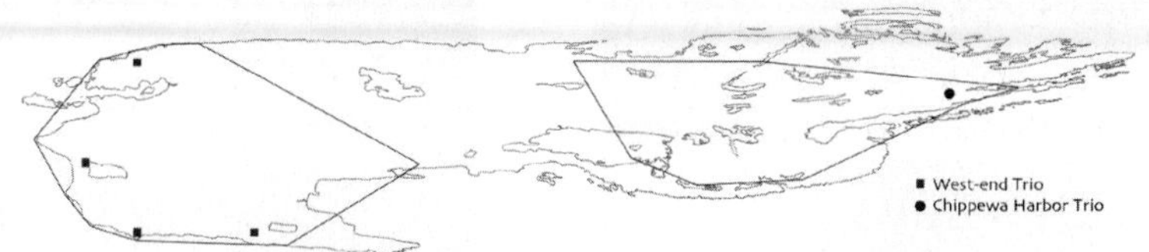

Moose are largely solitary creatures and are distributed more randomly on the island. Counting them requires a different strategy.

2013 Moose Distribution

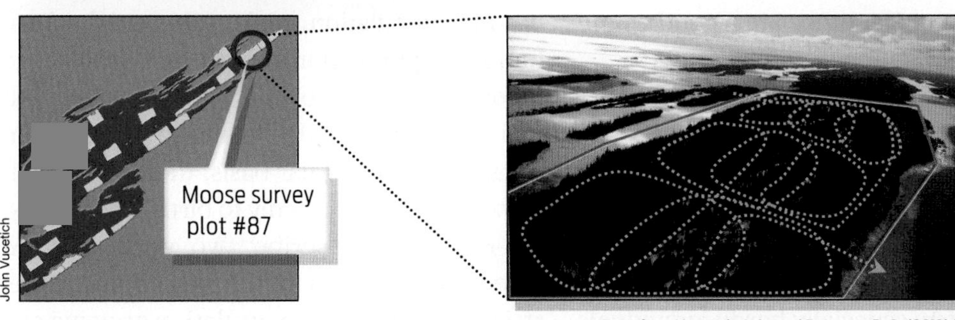

This is an aerial view of plot #87, outlined in orange. The green dotted line defines the 9 overlapping circles flown as moose are counted. This sampling is done for 91 plots across 20% of the island to gather enough data for an accurate estimate of moose population size and density on the island.

Data from Vucetich, J. A., and Peterson, R. O. (2013) Ecological studies of wolves on Isle Royale, Annual Report 2012–2013.

? In order to establish the population size of the moose on Isle Royale, do the researchers try to count all the moose?

and randomly dispersed may help protect moose from predation, since single moose are harder to spot in the forest than a large group would be. For some plants, a random dispersion may result from the way seeds are dispersed. Pine trees, for example, have air-blown seeds that are spread far and wide by gusty winds, resulting in a random dispersion of trees in the forests on the island.

A truly random dispersion is rare in nature; even wind-blown seeds must fall on fertile soil to grow, and this does not always happen. More common is a clustered, or clumped, dispersion pattern, which results when resources are unevenly distributed across the landscape, or when social behavior dictates grouping, as it does with the highly social wolf. Clumping has its advantages: for wolves, clumping helps them to gang up on moose; they circle their prey and close in for the kill. Clumping can also be a defense against predation, as it is for a school of fish.

A third dispersion pattern found in nature is uniform dispersion. In this case, individuals keep apart from one another at regular distances, usually because of some kind of territorial behavior. Birds such as

INFOGRAPHIC 20.3 Population Dispersion Patterns

Different organisms have different distribution patterns. There are three main types, but few organisms in nature fall into strictly one category.

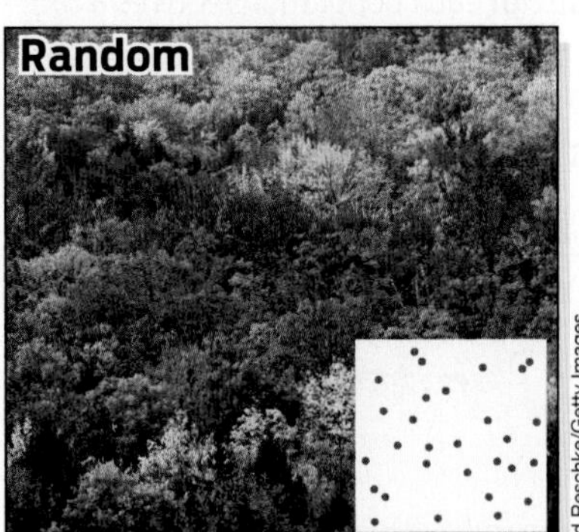

Ed Reschke/Getty Images

Individuals are equally likely to be anywhere within the area.

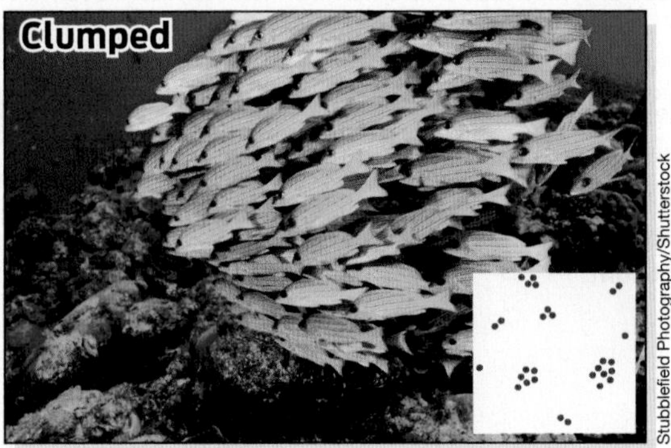

Stubblefield Photography/Shutterstock

High-density clumps are separated by areas of low abundance.

Marcel Krol/Dreamstime.com

Individuals maximize space between them by being uniformly spaced.

? If plant seeds are carried by animals that congregate at watering holes, what type of dispersion would you predict for the plants?

penguins that nest in defined spaces a few feet away from one another are a good example (**INFOGRAPHIC 20.3**).

Population Boom and Bust

Moose have not always roamed Isle Royale. The first antlered settlers likely arrived around 1900, when a few especially hardy individuals swam across the 15-mile-wide channel from Canada. With an abundant food supply and no natural predators on the island, the moose population exploded, growing from a handful of individuals around the turn of the century to more than a thousand by 1920.

This rapid increase in numbers reflected the population's high **growth rate,** defined as the birth rate minus the death rate. Because it denotes the simple balance between birth and death, the growth rate is also known as the rate of natural increase. When the birth rate of a population is greater than the death rate, the population grows; when the death rate is greater than the birth rate, the population declines; and when the two rates are equal, the result is zero population growth.

In many populations, **immigration,** the movement of individuals into a population, and **emigration,** the movement of individuals out of a population, make substantial contributions to population growth. But because the moose and wolves of Isle Royale are isolated, and individuals neither come to nor go from the island on a regular basis, their population growth rates are due only to births and deaths.

Ecologists describe two general types of population growth. The rapid and unrestricted increase of a population growing at a constant rate is called **exponential growth.** When a population is growing exponentially, it increases by a certain fixed percentage every generation. Thus, instead of a constant number of individuals being added at each generation—say, the population going from 100 to 120 to 140 to 160—the increase is more like credit card interest, with each increase added to the principal (the population) before the percentage is applied. And so, with an exponential growth rate of 20%, a population of 100 would increase at each generation from 100 to 120 to 144 to 173 to 207. If the population continued to grow exponentially, it would quickly get out of control, not unlike a credit card bill you don't pay on time.

Such unrestricted growth is rarely, if ever, found unchecked in nature. As populations

GROWTH RATE
The difference between the birth rate and the death rate of a given population; also known as the rate of natural increase.

IMMIGRATION
The movement of individuals into a population.

EMIGRATION
The movement of individuals out of a population.

EXPONENTIAL GROWTH
The unrestricted growth of a population increasing at a constant growth rate.

increase, various environmental factors such as food availability and access to **habitat,** the physical environment where an organism lives and to which it is adapted, limit an organism's ability to reproduce. When population-limiting factors slow the growth rate, the result is **logistic growth,** a pattern of growth that starts rapidly and then slows.

Eventually, after a period of rapid growth, the size of the population may level off and stop growing as the population reaches the environment's **carrying capacity**–the maximum number of individuals that an environment can support given its space and resources. Carrying capacity places an upper limit on the size of any population; no natural population can grow exponentially forever without eventually reaching a point at which resource scarcity and other factors limit population growth. This is true even of the human population (see Chapter 23).

The size of a population may fluctuate around the environment's carrying capacity, briefly exceeding it and then dropping back. After an initial overshoot of carrying capacity, factors like disease or food shortage will cause the population to shrink. This drop in turn may allow the environment time to recover its food supply, at which point the population may begin to grow again, briefly exceeding carrying capacity, and so on, in a cycle of "boom and bust" **(INFOGRAPHIC 20.4)**.

HABITAT
The physical environment where an organism lives and to which it is adapted.

LOGISTIC GROWTH
A pattern of growth that starts off fast and then levels off as the population reaches the carrying capacity of the environment.

CARRYING CAPACITY
The maximum population size that a given environment or habitat can support given its food supply and other natural resources.

INFOGRAPHIC 20.4 Population Growth and Carrying Capacity

A population growing at a constant rate without checks will increase exponentially. However, for most populations found in nature, as the population reaches its carrying capacity, the growth rate slows and eventually remains at or near zero.

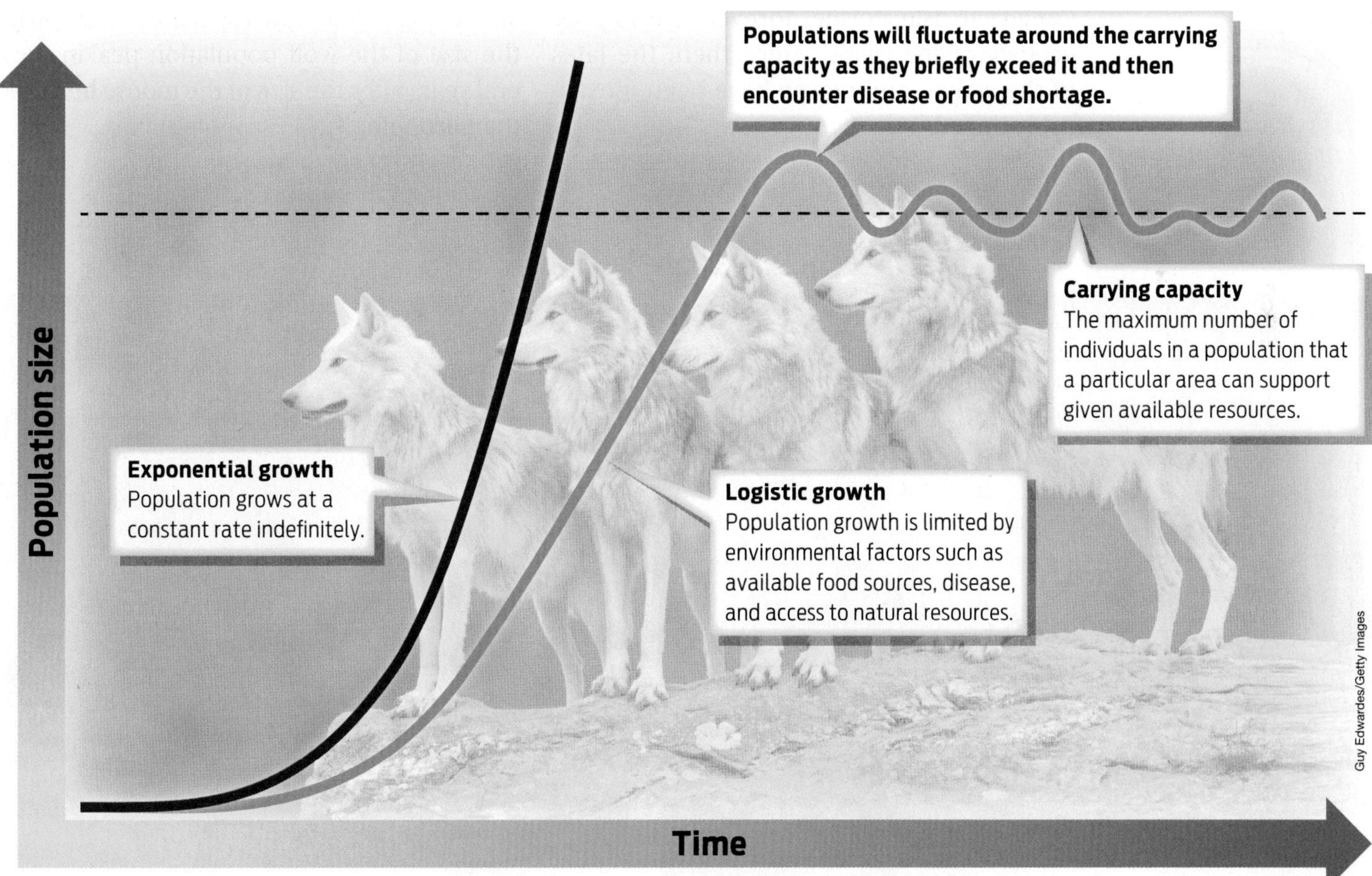
Populations will fluctuate around the carrying capacity as they briefly exceed it and then encounter disease or food shortage.
Population size
Carrying capacity
The maximum number of individuals in a population that a particular area can support given available resources.
Exponential growth
Population grows at a constant rate indefinitely.
Logistic growth
Population growth is limited by environmental factors such as available food sources, disease, and access to natural resources.
Time
Guy Edwardes/Getty Images

? What can cause an exponentially growing population to switch to logistic growth?

When moose first arrived on Isle Royale, their population grew exponentially. This unchecked proliferation of hungry mouths took a severe toll on the island; by 1929, the moose had munched their way through most of its vegetation. In turn, the reduction of the island's food supply caused the moose population to crash. The moose population had exceeded the island's carrying capacity, and by 1935 it had dwindled to a few hundred starving individuals.

The herd got lucky, though. The next summer, fire consumed 20% of the island, and the scorched areas provided space for new trees to grow. But as soon as the forest recovered, moose numbers again began to explode, ravaging the forests once more.

Then, around 1950, everything changed. One especially cold winter, a pair of gray wolves crossed an ice bridge connecting Canada to Isle Royale, forever altering the ecology of the island. Since then, the fates of the wolves and moose have been inextricably linked.

At the beginning of the Isle Royale study, in 1959, there were about 550 moose and 20 wolves on the island. Moose numbers climbed for about 15 years, reaching a peak of approximately 1,200 animals in 1972, and then declined rapidly, to a low of approximately 700 moose in 1980. As moose numbers fell, wolf numbers rose–from a low of 17 wolves in 1969 to a high of 50 animals in 1980. These two trends were linked: the wolves were feeding well enough to increase their own population, but by hunting and killing so many moose they caused the moose death rate to exceed the birth rate. With a negative growth rate, the moose population shrank.

What would happen next? Would the wolf predators simply drive their moose prey to extinction? No one knew. The only thing to do was watch and wait. Eventually, it became clear that the two populations were rising and falling together in a specific pattern, with the size of the wolf population peaking several years after the size of the moose herd and then dropping.

John Vucetich

A pack of wolves moving in for the kill.

John Vucetich

John Vucetich examining a kill site.

Why does the wolf population fall? Because even for wolves, there's no such thing as a free lunch: they pay a price for predation in the form of a declining food supply. The result is a repeating cycle in the numbers of predator and prey. Rather than growing exponentially and leveling off, the populations cycle through repeated rounds of boom and bust (**INFOGRAPHIC 20.5**).

Ecological Detectives

The size of the wolf population affects more than just the size of the moose population. Another pattern that has emerged in the decades of data collected on Isle Royale is a correlation between a large wolf population and vigorous tree growth. When wolves are plentiful, they keep the moose population in check. Because trees are the primary food source for moose, they grow more when fewer moose are eating them. It's therefore possible to follow the rise and fall of the wolf population by monitoring the state of the forest.

One way ecologists can determine forest growth and health is to count and measure the width of tree rings, which reflect how much trees have grown season by season. They also measure the height of the trees. Taller and bigger trees mean that fewer moose have been foraging on them, which in turn indicates that more wolves have been keeping the moose population in check.

Wolves affect tree growth in another, and indirect, way as well. Because the wolves don't always consume the entire carcass of a moose they kill, the remains decay and fertilize the ground where they lie, enriching the soil with nutrients for plant growth. Researchers have found that nitrogen levels are between 25% and 50% higher in these hot spots compared to control sites. This work shows that predators–in this case, wolves–are an important component of a balanced and healthy ecosystem and illustrates what

INFOGRAPHIC 20.5 Population Cycles of Predator and Prey

The wolf and moose populations are intimately linked and inversely proportional. When wolf populations are high, moose populations are low. When wolf populations are low, moose populations are high.

Moose–Wolf Populations 1959–2016

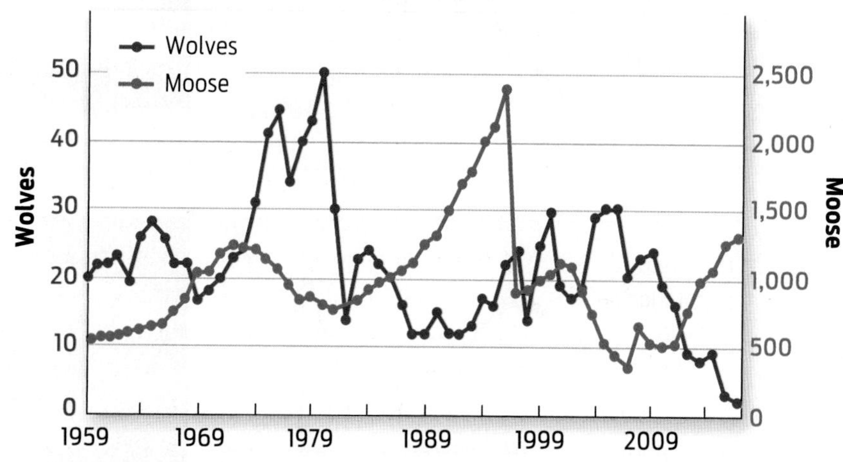

Wolf and moose populations, 1959–2016. Moose population estimates for 1959 to 2001 are based on population reconstruction from recoveries of dead moose. Estimates for 2002 to 2016 are based on aerial surveys.

Data from Vucetich, J. A., and Peterson, R. O. (2016) Ecological studies of wolves on Isle Royale, Annual Report 2015-2016.

? Why does the population of moose (prey) decline as the population of wolves (predator) increases?

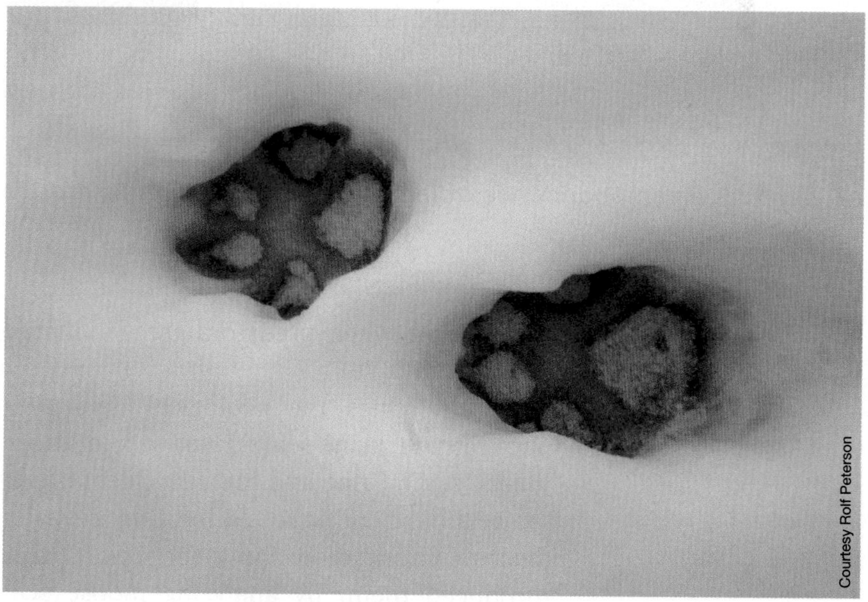

Wolf tracks in the snow on Isle Royale.

INFOGRAPHIC 20.6 Patterns of Population Growth

Wolf, moose, and tree populations are all interconnected. Trees provide food for moose, and moose provide food for wolves. Anything that affects the size of one population will affect the size of the others.

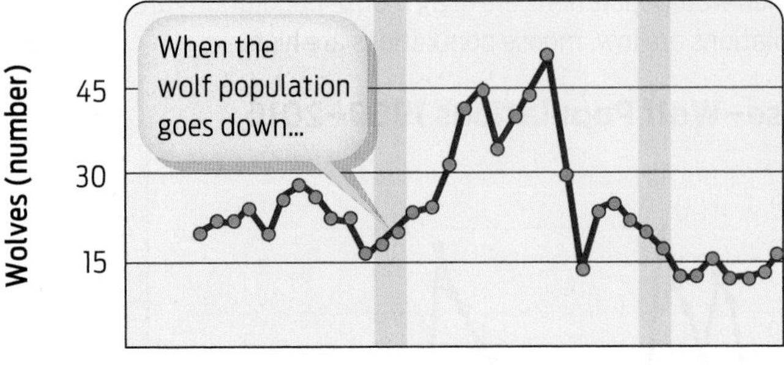

The main diet of Isle Royale wolves is moose. The wolf population grows and diminishes in response to the availability of this food resource.

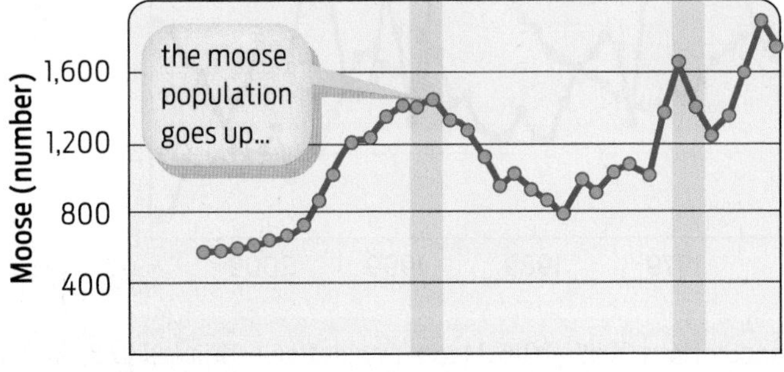

The main diet of moose is trees. A larger moose population means that more tree material is eaten, so tree growth slows.

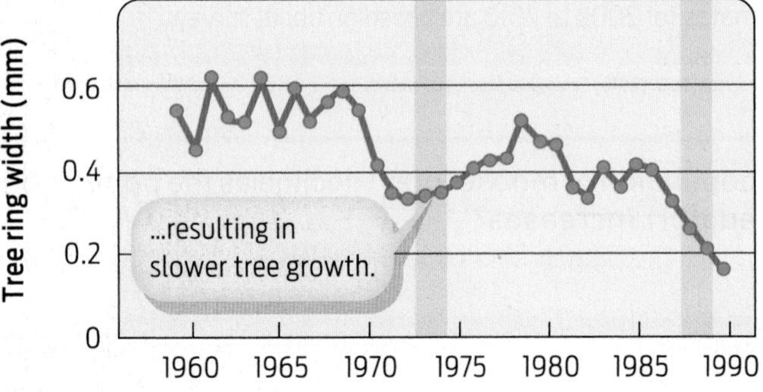

Tree growth can be measured by the width of each tree ring. One ring represents the amount of growth in 1 year: the wider the ring, the more the growth in that year.

Data from McLaren, B.E., and Peterson, R.O. (1994) Wolves, moose, and tree rings on Isle Royale. *Science* 266(5190):1555–1558.

? What does an abundance of large trees suggest about moose and wolf populations?

can be lost when predators are eliminated **(INFOGRAPHIC 20.6)**.

Other clues the ecological detectives look at are urine-soaked snow and droppings (scat). Urine and scat may seem crude objects of scientific study, but they reveal a host of information about the animal that produced them. By analyzing moose scat samples under the microscope, researchers can tell exactly what moose have been eating. During the winter months, for example, moose eat mostly twigs from deciduous (leaf-shedding) plants and needles from balsam fir and cedar trees. Knowing a moose's diet is important because the supply of balsam fir on the island has been steadily dwindling over the years. By monitoring moose scat, researchers will be able to tell if moose can switch to other, more plentiful food items.

Scat also provides important information about an animal's genetics. Scat can be used as a source of DNA for that animal. DNA profiles of scat samples can be used to confirm population counts and to track which wolves were involved in killing which moose. DNA can also be used to look for diseases or signs of inbreeding. "Through the DNA we can get a good sense of individual wolves–how they live and how they die," says Vucetich. (For more information on DNA profiling, see Chapter 7.)

Still more clues can come from studying a moose kill site, which is a bit like analyzing a crime scene. Researchers can tell if a moose was killed by wolves: in that case there will often be blood spattered on nearby trees and signs of struggle in the form of broken branches. Wolves also typically scatter bones as they feast, whereas the carcasses of moose that die of starvation may be relatively intact.

At the kill site, researchers gather moose bones. From these bones, the researchers can tell how old a moose was when it died, as well as learn about other aspects of the animal's health, such as whether it had arthritis or osteoporosis. The value of this information goes beyond understanding individual animals. It allows researchers to know whether wolves are targeting healthy moose or sickly ones. Killing a healthy moose has a bigger effect on moose population dynamics than killing one that is already near death, because a young, healthy moose might have gone on to reproduce had it lived (**INFOGRAPHIC 20.7**).

INFOGRAPHIC 20.7 A Variety of Methods Can Be Used to Monitor Populations

In addition to information about population size, researchers collect other data that are essential for monitoring the physical health of moose and wolf populations.

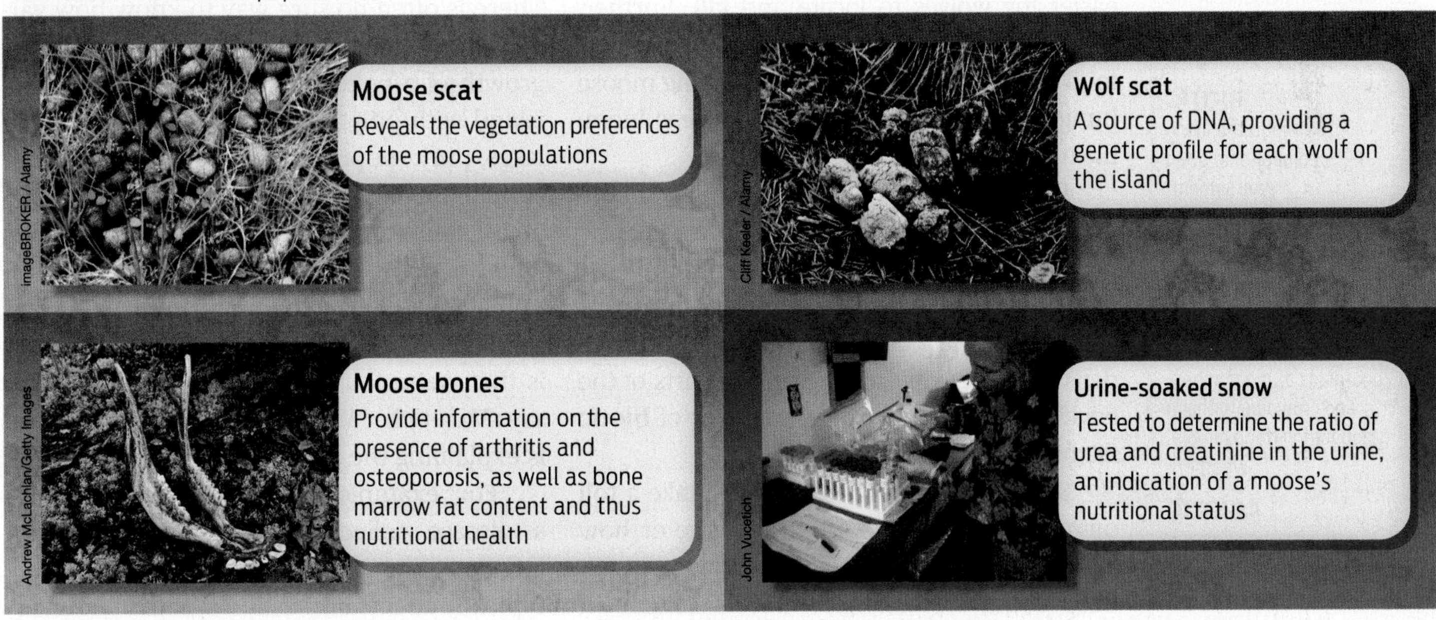

imageBROKER / Alamy
Moose scat
Reveals the vegetation preferences of the moose populations
Cliff Keeler / Alamy
Wolf scat
A source of DNA, providing a genetic profile for each wolf on the island
Andrew McLachlan/Getty Images
Moose bones
Provide information on the presence of arthritis and osteoporosis, as well as bone marrow fat content and thus nutritional health
John Vucetich
Urine-soaked snow
Tested to determine the ratio of urea and creatinine in the urine, an indication of a moose's nutritional status

? What can researchers learn about moose from studying their scat and urine?

Too Close for Comfort?

Moose are formidable foes of their wolf predators. At 900 pounds, 10 times the weight of a wolf, an adult moose can successfully defend itself against an aggressive pack of wolves with its powerful front legs. For that reason, wolves often attack older and weaker moose, or younger and smaller moose. They typically target the nose and hindquarters, where they bite and latch onto the flesh like a steel trap. When enough wolves are attached, their collective weight brings down the moose, and the feeding begins.

"Isle Royale is not too big and it's not too small and it's not too close and not too far."

—John Vucetich

A number of factors can influence the likelihood that wolves will kill moose. One of the simplest is **population density,** the number of organisms per unit area. Because the total area of Isle Royale stays the same, as the size of the moose population increases so does its density. At high population density, moose are easier for wolves to locate and kill. Further, when the moose population is at high density, food scarcity can be a problem, leaving moose hungry and weak and therefore more vulnerable to attack.

Because wolf predation and plant abundance have a greater effect on moose when the moose population is large, these are examples of **density-dependent factors**–factors that exert different degrees of influence depending on the density of the population. As living parts of the environment, they are also examples of **biotic** factors influencing growth.

Some environmental pressures take a toll on a population no matter how large or how small it is. In an exceptionally cold winter with deep snow, for example, moose can die of cold or starvation, or the weather can weaken them so they are easier to hunt and kill. Since cold weather affects moose regardless of population size, it is considered a **density-independent factor:** whether there are 10 moose or 1,000, a harsh winter affects them all.

The wolf population is also affected by density-independent factors such as weather. "A mild winter is always tough on the wolves," says Peterson, who notes that moose can more easily escape wolves when snow cover is light. Density-independent factors like snowfall can be considered nature's form of bad luck. Most density-independent factors are **abiotic,** or nonliving, and include things like temperature, precipitation, and fire (**INFOGRAPHIC 20.8**).

POPULATION DENSITY
The number of organisms per unit area.

DENSITY-DEPENDENT FACTOR
A factor whose influence on population size and growth depends on the number and crowding of individuals in the population (for example, predation).

BIOTIC
Refers to the living components of an environment.

DENSITY-INDEPENDENT FACTOR
A factor that can influence population size and growth regardless of the numbers and crowding within a population (for example, weather).

ABIOTIC
Refers to the nonliving components of an environment, such as temperature and precipitation.

Watching and Waiting

For the scientists on Isle Royale, population ecology is full of unexpected twists and turns. There is often no sure way to know how various environmental factors will influence the growth of a population. Even on an isolated island with only one large predator and one large prey, population dynamics are never simple. Scientists gather data, look for patterns, and form hypotheses, but predicting what will happen next is much more difficult. "What Isle Royale has shown us–and has shown us convincingly for the past 50 years–is that we're lousy at predicting the future," says Vucetich. "What we're a fair bit better at is explaining the past."

For example, beginning around 1980, a disease called canine parvovirus (CPV) infected Isle Royale's wolves. The disease typically affects domestic dogs and was likely brought unintentionally to the island on the boots of hikers. The disease killed all but

INFOGRAPHIC 20.8 Abiotic and Biotic Influences on Population Growth

Both nonliving (abiotic) and living (biotic) environmental factors influence the size and growth of populations. Some factors influence the population regardless of population density, whereas others are density-dependent.

Abiotic Factors

Photobac/Dreamstime.com

Precipitation (density-independent)
Snowfall stresses moose, making it hard for them to find food and avoid predators.

Fire (density-independent)
Fire is an immediate danger to moose; it can destroy their habitat and make food scarce.

Peter J. Wilson/Shutterstock

terrymorris/iStockphoto.com

Temperature (density-independent)
High summer temperatures can cause heat stress in moose and increase insect parasitism.

James Mattil/AGE Fotostock

Biotic Factors

Predators (density-dependent)
The presence of a wolf predator limits moose population growth, especially when moose density is high.

jim kruger/istockphoto

Oksana Churakova/Dreamstime.com

Food (density-dependent)
The availability of trees as a food source influences moose population growth, especially when moose density is high.

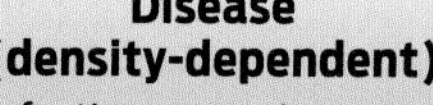

Disease (density-dependent)
Infectious organisms and parasites such as ticks can reproduce more and spread more easily between moose when moose density is higher. Tick infestations weaken moose, limiting their reproduction.

John Vucetich

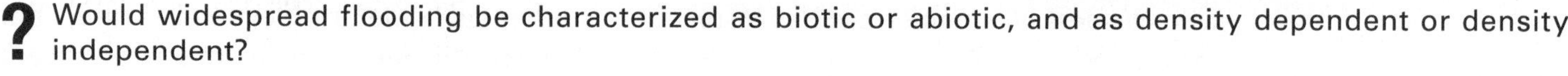

? Would widespread flooding be characterized as biotic or abiotic, and as density dependent or density independent?

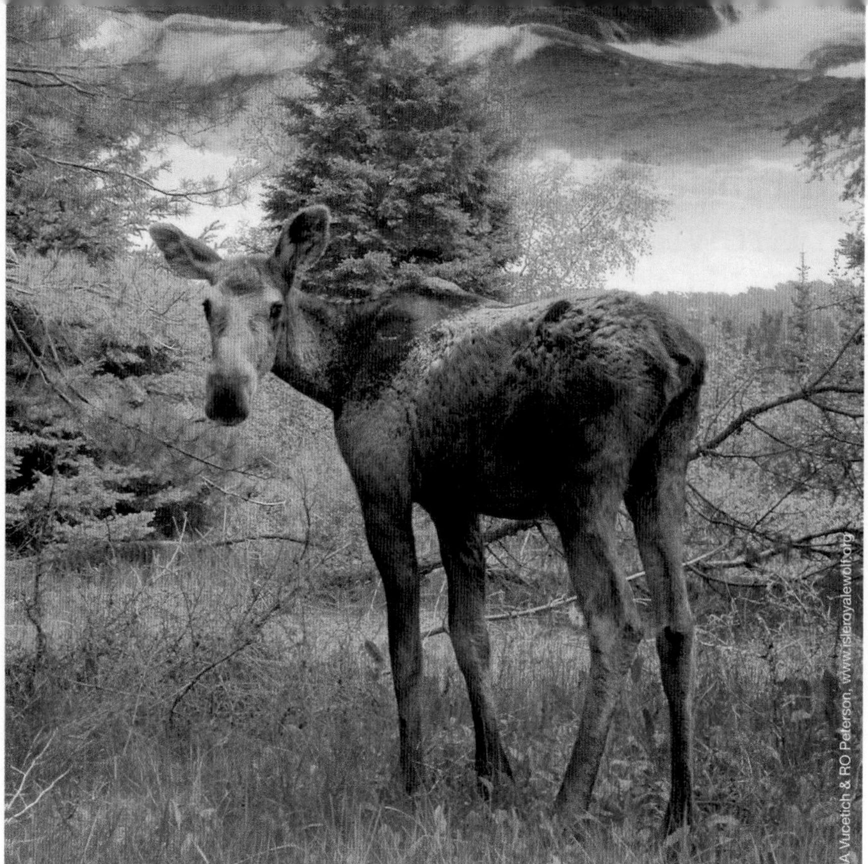

A young moose on Isle Royale (note the patchy fur, a sign of tick infestation).

14 of the island's wolves, and over the next 10 years the moose population skyrocketed, suggesting that wolves exert a strong influence on the abundance of their prey. The event was useful from a scientific standpoint–but entirely unexpected. "There's no way that anyone could have predicted that. Not in a million years," says Vucetich.

That wasn't the end of the surprises. In the last 15 years, it's become apparent that a warming climate, not just predation by wolves, is influencing moose population size. The first decade of the 21st century was one of the hottest on record. Sweltering summer temperatures hit moose especially hard. The large herbivores get hot easily, and they don't perspire; they escape the heat by resting in the shade. A lot of time spent resting means less time for eating, and a moose who eats less all summer has less insulation for winter.

Warmer temperatures have affected moose in a more insidious way as well. About 10 years ago, Vucetich and his colleagues began to notice that a tick parasite was bothering the moose, and that warm weather seemed to favor ticks. Ticks suck the moose's blood and cause them to itch. The moose scratch themselves against trees and chew their hair out trying to rid themselves of the itchy freeloaders. Since a single moose may host many thousands of ticks, the combination of tick-related blood loss and heat-induced weight loss can be deadly. In 2004, the average moose had lost more than 70% of its body hair, the result of carrying more than 70,000 ticks.

By 2007, the deadly combination of blood-sucking ticks, hot summers, and relentless predation from wolves had driven the moose population to its lowest point in at least 50 years–385, down from 1,100 in 2002. Predictably, the wolf population followed suit, declining from 30 individuals in 2005 to 21 in 2007. In recent years, winters have been colder and more long-lasting, which has led to fewer tick infestations and healthier moose **(INFOGRAPHIC 20.9)**.

Hunted by wolves, preyed on by ticks, dogged by oppressive heat, moose certainly do not have it easy. They can live to be 17 years old, but most moose on Isle Royale die before reaching their tenth birthday.

Life is no picnic for Isle Royale's wolves, either. Although they can live to be 12 years old, most die by age 4. The most common cause of death is starvation. With few available food sources, a wolf may go 10 days without eating. Obtaining a meal on the eleventh day may mean having to wrestle a 900-pound moose on an empty stomach.

The difficulty of finding food is just one obstacle for wolves. They also have a high incidence of bone deformities, which cause back pain and partial paralysis of the hind legs. In the early years of the Isle Royale study, such deformities were rare, but they've become more common in recent years, almost certainly a result of inbreeding (see Chapter 14). For the last 15 years, every

INFOGRAPHIC 20.9 A Warming Climate Influences Wolf and Moose Population Size

In recent years, climate change has become a significant influence on moose and wolf populations on Isle Royale. Warmer temperatures lead to increased tick infestations of moose, resulting in loss of hair and a weakened and depleted population. In cooler years, moose experience less hair loss and their populations grow.

One moose may be home to tens of thousands of ticks at a time.

Ticks cause moose to lose their hair, their appetite, and a good deal of blood.

Ticks make moose weak and vulnerable to predation and starvation. So, when ticks increase...

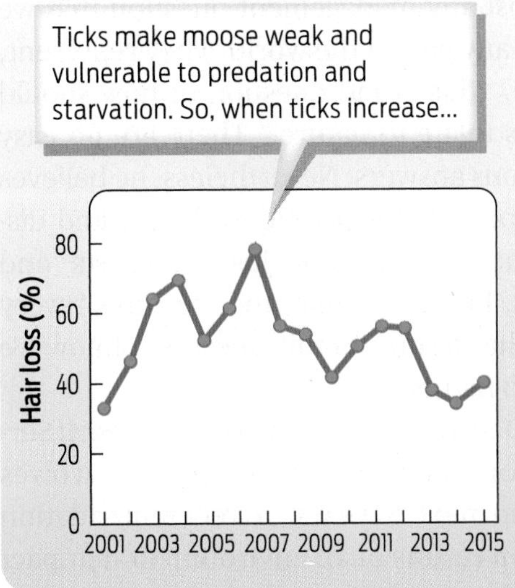

...the moose population decreases.

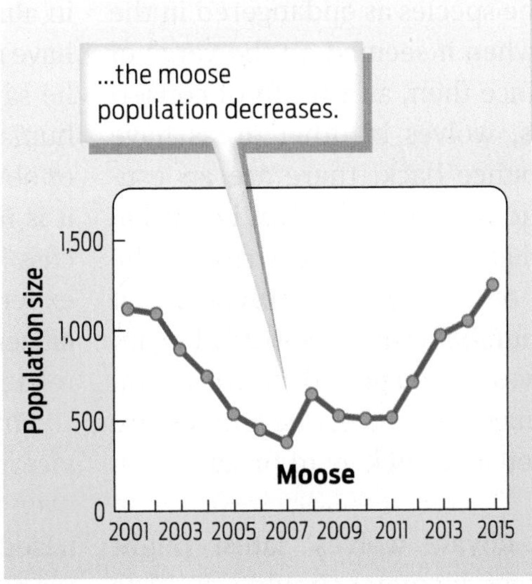

Moose weakened by ticks are easier for wolves to catch. After an initial population increase in response to an abundance of moose, the wolf population begins to suffer.

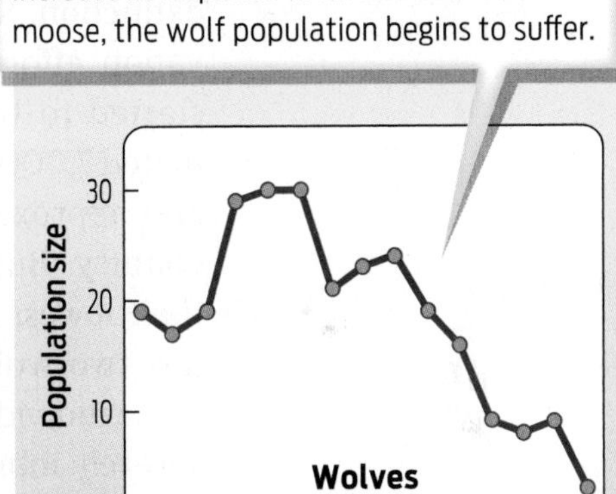

? If both summers and winters were to be warmer than usual in the next few years, what do you predict about the numbers of ticks and moose?

dead wolf on Isle Royale has had such deformities. Vucetich thinks that climate change may be an indirect cause of the inbreeding by lessening the frequency that ice bridges form between Canada and Isle Royale, bridges over which new wolves could reach the island and diversify the wolf gene pool.

It's not the first time wolf populations have been in trouble. When European settlers first arrived in North America, the gray wolf roamed throughout all of the future 48 contiguous U.S. states. By 1914, hunting and trapping had greatly reduced the population, and survivors were limited to remote

Wolves on Isle Royale are facing extinction.

wooded regions of Michigan, Wisconsin, and Minnesota. The federal government officially listed the species as endangered in the early 1970s, when it seemed on the verge of extinction. Since then, as a result of conservation efforts, wolves in other areas have started to bounce back: there are an estimated 7,000 to 11,000 gray wolves in Alaska, and approximately 5,000 in the rest of the country. But on Isle Royale, wolves are at their lowest numbers since the study began: just two wolves were spotted in 2016. And these individuals—one male, one female—are severely inbred and unlikely to produce viable offspring.

The Isle Royale wolves' latest plight poses an ethical dilemma: should scientists intervene on their behalf—say, by importing wolves from another population to reintroduce genetic diversity—or let nature take its course? It's a question that Vucetich thinks about a lot. The answer, he says, will require balancing a number of competing values—not just the value of individual populations, but the values of ecosystem health, scientific knowledge, and wilderness. Without wolves, for example, would the moose population once again explode and decimate the island's forest? Would healthier wolves be able to completely overwhelm moose, and drive them to extinction on the island? Does the fact that human-caused climate change has worsened these problems mean that we have a moral obligation to alleviate them? These are some of the difficult questions that wildlife managers will need to consider when debating whether and how to intervene.

The dilemma is a familiar one to conservation biologists. According to Vucetich, these competing values show up in varying degrees in almost any management question that we have in any part of the world. They represent, he says, "this grand question of how should humans relate to nature." There are no easy or obvious answers. Nevertheless, he believes it is important for people to debate and discuss these issues—not just scientists and experts, but lay people, too, because "every citizen has a stake in this question of how we relate to nature."

In December 2016, the National Park Service proposed introducing up to 30 wolves over the next 3 years, a recommendation based on results of an environmental impact report that states this is the only way to save the wolf population on the island and to preserve the natural laboratory. As of early 2017, public comments on that plan from citizens were being actively solicited. ■

CHAPTER 20 SUMMARY

- Ecology is the study of the interactions among organisms and between organisms and their nonliving environment.
- Ecologists study these interactions at a number of levels, including individual, population, community, and ecosystem.
- Living organisms may have a clumped, random, or uniform dispersion pattern, depending on ecological and behavioral adaptations. Few organisms fall into strictly one category.
- Population growth is an increase in the number of individuals in a population. The growth rate of a population, also known as the rate of natural increase, is defined as the birth rate minus the death rate.
- Exponential growth is the unrestricted growth experienced by a population growing at a constant rate. Logistic growth is the slowing of the growth of a population because of environmental factors such as crowding and lack of food.
- Carrying capacity is the maximum population size that an area can support, given its food supply and other life-sustaining resources. Populations cannot grow exponentially forever; eventually, they hit the carrying capacity for the region and stop growing.
- Population growth can be limited by a variety of factors, including biotic (living) and abiotic (nonliving) parts of the environment.
- Density-independent factors, such as a severely cold winter, can affect a population of any size.
- Density-dependent factors, such as the presence of predators, have different impacts on the population depending on the population size and crowding of individuals.
- Populations in a community are interconnected, the fate of one often influencing the fate of the others.

MORE TO EXPLORE

- **Wolves and Moose on Isle Royale: http://isleroyalewolf.org**
- **Vucetich, J. A., and Peterson, R. O. (2016) Ecological studies of wolves on Isle Royale, Annual Report 2015–2016.**
- **PBS (2007) *Nature*: In the Valley of the Wolves**
- **Mlot, C. (2017) Two wolves survive in world's longest running predator-prey study. *Science*: www.sciencemag.org/news/2017/04/two-wolves-survive-world-s-longest-running-predator-prey-study**
- **Seabird Monitoring on the Isle of May, Scotland: www.ceh.ac.uk/sci_programmes/IsleofMayLong-TermStudy.html**

CHAPTER 20 Test Your Knowledge

DRIVING QUESTION 1 What is ecology, and what do ecologists study?

By answering the questions below and studying Infographics 20.1, 20.2, and 20.3, you should be able to generate an answer for the broader Driving Question above.

KNOW IT

1 **What is the difference between a community and a population?**

2 **An ecosystem ecologist might study**

a. plant populations.
b. herbivores that eat the plants.
c. predators in the population.
d. the impact of precipitation patterns on the plant populations.
e. all of the above

3 **Palm trees can be found at oases in deserts. How would the dispersion pattern of desert palm trees be described?**

USE IT

4 How would you explain to a 10-year-old what ecologists do?

5 Your local environmental group wants to determine the population size of squirrels in a nearby nature preserve. What are some methods you could use to estimate the size of the squirrel population? Would the same approaches be as useful in determining the population size of maple trees in the same area? Why or why not?

6 An invasive plant species has been introduced into a particular region. It is displacing native plants in the area. What kinds of impact might you predict on other species in the community? Explain your answer.

DRIVING QUESTION 2 What are the different types of population growth?

By answering the questions below and studying Infographic 20.4, you should be able to generate an answer for the broader Driving Question above.

KNOW IT

7 Which of the following would cause a population to grow?

- **a.** identical increases in both the birth rate and the death rate of a population
- **b.** a decrease in the birth rate and an increase in the death rate of a population
- **c.** an increase in the birth rate and a decrease in the death rate of a population
- **d.** an increase in the birth rate and a larger increase in the death rate of a population
- **e.** identical decreases in the birth rate and the death rate of the population

8 Which of the following statements describes an example of population growth?

- **a.** The average weight of Americans has increased substantially in the past decade.
- **b.** Tropical fish have been found in waters more northerly than their usual habitat.
- **c.** The number of people in a town has increased by 25% in the past 5 years.
- **d.** The number of butterflies in a region has stayed the same from 1950 to 2010.
- **e.** all of the above

9 When a population reaches its carrying capacity, what happens to its growth rate?

USE IT

10 The wolf population on Isle Royale was nine in 2014, three in 2015 and two in 2016.

- **a.** What must have happened to wolf birth and/or death rates between 2013 and 2016?
- **b.** If the winter of 2017 is very cold and an ice bridge linking Isle Royale and the mainland forms, what factor could contribute to growth of the wolf population? What impact would this have on the moose population?

apply YOUR KNOWLEDGE

INTERPRETING DATA

11 Population Q has 100 members. Population R has 10,000 members. Both are growing exponentially at a 5% annual growth rate.

- **a.** Which population will add more individuals in 1 year? Explain your answer.
- **b.** After 5 years, what will be the size of each population?
- **c.** If the larger population reaches its carrying capacity at the end of the third year, what will its size be after 5 years?

DRIVING QUESTION 3 What factors influence population growth and population size?

By answering the questions below and studying Infographics 20.5, 20.6, 20.7, 20.8, and 20.9, you should be able to generate an answer for the broader Driving Question above.

KNOW IT

12 The Mexican gray wolf has been reintroduced into parts of New Mexico and Arizona. There have been several wolf deaths due to shootings and traffic kills. Are these influences on the wolf population biotic or abiotic factors?

13 Drought causes a pond to dry up almost completely. The pond's frog population dropped to 10% of its initial size. What kind of factor—biotic or abiotic—influenced the frog population?

14 Which of the following is a density-dependent factor influencing population growth?

- **a.** elevated temperature
- **b.** prolonged winters
- **c.** a viral disease
- **d.** a devastating forest fire
- **e.** all of the above

USE IT

15 **Why is it important for researchers to determine the cause of death of moose on Isle Royale? Can this information be used to help make predictions about moose and wolf populations? Explain your answer.**

16 **A group of predatory fish lives in a school in a large lake. If a parasite were introduced to the lake—for example, by a vacationing fisherman—would you expect it to have a greater impact on the population if the fish were at high density or at low density? (Assume the parasite is passed from one fish to another through the water but can remain alive in the water only for a very short period of time.) What would happen to this same population if there were a severe drought and very hot summer?**

17 **Classify each of the following as a biotic or an abiotic factor in an ecosystem. Then predict the impact of each factor on the moose population of Isle Royale. Explain your answers, keeping in mind possible interactions between the various factors and between the moose and wolf populations.**

a. hot summer temperatures
b. ticks that parasitize moose
c. declining numbers of balsam fir trees
d. a parvovirus in wolves
e. deep winter snowfall

18 **Assume that a new herbivore is added to Isle Royale that is not a prey for wolves. Predict the effect of this introduction on**

a. the populations of trees.
b. the moose population.
c. the wolf population.

19 **Using the estimates provided below, graph the moose population size on Isle Royale since 2009. Given the trend in the moose population, how can you explain the reduction in wolves from 24 in 2009 to 2 in 2016?**

Year	Estimated Moose Population
2009	530
2010	510
2011	515
2012	750
2013	975
2014	1,050
2015	1,250
2016	1,300

MINI CASE

apply YOUR KNOWLEDGE

20 **The wolves of Isle Royale are suffering from bone deformities, probably as a result of inbreeding in their small population.**

a. Do you think that humans should intervene to save the wolves? Would your answer be different if the wolves were near human populations or agricultural centers?
b. If humans were to intervene, what kinds of strategies might help stabilize or increase the wolf population? Explain your answer.

BRING IT HOME

apply YOUR KNOWLEDGE

21 **Use your knowledge of ecology to plan a home saltwater aquarium. Think about the factors you need to consider in order to have a healthy community of fish and plants in your aquarium. Consider both biotic and abiotic factors and the population densities of the organisms in your aquarium.**

21 Community Ecology

bo1982/Getty Images

What's Happening to Honey Bees?

A mysterious ailment threatens a vital link in the food chain

Dave Hackenberg has been keeping bees for more than 40 years. Every spring, as flowering plants start to bloom, he trucks bees from his home in central Pennsylvania to farms around the country, where they help farmers pollinate local crops—everything from California almonds to Florida melons. In November 2006, as he had done for years, Hackenberg brought his buzzing cargo to his winter base in central Florida. When he dropped them off, his 400 healthy hives were "boiling over" with bees, he says. Three weeks later, when he returned to check on them, the bees had essentially vanished; only 40 healthy hives remained.

DRIVING QUESTIONS

1. What are keystone species in a community, and why are pollinators considered keystone species?
2. What are food chains and food webs, and how does energy flow through them?
3. What kinds of interaction occur among members of a community, and how do these influence the structure of the community?

COMMUNITY A group of interacting populations of different species living in the same area.

POLLEN Small, thick-walled plant structures that contain cells that develop into sperm.

POLLINATION The transfer of pollen from male to female plant structures so that fertilization can occur.

Mysteriously, there were no dead bees lying in or near the hives. Nor were there any signs of intruders who might have destroyed the hives in search of honey. The bees were simply gone. It was, as Hackenberg said, a bee ghost town.

"I literally got down on my hands and knees and looked between the stones for dead bees," says Hackenberg, but the beekeeper found none. "I was kind of speechless. And people know I'm not speechless."

Hackenberg's was the first reported case of what has since become known as colony collapse disorder, or CCD. But Hackenberg was not alone. Surveys conducted in 2007 and 2008 by the U.S. Department of Agriculture (USDA) and the Apiary Inspectors of America found that beekeepers all across the United States had suffered similar unexplained devastation, losing anywhere from 30% to 90% of their colonies.

Since that first case, more than 10 million honey bee colonies across the United States have been wiped out, with American beekeepers losing an average of 30% to 40% of their colonies every year since 2006. In 2016, losses reached 44%. While some annual loss of bee colonies is considered normal, these numbers are more than double historical averages. To date, CCD has been documented in 35 states and in Canada, as well as in Europe and Asia.

What's happening to honey bees? No one knows for sure, but it's a predicament that has beekeepers, farmers, and scientists racing to understand and combat the plight of the precious pollinator. At stake are not only billions of dollars' worth of agricultural crops but also the health and diversity of natural **communities** that rely on the valuable services of bees.

Pollination Partners

A hundred million years ago, when dinosaurs roamed Earth, the plant world was dominated by cone-bearing conifers such as pine and redwood trees, which spread their **pollen** by the wind; when the pollen reaches another plant, fertilization of an egg can occur. But a new type of **pollination** was evolving at this time, one that would forever change the terrestrial landscape: pollination by insects.

2014–2015 Annual Losses of Managed Honey Bee Colonies by State

Few bees still remain in this hive displaying colony collapse disorder.

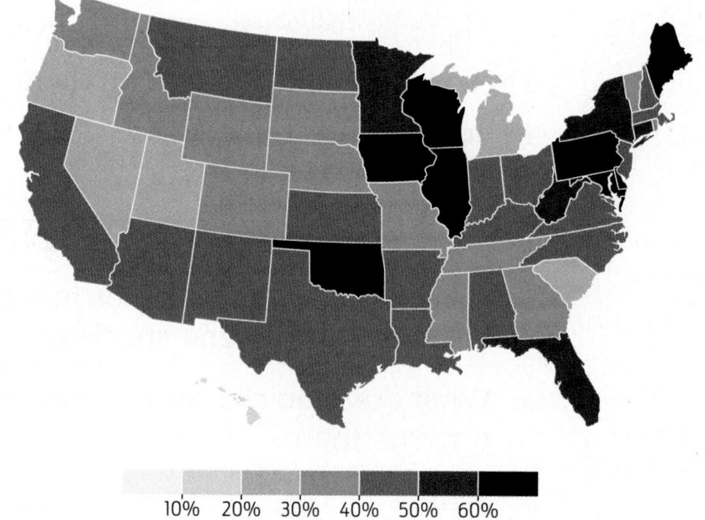

Data from the Colony Loss survey of the Bee Informed Partnership. USDA.

Pollinator Diversity

Flowering plants attract pollinators with their colorful flowers and fragrant nectar. When a pollinator such as a bee visits a flower to collect energy-rich nectar, it also picks up pollen. When it visits another flower, the pollen is transferred to that flower, helping the plant to reproduce. A variety of animals pollinate plants that are important to agriculture and to natural communities.

Silvrshoot/Getty Images

Monarch butterfly
Danaus plexippus

Alexander Gabrysch/Getty Images

Green-eyed hook-tailed dragonfly
Onychogomphus forcipatus

Merlin Tuttle/Getty Images

Long-nosed bat
Leptonycteris curasoae

Claudius Thiriet/Getty Images

Bumble bee
Bombus

Glenn Bartley/AGE Fotostock

Fork-tailed woodnymph hummingbird
Thalurania furcata

With the arrival on the evolutionary scene of pollinating insects like bees and butterflies, flowering plants multiplied, diversified, and radiated around the globe. Their great success owes everything to the reproductive advantage of relying on insects, rather than wind, to deliver pollen.

Wind pollination is like junk mail: you need to send thousands of letters to hook just one receptive customer—or in this case, to fertilize just one egg. Pollinating insects, on the other hand, are like FedEx: they deliver a pollen package directly to the appropriate recipient, contributing to the reproductive success of the plants.

Of the 250,000 species of flowering plants that exist worldwide, 90% are dependent on animal pollinators to reproduce. The vast majority of these pollinators are insects—beetles, bees, butterflies, and moths, mostly. By far the most important ecologically are bees. In fact, in many natural environments, bees represent **keystone species,** those that play a central role in holding the community together.

You can think of a keystone species as analogous to the keystone in an archway—it doesn't support as much weight as the other stones, but if it is removed, the archway collapses. Similarly, keystone species play a crucial role in supporting a functioning community. While not necessarily—or even frequently—the most abundant members of a community, they exert an outsized influence because of their important ecological services. Loss of a keystone

KEYSTONE SPECIES
Species upon which other species depend and whose removal has a dramatic impact on the community.

species from a community can spell disaster (**INFOGRAPHIC 21.1**).

The majority of beekeepers in the United States and Europe cultivate the bee species *Apis mellifera*, the Western honey bee. It is hard to overestimate the importance of this tiny pollinator to modern agriculture. In the United States alone, at least 90 different commercial crops–worth an estimated $15 billion annually–are dependent on honey bee pollination, including apples, oranges, blueberries, melons, pears, pumpkins, cucumbers, cherries, raspberries, broccoli, avocados, asparagus, clover, alfalfa, and almonds. Globally, 87 of the 115 leading food crops are dependent upon pollinators, the most important of which are honey bees. Pollinator-dependent crops represent about 35% of global crop production.

"One in every three bites of food we eat is pollinated directly or indirectly by honey bees," says Dennis vanEngelsdorp, an apiarist at the University of Maryland and project director for the Bee Informed Partnership, which has been monitoring bee losses. Without honey bees, he says, we wouldn't starve–we would still have wheat, rice, corn, and other crops that are either wind- or self-pollinated–but many of our favorite foods might no longer grace our tables (**INFOGRAPHIC 21.2**).

For bees, flowers provide food: they contain the protein-rich pollen and sugary nectar that bees need to nourish themselves and their hives. With their long tonguelike proboscis, bees are able to reach deep into a flower to draw out the nectar. Being fuzzy and having a slight electrical charge, bees attract pollen as they snuggle up to a flower the way warm socks attract other

INFOGRAPHIC 21.1 Keystone Species Hold Communities Together

As keystone species, bees play a fundamental role in supporting the entire community. Whether or not a keystone species is the most abundant member of its community, its loss has a huge impact on the community.

Bees are keystone species:
The presence of a keystone species supports the success of many other species in a community, just as a keystone supports the weight of other stones in an arch.

Loss of a keystone species:
The loss of a keystone species devastates a community, just as the removal of the keystone from an arch causes it to collapse.

? Does abundance or impact define a species as a keystone species in a community?

INFOGRAPHIC 21.2 Commercial Crops Require Bees

Many of the crops that we rely on for food, fuel, and fiber rely on bees for their pollination and reproduction.

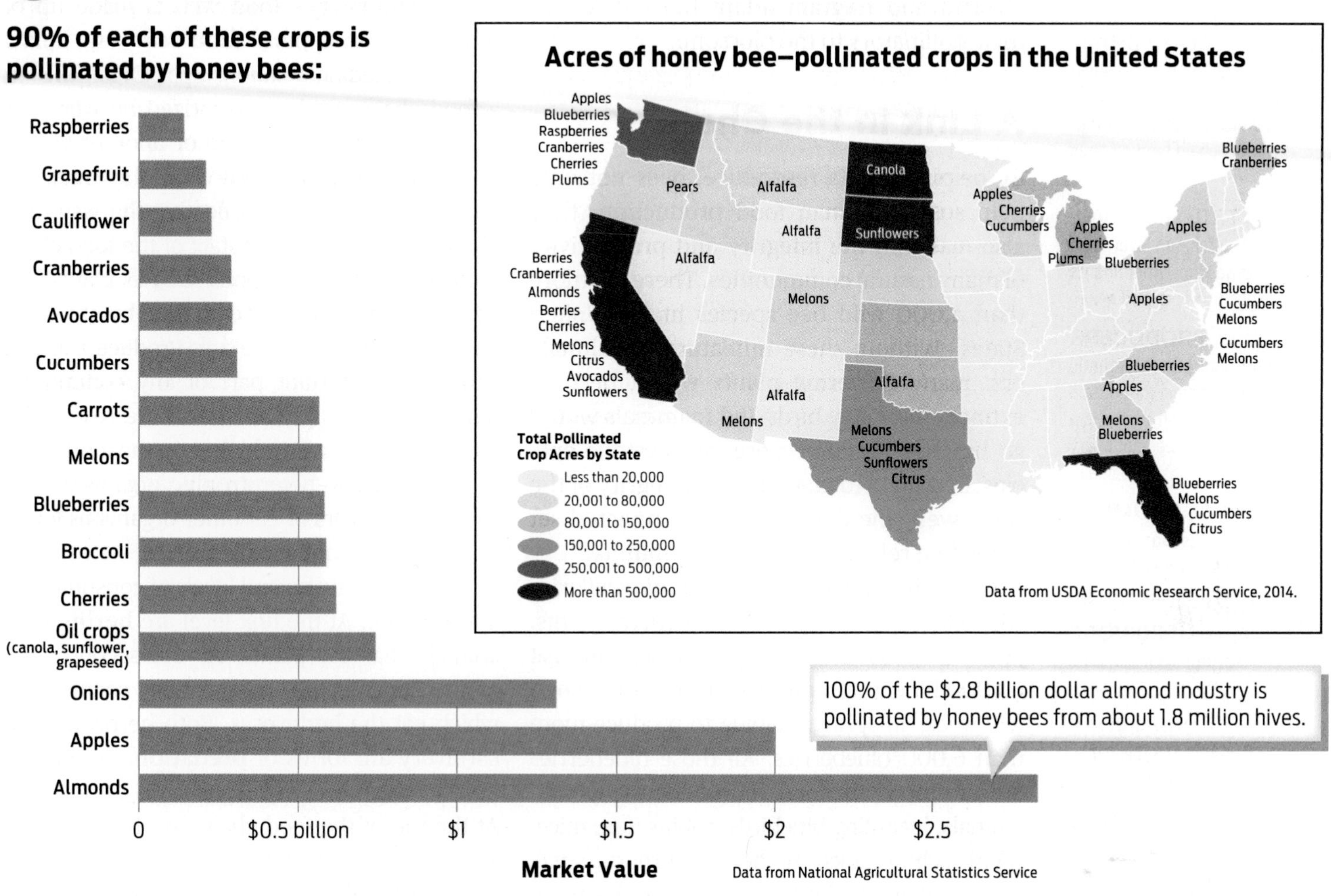

? From the data presented, what is the total market value of honey bee pollination of the crops shown?

clothes as they come out of the dryer. The bees then transfer this pollen to other plants as they continue their hunt for food.

Honey bees are more efficient at pollination than many other types of pollinators, which is why farmers have come to depend on them. The average honey bee will make 12 or more foraging trips a day, visiting several thousand flowers. On each trip, she (the foragers are all female) will confine herself to flowers from a single plant species, thus ensuring delivery of the proper pollen. Because they come and go from a central home base—their hive—honey bees can be counted on to stay in a fixed area around a crop. And with roughly 40,000 individual bees per hive, this is a versatile workforce that's also easy for beekeepers to transport from crop to crop.

Of course, pollination isn't just about feeding bees and humans; it's also how flowering plants have sex. A flower is the reproductive hub of the plant—it is where its reproductive organs are located, male and female in the same flower in some species, separate flowers in others. In order for a flowering plant to reproduce, male pollen, containing sperm, must find its way to the female eggs of a plant of the same species. Pollinators make this possible.

FOOD WEB
A complex interconnection of feeding relationships in a community.

FOOD CHAIN
A linked series of feeding relationships in a community in which organisms further up the chain feed on ones below.

TROPHIC LEVEL
The feeding level of an organism, reflecting its position in a food chain.

PRODUCERS
Autotrophs (photosynthetic organisms) that obtain energy directly from the sun and form the base of every food chain.

CONSUMERS
Heterotrophs that eat other organisms lower on the food chain to obtain energy.

HERBIVORE
An organism that eats plants.

CARNIVORE
An organism (typically an animal) that eats animals.

PREDATION
An interaction between two organisms in which one organism (the predator) feeds on the other (the prey).

Plants don't just passively wait for pollinators; they actively court them. Masters of seduction, flowers have evolved countless colorful and fragrant adaptations that lure their pollinators to the blossom.

A Link in the Chain

By helping plants reproduce, bees not only help sustain human food production, they also maintain the integrity and productivity of many natural communities. There are more than 4,000 wild bee species in the United States. Without these miniature matchmakers, many flowering plants would become extinct, and many birds and mammals would go hungry. That's because, as keystone species, bees help to maintain the integrity of the **food web,** the complex, interconnected set of feeding relationships in a community.

Take the tiny blueberry bee (*Osmia ribifloris*), which feeds on the flowers of blueberry plants. Just one of these speedy pollinators can visit 50,000 blueberry flowers over the course of a few weeks in spring, helping to produce more than 6,000 blueberries. All those blueberries are an important food source for many wild animals, including bluebirds, robins, fox, mice, rabbits, chipmunks, and deer. In turn, the small animals that eat the blueberries are fed upon by larger carnivorous animals, such as hawks and coyotes. Finally, mushrooms and other decomposers (see Chapter 18) feed on these organisms when they die. Without the pollinating bees—the keystone in this community—the web unravels.

A community's food web is made up of multiple, overlapping **food chains,** linked series of feeding relationships. Organisms in a food chain can be categorized by who eats whom and assigned to one or another of several distinct **trophic levels** (from the Greek *trophé,* meaning "food"), reflecting the way they obtain their food. At the start of the food chain are **producers**—autotrophs such as plants and algae, which make their own food by obtaining energy directly from the sun. Producers are an extremely important part of any community because they supply energy to the rest of the food chain. Organisms higher up the food chain are **consumers**—heterotrophic organisms that eat the producers or eat other organisms lower on the chain to obtain energy.

There can be several levels of consumers in a food chain. At the first level are **herbivores,** animals that eat plants. At the second level of consumers are meat-eating **carnivores,** which eat the herbivores. Both herbivory and carnivory are forms of **predation,** an interaction in which one organism feeds on another. At the top of the food chain are those animals known as top consumers—animals such as coyotes, hawks, and wolves (as well as meat-eating humans), who have no natural predators and are not generally eaten by anything else in the community **(INFOGRAPHIC 21.3).**

Carnivory

Gail Shumway/Getty Images

Herbivory

John Serrao/Science Source

A wolf hunting a mouse and a deer nibbling on a bush are both examples of predation.

INFOGRAPHIC 21.3 Who Eats Whom: Food Webs and Food Chains

In a community, organisms are organized into complex food webs, which contain intersecting and overlapping food chains. The energy in a food web is captured from the sun by producers. Energy and nutrients are then transferred to other organisms in other trophic levels as they feed on producers (herbivores) and other consumers (carnivores).

Food Web

Individuals in a community form complex relationships based on their eating behaviors.

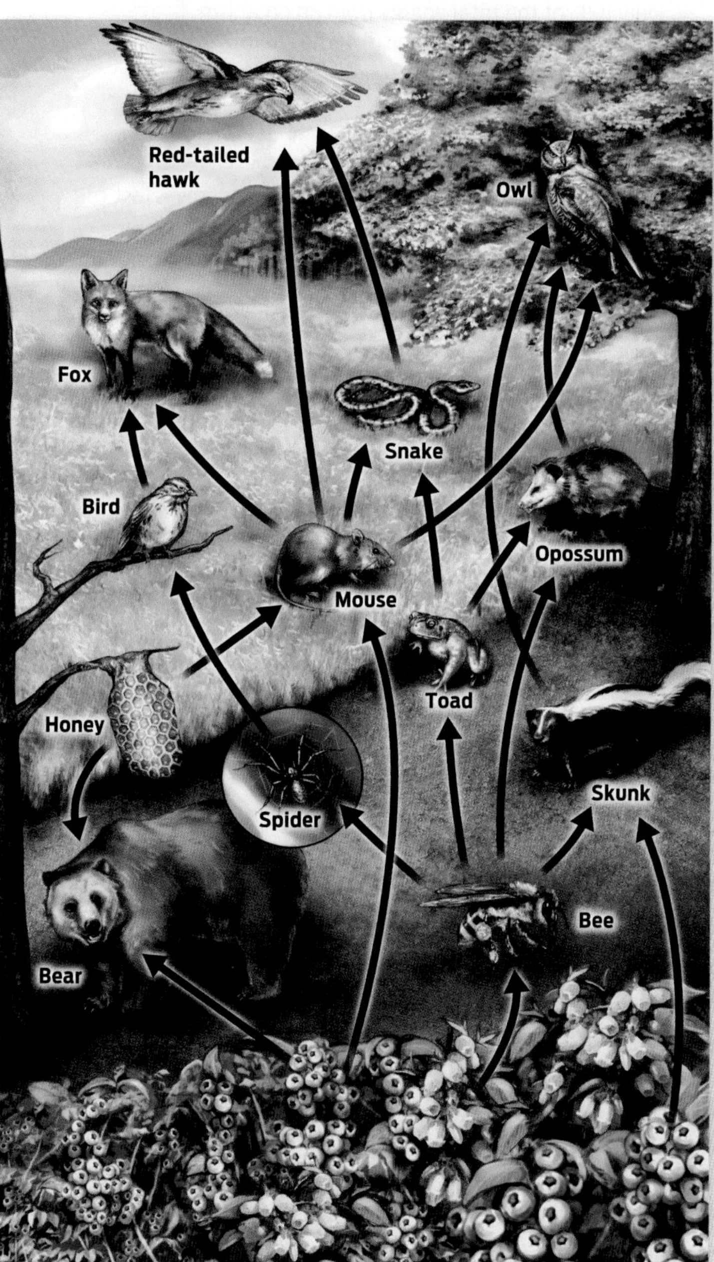

Food Chain

A food chain is a linked sequence of feeding events within a food web. Individuals in a food chain are classified into trophic levels by how they obtain food.

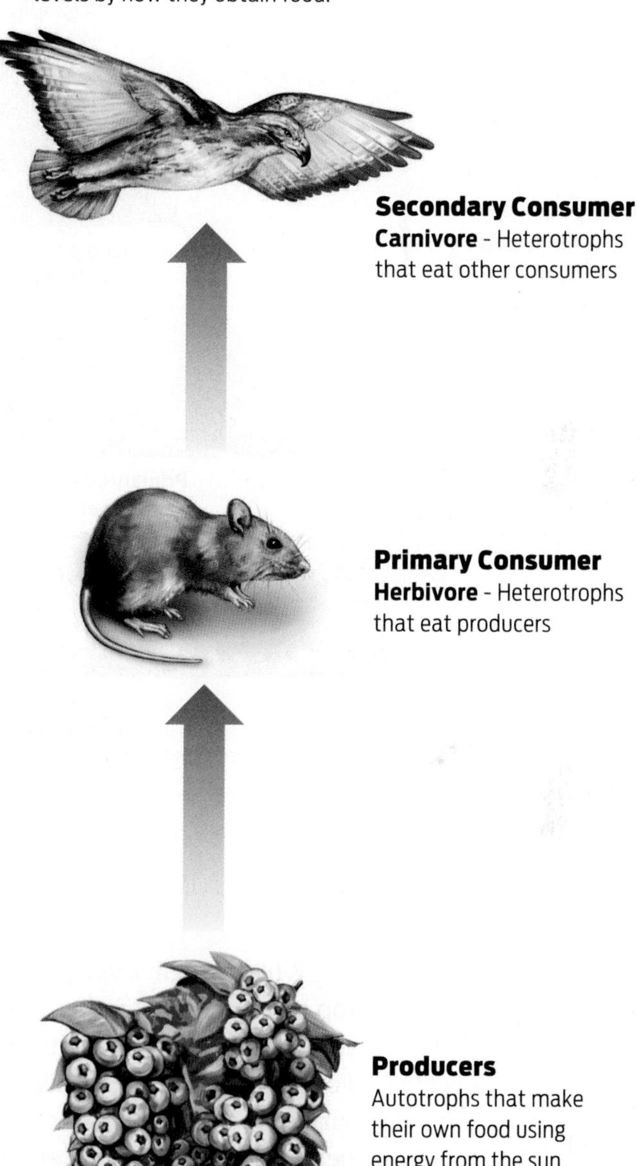

? Bears eat plants and berries, and also eat other animals (from insects to fish). How can bears be characterized: as producers? As consumers? As herbivores? As carnivores?

INFOGRAPHIC 21.4 Energy Is Lost as It Flows Through a Food Chain

In a food chain, energy flows in one direction: from producers to consumers. The passage of energy is not efficient, however, as only 10% of energy from one trophic level ends up stored in the bodies of the next. The result is an energy pyramid, which shows that fewer top consumers can be supported in a community compared to organisms lower down the food chain.

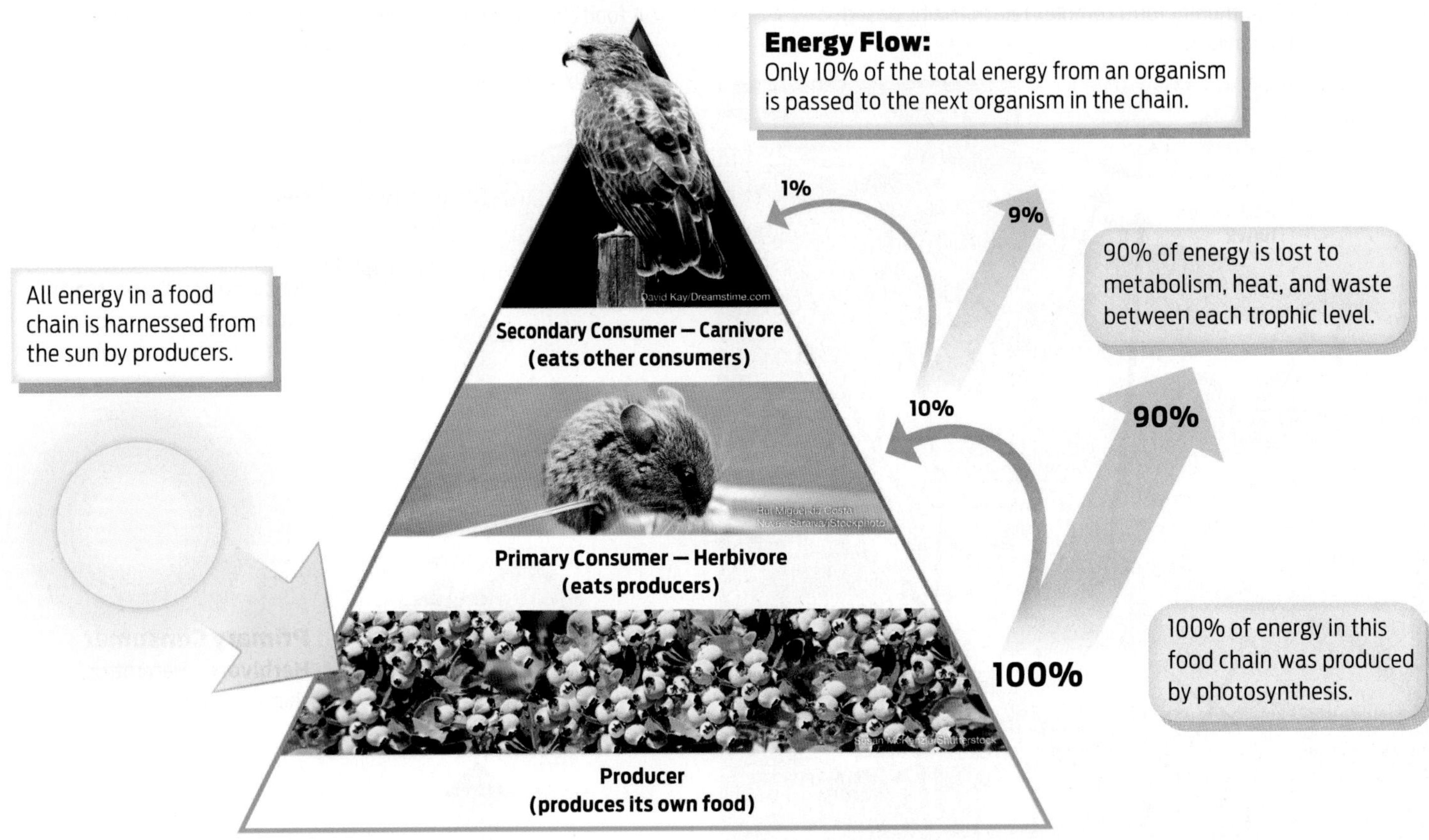
Energy Flow:
Only 10% of the total energy from an organism is passed to the next organism in the chain.
1%
9%
90% of energy is lost to metabolism, heat, and waste between each trophic level.
All energy in a food chain is harnessed from the sun by producers.
David Kay/Dreamstime.com
Secondary Consumer — Carnivore
(eats other consumers)
10%
90%
Primary Consumer — Herbivore
(eats producers)
100%
100% of energy in this food chain was produced by photosynthesis.
Producer
(produces its own food)

? If a predator were to eat a hawk, how much of the energy stored in the hawk's body would actually be stored in the top predator?

While it might seem preferable to be at the top of a food chain rather than at the bottom, there are downsides to being last in line to eat. For one, it's harder to obtain the energy necessary to live. As consumers prey on organisms below them in the chain, energy is transferred up the chain through each trophic level. But not all the energy stored in a lower level makes it to the level above it; at each step, much of this energy is lost from the chain.

When a mouse feeds on a blueberry bush, for example, most of the energy in the blueberries is either burned (in aerobic respiration) to fuel the mouse's activities and given off as heat, or passed through the mouse as indigestible plant fiber. Only a very small portion–about 10%–of the energy stored in the blueberries goes to putting weight on the mouse. The same is true for organisms, like birds, that eat the mouse that eat the blueberries–only about 10% of the energy in the mouse makes it into the body mass of the bird (**INFOGRAPHIC 21.4**).

This is the main reason top carnivores like coyotes, hawks, and wolves are scarce on Earth, and why there are no predators of these creatures: there's simply not enough energy left in the chain to sustain more of them. (It's also why human vegetarianism is more energetically efficient than meat-eating: the same amount of a crop can feed many more vegetarians than meat-eaters who eat the animals that eat the crop.)

A Swarm of Problems

The United States is home to approximately 1,000 commercial beekeepers, who together cultivate about 2.5 million bee colonies. To a beekeeper–a small business owner–losing 40% of his or her colonies represents an unsustainable financial loss. The sudden bee disappearances that have been occurring since 2006 are a serious worry for beekeepers.

> "One in every three bites of food we eat is pollinated directly or indirectly by honey bees."
>
> —Dennis vanEngelsdorp

While the losses from CCD have indeed been considerable, this is actually not the first time that beekeepers' livelihoods have been hit hard. Since 1987, beekeepers have had to battle significant annual losses from an aggressive pest: the blood-sucking varroa mite.

A non-native species, the varroa mite was likely introduced into the United States on the backs of imported bees. The sesame seed-size freeloader is a parasite that feeds on bees' blood, weakening the bees' immune systems and spreading viruses. **Parasitism** is another important relationship in a community, a type of **symbiosis**–a close relationship between two species–in which one species (in this case, the mite) clearly benefits, and one species (in this case, the honey bee) clearly loses. Because it involves one species feeding on another, parasitism is also a form of predation.

Not all symbiotic relationships are harmful to one of the partners; they can sometimes be mutually beneficial. Bees and flowering plants are a perfect example of one such **mutualism.** Bees can't survive without the flowers, which provide food, and plants depend on the bees to help them reproduce. Honey bees have other mutualistic symbioses, as well, including with bacteria that live safely inside the bees and benefit their hosts by helping them combat disease.

A third type of symbiotic relationship is **commensalism,** a relationship in which one species benefits while the other is unaffected or unharmed–bees living in a hollowed-out oak tree, for example **(INFOGRAPHIC 21.5)**.

As devastating as the parasitic varroa mite infestation has been, it is unlikely to be the sole or even the primary factor responsible for the most recent colony collapses. Research by apiarist vanEngelsdorp and others has shown that levels of mite infections in collapsing colonies are no higher than they had been in previous years. Moreover, a mite infestation does not explain the most curious aspect of the condition: the sudden disappearance of entire hives.

Not surprisingly, the sudden disappearances have fueled intense speculation among beekeepers and laypeople alike about what's going on. Hypotheses have included everything from pesticides, viruses, and genetically modified crops to cell phone radiation, global warming, and even alien abduction. It's a baffling who-done-it with many suspects but no smoking gun.

PARASITISM
A type of symbiotic relationship in which one member benefits at the expense of the other.

SYMBIOSIS
A relationship in which two different organisms live together, often interdependently.

MUTUALISM
A type of symbiotic relationship in which both members benefit; a "win–win" relationship.

COMMENSALISM
A type of symbiotic relationship in which one member benefits and the other is unharmed.

INFOGRAPHIC 21.5 Organisms May Live Together in Symbioses

Symbioses are relationships in which different species live together in close association. These associations can provide benefits, harm, or have no effect on the partners involved.

Mutualism: Both species benefit from the relationship

Bees and flowers
Bees get nectar and pollen for food. Flowers benefit from successful reproduction.

Anand Varma/Getty Images

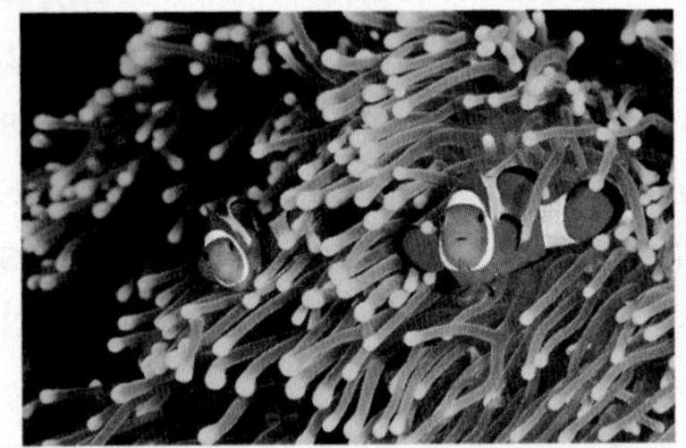

ullstein bild/Getty Images

Clownfish and anemone
Clownfish nestle in the tentacles of a poisonous anemone. The clownfish is protected from predators, and the anemone feeds on other fish attracted by the clownfish.

Parasitism: One species benefits and the other is harmed

Varroa mites and bees
The varroa mite feeds on bee adults and larvae. The bee's immune system is compromised and it becomes susceptible to disease.

Paul Tessier/Getty Images

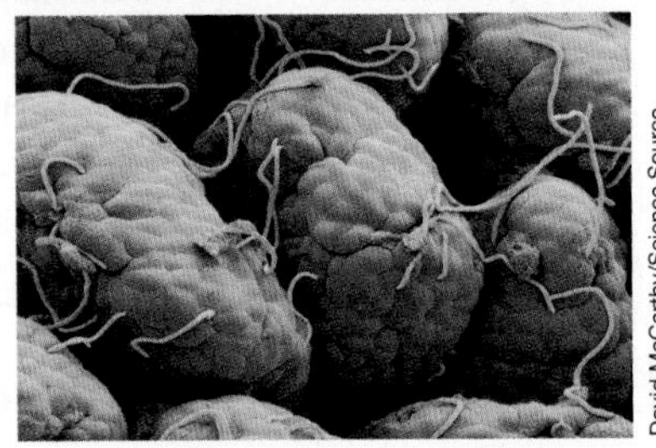

David McCarthy/Science Source

Pinworms and humans
The pinworm infects humans and feeds on food in the human gut. The human suffers discomfort.

Commensalism: One species benefits and the other is unharmed

Bees and trees
Natural bee swarms make safe hives on trees. The tree is not harmed.

Tim Graham/Getty Images

John Elk III/Getty Images

Egrets and water buffalos
The egret feeds on bugs that are stirred up by the African water buffalo. The water buffalo neither benefits nor is bothered.

What is the difference between commensalism and mutualism?

Honey Bee Forensics

Among the first to investigate the die-offs was a team of Pennsylvania State University biologists headed by vanEngelsdorp and Diana Cox-Foster. It was Cox-Foster whom beekeeper Hackenberg called the day in 2006 he saw that his bees had gone missing.

The team started their investigation by performing autopsies on the few remaining bees in Hackenberg's colonies. When vanEngelsdorp looked through his microscope, he was shocked: "I found a lot of different scar tissue, and [what] looked like foreign organs," he says. There were also signs of multiple infections, including a parasitic fungus called *Nosema ceranae*. The bees' insides were overrun with pathogens.

Though the bees were clearly sick, each collapsed colony seemed to suffer from a different spectrum of ailments. The researchers hypothesized that something had compromised the bees' immune system, making them vulnerable to infections that a healthy colony could normally fend off. Some observers have even likened the condition to "bee AIDS."

Hoping to isolate a previously unidentified culprit, in 2007 Cox-Foster and her colleagues enlisted genomics experts from Columbia University to examine genetic material from the hives for evidence of a new invader. After months of intensive work, their efforts seemed to pay off: genetic tests revealed that a virus called Israeli acute paralysis virus (IAPV) was present in 96% of the hives affected with CCD. The researchers thought they had found the smoking gun. Subsequent research, however, showed that not all honey bee colonies that are infected with IAPV have symptoms of CCD, suggesting that the virus alone is not the source of the problem.

There may in fact be no single cause of CCD, but rather a complex combination of causes. "All the evidence so far has really supported the idea that it's likely a combination of factors that are stressing the bees beyond their ability to cope," says Maryann Frazier, a bee researcher at Penn State who is part of Cox-Foster's team.

"All the evidence so far has really supported the idea that it's likely a combination of factors that are stressing the bees beyond their ability to cope."

—Maryann Frazier

One factor that almost certainly plays a role in exacerbating the condition is poor nutrition. Just like humans, bees need a well-balanced diet that contains all the essential nutrients to remain healthy. For a number of reasons, honey bees are finding it harder and harder to obtain a nutritious diet.

Competing for Resources

The Western honey bee is native to Europe, the Middle East, and Africa, but not to the United States, making it an **introduced species** here. When European settlers first brought the Western honey bee to the United States in the 1600s, the bees quickly spread from managed colonies into the wild, in some cases displacing native bee species.

Honey bees are excellent at colonizing new habitats because they are largely generalists when it comes to flower choice. Though they tend to visit a single species of flower on each foraging trip, honey bees may visit more than 100 species of flowers within a single geographic region over the course of a season. In warm climates, they are active year round, and tend to feed throughout the day and start foraging earlier in the morning than many native bee species. In other words, honey bees have a broad ecological **niche**—the space, environmental conditions, and resources (including other living species) that a species needs in order to survive and reproduce.

Different pollinators generally have different niches, owing to their varying sizes and preferences for different flower types. Bees, for instance, are attracted to brightly colored blossoms with yellow, blue, or purple petals—but not red ones. Butterflies, by contrast, are commonly attracted to red flowers that are large and easy to land on. Moth-pollinated flowers tend to have pale or white petals with no distinctive color pattern but with strong fragrance. These different adaptations allow multiple pollinators to partition available resources and exist in the same area (**INFOGRAPHIC 21.6**).

When two or more species rely on the same limited resources—that is, when their

INTRODUCED SPECIES
Species that are not native to a particular environment and which have arrived as a result of human activity.

NICHE
The space, environmental conditions, and resources that a species needs in order to survive and reproduce.

INFOGRAPHIC 21.6 Organisms Share Similar Niches by Resource Partitioning

Pollinators have different ecological niches, even within the same general space. They may require the same seasonal temperatures, structures for shelter, yearly rainfall, and food from the nectar and pollen of flowers, but prefer flowers of different size, color, and shape.

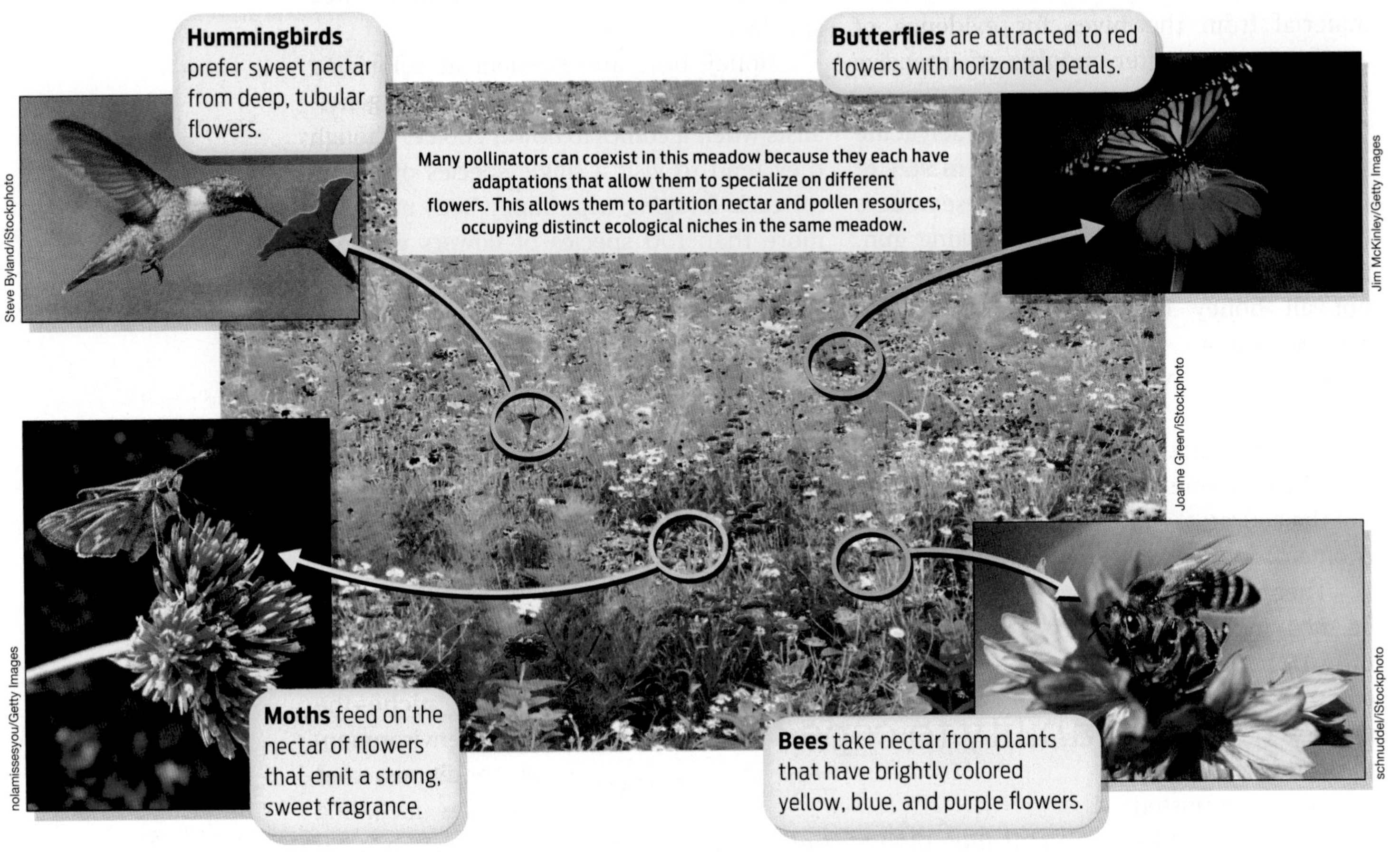

Hummingbirds prefer sweet nectar from deep, tubular flowers.
Butterflies are attracted to red flowers with horizontal petals.
Many pollinators can coexist in this meadow because they each have adaptations that allow them to specialize on different flowers. This allows them to partition nectar and pollen resources, occupying distinct ecological niches in the same meadow.
Moths feed on the nectar of flowers that emit a strong, sweet fragrance.
Bees take nectar from plants that have brightly colored yellow, blue, and purple flowers.
Steve Byland/iStockphoto
Jim McKinley/Getty Images
Joanne Green/iStockphoto
nolamissesyou/Getty Images
schnuddel/iStockphoto

? If bees, butterflies, and hummingbirds all rely on nectar, how can they coexist in a meadow of wildflowers?

COMPETITION
An interaction between two or more organisms that rely on a common resource that is not available in sufficient quantities.

COMPETITIVE EXCLUSION PRINCIPLE
The concept that when two species compete for resources in an identical niche, one is inevitably eliminated from that niche.

INVASIVE SPECIES
Introduced species that do harm in their new environment.

niches overlap—the result is **competition.** Competition tends to limit the size of competing populations and may even drive one out. The **competitive exclusion principle** holds that no two species can successfully coexist in identical niches in a community because one will eventually out-compete the other. Competition puts selection pressure on species to evolve adaptations to carve out a new niche. The different preferences of pollinators for different flowers, in fact, may be a result of this sort of selection pressure.

Some species compete through behavior. African honey bees (*Apis mellifera scutellata*), for example, are an **invasive species** that was brought from Africa to Brazil in 1956 and which quickly expanded their range to include Central America and the southern United States. When African bees mate with local varieties they produce hybrid species with a blend of traits: these so-called Africanized honey bees are much more aggressive than Western honey bees. They will chase away other pollinators from food sources, and even swarm and sting animals that get near their hives–behaviors earning them the colloquial name "killer bees" (**INFOGRAPHIC 21.7**).

Some scientists are worried that killer bees from Central America may displace or interbreed with U.S. populations of Western honey bees, with potentially disastrous results for beekeeping and agriculture. This hasn't happened–yet. The much more serious problem for the Western honey bee, it seems, is competition from an even more dangerous species: humans. Humans and their activities have limited the resources used by bees and other pollinators. Agriculture, suburban sprawl, and development, for example, have all decreased bees' natural forage areas and fragmented their habitat into nonoverlapping zones. Unable to access as many resources in a single foraging trip, bees must compete with each other in the patches that remain.

The ubiquitous well-manicured lawn is particularly problematic. An immense stretch of green grass and no flowers, a lawn is "basically a desert to pollinators," says Frazier. There is literally nothing for them to eat. Likewise, many agricultural areas are planted with monocultures (that is, single crops), which all bloom at the same time, leaving no flowers for the rest of the year. Worse yet, certain genetically engineered pollen-free crops trick bees into thinking they'll find food only to leave them hungry. And certain non-native plants have floral structures that are inaccessible to indigenous pollinating insects.

As a consequence of these and other human actions, in some geographic regions bees must compete both with one another and with other pollinators for a dwindling supply of food-providing flowers. Many are going hungry, and some are in danger of being forced out completely.

INFOGRAPHIC 21.7 Organisms Compete for Resources

Species relying on the same limited resource may display competitive behaviors. Species may out-compete one another for the resource, or find a balance between one another, depending on their abilities and behaviors.

H.S. Photos / Alamy

Blueberry bee
(*Osmia ribifloris*)

Food Partitioning
Native bees like the blueberry bee successfully coexist with non-native bees by specializing on one food (blueberry flowers) that the honey bees can't use efficiently.

ElementalImaging/iStockphoto

Honey bee
(*Apis mellifera*)

Generalist Foraging Patterns
Imported honey bees can forage over great distances and feed on a wide variety of flowers. As they feed year round in warmer climates, they are very successful in their competition with other bee species, which may have more limited niches.

NIC BOTHMA/European Pressphoto Agency/Newscom

Africanized honey bee
(*Apis mellifera scutellata*)

Defensive Behavior
Killer bees compete successfully because of their aggressive defense of food resources. They chase other pollinators away from available food.

? Are killer bees relying on resource partitioning or on competition for success in their niche?

Honey Bee in the Coal Mine?

Honey bees aren't the only pollinator in peril. According to a 2016 report issued by the Intergovernmental Science-Policy Platform on Biodiversity and Ecosystem Services, the number and abundance of pollinator species have declined greatly over the last several years. In fact, several bumble bee species are becoming or have become extinct in North America. The concern among researchers and beekeepers is that honey bees may be the "canary in the coal

Curt Pickens/iStockphoto

mrolands/Featurepics.com

Maigi/Dreamstime.com

Well-manicured lawns (top) and farms planted with monocrops (center) are "a desert to pollinators." Yards with wildflowers and diverse plants (bottom) are more welcoming to pollinating creatures.

mine," forecasting what's in store for other pollinators. "It's not only the honey bees that are in trouble," says Hackenberg. "All the beneficial insects are in a bad situation."

What's ailing these insects? In addition to a shrinking and fragmented habitat, a disquieting possibility is that they are being poisoned by pesticides. Penn State researcher Frazier and her colleagues have looked at pollen and wax from beehives and found large amounts of many different kinds of pesticides, some of which are approaching toxic levels for the bees. "Pesticides are definitely in the mix and we think they are definitely a player in the stresses that bees are experiencing," says Frazier.

Of particular concern to beekeepers is a class of pesticides known as neonicotinoids, or "neonics" for short. Neonics are an artificial form of nicotine heavily used in commercial agriculture all over the world. (Nicotine, produced by tobacco plants, is a natural deterrent to plant-eating insects.) Virtually all corn and most soybean seeds in the United States are treated with neonics. Research has shown that neonics can impair honey bees' ability to find and return to their hives, and the U.S. Environmental Protection Agency (EPA) acknowledges that neonics are "highly toxic to honey bees." The agency has begun a detailed review of neonic toxicity, yet so far it has not restricted use of these pesticides in the United States. A recent report produced by the USDA acknowledged that neonics and other pesticides may be a contributing factor to CCD but stopped short of singling them out as a primary cause. "The bottom line is we have not been able to put together, here in the United States, the evidence that neonicotinoids are causing the decline in honeybees," says Frazier. "But we are not convinced that they are *not* playing a role."

The situation is a bit different in Europe. In 2013, the European Union imposed a temporary continent-wide ban on neonicotinoids on flowering crops such as corn, rapeseed, and sunflowers that are attractive to bees after a report by the European Food Safety Authority identified a number of risks posed to bees from the pesticides. The provisional ban, which took effect in 2014, will allow researchers the chance to study the effects of neonics more thoroughly–especially at the sublethal doses bees are likely to encounter. The ban was set to be reviewed in 2017.

Although researchers have not yet been able to prove that neonics are playing a role in CCD–"the jury is still out," says Frazier–beekeepers like Hackenberg are understandably cautious about what pesticides they expose their colonies to (**INFOGRAPHIC 21.8**).

Because it may involve a complex combination of triggers, which may interrupt important community interactions, there is no easy remedy for CCD and other causes of bee losses. It may require making fundamental changes to our beekeeping and agricultural practices. In particular, we could break up fields of monocultures with varied bee-friendly plants: red clover, foxglove, and bee balm, for example. This would help diversify honey bee food webs. We could also use pesticides sparingly and avoid spraying at times of day when bees are actively foraging (although this won't necessarily help against neonics, which

INFOGRAPHIC 21.8 What Is Causing Colony Collapse Disorder?

The collapse of colonies all over the world is of great concern. The cause of this disorder is likely to be complex and to involve an interplay of several factors.

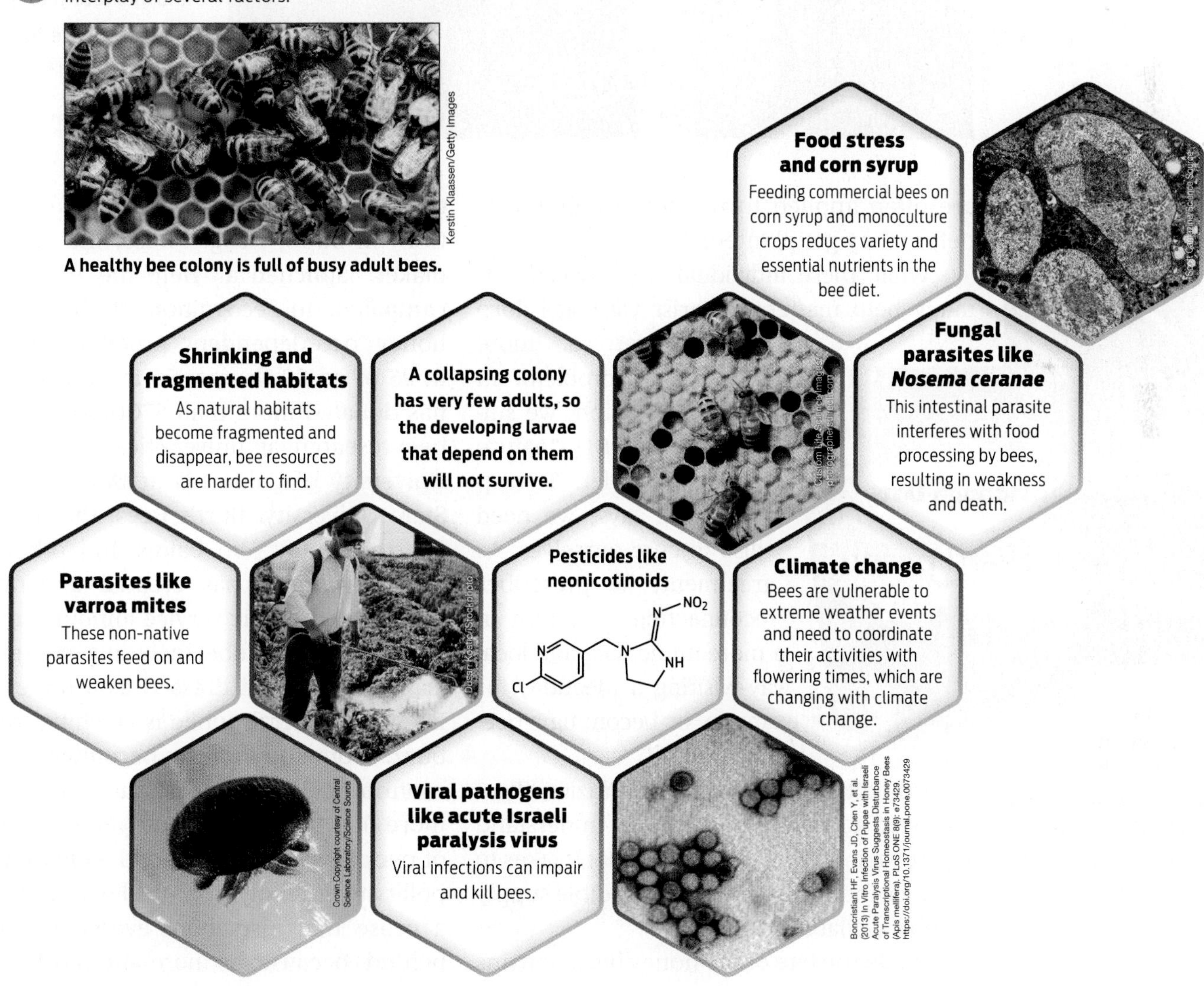

? What ecological role do honey bees play that makes their fate so important?

Pesticides May Contribute to the Decline of Bees

Pesticides are sprayed on crops to reduce predation by herbivorous pests. Bees have mutualistic relationships with many plants exposed to these pesticides. There is growing concern about the effects of pesticides on these natural keystone species, which are also important for agricultural success.

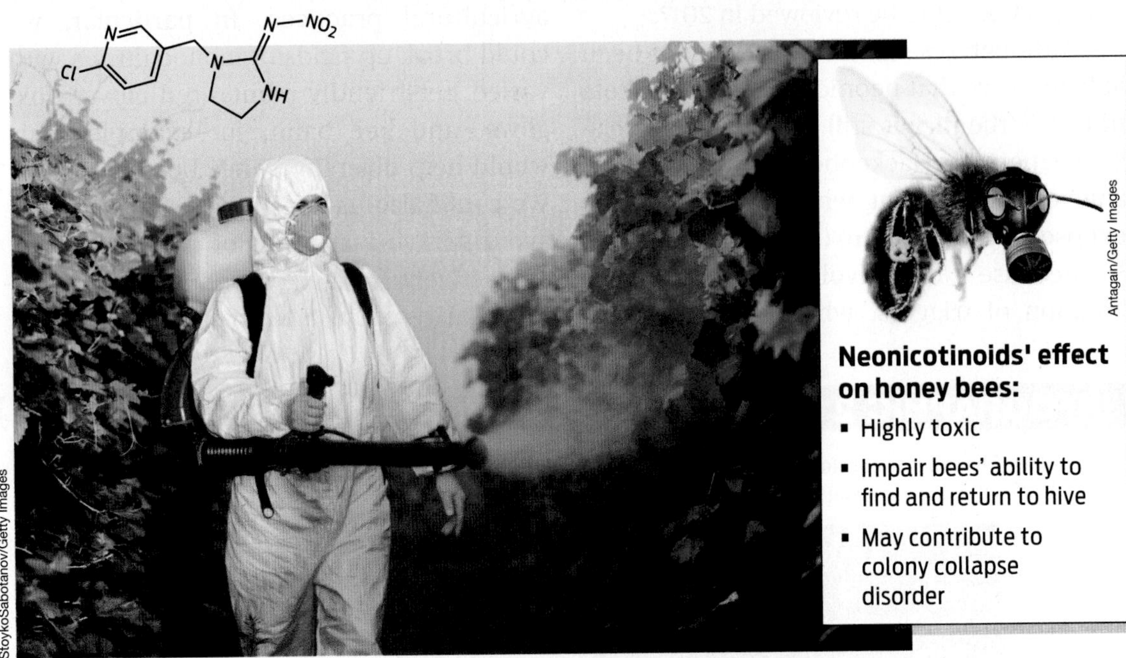

are often applied to the seeds themselves before they get planted).

While these individual steps would certainly help matters, apiarist vanEngelsdorp diagnoses a more systemic problem. In his estimation, we suffer from NDD—"nature deficit disorder." To help bees, he says, we need also to cure ourselves. As treatment, he prescribes reconnecting to nature in a more immediate and local way—visiting a meadow, for example, or becoming a beekeeper oneself.

Häagen-Dazs donates funds to support honey bee research, acknowledging the essential role that bees play in 26 of its 60 ice cream flavors.

In addition, says bee expert Frazier, "People need to take more time to understand where their food comes from, what it takes to produce food and have this incredible supply of food available to us."

While the fate of the honey bees remains uncertain, there are signs that a more bee-friendly awareness is beginning to emerge, thanks in part to the concerns raised by CCD. In 2008, Häagen-Dazs, the ice-cream maker, launched a "Help the Honey Bee" campaign, in recognition of the fact that honey bee-dependent products are used in 25 of its 60 flavors. So far, the company has donated more than $700,000 to honey bee research at universities including the University of California at Davis and Penn State University. Burt's Bees, the maker of "Earth-friendly" lip balms and other personal products, has created a series of CCD-related public service announcements, viewable on YouTube, including one starring Isabella Rossellini dressed as a honey bee. Even the general public is catching the bee buzz. From city-dwellers becoming amateur rooftop beekeepers to suburbanites letting more flowers grow in their yards, the ranks of people wanting to make the environment pollinator-friendly has swelled. And that's a cause that just about everyone can get behind—because, as more and more people are coming to realize, a world without honey bees just wouldn't be as sweet. ■

CHAPTER 21 SUMMARY

- An ecological community is made up of interacting populations of different species.
- Bees are keystone species because they play a fundamental role in supporting an entire community, much like the keystone in an arch.
- Bees are the primary pollinators for many species of flowering plants, which depend on the pollinators to transfer pollen between plants of the same species.
- The complex set of interconnected feeding relationships in a community makes up a food web. The food web is itself made up of linked series of feeding relationships called food chains.
- Organisms in a food chain can be placed into different trophic levels, based on how they obtain their nourishment. Organisms at the base of a food chain are producers—they obtain energy directly from the sun and supply it to the rest of the food chain; organisms higher up a food chain are consumers—they obtain energy by eating organisms lower on the chain.
- Energy flows in one direction through a food chain, from producers to consumers. As energy flows up the trophic levels in the food chain, most of it is lost to the environment, which explains why top consumers are rare in a community.
- Organisms can have different types of symbiotic relationships. In mutualistic symbioses, both members benefit; in parasitism, one member benefits while the other suffers; in commensalism, one member benefits while the other is unharmed.
- The space and resources that a species uses to survive and reproduce define its ecological niche. Some species have overlapping niches, leading to competition for resources.
- Colony collapse disorder may have no one single cause but may result from many interacting factors affecting bees, including poor nutrition, pathogens, and pesticides.
- Bees are not the only pollinators in peril. Human development and agriculture have decreased habitat and foraging areas for many natural pollinators, resulting in increased competition among them.

MORE TO EXPLORE

- **Bee Informed: http://beeinformed.org**
- **Woodcock, B.A., et al. (2016) Impacts of neonicotinoid use on long-term population changes in wild bees in England. *Nature Communications* 7: 12459.**
- **Dennis vanEngelsdorp (2009) Colony collapse disorder: A descriptive study. *PLoS ONE* 4(8):e6481.**
- **Cox-Foster, D. L. (2007) A metagenomic survey of microbes in honey bee colony collapse disorder. *Science* 318(12):283–286.**
- **Roberts, S. (June 17, 2016) Robert Paine, Ecologist Who Found 'Keystone Species,' Dies at 83. *New York Times*, p. B8.**

CHAPTER 21 Test Your Knowledge

DRIVING QUESTION 1 What are keystone species in a community, and why are pollinators considered keystone species?

By answering the questions below and studying Infographics 21.1 and 21.2, you should be able to generate an answer for the broader Driving Question above.

KNOW IT

1 **How does a community differ from a population?**

2 **What are keystone species?**

3 A rocky shoreline that is covered at high tide but exposed at low tide supports a community of mussels, algae, barnacles, and starfish. An ecologist systematically removes species from different areas of the beach. Removing the mussels or the barnacles doesn't substantially change the community, but removing the starfish dramatically changes the mix of species in the area. Which is the keystone species?

a. mussels
b. barnacles
c. algae
d. starfish
e. all of the above

USE IT

4 Think about a community of organisms that you are familiar with. From what you know about this community, choose what you think might be a keystone species and defend your choice.

5 If you have pollen allergies, are you more likely to be suffering from the effects of bee-carried pollen or wind-carried pollen? Explain your answer.

6 Sugar maple trees occur in forests in the northeastern United States and southeastern parts of Canada. Maple trees are wind pollinated. Do you predict bees to be a keystone species in a forest community dominated by maple trees? If not, what kinds of species might be a keystone species in this community?

DRIVING QUESTION 2 What are food chains and food webs, and how does energy flow through them?

By answering the questions below and studying Infographics 21.3 and 21.4, you should be able to generate an answer for the broader Driving Question above.

KNOW IT

7 In relation to a food chain, what do plants and photosynthetic algae have in common?

a. nothing
b. They are both producers.
c. They are both first-level consumers.
d. They are both top-level consumers.
e. Their numbers are limited by the energy they take in from heterotrophic food sources.

8 A bear that eats both blueberries and fish from a river is

a. an omnivore.
b. a heterotroph.
c. a consumer.
d. a producer.
e. all of the above
f. a, b, and c
g. a and c

USE IT

9 Describe a natural food web that includes a terrestrial food chain (including bees) and also at least one aquatic organism (that is, an organism from a lake, river, or ocean).

10 Explain why the kilograms of grass a cow eats do not produce the equivalent amount of energy in the form of meat. What happens to the energy stored in the grass once it is ingested by the cow?

11 Compare the diet of a human who is a herbivore with that of a human who is a top consumer. Consider what each might actually eat; how much energy from a producer is captured in the herbivore human; and how much energy from a producer is captured in the top consumer human.

DRIVING QUESTION 3 What kinds of interactions occur among members of a community, and how do these influence the structure of the community?

By answering the questions below and studying Infographics 21.5, 21.6, 21.7, and 21.8, you should be able to generate an answer for the broader Driving Question above.

KNOW IT

12 What are some important features of a honey bee niche? How is it that other nectar-feeding organisms can coexist with bees as part of a community?

13 Competition is most likely to occur

a. when one species eats another.
b. when two species occupy different niches.
c. when one species helps another.
d. when two species occupy overlapping niches.
e. when two species help each other.

14 Which of the following characterizations best describes a symbiotic relationship?

a. Both organisms benefit.
b. The organisms live in close association.
c. Only one organism benefits.
d. The relationship is mutually harmful.
e. Neither organism benefits.

15 Would you characterize the relationship between the bacteria that live symbiotically within bees and their bee hosts as competition, parasitism, mutualism, or commensalism? Explain your answer.

USE IT

16 **On a rocky intertidal shoreline (the area between the highest and lowest tidelines, so the intertidal zone is alternately exposed and covered by seawater), mussels and barnacles live together attached to rocks where they obtain food by filtering it from ocean water. Since these two species coexist in the same habitat, we predict that they do not have identical niches. What might be separating their niches enough to allow them to occupy the same rocky intertidal zone?**

17 **If a meadow of wildflowers were converted to a field of corn, would you predict the number and diversity of bees in the community to increase or decrease? Explain your answer.**

18 **What is the evidence for and against each of the following being responsible for colony collapse disorder (CCD)?**

a. varroa mites
b. IAPV
c. neonicotinoids

19 **We all have *E. coli* bacteria living in our intestinal tracts. Occasionally these *E. coli* can cause urinary tract infections. From this information, which of the following terms would you say describe(s) the relationship between us and our intestinal *E. coli*? Why did you choose the term(s) you did?**

a. competition
b. mutualism
c. parasitism
d. symbiosis
e. predator–prey

BRING IT HOME

apply YOUR KNOWLEDGE

20 **Many people consider bees a stinging nuisance. What could you say to such people to dissuade them from killing all the bees in their backyards?**

MINI CASE

apply YOUR KNOWLEDGE

21 **Farmers often plant large acreage of a single crop in order to maximize yield and simplify harvesting. This is true of almonds in the Central Valley of California.**

a. From what you have read in this chapter, what are some of the pros and cons of monoculture?
b. Do some online research to develop a specific model for an alternative to monoculture that addresses at least one of the issues you have identified.

MINI CASE

apply YOUR KNOWLEDGE

22 **Scientists carried out an experiment to test the hypothesis that a neonicotinoid pesticide called imidacloprid could cause colony collapse disorder (CCD). They had a total of 20 hives (colonies) that were broken into five groups (with 4 hives per group). Four groups received imidacloprid at different dosages (400 µg/kg; 200 µg/kg; 40 µg/kg and 20 µg/kg). One group did not receive imidacloprid. Hives were monitored for 23 weeks after the initial dose. The data are summarized in the table below.**

a. Graph these data.
b. What patterns do you observe?
c. Do you think that the data support the hypothesis? Why or why not?

Dose of imidacloprid (µg/kg)	No. of Dead Hives at 12 Weeks	No. of Dead Hives at 14 Weeks	No. of Dead Hives at 16 Weeks	No. of Dead Hives at 18 Weeks	No. of Dead Hives at 21 Weeks	No. of Dead Hives at 23 Weeks
400	0	2	2	4	4	4
200	0	0	2	2	3	4
40	0	1	1	2	3	3
20	0	0	0	0	2	4
0	0	0	0	0	1	1

Data from Lu, C., et al. 2012. In situ replication of honey bee colony collapse disorder. *Bulletin of Insectology* 65:99–106.

22

Ecosystem Ecology

DRIVING QUESTIONS

1. What are ecosystems, and how are ecosystems being affected by climate change?
2. What is the greenhouse effect, and what does it have to do with global warming?
3. How does carbon cycle through ecosystems, and how do scientists compare present and past atmospheric CO_2 levels and what can they learn from such comparisons?

The HEAT Is On

From migrating maples to shrinking sea ice, signs of a warming planet

For more than two centuries, Burr Morse's family has collected sap from Vermont's maple trees and boiled it to sweetened perfection. If you pour maple syrup over your breakfast pancakes or eat maple-cured ham, you've likely enjoyed the results of their careful craft, or that of other Vermont maple syrup farmers. About one in four trees in the state of Vermont is a sugar maple (*Acer saccharum*), and each year the state produces roughly a million gallons of syrup, making Vermont the number one maple syrup producer in the United States. But what has been a proud family tradition and the economic lifeblood for generations of maple syrup farmers could now very well be in jeopardy.

"In the last 20 years we have had a number of bad seasons and most of those I would attribute to temperature that is a little too warm," says Morse. "For maple sugaring to work right, the nights have to freeze down into the mid-20s, and the days have to thaw up into the 40s. And the nights for those 20 years, it seemed, were not quite getting cold enough."

Morse isn't the only one to notice the shift. Maple syrup farmers across New England have noted the changes in temperature and are concerned about their long-term effects.

Warmer winters in New England could have a large economic impact on the region. As ecologist Tim Perkins, director of the Proctor Maple Research Center at the University of Vermont, testified to Congress in 2007, "If

Education Images/Getty Images

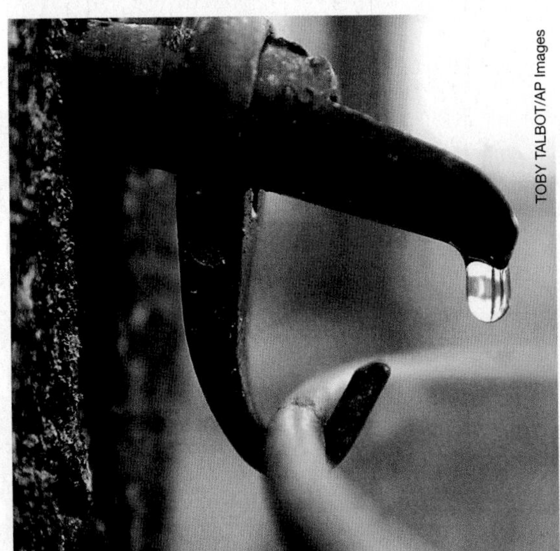

TOBY TALBOT/AP Images

Sap is collected from taps placed in sugar maple trees.

2016 Maple Syrup Production Values

In 2016, Canada out-produced the United States by a little over 9 million gallons. Within the United States, Vermont, New York, and Maine are the largest producers.

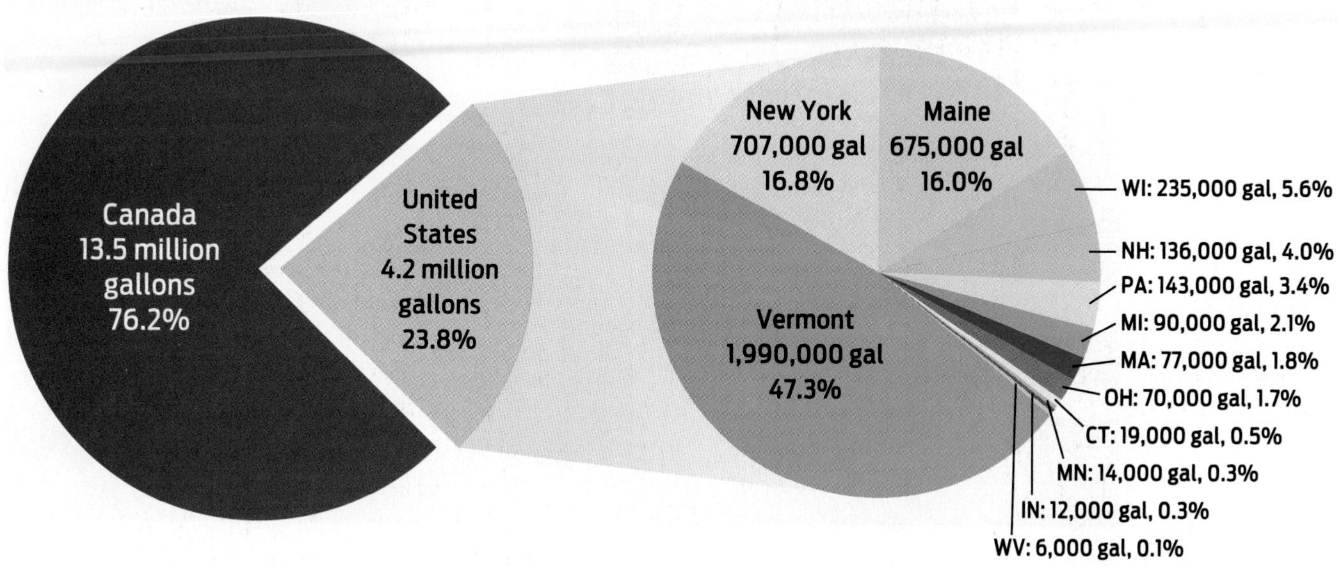

Data from USDA, National Agricultural Statistics Service

the northeast regional climate continues to warm as projected, we expect that the maple industry in the U.S. will become economically untenable during the next 50 to 100 years." This is not just a drop in the bucket: according to Perkins, "The total economic impact of maple in Vermont alone is nearly $200 million each year."

Before 1900, 80% of the world's maple syrup came from trees in the United States, the rest from trees in Canada. Today, the pattern is reversed, with Canada greatly out-producing the United States. Canada now accounts for about 80% of world maple syrup production. While part of this reversal has to do with marketing, Canadian government subsidies, and improved technologies, Perkins believes that climate change is a significant contributing factor, putting New England maple syrup farmers at a competitive disadvantage.

Note that climate is not the same as weather. The **climate** in any given area refers to the long-term average of atmospheric conditions. **Weather** describes local atmospheric conditions over a short interval–the sun or clouds, wind or rain predicted in your weekly forecast. If the weather demonstrates a consistent change over a long period of time–consistently warmer winters, for example–that can indicate a change in the climate. **Climate change** is defined as any substantial change in climate that lasts for an extended period of time (decades or more). One contributor to current climate change is **global warming,** a recent and continuing increase in the average global temperature.

New England's maples are not the only ones feeling the heat. Plant and animal species throughout the world–from herbs in Switzerland to starfish in California–are being affected by rising temperatures. Some are shifting their geographic ranges as a result. Many historically subtropical aquatic animals, such as seahorses and turtles, for

CLIMATE
The long-term average of atmospheric conditions.

WEATHER
Local atmospheric conditions over a short period of time.

CLIMATE CHANGE
Any substantial change in climate that lasts for an extended period of time (decades or more).

GLOBAL WARMING
An increase in Earth's average temperature.

example, are moving toward the coasts of northern England and Scotland, where ocean temperatures are warmer than they used to be. And fish that were once wholly tropical are turning up in North Atlantic waters. Other organisms that cannot easily relocate—like some plants and mountain-dwelling animals—are being driven to extinction.

Weather
Local atmospheric conditions over a short period of time, including temperature, precipitation, wind, etc.
Climate
Regional average trends of atmospheric conditions over a long period of time
Mike Hill/Getty Images
NASA

Earth's climate changes naturally and has done so many times in its long history. But scientists now believe that humans are accelerating the pace of change, with potentially dire consequences for life on our planet.

To Everything a Season

In nature, timing is everything. And for many species temperature is nature's clock, cueing seasonally appropriate tasks like mating or producing flowers. Rising temperatures around the globe are interfering with these natural rhythms. Many plants are flowering earlier now than they once did; animals are emerging from hibernation earlier; and many bird and butterfly species are migrating north and breeding earlier in the spring than they did a few decades ago—all because of slight changes in temperature cues. It's a pattern of change that scientists are seeing around the globe **(INFOGRAPHIC 22.1)**.

ECOSYSTEM
The living and nonliving components of an environment, including the communities of organisms present and the physical and chemical environment with which they interact.

Who cares if flowers bloom earlier or animals come out of hibernation sooner? By themselves, these changes wouldn't necessarily be a big deal. But because living things are exquisitely adapted to their environments and are interdependent, a change in one part of an ecosystem may upset others.

As the name suggests, an **ecosystem** is a complex, interwoven network of interacting components. It includes both the community of living organisms present in an area and the features of the nonliving environment—physical conditions such as temperature, moisture, light, and the chemical resources found in soil, water, and air. Because the biotic and abiotic parts of an ecosystem can and do change, ecosystems are not static entities but dynamic systems. And because the parts of an ecosystem are so interconnected, a small change in one

INFOGRAPHIC 22.1 Changing Temperatures Affect Plant Behavior

Changes in average temperatures are changing the seasonal behavior of plants and animals. Changes in average spring temperatures result in earlier flowering dates in some locations, and later flowering dates in other locations.

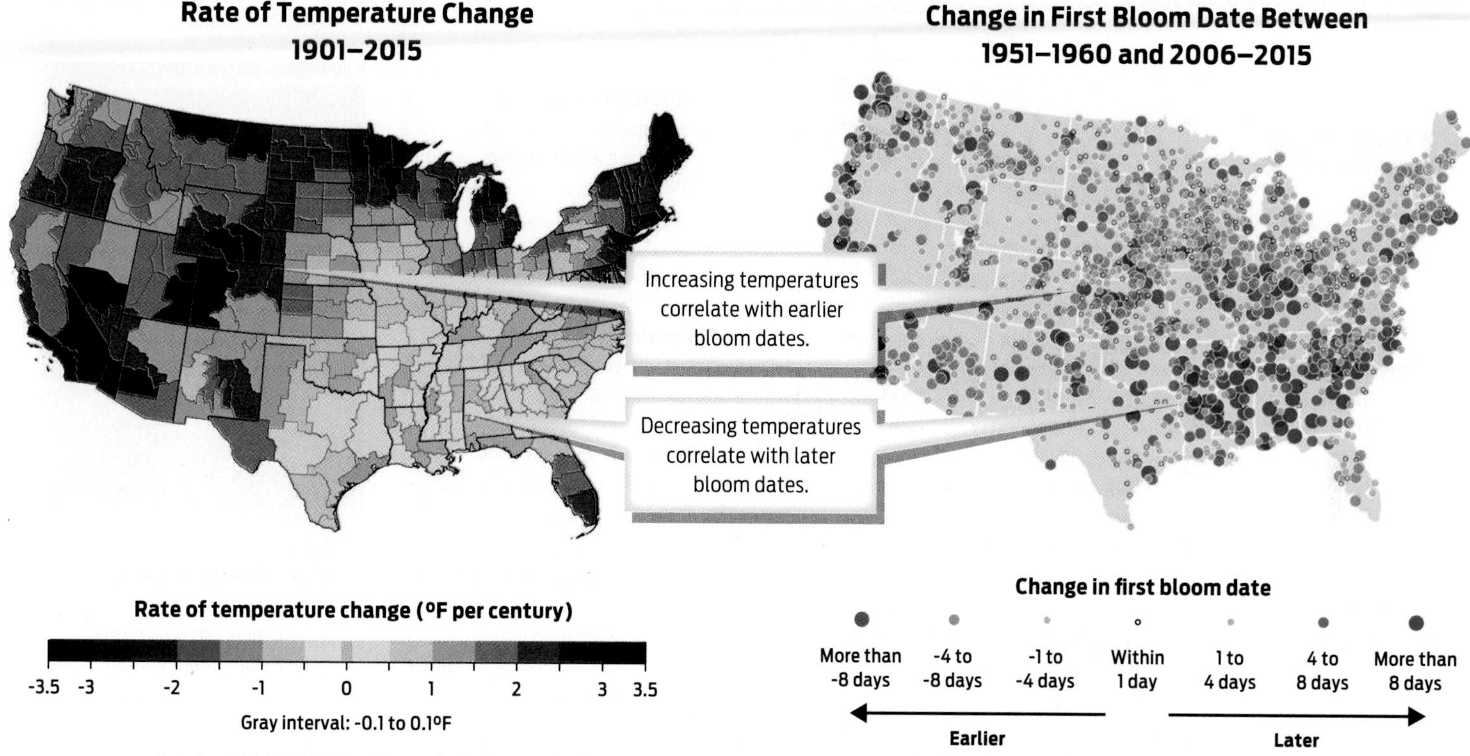

? Why are bloom dates in the southeastern United States occurring later and bloom dates in New England occurring earlier?

part of an ecosystem can have a domino effect (**INFOGRAPHIC 22.2**).

No one knows this better than maple syrup farmers. "The flow of sap from maple trees during the spring season is controlled almost entirely by the daily fluctuation in temperature," explains ecologist Perkins. "Small changes in the day-to-day temperature pattern will have large consequences on sap flow."

Historically, trees were tapped in early March when the sap began to flow; the sap was then collected for the next 4 weeks. But about 10 years ago, Perkins started getting calls from syrup producers saying that they were tapping earlier and making syrup earlier. Curious, he and his colleagues decided to investigate. They examined historical records and surveyed hundreds of maple syrup producers in New England and New York. Their results were startling: over a mere 40 years, between 1963 and 2003, the start of the tapping season had moved up by about 8 days. Even more significant, the end of the season, when maples begin to leaf out and the sap is no longer good for syrup, now comes 11 days earlier.

"Over that 40-year time period we've lost about 3 days of the season," Perkins told Vermont Public Radio in 2009. "That doesn't seem like a lot until you realize that the maple

INFOGRAPHIC 22.2 Ecosystem Interactions Can Be Altered by Warming Temperatures

Scientists studied the roe deer and their ecosystem for nearly 40 years. Warmer springs have made the preferred food for deer available earlier, but the deer have not shifted their birthing dates to match the food availability. In this case, the deer population is being negatively affected.

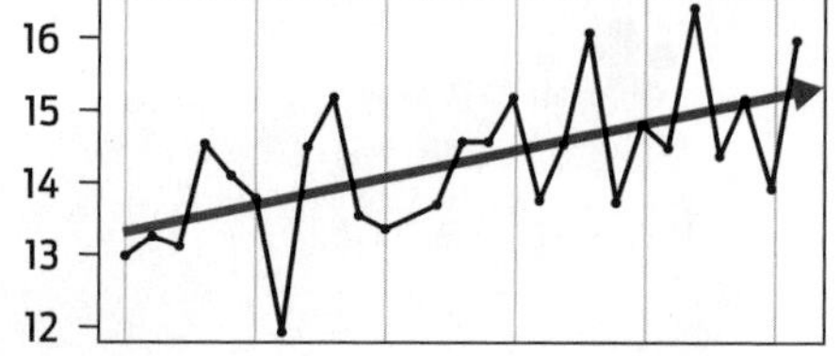

Temperature goes up:
Annual spring (April to June) temperature increased by 0.07°C per year.

Flowering Date (day of the year Jan 1 = 1)

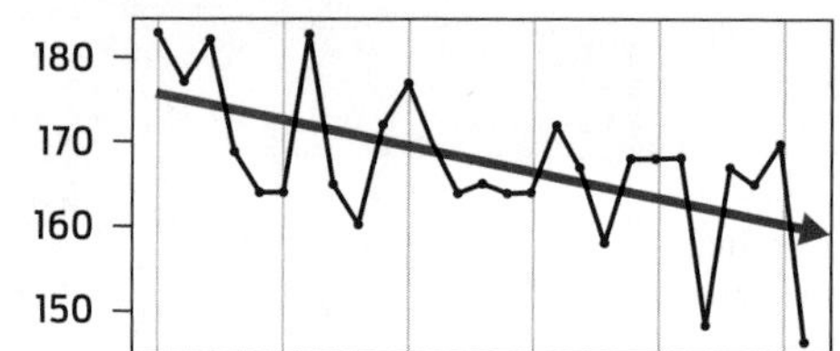

Flowering date comes early:
Annual flowering date was 0.6 days earlier per year, so the preferred food for roe deer was available earlier every year.

Mismatch (No. of days between flowering and birth)

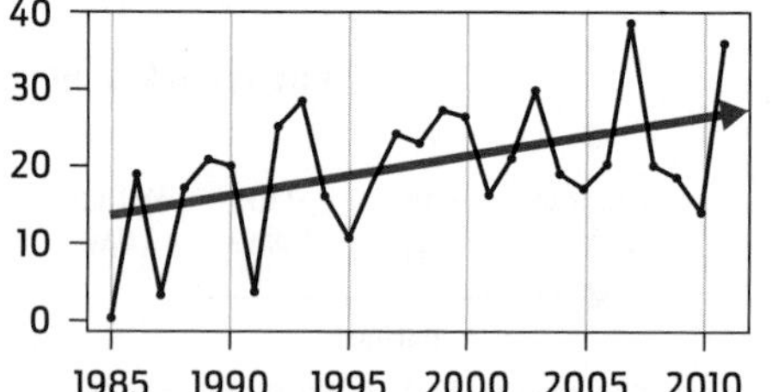

Mismatch between deer birth and food availability increases:
Roe deer historically gave birth about 1 month before the onset of flowering when food was most available. Since roe deer did not change their birthing date as the flowering date became earlier, there was an increasing mismatch between the two.

Juvenile Survival (Proportion of each fawn population that survived to the onset of winter)

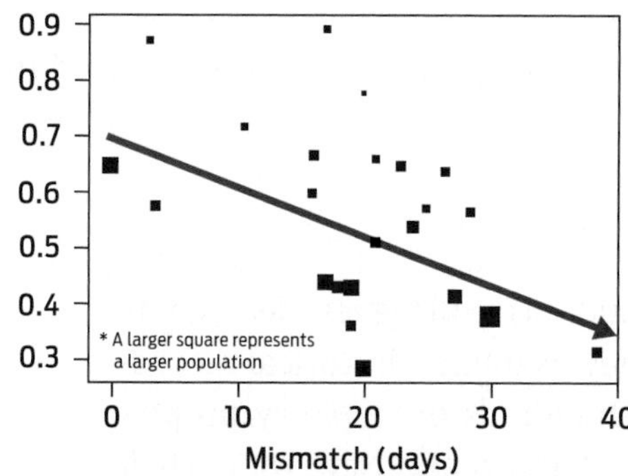

Data from Plard, F. et al. (2014) *PLoS Biol.* 12(4): e1001828

Young roe deer survival decreases:
As a result of the increasing food/birthing mismatch, survival of young roe deer in each population decreased, as newborn calves emerged with less food available to them. Survival of young deer decreased by 40% when the food/birthing mismatch was 1 month.

? Why is increasing spring temperature having a negative impact on deer populations in this ecosystem?

production season averages about 30 days in length. So we've lost about 10% of the season."

To some extent, losses from a shortened tapping season have been offset by improved sap-removal technologies that make it possible to extract sap even under poor conditions. "[The shortened season] hasn't yet impacted yields, because the technology of sap extraction has improved," says Perkins, who points in particular to the use of vacuum tubes to suck out sap. The bigger problem is what will happen if the climate changes so much that New England no longer provides a suitable **habitat** for maple trees.

Current climate computer models predict that by the end of the century New England's forests will more closely resemble those of present-day Virginia, North Carolina, and Tennessee, which are dominated by hickory, oak, and pine rather than by maple, beech, and birch. If that happens, not only maple syrup, but also the brilliant fall foliage New England is famous for will be a thing of the past (**INFOGRAPHIC 22.3**).

New England's colorful foliage is part of a distinct biome known as temperate deciduous forest. **Biomes** are large, geographically cohesive regions whose defining vegetation–their plant life–is determined principally by climatic factors like temperature and rainfall. The temperate deciduous forest biome is characterized by distinct seasons, trees that lose their leaves in the fall, and a total annual precipitation of between 750 and 1,500 mm that occurs evenly throughout the year. Losing their leaves in the fall–that is, being

HABITAT
The physical environment where an organism lives and to which it is adapted.

BIOME
A large geographic area defined by its characteristic plant life, which in turn is determined by temperature and levels of moisture.

INFOGRAPHIC 22.3 Plant Growing Range Is Affected by Rising Temperatures

Computer simulation models predict major changes in forest makeup in the northeast United States over the next century. Predictions suggest that the currently dominant maple, beech, and birch forests will be replaced with hickory, oak, and pine forests.

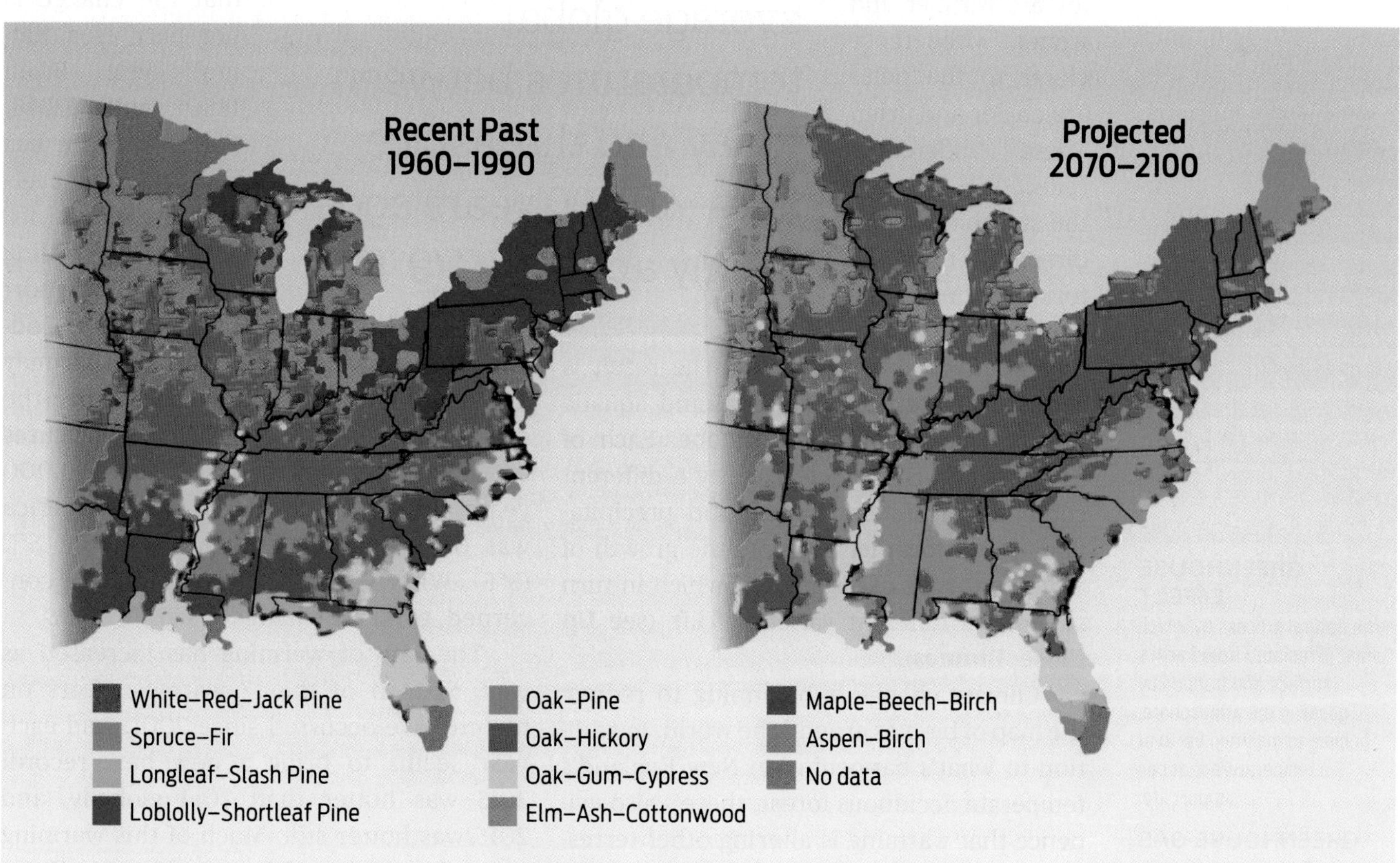

Data from Team, National Assessment Synthesis. *Climate Change Impacts on the United States: The Potential Consequences of Climate Variability and Change*. Cambridge, UK and New York, NY, 2001. Map: U.S. Global Change Research Program

? Describe what is predicted to happen to the vegetation in Florida in the next 45 to 75 years.

deciduous–is an adaptation of trees to freezing temperatures that occur over winter.

During this period, the trees are dormant and do not photosynthesize. Because the vegetation in a biome forms the base of the food chain (see Chapter 21), and because winter cuts down on the available food, animals that live here have adapted to the winter, too–for example, by hibernating, storing nuts, or migrating to warmer places. During the summer, trees leaf out again, providing food for the animals that thrive here (**INFOGRAPHIC 22.4**).

The climatic factors–temperature and rainfall–that shape life in each biome correlate with a region's position on the globe with respect to the equator. Biomes closer to the equator are warmer and wetter, while those closer to the poles are colder and drier. These differences reflect the fact that the sun shines more directly on the equatorial regions of the planet, whereas less direct sunlight hits the poles. The result is a distinct pattern of terrestrial and aquatic biomes located around the globe. Each of these biomes is characterized by a different combination of temperature and precipitation, which together promote the growth of a different type of vegetation, which in turn supports a different variety of life (see **Up Close: Biomes**).

Climate change is beginning to redraw the map of biomes around the world. In addition to what's happening in New England's temperate deciduous forest, there's also evidence that warming is altering other terrestrial and aquatic biomes. Some of the most dramatic changes are occurring in biomes that are historically cold, like tundra. In northern Alaska, where once there was only sparsely vegetated tundra, woody shrubs now grow. When Montana's Glacier National Park was opened in 1910, it held approximately 150 large glaciers; in 2013, there were only 25, and scientists predict that by 2030 there will be no glaciers left. As the vegetation in these landscapes changes, so does the community of organisms that rely on it for food and habitat, thereby changing the ecosystem as a whole.

Warming Planet, Diminishing Biodiversity

Although temperature swings and shifts in the ranges of organisms are natural phenomena, the amount of warming in recent years is unprecedented, and evidence suggests that the change is not part of a natural cycle. From 1880 until 2016, Earth's surface has warmed, on average, by about 1.1°C (2.0°F), according to a 2017 report by NASA's Goddard Institute for Space Studies. That may not sound like a lot. But consider this: the difference in average global temperatures between today and the last ice age–10,000 years ago, when much of North America was buried under ice–is only about 5°C (9°F). Where global temperatures are concerned, even a 1° change is significant.

The difference in average global temperatures between today and the last ice age—10,000 years ago—is only about 5°C (9°F).

The rate of warming has increased as well. Sixteen of the 17 warmest years on record have occurred since 2001. And each year seems to bring a new heat record: 2015 was hotter than 2014 globally, and 2016 was hotter still. Much of this warming is attributable to the **greenhouse effect,** the trapping of heat in Earth's atmosphere. As sunlight shines on our planet, it warms Earth's surface. This heat radiates back to the atmosphere, where it is absorbed by **greenhouse gases** such as carbon dioxide.

GREENHOUSE EFFECT
The natural process by which heat is radiated from Earth's surface and trapped by gases in the atmosphere, helping to maintain Earth at a temperature that can support life.

GREENHOUSE GAS
Any of the gases in Earth's atmosphere that absorb heat radiated from Earth's surface and contribute to the greenhouse effect, for example carbon dioxide and methane.

INFOGRAPHIC 22.4 The Temperate Deciduous Forest Biome

Maple trees live in the temperate deciduous forest biome. This biome is characterized by four distinct seasons.

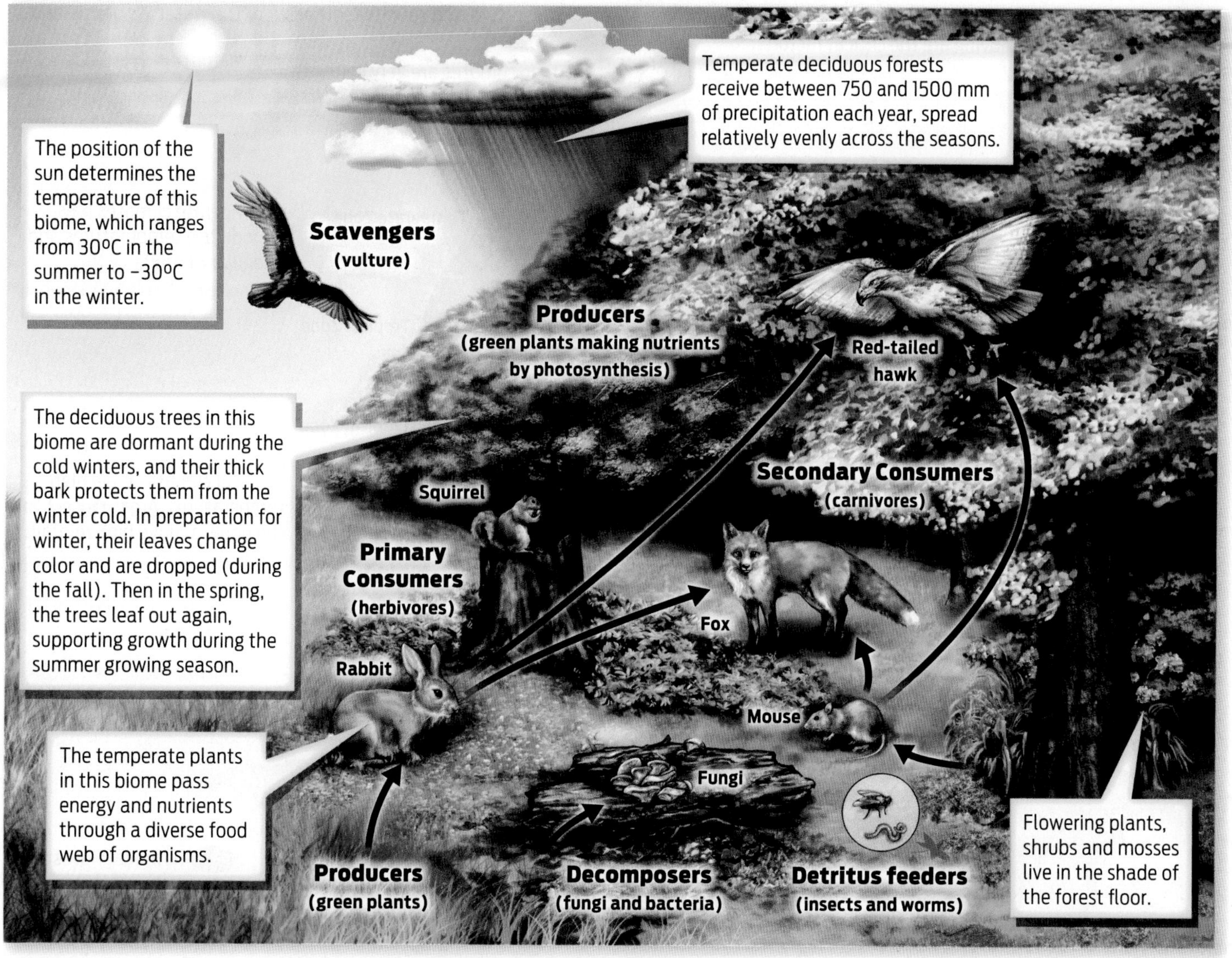

The position of the sun determines the temperature of this biome, which ranges from 30°C in the summer to −30°C in the winter.
Temperate deciduous forests receive between 750 and 1500 mm of precipitation each year, spread relatively evenly across the seasons.
Scavengers (vulture)
Producers (green plants making nutrients by photosynthesis)
Red-tailed hawk
The deciduous trees in this biome are dormant during the cold winters, and their thick bark protects them from the winter cold. In preparation for winter, their leaves change color and are dropped (during the fall). Then in the spring, the trees leaf out again, supporting growth during the summer growing season.
Squirrel
Secondary Consumers (carnivores)
Primary Consumers (herbivores)
Fox
Rabbit
Mouse
Fungi
The temperate plants in this biome pass energy and nutrients through a diverse food web of organisms.
Producers (green plants)
Decomposers (fungi and bacteria)
Detritus feeders (insects and worms)
Flowering plants, shrubs and mosses live in the shade of the forest floor.

Climate Data for a Typical Temperate Deciduous Forest

Temperature (°C)
40
30
20
10
0
-10
-20
-30
J F M A M J J A S O N D
Month
Precipitation (mm)
140
120
100
80
60
40
20
0
J F M A M J J A S O N D
Month

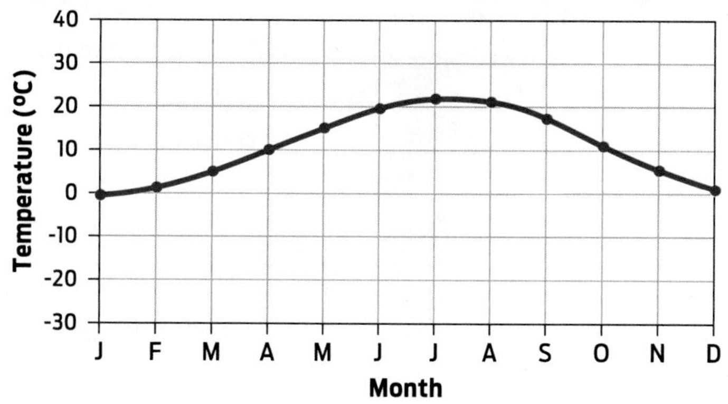

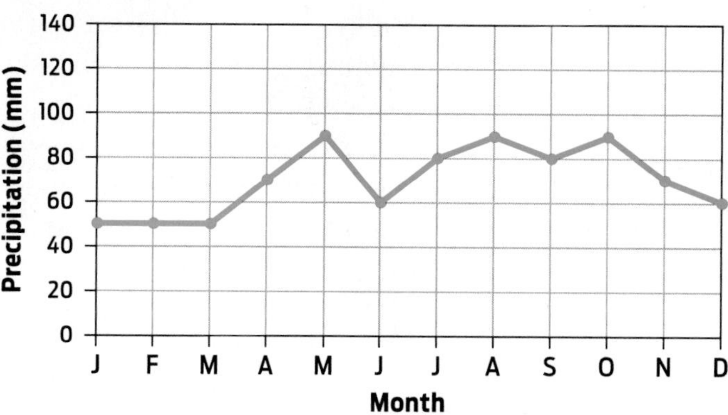

? In a temperate deciduous forest in the United States, how do the temperature and precipitation in summer (July) compare with temperature and precipitation in winter (January)?

UP CLOSE Biomes

Biomes are large geographic areas defined by their characteristic plant life. The characteristic vegetation is, in turn, determined by patterns of temperature and rainfall. For terrestrial biomes, average temperature and rainfall as well as seasonal variations in both are critical to determining the types of vegetation. For aquatic biomes, the temperature and salinity (saltiness) of the water, its depth, and whether it is still or moving are the critical factors.

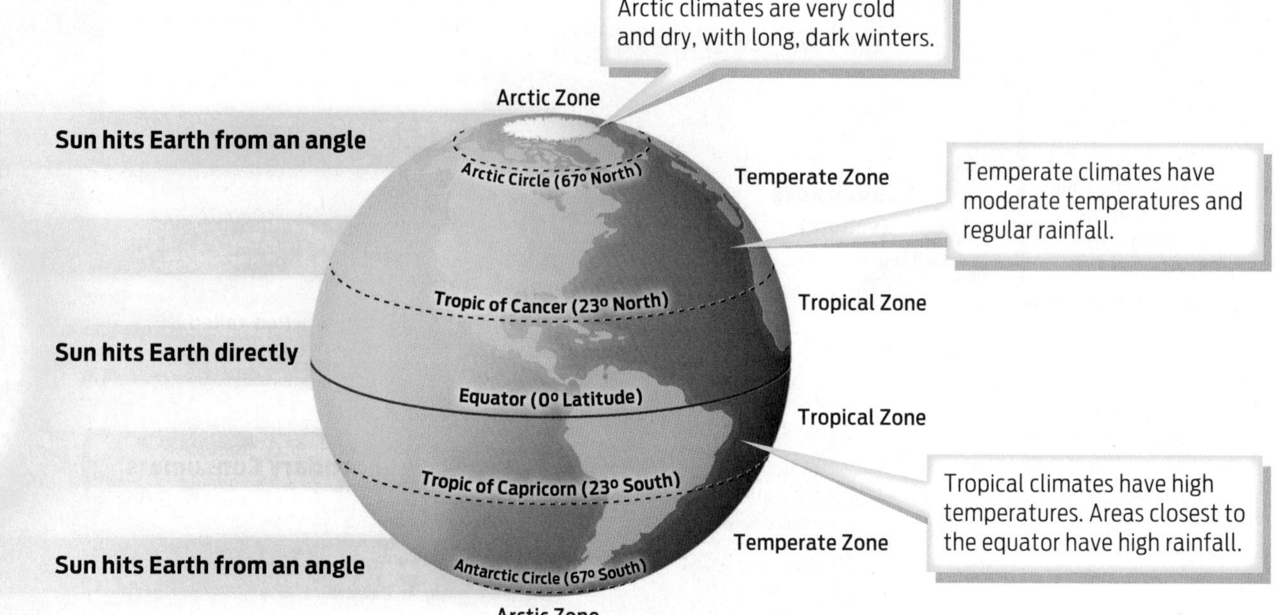

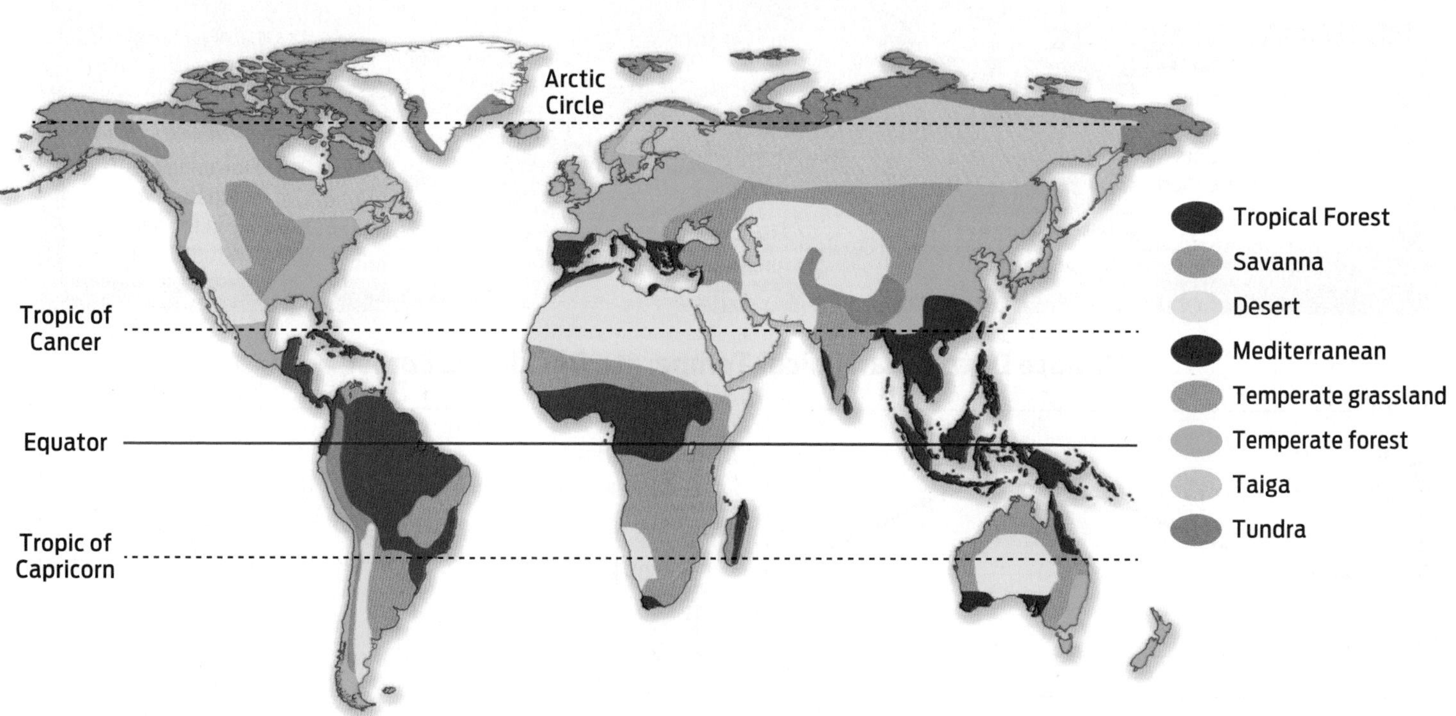

Taiga
A biome characterized by evergreen trees, with long and cold winters and only short summers.

Tundra
A biome that occurs in the Arctic and mountain regions. Tundra is characterized by low-growing vegetation and a layer of permafrost soil (frozen all year long) very close to the surface of the soil.

Tropical Forest
Tropical forests are biomes characterized by warm temperatures and sufficient rainfall to support the growth of trees. Tropical forests may be deciduous or evergreen, depending on the presence or absence of a dry season.

Temperate Deciduous Forest
Temperate forests are characterized by moderate winters and rainfall. Trees are mostly evergreen or deciduous, dropping their leaves in winter.

Savanna
A biome characterized by warm temperatures and two seasons (dry and rainy). The primary vegetation is grasses, with some shrubs and rare trees.

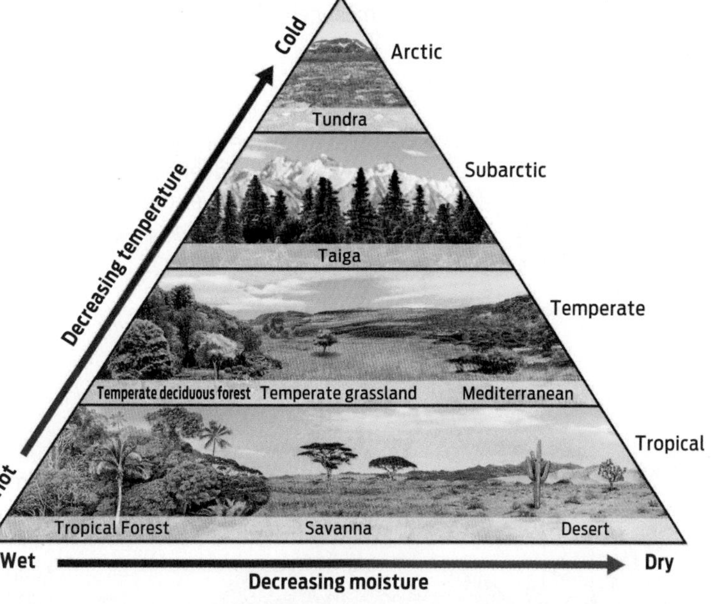

Mediterranean
A biome characterized by long, hot, and dry summers and cool, damp winters. The plant life includes characteristic short evergreen trees and shrubs with leathery leaves.

Temperate Grassland
A biome characterized by perennial grasses and other nonwoody plants. In North America, the prairies are examples of grasslands.

Desert
A biome characterized by extreme dryness. Cold deserts experience cold winters and hot summers, while hot deserts are uniformly warm throughout the year.

Aquatic: Marine
This biome covers about three-fourths of Earth and includes the oceans, coral reefs, and estuaries (where rivers meet the sea).

Aquatic: Freshwater
A biome characterized by a low salt concentration. Fresh-water biomes include ponds and lakes, rivers and streams, and wetlands.

The heat trapped by greenhouse gases raises the temperature of the atmosphere, and in turn, Earth's surface (**INFOGRAPHIC 22.5**).

The greenhouse effect is a natural process that helps maintain life-supporting temperatures on Earth. Without this greenhouse effect, the average surface temperature of the planet would be a frigid -18°C (0°F). In recent years, however, rising levels of greenhouse gases have increased the strength of the greenhouse effect, a phenomenon known as the enhanced greenhouse effect. As the amount of greenhouse gases in the atmosphere has increased, so have temperatures. The result is global warming, an overall increase in Earth's average temperature (**INFOGRAPHIC 22.6**).

For ecologist Hector Galbraith, director of the Climate Change Initiative at the

INFOGRAPHIC 22.5 The Greenhouse Effect

The greenhouse effect is a natural process that helps maintain steady and life-sustaining surface temperatures on Earth. Sunlight heats Earth's surface and that heat radiates back to the atmosphere. While some of the heat escapes to space, certain gases in Earth's atmosphere, known as greenhouse gases, trap heat within the atmosphere. This trapped heat warms the atmosphere and Earth's surface.

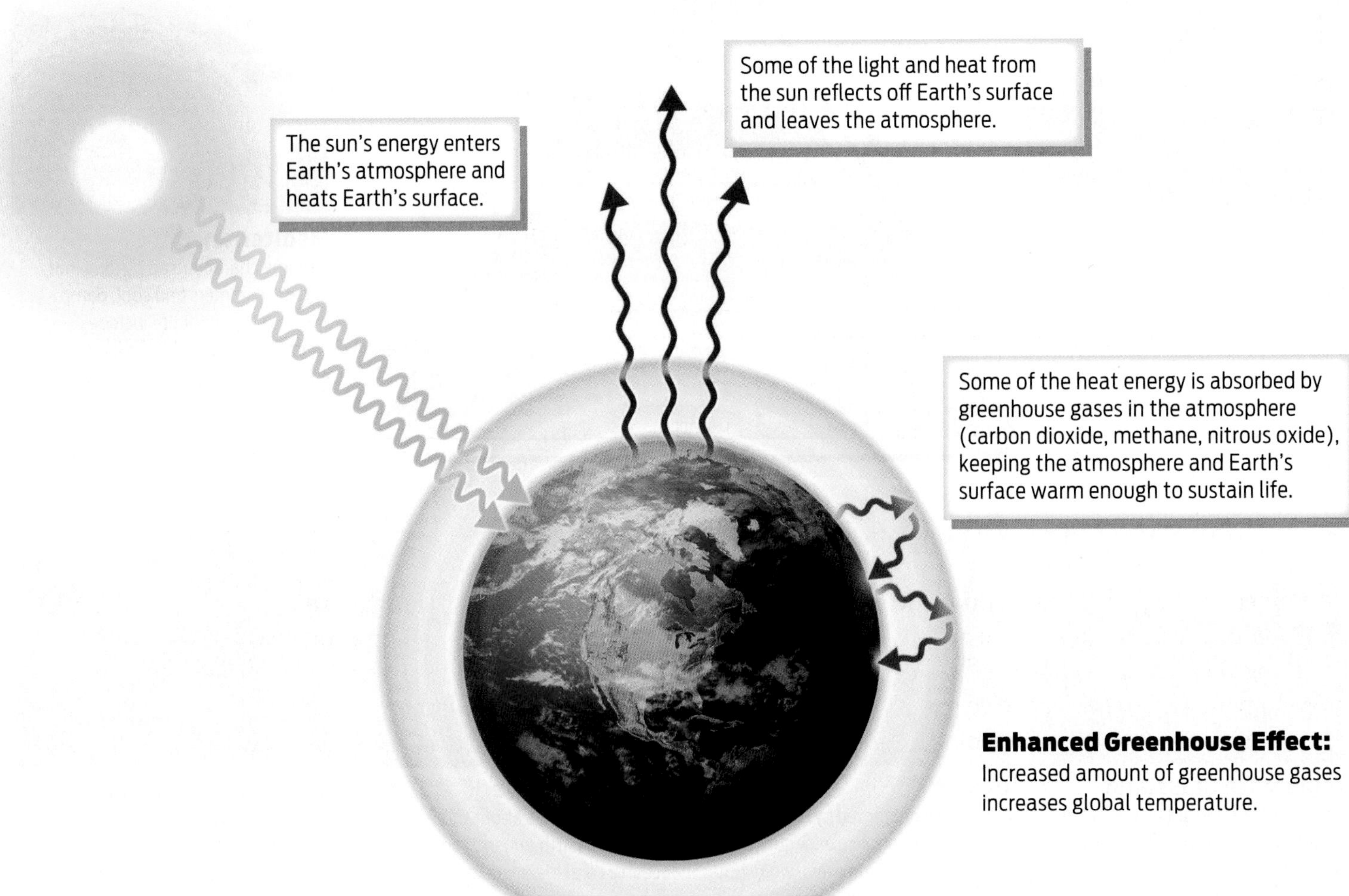

The sun's energy enters Earth's atmosphere and heats Earth's surface.
Some of the light and heat from the sun reflects off Earth's surface and leaves the atmosphere.
Some of the heat energy is absorbed by greenhouse gases in the atmosphere (carbon dioxide, methane, nitrous oxide), keeping the atmosphere and Earth's surface warm enough to sustain life.
Enhanced Greenhouse Effect:
Increased amount of greenhouse gases increases global temperature.

? How do greenhouse gases contribute to keeping Earth warm?

INFOGRAPHIC 22.6 Earth's Surface Temperature Is Rising with Increases in Greenhouse Gases Like Carbon Dioxide

Long-term measurements of global temperature and atmospheric CO_2 concentration show a clear, positive correlation.

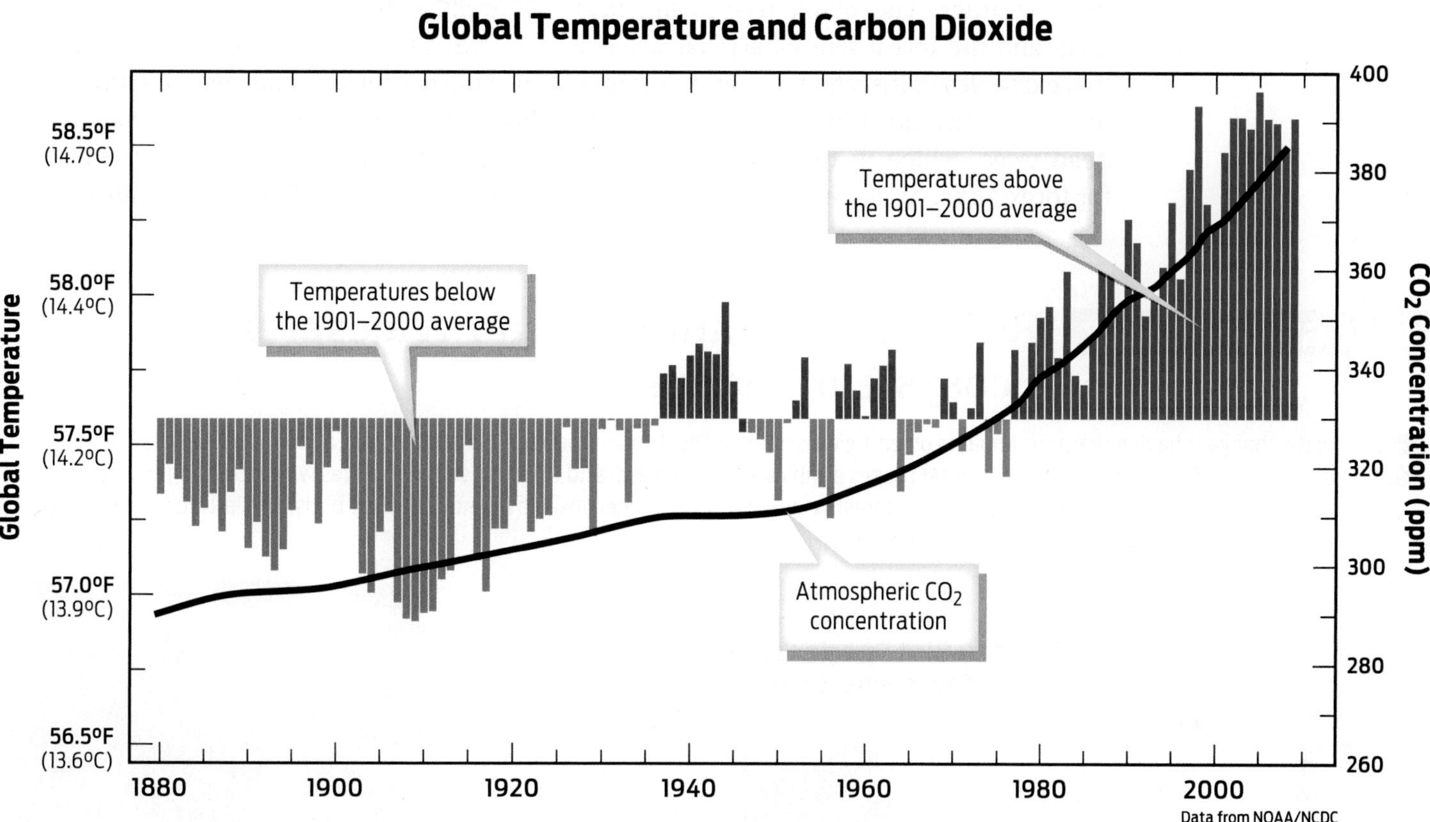

? From the most current data shown in the graph, how many degrees warmer is Earth than the 1901–2000 average? When did Earth's temperature start to be consistently higher than the 1901–2000 average?

Manomet Center for Conservation Sciences, in Plymouth, Massachusetts, the most worrying thing about climate change is how quickly it is happening and how sensitive species are to the changes. "Most people think of climate change as something that's 30 years out," says Galbraith. But that's simply not true, he notes. "We began seeing responses in ecosystems 20 years ago. The ecosystems knew about it before we did."

Plants, of course, are slower to migrate than animals; they cannot simply get up and move (although they may change their range over time by dispersing seeds into more favorable habitats). But some animals can change their ranges quite quickly. "A bird can simply open its wings, and within 2 hours it's 50 miles farther north," says Galbraith.

What will be the outcome of all these changes? Scientists don't really know. "We're seeing changes to systems that have been relatively stable for thousands of years," says Galbraith. "The really scary thing about climate change is it's very difficult to predict the ecosystem effects of these changes."

Nevertheless, there are disturbing scenarios. Take the relationship between birds and insects. Many forests are susceptible to insect attacks. Given their insect-rich diet, birds are a natural form of pest control in these forests. If the birds move north, as evidence suggests they are doing, the forests they leave behind will be more vulnerable to threats from insects that thrive in warmer, drier weather. The maple-tree-loving pear thrip and the forest tent caterpillar are just two examples of insects that might be happy to see the birds go. More insects mean more dead trees, which in turn mean more fuel for forest fires. As a result of climate change, many areas will also experience increased periods of drought. Reduced moisture in combination with hotter temperatures and trees devastated by insects creates a "perfect storm" of conditions for devastating fires in regions not historically swept by fire (**INFOGRAPHIC 22.7**).

Not all species will be negatively affected by climate change–some may actually benefit. But one species' success in coping with climate change may contribute to another's demise. For example, the adaptable red fox (*Vulpes vulpes*),

INFOGRAPHIC 22.7 Rising Temperatures Mean Widespread Ecosystem Change

Climate change is having dramatic impacts on entire ecosystems. The Annual Audubon Christmas Bird Count has shown that 58% of 305 bird species studied have shifted their winter ranges significantly northward, and that this shift correlates with increasing temperature. This has implications for insect control, leaving forests vulnerable to fire when insects are not eaten by birds. Similarly, climate change causes periods of drought, increasing the risk of fires.

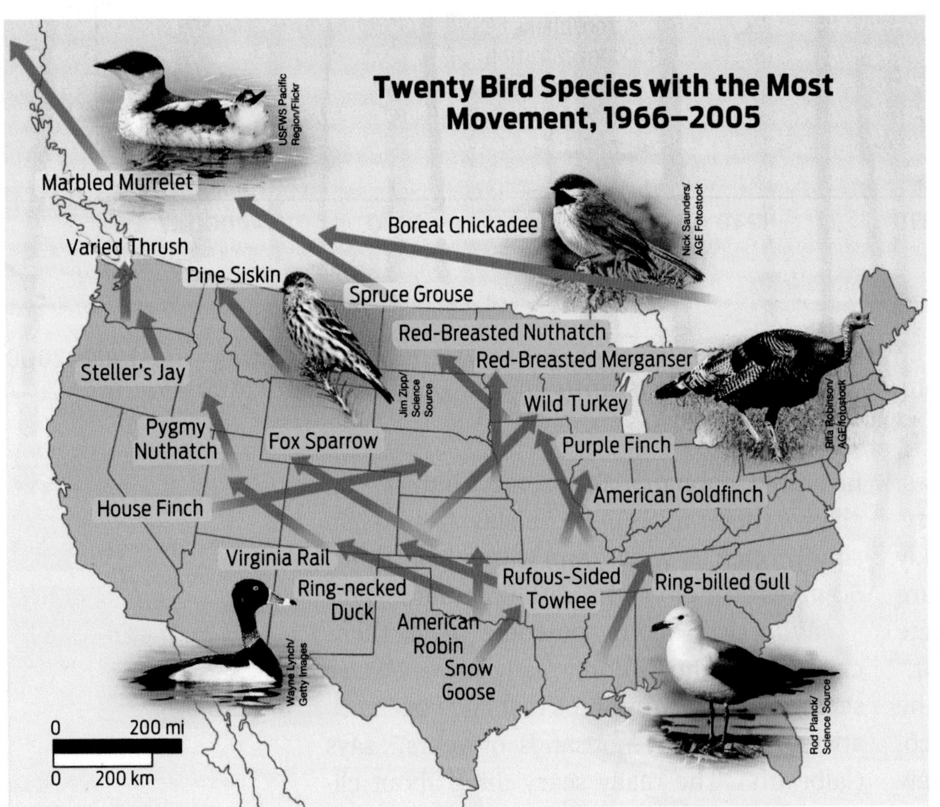

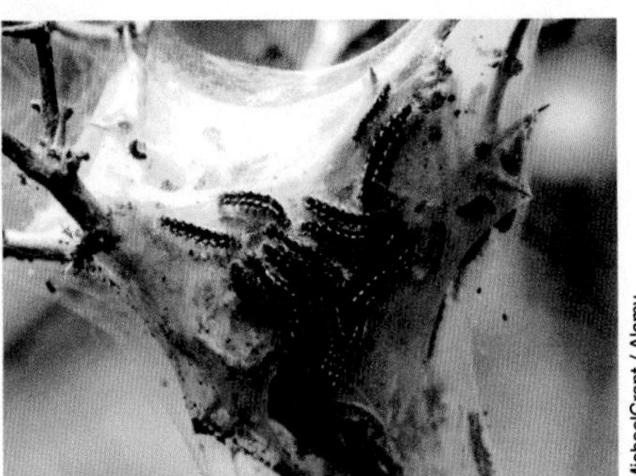

The migration of birds north leaves forests vulnerable to insects like the tent caterpillar that voraciously eat, and eventually kill, trees. Dead trees lead to increased forest fires.

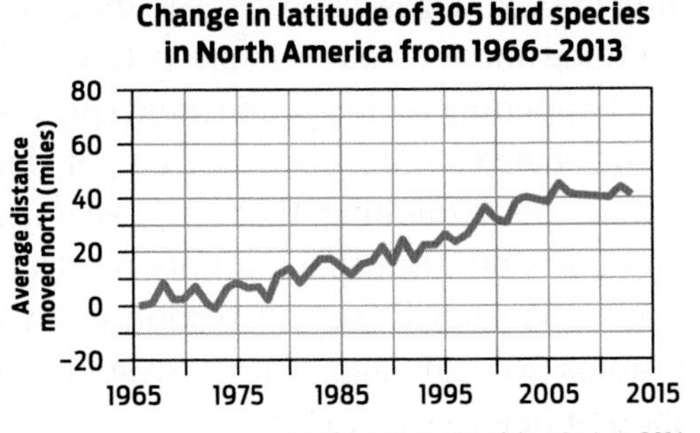

Data from National Audubon Society, 2014

? How can the migration of birds out of a particular area affect trees in that area?

found throughout the northern hemisphere, is venturing into the range of the endangered Arctic fox (*Vulpes lagopus*), whose habitat–the Arctic tundra–has become warmer. When the two species share a range, the Arctic fox inevitably suffers because the red fox out-competes it for food and preys on Arctic fox pups.

While some species can adapt to a changing climate by shifting range, future climate change will likely exceed the ability of many species to adapt, as hospitable habitats can no longer be found or accessed. According to a 2011 study published in *Proceedings of the National Academy of Sciences*, 1 in 10 species could be driven to extinction by 2100 because of climate change. The natural residents of mountaintops are especially vulnerable: as temperatures rise, species may move up to higher, colder elevations, but eventually they will have nowhere left to go.

> “If the northeast regional climate continues to warm as projected, we expect that the maple industry in the U.S. will become economically untenable during the next 50 to 100 years.”
>
> —Tim Perkins

Arctic Meltdown

Predictably, snow- and ice-covered regions such as the Arctic stand to suffer most immediately from a warming climate as frozen habitats start to melt. But the situation is worse than one might imagine. As Mark Serreze, director of the National Snow and Ice Data Center at the University of Colorado, Boulder, notes, the Arctic has warmed, on average, twice as much as the rest of the planet. This phenomenon is known among climate scientists as Arctic amplification, and it has to do with the way sea ice affects temperature. As Serreze explains, sea ice both reflects solar radiation and insulates the ocean. As global temperatures rise, ice begins to melt. With less sea ice, more solar radiation is absorbed by the ocean and more of the relatively warm ocean is exposed to air, raising the air temperature even more. It's a positive feedback loop: as additional ice is lost, temperatures rise at an accelerated pace.

According to the extensive Arctic Climate Impact Assessment, the result of 4 years' work by more than 300 scientists around the world, Arctic temperatures are projected to rise by an additional 4°-7°C (7°-13°F) over the next 100 years.

Warming temperatures could spell disaster for species that call the Arctic their home. Polar bears, for example, spend most of the year roaming the Arctic on large swaths of floating sea ice that blanket a good portion of the Arctic Ocean from September through March. The massive mammals travel on sea ice to hunt for seals, which periodically pop up through "whack-a-mole"-like breathing holes in the ice and are nabbed by the bears. The size of this frozen habitat has been shrinking, greatly reducing the bears' ability to obtain food.

Moreover, over the past few decades the ice has been breaking up earlier and earlier in spring. The sea ice in Hudson Bay, Canada, for example, now breaks up nearly 3 weeks earlier than it did in the 1970s. In the absence of sufficient summer sea ice, the polar bears are stuck on land (where there are no seals), or are forced to swim long distances to reach sea ice. Some, exhausted by the journey, drown. Those that survive have fewer opportunities to hunt. Canadian polar bears now weigh on average 55 pounds less than they did 30 years ago, a loss that seriously compromises their reproductive ability.

Scientists have monitored sea ice daily by satellite since 1979. Over the past three decades, the area of Arctic sea ice has shrunk by more than 1 million square miles, an area roughly four times the size of Texas, according to Walt Meier, a research scientist with the National Snow and Ice Data Center. Arctic sea ice hit a record low in September 2012, at the end of the summer melt season, shrinking to a level that climate change models had predicted wouldn't happen until at least 2050; September 2016 was the second lowest on record. Scientists now fear that nearly all of the polar bear's summer sea ice could vanish by 2040—possibly sooner **(INFOGRAPHIC 22.8)**.

Warming temperatures are also causing glaciers and ice caps on land to melt. Unlike sea ice, which, like an ice cube in a glass of water, doesn't raise the water level as it melts, melting glaciers and ice caps do. How much will seas rise? "By 2100, you're looking at probably about a meter," says Serreze. "Here in Boulder we're at 5,400 feet—we're not worried about that. But if you're living in Miami, this is something that should concern you."

It's important to note that much of the data we have on climate change relates

INFOGRAPHIC 22.8 Arctic Temperatures Are Rising and Sea Ice Is Melting

Current measurements suggest that the Arctic is warming faster than other parts of Earth. 2016 was the warmest year on record (since measurements began in 1880). Rising temperatures have caused Arctic sea ice to melt and break apart earlier in the season. Arctic sea ice extent in 2016 was the second lowest on record, behind 2012.

Global Mean Surface Temperature January–June, 2016

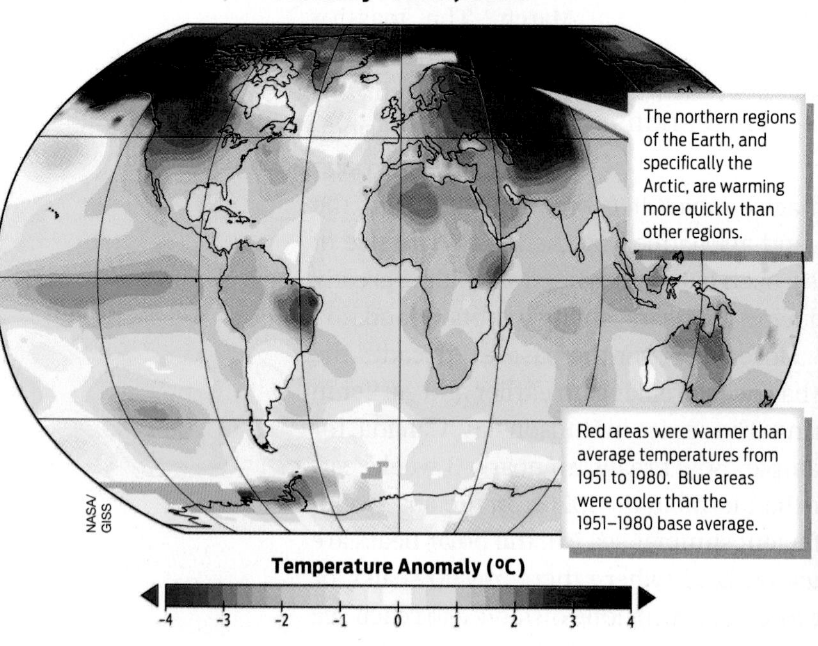

The northern regions of the Earth, and specifically the Arctic, are warming more quickly than other regions.
Red areas were warmer than average temperatures from 1951 to 1980. Blue areas were cooler than the 1951–1980 base average.
NASA/GISS
Temperature Anomaly (°C)
-4
-3
-2
-1
0
1
2
3
4

Arctic Sea Ice Extent, September 10, 2016

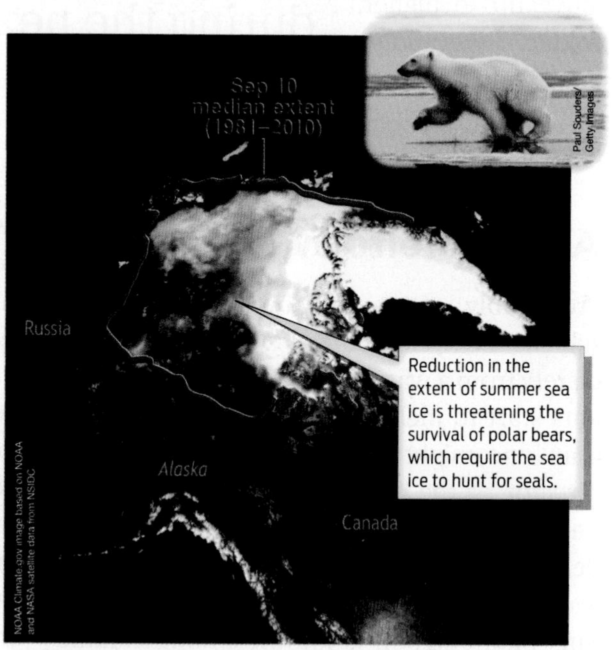

? Based on the map of global surface temperatures, which areas of Earth have higher than average temperatures, and which have lower than average temperatures?

to global, long-term trends. From year to year, there may be slight variations–slightly warmer summers and less sea ice one year, slightly cooler summers and more sea ice the next. And indeed, from a low in 2007, sea ice did indeed bounce back a bit in 2008 and 2009. But the overall trend is still unmistakably downward, toward less sea ice. By 2030 or 2040, says Serreze, there could be no summer ice to speak of. "You could take a ship across the north pole."

There are also less obvious, but no less dire consequences of Arctic warming. The Arctic tundra biome is characterized by very low temperatures and small amounts of annual precipitation, historically in the form of snow. In many ways it is like a cold, windy desert. The soil is permafrost (that is, permanently frozen soil) between 10 and 36 inches beneath the surface–at least it used to be. With global warming, the permafrost is beginning to melt for parts of the year. As the permafrost melts, organic material such as dead plants and animal remains can start to decay, releasing enormous amounts of carbon dioxide as a result. In other words, the Arctic is an important storage reservoir of carbon, whose disturbance can lead to negative consequences for the planet as a whole.

Follow the Carbon

The immediate cause of this planetary warming is a fired-up greenhouse effect. And that, scientists argue, is the result of human activity. As a 2010 statement from the National Academy of Sciences puts it, "There is compelling, comprehensive, and consistent objective evidence that humans are changing the climate in ways that threaten our societies and the ecosystems on which we depend." How did we get to be the culprits in this situation? In short, by pumping more carbon dioxide into the atmosphere.

Carbon dioxide is the most notorious player in the greenhouse effect, and scientists believe it is responsible for most of the warming. In fact, atmospheric carbon dioxide concentrations are higher now than they have been in more than 800,000 years.

As discussed in Chapter 2, carbon is a natural ingredient in every living organism, part of the backbone of all organic molecules. Carbon also exists in inorganic forms: as carbon dioxide in the atmosphere, as carbonic acid dissolved in water, as calcium carbonate in limestone rocks. If dead organisms are fossilized before being digested by decomposers, the organic molecules contained within their bodies become trapped below Earth's surface or under the seas. Over time, these compressed organic molecules become **fossil fuels**–coal, oil, and natural gas.

Like other chemical elements, the total amount of carbon on Earth remains essentially constant. In contrast to the way energy flows through an ecosystem in one direction (from the sun to producers to consumers and out to the universe as heat; see Chapter 21), elements such as carbon, nitrogen, and phosphorus move in cycles. The movement of carbon through the environment follows a predictable pattern called the **carbon cycle.**

As it cycles through the environment, carbon moves between organic and inorganic forms. For example, animals take in organic carbon when they eat other organisms and release inorganic gaseous CO_2 into the atmosphere as a by-product of cellular respiration. Similarly, when organisms die, decomposers in the soil use the dead organic material for food and energy, releasing some of the carbon during respiration as CO_2.

Plants, photosynthetic bacteria, and algae take up CO_2 during photosynthesis and convert it into organic sugar molecules, thus reducing atmospheric CO_2 levels. Photosynthesis, respiration, and decomposition form a cycle that keeps carbon dioxide at a relatively stable level in the atmosphere. But human actions, like deforestation and burning fossil fuels, inject

FOSSIL FUEL
A carbon-rich energy source, such as coal, petroleum, or natural gas, formed from the compressed, fossilized remains of once-living organisms.

CARBON CYCLE
The movement of carbon atoms as they cycle between organic molecules and inorganic CO_2.

INFOGRAPHIC 22.9 The Carbon Cycle

The carbon cycle involves the movement of carbon atoms as they cycle between organic molecules and inorganic CO_2. Natural processes such as photosynthesis, respiration, and decomposition are responsible for most carbon cycling. Since the late 1700s, human activities, including burning fossil fuels and deforestation, have made significant contributions to the carbon cycle, primarily by increasing the amount of carbon in the cycle in the form of CO_2 (measured in tons of carbon).

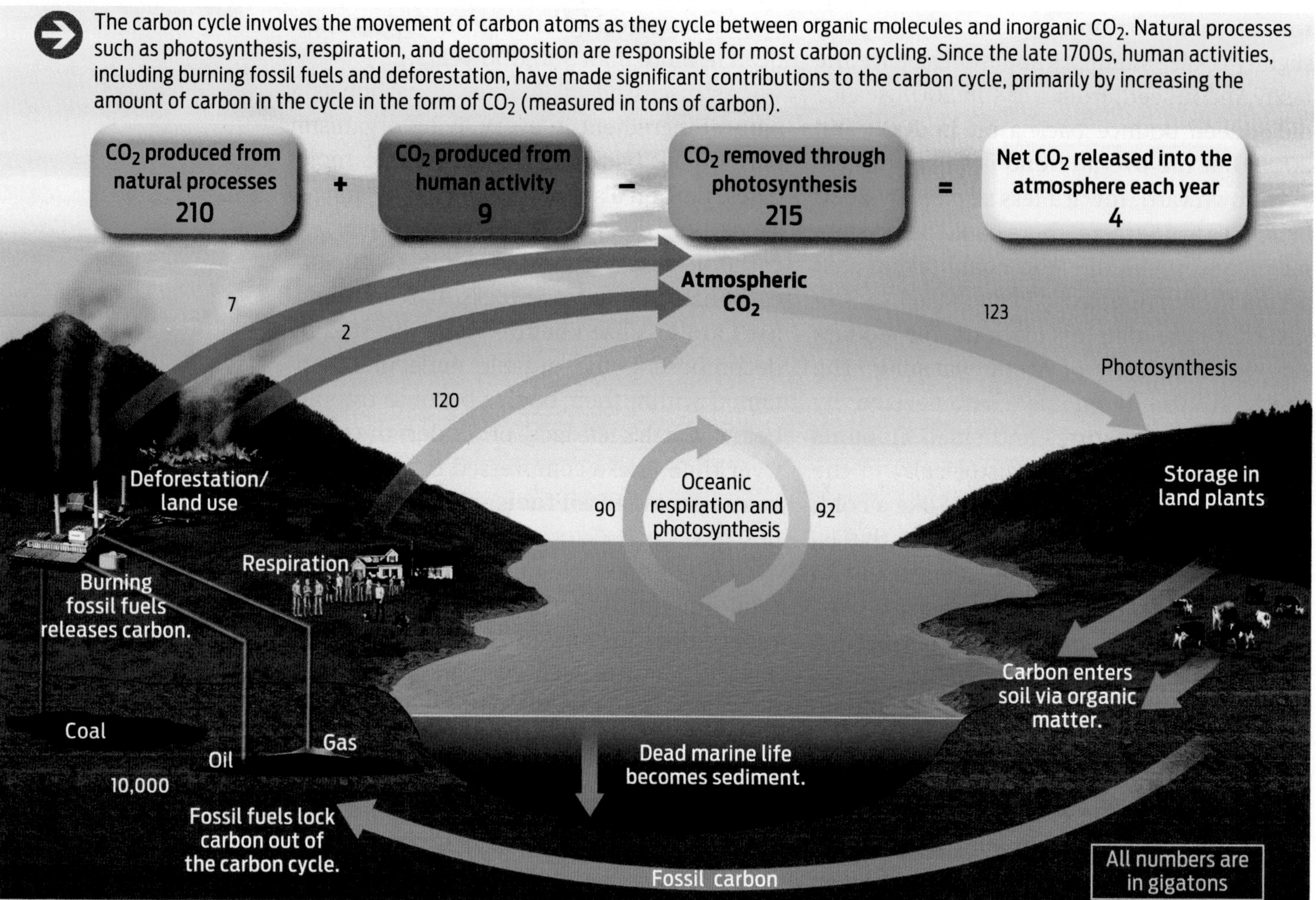

? If human activities had no effect on the carbon cycle, what would happen to carbon balance?

carbon dioxide that was not otherwise moving into the cycle (**INFOGRAPHIC 22.9**).

For most of human history, the amount of carbon present in the atmosphere as carbon dioxide has remained fairly constant. But since the industrial revolution, beginning in the late 1700s, humans have begun to alter the carbon cycle, adding increasing amounts of CO_2 to the atmosphere.

Before the industrial revolution, the carbon trapped in fossil fuels was not easily accessible and therefore not part of the carbon cycle. But modern drilling and mining methods have unlocked the deep reserves of this ancient planetary energy. The CO_2 released when humans burn fossil fuels is added to the carbon cycle and is a major contributor to the enhanced greenhouse effect.

How do scientists know that carbon dioxide levels are much higher now than in the past? There are two main sources of evidence. Air bubbles trapped in glacial ice from Greenland and Antarctica provide a historical measure of carbon dioxide in the atmosphere. Ice cores drilled at these sites provide data on very long term changes in CO_2 levels (and temperature), going back hundreds of thousands of years. These data show, for example, that levels of CO_2 have cycled in patterns that correlate with major ice ages. Since 1958, scientists have also directly measured CO_2 in the atmosphere—for example,

INFOGRAPHIC 22.10 Measuring Atmospheric Carbon Dioxide Levels

Examining data from ancient air bubbles and present-day air measurements, scientists have recorded an approximately 40% increase in atmospheric CO_2 levels since 1800.

Historical carbon dioxide levels are measured in glacial ice cores

As snow compacts into ice cores, bubbles of atmospheric gas remain trapped in the ice. The deepest ice is the oldest, and the air bubbles in those old layers contain levels of CO_2 that were present in the atmosphere at that time.

VIN MORGAN/AFP/Getty Images

Gas bubbles trapped in ice cores can be analyzed for carbon dioxide levels.

British Antarctic Survey/Science Source

Present-day carbon dioxide levels are measured directly from the air

Direct measurements of carbon dioxide are currently taken from the Mauna Loa Research Station in Hawaii.

U.S. Dept. of Commerce, NOAA, Earth System Research Laboratory

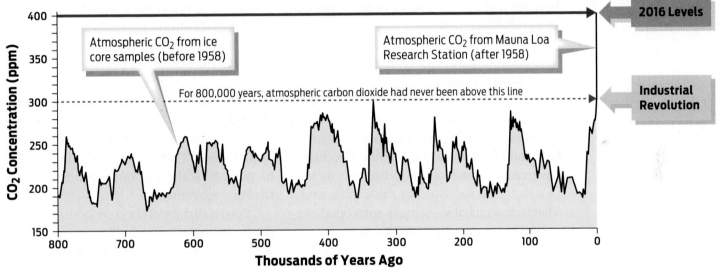

CO2 Concentration (ppm)
400
350
300
250
200
150
800
700
600
500
400
300
200
100
0
Thousands of Years Ago
Atmospheric CO2 from ice core samples (before 1958)
Atmospheric CO2 from Mauna Loa Research Station (after 1958)
For 800,000 years, atmospheric carbon dioxide had never been above this line
2016 Levels
Industrial Revolution

Data from NOAA and Scripps institution of Oceanography

? What is the value (in ppm) that CO_2 had not exceeded for 800,000 years? What is the current value of CO_2 concentration in the atmosphere? What percent increase does this represent?

at the Mauna Loa Research Station, which sits atop an inactive volcano in Hawaii. When combined, these data show that atmospheric CO_2 has been rising steadily since the industrial revolution—increasing from about 280 parts per million (ppm) in 1800 to more than 400 ppm in 2017—or more than 40% (**INFOGRAPHIC 22.10**).

Activities that decrease the number of photosynthetic organisms also increase global CO_2 levels. Since photosynthesizers are the only consumers of carbon dioxide in the carbon cycle, removing them not only reduces the amount of carbon dioxide they might have consumed, but also—in the case of large trees and stable populations of algae—eliminates what are in essence long-term storage vessels of carbon. Human activities that reduce the number of photosynthetic organisms on the planet include large-scale slash-and-burn agriculture, development that leads to deforestation, and various forms of pollution. Together,

CARBON FOOTPRINT A measure of the total greenhouse gases produced by human activities.

these activities contribute to our **carbon footprint,** a subset of our total ecological footprint.

Though CO_2 is the major player in global warming, another important greenhouse gas is methane (CH_4). Methane is produced by natural processes, such as the decomposition of organic material in swamps by microbes. However, agriculture, including raising cattle and growing rice in paddies, now accounts for more than half the total methane being pumped into the atmosphere. One of the main sources of methane is the gas produced by archaea that live in the digestive systems of cattle. Emitted as flatulence, it adds an estimated 100 million tons of methane a year to the atmosphere. Although the atmospheric concentration of methane is far less than that of CO_2, methane is particularly worrisome because it is 30 times more potent than CO_2 as a greenhouse gas.

With this steep rise in greenhouse gases have come steadily rising temperatures around the globe, with most of that warming occurring since the 1970s. Virtually all climate scientists agree that greenhouse gases emitted by human activities—primarily driving gasoline-powered cars and burning coal to generate electricity—have caused most of the global rise in temperature observed since the mid-20th century. That is, this global warming is anthropogenic—caused by humans **(INFOGRAPHIC 22.11)**.

And increased atmospheric carbon dioxide does more than just enhance the greenhouse effect; it also changes the chemical composition of the oceans, making them more acidic (because of the production of carbonic acid), which in turn harms life in the oceans. Many have called this change the "evil twin of global warming."

No Time for Fatalism

The United States is among the world's biggest emitters of greenhouse gases, yet for political reasons it has been reluctant to make significant reductions. It was one of the few countries that refused to ratify the Kyoto Protocol, a United Nations agreement adopted in 1997 that obligates endorsing countries to reduce carbon dioxide emissions. In 2015, the United States joined 194 other nations in signing the Paris Agreement, which commits nearly every country on Earth to reducing its greenhouse emissions through concrete plans. Together, these plans—if implemented—would mean a reduction in the level of carbon emissions by about half as is required to avoid the 2°C (3.6°F) of warming that many scientists fear is a point of no return for the climate. But many of the other concrete measures the Obama administration wanted to take to reduce CO_2 emissions in the United States were opposed by majorities in the Congress. The Environmental Protection Agency (EPA) could attempt to regulate CO_2 levels on its own; in 2007, the Supreme Court ruled that CO_2 qualifies as a pollutant and so the EPA is legally required to regulate it under the Clean Air Act. But so far, the EPA's efforts to regulate carbon emissions have been met with legal challenges. And with the election of Donald Trump to the presidency in 2016 and his appointment of someone who has disputed widely accepted climate science to head the EPA, the prospects for U.S. leadership on climate change seem bleak indeed. Trump announced in 2017 that he plans to withdraw the United States from the Paris Agreement.

Even if all the world's greenhouse gas emissions were turned off today like closing a faucet, we would still face decades of warming and its consequences because of past emissions—what climate scientists refer to as heat in the pipeline. Basically, the climate has not yet caught up with the effects of burning fossil fuels in past decades because the oceans are slower to heat than the land (as any beachgoer knows). But eventually, the oceans will catch up, leading to more warming of the atmosphere.

This grim reality could easily lead some to take a fatalistic attitude, but that would be a dangerous mistake, says Hector Galbraith. "We've got to get beyond the deer in the headlights stage and begin to think as conservation biologists about what we're going to do about this to help to mitigate the impact." These preventative measures are what he and other climate experts call "adaptation."

INFOGRAPHIC 22.11 Anthropogenic Production of Greenhouse Gases

A variety of human, or anthropogenic, activities are increasing the levels of greenhouse gases in the atmosphere. Power generation and transportation account for the largest proportion of greenhouse gas emission.

U.S. Greenhouse Gas Emissions in 2014
Total U.S. emissions = 6,870 million metric tons

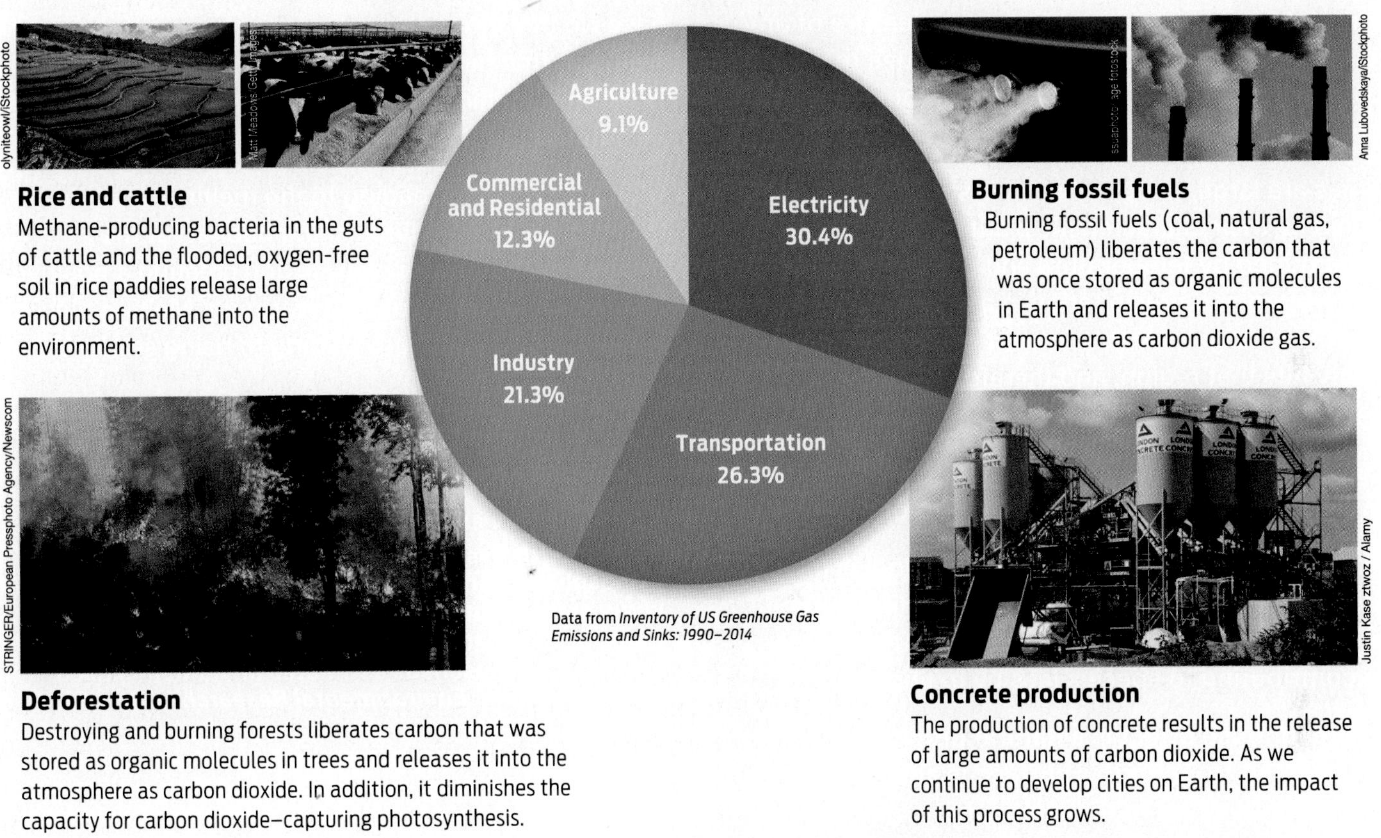

Rice and cattle
Methane-producing bacteria in the guts of cattle and the flooded, oxygen-free soil in rice paddies release large amounts of methane into the environment.

Burning fossil fuels
Burning fossil fuels (coal, natural gas, petroleum) liberates the carbon that was once stored as organic molecules in Earth and releases it into the atmosphere as carbon dioxide gas.

Deforestation
Destroying and burning forests liberates carbon that was stored as organic molecules in trees and releases it into the atmosphere as carbon dioxide. In addition, it diminishes the capacity for carbon dioxide–capturing photosynthesis.

Concrete production
The production of concrete results in the release of large amounts of carbon dioxide. As we continue to develop cities on Earth, the impact of this process grows.

Data from *Inventory of US Greenhouse Gas Emissions and Sinks: 1990–2014*

? If substantially more people drove electric cars with rechargeable batteries, what would happen to the distribution of emissions illustrated here?

Adaptation will not be easy. For many species, like Vermont's maples, it may already be too late. But doing nothing, say scientists, risks turning a bad problem into a catastrophic one.

In concrete terms, adaptation means planning for the inevitable: more frequent droughts, heat waves, and severe storms, as well as a rise in sea level. Increasingly, cities and towns all across the country are taking steps to incorporate adaptation into urban and land-use policies–restricting new construction in flood zones, for example, and working to conserve water in drought-prone areas, as well as protecting wetlands.

But these efforts won't do much to stem the tide unless we deal with the underlying cause of rapid climate change: runaway emissions of greenhouse gases. And that means dealing with the practices that collectively produce more than 90% of greenhouse gases in the United States today: burning fossil fuels for electricity, heat, transportation, and industry.

"Fossil fuels are incredibly efficient sources of energy," says Serreze, of the University of Colorado. "We've built our whole infrastructure around that. But what we didn't realize is that it's a trap, and that's what we're coming to grips with now." ■

CHAPTER 22 SUMMARY

- Ecosystems are made up of the living and nonliving components of an environment, including the communities of organisms present and the physical and chemical environment with which they interact.
- Temperature is an important physical feature of any ecosystem and serves as a clock that cues many biological events, such as breeding, blooming, and hibernation.
- Biomes are large, geographically distinct ecosystems, defined by their characteristic plant life, which in turn is determined by temperature and precipitation. Temperature and precipitation in a region are determined largely by the amount of sunlight that directly hits that region, and therefore by latitude.
- The amount and type of vegetation in a biome provide the base of the food chain and determine the variety of life that can thrive in that biome.
- Climate change, especially global warming, is having widespread effects on plant and animal life on the planet—altering seasonal life cycles, shifting ranges, and contributing to species loss by extinction.
- The greenhouse effect is a natural process by which heat radiated from Earth's surface is absorbed by heat-trapping gases in the atmosphere, maintaining a global temperature that can support life. Rising levels of greenhouse gases have led to the enhanced greenhouse effect.
- Elements cycle through ecosystems. The carbon cycle is the movement of carbon atoms through living and nonliving components of the environment by the biotic processes of photosynthesis, cellular respiration, and decomposition, as well as by long-term geological processes.
- Global warming results from an increase in the amount of carbon dioxide and other greenhouse gases in the atmosphere. Scientists agree that the warming that has occurred since the mid-20th century is anthropogenic, due largely to human activities such as burning fossil fuels and deforestation.
- Global warming is leading to the melting of sea ice in the Arctic, which is diminishing habitat for the organisms that rely on sea ice and creating a positive feedback loop for increased warming. Melting glaciers and ice caps on land are leading to rising sea levels.

MORE TO EXPLORE

- **National Snow and Ice Data Center: http://nsidc.org/**
- **NOAA, Earth System Research Laboratory, Trends in Atmospheric Carbon Dioxide (animation): www.esrl.noaa.gov/gmd/ccgg/trends/history.html**
- **Cook, J., et al. (2016). Consensus on consensus: a synthesis of consensus estimates on human-caused global warming. *Environmental Research Letters* 11(4), 048002.**
- **Ilya M. D., et al. (2011) Recent ecological responses to climate change support predictions of high extinction risk. *Proceedings of the National Academy of Sciences* 108:12337–12342.**
- **Oreskes, N., and Conway, E. M. (2010) *Merchants of Doubt: How a Handful of Scientists Obscured the Truth on Issues from Tobacco Smoke to Global Warming*. New York: Bloomsbury Press.**
- **Katz. J., and Daniel, J. (December 2, 2015) What You Can Do About Climate Change. Seven Simple Guidelines for Thinking About Carbon Emissions. *New York Times*.**

CHAPTER 22 Test Your Knowledge

DRIVING QUESTION 1 What are ecosystems, and how are ecosystems being affected by climate change?

By answering the questions below and studying Infographics 22.1, 22.2, 22.3, 22.4, and 22.7 and Up Close: Biomes, you should be able to generate an answer for the broader Driving Question above.

KNOW IT

1 **Which of the following are parts of an ecosystem?**

a. the plant life present in a given area
b. the animals living in a given area
c. the amount of annual rainfall in a given area
d. the soil chemistry in a given area
e. none of the above
f. all of the above

2 **List several examples of species discussed in this chapter that have changed their geographic distributions or the timing of events in their life cycles as a result of global climate change.**

3 **In identifying a biome, for which of the characteristics below would it be important to have data? (Select all that apply.)**

a. monthly rainfall
b. temperatures throughout the year
c. plant life
d. animal life
e. size of the human population in the area

4 **What plant life is abundant in the temperate forest biome?**

5 **Look at Up Close: Biomes. Where in North and South America do you find temperate forest? Tropical forest?**

6 **Distinguish between climate and weather.**

USE IT

7 **Although trees may not be able to walk away from increasingly warm regions, evolutionary adaptations may allow trees to survive in warmer regions. Discuss each of the adaptations listed below and decide if it is likely to be helpful or harmful in a warming environment. (Think about water—water is taken up by the roots of plants and lost through pores in the leaves; CO_2 levels—CO_2 is taken up by plants through pores in leaves, then used by leaves for photosynthesis; and the movement of other species, for example insects, in response to global warming.)**

a. having smaller leaves
b. having a larger number of pores on each leaf
c. having thicker and waxier bark

8 **What is a possible risk for humans if insects that carry pathogenic bacteria or viruses expand their range northward?**

9 **How can a change in spring flowering time in a particular region affect the animals of that particular ecosystem?**

10 **Why are tropical rain forests found at approximately the same latitude in Asia and in South America?**

DRIVING QUESTION 2 What is the greenhouse effect, and what does it have to do with global warming?

By answering the questions below and studying Infographics 22.5, 22.6, and 22.8, you should be able to generate an answer for the broader Driving Question above.

KNOW IT

11 **Which greenhouse gas is emitted every time you breathe out?**

a. oxygen
b. carbon dioxide
c. methane
d. nitrogen
e. water vapor

12 **Which of the following organisms contributes to reducing atmospheric CO_2 levels?**

a. maple trees
b. most algae
c. polar bears
d. Arctic fox
e. a and b
f. a, b, and d

13 **Could we live in the absence of the greenhouse effect? Explain your answer.**

14 **If global warming causes Arctic sea ice to melt, what will be the effect on sea levels in a low-lying region like Miami? If large parts of the Antarctic polar ice cap should melt, what would be the effect on sea level?**

USE IT

15 Explain how each of the following contributes to an elevation of levels of greenhouse gases.

a. engaging in large-scale slash-and-burn agriculture
b. driving gasoline-fueled cars
c. producing cattle for beef and dairy products
d. producing rice

16 Look at Infographic 22.8. From the data presented, how much warmer than the 30-year average was most of the continental United States in 2016? Which part of the United States had the greatest degree of warming? How much warmer was this region than the 30-year average?

DRIVING QUESTION 3 How does carbon cycle through ecosystems, and how do scientists compare present and past atmospheric CO_2 levels and what can they learn from such comparisons?

By answering the questions below and studying Infographic 22.9, 22.10, and 22.11, you should be able to generate an answer for the broader Driving Question above.

KNOW IT

17 Fill in the blanks in the diagram below.

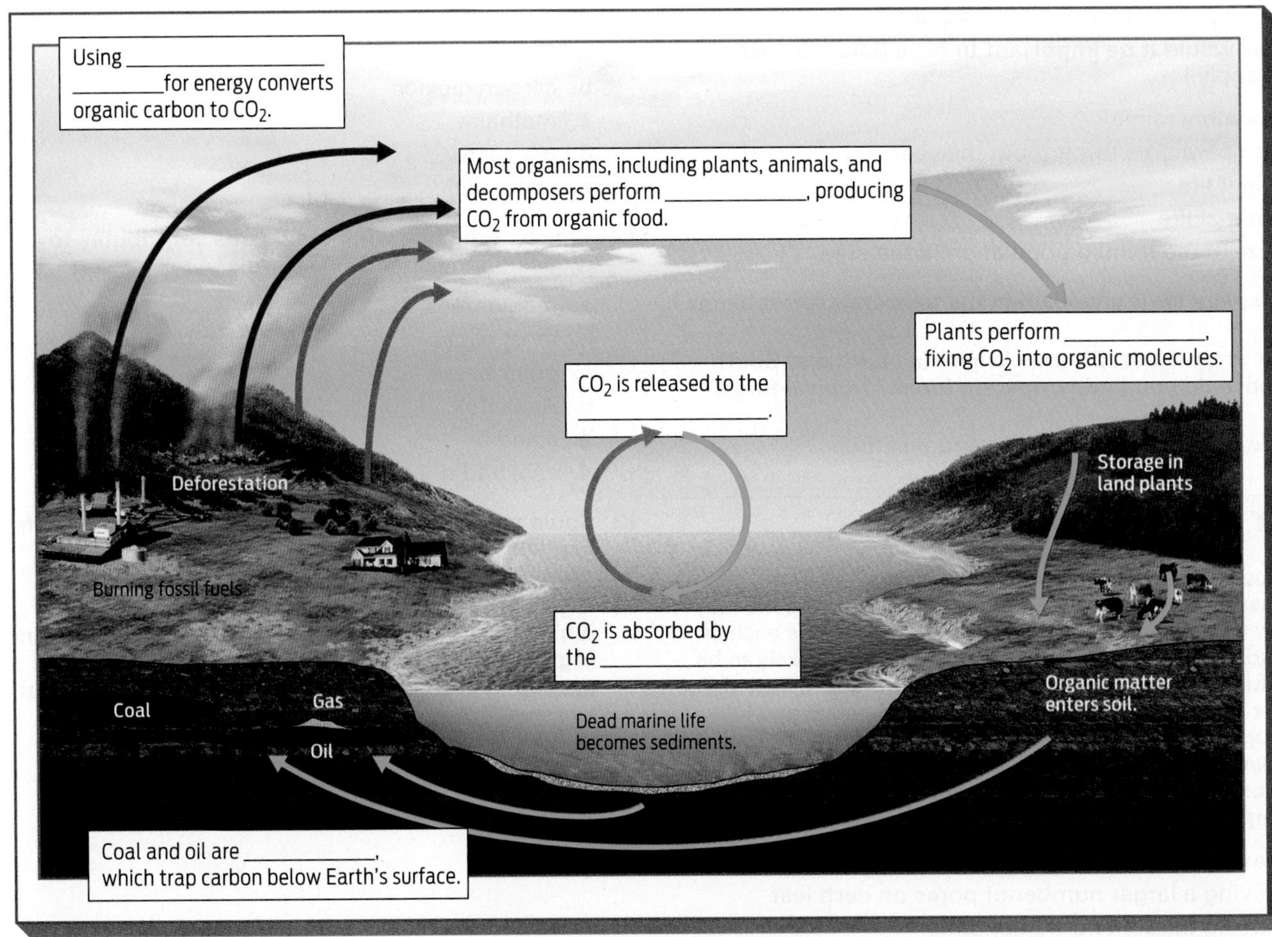
Using ____________
________for energy converts organic carbon to CO_2.
Most organisms, including plants, animals, and decomposers perform ____________, producing CO_2 from organic food.
Plants perform ____________, fixing CO_2 into organic molecules.
CO_2 is released to the ____________.
Storage in land plants
Deforestation
Burning fossil fuels
CO_2 is absorbed by the ____________.
Organic matter enters soil.
Coal
Gas
Oil
Dead marine life becomes sediments.
Coal and oil are ____________, which trap carbon below Earth's surface.

18 Plants like maple trees____________ CO_2 by ____________.

a. emit; photosynthesis
b. take up; photosynthesis
c. emit; cellular respiration
d. take up; cellular respiration
e. store; cellular respiration

19 Fossil fuels are most immediately derived from

a. organic molecules.
b. CO_2.
c. methane.
d. melting ice caps.
e. photosynthesis.

20 Name at least two human activities that increase CO_2 levels in the atmosphere and two natural processes that contribute CO_2 to the atmosphere.

USE IT

21 How is ice useful in the measurement of atmospheric levels of CO_2?

22 Describe the evidence that increasing levels of greenhouse gases are responsible for global climate change. What if someone suggested to you that global climate change is due to increased intensity of solar radiation (that is, the amount of sunlight reaching Earth)? What evidence would you ask for in support of this hypothesis?

23 Which of the following data would you use to determine the levels of atmospheric CO_2 in 1750? Justify your choice, including an explanation of why the other choices would not be as effective.

a. historical weather records of daily temperatures

b. records for 1750 from the archives of the Mauna Loa Research Station

c. tree-ring analysis (to look for evidence of extreme fires)

d. ice cores from ice formed in 1750

MINI CASE

apply YOUR KNOWLEDGE

24 Annie is considering a new car. She is contemplating several options, and has decided that environmental impact (particularly greenhouse gas emissions) is her top priority. After all, she tries to eat primarily grains, fruits and vegetables, and regularly embraces "meatless Mondays." She is looking at a fuel-efficient gasoline powered car, a traditional hybrid (that uses gasoline, but can also run on a battery that recharges while the car is being driven), and a fully electric plug-in car. Discuss the pros and cons of her three vehicle options. What other factors (which may be beyond her control) could affect the emissions associated with owning and driving some of the vehicles she is contemplating?

BRING IT HOME

25 Visit an online carbon footprint or carbon emissions calculator (for example, http://www.epa.gov/climatechange/ghgemissions/ind-calculator.html) and calculate your total carbon emissions.

a. What is your largest source of emissions?

b. What steps can you take to decrease your carbon emissions?

c. Explain how drying your laundry on a clothesline rather than in the dryer can decrease your carbon emissions.

INTERPRETING DATA

apply YOUR KNOWLEDGE

26 A 2010 study compared the amount of CO_2 emitted when locally grown broccoli was delivered to Virginia Tech University with the amount emitted when broccoli grown in California was delivered to Virginia Tech. The California broccoli was delivered in shipments containing 768 lbs of broccoli in a tractor-trailer that traveled 2,786 miles. Tractor-trailer fuel efficiency is 5 miles per gallon, and 20 lbs of CO_2 are released per gallon of fuel burned. The local broccoli was delivered in shipments of 587 lbs of broccoli in a cargo van that traveled 19.1 miles. Cargo van fuel efficiency is 16 miles per gallon, and 20 lbs of CO_2 are released per gallon of fuel burned.

a. Complete the table below to determine the CO_2 emissions associated with delivering 1 lb of local and 1 lb of nonlocal broccoli.

b. Is locally sourced fresh broccoli a year-round option at Virginia Tech?

c. Do some online research to determine approximately what proportion of CO_2 emissions are associated with food delivery vs. food production.

Source of Broccoli	Miles per Shipment	Gallons of Fuel Burned per Shipment	CO_2 Released per Shipment (lbs)	CO_2 Released per lb of Broccoli Delivered (lbs)
Nonlocal	2,786			
Local	19.1			

Data from Schultz, J., and Clark, S. (2010) Foodprint comparison of local vs. nonlocal produce: http://www.blacksburgfarmersmarket.com/docs/Schultz_Foodprint_Comparison_of_Local_vs_Nonlocal.pdf (accessed 5/4/2013)

7 Milestones in Biology

Progress or POISON?

Rachel Carson, pesticides, and the birth of the environmental movement

D.D.T.
Powerful Insecticide
Harmless to Humans
Applied by
TODD INSECT FOG APPLICATOR

DRIVING QUESTIONS

1. Why can carefully considered environmental policies still have unintended consequences?
2. What is biomagnification, and how does it occur?
3. Why was DDT widely adopted? What properties of DDT permit it to negatively affect organisms at a variety of levels in a food chain?

In 1958, the biologist and science writer Rachel Carson received a disturbing letter from a friend in Massachusetts. The letter described an event that had recently taken place on the friend's property near Cape Cod: an airplane had sprayed a thick cloud of a pesticide called DDT as part of a coordinated campaign to eradicate mosquitoes. In the days following the spraying, the friend noticed many dead songbirds in the area. "Their bills were gaping open, and their splayed claws were drawn up to their breasts in agony," she wrote. Could anything be done to stop these aerial sprayings, the friend wanted to know?

At the time she received the letter, Carson was a well-known science writer who had written some widely popular books about the sea. She cared deeply about nature and was horrified by her friend's report—especially since it wasn't the first time she'd heard about DDT toxicity. Carson had been concerned for a number of years about the effects DDT might be having on beneficial insects, birds, and fish, ever since she had worked as a marine biologist with the U.S. Fish and Wildlife Service, which had conducted studies on the pesticide in 1945. But the letter from her friend was a tipping point.

Carson decided that someone should research and write an article about the dangers of pesticides. She tried at first to persuade other writers to take on the topic. When no one would, she realized she had to tackle it herself. Four years later, the outcome was *Silent Spring*, a book that sparked a national conversation about pesticides and ushered in a new way of thinking about human impacts on the environment.

Bettmann/Getty Images

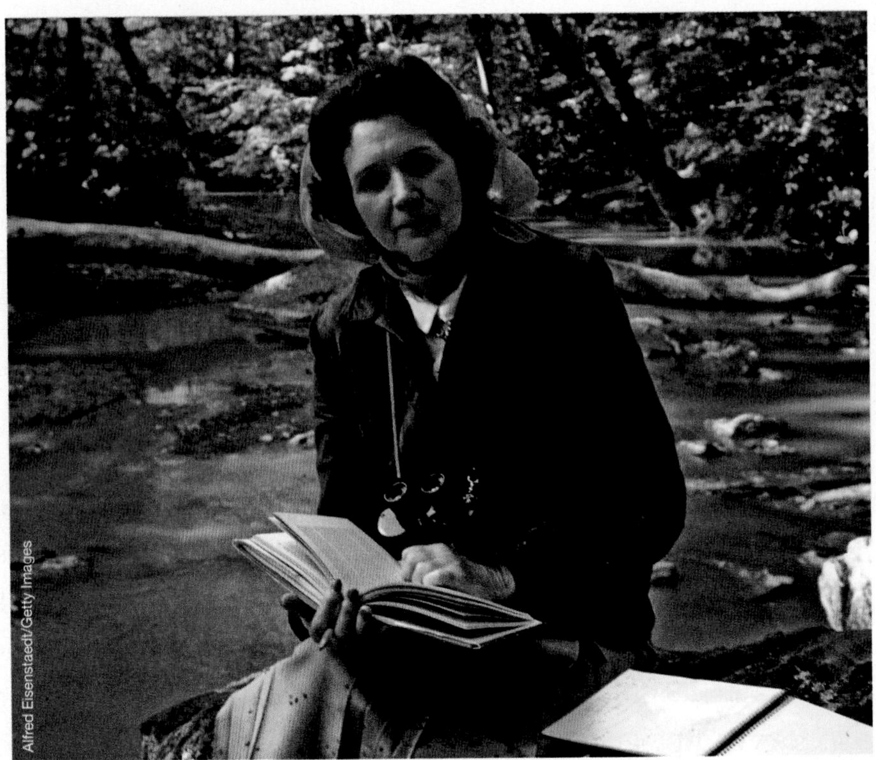
Alfred Eisenstaedt/Getty Images

Rachel Carson was a best-selling author and marine biologist before she wrote *Silent Spring*.

In the book's famous opening chapter, "A Fable for Tomorrow," Carson asked readers to imagine a town "in the heart of America" where "a strange blight" had crept over the land. The birds that had once greeted the coming of spring with a chorus of song now lay sick, dying, and silent. "It was a spring without voices," she wrote. What had caused this strange blight? "No witchcraft, no enemy action had silenced the rebirth of new life in this stricken world," Carson wrote. "The people had done it themselves."

Paul Müller, a Swiss chemist, showed that DDT was an effective insecticide. As a result of his work, DDT became a widely used pesticide.

The chemical formula for the synthetic chemical DDT

Basement Stock / Alamy

Bettmann/Getty Images

I Want My DDT

DDT, a synthetic chlorinated hydrocarbon, was discovered to be a potent insecticide in 1939 by the Swiss chemist Paul Müller. DDT poisons the nervous system of insects and other animals. It was first widely used during World War II to combat insect-borne diseases, such as typhus and malaria, among U.S. soldiers. Typhus is spread by lice, which infested many soldiers living in cramped quarters. Malaria is transmitted by mosquitoes, which live in tropical regions of the globe. The disease plagued soldiers fighting in the islands of the South Pacific, and was also a problem in the American South, where many U.S. military training bases were located. DDT was mixed with powder and dusted directly on clothing and was sprayed on walls in barracks. As a public health measure, DDT succeeded marvelously, saving countless lives during the war. For his work Müller was awarded a Nobel prize in 1948.

After the war, DDT became available to the public as a commercial pesticide and was widely used by farmers to protect their crops and by public health officials to contain insect-borne disease. The chemical companies that manufactured the potent bug-killer advertised it as an aid to domestic comfort and tranquility. Photos and newsreels from the era document mass spraying of suburban neighborhoods as children played happily in the chemical clouds (**INFOGRAPHIC M7.1**).

But by the late 1950s, a number of scientists and citizens had become concerned that indiscriminate spraying of pesticides like DDT was doing more harm than good. Efforts by the U.S. Department of Agriculture (USDA) to eliminate gypsy moths and fire ants in a number of states had reportedly wiped out scores of other insects, fish, and birds. By 1958, when Carson received the letter from her friend, several court cases had been filed against the USDA in an effort to stop the indiscriminate spraying.

Carson began approaching scientists around the country, including many of her former colleagues in the Fish and Wildlife Service, for information about pesticide use and toxicity. "The more I learned about the use

INFOGRAPHIC M7.1 Widespread Use of DDT to Kill Mosquitos and Lice

As an insect neurotoxin, DDT was used to combat a number of diseases associated with insects, including those borne by lice and mosquitoes.

DDT was applied directly to human hair, skin, and clothing to eliminate lice that spread typhus, a microbial disease.

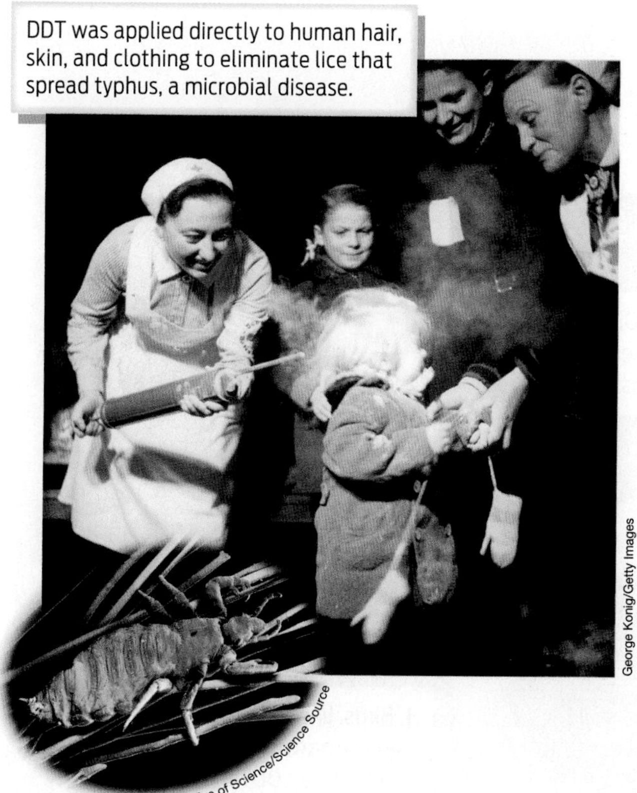

George Konig/Getty Images

Eye of Science/Science Source

DDT was sprayed freely throughout neighborhoods to eradicate mosquitoes and other insects that carry human pathogens like the one that causes malaria.

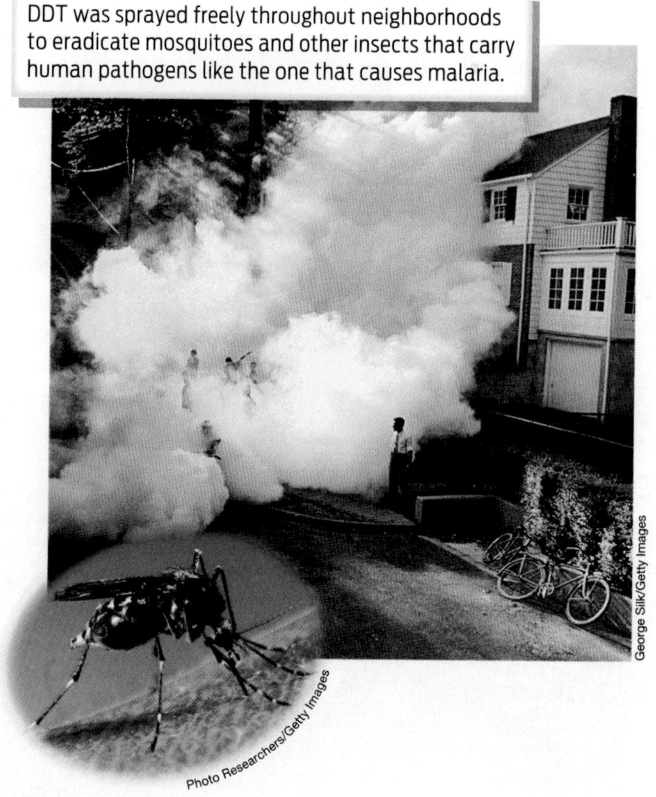

George Silk/Getty Images

Photo Researchers/Getty Images

? Name at least one disease that is transmitted by insects.

of pesticides, the more appalled I became," she later said. Over the next 4 years, she interviewed more scientists, scoured the research literature, and culled through newspaper reports to produce *Silent Spring*, a comprehensive treatise with 50 pages of footnotes.

Carson's book was filled with example after depressing example of how synthetic pesticides were wreaking destruction on natural populations of animals, most famously birds. For example, in the mid-1950s, government officials in states across the Midwest attempted to deal with the growing problem of Dutch elm disease, which was killing off the stately trees, by using DDT. Dutch elm disease is caused by a fungus, but it is spread from tree to tree by beetles that feed on the trees' leaves. To combat the beetles, scientists sprayed the trees with DDT. In the autumn, leaves coated in DDT fell to the ground, then earthworms feeding on the leaves took up the chemical and accumulated it inside their bodies. In the spring, robins migrated to these areas, fed voraciously on the earthworms, and were poisoned by the high amounts of DDT they ingested (**INFOGRAPHIC M7.2**).

Carson presented evidence that aquatic ecosystems are also affected by DDT. Since 1945, coastal waters around the United States had been sprayed with DDT to combat the salt-marsh mosquito. DDT entered the water supply, where it and its breakdown product, DDE, were taken up into the bodies of small fish and crabs. Larger fish would eat the smaller fish, thus taking in higher quantities of stored DDT and DDE, until finally

INFOGRAPHIC M7.2 Unintended Consequences of Using DDT

While DDT was effective against its insect targets, it also made its way into the food chain of many organisms, with unintended consequences.

1. Trees were sprayed with DDT to kill the beetle that spread the fungus that causes Dutch elm disease.
2. Leaves coated in DDT fell to the ground and were eaten by earthworms.
3. Birds, like the American robin, that ate earthworms died from the accumulated DDT in their systems.
Courtesy of the Greenwich Historical Society
Nature Graphic Science/Alamy
Valerie Giles/Getty Images
Kenneth C. Zirkel/Getty Images

? Were the robins harmed by being sprayed by DDT, by contacting DDT-coated leaves, or by eating earthworms?

the larger fish were eaten by predatory birds such as eagles, pelicans, and falcons, which ingested the highest quantities of all. Because of the interactions of organisms at different trophic levels in a food chain (see Chapter 21), organisms at the highest trophic levels can have high concentrations of DDT in their tissues even if they were not directly exposed to DDT. The chemical is fat soluble and not easily excreted in urine; therefore it accumulates in animal bodies. When other animals eat these animals, they eat their stored DDT, too. The process by which environmental toxins accumulate as they move up the food chain is called biomagnification (**INFOGRAPHIC M7.3**).

The accumulated DDE impaired reproduction in birds of prey, especially pelicans, eagles, and falcons: their eggshells became so thin that mothers would crush their eggs when they sat on their nests to incubate them. This, combined with overhunting and habitat destruction, led to a steep drop in predatory bird populations in the United States. From numbers in the hundreds, for example, the bald eagle population in the United States plummeted in the 1950s and appeared on the verge of extinction—an ominous fate for our national emblem.

Besides the threat to wild animals, Carson also pointed to possible dangers to humans. Though DDT was deemed safe for use as an insecticide on humans, there were concerns about long-term effects. "We have to remember that children born today are exposed to these chemicals from birth, perhaps even before birth," Carson said in a 1962 interview on CBS News. "Now what is going to happen to them in adult life as a result of that exposure? We simply don't know." These uncertainties were made more unnerving when DDT was shown to persist in the environment for many years, long after its application. We now know that DDT stays in soil for at least 15 years, and

? Explain how a chemical like DDT when used to treat water can eventually be detected at high concentrations in the tissues of birds like eagles.

INFOGRAPHIC M7.3 Biomagnification

Chemicals such as DDT are retained in the bodies of organisms that take them up. When those organisms are eaten by others, the concentration of chemicals increases at each level of the food chain.

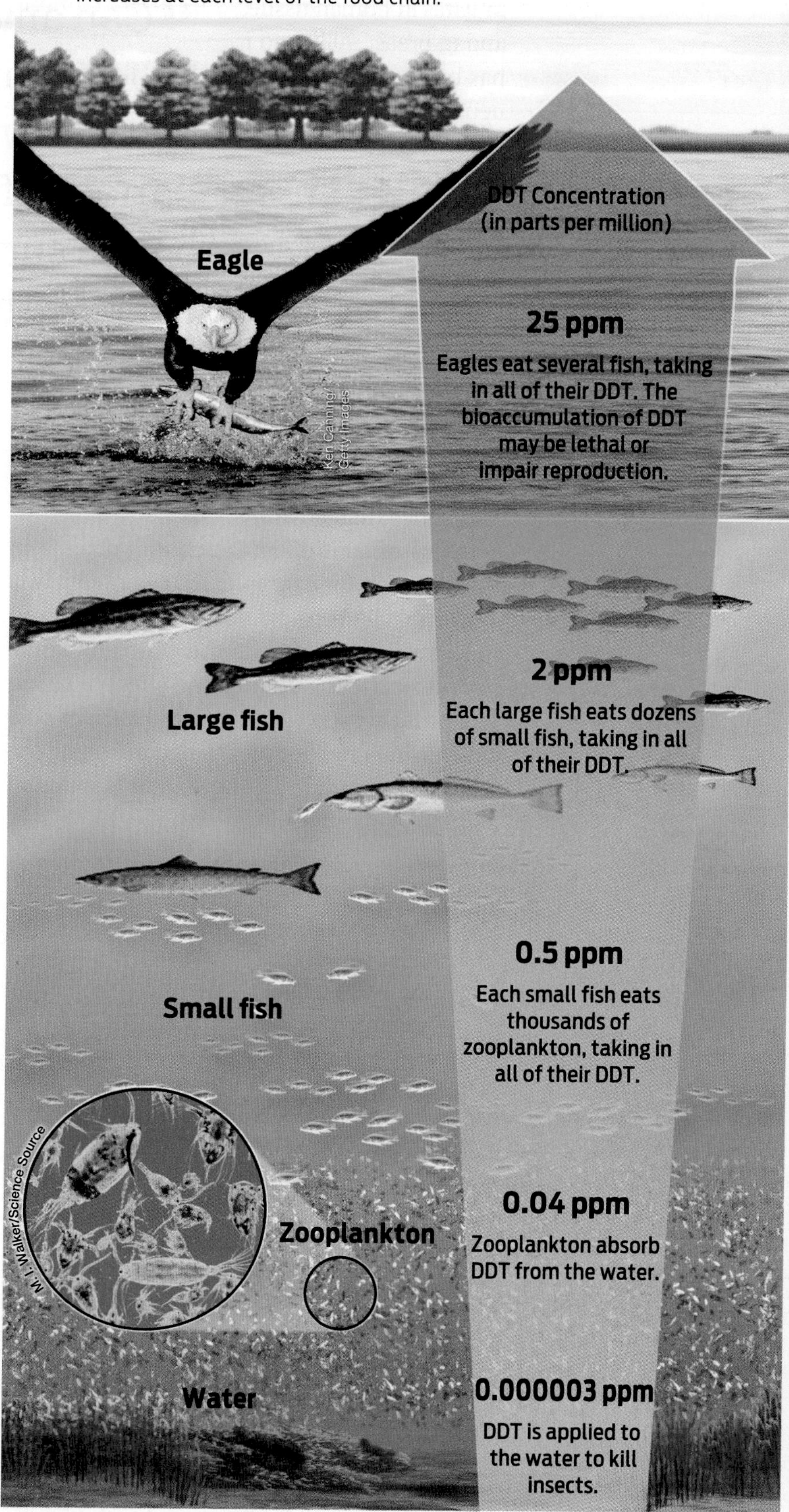
DDT Concentration (in parts per million)
Eagle
25 ppm
Eagles eat several fish, taking in all of their DDT. The bioaccumulation of DDT may be lethal or impair reproduction.
Ken Canning/Getty Images
Large fish
2 ppm
Each large fish eats dozens of small fish, taking in all of their DDT.
Small fish
0.5 ppm
Each small fish eats thousands of zooplankton, taking in all of their DDT.
M. I. Walker/Science Source
Zooplankton
0.04 ppm
Zooplankton absorb DDT from the water.
Water
0.000003 ppm
DDT is applied to the water to kill insects.

much longer in aquatic environments, possibly for hundreds of years. (The EPA currently classifies DDT as a "probable carcinogen"; because it is fat soluble, DDT accumulates in human tissues and in breast milk, and it has been associated with neurological disorders in places where it has been used heavily to prevent malaria.)

> "Can anyone believe it is possible to lay down such a barrage of poisons on the surface of the earth without making it unfit for all life?"
>
> —Rachel Carson

To Carson, the most galling and stupefying aspect of all these stories was the blind faith people put in pesticides—using them even when they had not been tested and when other methods had proved effective. As a means of controlling Dutch elm disease in the Midwest, for example, DDT spraying was a colossal failure; it did not save the trees, and may have actually made the problem worse because it had the unintended effect of killing the birds that served as a natural form of pest control. Yet effective means of controlling the disease—through the removal and burning of contaminated wood—had been used successfully for decades in other parts of the country, for example in New York.

How to account for these lapses of scientific judgment? A big part of the problem, Carson argued, was an unquestioned faith in the power of experts to control nature, to bend it to their whim with technology. In other words, it was as much a mind-set as a specific technology that was the source of the trouble.

"Who has made the decision that sets in motion these chains of poisonings, this ever-widening wave of death that spreads out, like ripples when a pebble is dropped into a still pond?" she asked in her book. Carson minced no words with her answer: it was shortsighted individuals in positions of power whose capacity for destruction was not matched by an informed awareness of the associated costs.

Rachel Carson testifying before Congress in 1963 on the dangers of pesticides.

From Silent Spring to Noisy Summer

Silent Spring was serialized in *The New Yorker* in the spring of 1962 and published as a book by Houghton Mifflin that fall. It caused an immediate sensation, as people around the country began to question the safety of these omnipresent and unseen chemicals. "'Silent Spring' Is Now Noisy Summer," the *New York Times* headlined a story documenting the furor that erupted in the wake of the book's publication. Though Carson found many supporters and admirers, among them President John F. Kennedy, she also found herself attacked by angry representatives of the chemical industry.

Velsicol Chemical, a maker of DDT and other pesticides, threatened to sue both

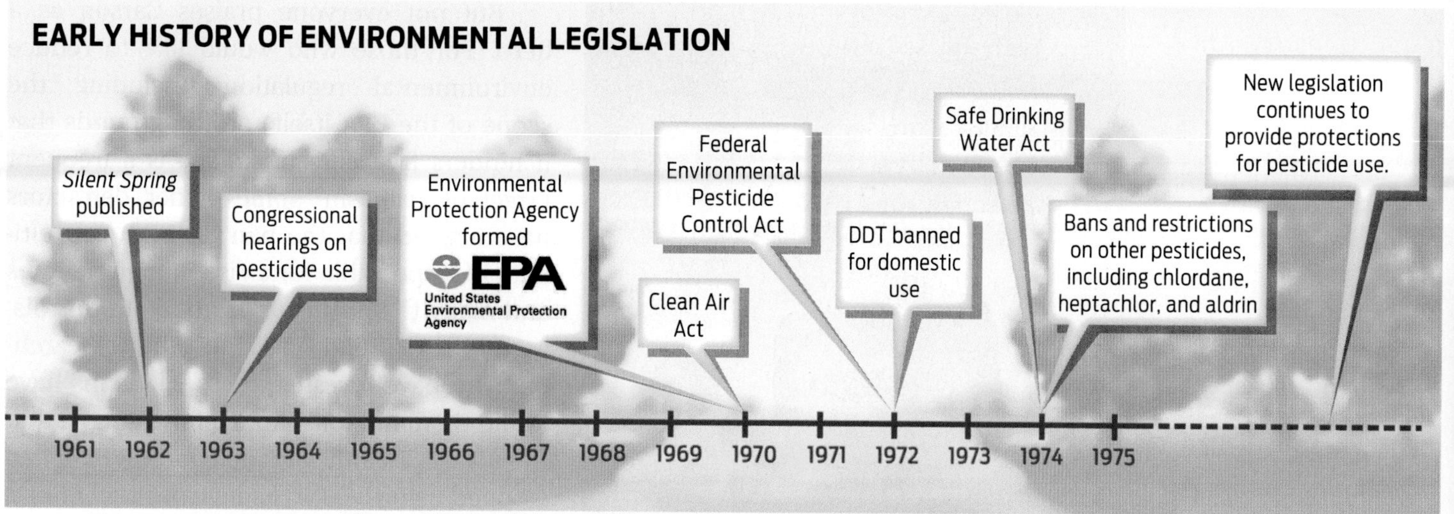

Houghton Mifflin and *The New Yorker*. A lawyer for the company accused Carson of being a Communist sympathizer who wanted to shrink the American food supply. Monsanto, another pesticide producer, published and distributed 5,000 copies of a brochure parodying Carson's book titled "The Desolate Year," which argued that without pesticides to help agriculture, food supplies would plummet and millions around the world would suffer from hunger and starvation. "If man were to faithfully follow the teachings of Miss Carson," scoffed an executive of American Cyanamid, "we would return to the Dark Ages, and the insects and diseases and vermin would once again inherit the earth."

Some of the criticism Carson faced was clearly because she was a woman. Her credentials as a scientist were routinely attacked (although she had a master's degree in zoology from Johns Hopkins), and she was often referred to as "hysterical." A former U.S. Secretary of Agriculture wondered aloud to the press "why a spinster with no children was so interested in genetics." A physician, writing in a medical journal, stated that reading *Silent Spring* "kept reminding me of trying to win an argument with a woman. It can't be done."

But Carson had done her homework, and her view ultimately carried the day. A commission convened by President Kennedy to investigate the claims made by Carson in her book found them to be sound. Soon thereafter, in 1963, congressional hearings were held at which Carson testified. She argued that a commission should be established to review pesticide issues and make decisions on the basis of broad public interest rather than the profit motives of a few.

Silent Spring is universally credited as the spark that ignited the modern environmental movement. In less than 10 years, what began as one woman's fight grew to enlist an army of activists, including many college students, who together changed the national conversation. The first national Earth Day celebration was held on April 22, 1970. Later that year, President Richard Nixon created the Environmental Protection Agency (EPA) and gave it the authority to set safe levels for chemicals. Shortly thereafter came several pieces of legislation including the Clean Air Act, the Federal Environmental Pesticide Control Act, and the Safe Drinking Water Act. The EPA banned DDT for domestic use in 1972, except for cases of public health (for example, to prevent malaria). In 1974, it further banned or severely restricted the other pesticides that Carson had written about in her book: chlordane, heptachlor, dieldrin, aldrin, and endrin.

Carson would surely have been gratified to see the enactment of these milestones,

George Silk/Getty Images

The beliefs of the time: An ad from *Life* magazine in 1948 featured model Kay Heffernon at Jones Beach, New York, apparently demonstrating that DDT was safe.

but she didn't live to witness them. She died of breast cancer in 1964 at the age of 56, less than 2 years after *Silent Spring* was published. Though she had been sick for much of the period when she wrote her book, she never mentioned her illness publicly, fearing that her analysis of the human dangers of pesticides–cancer among them–might be labeled as biased. But her legacy lives on in the many people who credit her with changing their worldview and relationship to nature.

"For me personally," former vice president Al Gore wrote in an introduction to the 1992 edition of her book, "*Silent Spring* had a profound impact. . . . Indeed, Rachel Carson was one of the reasons that I became so conscious of the environment and so involved with environmental issues."

But not everyone praises Carson as a hero. For those who would like to reduce environmental regulation (including the scope of the EPA itself) on the grounds that it hinders free enterprise, she is a frequent target of criticism. Some of her detractors have argued that the ban on DDT was ultimately responsible for a rise in malaria rates and many thousands of deaths from this disease in Africa and Asia. But other researchers who have investigated these claims have found them dubious, noting for example that the World Health Organization (WHO) had stopped using DDT in its international efforts to control malaria even before it was banned in the United States, principally because mosquitoes had grown resistant to it. (Today, WHO permits indoor spraying of DDT as one way to combat malaria, but given the problems of resistance as well as ecological and human dangers, it is preparing to phase it out by 2020.)

Carson herself never advocated the complete elimination of chemical pesticides. Rather, she opposed their indiscriminate use without conducting studies of their ecological effects. "It is not my contention that chemical insecticides must never be used. I do contend that we have put poisonous and biologically potent chemicals indiscriminately into the hands of persons largely or wholly ignorant of their potentials for harm," she wrote.

It was the height of human arrogance and stupidity, she argued, to think that we could inflict great damage on one part of the environment without harming others. The issues she raised are as applicable today as they were then, foreshadowing our current debates over greenhouse gases and global warming, as well as a new crop of synthetic pesticides (see Chapters 21 and 22).

"Can anyone believe it is possible to lay down such a barrage of poisons on the surface of the earth without making it unfit for all life?" she wrote. ■

MORE TO EXPLORE

- PBS, American Experience (2017) *Rachel Carson* (documentary): www.pbs.org/wgbh/americanexperience/films/rachel-carson/
- Haberman, C. (January 22, 2017) Retro Report: Rachel Carson, DDT and the Fight Against Malaria. *New York Times* (includes video)
- Conniff, R. (2015) Rachel Carson's critics keep on, but she told truth about DDT. *YaleEnvironment360.*
- Griswold, E. (September 23, 2012) How *'Silent Spring'* Ignited the Environmental Movement. *New York Times Magazine.*
- Oreskes, N., and Conway, E. (2010) *Merchants of Doubt: How a Handful of Scientists Obscured the Truth on Issues from Tobacco Smoke to Global Warming.* New York: Bloomsbury.

MILESTONES IN BIOLOGY 7 Test Your Knowledge

1 On the origins of DDT:

a. When was the chemical DDT first widely used, and what was its intended purpose?

b. How effective was it for that purpose?

2 On the mechanism of DDT:

a. How does DDT kill insects?

b. How does DDT harm top predator birds?

3 What is biomagnification?

4 PCBs (polychlorinated biphenyls), a type of chlorinated hydrocarbon, were used for a variety of purposes (including electrical insulation) until their use was banned in 1979. A 2000 survey of top predator fish in the Great Lakes showed that the concentration of PCBs in these fish ranged from 0.8 to 1.6 ppm. The wildlife protection value (the concentration that should not be exceeded in order to protect the safety of wildlife) is 0.16 ppm. How could there be such high levels of PCBs in top predator fish 21 years after PCBs were banned?

5 In 2013, a group of beekeepers launched a lawsuit against the EPA concerning the use of neonicotinoid pesticides and possible unintended impacts on honey bees (see Chapter 21). From what you have read here, what kinds of testing would you want the EPA to require before approving a pesticide applied widely to crops such as corn or soybeans?

23 Sustainability

Wichita Eagle/Getty Images

The Makings of a GREEN CITY

A Kansas town reinvents itself sustainable

On the night of May 4, 2007, at 9:50 P.M., a mammoth EF-5 tornado roared through the small town of Greensburg, Kansas. Nearly 2 miles wide–wider than the town itself–with winds topping 200 mph, the twister was one of the largest ever recorded. Residents had about 20 minutes from the time warning sirens sounded to retreat to their storm cellars before the monster hit.

DRIVING QUESTIONS

1. What human activities are considered when determining ecological footprint, and how does human population size influence our impact on Earth?
2. What resources do humans rely on, and what are the benefits and costs associated with different resources?
3. What is sustainability, and how can we live more sustainably?

"The sound was like a jet engine going right over us, about to take off," recalled Megan Gardiner, then a high school senior. She and her family huddled in their basement, ears popping, listening as first windows exploded and then nails were sucked out of walls with a high-pitched screech. "Just hearing the house rip into shreds was horrible."

The ordeal lasted 10 long minutes. By the time it was over, 11 people had died. People emerged from their storm cellars to find the homes above them gone. Fully 95% of the buildings in town—961 homes and 110 businesses—were leveled. The town of Greensburg had been wiped off the map.

Before the tornado hit, Greensburg was a struggling farming town of about 1,300, its population shrinking every year as people left the town for brighter horizons. After the disaster, the people of Greensburg were faced with a stark choice: abandon the town or rebuild?

Many residents, unable to face the prospect of another tornado, chose to leave. Others took a wait-and-see attitude. But for town administrators, who had for years been searching for ways to breathe life back into Greensburg, the tornado offered them a rare and unlikely opportunity to start over from scratch.

"We can paint this community any way we want," City Administrator Steve Hewitt, whose own home was destroyed, told the mayor the day after the storm. "It's a blank canvas."

Going Green

An ambitious young professional with a wife and son, Hewitt was one of the first to start thinking about rebuilding in a fundamentally different way. He saw in the catastrophe an opportunity to fix problems that had plagued Greensburg before the storm—crumbling infrastructure, a lackluster economy and jobs base, and dim prospects for future growth.

In the days immediately following the storm, he met with Kansas Governor Kathleen Sebelius to discuss ideas for improving on the old ways. Sebelius listened patiently to what Hewitt was saying about rebuilding smarter, with an eye toward economic growth, and

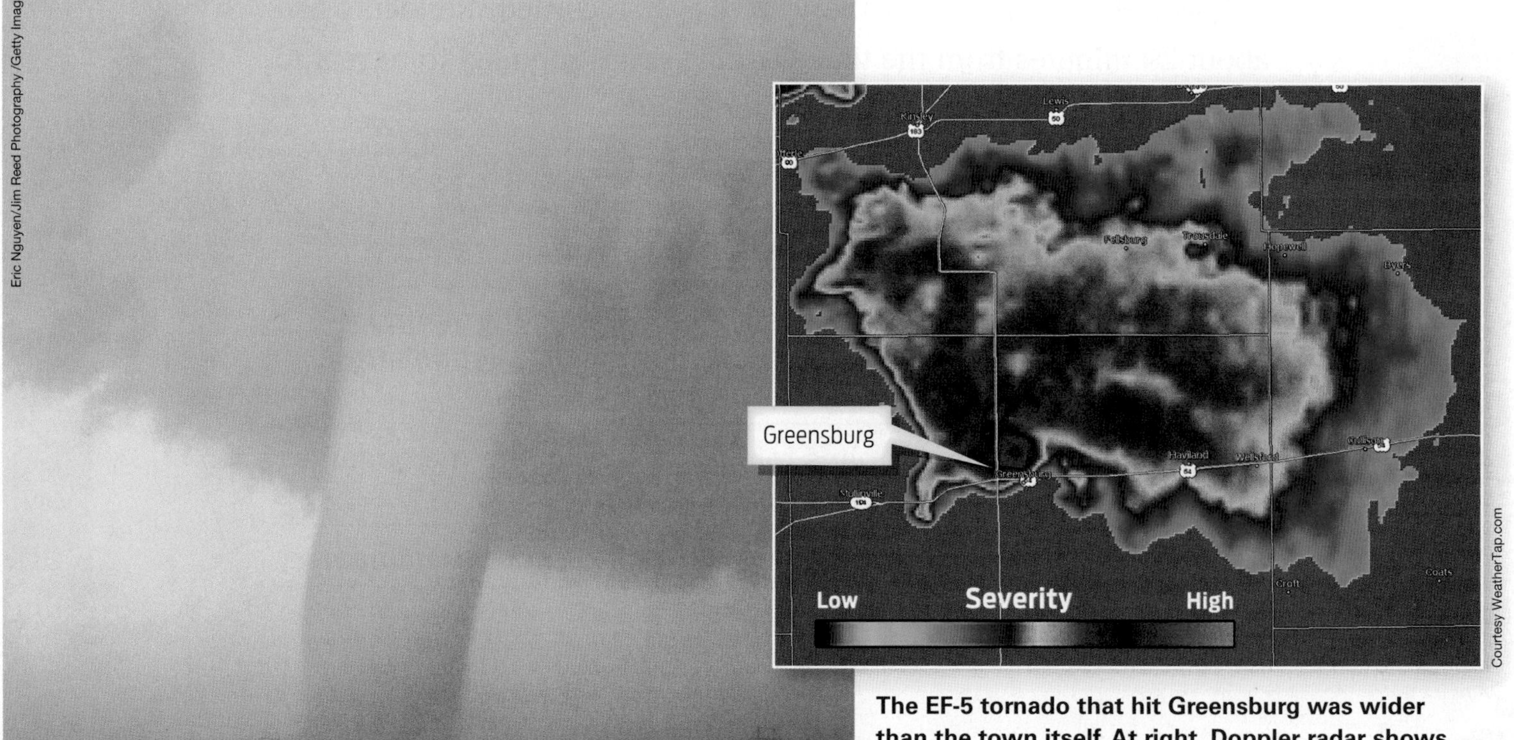

The EF-5 tornado that hit Greensburg was wider than the town itself. At right, Doppler radar shows the position and severity of the storm.

The Greensburg grain elevator was one of the few buildings left standing after the tornado.

then summarized what she heard in one simple statement: "Steve, you're talking about green."

Hewitt hadn't immediately connected his ideas to a green agenda, but the conceptual beauty was not lost on him. This was *Green*sburg, after all. "You're right," he said. "That is what we're talking about." From there, the idea took off.

A small town in Kansas might seem like a strange place for an environmental movement to emerge. After all, this is a politically conservative area in a staunchly red state, where climate change is still viewed skeptically by a large proportion of the population. "Green, I thought that was just for people who hug the trees," said one resident at an early town meeting called to discuss the idea of rebuilding green. But Greensburg residents–even those who did not see themselves as environmentalists–eventually warmed to the idea of putting the green in Greensburg. What sold them was the concept of living more sustainably.

Sustainability means different things to different people, and it can be a difficult thing to define and measure. "In its simplest sense, sustainability is just doing things today to ensure a vibrant successful future for others," says Stacey Swearingen White, a professor of environmental planning at the University of Kansas.

To ecologists, sustainability has to do with the way humans use Earth's resources. We require a variety of **natural resources** to live: farmland to grow crops and raise cattle, fossil fuels to power cars and planes, oxygen to fill our lungs, to name just a few. In addition to providing us with natural resources, Earth acts like a sponge, absorbing our wastes. The carbon dioxide we emit, for example, is soaked up and stored by plants; the garbage we produce is decomposed in landfills.

The collective demand we place on these resources is our **ecological footprint.** This measure is a tally of the land and water area needed both to supply a given population with resources and to absorb its wastes. By quantifying the amount of biologically productive area it takes to sustain our lifestyles, the ecological footprint puts

SUSTAINABILITY
The use of Earth's resources in a way that will not permanently destroy or deplete them.

NATURAL RESOURCES
Raw materials that are obtained from Earth and are considered valuable even in their relatively unmodified, natural form.

ECOLOGICAL FOOTPRINT
A measure of how much land and water area is required to both supply the resources an individual or a population consumes and also absorb the wastes it produces.

BIOCAPACITY The amount of Earth's biologically productive area—cropland, pasture, forest, and fisheries—that is available to provide resources and absorb wastes to support life.

a number on our environmental impact (**INFOGRAPHIC 23.1**).

The human ecological footprint is often compared with Earth's **biocapacity**–its ability to sustain human demand given its available natural resources and its ability to absorb waste. If we think of the footprint as our demand on Earth, the biocapacity is the amount of supplies that Earth can produce to meet that demand. The question is, does Earth have a sufficient supply of resources to meet our demands?

Earth's biocapacity is measured in units called **global hectares (gha).** One global hectare represents the biological productivity–both the resource-providing and the waste-absorbing capacity–of an average hectare of Earth's surface (including cropland, pasture, forest, and fisheries). A hectare is 10,000 square meters–about the

INFOGRAPHIC 23.1 The Human Ecological Footprint

How much of Earth's resources does your lifestyle require? The ecological footprint is a measure of our demand on nature. Ecologists use 5,400 different measures gathered from government agencies and scientific publications to calculate a footprint, a measure of how much biologically productive land and water area (cropland, forests, grazing lands, fishing area, and built-up land) a human population requires to produce the resources it consumes and to absorb the waste it produces.

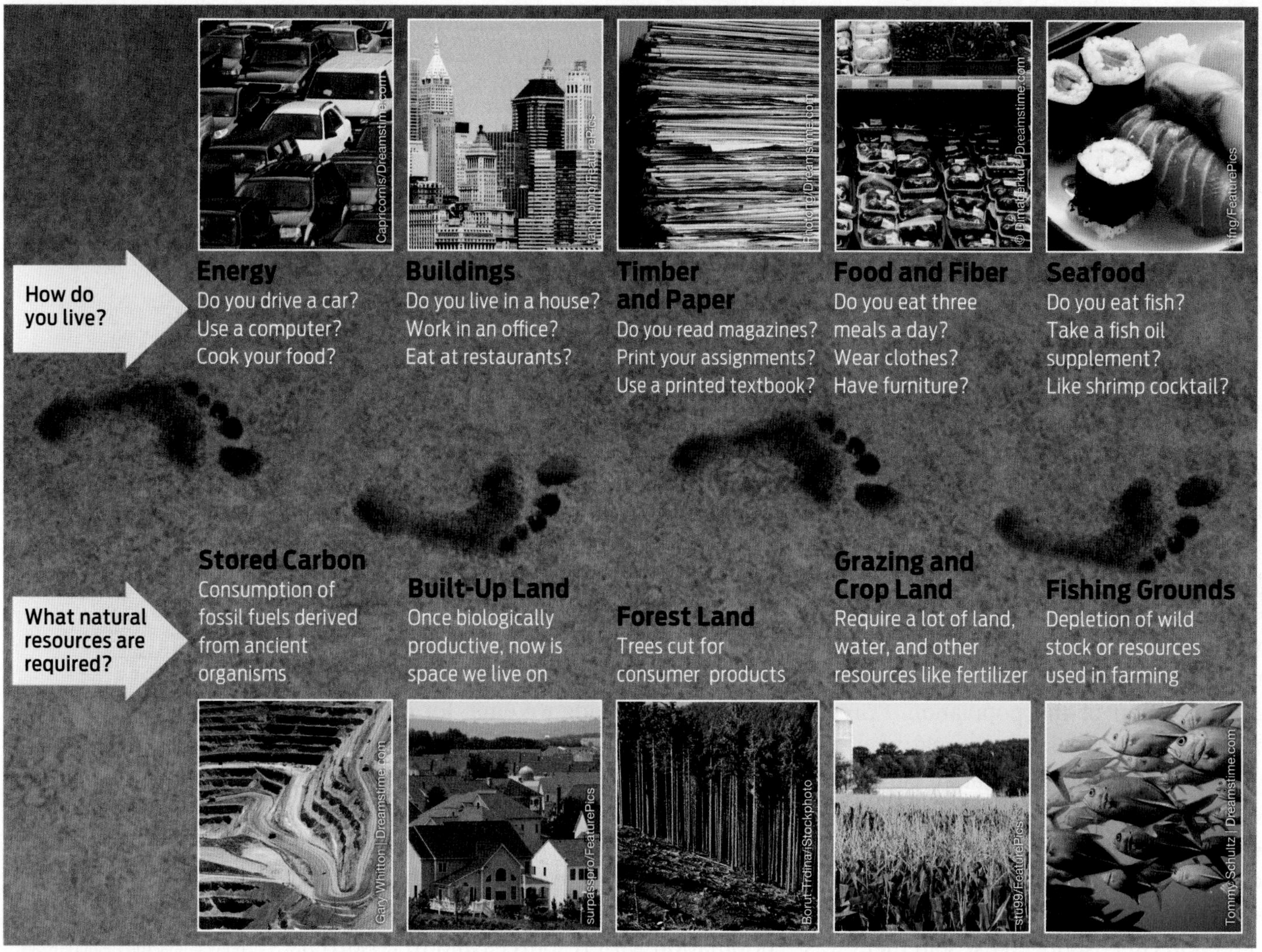

How do you live?
Energy
Do you drive a car?
Use a computer?
Cook your food?
Buildings
Do you live in a house?
Work in an office?
Eat at restaurants?
Timber and Paper
Do you read magazines?
Print your assignments?
Use a printed textbook?
Food and Fiber
Do you eat three meals a day?
Wear clothes?
Have furniture?
Seafood
Do you eat fish?
Take a fish oil supplement?
Like shrimp cocktail?
What natural resources are required?
Stored Carbon
Consumption of fossil fuels derived from ancient organisms
Built-Up Land
Once biologically productive, now is space we live on
Forest Land
Trees cut for consumer products
Grazing and Crop Land
Require a lot of land, water, and other resources like fertilizer
Fishing Grounds
Depletion of wild stock or resources used in farming
Capricornis/Dreamstime.com
Gary Whitton/Dreamstime.com
surpasspro/FeaturePics
Borut Trdina/iStockphoto
stu99/FeaturePics
Tommy Schultz/Dreamstime.com

? If you used a solar oven to cook your food, which natural resource would you be conserving?

size of a soccer field. In 2012 (the last year for which complete data are available), Earth's total biocapacity was 12.2 billion gha, or 1.7 gha per person. That same year, the average ecological footprint was 2.8 gha per person per year. In other words, it takes more than two and a half soccer fields of land and water area to support one average human for 1 year. Since the average human ecological footprint is greater than Earth's total biocapacity, our demand on Earth's resources is currently outstripping its supply (by 1.1 gha per person).

How is this possible when there is only one Earth? You can think of Earth as being like an interest-earning bank account: it has a certain amount of natural capital that is continually renewing itself. Just as it is possible to overdraw a bank account faster than the interest can regenerate the principal, our demands on Earth's resources are diminishing the natural capital faster than it can be renewed. We're living in the red **(INFOGRAPHIC 23.2)**.

These footprint numbers are averages, tallied across the whole globe. But, of course, not everyone on the planet uses resources to the same extent. An average American, for instance, has an ecological footprint of about 8.2 global hectares, while the average Haitian uses just 0.6 global hectares. These are per capita figures, averages for one resident in each of those countries. It's also possible to calculate the ecological footprint

GLOBAL HECTARE (gha)
A unit of measurement representing the biological productivity (both resource-providing and waste-absorbing) of an average hectare of Earth.

INFOGRAPHIC 23.2 The Human Ecological Footprint Is Greater than Earth's Biocapacity

When comparing our biological demand, or ecological footprint, with Earth's biocapacity, it is clear that our footprint has been exceeding biocapacity since the mid-1970s. Our greatest demand is energy, indicated by our large carbon footprint.

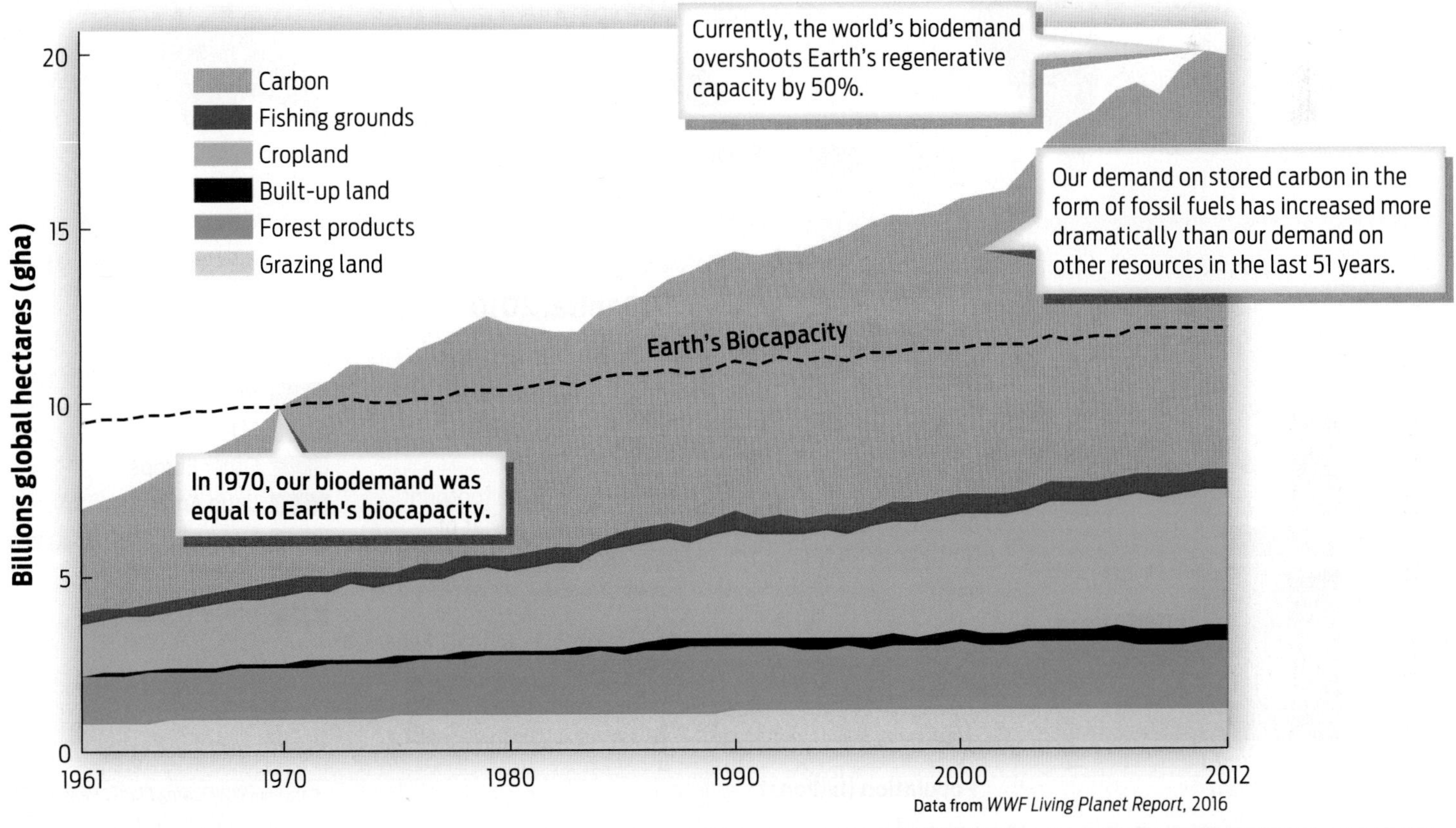

Data from *WWF Living Planet Report*, 2016

? In what year did our ecological footprint first exceed Earth's biocapacity?

of a whole country. China, for example, has a per capita footprint of 3.4 global hectares, much smaller than that of the United States, but because China's population is so large, its total footprint is larger than that of the United States, which has fewer people. In fact, China has the largest total footprint of any country in the world **(INFOGRAPHIC 23.3)**.

INFOGRAPHIC 23.3 Countries Differ in Their Ecological Footprint

Per capita, the United States has a larger footprint than most other world populations. However, the enormous size of some populations like China's, means that their total footprints are greater than that of the United States–in fact, China has the largest footprint of any country in the world.

Total Ecological Footprint by Country, 2012

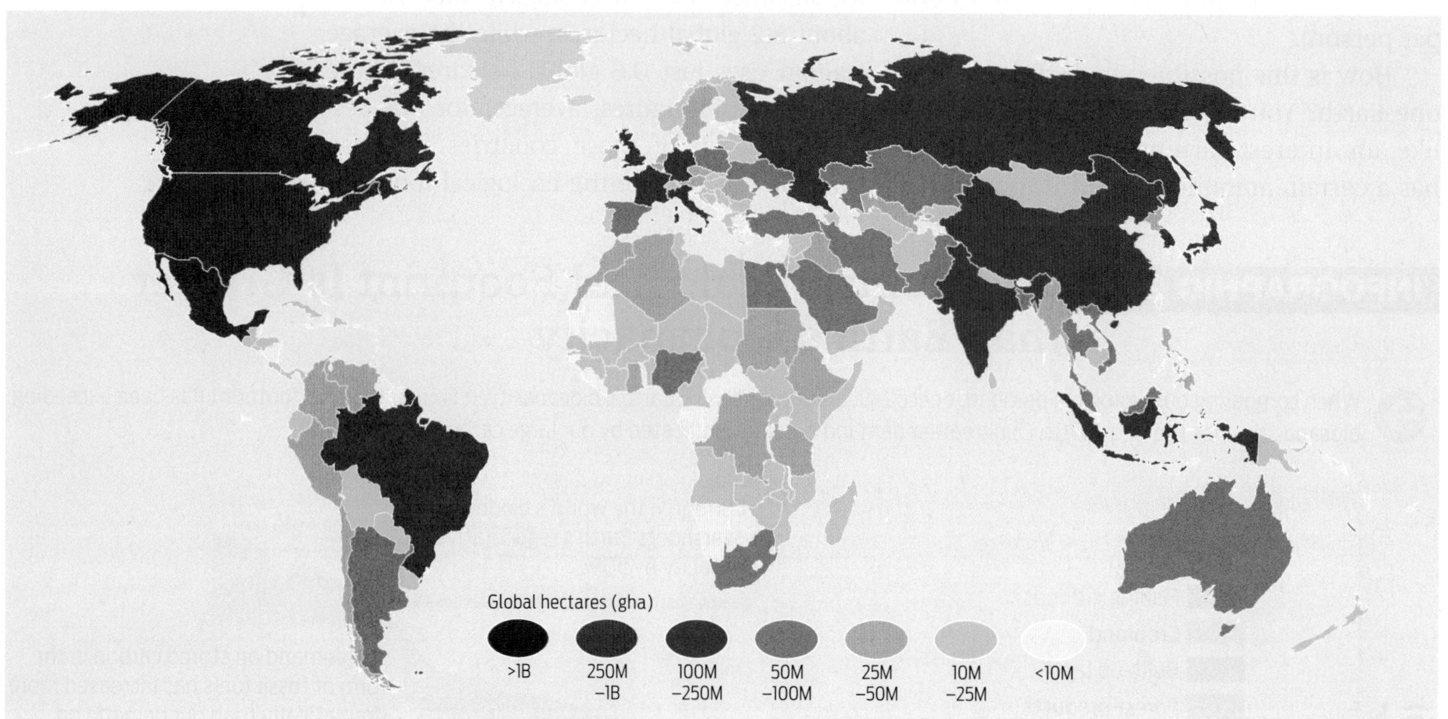

Data from *Global Footprint Network, 2014*

Ecological Footprint per Capita, 2010

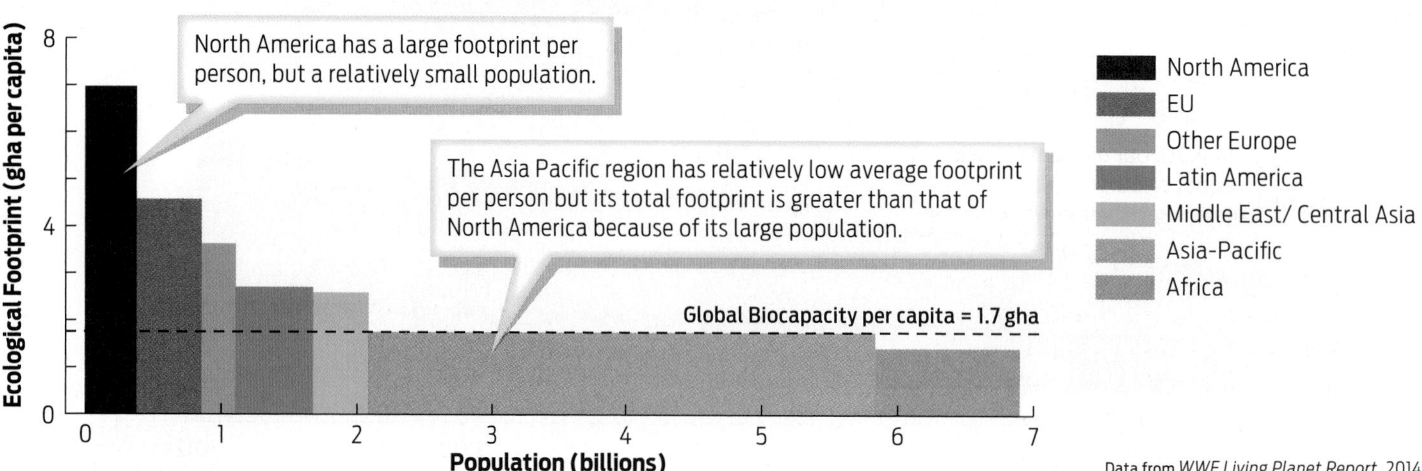

Data from *WWF Living Planet Report*, 2014

? If the per capita footprint in the Asia Pacific region (including China) is low, why is China's total footprint so high?

Greensburg Style: LEED Platinum Buildings

Stefan Falke/laif/Redux

Greensburg City Hall

LEED Platinum Features

- Solar panels
- Geothermal heating and cooling
- Reclaimed brick and wood from buildings destroyed in the tornado
- Green roof with vegetation
- Rainwater collection and irrigation
- 38% more energy efficient

Kiowa, KS, County Schools

LEED Platinum Features

- Natural daylight and ventilation
- Wind energy generator
- Recycling center
- Cabinets and lockers made from natural and recycled materials
- Spaces for community sharing
- Rainwater collection and irrigation

Farshid Assassi/Courtesy BNIM

According to ecological footprint analysis, if everyone on the planet were to live like the average resident of the United States, it would take about four Earths to support all of us. By contrast, if everyone in the world lived like the average person in Haiti, we would need only one-third of an Earth to sustainably satisfy our demands.

What is it about the U.S. lifestyle that leaves such a heavy footprint compared to that of other countries? Energy consumption, by far, is the largest culprit. The cars and SUVs we drive, the computers we work and play with, the washers and dryers that clean our clothes, the air conditioners that cool our homes, the food we truck across the country or fly around the world—all these require energy.

In the United States, as in most parts of the world, most of this energy comes from fossil fuels—oil, coal, and natural gas. As discussed in Chapter 22, burning fossil fuels releases carbon dioxide to the atmosphere, contributing to global warming. Therefore, the ecological footprint takes into account the amount of land and water area needed to absorb CO_2. This, combined with increased consumption, is what makes our ecological footprint so large.

Because they take millions of years to form naturally, fossil fuels are considered **nonrenewable resources:** once depleted, they are essentially gone for good. There's a good reason we have historically relied on these fuels so heavily to meet our energy demands: they are plentiful and relatively cheap to obtain. But they come with environmental costs. Extracting these resources from the ground can damage both terrestrial and marine habitats. Burning these nonrenewable

NONRENEWABLE RESOURCES
Natural resources that cannot easily be replaced; fossil fuels are an example.

INFOGRAPHIC 23.4 Fossil Fuels Are Nonrenewable

Most of the natural resources we use to supply our energy needs are nonrenewable. Coal, oil, and natural gas are fossil fuels that take millions of years to form as organic material is compressed by layers of sedimentary rock. While plentiful today and relatively cheap to obtain, fossil fuels come with significant environmental and human costs.

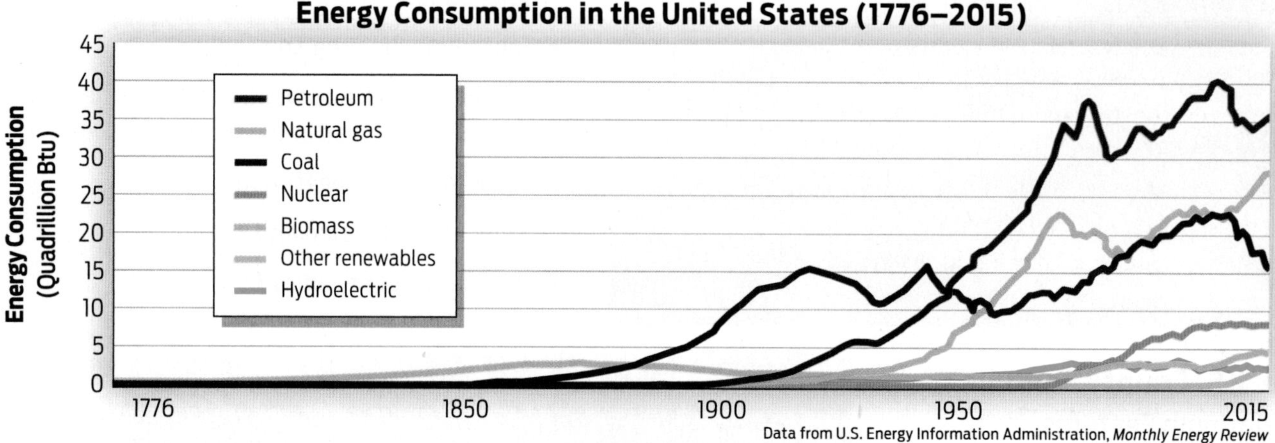

Data from U.S. Energy Information Administration, *Monthly Energy Review*

Coal: Supplies 16% of U.S. Energy

Why Do We Use It?

Coal is burned in power plants to produce steam to turn turbines that generate electricity. Coal is relatively cheap to mine, and there is currently an abundance of it in Earth.

Environmental Impact

Mining coal from Earth often damages the habitat on large tracts of land. Greenhouse gases and pollutants like arsenic, nitrogen dioxide, and sulfur dioxide are released when coal is burned to make electricity. Coal miners have increased risk of respiratory illness. Burning of coal and other fossil fuels can contribute to acid rain, which can kill fish and leach nutrients from soils.

Natural Gas: Supplies 29% of U.S. Energy

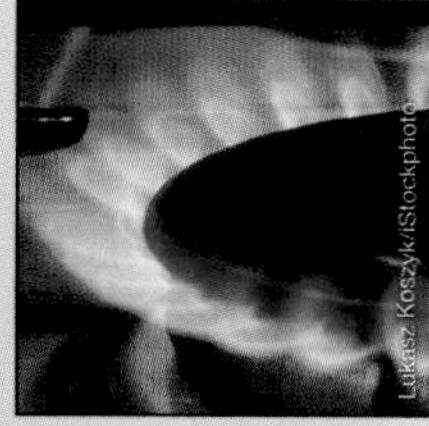

Why Do We Use It?

Natural gas is burned to heat buildings and water. It is relatively cheap to extract, and there are currently large reservoirs of it deep in Earth.

Environmental Impact

Natural gas is extracted from underground and offshore reservoirs. Drilling platforms can disrupt ocean habitat. Burning natural gas releases greenhouse gases to the atmosphere. Extraction by fracking requires large amounts of water, as well as wells to contain contaminated wastewater. Such wells have been implicated in contributing to earthquakes.

Petroleum Oil: Supplies 37% of U.S. Energy

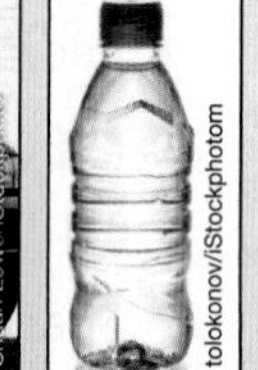

Why Do We Use It?

Oil is used to produce gasoline, petroleum products, and plastics. It is relatively cheap to extract, and there are currently large reservoirs of it deep in Earth.

Environmental Impact

Oil is drilled from underground and offshore reservoirs. Drilling platforms can disrupt ocean habitat. Oil spills can devastate ocean ecology and the seafood economy. Burning products made from oil produces pollution and emits greenhouse gases. Plastics do not biodegrade and therefore create a huge amount of landfill waste.

? Rank the consumptions of the three fossil fuels shown in 1950 and in 2015.

resources releases not only copious amounts of greenhouse gases but also pollutants such as sulfur dioxide and nitrogen dioxide, which contribute to acid rain. Acid rain can acidify aquatic ecosystems, leading to death of fish and some insects that fish and amphibians rely on for food. In terrestrial ecosystems, acid rain leaches important nutrients from soils, impairing tree growth. For many reasons, urban planners are eager to wean people off fossil fuels (**INFOGRAPHIC 23.4**).

Hewitt didn't talk about ecological footprints at the first town meeting he called. But he knew that the idea of sustainability would resonate with this group of midwesterners whose pioneering ancestors had settled the area generations ago. As farmers, many of them had an intrinsic connection to nature and an awareness of the goods it can provide if managed properly.

Environmental concerns are crucial to the concept of sustainability, but they aren't the whole story. "The term sustainability often gets focused exclusively on environmental aspects," says White. "I'm an environmental planner so those are near and dear to my heart, but I think true sustainability also has to bring in the social or the equity concerns as well as economic concerns." People have to be able to work and support themselves, for example, and can't be forced to choose between providing for their families and protecting the environment.

Greensburg residents recognized the importance of these factors, too. As the building plan the town eventually drew up stated, "A truly sustainable community is one that balances the economic, ecological, and social impacts of development." What many did not realize at first was that going green can be good for business.

Greensburg Style: Net-Zero Energy

Native Energy, Inc./ NREL

Greensburg Wind Farm

- 100% renewable energy
- Powers the entire city
- Greensburg uses 25%–33% of the power generated from 10 turbines. Remaining energy goes to powering neighboring towns

100% LED Streetlights

- 40% more efficient than traditional fixtures
- 70% total energy and maintenance savings
- Reduce nighttime light pollution by pointing straight downward

ZUMA Press Inc / Alamy

Solar Energy

- 100% renewable energy
- Many individual homes and buildings have a collection of solar panels.
- Additional panels on roofs and walls or in vacant fields

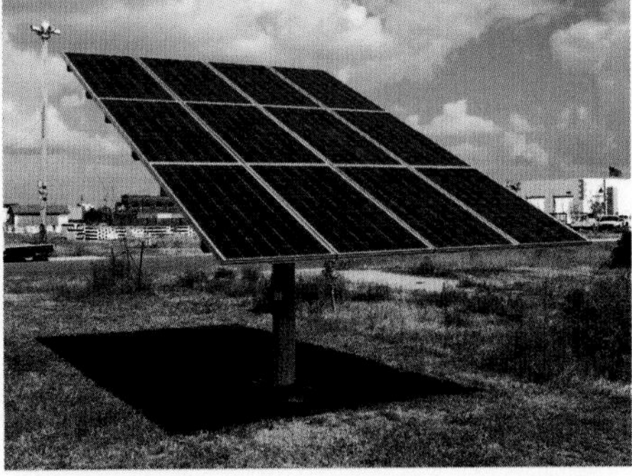

Stefan Falke/laif/Redux

Platinum Rated

Mike Estes, owner of the local John Deere dealership and service shop, was one of the Greensburg residents who initially resisted the notion of rebuilding green. The 63-year-old Kansas native had "only minimal interest in green building" at first. Then he realized how much money he'd save in energy costs over the years and was ultimately persuaded. With technical assistance provided by the National Renewable Energy Laboratory (NREL) of the U.S. Department of Energy, he rebuilt his dealership to include a suite of green technologies, including solar panels and on-site wind turbines. Since rebuilding, he says, he's saved $25,000 to $30,000 a year in energy costs.

From the beginning, going green was presented to the residents of Greensburg as an opportunity for economic success rather than a financial burden. "We're very honest and straightforward about getting people the best economic solutions," says Lynn Billman, a research analyst with NREL who worked closely with residents of Greensburg on their building plans. "We really believe that the best economic solutions can usually be, and often are, green solutions."

RENEWABLE RESOURCES
Natural resources that are replenished after use as long as the rate of consumption does not exceed the rate of replacement.

Yes, rebuilding green would cost more in the short term, but in the long term the savings in energy costs would more than make up for it. Moreover, becoming a pioneer in the green energy movement promised to attract national attention as well as new business to the area, putting Greensburg back on the map with a claim to fame (besides having the world's largest hand-dug well).

Before the storm, City Administrator Hewitt couldn't get businesses to talk to him about relocating to Greensburg. After they went green, he says, his phone started ringing off the hook.

One of the first green things the city council did was pass an ordinance requiring all new public buildings larger than 4,000 square feet to be built to LEED platinum standards. "LEED" stands for "Leadership in Energy and Environmental Design," a certification system run by the U.S. Green Building Council, a nonprofit organization that supports sustainable building practices. Platinum is the highest rating a building can achieve, above gold and silver.

LEED-certified buildings must be shown to conserve energy and water, reduce waste sent to landfills, lower greenhouse gas emissions, and be healthier and safer for occupants. By committing to platinum standards, the town agreed to design buildings that would reduce energy consumption by more than 40% over current building code requirements.

Greensburg now has the most LEED-certified buildings per capita in the world. Among the more than 15 LEED platinum buildings that have been constructed are a new city hall, public school, hospital, art center, and business incubator.

The city ordinance for platinum buildings applies only to public buildings. On the private side, more than half the residents of Greensburg chose to rebuild their homes using green technologies to reduce energy use. An analysis by NREL revealed that more than 100 of the newly built homes would, according to their design plans, use an average of 40% less energy than before, for an average saving of about $512 per year.

In addition to using less energy overall, the residents of Greensburg also chose to have their electricity be "100% renewable, 100% of the time." They decided to take advantage of a **renewable resource** that Kansas has in abundance: wind.

On the outskirts of town, a stately row of 10 wind turbines, each more than 200 feet tall, turns slowly and reliably in the breeze. Completed in 2010, the wind farm supplies enough electricity to power about 4,000 households–enough for all of Greensburg and dozens of neighboring towns.

Other initiatives to reduce Greensburg's footprint include LED (light-emitting diode) lamps that illuminate city streets. In 2010, Greensburg became the first city in the United States to use LEDs for 100% of its street lighting. LED lamps are 40% more efficient than sodium vapor streetlamps, and help avoid

INFOGRAPHIC 23.5 Human Population Growth

Since the advent of agriculture, the human population has been following an exponential growth pattern, reaching 7 billion people in 2011. Some estimate that the human population will reach 9 billion by 2050.

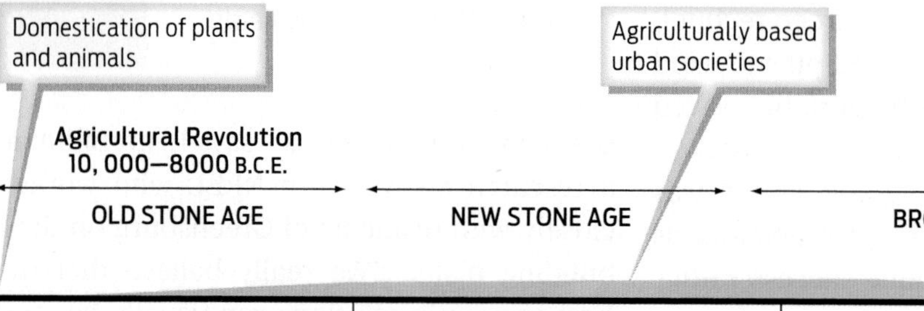

light pollution because they are easier to direct in only one direction: down.

With all these changes, Greensburg has effectively achieved its energy goal. It is a net-zero energy city, generating as much electricity from renewable energy as it uses. "Greensburg has the best infrastructure of any community in the world right now," Hewitt says proudly.

Green Like Us

With a population of about 800 people, what this small town does to conserve resources may seem insignificant. But if Greensburg serves as a compelling model for other cities and states to emulate, the gains could begin to add up. And that's important because the number of people on the planet is likely to keep going up as well.

Two thousand years ago, there were roughly 300 million people on Earth, less than the current population of the United States alone. In 2017, there were 7.5 billion people. Much of that growth has occurred since 1950, thanks in large part to antibiotics and advances in public health that have allowed many of us to live longer. And each hour more than 10,000 new people are added to the planet. By 2050, demographers estimate, we'll hit the 9 billion mark (**INFOGRAPHIC 23.5**).

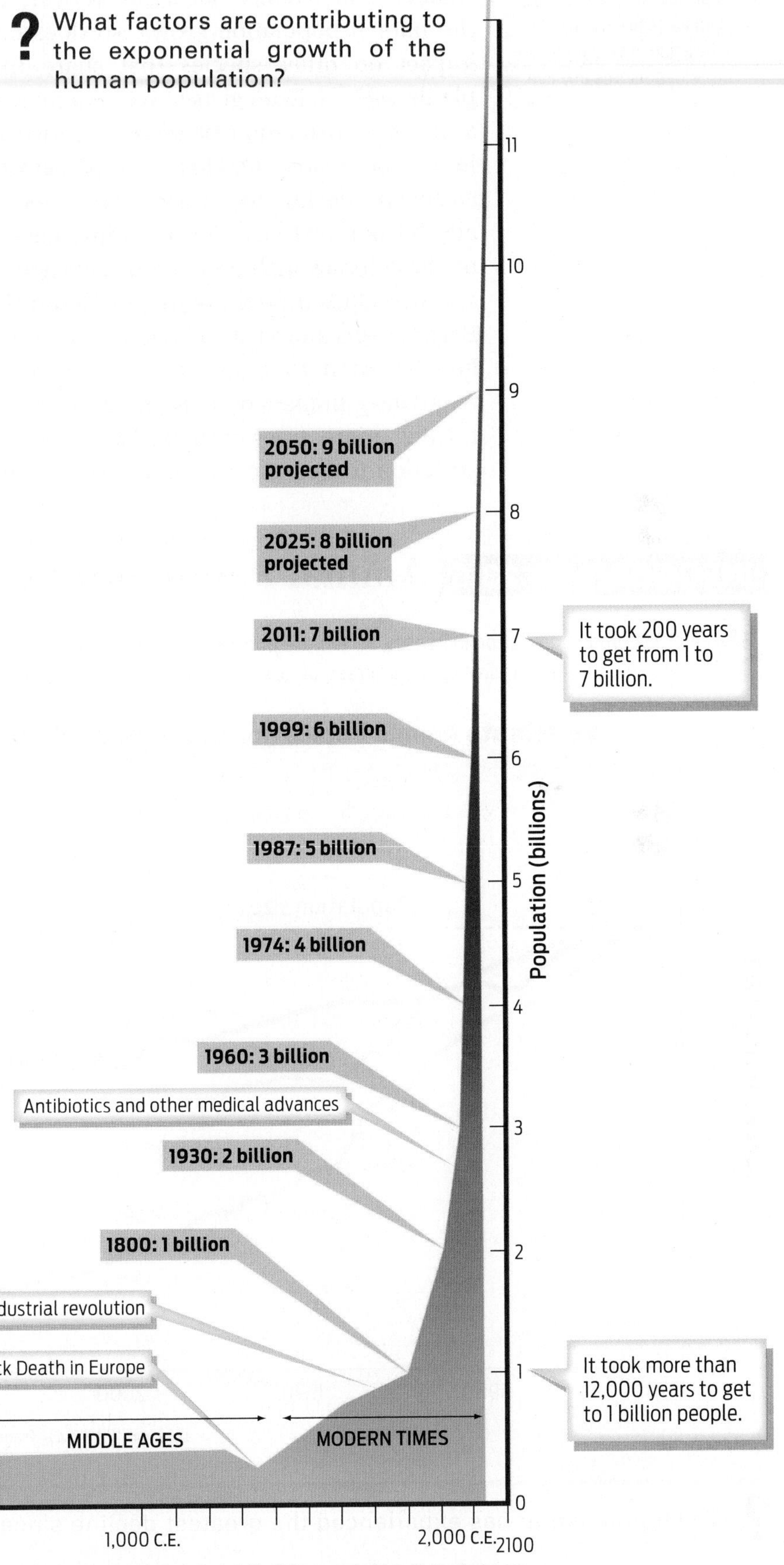

Data from United Nations; Sustainable Scale Project; World Resources Institute; NationMaster.com.

BIODIVERSITY
The number of different species and their relative abundances in a specific region or on the planet as a whole.

With every person added to the planet, Earth's pie of resources is divided into smaller and smaller slices. Moreover, as the human population grows, so does our impact on other species that share this planet with us. Ecologists have documented a striking correlation between the rise in human population and loss of **biodiversity** as species decline in numbers and eventually become extinct. The principal causes of these losses include habitat destruction and overconsumption. More people on the Earth means more natural habitat that must be converted to cropland to grow food. A growing human population also means turning more trees into lumber for construction to house these new arrivals, and more pollution sullying natural habitats as a result of our activities. Faced with the prospect of 2 billion more people in the coming years, the world is badly in need of new ways of living within Earth's biocapacity **(INFOGRAPHIC 23.6)**.

What Greensburg lacks in size, it is hoping to make up for in visibility and daring. It seeks to be a model for other towns and cities across the country, and indeed the world, to emulate.

In a speech to Congress in 2009, President Barack Obama echoed this view, describing Greensburg as "a global example of how clean energy can power an entire community, how it can bring jobs and businesses to a place where piles of bricks and rubble once lay."

INFOGRAPHIC 23.6 Animal Populations Are Declining

As the human population increases, vertebrate populations are declining. The causes include climate change, habitat destruction, overconsumption, invasive species, spread of disease, and pollution.

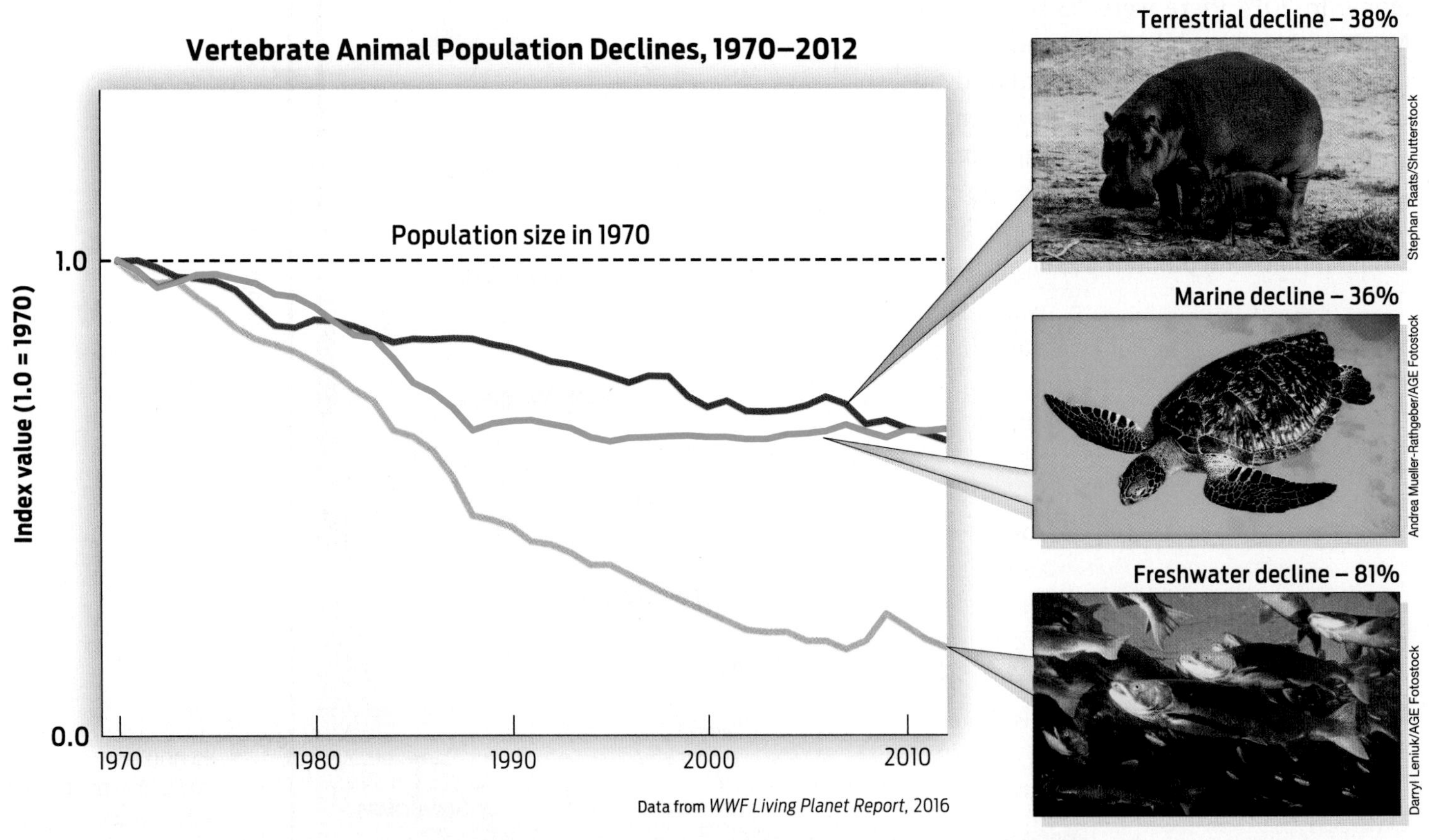

? Which population has experienced the greatest decline since 1970: terrestrial, marine, or freshwater?

Already, the city is attracting international attention. In 2012, a group of architects from tsunami-ravaged areas of Japan came to Greensburg to learn about building sustainable infrastructure. And many of the strategies enacted by Greensburg are being used to design and build ecocities around the world in countries as diverse as Argentina, Australia, Finland, and Vietnam.

> "In its simplest sense, sustainability is just doing things today to ensure a vibrant successful future for others."
> — Stacey Swearingen White

A revitalized Greensburg suddenly finds itself serving as the poster child for a sustainability movement that seems at times to be facing an uphill battle.

"It's Not Easy Bein' Green"

From an ecological standpoint, the appeal of renewable sources of energy such as wind and solar is undeniable: they are plentiful, powerful, and environmentally neutral in their carbon emissions. Solar power alone could theoretically provide more than enough clean energy to supply the needs of everyone on the planet many times over–assuming we could adequately and inexpensively harvest it.

The major problem with wind and solar power is that the technologies to harness them are currently much more expensive to build and operate than ones based on oil, coal, and natural gas. What makes fossil fuels such convenient and inexpensive sources of energy is that the difficult work of harvesting the energy of sunlight has already been done by the photosynthetic organisms that were compressed over millions of years into oil, coal, and gas. (So in a way, we are already, indirectly, using the energy of sunlight to power our lifestyles.) New natural gas and oil harvesting technologies, like hydraulic fracturing ("fracking"), have upped the ante even more because they deliver products that are cheaper than coal.

Of course, when you consider the environmental and human costs of obtaining and burning fossil fuels–from coal-mine explosions to air and water pollution to oil spills–they aren't actually that cheap. Think of the 2010 disaster in the Gulf of Mexico: an oil rig exploded and sank, spewing 4.9 million barrels of oil into the Gulf and costing 11 lives and more than $40 billion in cleanup. Or consider that hydraulic fracturing can lead to water pollution and, increasing evidence suggests, earthquakes. These downstream costs of fossil fuels, which are not reflected in their market price, are known among economists as externalities. If externalities were included in the price, as some economists and environmentalists suggest they should be, then the playing field with other forms of energy would be more level.

Stefan Falke/laif/Redux

City Administrator Steve Hewitt, who led Greensburg's efforts to go green.

AQUIFER
An underground layer of porous rock from which water can be drawn for use.

Moreover, as NREL's Billman points out, the government subsidizes fossil fuel use in lots of ways that aren't as obvious to the public. According to a 2009 study by the nonpartisan Environmental Law Institute, the U.S. government provided $72 billion in tax breaks and subsidies to the fossil fuel industry over the years 2002-2008 versus $29 billion to renewable energy companies over the same period. Such hidden subsidies can create disincentives to change.

But as Greensburg shows, there are ways to make renewables more competitive. First and foremost, there's the approach of emphasizing cost savings: when energy savings over time are considered in the math, the economic calculations look pretty good. As Mike Estes, owner of the Greensburg John Deere dealership, experienced firsthand, renewables make up for their initial higher price by saving money in the long term.

There are also ways to capitalize on the desire of companies that want to become better stewards of the environment. The Greensburg wind farm, for example, was funded in part with money provided by Vermont-based *Native*Energy, a company that sells carbon offsets to businesses that want to counterbalance their fossil fuel use but may not be able to implement substantial carbon reductions of their own. Current purchasers of *Native*Energy carbon offsets include Ben & Jerry's, Stonyfield Farm, Clif Bar, and Keurig Green Mountain Coffee Roasters.

> "We really believe that the best economic solutions can usually be, and often are, green solutions."
>
> — Lynn Billman

Yet there are factors besides cost that limit how much we can rely on renewables at present. The most significant are technological problems of storing and transmitting energy produced from wind and solar. Threats to native animal habitat posed by wind farms and solar panels also complicate the picture. Given these limitations, says Billman, "wind and solar still are a ways from sweeping over the country." The same goes for algae-based biofuels (see Chapter 5) and other renewables. At least for the next decade, we cannot take fossil fuels out of our energy mix (**INFOGRAPHIC 23.7**).

"Nor Any Drop to Drink"

As any Kansas farmer will tell you, fossil fuels aren't the only natural resources being overdrawn by a growing human population—water is, too. "We are *so dry* right now," says urban planner White.

For residents of Kansas and other states of the Great Plains, water for drinking and irrigation comes from the large Ogallala Aquifer, a deep underground source of water spanning eight states in the middle of the country. Rain is so scarce in this part of the United States that, were it not for the **aquifer,** it would be a desert. But thanks to the aquifer, it is one of the most fertile agricultural regions in the world.

Studies by the U.S. Geological Survey and others suggest that the Ogallala, also known as the High Plains Aquifer, is being depleted faster than it is being replenished, and some experts predict that it could dry up by 2050. The aquifer is at "historical lows right now," White says.

To help conserve water, many buildings in Greensburg have rooftop rain filtration and storage systems, which collect rainwater and use it to flush toilets and irrigate the grounds. Rain barrels to collect water are also becoming quite common in many parts of Kansas.

The very fact of water shortages may seem counterintuitive. After all, 70% of the globe's surface is covered with water. But of this vast supply, only 2.5% is freshwater,

INFOGRAPHIC 23.7 Nonfossil Fuel Resources Reduce Our Ecological Footprint

Although many renewable resources are available, a variety of economic, technological, and environmental considerations currently limit their use as alternatives to fossil fuels.

	How It Reduces the Ecological Footprint	Why Don't We Use It More?
Solar: Supplies 0.04% of total U.S. Energy  Aerial Archives / Alamy	Solar energy traps energy from the sun and converts it into electricity and heat with little impact on the environment. As nothing is burned to produce the electricity, there are zero polluting emissions from this process.	Solar power is currently much more expensive to produce than nonrenewable options. The mining and processing of the rare earth metals used in the solar panels use hazardous chemicals, and the resulting waste must be properly disposed of.
Wind: Supplies 1.8% of total U.S. Energy  Ron Thomas/iStockphoto	Wind energy is used to turn turbines, producing electricity with little impact on the environment. In the absence of combustion, no pollutants are released to the environment.	Wind power is currently much more expensive to produce than nonrenewable options. Wind generators take up space, either on land or in the water, and must be located in windy areas. Some people don't want a visible wind farm near their homes. The mining and processing of the rare earth metals used in the turbines uses hazardous chemicals. Bird species may be affected as turbines encroach on their air space.
Nuclear: Supplies 8% of total U.S. Energy  Petr Nad/iStockphoto	Nuclear energy uses concentrated radioactive elements harvested from Earth. As these elements decay, they give off tremendous heat, which is used to produce electricity. As nothing is burned in the process, there are no polluting emissions.	Nuclear reactors are expensive to design and build, and a reactor has a limited life span. Extracting uranium from mines has an environmental impact, and the waste from uranium mines is radioactive. Most important, the waste from nuclear reactors is highly radioactive, making storage complicated. Weapons-grade plutonium can be made from reactor waste, posing a security threat.
Biomass: Supplies 4.5% of total U.S. Energy  Chine Nouvelle/SIPA/Newscom	Biomass includes biofuels, biomass waste, and wood. When biomass is burned, the only CO_2 released to the atmosphere is what the plants and algae took in through photosynthesis, so fossil deposits of carbon are not used.	While being intensively researched, biofuels have not yet become feasible replacements for fossil fuels. In some cases, significant emissions are associated with their production. In other cases, more research and investment are required to optimize production. Growing plants for biofuels can also compete with growing crops.
Hydroelectric: Supplies 2.6% of total U.S. Energy  Pavle Marjanovic/Dreamstime.com	Hydroelectric power relies on the conversion of potential energy (stored in the position of accumulated water behind a dam) to kinetic energy, which can turn a generator. There are no emissions associated with hydroelectric power. Hydro plants have long life spans, and hydro power can potentially power half the projected energy demands of the planet.	To produce hydroelectric power, dams are built to block rivers, creating lakes with immense amounts of potential energy. Building dams destroys habitat, affects local fish populations, and can force human populations to relocate.
Geothermal: Supplies 0.2% of total U.S. Energy  Rob Broek/iStockphoto	Geothermal energy relies on naturally occurring heat from the magma layer beneath Earth's crust. This is a sizable and sustainable resource that can be tapped to drive generators or directly heat homes and businesses.	Geothermal energy is used extensively in Iceland and in some areas of California. However, it has yet to be fully developed in other areas, primarily because optimal technologies require further development. It can also be very expensive, and in some cases, noxious pollutants are released with the steam from geothermal resources.

? Rank the contributions (highest to lowest) of nonfossil fuels to U.S. energy.

and most of that is locked up in ice caps and glaciers. A meager 1% of the total water on Earth is available for human consumption.

Nevertheless, freshwater is considered a renewable resource because the supply in lakes, rivers, reservoirs, and underground aquifers is continually being replenished in the water cycle. As long as the rate of water withdrawal from these sources is less than the rate of replacement, the supply of freshwater remains relatively constant **(INFOGRAPHIC 23.8)**.

Although water is not consumed in the same way as coal or oil, our supply of freshwater is being divided among more and more people, so there is less available, on average, for each of us. According to the United Nations, water use increased sixfold during the 20th century, more than twice the rate of population increase. Today, more than half of all the accessible freshwater contained in rivers, lakes, and aquifers is appropriated by humans, most of it for irrigation in agriculture. A striking example of the consequences of increased

INFOGRAPHIC 23.8 Water Is a Renewable Resource

Freshwater is a valuable resource. In addition to its role in keeping us hydrated, it irrigates crops, sustains fisheries, and provides recreational opportunities. Although water is "used," it is not "used up." It is ultimately returned to the global ecosystem as it evaporates to the atmosphere, flows into rivers and streams, or enters underground aquifers.

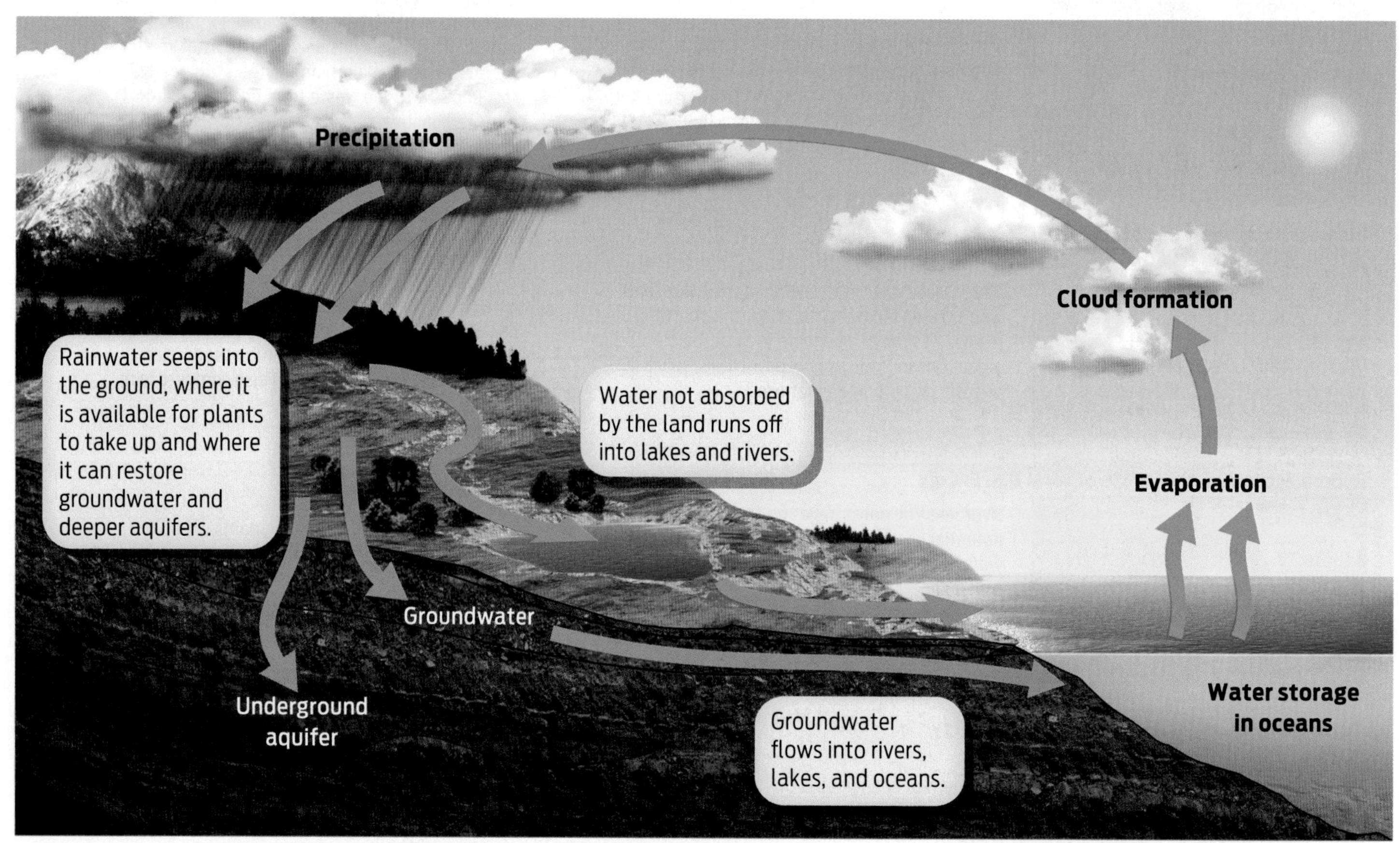

Precipitation
Cloud formation
Rainwater seeps into the ground, where it is available for plants to take up and where it can restore groundwater and deeper aquifers.
Water not absorbed by the land runs off into lakes and rivers.
Evaporation
Groundwater
Underground aquifer
Groundwater flows into rivers, lakes, and oceans.
Water storage in oceans

? Describe the roles of oceans, groundwater, and aquifers in the water cycle.

INFOGRAPHIC 23.9 Depletion of Freshwater by a Growing Population

Water is pumped from rivers and lakes to irrigate dry land for agriculture. More than 90% of the world's water usage is for agriculture. The demand for water can exceed the replenishment of rivers, lakes, and aquifers. During drought years, irrigation sources dry up, affecting aquatic species and human livelihood.

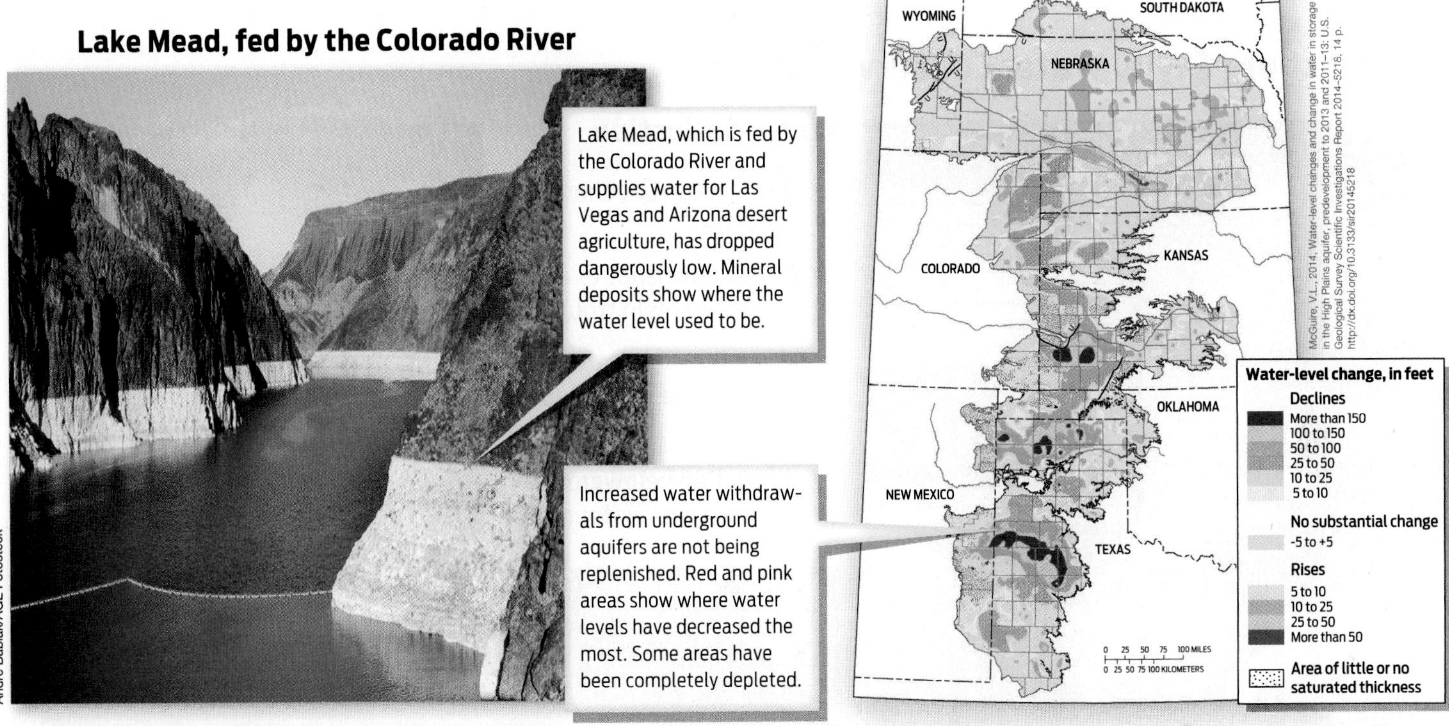

Lake Mead, fed by the Colorado River
Lake Mead, which is fed by the Colorado River and supplies water for Las Vegas and Arizona desert agriculture, has dropped dangerously low. Mineral deposits show where the water level used to be.
Increased water withdrawals from underground aquifers are not being replenished. Red and pink areas show where water levels have decreased the most. Some areas have been completely depleted.
André Babiak/AGE Fotostock
WYOMING
SOUTH DAKOTA
NEBRASKA
COLORADO
KANSAS
OKLAHOMA
NEW MEXICO
TEXAS
0 25 50 75 100 MILES
0 25 50 75 100 KILOMETERS
Water-level change, in feet
Declines
More than 150
100 to 150
50 to 100
25 to 50
10 to 25
5 to 10
No substantial change
-5 to +5
Rises
5 to 10
10 to 25
25 to 50
More than 50
Area of little or no saturated thickness
McGuire, V.L., 2014, Water-level changes and change in water in storage in the High Plains aquifer, predevelopment to 2013 and 2011–13: U.S. Geological Survey Scientific Investigations Report 2014–5218, 14 p. http://dx.doi.org/10.3133/sir20145218

? If per capita withdrawals are relatively stable, why are total withdrawals increasing?

water use can be seen in Lake Mead, which is fed by the Colorado River. This reservoir, which provides freshwater to much of the southwestern United States, is at historic lows today (**INFOGRAPHIC 23.9**).

Pollution also shrinks the total amount of available clean freshwater on the planet. Agriculture, industry, and cities all play a role. Runoff from streets carries pollutants such as motor oil and sewage, and fertilizers, pesticides, and toxic chemicals leach from fields and factories. These substances can eventually reach aquifers, rivers, and oceans, contaminating the water that both humans and wildlife depend on.

Food and lifestyle choices also affect water availability. According to environmental scientist Arjen Y. Hoekstra, it takes 900 liters of water to produce 1 kilogram of corn, but more than 15 times that to produce 1 kilogram of beef.

Some countries experience water scarcity more acutely than others. That's because the geographic distribution of freshwater does not match the distribution of the world's population. Canada has just 0.5% of the world's population but 20% of the global freshwater supply. The United Nations estimates that at least a billion people in the world currently lack access to clean and safe drinking water, and by 2025, two-thirds of the world's population will live in areas of moderate to severe water stress. Climate change, if it changes precipitation patterns, may also affect the global availability of water in unpredictable ways (**INFOGRAPHIC 23.10**).

In their sustainable building plan, Greensburg pledged to "treat each drop of water as a precious resource."

Greensburg Style: Water Conservation

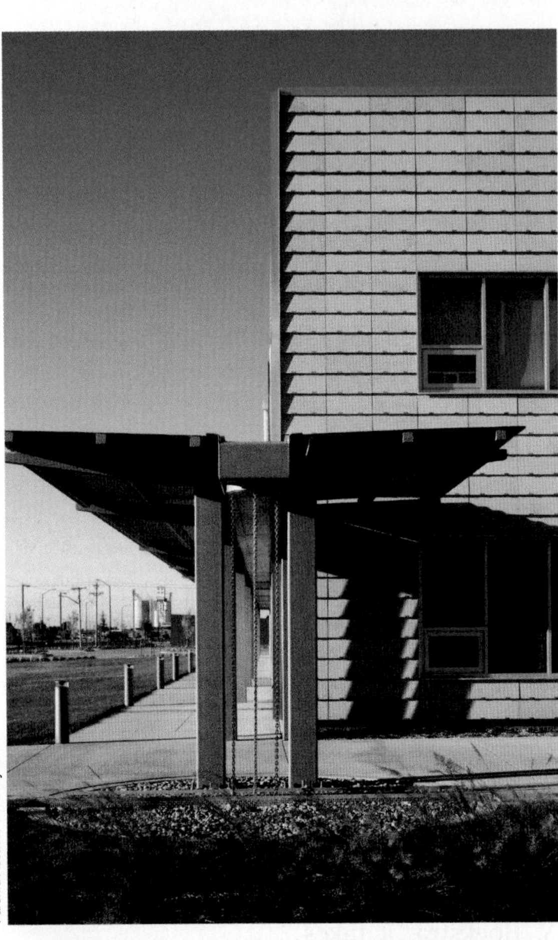

Farshid Assassi/Courtesy BNIM

Water Collection

- Roofs tilted to channel rainwater
- Gutter downspouts collect water in troughs
- 10k gallon aboveground cistern stores collected rainwater
- 50k gallon belowground cistern stores water runoff

© conflictedracer

© conflictedracer

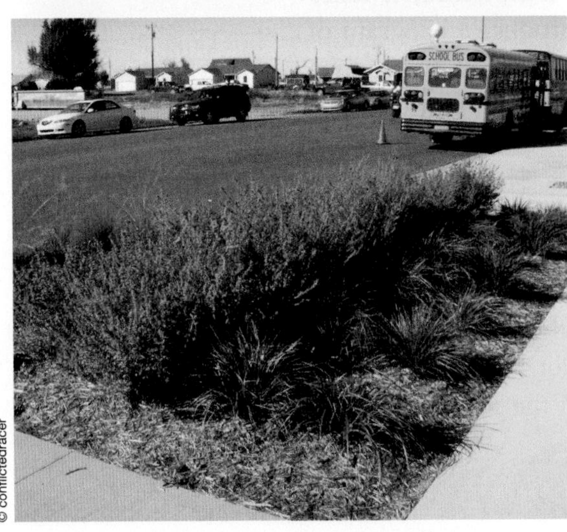

© conflictedracer

Irrigation

- Collected water used for irrigation of landscaping
- Collected water used to flush water-conserving toilets
- Waterless urinals save 0.5 gallons per use

Plants

- Drought-tolerant native plants require less watering
- Landscaping minimizes rainwater runoff into city sewers —it filters and channels it for capture
- Artificial turf athletic fields require no watering

No Place Like Home

When Greensburg decided to go green, its leaders hoped to capitalize on the green message to lure new business to the area. Mayor Bob Dixson said he wanted to see Greensburg become the "ecotourism capital of the world."

Unfortunately, the town's green building project was just getting under way when the 2008 financial crisis hit and made businesses wary about making new investments. But town leaders haven't given up hope. They believe more companies will eventually flock to Greensburg. And they are happy and proud of what they have been able to build already: a community that has a future.

City Administrator Hewitt frames the discussion of sustainability in terms of parenting. As citizens and community members, he says, we have the same

INFOGRAPHIC 23.10 Water Availability Is Not Equally Distributed

Freshwater is not evenly distributed across the globe, and its availability does not always follow international borders. In addition, access to even a sufficient water supply may be limited by economic, social, and political circumstances, such as war and ethnic conflict. As the human population continues to grow and access to clean freshwater continues to decline, these problems are likely to intensify, particularly in areas with existing scarcities of water.

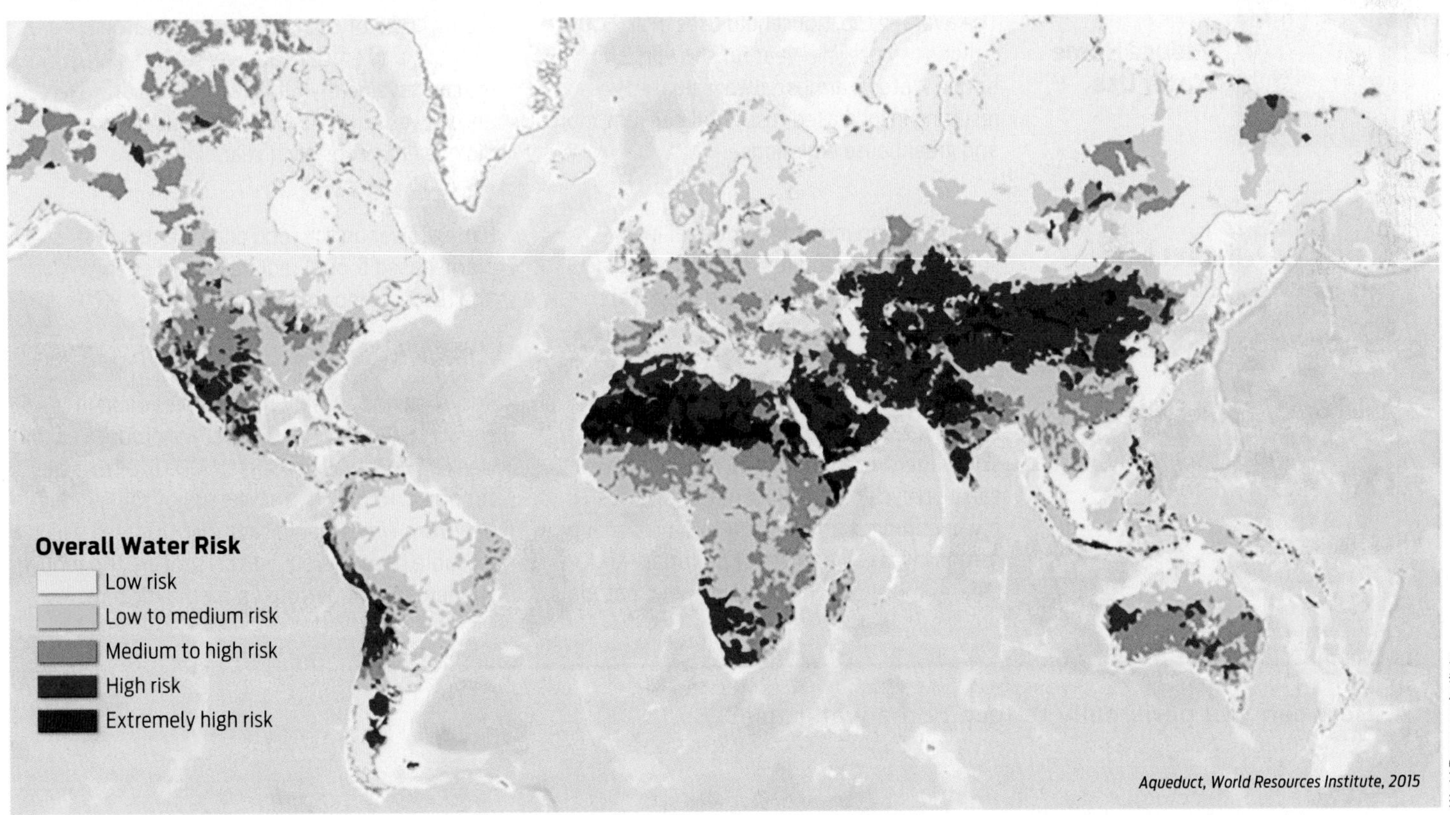

World Resources Institute

? Find the place where you grew up on this map. What is its current overall water risk?

INFOGRAPHIC 23.11 What You Can Do to Live More Sustainably

Take Action	Why?	Your Impact!
Dimitri Otis/Getty Images; Damian Dovarganes/AP Images **Drive Less and Invest in Fuel Efficiency**	With less than 5% of the world's population, the United States consumes a quarter of the world's oil and emits a quarter of its greenhouse gases, largely from automobiles.	Driving smaller vehicles and those with more fuel efficiency cuts carbon dioxide emissions and reduces dependence on nonrenewable fossil fuels.
rockphoto/FeaturePics **Install Compact Fluorescent Lightbulbs (CFLs)**	Electricity production is the largest source of greenhouse gas emissions in the United States, and lighting accounts for about 25% of American electricity consumption. On the other hand, CFLs contain mercury, and require proper disposal.	By replacing just four standard bulbs with CFLs, you can prevent the emission of 5,000 pounds of carbon dioxide and reduce your electricity bill by more than $100 over the lives of those bulbs.
Don Nichols/iStockphoto **Reduce Vampire Energy Waste**	Electronics use energy even when they are turned off. This standby "vampire energy" accounts for 5% to 8% of a single family's home electricity use per year.	When you plug your electronics into a power cord that you turn off each night, you will save the equivalent of 1 month's electric bill each year.
stuartbur/iStockphoto **Reduce Home Water Use**	The average U.S. household uses over 22,000 gallons of water per year for showers and baths. Water is almost always heated, resulting in increased fossil fuel consumption and greenhouse emissions.	If only 1,000 of us install faucet aerators ($2–$5) and efficient showerheads (<$20), we can save nearly 8 million gallons of water and prevent over 450,000 pounds of carbon dioxide emissions each year.
okea /FeaturePics **Eat Less Feedlot Beef**	Producing 1 pound of feedlot beef in some cases can require 5 pounds of grain and over 2,400 gallons of irrigation water.	Eating lower on the food chain (eating plant-based foods) requires fewer energy and resource inputs.
tolokonov/iStockphoto **Recycle**	Each year, the U.S. population discards enough glass bottles and jars to fill 12 giant skyscrapers, even though 75% of our trash can be recycled. Recycling materials uses fewer nonrenewable resources, saves energy, results in less air and water pollution, and creates more jobs than making new materials.	Recycling one aluminum can saves enough energy to run a TV for 3 hours. To produce each week's Sunday newspapers, 500,000 trees are cut down. Recycling a single run of the Sunday *New York Times* would save 75,000 trees. Taking reusable bags on your weekly grocery trip reduces our demand for petroleum for plastic bags.

? How can you personally reduce home water use?

obligation that we do as parents: to recognize that the choices we make today have consequences, and to make decisions that are going to be good for the long term (**INFOGRAPHIC 23.11**).

Hewitt says his 4-year-old son, Gunner, was often on his mind while he was working to transform Greensburg into a sustainable city. "By doing it for Gunner," he says, "I'm taking care of a lot of people." ■

CHAPTER 23 SUMMARY

- Sustainability refers to the ability of humans to live within Earth's biocapacity—its ability to provide current and future generations with natural resources and to absorb our wastes.
- Natural resources include nonrenewable resources such as fossil fuels (oil, coal, and gas) and renewable resources such as sunlight, wind, and water.
- Ecologists measure human demand on Earth's resources using ecological footprint analysis, which quantifies the amount of biologically productive land and water needed to support our lifestyles.
- The ecological footprint of the current human population is greater than Earth's biocapacity, which means that we are living unsustainably.
- The largest component of our footprint is the burning of fossil fuels, which generates harmful wastes, including greenhouse gases and pollutants.
- As the human population grows, so does our ecological footprint. As of 2017, the human population totaled 7.5 billion people. Some demographers predict the number could hit 9 billion by 2050.
- Freshwater is a renewable resource, but the world's supply is not distributed equally, and many people around the world suffer from water scarcity, a problem exacerbated by a rising population, the demands of agriculture, and socioeconomic challenges.
- Sustainable practices minimize the consumption of nonrenewable resources by using renewable resources like wind and solar power instead of fossil fuels to generate electricity and heat.
- At their current level of development, technologies to harvest renewable energy cannot meet our total energy demands. Fossil fuels cannot yet be taken out of our energy mix.
- Individually, we can decrease our ecological footprint by driving less, reducing water and electricity use, consuming less meat, and recycling.

MORE TO EXPLORE

- **National Renewable Energy Laboratory, Greensburg Deployment Project: eere.energy.gov/deployment/greensburg.html**
- **Global Footprint Network, Footprint Calculator: www.footprintnetwork.org/en/index.php/GFN/page/calculators/**
- **World Wildlife Fund (2016) Living Planet Report.**
- **White, S. S. (2010) Out of the rubble and towards a sustainable future: the "greening" of Greensburg, Kansas. *Sustainability* 2:2302–2319.**
- **Planet Green (Discovery Communications) (2010) *Greensburg* (documentary series).**

CHAPTER 23 Test Your Knowledge

DRIVING QUESTION 1 What human activities are considered when determining ecological footprint, and how does human population size influence our impact on Earth?

By answering the questions below and studying Infographics 23.1, 23.2, 23.3, 23.5, and 23.6, you should be able to generate an answer for the broader Driving Question above.

KNOW IT

1 From what you have read in this chapter, explain some of the developments that have permitted the human population to grow exponentially.

2 What is an ecological footprint?

USE IT

3 From your reading of this chapter, how do you think that the ecological footprint of Greensburg compares to that of a small tribal village without electricity in the northern hill regions of Thailand?

4 On the outskirts of a small town, a farmer has just sold his 5 acres of cropland to a developer who is planning to build 20 single-family condominium units on that land. Discuss the ways in which this transaction will affect the size of the nearby town's population and the ecological footprint of the residents of the town and its outskirts.

5 What building considerations could the developer in Question 4 take into account to minimize the impact of this development on the ecological footprint of the town and its outskirts?

DRIVING QUESTION 2 What resources do humans rely on, and what are the benefits and costs associated with different resources?

By answering the questions below and studying Infographics 23.4, 23.7, 23.8, 23.9, and 23.10, you should be able to generate an answer for the broader Driving Question above.

KNOW IT

6 Which of the following waste products is/are associated with the burning of fossil fuels?

a. water
b. carbon dioxide
c. nitrogen dioxide
d. all of the above
e. b and c

7 Mark each of the following natural resources as renewable (R) or nonrenewable (N).

_____ Freshwater
_____ Coal
_____ Codfish populations in the North Atlantic
_____ Wind
_____ Sunlight

8 If oil is formed from fossilized remains of once-living organisms, and if organisms keep dying, why is oil considered to be a nonrenewable resource?

USE IT

9 The renewability of some resources can depend on human choices and activities. List some such resources, and explain how human activities may lead a renewable resource to become essentially nonrenewable.

10 Think about your local region—for example, do you live in the southwestern desert or on the northeastern ocean shore? Describe the nonrenewable and renewable energy resources that are available in your region or that your region can harvest. What are some of the challenges that must be overcome in order to tap into the renewable energy resources in your region?

apply YOUR KNOWLEDGE

INTERPRETING DATA

11 The levelized cost of energy is the overall cost to utilities to generate electricity from various power sources. It includes the cost to finance, build, and operate a generating plant over its lifetime. The levelized costs of energy from two sustainable sources, wind power and solar photovoltaic, are decreasing:

Wind Energy
2009: $135 per MWh
2014: $59 per MWh

Utility-Scale Solar Photovoltaic
2009: $359 per MWh
2014: $79 per MWh

a. As a percent, what is the 2014 wind energy levelized cost relative to the 2009 cost?

b. As a percent, what is the 2014 solar photovoltaic levelized cost relative to the 2009 cost?

In 2014, the levelized cost of coal was $110 per MWh, and the levelized cost of residential rooftop solar was $215 per MWh.

c. Arguments are often made that while sustainable resources have many benefits, the biggest challenge to large-scale adoption is the cost of the new technologies required. From all the data presented, what energy sources appear to be the most financially viable for electricity generation? What are the environmental impacts of using these resources?

DRIVING QUESTION 3 What is sustainability, and how can we live more sustainably?

By answering the questions below and studying Infographics 23.6, 23.7, and 23.11, you should be able to generate an answer for the broader Driving Question above.

KNOW IT

12 How would you define "sustainability"? Highlight several features of sustainability in your answer.

13 What aspects of LEED-certified buildings contribute to sustainability?

USE IT

14 What practices can you adopt where you live to reduce your ecological footprint and embrace the philosophy of sustainable living? For each practice that you think of, explain how it would contribute to sustainability and the reduction of your ecological footprint.

15 Many cities have been developed in the hot and dry southwestern states of the United States. What are some of the sustainability implications of living in the desert?

apply YOUR KNOWLEDGE

MINI CASE

16 A single mother of two children living in Boston (a city with a decent public transportation system) asks you to help her reduce both her day-to-day living expenses and her ecological footprint. What specific recommendations can you make? What evidence do you have to back up your recommendations? Are any of your recommendations also consistent with a healthier lifestyle? Explain your answer.

apply YOUR KNOWLEDGE

BRING IT HOME

17 Explore your school's website to find out about your school's commitment to sustainability. Are there plans for LEED-certified buildings? Are there bike lanes? Bike-parking spots? What about a recycling or composting program? From what you learn, assign a letter grade (A–F) for your school's sustainability plan, and write a short report explaining the reasons for your grade. If you assigned a high grade, what factors contributed to it? If you assigned a low grade, what could be improved at your school?

24 Plant Growth and Reproduction

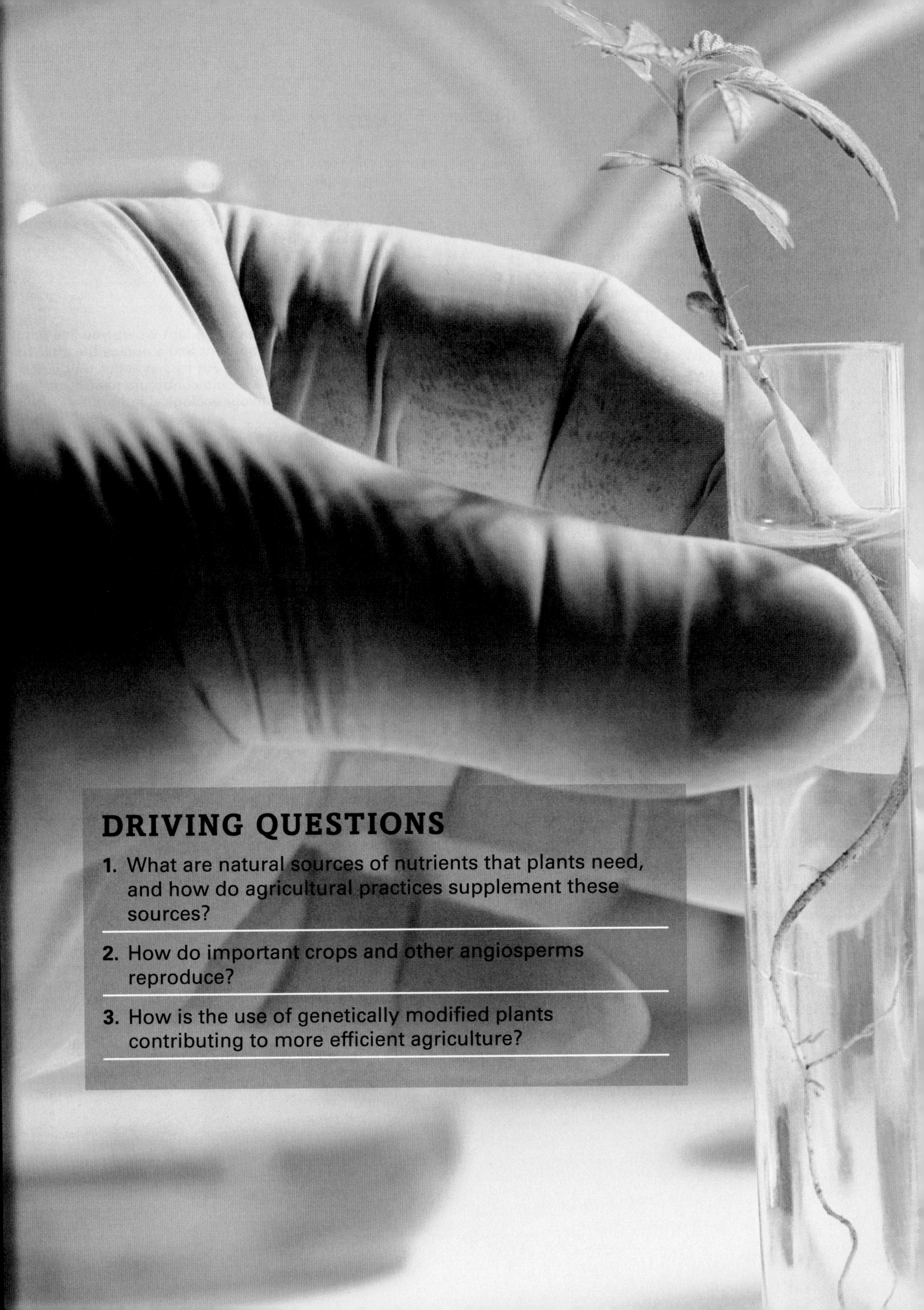

DRIVING QUESTIONS

1. What are natural sources of nutrients that plants need, and how do agricultural practices supplement these sources?
2. How do important crops and other angiosperms reproduce?
3. How is the use of genetically modified plants contributing to more efficient agriculture?

Plants 2.0

Is genetic engineering the solution to world hunger?

In 1856, the U.S. government tried to corner the market on bird poop. That was the year Congress passed the Guano Islands Act, which led to the country's acquiring several small islands in the South Pacific–along with their stockpile of bird droppings.

Not just any bird droppings, but those of seabirds such as boobies, pelicans, and cormorants. Thanks to the fish these birds eat, their poop–that is, guano–is rich in nutrients, making it an excellent fertilizer.

In the mid-19th century, farmers around the world were desperate for ways to boost crop production. Cities were growing and populations were rising fast. But manure from cattle was pricey and hard to transport. Guano seemed a perfect solution, leading to what contemporaries called "guano mania." The United States imported nearly 200,000 tons of guano in 1856 for use on cropland.

REB Images/Getty Images

Today, farmers rely on industrially produced fertilizer rather than guano as a source of nutrients for their crops. But the need to improve crop yields continues unabated. If anything, the need is greater now than ever.

Tomas Munita/The New York Times/Redux Pictures

Workers collecting guano on Isla Guañape off the coast of Peru.

By 2050, there will be an estimated 2 billion more people on the planet, bringing the total population up to 9.5 billion. That's 2 billion more people who will need food, clothing, fuel–not to mention jobs and health care. The United Nations predicts that to accommodate them, by 2050 we will need to produce 70% more food globally than we currently do. In developing countries, where most of the population growth will take place, food production will need to double. How to do that sustainably is a challenge with no clear answer.

"You only have a few options for how to deal with food production," says Joseph Jez, a plant biologist at Washington University in St. Louis, Missouri, whose lab focuses on engineering plants to help solve ecological problems. "You either boost productivity or you use more land to grow more crops."

But agricultural lands in many parts of the world are already stressed and at near-maximum capacity. Converting more land to farms means sacrificing other valuable resources, like rain forest. For this reason, scientists are increasingly looking to genetic

Human Population and Food Production

In order to keep up with a growing human population, global food production will need to increase substantially by 2050. This increase is projected to come from enhanced crop yields, more frequent crop harvests, and increases in the amount of land used to grow crops.

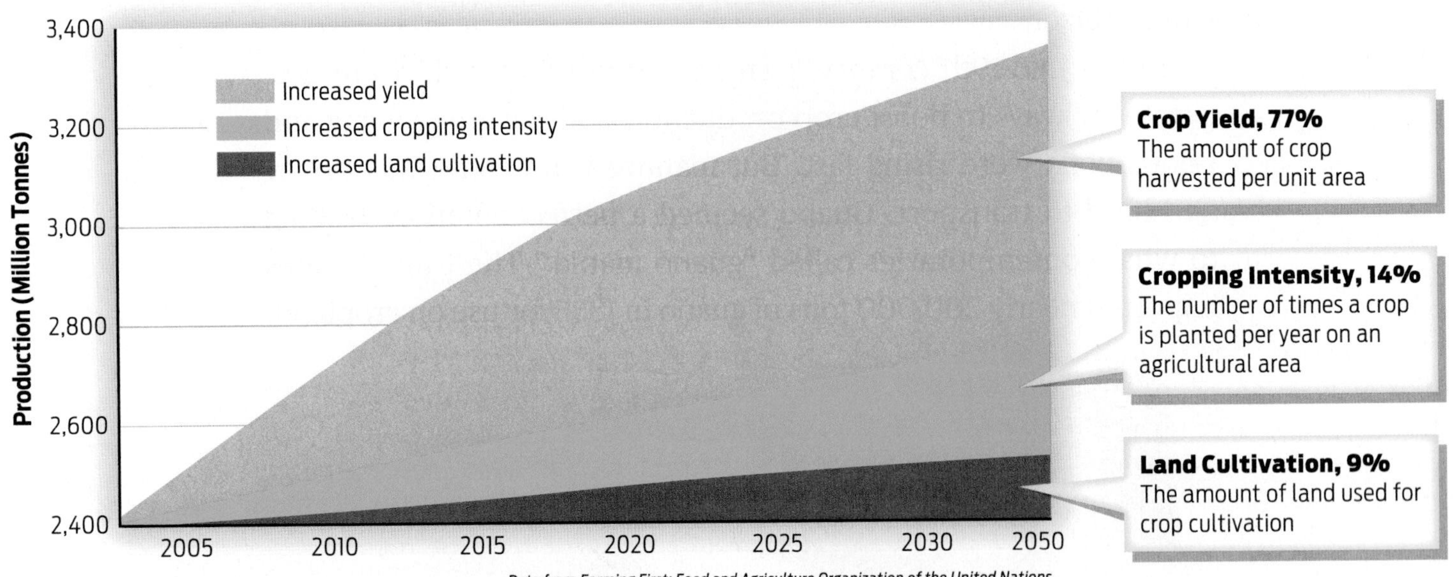

Data from *Farming First; Food and Agriculture Organization of the United Nations*

engineering to help improve yields. Think of it as building better plants.

Nutrient Fix

Farming is one long experiment in coaxing better harvests out of crops. Ancient Egyptian, Roman, and Babylonian civilizations all used some form of fertilizer—typically manure—to enhance their crop yields. But it was not until the 19th century that scientists fully understood why fertilizer helps plants grow.

The reason is nitrogen. Like all organisms, plants need nitrogen to build important biological molecules, like proteins and DNA. Though 80% of the atmosphere is made up of nitrogen gas (N_2), and this gas is also present in air pockets in the soil, this form of nitrogen is not usable by plants.

To obtain usable nitrogen, plants rely on **nitrogen-fixing bacteria** that live in the soil. These organisms—among the most important on Earth—convert gaseous nitrogen into a form that plants can take up and use. That form is ammonia (NH_3). Only certain types of bacteria can perform this important feat of nitrogen fixation, and no nonmicrobial organisms have this ability. Without these bacteria, the planet's eukaryotic life simply wouldn't exist.

Plants have a structure that is well adapted to obtaining nitrogen as well as the other aboveground and belowground resources they need to grow. The plant **shoot** consists of a stem and leaves that extend skyward, giving the plant access to the sunlight and carbon dioxide it needs to make carbohydrates by photosynthesis (see Chapter 5). Plant **roots** extend downward into the soil, giving the plant access to water and the nitrogen and other nutrients it needs to make proteins and other molecules (**INFOGRAPHIC 24.1**).

Studies have shown that what most limits plant growth in many terrestrial environments is the amount of available nitrogen in the soil.

Though usable nitrogen is naturally found in soil, the amount is often limiting. And every time crops are harvested, they take a portion of the nitrogen with them, leaving the soil less fertile. Studies have shown that what most limits plant growth in many terrestrial environments is the amount of available nitrogen in the soil. This is why fertilizer is such an important component of modern agriculture—it's the fastest and surest way to boost yields.

Today, most of the nitrogen in fertilizer is produced industrially by the Haber-Bosch process, in which high heat and pressure convert gaseous nitrogen and hydrogen into ammonia. By some analyses, the growth of the human population from about 1 billion in 1900 to more than 7 billion today would not have been possible without the ammonia produced through the Haber-Bosch process.

Nitrogen from fertilizer is taken up into plants, which are eaten by animals, which thus obtain the plant's nitrogen. In this way, nitrogen from fertilizer literally becomes a part of our bodies. "There's the old saying you are what you eat," Jez says. "Well, depending on where you live in the world, between 30% and 70% of the nitrogen in your body actually came from a chemical factory."

The nitrogen stored in animal bodies eventually returns to the soil through feces and urine, which are rich in ammonia, and ultimately through decomposition when animals die. Bacteria and fungi decompose proteins and other nitrogen-rich organic molecules into ammonia, replenishing usable nitrogen in the soil. Nitrogen gas is returned to the air from the soil through the action of other bacteria. The movement of nitrogen through the environment is called the **nitrogen cycle.**

NITROGEN-FIXING BACTERIA
Bacteria that convert gaseous nitrogen to ammonia, a form of nitrogen usable by plants.

SHOOT
The aboveground parts of a plant: the stem and photosynthetic leaves.

ROOT
The belowground parts of a plant, which anchor it in the soil and absorb water and nutrients.

NITROGEN CYCLE
The movement of nitrogen atoms as they cycle between different molecules in living organisms and the environment.

INFOGRAPHIC 24.1 What a Plant Needs to Grow

Plants access sunlight (energy) and carbon (CO_2) through their aboveground shoot system. The root system extends into the soil to take up water and nutrients such as nitrogen and phosphorus, which are transported through the stem to the leaves.

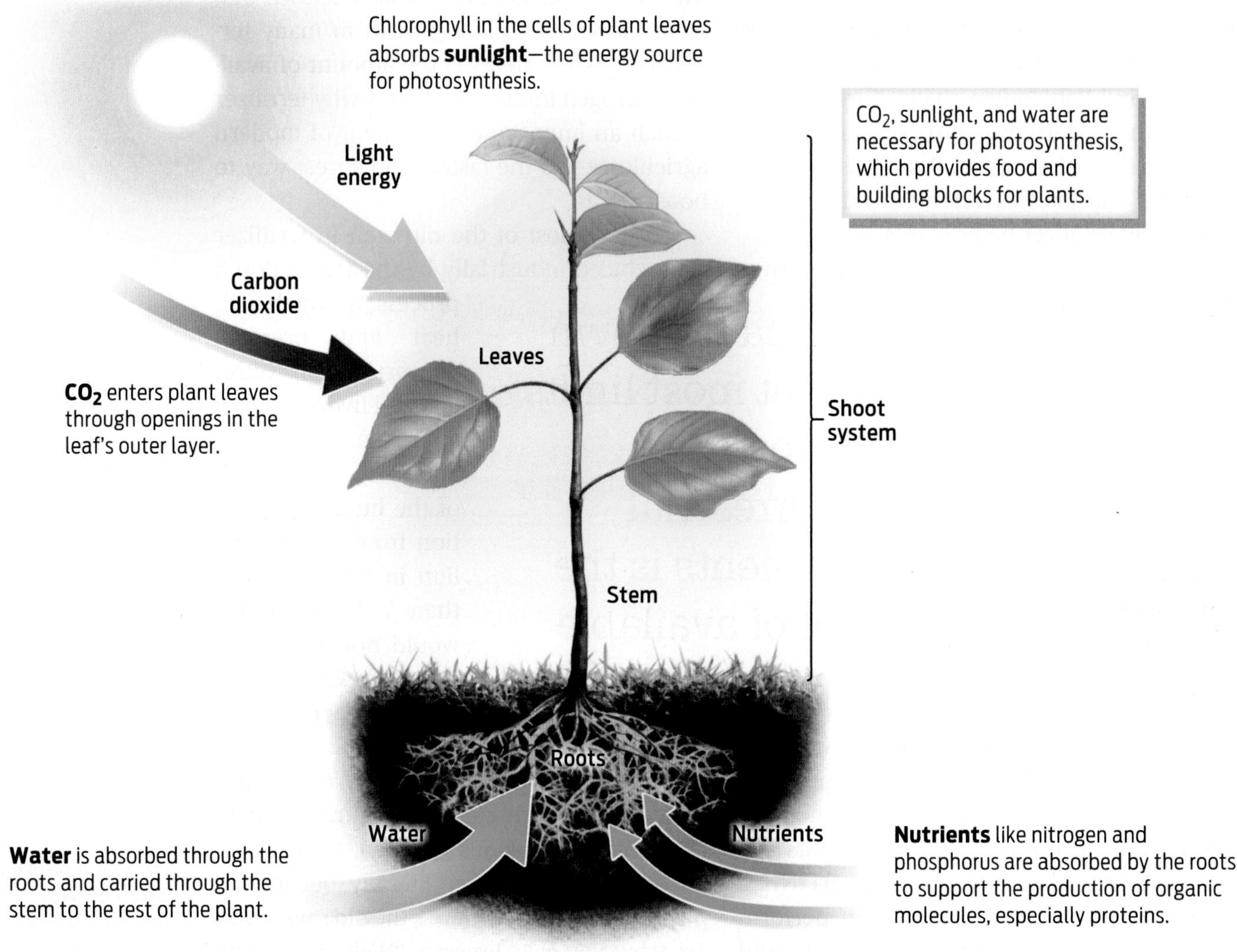

Chlorophyll in the cells of plant leaves absorbs **sunlight**—the energy source for photosynthesis.
CO_2, sunlight, and water are necessary for photosynthesis, which provides food and building blocks for plants.
Light energy
Carbon dioxide
Leaves
CO_2 enters plant leaves through openings in the leaf's outer layer.
Shoot system
Stem
Roots
Water
Nutrients
Water is absorbed through the roots and carried through the stem to the rest of the plant.
Nutrients like nitrogen and phosphorus are absorbed by the roots to support the production of organic molecules, especially proteins.

? How do plants acquire water; CO_2; nitrogen; phosphorus—through leaves or through roots?

PHOSPHORUS CYCLE The movement of phosphorus atoms as they cycle between different molecules in living organisms and the environment.

In addition to nitrogen, commercial fertilizer also contains plentiful phosphate, which provides plants with the phosphorus they need to grow—phosphorus is a crucial component of nucleotides and cell membranes. Phosphorus exists naturally in rocks and enters soil through the weathering of rocks. Plants take up phosphorus through their roots. Animals, in turn, eat these plants and so obtain their phosphorus. When plants and animals die, phosphorus is returned to the soil in the **phosphorus cycle.** As with nitrogen, harvesting crops removes phosphorus from the land, making it less fertile. In some regions, phosphorus is more limiting than nitrogen **(INFOGRAPHIC 24.2)**.

Though we cannot feed the world without it, industrial fertilizer has unwanted environmental side effects. When too much fertilizer is used, the excess washes out of fields and into lakes and rivers, leading to a damaging process called **eutrophication**. This enrichment of nutrients permits algae to bloom out of control. The rapidly growing algae eventually block sunlight from reaching lower regions of the lake or river, causing aquatic plants along the bottom to die. Bacteria decompose the dead plants, using oxygen in the process. Eventually, the bacteria deplete the water of this important gas, creating an

EUTROPHICATION
An excessive enrichment of nutrients in bodies of water that can cause ecological problems.

INFOGRAPHIC 24.2 Nutrient Cycles

Nutrients such as nitrogen and phosphorus move into and out of the bodies of living organisms and through the environment. Nutrients are taken up and incorporated into molecules of living organisms. When organisms and their waste products decompose, the nutrients are returned to the environment. These movements make up a cycle for each nutrient.

Nitrogen Cycle
Nitrogen is a critical component of the amino acids that make up proteins and the nucleotides of DNA and RNA. Nitrogen atoms cycle between different chemical and biochemical compounds as they move from organisms to the soil, water, and air and back to organisms

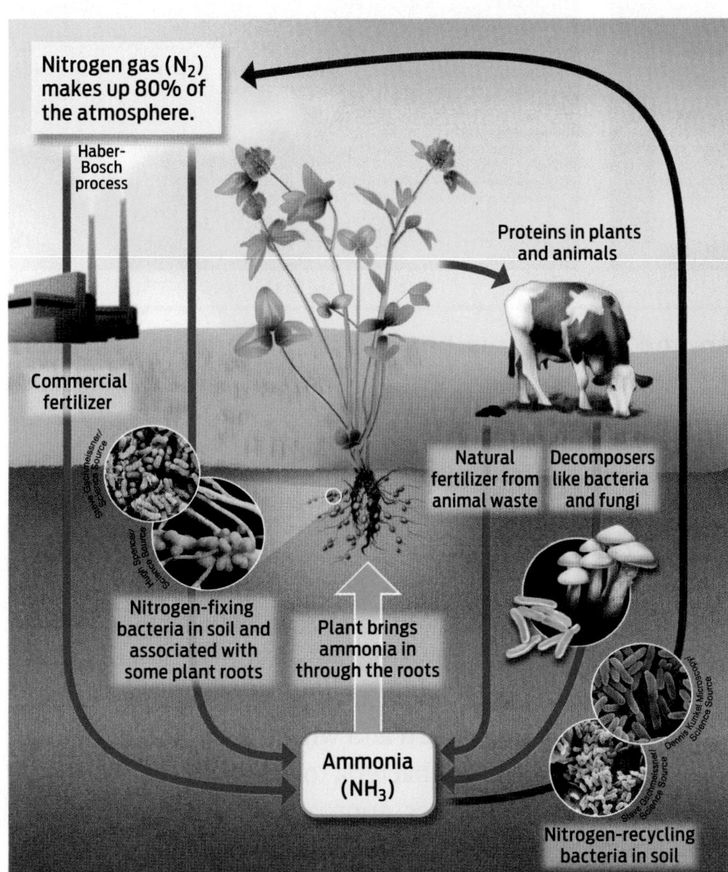

Phosphorus Cycle
Phosphorus is critical for the structure of nucleic acids and phospholipids. In animals, it is critical for bones and teeth. Phosphorus is added to an ecosystem by the weathering of rocks, or through human activities.

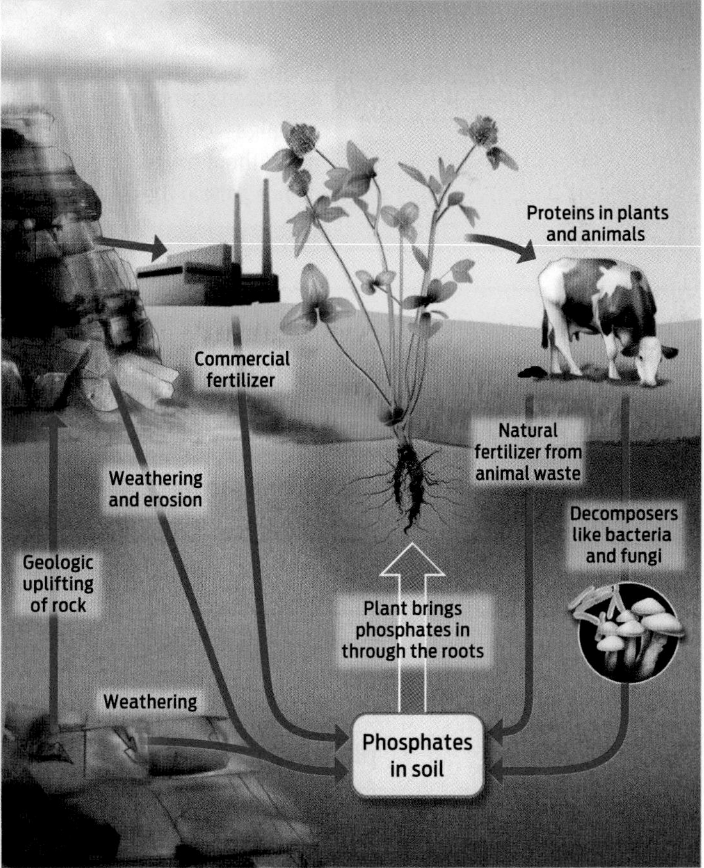

? Identify one human-produced source and one natural source for soil nitrogen and for soil phosphorus.

INFOGRAPHIC 24.3 Eutrophication Is a Danger of Modern Agriculture

Addition of nutrients to aquatic environments stimulates excessive growth of algae. This excessive growth initiates a chain of events that culminates in oxygen-depleted dead zones that cannot support life.

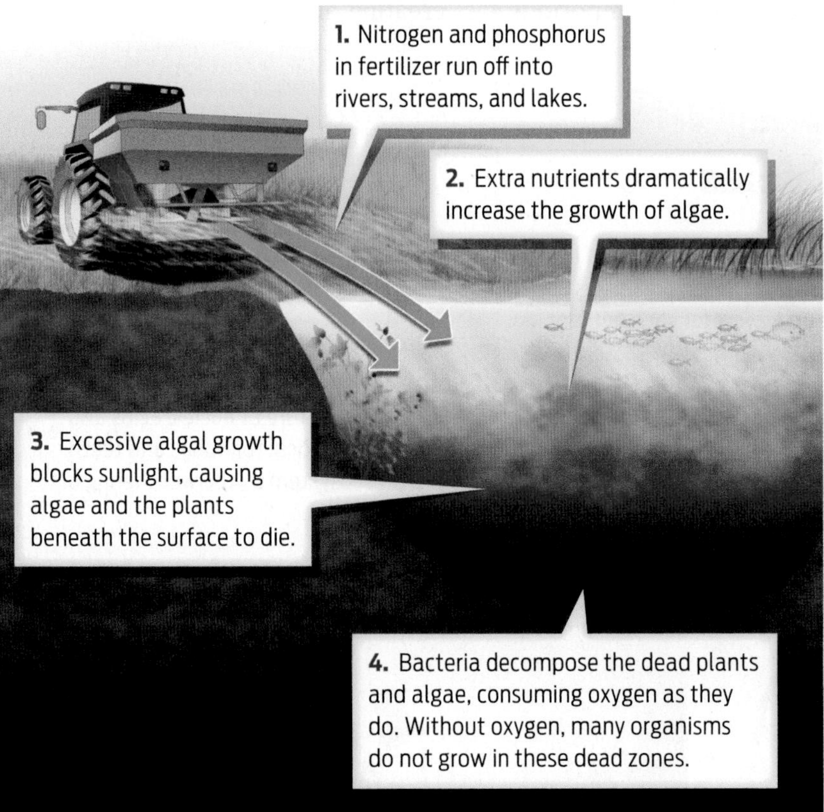

1. Nitrogen and phosphorus in fertilizer run off into rivers, streams, and lakes.
2. Extra nutrients dramatically increase the growth of algae.
3. Excessive algal growth blocks sunlight, causing algae and the plants beneath the surface to die.
4. Bacteria decompose the dead plants and algae, consuming oxygen as they do. Without oxygen, many organisms do not grow in these dead zones.

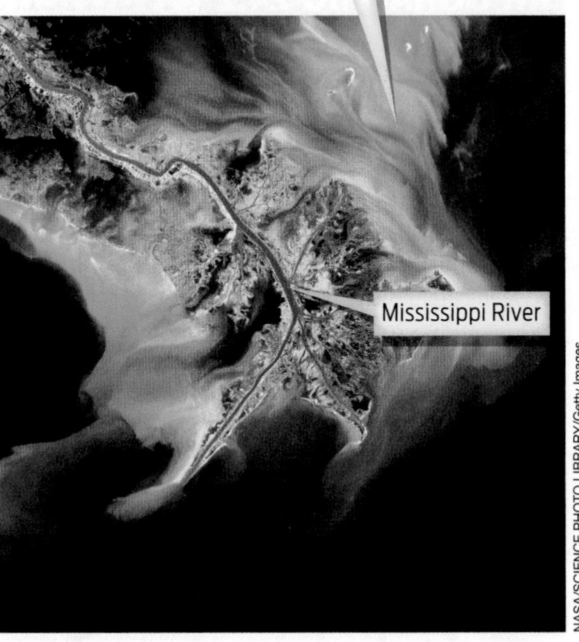

Massive algae blooms (light blue areas) grow in areas where fertilizer runoff in the Mississippi River flows into the Gulf of Mexico. These blooms have caused dead zones through eutrophication.
Mississippi River

NASA/SCIENCE PHOTO LIBRARY/Getty Images

? Why are organisms that, like fish, require nitrogen and phosphorus harmed by the addition of these nutrients to a body of water?

oxygen-free dead zone where nothing can live (**INFOGRAPHIC 24.3**).

Producing fertilizer on an industrial scale also uses a lot of energy, primarily in the form of fossil fuels. Burning these fuels adds greenhouse gases to the atmosphere, contributing to climate change (see Chapter 22).

But what if there were a way to engineer plants to make their own fertilizer so that farmers didn't need to provide it? That's an approach that scientists are actively pursuing.

Some plants are already halfway there. For example, soybeans, peanuts, alfalfa, and other legumes partner with nitrogen-fixing bacteria living in lumpy structures called **root nodules** on the plant roots. Plants provide the bacteria with sugars to eat, and the bacteria provide the plants with usable nitrogen.

According to Jez, that has given scientists an idea: "Can you take what a legume does all the time and move those processes into plants like corn or wheat?" This would mean engineering corn and wheat to partner with nitrogen-fixing bacteria to produce root nodules.

Another approach might be to transfer the nitrogen-fixing abilities of a type of bacteria called *Rhizobium* into other bacteria and fungi that already associate with plants—"to provide an analogous nitrogen source that doesn't require the *Rhizobium*," Jez explains.

ROOT NODULE
An enlargement on the root of a plant that contains nitrogen-fixing bacteria.

These are still pie-in-the-sky solutions at the moment, he says, but they're important to pursue. The primary reason for low crop yields in sub-Saharan Africa, for example, is related to limited nitrogen. "Any way you can make nitrogen use more efficient is critical," Jez says.

Seeds of Plenty

This isn't the first time that concerns over feeding a growing human population have fueled innovations in plant science. In the 1950s and 1960s, advances in public health spurred rapid population growth in developing countries around the globe. Yet agricultural gains lagged behind, threatening food shortages and mass hunger.

In some crops, fertilizer itself caused unforeseen trouble. For example, fertilized wheat (*Triticum aestivum*) grows rapidly, and the tall plants that result have thin stems that fall over easily in wind, decreasing yields. In some tropical regions, such as Mexico, a fungal disease called stem rust reduced yields still further.

Norman Borlaug, a plant biologist, eventually came up with a solution. He bred strains of high-yielding American wheat plants to a dwarf variety that grew in Japan, yielding a "semidwarf" strain that could stay upright when fertilized. He then crossed this semidwarf variety with strains resistant to stem rust, producing plants that were well suited to tropical climates like Mexico's. Yields nearly doubled.

Borlaug's methods quickly caught the attention of leaders in India and Pakistan, who enlisted him to help breed semidwarf strains of wheat suitable to grow in that part of the world. Semidwarf varieties of rice (*Oryza sativa*) came next and helped increase rice yields in the Philippines, China, Taiwan, and other East Asian countries.

For his work, Borlaug has come to be known as the father of the green revolution. By some estimates, he saved a billion people from starvation. Borlaug won the Nobel Peace Prize in 1970.

Borlaug focused on wheat and rice because these are two of the world's most important **staple crops,** providing nourishment to billions of people every day. Wheat and rice—as well as corn, barley, oats, and rye—are cereals (the name is from Ceres, the Roman goddess of harvest and agriculture). Cereals are members of the grass family of plants, known for their edible, starchy seeds.

STAPLE CROP
A crop that is eaten in large quantities and provides most of the energy and nutrients in the human diet.

Root nodules on the roots of a pea plant (*Pisum sativum*). Nitrogen-fixing bacteria living in the root nodules convert nitrogen gas to ammonia, which the plant can use for growth.

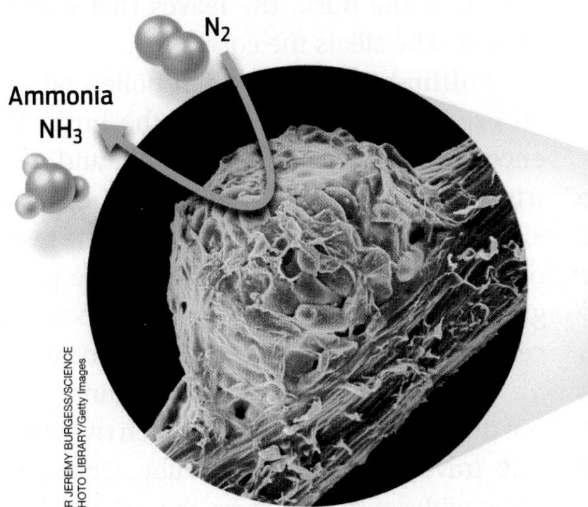

STAMEN
The male reproductive structure of a flower, made up of a filament and an anther.

ANTHER
The part of the stamen that produces pollen.

PISTIL
The female reproductive structure of a flower, made up of a stigma, style, and ovary.

STIGMA
The sticky "landing pad" for pollen on the pistil.

OVARY (PLANT)
The structure at the base of the pistil that contains the ovules.

OVULE
The part of a flower that develops into a seed after fertilization.

STYLE
The tubelike structure that leads from the stigma to the ovary.

POLLINATION
The transfer of pollen from male to female plant structures so that fertilization can occur.

POLLEN TUBE
A hollow tube that grows from a pollen grain after pollination and transports the male gametes to the egg.

DOUBLE FERTILIZATION
The process in angiosperms in which one sperm fertilizes the egg and one sperm fuses with other cells in the ovule.

As discussed in Chapter 18, seeds are protective structures containing the embryo of a plant together with a starting supply of food for that embryo. Cereal crops are valuable because they are nutritious and, because they are dry, can be stored easily without rotting.

Where do seeds come from? Seeds are products of plant reproduction–plant sex. Just as human embryos result from the union of egg and sperm during fertilization, plant embryos result from plant sperm meeting and fusing with plant eggs. But unlike humans and many other animals, which raise their embryos in a protective womb, seed plants (including many crop plants) encase their newly formed embryos in a protective seed. The seed provides the embryo with food and shelter as it develops.

Like other sexually reproducing creatures, plants have distinct male and female reproductive organs that produce haploid gametes: sperm and eggs. In angiosperms, or flowering plants, male and female sex organs are housed within a plant's flowers. Though they might not look like it, grasses like corn and wheat are types of flowering plants.

Though they might not look like it, grasses like corn and wheat are types of flowering plants.

Many flowering plants have "perfect" flowers, meaning that male and female reproductive parts are found within the same flower. The male part of the flower is called the **stamen.** It consists of a stemlike filament topped with a pollen-producing **anther.** Pollen grains contain the male gametes.

The female part of the flower is called the **pistil.** The top of the pistil is a sticky "landing pad" for pollen called the **stigma.** At the base of the pistil is the **ovary,** a protective structure that holds the egg-producing **ovules.** In angiosperms, the ovary develops into a seed-containing fruit. The tube connecting the stigma to the ovary is the **style.** For seed-producing plants to reproduce, sperm from pollen produced by the male anther must join with the eggs in a plant ovule. How that happens takes different forms, depending on the plant **(INFOGRAPHIC 24.4)**.

Take corn as an example. Corn is the most common crop grown in the United States–it is fed to both people and livestock and is used to produce corn ethanol, a common gasoline additive and fuel for flex-fuel vehicles. Every kernel on an ear of corn is a tiny fruit that contains a seed. And each of these seeds contains an embryo, the product of fertilization between a sperm and an egg. The male organ of a corn plant, called the tassel, emerges from the top of the plant stalk. It contains many small flowers that produce pollen grains. The tassel is the corn stamen.

The female reproductive organ is the ear, which develops on the plant stalk about halfway between the ground and the tassel. (Note that in corn, male and female reproductive organs are housed in different flowers on the same plant, making the flowers "imperfect.") The ear consists of a cob studded with egg-containing ovules that, when fertilized, will develop into seed-containing kernels. Each ovule produces a hairlike structure called a silk that grows along the ear and eventually emerges from the tip of the husk, the leaves that enclose the ear. The silk is the corn style.

Pollination occurs when pollen falls on an exposed silk at the top of the husk. The encounter between a pollen grain and a silk triggers the growth of a **pollen tube,** which extends from the pollen grain down the inside of the silk. Male sperm cells from the pollen grain travel through the pollen tube to the egg cell on the cob, where fertilization occurs.

Like all angiosperms, corn plants carry out **double fertilization,** in which two sperm cells travel down the pollen tube. One sperm fuses with an egg to form the embryo. The

INFOGRAPHIC 24.4 Sexual Anatomy of a Flower

Flowers are the reproductive structures in angiosperms. The anthers at the top of stamens produce pollen, which lands on the stigma during pollination. Fertilization occurs when a male gamete reaches an egg within an ovule within the ovary.

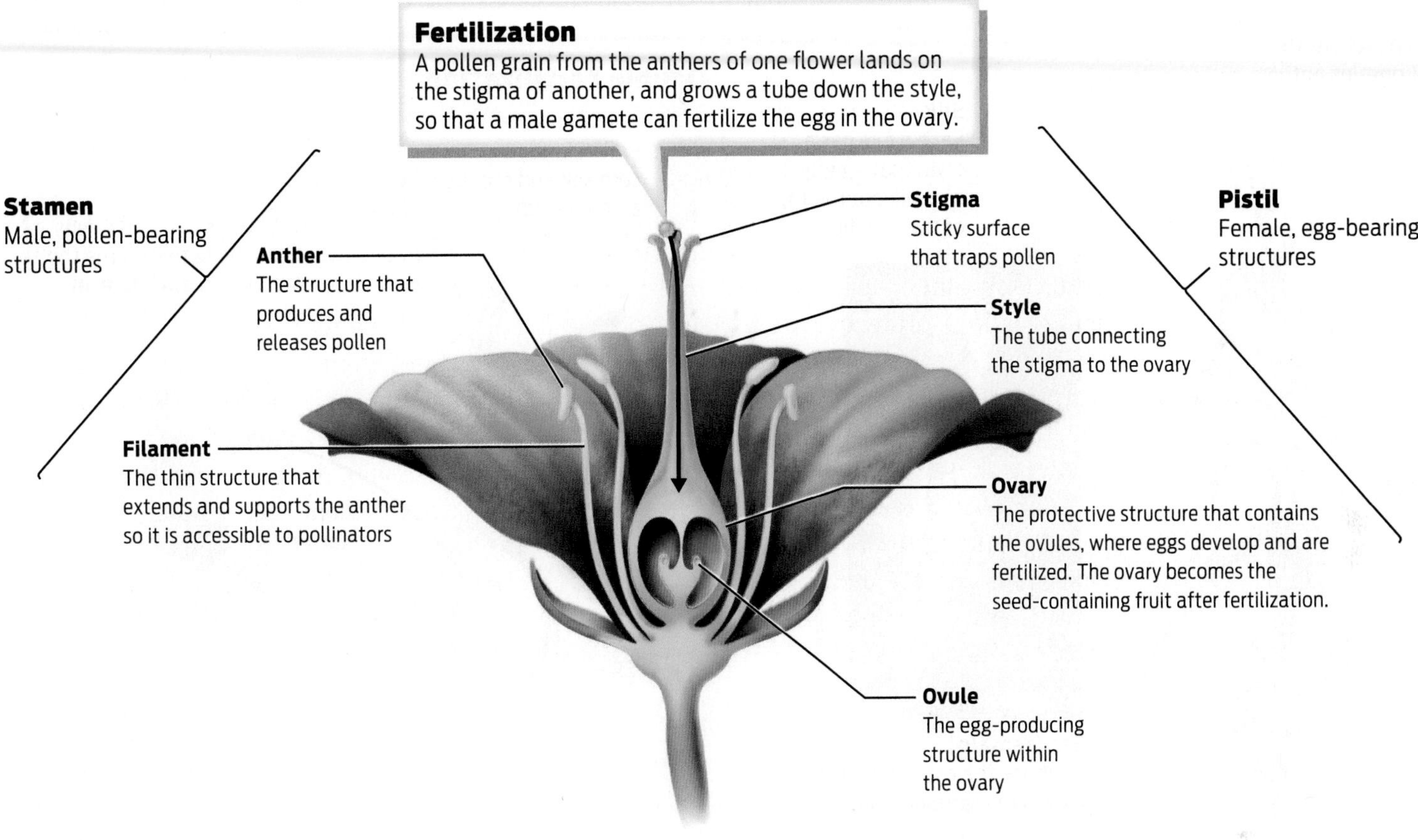
Fertilization
A pollen grain from the anthers of one flower lands on the stigma of another, and grows a tube down the style, so that a male gamete can fertilize the egg in the ovary.
Stamen
Male, pollen-bearing structures
Anther
The structure that produces and releases pollen
Filament
The thin structure that extends and supports the anther so it is accessible to pollinators
Stigma
Sticky surface that traps pollen
Style
The tube connecting the stigma to the ovary
Pistil
Female, egg-bearing structures
Ovary
The protective structure that contains the ovules, where eggs develop and are fertilized. The ovary becomes the seed-containing fruit after fertilization.
Ovule
The egg-producing structure within the ovary

? Where in the plant does pollination occur, and where does fertilization occur?

other sperm cell joins with other ovule cells to form the **endosperm** of the seed. The endosperm is the source of stored nutrients and energy for the growing embryo; it is also what makes cereals an energy-rich food source for humans. The hearty outer covering of the seed, the **seed coat,** protects the developing embryo. For humans who consume whole grain foods, the seed coat is a good source of **dietary fiber,** a largely indigestible complex carbohydrate that aids digestion and regularity (**INFOGRAPHIC 24.5**).

In addition to the edible and nutritious seeds of cereals, humans eat many other plant parts. Apples, tomatoes, eggplants, and other fruits are the developed ovaries of plants containing one or more seeds. Fruit serves as a delicious incentive for animals to eat and disperse the plant's seeds. Carrots, radishes, beets, and sweet potatoes are plant roots. Broccoli is a flower, and asparagus is a stem.

From Seed to Crop

For those of us who don't live on farms, it's easy to forget just how central agriculture is to our modern way of life. But a plant biologist will tell you that crop plants are the very foundation of civilization. "Think back to when

ENDOSPERM
A part of the seed containing nutrients that is the seed's source of energy, produced when a sperm fuses with cells in the ovule during double fertilization.

SEED COAT
The hearty outer covering of a seed that protects the developing embryo.

DIETARY FIBER
Plant-based carbohydrates that contribute to proper digestion.

INFOGRAPHIC 24.5 Reproduction in Corn

Corn is an angiosperm. Male flowers produce pollen, which lands on a female stigma at the tip of a silk. A pollen grain produces a pollen tube through which sperm travel to the ovule. One sperm fertilizes the egg to produce the embryo, and the other fuses with cells in the ovule to produce the endosperm.

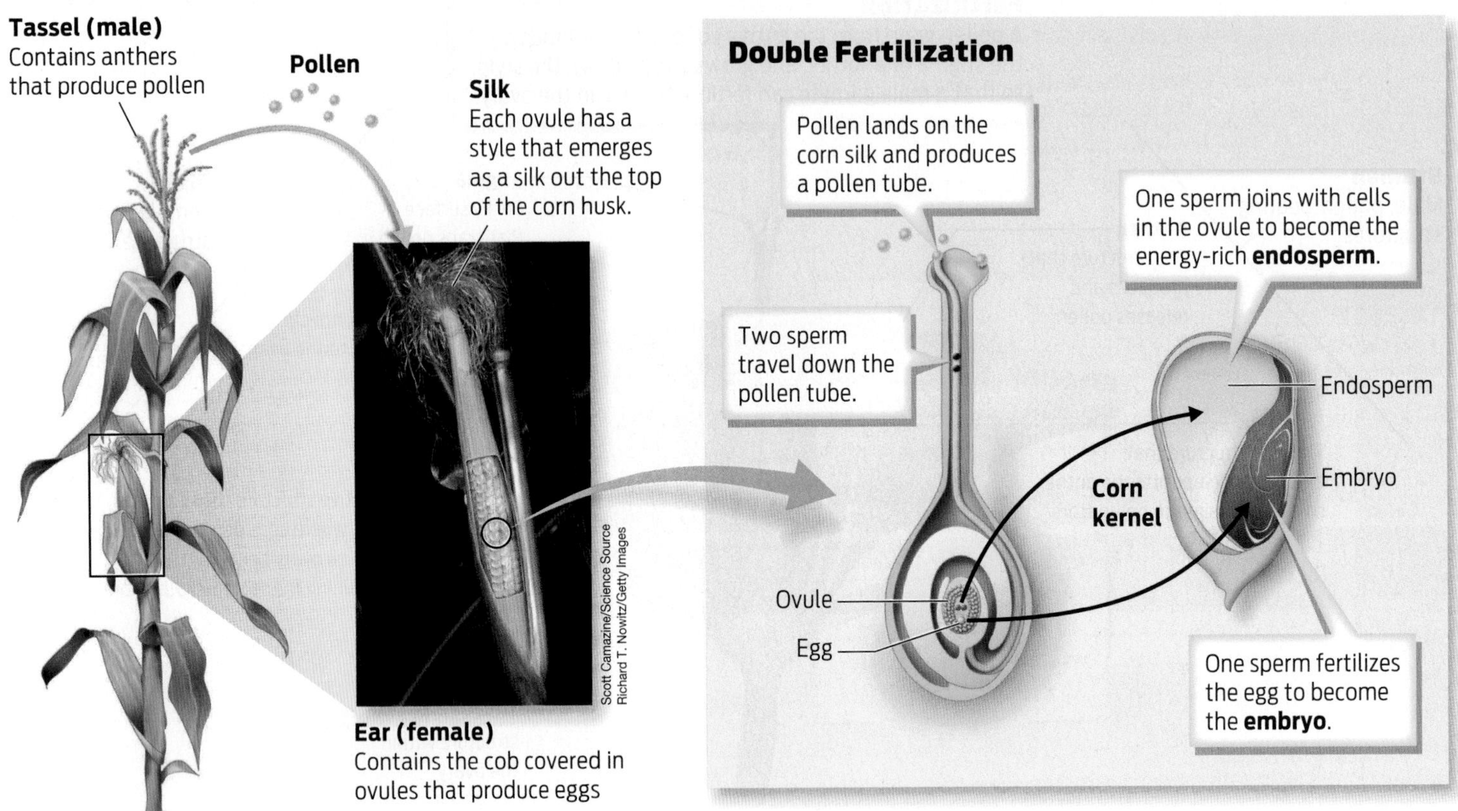
Tassel (male)
Contains anthers that produce pollen
Pollen
Silk
Each ovule has a style that emerges as a silk out the top of the corn husk.
Ear (female)
Contains the cob covered in ovules that produce eggs
Scott Camazine/Science Source
Richard T. Nowitz/Getty Images
Double Fertilization
Pollen lands on the corn silk and produces a pollen tube.
Two sperm travel down the pollen tube.
One sperm joins with cells in the ovule to become the energy-rich endosperm.
Endosperm
Embryo
Corn kernel
Ovule
Egg
One sperm fertilizes the egg to become the embryo.

? What are the two events of double fertilization?

GERMINATION
The process by which a new plant begins to grow from a seed.

COTYLEDON
The part of a plant embryo that forms the first leaf of the seedling.

MERISTEM
A plant tissue containing cells that can divide and contribute to the growth of the plant.

society first began," Jez explained in a recent TED Talk. "Where did cities come from? It was when we began to plant crops, to create agriculture." Planting crops allowed humans to settle in one place rather than live as nomads who travel continuously to find food.

Today, most of us buy our food from supermarkets, with little understanding of the back story–what it takes to get it there.

It starts with a seed. Seeds grow into new crop plants when they are planted in fertile soil and conditions are right for **germination,** the process in which a new plant begins to grow. When there is enough moisture and warm enough temperatures, a seed begins to sprout, sending out both a young root and a shoot. The embryo within the seed contains at least one **cotyledon,** an embryonic leaf that emerges with the shoot when the seed germinates. The cotyledon is ready to begin photosynthesizing as soon as the young shoot emerges from soil into sunlight. But if seeds are planted too deep, and the young seedling runs out of stored energy before a cotyledon reaches sunlight, the plant will die.

As in humans and other animals, growth in plants occurs as a result of cell division of stem cells (see Chapter 9). In plants, the regions of a developing plant that contain these stem cells are called **meristems.**

The root apical meristem is the region from which new roots grow. Likewise, the shoot apical meristem is the region from which new shoots emerge. Plant meristems are active throughout a plant's life, and so plant growth never stops **(INFOGRAPHIC 24.6)**.

Controlling Pests

Historically, one of the biggest threats to crop yields is pests–insects, fungi, and bacteria that eat plant parts, sometimes sickening or killing the plant. Almost as soon as humans started planting large fields of a single crop, so-called **monocrops,** they had to confront pests attracted to this easy-to-access bounty. And pests continue to be a problem: The Food and Agriculture Organization of the United Nations estimates that 20%-40% of global crop yields are lost each year because of pests and the diseases they cause.

To combat pests, farmers rely on a variety of natural and synthetic **pesticides.** Some of the earliest pesticides were toxic elements such as arsenic, mercury, and lead. These were toxic not only to the intended pest targets but also to people, creating health risks for farmers who applied these elements to their crops. Today, farmers use a wide variety of chemicals as pesticides, including ones made by other living organisms. One particular chemical, made by a bacterium, has had far-reaching effects.

MONOCROP
A single crop species grown in the same field over many seasons.

PESTICIDE
A substance that is toxic to pests, organisms that can damage crops or farm animals.

INFOGRAPHIC 24.6 Seed Germination and Growth

Seeds contain meristems that divide to produce the roots and the shoot as the seed germinates. They also contain embryonic leaves known as cotyledons. Plant growth occurs as a result of the division of cells located at meristems. Root and shoot meristems produce cells that contribute to the growth of roots and shoots, respectively.

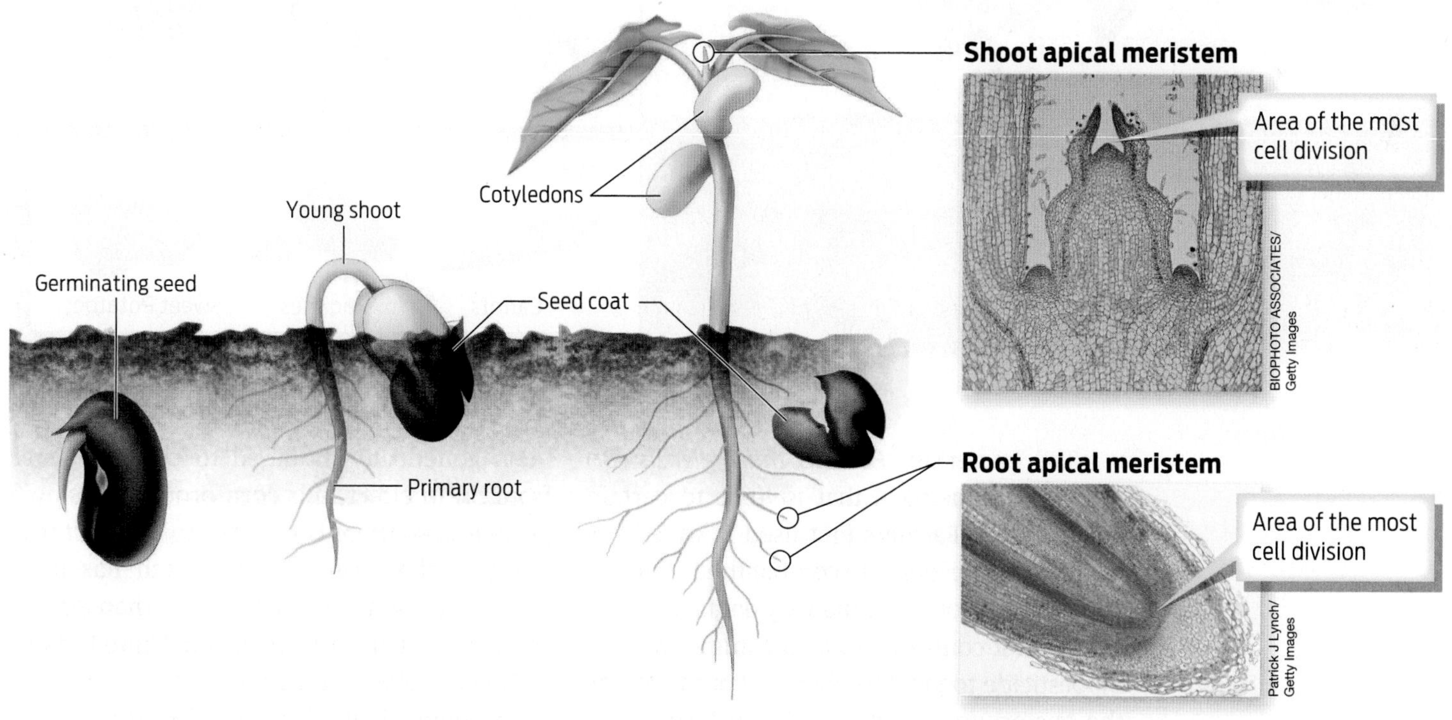

? If the end—the very tip—of the root shown above is cut off, can there be any further root growth in this plant? Explain your answer.

Edible Parts of a Flowering Plant

Many parts of flowering plants are edible and make up important parts of the human diet.

The bacterium *Bacillus thuringiensis* (*Bt*) produces a chemical that is toxic to certain insect larvae. Farmers first used *Bt* in 1928 as a natural biopesticide to combat the European corn borer, a common and very damaging corn pest, and it continues to be a commonly used biopesticide today. The chemical produced by the bacterium is toxic to certain insects but harmless to humans and other mammals.

A major innovation came in 1995 when scientists unveiled a variety of corn that had been genetically modified to contain genes from *Bt*. In effect, this corn produces its own pesticide, so there is no need to apply *Bt* pesticide to the crop. The approach has been wildly successful and today more than 80% of corn and cotton planted in the United States is the *Bt*-engineered variety.

Genetically modified (GM) crops are created by variations of the genetic engineering techniques we encountered in Chapter 8. First, the desired gene–in this case the one for the *Bt* pesticide–is cut out of the source organism, the bacterium. It is then incorporated into a circular piece of DNA called a vector, which will deliver the bacterial gene to the plant nucleus. Because plant cells are surrounded by a sturdy cell wall, it's harder to introduce the DNA vector into their cells than into animal cells. Therefore, scientists often use what's called a gene gun to bombard plant cells with the DNA vector, penetrating the cell wall. The vector enters the cell and the transgene is incorporated at random into the plant genome **(INFOGRAPHIC 24.7)**.

INFOGRAPHIC 24.7 Genetic Modification Has Lowered Pesticide Use

Genetically modified crops have been engineered to have a variety of traits, including resistance to pests. *Bt* corn is widely planted, reducing the need to apply pesticides to millions of acres of corn.

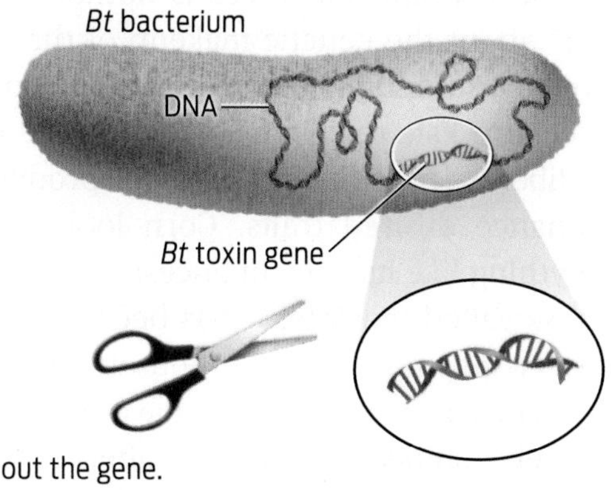

1. Cut out the gene.

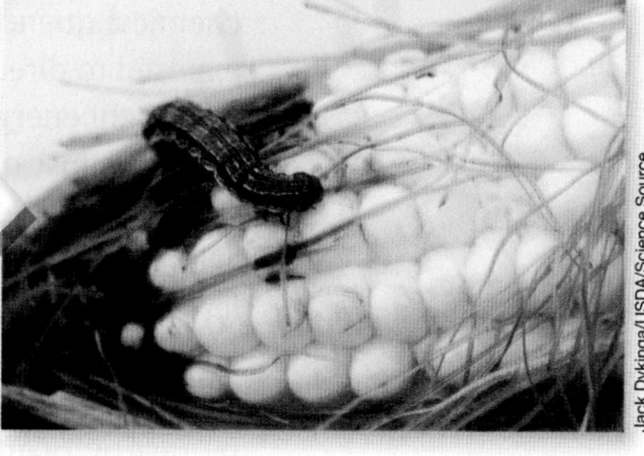

Bt corn prevents attack by pests like the corn borer without the need for pesticides.

2. Insert the gene into a vector.

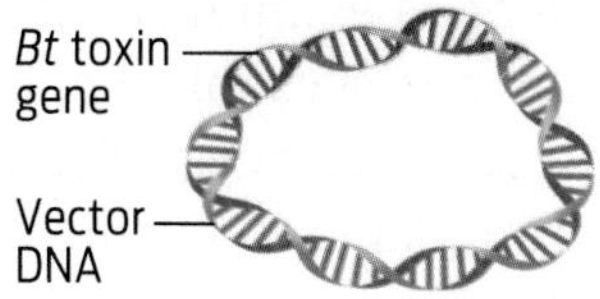

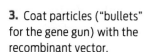

3. Coat particles ("bullets" for the gene gun) with the recombinant vector.

4. Shoot the coated particles at plant cells, allowing the recombinant vector to enter the nucleus.

5. Grow the engineered cells into small plants that can then be grown to maturity in soil.

Insecticide use on U.S. farms has dropped dramatically with increased use of genetically modified *Bt* corn varieties. However, insecticide use may rise in the future due to the emergence of *Bt*-resistant insects.

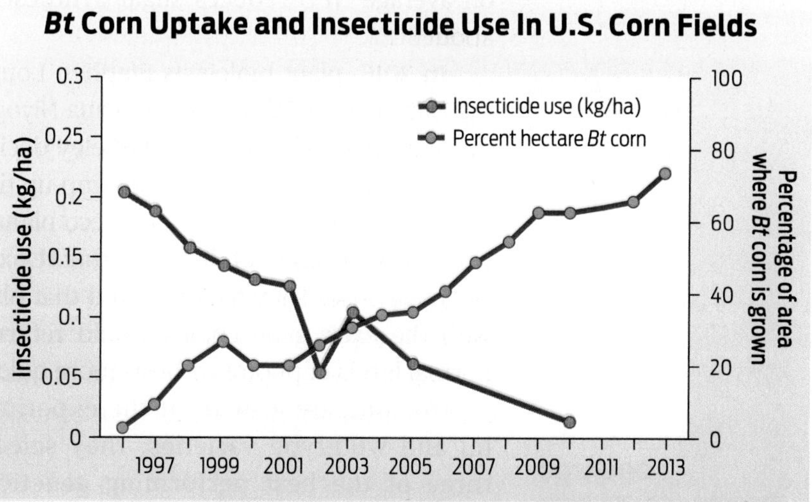

Data from EPA Pesticides Industry Sales and Usage 2006 and 2007 Market Estimates, February 2011

? Why is reducing pesticide use important?

Made in the Shade

The effort to build better plants through genetic engineering doesn't stop with pest control. Scientists have used GM technologies to develop drought-resistant varieties of corn and tomatoes that ripen more slowly. They've designed plants that can grow on land contaminated with toxic metals to help bioremediation projects. Recently, scientists have turned to using GM technologies to improve the efficiency of photosynthesis itself.

As miraculous as the ability to capture energy from sunlight seems, photosynthesis is actually not very efficient. Partly, that's because of a phenomenon called nonphotochemical quenching (NPQ). When plants are exposed to direct sunlight, they often receive more light energy than they can safely use. To guard against damage to the photosynthetic machinery in leaves, plants tamp down the activity of this machinery, as a result turning less carbon dioxide into carbohydrate sugars. When plants are exposed to shade–because of clouds moving overhead, or shade cast by nearby plants–they ramp photosynthesis back up. While the tamping down of photosynthesis happens quickly, the ramping back up takes time, resulting in a loss of total photosynthesis time. Scientists estimate that on average NPQ reduces sugar synthesis by about 30%.

In 2016, plant biologists Stephen Long of the University of Illinois and Krishna Niyogi of the University of California, Berkeley decided to see if they could shorten the ramp-up time. They genetically engineered tobacco plants to contain extra copies of the genes that relax the NPQ response. They hypothesized that plants with the extra gene copies would return to normal levels of photosynthesis more quickly.

To compare growth of the experimental and wild-type varieties, they selected three of the best performing genetically modified tobacco plants. Then they grew these three varieties in plots of land alongside unmodified tobacco plants. They found that after 22 days the GM versions weighed 15%-20% more than their unmodified neighbors (**INFOGRAPHIC 24.8**).

Tobacco isn't a food crop, of course, but the researchers hope that the improvements seen in this plant can be reproduced in food crops like corn and rice.

ORGANIC
Describes a way of growing crops that conforms to several regulations, among them that synthetic pesticides must not be used.

Frankenfoods?

Using genetic engineering technologies to "improve" on nature is not without controversy. Some critics object to scientists' fiddling with the genetic blueprints of plants and other organisms that have evolved naturally. Yet, as farmers and plant breeders are quick to point out, there is nothing "natural" about the genetic makeup of the crops we eat. Essentially every crop–from corn to wheat to kale to kiwi fruit–is the product of deliberate breeding designed to produce or enhance desired traits. Corn looks hardly anything like its ancient ancestor, a Mexican grass called teosinte; it has been artificially selected by humans over thousands of years.

Moreover, many crops have been deliberately subjected to DNA-altering treatments. Beginning after World War II, under the auspices of the International Atomic Energy Agency, scientists began exposing crop plants to atomic radiation in the hope of increasing the frequency of useful mutations that could be artificially selected. This "radiation breeding" has produced many of the most common varieties of a large fraction of the world's crops, including rice, wheat, barley, pears, peas, cotton, peppermint, sunflowers, peanuts, grapefruit, sesame, and bananas. These crops are sold in grocery stores without any special labeling–and include some **organic** crops raised without the use of pesticides.

The main difference between these older forms of genetic modification and modern genetic engineering approaches is that today the genetic changes can be made more precisely. Whereas earlier techniques might alter or delete surrounding genes in addition to the desired gene, now specific genes can be targeted.

INFOGRAPHIC 24.8 Making Photosynthesis More Efficient

Nonphotochemical quenching (NPQ) is a defense mechanism that protects plants from excess sunlight. Genetically engineered tobacco plants contain extra copies of genes that help reverse nonphotochemical quenching. After exposure to bright light that inhibits photosynthesis, these plants resume photosynthesis faster than normal tobacco plants. As a result, the engineered plants are significantly larger.

Nonphotochemical Quenching

Protection in Intense Light
In intense sunlight, a plant cannot use all available light for photosynthesis. To avoid being damaged, it quenches the activity of some chlorophyll molecules, converting the extra energy to heat that can be easily removed from the plant.
Recovery in Shade
When a plant becomes shaded or protected from high-intensity light, it reverses its quenching and uses all available light for photosynthesis. Plants that recover faster get the benefits of more photosynthesis and increased growth.
Quenched Photosynthesis
Unused energy lost as heat
Full Capacity Photosynthesis
Faster recovery leads to more plant growth

Genetically Modified Plants Recover Faster from Quenching

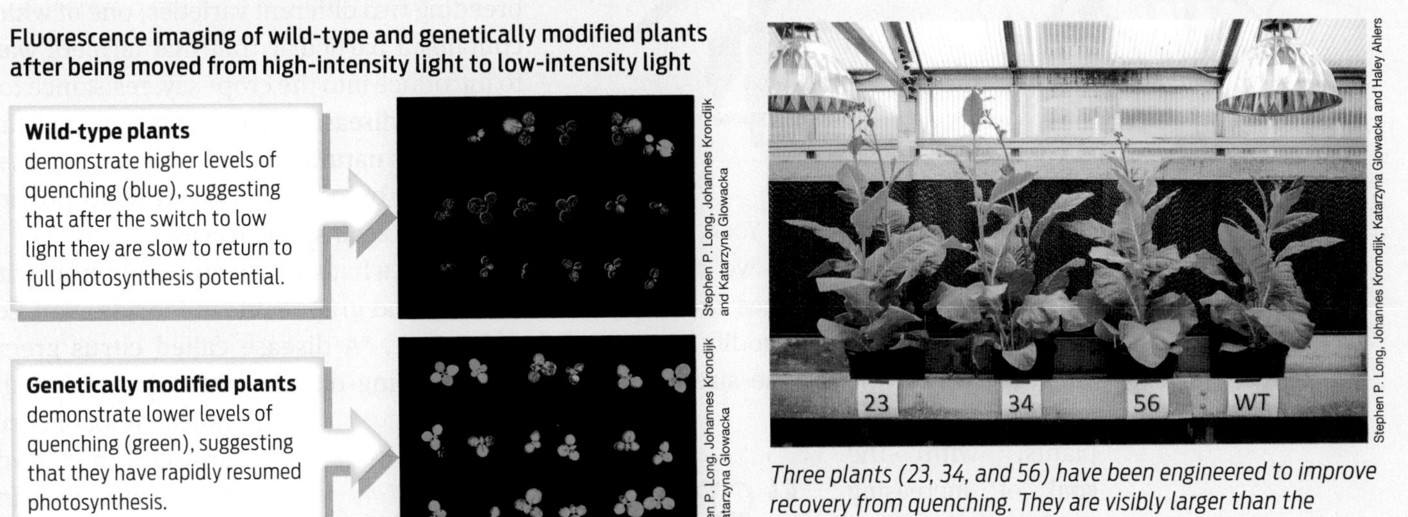

Three plants (23, 34, and 56) have been engineered to improve recovery from quenching. They are visibly larger than the unmodified (WT) tobacco plant.

? Explain why these plants were engineered to recover quickly after nonphotochemical quenching rather than to reduce nonphotochemical quenching.

One new technique in particular has transformed the genetic engineering landscape. Called CRISPR, it allows scientists to precisely edit DNA at very specific regions of the genome. CRISPR (which stands for "clustered regularly interspaced short palindromic repeats") is a kind of molecular scissors that defends bacteria against viruses. Scientists have adapted this "scissors" to make it an exquisite tool for genetic engineering.

"CRISPR is one of those things I feel like I've been waiting for my entire career," says Joyce Van Eck, a plant biologist at the Boyce Thompson Institute, affiliated with Cornell

Artificially Selected Crops: From Teosinte to Corn

The ancestor and closest relative of corn, teosinte, is a wild grass that grows in parts of Central America and Mexico. Over time, ancestral farmers bred teosinte to have the desired features of corn. These include taller, unbranched stalks that are easier to harvest and cobs with hundreds of soft kernels.

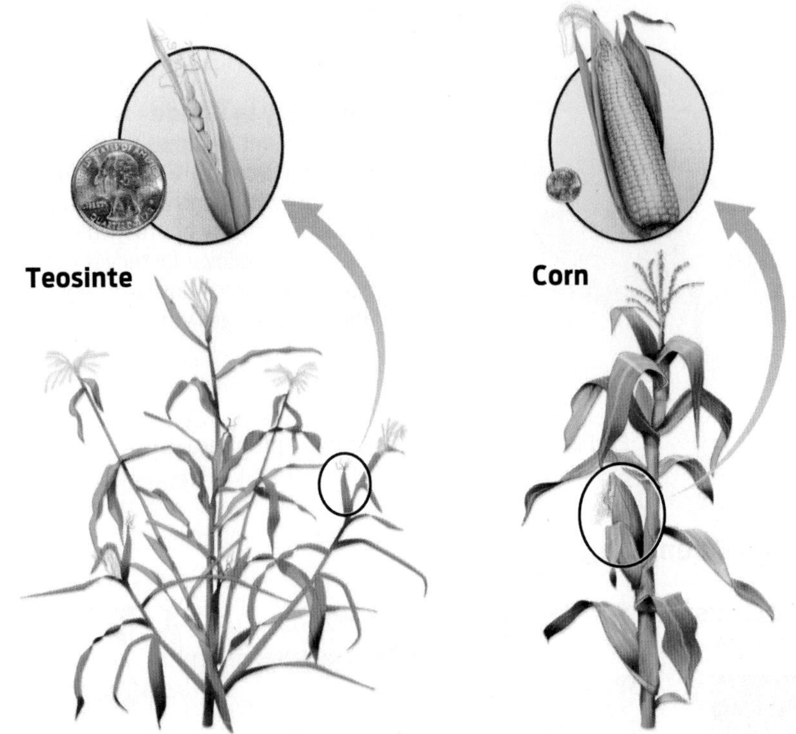

University in Ithaca, New York. "It's really transformed the way that we work in my lab." She and her colleagues are using the technique to study and modify genes that affect plant growth and the size of flowers and fruit in tomato plants, with the goal of increasing productivity.

> "CRISPR is one of those things I feel like I've been waiting for my entire career."
>
> –Joyce Van Eck

CRISPR has two main components: an enzyme that cuts DNA and a small piece of RNA that serves as a guide for the enzyme, directing it to a specific sequence of DNA. The RNA binds to DNA through complementary base pairing, as we've seen in other chapters. Once the complementary match is found, the enzyme cuts the DNA in that precise location. The approach can be used to insert a new gene at the cut site, alter the sequence of a gene, or remove the gene altogether.

The biggest difference between CRISPR and previous genetic engineering methods is that with CRISPR changes can be directed to very precise locations in the genome. Other methods, by contrast, result in genes being inserted randomly into the genome, which can result in unpredictable effects.

The first CRISPR-modified food to hit supermarket shelves will likely not be a plant but a fungus. Scientists at Penn State University used CRISPR to inactivate a gene for a particular enzyme in mushrooms that causes browning. Without this enzyme, the mushrooms have a longer shelf life **(INFOGRAPHIC 24.9)**.

Though CRISPR is revolutionizing many biological fields, it will not replace traditional breeding methods. The fastest and surest way to improve a crop is still through cross breeding two different varieties, one of which contains a particular trait that farmers want to introduce into the crop–say, resistance to a particular disease. But sometimes no existing variety has natural disease resistance, and so traditional breeding is powerless to provide protection against the disease.

That's what's happening right now with oranges and grapefruits in Florida, Van Eck points out. "A disease called citrus greening is wiping out the citrus industry," she notes. "A lot of citrus growers in Florida are now growing peaches instead."

Citrus greening is caused by a bacterium that is spread by an insect. The bacteria choke the plant's vascular system, causing fruit to shrivel and, eventually, the plant to die.

Similar problems confront certain varieties of bananas, coffee, and chocolate–the plants lack natural resistance to diseases that are wreaking havoc on the populations. But new varieties of CRISPR-modified plants may

INFOGRAPHIC 24.9 Precision Modification with CRISPR

CRISPR is a genetic engineering method that can precisely modify specific gene sequences. Molecular tools target a specific gene, which can then be altered, removed, or used as a site to introduce a new gene. The resulting genetic changes confer new phenotypes on the modified organism.

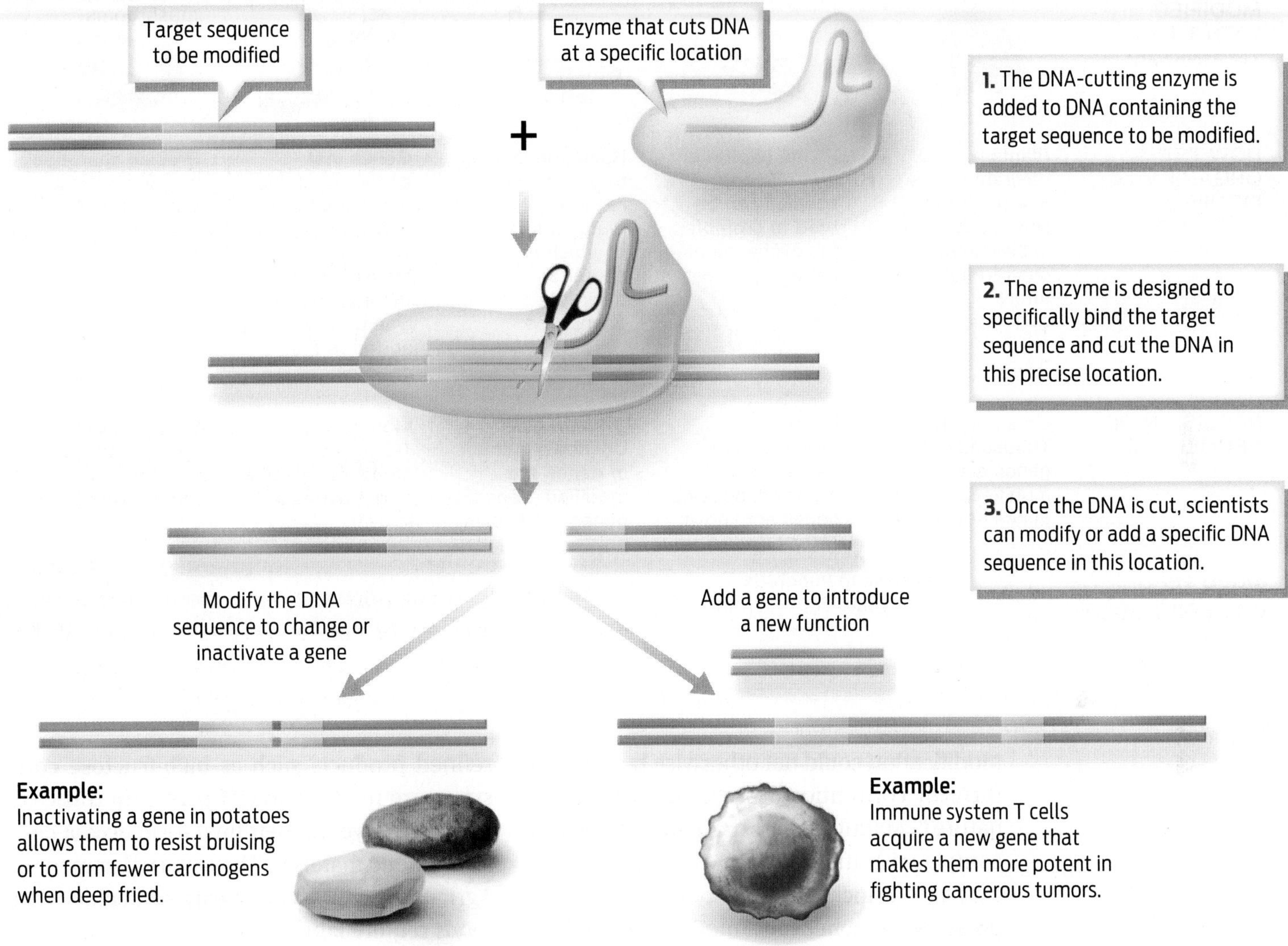

Target sequence to be modified
Enzyme that cuts DNA at a specific location
+
1. The DNA-cutting enzyme is added to DNA containing the target sequence to be modified.
2. The enzyme is designed to specifically bind the target sequence and cut the DNA in this precise location.
3. Once the DNA is cut, scientists can modify or add a specific DNA sequence in this location.
Modify the DNA sequence to change or inactivate a gene
Add a gene to introduce a new function
Example:
Inactivating a gene in potatoes allows them to resist bruising or to form fewer carcinogens when deep fried.
Example:
Immune system T cells acquire a new gene that makes them more potent in fighting cancerous tumors.

? Why is the precision of CRISPR so valuable?

offer hope. "Genetic engineering might be the only solution at this point," Van Eck says **(TABLE 24.1)**.

Many studies–including a 2016 report by the U.S. National Academies of Sciences, Engineering, and Medicine–have shown that GM foods are safe to eat, and the Food and Drug Administration (FDA) has concluded the same. Nonetheless, the subject remains controversial to some consumers and food safety advocates, who pressured lawmakers to address their concerns. In 2016, Congress passed a bill–signed into law by President Barack Obama–that requires food manufacturers to label foods that "contain genetic material that has been modified through in vitro recombinant deoxyribonucleic acid (DNA) techniques; and for which the

TABLE 24.1 Genetic Modification of Crops

	DELIBERATE BREEDING	CROSSES BETWEEN SPECIES	RADIATION BREEDING	GMOs	CRISPR
FOODS MODIFIED WITH THIS TECHNIQUE:	Many crops	Many crops	Many crops	Corn and soybeans	Potatoes and mushrooms
HOW THE ORGANISM IS MODIFIED:	Plants with desirable traits can be bred to select and enhance the desired trait in future generations.	Plants of different (but related) species can be bred to promote the exchange of genes.	Radiation is used to generate random mutations, some of which may confer a desirable phenotype.	Genes that encode proteins that confer a desired trait are added to different species. Little control of where the gene is added to the genome.	Precise alteration or addition of a gene
PRECISION OF PROCEDURE:	Less precise: Thousands of genes are modified; specific genes not known	Less precise: Thousands of genes are modified; specific genes not known	Less precise: Unknown number of genes are modified; specific genes not known	More precise: 1-3 genes modified; specific target genes are known	Very precise: 1 specific gene targeted for modification
WHEN TECHNIQUE WAS FIRST USED:	Thousands to hundreds of years ago		Mid 1950s	Early/Mid 1980s	~2016

modification could not otherwise be obtained through conventional breeding or found in nature." Manufacturers have three ways to comply with the new law. They can provide a label that includes a symbol indicating the presence of genetically modified organisms (GMOs); print a label with GMO information in words; or add a Quick Response (QR) code, readable by a smartphone, that links to ingredient information.

The new law will likely not end controversy since the wording of the law leaves room for debate about what foods will require a label. Historically, genetically modified organisms–GMOs–contained one or more genes from another species, making them transgenic (see Chapter 8). But with newer tools like CRISPR, organisms can be modified without adding foreign genes, making them not much different from natural varieties. And refined products such as high-fructose corn syrup are made from GM plants (in this case from corn), yet in their final form do not contain genetic material. The U.S. Department of Agriculture will have 2 years to decide what foods require labels.

The protection of human health is only one consideration that motivates opposition to GMOs. For example, many environmental activists worry about what would happen if a genetically modified trait, such as the production of a natural pesticide, were to "escape" from the modified plant into wild populations. Would beneficial insects like bees and butterflies be harmed, for example? It's not possible to say with certainty, though available evidence suggests this has not happened.

A further critique concerns economics. Monsanto, for example, is a company that sells genetically engineered seeds that are resistant to a commonly used herbicide called Roundup, which the company also produces. These seeds, called Roundup Ready, allow farmers to spray Roundup on their fields to control weeds without at the same time killing their crops. But because Monsanto holds a patent on the seeds, farmers are not legally allowed to re-plant seeds from their harvest and must buy them each year from the company. Though many farmers don't mind this obligation, given the benefits of the technology, others find the increasing reliance on GM products to be a threat to traditional ways of farming.

The Gates Foundation, which sponsored the research on improving photosynthesis efficiency, wants to make sure that any resulting technologies reach farmers in need at low cost.

New legislation requires food manufacturers to label GMOs.

No Simple Solution

As important as advances in biotechnology are likely to be to improving crop yields, feeding the planet in the next 50 years will require more than a technological fix. Consider that, right now, the world produces enough food to theoretically feed everyone on the planet. And yet, 1 in 8 individuals–about a billion people worldwide–go hungry each day. That has more to do with economics and politics than with biology–many people in the world are too poor to afford to buy much food.

Adding to the issue of cost, only about 55% of the world's crop calories go directly to feeding people. The rest goes to feed livestock (about 36%) or is turned into biofuels or other products (about 9%). When droughts hit or other weather conditions unexpectedly reduce harvests, these competing uses of grain can lead to price spikes, food shortages, and riots–such as happened around the world in 2008 in Haiti, Egypt, Bangladesh, and Mozambique.

The UN's call to double food production in the developing world by 2050 assumes that as lifestyles and wages improve in developing countries, more people will be eating meat. But many experts believe that richer countries must eat less meat if we are going to feed the world sustainably.

There is also the problem that climate change is likely to affect the distribution of farmland and water resources, making agriculture as a whole less predictable.

"It's basically a swirling mass of all sorts of issues," says Jez. "I don't think any one person actually has their mind wrapped around all of the issues completely."

That's why, he says, it's important to have people from different areas of policy and science and government and economics talking to one another. The next generation is going to have to weigh all these different issues to solve this problem.

But given their intimate connection to human livelihoods, plants are a good place to start. "Plants are cool," Jez says. "When you actually stop and look at what they're able to do, you realize they're worth paying attention to." ■

CHAPTER 24 SUMMARY

- Plants have a structure that is well suited to obtaining the materials they need to grow. Through their aboveground shoot system they absorb sunlight and take in carbon dioxide. Through their belowground root system they absorb water and nutrients like nitrogen and phosphorus.
- Plants use sunlight, carbon dioxide, and water in photosynthesis to make sugar. Nitrogen and phosphorus are important components of plant molecules such as proteins and the phospholipids that make up cell membranes.
- Although nitrogen gas makes up 80% of the atmosphere, plants cannot use nitrogen in this form. They require the assistance of nitrogen-fixing bacteria in the soil to convert gaseous nitrogen (N_2) into ammonia (NH_3).
- Nutrients like nitrogen and phosphorus cycle through the environment as they move through the bodies of living organisms and are returned to the soil through decomposition.
- Fertilizer helps plants grow by providing supplemental nutrients that may be limiting in soils.
- A danger of fertilizer use is eutrophication: the enrichment of nutrients in aquatic environments causes overgrowth of algae, leading ultimately to oxygen depletion and the creation of a dead zone where nothing can grow.
- Staple crops are those eaten in large quantities that provide most of the energy and nutrients in the human diet. Many staple crops are grasses with edible seeds. Seeds are a product of plant reproduction.
- Many crop plants are angiosperms, or flowering plants, which reproduce sexually. Some angiosperms have perfect flowers: male and female sex organs are present in the same flower. Others have imperfect flowers: male and female sex organs are present in separate flowers.
- Fertilization of female eggs by male sperm occurs when pollen (which contains sperm) travels from a male sex organ (the stamen) to a female sex organ (the pistil).
- Angiosperms perform double fertilization: one sperm fertilizes the egg to form the plant embryo, and the other joins with other cells in the ovule to produce the endosperm, a food source in the seed for the developing embryo.
- Seeds planted in fertile soil will germinate—producing a young seedling with at least one photosynthetic leaf, called a cotyledon. Plants grow through division of cells in tissues called meristems located in specific regions of the plant.
- Through genetic engineering techniques, scientists can modify plants to give them unique properties, such as resistance to pests or more efficient photosynthesis.
- A new genetic engineering technique called CRISPR allows scientists to easily make very precise changes to specific genes in cells.
- Genetically engineering better crops may be an important way to address the needs of a growing human population.

MORE TO EXPLORE

- **Jez, J. (2016) TEDx Talks: The once and future green factory: www.youtube.com/watch?v=7NCOKLqdEHE**
- **Jez, J. M., et al. (2016). The next green movement: Plant biology for the environment and sustainability. *Science* 353.6305:1241–1244.**
- **Stokstad, E. (2016) How turning off a plant's sunshield can grow bigger crops. *Science*. www.sciencemag.org/news/2016/11/how-turning-plants-sunshield-can-grow-bigger-crops (video)**
- **National Academy of Sciences (2016) *Genetically Engineered Crops: Experiences and Prospects:* www.nas-sites.org/ge-crops/**
- **HHMI BioInteractive (2015) Popped Secret: The Mysterious Origin of Corn (film): www.hhmi.org/biointeractive/popped-secret-mysterious-origin-corn.**

CHAPTER 24 Test Your Knowledge

DRIVING QUESTION 1 What are natural sources of nutrients that plants need, and how do agricultural practices supplement these sources?

By answering the questions below and studying Infographics 24.1, 24.2, and 24.3, you should be able to generate an answer for the broader Driving Question above.

KNOW IT

1 Which form of nitrogen is abundant in the atmosphere? Which form of nitrogen can be easily taken up and used by plants?

2 Match each nitrogen conversion process with the organism that can carry out that process.

Nitrogen Conversion Process	Organism
___ $N_2 \rightarrow NH_3$	a. humans
___ proteins → NH_3	b. nitrogen-fixing bacteria
___ N_2 → chemical fertilizer	c. decomposers

3 Which of the following is *not* a reservoir of phosphorus?

a. rocks
b. bodies of plants and animals
c. atmospheric gases
d. the soil
e. bodies of water

4 If phosphorus is important for bones and teeth, why do plants need phosphorus?

USE IT

5 Why is carbon not included in industrial fertilizers?

6 If you applied to the soil around your plants a chemical that kills bacteria (but not plants), why might your plants die?

7 You have five containers with specific soils and for each container you can control the composition of the surrounding air. The setup of each container is as follows:

A: sterile soil with no ammonia; limited N_2 in the air
B: sterile soil with no ammonia; abundant N_2 in the air
C: soil with no ammonia but with Rhizobium bacteria; limited N_2 in the air
D: soil with no ammonia but with Rhizobium bacteria; abundant N_2 in the air
E: sterile soil with ammonia added; abundant N_2 in the air

Which container do you predict will support the most robust growth of plants? Explain your answer.

DRIVING QUESTION 2 How do important crops and other angiosperms reproduce?

By answering the questions below and studying Infographics 24.4, 24.5, and 24.6, you should be able to generate an answer for the broader Driving Question above.

KNOW IT

8 Match the parts of a flower with the corresponding letter from the diagram below:

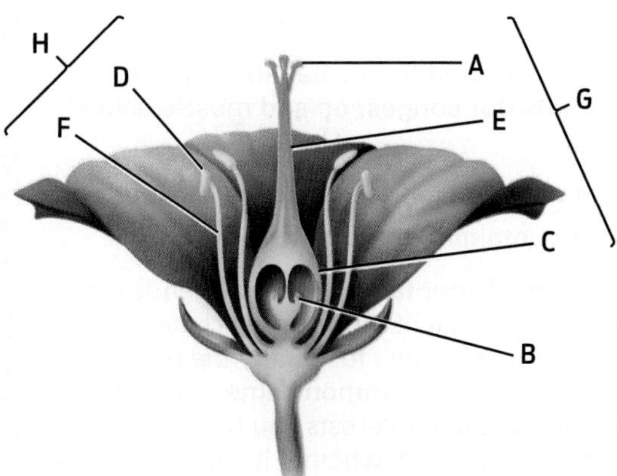

___ Anther
___ Filament
___ Ovary
___ Ovule
___ Pistil
___ Stamen
___ Style
___ Stigma

9 Mark each of the flower parts listed below as a male reproductive structure (M) or a female reproductive structure (F).

___ Stamen
___ Pistil
___ Ovary
___ Anther
___ Filament
___ Ovule
___ Stigma
___ Style

10 What is the role of endosperm in a seed?

11 Complete the sentence below using the terms *ovary, egg,* and *ovule*.

The ____ is the female gamete. In angiosperms, it is found in an ________, which in turn is found in the ________.

USE IT

12 A plant is producing seeds that won't germinate. Examination of the seeds reveals endosperm with an extra set of chromosomes (relative to normal endosperm) and no embryo. What process is likely to be failing in this case? Explain your answer.

apply YOUR KNOWLEDGE

BRING IT HOME

13 Many everyday products contain compounds derived from plants. A few of these are:

- Caffeinated sodas or energy drinks
- Camphor and menthol-containing vapor rubs and steams (for congestion and muscle aches)
- Nicotine
- Tea tree oil for minor skin irritations and infections
- Quinine-containing tonic water

You, members of your family, or your friends may have used some or all of these. For each item listed, do some internet research to identify the plant (find both the scientific and common name) and something else about it that interests you (for example, where the plant is found, whether it has historically been used for medicinal purposes, other sources of the compound).

DRIVING QUESTION 3 How is the use of GM plants contributing to more efficient agriculture?

By answering the questions below and studying Infographics 24.7, 24.8, and 24.9, you should be able to generate an answer for the broader Driving Question above.

KNOW IT

14 Compare and contrast the genetic engineering of a plant with the genetic engineering of a yeast (see Chapter 8).

15 Compare and contrast *Bt* and Roundup Ready GM plants. Discuss purpose, benefits and potential risks or unintended consequences.

16 Which of the following is consistent with organic farming practices?

a. using a synthetic chemical fertilizer
b. using *Bt* toxin proteins as an insecticide
c. using genetically modified *Bt* plants
d. using a synthetic pesticide

USE IT

17 Growing plants in the shade reduces NPQ. Why is shading plants not an effective strategy for increasing plant biomass production?

apply YOUR KNOWLEDGE

MINI CASE

18 There are many strategies that can help increase food availability for a growing global population. If you were the head of a major funding agency, which of the following strategies would you prioritize for funding? Justify your decision.

a. reducing NPQ
b. designing corn and wheat to form root nodules
c. public campaigns to promote reduced amounts of meat consumption

apply YOUR KNOWLEDGE

INTERPRETING DATA

19 The figure below shows data about quenching (NPQ) and plant growth for GM plants that have been engineered to speed up the rate at which plants recover from quenching. Panel A shows the amount of quenching in wild-type and GM plants under conditions of steady light and variable (alternating bright and dim) light. Panel B shows the increase as a percent of the weight, leaf area and height of GM plants relative to wild type for plants grown out in the field.

a. Does the genetic modification appear to be important for NPQ when plants are grown in steady light conditions?

b. Would you predict the GM plants to grow significantly larger than wild-type plants if the plants were grown in a greenhouse under conditions of constant illumination? Explain your answer.

c. If you were a radish or carrot farmer, what data would you like to see to persuade you to grow a GM plant to increase your crop yield?

Quenching (NPQ) and Plant Growth in Wild-Type and Genetically Modified Tobacco

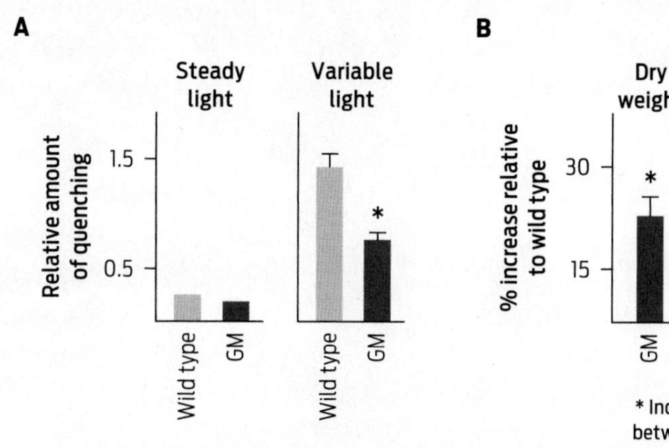

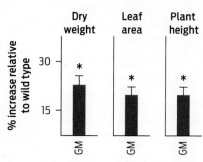

25

Plant Physiology

A mosquito caught on a carnivorous sundew plant.

Julie DeRoche/Getty Images

DRIVING QUESTIONS

1. What are key structural features of plants?
2. How do plants respond to limiting nutrients and harsh conditions?
3. What strategies do plants use to reproduce?
4. How do plants respond to stimuli and protect themselves?

Plants

Q&A

Exploding seeds, carnivorous flowers, and other colorful adaptations of the plant world

In Chapter 24, we learned about basic plant structure, growth, and reproduction. In this chapter, we consider these topics in more detail, focusing on how plant tissues transport water and nutrients; how plants obtain nutrients for photosynthesis and growth in limiting or harsh conditions; and the various adaptations that assist plant reproduction. We also consider plant hormones and plant defenses.

PLANT STRUCTURE

Q: Plants lack bones, so how do they stand up?

A: From tiny 3-inch-tall crocus flowers to massive redwood trees standing 350 feet, plants share the same basic design: a belowground root system for absorbing water and nutrients and an aboveground shoot system made up of stems and leaves. This body arrangement works well for plants, allowing them to simultaneously reach for the sky and anchor themselves safely in the soil–all without the help of bones, tendons, or muscles. In the absence of such a musculoskeletal system, what keeps a plant from flopping over?

Like animals and other eukaryotes, plants are made of cells packed with organelles, including a nucleus, endoplasmic reticulum, and mitochondria. Plant cells contain a few plant-specific parts as well, including a supportive cell wall made of a complex carbohydrate called cellulose; chloroplasts, the sites of photosynthesis; and a **central vacuole**, essentially a large water balloon occupying the center of the cell. When filled with water, vacuoles create **turgor pressure** against the cell wall, keeping a plant body rigid and upright. When the vacuoles are less than full, turgor pressure is reduced, and the plant wilts.

Water flows into and out of the vacuole by osmosis, the movement of water across a cell membrane from an area of lower solute concentration to an area of higher concentration (see Chapter 3). Vacuoles have a relatively high solute concentration and tend to draw water in. When a plant dries out, the relative concentration of solutes outside the cell increases, and water moves out of the cell, decreasing turgor pressure against the cell wall.

CENTRAL VACUOLE
A fluid-filled compartment in a plant cell that contributes to cell rigidity by exerting turgor pressure against the cell wall.

TURGOR PRESSURE
The pressure exerted by the water-filled central vacuole against the plant cell wall, giving a stem its rigidity.

LIGNIN
A stiff strengthening agent found in the inner cell wall of plants.

Cell walls contribute to a plant's stiffness in another way. Plant cells are packed tightly together, much like bricks in a wall; carbohydrate molecules associated with the cell walls act like glue, helping adjacent cells stick together. With many cells held together in this way, plant tissues are exceptionally strong, which is why they make such durable ropes and fabrics.

Some plant tissues, such as those that make up the stem, are made of cells with two cell walls: an outer cell wall containing cellulose, and a second, inner cell wall containing cellulose and **lignin.** Lignin is a hard, durable material, which lends added support and strength to plant tissues; it is what makes them "woody." A plant with a thick, woody stem (or trunk, in the case of a tree) can grow tall and still not topple over. The tallest plants on Earth are the California redwoods (*Sequoia sempervirens*), which can reach heights of more than 350 feet (**INFOGRAPHIC 25.1**).

INFOGRAPHIC 25.1 How Plants Stand Up

Plants have water-filled vacuoles that prevent wilting. The pressure in these vacuoles keeps cells rigid and the plant standing upright. Woody stems have a second layer of lignin, in addition to the fibrous cellulose in their cell walls. Lignin provides additional strength to stems.

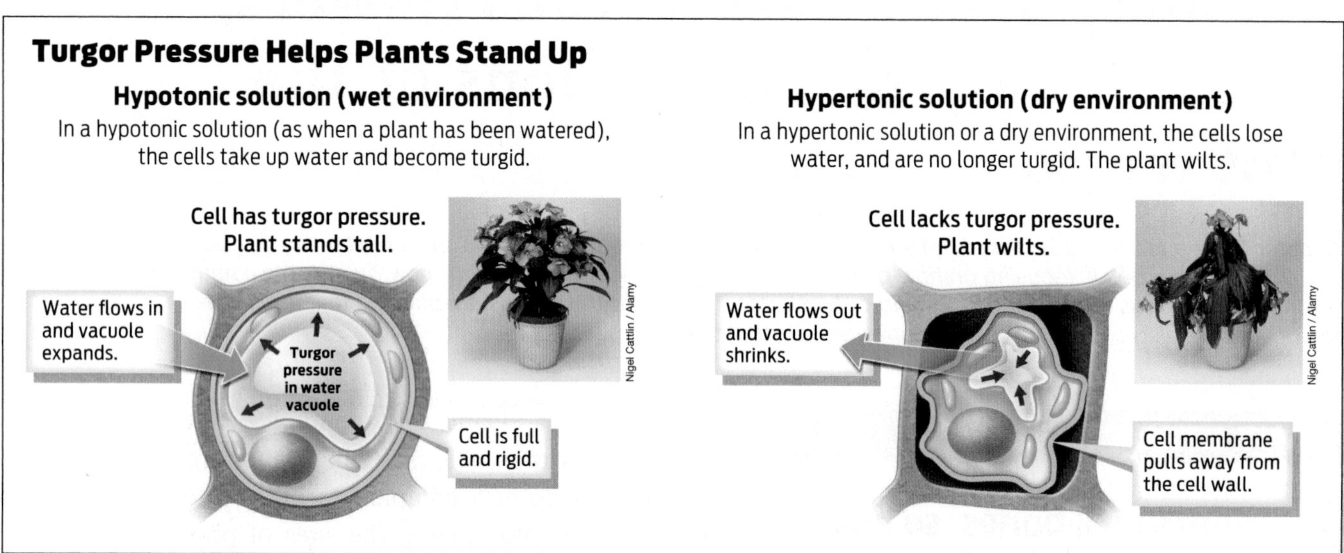

Turgor Pressure Helps Plants Stand Up
Hypotonic solution (wet environment)
In a hypotonic solution (as when a plant has been watered), the cells take up water and become turgid.
Cell has turgor pressure. Plant stands tall.
Water flows in and vacuole expands.
Turgor pressure in water vacuole
Cell is full and rigid.
Nigel Cattlin / Alamy
Hypertonic solution (dry environment)
In a hypertonic solution or a dry environment, the cells lose water, and are no longer turgid. The plant wilts.
Cell lacks turgor pressure. Plant wilts.
Water flows out and vacuole shrinks.
Cell membrane pulls away from the cell wall.
Nigel Cattlin / Alamy

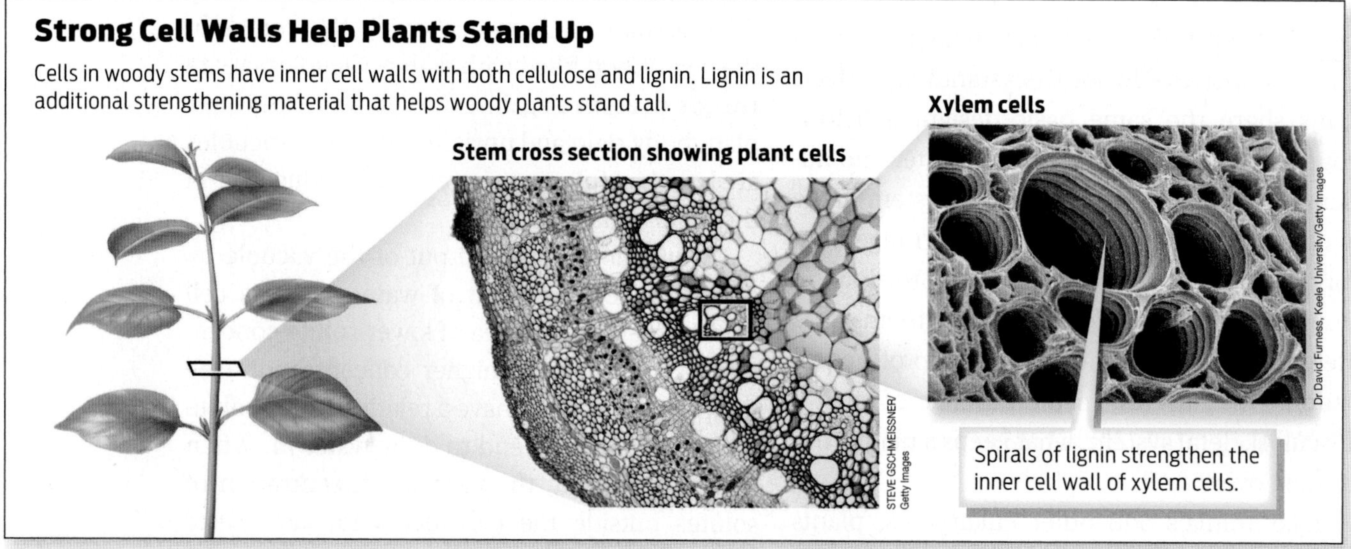

Strong Cell Walls Help Plants Stand Up
Cells in woody stems have inner cell walls with both cellulose and lignin. Lignin is an additional strengthening material that helps woody plants stand tall.
Stem cross section showing plant cells
STEVE GSCHMEISSNER/ Getty Images
Xylem cells
Dr David Furness, Keele University/Getty Images
Spirals of lignin strengthen the inner cell wall of xylem cells.

? Why does watering plants with an excessive amount of fertilizer (which contains a high concentration of solutes) cause the plant to wilt?

Q: What travels through a plant's veins?

A: Like animals, plants have a **vascular system** for transporting fluids. Instead of blood, however, a plant's vascular system transports water and nutrients throughout the plant body. Plants need water to live and grow (you may have learned this the hard way if you've ever killed a plant with neglect). Among other things, water is crucial for photosynthesis in leaves. But you don't water a plant's leaves; you water the roots by pouring water in the soil. How does water get from the roots to the rest of the plant?

A plant's water-carrying tissue is called **xylem** (pronounced ZYE-lum). Xylem tissue is made of cells arranged into long, stiff tubes; the cells have holes in each end and are stacked one on top of the next. Water moves up from the roots through these tubes to the aboveground stems, and eventually into the leaves, where it is used during photosynthesis to make sugar. Plant cells use the newly synthesized sugar as food, so the sugar must be transported out of the leaves and back down through the plant. Another series of tubes, called **phloem** (pronounced FLO-um; think "f" for "food"), transports sugar through the plant. Phloem supports two-way transport: sugar moves down to the roots, where some of it is stored; later, sugar moves up to the shoot system, where it provides nutrition and energy for growing fruit, buds, and leaves. The veins you see in a plant's leaf are bundles of both xylem and phloem tissue.

Moving water and other nutrients up to the top of a 300-foot tree against the force of gravity is no easy task—hundreds of pounds of water must be lifted through what is essentially a long water pipe. And unlike animals, plants have no heart to pump fluid along. Instead, plants rely on evaporation of water from the leaves to draw water up the plant, a process called **transpiration.** Because water is a polar molecule that can form hydrogen bonds with other water molecules, it has great cohesive strength (see Chapter 2). As water evaporates from leaves into the air, water in the xylem is pulled up to replace it. The cohesive strength of water is enough to counteract the force of gravity and pull water up to astonishing heights.

For a plant, transpiration is life sustaining: it is the force that carries water through the plant. But losing too much water can be dangerous. On a hot, dry, or windy day, a large tree can lose hundreds of gallons of water vapor from its leaves. To control the amount of water lost by transpiration, a plant's leaves are coated with a waxy layer called the **cuticle** that functions somewhat like a rubber suit, sealing in moisture. At regular intervals, the cuticle is punctuated by pores, called **stomata,** which open and close. When the stomata are open, water vapor leaves freely and other gases enter and exit–specifically, carbon dioxide enters and oxygen exits. When the stomata are closed, water and gases are sealed in. Many plants keep their stomata open during the day, letting in carbon dioxide for photosynthesis.

VASCULAR SYSTEM
A system of tube-shaped vessels and tissues that transports water and nutrients throughout an organism's body.

XYLEM
Plant vascular tissue that transports water from the roots to the shoots.

PHLOEM
Plant vascular tissue that transports sugars throughout the plant.

TRANSPIRATION
The loss of water from plants by evaporation, which powers the transport of water and nutrients through a plant's vascular system.

CUTICLE
The waxy coating on leaves and stems that prevents water loss.

STOMATA (singular: STOMA)
Pores on leaves that permit the exchange of oxygen and carbon dioxide with the air and allow water loss.

Giant sequoias are the world's tallest trees. They can grow to more than 350 feet in height and more than 25 feet in diameter.

At night, they close the stomata, conserving water (**INFOGRAPHIC 25.2**).

The presence of specialized tissues for transporting water is what distinguishes plants from their water-dwelling ancestors, the algae. The evolution of this vascular tissue is what allowed plants to colonize nearly every part of the land, from valley to mountaintop. A few primitive land plants, such as mosses and liverworts, also lack true roots and shoots containing vascular tissue (see Chapter 18). Without specialized tubes for transporting water, these nonvascular plants are limited to environments that are saturated with water, where they grow close to the ground in squat, spongy mats.

Some vascular plants have made the evolutionary journey in the other direction,

INFOGRAPHIC 25.2 A Plant's Vascular System Transports Water and Sugar

Plants have a vascular system that transports specific substances throughout the plant body. Transport of water relies on a type of evaporation from leaves called transpiration.

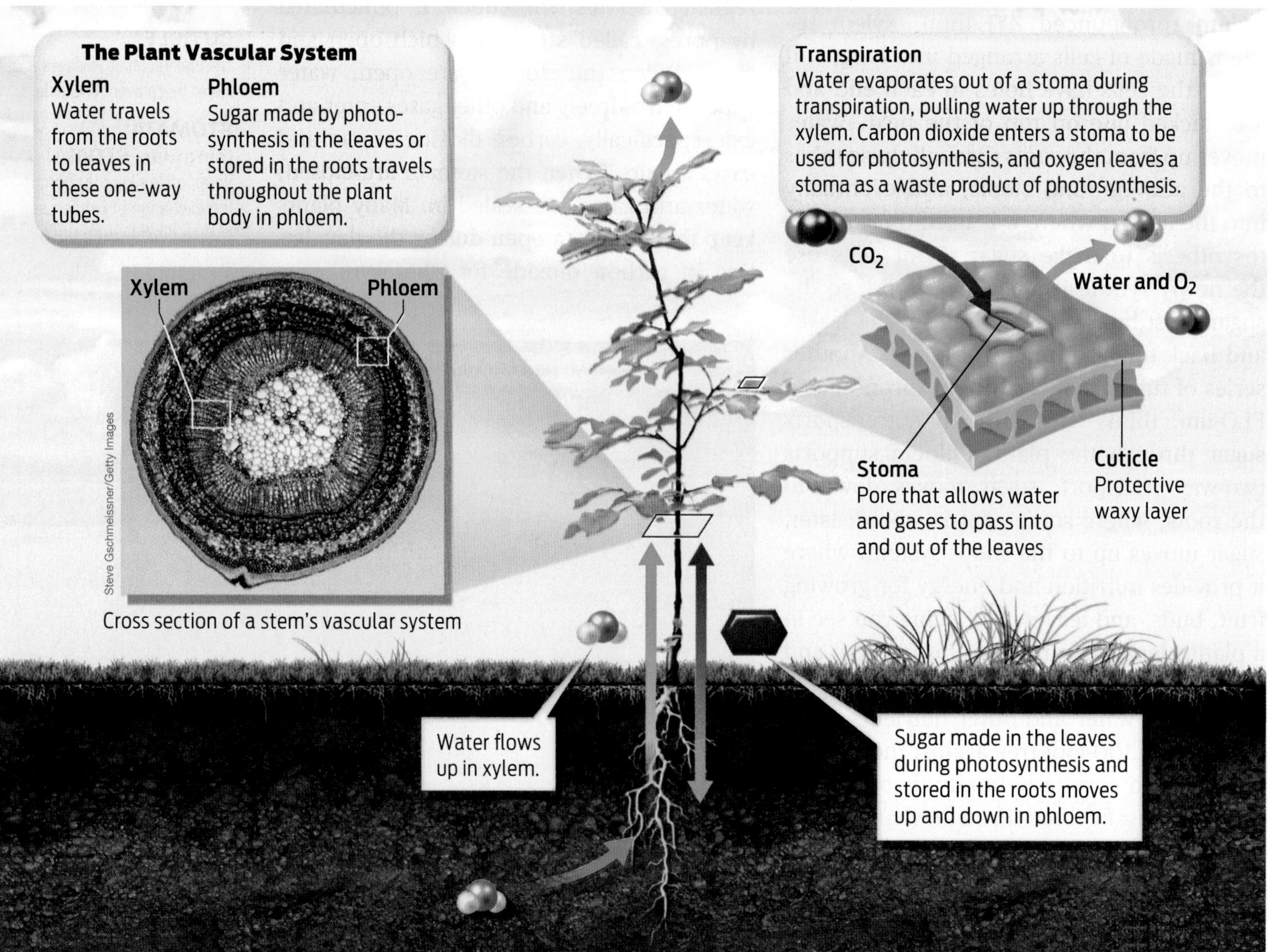

The Plant Vascular System
Xylem
Water travels from the roots to leaves in these one-way tubes.
Phloem
Sugar made by photosynthesis in the leaves or stored in the roots travels throughout the plant body in phloem.
Xylem
Phloem
Steve Gschmeissner/Getty Images
Cross section of a stem's vascular system
Transpiration
Water evaporates out of a stoma during transpiration, pulling water up through the xylem. Carbon dioxide enters a stoma to be used for photosynthesis, and oxygen leaves a stoma as a waste product of photosynthesis.
CO_2
Water and O_2
Stoma
Pore that allows water and gases to pass into and out of the leaves
Cuticle
Protective waxy layer
Water flows up in xylem.
Sugar made in the leaves during photosynthesis and stored in the roots moves up and down in phloem.

? Compare and contrast xylem and phloem.

to life in the water. These aquatic plants include water lilies, cattails, and the *Elodea* popular in fish tanks. All have unique adaptations that permit them to thrive in wetlands and other environments where they are partially or completely submerged. For example, many aquatic plants lack the external protective cuticle layer that prevents water loss; carbon dioxide and oxygen can therefore pass in and out of any cell on the surface of the aquatic plant body. In addition, aquatic plants have no trouble obtaining water directly from their surroundings. And so the xylem tissue, which normally transports water from the roots to stems, is often greatly reduced or absent in aquatic plants. Since water provides natural buoyancy, aquatic plants invest little energy making lignin-based supportive tissues to help them resist gravity.

Q: What are tree rings?

A: We discussed in Chapter 24 how plants grow through cell division in meristems, which are present in specific regions of the plant body. Growth from the apical meristem in the tip of the shoot results in a plant's getting taller. This is primary growth. Many plants also grow wider as well, through cells dividing in meristems located in a ring around the stem or trunk. This is secondary growth. In trees, secondary growth produces a thickened, woody trunk. **Wood** is hardened xylem tissue (the word "xylem" derives from the Greek *xylon,* meaning "wood"). Wood allows trees to grow tall and therefore provides an evolutionary advantage over nonwoody plants in the competition for sunlight.

In temperate regions–as in most of the United States–xylem growth is dormant, or inactive, in the winter. In the spring, xylem growth starts again. The first xylem cells to grow in the spring are usually larger in diameter and thinner-walled than xylem cells produced later in the summer. The boundary between the smaller xylem cells from one summer and the larger xylem cells from the next spring appears as a ring in the cross section of a tree trunk. You can determine the age of a tree by counting the number of rings. Tree rings vary in width from year to year because differences in temperature and rainfall affect the amount of xylem tissue that is produced in any one season.

Not all plants display secondary growth. In fact, the presence of secondary growth is one feature distinguishing two large classes of flowering plants: monocots and dicots. **Monocots** are defined as plants that contain one cotyledon, or seed leaf, in their seeds, whereas **dicots** have two seed leaves. But anywhere you look in a plant body there are differences between these two groups of plants. The leaves of monocots typically have parallel veins, whereas the leaves of dicots have branching veins. In the stems of monocots, the vascular bundles of xylem and phloem appear to be scattered, whereas the vascular bundles of dicots are arranged in a ring. Even the flowers of monocots and dicots differ: monocots have flower parts like petals and stamens that are multiples of threes, and dicots have floral parts in multiples of four or five.

It is the presence of vascular bundles arranged in a ring along the stem's outer edge that permits the secondary growth (and wood generation) in dicots; this is where new xylem tissue grows every season. For that reason, among the flowering plants, most trees are dicots. Although palm trees and some other monocots have hardened stem tissues, this material is not technically wood since it is not made of xylem. Nor do palm trees experience secondary growth resulting from the addition of new xylem every

WOOD
Hard, secondary xylem tissue found in the stem of a plant.

MONOCOT
Flowering plants with one cotyledon (seed leaf) in their seeds.

DICOT
Flowering plants with two cotyledons (seed leaves) in their seeds.

Gyro Photography/amanaimagesRF/Getty Images

Each dark ring represents the end of one year of growth. This tree was 37 years old.

INFOGRAPHIC 25.3 Differences between Monocots and Dicots

Monocots and dicots are two major groups of angiosperms. They are distinguished on the basis of the number of cotyledons (embryonic leaves), the arrangement of vascular tissue, patterns of veins in leaves, and the arrangement of floral parts such as petals.

	Seed	Stem	Leaf	Flower
Monocots	One cotyledon	Vascular bundles scattered in stem	Leaf veins form a parallel pattern	Flower parts in threes
Dicots	Two cotyledons	Vascular bundles in a distinct ring	Leaf veins form a net pattern	Flower parts in fours and fives

? Consider the famous cherry blossoms in Washington, DC. Find some detailed photographs of the blossoms, the leaves, and the trees. From your observations, are the cherry trees monocots or dicots?

year—which is why palm trees may be quite tall, but are also usually very thin. It's also why, if you cut a palm tree in half, you won't find any tree rings (**INFOGRAPHIC 25.3**).

PLANT NUTRITION

Q: Why are some plants carnivorous?

A: Plants are autotrophs: they make their own food through photosynthesis (see Chapter 5). Give a plant some sunshine, carbon dioxide, and water—plus a few soil nutrients—and you will have a happy plant, capable of feeding and nourishing itself. So then why do some plants prey on animals, earning them the title "carnivorous"?

Plants rely on soil bacteria to fix nitrogen into a form that is usable by the plant (see Chapter 24). Fertile soil naturally contains these nitrogen-fixing bacteria and therefore provides adequate supplies of usable nitrogen that plants can absorb and use to grow. And soil can be supplemented with fertilizer, giving plants a boost of artificial nitrogen and other nutrients. But in certain natural environments, such as bogs or rock outcroppings, it's hard for plants to obtain the nitrogen they need. The acidity of a bog, for example, prevents organic matter from breaking down, so nutrients are recycled more slowly. In these environments, plants have evolved novel ways to obtain scarce nitrogen—some of which would put animal carnivores to shame.

Trumpet pitchers (*Sarracenia*), for example, lure insects with brightly colored flowers and nectar "bribes." But the rim of the plant's trumpet-shaped flower is slippery. Unsuspecting trespassers climb onto the rim, lose their grip, and tumble into a deep cavity filled with digestive juices. Prevented from escape by downward-pointing spikes, the tiny prisoners drown and are slowly dissolved. The resulting insect soup—a rich source of nitrogen—is absorbed by the plant.

The Venus flytrap (*Dionaea muscipula*) takes an even more dramatic approach. The plant's "flower" is actually a spring-loaded trap that snaps shut around unsuspecting prey. Tiny hairs inside the flower act as sensors; when the sensors are tripped by a moving insect, the trap slams shut and the feasting begins.

Even when plants obtain nitrogen in this carnivorous way, they must still perform photosynthesis to make sugar. The plant body is composed of complex carbohydrates, such as cellulose, which the plant makes by stringing sugar molecules together. And the starting material for photosynthesis is carbon dioxide gas. Thus the air, rather than the soil, is where a plant obtains the material to put on weight (**INFOGRAPHIC 25.4**).

INFOGRAPHIC 25.4 Carnivorous Plants Have Unique Adaptations for Acquiring Nutrients

Carnivorous plants perform photosynthesis to make carbohydrates, but they acquire scarce nutrients like nitrogen by digesting insects and other small animals, instead of pulling them from the soil.

The soil in acidic bog habitats recycles organic matter slowly, making the soils low in nutrients like nitrogen. Instead of pulling them from the soil, carnivorous plants acquire scarce nutrients by digesting insects and other small animals.

The Venus flytrap (*Dionaea muscipula*) attracts insects into its spring-loaded trap. Projections interlock at the opening, preventing the insect's escape.

? Are insects primarily a source of carbon or of nitrogen for carnivorous plants?

Q: Can plants photosynthesize at night?

A: By definition, there is a crucial part of photosynthesis that can occur only during daylight hours—the light-absorbing "photo" part. So technically, the answer is, no—plants can't photosynthesize at night. However, some plants are still hard at work at night, especially those living in hot, dry climates.

Because daylight hours tend to be the time when it's hottest and driest, plants lose a lot of water during the day. To conserve water, plants can close their stomata. But closing stomata also prevents carbon dioxide—the raw material of photosynthesis—from entering the leaf. In many plants, such as wheat and rice, the result of this trade-off is reduced output of plant food—that is, sugar.

Some plants have adapted to sun-scorched surroundings. For example, corn and sugar cane can thrive in hot, sunny climates by keeping their stomata closed as much as possible. They have scavenging molecules that can capture CO_2 even when it is present only at low concentrations in air pockets in the leaf when the stomata are closed.

Still other plants have adapted to hot, dry conditions by splitting up two parts of photosynthesis that usually occur together: taking in CO_2 and making sugar. Pineapples as well as cacti and other succulent plants conserve water by keeping their stomata closed during the day when it's hot. But at night, when it's cooler, they open their stomata, allowing carbon dioxide in. The CO_2 that is captured at night isn't used right away but instead is stored. During the day, while their stomata are closed, these plants use the stored CO_2 to complete photosynthesis and make sugar. By separating in time the steps of photosynthesis, these well-adapted plants can thrive in conditions that would wither their less physiologically adapted cousins (**INFOGRAPHIC 25.5**).

Like other eukaryotic organisms, plants use sugar to perform aerobic respiration to make ATP (see Chapter 6). All plants respire both during the day and at night. Because the total amount of carbon dioxide given off by plants during cellular respiration is less than the total amount taken in for photosynthesis, plants are carbon dioxide sinks that ultimately lower the amount of carbon dioxide in the atmosphere.

The tallest plants on Earth are the California redwoods (*Sequoia sempervirens*), which can reach heights of more than 350 feet.

Q: Why do leaves change color in the fall?

A: For most of the year, leaves are photosynthesis factories, churning out sugar. A key component of the photosynthetic machinery is the pigment **chlorophyll,** which absorbs red and blue wavelengths of light and reflects green wavelengths. Chlorophyll is the reason plants appear green. To keep photosynthesis running, plants make abundant chlorophyll in spring and summer. But they also produce a smaller amount of other pigments—yellow-reflecting xanthophyll and orange-reflecting carotene. In the leaves, these pigments capture additional wavelengths of light and therefore expand the range of light that is useful for photosynthesis. You can't see these other pigments in leaves during spring and summer because leaves are chock full of green chlorophyll, camouflaging the other hues (although

CHLOROPHYLL
The dominant pigment in photosynthesis, which makes plants appear green.

INFOGRAPHIC 25.5 Beating the Heat: Photosynthesizing without Drying Out

All plants must perform the "photo" reactions of photosynthesis during the day. But many plants experience reduced levels of photosynthesis in hot and dry conditions. Other plants are adapted to live in hot and dry climates, and have different strategies to take in CO_2 while minimizing water loss.

Stomata open all day: too much water loss

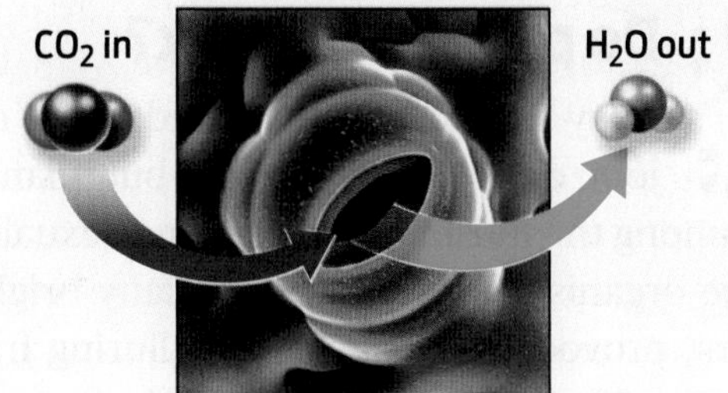

Dr Jeremy Burgess/ Science Source

When stomata are open, carbon dioxide can enter the leaves for photosynthesis. However, this increases transpiration and therefore water loss.

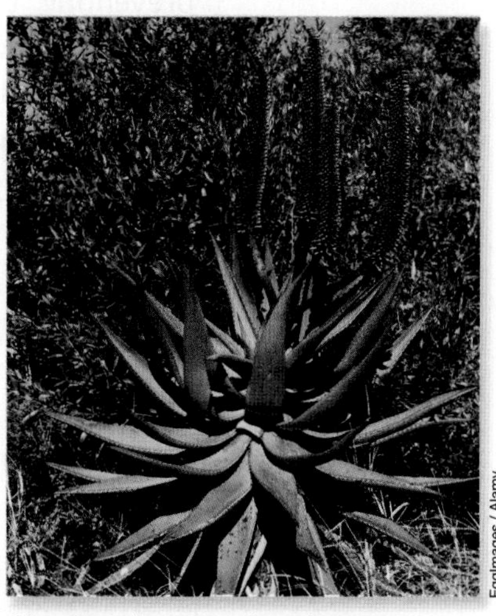

Egglmages / Alamy

Aloe ferox has succulent leaves.

Stomata closed all day: too little sugar produced

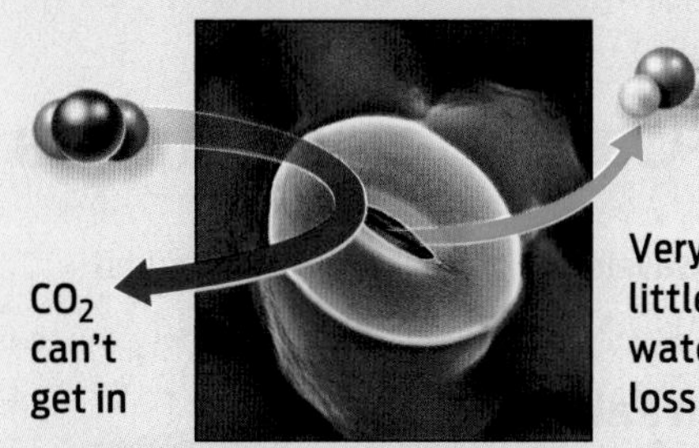

Dr Jeremy Burgess/ Science Source

In dry climates, some plants keep stomata closed during the hot hours of the day. This conserves water, but inhibits the uptake of carbon dioxide for photosynthesis. These plants make less sugar food.

Strategies to combat water loss in dry climates:

- Thick, waxy succulent leaves
- Extract CO_2 from airspaces even when stomata are mostly closed
- Open stomata at night to capture CO_2 in cooler temperatures

? Why do many plants make less sugar on very hot, sunny, dry days?

you can see them elsewhere in some plants: in the flesh of a pineapple, for example, or the taproot that is a carrot).

After a summer of intense sugar stockpiling, trees, bushes, and other deciduous plants that seasonally drop their leaves start to settle in for the winter and begin to shut down their photosynthesis machinery. During the winter months in temperate regions, there isn't enough water to drive photosynthesis; water in the ground is frozen and can't be absorbed. As temperatures cool, plants turn off the production of their light-absorbing pigments. Of all the pigments, chlorophyll is the most unstable chemically and therefore the shortest lived: levels fall quickly once production stops. By contrast, xanthophyll and carotene linger. As green fades, the other colors peek through. These mostly yellow and orange colors have been there all along, but hidden.

The really intense colors—the fiery reds and deep purples—that some trees and bushes turn in autumn are the result of a fourth pigment, called anthocyanin. This distinctive chemical, which is the same one that gives apples and acai berries their color, is produced in leaves in response to high sugar concentration. Why does sugar concentration rise in the fall? To conserve water in winter, a tree develops a corklike membrane between the leaf stem and the branch, sealing off the leaf and preventing transpiration. This corky membrane also prevents phloem from transporting sugar out of the leaf. Unable to move out of the leaf, sugar begins to collect. More sugar means more anthocyanin is produced, yielding the bright colors we associate with fall.

Trees such as maple and oak produce brilliant colors because they naturally produce lots of anthocyanin. Others, such as aspen and poplar, produce less anthocyanin and turn yellow or orange (**INFOGRAPHIC 25.6**).

Eventually the corky membrane that seals off the leaf from the tree causes the leaf to dry out. With a slight gust of wind, the leaf flutters to the ground. The tree is now prepared for winter.

PLANT REPRODUCTION

Q: Do plants have sex?

A: They may not go on blind dates or join online dating sites, but plants are among the most "flirtatious" and sexually active organisms on earth. The many bright colors, provocative shapes, and alluring fragrances of flowers are the plant equivalents of lipstick, muscle shirts, and perfume–tempting

INFOGRAPHIC 25.6 Plants Produce Multiple Light-Capturing Pigments

Leaves contain chlorophyll and other pigments. These pigments help capture a wide range of wavelengths of light to maximize the efficiency of photosynthesis.

More Chlorophyll in Summer

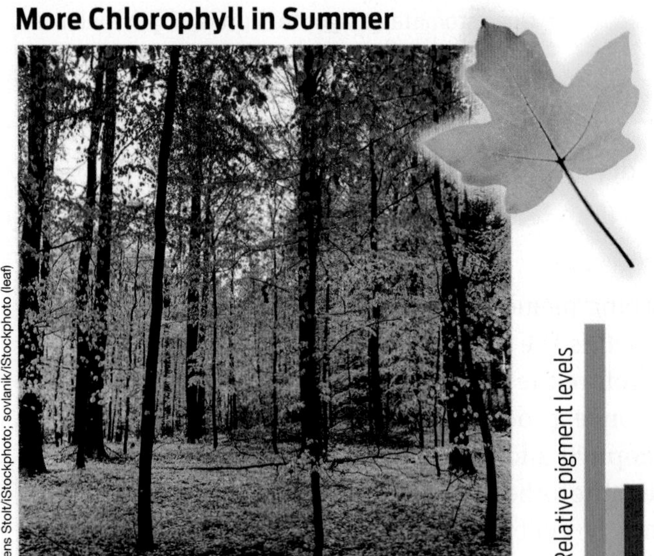

Jens Stolt/iStockphoto; sovlanik/iStockphoto (leaf)

Green leaves produce an abundance of the green pigment chlorophyll during the warm, sunny months, and smaller amounts of red and yellow pigments.

Less Chlorophyll in Fall

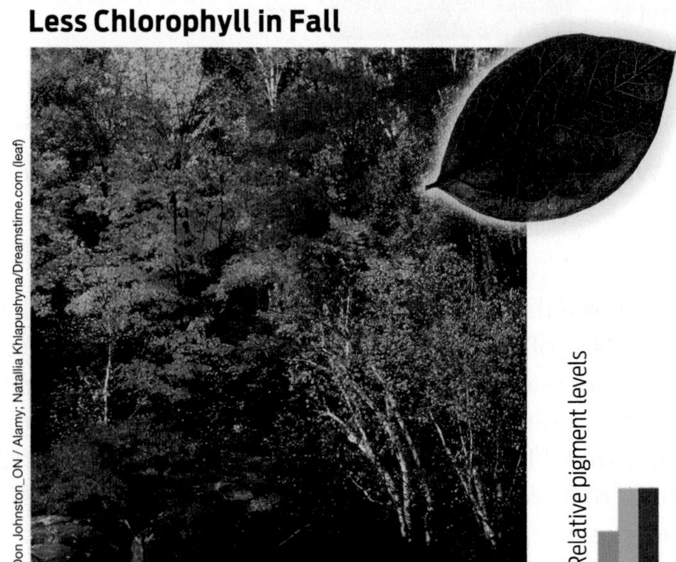

Don Johnston_ON / Alamy; Natallia Khlapushyna/Dreamstime.com (leaf)

When the hours of daylight are shorter and the temperature is cooler, the trees perform less photosynthesis. Chlorophyll is no longer made and levels quickly decrease. The red and yellow pigments produced by the leaves are revealed.

? What is the function of the red and yellow pigments that become visible during the fall?

novelties designed to attract mates (or pollinating partners).

Like other sexually reproducing creatures, plants have distinct male and female reproductive organs that produce sperm and egg (see Chapter 24). In flowering plants, male and female sexual organs are housed within a plant's flowers. The male organ, the stamen, produces pollen that contains sperm. The female structure, the pistil, produces eggs. Transfer of pollen from the stamen to the pistil results in pollination, which may then lead to fertilization–the joining of sperm and egg.

In some flowering plant species, male and female organs are found on separate male and female plants. Ginkgo trees and holly bushes are examples of such unisexual plants. More commonly, flowering plants–everything from dandelions to lilies–are hermaphrodites: their flowers contain both male and female reproductive parts. A few such hermaphroditic plants can self-pollinate–transferring pollen from stamen to pistil within the same flower or plant. But most hermaphroditic plants have evolved ways to prevent self-pollination and to encourage matings between different individuals. They may have pistils and stamens of different heights, for example, or their pollen may be chemically rejected by eggs of the same plant.

Plants have evolved many elaborate ways to spread their pollen from plant to plant. For many flowering plants, pollinators such as bees and hummingbirds transfer pollen grains from the anther of one flower to the stigma of another. To these animal pollinators, a flower's scent is a potent draw, irresistibly attracting them to the blossom as they search for nectar to eat. Some flowers practice sexual deception, mimicking the shape or smell of a female insect in order to lure a male to copulate. Other plants rely on wind to transfer pollen.

Not all plants are so sexually adventurous. Some are asexual, or have asexual phases, and can produce new individuals by extending runners or producing bulbs that develop into whole new plants. Asexual reproduction, in the form of taking cuttings and grafting, is especially important in agriculture. A cutting taken from one plant can be planted in soil to form a new plant; sugar cane and pineapples are often reproduced this way. A cutting taken from a plant can also be surgically grafted to the root system of a different plant–a procedure commonly used to perpetuate vineyard grapes (**INFOGRAPHIC 25.7**).

Q: What spins like a helicopter, shoots like a rocket, and contains its own parachute?

A: The answer to this biological riddle: a seed. Unlike animals, plants cannot move to seek out more hospitable living conditions for their offspring when resources are scarce or the neighborhood gets too crowded. Forest or desert, valley or mountaintop, plants are stuck where nature put them. Their solution to this enforced sedentary existence is to disperse their offspring far and wide through seeds. A seed is a small embryonic plant contained within a sac of stored nutrients–essentially, a plant starter kit (see Chapter 24). It develops from a fertilized egg and is a perfect package for delivering an immature plant to its new home and protecting it from harsh conditions.

> On a hot, dry, or windy day, a large tree can lose hundreds of gallons of water vapor from its leaves.

Seeds come in many shapes and sizes, and they are dispersed in many ways. Some are small and hitch a ride on fur or clothing by means of tiny hooks or burrs. Some, like coconuts, are large and buoyant and can float across oceans to reach distant beaches. The delicate parachutes of dandelions, the spinning helicopters of maples and pine

INFOGRAPHIC 25.7 Plants Reproduce Sexually and Asexually

In sexual reproduction, pollen must make its way to female sexual organs. Flowering plants use their flowers to attract pollinators. Other plants rely on wind to transfer pollen. Plants can also reproduce asexually. Some plants extend runners that can generate new plants.

Sexual Attraction
Flowers of the fly orchid (*Ophrys insectifera*) imitate the shape, color, and smell of a female insect. Male bees and wasps transfer pollen from plant to plant as they attempt to mate with the flowers.

Wind Pollination
Windborne pollen is released in clouds from male pinecones to neighboring female plant structures. The plant must release huge amounts of pollen into the wind, to increase the chance of at least one pollen grain finding its target. These clouds of pollen plague people with allergies during pollen and allergy season.

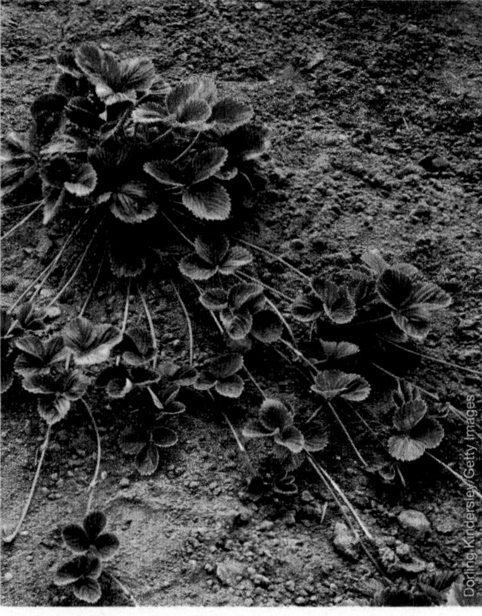

Asexual Runners
Plants like this wild strawberry can extend long runners away from the parent plant. These runners can generate new plant clones, reproducing without the need for male and female gametes.

? A plant has flowers with a distinctive fragrance, and always has flies buzzing around it. Which reproductive strategy is it using?

trees, and other lightweight seeds sail on the wind. Cottonseeds are little more than hairy specks of dirt, but on a steady breeze they can windsurf for miles. Other seeds are packaged inside fruit. Tempted by the fruit's bright color and sugary content, animals eat the fruit and then deposit the seeds in feces some distance away from the original plant.

Some seeds are dispersed through a ballistic mechanism. The seedpod of the squirting cucumber (*Ecballium elaterium*), for example, fills with slimy juice as it ripens. Eventually, the mounting pressure of the increased volume of juice causes the cucumber to shoot off the plant like a rocket, trailing a plume of seeds and slime in its wake. At the slightest touch, the bulging seedpods of the aptly named touch-me-not plant explode, spraying seeds like bullets. When a seed lands in favorable conditions, with enough water, it will germinate and grow into a young plant, or seedling (**INFOGRAPHIC 25.8**).

PLANT HORMONES

Q: Can plants see?

A: Observe an old building and you'll likely see ivy scaling up its walls. Keep houseplants next to a window and you'll find them bending toward the sunlight. How does a plant know where it's going? While plants don't have eyes and therefore cannot see, they are quite adept at sensing and responding to their environment, which they do through various kinds of tropism (from the Greek *tropos*, meaning "turn"). A tropism is a turning in a

INFOGRAPHIC 25.8 Seeds Carry a Young Plant to a New Destination

Seeds are important for the dispersal of plant populations. As seeds are dispersed, offspring reach new locations, often far from their parents.

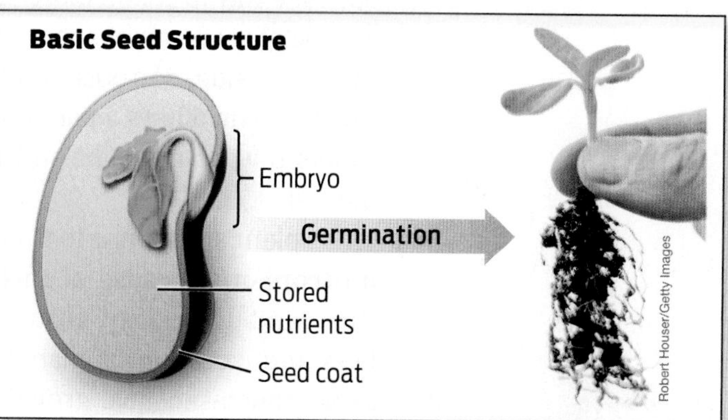

Seeds Disperse in a Variety of Ways

Burrs hitch a ride on animal fur.

Small dandelion seeds fly for miles in the wind.

Coconuts float, relocating seeds from one shore to another.

Animals eat fruits, dispersing the seeds in their feces.

The squirting cucumber shoots its seeds out with water.

? Do burrs need to have a tasty interior to facilitate their dispersal? Why or why not?

specific direction in response to an external stimulus. This type of response is triggered by plant hormones, which are signaling molecules that regulate growth and development.

The growth of a plant shoot toward light is called **phototropism.** Through phototropism, leaves get the sunlight they need for photosynthesis, and it's why ivy climbs up walls and houseplants bend toward the window. Shoots grow toward the light because of **auxin,** a plant hormone that promotes cell elongation as one of its effects. Auxin is produced in the tip and travels continuously down the stem. When light hits one side of a plant shoot, auxin moves to the shaded side, creating a gradient of the hormone in the stem. The side receiving the most direct sunlight contains the least auxin, while the shaded side contains the most. Auxin promotes elongation of cells on the shady side of the stem. The shaded side thus elongates faster than the sunny side, pushing the stem toward the sun. The whole stem doesn't sense light, though–just the tip. Cover the tip of a young plant and it won't turn toward the light, but will instead grow straight up because, in the absence of sunlight, auxin does not accumulate more on one side of the stem than the other.

Other mechanisms help a plant sense where it is and orient itself in space. **Gravitropism** is the growth of plants in response to gravity–roots grow downward, with the force of gravity, and shoots grow upward, against it. Auxin is again the main player in this mechanism. When

PHOTOTROPISM
The growth of the stem of a plant toward light.

AUXIN
A plant hormone that causes elongation of cells as one of its effects.

GRAVITROPISM
The growth of plants in response to gravity. Roots grow downward, with gravity; shoots grow upward, against gravity.

THIGMOTROPISM
The response of plants to touch and wind.

a plant is placed on its side, more auxin is sent to the down side of the stem, in the direction of gravity. This causes the cells on the down side to elongate more. Stems begin to curve away from gravity. Root cells, however, respond the opposite way to auxin: more auxin on the gravity side of roots inhibits root cell elongation on the down side, so roots bend toward gravity. Gravitropism allows a planted seed to send its shoots toward the light and its roots toward the soil.

A plant's response to touch is called **thigmotropism;** because of this response vines can wind their way around poles or trellises and carnivorous plants can sense their prey. Touch-sensitive growth, like the other tropisms, is controlled by auxin. Vine cells touching a pole, for example, get less auxin and consequently elongate less, while cells on the other side elongate more. The result is another kind of lopsided growth in the shoot, which eventually causes the vine to coil around whatever it's touching. A plant's sense of touch can be exquisitely sensitive—more sensitive than a human's. A human can detect the presence of a thread weighing 0.002 mg laid across the arm. By contrast, the feeding tentacle of the insectivorous sundew plant can sense a thread of less than half that weight. The legs of a single gnat are enough to trigger the tentacle into swift action (**INFOGRAPHIC 25.9**).

INFOGRAPHIC 25.9 Plants Sense and Respond to Their Environment

Plants can respond to a variety of stimuli, including light, touch, and gravity. Many of these responses are triggered by a plant hormone called auxin. Unequal distribution of auxin leads to a corresponding unequal pattern of cell elongation in stems and roots, and thus growth in a particular direction.

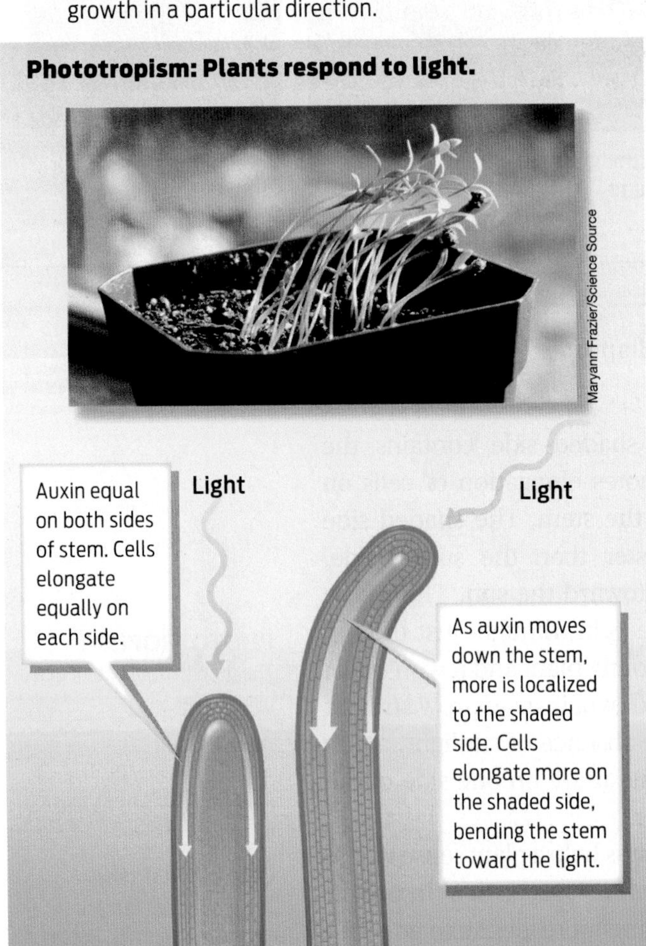

Phototropism: Plants respond to light.
Maryann Frazier/Science Source
Auxin equal on both sides of stem. Cells elongate equally on each side.
Light
Light
As auxin moves down the stem, more is localized to the shaded side. Cells elongate more on the shaded side, bending the stem toward the light.

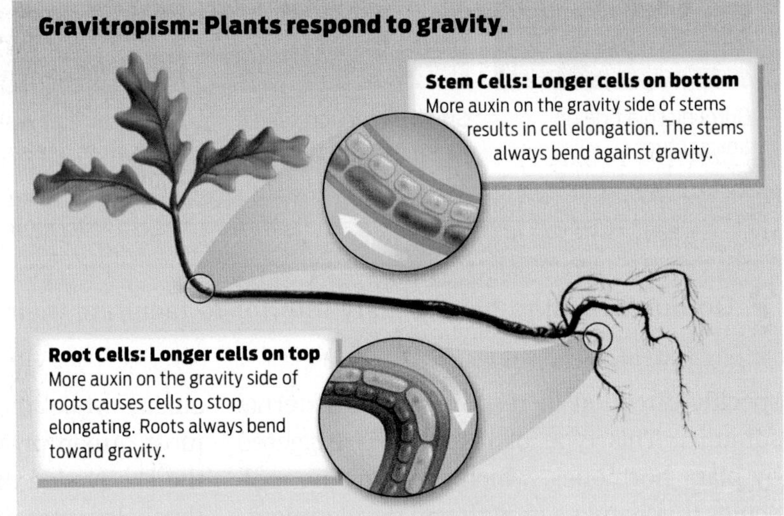

Gravitropism: Plants respond to gravity.
Stem Cells: Longer cells on bottom
More auxin on the gravity side of stems results in cell elongation. The stems always bend against gravity.
Root Cells: Longer cells on top
More auxin on the gravity side of roots causes cells to stop elongating. Roots always bend toward gravity.

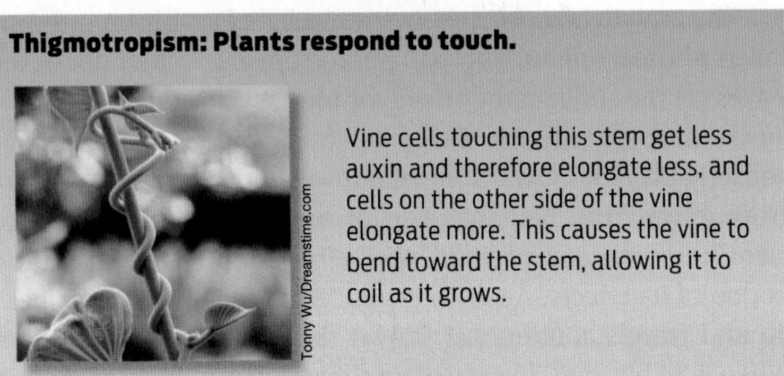

Thigmotropism: Plants respond to touch.
Tonny Wu/Dreamstime.com
Vine cells touching this stem get less auxin and therefore elongate less, and cells on the other side of the vine elongate more. This causes the vine to bend toward the stem, allowing it to coil as it grows.

? What happens to cells of stems and roots as a result of auxin action?

Q: Does one bad apple really spoil the bunch?

A: This old adage is true, and its truth can be shown empirically: put a ripe apple in a bowl of unripe ones and the unripe neighbors will quickly ripen. That's because ripe fruit–bananas and apples, especially–emit **ethylene,** a gaseous plant hormone that promotes ripening. In a confined space, the ethylene gas collects and causes nearby fruit to ripen through the loss of chlorophyll and the breakdown of cell walls. The result is the conversion of a hard, green fruit to a soft, ripe one.

Commercial fruit growers take advantage of the action of ethylene when they ship fruit to distributors. Often, fruit is picked while still green and then exposed to natural or synthetic ethylene just before arrival in grocery stores to induce ripening at the preferred time.

An example of gravitropism: leaf-bearing stems bend upward, away from gravity.

Q: Can plants take hormones to improve their performance?

A: In a manner of speaking, yes. **Gibberellins** are plant hormones that promote growth. Scientists have identified more than 100 different types of gibberellin. One effect of gibberellins is to provide the chemical cue for seeds to germinate and grow. When conditions are right for a seed to germinate–when rising temperatures begin to melt frost, for example–these growth-promoting hormones give the green light for a seedling to grow.

Applying gibberellins to a young plant can increase the length of its stem, which also indirectly increases the size of its fruits–a fact that makes them very useful in agriculture. Gibberellins are commonly used to increase the size of seedless grapes, for example. On an untreated seedless grape plant, the stem remains relatively short, so the bunches of grapes growing on the stem are clustered densely together, resulting in small grapes. When sprayed with gibberellins, the stems grow longer, giving the grapes more room to grow. Because seeds are the normal source of gibberellins, seedless grapes have no source of gibberellins to help them grow naturally, which is why farmers need to spray them in the first place.

In many species of plant, seed germination is controlled by the balance between the growth-promoting properties of gibberellins and the growth-inhibiting effects of another

Ethylene is used commercially to ripen fruit that is picked and transported while still green.

ETHYLENE
A gaseous plant hormone that promotes fruit ripening as one of its effects.

GIBBERELLINS
Plant hormones that cause cell division and stem elongation.

ABSCISIC ACID (ABA)
A plant hormone that helps seeds remain dormant.

hormone, **abscisic acid (ABA).** ABA is the plant hormone that keeps seeds from growing, that is, keeps them dormant. For example, seeds are dormant over the cold winter months, when conditions for growth are not ideal. With appropriate environmental cues–the right combination of temperature, moisture, and light–ABA is degraded and gibberellins are synthesized, leading to the end to dormancy and the start of germination. You can think of ABA as the brakes and gibberellins as the gas pedal: as the brakes on growth are lifted, the gas pedal is pressed, ushering in a new season's blooms **(INFOGRAPHIC 25.10)**.

PLANT DEFENSES

Q: Why are some plants poisonous?

A: From juicy peaches and succulent strawberries to zesty basil and peppery arugula, many plants are incontestably delicious. Humans aren't the only ones who think so. Many herbivores such as insects, birds, rodents, and other small mammals find plant parts tasty and irresistible. This is both helpful and harmful to a plant. On the one hand, plants rely on animals to eat their fruits and disperse their seeds. On the other hand, plants must ensure that other organisms eat only noncrucial parts of the plants. Eating a plant's fruit is one thing; eating all of its leaves is quite another.

Plants have evolved many defenses to protect their important parts from an herbivore's chomping. Some defenses are mechanical: the stems of a raspberry plant are covered in prickly spines to prevent unwanted chewing; holly leaves are waxy and difficult for insect jaws to grasp; and cacti have sharp needles that deter all but the most fearless animals, who would otherwise be attracted to these water-hoarding succulent plants to quench their thirst in the desert heat. Other defenses are chemical: leaves of the tobacco plant produce nicotine, which is toxic to insects; the bark of the South American evergreen cinchona tree produces quinine, an

INFOGRAPHIC 25.10 Hormones Trigger Plant Growth and Development

Certain plant hormones regulate growth and development of plants. When humans apply these hormones to crops, plant growth can be dramatically enhanced.

Gibberellin hormones enhance plant growth

Spraying plants with gibberellins enhances the number and size of the fruit produced.

Treated with gibberellins

Not treated with gibberellins

Photo by Stephen Vasquez, used with the permission of UC Regents.

The relative amounts of ABA and gibberellins trigger germination

Environmental cues cause inhibitory ABA to be degraded and growth-promoting gibberellins to be produced. This results in germination and plant growth.

JTB MEDIA CREATION, Inc. / Alamy

? If you wanted to stimulate a seed to germinate, which hormone would be more helpful, ABA or gibberellins?

INFOGRAPHIC 25.11 Plants Defend Themselves

Plants use both physical and chemical defenses to protect themselves against a variety of herbivores.

Physical Defenses

Cathleen Abers-Kimball/iStockphoto

Raspberry thorns can impale hungry insects.

histhorhome/iStockphoto

The jaws of insects that try to feed on holly leaves may not be strong enough to penetrate their waxy coat.

Chemical Defenses

Serghei Piletchi/Dreamstime.com

Peach pits have a tough exterior that protects the embryo inside. If an animal is successful in cracking the seed open, it is met with cyanide, a lethal toxin.

timsa/iStockphoto

When a hornworm feeds on a wild tobacco plant, the plant releases chemicals that both repel the worm and attract the worm's predators.

? Which defense(s), chemical and/or physical, protect a peach seed?

extremely bitter substance that many animals find distasteful (except certain humans, who use it in their gin and tonics). Such antiherbivory chemicals are a highly effective way to deter pests from eating a plant's leaves.

While a plant's fruits are often tasty and meant to be digested, seeds generally are not. The seed contains a new plant, and therefore it must be protected. Many seeds are encased within an indigestible shell that keeps them from being destroyed by an animal's stomach juices. The unlucky animal that succeeds in breaking open the shell and eating the seed is in for an unwelcome surprise. Seeds are sources of some of the most potent poisons on Earth, including ricin, cyanide, and strychnine. Ricin, found in castor beans, can be lethal to animals in quantities as little as two beans. Cyanide, which is found in small doses in the seeds of peaches, apricots, and apples, kills by interrupting cellular respiration in mitochondria; unable to make ample amounts of the short-term energy-storage molecule ATP, nerve and muscle tissues quickly shut down (**INFOGRAPHIC 25.11**).

Although antiherbivory chemicals complicate an herbivore's life, they are often quite useful to humans. Some of our most important medicines, including aspirin, morphine, digitalis, quinine, and the anticancer drug Taxol, which was originally obtained from the bark of the Pacific yew tree, are extracts of plant chemical defenses. ■

CHAPTER 25 SUMMARY

- All plant cells are surrounded by a cell wall made of the complex carbohydrate cellulose. A central water-filled vacuole creates turgor pressure against the cell wall and helps a plant stand up.
- Some plant tissues have cells with an additional inner cell wall made of cellulose and lignin. Lignin is extremely durable and is what makes plant tissues woody.
- Plants have a vascular system. Xylem transports water and nutrients from the roots to the shoots. Phloem transports sugars throughout the plant body.
- Water transport is powered by evaporation through stomata in leaves, a process called transpiration. Plants have mechanisms that control the amount of water lost by transpiration.
- Plants undergo primary (vertical) and secondary (horizontal) growth. Secondary growth is mostly growth of xylem tissue. Tree rings reflect the growth of xylem tissue from season to season.
- Two major groups of flowering plants are the monocots and dicots, which have characteristic anatomical differences that can be seen throughout their bodies.
- Plants are autotrophs, able to make their food through photosynthesis using sunlight, carbon dioxide, and water. They also require nutrients from the soil, including nitrogen, phosphorus, and sulfur. In nutrient-poor soils, some plants obtain nitrogen by trapping and digesting insects.
- Some plants have adapted to hot, dry climates, keeping their stomata closed during the day and collecting carbon dioxide for photosynthesis primarily at night.
- A variety of pigments assist with photosynthesis. Chlorophyll is dominant and is responsible for the green color of plant leaves. Other pigments contribute to photosynthesis and are visible in the fall, after chlorophyll has broken down.
- Many plants reproduce sexually and have male and female sexual organs that produce haploid gametes, sperm and egg. Sperm is contained within pollen.
- Flowering plants rely on pollinators such as insects to deliver male pollen to female reproductive structures. Other seed plants depend on the wind to deliver pollen to the female structures.
- Asexual reproduction, for example through runners or bulbs, is possible is some plants.
- Plants respond to their environment by various tropisms. Phototropism is growth toward light; gravitropism is growth in response to gravity; thigmotropism is growth in response to touch.
- Plants produce hormones that contribute to growth and development: ethylene contributes to fruit ripening; gibberellins and ABA regulate germination and stem growth; auxin controls cell elongation.
- Plants have physical defenses (for example, thorns) and chemical defenses (for example, toxins) that fend off herbivores. Humans make use of chemical defenses in developing pharmaceuticals.

MORE TO EXPLORE

- **Chamovitz, D. (2013) *What a Plant Knows: A Field Guide the Senses.* New York: Scientific American/Farrar, Straus, Giroux.**
- **U.S. Forest Service, Ethnobotany: www.fs.fed.us/wildflowers/ethnobotany/**
- **PBS (2013) *Nature*: What Plants Talk About: www.pbs.org/wnet/nature/episodes/what-plants-talk-about/video-full-episode/8243/**
- **PBS (1999) *Nature*: The Seedy Side of Plants: www.pbs.org/wnet/nature/episodes/the-seedy-side-of-plants/introduction/1268/**
- **Walters, D. (2017) *Fortress Plant: How to Survive When Everything Wants to Eat You.* Oxford University Press.**

CHAPTER 25 Test Your Knowledge

DRIVING QUESTION 1 What are key structural features of plants?

By answering the questions below and studying Infographics 25.1, 25.2, and 25.3, you should be able to generate an answer for the broader Driving Question above.

KNOW IT

1 Which organelle generates turgor pressure?

a. the chloroplast
b. the central vacuole
c. the nucleus
d. the mitochondrion
e. the endoplasmic reticulum

2 Which of the following statements represents a true distinction between xylem and phloem?

a. Xylem provides support only; phloem provides transport.
b. Xylem provides water and nutrient transport; phloem provides sugar transport.
c. Xylem transports materials from shoots to roots; phloem transports materials in either direction.
d. Xylem transports sugars in either direction; phloem transports water from roots to shoots.
e. all of the above

3 What is the function of the cuticle?

a. It enables neighboring cells to stick together.
b. It provides rigidity to the cell wall.
c. It is toxic to many herbivorous insects.
d. It prevents water loss.
e. It is sticky and helps pollen adhere to a plant during pollination.

4 Mark each of the following features as a monocot trait (M) or a dicot trait (D).

_____ Secondary growth
_____ Parallel leaf veins
_____ Five-petaled flowers
_____ Tree rings
_____ One seed leaf

5 Which cross section below (left or right) shows the stem of a dicot? How do you know?

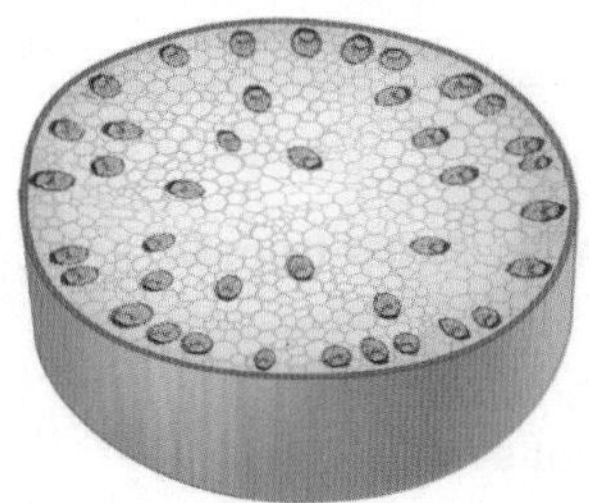

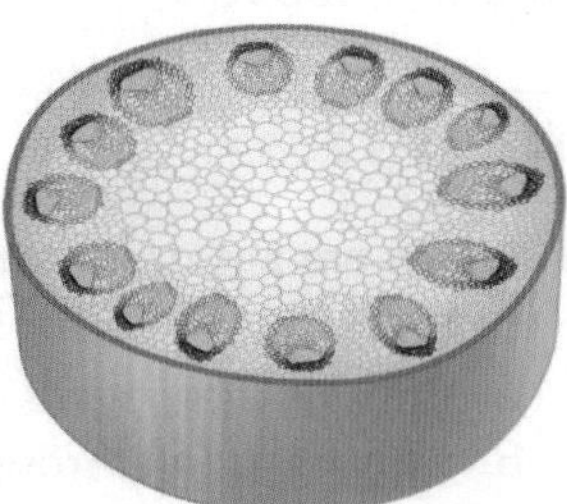

USE IT

6 Paper is made from wood that is broken down to pulp. Why are lignin-digesting enzymes included in the pulping process? Would these enzymes have to be included in the pulping process if paper were made from green leaves? Explain your answer.

7 In order to crisp up some wilted celery, what should you soak it in?

a. pure water
b. a solution with the same solute concentration as the celery
c. a solution with a higher solute concentration than the celery
d. you should let it air dry for a few minutes
e. any of the three solutions (a, b, or c) will be equally effective

8 The herbicide 2,4-D kills dicots but not monocots. Which of the following are uses of 2,4-D? Explain your answer.

a. to kill invasive grass that is growing into and taking over a flower garden
b. to kill dandelion weeds in an otherwise perfect lawn of grass
c. to kill "broadleaf" weeds in a cornfield

DRIVING QUESTION 2 How do plants respond to limiting nutrients and harsh conditions?

By answering the questions below and studying Infographics 25.4, 25.5, and 25.6, you should be able to generate an answer for the broader Driving Question above.

KNOW IT

9 Plants are autotrophs and can make sugar from CO_2. How do they obtain CO_2?

a. through stomata
b. by absorption through the root system
c. by digesting insects
d. by breaking down carbon-rich carbohydrates stored in roots
e. a and b

10 When stomata are open, what is happening?

a. O_2 is entering the plant for photosynthesis.
b. CO_2 is entering the plant for photosynthesis.
c. H_2O is entering the plant for photosynthesis.
d. H_2O is leaving the plant.
e. a, b, and c
f. b and d

11 Which pigments are present in green leaves in midsummer?

a. chlorophyll
b. xanthophyll
c. carotene
d. anthocyanin
e. all of the above
f. a, b, and c

12 Why might the bright coloration of a trumpet pitcher (a carnivorous plant) have a different function from that of bright yellow or orange squash blossoms? (Hint: Squash are pollinated by bees).

USE IT

13 Describe the "conflict" that plants face with respect to opening and closing their stomata.

14 If a plant could not make chlorophyll, would you expect it to survive? Why or why not?

apply YOUR KNOWLEDGE

INTERPRETING DATA

15 Scientists carried out an experiment to examine the effect of CO_2 concentrations on plant growth in a semiarid (that is, a dry) grassland environment in Colorado. They set up several plots in the field, consisting of chambers that allowed the concentration of CO_2 to be controlled. One set of plots (A) was kept at ambient CO_2 concentration, and one set (B) was kept at elevated CO_2 concentration (two times ambient concentration). In mid-July the total plant mass in each plot set was recorded. The data for three consecutive years are shown in the table. (In order to establish a baseline, the CO_2 levels in 1996 were not manipulated.)

Plot Set	Average Plant Mass (g/m^2), 1996	Average Plant Mass (g/m^2), 1997	Average Plant Mass (g/m^2), 1998
A (ambient CO_2)	110	108	145
B (elevated CO_2)	112	145	205

Data from Morgan, J. A., et al. (2001) Elevated CO_2 enhances water relations and productivity and affects gas exchange in C_3 and C_4 grasses of the Colorado shortgrass steppe. *Global Change Biology* 7:451–466.

a. Graph these data.
b. Are there any differences between different plot sets in any given year? If so, describe the differences observed.
c. Are there any differences in the same plot set between years? If so, describe the differences and propose an explanation.
d. What are the implications of this study for grassland productivity (at least in Colorado) with rising CO_2 levels?

DRIVING QUESTION 3 What strategies do plants use to reproduce?

By answering the questions below and studying Infographics 25.7 and 25.8, you should be able to generate an answer for the broader Driving Question above.

KNOW IT

16 What is the function of endosperm in a seed?

17 List at least three ways by which seeds can be dispersed.

USE IT

18 Imagine an island that has no animals, but does have seed-bearing plants. What seed dispersal mechanisms would most likely be found on this island? Explain your answer.

DRIVING QUESTION 4 How do plants respond to stimuli and protect themselves?

By answering the questions below and studying Infographics 25.9, 29.10, and 25.11, you should be able to generate an answer for the broader Driving Question above.

KNOW IT

19 If you wanted a plant to grow very tall, which hormone should you apply?

a. auxin
b. ethylene
c. gibberellins
d. anthocyanin
e. ABA

20 Which plant hormone is responsible for a plant's bending toward light?

a. auxin
b. ethylene
c. gibberellins
d. anthocyanin
e. ABA

USE IT

21 Why do seedless grapes need hormone treatment to develop big clusters of big grapes, whereas seeded varieties can develop large fruits without the application of hormones?

22 Nopales are cactus pads (the large, thick "leaves" of the prickly pear cactus) and make a delicious salad. What antiherbivory mechanism fails when we succeed in making ensalada de nopales—prickly pear salad?

MINI CASE

apply YOUR KNOWLEDGE

23 The Natural Products Branch of the National Cancer Institute looks for defensive compounds produced by plants, microbes, and marine organisms that may have anticancer activity. Once compounds have been isolated, they can be chemically modified to enhance their activity. Several drugs have come out of this program, including eribulin mesylate, a chemically modified compound originally purified from a sea sponge. This drug has been approved for women with metastatic breast cancer whose disease has not been responsive to previous treatment. In a clinical trial, patients taking eribulin mesylate had a significantly longer survival time (13.1 months) than did patients in chemotherapy regimens prescribed by their oncologists (10.6 months). (Data are from http://www.cancer.gov/ncicancerbulletin/041911/page5.)

The first step in developing drugs such as these is to prepare extracts by grinding up the natural product. Design a procedure to test such extracts for anticancer activity. Consider what tests you will use to determine if an extract has an anticancer activity, and what variables you will measure and manipulate in order to find promising candidates to advance along the drug discovery pipeline.

BRING IT HOME

apply YOUR KNOWLEDGE

24 Many household and garden plants are toxic to household pets. For example, rhododendrons are toxic to dogs and cats, and oleanders are toxic to cats, dogs, horses, birds, and cows. Do some Internet research to identify at least three other plants that are toxic to household pets or farm animals. For each plant, find both the scientific and common name and which part(s) of the plant is/are toxic to animals. Do any of the plants you identified have medicinal uses for humans?

MAN *vs.* MOUNTAIN

Physiology explains a 1996 disaster on Everest

Harry Kikstra/Getty Images

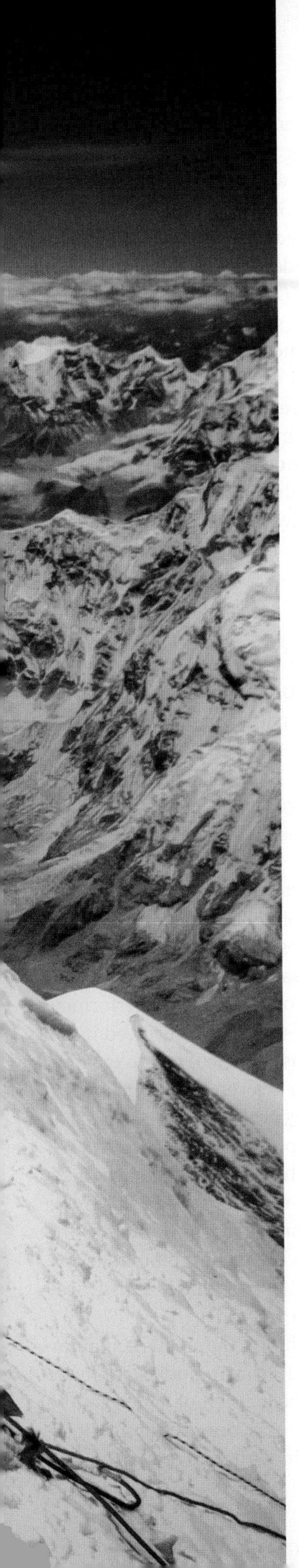

DRIVING QUESTIONS

1. How are the bodies of living organisms organized?
2. How do humans and other organisms maintain homeostasis?
3. What are homeostatic feedback loops, and which physiological systems are regulated by them?

At 1:17 P.M. on May 10, 1996, Jon Krakauer planted one foot in China, the other in Nepal, and stood on the roof of the world. He was at the highest point above sea level that any human has ever reached–short of standing on the moon. Yet he didn't feel like celebrating. It had taken him 6 long weeks to climb to the top of Mount Everest, and now that he was here his toes ached in the subzero cold, his breath came in short, painful bursts, and his head pounded from the altitude. It was a struggle just to stay upright. "I cleared the ice from my oxygen mask, hunched a shoulder against the wind, and stared absently at the vast sweep of earth below," wrote Krakauer in an account of his climb for *Outside* magazine later that year.

Krakauer's journey to Everest was a lifetime in the making. While other kids were idolizing astronaut John Glenn and baseball pitcher Sandy Koufax, Krakauer's childhood heroes were Tom Hornbein and Willi Unsoeld–two men from his hometown in Oregon who, in 1963, became the first climbers to scale the daunting western ridge of Everest. As a teenager, Krakauer became a skilled climber, vanquishing many of the world's most difficult peaks, and he dreamed of one day climbing Everest himself. By his mid-twenties, though, he had largely abandoned the idea as a boyhood fantasy. But old dreams die hard.

ANATOMY
The study of the physical structures that make up an organism.

In 1995, Krakauer was working as a journalist when the opportunity came to shadow an Everest climb and report on it for *Outside* magazine. The 42-year-old writer-adventurer jumped at the chance. He would join a team headed by the celebrated climbing guide Rob Hall, whose company, Adventure Consultants, had successfully put 39 amateur climbers on top of Everest. Reaching the summit himself would mean enduring a weeks-long ascent from Base Camp, giving his body time to adjust to the high altitude.

It would also mean risking his life on a daily basis.

The icy tip of Mount Everest sits at 29,035 feet above sea level. For perspective, consider that the cruising altitude of most commercial jetliners is 30,000 feet. A human plucked from sea level and deposited at this altitude would quickly lose consciousness and die. A climber who has spent weeks adjusting to the altitude can function better at the summit, but not very well, and not for very long. Everest is not only the highest place on Earth, it is also one of the coldest. At the summit, where windchill temperatures average −53°C (−63°F), freezing to death is a real possibility.

Despite these dangers—or perhaps because of them—about 150 fearless men and women try to climb Everest every season. And every season, some of them don't come back. Nineteen people died on Everest in 2015 alone. There are many reasons for these disasters—poor training, unforeseen accidents, raw egotism—but among the most important is basic biology: the human body is not equipped to survive at such extreme altitude, and in such extreme temperatures, for long.

Andrew Bardon/Getty Images

Climbers using aluminum ladders to bridge dangerous crevasses in the Khumbu Icefall region of their ascent.

The Body as Machine

Like a car or a computer, a human body is made up of many parts working together in a coordinated fashion. The parts are organized hierarchically, so that smaller components are organized into increasingly larger units, which are themselves organized into more complex systems. The study of all this intricate hardware is called **anatomy.**

The product of millions of years of evolution, human bodies have an anatomical structure that is impressively well adapted to living in certain environments and performing certain functions. Our species evolved in the hot, flat savannahs of Africa, where environmental conditions favored big brains, opposable thumbs, and bipedal posture—as well as the ability to keep cool (see Chapter 19). As a result, modern humans excel at grasping a pencil or looking through a microscope; we do less well swimming at the bottom of the ocean or living on the highest mountaintops. Fundamentally, that's because of the way we're put together.

For all living things, the smallest anatomical unit is the cell. Human bodies are made up of trillions of cells, each of which can be classified as one of a few hundred different types. Cells, in turn, are organized into

tissues–groups of specialized cells working together to execute a particular function. Humans and other animals have four different types of tissue–epithelial, connective, muscle, and nervous–which carry out specific tasks in the **organs** of which they are a part. The stomach, for example, is an organ composed of the four types of tissue organized into a compartment for churning and digesting food. At the highest level of organization, organs interact chemically and physically as part of **organ systems.** For instance, as part of the digestive system, the stomach works with the esophagus, small intestine, and liver to digest and absorb food (**INFOGRAPHIC 26.1**).

If the body is like a machine, then physiologists–the scientists who study **physiology**, the exploration of how the body functions–are interested in how this machine keeps running smoothly. Physiologists want to understand how organ systems cooperate to accomplish basic tasks, such as obtaining energy from food, taking in nutrients to build new molecules during growth and repair, and ridding the body of

TISSUE
An organized collection of cells working to carry out a specific function.

ORGAN
A structure made up of different tissue types working together to carry out a common function.

ORGAN SYSTEM
A set of cooperating organs within the body.

PHYSIOLOGY
The study of the way a living organism's physical parts function.

INFOGRAPHIC 26.1 How the Human Body Is Organized

In the human body, the smallest anatomical units, cells, are organized into increasingly complex units, including tissues, organs, and organ systems. These organ systems then work together to allow the organism to function.

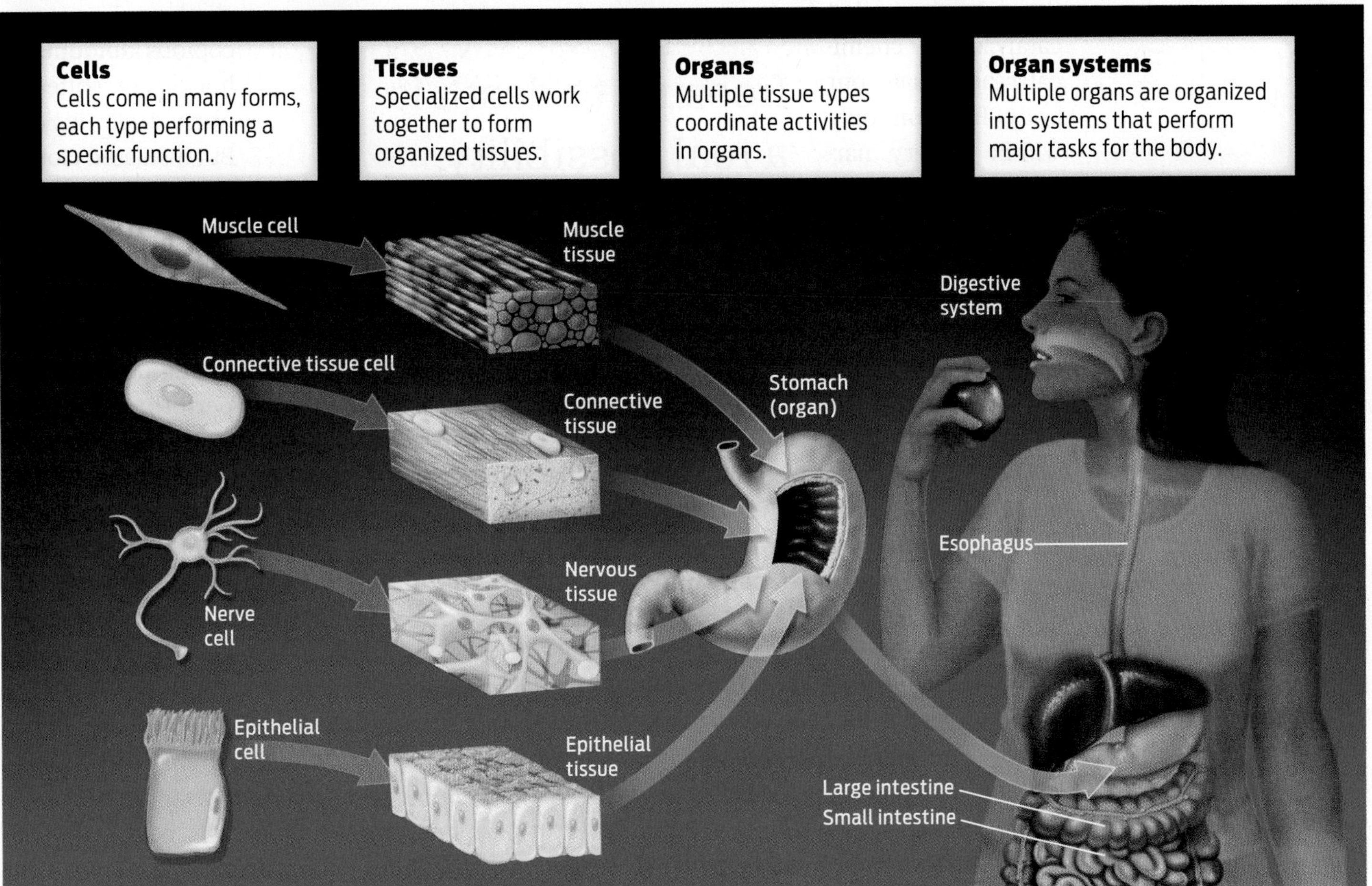

? What is a primary difference between a tissue and an organ?

wastes. To the physiologist, the body is an integrated system for processing inputs and outputs and maintaining **homeostasis**—an internal environment that remains relatively stable even when the external environment changes.

Thermoregulation: The Physiology of Staying Warm

Like many other animals, humans have an optimal operating temperature and are exquisitely sensitive to temperature changes. Although we can tolerate a wide range of external temperatures in our daily lives, we cannot tolerate large fluctuations in our internal temperature. That's because the enzymes that catalyze the chemical reactions in our body function only within a very narrow temperature range. The body works hard to maintain a relatively constant internal temperature compatible with life, fluctuating only about 0.5°C. Through this process of **thermoregulation,** our body temperature is kept at a consistent—and toasty—37°C (98.6°F).

Keeping a consistent body temperature is just one example of how the body maintains homeostasis. "What we're really thinking about with homeostasis are certain set points that your body needs to maintain," explains Robert Kenefick, an exercise physiologist with the U.S. Army Research Institute of Environmental Medicine in Natick, Massachusetts. The body has a number of such set points, he says, for things like temperature, blood pH, and blood pressure, and it works hard to keep these factors balanced within a very narrow range, even in the face of a changing external environment. The consequences of not maintaining this balance can be deadly: a body temperature increase or decrease of just a few degrees, for example, can be lethal **(INFOGRAPHIC 26.2)**.

As an exercise physiologist, Kenefick has spent his career trying to understand how the human body maintains homeostasis during strenuous activities like hiking and running marathons. He works in the Army Research Institute's Thermal and Mountain Medicine Division, where a main focus of his research is understanding how the body performs in extreme cold.

Staying warm is hard to do when ambient temperatures drop below −50°C (−58°F), as they routinely do on Everest. To seal in heat, mountain climbers wear layers of protective gear designed to trap heat, wick away moisture, and insulate their bodies from the wind and cold. When not hiking, climbers consume copious amounts of hot tea or coffee to warm their insides. But insulated clothes and hot beverages would be of little help without the body's natural way of keeping warm.

> At the summit, where windchill temperatures average −53°C (−63°F), freezing to death is a real possibility.

Human bodies respond to cold in two main ways: by conserving the heat they already have, and by generating more. To conserve heat, the body performs peripheral **vasoconstriction**—a decrease in the diameter of blood vessels just below the surface of the skin. By constricting blood vessels in the skin, peripheral vasoconstriction pushes blood from the skin to the body core, where the internal organs are.

"A lot of people believe this is done to increase the amount of blood that goes to your core to help protect those organs," says Kenefick. There's some truth in that, Kenefick notes, but a more important reason for peripheral vasoconstriction is, he explains, "to decrease the amount of heat loss from your skin to the environment." Like most things in the universe, heat moves along a gradient from higher to lower. "If the temperature is higher

HOMEOSTASIS
The maintenance of a relatively stable internal environment even when the external environment changes.

THERMOREGULATION
The maintenance of a relatively stable internal body temperature.

VASOCONSTRICTION
The reduction in diameter of blood vessels, which helps to retain heat.

INFOGRAPHIC 26.2 The Body Works to Maintain Homeostasis

The body expends a great deal of energy to maintain a constant internal environment. Only small fluctuations are tolerated, even in extreme external conditions.

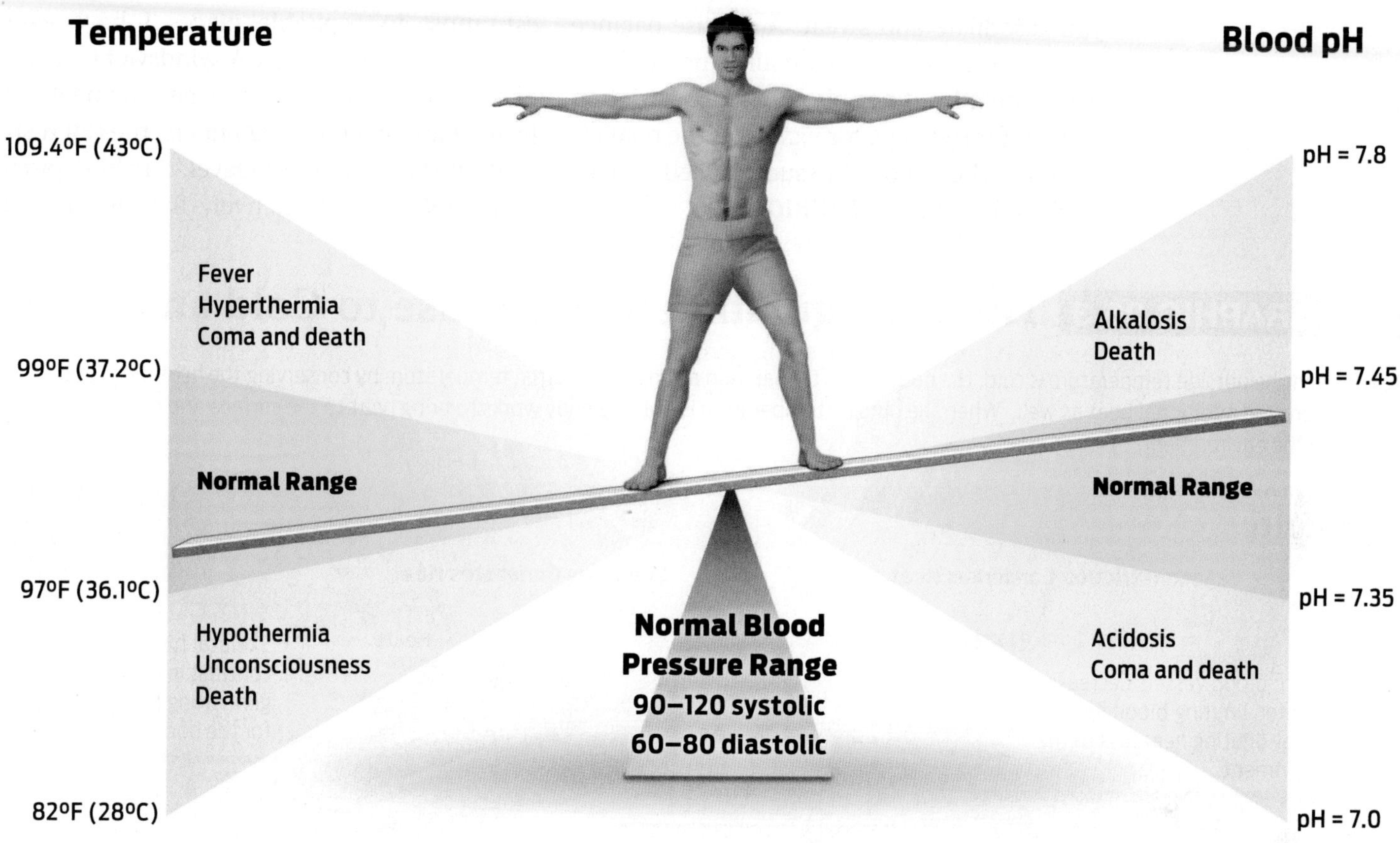

? Is a body temperature that drops to 36.5°C (97.7°) within the normal range, or will it trigger a response to restore homeostasis?

in your skin and lower in the air, then you're going to lose heat to the air. By bringing [these temperatures] closer together, you lose much less heat to the environment." The clamping down of blood flow near the skin surface is the reason hands, feet, and noses are the first to feel cold on a cold day, and it's a sign that your body is trying to retain heat.

The second way the body responds to cold is by trying to generate more heat. It does this by shivering, the involuntary contraction of normally voluntary muscles. "We know that the by-products of cellular respiration–any time cells work, and that includes your muscle cells–are CO_2, heat, and water," says Kenefick. "So when your muscles contract through shivering, they [generate] heat, and that heat helps to warm up the core of your body."

Of course, to maintain a constant temperature, our bodies must not only keep from getting too cold, they must also keep from getting too hot. Two main physiological responses help prevent overheating: peripheral **vasodilation,** the expansion of the diameter of blood vessels, which increases blood flow to the skin; and evaporative cooling, or

VASODILATION
The expansion in diameter of blood vessels, which helps to release heat.

sweating, which cools the body by releasing heat to the air.

In other words, you have a set point for body temperature: if you get too cold, you vasoconstrict and shiver; if you get too hot, you vasodilate and sweat. A precise balance between the two must be maintained to keep tissues healthy. If peripheral vasoconstriction goes on for too long, for example, the result is frostbite–the death of tissues caused by lack of blood flow (**INFOGRAPHIC 26.3**).

Krakauer and his teammates were no strangers to the cold. After weeks of slowly ascending from camp to camp, they reached the launching pad for the summit, the South Col, at 1 P.M. on May 9. "It is one of the coldest, most inhospitable places I have ever been," Krakauer wrote. A windswept saddle of rock and ice that sits between the peaks of Everest and neighboring Lhotse, the Col rests at 26,000 feet above sea level. Climbers pitch their tents on the relatively flat terrain and

INFOGRAPHIC 26.3 Thermoregulation in Response to Cold and Heat

When the outside temperature is cold, the body works to maintain a constant internal temperature by conserving the heat it has and generating additional heat as well. When the outside temperature is hot, the body works to bring heat to the surface and release it into the environment.

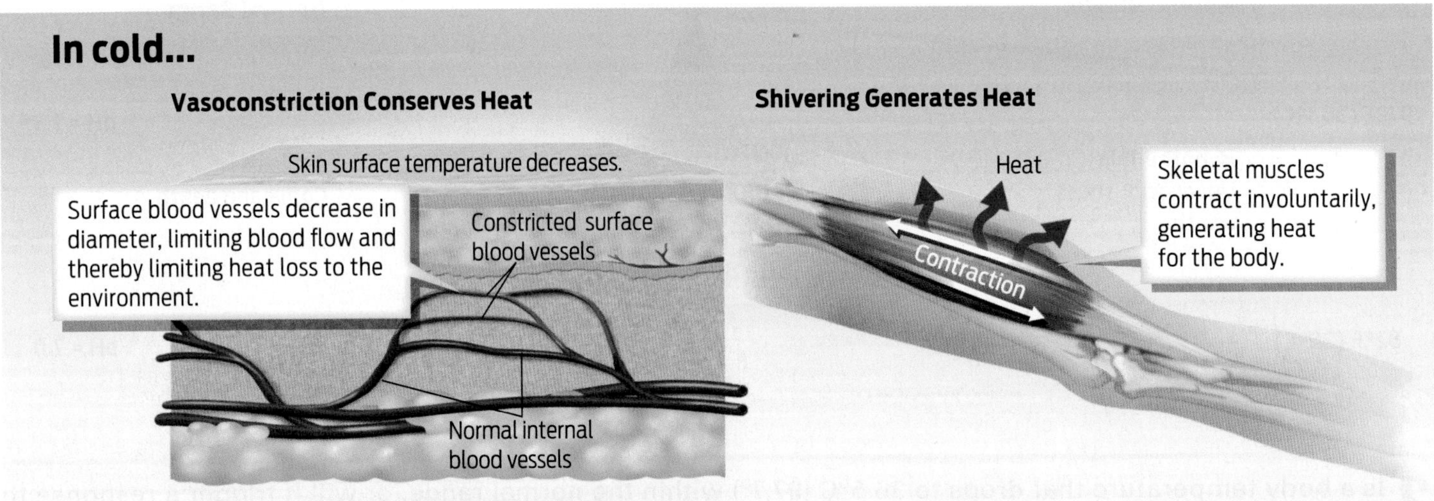
In cold...
Vasoconstriction Conserves Heat
Skin surface temperature decreases.
Surface blood vessels decrease in diameter, limiting blood flow and thereby limiting heat loss to the environment.
Constricted surface blood vessels
Normal internal blood vessels
Shivering Generates Heat
Heat
Contraction
Skeletal muscles contract involuntarily, generating heat for the body.

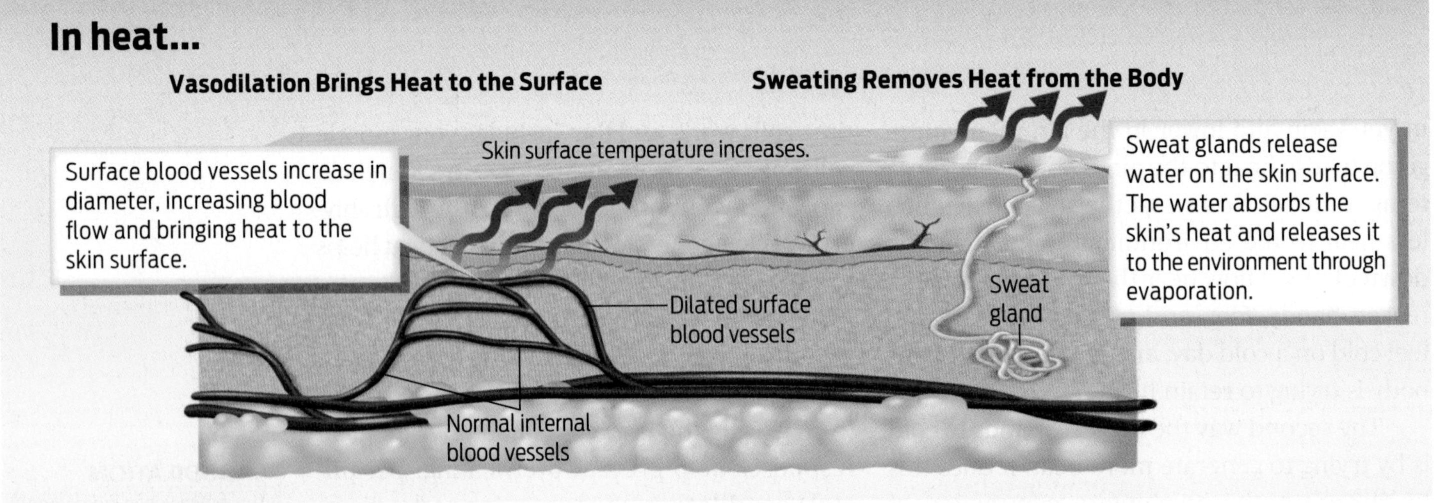
In heat...
Vasodilation Brings Heat to the Surface
Skin surface temperature increases.
Surface blood vessels increase in diameter, increasing blood flow and bringing heat to the skin surface.
Dilated surface blood vessels
Normal internal blood vessels
Sweating Removes Heat from the Body
Sweat gland
Sweat glands release water on the skin surface. The water absorbs the skin's heat and releases it to the environment through evaporation.

? How does the body generate heat? Is there a way to generate "cool," or do we rely on dissipating heat to reduce body temperature?

try not to think about the fact that they have entered what's known as the death zone.

Conditions were particularly bad on the Col that day. Gale-force winds blew through the camp, limiting visibility. As Krakauer's teammate, Beck Weathers, later recalled, "The weather was so crummy that when we first got in there, I didn't think there was any chance that we were going to climb that night."

But at 7 P.M., conditions improved markedly. It was still cold –26°C (–15°F)–but the wind had abruptly ceased, and by 11 P.M., above their heads, the stars appeared, while a gibbous moon reflected off the mountain snow. It was the perfect night for a climb.

"Into Thin Air"

The 15-member team left camp shortly after 11 P.M. Night climbing is necessary on Everest in order to arrive at the most difficult parts of the climb during daylight hours and still have enough time to get back down to camp before nightfall. Krakauer led the pack that night, along with the team's head Sherpa, Ang Dorjee.

The pair reached the Southeast Ridge, the second to last stop before the summit, at 5:30 A.M., just as the sun was peering over the eastern peaks. By this time, Krakauer's hands and feet felt like unwieldy blocks of ice, nearly useless in performing the careful work of laying ropes and scaling ice. But it wasn't just the cold he had to deal with. His brain and body were also showing the effects of altitude: "Plodding slowly up the last few steps to the summit," wrote Krakauer in *Into Thin Air,* his 1997 book about the expedition, "I had the sensation of being underwater, of life moving at quarter speed."

At high altitudes, there is lower barometric pressure, which means there are fewer air molecules banging around in the atmosphere–including fewer molecules of oxygen. The percentage of oxygen in the air is the same as at sea level (about 20%), but since there are many fewer air molecules overall, the pressure of oxygen is much less, and therefore fewer oxygen molecules are taken in per breath of air (see Chapter 29). The lower pressure means that fewer oxygen molecules bind to the hemoglobin in blood, which means that blood is less saturated with oxygen, a condition called **hypoxia.** Since all cells require oxygen to function, hypoxia has many bodily consequences. The most serious and immediate occur in the brain. "I've been at 19- and 20,000 feet climbing myself," says physiologist Kenefick, "and I can tell you that doing simple tasks like tying your shoes–even though you've tied your shoes many times–is much more difficult." For the climbers on Everest, he says, each step would have been a struggle.

To help cope with conditions of low oxygen, climbers spend about 6 weeks **acclimatizing** their bodies to the conditions, spending a few nights at progressively higher elevations. Their bodies respond by increasing the production of red blood cells, the cells that contain hemoglobin and carry oxygen.

The physiological adjustment of acclimatization allows a climber to carry more oxygen than someone coming straight from sea

HYPOXIA
A state of low oxygen concentration in the blood.

ACCLIMATIZATION
The process of physiologically adjusting to an environmental change over a period of time. Acclimatization is generally reversible.

Climbers spend weeks moving from camp to camp, acclimatizing to increasingly higher elevations before attempting to reach the summit from Camp IV (C4) on the South Col.

level could. But even well-acclimatized climbers usually need bottled oxygen to reach the summit successfully.

At 1:17 P.M., after more than 12 hours of climbing, Krakauer finally reached the summit. It was smaller than he expected—a patch of ice the size of a picnic table, with Buddhist prayer flags tied to a pole and flapping noisily in the wind. He stood and took in the 360° view. The towering peaks of the surrounding Himalayas, which had once loomed so large, now sat below him, draped in low-lying clouds. Beyond the mountain range, endless miles of continent stretched to the horizon.

Standing on top of the world, Krakauer was surprised by his lack of elation. He had just cleared a huge personal hurdle, yet the victory felt hollow. Partly, he was too exhausted to truly care: he hadn't slept soundly in more than 50 hours, and the only food he had been able to choke down in 3 days was a bowl of ramen soup and some peanut M&Ms (sleep disturbances and digestive difficulties are additional side effects of high elevation). But another thought lurked in his brain: the oxygen tank he had slung on his back to help him breathe was running low, and he still had to get down the mountain.

STR/Getty Images

High traffic on the Hilary Step.

"With enough determination, any bloody idiot can get up this hill," guide Rob Hall had famously said. "The trick is to get back down alive." Keenly aware of the clock, Krakauer snapped a few perfunctory photos, and within 5 minutes was headed back down the mountain toward Camp IV on the Col.

Fifteen minutes later, after scaling the steep ice fin of the Southeast Ridge, he arrived at the pronounced notch in the mountain known as the South Summit, just below the main peak. As he prepared to rappel over the edge, he caught a glimpse of an alarming sight: a queue of 20 climbers, from three separate expeditions, waiting to come up. They were backed up at the notorious Hillary Step—a 40-foot wall of rock and ice named for Sir Edmund Hillary, who, with Tenzing Norgay, was the first to scale it successfully, in 1953. Getting up the Step requires ropes, so climbers must go up one by one, and on this day there was a traffic jam.

While waiting for his turn to get down the Step, Krakauer peered into the distance and saw something he hadn't noticed before: on the horizon, dark clouds were sweeping in from the south, filling up a corner of what had been a clear blue sky. A storm was brewing.

By this point, it was well past the agreed-upon turnaround time of 1 P.M. set by Hall. The climbers who were still headed up the mountain at this hour were willfully flouting safety rules. Not only that, weather conditions were getting worse. Snow had started to fall, and it had become hard to see where mountain ended and sky began. The lower Krakauer got on the mountain, the worse the weather became.

Krakauer made it back to Camp IV on the Col just before 6 P.M. The bedraggled climber fell into his tent and quickly passed out. He was delirious, shivering uncontrollably, and exhausted. But he was alive.

Sensors Working Overtime

Even as he slept, Krakauer's body was working hard to thermoregulate. Like many physiological processes, thermoregulation is not something that requires conscious thought. It is more like the automated response of a home heating system, triggered when the thermostat is tripped.

The body's thermostat is the **hypothalamus,** a grape-size structure that sits at the base of the brain, right above the brain stem. The hypothalamus receives signals from many different **sensors,** specialized cells in the body that detect changes in both the internal and external environment. For cold, the major sensors are thermoreceptors in the skin and in the hypothalamus itself. Information from various sensors is fed to the hypothalamus, which then integrates the information and directs an appropriate response.

Acting as a thermostat, the hypothalamus has a specific temperature set point below which a warning message is triggered that body temperature is dropping. When that happens, the hypothalamus "tells" the body to take corrective action. For example, it can send a signal to blood vessels in the skin, causing them to constrict in peripheral vasoconstriction. It can also send a signal to muscles to start shivering. Both signals are sent from the hypothalamus to their target tissues by nerve fibers running from the brain to the rest of the body. The cells, tissues, or organs that respond to such signals are known as **effectors:** they act to cause a change in the internal environment. Once the effectors have raised the body temperature, the sensors detect the changed conditions and the signals are turned off.

This circuit of sensing, processing, and responding is an example of a homeostatic **feedback loop.** In this case, the loop is a negative feedback loop because the output of the loop–an increase in body temperature–feeds back to the sensors to *decrease* the response. As body temperature rises, the responses that would further increase body temperature–shivering and vasoconstriction–are no longer needed and so they are turned off. This loop helps keep the system at the set point **(INFOGRAPHIC 26.4)**.

Not all feedback loops act in a negative fashion. Positive feedback loops occur when the output of a system acts to *increase* the response of the system. An example is the formation of a blood clot when you cut yourself, a response critical to preventing blood loss. Blood platelets stick to damaged blood vessels and release molecules that attract even more platelets to the area, which in turn attract even more platelets, and eventually a blood clot forms. Positive feedback loops are effective at rapidly amplifying a response, but negative feedback loops tend to be more common in physiology because they help return the body to its set point and ensure that homeostasis is maintained.

> "Within the space of 5 minutes, it changed from really a good day with a little bit of wind to desperate conditions, something I'd never experienced the ferocity of before."
>
> —John Taske

The hypothalamus does more than regulate body temperature. In fact, it is the body's main homeostasis control center, regulating many bodily states including hunger, thirst, sexual arousal, and sleep. The hypothalamus is part of what Kenefick calls our "lizard brain"–the evolutionarily ancient parts of the brain, which control our most basic physiological responses through unconscious reflexes.

The hypothalamus is able to play such an important role in homeostasis because it is so well connected to sensors and effectors. The hypothalamus is not only a key part of the **nervous system,** connected to parts of the body through nerves, it is also intimately associated with the **endocrine system,** which produces changes in the body through the action of chemical signaling molecules

HYPOTHALAMUS
A master coordinator region of the brain responsible for a variety of physiological functions.

SENSOR
A specialized cell that detects specific sensory input like temperature, pressure, or solute concentration.

EFFECTOR
A cell or tissue that responds to information relayed from a sensor.

FEEDBACK LOOP
A pathway in which the output from an effector feeds back to a sensor and changes further output.

NERVOUS SYSTEM
The collection of organs that sense and respond to information, including the brain, spinal cord, and nerves.

ENDOCRINE SYSTEM
The collection of hormone-secreting glands and organs with hormone-secreting cells.

INFOGRAPHIC 26.4 Homeostasis Feedback Loops Require Sensors and Effectors

By means of sensors, the body constantly monitors factors like body temperature. The sensors relay temperature information to the hypothalamus. If the temperature is too hot or too cold, the hypothalamus sends signals to effector tissues and organs that work to return the temperature to homeostasis levels.

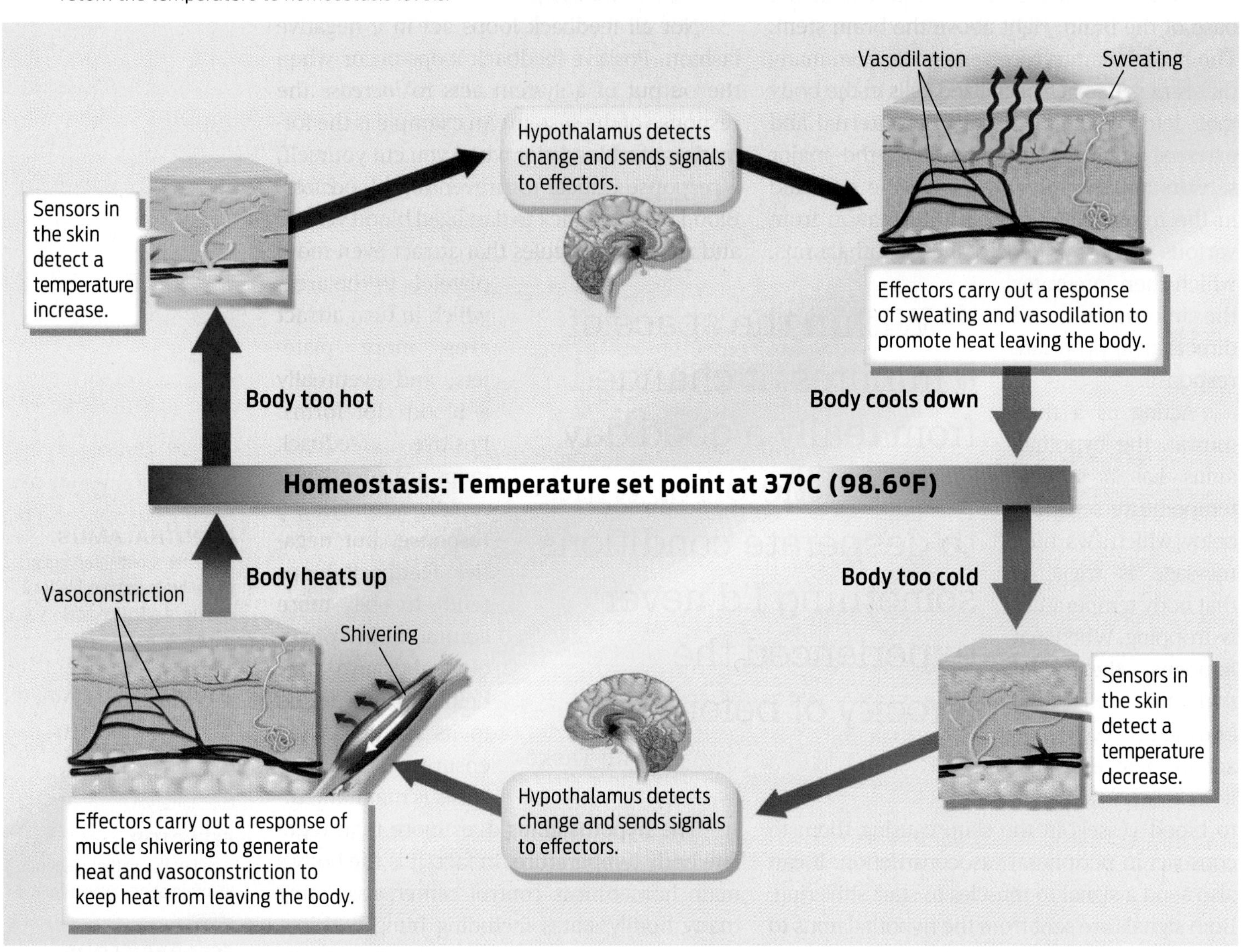

Sensors in the skin detect a temperature increase.
Hypothalamus detects change and sends signals to effectors.
Vasodilation
Sweating
Effectors carry out a response of sweating and vasodilation to promote heat leaving the body.
Body too hot
Body cools down
Homeostasis: Temperature set point at 37°C (98.6°F)
Body heats up
Body too cold
Vasoconstriction
Shivering
Sensors in the skin detect a temperature decrease.
Hypothalamus detects change and sends signals to effectors.
Effectors carry out a response of muscle shivering to generate heat and vasoconstriction to keep heat from leaving the body.

? Identify at least two effectors involved in homeostasis for body temperature.

HORMONE A chemical signaling molecule that is released by a cell or gland and travels through the bloodstream to exert an effect on target cells.

PITUITARY GLAND An endocrine gland in the brain that secretes many important hormones.

called **hormones,** which travel through the blood. The hypothalamus connects to the endocrine system via a direct circulatory connection to the **pituitary gland,** a pea-size structure that sits right below the hypothalamus. Hormones released by the hypothalamus travel directly to the pituitary gland, signaling it to release more hormones, which in turn travel through the bloodstream and act on many tissues in the body, including other glands **(INFOGRAPHIC 26.5)**.

The endocrine system, with its many hormone-secreting glands, is just one of 11 organ systems in the human body that work together to perform the tasks necessary for survival–from food intake and waste removal to self defense and reproduction

INFOGRAPHIC 26.5 The Endocrine System

The endocrine system, which helps us respond to changes in environment, relies on the activity of hormones—chemical signaling molecules that travel to target tissues via the circulation. The hypothalamus releases hormones that act on the pituitary gland, signaling the release of hormones that act on many tissues in the body.

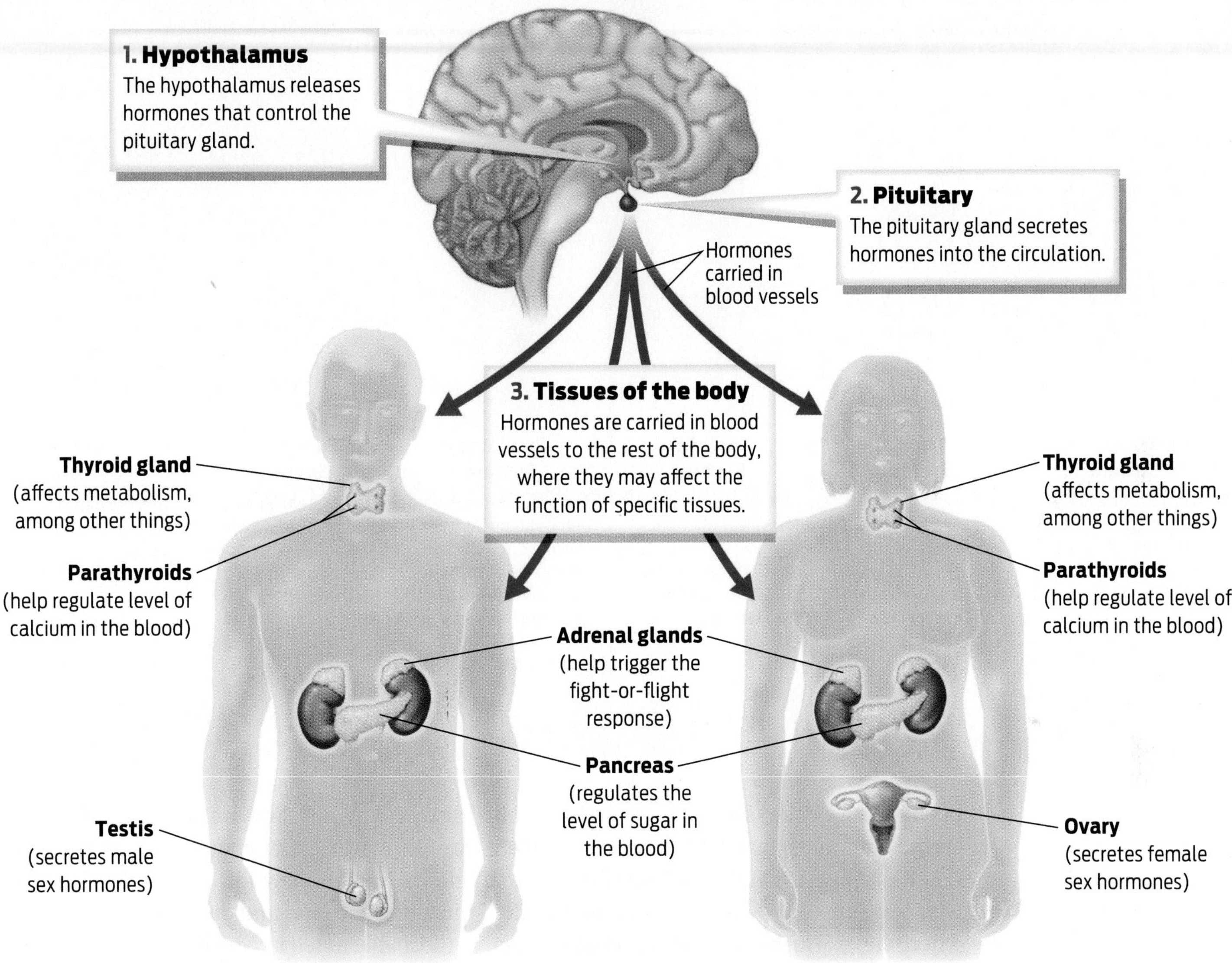
1. Hypothalamus
The hypothalamus releases hormones that control the pituitary gland.
2. Pituitary
The pituitary gland secretes hormones into the circulation.
Hormones carried in blood vessels
3. Tissues of the body
Hormones are carried in blood vessels to the rest of the body, where they may affect the function of specific tissues.
Thyroid gland
(affects metabolism, among other things)
Parathyroids
(help regulate level of calcium in the blood)
Adrenal glands
(help trigger the fight-or-flight response)
Pancreas
(regulates the level of sugar in the blood)
Testis
(secretes male sex hormones)
Thyroid gland
(affects metabolism, among other things)
Parathyroids
(help regulate level of calcium in the blood)
Ovary
(secretes female sex hormones)

? When hormones are released from the hypothalamus, what organ to do they tend to act on?

(see **UP CLOSE**: Organ Systems and subsequent chapters in this unit).

During the night, Krakauer was awakened by a teammate who gave him grave news: a number of his teammates, including Rob Hall, had not yet returned to Camp IV. They were still out in the blistering subzero cold somewhere above 26,000 feet. Krakauer's heart sank. He knew the chances of surviving in the cold for that long were slim. By 5 P.M., everyone's oxygen tank would have been empty. It was now midnight. Krakauer feared for the others' lives. But he was also dumbfounded: Hall and the rest of his team had not been far behind him on the mountain. What had gone wrong?

The storm he had spotted on the horizon began as a cyclone in the Bay of Bengal. It came in low from the valley and then rose up the mountain, gaining in ferocity and strength as it climbed. "One minute, we could look

UP CLOSE Organ Systems

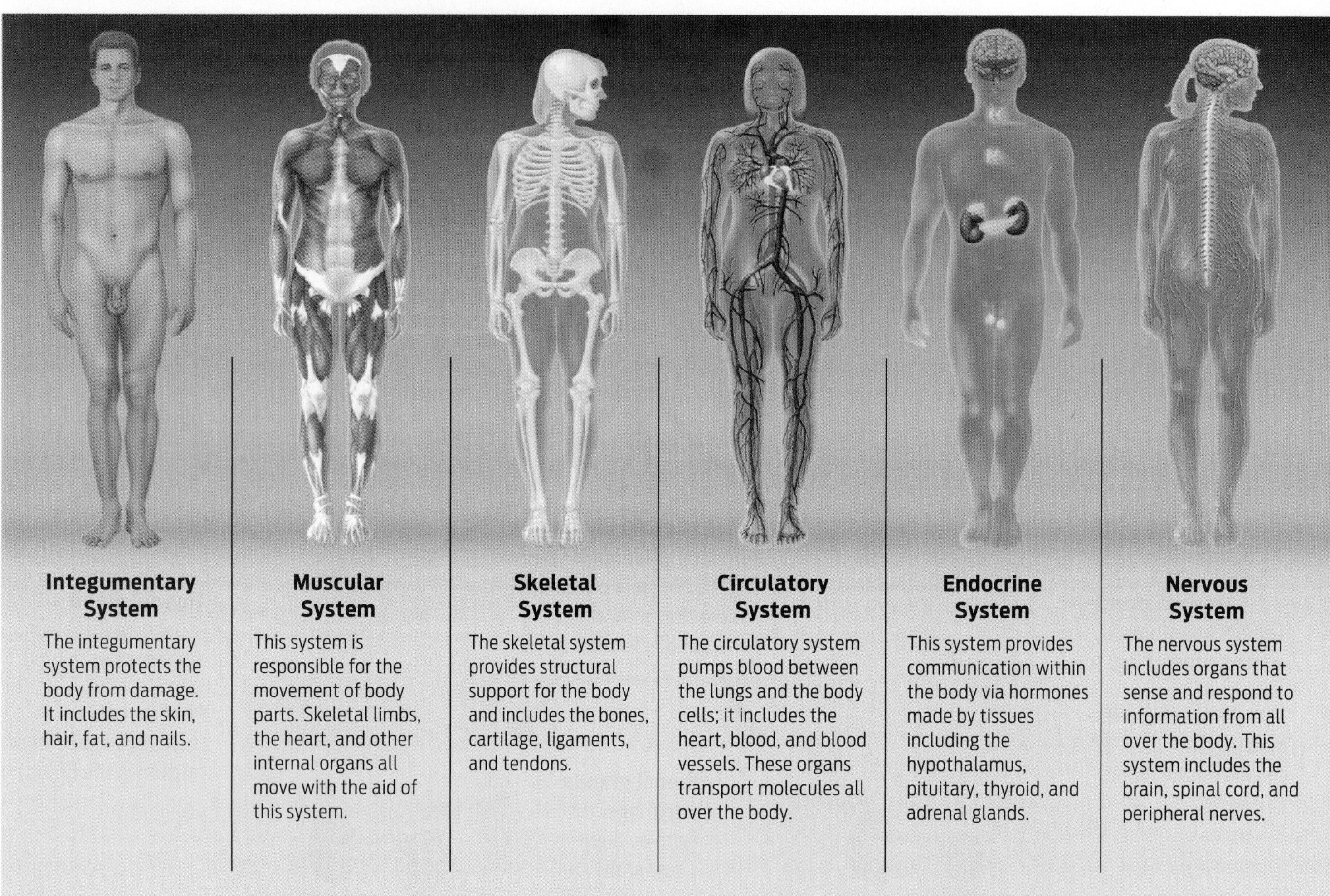

Integumentary System
The integumentary system protects the body from damage. It includes the skin, hair, fat, and nails.
Muscular System
This system is responsible for the movement of body parts. Skeletal limbs, the heart, and other internal organs all move with the aid of this system.
Skeletal System
The skeletal system provides structural support for the body and includes the bones, cartilage, ligaments, and tendons.
Circulatory System
The circulatory system pumps blood between the lungs and the body cells; it includes the heart, blood, and blood vessels. These organs transport molecules all over the body.
Endocrine System
This system provides communication within the body via hormones made by tissues including the hypothalamus, pituitary, thyroid, and adrenal glands.
Nervous System
The nervous system includes organs that sense and respond to information from all over the body. This system includes the brain, spinal cord, and peripheral nerves.

down and we could see the camp below. And the next minute, you couldn't see it," recalled Lou Kasischke, a member of Krakauer's team, who was one of 11 people trapped on the Col when the storm hit and who recounted his experience in the PBS documentary *Storm over Everest*. "Within the space of 5 minutes, it changed from really a good day with a little bit of wind to desperate conditions, something I'd never experienced the ferocity of before," said John Taske, another member of Krakauer's team, on the same PBS program.

According to Kent Moore, a physics professor at the University of Toronto who has studied the disaster, the storm that hit Everest that day also caused a particularly severe drop in barometric pressure, greatly reducing the availability of oxygen. The sudden drop in pressure may have caused severe physiological distress in the climbers. In particular, they would have experienced the mental side effects of extremely low oxygen levels, which include confusion and disorientation.

Unable to tell in which direction they were going, and not wanting to take a wrong turn and step off a cliff, the climbers were forced to hunker down in the hurricane-force winds and wait for the storm to abate. Eventually, after 4 long hours, the clouds parted long enough for one of them to see where they were. Six climbers who were able to walk made it back to camp during this lull. An additional three were brought back safely by the efforts of Anatoli Boukreev, a Russian guide who, having descended to Camp IV, went back to search for them.

But others were not so lucky. Two climbers, too weak to make it back to camp, suffered severe frostbite before being rescued. One

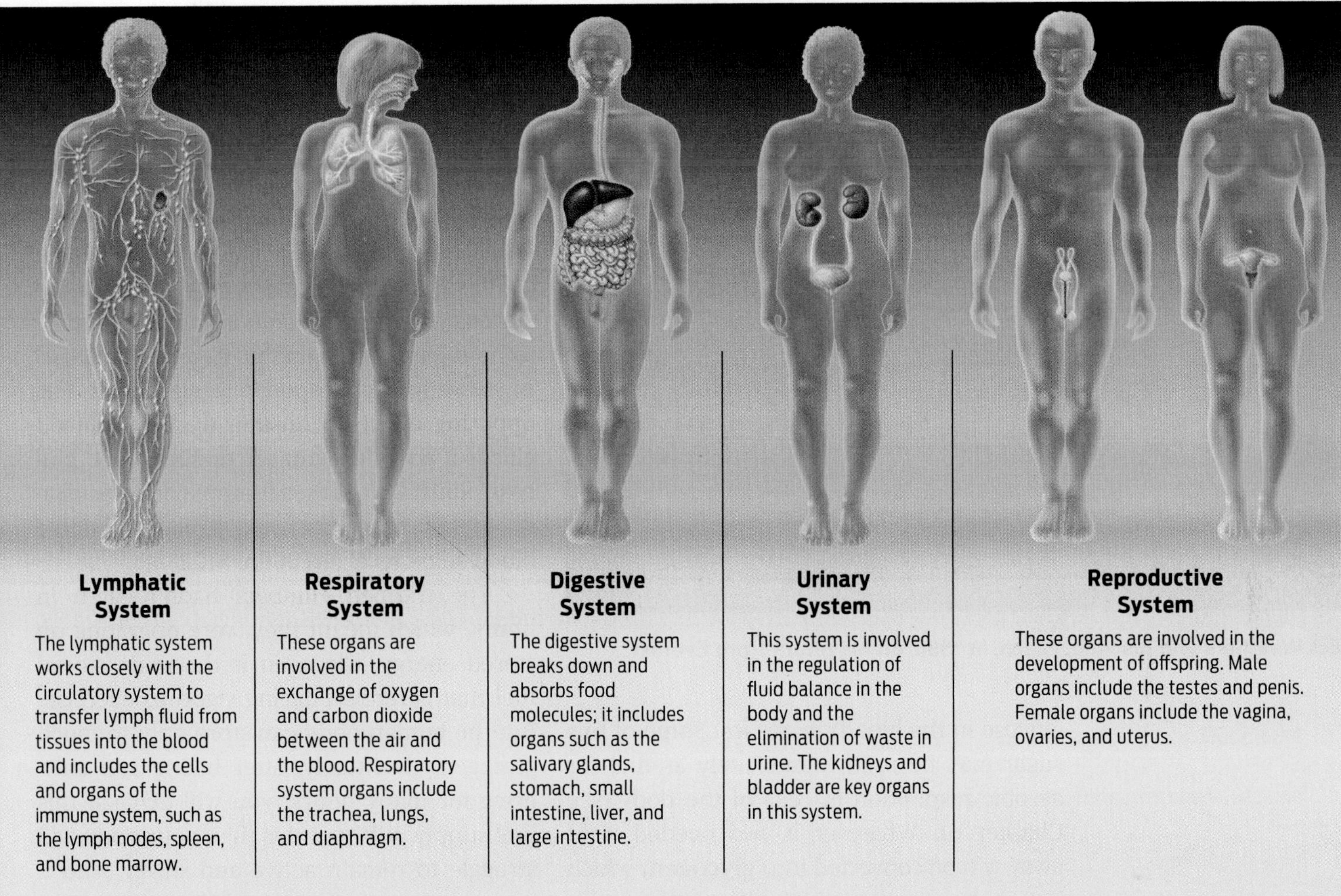

Lymphatic System
The lymphatic system works closely with the circulatory system to transfer lymph fluid from tissues into the blood and includes the cells and organs of the immune system, such as the lymph nodes, spleen, and bone marrow.
Respiratory System
These organs are involved in the exchange of oxygen and carbon dioxide between the air and the blood. Respiratory system organs include the trachea, lungs, and diaphragm.
Digestive System
The digestive system breaks down and absorbs food molecules; it includes organs such as the salivary glands, stomach, small intestine, liver, and large intestine.
Urinary System
This system is involved in the regulation of fluid balance in the body and the elimination of waste in urine. The kidneys and bladder are key organs in this system.
Reproductive System
These organs are involved in the development of offspring. Male organs include the testes and penis. Female organs include the vagina, ovaries, and uterus.

lost all his fingers and toes; the other had to have his right hand amputated. Those climbers stuck higher on the mountain–including Hall–could not be rescued. Trapped without shelter in the subzero temperatures all night, their supplemental oxygen and food long gone, the hikers eventually lost their ability to cope with the cold and succumbed to **hypothermia,** a drop in body temperature below 35°C (95°F). In all, eight climbers died on Everest that day.

This was not the first time that disaster had struck the summit. A 2008 study of all reported Everest deaths from 1921 to 2006 led by researchers at Massachusetts General Hospital found that more than 80% occurred above 26,000 feet, either during a summit attempt or the day after. While many of these deaths were attributable to traumatic injuries resulting from falls and avalanches, nearly as many were caused by hypoxia and hypothermia.

No Fuel Left to Burn

Although the body is able to cope with cold temperatures for some time through vasoconstriction and shivering, it cannot do so indefinitely. Thermoregulation is work, and work takes energy–roughly 150-300 kilocalories per hour for a 150-pound man. Eventually, if the body is not consuming food, it will run out of fuel.

The main fuel the body uses in times of intense activity is the sugar glucose, a breakdown product of carbohydrate digestion (see Chapter 4). When we eat carbohydrates, sugars are released and absorbed into the circulation, and blood sugar (the concentration of

HYPOTHERMIA
A drop of body temperature below 35°C (95°F), which causes enzyme malfunction and eventually death.

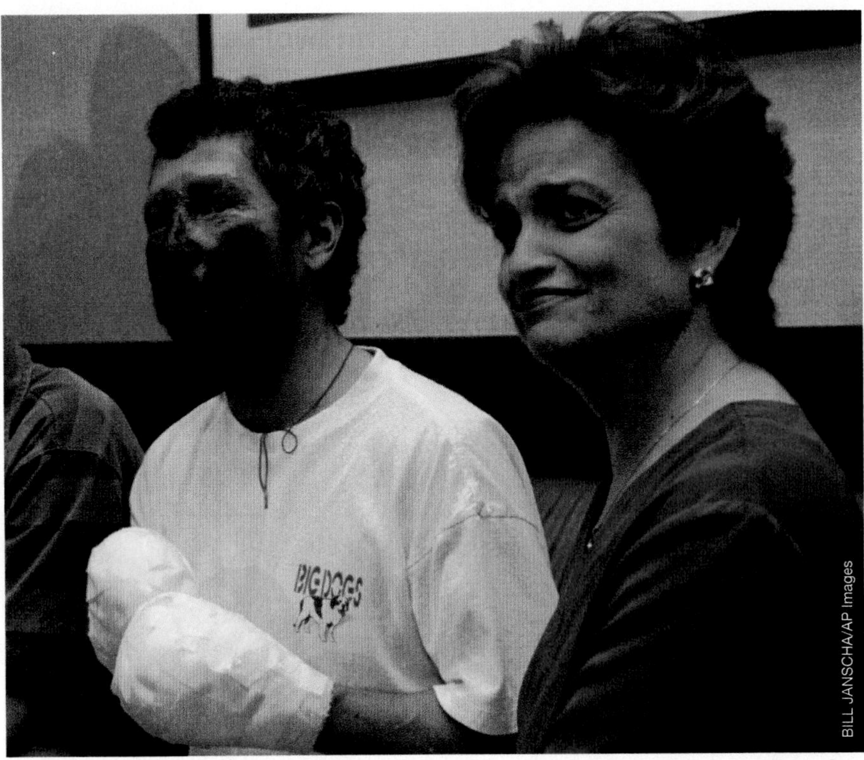

BILL JANSCHA/AP Images

Beck Weathers and his wife, Peach, in 1996, on his return from Everest.

glucose in the blood) increases. Some of this sugar may be used immediately as fuel for aerobic respiration in cells of the body (see Chapter 6). Whatever is not needed right away will be converted into **glycogen,** which is stored in muscles and the liver.

Blood sugar is monitored and controlled by the **pancreas,** a small organ near the stomach that functions in both the endocrine and digestive systems (see **Milestone 8: Stumbling on a Cure**). In response to high blood sugar, endocrine tissue in the pancreas produces the hormone **insulin,** which acts on liver and muscle cells, signaling them to remove sugar from the blood. The glucose taken up from the blood will either be used immediately for cellular respiration or stored as glycogen. Like many hormones, insulin acts by binding to receptors on target cells, much as a key fits into a lock.

When blood sugar is low, the body first prompts us to eat by sending a signal to the hypothalamus, which cues hunger. If eating isn't an option, the body begins to break down its stored glycogen. The key signal here is the hormone **glucagon,** which is released by the pancreas in response to low blood sugar and triggers liver and muscle cells to break down their stored glycogen into glucose. The liver can release this glucose into the blood, providing fuel for other tissues, particularly the brain. The breakdown of glycogen in skeletal muscle helps provide energy for the muscle itself.

The amount of glucose in the blood is therefore tightly regulated: when blood-sugar levels are high, excess glucose is stored in cells as glycogen (in response to insulin). The stored glycogen represents a source of glucose to be released into the blood during periods of starvation (in response to glucagon). The opposing effects of insulin (reducing blood glucose) and glucagon (increasing blood glucose) illustrate another example of homeostasis, in this case maintaining a relatively stable blood-sugar level (**INFOGRAPHIC 26.6**).

The trapped climbers hadn't eaten in hours, which meant they were operating on stored energy. Glycogen is the main stored fuel that is tapped during vigorous exercise. But the human body can store only so much glycogen. Eventually, after hiking and shivering for many hours, you will exhaust this fuel supply. Without this fuel, your body will struggle to remain active and shiver. And if you can't remain active and shiver, then you can't generate heat and your body temperature will fall. That's when hypothermia can set in.

The average adult has enough stored glycogen to power about 12 to 14 hours of routine activity. When a person is exercising strenuously–say, running or hiking–glycogen stores can be depleted in as little as 2 hours. Marathon runners often refer to this point, which occurs at about mile 20, as "hitting the wall." Then, in order to continue exercising, you must eat something–preferably something with carbohydrates.

"A lot of mountaineering communities think you need fat," says Kenefick. "And that's true–fat has more calories per gram–9 kcals per gram compared to 4 kcals per gram of protein or carbohydrate. But when you're doing things like shivering, those types of contractions, especially, use a lot of glucose." Fats–though a good source

GLYCOGEN
An energy-storing carbohydrate found in liver and muscle.

PANCREAS
An organ that secretes the hormones insulin and glucagon as well as digestive enzymes.

INSULIN
A hormone secreted by the pancreas that causes a decrease in blood sugar.

GLUCAGON
A hormone produced by the pancreas that causes an increase in blood sugar.

INFOGRAPHIC 26.6 The Pancreas Regulates Blood-Glucose Levels

The pancreas responds to variation in blood-glucose levels by secreting insulin, inducing cells to take in excess glucose, or by secreting glucagon, inducing liver and muscle cells to release stored glucose.

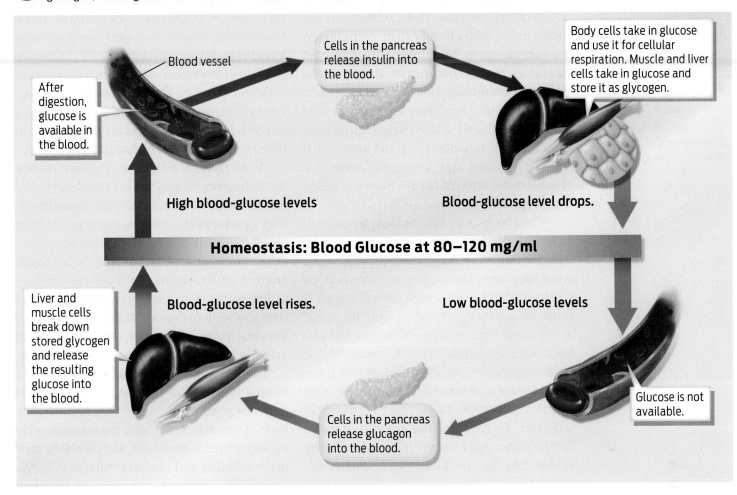

? What effects do insulin and glucagon have on blood sugar?

of stored energy–are not as readily available for immediate use. And glucose is the primary fuel for the brain.

Fitness also likely played a role in how the Everest climbers fared. Being fit means having more muscle mass relative to fat for a given body weight. Having more muscle mass means you have more glycogen and can exercise longer and generate more heat through cellular respiration. Someone who is less fit, or who simply has less muscle mass, will tire sooner, need to sit down and rest, and continue to lose heat to the environment. This is likely what happened to the climbers who were too weak to hike back to camp: they ran out of their glycogen stores sooner than other climbers.

Triggering Thirst

Another exacerbating factor would have been dehydration–a little-known cold-weather risk. In cold conditions, our bodies must work harder under the extra weight of heavy clothing, and sweat evaporates quickly in cold, dry air. We also lose a significant amount of water as water vapor when we exhale. The human body is about 65% water by weight, and when the total amount of water drops by only a few percent, we become dehydrated–which can cause dangerous side effects like delirium, confusion, and convulsions. Kenefick points out that people do not feel as thirsty when it's cold, and thus become even more

OSMOLARITY
The concentration of solutes in blood and other bodily fluids.

OSMOREGULATION
The maintenance of relatively stable volume, pressure, and solute concentration of bodily fluids, especially blood.

KIDNEY
An organ involved in osmoregulation, filtration of blood to remove wastes, and production of several important hormones.

dehydrated. "We're really tropical animals," says Kenefick. "We came from the sub-Sahara. We do much better in the heat."

Our body's sense of thirst is another physiological response that depends on the endocrine system. It relies on **osmolarity,** the concentration of solutes (the dissolved substances in a fluid) in the blood. Among the solutes in blood are electrolytes–ions such as sodium and potassium that are critical for nerve signaling and muscle contraction. Osmolarity is monitored by the hypothalamus, as is volume and pressure of bodily fluids; this monitoring of water balance is called **osmoregulation.**

When you are dehydrated–when you have less fluid in your blood–the concentration of solutes is higher. If the hypothalamus registers that the concentration of solutes in the blood is high, it will trigger a sense of thirst, encouraging you to drink. At the same time, it triggers the release of a hormone called antidiuretic hormone (ADH) from the pituitary, which travels through the bloodstream and acts on the **kidneys.** ADH signals the kidneys to excrete less water in the urine. By reducing the amount of water lost in urine, ADH causes more water to be reabsorbed by the kidneys back into the bloodstream. Water in the bloodstream dilutes solutes and lowers osmolarity. That's why people who are dehydrated have darker urine–it contains less water and so is more highly concentrated.

Osmoregulation depends on sensors that detect changes in blood volume and pressure. Sensors in the heart, for example, sense how full the heart's chambers are; sensors in blood vessels sense how stretched the vessels are. When low blood volume and pressure are detected, the hypothalamus responds by triggering the release of ADH from the pituitary into the blood, which acts on the kidneys to help retain water **(INFOGRAPHIC 26.7).**

With these multiple sensors for detecting dehydration, why do we feel less thirsty in the cold? The reason, says Kenefick, is that peripheral vasoconstriction pushes blood toward the core. All that blood pushed centrally is sensed by the body as a normal amount of hydration. As a result, the sensation of thirst is reduced, despite the fact that you're dehydrated. This is why it's very important to drink adequate amounts of water in winter, even when you aren't thirsty.

"Because water plays such a large role in cellular function," says Kenefick, "being dehydrated is going to put a greater stress on your body." Dehydration can alter the concentration of electrolytes in the blood, and therefore alter nerve function and muscle contraction. Dehydration also lowers blood pressure and thus makes the heart work harder. Together, these effects can have dangerous consequences, impairing thinking and coordination–two things that matter a great deal when you're navigating the treacherous terrain of the world's tallest mountain during a blizzard.

DEVENDRA SINGH/Getty Images

Jon Krakauer being interviewed by reporters after returning from Everest.

Warning Signs

Hypothermia isn't only a danger for death-defying mountain climbers: it's a leading cause of death during outdoor recreation like rafting and skiing, and is the number one way to lose your life while outdoors in cold weather. The Centers for Disease Control and Prevention estimate that hypothermia causes more than 1,000 deaths each year in the United States.

Wilderness-medicine experts say the best way to prevent hypothermia–in addition to dressing appropriately and carrying plenty of

INFOGRAPHIC 26.7 The Kidneys Respond to Changes in Water Balance

The amount of water in the circulation controls the concentration of dissolved molecules in the blood and also determines blood volume and blood pressure. The kidneys control water availability by responding to a variety of signals.

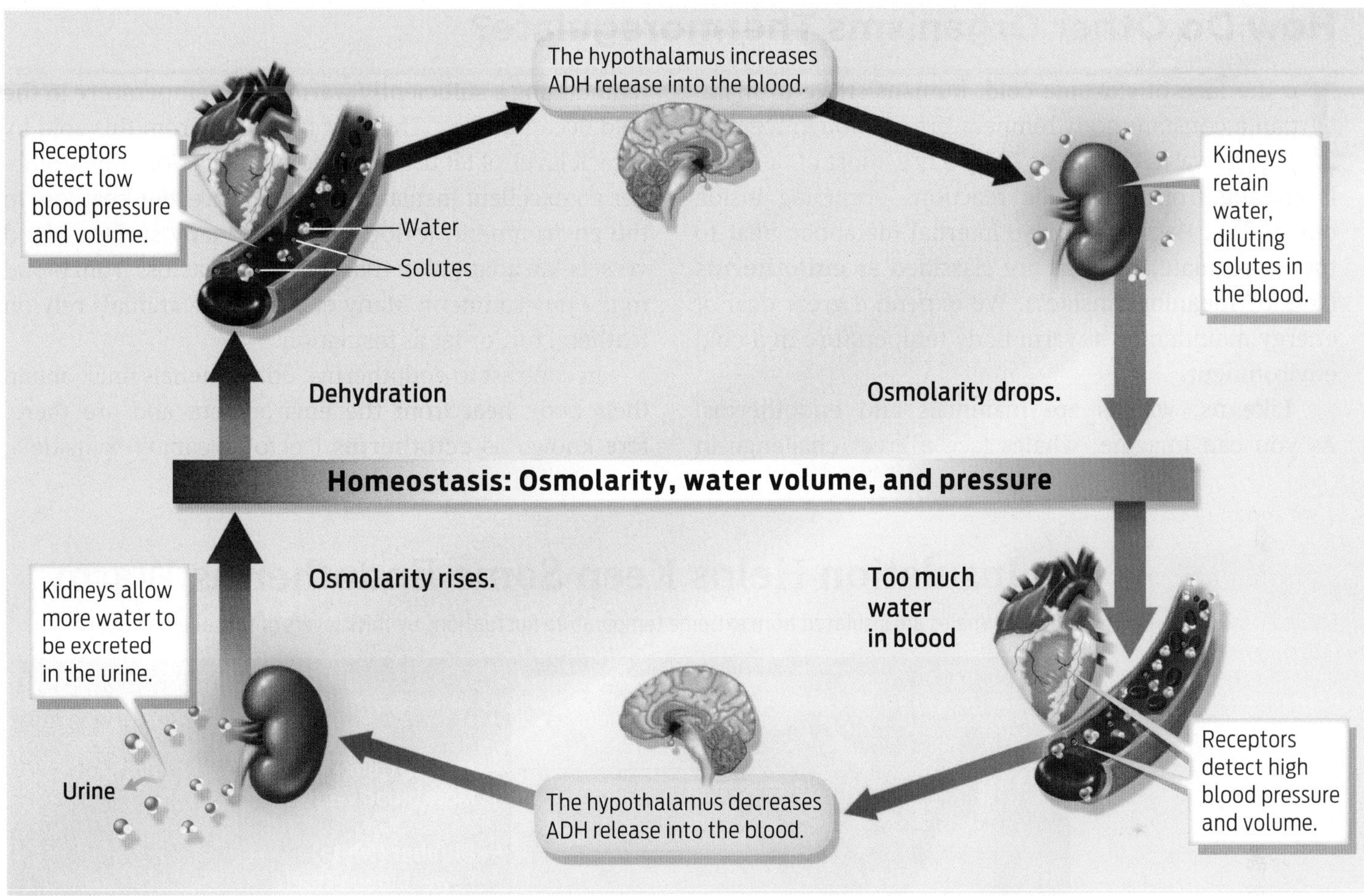

The hypothalamus increases ADH release into the blood.
Receptors detect low blood pressure and volume.
Water
Solutes
Kidneys retain water, diluting solutes in the blood.
Dehydration
Osmolarity drops.
Homeostasis: Osmolarity, water volume, and pressure
Osmolarity rises.
Too much water in blood
Kidneys allow more water to be excreted in the urine.
Urine
The hypothalamus decreases ADH release into the blood.
Receptors detect high blood pressure and volume.

? Where is ADH made, what organ does it act on, and what impact does it have on its target organ?

food and water–is to be aware of its signs. In particular, watch for the "umbles": stumbles, mumbles, fumbles, and grumbles, which show changes in motor coordination and altered brain function. If you experience any of these signs in cold conditions, it's time to seek shelter.

None of the climbers who died on Everest in 1996 was an inexperienced climber–three of them, in fact, were professional guides. Why didn't they heed these physiological warning signs? Part of the reason is that there was simply no time. The swift-moving storm made the decision for them. But the climbers had also earlier made questionable choices that affected their fate. Whether from overconfidence or brain-addled thinking, they continued climbing toward the summit even when the hour was late. In the end, the climbers made a fatal wager with biology: in their race to the summit, they pushed themselves beyond the breaking point, overestimating, in Krakauer's words, "the thinness of the margin by which human life is sustained above 25,000 feet."

Not long after the disaster, Krakauer returned to climbing mountains. In a 1997 interview with *Bold Type* magazine, he was asked whether he was fearful of climbing again after the trauma he experienced on Everest. The chastened climber replied: "It wasn't like 'Am I afraid of this?' It was more like 'Is this right? Is it too selfish?' I won't go back to Everest–I'm afraid of that." ■

FOR **COMPARISON**

How Do Other Organisms Thermoregulate?

In the face of extreme cold, humans strive to maintain a constant body temperature. Although we may pile on warm clothes or sit by a fire, most of our heat is coming from metabolic reactions occurring inside our bodies. Because we use internal metabolic heat to thermoregulate, humans are classified as **endotherms** ("endo" meaning "inside"). We expend a great deal of energy maintaining a warm body temperature in a cold environment.

Like us, whales are mammals and endotherms. As you can imagine, whales face a great challenge in maintaining a sufficiently warm body temperature in the cold ocean depths. They are protected from this cold by a thick layer of fat tissue called blubber. Not only is blubber an excellent insulator that helps prevent heat loss to the environment, it does not have many surface blood vessels—an adaptation that prevents heat loss from blood to the environment. Many endothermic animals rely on feathers, fur, or fat as insulation.

In contrast to endotherms, other animals must obtain their body heat from the environment and are therefore known as **ectotherms** ("ecto" meaning "outside").

Insulation Helps Keep Some Endotherms Warm

Whales are insulated from extreme temperature fluctuations by thick layers of fat called blubber.

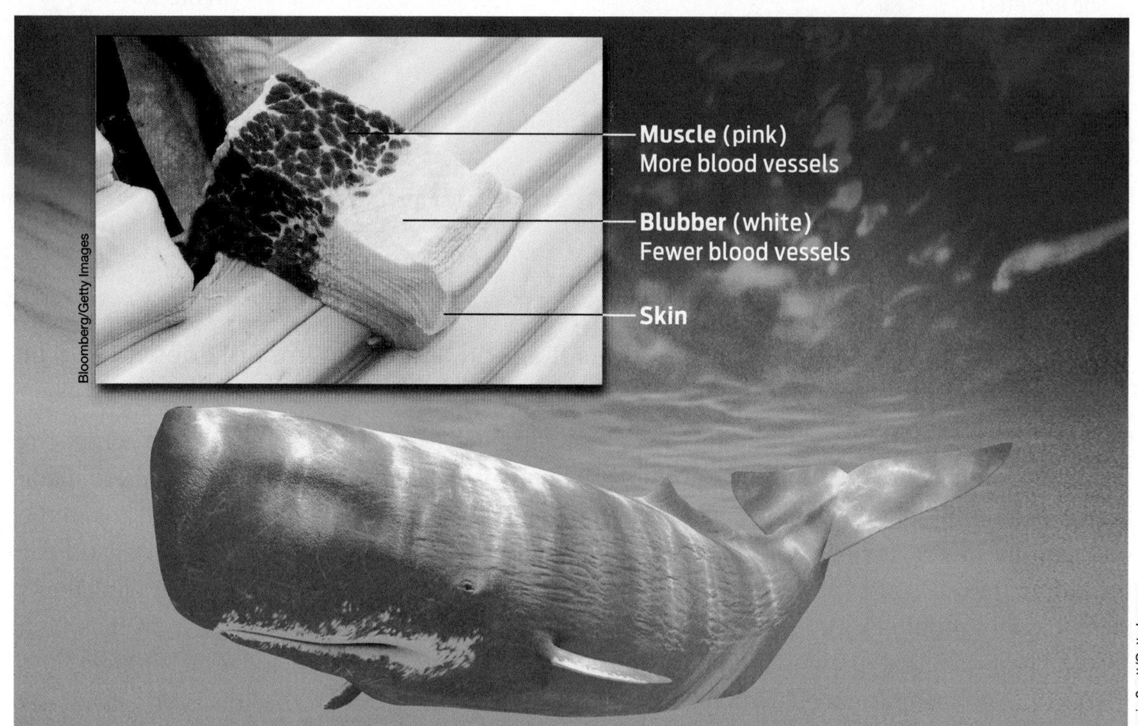

ENDOTHERM
An animal that can generate body heat internally to maintain its body temperature.

ECTOTHERM
An animal that relies on environmental sources of heat, such as sunlight, to maintain its body temperature.

Fish, reptiles, and amphibians are all ectotherms. While these animals are commonly referred to as "cold-blooded," their body temperature actually mimics that of their environment. Through behavioral adaptations, many ectotherms also maintain a relatively stable body temperature (although by definition not as stable as that of endotherms). For example, lizards bask in the sun to warm up and seek out shade or a protected burrow to prevent overheating. Through these behaviors, lizards keep their bodies in a temperature range compatible with their metabolism.

Most fish are ectotherms and can do something similar—swim to warmer or cooler water to maintain their body temperature. But some marine-dwelling fish are exceptions. Fish such as swordfish and marlin that dive to great depths to hunt cannot rely solely on external sources of heat to regulate their temperature. Instead, they rely on a type of regional endothermy in the form of specialized "heater tissue" that generates heat to keep their brains warm. The heater tissue is modified eye muscle that acts to generate heat rather than force or movement. This form of heating is a type of nonshivering thermogenesis.

Ectotherms Rely on the Environment to Regulate Body Temperature

Ectotherms can increase their body temperature by basking in the sun and avoid overheating by moving into the shade.

jimbo /FeaturePics

Clément Philippe/AGE Fotostock

Another form of nonshivering thermogenesis, seen in bats (and human babies), occurs in a tissue called brown fat. Brown fat is located in the neck and shoulders areas and is several degrees Celsius warmer than the rest of the body. In brown fat, specialized mitochondria convert energy to heat rather than to ATP, and the many blood vessels in brown fat deliver that heat to other parts of the body.

Some Organisms Generate Heat from Nonshivering Thermogenesis

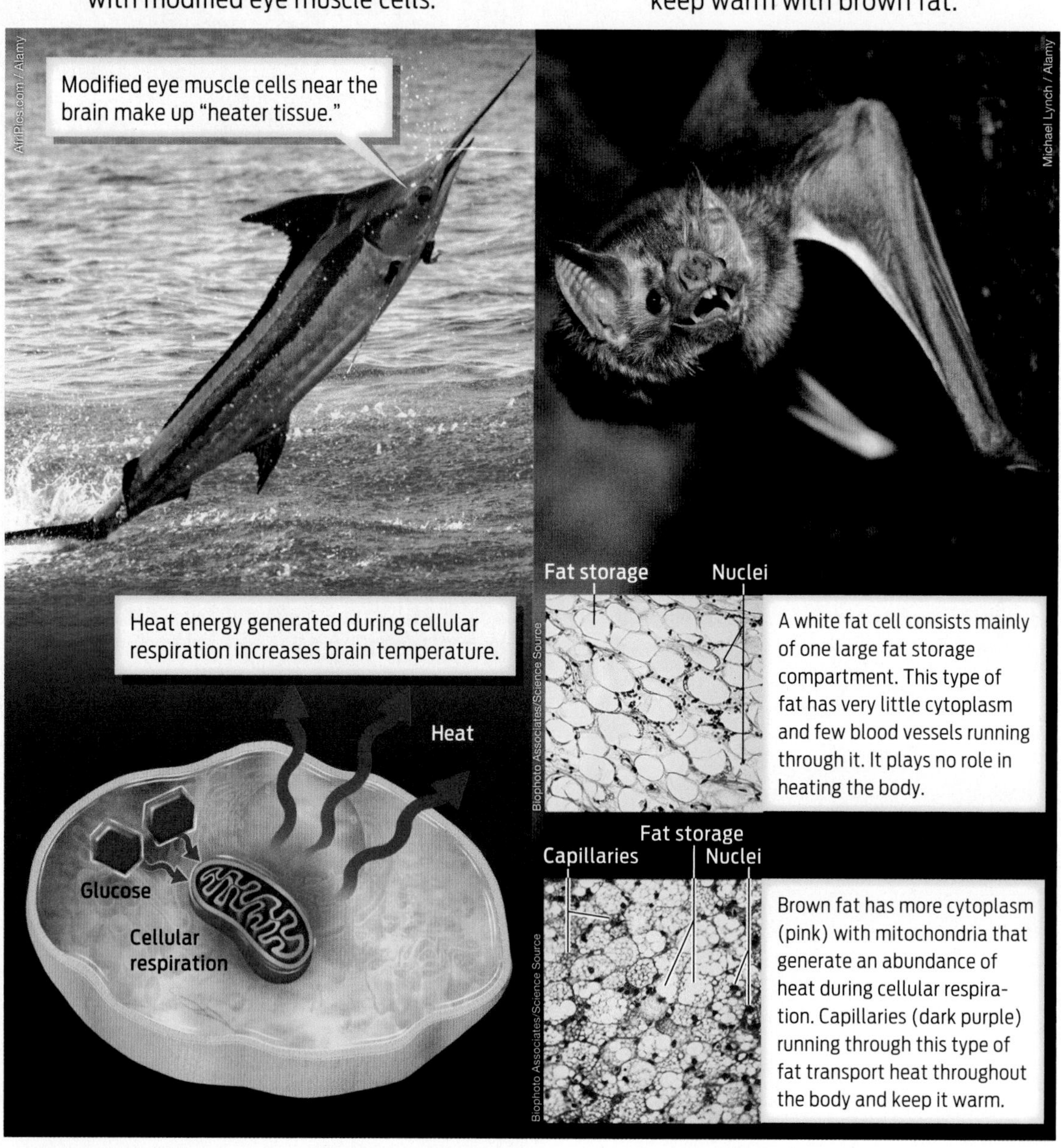

Marlins heat their brains with modified eye muscle cells.
Small hibernating endotherms keep warm with brown fat.
AriPics.com / Alamy
Michael Lynch / Alamy
Modified eye muscle cells near the brain make up "heater tissue."
Heat energy generated during cellular respiration increases brain temperature.
Heat
Glucose
Cellular respiration
Fat storage
Nuclei
Biophoto Associates/Science Source
A white fat cell consists mainly of one large fat storage compartment. This type of fat has very little cytoplasm and few blood vessels running through it. It plays no role in heating the body.
Fat storage
Capillaries
Nuclei
Biophoto Associates/Science Source
Brown fat has more cytoplasm (pink) with mitochondria that generate an abundance of heat during cellular respiration. Capillaries (dark purple) running through this type of fat transport heat throughout the body and keep it warm.

CHAPTER 26 SUMMARY

- Living organisms have an anatomical structure that is adapted to suit their physiological functions.
- Humans and other multicellular organisms are organized hierarchically: cells assemble to make up tissues; tissues congregate to form organs; organs work together as part of organ systems.
- Humans have many different organ systems that cooperate to accomplish basic physiological tasks, such as obtaining energy, taking in nutrients to build new molecules during growth and repair, and ridding themselves of wastes.
- Most organisms cannot tolerate wide fluctuations in their internal environment; their bodies work to maintain homeostasis, a stable internal environment.
- The process whereby organisms maintain a relatively constant internal temperature is called thermoregulation.
- The body responds to cold temperatures in two main ways: by conserving the heat it has through vasoconstriction, and by generating more heat through shivering. When overheated, the body releases heat by vasodilation and sweating.
- Maintaining homeostasis requires both sensors and effectors. Sensors include nervous system receptors that detect changes in a variety of internal states (for example, temperature and blood pressure). Effectors include the endocrine glands and muscles that respond to an abnormal state in an effort to correct it.
- Sensors and effectors work together as part of a circuit or feedback loop. Negative feedback loops are important in homeostasis.
- The endocrine system produces hormones—chemical messengers that travel through the bloodstream, bind to receptors on a target cell, and effect a change in that cell. Insulin and glucagon are hormones that regulate blood-glucose levels.
- Osmoregulation is the control of water balance in the body. Sensors detect blood pressure, blood volume, and solute concentration. Kidneys are important effectors in maintaining water balance.
- Maintaining homeostasis is work and requires adequate energy and oxygen to power cellular respiration.
- Humans (and other mammals) are endotherms: we generate heat internally. Many other organisms, such as reptiles and fish, are ectotherms: they rely on behavior and the environment to maintain a temperature compatible with life.

MORE TO EXPLORE

- Krakauer, J. (1997) *Into Thin Air: A Personal Account of the Mt. Everest Disaster*. New York: Random House.
- *Everest* (2015). Film, Universal Pictures. Based on Weathers, B. (2002) *Left for Dead: My Journey Home from Everest*. New York: Villard Books.
- PBS (2008) Frontline: *Storm over Everest*: www.pbs.org/wgbh/pages/frontline/everest/
- Firth, P. G., et al. (2008) Mortality on Mount Everest, 1921–2006: descriptive study. *British Medical Journal* 337:a2654.
- Moore, G. W. K., et al. (2006) Weather and death on Mount Everest: an analysis of the *Into Thin Air* storm. *Bulletin of the American Meteorological Society* 87:465–480.

CHAPTER 26 Test Your Knowledge

DRIVING QUESTION 1 How are the bodies of living organisms organized?

By answering the questions below and studying Infographic 26.1, you should be able to generate an answer for the broader Driving Question above.

KNOW IT

1 **Compare and contrast anatomy and physiology.**

2 **Organize the following terms on the basis of level of structure, from the simplest (1) to the most complex (4).**

_____ Small intestine
_____ Mucus-secreting cell of the small intestine
_____ Digestive system
_____ Layer of muscle that contributes to the function of the small intestine

3 **Which of the following groups is in the correct order of organization from most inclusive level to lowest level?**

a. tissues, cells, organ systems, organs
b. organ systems, organs, tissues, cells
c. cells, organ systems, tissues, organs
d. cells, tissues, organs, organ systems
e. cells, organs, organ systems, tissues

USE IT

4 **An emergency room doctor setting a complex bone fracture is relying primarily on knowledge of**

a. anatomy.
b. physiology.
c. thermoregulation.
d. homeostasis.
e. osmoregulation.

5 **Is a personl trainer who works with clients to help them lose weight through a combination of diet and exercise focusing primarily on anatomy or physiology? Explain your answer.**

6 **Why is the heart considered an organ and not a tissue?**

DRIVING QUESTION 2 How do humans and other organisms maintain homeostasis?

By answering the questions below and studying Infographics 26.2, 26.3, 26.4, 26.6, 26.7 and For Comparison: How Do Other Organisms Thermoregulate? you should be able to generate an answer for the broader Driving Question above.

KNOW IT

7 **What is homeostasis?**

8 **How does brown fat contribute to thermoregulation in bats?**

a. by providing insulation to retain heat
b. by providing a highly vascularized tissue to release heat to the environment
c. by generating heat through shivering
d. by generating heat via cellular respiration in specialized mitochondria
e. b and d

9 **Describe the feedback loop involved in thermoregulation in cold conditions. Use the following terms in your answer: *hypothalamus, sensor, muscle, effector, low body temperature, normal body temperature*.**

10 **Which hormone regulates water conservation by the kidneys?**

a. insulin
b. ADH
c. glucagon
d. ADH and insulin work together

USE IT

11 **How could damage to the hypothalamus prevent shivering even if the core body temperature drops dramatically?**

12 **Glucagon is released as part of the response to a drop in body temperature. Why do you think this happens?**

13 **Name two or three physiological responses that could help the body dissipate heat during exertion on a hot day. For each mechanism that you propose, explain how it would dissipate heat.**

DRIVING QUESTION 3 What are homeostatic feedback loops, and which physiological systems are regulated by them?

By answering the questions below and studying Infographics 26.6 and 26.7, you should be able to generate an answer for the broader Driving Question above.

KNOW IT

14 What internal signals are associated with dehydration?

- **a.** high osmolarity
- **b.** low blood volume
- **c.** low blood pressure
- **d.** all of the above
- **e.** a and b

15 People who are severely dehydrated produce ____________ of urine that is ____________.

- **a.** a high volume; highly concentrated and dark in color
- **b.** a high volume; dilute and light in color
- **c.** a low volume; highly concentrated and dark in color
- **d.** a low volume; dilute and light in color
- **e.** a normal volume; a normal color (neither very light nor very dark)

16 Insulin is released from the ________ in response to ________.

- **a.** pancreas; elevated blood sugar
- **b.** pancreas; low levels of blood sugar
- **c.** hypothalamus; elevated blood sugar
- **d.** hypothalamus; low levels of blood sugar
- **e.** pituitary; signals from the hypothalamus

USE IT

17 What conditions might cause high levels of insulin in the bloodstream? What events would follow?

18 Tibetan Sherpas, many of whom serve as guides and rescuers on Everest, often do not require bottled oxygen to reach the summit. Why might Tibetans, who have lived at high elevations for many generations, have an easier time than others with hypoxia? (Think about both short-term and long-term changes.)

MINI CASE

apply YOUR KNOWLEDGE

19 Hypertension (high blood pressure) has been diagnosed in a 65-year-old woman. She needs medication that will return her blood pressure to normal (and safe) levels. Her doctor tells her that there are two main categories of drugs for hypertension: thiazides, a type of diuretic, and ACE inhibitors, which help relax blood vessels and prevent their constriction. Explain how both of these could reduce blood pressure.

INTERPRETING DATA

apply YOUR KNOWLEDGE

20 Jonas and Jennifer have abnormal fasting glucose levels in their blood. They each had a blood test to measure the levels of insulin in their blood (also measured after fasting). Their blood values for both glucose and insulin are shown in the table below. From the data shown, is Jonas or Jennifer more likely to have type 2 diabetes (which is characterized by an inability of cells to respond to insulin)? Which of them is more likely to have type 1 diabetes (which results from a failure of insulin production)? Explain all their blood test results.

	Fasting Glucose Level (mg/dL)	Fasting Insulin Level (international units, μIU/mL)
Jonas	115	1
Jennifer	119	35
Normal Range	70–100	5–25

BRING IT HOME

apply YOUR KNOWLEDGE

21 The U.S. National Park Service has to rescue stranded hikers, often at great expense. Do you think that hikers' level of preparation should be a factor in determining whether or not they should bear the cost of their rescue? What factors would you consider to determine whether or not a hiker was adequately prepared? Give a physiological reason for each factor that you propose.

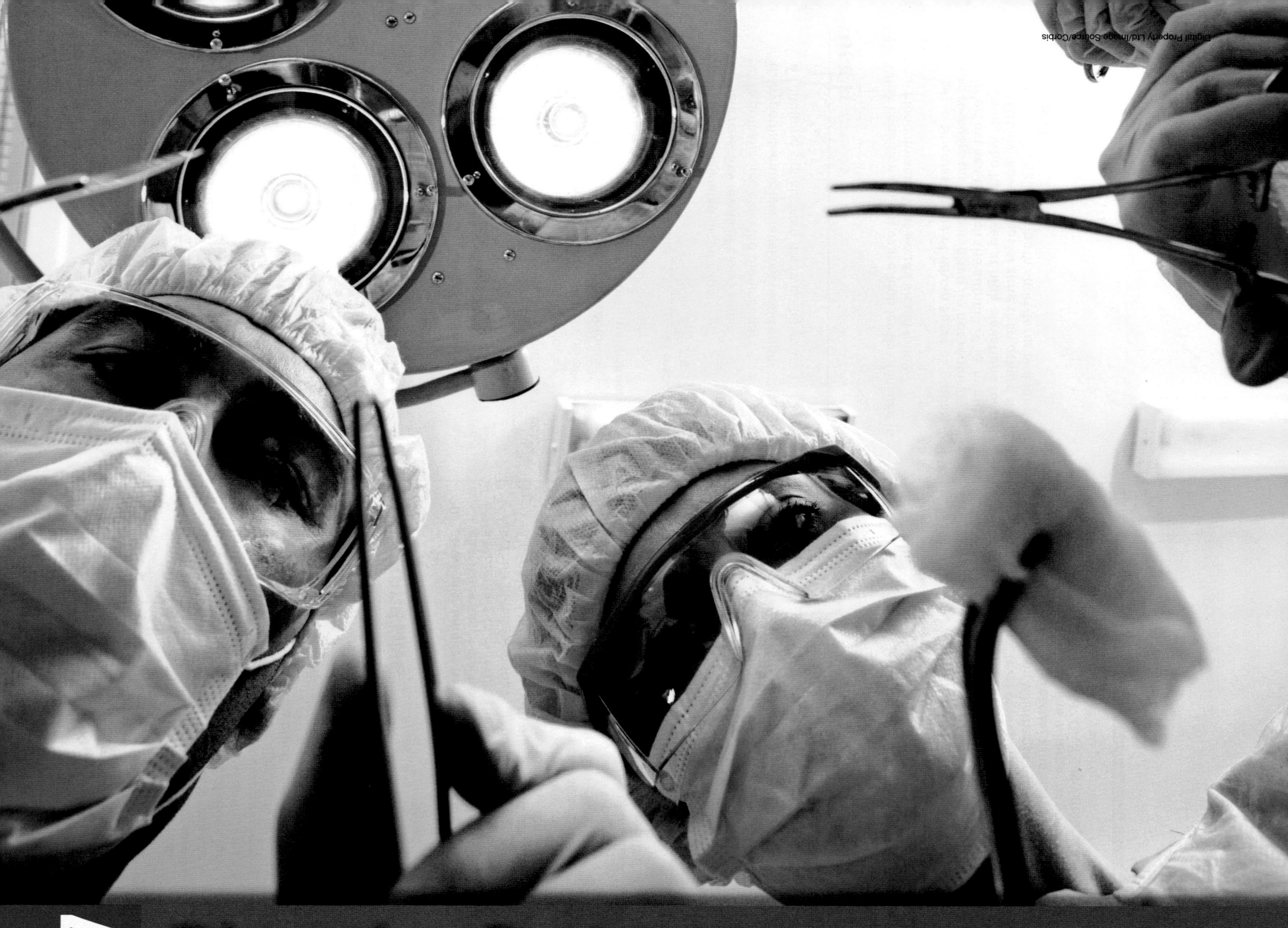

Digital Property Ltd/Image Source/Corbis

27 Digestive System

DRASTIC MEASURES

For the morbidly obese, stomach-shrinking surgery is a last resort

Amy Jo Smith hardly recalls a time growing up when her family wasn't dieting. Her parents were both obese, and they were always trying to lose weight.

Smith herself was relatively slender until her senior year in high school, when her weight began to creep up. She grew up on a horse farm in northeast Maryland, and as a teenager spent much of her spare time on the road, taking her horses to shows. She attributes her weight gain to a diet that consisted primarily of fast food. "I was always eating on the run," says Smith, now 38 years old and a computer literacy teacher. But, she says, her growing girth "never stopped me from doing the things I wanted to do." The extra weight did bother her, though, and she tried several diets and diet pills, only to see her weight yo-yo up and down.

DRIVING QUESTIONS

1. What is the anatomy of the digestive system?
2. How is food broken down and utilized as it moves through the digestive tract?
3. How does bariatric surgery change the digestive tract and digestion, and what are the risks and benefits of bariatric surgery?

In 2004, at a routine checkup, Smith's doctor noticed that Amy Jo was suffering a number of ills that were likely caused by Smith's 264-pound weight. For one thing, she had been experiencing migraines. She also had stress incontinence: her bladder would leak when she coughed or laughed. "I thought that was just normal," she says. And she went months at a time without having a period—a telltale sign of a hormonal imbalance often associated with obesity. Smith's physician suggested that, to lose weight, Smith consider having a surgical procedure that would shrink her stomach to the size of a golf ball—a drastic measure, but one that her doctor believed was warranted given the health complications caused by Smith's weight.

Studies show that even 10 years after surgery, most patients still weigh 25% to 30% less than they did before the surgery.

Obesity is a medical condition defined as weighing 20% or more than is recommended for one's height or having a body mass index (BMI) of 30 or higher (see Chapter 6). Morbid obesity—sometimes called clinically severe obesity—is defined as being 100 pounds or more overweight or having a BMI of 40 or higher. Obesity becomes "morbid" when it significantly increases the risk of one or more obesity-related health conditions or serious diseases, such as heart disease and diabetes. At 5′ 2″ and with a BMI of 48, Smith had become morbidly obese.

Amy Jo Smith in 2004.

Even then, Smith had a hard time accepting that she needed such a drastic method to lose weight. "At first I thought he was a quack," she says of her doctor. But she began to think more about the surgery after a friend underwent the stomach-shrinking surgery with dramatic results.

There are several types of bariatric, or weight-loss, surgery. (The word "bariatric" comes from the Greek *báros,* meaning "heaviness.") All involve surgically reducing the size of the stomach, either temporarily or permanently. The specific type of surgery recommended depends on the individual patient's medical history and weight-loss goal. Because there are many associated risks, a National Institutes of Health panel of experts has recommended surgery only for people considered morbidly obese—people whose risk of death from diabetes or heart disease because of excess weight is five to seven times greater than for those of average weight.

But bariatric surgery is no miracle cure. "It's sort of barbaric," says Monica Skarulis, director of the Metabolic Clinical Research Unit at the National Institutes of Health. Because the surgery so drastically reduces the size of the stomach and restricts how much a person can eat, it amounts to "forced behavior control," since patients must live on a strict diet, she says. If they overeat, they suffer nasty side effects such as vomiting

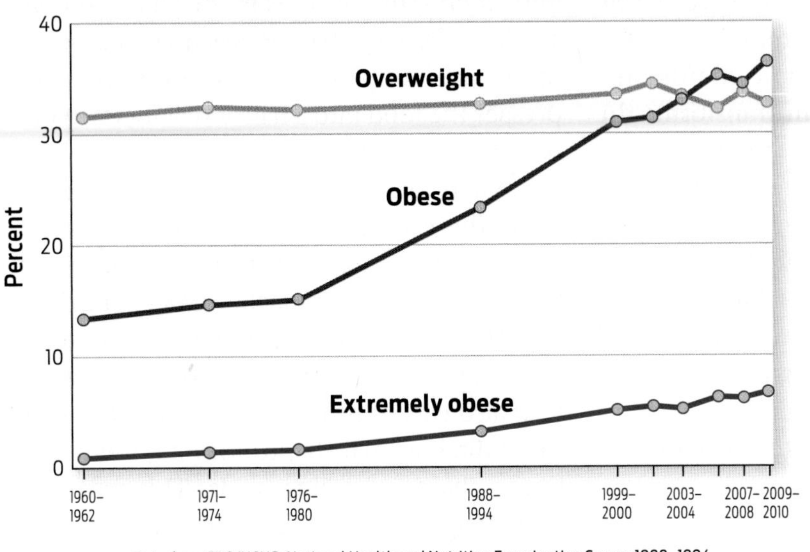

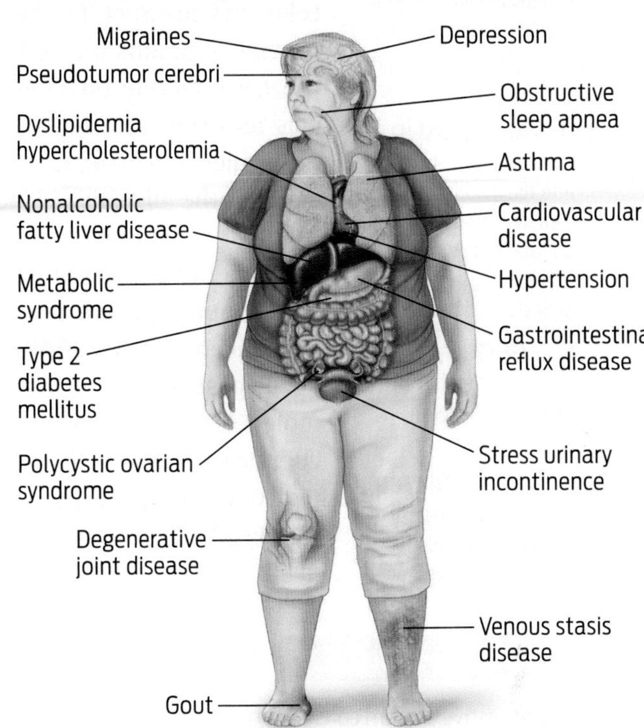

and diarrhea. On top of that, some bariatric surgery patients suffer mineral and vitamin deficiencies over the long term that cause bone loss and potentially other health impairments. And the surgery itself is risky: as many as 20% of patients suffer complications a year after the surgery that are severe enough to put them back in the hospital.

For some morbidly obese people, however, the risk of dying from obesity-related diseases is higher than the risk of surgical complications. And for weight reduction, the surgery is more effective than lifestyle changes alone. Almost all patients lose 30% to 50% of their excess weight in the first 6 months and 77% of their excess weight after about a year. Studies show that even 10 years after surgery, most patients still weigh 25% to 30% less than they did before the surgery. Consequently, demand for the surgery has soared in the United States since the early 2000s, peaking at about 220,000 surgeries a year in 2009, and today hovering around 180,000 per year.

Food Fix

The rationale behind bariatric surgery is simple: by reducing the amount of food the **stomach** can hold, the surgeries prevent overeating. Reducing the amount of food taken in means less food is digested, fewer calories are absorbed into the body, and eventually weight is lost.

The stomach is an easy target for a surgical fix to overeating because it's relatively simple to reduce its size surgically. But care must be taken to make sure that the rest of the **digestive system** still works as it's supposed to—breaking down food molecules into smaller subunits, absorbing nutrients, and eliminating waste.

The digestive system can be thought of as having two main components: a central digestive tract—essentially a long tube lined with muscles that extends from the mouth to the anus; and accessory organs that flank the tract and assist in digestion. (The stomach is one part of the central digestive tract.) As

STOMACH
An expandable muscular organ that stores and mechanically breaks down food. Specific enzymes in the stomach digest proteins.

DIGESTIVE SYSTEM
The organ system that breaks down food molecules into smaller subunits, absorbs nutrients, and eliminates waste; it is composed of the digestive tract and accessory organs.

DIGESTION
The mechanical and chemical breakdown of food into subunits, enabling the absorption of nutrients.

SALIVARY GLANDS
Glands that secrete enzymes into the mouth to break down macromolecules in food. One such enzyme is salivary amylase, which digests carbohydrates.

ESOPHAGUS
The section of the digestive tract between the mouth and the stomach.

PERISTALSIS
Coordinated muscular contractions that force food down the digestive tract.

the muscles of the digestive tract alternately relax and contract, food is pushed along. The accessory organs located along the length of the tract secrete enzymes and other chemicals into the tract to help break down food molecules. Through these coordinated actions, the digestive system transforms the food we eat into a form our bodies can use and rids the body of the waste left over once usable nutrients are removed from food we have taken in (**INFOGRAPHIC 27.1**).

Digestion–the breaking down of food molecules–relies on both mechanical and chemical processes. These begin as soon as we put food into our mouths–that is, as soon as we ingest it. The act of chewing mechanically breaks food down into smaller pieces, while **salivary glands** secrete enzymes into saliva that chemically dismantle macromolecules into their subunits. The enzyme salivary amylase, for example, breaks down carbohydrates into simpler sugars. The tongue compresses the food into a ball and works it to the back of the mouth.

When we swallow, food is propelled along the **esophagus** by rhythmic waves of muscle contractions called **peristalsis.** Food then enters the stomach, where stomach acid destroys harmful bacteria and protects us against food-borne diseases. Stomach acid also causes proteins in food to lose their three-dimensional shapes, turning them into linear chains of amino acids. This makes it easier for the enzyme **pepsin,** which is produced in the stomach, to chemically break proteins apart into individual amino acids.

Like the esophagus, the stomach is muscular, expanding and contracting as it accepts food and churns it. Each time it contracts, stomach acid mixes with food, producing a soupy mixture called **chyme** (**INFOGRAPHIC 27.2**).

Despite the powerful acid churning inside it, the stomach remains intact. This is possible because the stomach is lined with a thick layer of protective mucus. Occasionally, this mucus layer is damaged–by a bacterial infection, for example–and the stomach lining becomes more vulnerable to gastric juices; the result is a painful sore called an ulcer.

While the stomach can absorb some substances directly into the bloodstream–water, ethanol, and certain drugs, for example–most of the chyme is pushed farther down the digestive tract, where it is further processed.

Although the stomach is only a small part of the upper digestive tract, it plays a large part in weight gain. Evolutionarily speaking, the reason we have a stomach in the first place is to enable us to temporarily store the food we eat. Without a stomach, we would have to eat constantly to fuel our activities. When we eat a large meal, the stomach expands greatly

INFOGRAPHIC 27.1 The Digestive System

The digestive system consists of a long tube with specialized sections and accessory organs that secrete enzymes and other chemicals into the digestive tract. As food travels down the digestive tract, macro-molecules are broken into subunits, nutrients are absorbed, and waste is eliminated.

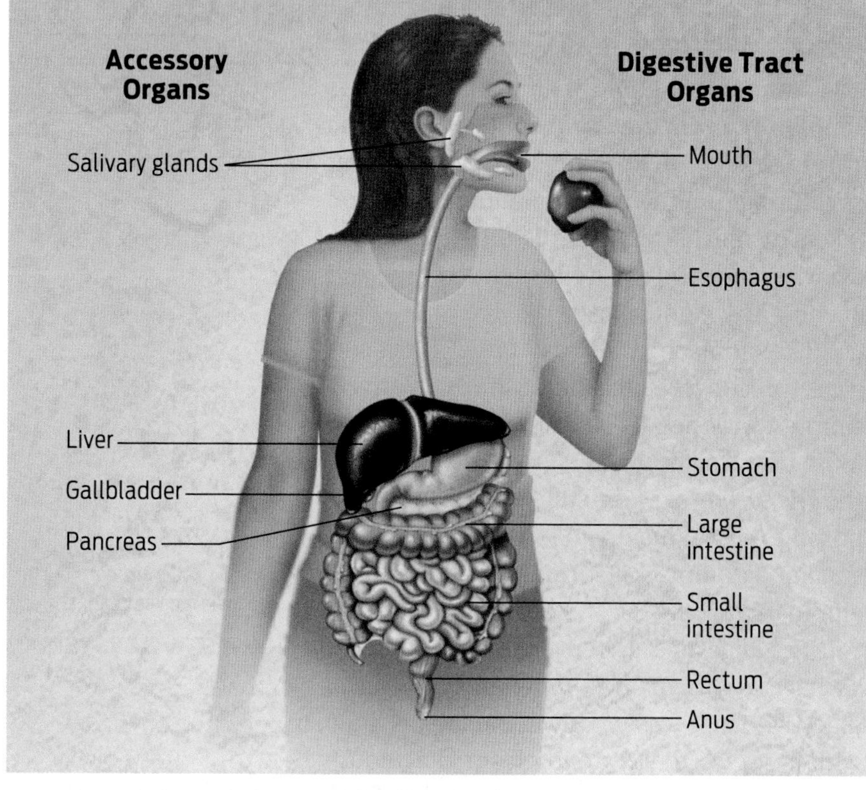

? List the accessory organs in the digestive tract.

INFOGRAPHIC 27.2 The Upper Digestive System

 The upper digestive tract includes the mouth, esophagus, and stomach, as well as enzymes and other chemicals secreted by the salivary glands and the stomach.

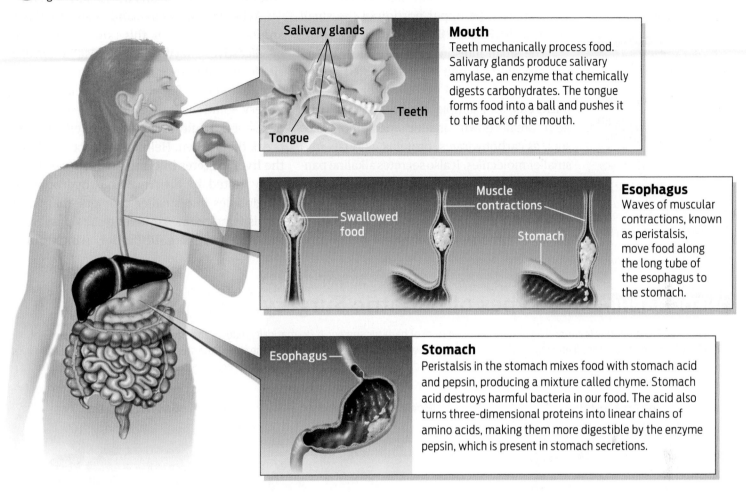

Mouth
Teeth mechanically process food. Salivary glands produce salivary amylase, an enzyme that chemically digests carbohydrates. The tongue forms food into a ball and pushes it to the back of the mouth.

Esophagus
Waves of muscular contractions, known as peristalsis, move food along the long tube of the esophagus to the stomach.

Stomach
Peristalsis in the stomach mixes food with stomach acid and pepsin, producing a mixture called chyme. Stomach acid destroys harmful bacteria in our food. The acid also turns three-dimensional proteins into linear chains of amino acids, making them more digestible by the enzyme pepsin, which is present in stomach secretions.

? Name two key events that occur in the stomach.

to accommodate and store all that food. It's partly because of this elasticity that we can eat enough to sustain us for hours. But this elasticity also means that we can eat more than our bodies need.

To Sleeve or Bypass

Because of Smith's weight-related illnesses, her doctor recommended that she have bariatric surgery. The doctor explained that bariatric surgery changes the anatomy of the digestive system to limit the amount of food a person can eat and digest before feeling full.

Of the different types of bariatric surgery, sleeve gastrectomy is currently the most common, followed by gastric bypass. Both of these dramatically reduce the size of the stomach. In the gastric sleeve, the reduced stomach still "feeds" into the small intestine, and there are no other changes to the digestive tract or digestion. In gastric bypass, food from the reduced stomach is redirected to the lower portion of the **small intestine,** bypassing part of the stomach and the upper small intestine. In order to ensure that the food can still mix with digestive enzymes in the stomach and small intestine, the portion of the

PEPSIN
A protein-digesting enzyme that is active in the stomach.

CHYME
The acidic "soup" of partially digested food that leaves the stomach and enters the small intestine.

SMALL INTESTINE
The organ in which the bulk of chemical digestion and absorption of food occurs.

PANCREAS
An organ that helps digestion by producing enzymes (such as lipase) that act in the small intestine, and by secreting a juice that neutralizes acidic chyme.

DUODENUM
The first portion of the small intestine, where mixing of chyme and digestive enzymes occurs.

LIVER
An organ that aids digestion by producing bile salts that emulsify fats.

BILE SALTS
Chemicals produced by the liver and stored by the gallbladder that emulsify fats so that they can be chemically digested by enzymes.

stomach and small intestine that is bypassed is surgically connected to the lower small intestine. Digestive tract secretions still mix with food, but after food has bypassed the stomach and the upper portion of the small intestine (**INFOGRAPHIC 27.3**).

The small intestine is where most chemical digestion of food occurs, assisted by the secretions of several accessory organs. The **pancreas,** for example, secretes enzymes that help break down organic macromolecules such as carbohydrates, proteins, and fats into smaller molecules. It also secretes alkaline pancreatic juice into the small intestine to neutralize the acidic chyme, which would otherwise damage the small intestine. The first part of the small intestine, where this mixing occurs, is called the **duodenum**. Enzymes secreted by the small intestine itself further break down macromolecules into building blocks such as amino acids, sugars, fatty acids, and glycerol.

Whereas proteins and carbohydrates are easily digested by this powerful mixture of digestive enzymes, fats pose a special challenge. Because they are hydrophobic (see Chapter 2), fats don't mix well with the watery solutions in the small intestine. This makes it difficult for fat-digesting enzymes to break them down. Helping the process along, the **liver** secretes **bile salts,** which are chemically suited to dividing large hydrophobic fat globules into smaller droplets—that is, to emulsifying them. These bile salts pass from the liver into the **gallbladder,** which stores

INFOGRAPHIC 27.3 Types of Bariatric Surgery

Sleeve gastrectomy is currently the most common form of bariatric surgery. It dramatically reduces the size of the stomach but does not redirect the flow of food and chyme through the digestive tract. Gastric bypass reduces the size of the stomach and redirects food from the upper stomach pouch to the lower small intestine. Additional anatomical rearrangements are necessary in order to expose food to digestive secretions in gastric bypass.

Before Surgery

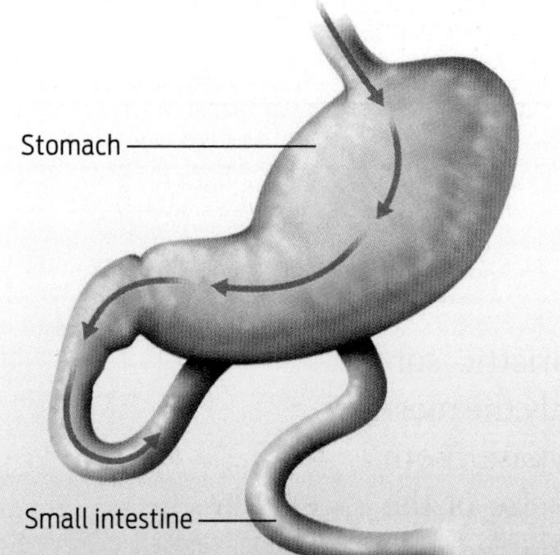

Food passes into the stomach, which can stretch from the size of a large sausage when empty to hold 3–4 liters of food. Chyme made in the stomach enters the small intestine, where digestion releases food nutrients.

Sleeve Gastrectomy

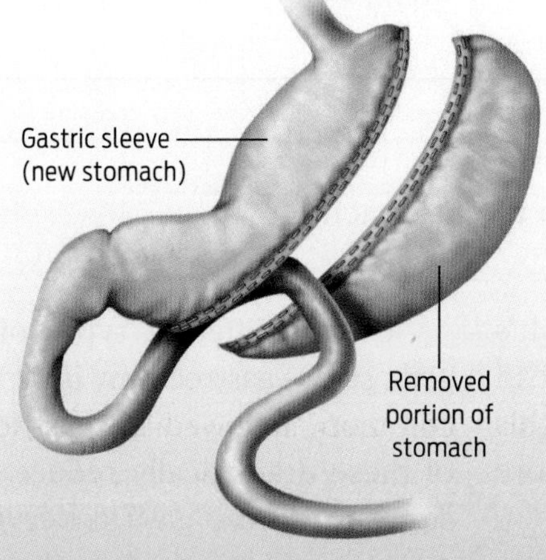

The stomach is reduced in size by up to 80%. Its small size limits the amount of food (and therefore Calories) that can be consumed. Stomach contents are still released into the upper small intestine for further processing.

Gastric Bypass

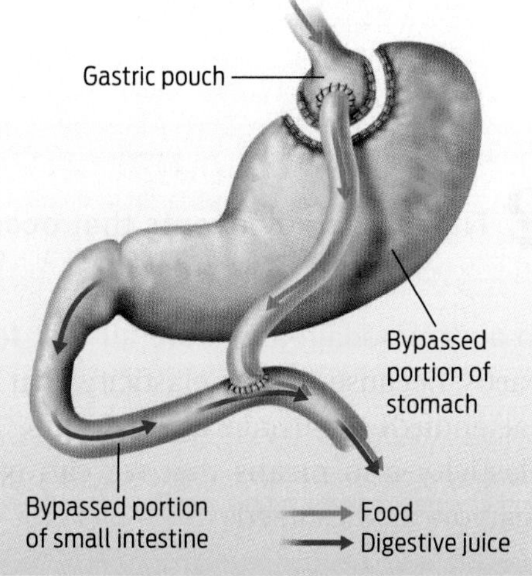

The stomach is reduced to the size of a golf ball, diminishing drastically the amount of food it can hold. Also, stomach contents bypass the upper part of the small intestine, reducing the amount of nutrients (and therefore of Calories) that can be absorbed.

? Compare and contrast sleeve gastrectomy and gastric bypass.

INFOGRAPHIC 27.4 Accessory Organs and the Small Intestine Work Together to Digest Food

The small intestine is the major organ that digests food. Accessory organs secrete enzymes and other substances into the small intestine, which itself also produces digestive enzymes.

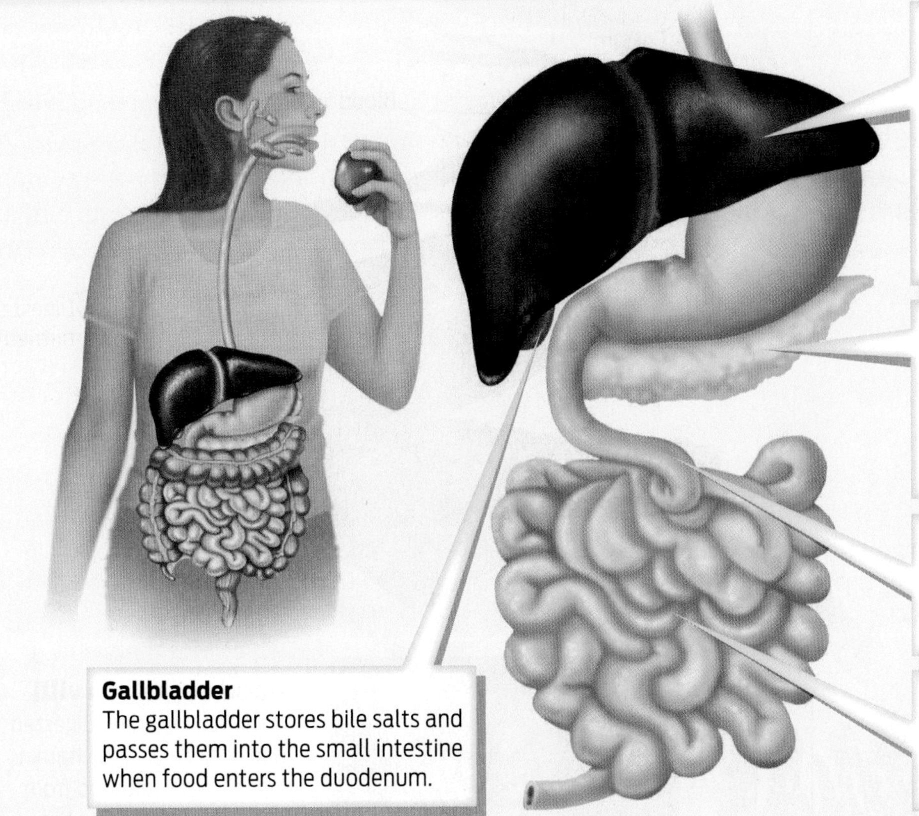

Liver
The liver makes bile salts and secretes them into the gallbladder. When released into the small intestine, bile salts emulsify fats, breaking them up into smaller droplets. This allows the enzyme lipase to more efficiently break down fat molecules. Lipase is made by the pancreas and secreted into the small intestine.
Pancreas
Pancreatic juice secreted into the small intestine neutralizes acids in chyme. Enzymes in the pancreatic juice also break down carbohydrates, proteins, fats, and nucleic acids into their smallest subunits.
Duodenum
Food from the stomach is mixed with digestive secretions in the duodenum, the first portion of the small intestine.
Small intestine
The small intestine produces some digestive enzymes and is the site of most chemical digestion of food.
Gallbladder
The gallbladder stores bile salts and passes them into the small intestine when food enters the duodenum.

? What do the liver, gallbladder, and pancreas have in common?

them for future use. When we eat a high-fat meal, bile salts pass from the gallbladder into the duodenum, where they help emulsify the fats. Once the fats are emulsified, the enzyme **lipase,** secreted by the pancreas, chemically breaks them down to release their constituent fatty acids and glycerol (**INFOGRAPHIC 27.4**).

Once digested into their smallest subunits, food molecules are taken up–absorbed–by cells lining the small intestine. The lining of the small intestine is folded into fingerlike projections called villi that greatly increase the surface area through which the intestine can absorb nutrients. The food molecules then pass into blood vessels of the circulatory system, and the bloodstream transports them throughout the body, where they are a source of nutrients to build and maintain cells (**INFOGRAPHIC 27.5**).

Once chyme passes through the small intestine, it moves on to the **large intestine,** or colon, which functions like a trash compactor–holding and compressing material that the body can't use or digest, such as plant fiber. Within the large intestine, fiber, small amounts of water, vitamins, and other substances mix with mucus and bacteria that normally live in the large intestine. As this

GALLBLADDER
An organ that stores bile salts and releases them as needed into the small intestine.

LIPASE
A fat-digesting enzyme active in the small intestine.

LARGE INTESTINE
The last organ of the digestive tract, in which remaining water is absorbed and solid stool is formed.

INFOGRAPHIC 27.5 The Small Intestine Absorbs Nutrients

The small intestine is the primary organ that absorbs nutrients from food. Nutrients enter the circulatory system via blood vessels connected to the small intestine.

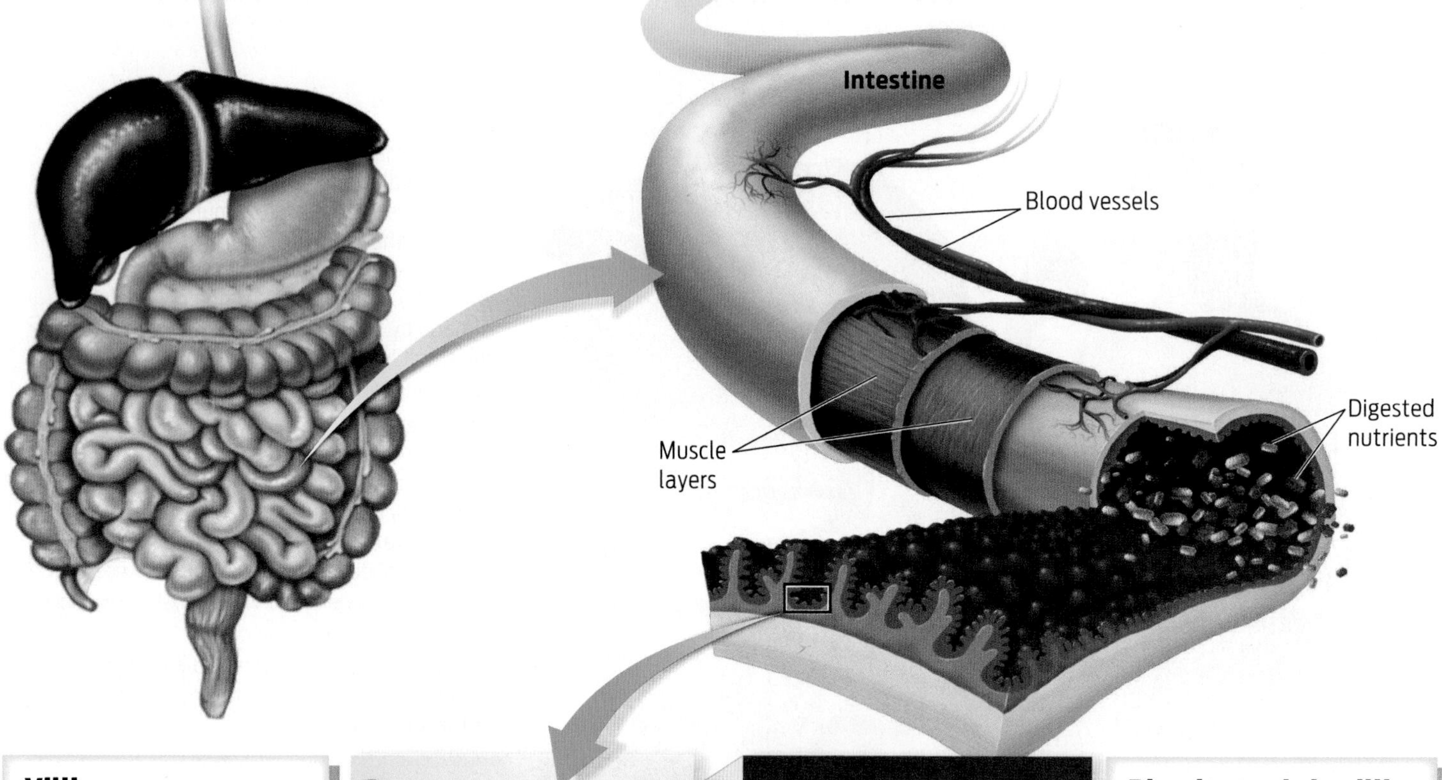

Villi
The inner surface of the small intestine is folded into fingerlike projections called villi, which are composed of many densely packed epithelial cells. Villi increase the surface area of the small intestine, enabling more nutrients to be absorbed.

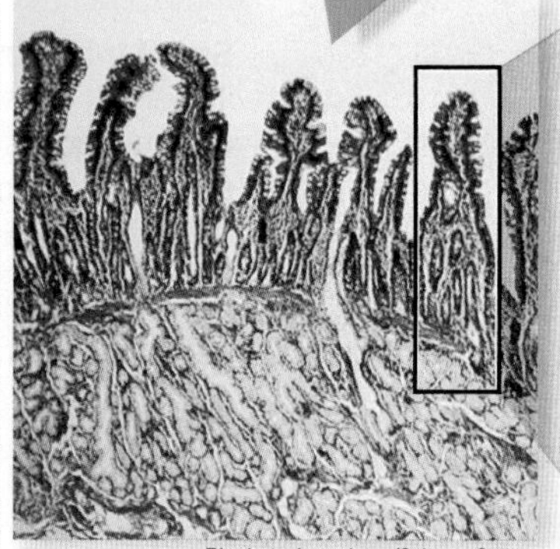

Biophoto Associates/Science Source

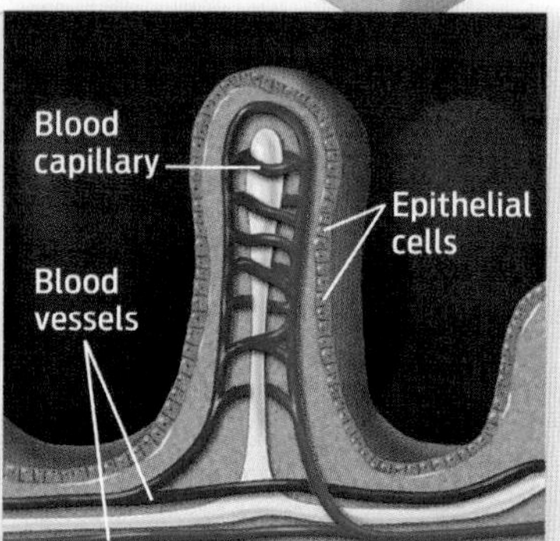

Blood vessels in villi
Nutrients, including digested food molecules and vitamins and minerals released from food, are absorbed by the lining of the small intestine. On the other side of the lining, nutrients enter blood vessels which transport nutrients to the rest of the body.

? What would you predict about the ability to absorb nutrients in someone who lacked normal villi?

STOOL
Solid waste material eliminated from the digestive tract.

waste travels through the large intestine, most of the water and some vitamins and minerals are reabsorbed into the body through the intestinal lining. Bacteria chemically break down some of the fiber to produce nutrients for their own survival and also to provide valuable vitamins, which is one reason fiber is an important dietary nutrient. As the large intestine expands and contracts, it creates **stool,** which is pushed into the rectum and eliminated from the body through the anus (**INFOGRAPHIC 27.6**).

Costs and Benefits of Surgery

Smith had her surgery in August 2009, opting for gastric bypass. After the surgery, she lost

INFOGRAPHIC 27.6 The Large Intestine

The large intestine absorbs water and some nutrients. It also packages waste material into stool.

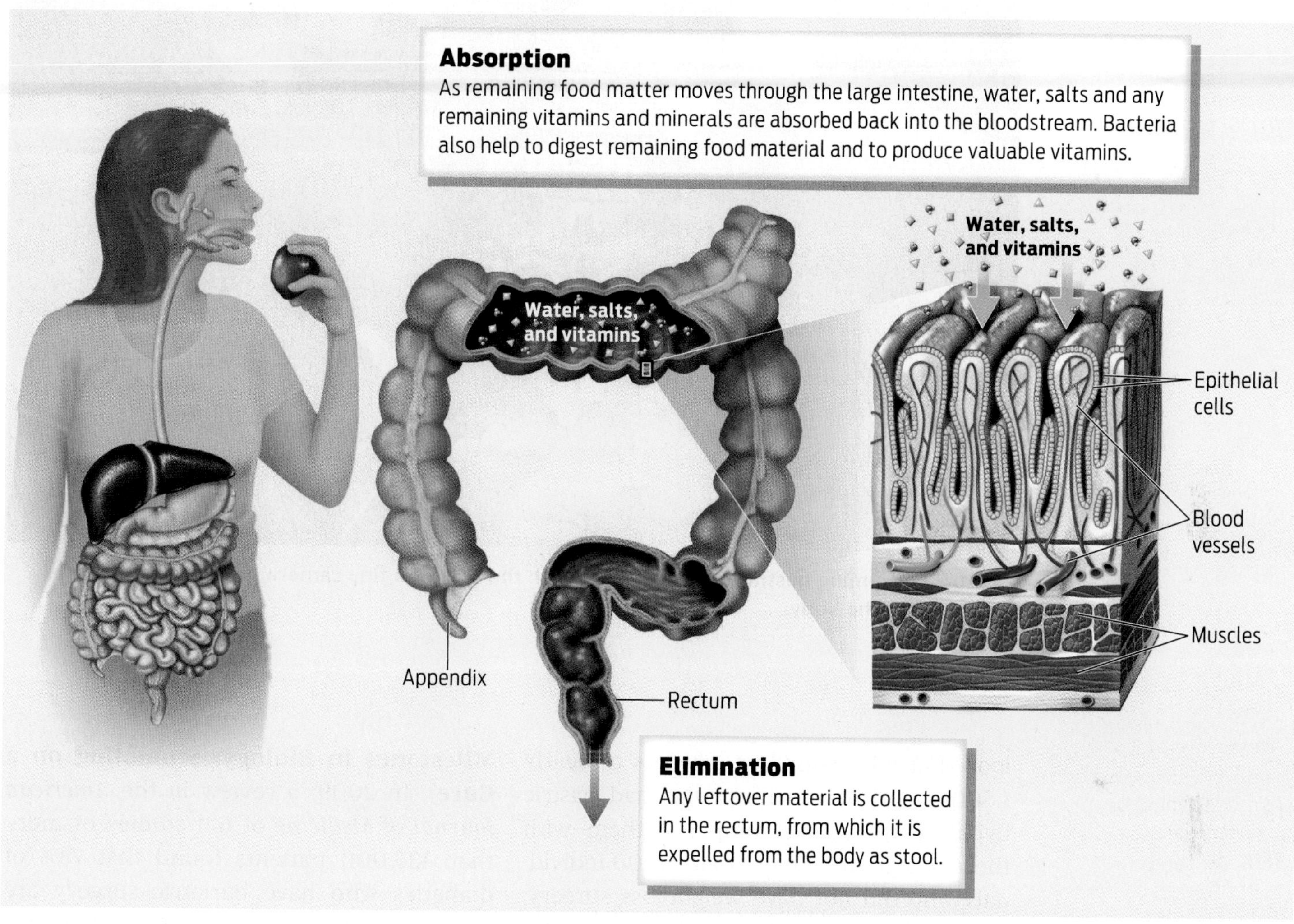

Absorption
As remaining food matter moves through the large intestine, water, salts and any remaining vitamins and minerals are absorbed back into the bloodstream. Bacteria also help to digest remaining food material and to produce valuable vitamins.
Water, salts, and vitamins
Water, salts, and vitamins
Epithelial cells
Blood vessels
Muscles
Appendix
Rectum
Elimination
Any leftover material is collected in the rectum, from which it is expelled from the body as stool.

? Should undigested fats normally be found in stool? Why or why not?

weight, but it wasn't a smooth ride. The stomach takes time to heal, and so doctors advised her to ingest only liquids for the first few weeks, puréed foods for the next few weeks, and then gradually progress to solid foods. Because the stomach is made so small, it can carry only about an ounce of food at a time–a handful of crackers or a few broccoli florets. Eating too much at once can cause vomiting or intense stomach pain.

Moreover, patients must stick to a special diet after the surgery or suffer other unpleasant consequences. Patients must introduce carbohydrates like breads and pasta into their diet very slowly, says Skarulis, of the National Institutes of Health. If they eat too many simple carbohydrates, the carbs enter the small intestine too quickly. This effect, called gastric dumping, causes nausea and diarrhea.

There are financial costs, too: the surgery runs anywhere from $11,000 to $26,000, and it's not always covered by insurance.

But the surgery does lead to weight loss. A 2016 study published in *JAMA Surgery*

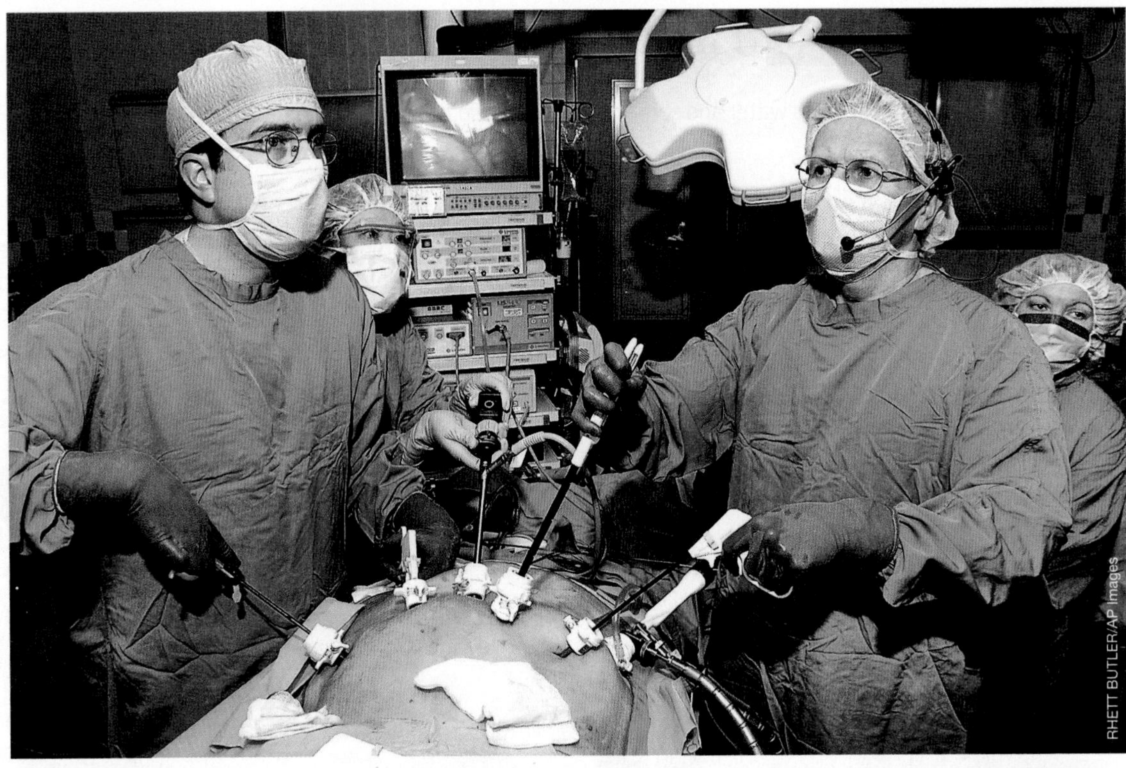

RHETT BUTLER/AP Images

Doctors performing gastric bypass surgery with the help of a tiny camera, called a laparoscope, inserted into the body.

looked at the weight-loss patterns of nearly 1,800 American veterans who had gastric bypass surgery and compared them with those of a group of more than 5,000 individuals who did not have weight-loss surgery. The results were striking: at 1 year postsurgery, patients undergoing gastric bypass had lost 31% of their baseline weight; 10 years out, they still weighed 21% less than the control group. Another study, published in 2007 in the *New England Journal of Medicine*, found that bariatric surgery also saves lives: among 2,000 obese patients who underwent weight-loss surgery, over the course of 16 years of follow-up, there were 129 deaths in the diet-only group, mostly from weight-related heart disease and cancer, and 101 deaths in the surgery group–a large difference statistically (**INFOGRAPHIC 27.7**).

Also encouraging to some doctors is the finding that weight-loss surgery can reverse or prevent type 2 diabetes (see **Milestones in Biology: Stumbling on a Cure**). In 2009, a review in the *American Journal of Medicine* of 621 studies of more than 135,000 patients found that 78% of diabetics who have bariatric surgery are cured of diabetes and that 87% are either cured or have their symptoms lessen. A 2012 study published in the *New England Journal of Medicine* showed that bariatric surgery could prevent the development of diabetes in obese patients, reducing the incidence of the disease in this patient population by more than 75% over a period of 15 years of follow-up.

As with any major surgery, there are risks. Complications of surgery include blood clots, hernias, bowel obstructions, and–rarely–death. Patients can also end up back in the hospital to repair intestinal leaks that can lead to serious infection. For these and other reasons, some obesity and diabetes experts think that patients

INFOGRAPHIC 27.7 Weight-Loss Surgery Is Effective and Saves Lives

A large study of weight-loss surgery showed that people who had gastric bypass lost substantially more weight than people who attempted to lose weight using nonsurgical methods. People who had gastric bypass also experienced fewer deaths in the 16 years following surgery compared to people who relied on nonsurgical methods for weight loss.

Difference in Weight Loss among Patients Undergoing Gastric Bypass and Nonsurgical Methods

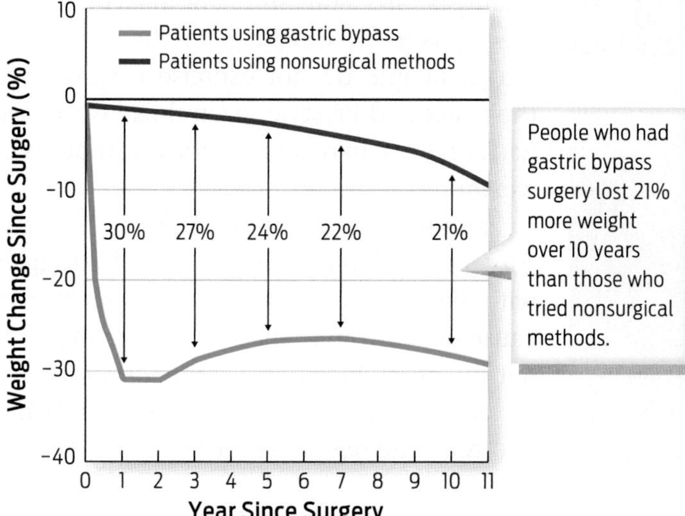

Data from Maciejewski, M.L., et al. JAMA Surg Published online August 31, 2016

Difference in Mortality among Patients Undergoing Gastric Bypass and Nonsurgical Methods

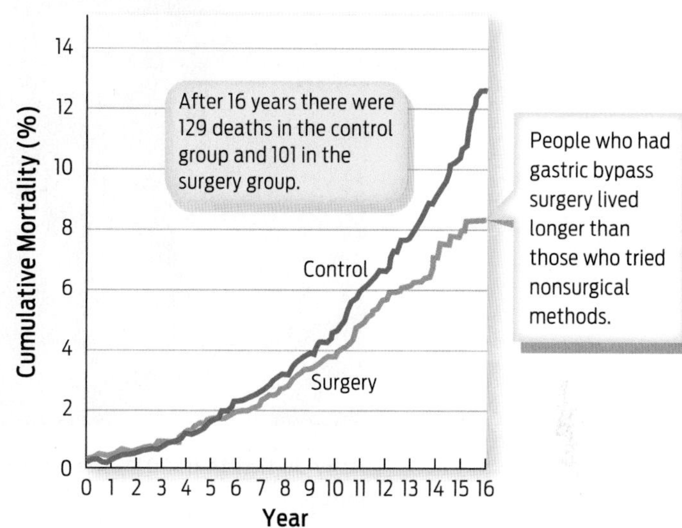

Data from Sjostrom L., et al. (2007) New England Journal of Medicine 357:741–752.

? A patient weighs 300 pounds at the time of surgery. What will the patient's weight be in 10 years if 28% of body weight is lost (as happened with the people in the 2016 study plotted above)?

should consider less radical options first, such as a diet that sharply restricts carbohydrates.

Smith has had a host of complications that have put her back in the hospital. A few months after her surgery, she felt terrible cramping in her side. Tests showed that scar tissue had formed at the site where her small intestine had been cut from her stomach. Surgery to remove the tissue revealed that part of her intestine and stomach had twisted and anchored onto this scar tissue, causing her pain. Even after the surgery, she was still having stomach pains when she ate, so doctors temporarily put a feeding tube into her stomach and a catheter into a vein in her arm through which she could take nutrients directly into her bloodstream. Smith spent weeks in and out of the hospital between January and April of 2010.

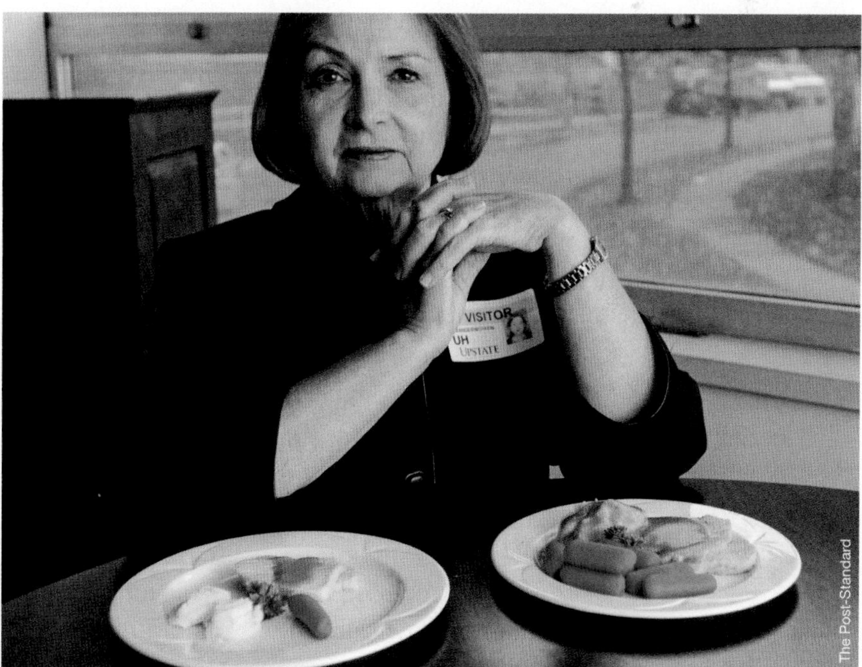

After having gastric bypass, patients must reduce food portion size. Since her surgery, this patient can comfortably eat only about a half cup of food at a time—even at Thanksgiving (plate on the left).

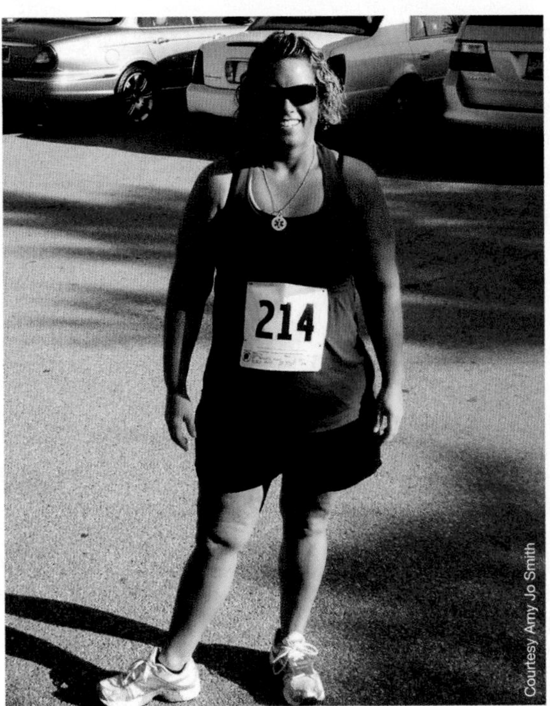

Amy Jo Smith in 2011 after her weight-loss surgery.

Smith's health has improved since then, but she is still at risk. Since people who have gastric bypass surgery end up with part of the small intestine bypassed, they absorb fewer of the micronutrients they eat. Patients must take vitamin supplements for the rest of their lives. There may be additional micronutrient deficiencies that scientists haven't yet recognized; only long-term follow-up of these patients will reveal how serious a problem this is. To monitor her micronutrient levels, Smith has a blood test every 3 months.

What's more, the surgery may not be a permanent cure for obesity. Most people who have the surgery regain some weight over time. The stomach and intestines can sometimes expand to allow greater ingestion of food. And hormonal changes may occur after surgery that alter body metabolism. While scientists are still studying the hormonal mechanisms of weight gain, they do know that if people do not exercise control over their diet and lifestyle, even those who have had surgery can regain significant amounts of weight.

Smith's own experience with bariatric surgery has been mixed. She still struggles with nausea every day; strong smells can cause her to vomit. She also feels pain in her left side, for which she takes medication.

Despite these complications, Smith has had not "one day of regret," she says. On her first "surgiversary"—her surgical anniversary—she wrote a letter to her surgical team in which she said: "I have been blessed with 35 birthdays but none can compare to my surgiversary. I never imagined in a year that I would lose over 100 lbs, run a 5K the day of my surgiversary . . . sit sideways in a student desk, wear a size 12 pants from a 24-26 . . . be able to sit comfortably in a restaurant booth, and be able to stand on a table or chair without thinking, 'My gosh, am I going to break this?' "

Within a year and half of her surgery, Smith had dropped down to 146 pounds. "I don't recognize myself anymore," she says. ■

How Do Other Organisms Digest Food?

Humans and many other animals have what's known as a complete digestive tract—one shaped like a tube, with two openings: a mouth and an anus. Not all organisms have such a tubelike digestive tract. In fact, many organisms have no digestive tract at all and yet are still able to obtain and process food from their environment.

Take fungi, for example. Fungi are eukaryotic organisms. Some, such as yeasts, are unicellular, while the majority are multicellular. Multicellular fungi have bodies that are made up of microscopic filaments called hyphae. Their bodies do not have distinct organs or organ systems, and they do not have digestive tracts. To obtain nutrients, fungi extend hyphae into food—a piece of bread, say, or

Fungi Digest Food Externally

Fungi do not have digestive tracts. Their cells secrete enzymes into the food source and food digestion occurs externally.

Gregory G. Dimijian/Science Source
Bread mold
(*Rhizopus stolonifer*)
Hyphae
These filaments of fungal cells grow into the food they land on and secrete digestive enzymes.
Digestive enzymes secreted into bread
External digestion
Digested nutrients absorbed by hyphae

leaf litter in a forest. Individual hypha cells then release digestive enzymes directly into the food and absorb the digested nutrients directly. Since digestion occurs outside their bodies, fungi do not need a stomach or a mouth, or even a digestive tract. Rather, each cell can absorb the products of this external digestion directly.

As another example, consider sea anemones, invertebrate animals that live in the oceans, attached to surfaces such as rocks. Sea anemones digest their food internally, but they don't have a digestive tract like humans. Rather, they have a single, multifunctional digestive cavity where digestion and absorption of nutrients take place. Sea anemones capture food and shove it into this cavity using the tentacles that surround their mouths. The cavity has only one opening–the mouth–through which food enters and wastes exit. This is an example of an incomplete digestive tract, in contrast to our own complete digestive tract in which food flows one way from the mouth to the anus.

Sea Anemones Have an Incomplete Digestive Tract

While digestion occurs internally in sea anemones, their digestive tracts have only one opening through which food enters and waste exits—the mouth.

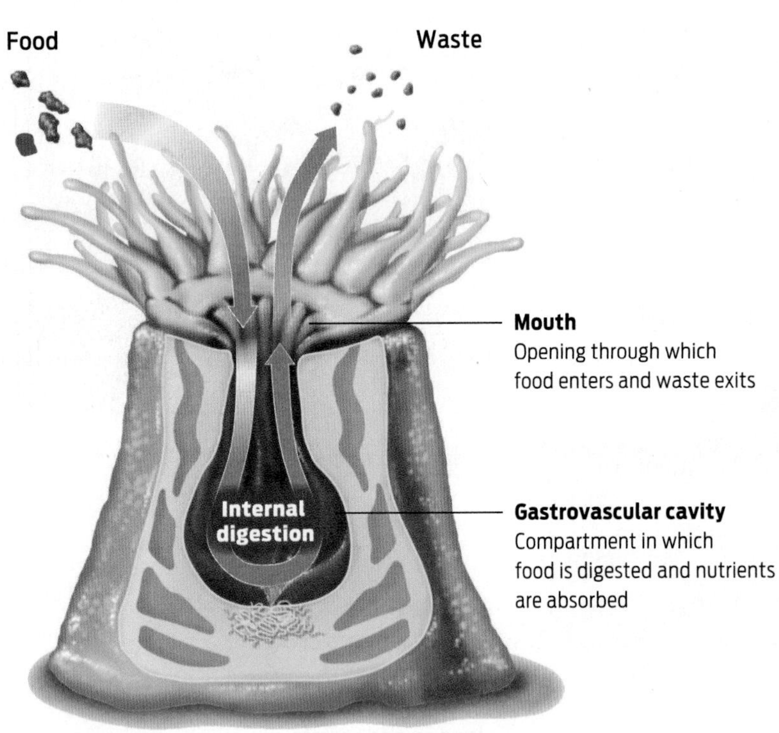
Food
Waste
Mouth
Opening through which food enters and waste exits
Internal digestion
Gastrovascular cavity
Compartment in which food is digested and nutrients are absorbed

ullstein bild/Getty Images

Yellow cluster sea anemone (*Parazoanthus axinellae*)

CHAPTER 27 SUMMARY

- The digestive system is composed of a central digestive tract and accessory organs. Its function is to break down food molecules into smaller subunits, absorb nutrients, and eliminate waste.
- Digestion begins in the mouth, where teeth chew food and the tongue compresses food and pushes it to the back of the mouth (mechanical digestion) and salivary enzymes begin breaking down carbohydrates (chemical digestion).
- Food passes from the mouth into the stomach through the esophagus, propelled by waves of muscular contractions called peristalsis.
- The stomach is muscular and acidic and contains pepsin, a protein-digesting enzyme. It is elastic and can expand after a large meal to store food for a few hours.
- Food processed in the stomach is called chyme. Chyme passes into the small intestine, where enzymes further digest it.
- Enzymes from the pancreas help to digest organic molecules in the small intestine.
- Bile salts, produced in the liver and stored in the gallbladder, emulsify fats and help the body digest them.
- The small intestine absorbs the broken-down products of food; once absorbed, food molecules enter the bloodstream and are transported throughout the body.
- The large intestine absorbs water and forms solid stool from indigestible matter in food such as fiber.
- Humans and many other animals have a complete digestive tract—one with a mouth and an anus. Not all organisms have a complete digestive tract; many have no digestive tract at all.

MORE TO EXPLORE

- **Centers for Disease Control and Prevention: Obesity www.cdc.gov/obesity**
- **Maciejewski, Matthew L., et al. (2016) "Bariatric Surgery and Long-term Durability of Weight Loss."** ***JAMA Surgery.***
- **Sjöström L., et al. (2007) Effects of bariatric surgery on mortality in Swedish obese subjects.** ***New England Journal of Medicine*** **357:741–752.**
- **Taubes, G. (2011)** ***Why We Get Fat: And What to Do about It.*** **New York: Anchor Books.**
- **Sara Hallberg and Osama Hamdy (September 2, 2016) Before You Spend $26,000 on Weight-Loss Surgery, Do This.** ***New York Times.***

CHAPTER 27 Test Your Knowledge

DRIVING QUESTION 1 What is the anatomy of the digestive system?

By answering the questions below and studying Infographics 27.1, 27.2, 27.4, 27.5, and 27.6, you should be able to generate an answer for the broader Driving Question above.

KNOW IT

1 Place the following structures of the digestive system in order from the entry of food (1) to the exit of waste (6).

_____ Esophagus
_____ Large intestine
_____ Stomach
_____ Mouth
_____ Small intestine
_____ Anus

2 Which part of the digestive tract has the most acidic pH?

a. esophagus
b. colon
c. small intestine
d. stomach
e. mouth

3 Why is it helpful to have an expandable stomach?

USE IT

4 What do the gallbladder, liver, and pancreas have in common with respect to the digestive system? How do they differ from the mouth, stomach, and small intestine?

5 Muscle paralysis in the digestive tract would compromise which digestive function?

a. chemical digestion in the stomach
b. chemical digestion in the small intestine
c. absorption in the small intestine
d. chemical digestion in the mouth
e. movement of food from the mouth to the stomach

6 Pepsin is most effective at a pH of about 2. Digestive enzymes in the small intestine are most effective at a pH of or near 7. If the pancreas were unable to secrete bicarbonate (the basic component of pancreatic juice that neutralizes acid), what would you predict about the waste eliminated from the large intestine?

DRIVING QUESTION 2 How is food broken down and utilized as it moves through the digestive tract?

By answering the questions below and studying Infographics 27.4 and 27.5, you should be able to generate an answer for the broader Driving Question above.

KNOW IT

7 Where does the majority of chemical digestion take place?

a. small intestine
b. esophagus
c. mouth
d. stomach
e. colon

8 What do pepsin and salivary amylase have in common? How do their activities differ?

9 What organ produces lipase?

USE IT

10 Someone whose gallbladder has been surgically removed will have trouble processing

a. fats.
b. carbohydrates.
c. minerals.
d. vitamins.
e. proteins.

11 Compare and contrast the functions of bile salts and lipase.

12 Why would someone with a blocked duct between the pancreas and the small intestine experience pancreatic inflammation (pancreatitis)? Note that in this case inflammation is a response to tissue damage.

13 If you stand on your head, can processed food still pass from your small intestine into your large intestine? Explain your answer.

MINI CASE

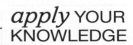

apply YOUR KNOWLEDGE

14 Alicia has had her gallbladder removed. She must be careful not to eat high-fat meals, or else she is likely to experience greasy diarrhea. Her friend Tammy is taking Alli, a weight-loss drug that inhibits the fat-digesting enzyme lipase. Tammy must also avoid high-fat meals, in order to avoid oily stools, a possible side effect of the drug. Why are both Alicia and Tammy at risk for similar digestive upsets when their situations are different (gallbladder removal, taking Alli)?

DRIVING QUESTION 3 How does bariatric surgery change the digestive tract and digestion, and what are the risks and benefits of bariatric surgery?

By answering the questions below and studying Infographics 27.3 and 27.7, you should be able to generate an answer for the broader Driving Question above.

KNOW IT

15 Gastric bypass surgery causes the __________ to become __________.

a. stomach; smaller
b. small intestine; larger
c. stomach; less acidic
d. small intestine; less acidic
e. stomach; larger

16 What is one similarity and one key difference between sleeve gastrectomy and gastric bypass?

17 From what you've read in this chapter, are there any other benefits to gastric bypass besides weight loss? Explain your answer.

USE IT

18 Consider this patient:

a. A 5'11" man weighs 320 pounds. What is his BMI? (See Infographic 6.1.)

b. If he has gastric bypass, and if his weight-loss trajectory is exactly that shown in the left panel of Infographic 27.7, how much will he weigh in 10 years? What will be his BMI at that point? Will he still be considered obese?

c. What health outcomes might he experience as a result of his weight-loss surgery?

apply YOUR KNOWLEDGE

BRING IT HOME

19 If a morbidly obese person who is considering gastric bypass surgery asked for your opinion on the procedure, what would you say about its known risks, benefits, and any unknowns? Would you say the same to someone considering the surgery who is simply overweight, not morbidly obese? Explain your answer.

apply YOUR KNOWLEDGE

INTERPRETING DATA

20 A 2012 study compared the impacts of medical therapy and bariatric surgery in obese people with uncontrolled type 2 diabetes. Patients were randomly assigned to receive aggressive medical therapy for their diabetes (including medications and diet and lifestyle modifications) or bariatric surgery. Several dependent variables were measured for 1 year. Two of these variables—average BMI and average number of diabetes medications—are shown in the table below.

a. Draw two graphs, one plotting the average diabetes medications over time for the group receiving medical therapy and for the group receiving surgery, and one plotting the change in BMI from baseline for the two groups (set the baseline values at 0 on the time axis because by definition no change can have taken place yet).

b. From these graphs, how does gastric bypass compare to medical therapy for diabetes management in obese patients with type 2 diabetes?

c. The data shown in the table are for 41 patients who had medical therapy and 50 patients who had gastric bypass. From this information and any other limitations you can identify, are these data sufficient to make a recommendation of surgery for diabetes management? Why or why not?

Treatment	Variable	Baseline	Time after Treatment 3 Mos.	6 Mos.	9 Mos.	12 Mos.
Medical therapy	Average BMI	36.3	35.4	34.8	34.5	34.4
	Average no. of diabetes medications	2.8	3.1	3.1	3.0	3.0
Gastric bypass	Average BMI	37.0	31.8	28.2	26.9	26.8
	Average no. of diabetes medications	2.6	1.1	0.6	0.4	0.3

Data from Schauer, P. R., et al. (2012) Bariatric surgery versus intensive medical therapy in obese patients with diabetes. *New England Journal of Medicine* 366(17):1567–1576.

8 Milestones in Biology

Stumbling on a CURE

Banting, Best, and the discovery of insulin

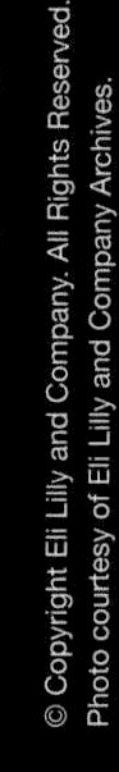

DRIVING QUESTIONS

1. What is the role of insulin in blood-sugar regulation and diabetes?
2. In general terms, what is a hormone?
3. What features are shared by type 1 diabetes and type 2 diabetes, and what features are unique to each type?

In December 1921, a 14-year-old boy named Leonard Thompson lay sick and dying in a Canadian hospital bed. He weighed just 65 pounds and was lapsing in and out of consciousness. Elevated levels of sugar in his blood, caused by the disease diabetes mellitus, were wreaking havoc on his internal organs, and doctors told his parents he would likely not survive more than a month. With nothing to lose, Thompson's parents agreed to let the doctors try something unusual: they injected the boy with a chemical that scientists had recently isolated from the pancreas of a dog. The chemical radically altered Thompson's fate and that of countless others since.

For most of human history, diabetes was a dreaded and deadly disease. People with diabetes have unusually high blood-sugar levels–a sign that the body's cells are not taking up sugar from the blood. Since cells require sugar (principally glucose) as fuel to power their activities, this lack of uptake eventually starves the body of nourishment, while the high blood-sugar levels cause a host of problems of their own.

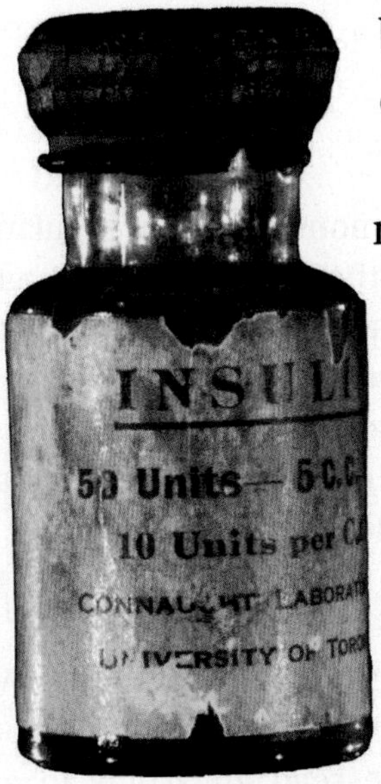

The word "diabetes" comes from the Greek word meaning "to pass through," and refers to the fact that diabetics tend to urinate excessively. By eliminating excess sugar, urinating restores normal blood-sugar levels. Before there were blood tests, doctors diagnosed diabetes by testing for the presence of sugar in a patient's urine, which would often attract flies because it was so sweet (the word "mellitus" means "like honey"). Until 1921, there was no effective treatment for diabetes, and a diagnosis was in essence a death sentence.

Sanofi Pasteur Limited (Connaught Campus) Archives, Toronto

The breakthrough came in a laboratory located just footsteps from where Thompson was being treated at the University of Toronto. A team of ragtag researchers, working on a shoestring budget, succeeded where many other scientists had failed. With nothing more than a pack of stray dogs and some borrowed laboratory equipment, they discovered a treatment for millions.

Hunting Down a Mystery Chemical

In the fall of 1920, a young Canadian surgeon named Frederick Banting was preparing a lecture on the physiology of the pancreas, the carrot-shaped organ that sits next to the stomach. At the time, the pancreas was known to be important for the digestion of food molecules. Digestive juices secreted from the pancreas into the small intestine contain enzymes that help break down carbohydrates, proteins, and fats. But scientists were learning that the pancreas has other functions as well. Experiments in 1890 had shown that surgical removal of the pancreas in dogs led quickly and inevitably to all the symptoms of severe diabetes: high blood-sugar levels, sugar in the urine, coma, and death. This finding suggested that the pancreas plays a critical role in the regulation of blood sugar.

Until 1921, there was no effective treatment for diabetes, and a diagnosis was in essence a death sentence.

Scientists were beginning to suspect that the pancreas was essentially two organs in one. One part of the pancreas released digestive enzymes into the intestines through ducts, while another part, made up of discrete islands of cells, released an unidentified substance into the blood that regulated blood-sugar levels. A number of researchers had tried, unsuccessfully, to isolate this mystery chemical from these groups of cells (called islets of Langerhans after Paul Langerhans, who first noticed them under a microscope in 1869). So convinced were researchers that the substance existed that they gave it a name, "insuline"–from the Latin *insula,* "island" (**INFOGRAPHIC M8.1**).

The notion that certain "internal secretions" produced by glands in the body and released into the bloodstream could influence physiology was a relatively new idea at the beginning of the 20th century–the term "hormone" was first used in 1905. Before that time, conventional wisdom held that physiological processes, like the regulation of blood sugar, were controlled primarily by the brain through nerves. But evidence was accumulating that blood-borne chemicals–hormones–controlled a number of different physiological processes, including blood-sugar regulation. The study of hormones and the glands that produce them would come to be known as endocrinology.

Today we know that the body has many hormone-secreting glands, which together coordinate many physiological processes. For example, the adrenal glands (located above the kidneys) secrete the hormone adrenaline, which controls the "fight or flight" response. The ovaries and testes produce sex hormones that control reproductive physiology. The thyroid gland (located in the neck) produces thyroid hormones, which regulate metabolism. And the brain's pituitary gland produces several hormones that control glands and tissues throughout the body. All these hormones work in a similar fashion: they are released by glands into the circulation and carried in the blood to their target cells (see Chapter 26).

Banting knew about previous attempts to isolate a pancreatic hormone. While preparing his lecture on the physiology of the pancreas, he read an article in a medical journal describing how surgically tying off

INFOGRAPHIC M8.1 The Pancreas Produces Digestive Enzymes and Insulin

The pancreas plays two distinct roles. Some pancreatic cells secrete digestive enzymes into the small intestine, where those enzymes help break down organic molecules, including carbohydrates, into smaller components. Pancreatic cells in the islets of Langerhans secrete hormones—including insulin—into the circulatory system.

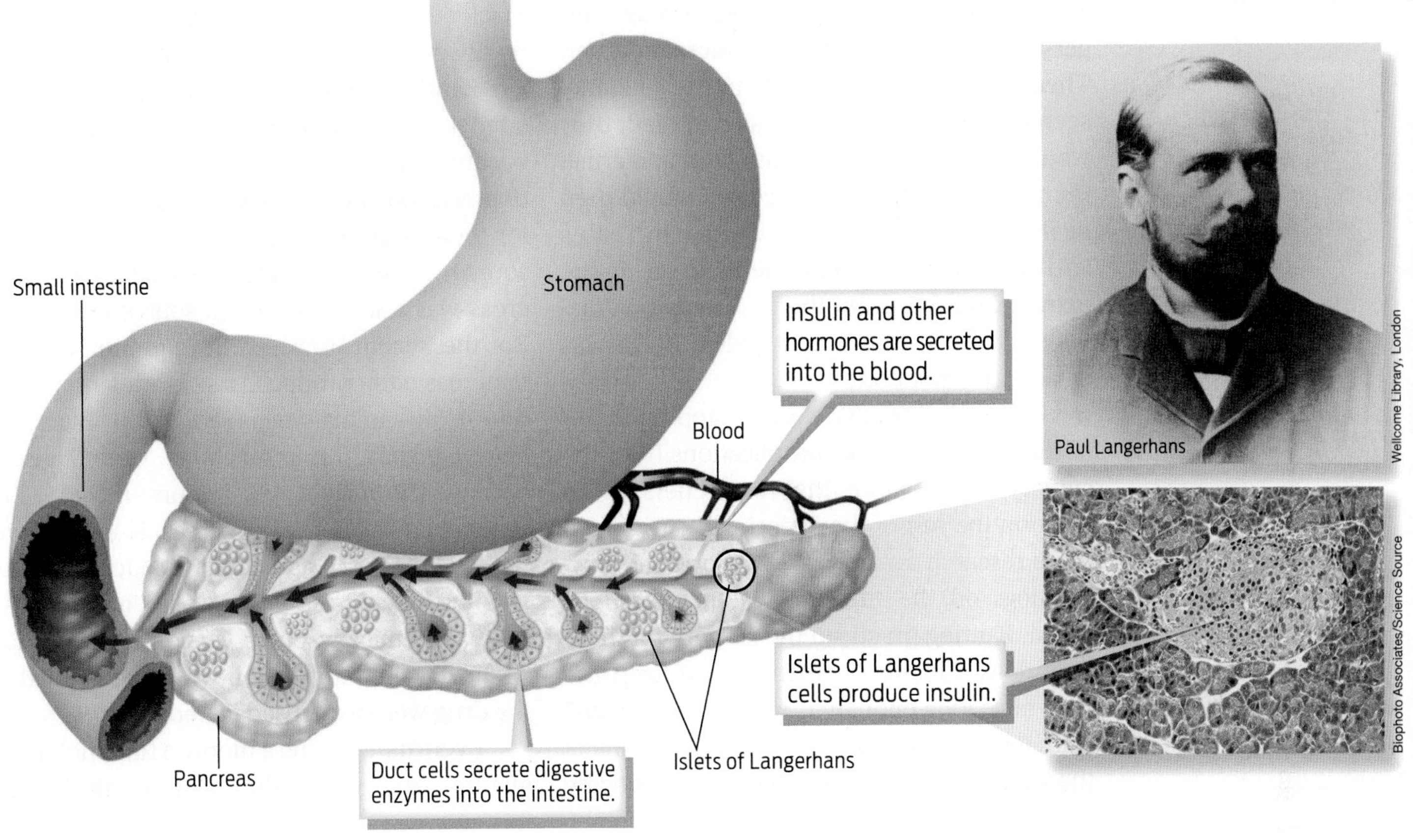
Small intestine
Stomach
Insulin and other hormones are secreted into the blood.
Blood
Paul Langerhans
Wellcome Library, London
Islets of Langerhans cells produce insulin.
Biophoto Associates/Science Source
Pancreas
Duct cells secrete digestive enzymes into the intestine.
Islets of Langerhans

? Which pancreatic cells (duct cells or islet of Langerhans cells) produce insulin? Which produce digestive enzymes?

the main duct leading from the pancreas to the intestine led to the death of the enzyme-secreting cells of the organ but left the "islands" unharmed. That's when an idea came to him: perhaps other researchers had been unable to isolate insulin because when they ground up the pancreas to perform the isolation procedure, the digestive enzymes that the pancreas produces destroyed the insulin. Perhaps he could avoid that problem by first performing surgery on an animal to tie off the pancreatic duct and then waiting for the enzyme-secreting part of the organ to die. He jotted down some hasty notes, then went to bed, but was so excited he was unable to sleep.

A few days later, Banting presented his idea to John Macleod, the head of the physiology department at the University of Toronto. Macleod listened patiently to Banting but was skeptical of his ideas. Macleod was aware that other researchers had tried and failed to isolate insulin and was doubtful about the new approach. But Banting was enthusiastic and persistent; eventually, he persuaded Macleod to provide him with some lab space and

10 dogs on which to test his hypothesis. Macleod also agreed to appoint a graduate student, Charles Best, to assist him.

A Rocky Start

Banting and Best began by performing surgery on dogs to tie off the pancreatic duct. Then, in a few weeks, when the enzyme-secreting part of the pancreas had died, they would remove the Langerhans cells and crush them into an extract. This extract would then be injected into other dogs whose pancreases had been surgically removed, to see if the extract would prevent them from getting diabetes. At least, that was the idea. But it was easier said than done.

The initial results were disappointing: 7 of the 10 dogs died of complications from the surgery even before the researchers could perform the rest of the experiment. To obtain more research animals, they resorted to buying dogs off the street for $1, no questions asked. Banting practiced the surgery on these new dogs. Finally, he managed to keep the dogs alive long enough to extract the "island" cells. Banting and Best were now ready to inject the extract into another dog made diabetic by removing its pancreas. The procedure worked beautifully. The extract significantly lowered the dog's blood-sugar levels and also reduced the amount of sugar in the dog's urine. A pancreatic hormone did indeed seem to regulate blood sugar. They reported these results to Macleod, who realized a major breakthrough had been made **(INFOGRAPHIC M8.2)**.

Banting and Best were excited to have isolated the elusive hormone insulin, but it quickly became clear that producing insulin from stray dogs was not a viable way to create a treatment for thousands, and they kept looking for other sources. They discovered that fetal calf pancreas was also an abundant source of insulin. But the insulin prepared from ground-up calf pancreas often produced a severe allergic reaction in the animals who received it. Macleod believed that these reactions were caused because the insulin was impure—mixed with other chemicals of the pancreas. Since neither Banting nor Best was a chemist, Macleod hired a young biochemist, James Collip, to help prepare a chemically purer version of the drug. Such a cleaned-up version would be needed if the drug were to be used in people.

Eventually, using a method that included alcohol and evaporation, Collip was able to

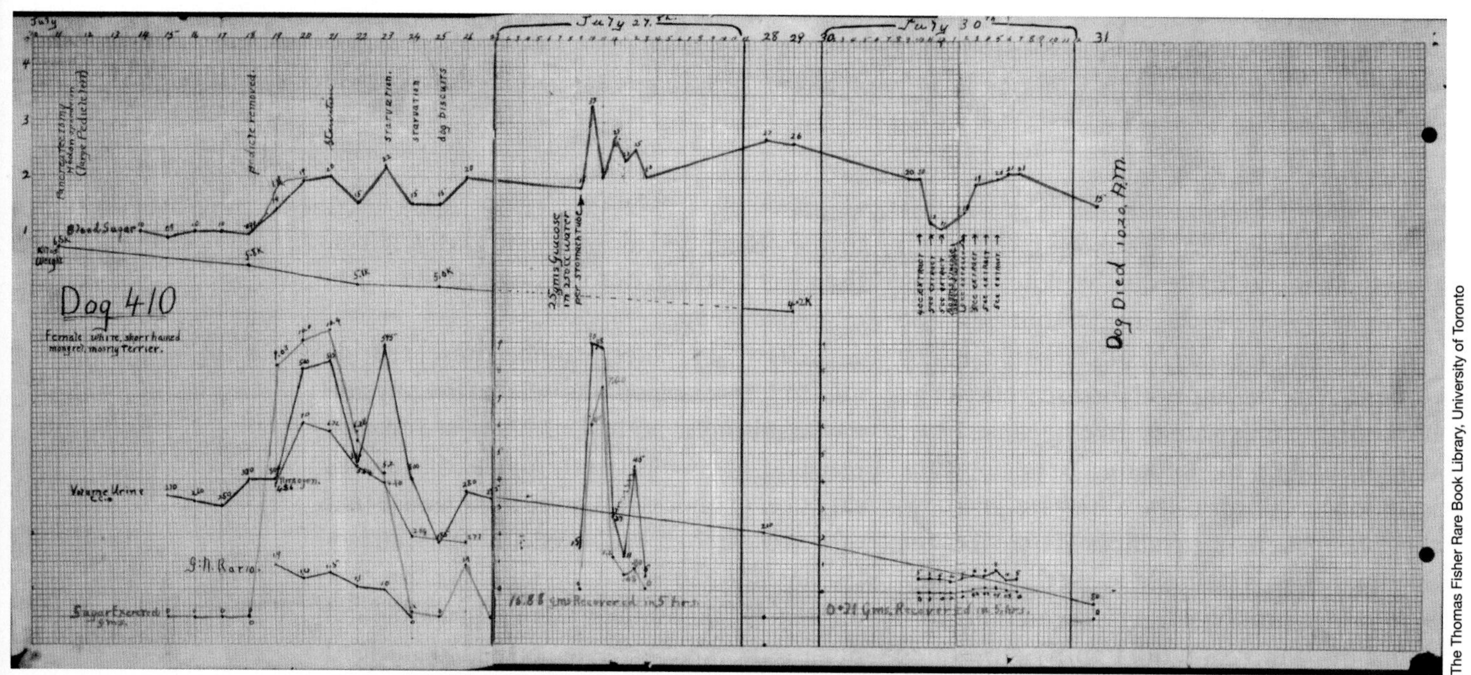

The Thomas Fisher Rare Book Library, University of Toronto

A page from Banting's laboratory notebook documenting glucose levels in dogs.

INFOGRAPHIC M8.2 Banting and Best Isolated Insulin from Dog Pancreases

Banting and Best developed a method of extracting insulin from dog pancreases. The extracted insulin reversed the symptoms of diabetes in diabetic dogs.

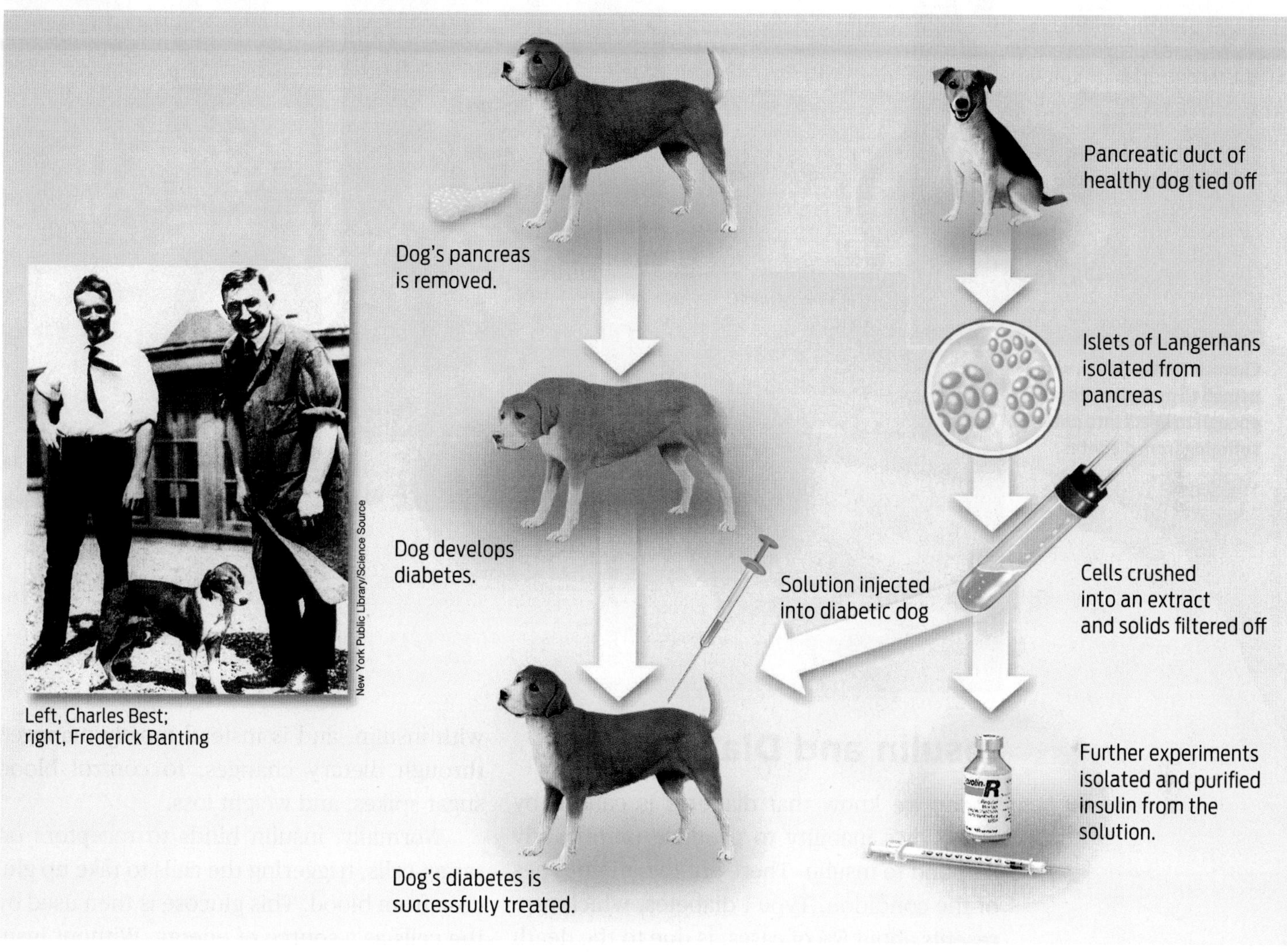
Dog's pancreas is removed.
Dog develops diabetes.
Dog's diabetes is successfully treated.
Pancreatic duct of healthy dog tied off
Islets of Langerhans isolated from pancreas
Cells crushed into an extract and solids filtered off
Solution injected into diabetic dog
Further experiments isolated and purified insulin from the solution.
New York Public Library/Science Source

Left, Charles Best; right, Frederick Banting

? Why did removal of the pancreas cause the dog on the left to develop diabetes?

produce a pure form of the drug, and the team was ready to test insulin on humans. They talked to doctors at the hospital across the street at the University of Toronto, who were desperate for a new way to treat patients with diabetes. Leonard Thompson, the 14-year-old diabetic patient who was falling in and out of consciousness, received this purified version on January 23, 1922. Within moments, he was alert and smiling, to the astonishment of his parents. Encouraged by these results, the doctors next went from bed to bed in the diabetes ward, injecting one patient after another. Before they reached the last patient, the first patients to be injected were regaining consciousness. Insulin as a treatment for diabetes was born.

The Thomas Fisher Rare Book Library, University of Toronto

Chemist James Collip, who helped to make insulin pure enough to inject into patients suffering from diabetes.

SSPL/The Image Works

A diabetic child before (left) and after (right) insulin treatment.

The Thomas Fisher Rare Book Library, University of Toronto

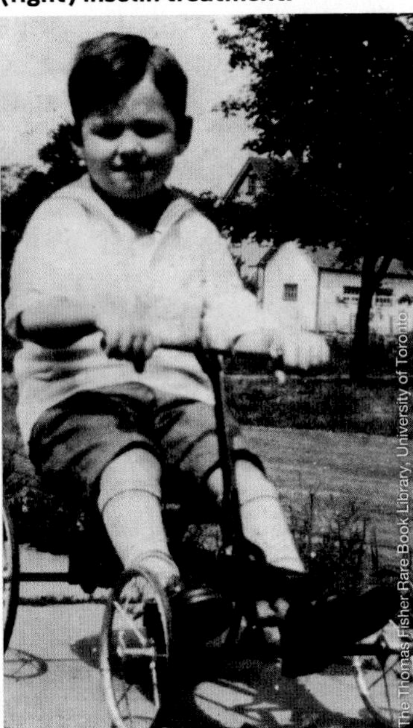
The Thomas Fisher Rare Book Library, University of Toronto

Insulin and Diabetes

Today we know that diabetes is caused by the body's inability to produce or properly respond to insulin. There are two main types of the condition. Type 1 diabetes, which represents about 5% of cases, is due to the death of the insulin-producing cells in the islets of Langerhans. (The cells die when they are attacked by the immune system, in a form of autoimmunity; see Chapter 32.) People with type 1 diabetes require daily injections of insulin to survive. Type 2 diabetes, which accounts for about 95% of cases, begins when cells in the body don't respond to insulin. Scientists don't fully understand why this happens, but they think that inflammation resulting from excess body fat may play an important role. Having cells that are unresponsive to insulin is as bad as not making insulin at all. Type 2 diabetes is not treated with insulin, and is instead mostly managed through dietary changes, to control blood sugar spikes, and weight loss.

Normally, insulin binds to receptors on target cells, triggering the cells to take up glucose from blood. This glucose is then used by the cells as a source of energy. Without insulin or properly functioning insulin receptors, cells cannot take up sugar from the blood and so they starve (**INFOGRAPHIC M8.3**).

The pancreas produces other hormones as well, including glucagon, which raises blood sugar. Glucagon and insulin work together to regulate blood sugar through a classic negative feedback loop. (For more on glucose metabolism, including the role of glucagon, see Chapter 26.)

Left untreated or unmanaged, chronic diabetes (both type 1 and type 2) can cause a suite of health problems, including increased risk of heart disease and stroke, loss of vision,

INFOGRAPHIC M8.3 Insulin and Diabetes

Insulin normally signals cells to take up sugar from the blood, thereby reducing blood-sugar levels. In diabetes, insulin is either not produced or cells don't respond to it, leading to elevated blood-sugar levels.

Normal Insulin Response

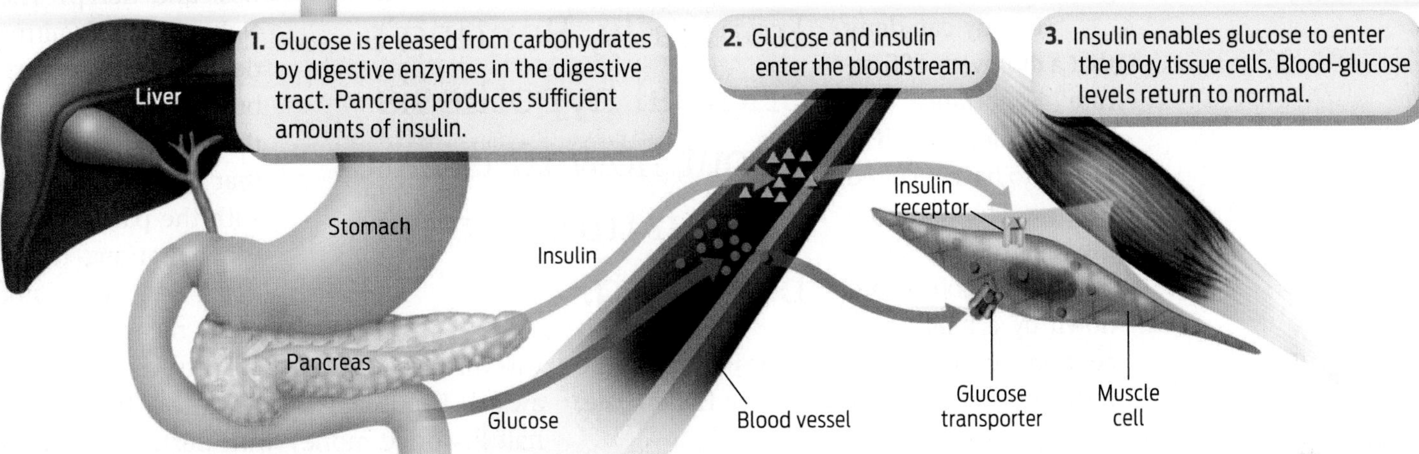
1. Glucose is released from carbohydrates by digestive enzymes in the digestive tract. Pancreas produces sufficient amounts of insulin.
2. Glucose and insulin enter the bloodstream.
3. Insulin enables glucose to enter the body tissue cells. Blood-glucose levels return to normal.
Liver
Stomach
Pancreas
Insulin
Glucose
Blood vessel
Insulin receptor
Glucose transporter
Muscle cell

	Digestive System	Insulin	Body Tissue Cells	Bloodstream Glucose
Normal Insulin Response	Digestive enzymes release high levels of glucose into the blood ↑	Pancreas produces sufficient amount of insulin. ↑	Insulin enables glucose to enter cells.	Blood glucose levels return to normal.
Type 1 Diabetes	Digestive enzymes release high levels of glucose into the blood ↑	Pancreas produces little or no insulin. ↑	Very little insulin allows only small amount of glucose to enter cells.	Blood glucose levels remain high. ↑
Type 2 Diabetes	Digestive enzymes release high levels of glucose into the blood ↑	Pancreas produces sufficient amount of insulin. ↑	Cells are unresponsive to insulin.	Blood glucose levels remain high. ↑

? What is the primary difference between type 1 and type 2 diabetes?

kidney disease, nerve damage, even death. An estimated 29 million Americans (roughly 8% of the population) suffer from diabetes, making it the most common endocrine disorder in the United States.

Reception and Controversy

Overnight, Banting went from being an unknown surgeon to a world-famous scientist, receiving many accolades. Because of his efforts, Leonard Thompson would live 13 more years, to age 27, when he died of pneumonia.

Ironically, Banting's original hypothesis about how to isolate insulin turned out to be wrong. Contrary to what Banting believed, it was not necessary to remove the digestive enzymes from the pancreas in order to isolate insulin. With Collip's purification method, the researchers were able to extract insulin from whole pancreas, without first performing Banting's surgery.

Some scientists thought Banting's fame and status were undeserved because his initial premise was wrong. But these detractors, others said, misunderstood the nature of the scientific method, which often involves a good deal of luck.

Ironically, Banting's original hypothesis about how to isolate insulin turned out to be wrong.

"Nobody can deny that a discovery of first-rate importance has been made, and, if it proves to have resulted from a stumble into the right road, where it crossed the course laid down by a faulty conception, surely the case is not unique in the history of science," wrote Henry Hallet Dale, a British physiologist, in a letter to the *British Medical Journal* in 1922.

The Granger Collection, New York

Frederick Banting.

For their work on developing insulin, Banting and Macleod shared the 1923 Nobel Prize in Physiology or Medicine. What about Best and Collip? The Nobel committee deemed their contributions less critical to the discovery, a view that didn't sit well with the participants. When Banting heard that Best had not been recognized, he was furious and refused at first to accept the award. Eventually, he accepted, but shared half his prize money with Best. Macleod, in turn, shared half his prize money with Collip. The earlier researchers who had determined the role of the pancreas in diabetes and deduced the existence of insulin–even gave it a name–got nothing.

Since Banting and his team made their important discoveries, a number of other scientific milestones have occurred. Researchers learned that insulin was a protein and, in 1952, the British biochemist Frederick Sanger determined its amino acid sequence–the first such protein sequence to be described. Today, insulin is no longer obtained from pancreas harvested from animal tissues but is synthesized by genetically engineered bacterial cells modified to contain the human insulin gene (see Chapter 8 for a discussion of genetic engineering). Insulin remains one of the most important drugs of modern medicine. ■

MORE TO EXPLORE

- American Diabetes Association: www.diabetes.org
- Rosenfeld, L. (2002) Insulin: discovery and controversy. *Clinical Chemistry* 48:2270–2288.
- Henderson, J. (2005) Ernest Starling and 'hormones': an historical commentary. *Journal of Endocrinology* 184:5–10.
- Taubes, G. (2016). *The Case Against Sugar.* New York: Knopf.

MILESTONES IN BIOLOGY 8 Test Your Knowledge

1 What organ produces insulin?

a. the liver
b. the stomach
c. the pancreas
d. blood
e. skeletal muscle

2 Someone who cannot produce insulin will likely have blood-sugar levels that are

a. normal.
b. lower than normal.
c. higher than normal.

3 In general terms, what is a hormone?

a. A signaling molecule that is transported by the circulatory system and acts on target cells.
b. A signaling molecule that regulates blood sugar.
c. A signaling molecule that is produced by the pancreas.
d. A signaling molecule that increases the concentration of blood sugar.
e. A protein that is released by one cell and acts on an adjacent cell.

4 Knowing that someone has chronically elevated blood-sugar levels, are you able to say if this person has type 1 diabetes or type 2 diabetes? Why or why not?

5 Insulin is a/an

a. monosaccharide.
b. protein.
c. triglyceride (fat).
d. phospholipid.
e. any of the above, depending on the diet

6 What are the key differences in glucose levels and insulin levels between type 1 and type 2 diabetes? Consider levels during periods of fasting and levels after a carbohydrate-rich meal.

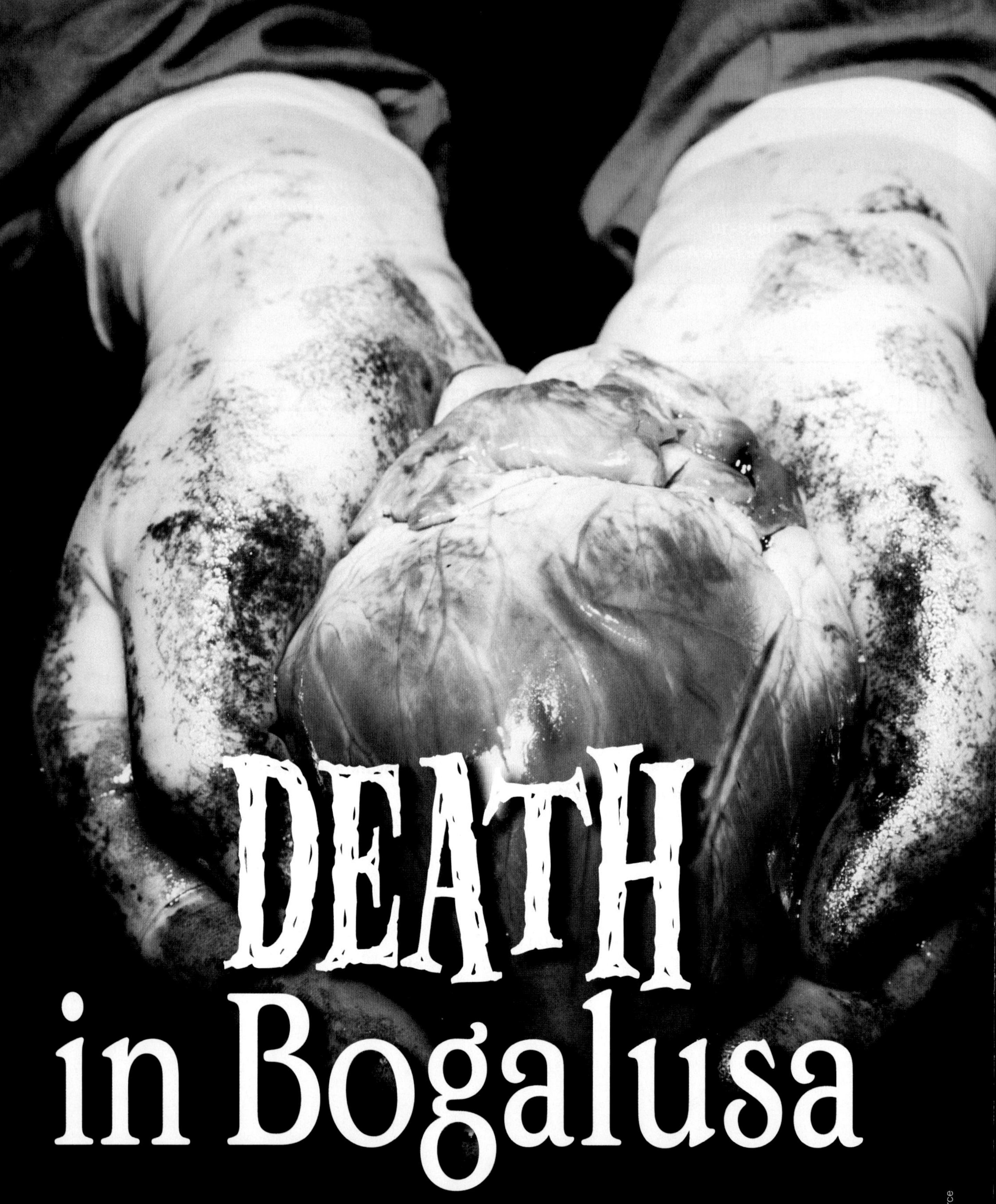

DEATH in Bogalusa

From tragic deaths in a southern town, insight into heart disease

Kevin Curtis/Science Source

DRIVING QUESTIONS

1. What structures make up the cardiovascular system, and how does blood flow through the system?
2. What is the structure of the heart and of the different types of blood vessel?
3. What is the composition of blood, and what does blood do?
4. What is cardiovascular disease, and what are some of the risk factors for developing cardiovascular disease?

Some scientific discoveries begin in a laboratory. This one begins in a funeral home, in 1978. Under glaring lights, in the back room of the Cook-Richmond Funeral home in Bogalusa, Louisiana, two pathologists hover over the lifeless body of a young African American male. There is no morgue in Bogalusa, so an autopsy is being conducted here, with newspapers spread beneath the body. The pathologists slice through skin and muscle with scalpels, looking for the usual suspects–internal bleeding, broken bones–then write up their report for the coroner.

The pathologists aren't quite done, however. Before closing up, they carefully remove the victim's heart, pack it in saline, and prepare it for a trip to a medical school in New Orleans. It was an unusual step, not standard for an autopsy. But the pathologists were heeding the instructions of a prominent local physician, who had the blessing of the family.

Gerald Berenson had a passionate interest in heart disease. Since 1972, as a cardiologist at the Louisiana State University School of Medicine, he had spearheaded what was then a novel study: an epidemiological study of heart disease in Bogalusa, a rural town of about 16,000 people 60 miles north of New Orleans.

CARDIOVASCULAR DISEASE (CVD) A disease of the heart or blood vessels or both.

CARDIOVASCULAR SYSTEM The system that transports nutrients, gases, and other critical molecules throughout the body. It consists of the heart, blood vessels, and blood.

HEART The muscular pump that generates force to move blood throughout the body.

The study began with a pretty simple idea: follow a large group of children over a period of time and correlate their physical and lifestyle characteristics with the development of heart disease in adulthood. The biggest hurdle was a logistical one–how to enlist the thousands of children necessary to produce a robust data set and keep them coming back for evaluation year after year.

But Berenson wanted to do more than make statistical correlations. He also wanted to document the progression of heart disease directly. And for that he needed a different type of evidence.

The heart that Berenson obtained in 1978 was one of more than 200 such organs collected from young people in Bogalusa over the next 20 years. In fact, nearly every young person who died, of whatever cause, was autopsied and had the heart removed. From this medical detective work has emerged a detailed understanding of heart disease in young people and the evidence needed to clinch the case against a cold-blooded killer.

A Silent Epidemic

Cardiovascular disease (CVD), which includes heart disease and stroke, is the number one killer of men and women in the United States and the developed world. According to the American Heart Association, one in three deaths in the United States is caused by cardiovascular disease–roughly 2,200 deaths per day (**INFOGRAPHIC 28.1**).

CVD claims so many lives each year for two main reasons. The disease is largely "silent"–people don't know they're sick until it's too late. And the system it affects is one of the most important to our survival, so when things go wrong, the result tends to be fatal.

The **cardiovascular system** transports nutrients, gases, hormones, and other critical molecules throughout the body. It consists of the heart, the blood vessels, and the blood. The **heart** is essentially a pump. It is about the size of a fist and very muscular. By repeatedly contracting and relaxing, it pumps blood throughout the body.

INFOGRAPHIC 28.1 Cardiovascular Disease Is the Major Killer in the United States

Of the top ten causes of death, cardiovascular disease (including heart disease and stroke) continues to be the leading cause of death in both men and women in the United States.

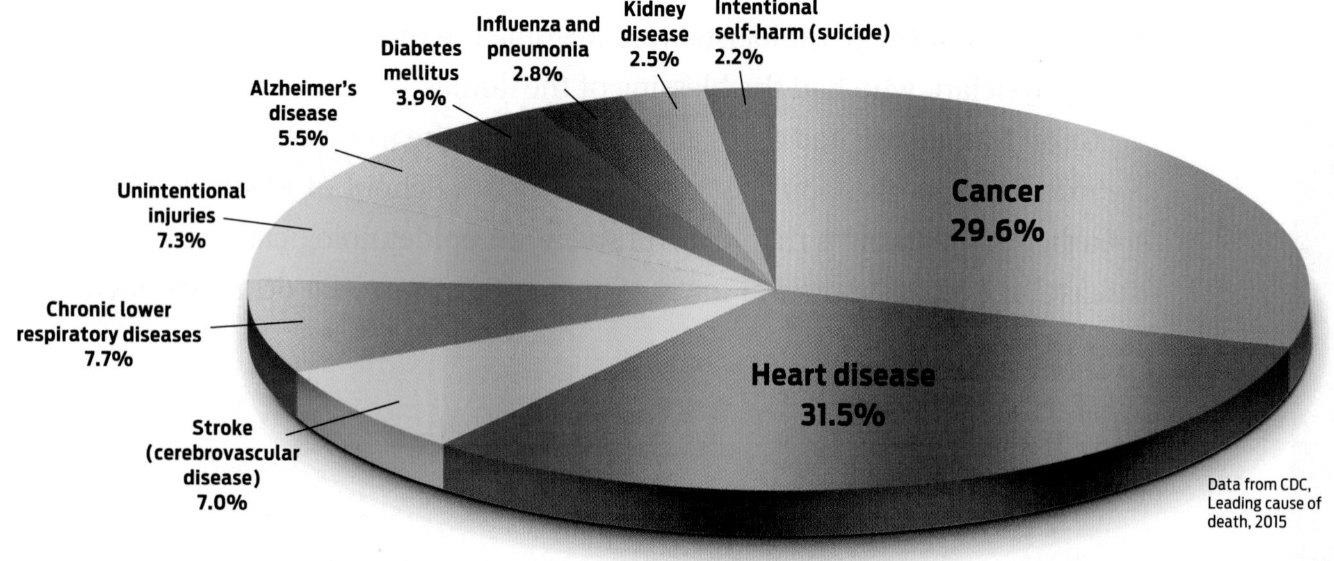

? Of the top ten causes of death in the United States, what percent is attributable to heart disease and cerebrovascular disease (stroke) combined?

Blood travels through the body in different types of vessel. Blood vessels that carry blood away from the heart are the **arteries,** and blood vessels that return blood to the heart are the **veins.** Blood is composed of cells and liquid. One of its important roles is to deliver oxygen to tissues of the body and carry away carbon dioxide waste (**INFOGRAPHIC 28.2**).

For decades, doctors have known that cardiovascular disease is essentially a matter of bad plumbing. Fatty deposits develop in the arteries that deliver blood to the body or to the heart muscle itself, cutting off or reducing blood flow to these tissues. This is dangerous because the oxygen traveling in the blood is needed by the tissues in order to carry out aerobic respiration (see Chapter 6). When the arteries that supply blood directly to the heart muscle–the **coronary arteries**–are blocked, the cells of the heart begin to die. This is a **heart attack.**

For the most part, young people do not have heart attacks. But just because teenagers do not die from heart disease does not mean they can ignore heart health. That's because heart disease can be insidiously gaining a foothold long before we have any obvious symptoms.

ARTERIES
Blood vessels that carry blood away from the heart.

VEINS
Blood vessels that carry blood toward the heart.

CORONARY ARTERIES
The blood vessels that deliver oxygenated blood to the heart muscle.

HEART ATTACK
Damage to the heart muscle resulting from the restriction of blood flow to heart tissue.

INFOGRAPHIC 28.2 The Cardiovascular System

The cardiovascular system transports nutrients and gases throughout the body. It consists of the heart, the blood vessels, and the blood.

The Blood Vessels
Blood vessels are the tubes that carry blood throughout the body.

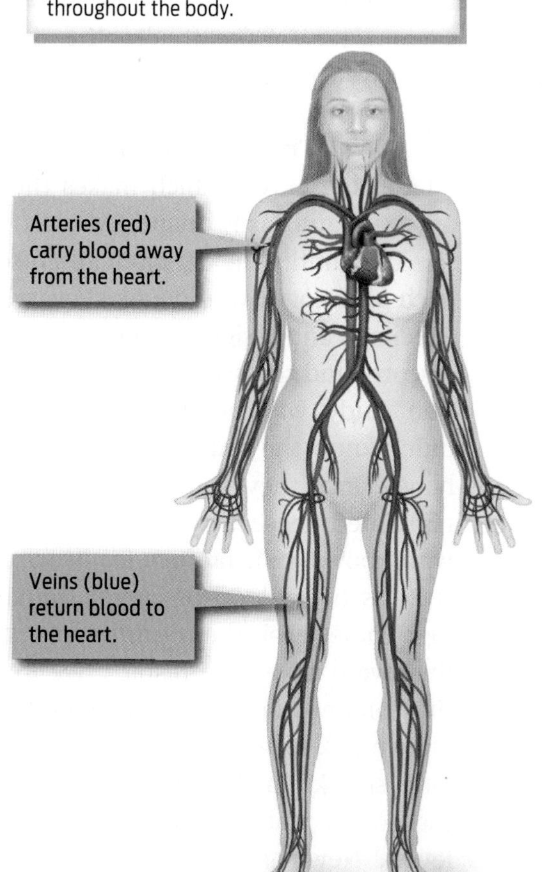

The Heart
The heart is a muscular pump that contracts to move blood through the blood vessels. It has four chambers that coordinate to receive and direct blood to the lungs and body.

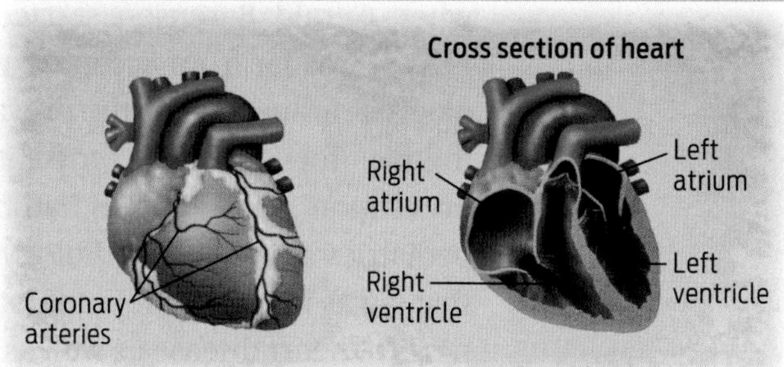

The Blood
Blood is composed of a variety of cell types, molecules, and gases suspended in liquid.

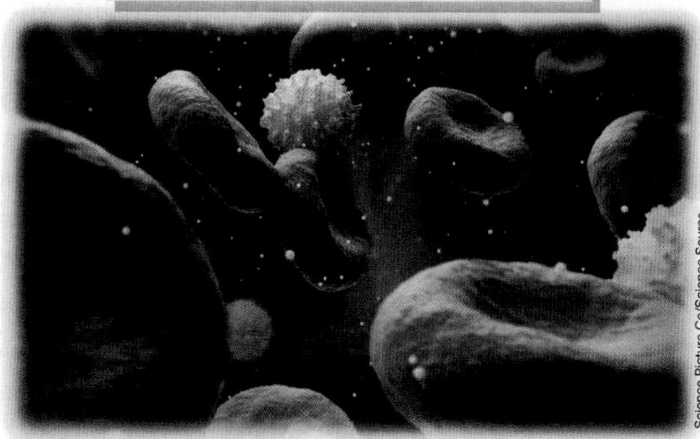

? Is blood traveling from your heart toward your big toe traveling in an artery or a vein?

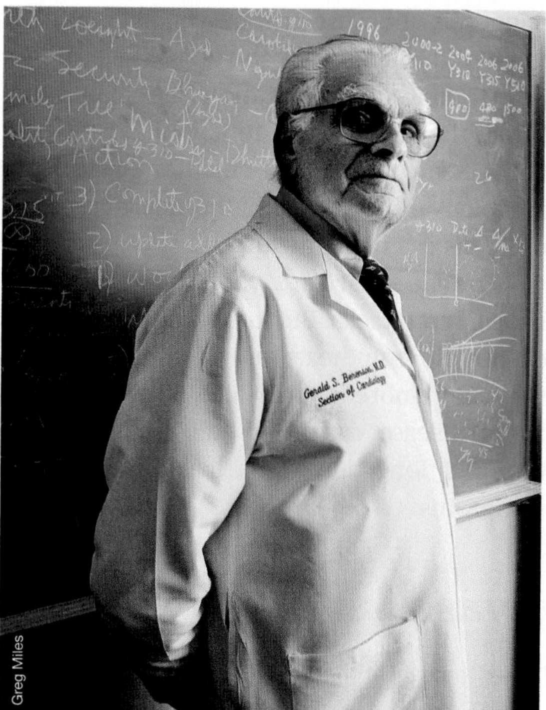

Gerald Berenson.

This is common knowledge now, but when Gerald Berenson started his study in 1972, it was far from accepted wisdom. Cardiologists–influenced by the prevailing beliefs of the day–focused more on treatment than on prevention. But Berenson had been trained in pediatrics as well as cardiology, and he knew that it was important to understand the beginnings of heart disease as well as its endings.

AORTA The large artery that receives oxygenated blood from the left ventricle.

ATHEROSCLEROSIS A condition in which fatty deposits build up in the lining of arteries, restricting blood flow; also known as hardening of the arteries.

CHOLESTEROL A lipid that is an important component of cell structures; it is used to make important molecules and also plays a role in heart disease.

RISK FACTOR A behavior, exposure, or other factor that increases the probability of developing a disease.

Under the Knife

William Newman was one of three pathologists on Berenson's team who analyzed the tissues collected from the autopsies. He was also directly involved in harvesting the specimens. Whenever a young person in Bogalusa died, Newman and a colleague would drive the 66 miles up from Louisiana State University School of Medicine in New Orleans to Bogalusa to perform the autopsy for the coroner.

What Newman and his colleagues saw when they opened the hearts of the young people shocked them: fatty streaks lining both the coronary arteries that supply the heart muscle with blood and the **aorta**, the major artery leaving the heart that ultimately supplies blood to the rest of the body. Virtually all the individuals autopsied had these fatty streaks, and the extent of the streaking increased with age.

As Newman explains, a fatty streak is an early form of **atherosclerosis,** a condition in which fatty deposits and other substances build up in the lining of blood vessels and restrict blood flow in those vessels. Atherosclerosis, sometimes called hardening of the arteries, is a common cause of heart disease. It's a complex process, and scientists aren't entirely sure what triggers it. One hypothesis is that it begins when blood vessels become damaged in some way. Once a vessel is damaged, a waxy lipid called **cholesterol** begins to accumulate in the lining of the vessel. White blood cells that normally help to heal injuries collect at the site of cholesterol deposition, but instead of helping the situation, they make it worse, causing more cholesterol to accumulate. A fatty streak is the result. Newman says you can identify these streaks when you cut open a vessel because they "look a bit yellow." They will also take up a lipophilic ("fat-loving") dye, which makes them more apparent and easily quantified.

To quantify the extent of atherosclerosis, each pathologist would look at every autopsy specimen and assign the stained fatty streaks a score, and the scores were then averaged. The study was blind, so the pathologists didn't know the source of the tissues beforehand. Once they had the data, statisticians correlated these anatomical measurements with known **risk factors** for heart disease–such as high blood pressure, smoking, and high cholesterol–the young people had when they were alive. The result? "We found that those individuals who had higher levels of known risk factors on the average had more fatty streaks in their coronary arteries and the aortas than individuals who had lower levels," Newman says.

In other words, though many of them were not even old enough to vote, these young people already had telltale signs of heart disease. Some even had evidence of more severe atherosclerosis in the form of thick and hardened deposits of cholesterol, fat, calcium, and other materials collectively called **plaque** that had begun to obstruct blood flow. Had these young people lived, those with the higher levels of atherosclerosis would have been at risk for a heart attack or other complications of heart disease (**INFOGRAPHIC 28.3**).

A Series of Tubes

The evidence that Newman and his colleagues found of atherosclerosis in the heart's coronary arteries in young people was particularly troubling. The coronary arteries branch off the aorta and supply the heart muscle with blood. It is these small arteries that are the usual sites of blockages that lead to a heart attack.

If the heart is full of blood, why does it need its own blood supply from arteries leading from the aorta back into the heart muscle? The answer has to do with how gases like oxygen and carbon dioxide, and nutrients like sugar, are exchanged between blood vessels and tissues.

Arteries and veins are like major highways for blood transport–their main responsibility is bringing blood to and from the heart. Situated between these two large

PLAQUE
Deposits of cholesterol, other fatty substances, calcium, blood clotting proteins and cellular waste that accumulate inside arteries, limiting the flow of blood.

INFOGRAPHIC 28.3 Atherosclerosis: A Common Cause of Cardiovascular Disease

Atherosclerosis begins with fatty streaks that thicken the wall of a blood vessel. A fatty streak is composed of lipids, such as cholesterol, and white blood cells. These fatty streaks may, over time, increase in size and harden, developing into plaques. Plaques obstruct blood flow. When blood flow in the coronary arteries is blocked, the result is a heart attack.

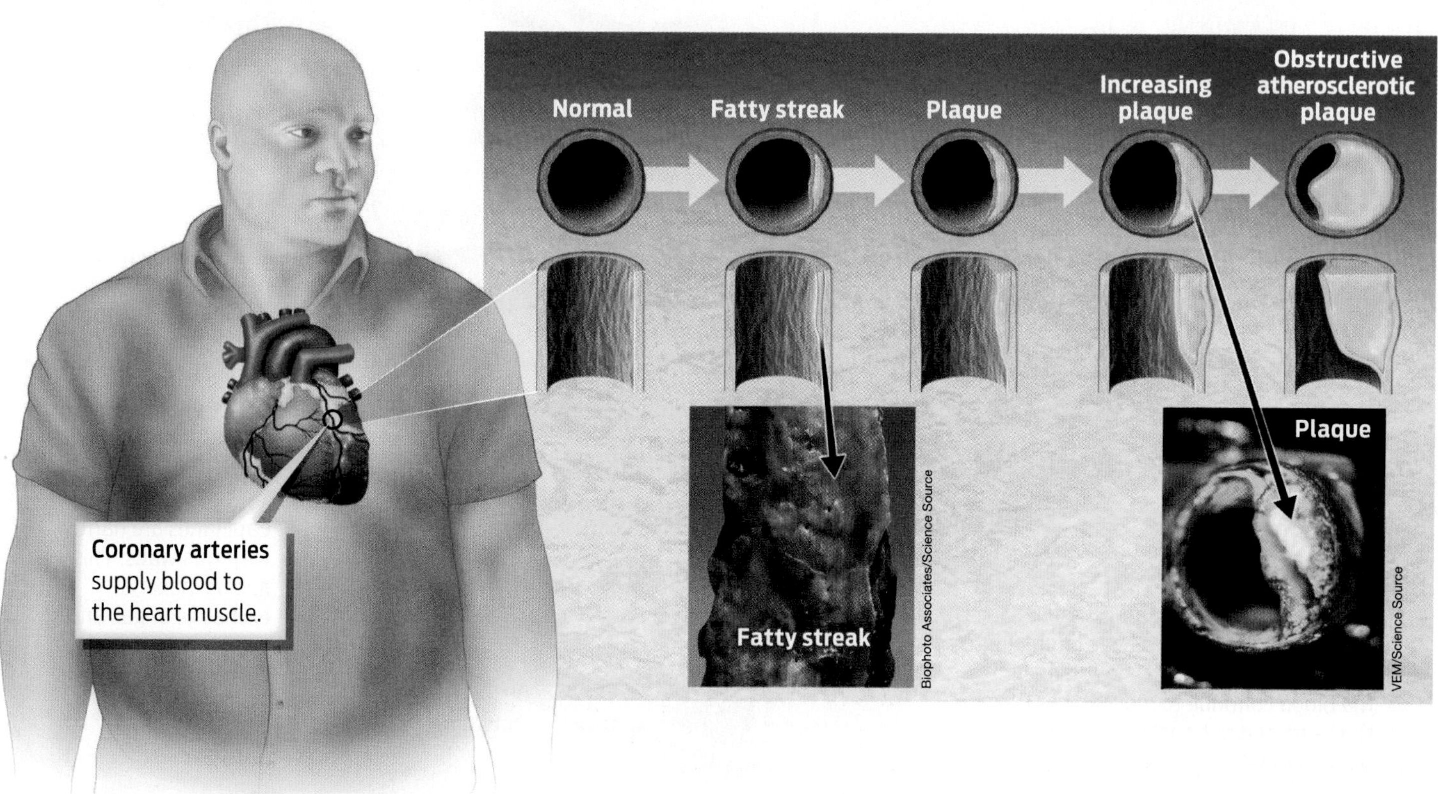

? If fatty streaks don't obstruct blood flow, are they still a cause for concern? Why?

CAPILLARIES
The smallest blood vessels. Capillaries are the sites of gas, nutrient, and waste exchange between the blood and tissue cells.

types of vessel, however, are capillaries. **Capillaries** are tiny blood vessels, located in tissues, where gas and nutrient exchange occurs. They are where the action is. Oxygen and nutrients diffuse *out* of capillaries into tissues, and carbon dioxide and wastes diffuse from tissues *into* capillaries. Capillaries are ideally suited to their job because of their thin walls and their slow rate of blood flow, which allows ample opportunity for gas, nutrient, and waste exchange between blood and tissues. Every tissue in the body, including the heart, is infiltrated with capillaries (**INFOGRAPHIC 28.4**).

INFOGRAPHIC 28.4 Capillaries Are Sites of Nutrient and Gas Exchange

Blood is carried by three different types of blood vessel. Arteries are large vessels that carry blood away from the heart. Arteries branch into smaller vessels in body tissues. The smallest blood vessels are the capillaries. The slow flow of blood in capillaries allows plenty of time for the exchange of nutrients and gases between the blood and tissue cells. Veins collect blood from the capillaries and deliver it back to the heart.

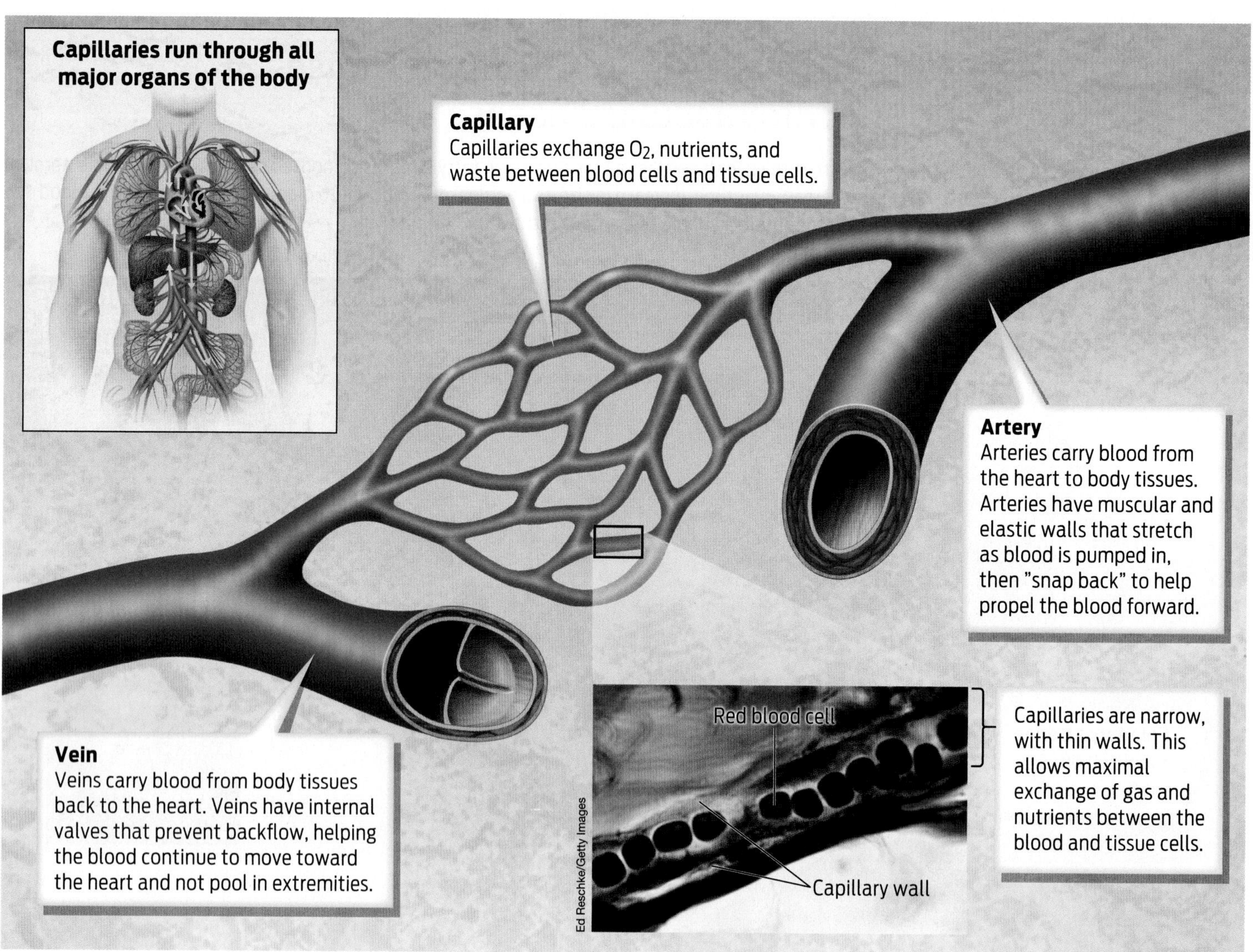

Capillaries run through all major organs of the body
Capillary
Capillaries exchange O2, nutrients, and waste between blood cells and tissue cells.
Artery
Arteries carry blood from the heart to body tissues. Arteries have muscular and elastic walls that stretch as blood is pumped in, then "snap back" to help propel the blood forward.
Vein
Veins carry blood from body tissues back to the heart. Veins have internal valves that prevent backflow, helping the blood continue to move toward the heart and not pool in extremities.
Red blood cell
Capillary wall
Capillaries are narrow, with thin walls. This allows maximal exchange of gas and nutrients between the blood and tissue cells.
Ed Reschke/Getty Images

? When a capillary enters a tissue (like the muscle tissue in your leg), what happens to the oxygen and carbon dioxide carried by the capillary as it passes through the tissue?

Arteries are high-pressure vessels; the blood that flows through them is propelled by the force exerted by the heart muscle. As blood slows down and loses pressure in the capillaries, it enters the veins at low pressure. As low-pressure vessels, veins rely on contractions of the skeletal muscles in which they are found, and on valves located in these vessels, to prevent the backflow of blood and return it to the heart.

The function of this interconnected system of blood vessels is to bring blood in proximity to nearly every cell in the body. What makes blood so vital? **Blood** is a complex tissue made up of cells suspended in a liquid known as plasma. For cardiovascular purposes, the most important cells are the **red blood cells** (or erythrocytes), which are specialized for carrying oxygen. Red blood cells lack a nucleus and have a flexible, concave shape that allows them to squeeze through capillaries in single file. Oxygen diffuses directly across the membranes of these cells into tissues.

Also present in blood are **white blood cells** (or leukocytes), which protect us from infection and respond to injuries as part of the immune response (see Chapter 32), and **platelets,** which are cell fragments that play a critical role in blood clotting. Blood also carries dissolved carbon dioxide, the waste product of cellular respiration (see Chapter 6). Finally, blood contains traces of whatever else we put into our bodies–from the food we eat to the drugs we take to the cigarette smoke we may inhale. Some of the things that circulate in our blood may contribute, over time, to the development of atherosclerosis (**INFOGRAPHIC 28.5**).

BLOOD
A circulating fluid that contains several types of cell and transports substances, including nutrients, gases, and hormones.

RED BLOOD CELLS
Cells specialized for carrying oxygen.

WHITE BLOOD CELLS
Cells involved in the body's immune response.

PLATELETS
Fragments of cells involved in blood clotting.

INFOGRAPHIC 28.5 The Components of Blood

Blood contains a variety of cells in a liquid called plasma. Red blood cells pick up oxygen in the lungs and deliver it to tissues in the body. White blood cells are critical for immune defense. Platelets are cell fragments important for blood clotting in areas of injury. The plasma contains many types of dissolved proteins, gases, nutrients, and other important molecules.

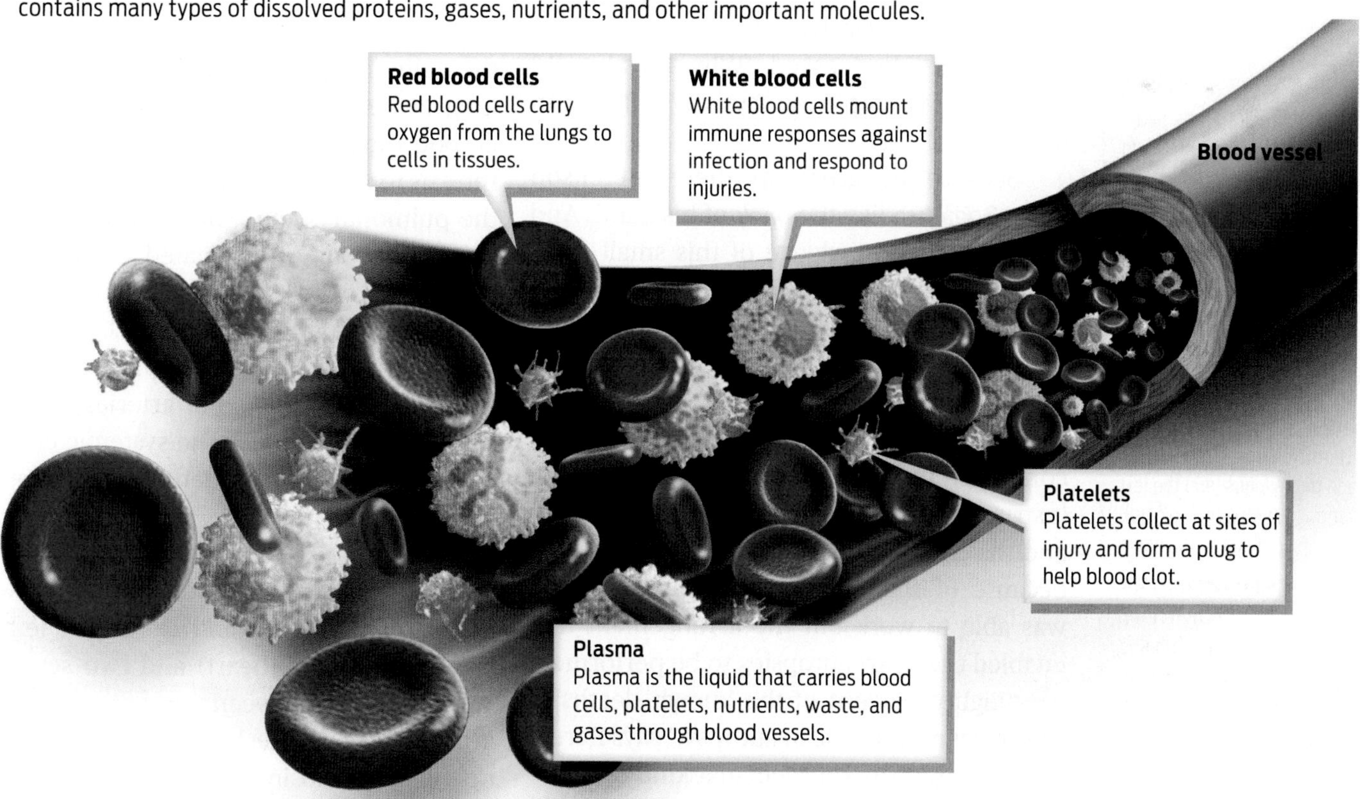
Red blood cells
Red blood cells carry oxygen from the lungs to cells in tissues.
White blood cells
White blood cells mount immune responses against infection and respond to injuries.
Blood vessel
Platelets
Platelets collect at sites of injury and form a plug to help blood clot.
Plasma
Plasma is the liquid that carries blood cells, platelets, nutrients, waste, and gases through blood vessels.

? What are the functions of red blood cells, white blood cells, and platelets?

Matters of the Heart

At 90, Gerald Berenson has been around long enough to see many changes in the practice of medicine. He is currently Director of the Tulane Center of Cardiovascular Health and Principal Investigator of the Bogalusa Heart Study, yet he still retains an air of the old-fashioned family doctor. Over the years, he has even been known to make house calls to many of the participants in the study.

"If there was any kind of medical problems I went and examined them myself and took care of them. I had 4,500 patients to look at initially," Berenson recalled in an interview with Tulane's *Global Health News* in 2012.

It's this personalized approach to medicine, combined with Berenson's brand of southern tenacity, that has ensured success of the study over the years. "I'm often asked 'Why Bogalusa?'" The answer is simple, he says: "It's where I'm from."

> One in three deaths in the United States is caused by cardiovascular disease—roughly 2,200 deaths per day.

Because he grew up there, Berenson knew Bogalusa like the back of his hand. And that firsthand knowledge of this small Louisiana town proved crucial in solidifying support for the study. In fact, nearly everyone in the town has participated in the study in some way. Teachers and nurses at local schools serve as study liaisons. The pathologists who conducted the autopsies had Berenson as an instructor in medical school. Even Berenson and the coroner were old friends, and it was because of that relationship that Berenson was able to work out the arrangement that enabled the heart autopsies to be performed.

"Eighty percent of the known deaths in the area we were able to autopsy," says Berenson proudly. "Nobody gets that kind of rate."

The heart pump has a distinctive structure that underlies its important function. It has four chambers that work together to deliver blood: two paired **atria** above two paired **ventricles.** The atria receive blood into the heart, and the ventricles pump blood out of it.

Key to understanding heart function is the fact that blood passes through the heart twice for every trip through the body–it follows a double circuit. One circuit involves the trip of blood between the heart and the lungs, known as the **pulmonary circuit;** the other involves the trip of blood between the heart and the rest of the body, known as the **systemic circuit.**

The two circuits can best be understood by tracing the path of blood as it makes one complete trip through the body. Deoxygenated blood enters the heart through the right atrium. The deoxygenated blood is then pumped into the right ventricle. From there, the pulmonary artery carries the blood to the lungs, where it picks up oxygen and drops off carbon dioxide. The oxygenated blood then returns to the left atrium of the heart through the pulmonary vein. This completes the pulmonary circuit. Oxygenated blood is then pumped into the left ventricle, and from there into the **aorta,** the major artery leading to the rest of the body. The aorta carries oxygenated blood to the body's other arteries, including the coronary arteries. The systemic circuit is completed when deoxygenated blood returns to the right atrium.

The heart is equipped with four valves–two atrioventricular valves (located between the atrium and the ventricle on the left and right sides of the heart) and two semilunar valves (located between the ventricles and the arteries leaving the heart). The coordinated opening and closing of these valves ensures a one-way flow of blood through the heart, with no backflow. The "lub dub" sound the heart

ATRIA
The chambers of the heart that receive blood. In humans, the right atrium receives low-oxygen blood from the body, and the left atrium receives high-oxygen blood from the lungs.

VENTRICLES
The chambers of the heart that pump blood away from the heart. In humans, the right ventricle pumps low-oxygen blood to the lungs, and the left ventricle pumps high-oxygen blood to the body.

PULMONARY CIRCUIT
The circulation of blood between the heart and the lungs.

SYSTEMIC CIRCUIT
The circulation of blood between the heart and the rest of the body.

makes as it beats is actually the sound of these valves closing in pairs; first the two atrioventricular valves snap shut ("lub"), then the two semilunar valves close ("dub"). On average, it takes about 1 minute for blood to make a complete trip around the body **(INFOGRAPHIC 28.6)**.

Obviously, for the heart to do its job properly, it has to keep pumping. A group of cells located in the right atrium is the heart's natural pacemaker, setting the heart's tempo—60–90 beats per minute when we are at rest. In fact, for as long as you are alive the heart muscle never stops beating, and in that sense it is the only muscle in the body that never gets a rest. The human heart beats more than 2.5 billion times in an average lifetime.

INFOGRAPHIC 28.6 Blood Flow Follows a Double Circuit

For every trip through the entire body, blood makes two passes through the heart. In one of these passes, the pulmonary circuit, blood is pumped from the right ventricle of the heart to the lungs, where it picks up oxygen. Blood then returns to the left atrium of the heart. The systemic circuit then pumps blood from the left ventricle to the rest of the body's tissues, where it delivers oxygen. Blood then returns to the right atrium of the heart.

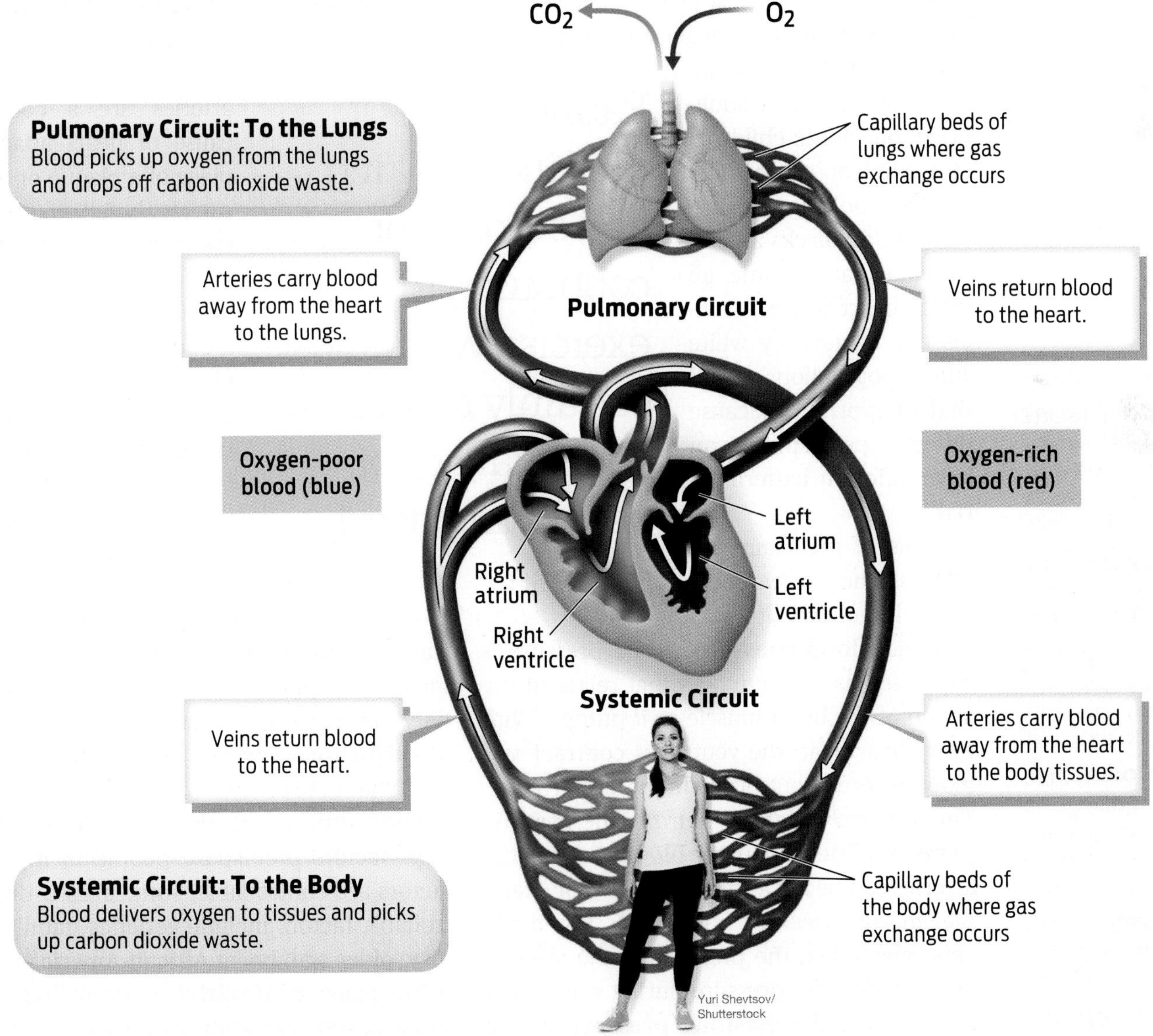

CO2
O2
Pulmonary Circuit: To the Lungs
Blood picks up oxygen from the lungs and drops off carbon dioxide waste.
Capillary beds of lungs where gas exchange occurs
Arteries carry blood away from the heart to the lungs.
Pulmonary Circuit
Veins return blood to the heart.
Oxygen-poor blood (blue)
Oxygen-rich blood (red)
Left atrium
Right atrium
Left ventricle
Right ventricle
Systemic Circuit
Veins return blood to the heart.
Arteries carry blood away from the heart to the body tissues.
Systemic Circuit: To the Body
Blood delivers oxygen to tissues and picks up carbon dioxide waste.
Capillary beds of the body where gas exchange occurs
Yuri Shevtsov/ Shutterstock

? Is blood in a vein in the pulmonary circuit traveling to the heart or to the lungs? Is it oxygenated or deoxygenated?

Like any muscle, the heart requires oxygenated blood to power the work of contraction. If a part of the heart loses its blood supply, that portion of the muscle begins to die, and the heart as a whole can begin to beat irregularly, or stop beating altogether. At that point, blood circulation stops. And very quickly, so do you. Understanding how blood flow becomes obstructed, then, is key to combating heart disease.

Under Pressure

When the Bogalusa Heart Study began, in 1972, Berenson and his colleagues were interested in identifying risk factors that contribute to heart disease, specifically in children. Other studies had identified risk factors in adulthood, but not in children. Moreover, no other study had ever looked at risk factors among blacks as well as whites (at the time, all major heart disease studies had exclusively white study populations). And that's important, because different populations can have different patterns of risk.

Eating better, quitting smoking, and starting or continuing to exercise will almost certainly reduce your chances of developing heart disease.

One of the main risk factors the researchers were interested in was elevated **blood pressure.** Blood pressure is expressed as two numbers that relate to the action of the heart muscle as it pumps. With every heartbeat, the ventricles contract and blood is forced out of the heart into arteries. This relatively high pressure is the **systolic pressure.** The force of the blood entering the arteries can be felt as the **pulse**—as, for example, in your neck or your wrist. When the ventricles relax, the pressure in the arteries drops; this lower pressure during ventricular relaxation is the **diastolic pressure.** Blood pressure is expressed as systolic pressure over diastolic pressure. The normal measure for blood pressure is systolic pressure less than 120 mmHg (millimeters of mercury, a unit of pressure) and diastolic pressure less than 80 mmHg, or 120/80. Chronic elevation of either number increases the risk of cardiovascular disease, including atherosclerosis, heart attack, and stroke.

High blood pressure, or **hypertension,** is dangerous because it can put stress on the walls of arteries, causing microscopic tears. These tears provide sites for the buildup of cholesterol, fats, and other substances, ultimately forming plaques.

When a plaque in a vessel grows large enough, it can rupture, exposing its contents to the bloodstream. This can cause blood clots to form on the plaque, obstructing blood flow. Blood clots in the coronary arteries are a common cause of heart attack. A piece of a plaque or a clot can also break off, travel to another spot in the body, and clog other vessels. If blood vessels in the brain are obstructed, the supply of oxygen to the brain is disrupted, causing a **stroke** (**INFOGRAPHIC 28.7**).

Hypertension is particularly dangerous because although it can have deadly consequences, there are typically no symptoms associated with it. This means that people have no way of knowing they are hypertensive unless they have their blood pressure checked and monitored.

What causes hypertension? Researchers are not entirely sure. However, they do know what factors predispose people to it. Some factors are controllable, some aren't. Uncontrollable factors include genetics, family history, older age, being African American, and being male; controllable ones include low physical activity, eating a high-salt diet, being overweight or obese, drinking too much alcohol, and stress.

BLOOD PRESSURE
The overall pressure in blood vessels, expressed as the systolic pressure over the diastolic pressure.

SYSTOLIC PRESSURE
The pressure in arteries at the time the ventricle contracts.

PULSE
The detectable force of blood entering arteries, which can be felt in the neck or wrist.

DIASTOLIC PRESSURE
The pressure in arteries when the ventricle is relaxed.

HYPERTENSION
Elevated (high) blood pressure.

STROKE
A disruption in blood supply to the brain.

INFOGRAPHIC 28.7 High Blood Pressure Can Result in Atherosclerosis

When the heart ventricles contract, they force blood out of the heart. The force of the blood leaving the heart puts pressure on the walls of the arteries—the systolic pressure. When the ventricles relax, there is less pressure on the arteries—the diastolic pressure. Chronically elevated blood pressure can cause small tears in the artery walls. These are sites where plaque can build up.

Blood pressure

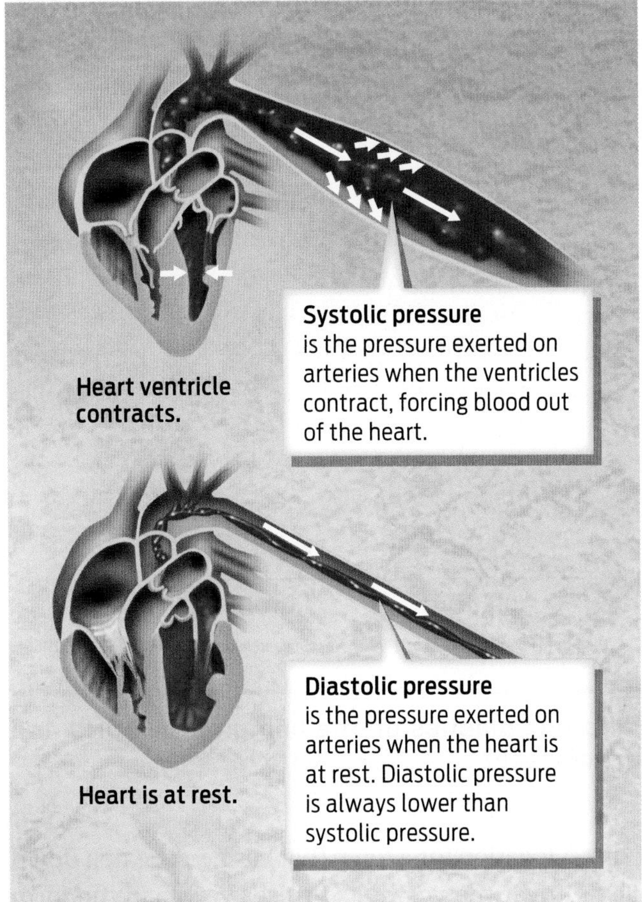

Systolic pressure
is the pressure exerted on arteries when the ventricles contract, forcing blood out of the heart.
Heart ventricle contracts.
Diastolic pressure
is the pressure exerted on arteries when the heart is at rest. Diastolic pressure is always lower than systolic pressure.
Heart is at rest.

Lengthwise section of an artery and its wall

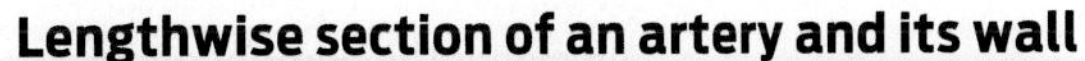

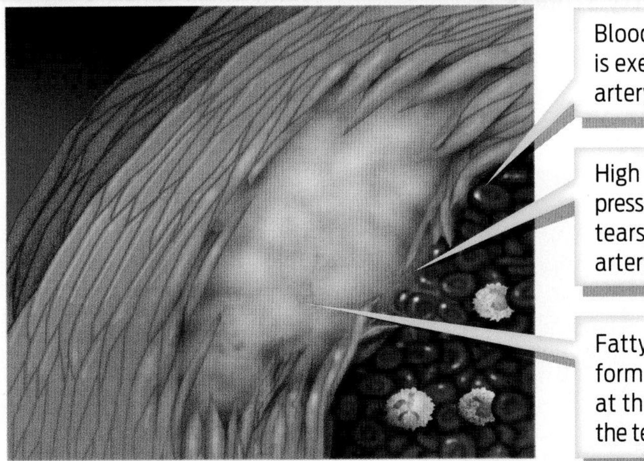

Blood pressure is exerted on artery walls.
High blood pressure causes tears in the artery wall.
Fatty deposits form plaques at the site of the tears.

Blood clots can form on the surface of a plaque and obstruct blood flow. High blood pressure can also cause plaques to rupture, sending small fragments to other vessels where they may form clots. In coronary arteries, blood clots cause a heart attack. In the brain, they cause a stroke.

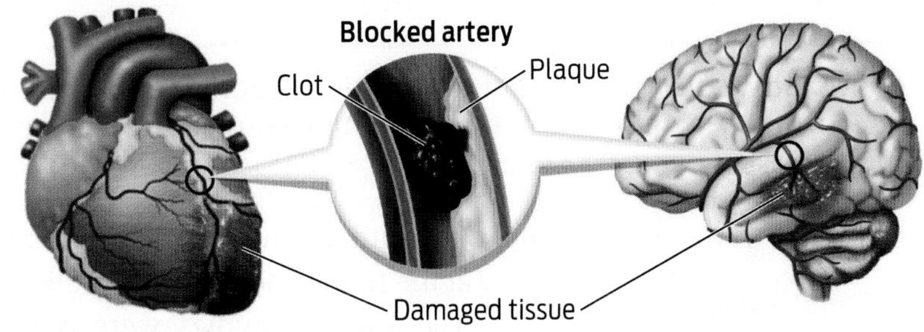

? If an artery can sustain high blood pressure without bursting, why is high blood pressure dangerous?

Obviously, when the bodies of the young people arrived at the funeral home in Bogalusa, they no longer had any vital signs, so it was not possible to take their blood pressure. However, because nearly every child in the town had been followed as part of the heart study, researchers were able to compare the victims' blood pressure when alive with the autopsy results. They found that having had high blood pressure correlated closely with the extent of atherosclerosis found in the arteries of the victims.

High blood pressure wasn't the only risk factor that researchers were able to link with the autopsy results. Other strong links were high body mass index (BMI), high blood-cholesterol levels, and smoking cigarettes. When all the known risk factors were considered, a clear pattern emerged: the amount and extent of atherosclerosis increased with age and was strongly correlated with the known risk factors for heart disease. The young people with the most risk factors when alive were

INFOGRAPHIC 28.8 Risk Factors and Heart Disease

Researchers measured the percentage of the inside lining of the aorta and coronary arteries covered in fatty streaks and fibrous plaques. These percentages were compared with the number of risk factors for heart disease a person had. The study found a direct correlation between the two. On average, young people with more risk factors had more evidence of heart disease.

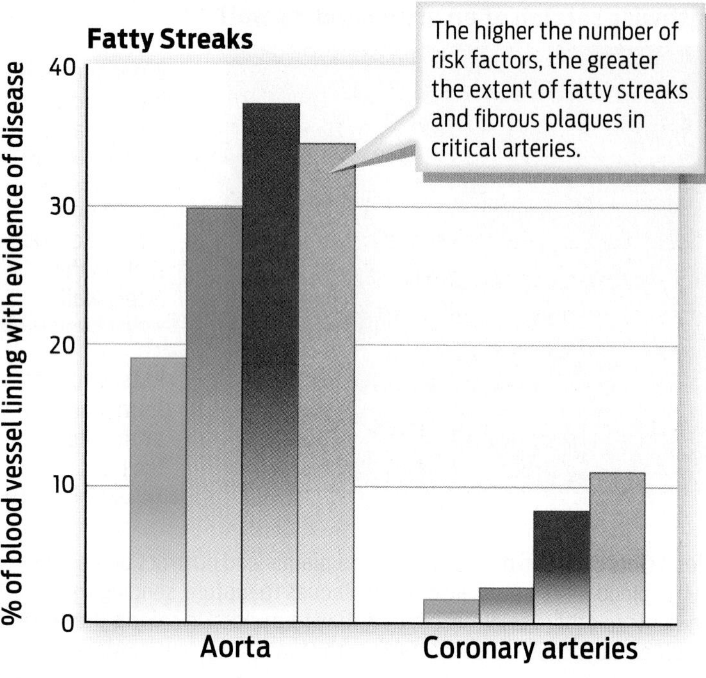

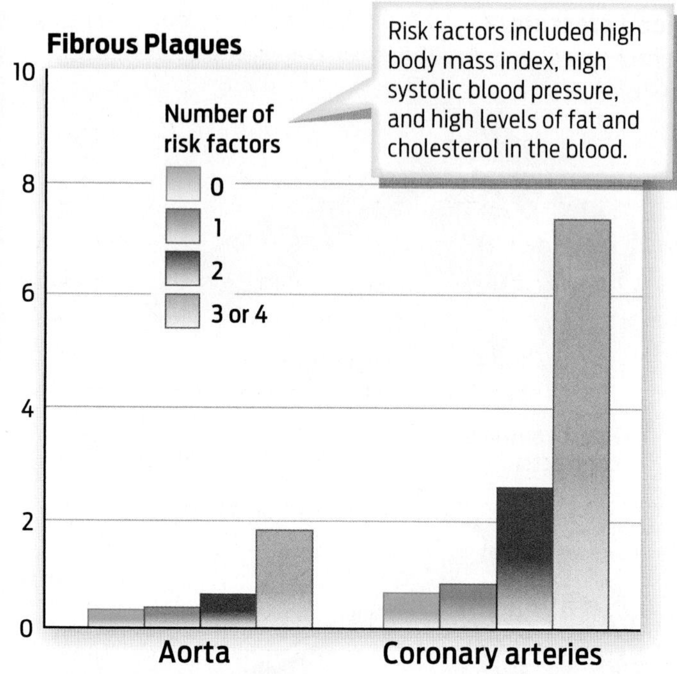

Data from Borenson G.S. et.al. (1998). *New England Journal of Medicine.* 338:1650-1656.

? What would you predict about the presence and extent of fibrous plaques on the lining of the aorta and coronary arteries in someone with two risk factors for heart disease?

the ones with the most evidence of heart disease at the time of death (**INFOGRAPHIC 28.8**).

"The autopsy studies were really landmark studies," says Peter Katzmarzyk, an epidemiologist at the Pennington Biomedical Research Center in Baton Rouge. Before them, he says, scientists didn't really understand that things like heart attack and stroke have their genesis in childhood. "The Bogalusa Heart Study really put that on the map," he says. And it made a strong case that averting severe consequences of heart disease later in life would mean addressing the lifestyle choices we make when we are young.

Southern Discomfort

If Louisiana is shaped like a boot, then the town of Bogalusa sits in ideal kicking position—right at the easternmost toe. It's a fitting location for a town that has been beaten down over the years, its bruises visible in the form of crumbling playgrounds, dilapidated storefronts, and damage from Hurricane Katrina in 2005. Once known as the Magic City because of its rapid growth around the burgeoning timber industry, Bogalusa has since lost some of its luster, the acrid smell of exhaust from the local paper mill the sole reminder of its former glory days.

Bogalusa is typical of many rural southern towns in which poverty and poor education have exacerbated problems of public health. More than 40% of individuals in Bogalusa live below the poverty line (contrasting with 23% for Louisiana as a whole and 15% nationally). And CVD disproportionately affects people of lower socioeconomic status.

Partly, the issue is access to health care. "One of the number one things related to health is socioeconomic status," says epidemiologist Katzmarzyk. "If you're wealthy and you have access to health care, you will be healthier." But there are likely cultural and behavioral factors at work, too.

Take diet, for example. Southern cuisine is famous for its fried foods, including fried chicken, fried fish, and fried potatoes. Also popular are bacon, ham, and sugar-sweetened beverages such as sweet tea and soda. In short, there's no shortage of salt, fat, and sugar in the southern diet.

Unhealthful diet choices can directly contribute to heart disease. Consuming large amounts of salt, for example, is known to raise blood pressure, contributing to hypertension. And many studies suggest that eating lots of certain fats—particularly **saturated fat** and **trans fat**—can elevate the level of cholesterol in the blood, contributing to atherosclerosis.

Cholesterol is not inherently harmful. In fact, it is a critical constituent of cell membranes as well as a precursor to the synthesis of many hormones. However, certain foods can increase the amount of cholesterol that our bodies produce, which can do more damage than good—especially if the result is more cholesterol hanging around in our arteries.

Whether cholesterol stays in our arteries or not depends upon the specific form it takes in the blood. Because cholesterol is a lipid, it is hydrophobic (see Chapter 2), and so it is not soluble in water-based solutions such as blood. It therefore is packaged along with other lipids in specialized particles called lipoproteins that allow it to be carried in the blood. Lipoproteins are spherical particles made of protein and lipids that contain a cargo of cholesterol (and triglycerides) in their hydrophobic center. There are two main varieties. **Low-density lipoprotein (LDL)** carries cholesterol to body cells and can accumulate in the walls of blood vessels, contributing to atherosclerosis. **High-density lipoprotein (HDL)** carries cholesterol out of blood vessels and delivers it to the liver, where it can be processed into a form that permits its elimination from the body, for example as bile salts (see Chapter 27).

In general, HDL is "good" and LDL is "bad" (think "H" for "healthy" and "L" for "lethal"). Too much LDL relative to the amount of HDL in the blood can lead to cholesterol becoming deposited in a blood vessel wall. This is the start of atherosclerosis. Over time, as atherosclerosis progresses, the affected area can

SATURATED FAT
A fat, such as butter, that is solid at room temperature; often found in animal products.

TRANS FAT
An unhealthful fat made from vegetable oil that has been chemically altered to make it solid at room temperature; often found in snack foods.

LOW-DENSITY LIPOPROTEIN (LDL)
A form of cholesterol and protein in the blood that contributes to CVD.

HIGH-DENSITY LIPOPROTEIN (HDL)
A form of cholesterol and protein that is protective against CVD.

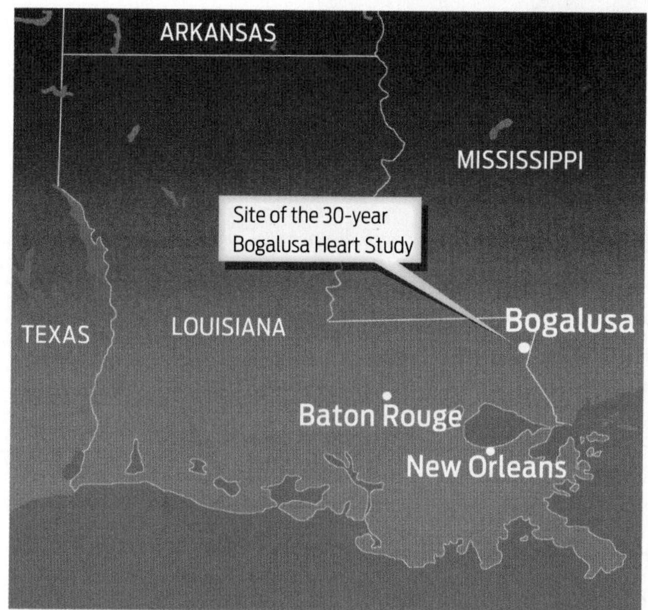

The Daily News, Bogalusa, LA

Once known as the Magic City, Bogalusa has since lost some of its luster.

UNSATURATED FAT
A fat, such as olive oil, that is liquid at room temperature; often found in plants and fish.

form into a thickened plaque that may eventually obstruct blood flow (**INFOGRAPHIC 28.9**).

Eating saturated fats (found in meats, butter, and cheeses) tends to raise LDL, whereas **unsaturated fats** (from olive oil, fish, and nuts) tend to raise HDL; doctors therefore recommend that people choose unsaturated fats over saturated ones (and avoid trans fats altogether).

Unfortunately, it's not just eating lots of unhealthful fat that is the problem. Accumulating evidence suggests that consuming lots of added sugar may be just as bad for you. Sugars in foods like white bread, candy, fruit drinks, and soda can elevate blood triglycerides and lower HDL, worsening atherosclerosis. A 2014 study published in *JAMA Internal Medicine* found that the more sugar a person eats as a percentage of the total diet, the higher the risk of dying from heart disease.

By making it easy for people to eat more calories than they need to maintain their weight, the abundance of added sugar in the modern diet may also contribute to obesity—an independent risk factor for cardiovascular disease. Being obese makes the heart work

INFOGRAPHIC 28.9 Cholesterol: Good and Bad

Cholesterol is an important lipid that the body makes in order to provide building materials for cell membranes, hormones, and vitamins. It is carried in the blood in different kinds of lipoprotein particles – HDL and LDL. Cholesterol in HDL ("good cholesterol") is easily removed from the circulation and transported to the liver for processing. Cholesterol in LDL ("bad cholesterol") tends to build up in the walls of arteries, contributing to atherosclerosis.

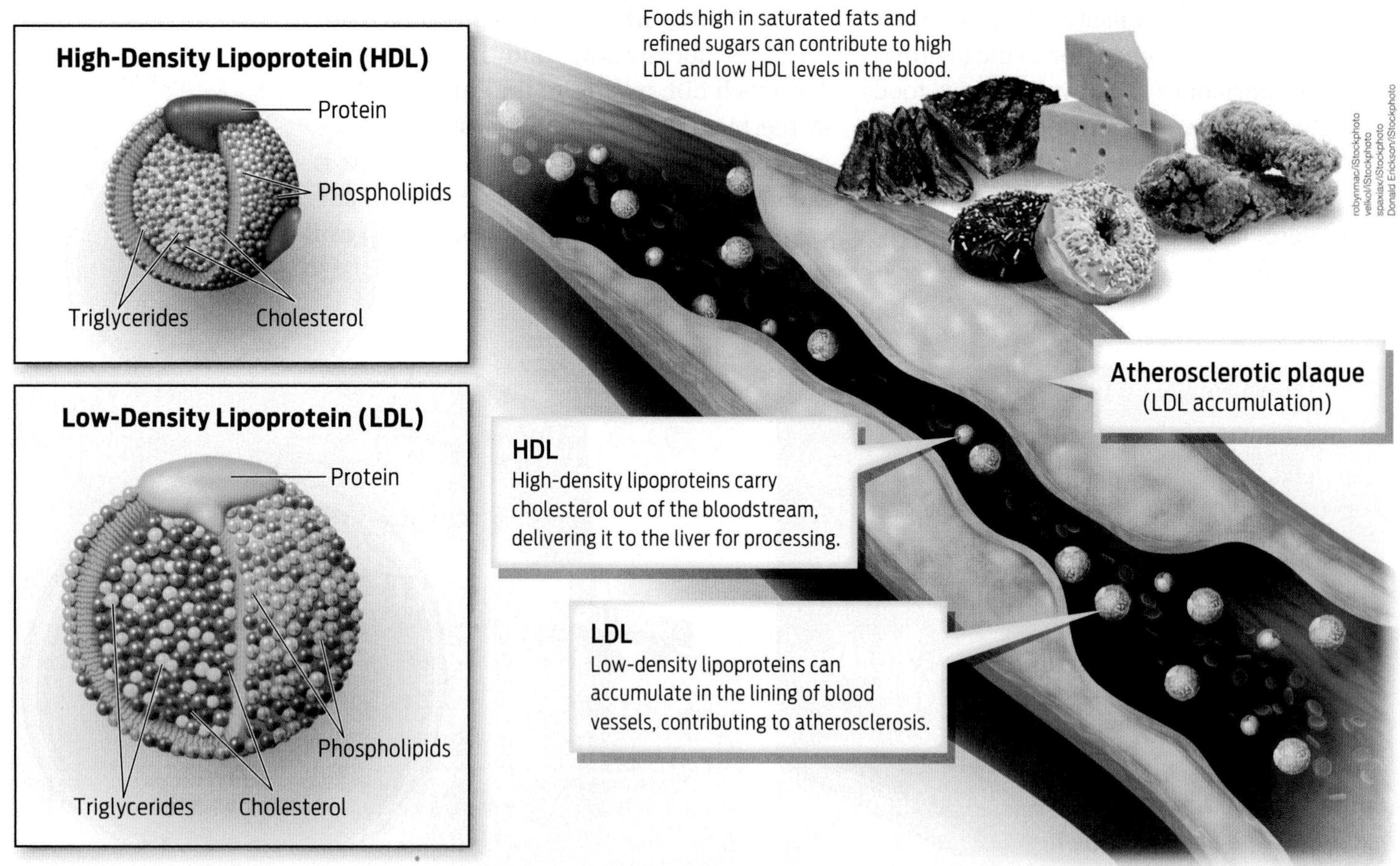
High-Density Lipoprotein (HDL)
Protein
Phospholipids
Triglycerides
Cholesterol
Low-Density Lipoprotein (LDL)
Protein
Phospholipids
Triglycerides
Cholesterol
Foods high in saturated fats and refined sugars can contribute to high LDL and low HDL levels in the blood.
Atherosclerotic plaque
(LDL accumulation)
HDL
High-density lipoproteins carry cholesterol out of the bloodstream, delivering it to the liver for processing.
LDL
Low-density lipoproteins can accumulate in the lining of blood vessels, contributing to atherosclerosis.
robynmac/iStockphoto
velkol/iStockphoto
spaxiax/iStockphoto
Donald Erickson/iStockphoto

? Why is cholesterol in the circulation found packaged in HDL and LDL particles and not as free cholesterol?

harder, and can lead to high blood pressure. The extra stored fat in obese individuals also secretes hormones that can worsen atherosclerosis. And obesity is the number one risk factor for type 2 **diabetes,** which, in turn, is a leading cause of CVD. In fact, 65% of people with diabetes die of cardiovascular disease–heart attack or stroke–according to the American Heart Association.

One of the most alarming trends revealed by recent epidemiological research is that obesity rates in the United States have skyrocketed since the 1980s. Nearly 36% of adults and 17% of children in the United States are obese (BMI ≥ 30), according to the CDC (**INFOGRAPHIC 28.10**).

In some parts of the country, such as in Bogalusa, the problem is even worse. Since 1973, when the Bogalusa study began, the proportion of children and young people ages 5 to 17 who are obese has increased more than fivefold, from 6% in 1973 to 31% today.

Louisiana is one of a series of states–those that lie between Texas and Florida–that epidemiologists sometimes refer to as the "diabetes belt," "stroke belt," or "obesity belt," because cardiovascular diseases seem to cluster in this region. What's behind this regional concentration? No one knows for sure. In addition to diet and other lifestyle factors, like smoking, it may have something to do with climate. "It's so hot for large parts of the year, people tend not to

DIABETES
A disease characterized by chronically elevated levels of blood sugar.

INFOGRAPHIC 28.10 Obesity in the United States Continues to Rise

The percentage of individuals in the United States who are overweight (BMI≥25) and obese (BMI≥30) has been steadily increasing. Currently, there are no states that report <20% obesity, with the highest levels of obesity found in the southern and midwestern states.

Historical Record of Obesity (BMI≥30) Among U.S. Adults

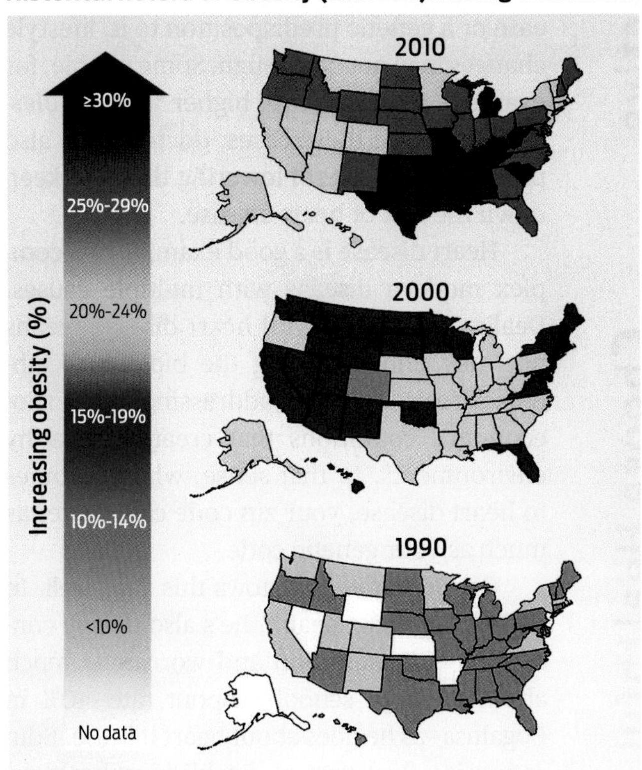

Percentage of Self-Reported Obesity (BMI≥30) Among U.S. Adults, 2013–2015

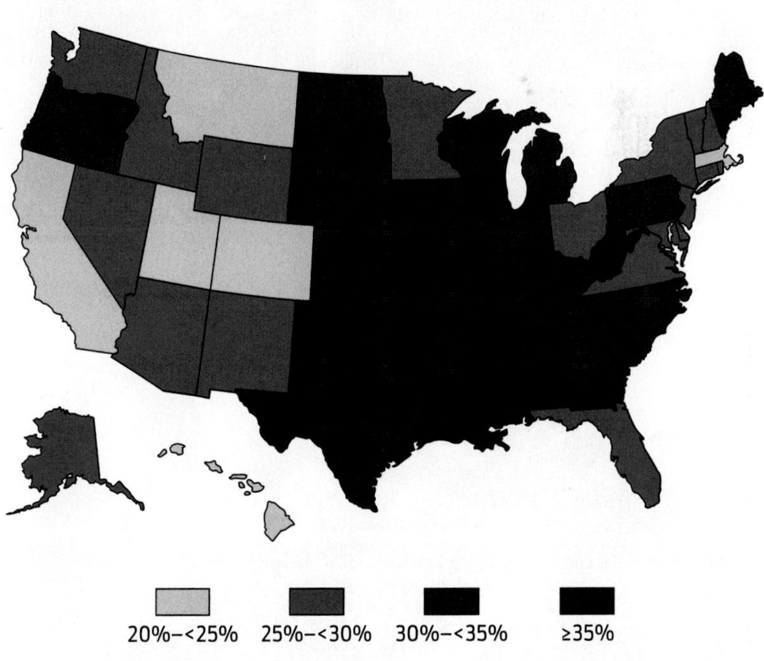

Data from Centers for Disease Control and Prevention. Data source: Behavioral Risk Factor Surveillance System.

? Which state(s) has/have consistently had the highest proportion of obese people, and which state(s) has/have consistently had the lowest proportion of obese people?

To reduce your risk of heart disease, follow the ABCs of heart health.

Avoid Tobacco

alptraum/iStockphoto

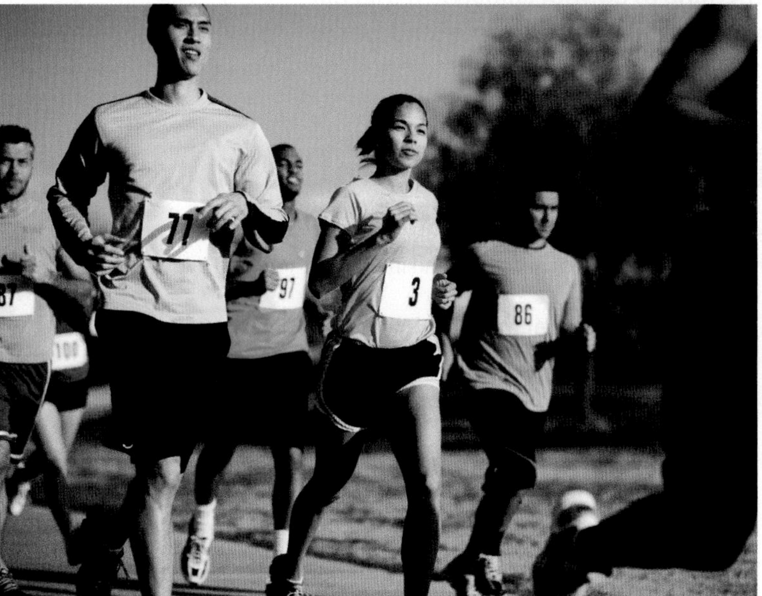

Be More Active

FatCamera/iStockphoto

Choose Healthful Foods

NightAndDayImages/iStockphoto

exercise outside as they would in the north," says Katzmarzyk. It could be that a confluence of factors has collectively created "a perfect storm" for cardiovascular disease in this part of the country (**INFOGRAPHIC 28.11**).

An Ounce of Prevention

Given the observed link between risk factors for heart disease and the extent of atherosclerosis in young people, it's obvious that reducing one's risk factors will reduce one's risk for heart disease. Eating better, quitting smoking, and starting or continuing to exercise will almost certainly reduce your chances of developing heart disease. In fact, a 2012 study found that eating more fruits and increasing one's activity level can positively affect blood cholesterol levels and reduce BMI. So that is the good news—there are things you can do to reduce your risk of heart disease: avoid tobacco, get more physical activity, and choose healthful foods.

Unfortunately, for some people, especially those with a family history of heart disease or a genetic predisposition to it, lifestyle changes may not be enough. Some people, for example, naturally have higher "bad" cholesterol levels. In these cases, doctors may also prescribe cholesterol-lowering drugs to keep down the risk of heart disease.

Heart disease is a good example of a complex modern disease with multiple causes. Dealing effectively with heart disease means not only understanding the biology of atherosclerosis but also addressing the socioeconomic conditions that create unhealthy environments. In that sense, when it comes to heart disease, your zip code can matter as much as your genetic code.

Gerald Berenson knows this very well. In addition to heart health, he's also deeply concerned with education and worries as much about the high school dropout rate—40% in Bogalusa—as he does about heart disease. Education is a key part of establishing healthful lifestyles, Berenson says, and it starts young. Likewise, the bad habits that lead to negative health consequences down the line start in childhood. In one very depressing statistic,

INFOGRAPHIC 28.11 Southern States Have the Highest Average Incidence of Cardiovascular Disease

States in the Southeast and the Southwest have the highest levels of heart disease in the United States. Societal factors such as diet, activity level, and smoking can influence the risk for developing cardiovascular disease.

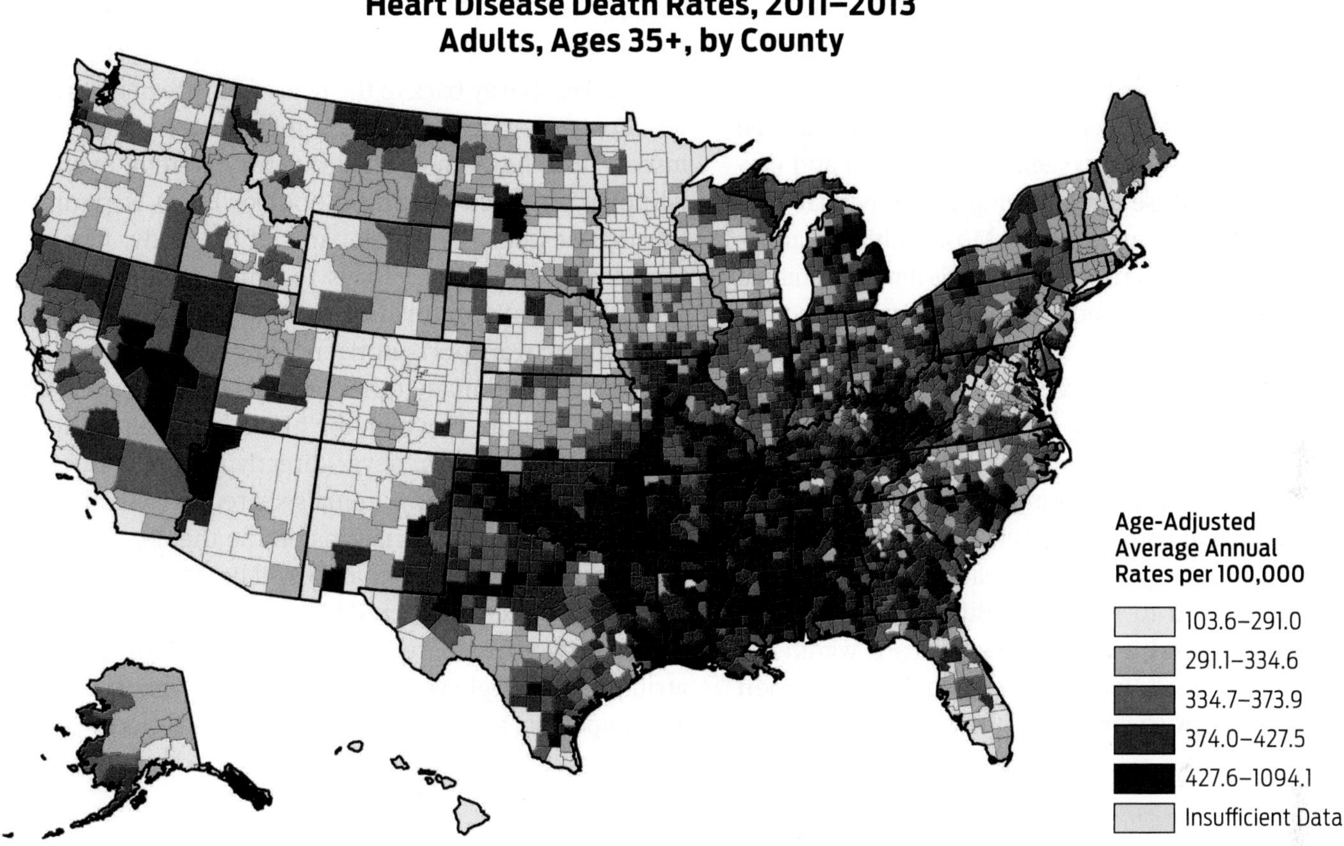

Data from "Heart Disease Death Rates, 2011-2013, Adults Ages 35+, by County." CDC. June 24, 2015.

? When comparing states that have high rates of obesity (see Infographic 28.10) with states that have high rates of death due to heart disease, is there any overlap?

Heart health starts young.

Berenson found that many kids in Bogalusa start smoking as early as the third grade.

That's why he's on a new mission: to get heart disease prevention taught in elementary schools alongside standard subjects like reading and math. He and his colleagues have developed a curriculum called Health Ahead/Heart Smart that builds on the lessons of the Bogalusa Heart Study and attempts to apply them in practical ways to help *prevent* heart disease.

More than 40 years later, the heart study that Berenson created continues to yield important insights. "There's no other study in the world like Bogalusa," he says. ■

Open vs. Closed Circulatory Systems (and Other Variations)

All organisms must exchange gases, nutrients, and wastes with their environment. For unicellular organisms, such as bacteria or yeast, this exchange occurs directly across the cell membrane by **diffusion** (see Chapter 3). For example, oxygen can diffuse in and carbon dioxide can diffuse out, allowing the individual cells to meet their gas exchange needs. Most multicellular organisms are too large to rely solely on this direct exchange with their environment, and so require a cardiovascular system to deliver nutrients and gases to body cells. All cardiovascular systems have three main common elements: a pump, a series of tubes, and a circulating fluid. But there are also important differences among different organisms.

Take the sea squirt, for example. The sea squirt is an invertebrate animal with a single-chambered heart that is surprisingly similar to the human heart in the way it contracts from one end to the other. But, in an interesting twist, sea squirt blood is not entirely enclosed within the heart and blood vessels. Instead, the sea squirt has an **open circulatory system,** in which the heart pumps blood into vessels, which then empty into its central body cavity, where exchange of gases and nutrients occurs. The blood then makes its way back to the heart from the body cavity.

Most other invertebrates, including mollusks and insects, also have an open circulatory system, in which hemolymph (a circulating bloodlike fluid) is pumped through vessels to various parts of the body, where it then leaves the open vessels to mingle with the body cells, eventually returning to the heart. In insects, hemolymph transports nutrients but not oxygen.

Humans and many other animals have a **closed circulatory system,** in which the blood is contained in the heart or blood vessels, never directly mixing with other body fluids. While many animals share this overall structure, there are many variations on the theme. Most fish, for example, have a two-chambered heart, and amphibians have a three-chambered heart.

The two-chambered fish heart contains a single atrium and a single ventricle. This means that the blood is pumped in a single circuit around the body. Blood

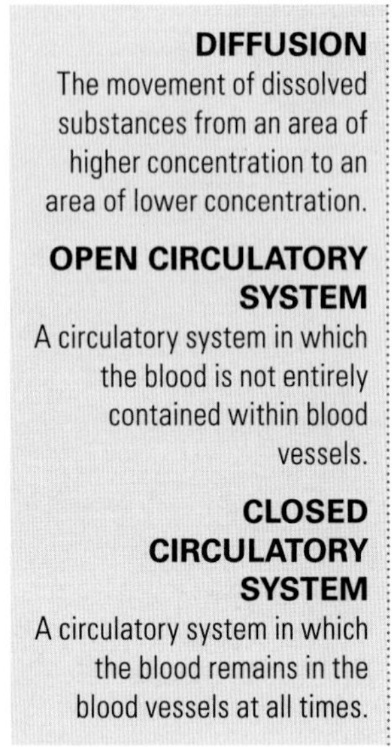

DIFFUSION
The movement of dissolved substances from an area of higher concentration to an area of lower concentration.
OPEN CIRCULATORY SYSTEM
A circulatory system in which the blood is not entirely contained within blood vessels.
CLOSED CIRCULATORY SYSTEM
A circulatory system in which the blood remains in the blood vessels at all times.

Sea Squirts Have an Open Circulatory System

In an open circulatory system, a simple tubular heart pumps blood into open-ended vessels, delivering blood to tissues in the body cavity (where gas and nutrient exchange occur).

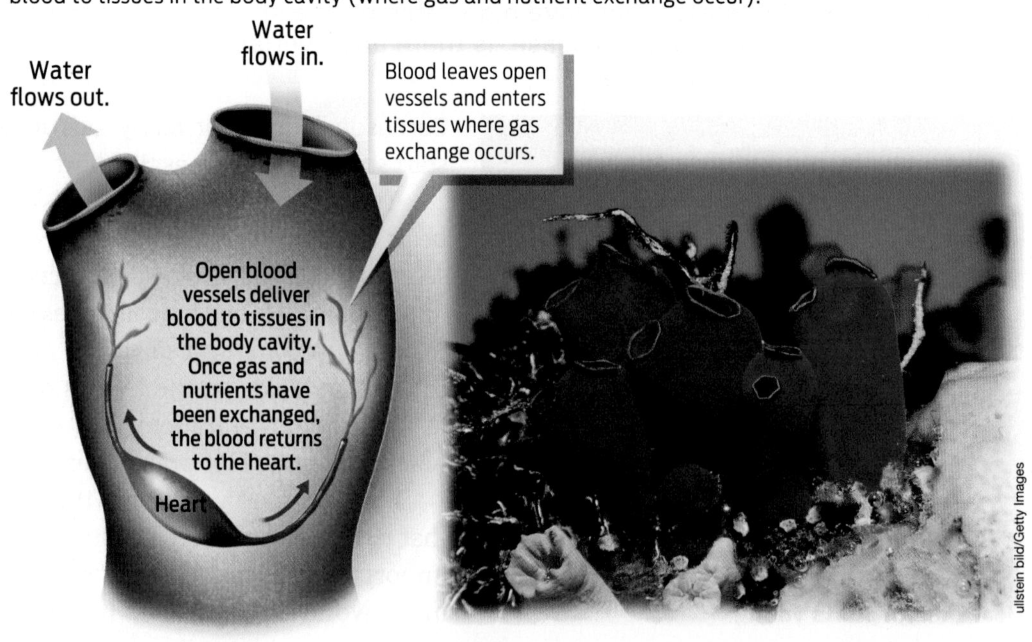

returns from the body and enters the atrium, then is pumped to the ventricle. The ventricle pumps the blood to the gills (the fish equivalent of lungs), where it picks up oxygen, and from there to body tissues. Deoxygenated blood then returns to the heart atrium.

Amphibians, which exchange gases through both their lungs and their skin, have a three-chambered heart with two atria and a single ventricle. One atrium receives oxygenated blood from the lungs and skin, and one receives deoxygenated blood from the body. Both then pour into the ventricle, which pumps blood to the lungs and skin. You might think that this would be disastrous, having oxygenated and deoxygenated flowing into the same heart chamber. But anatomical features of the vessel leaving the single ventricle help direct deoxygenated blood from the right atrium preferentially to the pulmonary circuit, and similarly preferentially direct oxygenated blood from the left atrium to the systemic (that is, the body) circuit.

When an amphibian is submerged, it cannot use its lungs to obtain oxygen. Rather, oxygen diffuses from the water through the skin directly into the blood. The three-chambered system allows the blood to bypass the lungs and be redirected to the body.

Fish and Amphibians Have Closed Circulatory Systems

Fish
Blood is pumped from the ventricle to the gills (where blood is oxygenated), then to body tissue capillaries (where oxygen and nutrients are delivered to the tissues). Finally, blood flows to the atrium of the heart, which pumps it back into the ventricle.

Frogs
The single ventricle pumps to both the lungs and the skin (where blood is oxygenated), and also to the body tissue capillaries (where oxygen and nutrients are delivered to the tissues). Blood returns from the skin/lungs and the body into two different atria, which pump the blood back into the single ventricle.

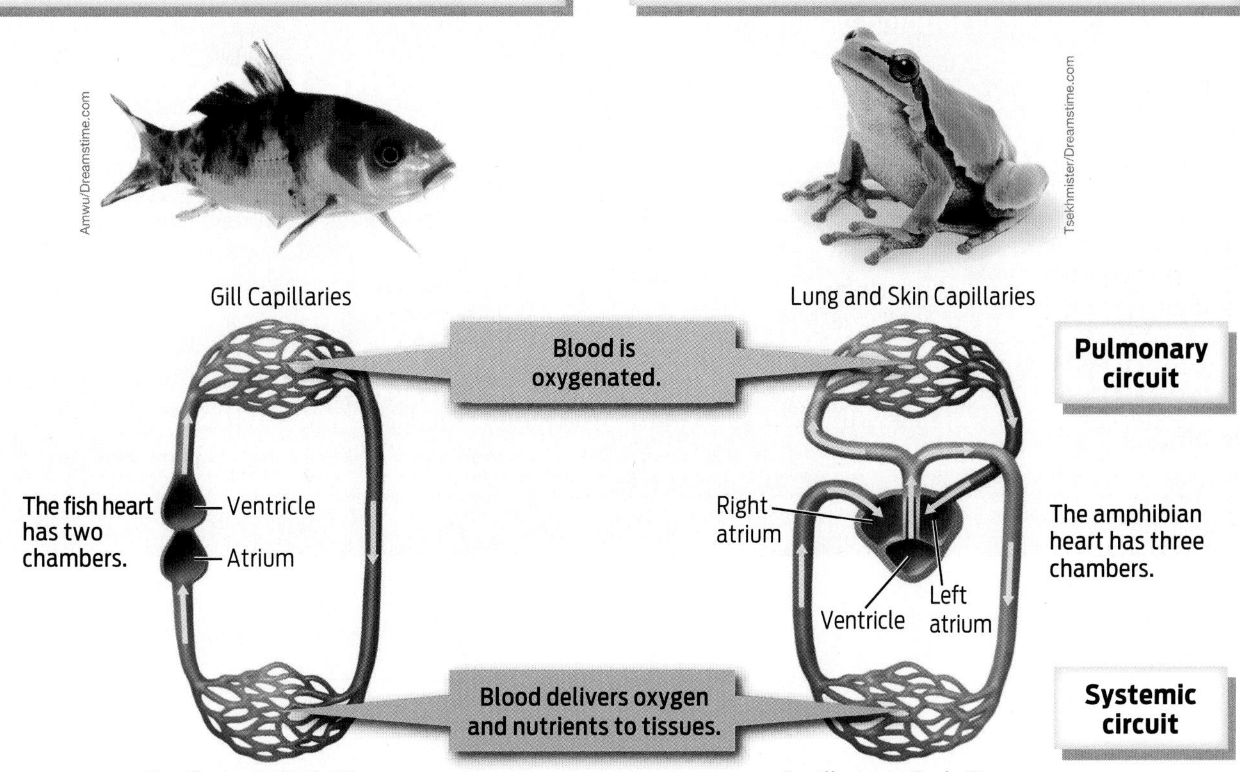

CHAPTER 28 SUMMARY

- The cardiovascular system transports nutrients, oxygen, hormones, and other substances throughout the body. It consists of the heart, blood vessels, and blood.
- The heart acts as a pump, moving blood in two circuits between the heart and lungs (the pulmonary circuit) and between the heart and the rest of the body (the systemic circuit).
- Different types of blood vessel have different structures and functions. Arteries transport blood away from the heart, veins transport blood to the heart, and capillaries are the sites of nutrient and gas exchange in tissues.
- Blood is composed of cells, water, gases, hormones, nutrients, and traces of whatever else we put in our body. Red blood cells carry oxygen, white blood cells participate in immune defenses, and platelets help blood clot.
- Atherosclerosis, or hardening of the arteries, is a common cause of cardiovascular disease. Cholesterol, fats, and other substances build up in the wall of a blood vessel, forming plaques that can obstruct the flow of blood.
- A heart attack results when the blood supply to the heart muscle is cut off and the muscle begins to die; atherosclerosis in coronary arteries is a common cause of a heart attack.
- High blood pressure (hypertension), obesity, high cholesterol, and smoking are known risk factors for atherosclerosis.
- Blood pressure is expressed as systolic pressure/diastolic pressure. Normal blood pressure is ~120/80 mmHg.
- Hypertension is a dangerous condition if left untreated. There are both controllable and uncontrollable causes of hypertension.
- Cholesterol is a normal constituent of cells and is transported through the body in the form of lipoproteins, LDL or HDL. Too much LDL cholesterol relative to HDL in the blood can lead to the accumulation of plaque in arteries.
- Unhealthful diet choices can raise LDL, lower HDL, and contribute to atherosclerosis.
- Cardiovascular disease begins early in life, long before symptoms occur. Autopsy studies have documented that atherosclerosis can occur in the arteries of young people and correlates with risk factors during life.

MORE TO EXPLORE

- **American Heart Association: www.heart.org**
- **Berenson, G. S., et al., for the Bogalusa Heart Study. (1998) Association between multiple cardiovascular risk factors and atherosclerosis in children and young adults. *New England Journal of Medicine* 338:1650–1656.**
- **Broyles, S., et al. (2010) The pediatric obesity epidemic continues unabated in Bogalusa, Louisiana. *Pediatrics* 125:900–905.**
- **HBO (2012) *Weight of the Nation* (film).**
- **Yang, Q., et al. (2014) Added sugar intake and cardiovascular diseases mortality among US adults. *JAMA Internal Medicine* 174(4):516–524.**

CHAPTER 28 Test Your Knowledge

DRIVING QUESTION 1 What structures make up the cardiovascular system, and how does blood flow through the system?

By answering the questions below and studying Infographics 28.2, 28.4, 28.5 and 28.6, you should be able to generate an answer for the broader Driving Question above.

KNOW IT

1 List the three components of the cardiovascular system.

2 Blood enters the systemic circuit from

a. the lungs.
b. the left side of the heart.
c. the right side of the heart.
d. the body.
e. both b and c

3 Which of the following is the correct order of structures through which blood flows as it passes through the cardiovascular system?

a. right side of heart→body→left side of heart→lungs
b. left side of heart→body→right side of heart→lungs
c. left side of heart→right side of heart→body→lungs
d. right side of heart→left side of heart→lungs→body

USE IT

4 In which circuit (pulmonary or systemic) does blood pick up nutrients such as sugar?

5 Are the coronary arteries part of the pulmonary or the systemic circuit? Explain your answer.

6 During heart surgery, patients are often placed on a heart–lung machine (known as a cardiopulmonary bypass). This machine circulates and oxygenates blood during the operation, taking over from the patient's heart and lungs. At what points would the patient's blood have to be diverted into the machine and back into the body in order to properly bypass the flow of blood through the heart and lungs?

DRIVING QUESTION 2 What is the structure of the heart and of the different types of blood vessel?

By answering the questions below and studying Infographics 28.4 and 28.6, you should be able to generate an answer for the broader Driving Question above.

KNOW IT

7 Blood returning to the heart from the lungs enters the

a. right atrium.
b. right ventricle.
c. left atrium.
d. left ventricle.
e. coronary arteries.

8 Which chamber of the heart pumps oxygen-rich blood to the body?

a. the right atrium
b. the right ventricle
c. the left atrium
d. the left ventricle
e. both the right atrium and the left atrium (to the right and left sides of the body, respectively)

9 The heart obtains oxygen from

a. the blood in its chambers.
b. the coronary arteries.
c. the systemic arteries.
d. the coronary veins.
e. the systemic veins.

10 What is the defining feature of an artery?

a. It carries blood away from the heart.
b. It has very thin walls.
c. It has valves.
d. It carries oxygenated blood.
e. both a and d

USE IT

11 Atrial fibrillation is a condition in which the atria contract very rapidly and irregularly. The ventricles also contract faster than normally, but their contraction is not coordinated with the atrial contraction. Explain how this condition can disrupt blood supply to the body.

12 Which of the following is most likely to be associated with left-side heart failure?

- **a.** fluid backing up in the feet and legs, causing swelling
- **b.** increased systolic pressure in the aorta
- **c.** reduced oxygenation of the blood leaving the left ventricle
- **d.** fluid backing up in the lungs, causing shortness of breath
- **e.** increased oxygenation of the blood returning to the right atrium

13 Some babies are born with a congenital heart defect in which the septum, or wall, between the left and right ventricles has an opening in it, and the aorta therefore can accept blood from both the left and right ventricles. These babies are known as "blue babies." Why?

DRIVING QUESTION 3 What is the composition of blood, and what does blood do?

By answering the questions below and studying Infographic 28.5, you should be able to generate an answer for the broader Driving Question above.

KNOW IT

14 Describe the function of each of the following cellular components of blood: white blood cells; red blood cells; platelets.

15 Which of the following statements is *not* true of capillaries?

- **a.** Blood moves through them slowly.
- **b.** They have a very small diameter.
- **c.** They have very thin walls.
- **d.** They occur between arteries and veins.
- **e.** They have valves.

16 What is happening in a capillary in your big toe?

- **a.** Oxygen is diffusing from the blood into your toe tissue.
- **b.** Carbon dioxide is diffusing from your toe tissue into the blood.
- **c.** Nutrients are diffusing from the blood into your toe tissue.
- **d.** all of the above
- **e.** a and c

USE IT

17 TV crime dramas often analyze arterial spray patterns of blood at crime scenes. Why do you never hear about venous spray patterns?

18 Why are soldiers standing at attention for long periods of time advised to contract their calf muscles to avoid becoming light-headed?

DRIVING QUESTION 4 What is cardiovascular disease, and what are some of the risk factors for developing cardiovascular disease?

By answering the questions below and studying Infographics 28.1, 28.3, 28.7, 28.8, 28.9, 28.10, and 28.11, you should be able to generate an answer for the broader Driving Question above.

KNOW IT

19 Match each disease in the left column with the letter corresponding to its underlying cause:

___ heart attack	a. elevated systolic and diastolic pressure
___ hypertension	b. fatty streaks in arteries
___ stroke	c. blockage of a coronary artery
___ atherosclerosis	d. blockage of blood vessels in the brain

20 List at least four risk factors for cardiovascular disease. For each, state whether or not it is modifiable.

21 How does atherosclerosis increase the risk of stroke?

22 Which type of cholesterol is dangerous with respect to cardiovascular disease?

- **a.** HDL
- **b.** LDL
- **c.** both
- **d.** neither; cholesterol is a healthy lipid
- **e.** HDL is a risk for hypertension and LDL is a risk for atherosclerosis.

USE IT

23 The data generated from the Bogalusa study show a correlation between atherosclerosis and risk factors for cardiovascular disease. Does this mean that everyone with a risk factor (for example, high BMI) will develop atherosclerosis? Explain your answer.

MINI CASE

apply YOUR KNOWLEDGE

24 Steven is 14 years old. He is an ace goalie for his soccer team, which practices twice during the week and plays a game every weekend. He is also an ace online gamer, holding the highest player skill level in World of Warcraft. After many long afternoons of playing, Steven has become a very good graphic designer and his mother pays him to design computer-generated fliers for her business. She prepares healthful meals with lots of whole grains, fruits, and dairy (milk and cheese), and she gets regular checkups to monitor her elevated blood pressure. Steven spends some of his money on chocolate bars and potato chips. He is a big guy, and in the heaviest 5% of his age group. From what you've read in this chapter, what cardiovascular risk factors do you identify in Steven? What could you say to Steven and his mother about Steven's potential risk and reducing that risk?

BRING IT HOME

apply YOUR KNOWLEDGE

26 Evaluate your own health and lifestyle with respect to risk for cardiovascular disease. Consider what kinds of changes you might want to talk about with your physician.

INTERPRETING DATA

apply YOUR KNOWLEDGE

25 Refer to Infographic 28.10 and record your answers to the following questions in the table below.

a. Look at the 2013–2015 data for Texas, Arizona, California, South Dakota and your home state. For each state, record the average percentage of adults with BMI ≥30.

b. Determine the total population of each state (you can find census data at https://www.census.gov/quickfacts/table/PST045216/00).

c. Determine the total number of people in each of those states with a BMI of at least 30.

State	Adults with BMI ≥30 (%)	Total State Population	No. of Adults with BMI ≥30
Texas			
Arizona			
California			
South Dakota			
Your home state			

29 Respiratory System

Michael Phelps taking a breath.

PEAK PERFORMANCE

An inside look at altitude training among elite athletes

DRIVING QUESTIONS

1. What structures make up the respiratory system?
2. How do the respiratory and cardiovascular systems cooperate to deliver oxygen to body cells and remove carbon dioxide from tissues?
3. What factors influence the oxygen-carrying capacity of blood and breathing rate?
4. How can scientific knowledge of the respiratory system be used to design training regimens for elite athletes?

With a wingspan of 6 feet 7 inches, feet the size of flippers, and just 4% body fat, swimmer Michael Phelps is built for speed. He holds seven current world record titles, is a 23-time Olympic gold medalist, and is universally recognized as the greatest swimmer of all time. With stats like these, it's hard to imagine how he could possibly improve his performance—short of growing gills. Yet the then 27-year-old athlete proved he still has a few tricks up his swimsuit when he announced in June 2012 that he sleeps in a "contraption" designed to give him an athletic boost.

"Once I'm already in my room, I still have to open a door to get into my bed," explained the swimmer to Anderson Cooper on *60 Minutes*. "It's like the boy in the bubble."

Phelps wouldn't show the device on TV, but he later tweeted a picture of the glass-walled structure to his followers. Known as a hypoxic chamber, the device creates an artificial low-oxygen environment, equivalent to that at 8,500 feet above sea level. The swimmer claims that sleeping in the chamber every night while he's training helps his body heal faster. In fact, it was a central part of his training regimen as he prepared for the summer 2012 Olympics in London, at which he broke the world record for most medals won by an Olympian at the time—22 in all; his current record is 28.

The hypoxic chamber in Michael Phelps's bedroom.

Michael Phelps/Twitter

Phelps isn't the only elite athlete to resort to a technical fix for improved performance. Manufacturers of hypoxic chambers have seen an increased interest in their product from athletes such as tennis pro Novak Djokovic, cyclist Lance Armstrong, and triathlete Jonathan Brownlee. Even entire Olympic sports teams, like the U.S. men's rowing team, are getting in on the action.

But what does sleeping in a low-oxygen chamber have to do with athletic performance? Does it work? And is it ethical?

According to Randall Wilber, a sports physiologist with the U.S. Olympic Training Center in Colorado Springs, Colorado, the use of hypoxic chambers is a variation on the theme of natural altitude training. Athletes often spend several weeks living at high altitude, typically defined as 6,000 feet or more above sea level, and then return to sea level to train or compete. The widely used practice is based on the idea that the body will acclimatize to higher elevation by becoming better able to take up and transport oxygen, ultimately improving performance.

"If you look at the number of athletes who win medals in winter sport, summer sport, across many endurance-based sports, I would say in my estimation 90 to 95% of those athletes at minimum are using altitude training in some way or another," says Wilber.

But with more and more athletes using technical means to enhance performance, many in the sports field are taking a closer look at the practice, and some are asking questions about fairness.

Live High, Train Low

Altitude training isn't new. According to Wilber, altitude training first became popular after the 1968 Olympic Games, held in Mexico City, which sits at a relatively high elevation of 7,350 feet above sea level. A number of strange things happened at those Games that drew the attention of scientists and athletes alike. World records were set in a number of short-distance track and field events–like the 100m and 200m dash–but athletes performing in endurance events, like the 1,500m run, fared worse than usual (**TABLE 29.1**).

"From 400 meters on down, including the long jump–think of Bob Beamon–all those people were just smashing world records left and right," says Wilber. "But from 800 meters on up, it went the other way: people were running slower than the world records." These results make perfect sense, he says. At altitude, the air is thinner, its molecules less

TABLE 29.1 Comparison of Performances in Short Races in the 1964 and 1968 Olympic Games

OLYMPIC GAMES	SHORT RACES: MEN				SHORT RACES: WOMEN			
	100 m	*200 m*	*400 m*	*800 m*	*100 m*	*200 m*	*400 m*	*800 m*
1964 (Tokyo)	10.0 s	20.3 s	45.1 s	1 m 45.1 s	11.4 s	23.0 s	52.0 s	2 m 1.1 s
1968 (Mexico City)	9.9 s	19.8 s	43.8 s	1 m 44.3 s	11.0 s	22.5 s	52.0 s	2 m 0.9 s
% change*	+1.0	+2.5	+2.9	+0.8	+3.5	+2.2	0	+0.2

*+ sign indicates improvement over 1964 performance.
Data from E.T. Howley. "Effect of Altitude on physical performance," in G.A. Stull and T.K. Cureton, *Encyclopedia of Physical Education, Fitness and Sports: Training, Environment, Nutrition, and Fitness*. Copyright © 1980 American Alliance for Health, Physical Education, Recreation and Dance, Reston VA.

Agence France Presse/Getty Images

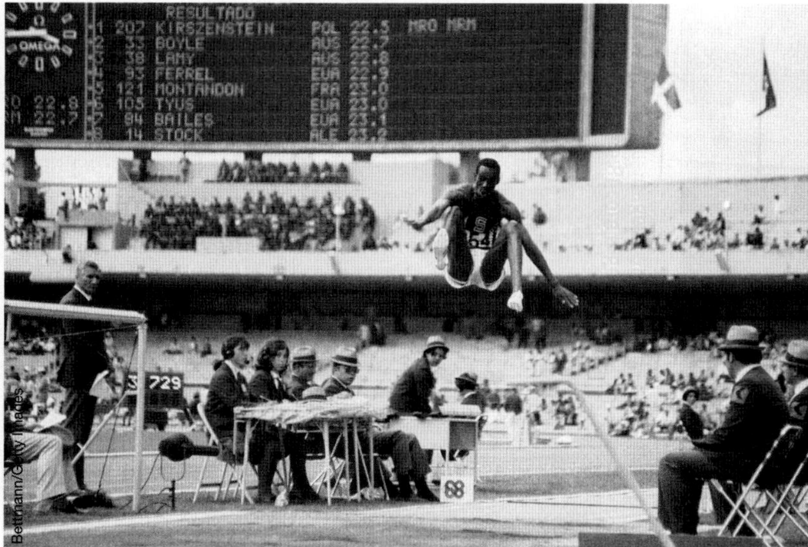

Bettmann/Getty Images

American long sprinter Tommie Smith (left) setting a World and Olympic record, and American long-jumper Bob Beamon (right) setting a World record, both at the 1968 Olympics in Mexico City.

RESPIRATORY SYSTEM
The organ system that allows us to take in oxygen and unload carbon dioxide.

LUNG
The major respiratory organ, the site of gas exchange between air and the blood.

PHARYNX
The throat.

LARYNX
The opening to the lower respiratory tract; also known as the voicebox.

TRACHEA
A large airway leading into the lower respiratory tract.

BRONCHI
Two airways that branch from the trachea; one bronchus leads into each lung.

densely packed, and so there is less air resistance, which explains why records were set in sprinting events. But at high altitude, there is also less oxygen, and so endurance events requiring prolonged exertion, like distance running, predictably suffered.

The question that emerged was, what would happen if you did the opposite–lived at high elevation and then competed at lower altitude? Would the body respond differently?

There was tantalizing evidence that it would. In Mexico City, athletes coming from higher-altitude countries like Kenya and Ethiopia in general performed better. They did not experience a dip in performance as other athletes did, and thus had a competitive advantage. The race was on to understand these results and incorporate the lessons into effective training regimens.

As a sports physiologist with the U.S. Olympic Committee, Wilber serves as a consultant to Olympic athletes, helping them to design appropriate training regimes using the latest science. He has worked with an impressive roster of elite athletes, including cyclist Kristin Armstrong and speed skaters Apolo Ohno and Christine Witty, as well as swimmer Michael Phelps. One of Wilber's main areas of research is the effect of altitude on the **respiratory system** of athletes.

The respiratory system is what allows us to take in oxygen from the air and release carbon dioxide. When we breathe, air first enters the nose and mouth and then moves through a series of branched tubes on its way to the **lungs.** It first passes through the throat, or **pharynx;** then through the **larynx,** which houses our vocal cords; then through the **trachea,** or windpipe. At that point, air is diverted into two major **bronchi,** one traveling to each lung. Much like an inverted tree, the tubes continue to branch and split into smaller and smaller

INFOGRAPHIC 29.1 The Human Respiratory System

The respiratory system allows us to take in oxygen from the air and eliminate carbon dioxide waste. Air travels through a series of tubes to small sacs in the lungs called alveoli. The alveoli have a close association with capillaries, permitting the uptake of oxygen into blood and the release of carbon dioxide from blood.

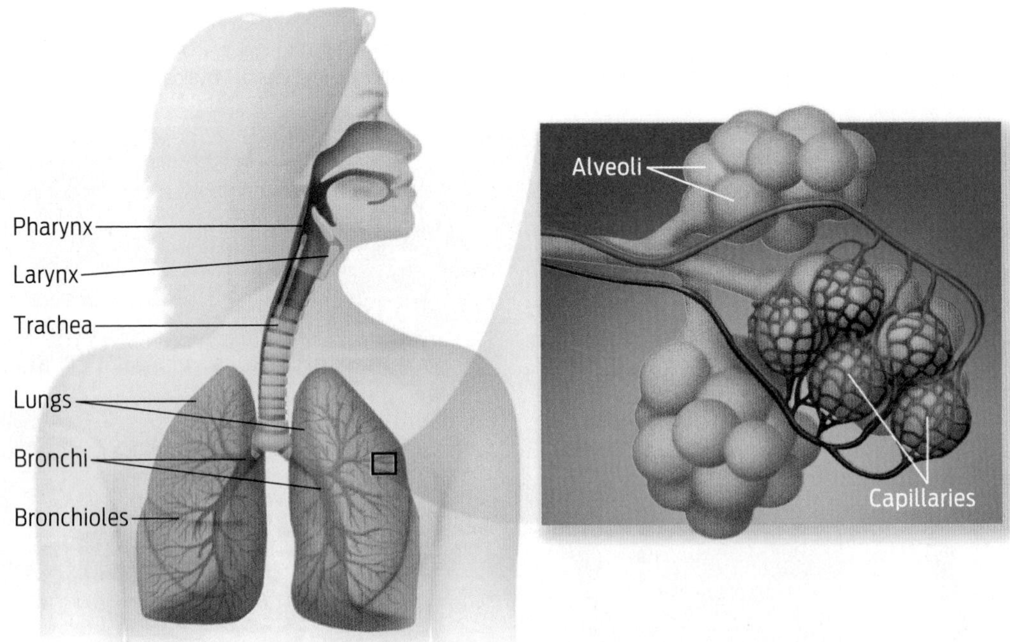

? Does oxygen move from the blood to alveoli, or from alveoli to the blood?

bronchioles until finally air hits the many tiny sacs called **alveoli,** located deep within the lungs. Alveoli are closely associated with capillaries (see Chapter 28), permitting oxygen and carbon dioxide to be exchanged between the air and the blood **(INFOGRAPHIC 29.1)**.

Each lung's **respiratory surface** is made up of millions of gas-exchanging alveoli. In an adult human, that respiratory surface is enormous: if all the alveoli were splayed out flat, they would occupy an area about the size of a tennis court (although the one-cell-thick layer would be so thin you wouldn't be able to see it). This large respiratory surface ensures that our bodies are able to take in enough oxygen to power the work of the trillions of cells in our body. This oxygen enables cells to perform the aerobic respiration that generates ATP (see Chapter 6).

The alveoli are the sites of **gas exchange** between the air and blood: they provide the means by which oxygen enters the blood. Once oxygen has diffused from the air into the blood, and carbon dioxide has diffused from the blood into the airspaces in the lungs, blood vessels carry the oxygenated blood to the heart, which will pump it around the body to various tissues **(INFOGRAPHIC 29.2)**.

BRONCHIOLES
Small airways that branch from the bronchi.

ALVEOLI
Air sacs in the lung across which gases diffuse between air and blood.

RESPIRATORY SURFACE
A surface across which oxygen enters the blood and carbon dioxide leaves the blood.

GAS EXCHANGE
In the lungs, the process of taking up and releasing oxygen and carbon dioxide across capillaries.

INFOGRAPHIC 29.2 Gas Exchange and Transport

Oxygen-rich air enters alveoli and oxygen enters the bloodstream. Carbon dioxide from the blood enters the alveoli and is breathed out. The oxygen-rich blood eventually reaches tissues, where oxygen is delivered, and carbon dioxide produced by aerobic respiration enters the blood to be carried back to the lungs.

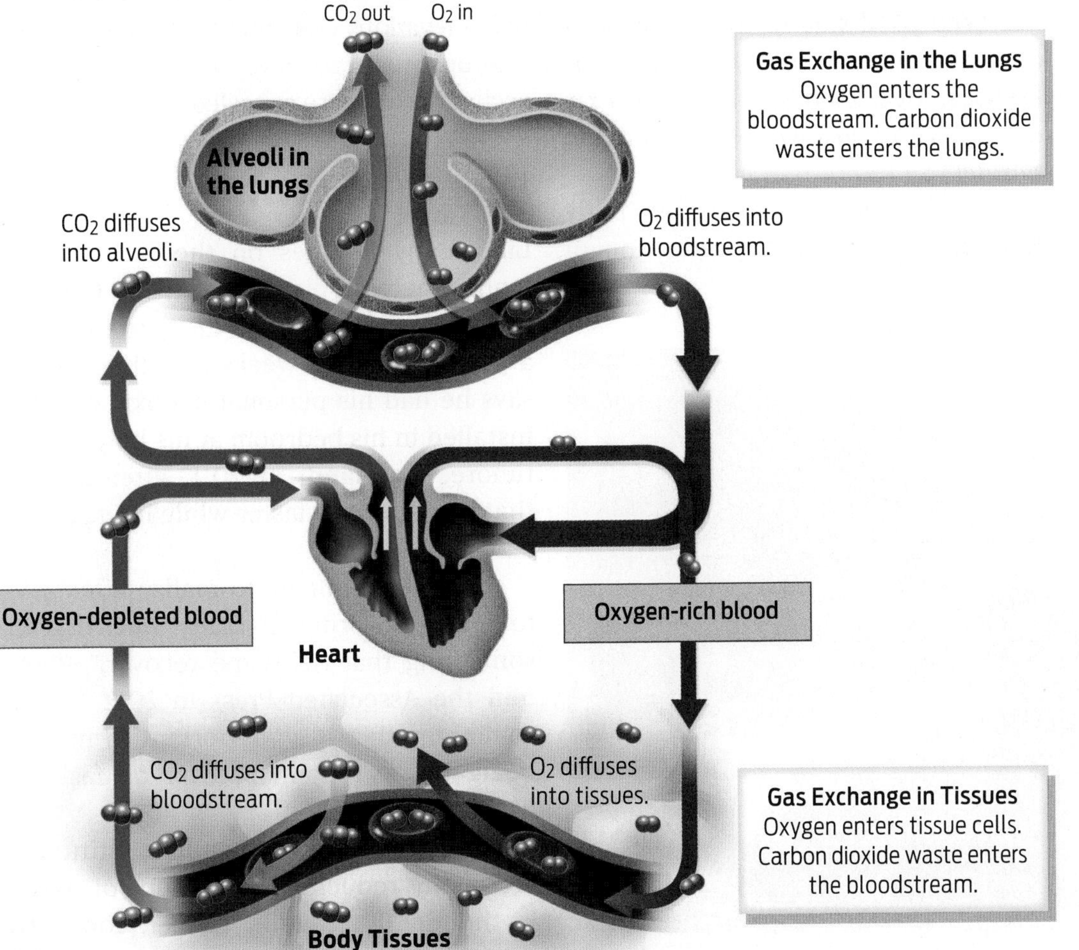
CO2 out
O2 in
Alveoli in the lungs
Gas Exchange in the Lungs
Oxygen enters the bloodstream. Carbon dioxide waste enters the lungs.
CO2 diffuses into alveoli.
O2 diffuses into bloodstream.
Oxygen-depleted blood
Heart
Oxygen-rich blood
CO2 diffuses into bloodstream.
O2 diffuses into tissues.
Gas Exchange in Tissues
Oxygen enters tissue cells. Carbon dioxide waste enters the bloodstream.
Body Tissues

? Compare and contrast the movement of the gases O_2 and CO_2 in the lungs and in a tissue (such as muscle).

RED BLOOD CELLS
Blood cells specialized for transporting oxygen throughout the body.

The respiratory and cardiovascular systems thus work together in close coordination. In fact, it's probably more useful to think of a cardiorespiratory system than of two separate systems acting individually.

If oxygen is so critical to athletic performance, why would athletes willingly deprive themselves of the needed gas by traveling to a place where oxygen is scarce? It may seem counterintuitive, but the lower oxygen levels actually help an athlete in the long run, because of the way the body responds to the altered conditions.

As Wilber explains, an athlete who travels to high altitude encounters less barometric air pressure, which means fewer total air molecules bouncing around in the atmosphere–including fewer molecules of oxygen. "Thus you have a tougher time running or walking or climbing or swimming," he says. "The good side of the situation is that when you come to [high] altitude your body senses, *Hey, there's not enough oxygen here, I need to do something about it to compensate!*"

As scientists have learned, the body indeed compensates for the reduction of oxygen molecules in the atmosphere by increasing the number of oxygen-carrying **red blood cells (RBCs)** in the blood. With more RBCs, the body is able to take in roughly the same amount of needed oxygen, despite the altered conditions, thus maintaining homeostasis. When the athlete returns to sea level, after a period of acclimatization to altitude, he or she will have a competitive advantage because of the enhanced ability of the blood to carry oxygen (because of the higher number of red blood cells) **(INFOGRAPHIC 29.3)**.

The most popular form of altitude training in use today is known as "live high, train low": athletes live for a period of time at higher elevations and then return to lower altitudes to train or compete. This approach is better than living and training full time at high elevation, say some experts, since at high elevations one can't exercise or train as rigorously as at sea level–thus the benefits of acclimatization are canceled out.

Boy in a Bubble

Like many athletes on the U.S. Olympic team, Phelps typically travels to Colorado Springs anywhere from three to five times a year, for 4 to 5 weeks at a time. Phelps says he had his personal hypoxic chamber installed in his bedroom at his home in Baltimore, Maryland, in 2011, after noticing that he recovered faster while living at high elevation.

"We've been able to realize after going to Colorado Springs so many times that it is something that helps me recover," Phelps told the Associated Press in 2012. "That's something that is so important to me now being older. I don't recover as fast as I used to."

Hypoxic chambers simulate altitude by artificially reducing the amount of oxygen in the air. By reducing oxygen concentration from about 21% (in untreated air) to about 15%, the chambers reduce oxygen's

ATS-Altitude Training Systems

Australian swimmer James Magnussen using the Simulated Altitude Training Pool at NSW Institute of Sport.

INFOGRAPHIC 29.3 The Effect of Altitude on Blood

The amount of available oxygen decreases with increasing altitude. Under these conditions, RBC production is stepped up to increase the oxygen-carrying capacity of the blood.

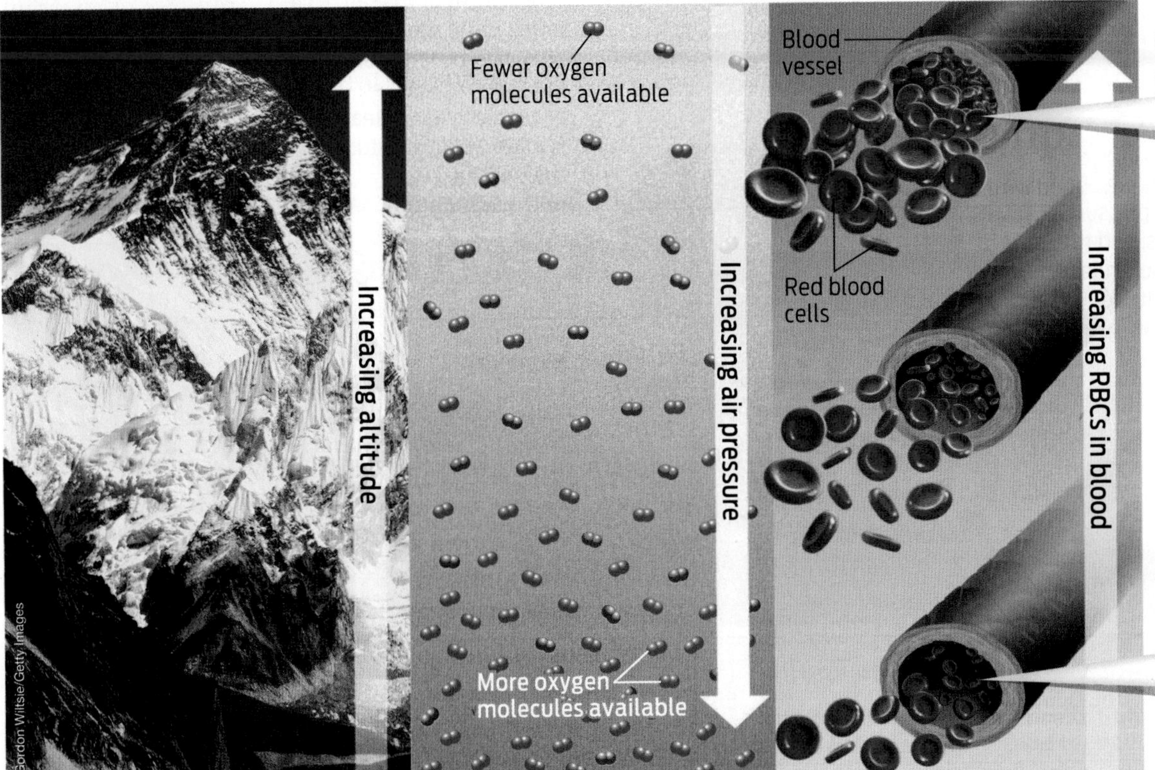

Fewer oxygen molecules available
Increasing altitude
Increasing air pressure
More oxygen molecules available
Blood vessel
Red blood cells
Increasing RBCs in blood
At higher altitudes there is lower air pressure, and therefore fewer available oxygen molecules. More RBCs are made to maintain adequate oxygen levels in the blood.
At lower altitudes there is higher air pressure, and therefore more available oxygen molecules. Adequate oxygen can be maintained in the blood with fewer RBCs.
Gordon Wiltsie/Getty Images

? Would you expect someone using supplemental oxygen while at altitude to have an increased RBC count? Explain your answer.

partial pressure, simulating an altitude of approximately 8,000 feet. Partial pressure and barometric pressure are measured in units called millimeters of mercury (mmHg) (**INFOGRAPHIC 29.4**).

The lower oxygen concentration Phelps experiences while he sleeps stimulates his body to produce RBCs, while the regular oxygen concentration during the day helps him train at his normal, vigorous levels. It's a simulated version of the "live high, train low" scenario.

Phelps may be unusual among athletes in the degree to which he has used a hypoxic chamber, but he is certainly not alone in using altitude training. In fact, according to Wilber, nearly all elite athletes do (**INFOGRAPHIC 29.5**).

Are there downsides to sleeping in hypoxic conditions? At least over the short term, there don't appear to be. In fact, many humans, including some Tibetan and Andean peoples, live easily at much higher elevations (above 10,000 feet) with even less oxygen. However, there is evidence that these populations have adapted over many years to the hypoxic conditions by becoming better able to absorb and transport oxygen. Someone coming from sea level, who lacked these adaptations, would have a hard time thriving at these elevations, and may in fact succumb

PARTIAL PRESSURE
The proportion of total air pressure contributed by a given gas.

INFOGRAPHIC 29.4 The Relationship between Altitude and Oxygen Pressure

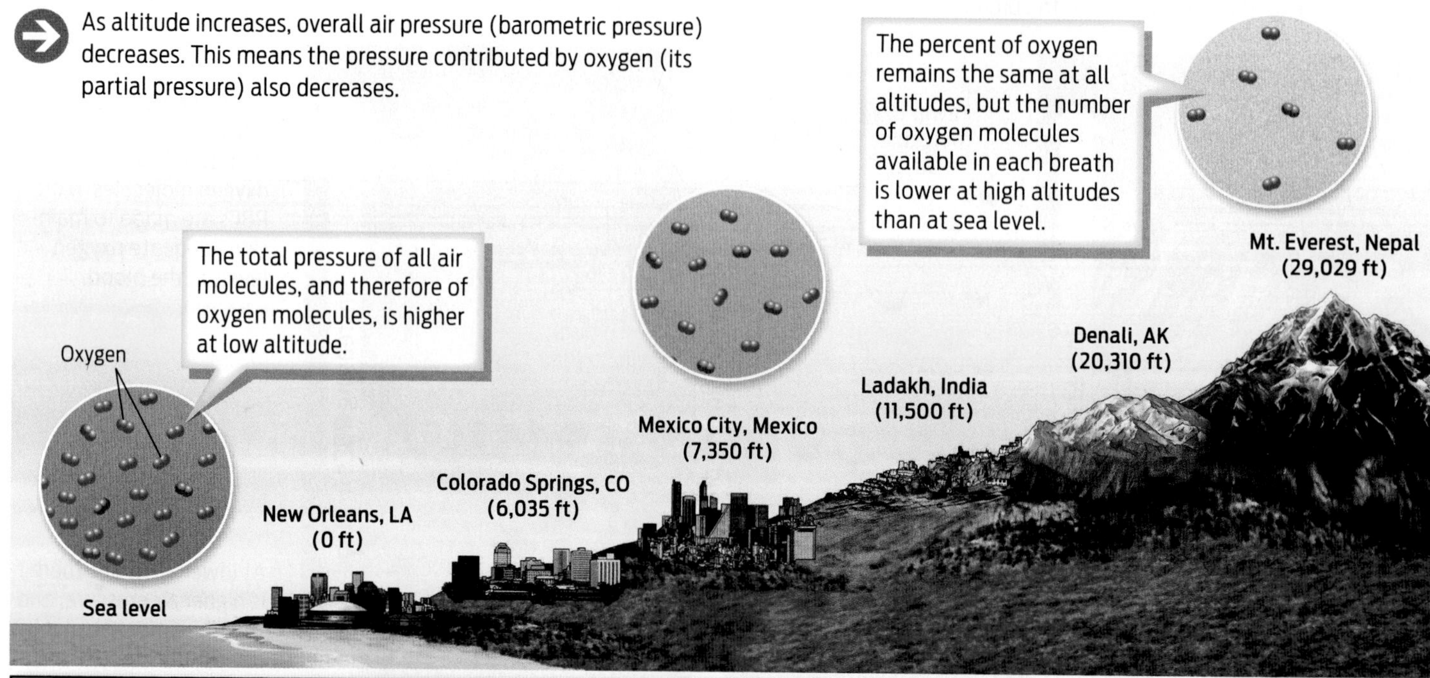

Barometric pressure (total air molecules, mmHg)	760	616	585	505	360	253
Partial pressure of oxygen (oxygen molecules, mmHg)	159	129	125	105	75	53
% O_2 in the air	20.93	20.93	20.93	20.93	20.93	20.93

Data from Altitude.org

? What is the partial pressure of oxygen at the peak of Mount Everest; in Colorado Springs, Colorado; and at sea level?

ALTITUDE SICKNESS
An illness that can occur as a result of an abrupt move to an altitude with a reduced partial pressure of oxygen.

to **altitude sickness,** a condition whose symptoms resemble those of an alcohol-induced hangover: headache, tiredness, and nausea. People can experience milder forms of altitude sickness when they ascend rapidly to altitudes above 8,000 feet—for example, on a ski trip in Colorado. Wilber notes that some athletes find that the chambers give them headaches or that they have trouble sleeping.

Phelps's biggest complaint is more mundane: "The worst thing is trying to watch TV in it. I've got to watch it through Plexiglas—it's blurry," Phelps told the Associated Press.

These are complaints that nonendurance athletes don't have to worry about. Typically, nonendurance athletes don't use altitude training. That's because sports such as weight-lifting and sprinting require only short bursts of energy, and thus do not rely as much on aerobic respiration, and so having extra RBCs doesn't really help these athletes (see the discussion of fermentation in Chapter 6). "If you wanted to, you could

INFOGRAPHIC 29.5 Altitude Training Hypothesis

Both hypoxic chambers and natural altitude training are based on the same principle. By living or sleeping at lower partial pressures of oxygen, athletes will elevate their RBC count. This enhances the oxygen-carrying capacity of the blood when competing at sea level.

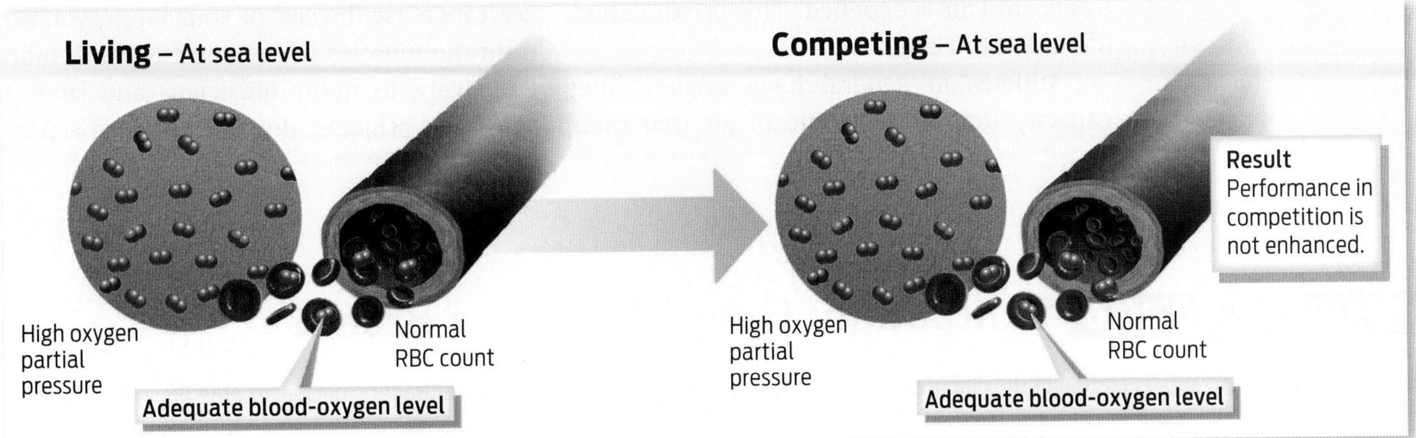

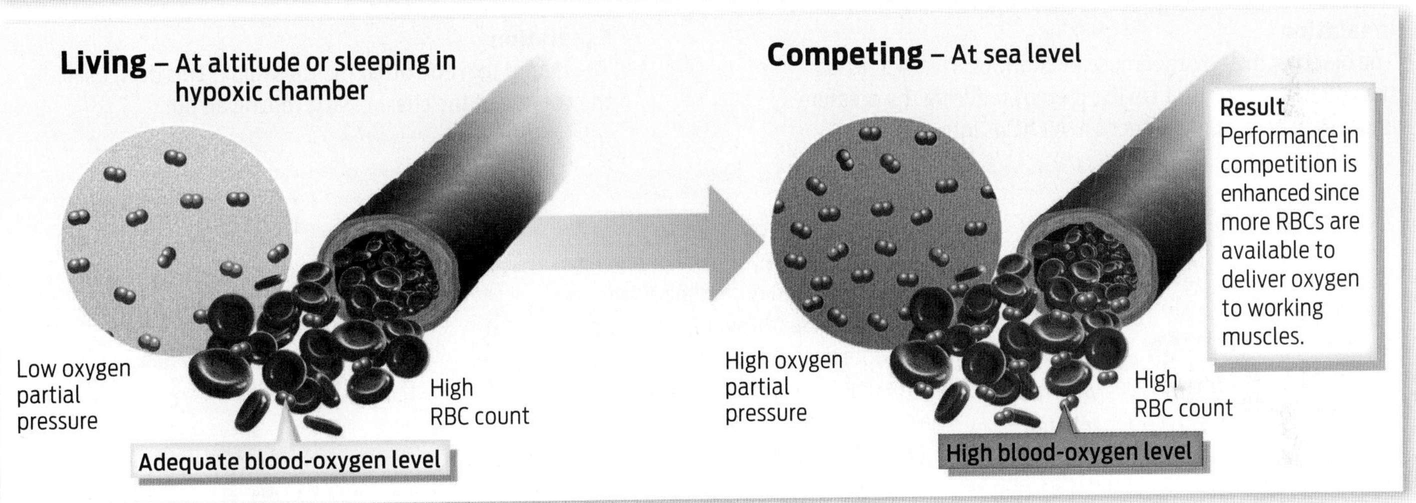

? What happens to the RBC count of an athlete living and sleeping in hypoxic conditions, and how is this advantageous when competing at sea level?

run a 100-meter dash holding your breath," notes Wilber.

Every Breath You Take

Obviously, to take in oxygen, we need to breathe. Breathing is the way we **ventilate** our lungs. With every breath, we bring in a new volume of air for gas exchange at the respiratory surface, and then expel the waste products. If our lungs didn't ventilate, the process of gas exchange would be much less efficient because there would be no way to move fresh air to the respiratory surface and get stale air out of the way.

On a mechanical level, breathing is the effect of a series of muscular movements of the **diaphragm,** a sheet of muscle working in coordination with other muscles connected to the rib cage. As these muscles contract, the diaphragm moves downward and the rib cage moves outward, thereby increasing the size of the chest cavity overall. This sudden increase in the volume of

VENTILATION
The process of moving air in and out of the lungs.

DIAPHRAGM
A sheet of muscle that contributes to breathing by contracting and relaxing.

the chest cavity has the effect of decreasing air pressure there, creating suction that draws air into the lungs. This is inhalation. As the muscles relax, the chest cavity contracts, and air is expelled. This is exhalation (**INFOGRAPHIC 29.6**).

Athletes in general have greater lung capacity (the total volume of air that their lungs can hold) than the average person, which is partly why they can exercise longer, and harder, without getting as winded as someone who is not as fit. And while you can't increase the size of your lungs, you can train the muscles of the chest to work more effectively, as many musicians and singers, as well as athletes, do. Exercising in a pool,

INFOGRAPHIC 29.6 Ventilation: How We Take a Breath

The contraction and relaxation of the diaphragm and rib cage muscles ventilate the lungs. Contraction decreases pressure in the lungs and air flows in. Relaxation of the muscles forces air out of the lungs.

Inhalation
The diaphragm and other muscles contract. This increases the volume of the chest cavity, thereby reducing the pressure and creating vacuumlike suction. Air flows into the lungs.

Exhalation
The diaphragm and other muscles relax. This decreases the volume of the chest cavity, forcing air out.

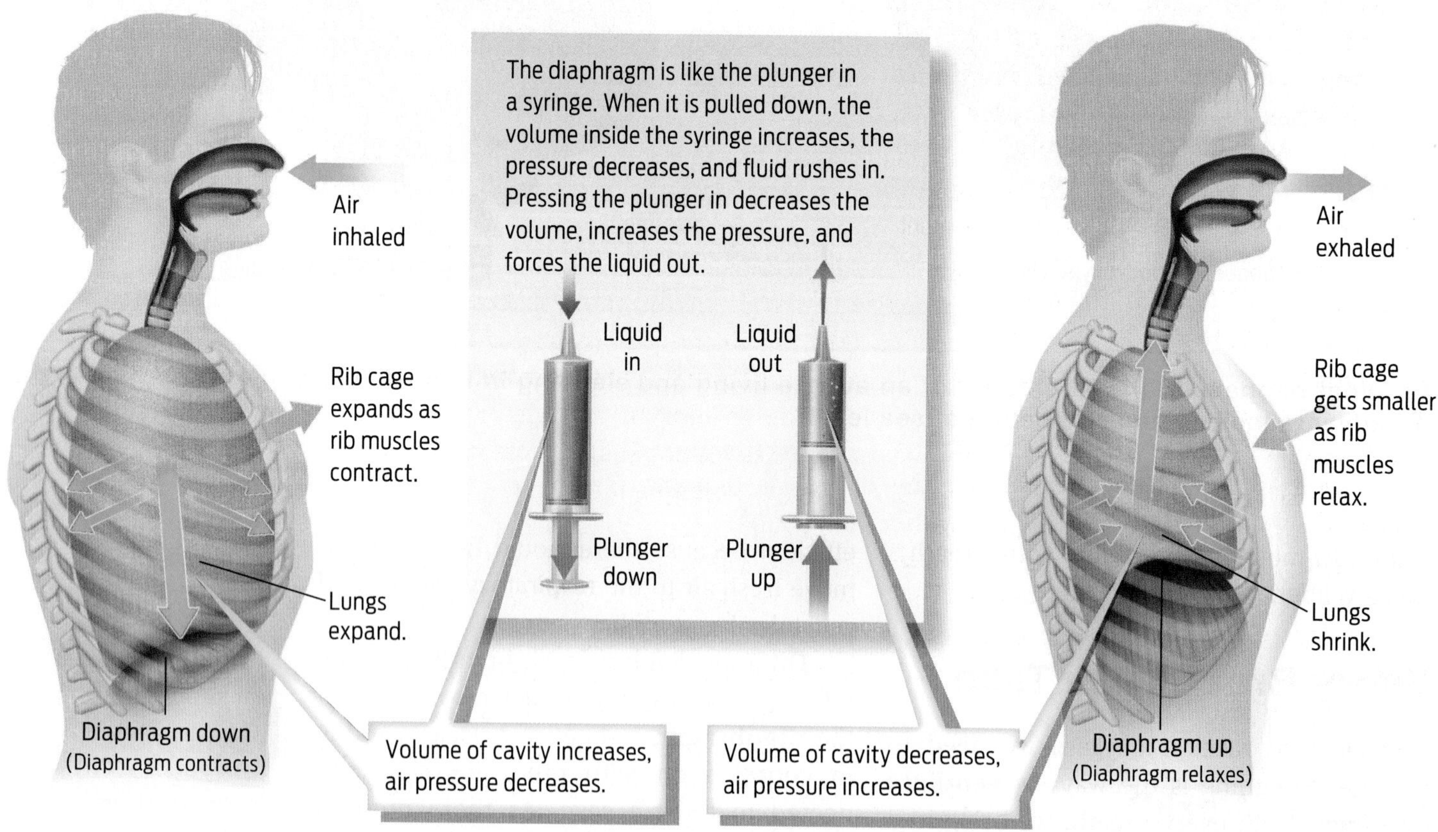
Air inhaled
Rib cage expands as rib muscles contract.
Lungs expand.
Diaphragm down
(Diaphragm contracts)
The diaphragm is like the plunger in a syringe. When it is pulled down, the volume inside the syringe increases, the pressure decreases, and fluid rushes in. Pressing the plunger in decreases the volume, increases the pressure, and forces the liquid out.
Liquid in
Liquid out
Plunger down
Plunger up
Volume of cavity increases, air pressure decreases.
Volume of cavity decreases, air pressure increases.
Air exhaled
Rib cage gets smaller as rib muscles relax.
Lungs shrink.
Diaphragm up
(Diaphragm relaxes)

? When the diaphragm and rib muscles contract, does air flow into or out of the lungs?

for example, makes your chest muscles work harder, which strengthens them, ultimately enhancing lung capacity. Some populations of humans who live at high elevations (like the Bods, whose home, Ladakh, is at 11,500 feet in the Himalaya mountains), have wider chests with stronger chest muscles, permitting them to draw in more air in order to get sufficient oxygen.

Although getting oxygen is an essential part of breathing, the rate of our breathing is not actually determined by the demand for oxygen in our body. It's actually carbon dioxide that is more important. When we exercise, we are using more oxygen and producing more carbon dioxide. The increased concentration of carbon dioxide in the blood causes a decrease in blood pH; when CO_2 dissolves in the water in the blood–the plasma–it produces carbonic acid, which has a low pH (see Chapter 2). The brain senses this drop in pH and responds by sending a signal to the lungs to increase breathing rate. The increased breathing rate increases the rate of gas exchange, bringing more oxygen into the body while unloading excess carbon dioxide, and thus raising blood pH back to normal.

If for some reason a person is unable to unload carbon dioxide from the lungs–as can happen when breathing is challenged by asthma or obesity, for example–the concentration of CO_2 in the blood can rise to dangerous levels, creating a potentially fatal condition known as **acidosis.** The opposite problem occurs when someone hyperventilates and releases too much carbon dioxide, which can cause dizziness and numbness; an easy fix is to breathe into a paper bag to re-inhale the exhaled carbon dioxide.

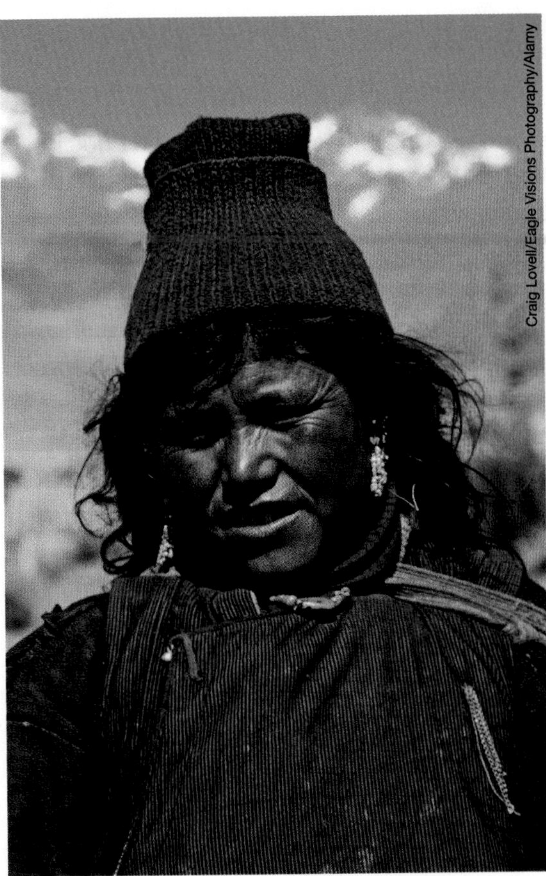
Craig Lovell/Eagle Visions Photography/Alamy

The Ladakhi people live at an elevation of more than 11,500 feet with no adverse health effects.

Does It Work?

The logic of altitude training seems airtight: increase the concentration of RBCs in the blood and you will have greater oxygen-carrying capacity, more aerobic respiration, and better performance. But actually proving that altitude training improves performance is harder than it may seem. Even Wilber, who literally wrote the book on the subject, acknowledges that there is still a fair amount of controversy surrounding the science of altitude training, with some studies concluding that it works, and others concluding it doesn't.

Partly, the issue is the difficulty of designing properly controlled studies. Any gains in performance are likely to be small, so how can you be sure the gains were due to high altitude and not some other aspect of training? Likewise, if athletes don't show consistent improvement, is it because altitude training doesn't work, or because not everyone responds to altitude in the same way?

For his part, Wilber is persuaded by the available evidence that, when pursued

ACIDOSIS
A dangerous condition in which blood is too acidic.

INFOGRAPHIC 29.7 Altitude Exposure Improves Performance

In a study of 22 elite distance runners, maximal oxygen uptake and 3,000m race performance were measured before and after a 4-week "live high, train low" altitude-training regimen (living at 2,500 m and training at 1,250 m). Performance and oxygen uptake were improved after living at altitude. Similar performance results were seen for collegiate athletes.

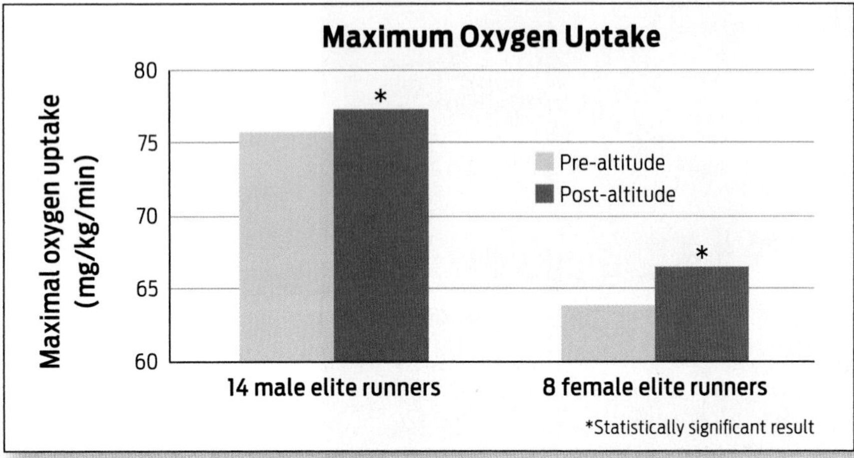

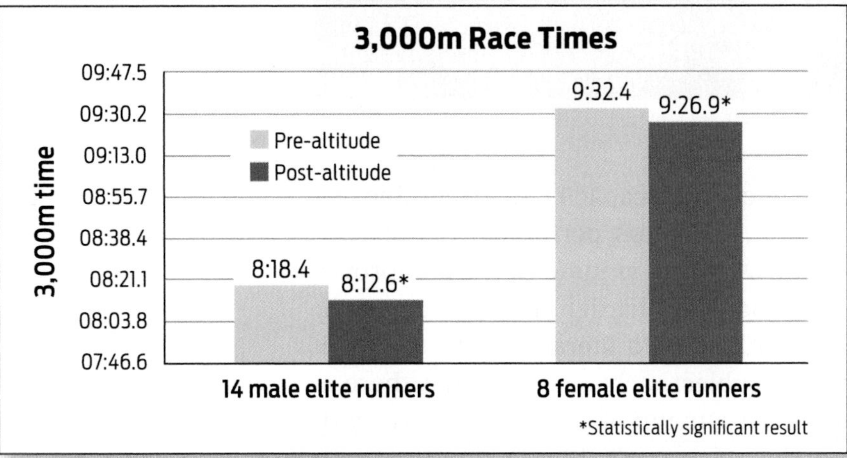

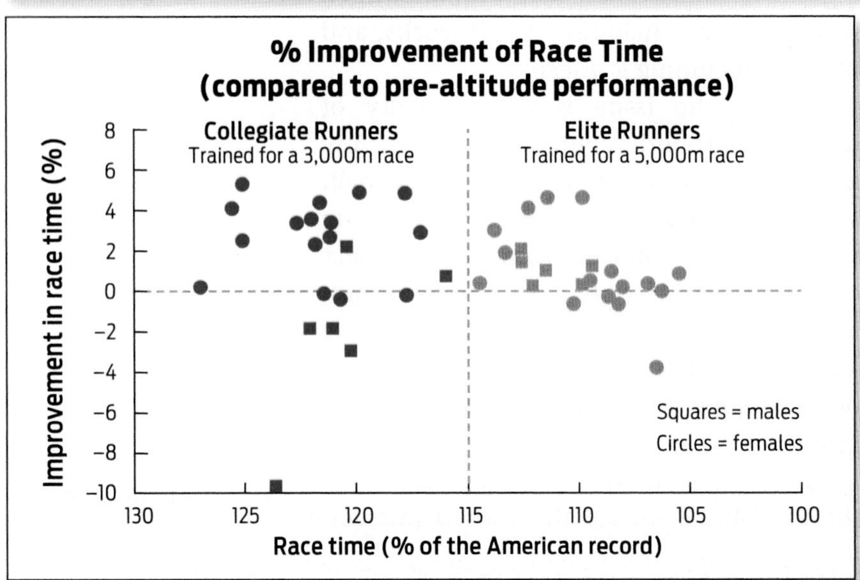

Data from Stray-Gunderson, *J Appl Physiol* 91:1113-1120,2001.

methodically, altitude training does indeed improve performance (**INFOGRAPHIC 29.7**).

Additional (but anecdotal) evidence that it works, he says, comes from the loyalty of athletes and coaches themselves. "When you see a Michael Phelps and his coach and other athletes coming back to altitude year after year after year after year, you know that it's working for them."

At the same time, Wilber cautions that altitude training requires consistent and long-term practice. For that reason, ordinary gymgoers are unlikely to see any significant results from periodic workouts in the numerous "hypoxic rooms" that have sprung up in commercial gyms all over the world.

Picking Up, Dropping Off

If an athlete experiences hypoxia regularly, for 12 to 20 hours a day, for at least 3 weeks, according to the research he or she will start to see benefits upon returning to sea level. The biggest benefit will be the ability to carry more oxygen in the blood. But unlike carbon dioxide, which travels easily in the blood plasma as carbonic acid, oxygen does not readily dissolve in the watery plasma and so is not transported this way. Instead, it's carried inside RBCs, bound to a protein called **hemoglobin.** RBCs are essentially bags of hemoglobin, with each red blood cell having between 250 and 300 million molecules of hemoglobin. RBCs lack a nucleus, and are streamlined for doing one thing very well: carrying oxygen.

Hemoglobin is a highly dexterous molecule, composed of multiple interworking parts. Each molecule of hemoglobin is made up of four interacting protein globin chains and four iron-containing **heme groups**. The heme groups are what bind oxygen. Since

? Look at the absolute change in maximal oxygen uptake in males and females, and in the 3,000m race times for males and females. Who had the larger increase in oxygen uptake and who had the larger change in race time?

INFOGRAPHIC 29.8 Hemoglobin Carries Oxygen in Red Blood Cells

Hemoglobin has four globin chains and four heme groups. Each heme group contains an iron atom and has the capacity to bind to one oxygen molecule.

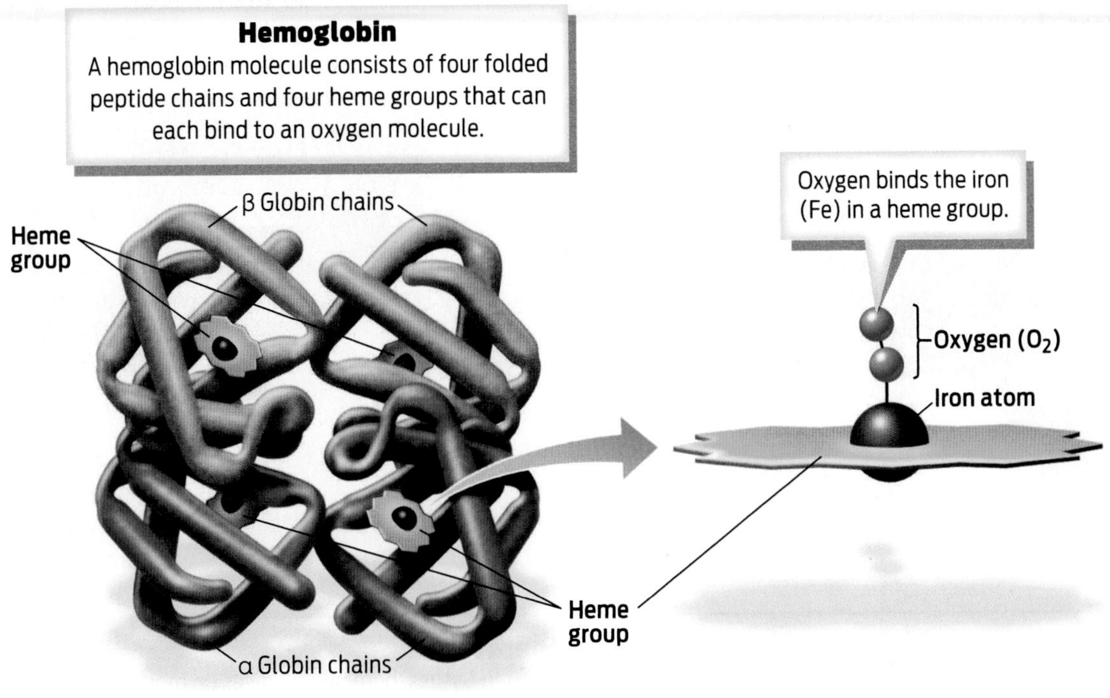
Hemoglobin
A hemoglobin molecule consists of four folded peptide chains and four heme groups that can each bind to an oxygen molecule.
β Globin chains
Heme group
Oxygen binds the iron (Fe) in a heme group.
Oxygen (O_2)
Iron atom
Heme group
α Globin chains

? How would a dietary deficiency of iron affect the transport of oxygen in red blood cells?

each hemoglobin molecule has four heme groups, each hemoglobin molecule can carry up to four molecules of O_2. Because the heme groups of hemoglobin contain iron, a dietary deficiency of iron can limit the production of hemoglobin, and therefore reduce the overall oxygen-carrying capacity of blood, leading to a condition known as iron-deficient anemia (**INFOGRAPHIC 29.8**).

Key to understanding how hemoglobin functions in oxygen transport is the fact that it binds oxygen reversibly. It can pick up (that is, bind) oxygen in the lungs and drop off (that is, release) oxygen in the tissues. Whether hemoglobin is picking up or dropping off oxygen depends mainly on the partial pressure of oxygen (P_{O_2}). Fresh air entering the lungs has a relatively high partial pressure of oxygen, so the hemoglobin will become saturated (that is, fully loaded) with oxygen. In tissues, the partial pressure of oxygen tends to be lower, because cells in tissues are consuming oxygen as they carry out aerobic respiration, so hemoglobin tends to give up some of its oxygen to tissues.

Another important factor affecting oxygen binding is temperature. Hemoglobin is more likely to release oxygen as temperature increases—as it does, for example, in an actively contracting muscle that is burning energy.

Lastly, pH plays a role. At lower pH, hemoglobin releases more oxygen. As we've seen, one of the main factors that can lower pH is the concentration of CO_2. Since CO_2 is produced during aerobic respiration, the pH in muscle tissue goes down, becoming more acidic during exercise. In turn, the lower pH causes the hemoglobin to give up more of its oxygen, thus ensuring a continuous

HEMOGLOBIN
A protein found in red blood cells specialized for transporting oxygen.

HEME GROUPS
Iron-containing structures on hemoglobin, the sites of oxygen binding.

INFOGRAPHIC 29.9 Oxygen Binding to Hemoglobin Is Reversible

Hemoglobin reversibly binds oxygen. When the partial pressure of oxygen (P_{O_2}) is high, hemoglobin will bind oxygen. At low P_{O_2}, hemoglobin will release oxygen. The binding is also influenced by temperature and pH.

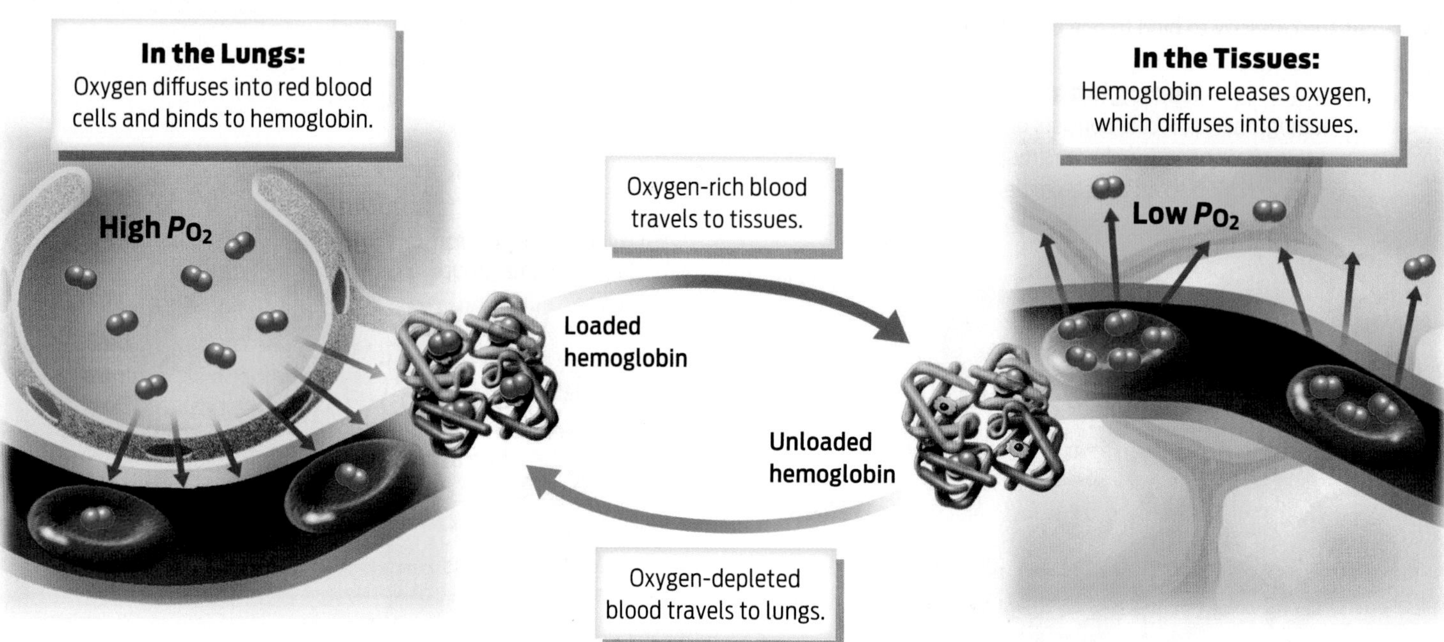
In the Lungs:
Oxygen diffuses into red blood cells and binds to hemoglobin.
High P_{O_2}
Oxygen-rich blood travels to tissues.
Loaded hemoglobin
Unloaded hemoglobin
Oxygen-depleted blood travels to lungs.
In the Tissues:
Hemoglobin releases oxygen, which diffuses into tissues.
Low P_{O_2}

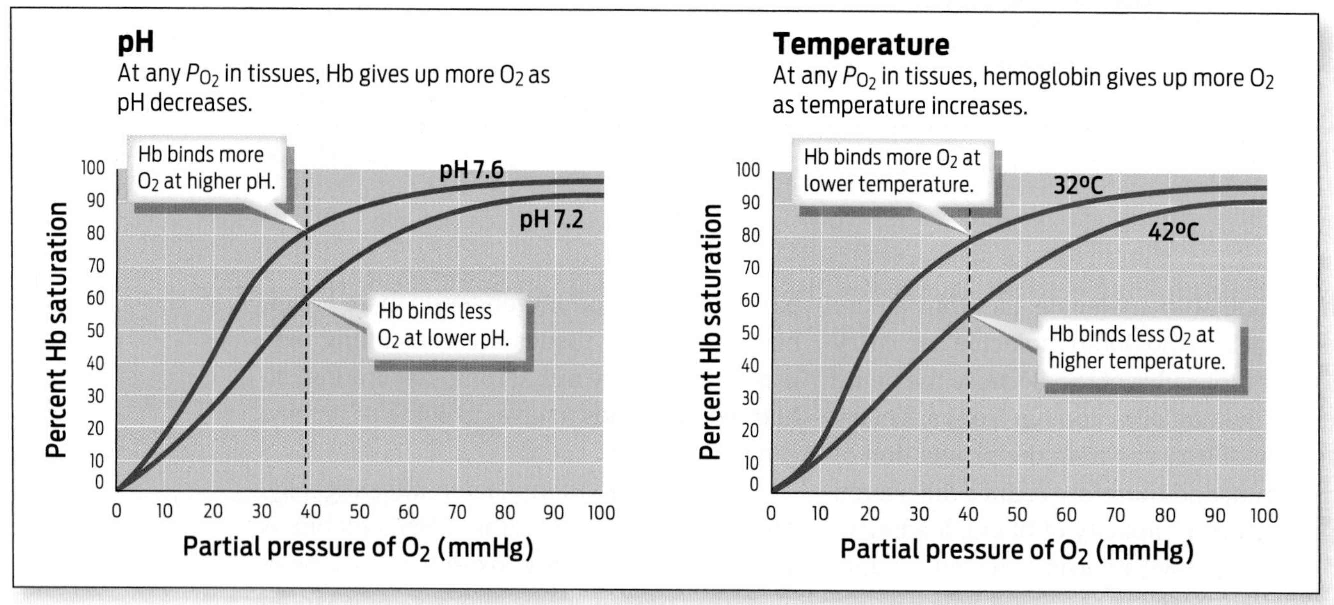
pH
At any P_{O_2} in tissues, Hb gives up more O_2 as pH decreases.
Hb binds more O_2 at higher pH.
pH 7.6
pH 7.2
Hb binds less O_2 at lower pH.
Percent Hb saturation
Partial pressure of O_2 (mmHg)
0 10 20 30 40 50 60 70 80 90 100
Temperature
At any P_{O_2} in tissues, hemoglobin gives up more O_2 as temperature increases.
Hb binds more O_2 at lower temperature.
32°C
42°C
Hb binds less O_2 at higher temperature.
Percent Hb saturation
Partial pressure of O_2 (mmHg)
0 10 20 30 40 50 60 70 80 90 100

? At pH 7.2 and P_{O_2} of 30 mmHg, what is the percent saturation of hemoglobin?

supply of this critical gas for the active muscle (**INFOGRAPHIC 29.9**).

So hemoglobin has an extremely important role to play in respiration. If there were some easy way to increase the amount of hemoglobin in blood, without sleeping in a hypoxic chamber or living at altitude, then an athlete would be at a competitive advantage in terms of conducting aerobic respiration.

High-Tech Doping?

When an athlete (or anyone) spends time at altitude, his or her kidneys respond by secreting a hormone called **erythropoietin (EPO),** which stimulates RBC production in the bone marrow. By producing more RBCs—each of which contains millions of hemoglobin molecules—the body increases its ability to absorb and transport oxygen. This is a natural process, which goes on anytime the body needs to make more RBCs—for example, after blood loss from a wound.

Because there is such a clear relationship between the concentration of RBCs in the blood and the ability to perform aerobic exercise, many athletes in elite sports in the past legally resorted to blood doping—artificially increasing the concentration of RBCs in their blood by undergoing blood transfusions or by injecting themselves with synthetic EPO (**INFOGRAPHIC 29.10**).

Blood doping with EPO can have dangerous side effects, including abnormal blot clots and stroke. Because EPO doping literally thickens the blood by increasing the concentration of red blood cells, the heart has to work harder to pump it, and the thickened blood is more likely to cause a blockage in a vessel. A few athlete deaths have been attributed to the practice.

The International Olympic Committee officially banned blood doping by transfusion in 1986 and by EPO injection in 1990.

ERYTHROPOIETIN (EPO)
A hormone produced by the kidneys that stimulates red blood cell production.

INFOGRAPHIC 29.10 EPO and Blood Doping

Erythropoietin (EPO) stimulates the production of red blood cells in the bone marrow. After EPO treatment, more immature red blood cells are detectable, and maximum oxygen uptake is enhanced.

Daily injection of recombinant human EPO over 25 days increased the number of immature red blood cells and maximum oxygen uptake, compared to a placebo.

	Placebo (7 subjects)	Human EPO (18 subjects)
Serum EPO	44 ± 44%	560 ± 316%
Immature RBCs	32 ± 22%	95 ± 45%
Maximal Blood Oxygen Uptake	0.4 ± 3.9%	6.5 ± 4.3%

Data are reported in % increase over baseline level.
Data from Ashenden, *J Sports Sciences*, 2001, 19, 831-837.

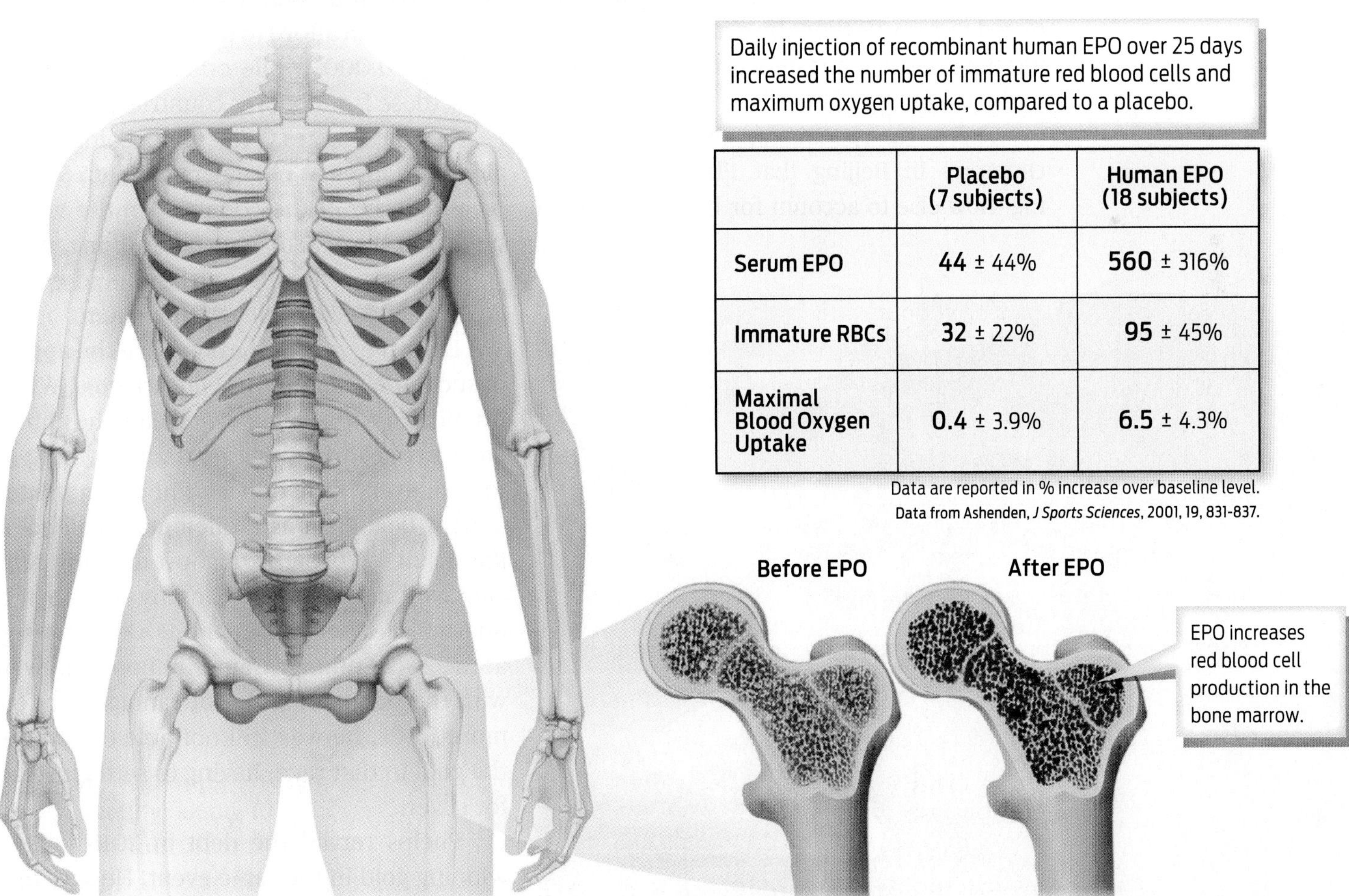

? What is the natural source of EPO?

However, the difficulty and expense of testing has made enforcement difficult.

Blood doping became endemic in cycling and caused a major scandal in 1998, when it was discovered that many cyclists were using EPO. That scandal led to the formation of the World Anti-Doping Agency (WADA) in 1999, which has worked to develop tighter and more effective testing programs, including urine and blood tests that can detect synthetic EPO.

Of course, for testing to be effective, it has to be done systematically and rigorously. Cyclist and seven-time Tour de France winner Lance Armstrong, who admitted to doping and was stripped of his titles in 2012, managed to avoid being tested numerous times by simply not answering the door when testers arrived, according to the testimony of his fellow cyclists.

There was speculation during the 2008 Olympics in Beijing that Phelps was doping–how else to account for his superhuman performance, eight gold medals in eight events?–but repeated tests proved that he was clean.

Unlike doping with EPO, the use of hypoxic chambers is not currently banned by WADA. Even though the underlying goals of blood doping and the use of hypoxic chambers are the same–increasing the production of RBCs–leaders in the sporting world have so far deemed the use of such chambers a variation on natural training regimens. So for now, it is not illegal to use them. However, it does raise the question of whether is it ethically proper or goes against the spirit of sport. Is it fair that some athletes–especially those with lucrative sponsorships, such as Phelps–can afford to buy a hypoxic tent for $7,000-$10,000, while other athletes, especially those from poorer countries, cannot?

Is it fair that some athletes can afford to buy a hypoxic tent for $7,000–$10,000, while other athletes cannot?

As Wilber notes, "You're not going to find altitude houses or chambers in South Sudan, or a majority of the countries in the world, but yet in the U.S. or Finland or China, yeah, you're going to find them there. And you're going to find elite athletes using them."

Ethical or not, it is easy to see the appeal of such practices among elite athletes. When fractions of a second can mean the difference between winning gold and coming in fourth, every little ounce of help can matter a great deal. This is a dilemma that Phelps knows all too well, having lost to South African swimmer Chad le Clos by less than a quarter of a second in the 200m butterfly at the 2012 Summer Games in London. Even with the potential added advantage of a few more RBCs, he was still not able to capture the gold in that race, having to settle instead for silver.

Phelps repaid the debt in 2016 in Rio, winning gold in the same event. He says that 2016 will be his last Olympic Games. Maybe now he can retire his bubble as well. ■

Michael Phelps celebrates after winning one of his 23 Olympic gold medals.

How Do Creatures without Lungs "Breathe"?

In humans and other mammals, the lungs are the primary sites of gas exchange–where oxygen is picked up and carbon dioxide is dropped off. But not all animals have lungs, or any kind of centrally located respiratory surface. Insects, for example, have a vast network of air-filled tubes that bring in air from the body surface to individual cells. This arrangement of air tubes is called a **tracheal system.** The individual tubes are called tracheae. Each trachea is an inner extension (that is, an invagination) of the body surface. The opening to the air (at the body surface) is called a spiracle. The spiracles are adjustable, to balance the intake of air and the loss of water, and to keep out dust. The tracheae branch into smaller and smaller tubes, resulting in a highly branched network that brings air very close to every cell in the body. Oxygen and carbon dioxide diffuse directly between the air tubes and the cells of the body, which means there is no need for a circulatory system to shuttle gases to and from cells.

In a process similar to lung ventilation, some insects generate air flow within their tracheal system by compressing and expanding their body wall. This action changes the volume and pressure of the air tubes and pushes air through the tracheal system.

Lungs are specialized structures for extracting gases from air. Not surprisingly, most fish do not have lungs. Instead, they have **gills**–delicate tissues that allow the fish to extract oxygen from water. As water flows over gills, oxygen diffuses directly into the blood and is carried

Insects "Breathe" through a Tracheal System

Insects have a series of air tubes (tracheae) that deliver gases directly to body cells. Air (and other gases) enter and leave the tracheae at spiracles.

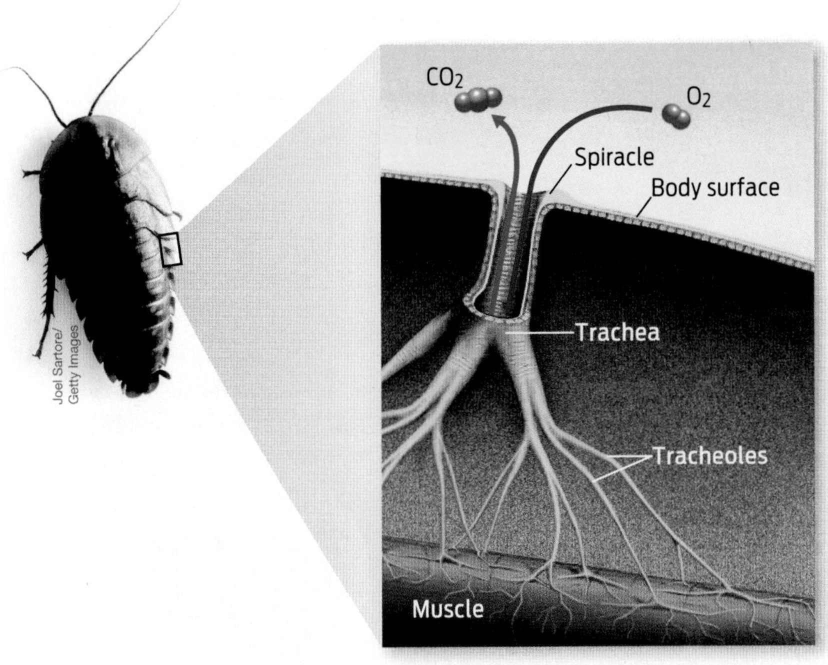

TRACHEAL SYSTEM
A series of air-filled tubes in insects that deliver air to cells.

GILLS
The gas exchange surface in fish.

through the body. Fish move water over their gills in a one-way direction as they swim. The more water they can move over their gills, the more oxygen they can extract from the water, and the more oxygen can be carried to cells for use in cellular respiration. Since water is the fish's source of oxygen, a fish out of water will die.

Fish "Breathe" with Gills

Most fish use gills for gas exchange between water and blood. As water flows over the gills, oxygen diffuses into the blood and carbon dioxide diffuses out of the blood.

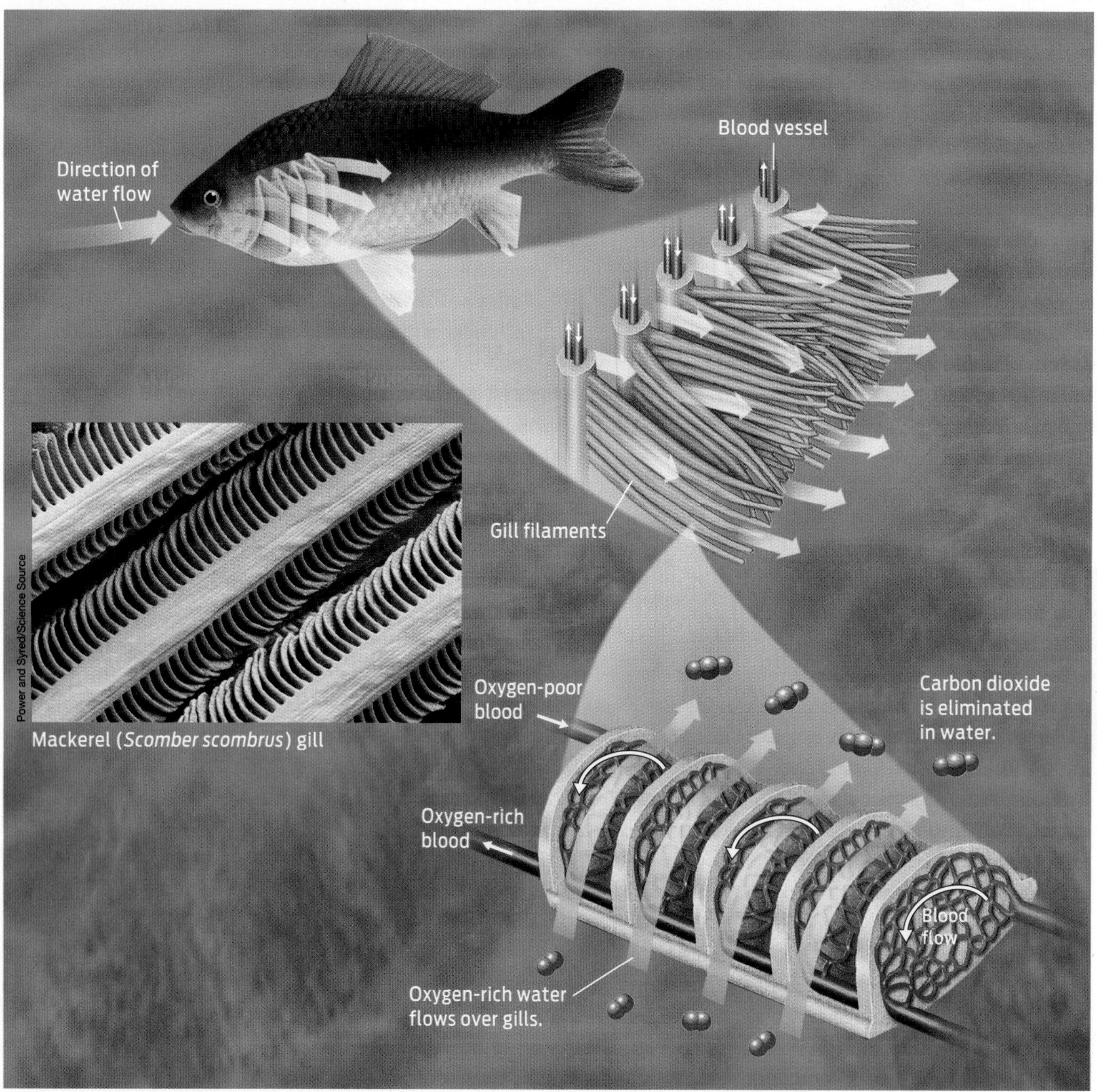

Mackerel (*Scomber scombrus*) gill

CHAPTER 29 SUMMARY

- The respiratory system takes in oxygen and eliminates carbon dioxide waste. It consists of the respiratory tract and the lungs.
- The respiratory system works in close conjunction with the cardiovascular system. Capillaries in the alveoli of the lungs are the sites of gas exchange between air and blood.
- Breathing is the way we ventilate our lungs and bring fresh air to the site of gas exchange. Ventilation requires a coordinated set of muscle movements that create negative pressure in the lungs to draw air in and positive pressure to expel air out.
- Breathing rate is controlled by the brain and relies on sensors that detect pH, which is directly related to the concentration of carbon dioxide in the blood.
- Oxygen is carried by hemoglobin in red blood cells. Hemoglobin binds reversibly to oxygen, picking it up from the lungs and dropping it off in tissues.
- Extended exposure to low oxygen pressure, such as occurs at high altitude, stimulates the production of red blood cells in the bone marrow. The growth signal comes from the hormone erythropoietin (EPO), which is secreted by the kidneys.
- Altitude training programs such as "live high, train low" or using a hypoxic chamber can increase the number of RBCs in the blood and improve athletic performance.

MORE TO EXPLORE

- **World Anti-Doping Agency: www.wada-ama.org/**
- **Wilber, R. (2004) *Altitude Training and Athletic Performance*. Champaign, IL: Human Kinetics Publishers.**
- **Stray-Gundersen, J., et al. (2001) "Living high-training low" altitude training improves sea level performance in male and female elite runners. *Journal of Applied Physiology* 91:1113–1120.**
- **Beall, C. M., et al. (2010) Natural selection on EPAS1 (HIF2a) associated with low hemoglobin concentration in Tibetan highlanders. *Proceedings of the National Academy of Sciences* 107:11459–11464.**
- **Matson, J. (July 30, 2012) Rope a dope: drug testing in sports enters a more aggressive era. *Scientific American*.**

CHAPTER 29 Test Your Knowledge

DRIVING QUESTION 1 What structures make up the respiratory system?

By answering the questions below and studying Infographics 29.1 and 29.6, you should be able to generate an answer for the broader Driving Question above.

KNOW IT

1 Add the names of the structures indicated by the labels A–F in the diagram below.

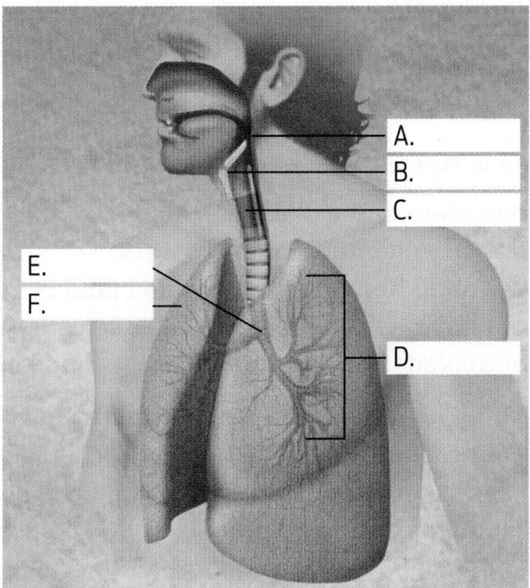

2 Which part of the respiratory system is the site of exchange of gases between blood and air?

a. alveoli
b. bronchioles
c. trachea
d. pharynx
e. bronchi

3 Inhalation is accompanied by

a. muscular relaxation and a decrease in lung volume.
b. muscular relaxation and an increase in lung volume.
c. muscular contraction and a decrease in lung volume.
d. muscular contraction and an increase in lung volume.
e. muscular contraction and no change in lung volume.

USE IT

4 Pneumonia tends to cause an accumulation of fluid in the lungs. This makes it harder for oxygen to diffuse between the air in alveoli and red blood cells in capillaries in the lungs.

a. From this description, explain why people with pneumonia experience shortness of breath.
b. Explain why supplemental oxygen is often a treatment for severe cases of pneumonia.

5 Asthma is a disease that causes swelling and constriction of the airways in the lungs. Compare the predicted symptoms of asthma with the symptoms of pneumonia. Can you think of any treatments for asthma that might be different from treatments for pneumonia?

DRIVING QUESTION 2 How do the respiratory and cardiovascular systems cooperate to deliver oxygen to body cells and remove carbon dioxide from tissues?

By answering the questions below and studying Infographics 29.2, 29.6, 29.8, and 29.9, you should be able to generate an answer for the broader Driving Question above.

KNOW IT

6 How is O_2 transported throughout the body?

a. dissolved in the plasma of blood
b. bound to hemoglobin in plasma
c. bound to hemoglobin in white blood cells
d. bound to hemoglobin in red blood cells
e. dissolved in the cytoplasm of red blood cells

7 What can cause a drop in blood pH?

a. a decrease in O_2
b. an increase in O_2
c. a decrease in CO_2
d. an increase in CO_2
e. b or d

USE IT

8 If blood pH drops, what happens to the breathing rate? Explain your answer.

9 Oxygen diffuses from the air in alveoli to the blood in lung capillaries. Diffusion occurs rapidly over short distances, but decreases dramatically with increases in distance. Pneumonia is an accumulation of fluid in the alveolar air spaces. Why does pneumonia cause shortness of breath and give a bluish tint to the skin and nails?

10 Breathing in and out of a paper bag will _____ pH and therefore _____ ventilation.

a. not change; not change
b. increase; increase
c. increase; decrease
d. decrease; decrease
e. decrease; increase

DRIVING QUESTION 3 What factors influence the oxygen-carrying capacity of blood and breathing rate?

By answering the questions below and studying Infographics 29.3, 29.4, 29.6, 29.8, and 29.9, you should be able to generate an answer for the broader Driving Question above.

KNOW IT

11 Relative to a tissue at rest, actively exercising tissues have

a. higher temperature, higher P_{O_2}, and higher pH.
b. higher temperature, lower P_{O_2} and lower pH.
c. higher temperature, higher P_{O_2} and lower pH.
d. lower temperature, higher P_{O_2} and higher pH.
e. lower temperature, lower P_{O_2} and lower pH.

12 What is the particular feature of altitude that increases the oxygen-carrying capacity of the blood?

a. the actual height (elevation)
b. the reduced atmospheric (barometric) pressure
c. the reduced partial pressure of oxygen
d. the increased atmospheric (barometric) pressure
e. the decreased relative humidity

USE IT

13 Hemoglobin releases O_2 at low pH. Give two reasons why a tissue might have a low pH.

14 Carbon monoxide (CO) binds to hemoglobin more tightly than does oxygen. In fact, CO can displace oxygen from hemoglobin. Predict the symptoms of CO poisoning, and provide an explanation.

15 Why do people "suck wind" (that is, breathe very heavily) with vigorous exercise?

DRIVING QUESTION 4 How can scientific knowledge of the respiratory system be used to design training regimens for elite athletes?

By answering the questions below and studying Infographics 29.5, 29.7, and 29.10, you should be able to generate an answer for the broader Driving Question above.

KNOW IT

16 Which of the following mimics high altitude?

a. sleeping in a high-O_2 chamber
b. sleeping in a low-O_2 chamber
c. transfusing RBCs into the circulation
d. a and c
e. b and c

17 What does EPO do?

a. stimulates RBCs to release stored O_2
b. stimulates RBC production
c. increases the number of heme groups per molecule of hemoglobin
d. increases ventilation rate.
e. b and c

18 Design an experiment to determine whether hypoxic chambers confer an advantage relative to altitude training. Consider how you will set up your experiment, including appropriate controls and the variables that you will consider and measure.

USE IT

19 WADA must be able to test for a variety of banned substances. How could WADA test for each of the following? Rank them from easiest to detect (1) to hardest to detect (4).

_____ EPO doping
_____ Transfusion of whole blood
_____ Transfusion of RBCs
_____ Use of a hypoxic chamber

INTERPRETING DATA

apply YOUR KNOWLEDGE

20 Look at Infographic 29.7.

a. Start with the 3000m performance data for the elite athletes. How many seconds did the men improve by? How many seconds did the women improve by? Now determine that improvement as a percent of the pre-altitude race time. What percent improvement did altitude training confer? In your opinion, is this method of training worth it for that percent improvement?

b. Now look at the graph with both the collegiate and the elite athletes. Even if the graph were not labeled, how could you know which group was which? (Hint: Look at their pre-altitude race times.) Did every athlete experience an improvement in race time? Was the change in race time identical for every athlete? Given the variability in results in this small sample size of conditioned athletes, what do you think you can extrapolate about altitude training in a larger population of active people?

BRING IT HOME

apply YOUR KNOWLEDGE

21 A friend wants to join a new gym. The gym has many amenities, including personal trainers (for an additional fee) and a hypoxic chamber for "performance training." There is a steep sign-up fee and high monthly dues at this gym. Another gym is offering a no-fee sign-up and lower monthly dues. This gym has the same cardiovascular equipment and access to personal trainers (at an additional fee comparable to that at the other gym). Your friend is fit. She runs local 10K races and often places in the top 10 in her age group in your small town. Given the costs of the two gyms and your friend's fitness level and goals, which gym would you advise her to join? Explain your reasoning.

30 Central Nervous System

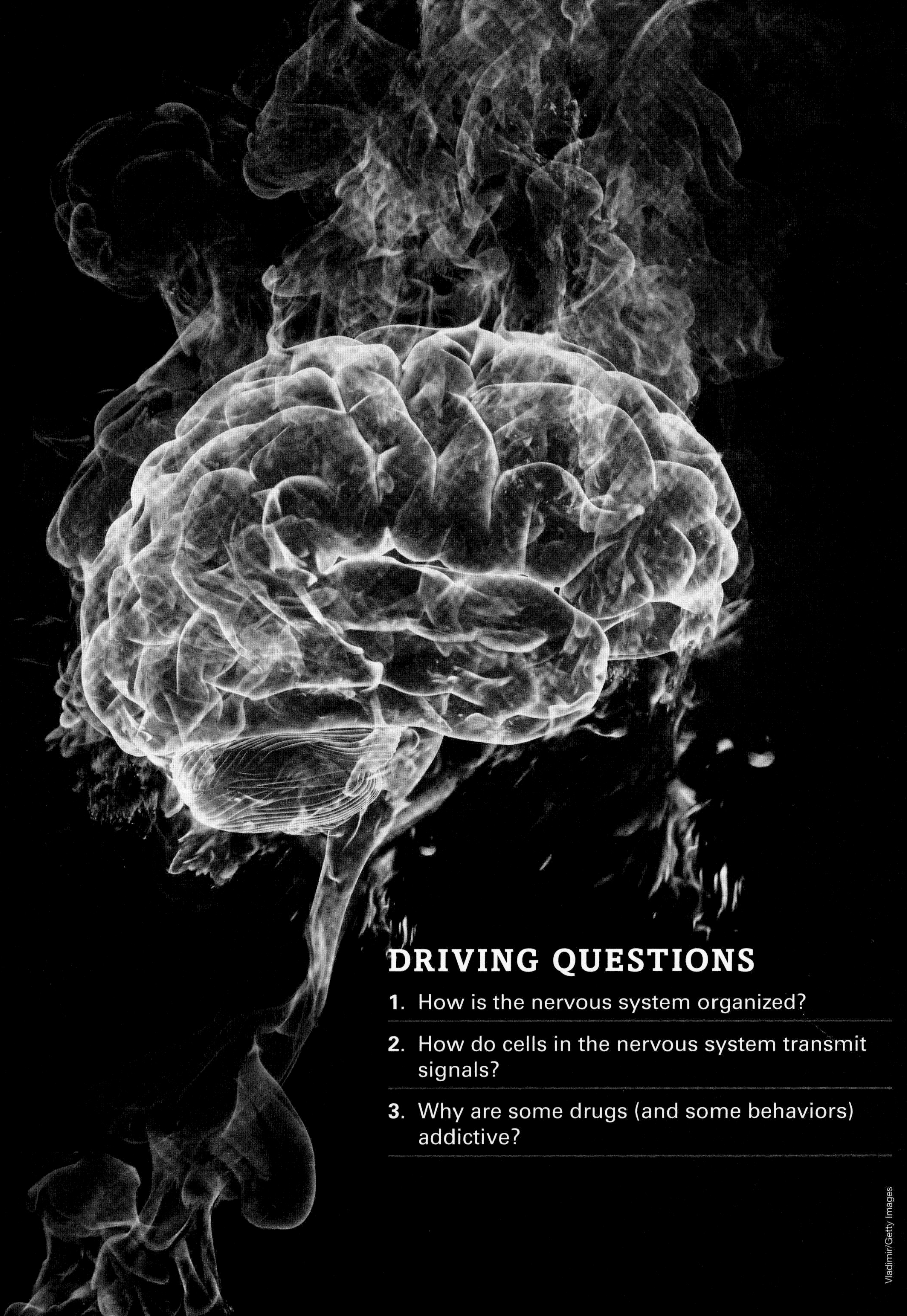

DRIVING QUESTIONS

1. How is the nervous system organized?
2. How do cells in the nervous system transmit signals?
3. Why are some drugs (and some behaviors) addictive?

SMOKE on the Brain

Neuroscience explains why nicotine and other drugs are hard to kick

Jack Ward thought he could resist picking up a cigarette, but the temptation was too strong. He had successfully quit smoking in 2006. But when he walked into a smoke-filled poker room 3 years later at a friend's house in Brooklyn, New York, the scene before him seemed to run in slow motion. He watched as smokers took deep satisfying drags on their cigarettes and exhaled wispy ringlets of contentment. He resisted that night. But poker night became a weekly event, and finally he gave in: he began smoking again.

Ward knew that smoking is risky. He'd had countless arguments with his wife, who urged him to stop smoking to protect his own health and hers as well, which is partly why he quit in the first place. But during poker night, none of that seemed to matter.

Ward lit his first cigarette in high school. By the time he went to graduate school, he was smoking a pack and a half a day. "I was surrounded by people who smoked," Ward recalls. "It never seemed unusual to smoke so much."

Cigarette smoking is highly addictive: smokers can develop a physical and psychological need to smoke. And while anyone might be able to smoke one cigarette or even several and not become addicted, most people find it extremely difficult to stop if they have smoked for an extended length of time.

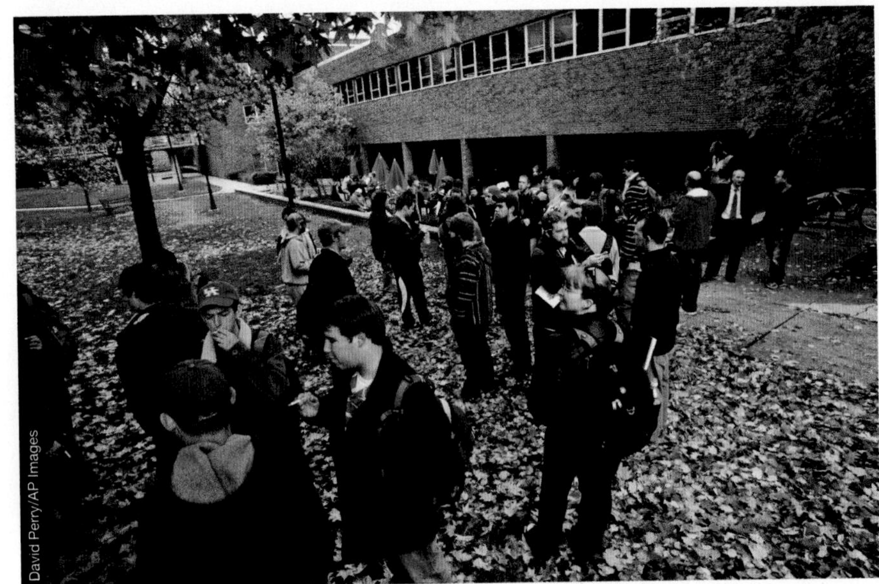

David Perry/AP Images

Despite public health campaigns, smoking remains a common practice.

For many years, addiction was seen as a personal or moral failing–the result of a weak will or a set of bad choices. Increasingly, however, researchers are looking to biology to explain addiction's awesome power. Brain scientists now know that cigarettes, alcohol, cocaine, and other drugs of abuse stimulate the brain's reward system: a complex circuit of brain cells that evolved to make us feel good after eating or having sex–activities we must engage in if we are to survive and pass along our genes. Without a feeling of pleasure from these activities that ensure our survival, we might never seek out food or sex, and our species would die out. Drugs of abuse stimulate this same reward system, which is why they make people feel so good, and why they can be so hard to resist.

Chemical substances like drugs and alcohol aren't the only things that can become addictive. Certain behaviors–gambling, shopping, sex, exercise–can start out as pleasurable habits but slide into addictions. Even eating can be addictive. Studies of obese people, for example, have shown that the brains of compulsive eaters are hyperactive in areas that respond to food. For these people, the mere thought of eating floods the brain with pleasure. In other words, almost anything deeply enjoyable can become an addiction.

But the pleasure comes at a massive cost. After prolonged use, drugs of abuse change the structure and function of the brain in ways that can wreak havoc on users' lives. Severely addicted people may stop eating, stop working, in fact stop all activities because nothing matters except the drug–and they will do anything to get it. Without it, they have intense cravings, obsessive thoughts, and are deeply depressed.

Once established, addiction is difficult to treat. The stories of celebrities who have been in and out of rehab and the deaths of superstars such as Michael Jackson, Heath Ledger, and Prince from drug overdoses are testament to just how difficult. But scientists today are gaining a better understanding of how physiological changes in the brain cause addiction, and that knowledge is leading to better treatments, which will help addicts of all kinds reclaim their brains and their lives.

BRAIN
The organ of the central nervous system that integrates information and coordinates virtually all functions of the body.

Addiction and the Brain

Despite a massive public health campaign aimed at curbing the dangerous habit, cigarette smoking remains a common practice in this country. Today, in 2017, about 20% of Americans smoke tobacco regularly, according to the National Institute on Drug Abuse (NIDA), down from about 45% in 1955. These numbers testify to tobacco's addictive potential.

The component in tobacco that makes cigarettes addictive is the chemical nicotine. When a person smokes, nicotine floods the lungs, where it is absorbed into the bloodstream. Once in the bloodstream, nicotine travels quickly throughout the body, reaching the brain within 8 seconds after tobacco smoke is inhaled. Chewing tobacco also contains nicotine, which the body absorbs into the bloodstream through the mucous membranes that line the mouth.

Nicotine is addictive because it stimulates feelings of pleasure in the brain. The **brain** is the master coordinator of the body, controlling virtually all of the body's activities, including sensation, movement, thinking, and just about all involuntary actions. The brain integrates information it receives from the internal and external environment and produces appropriate actions. The brain is also the seat of memory. Pleasurable experiences recorded in our memories serve as motivation; we tend to seek out the same pleasurable experiences over and over. Ultimately, the decision to smoke a cigarette, or to carry out any other behavior, is made in the brain (**INFOGRAPHIC 30.1**).

The brain doesn't act in isolation: it communicates with the rest of the body, sending out and receiving signals to coordinate our many activities. When we decide to reach for a cigarette, for example, our brain sends a signal to muscles in our arm. These signals travel along **nerves,** long fibers made up of specialized cells and supportive tissue that transmit signals through the nervous system. A major collection of nerves is the **spinal cord,** which extends from the base of the brain down to the lower back and is contained in and protected by our bony spinal column. Together, the brain and the spinal cord make up the **central nervous system (CNS).** Nerves that travel

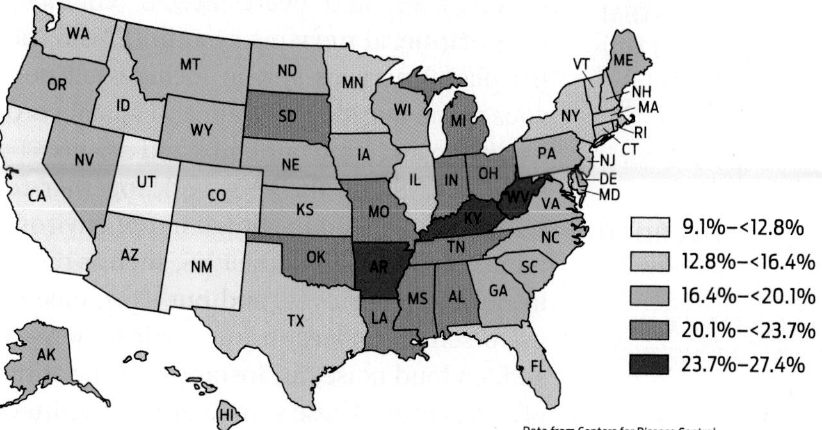
Cigarette Smoking Rate among Adults, 2015
9.1%–<12.8%
12.8%–<16.4%
16.4%–<20.1%
20.1%–<23.7%
23.7%–27.4%
Data from Centers for Disease Control

NERVE
A bundle of specialized cells that transmits information.

SPINAL CORD
A bundle of nerve fibers, contained within the bony spinal column, that transmits information between the brain and the rest of the body.

CENTRAL NERVOUS SYSTEM (CNS)
The brain and the spinal cord.

INFOGRAPHIC 30.1 Nicotine Stimulates Pleasure in the Brain

The brain coordinates the vast majority of all the body's activities. When nicotine is absorbed, it stimulates feelings of pleasure in the brain. The brain ultimately makes the decision to smoke a cigarette or not.

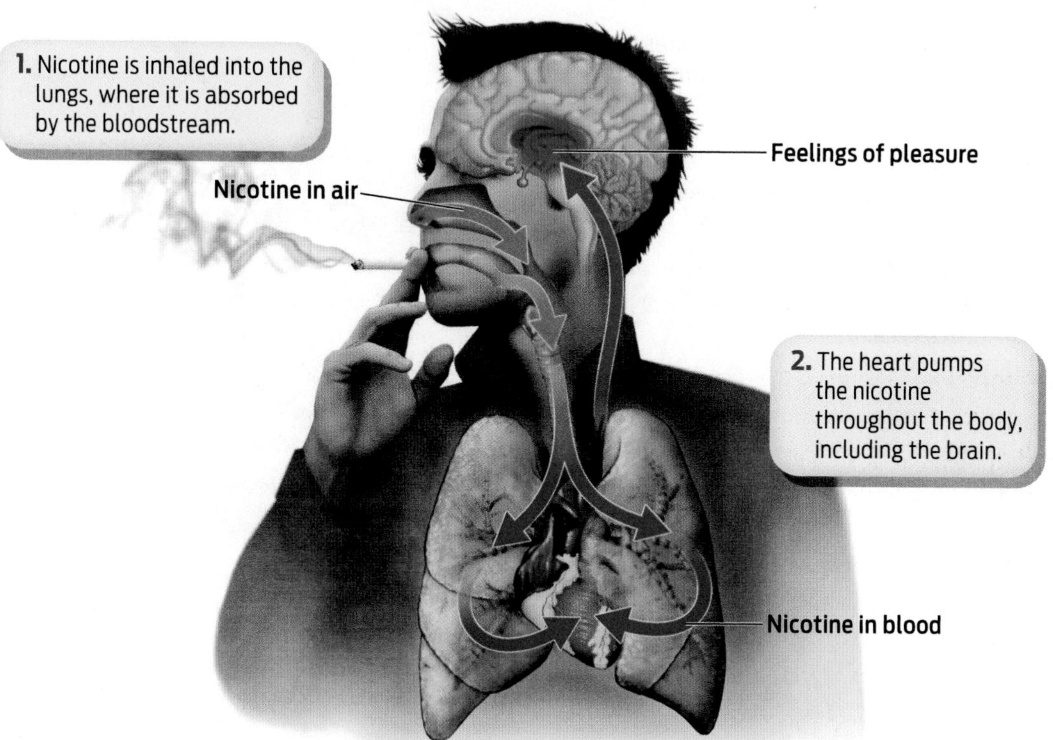
1. Nicotine is inhaled into the lungs, where it is absorbed by the bloodstream.
Nicotine in air
Feelings of pleasure
2. The heart pumps the nicotine throughout the body, including the brain.
Nicotine in blood

? Why can nicotine stimulate feelings of pleasure in the brain when it is chewed in chewing tobacco, rather than smoked?

PERIPHERAL NERVOUS SYSTEM (PNS)
All the nervous tissue outside the central nervous system that transmits signals from the CNS to the rest of the body.

NEURONS
Specialized cells of the nervous system that generate electrical signals in the form of action potentials.

from spinal cord to distant body sites, like fingers, toes, and heart muscle, compose the **peripheral nervous system (PNS).** The peripheral nervous system includes all nervous tissue outside the brain and spinal cord leading to and from our limbs and organs.

The CNS and the PNS work together to sense and respond to stimuli in the environment. Sensory cells in the PNS, such as those in our ears, our eyes, and our skin, enable us to sense changes in our environment—a sudden loud noise, for instance, or the sting of a mosquito. These stimuli are transmitted to the brain and spinal cord, which receive and integrate the incoming signals and trigger an appropriate response, such as a reflex. The reflex is carried out by effector tissues like muscles that allow us to respond, by jumping or swatting our arm. Peripheral nervous tissue in other organs, including the lungs, heart, and digestive organs, keeps our bodies operating without conscious thought. The nervous system allows us to perceive and understand the world around us and to translate thought into action (**INFOGRAPHIC 30.2**).

INFOGRAPHIC 30.2 The Central and Peripheral Nervous Systems Work Together

The nervous system has two main branches: the central nervous system (yellow) and the peripheral nervous system (orange). The central nervous system consists of the brain and spinal cord. The rest of the nervous system is the peripheral nervous system, which connects the CNS to the rest of the body.

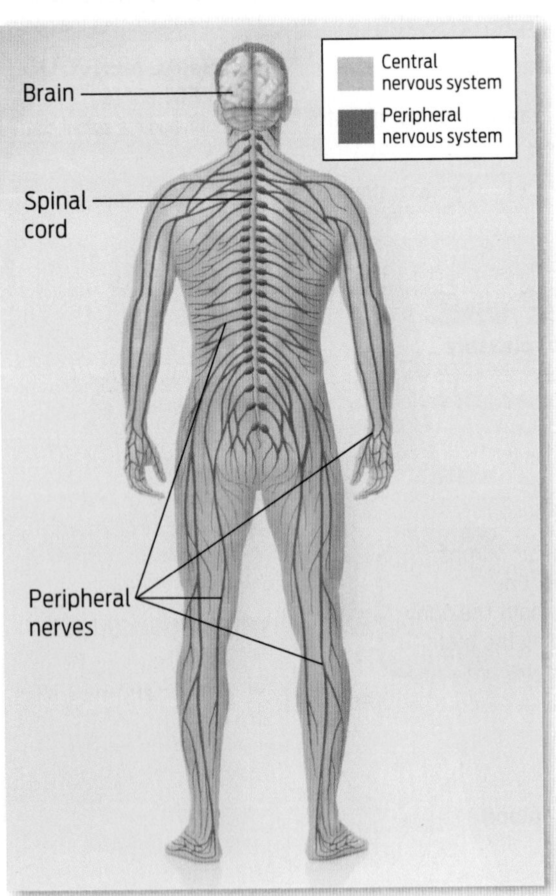

1. A mosquito bites the skin on your arm. Sensory receptors of the peripheral nervous system detect pain at the site.

2. The pain signal is transmitted to the spinal cord. The spinal cord can initiate a reflex and send a signal to the peripheral nerves of the other arm. The spinal cord can also send signals to the brain, where pain is detected.

3. The brain sends signals to the peripheral nerves of the other arm. Muscle tissues (the effector cells) in the arm and hand respond to slap the mosquito.

? Which function(s)—sensing, integrating signals, responding to the mosquito—is/are carried out by the peripheral nervous system?

Excitable Cells

Addiction is a complex psychological and physiological process that affects many parts of the nervous system, but it begins at the level of the cell. Fundamentally, the brain–like any organ–is a collection of cells. The cells of the brain are organized into networks that receive, process, and send information. Highly specialized cells called **neurons** are the individual units of this elaborate network.

Neurons have a unique structure that enables them to send and receive signals: a neuron consists of a large **cell body** with branched extensions called **dendrites** that receive signals and a single large **axon** that carries signals away from the neuronal cell body. The **axon terminals,** at the end of the axon, transmit the signals to the next cell or cells in the network; this could be another neuron, a muscle cell, or an endocrine cell **(INFOGRAPHIC 30.3)**.

There are different types of neuron. **Sensory neurons,** such as the receptor cells in the eyes and skin, receive information from the external world and transmit it to the CNS. **Motor neurons** transmit information from the CNS to muscle cells, signaling them to contract or relax.

The signals sent along neurons are electric currents. Dendrites and axons are like electrical cords that help collect and transmit information in the nervous system. In fact, a nerve is just a bundle of axons of neurons. The axon of a single motor neuron can be over a meter long.

Neurons can conduct electrical signals because of the special properties of their cell membranes (see Chapter 3). Like other cells in the body, neurons have a slight difference in electrical charge across the membrane: because of the balance of charged ions like sodium (Na^+) and potassium (K^+) on each side of the

CELL BODY
The part of a neuron that contains the nucleus and most of the cell's other organelles.

DENDRITES
Branched extensions from the cell body of a neuron, which receive incoming information.

AXON
The long extension of a neuron that conducts action potentials away from the cell body toward the axon terminal.

AXON TERMINALS
The tips of an axon, which communicate with the next cell or cells in the pathway.

SENSORY NEURONS
Cells that convey information from both inside and outside the body to the CNS.

MOTOR NEURONS
Neurons that control the contraction of skeletal muscle.

INFOGRAPHIC 30.3 Neurons Are Highly Specialized

Neurons, the highly specialized cells of the nervous system, have a structure that enables them to send and receive electrical signals quickly.

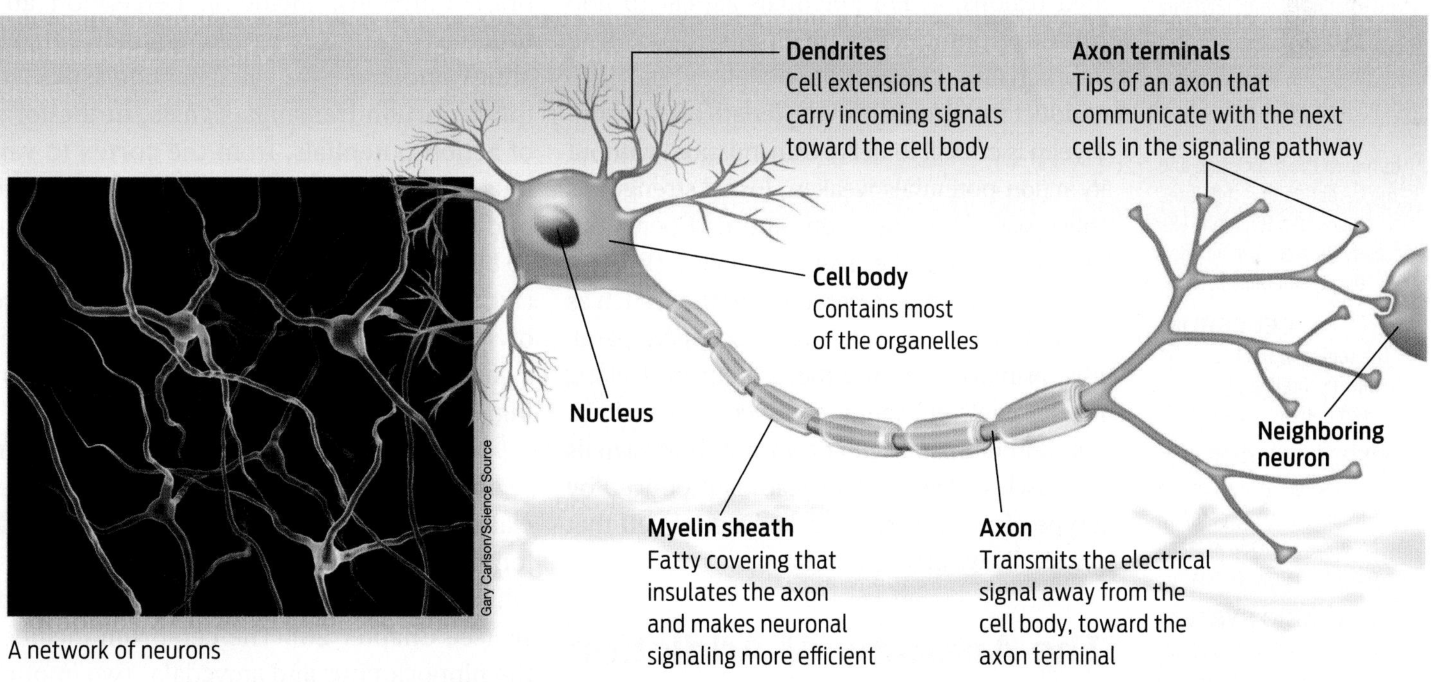
Dendrites
Cell extensions that carry incoming signals toward the cell body
Axon terminals
Tips of an axon that communicate with the next cells in the signaling pathway
Cell body
Contains most of the organelles
Nucleus
Neighboring neuron
Myelin sheath
Fatty covering that insulates the axon and makes neuronal signaling more efficient
Axon
Transmits the electrical signal away from the cell body, toward the axon terminal
Gary Carlson/Science Source

A network of neurons

? Which part of a neuron receives signals, and which part transmits them to the next cell in a signaling pathway?

ACTION POTENTIAL
An electrical signal within a neuron caused by ions moving across the cell membrane.

MYELIN
A fatty substance that insulates the axons of neurons and facilitates rapid conduction of action potentials.

GLIAL CELLS
Supporting cells of the nervous system, some of which produce myelin.

CEREBELLUM
The part of the brain that processes sensory information and is involved in movement, coordination, and balance.

BRAIN STEM
The part of the brain that is closest to the spinal cord and which controls vital functions such as heart rate, breathing, and blood pressure.

DIENCEPHALON
A brain region located between the brain stem and the cerebrum that regulates homeostatic functions like body temperature, hunger, thirst, and the sex drive.

CEREBRUM
The region of the brain that controls intellect, learning, perception, and emotion.

CEREBRAL CORTEX
The outer layer of the cerebrum, the cerebral cortex is involved in many advanced brain functions.

LIMBIC SYSTEM
A set of brain structures that is stimulated during pleasurable activities and which is involved in addiction.

membrane, the inside of the neuron has a negative charge relative to the charge outside the cell. But unlike most other cells, neurons can dramatically change the charge difference across the membrane, resulting in an electrical signal. Ion channels in the axon membrane open and close to allow specific ions to cross. When the axon of a neuron allows positive ions to cross the cell membrane, the charge across the membrane changes. This sudden change of charge initiates an **action potential,** a coordinated pattern of ion flow across the membrane that is conducted down the axon like a wave (**INFOGRAPHIC 30.4**).

Action potentials travel down the length of a neuron very quickly. Just think how quickly your arm moves in response to touching a hot stove—the movement is almost instantaneous. An action potential can travel down the length of an axon at a rate of 120 m/s—about 270 mph.

Electrical signals can travel so quickly down neurons because, much like electrical wires, neurons are insulated: they are coated with a sheath of **myelin** that wraps around the axon at repeating intervals and prevents the electrical charge from leaking across the membrane. Because ion channels open only at unmyelinated regions, action potentials appear to hop from one unmyelinated region of the axon to another, and it takes fewer action potentials to conduct the signal down the length of the axon. Myelin is critical to nerve transmission: without it, action potentials weaken, losing strength as they pass along the axon, and can peter out before reaching the axon terminals. Diseases in which myelin degenerates over time, such as multiple sclerosis, can cause progressive paralysis, primarily because motor neurons lacking myelin can't efficiently conduct electrical signals and ultimately can't transmit those signals to muscles. The myelin sheath is produced by a type of supportive cell called a **glial cell** that physically wraps around the axon.

The Anatomy of Addiction

Neurons communicate not only to move muscles that control limbs but also to solidify thoughts and lay down memories. This constant chatter within the nervous system is the way in which the brain grows and adapts to new environments and learns new tasks. The great capacity that humans have for conscious thought is a function of the sheer number of neurons found in the human brain: approximately 100 billion.

But the brain isn't just an undifferentiated mass of neurons; like any organ, it has specialized regions that perform distinct functions. For example, one part of the brain processes visual information, another specializes in hearing.

Scientists typically divide the brain into four major regions that orchestrate different functions. The **cerebellum,** located in the rearmost portion of the brain, controls movement, coordination, and balance. The **brain stem,** at the base of the brain, coordinates involuntary functions like reflexes, heart rate, and breathing. The **diencephalon,** located above the brain stem, regulates homeostatic functions like body temperature, hunger, thirst, and the sex drive. The **cerebrum** is the largest part of the brain, sitting right on top. Its outer layer, the **cerebral cortex,** is the seat of our more advanced brain functions, including perception and thinking, and gives us our distinct personalities and most human characteristics. Its inner portion transmits signals, in the form of action potentials, from the cortex to various brain regions and to other parts of the body. The cerebrum is made up of about 10 billion neurons and is divided into left and right hemispheres, with each hemisphere divided into four lobes. Each lobe processes a variety of functions, including smell, hearing, speech, and vision.

Within each of the four major brain regions are subregions that are further specialized for certain functions. The diencephalon, for example, contains the hypothalamus, which is important for maintaining homeostasis (see Chapter 26). The cerebrum includes the hippocampus and amygdala—two important components of what is sometimes called the **limbic system**, the primary seat of our emotions and memories. ("Limbic" comes

INFOGRAPHIC 30.4 Electrical Signals Are Transmitted Along Axons

Neurons conduct electrical signals called action potentials, which are generated as ions flow across the cell membrane. When a neuron "fires," sodium (Na^+) and potassium (K^+) ions enter and leave the axon in a characteristic pattern, creating an action potential. In an axon, the firing of an action potential at one location triggers the firing of another farther down the axon. In this way, signals travel down the axon.

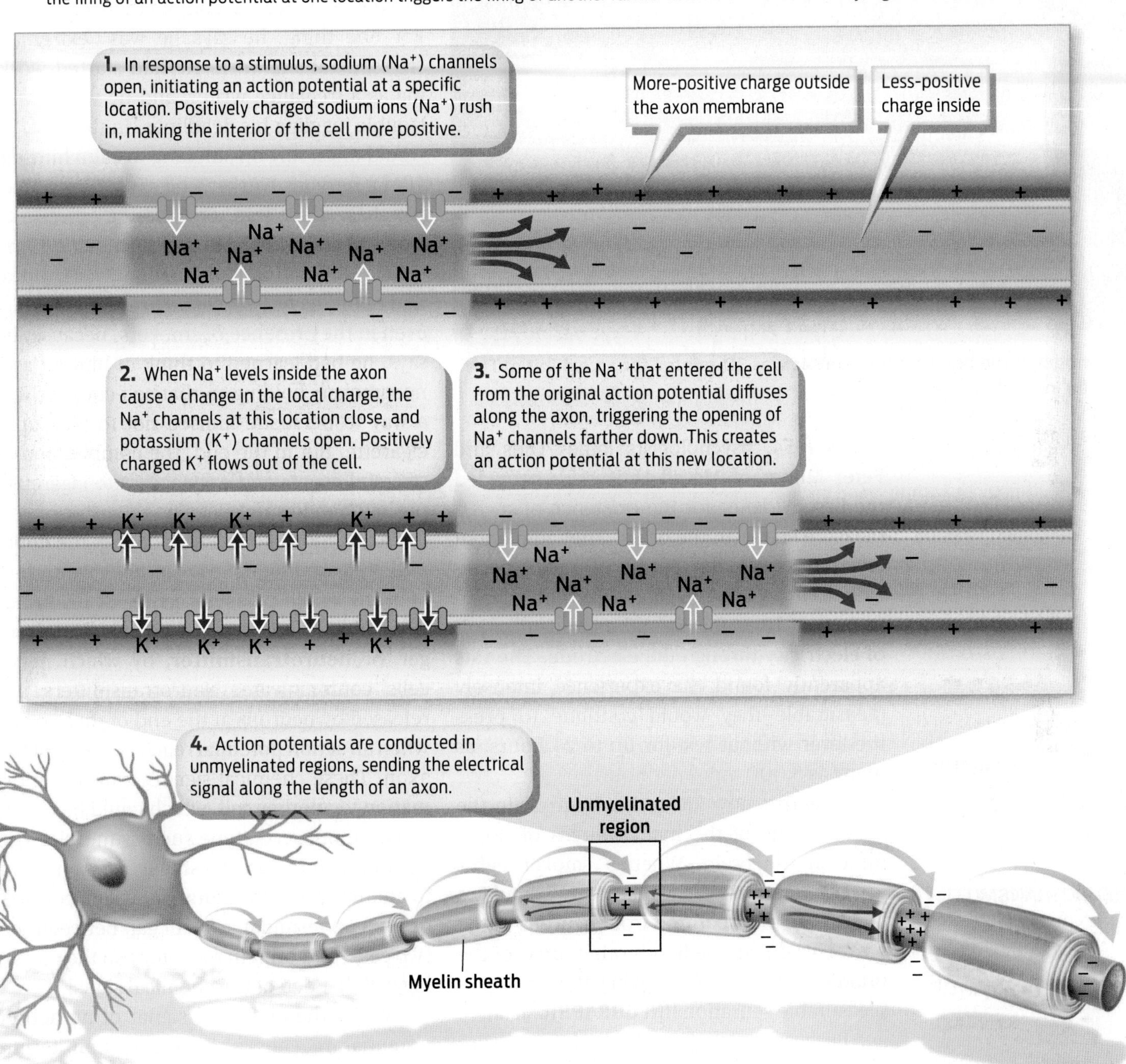

1. In response to a stimulus, sodium (Na⁺) channels open, initiating an action potential at a specific location. Positively charged sodium ions (Na⁺) rush in, making the interior of the cell more positive.
More-positive charge outside the axon membrane
Less-positive charge inside
Na⁺
2. When Na⁺ levels inside the axon cause a change in the local charge, the Na⁺ channels at this location close, and potassium (K⁺) channels open. Positively charged K⁺ flows out of the cell.
3. Some of the Na⁺ that entered the cell from the original action potential diffuses along the axon, triggering the opening of Na⁺ channels farther down. This creates an action potential at this new location.
K⁺
4. Action potentials are conducted in unmyelinated regions, sending the electrical signal along the length of an axon.
Unmyelinated region
Myelin sheath

? Which ion, flowing in which direction, triggers the initiation of an action potential at a given spot along a neuron?

from the Latin *limbus,* which means "border"; the structures in this system lie along the border of the cerebrum and diencephalon.) Along with two nearby regions, the nucleus accumbens and the ventral tegmental area (VTA), the limbic system is stimulated during pleasurable activities and is a major region involved in addiction. It is this system that researchers sometimes refer to as the "pleasure center" or "reward system" of the brain.

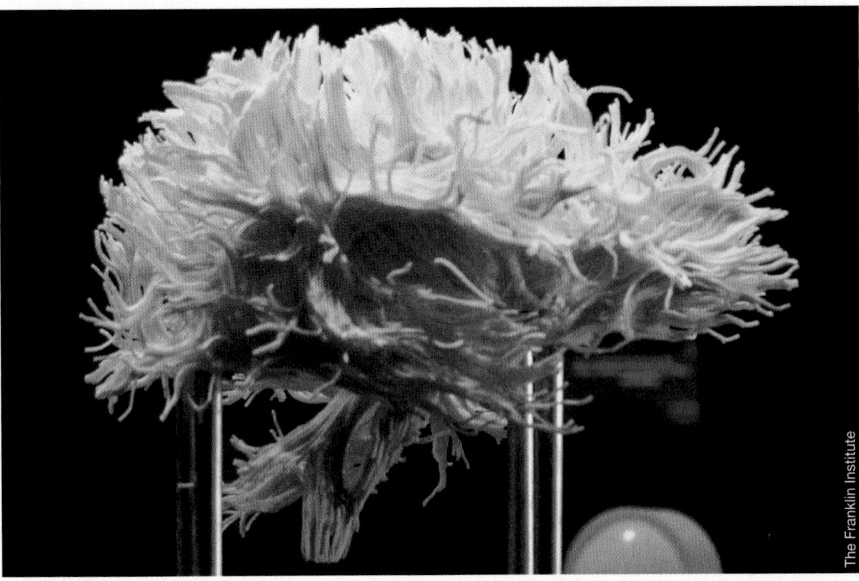

This model of the brain's white matter was 3D-printed for an exhibit at The Franklin Institute in Philadelphia.

In 1954, psychologists James Olds and Peter Milner of McGill University provided the first evidence for the existence of such a pleasure center in the brain. They implanted electrodes into the brains of rats, specifically in the VTA. The rats were then trained to press a lever that would administer a jolt of electricity through the electrode. The rats apparently found the experience intensely pleasurable—they would continue to press the lever without rest for up to 24 hours, as many as 5,000 times per hour!

Scientists now know that neurons in the VTA are one of the main sources of dopamine in the brain. When a smoker lights up, nicotine that reaches the brain stimulates the release of **dopamine** from neurons in the VTA. This dopamine floods other areas of the brain's reward system, giving the smoker a pleasurable sensation (**INFOGRAPHIC 30.5**).

Because drugs affect the brain's pleasure centers, they can be very hard to kick. This is also why even addicts who successfully stop taking drugs for decades can relapse simply by being in an environment that conjures up memories of the drug. The tinkling of a whiskey glass, meeting an old drinking buddy, being in a bar or a room filled with smokers puffing away can be reminders powerful enough to lure former addicts back to their old ways.

DOPAMINE
A chemical messenger that is involved in conveying a sense of pleasure in the brain.

NEUROTRANSMITTER
A chemical signaling molecule released by a neuron to transmit a signal to a neighboring cell.

SYNAPSE
The site of transmission of a signal between a neuron and another cell; includes the axon terminal of the signaling neuron, the space between the cells, and receptors on the receiving cell.

SYNAPTIC CLEFT
The physical space between a neuron and the cell with which it is communicating.

Chemical Messengers

For Ward, it wasn't just the smoking at the weekly poker game that rekindled his desire to smoke, it was a combination of events. For one thing, he says he was "extremely stressed" about work. He had started working as a freelance writer, and he was having trouble juggling his deadlines. In the past, he would often rely on smoking to calm himself. "I'm a nervous worker," he says. "Some people shake their legs, or pace, or exercise to work off stress. For me it was smoking." And then a difficult family situation arose. Up to that point, he had managed to avoid smoking, even in the presence of smokers, because, he says, he had learned to think of himself as a nonsmoker: holding to that self-concept was a way to persuade himself not to pick up a cigarette. But in the end, the combination of stressful events and easy access to cigarettes pushed him to smoke again.

As he took a long-awaited drag, the dopamine released by neurons in his VTA excited neurons in the pleasure center of his brain. Dopamine is one type of chemical messenger, or **neurotransmitter,** by which nerve cells communicate. Neurotransmitters are released by neurons at the end of their axons when an action potential reaches the tip of the axon. These chemical signals transmit information to another cell, which could be a neuron or another cell type such as muscle. The site of transmission of a signal from a neuron to another cell is the **synapse,** which consists of an axon terminal, a small gap between the two cells, and protein receptors on the receiving cell that detect the chemical signal.

When the electrical signal of an action potential reaches a neuron's axon terminal, it stimulates the neuron to release neurotransmitters from its axon terminal into the space between the neuron and the cell with which it is communicating. Neurotransmitters diffuse across this space, called the **synaptic cleft,** carrying messages from one cell to another. Each neurotransmitter molecule fits into a receptor on the surface of the cell receiving the signal as a key fits into a lock. The act of a neurotransmitter's binding to its receptor initiates a response in the receiving cell. If the receiving cell is another neuron, the binding of neurotransmitter may spark another action potential, thereby perpetuating the signal. Signaling between the two cells is terminated when the neurotransmitter is removed from the synaptic cleft. Neurotransmitter may be degraded by enzymes in the cleft, or taken back up into the signaling neuron via reuptake receptors (**INFOGRAPHIC 30.6**).

Cells in the nervous system use several different neurotransmitters to communicate different messages. For example, the neurotransmitter serotonin, which is active in the CNS and the gastrointestinal tract, regulates anxiety, appetite, and sleep. Another important neurotransmitter is acetylcholine, which is involved in learning, memory, and muscle contraction.

Many drugs–including both drugs of abuse and therapeutic medicines–interact with neurotransmitters. The class of antidepressant drugs known as selective serotonin reuptake inhibitors (SSRIs), which includes Prozac and Zoloft, helps to treat depression by preventing the reuptake of serotonin from the synaptic cleft, thus prolonging the activity of this mood-elevating neurotransmitter. Nicotine binds to receptors on neurons and triggers the release of multiple neurotransmitters in the brain, which accounts for the range of sensations that smokers experience: a reduction in anxiety, for instance, as well as an increased ability to concentrate. And as we saw, nicotine dramatically increases the release of dopamine from certain neurons, such as those in the VTA.

INFOGRAPHIC 30.5 Functional Regions of the Brain

The brain has specialized regions that carry out distinct functions. Several areas are stimulated by pleasurable activities and are important in addiction.

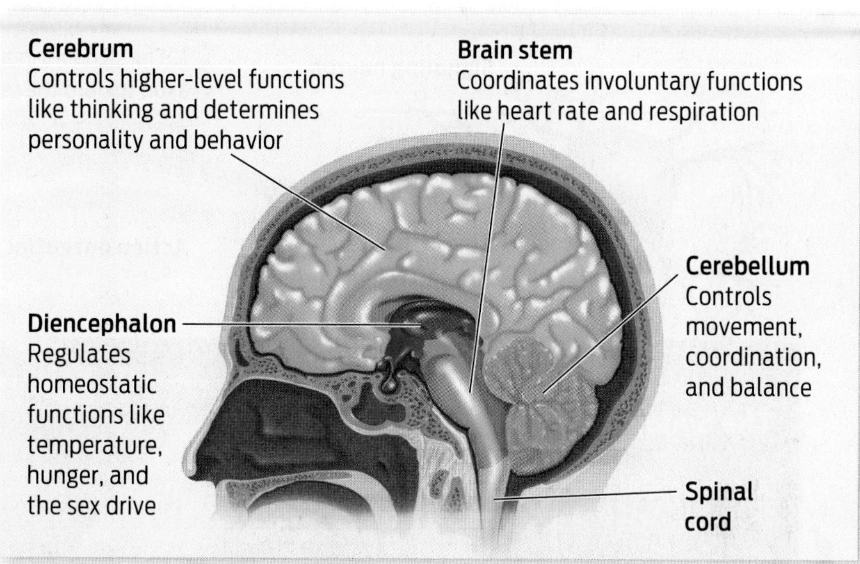

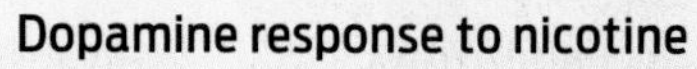

Dopamine response to nicotine
Neurons in the ventral tegmental area release dopamine to other parts of the brain (shown here in red), resulting in feelings of pleasure.

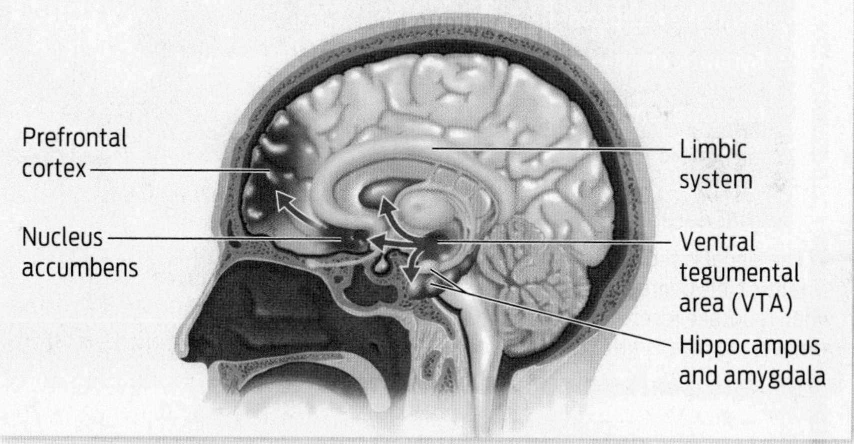

? What could be the consequences of an injury to the cerebellum?

Too Much of a Good Thing

In normal circumstances, dopamine's main job is to convey information related to elation and pain. The joy we get from a meal, sex, a winning poker hand, or indeed anything that gives us pleasure is conveyed in part by dopamine. Drugs of abuse stimulate dopamine production, which is why they can become so addictive.

But these drugs cause such a high that over time they can alter the dopamine signaling system so that the pleasures of everyday life become meaningless compared to the pleasure of the drug. Normally the brain produces dopamine at a relatively constant

INFOGRAPHIC 30.6 Neurons Communicate with Other Cells by Chemical Signals

Within a neuron, action potentials are electrical signals that carry information. When an action potential reaches the axon terminal, a neuron releases molecules called neurotransmitters that communicate with the next cell in the pathway at the synapse.

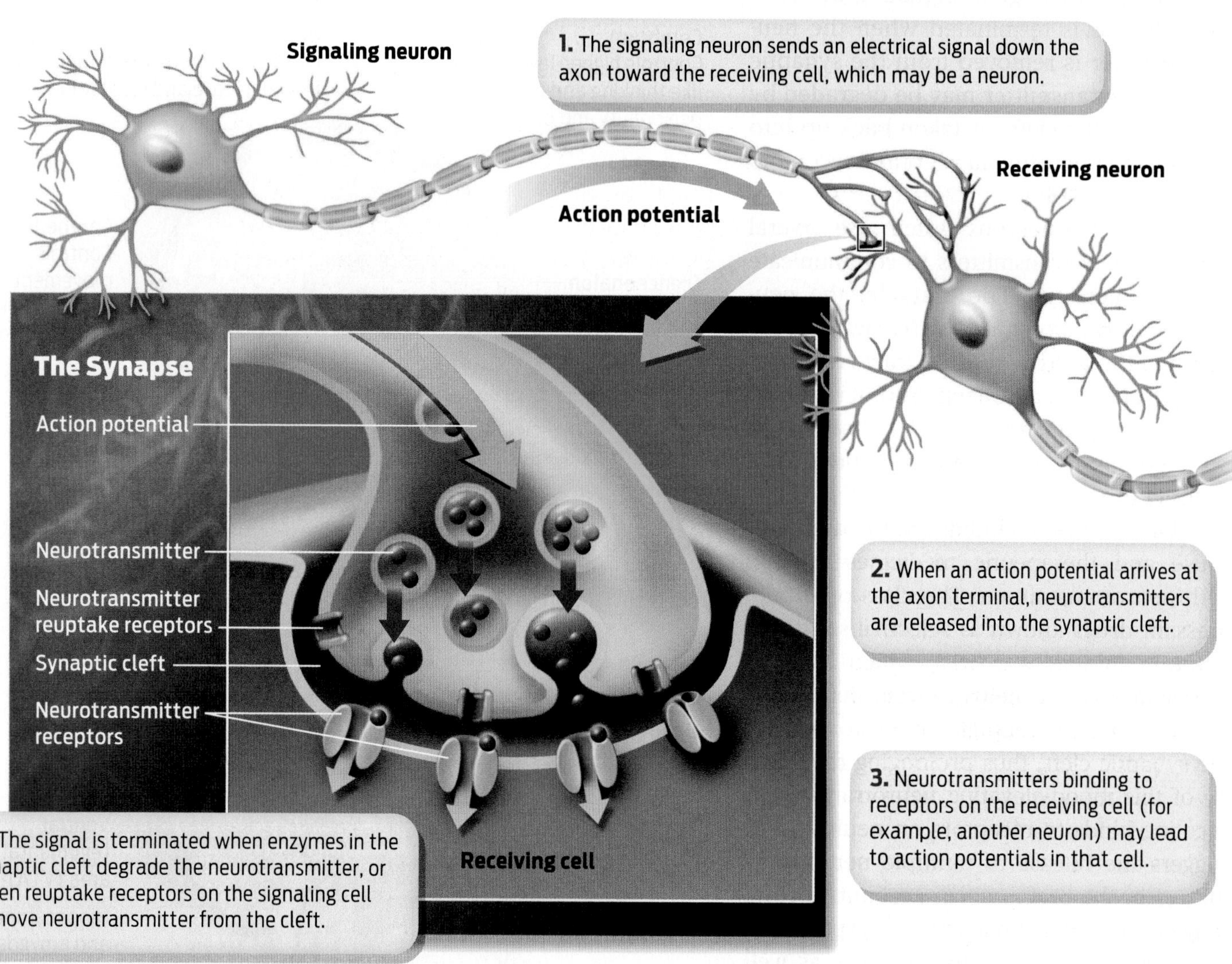

Signaling neuron
1. The signaling neuron sends an electrical signal down the axon toward the receiving cell, which may be a neuron.
Receiving neuron
Action potential
The Synapse
Action potential
Neurotransmitter
Neurotransmitter reuptake receptors
Synaptic cleft
Neurotransmitter receptors
2. When an action potential arrives at the axon terminal, neurotransmitters are released into the synaptic cleft.
3. Neurotransmitters binding to receptors on the receiving cell (for example, another neuron) may lead to action potentials in that cell.
4. The signal is terminated when enzymes in the synaptic cleft degrade the neurotransmitter, or when reuptake receptors on the signaling cell remove neurotransmitter from the cleft.
Receiving cell

? Which part of a neuron releases neurotransmitter molecules into the synaptic cleft?

rate, and dopamine occupies only a portion of dopamine receptors at any given time. But when a person smokes, snorts cocaine, or takes heroin, for example, dopamine levels in the synaptic cleft increase dramatically. With so much dopamine available, practically all of the brain's dopamine receptors become activated simultaneously.

The immediate effect is euphoria. But there is a downside. Because so much dopamine is produced, the brain becomes overwhelmed and tries to dampen the drug's effect by switching off some of its dopamine receptors. When the drug wears off, fewer receptors are functioning–bringing down mood.

The low can be so low that normal pleasures such as eating or socializing become dull and listless affairs. The user's mood may now be even lower than it was before taking the drug. As dopamine receptors shut down, ever-larger quantities of the drug are required to produce a high—and the high may never be as high as the user experienced the first time.

As they come down from a high, addicts will likely feel even more unhappy and depressed as the dopamine response system is dampened. Eventually, many addicts need to take drugs simply to feel normal. Without drugs, they suffer the physical symptoms of withdrawal, which may include depression, anxiety, and intense cravings for a dopamine fix. The specific symptoms and their intensity vary depending on the drug **(INFOGRAPHIC 30.7)**.

Altered dopamine signaling isn't the only brain change that scientists have observed in drug addicts. Researchers have also shown that drugs such as cocaine can change the shape of neurons in specific parts of the brain and consequently may impair their ability to transmit signals. Brain-imaging studies have also shown that addicts consistently have lower than normal levels of blood flow in the frontal regions of the cerebrum during withdrawal from cocaine, and higher than normal levels while they are on the drug. This region of the cerebrum is involved in decision making. In a variety of tests, drug addicts seem to make poorer decisions, with detrimental consequences.

As dopamine receptors shut down, ever-larger quantities of the drug are required to produce a high.

Even more troubling, adolescents who take drugs may be preventing their brains from developing normally. A 2004 study by researchers at University of California, Los Angeles, and the National Institute of Mental Health (NIMH) that imaged the brains of 13 children and young people age 4 to 21 over 10 years showed that that some parts of the brain, such as the prefrontal cortex, are not fully developed until the mid-twenties or so. Taking drugs at an early age may hinder normal development of this region.

And that's not the end of the addiction story. In addition to dopamine, there are several other neurotransmitters involved in addiction, says Joe Frascella, director of the division of Clinical Neuroscience and Behavioral Research at the National Institute on Drug Abuse. Scientists are just starting to study how drugs of abuse affect these other neurotransmitters. While scientists have shown that almost every drug of abuse affects the dopamine system in varying degrees, Frascella says, "It's certainly more complex than just dopamine." Scientists have only just scratched the surface when it comes to learning exactly how long-term drug use affects the brain **(TABLE 30.1)**.

Born Addicts?

Not everyone who takes drugs becomes addicted. Exactly why some people seem to be more at risk than others isn't clear. But researchers have a few hypotheses based on existing evidence. For some of us, it may be a matter of biology: a predisposition to addiction caused by a shortage of dopamine or other types of receptors. And some people may have been born with fewer receptors, or their brains may have lost receptors over time because of difficult life experiences. Consequently, drugs provide these people a high that they can't get from any other stimulus. And the drug feels too good to stop.

And just as some of us may be biologically predisposed to addiction, others may be predisposed to avoid it: some people's brains may simply be better at overriding the pleasure-seeking impulse.

INFOGRAPHIC 30.7 Addictive Drugs Alter Dopamine Signaling

Addictive drugs increase dopamine release, causing initial feelings of pleasure. Continued use of the drug alters dopamine signaling, requiring more drug to achieve the same high. Attempts to quit can be difficult because of symptoms of withdrawal, the result of diminished dopamine signaling.

a. Normal Dopamine Signaling

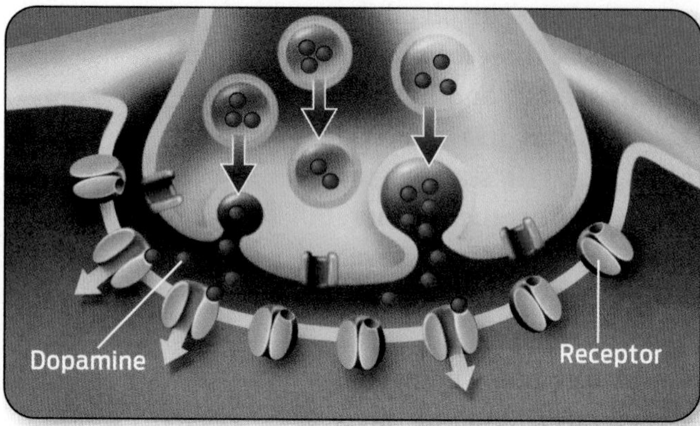

Cells release moderate amounts of dopamine into the synapse and not all dopamine receptors are occupied.

b. After Drug Use

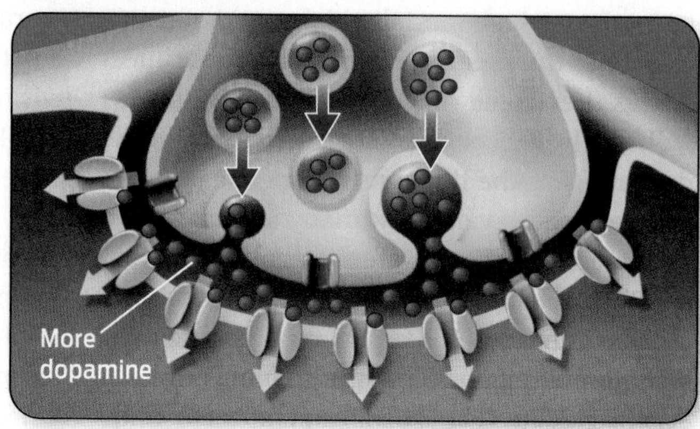

Certain drugs cause massive dopamine release into the synapse. Many more dopamine receptors become occupied. The result is an intense feeling of pleasure.

c. After Repeated Drug Use

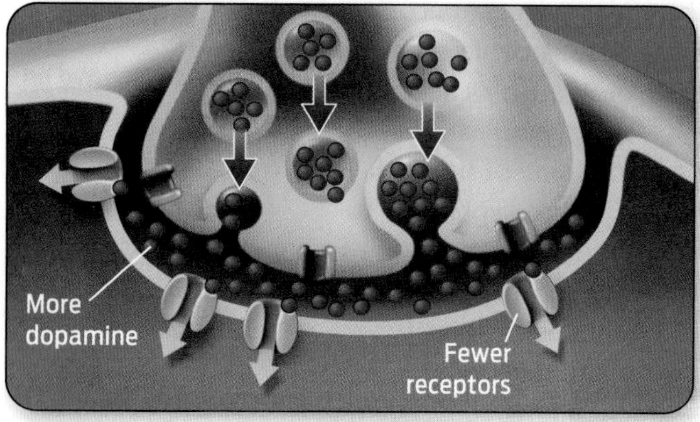

Dopamine overstimulation causes the receiving cell to down-regulate (that is, shut down) some dopamine receptors. Because the user now has fewer receptors available, more of the drug is required to feel high.

d. After Drug Withdrawal

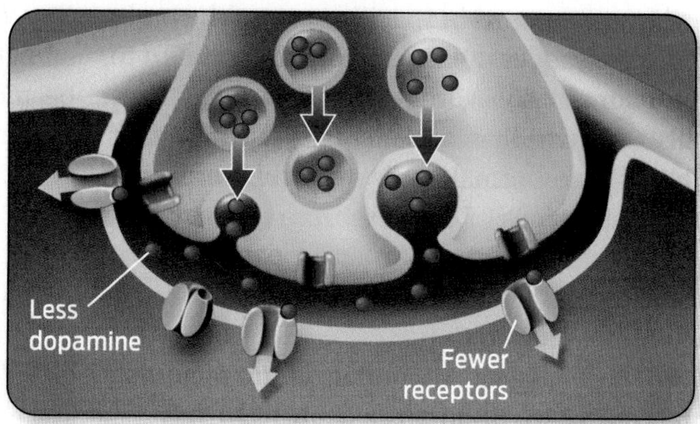

Removing the drug reduces the amount of dopamine released. In combination with fewer dopamine receptors, the user may feel sick and depressed.

? Why do people who are addicted to drugs need to take increasing amounts of the drug in order to experience the same high?

A 2011 study by researchers at the Scripps Research Institute Florida, for example, identified a brain pathway involved in nicotine addiction. The researchers found that the number of a nicotine receptor called alpha-5 influences how susceptible mice are to nicotine addiction. When given the opportunity to self-administer nicotine, normal mice will stop after reaching a certain dose. Mice with no alpha-5 receptors, however, won't

TABLE 30.1 Potentially Addictive Drugs and Their Effects

	MODE OF ACTION	EFFECT
COCAINE	• Causes a large release of dopamine into the synapse • Inhibits reuptake receptors • Causes an amplified signal between neurons	• Highly addictive • Causes powerful feelings of well-being and confidence • Users lose interest in life activities. • High doses lead to paranoia, anxiety, and increased blood pressure at high doses.
HEROIN	• Mimics natural endorphins • Binds opiate receptors in specific regions • Affects mood, respiration, and pain response	• Rush of euphoria followed by a foggy feeling • Powerful withdrawal symptoms make this drug extremely addictive. • Overdose slows breathing to dangerous levels.
CAFFEINE	• Mimics adenosine; blocks the natural sleep response • Causes release of adrenaline and dopamine; result is alertness and sense of pleasure	• Inhibits sleepiness and increases alertness • Withdrawal symptoms include headaches, jittery feelings, and increased anxiety. • Interferes with deep-sleep cycles, causing exhaustion and depression. • The adrenaline produced constricts blood vessels, affecting heart rate.
NICOTINE	• Increases levels of dopamine	• Enhanced short-term feelings of pleasure, relaxation • Increased concentration • Very addictive
ECSTASY	• Causes excessive release of serotonin • Destroys nerve cells that produce serotonin	• Intense euphoria, followed by depressive "crash" • Side effects include paranoia, anxiety, confusion, and difficulty concentrating. • Long-term users can't distinguish between reality and fantasy.
ALCOHOL	• A general depressant of the central nervous system • Changes communication patterns between neurons in specific brain regions	• Acts as an anesthetic • Influences breathing, motor and behavior control • Diminishes senses • Destroys cells in the brain and other organs
MARIJUANA	• Mimics the chemical anandamide • Stimulates anandamide receptors in areas of the brain that affect memory, emotion, and sensory perception	• Relaxation, mild euphoria, and appetite stimulation • Can cause paranoia • Weakens short-term memory and can block the production of long-term memory • Diminishes problem-solving ability and coordination
INHALANTS (GLUE, HAIR SPRAY, PAINT THINNER, ETC.)	• Vapors destroy the myelin sheath of axons. • Damages cells in the brain, lungs, heart, liver, kidneys, and bones	• Causes headaches, nausea, and disorientation • Can diminish the ability to learn, remember, and solve problems • May cause a rapid and irregular heartbeat
METHAMPHETAMINE	• Causes release of dopamine and norepinephrine into the synapse • Dopamine leads to feelings of pleasure. • Norepinephrine increases blood pressure and heart rate.	• Creates feelings of pleasure and euphoria, paranoia, and hallucinations • May alter dopamine-producing neurons connected with Parkinson's disease
RITALIN	• Causes increased levels of dopamine	• At lower doses increases the ability to focus and concentration • At higher doses can inhibit formation of new nerve pathways, interfering with cognition and brain development • Prescribed for attention deficit disorder, but has become a common street drug

stop until they've taken a much higher dose. Humans also have alpha-5 receptors in varying numbers, and scientists hypothesize that people with fewer receptors are less sensitive to nicotine and may become more easily addicted.

Regardless of the reason people become addicted, addiction is a serious public health problem. Tobacco, for example, is responsible in some way for one out of every five deaths in the United States, according to NIDA. Smokers have a higher incidence of both heart and lung disease. And smoking causes cancer. Tobacco use is the leading preventable cause of death and disease in the United States, killing some 440,000 people a year.

Tobacco is responsible in some way for one out of every five deaths in the United States.

Drug use also affects health in ways that diminish the quality of life. For example, deficits in the dopamine system caused by drug use weaken memory and motor skills. Nora Volkow, head of NIDA, has shown that methamphetamine (crystal meth) users have poorer short-term memory and score much lower on tests of motor skills, such as quickly walking a straight line (**INFOGRAPHIC 30.8**).

Kicking the Habit

Research on the neurobiology of drug addiction is informing new and better ways to treat addiction. Volkow and other researchers have shown that after a period of abstinence from drugs, the dopamine system can repair itself–but it generally takes longer than a year. And while studies have shown that functional skills such as short-term memory and motor control do come back, it's not clear whether they come back completely.

Given the nature of many of the drug-induced brain changes, experts now see addiction as a chronic disease. Like heart disease or diabetes, diseases for which patients require long-term treatment, patients with addictions require long-term treatment plans. And the occasional relapse is only a predictable setback, not a failure of the treatment, says Volkow.

Experts also now know that the best treatments should target addictive behaviors in several ways. They should, for example, decrease the reward value of the drug, according to Volkow, perhaps by counseling addicts to seek out other pleasurable experiences and to repeat those experiences to reinforce their value in the brain. Avoiding the drug and focusing on other pleasurable experiences will, over time, weaken conditioned memories of the drug and drug-related stimuli.

A better understanding of the effects of substance abuse on the brain is also informing efforts to develop medications that can help an addicted person kick the habit. Two popular antismoking drugs, Chantix and Zyban, for example, work by competing with nicotine for binding sites on neurons. By partially activating dopamine release, the drugs help reduce nicotine cravings in someone trying to quit smoking. They also make smoking less pleasurable because nicotine from cigarettes cannot bind to nicotine receptors while the drugs are present. As a result, smoking becomes much less enjoyable, and therefore easier to stop. Research has shown these medications to be two to three times more effective at aiding smoking cessation compared to placebo. However, there have also been reports of serious side effects of these drugs, such as suicidal thoughts, so use of these medications should be considered carefully. A combination of behavioral strategies and medications that target specific neurotransmitters or brain circuits will likely work the best to help addicts kick the habit, says Frascella, of NIDA.

Before Ward quit smoking in 2006, he was up to a pack and a half a day. Smoking had increasingly become a point of tension between him and his nonsmoking wife. She desperately

INFOGRAPHIC 30.8 Drugs Diminish Memory and Motor Skills

Researchers used a series of measurements to estimate the number of dopamine reuptake receptors in the brains of methamphetamine users. Meth users had lower numbers of dopamine reuptake receptors than nonusers, and this reduction was correlated with a reduction in motor skills and memory.

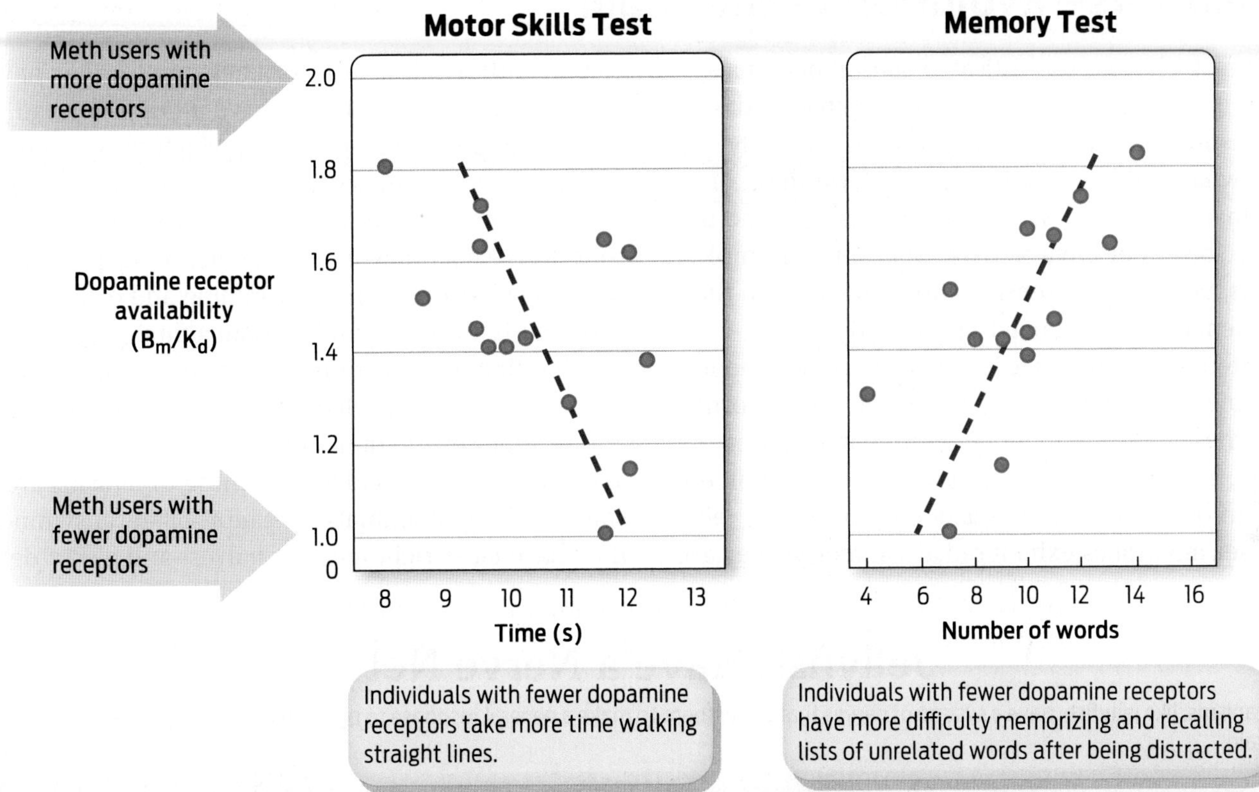

Data from Volkow, N., et al. (2001) *American Journal of Psychiatry*, 158:377-382.

? How does the number of dopamine reuptake receptors for the participant who recalled the greatest number of words compare to that of the participant who recalled the fewest number of words?

wanted him to quit, both for the sake of his health and because of the risks to herself of heart disease and cancer from secondhand smoke. A medical examination showed that Ward already had risk factors for heart disease, including high blood pressure. His doctor told him to stop smoking. "She didn't ask me to stop," Ward says. "She simply said, 'You are going to stop smoking in two weeks.'" She prescribed a nicotine patch–a skin patch that delivers nicotine to the bloodstream and eases nicotine withdrawal symptoms–and other medication, which helped him quit.

But after that fateful poker night in 2009, Ward caved into his cravings: he went back to smoking. This time, instead of his old pack-and-a-half-a-day habit, he managed to cut down to only a few cigarettes a week, all of them smoked on his weekly poker night.

Two years after Ward started smoking again, he moved. The weekly poker nights were gone, and gone, too, were the social cues that had tempted him in the first place. There are more antismoking social influences in his life now than ever. Most of his friends do not smoke, he exercises more, and perhaps most important of all, he wants to model healthy behavior for his 4-year-old daughter: "I don't want her to see me smoking." ■

Size Matters—Evolution of the Brain

Humans have a sophisticated central nervous system that consists of a brain and spinal cord. Our distinctively large brain relative to our body size is partly what endows us with the traits we think of as most human: language, self-consciousness, and the use of tools to shape our environment. But even the simplest organisms have ways of sensing and responding to their environments, and don't need a brain to do so.

Invertebrate animals like jellyfish and sea anemones, for example, have neurons that sense and send information. But these neurons are not organized into a brain. By definition, a brain is a grouping of neurons located in the head of an organism that acts as a control center. Jellyfish and sea anemones exhibit radial symmetry—they are round, with no clear front or back, and so they have no head. They therefore do not have a brain. Instead, they have what are called nerve nets, which are collections of sensory neurons, motor neurons that control muscles, and neurons that connect the sensory and motor neurons. Nerve nets enable these animals to respond to stimuli with behaviors such as swimming and eating—but the response is more like a reflex than a thought process.

The first animals to have primitive brains were flatworms, invertebrates that evolved 550 million years ago and include several human parasites such as tapeworms and flukes. Not coincidentally, these organisms are also some of the first animals to exhibit bilateral symmetry—they have clear right and left halves—and also a defined

Jellyfish Have a Nerve Net

Some organisms, like jellyfish, have a system of nerves that allow them to make a general response to physical contact, but have no central brain.

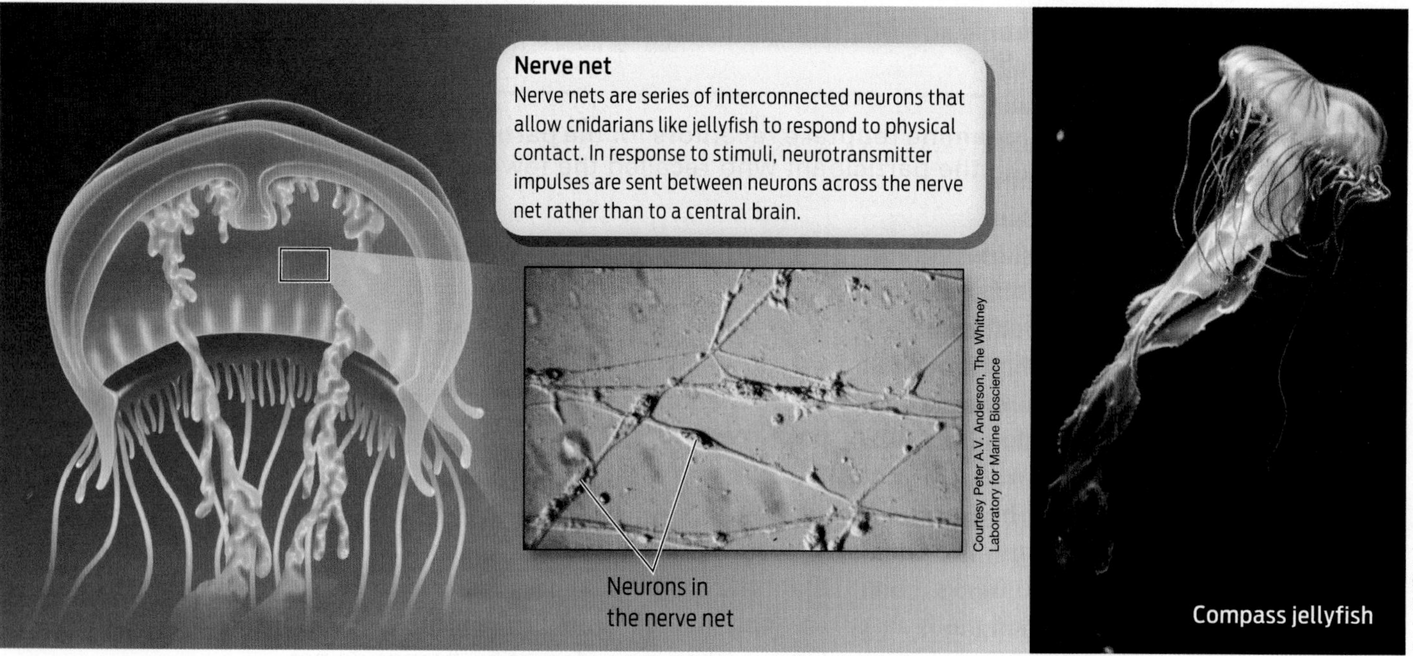
Nerve net
Nerve nets are series of interconnected neurons that allow cnidarians like jellyfish to respond to physical contact. In response to stimuli, neurotransmitter impulses are sent between neurons across the nerve net rather than to a central brain.
Neurons in the nerve net
Courtesy Peter A.V. Anderson, The Whitney Laboratory for Marine Bioscience
Compass jellyfish

ullstein bild/Getty Images

head end. The primitive brain of the flatworm consists of a collection of neurons located at the head end of the organism. The forward location of the flatworm brain allows the organism to sense its environment as it moves forward, and also permits the coordinated movement of its right and left halves. Flatworms also have two large nerves, called nerve cords, that extend along the body and sprout smaller nerves that reach the body tissues—a simple version of a peripheral nervous system. The brain and nerve cords work together to coordinate movement and other responses to the environment.

As animals evolved over time, their brains became larger and more complex, allowing them to perform more-complex functions. In vertebrate animals such as fish, frogs, birds, and humans, brains and peripheral nervous systems are larger and more complex than the primitive brains of flatworms and other invertebrates. But the brains of vertebrate organisms vary, too, which accounts for differences in computational and processing ability. The human cerebrum contains more folds than do the cerebrums of birds and amphibians, for example. The folded structure permits more tissue to

Flatworms Have a Primitive Brain

Flatworms were the first group of organisms to have both a central and a peripheral nervous system.

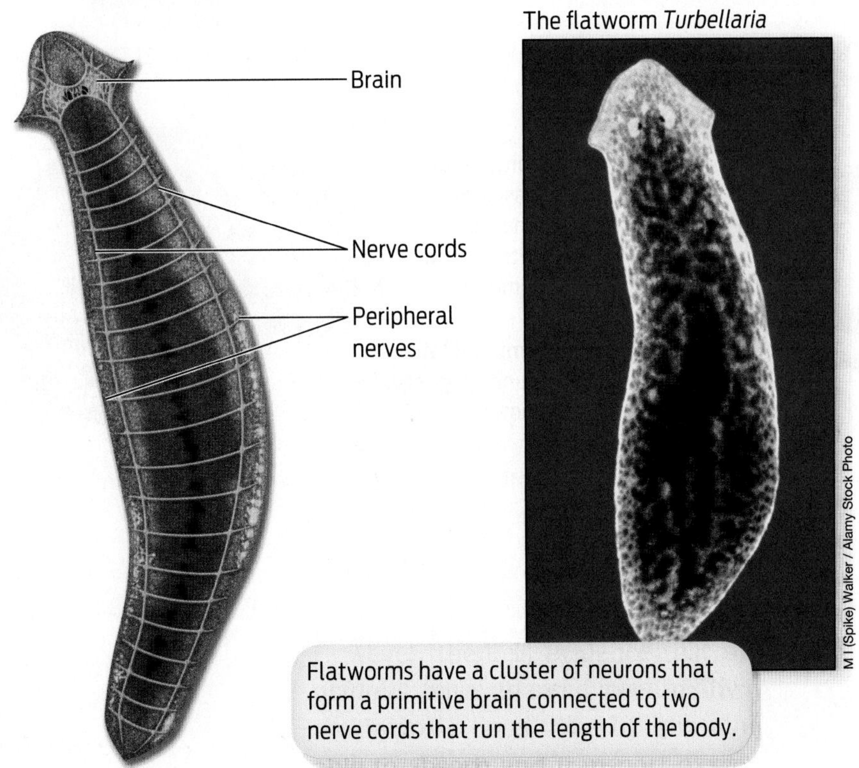
Brain
Nerve cords
Peripheral nerves
The flatworm *Turbellaria*
M I (Spike) Walker / Alamy Stock Photo
Flatworms have a cluster of neurons that form a primitive brain connected to two nerve cords that run the length of the body.

be packed into a given amount of space. In addition, the human brain is organized into hemispheres and lobes. This folded structure and complex organization account for the intellectual advantage that humans have over other organisms. In fact, the relative size and degree of folding of the cerebrum of any organism is associated with higher processing abilities. A fish cerebrum is small and has a relatively smooth surface, whereas mammalian cerebrums, such as those of humans, dolphins, and whales, are larger and contain many more folds.

Vertebrates Have Large and Complex Brains

All vertebrates have both a peripheral and a central nervous system. They have brains that vary in size and complexity.

Cerebrum size and complexity

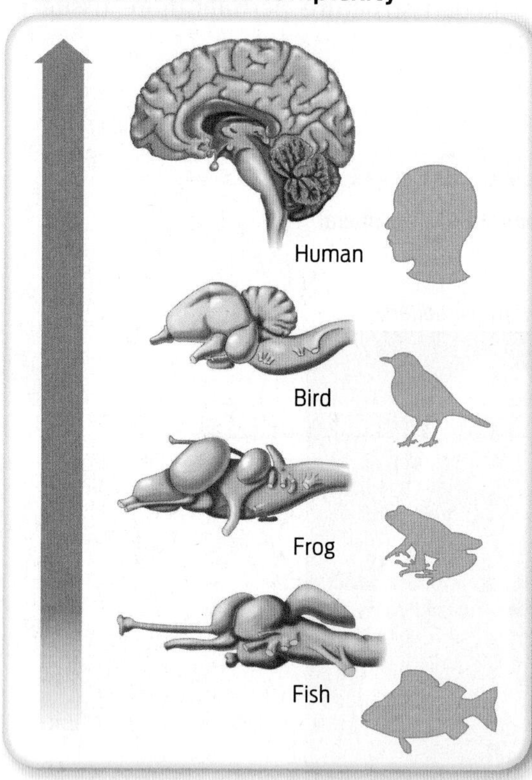

Mammal brains have complex organization

Left hemisphere
This hemisphere controls the right side of the body; it is responsible for linear, rational, and verbal processes.

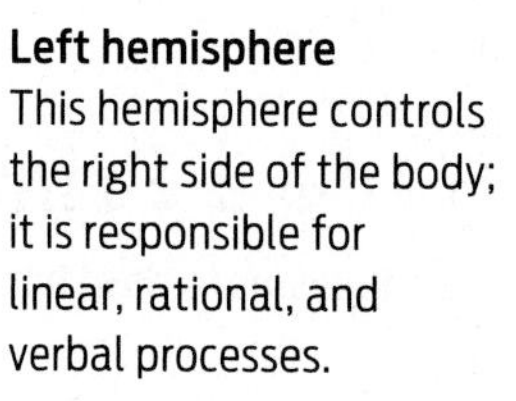

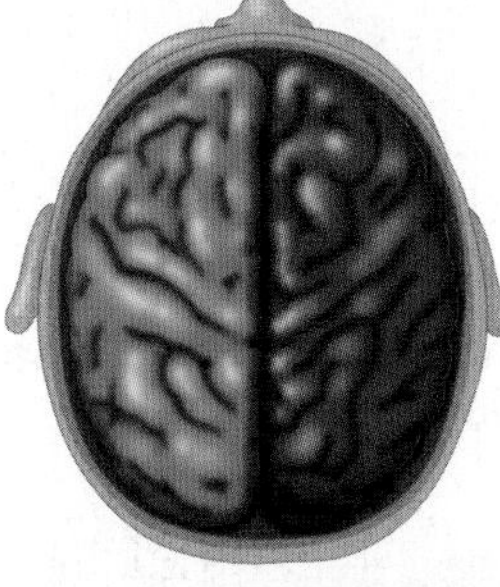

Right hemisphere
This hemisphere controls the left side of the body; it is responsible for artistic, spatial, and musical processes.

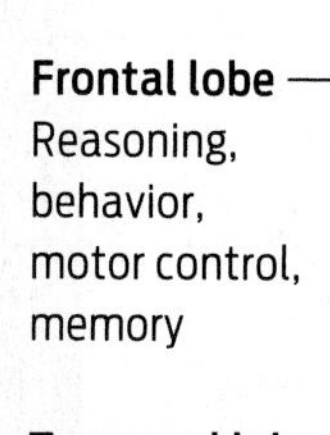

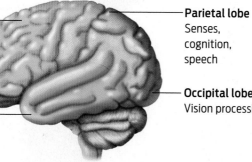

Frontal lobe
Reasoning, behavior, motor control, memory

Temporal lobe
Sound, language, emotion

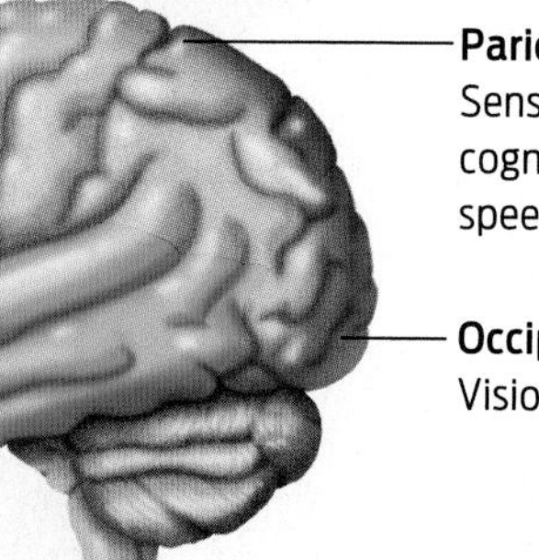

Parietal lobe
Senses, cognition, speech

Occipital lobe
Vision processing

CHAPTER 30 SUMMARY

- The nervous system senses and responds to signals from the environment and coordinates bodily functions, both voluntary ones like moving and thinking, and involuntary ones like breathing and heart rate.
- The central nervous system (CNS) consists of the brain and the spinal cord, a thick bundle of nerves extending from the base of the brain to the lower back; the peripheral nervous system (PNS) consists of all the nerves extending from the spinal cord to the limbs and internal organs.
- The PNS senses and responds to information both inside and outside our bodies. It includes the sensory receptors of our sensory organs, such as the eyes, ears, and skin, as well as the effectors, such as the muscles, that respond to signals sent from the CNS.
- Neurons are specialized cells that consist of a cell body, branched dendrites, a long axon, and axon terminals. Neurons conduct electrical signals called action potentials.
- The coordinated movement of positively charged ions across the axon membrane initiates an action potential. When an action potential reaches the end of a neuron, it causes the neuron to release neurotransmitters.
- Neurotransmitters are chemical signaling molecules released from neuron axon terminals into the synaptic cleft of a synapse and bind to the receptors of other cells—neurons, muscle cells, or endocrine gland cells. Important neurotransmitters include dopamine, serotonin, and acetylcholine.
- Dopamine is a neurotransmitter that produces feelings of pleasure. It is one of the primary neurotransmitters involved in addiction.
- Neurotransmitters bind to receptors on their target cells. The number of receptors can be down-regulated in response to persistently high levels of a neurotransmitter.
- Different parts of the brain coordinate different functions. The four main regions of the brain are the cerebellum, brain stem, diencephalon, and cerebrum. A subregion of the brain called the limbic system makes up the brain's "pleasure center" and has been implicated in addiction.
- All organisms have ways of sensing and responding to their environment, but not all have a central nervous system with a brain and spinal cord.

MORE TO EXPLORE

- **National Institute of Drug Abuse: www.drugabuse.gov/**
- **Volkow, N. D., et al. (2011) Addiction: beyond dopamine reward circuitry. *Proceedings of the National Academy of Sciences* 108(37):15037–15042.**
- **Kringelbach, M. L., and Berridge, K. C. (2010) The functional neuroanatomy of pleasure and happiness. *Discovery Medicine* 9(49):579–587.**
- **Proctor, R. N. (2011) *Golden Holocaust: Origins of the Cigarette Catastrophe and the Case for Abolition.* Berkeley: University of California Press.**
- **Olds J. (1956) Pleasure centers in the brain. *Scientific American* 195:105–116.**

CHAPTER 30 Test Your Knowledge

DRIVING QUESTION 1 How is the nervous system organized?

By answering the questions below and studying Infographics 30.2 and 30.5, you should be able to generate an answer for the broader Driving Question above.

KNOW IT

1 **Mark each of the following structures as a part of the central nervous system (CNS) or of the peripheral nervous system (PNS).**

_________ Light-detecting receptor in the eye

_________ Amygdala

_________ Pain receptor in the skin

_________ Spinal cord

_________ Thalamus

2 **Which part of the brain coordinates movement? Which part of the brain maintains body temperature?**

3 **Rank the complexity of the cerebrums of the following vertebrates from 1 (most complex) to 4 (least complex).**

_________ Bird

_________ Chimpanzee

_________ Salamander (an amphibian)

_________ Fish

USE IT

4 **Is information flow in the spinal cord one way or two way? Explain your answer.**

5 **How does multiple sclerosis cause muscle weakness? Does multiple sclerosis directly affect muscles? Which part of the nervous system is affected?**

6 **A brain injury (caused by a blow to the head, for example) that results in the loss of the ability to speak most likely affected the**

a. cerebellum.
b. cerebrum.
c. diencephalon.
d. brain stem.
e. hypothalamus.

DRIVING QUESTION 2 How do cells in the nervous system transmit signals?

By answering the questions below and studying Infographics 30.3, 30.4, and 30.6, you should be able to generate an answer for the broader Driving Question above.

KNOW IT

7 **Neurons receive information through their**

a. axons.
b. axon terminals.
c. cell bodies.
d. dendrites.
e. nuclei.

8 **Action potentials are a type of _________ signaling that relies on _________.**

a. electrical; neurotransmitters
b. electrical; charged ions
c. electrical; electrons
d. chemical; neurotransmitters
e. chemical; charged ions

9 **Neurons release neurotransmitters from their**

a. cell bodies.
b. dendrites.
c. axon terminals.
d. all of the above
e. b and c

10 **What happens when a neurotransmitter is released into a synaptic cleft?**

11 **Compare and contrast electrical and chemical signaling by neurons.**

USE IT

12 **Gatorade and other sports drinks contain replacement electrolytes (ions necessary to enable muscles to continue to contract, especially the ions lost during sweating). Gatorade contains sodium and potassium ions. Other than in the muscle, where else might these ions be crucial during sustained exercise?**

13 **Botox is a chemical treatment injected into skin to prevent wrinkling. It is a bacterial toxin that prevents certain neurons from releasing the neurotransmitter acetylcholine. Acetylcholine is normally released by motor neurons to signal muscles to contract. Does Botox paralyze muscles in a relaxed state or a contracted state?**

14 **Is more or less of the neurotransmitter acetylcholine released by the axon terminals of neurons in patients with multiple sclerosis compared to those in people who do not have multiple sclerosis? Explain your answer.**

DRIVING QUESTION 3 Why are some drugs (and some behaviors) addictive?

By answering the questions below and studying Infographics 30.1, 30.5, 30.7, and 30.8, you should be able to generate an answer for the broader Driving Question above.

KNOW IT

15 Addictive substances confer a sense of pleasure because they

a. decrease the amount of dopamine in synaptic clefts.
b. increase the amount of dopamine in synaptic clefts.
c. increase the number of dopamine receptors on the axon terminals of cells that release dopamine.
d. increase the number of dopamine receptors on dendrites of cells that release dopamine.
e. c and d

16 Why do drug users need to take ever-increasing amounts of drugs to get the same high?

USE IT

17 Cocaine prevents dopamine from being removed from the synapse. Why does this cause feelings of pleasure?

18 Would you expect a person born with a relatively low number of dopamine receptors to be happier or sadder than the average? Explain your answer.

MINI CASE

apply YOUR KNOWLEDGE

19 Parkinson's disease is caused primarily by a gradual loss of dopamine-producing neurons in the brain. Why is depression often among the debilitating symptoms of Parkinson's disease? There are a variety of medications available to treat people with Parkinson's disease. Do some Internet research to match each medication listed in the left column with its probable mechanism of action.

_____ Mirapex	a. can be used by neurons to make dopamine
_____ Eldepryl	b. binds to and activates the dopamine receptor
_____ Levodopa	c. inhibits an enzyme that breaks down dopamine

How can drugs with different mechanisms of action all help treat Parkinson's disease? (Hint: What do all the underlying mechanisms have in common in terms of their effect?)

INTERPRETING DATA

apply YOUR KNOWLEDGE

20 Study Infographic 30.8.

a. What are the independent and dependent variables in each experiment?
b. The performances of 14 participants are plotted in each graph. Calculate the average time (in seconds) these participants took to walk the straight line, and the average number of words recalled by these participants.
c. In the right-hand graph in Infographic 30.8, you will note that there are a few outliers (two are circled in red in the copy of the graph shown below). For the outlier in the lower left, is this participant recalling more or fewer words than would be predicted based on his or her availability of dopamine receptors? Explain your answer.
d. For the outlier on the top right, is this participant recalling more or fewer words than would be predicted based on his or her availability of dopamine receptors? Explain your answer.

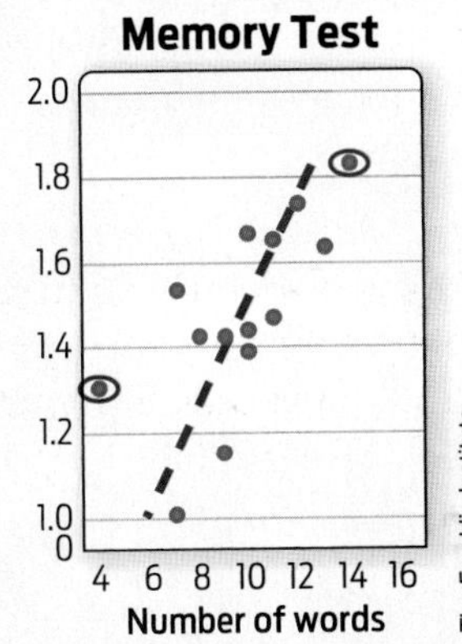

BRING IT HOME

apply YOUR KNOWLEDGE

21 Replicate the motor skills experiments shown in Infographic 30.8 with some students. On a flat surface, set up a start line and, 10 yards away, a finish line. Instruct each subject to walk in a straight line from the start to the finish and back as fast as they can without running. Start timing when they cross the start line, and stop timing when they cross the start line again at the end of the trial. Have each participant do the trial three times, and record the average of the three trials as the participant's final time. Now calculate the average time in your set of participants. How does this compare to the 14 methamphetamine users whose performances are plotted in Infographic 30.8?

Method from Robertson, K. R., et al. (2006) Timed gait test: normative data for the assessment of the AIDS dementia complex. *Journal of Clinical and Experimental Neuropsychology* 28:1053–1064.

Too Many Multiples?

The birth of octuplets raises questions about the fertility business

DRIVING QUESTIONS

1. What is the anatomy of the male and female reproductive tracts, and how does the anatomy contribute to the function of the reproductive system?
2. What hormones are involved in reproduction, and how do they work in the reproductive systems of males and females?
3. What are the different types of assisted reproduction, and how do they work?

The live birth of octuplets is an extremely rare event, having occurred only once in recorded U.S. history before the year 2000. So the arrival of a second set in a California hospital in January 2009 was greeted with fanfare. Headlines read "Octuplets Stun Doctors" and "Eight Babies!"

But days after news of the miracle multiple birth spread worldwide, the public reception turned sour when it came to light that the 33-year-old mother, Nadya Suleman, already had six children all under the age of 7 who were, like the octuplets, conceived using a form of assisted reproductive technology called in vitro fertilization.

Even more disturbing to some, Suleman was an unemployed single mother on welfare. The public outcry was fierce. How could she support her children? Was she psychologically disturbed? And why had her doctor agreed to give her fertility treatments when she already had six children?

The case cast a spotlight on the business of fertility treatment: in the United States, fertility clinics are largely unregulated. Although the American Society of Reproductive Medicine (ASRM) issues guidelines on how doctors should administer fertility services, in most states there are no laws regulating what doctors can and cannot do in this regard. Though critics called Suleman's doctor irresponsible, he had not violated any laws.

Nancy Pastor/Polaris Images

Multiple Births Have Become More Common

The rate of pregnancies resulting in multiple births has increased substantially since 1980. The increase is most dramatic in women over the age of 35.

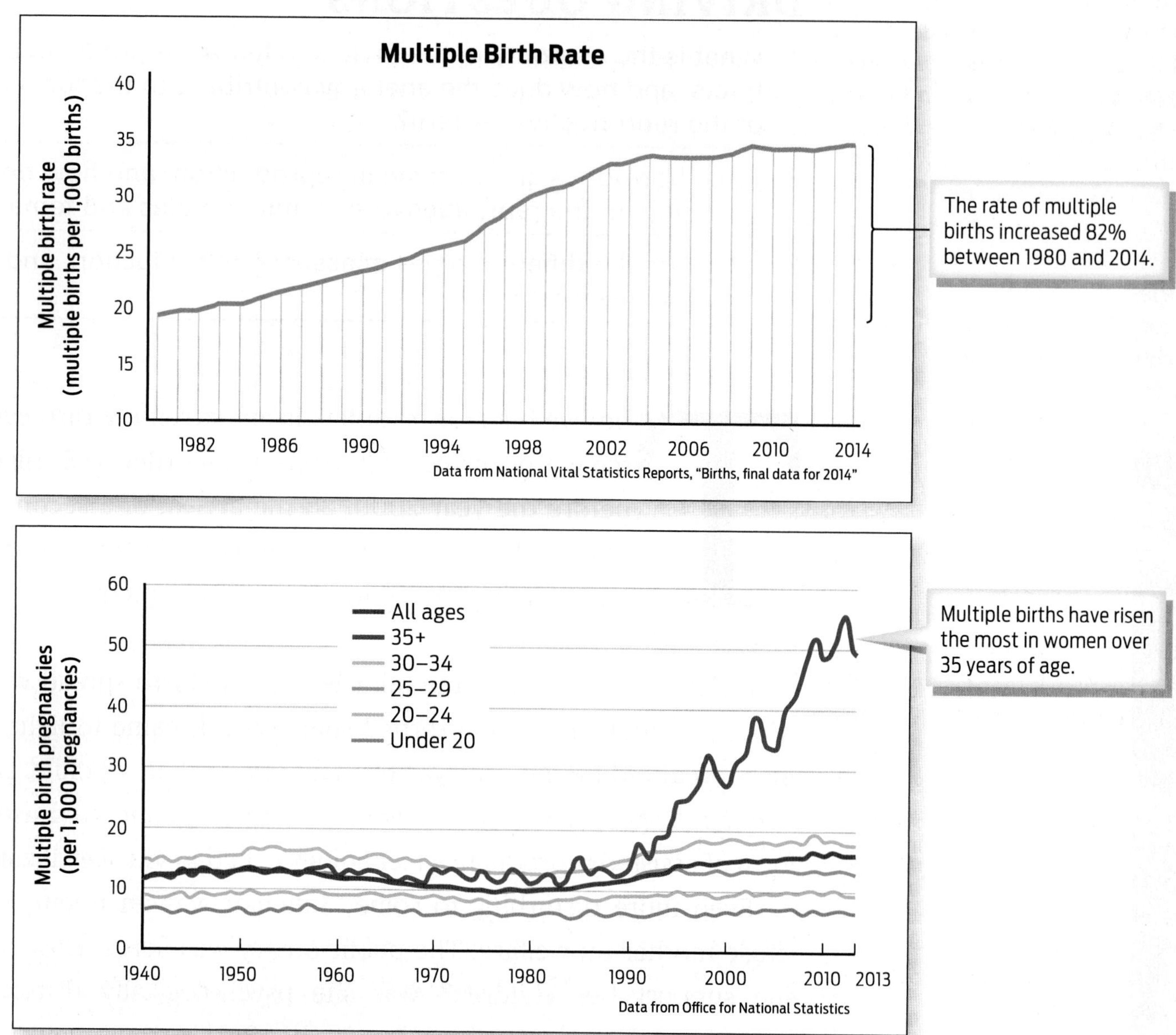

Infertility treatment wouldn't be nearly as controversial if it didn't increase the odds that a woman will conceive more than one child during the treatment. "Multiples," as these babies are called, are often born prematurely. Consequently, they are underweight and have underdeveloped organs, and so are at risk for birth defects such as cerebral palsy and infant respiratory distress syndrome. Carrying multiples also increases the risk that the mother will develop dangerously high blood pressure, diabetes, vitamin deficiencies, or other medical conditions during her pregnancy that may affect her health or that of her unborn children. Even women who have twins have a higher risk of medical complications during pregnancy, and their babies are at a higher risk of premature birth than are singletons. Suleman's eight children are the only surviving set of octuplets ever.

Multiple births without fertility treatment are extremely rare. Scientists estimate that the incidence of natural triplets is 1 in 6,000 to 8,000 births–of quadruplets, 1 in 500,000 births. The incidence of higher-order multiple births is even rarer. But because the use of assisted reproductive technologies has skyrocketed, so, too, have the number of multiple births. The number of triplet births rose 400% between 1980 and 1998, according to statistics compiled by the Centers for Disease Control and Prevention. The medical complications associated with such pregnancies place a large burden on the health care system.

Although there have been efforts to improve assisted reproductive technology and therefore reduce the likelihood of multiple births, extreme cases such as Suleman's have drawn lawmakers to the scene. Many states are considering legislation that would place restrictions on fertility doctors. Working from another perspective, some groups are fighting for legislation that would require health insurance providers to pay for fertility treatments. Because many patients pay out of pocket, patients with limited financial resources can, because of their circumstances, pressure doctors to be aggressive with treatment, despite the health risks associated with carrying multiples. The goal of this legislation is to make assisted reproduction safer for everyone, not just for those who can afford it.

Fertility and Infertility

Suleman, who grew up as an only child, said she had always wanted a large family to help make up for her lonely childhood. She told Ann Curry on NBC's *Today* show in 2009 that she had tried to become pregnant for several years but had been unable to. That's when she turned to a fertility doctor for help.

In the same interview, Suleman said she suffered from a medical condition that prevented her from conceiving a child naturally. If she did, she wasn't alone. In the United States, an estimated one out of eight couples–about 7.3 million women and their partners–experience infertility, which is defined as the inability to conceive within a year or to bring a pregnancy to term. Many things can cause infertility: advanced age, infections, hormonal imbalances, chromosomal abnormalities, and physical blockage of reproductive passages. Men can suffer from a low sperm count or have abnormal sperm. In many cases, the reason for a couple's infertility remains unknown, and most couples are unaware that they have a fertility problem until they begin trying to conceive a child and can't.

According to the ASRM, modern medicine can offer treatment to 90% of infertile

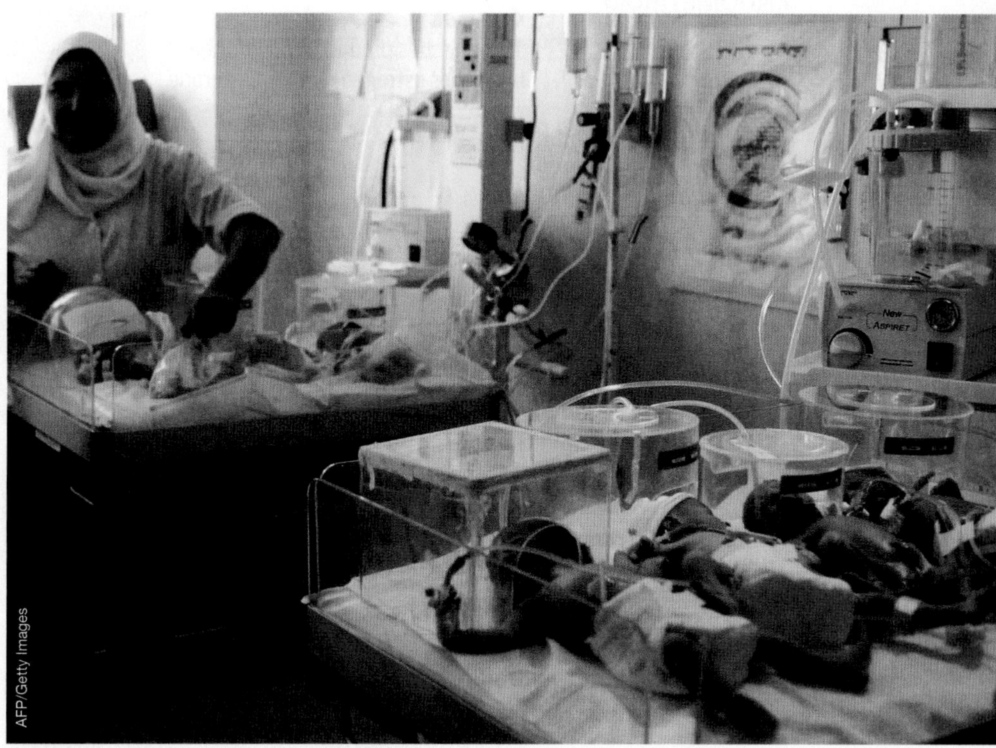

Multiples are often born prematurely and underweight, as were these septuplets born in Alexandria, Egypt, in 2008.

OVARIES
Paired female reproductive organs; the ovaries contain eggs and produce estrogen and progesterone.

ESTROGEN
A female sex hormone produced by the ovaries that supports female sexual development and function.

PROGESTERONE
A female sex hormone produced by the corpus luteum of the ovary that prepares and maintains the uterus for pregnancy.

OVIDUCT
The tube connecting an ovary and the uterus in females. Eggs are ovulated into and fertilized within the oviducts.

UTERUS
The muscular organ in females in which a fetus develops.

ENDOMETRIUM
The lining of the uterus.

TESTES
Paired male reproductive organs, which contain sperm and produce androgens (primarily testosterone).

SCROTUM
The sac in which the testes are held.

couples. But fertility isn't an exact science, and treatment isn't always effective. There are many organs and hormones involved in human reproduction, and communication among them is a highly orchestrated process; even in the best circumstances, successful pregnancies require a bit of lucky timing.

The female reproductive system consists of two **ovaries** and accessory structures that facilitate fertilization and pregnancy. The ovaries contain eggs and produce **estrogen** and **progesterone,** the major female sex hormones. Each month, one ovary typically releases one egg into an adjacent organ called the **oviduct**, also known as the fallopian tube. The egg travels via the oviduct to the **uterus,** an elastic muscular compartment where a fetus may develop should an egg be fertilized by sperm. The tissue that lines the uterus is called the **endometrium,** which becomes enriched with blood vessels to support a potential pregnancy **(INFOGRAPHIC 31.1).**

In males, paired glands called **testes** (or testicles) produce sperm. The testes are contained in a sac of skin called the **scrotum** that hangs outside the body (an arrangement that keeps the testes slightly cooler than body temperature, ensuring proper sperm development). The testes

INFOGRAPHIC 31.1 Female Reproductive System

The female reproductive system consists of the ovaries and accessory reproductive structures. The ovaries produce eggs and hormones. Accessory structures facilitate fertilization and pregnancy.

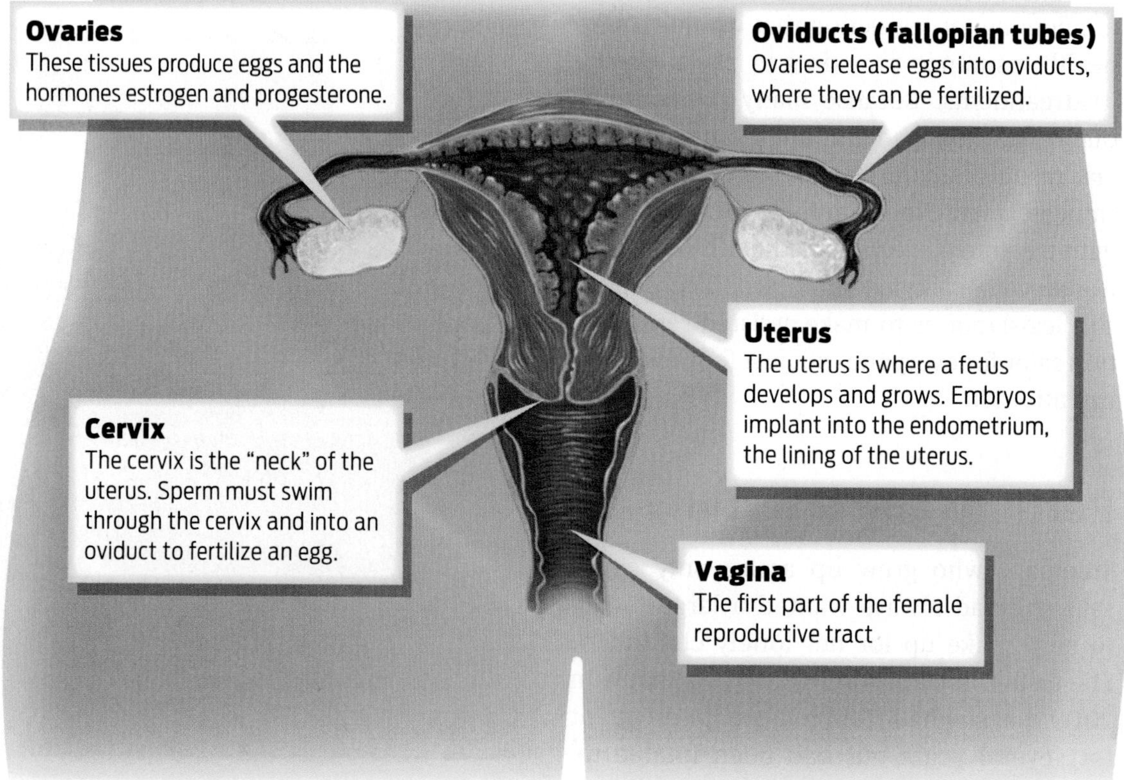
Ovaries
These tissues produce eggs and the hormones estrogen and progesterone.
Oviducts (fallopian tubes)
Ovaries release eggs into oviducts, where they can be fertilized.
Uterus
The uterus is where a fetus develops and grows. Embryos implant into the endometrium, the lining of the uterus.
Cervix
The cervix is the "neck" of the uterus. Sperm must swim through the cervix and into an oviduct to fertilize an egg.
Vagina
The first part of the female reproductive tract

? Where does fertilization, if it occurs, take place? Where will the embryo develop?

produce **testosterone,** the primary male sex hormone. Each testis contains tightly coiled **seminiferous tubules** within which sperm develop. Remarkably, each testis contains approximately 250 meters of seminiferous tubules. Sperm travel through the seminiferous tubules and enter the **epididymis,** where they mature and are stored until ejaculated. From the epididymis, sperm travel through paired tubes called the **vas deferens** and exit the body through the **urethra,** which ends at the tip of the penis. Along the way, the prostate and other glands add fluid to sperm that helps the sperm survive in the female reproductive tract. The fluid contains the sugar fructose as an energy source, bases that help buffer the acidic pH in the vagina and protect the sperm DNA from being damaged, and mucus that helps keep the sperm mobile in the female reproductive tract. The mixture of ejaculated sperm and accompanying fluid is called **semen** (**INFOGRAPHIC 31.2**).

During sex, when a man ejaculates through the penis into a woman's **vagina,** sperm swim up the reproductive tract, through the opening of the uterus called the **cervix,** and into the oviducts. The oviducts are where **fertilization**—the fusion of egg and

TESTOSTERONE
The primary male sex hormone, which stimulates the development of masculine features and plays a key role in sperm development.

SEMINIFEROUS TUBULES
Coiled structures that constitute the bulk of the testes and in which sperm develop.

EPIDIDYMIS
A system of tubes in which sperm mature and are stored before ejaculation.

VAS DEFERENS
Paired tubes that carry sperm from the testes to the urethra.

URETHRA
A tube that connects the bladder to the genitals and carries urine out of the body. In males, the urethra travels through the penis and also carries sperm.

SEMEN
The mixture of fluid and sperm that is ejaculated from the penis.

VAGINA
The first part of the female reproductive tract, extending to the cervix; also known as the birth canal.

CERVIX
The opening or "neck" of the uterus, where sperm enter and babies exit.

FERTILIZATION
The fusion of an egg and a sperm; the resulting cell is called a zygote.

INFOGRAPHIC 31.2 Male Reproductive System

Paired testes produce sperm and hormones. Sperm travel through a series of ducts and are ejaculated through the urethra.

Testes
Within the testes are seminiferous tubules, which produce sperm. Mature sperm travel through the epididymis to the vas deferens.

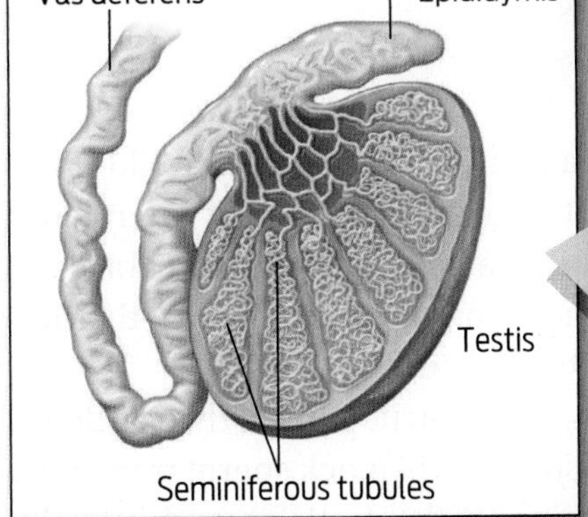

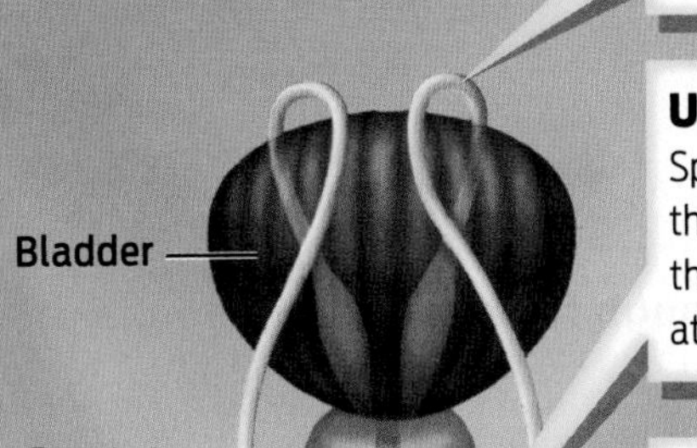

Vas deferens
Each vas deferens carries sperm from a testis to the urethra.

Urethra
Sperm leave the body through the urethra. Urine also passes through the urethra, but not at the same time.

Penis
This organ enables sperm to be delivered to the female vagina.

Scrotum
The testes are contained in a sac called the scrotum, which hangs outside the body.

? Trace the path of sperm from the seminiferous tubules to the end of the urethra.

INFOGRAPHIC 31.3 Fertilization Occurs in the Oviduct

During intercourse, sperm ejaculated from the penis enters the female reproductive tract. Sperm must swim through the tract into the oviducts to fertilize an ovulated egg. While many sperm may make it to an oviduct, only one will actually fertilize an egg. Blockages in the female reproductive tract can impede sperm passage to an egg, and consequently compromise fertilization.

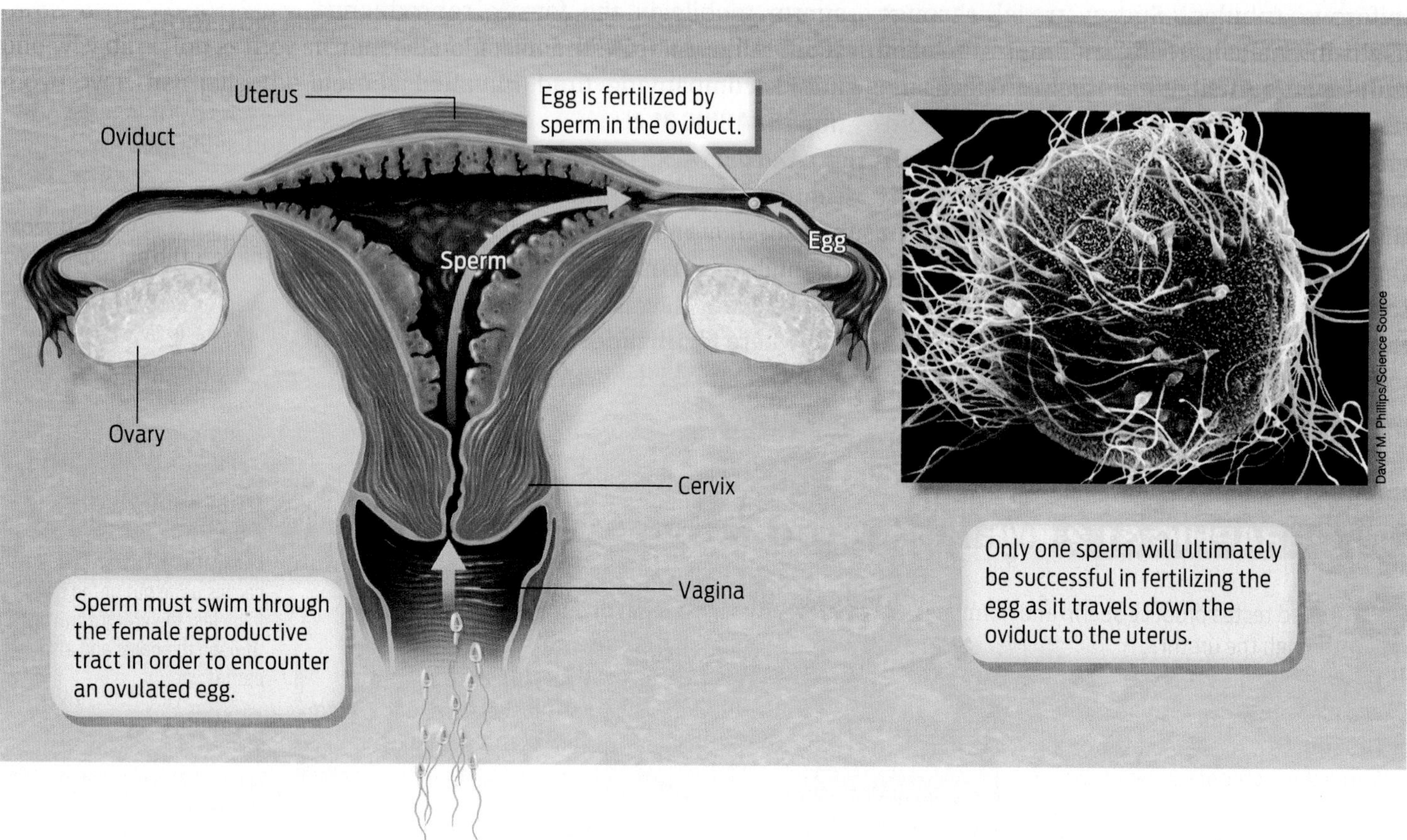
Uterus
Oviduct
Egg is fertilized by sperm in the oviduct.
Sperm
Egg
Ovary
Cervix
Vagina
Sperm must swim through the female reproductive tract in order to encounter an ovulated egg.
Only one sperm will ultimately be successful in fertilizing the egg as it travels down the oviduct to the uterus.
David M. Phillips/Science Source

? What structure do sperm have to pass through from the vagina to the uterus? Once sperm are in the uterus, is their journey complete?

sperm—occurs. Only a single sperm will be successful in fertilizing an egg (**INFOGRAPHIC 31.3**).

That's the way conception normally happens—but several things can go wrong along the way. Physical damage to the reproductive organs can prevent fertilization. In some cases, a woman's oviducts can be blocked or damaged, preventing eggs from entering the uterus or sperm from getting into the oviduct, where fertilization normally takes place. The most common cause of blocked tubes is pelvic inflammatory disease, which can be caused by sexually transmitted diseases such as chlamydia, a bacterial infection. An untreated infection can cause scar tissue to build up in the oviducts and block them. In interviews, Suleman said that she suffered from fibroids—benign tumors that grow in muscle—that left her oviducts scarred.

In men, any obstructions in the vas deferens or epididymis will block sperm transport. Varicose veins in the testicles and sexually transmitted bacterial infections such as chlamydia or gonorrhea can also block tubes.

Fertility specialists can test for physical blockages and, in some cases, surgically correct them. When surgery isn't an option, **in vitro fertilization (IVF)** is often

IN VITRO FERTILIZATION (IVF) A form of assisted reproduction in which eggs and sperm are brought together outside the body and the resulting embryos are inserted into a woman's uterus.

recommended. In IVF, hormones are administered to a woman to promote egg development, and then eggs are extracted from her ovaries through a needle inserted through the vagina. Sperm are extracted from ejaculate or, in cases of physical blockage, from the epididymis. The sperm and eggs are combined outside the body in a petri dish to allow the sperm to fertilize the eggs. The fertilized eggs begin to divide, and the resulting **embryos** are then inserted into the woman's uterus in hope that at least one will develop into a fetus. IVF has been used to help infertile couples conceive children since 1978, when Louise Brown, the first "test-tube baby" conceived with IVF, was born **(INFOGRAPHIC 31.4)**.

IVF may also be recommended in cases of low sperm number, abnormal sperm, and

EMBRYO
An early stage of development; an embryo forms when a zygote undergoes cell division.

INFOGRAPHIC 31.4 In Vitro Fertilization

In vitro fertilization involves extracting eggs and sperm and combining them outside the body to allow fertilization. The resulting embryos are then inserted back into a woman's uterus in hope that at least one of them will implant into the uterus and grow into a fetus.

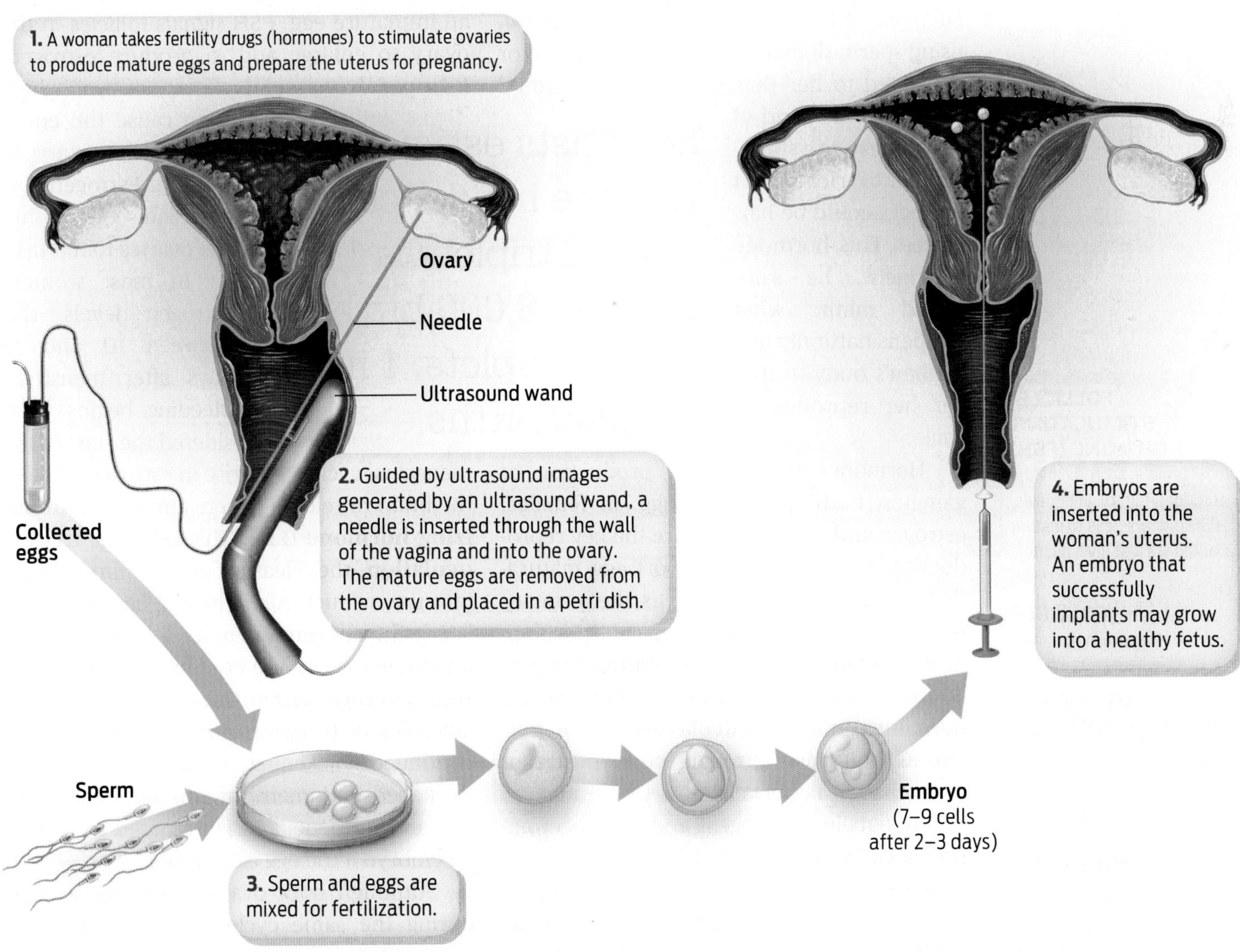

1. A woman takes fertility drugs (hormones) to stimulate ovaries to produce mature eggs and prepare the uterus for pregnancy.
Ovary
Needle
Ultrasound wand
Collected eggs
2. Guided by ultrasound images generated by an ultrasound wand, a needle is inserted through the wall of the vagina and into the ovary. The mature eggs are removed from the ovary and placed in a petri dish.
Sperm
3. Sperm and eggs are mixed for fertilization.
Embryo
(7–9 cells after 2–3 days)
4. Embryos are inserted into the woman's uterus. An embryo that successfully implants may grow into a healthy fetus.

? In IVF, which processes occur in their normal locations? Which occur in other places? (Hint: Think about egg production, fertilization, and embryo implantation.)

even in cases in which the cause of infertility can't be determined. In such cases the term "infertility" can actually be a misnomer. Many couples with defective sperm or unexplained infertility can still conceive a child naturally—it just may take longer. But because no one can predict how long it might take to achieve a successful pregnancy and because fertility decreases with age, IVF makes conception more likely by bringing sperm and egg together artificially.

Hormones and Pregnancy

Suleman had her first round of IVF in 2000, using sperm donated by a friend. Her doctor explained to her that the procedure would begin with a round of hormones to stimulate her ovaries so that her eggs could be harvested. This hormone treatment, he said, would mimic what happens naturally in a woman's body to trigger her reproductive cycle.

Scientists estimate that the incidence of natural triplets is 1 in 6,000 to 8,000 births—of quadruplets, 1 in 500,000 births.

Hormones regulate the production of gametes, both sperm and egg. In females, estrogen and progesterone are the key reproductive hormones that support egg maturation. In males, testosterone is the primary hormone that stimulates sperm to develop. As we saw in Chapter 26, hormones are produced by endocrine glands, which secrete hormones into the circulation. These hormones then travel through the bloodstream to reach their target cells.

In females, estrogen and progesterone produced by the ovaries drive the menstrual cycle, a reproductive cycle that repeats roughly once every 28 days after the onset of puberty. During each cycle, estrogen and progesterone levels rise and fall, triggering the ovaries to release an egg and prepare a woman's uterus for pregnancy should an egg be fertilized.

The brain's hypothalamus controls levels of estrogen and progesterone in the body and therefore is the ultimate regulator of fertility in females. The hypothalamus works closely with the anterior pituitary, which sits just below the hypothalamus in the brain. The hypothalamus secretes hormones that act on the anterior pituitary, causing it to produce two hormones of its own. These hormones, called follicle-stimulating hormone and luteinizing hormone, travel through the bloodstream and directly stimulate the ovaries.

In women, **follicle-stimulating hormone (FSH)** acts on structures in the ovaries called **follicles,** each of which contains an immature egg. FSH signals follicles in the ovary to enlarge and to produce estrogen. Estrogen has several effects, one of which is to cause the endometrium to start to thicken. Estrogen also stimulates eggs within the ovaries to mature.

In most women, estrogen levels rise between 10 and 14 days after menstrual bleeding begins (considered the start of the menstrual cycle). This rise in estrogen triggers the brain to release a large amount of **luteinizing hormone (LH).** This LH surge triggers **ovulation**—the release of an egg from a follicle into the oviduct. After the egg has been ovulated, the remaining empty follicle becomes a structure called the **corpus luteum,** which secretes progesterone. One of the most important roles of progesterone is to promote the continued thickening of endometrium. The thickened endometrium contains blood vessels and nutrients and is prepared to receive an embryo if the egg is fertilized.

Although both ovaries can release eggs during the same cycle, they typically take turns and only one egg is released per cycle. In about 1% of cycles, however, there are multiple ovulations, in which case fraternal twins, triplets, or higher multiples can develop. (Identical

FOLLICLE-STIMULATING HORMONE (FSH)
A hormone secreted by the anterior pituitary. In females, FSH triggers eggs to mature at the start of each monthly cycle.

FOLLICLE
A structure in the ovary where eggs mature.

LUTEINIZING HORMONE (LH)
A hormone secreted by the anterior pituitary. In females, a surge of LH triggers ovulation.

OVULATION
The release of an egg from an ovary into the oviduct.

CORPUS LUTEUM
The structure in the ovary that remains after ovulation. It secretes progesterone.

twins occur when a single egg is released and fertilized and then splits into two embryos early in embryonic development.)

Ovulation presents a crucial time window during which a woman can become pregnant. Sperm must swim through the cervix and uterus and into the oviduct containing the released egg in order to fertilize it. Because sperm can survive in the female reproductive tract anywhere from 3 to 7 days, a woman can become pregnant even if she has sex before she ovulates. Sperm can, in effect, wait in the oviduct for an egg to be ovulated. Once the egg leaves the oviduct and enters the uterus, however, the odds that it will be fertilized are extremely small.

If an egg is not fertilized within 24 hours of ovulation, it is no longer viable. The corpus luteum degenerates at about day 26 of the cycle, progesterone levels drop, and the uterine lining sloughs off in the **menstruation** that follows **(INFOGRAPHIC 31.5)**.

MENSTRUATION
The shedding of the uterine lining (the endometrium) that occurs when an embryo does not implant.

INFOGRAPHIC 31.5 Hormones Regulate the Menstrual Cycle

A complex interplay of hormones from the hypothalamus, anterior pituitary gland, and ovaries drives the monthly female reproductive cycle.

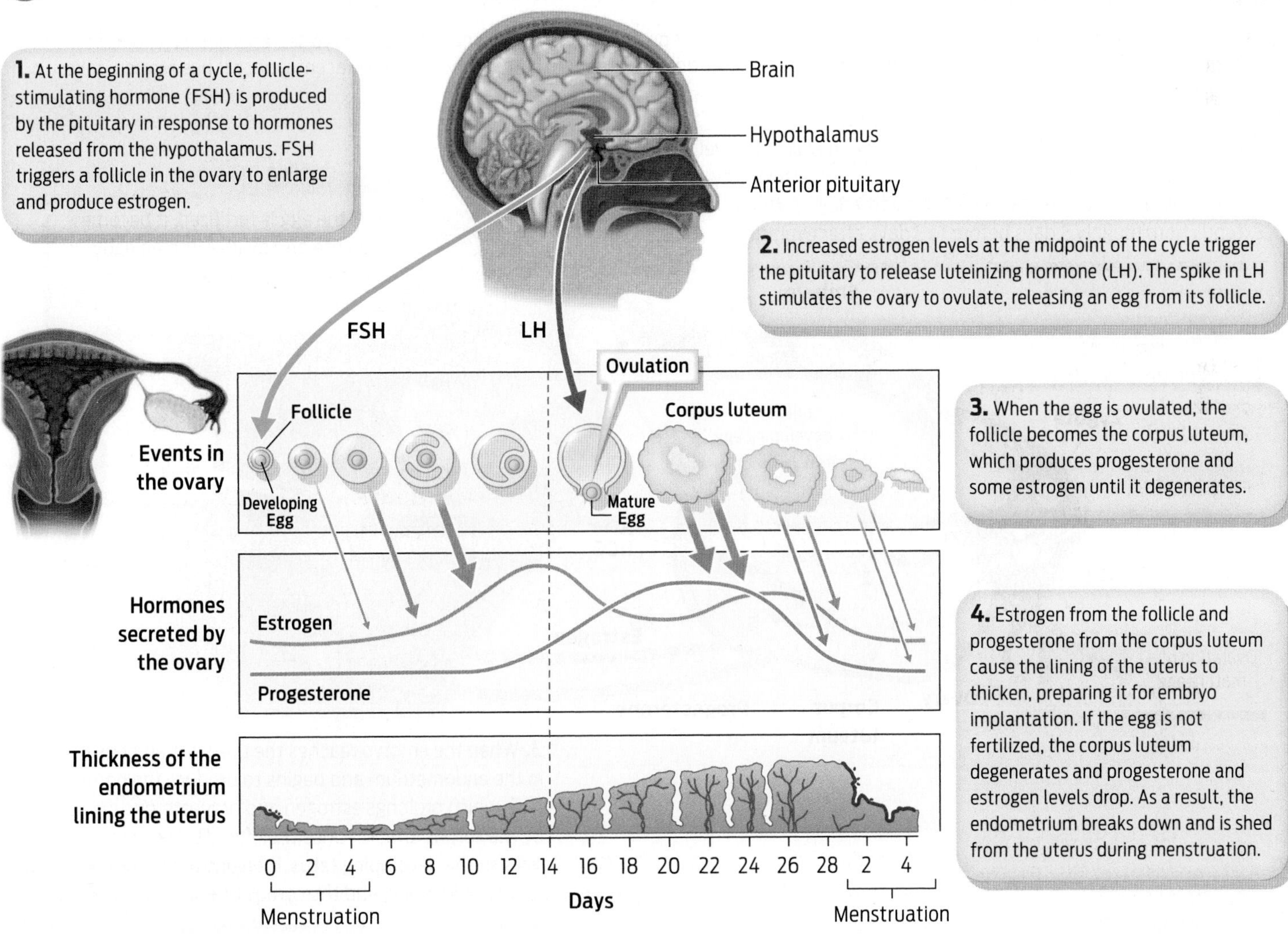

1. At the beginning of a cycle, follicle-stimulating hormone (FSH) is produced by the pituitary in response to hormones released from the hypothalamus. FSH triggers a follicle in the ovary to enlarge and produce estrogen.
Brain
Hypothalamus
Anterior pituitary
2. Increased estrogen levels at the midpoint of the cycle trigger the pituitary to release luteinizing hormone (LH). The spike in LH stimulates the ovary to ovulate, releasing an egg from its follicle.
FSH
LH
Ovulation
Follicle
Corpus luteum
Events in the ovary
Developing Egg
Mature Egg
3. When the egg is ovulated, the follicle becomes the corpus luteum, which produces progesterone and some estrogen until it degenerates.
Hormones secreted by the ovary
Estrogen
Progesterone
4. Estrogen from the follicle and progesterone from the corpus luteum cause the lining of the uterus to thicken, preparing it for embryo implantation. If the egg is not fertilized, the corpus luteum degenerates and progesterone and estrogen levels drop. As a result, the endometrium breaks down and is shed from the uterus during menstruation.
Thickness of the endometrium lining the uterus
0 2 4 6 8 10 12 14 16 18 20 22 24 26 28 2 4
Days
Menstruation
Menstruation

? For each of the four hormones (LSH, FH, estrogen, and progesterone), state where it is produced and what effect it has on its target.

ZYGOTE
A fertilized egg.

HUMAN CHORIONIC GONADOTROPIN (hCG)
A hormone produced by an early embryo that helps maintain the corpus luteum until the placenta develops.

PLACENTA
A structure made of fetal and maternal tissues that helps sustain and support the embryo and fetus.

A fertilized egg is called a **zygote**. As the zygote travels to the uterus it begins dividing and developing into an embryo. Typically, the embryo implants in the endometrium of the uterus about 1 week after the egg is fertilized. Once implanted in the uterus, the embryo secretes a hormone called **human chorionic gonadotropin (hCG),** which signals the corpus luteum to continue producing progesterone, which supports the thickening endometrium. This is the hormone that, in effect, tells the reproductive system that pregnancy has begun (hCG is the hormone that most pregnancy tests use as a marker to detect a pregnancy).

Once the embryo implants, embryonic and maternal endometrial tissues interact to form the **placenta,** a disc-shaped structure that provides nourishment and support to the developing **fetus,** as the embryo is now called. In addition to delivering oxygen, nutrients, and other key molecules like antibodies from the mother that help protect the embryo against infections, the placenta eventually takes over from the corpus luteum the task of producing estrogen and progesterone **(INFOGRAPHIC 31.6)**.

INFOGRAPHIC 31.6 Hormones Support Pregnancy

Estrogen and progesterone support the implanted embryo as it develops. Early in pregnancy, the embryo secretes human chorionic gonadatropin (hCG), which signals the corpus luteum to continue to produce these hormones. The placenta, once it has formed, takes over estrogen and progesterone production.

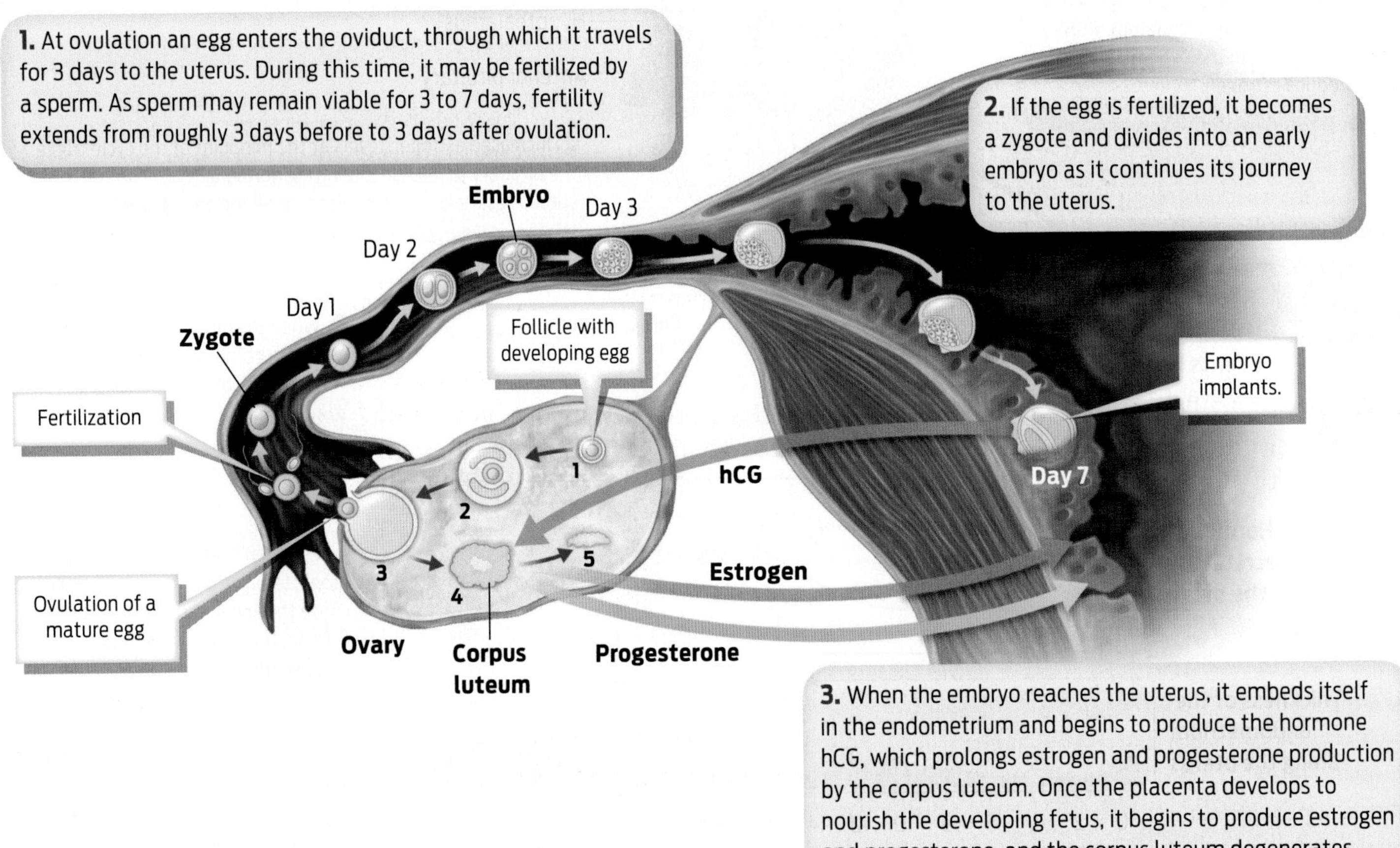

1. At ovulation an egg enters the oviduct, through which it travels for 3 days to the uterus. During this time, it may be fertilized by a sperm. As sperm may remain viable for 3 to 7 days, fertility extends from roughly 3 days before to 3 days after ovulation.
2. If the egg is fertilized, it becomes a zygote and divides into an early embryo as it continues its journey to the uterus.
Embryo
Day 3
Day 2
Day 1
Zygote
Follicle with developing egg
Fertilization
Embryo implants.
1
hCG
Day 7
2
3
5
Estrogen
4
Ovulation of a mature egg
Ovary
Corpus luteum
Progesterone
3. When the embryo reaches the uterus, it embeds itself in the endometrium and begins to produce the hormone hCG, which prolongs estrogen and progesterone production by the corpus luteum. Once the placenta develops to nourish the developing fetus, it begins to produce estrogen and progesterone, and the corpus luteum degenerates.

? What makes hCG? Why is it important during early pregnancy?

Even if a woman never becomes pregnant, her hormonal cycles will continue until the ovaries stop responding to follicle-stimulating hormone and luteinizing hormone at the time of menopause, which in American women occurs at about age 51; at this time women stop ovulating and stop having monthly reproductive cycles, and are therefore no longer able to conceive.

Because hormones play such a crucial role in pregnancy and reproduction, many types of **contraception** are designed to interfere with the normal female hormone cycle. Most birth control pills contain both estrogen and progesterone at levels that prevent the anterior pituitary from releasing follicle-stimulating and luteinizing hormones. This prevents ovulation and consequently a woman taking this pill does not release eggs. Progesterone in birth control pills prevents successful pregnancy in other ways, too: it thickens the cervical mucus, blocking sperm from entering the uterus and oviducts, and also reduces endometrial thickening—a process that is necessary to support an embryo (**TABLE 31.1**).

Men do not have a monthly hormone cycle, but the male gametes (sperm) develop under the influence of the same hormones that regulate the menstrual cycle in women. Beginning in puberty, hormones from the hypothalamus act on the anterior pituitary, stimulating it to release FSH and LH. FSH and LH act on specific cell types in the testes, resulting in the production of testosterone, which is essential for spermatogenesis. The seminiferous tubules house precursor sperm cells that go through meiosis (see Chapter 11) and specialization to produce mature sperm cells. It takes approximately 6 weeks for sperm to mature. Maturation is stimulated by testosterone, and although men produce testosterone continuously throughout their adult lives, they produce slightly less of the hormone as they age and consequently sperm production declines over time (**INFOGRAPHIC 31.7**).

The Many Causes of Infertility

Given how important hormones are to egg maturation and ovulation, it's not surprising that hormonal imbalances are a common cause of female infertility. Low levels of luteinizing hormone and follicle-stimulating hormone, for example, may prevent ovulation or make it erratic. Even slight irregularities in the hormone system can prevent the ovaries from releasing eggs. Specific causes of such hormonal imbalances include injury, tumors, excessive exercise, and starvation. Some medications are associated with ovulation disorders, and some studies have shown that stress can negatively affect fertility, as can poor nutrition.

> By age 40, nearly half of all women have difficulty conceiving.

Some women suffer from polycystic ovary syndrome, a condition defined by the presence of multiple ovarian cysts that impair ovulation and associated with the production of excessive amounts of **androgens** (the "male" sex hormones, typically present in lower amounts in females). This syndrome is one of the most common hormonal disorders, affecting an estimated 10% of women of reproductive age. In addition to impairing ovulation, the condition is associated with irregular menstrual cycles, diabetes, and obesity.

One of the most common causes of infertility in women, especially those living in Western industrialized societies, is advanced maternal age (typically defined as over 35). Women often have a harder and harder time becoming pregnant as they age, and after age 35 they may choose to consult a fertility specialist to assist in becoming pregnant. Age-related declines in fertility occur in part

FETUS
After the eighth week after fertilization, the embryo is referred to as a fetus.

CONTRACEPTION
The prevention of pregnancy through physical, surgical, or hormonal methods.

ANDROGEN
A class of sex hormones, including testosterone, that is present in higher levels in men than in women and causes male-associated traits like deep voice, growth of facial hair, and defined musculature.

TABLE 31.1 Contraception

There are many ways to prevent conception. Currently available contraceptives include behavioral methods, physical and chemical barriers, hormones, and surgery.

Pregnancies per 100 women in 1 year	METHOD	DESCRIPTION
<1	**Abstinence:** 0% failure rate Photomac/FeaturePics.com	No sexual intercourse
	Intrauterine device: 0.2%–0.8% failure rate Saturn Stills/Science Source	An intrauterine device (IUD) is a small T-shaped device that typically contains copper or plastic. It is inserted into the base of the uterus through the cervix. The IUD is a long-term contraceptive option, lasting from 3 to 10 years. It prevents pregnancy by thickening cervical mucus and consequently impeding sperm from entering the uterus. It also weakens the endometrium, making it less able to support an embryo.
	Sterilization surgery (tubal ligation, vasectomy): 0.5%–0.15% Failure rate Tek Image/Science Source	Surgical options include a vasectomy for men and tubal ligation for women. In a vasectomy, the vas deferens is cut, so sperm can no longer be ejaculated. In tubal ligation, the oviducts are cut and tied off, so sperm can no longer reach an ovulated egg. Both surgeries are essentially permanent, although they can be reversed with limited success. Those who do not want to have children, or any more children, typically choose surgical sterilization.
	Hormones (implant, shot, ring, patch, pill): 0.3%–8% failure rate crankyT/iStockphoto moodboard/Superstock	Female hormonal contraceptives contain a combination of synthetic estrogen and progesterone or progesterone only. Women can take these hormones in the form of combination pills, a skin patch, a cervical ring, a minipill, a regular injection, or an implant. All hormonal methods prevent pregnancy in three major ways: they thicken the cervical mucus, making it less likely that sperm will be able to swim through it and get into the oviducts; they prevent ovulation; and they thin the endometrium, so it cannot support the implantation of an embryo.
10 – 20	**Barriers** 2%–21% failure rate Superstock Jenny Swanson/iStockphoto	The male and female condom, the diaphragm, and the cervical cap. All of these prevent sperm from entering the uterus and are typically used with spermicidal foams or jellies that contain sperm-inactivating chemicals.
	Rhythm method and withdrawal: 3%–27% failure rate	The rhythm method (avoiding intercourse around the time a woman ovulates); withdrawal (the male withdraws his penis before ejaculation)
85	**No contraception** 85% failure rate	Sexual intercourse without any method of contraception

Data from Trussel, J. (2007) In Hatcher, R. A., et al., *Contraceptive Technology*, 19th revised ed. New York: Ardent Media.

INFOGRAPHIC 31.7 Sperm Develop in the Testes

Beginning at puberty the release of hypothalamic hormones triggers LH and FSH release from the anterior pituitary. LH and FSH act on the testes to trigger spermatogenesis and testosterone release. Precursor cells in the seminiferous tubules in the testes begin to divide by meiosis, and the haploid products differentiate into sperm. While the entire process takes approximately 6 weeks, because cells are in various stages of development a continuous supply of sperm is produced.

1. Luteinizing hormone (LH) and follicle-stimulating hormone (FSH) are produced by the pituitary in response to hormones released from the hypothalamus.
Brain
Hypothalamus
Anterior pituitary
FSH
LH
3. Precursor sperm cells divide by mitosis and then complete meiosis to form four haploid cells that develop into sperm.
2. LH and FSH stimulate testosterone production and spermatogenesis in the testes.
Mitosis
Meiosis I
Meiosis II
Lumen of the seminiferous tubule
Testis
Penis
Seminiferous tubule
Steve Gschmeissner/SPL/Getty Images
Inner space (lumen) of a seminiferous tubule
Precursor sperm cells (diploid)
Mature sperm cells (haploid)

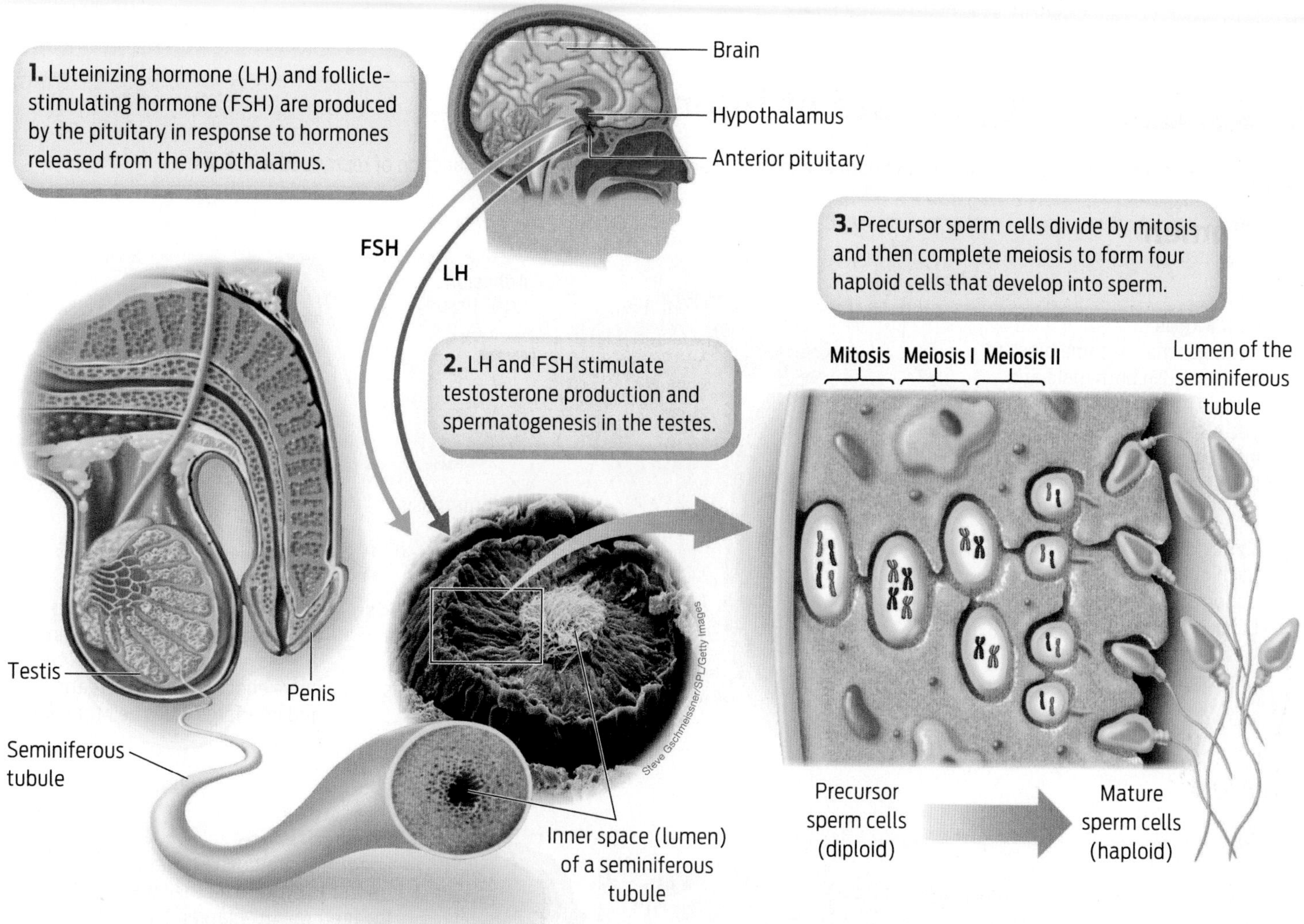

? Where are the hormones that trigger spermatogenesis produced?

because both the quantity and quality of a woman's eggs decline with age. At birth, a baby girl has about 1 million follicles in her ovaries. By puberty, she has about 300,000 follicles. Of these, only about 300 will ever ovulate an egg. Fertility in women peaks around age 25 and declines thereafter. As a result, according to the Centers for Disease Control, by age 40, nearly half of all women have difficulty conceiving.

Men, too, experience declines in fertility as they age and testosterone levels fall, but usually not as quickly or as dramatically as women. Men produce sperm throughout their lives and therefore can still father children in their 50s, 60s, and 70s. However, many men suffer from other problems, such as erectile dysfunction and testicular varicose veins, that can interfere with sperm reaching their target.

About a third of all cases of infertility are caused by reproductive impairments in women, and about a third are caused by impairments in men. In the remaining cases, the impairments affect both men and women or the cause of infertility cannot be determined **(INFOGRAPHIC 31.8)**.

INFOGRAPHIC 31.8 Causes of Infertility

There are multiple causes of infertility in both women and men that disrupt the normal function of reproductive tissues.

In Women

Blockages
Passages may become blocked or disabled in both male and female reproductive systems because of tissue scarring, infection, cancer, or abnormal tissue growth.

Cysts, fibroids, polyps
Each of these is a type of abnormal growth that may block passages or interfere with normal function of the tissue.

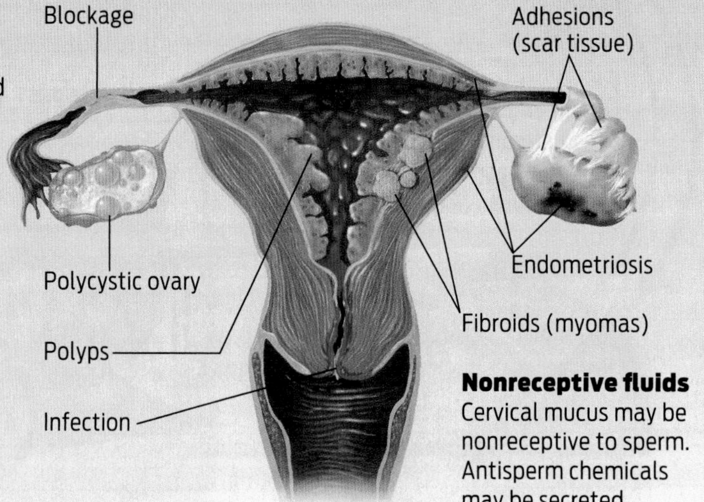

Nonreceptive fluids
Cervical mucus may be nonreceptive to sperm. Antisperm chemicals may be secreted.

Nonfunctional ovaries
Ovaries may fail to ovulate because of a variety of factors, including hormonal imbalances, genetic abnormalities, undeveloped ovarian tissue, endometriosis, and cancer.

Endometriosis
The tissue that lines the uterus grows abnormally and invades other tissues in the pelvic region. The wayward tissue irritates the nerve endings of these organs and interferes with their function.

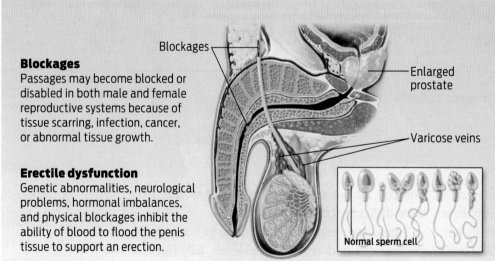

In Men

Blockages
Passages may become blocked or disabled in both male and female reproductive systems because of tissue scarring, infection, cancer, or abnormal tissue growth.

Erectile dysfunction
Genetic abnormalities, neurological problems, hormonal imbalances, and physical blockages inhibit the ability of blood to flood the penis tissue to support an erection.

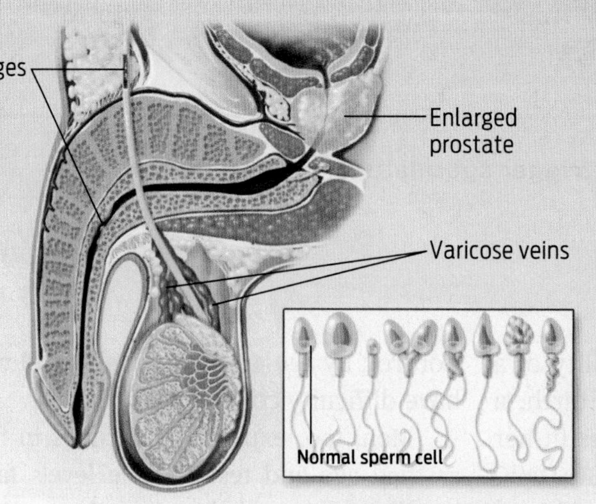

Prostatitis
Sperm pass through and receive fluid from the prostate on their way into the urethra. An enlarged prostate can block the passage of semen.

Testicular varicose veins
Valves in the veins that keep blood flowing in one direction deteriorate, causing blood to back up and pool. These enlarged veins can interfere with sperm production and transport.

Sperm abnormalities
Men may have low numbers of healthy sperm and/or physically abnormal sperm.

? Identify at least two specific places in the male and female reproductive tracts where blockages may lead to infertility.

The Problem of Multiples

While Suleman's case made numerous headlines and generated a strong backlash among the public, the lessons of the case extend well beyond this one woman's unusual story. Increasingly, people with all kinds of fertility difficulties are turning to assisted reproduction for help conceiving a child. Lesbian and gay couples, too, are seeking help from fertility specialists in their quest to become parents. Multiples are often an unintended consequence of the technologies used.

Assisted reproduction takes a number of forms. Besides IVF, another popular method is **intrauterine insemination (IUI),** in which sperm are injected directly into the uterus. This approach, also known as artificial insemination, is used most often when infertility can be traced to low sperm count, or to sperm that are "slow swimmers." It is also commonly used when a woman is trying to conceive with donor sperm or if the cause of infertility cannot be determined. Ultrasound technology is often used to monitor ovulation, so the insemination can be performed when chances are highest that one or more eggs have been released into the oviduct.

Commonly, both IVF and IUI start with a course of fertility drugs administered in the weeks before the procedure. These are usually FSH and LH, the hormones that stimulate ovarian follicles to develop and eggs to ovulate. The treatment works well for promoting follicle development and for controlling the precise time of ovulation, but there is a catch: fertility drugs almost always cause multiple eggs in multiple follicles to develop. This effect is desirable during IVF, in which the number of eggs that are fertilized outside the body and implanted into the uterus can be controlled. But this advantage turns into a liability with intrauterine insemination. Since millions of sperm are injected into the uterus during insemination, it's difficult to control the number of eggs that are fertilized. Multiple births result when more than one egg is fertilized (each by a different sperm), leading to more than one embryo implanting in the uterus and developing into a fetus.

This is what happened to Kate Gosselin, who with her husband, Jon, starred in the TLC television reality series *Jon & Kate Plus 8* from 2007 to 2011. The couple first had twins and then later sextuplets, all of whom were conceived through intrauterine insemination **(INFOGRAPHIC 31.9).**

INTRAUTERINE INSEMINATION (IUI)
A form of assisted reproduction in which sperm are injected directly into a woman's uterus.

Costly Care

In the United States, only 15 states currently mandate that health insurance policies must offer coverage for infertility treatment, and those that do often do not cover the more expensive treatments, like IVF. As a consequence, many patients must pay out of pocket. Because the costs of treatment are so high, many doctors find themselves under pressure to be aggressive with treatment with

Kate Gosselin and six of her children.

INFOGRAPHIC 31.9 Assisted Reproductive Technologies Can Result in Multiple Births

A hazard of assisted reproduction is a high probability of multiple births. Babies born as multiples are more likely to be born underweight and premature, putting them at risk for a variety of serious health conditions.

In Vitro Fertilization (IVF)

A woman is given fertility drugs to stimulate egg development and maturation. Eggs are then retrieved and fertilized in a dish. If multiple embryos are injected, more than one of them may implant in the uterus.

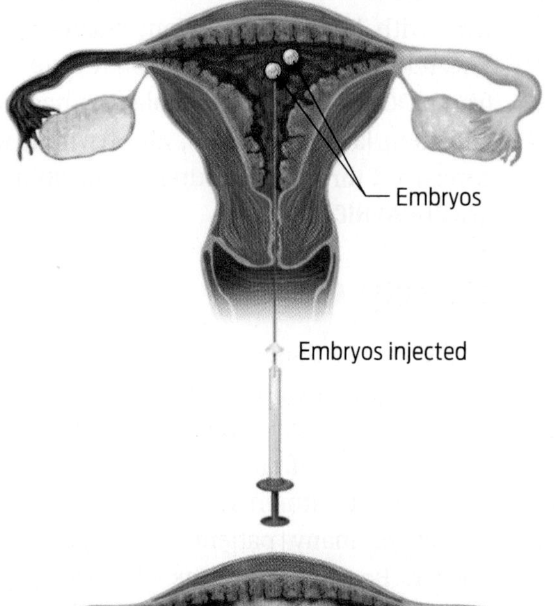

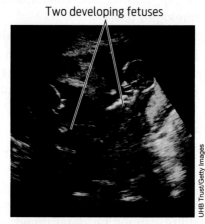

Two developing fetuses

UHB Trust/Getty Images

Multiple births: 41% of births by IVF are twins; 2.4% are triplets or more.

Intrauterine Insemination (IUI)

When a woman has been given fertility drugs to stimulate ovulation, several eggs may mature and ovulate simultaneously. Sperm injected into the uterus may fertilize multiple eggs and create multiple embryos.

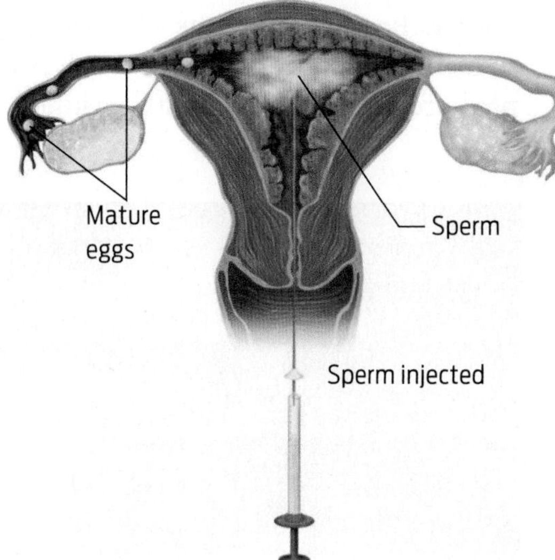

? Why is it easier to reduce the chance of multiples with IVF than with IUI?

patients who have limited resources. Therefore, financial considerations can dictate a course of treatment, even when the treatment is unlikely to be successful or may be too successful, as it proved to be in Suleman's case.

As of 2017, a single insemination treatment (including fertility drugs and monitoring) can cost as much as $1,200—steep, but barely one-tenth the cost of an average in vitro fertilization cycle. Many couples may choose insemination over in vitro fertilization simply because it is less expensive, even though the procedure has a low success rate. On average, only 5% to 15% of all insemination treatments result in a live birth. But younger women and women who take fertility drugs and, as a result, have multiple ovulations, tend to have higher success rates. The multiple ovulations increase the odds of a successful pregnancy by insemination.

IVF has a higher success rate—about 30% for women aged 35, according to ASRM—but if the woman must undergo multiple rounds before becoming pregnant, the $12,400 price tag for each round can be prohibitive. This financial burden can cause couples to hedge their bets by insisting that the doctor implant more embryos at one time than is typically recommended. The more embryos that are transferred into a woman's uterus during each round of IVF, the higher the odds of a pregnancy.

> "If nothing else, these high profile cases have served as a wake-up call."
>
> — Barbara Collura

Medical details revealed during a court case brought by the California Medical Board in 2009 against Suleman's doctor, Michael Kamrava, show that Kamrava created 14 embryos and implanted a dozen of them—six times the recommended number for a patient Suleman's age. Eight embryos survived, and the babies were delivered 9 weeks prematurely by C-section (cesarean section). The Associated Press reported that during the hearing Kamrava said he regretted implanting the 12 embryos and "would never do it again."

Kamrava further stated that Suleman was adamant about using all 12 embryos, even though he suggested implanting only 4. "She just wouldn't accept doing anything else with those embryos. She did not want them frozen, she did not want them transferred to another patient in the future," he said, according to the AP story. Kamrava said that he consented only after Suleman agreed to have a fetal abortion if necessary to reduce risk. After the implantation, however, he only heard from Suleman after the birth of her octuplets.

After a lengthy investigation, the California Medical Board revoked Kamrava's medical license. The Board claimed he was negligent not only in Suleman's case but also in the cases of two other women who suffered serious medical complications because of aggressive fertility treatments.

From a medical perspective, the birth of multiples from any assisted reproduction procedure carries risks. According to the March of Dimes, 50% of twins and 90% of triplets are born prematurely, as are virtually all quads and quintuplets. A premature baby is defined as one born before 37 completed weeks of pregnancy. When babies are born prematurely, their organs may not be fully developed, which can lead to medical problems. In particular, their lungs are often immature and so the babies must be hooked up to mechanical breathing ventilators, which sometimes scar the lungs so that these children will for the rest of their lives be prone to asthma, pneumonia, chronic lung disease, and other respiratory disorders. And because their brains aren't fully developed, premature babies are susceptible to brain hemorrhages and to developmental difficulties, including learning disabilities.

More than three-quarters of all triplets and higher-order multiples born in the United States are attributable to artificial reproductive technology. To reduce the birth of multiples and prevent the associated health problems, in 1998 the ASRM began recommending limits on the number of embryos transferred and has since adjusted those recommendations downward. Current guidelines recommend transferring no more than two embryos in women under 35 years of age and no more than five in women over 40, because as a woman ages the success rate tends to drop. These guidelines appear to have helped: in 2012, 2.4% of all infants born via assisted reproduction were triplets,

Prenatal Development

Pregnancy is divided into three trimesters. A normal pregnancy lasts 38 weeks from the time of fertilization, which is 40 weeks from the last menstrual period.

1st Trimester: Development of tissue layers and vital organs

The first trimester includes the embryonic stage of development and the early fetal stage. The embryo and fetus grow rapidly and critical organs develop.

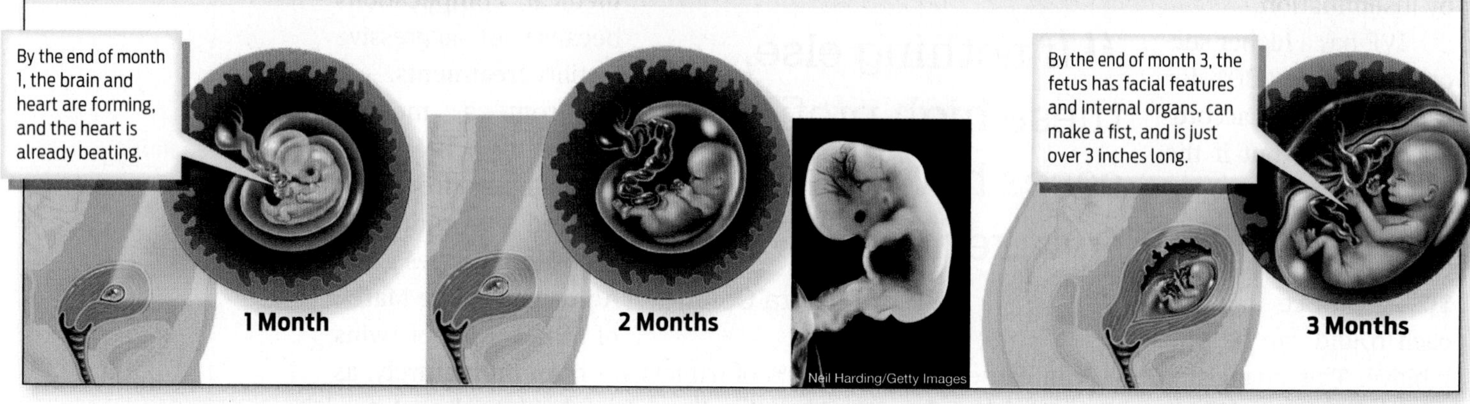

By the end of month 1, the brain and heart are forming, and the heart is already beating.
1 Month
2 Months
By the end of month 3, the fetus has facial features and internal organs, can make a fist, and is just over 3 inches long.
3 Months

Neil Harding/Getty Images

2nd Trimester: Growth and sex development

During the second trimester, the fetus continues to grow and develop features, including external genitalia.

6 Months

ninjaMonkeyStudio/Getty Images

3rd Trimester: Weight gain and organ system development

During the third trimester the fetus becomes fully developed and continues to gain weight. A fetus is technically "full term" at 35 weeks after fertilization, but most pregnancies last 38 weeks after fertilization (or 40 weeks after the last menstrual period).

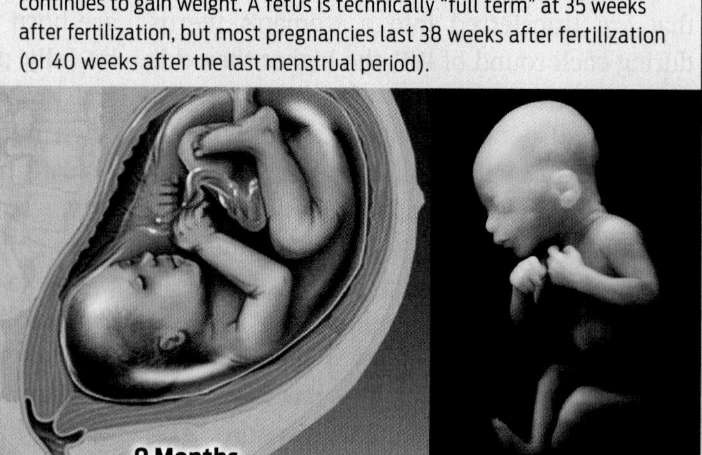

ninjaMonkeyStudio/Getty Images

as compared to 6.4% in 2003. But assisted reproduction still results in a high number of twins—in 2012, 41% percent of births from IVF alone were twins, compared to 3% of all births.

Despite the drop in multiple births, some lawmakers want more control. The risk that doctors may act against the ASRM guidelines, with negative consequences for women and their babies, remains. In the wake of the Suleman affair, in 2009 several states introduced legislation to limit the number of embryos that can be transferred during IVF. So far, no such federal bills have been enacted. Other countries, including Canada, the United Kingdom, Sweden, Germany, and Australia, heavily regulate the fertility business.

But some experts fear that regulation would hinder good care. Fertility doctors need flexibility to tailor treatment to a couple's individual condition of fertility, they argue. Such legislation "seeks to substitute the judgment of politicians for that of physicians

Nadya Suleman with some of her 14 children.

and their patients," ASRM president R. Dale McClure said in a statement.

Instead of putting controls on doctors, some groups favor regulation that would require health insurers to cover infertility diagnostics and treatment. The 2010 Affordable Care Act, passed by Congress and signed into law by President Barack Obama, does not include infertility treatments in the "Essential Health Benefits" that health insurers were required to cover beginning in 2014. However, it does make changes to the tax code that would allow individuals to write off a larger portion of the cost of these procedures on their income taxes.

As long as fertility treatment continues to be a financial burden on couples, fertility doctors will face pressure from patients to give them the most for their money, says Barbara Collura, executive director of RESOLVE, an advocacy group supporting infertile couples. Broad health insurance coverage would eliminate the cost factor. The result might be that couples would be able to forgo treatments such as insemination that have high risks with low success rates and skip directly to IVF when appropriate.

"If nothing else, these high profile cases have served as a wake-up call," says Collura. The community of health care workers is examining its own procedures and methods because it would rather self-regulate than have regulation imposed on it from the outside, she adds. "After Suleman, the community is really looking into how [the birth of octuplets] happened and how it can prevent it from happening again." ■

Why Do Organisms Have Sex?

Reproduction is a fundamental fact of life: all species must reproduce if they are to survive. But not all organisms reproduce in the same way, and not all have sex.

Sexual reproduction–reproduction that involves the combination of eggs and sperm–occurs in most eukaryotic organisms: animals, plants, fungi, and single-celled protists. In humans and other mammals, fertilization of an egg by sperm occurs internally, in an oviduct. This approach to sex allows the fertilized egg to be protected as it grows, but limits the number of eggs that can be fertilized at any one time. Internal fertilization also requires that the male and female physically copulate in order to achieve fertilization. In humans and other mammals, the penis serves as a sperm delivery system to place sperm in a location from which they can easily swim to the egg.

Many nonmammalian species also rely on internal fertilization but have a very different sexual anatomy. Internal fertilization in birds and reptiles, for example, occurs in an organ called the cloaca. The cloaca is the shared opening for solid waste, urine, and the reproductive system. When birds have sex, they briefly touch their cloacae together in

Birds Reproduce Sexually and Fertilize Eggs Internally

Birds and reptiles reproduce sexually and fertilize internally in the cloaca, following a "cloacal kiss."

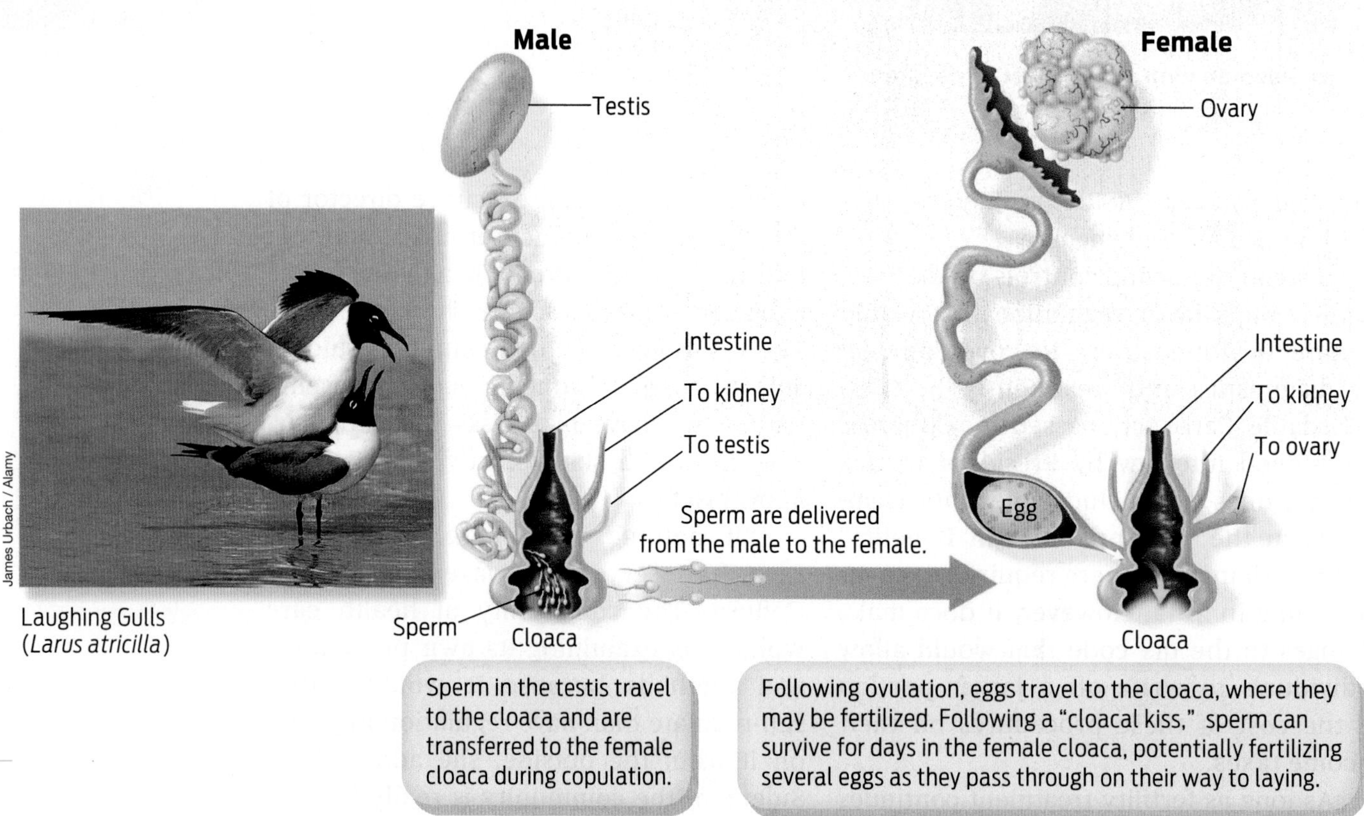

Male
Testis
Female
Ovary
Intestine
To kidney
To testis
Intestine
To kidney
To ovary
Egg
Sperm are delivered from the male to the female.
James Urbach / Alamy
Laughing Gulls (*Larus atricilla*)
Sperm
Cloaca
Cloaca
Sperm in the testis travel to the cloaca and are transferred to the female cloaca during copulation.
Following ovulation, eggs travel to the cloaca, where they may be fertilized. Following a "cloacal kiss," sperm can survive for days in the female cloaca, potentially fertilizing several eggs as they pass through on their way to laying.

what's called a cloacal kiss. This speedy kiss-and-run is sufficient to transfer sperm from the male to the female.

Not all sex takes place inside the body. Many fish and amphibians rely on external fertilization—their gametes fuse in the outside environment. Female salmon, for example, deposit their eggs in gravel nests in streambeds. Male salmon then swim over the eggs and release sperm, fertilizing thousands of eggs simultaneously. Other aquatic species such as coral and hydra also reproduce this way.

Many very successful creatures reproduce without having sex at all. Bacteria, for example, reproduce asexually, without additional genetic input from another individual. In asexual reproduction, a single parent cell simply divides to produce identical offspring. Some organisms, such as certain types of fungus, can reproduce both sexually and asexually. Baker's yeast (*Saccharomyces cerevisiae*) is a fungus that is commonly used to make bread rise. *S. cerevisiae* can produce gametes called spores, and these gametes can fuse to generate zygotes that develop into unicellular yeast.

Fish Reproduce Sexually and Most Fertilize Eggs Externally

Organisms like fish reproduce sexually, with two parents each contributing a gamete. Some fish fertilize internally, but most lay eggs that are fertilized externally.

Mark Conlin / Alamy
Natural Visions / Alamy
Female Sockeye Salmon
(*Oncorhynchus nerka*)
Mark Conlin / Alamy
Male Sockeye Salmon
(*Oncorhynchus nerka*)
Eggs are laid by the female in the river rock bed.
Sperm are deposited by the male over the eggs.
Genevieve Anderson
Eggs and sperm mix in the river gravel and thousands of eggs are fertilized externally.

But both the spores and the yeast can also make exact copies of themselves by mitotic cell division, resulting in identical populations of cells.

If some organisms can reproduce on their own, without the involvement of a partner, why bother having sex at all? Biologists don't fully understand why sex evolved, but they do have some good ideas. One hypothesis has to do with the evolutionary advantage of combining genetic information from two individuals. The resulting genetic diversity may enable offspring to better survive and adapt to changing environments—better than they would if each individual were identical to the parent.

But there are downsides to sex, too. Sexual reproduction generally takes longer than asexual reproduction. Certain bacteria, for example, can reproduce in as little as 20 minutes, which is a great evolutionary advantage in the race of natural selection. And sexual reproduction requires more energy and investment, especially on the part of the female. Internal fertilization, in particular, places a large demand on the mother. In the case of birds and reptiles, she has to produce an energy-rich egg in which the embryo will develop, and then in some cases protect it while it hatches. In most mammals, the maternal investment is even greater: she must supply all nutrients to a growing fetus and also provide a protective environment within her body during the long time it takes for a fetus to grow into a baby. However, this maternal investment, while leaving the mother vulnerable during the pregnancy, often results

Bacteria Reproduce Asexually

Bacteria reproduce asexually, producing offspring from only a single parent by binary fission. Offspring are genetically identical to each other, as well as to the single parent cell.

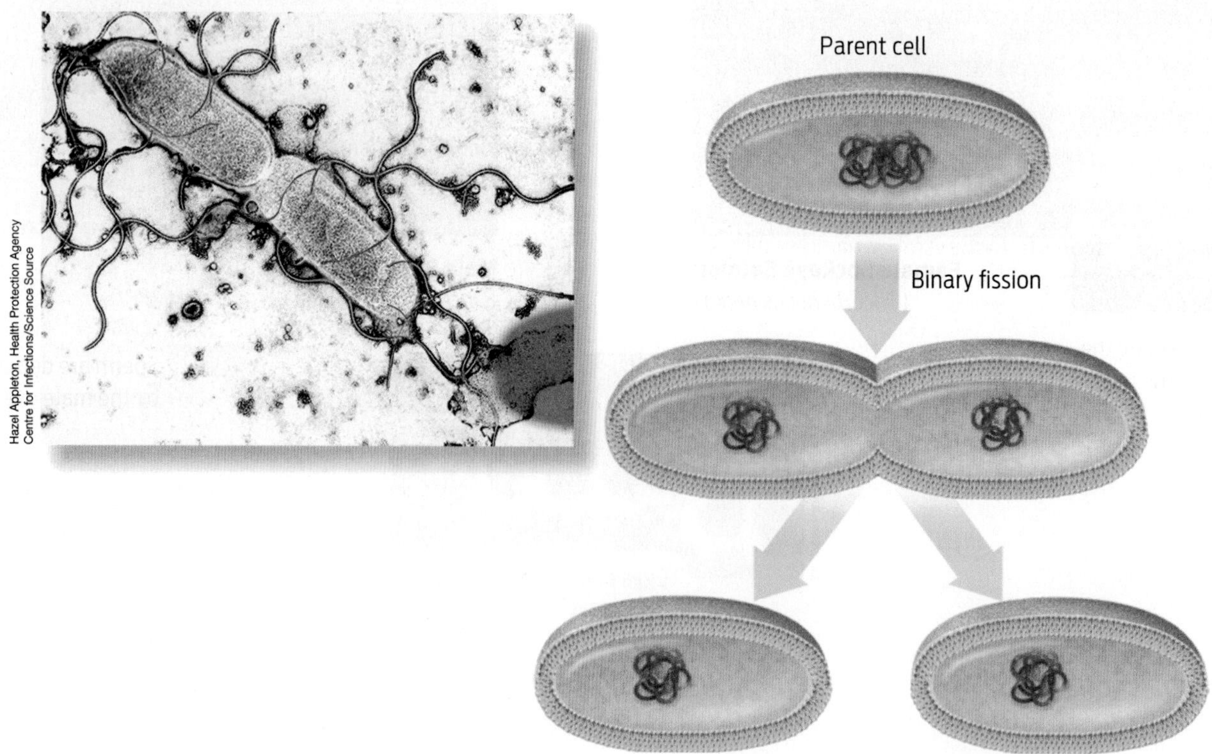

in the successful birth of a baby from each egg that is fertilized, compared to the exposed embryos that result from external fertilization, which are often eaten by predators before they ever mature.

Over time, different organisms have evolved different reproductive strategies that ensure that their offspring survive in the particular environment each organism occupies.

Yeast Reproduce Both Sexually and Asexually

Yeast can reproduce asexually by going through mitosis. They can also reproduce sexually by producing spores.

Asexual

A parent goes through mitosis, producing two genetically identical offspring. The parent cell forms a bud during this process, which will form one of the new daughter cells.

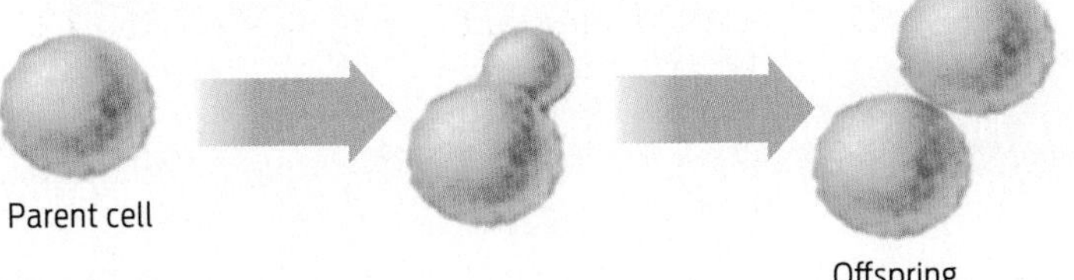

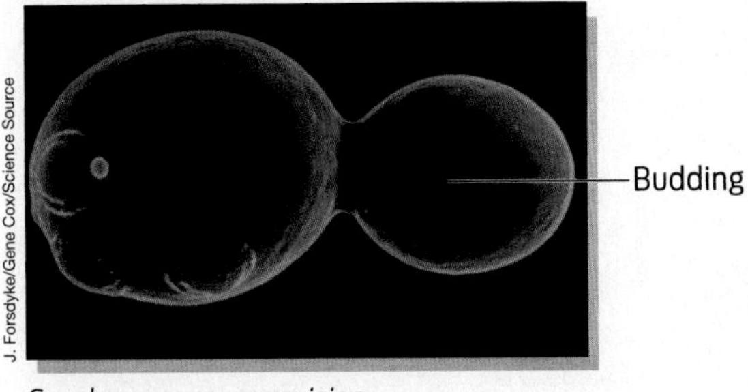

Saccharomyces cerevisiae

Sexual

First, a diploid parent cell forms four haploid spores during meiosis:

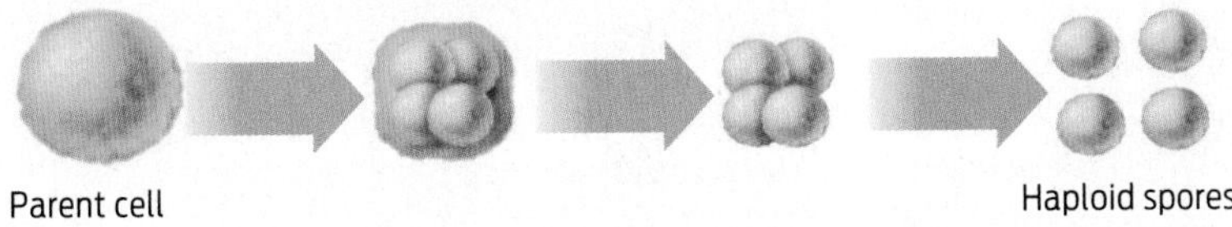

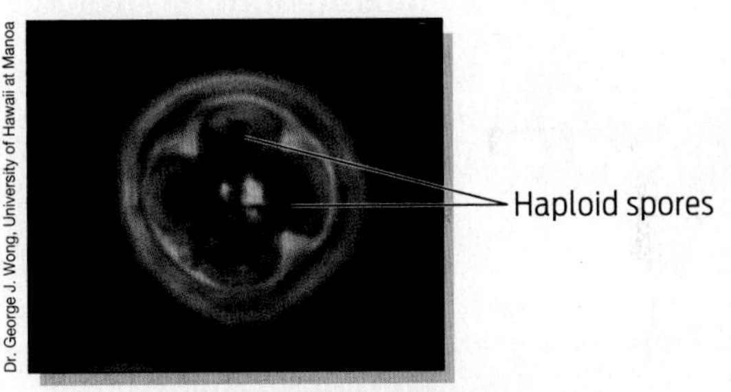

Then, two haploid spores from the same or different parent cells combine, producing a diploid offspring that reproduces asexually:

CHAPTER 31 SUMMARY

- The female reproductive system consists of paired ovaries, which produce eggs and the hormones estrogen and progesterone, and accessory structures that enable fertilization and support pregnancy.
- The male reproductive system consists of paired testes, which produce sperm and the hormone testosterone, and accessory structures that produce seminal fluids and permit the delivery of sperm to the egg.
- Fertilization of an egg by a sperm occurs in an oviduct. Only one sperm can fertilize an egg at one time.
- The monthly menstrual cycle in females and spermatogenesis in males are coordinated by a complex balance of hormones that is controlled by the brain.
- Hormones from the hypothalamus trigger the anterior pituitary to release follicle-stimulating hormone (FSH) and luteinizing hormone (LH), which in females stimulate eggs to develop and the ovary to secrete estrogen and progesterone. In males, FSH and LH trigger spermatogenesis and testosterone production.
- Ovulation, the monthly release of an egg from an ovarian follicle, is caused by a spike in production of LH. Estrogen and progesterone stimulate eggs to develop and the endometrium to thicken and prepare for a possible pregnancy.
- Upon fertilization, the zygote divides and travels to the uterus, where it implants in the nutrient-rich endometrium. If an egg is not fertilized, progesterone levels fall and the endometrium sloughs off during menstruation.
- There are many approaches to contraception, including barriers between sperm and eggs and manipulation of reproductive hormones.
- Infertility has many causes, including blocked passageways caused by infection and scar tissue, genetic abnormalities, hormonal deficiencies, and advanced age.
- Assisted reproduction involves artificially bringing sperm and egg together, either inside the body (in IUI) or outside the body (in IVF).
- Fertility drugs increase the number of eggs that mature and are ovulated by a female at one time. Multiple pregnancies result when sperm fertilize more than one available egg.
- Humans and other mammals, as well as birds and reptiles, reproduce sexually by internal fertilization. Other sexually reproducing animals, such as fish, reproduce by external fertilization. Some organisms are asexual and have no sex at all.

MORE TO EXPLORE

- **American Society for Reproductive Medicine: http://reproductivefacts.org**
- **Mundy, L. (2008) *Everything Conceivable: How the Science of Assisted Reproduction Is Changing Our World.* New York: Anchor Books.**
- **Hamzelou, J. (2016) World's first baby born with new "3 parent" technique. *New Scientist, 27.***
- **Maines, R. (2001) *The Technology of Orgasm: "Hysteria," the Vibrator, and Women's Sexual Satisfaction.* Baltimore: Johns Hopkins University Press.**
- **Roach, M. (2009) *Bonk: The Curious Coupling of Science and Sex.* New York: W. W. Norton.**

CHAPTER 31 Test Your Knowledge

DRIVING QUESTION 1 What is the anatomy of the male and female reproductive tracts, and how does the anatomy contribute to the function of the reproductive system?

By answering the questions below and studying Infographics 31.1, 31.2, 31.3, and 31.7, you should be able to generate an answer for the broader Driving Question above.

KNOW IT

1 Sperm develop in

a. the epididymis.
b. the vas deferens.
c. the seminiferous tubules.
d. the urethra.
e. the penis.

2 Why can untreated pelvic inflammatory disease lead to infertility?

a. because it prevents ovulation
b. because it scars and blocks the oviducts
c. because it scars and blocks the cervix
d. because it interferes with estrogen production by the ovaries
e. because it interferes with FSH and LH production by the anterior pituitary gland

3 Describe the relationship between the uterus and the cervix, and between the uterus and the endometrium.

USE IT

4 List the structures that sperm must pass through to reach and fertilize an egg. Begin with the seminiferous tubules.

5 A woman can become pregnant if she has intercourse 5 days before ovulating or on the day she ovulates, but not generally more than 24 hours after ovulation. What does this suggest about sperm and egg?

a. Sperm can survive for up to 6 days in the female reproductive tract.
b. Fertilization can occur in the ovary.
c. Human eggs cannot survive for very long after ovulation.
d. all of the above
e. a and c

6 A friend tells you that her boyfriend has received a diagnosis of gonorrhea, a sexually transmitted infection. She isn't worried for herself because she doesn't have any symptoms of infection. What can you tell her about the invisible risks of an untreated sexually transmitted bacterial infection?

By answering the questions below and studying Infographics 31.5 and 31.6, you should be able to generate an answer for the broader Driving Question above.

DRIVING QUESTION 2 What hormones are involved in reproduction, and how do they work in the reproductive systems of males and females?

KNOW IT

7 What is the source—testes, ovaries, anterior pituitary gland, or embryo—of each of the following hormones?

Luteinizing hormone (LH) ________________

Follicle-stimulating hormone (FSH) ________________

Testosterone ________________

Estrogen ________________

Progesterone ________________

hCG ________________

8 The hormone hCG is an indicator of pregnancy; it also

a. signals the corpus luteum to keep producing progesterone.
b. triggers ovulation.
c. acts on the anterior pituitary gland, causing it to release a surge of LH.
d. acts on the endometrium, causing it to thicken.
e. attracts sperm.

9 Which of the following hormones is/are produced by the anterior pituitary in males?

a. FSH
b. LH
c. testosterone
d. hCG
e. FSH and LH
f. FSH, LH, and testosterone

USE IT

10 Which of the following would most directly cause reduced levels of estrogen production?

a. an anterior pituitary tumor that increases secretion of LH

b. an increase in hypothalamus hormones that target the anterior pituitary

c. an anterior pituitary tumor that increases secretion of FSH

d. a decrease in hypothalamus hormones that target the anterior pituitary

e. anterior pituitary damage that prevents synthesis and release of FSH

11 In an episode of a popular "ripped from the headlines" TV crime series, a blood sample from a crime scene was found to have extremely low levels of FSH and LH. From this information, detectives determined that the blood came from a prepubescent girl, not a woman of reproductive age. Explain how they reached this conclusion.

12 As discussed in this chapter, oral contraceptives (such as the combination birth control pill, which contains both estrogen and progesterone) are designed to block ovulation in women. As males do not ovulate, a male hormonal contraceptive would have to target sperm development. Why would blocking testosterone secretion or action be effective in terms of contraception? What would be a likely undesired consequence of this type of male contraception?

INTERPRETING DATA

apply YOUR KNOWLEDGE

13 Emergency contraception—that is, contraception following intercourse—works by delaying or preventing ovulation. To be effective, it must be taken before LH levels start to rise.

a. Will emergency contraception be effective in preventing a pregnancy after unprotected intercourse on the day that a woman ovulates?

b. Different forms of emergency contraception have different efficacy rates, expressed as the percent reduction in the number of pregnancies that would have otherwise occurred. For example, if 80 women out of 1,000 become pregnant after unprotected intercourse, and an emergency contraceptive is 75% effective, then one would expect the use of the emergency contraceptive to reduce the number of pregnancies by 75% of 80 pregnancies, to 20 pregnancies. Predict the number of pregnancies with each of the following emergency contraceptives (given 80 of 1,000 pregnancies as a baseline): the pill marketed as Plan B (progestin—a synthetic progesterone—only), 89% effective; combined progestin and estrogen, 74% effective.

DRIVING QUESTION 3 What are the different types of assisted reproduction, and how do they work?

By answering the questions below and studying Infographics 31.4, 31.8, and 31.9, you should be able to generate an answer for the broader Driving Question above.

KNOW IT

14 Which of the following could interfere with ovulation?

a. blocked oviducts

b. chronically low levels of LH

c. excessive production of cervical mucus that blocks the cervix

d. presence of sperm in the oviduct

e. low levels of hCG

15 Compare and contrast in vitro fertilization (IVF) and intrauterine insemination (IUI).

USE IT

16 Assume that there is an array of diagnostic methods available to you, including blood tests to determine hormone levels and ultrasound to visualize internal structures. What results might confirm each of the following infertility-associated conditions? Be specific.

a. a blocked epididymis

b. polycystic ovary syndrome

c. menopause

d. oviduct scarring

17 Why does IUI have a higher risk of multiple births than IVF?

apply YOUR KNOWLEDGE

MINI CASE

18 A young couple has been trying to have a baby for over a year, but so far they have not had any luck. Analysis of the man's semen reveals a normal sperm count and no evidence of high rates of abnormal sperm. A physical exam and ultrasound revealed blockages in both of the woman's oviducts (fallopian tubes). From this information, describe two forms of fertility treatments (including assisted reproduction) that the couple could consider, and two that would not be helpful. For each, describe what is involved in the treatment, and why it would or would not be a helpful strategy for this couple.

apply YOUR KNOWLEDGE

BRING IT HOME

19 From the perspective of a fertility specialist, how would you respond to a congressional representative proposing increased regulation of fertility clinics? In order to make a convincing argument, include both pros and cons, medical and scientific considerations, and a description of the patient population that this specialist serves.

32 Immune System

DRIVING QUESTIONS

1. What is the structure of a virus, and how do viruses cause disease?
2. Why is nonspecific immunity important even though another system provides both specificity and memory?
3. What is adaptive immunity, and how does vaccination rely on adaptive immunity?
4. What are the features of the influenza virus that allow it to cause worldwide outbreaks?

THE FORGOTTEN PLAGUE

After nearly a century, scientists learn what made the 1918 influenza virus so deadly

In the fall of 1918, World War I was coming to an end—but another, insidious, threat was surfacing. In communities around the world, a deadly illness was gaining momentum, spreading like wildfire from person to person.

At first the disease seemed much like a bad case of influenza, or "flu." High fevers and body aches were common. But some patients developed more severe symptoms, such as coughing so intense that they spat up blood. Many died so quickly after falling ill that doctors began to suspect they were witnessing something far worse than ordinary flu.

Stories were told of victims who were completely healthy in the evening but were dead by morning. An army physician stationed at Camp Devens in Massachusetts wrote to a friend that patients would rapidly "develop the most vicious type of pneumonia that has ever been seen" and later would "struggle for air until they suffocated." Patients "died struggling to clear their airways of a blood-tinged froth that sometimes gushed from their nose and mouth," recalled another physician.

Between 1918 and 1920, the mysterious disease swept around the globe. Had the world not been at war, the scourge might never have happened. Troops carried the illness with them across Europe and back to the United States, infecting relatives, friends, and strangers along the way.

The result was one of the deadliest **pandemics** in human history. Even more frightening than an epidemic, a pandemic is an exceptionally high number of cases occurring worldwide. According to historian John Barry, the 1918 flu pandemic killed more people in 1 year than the Black Death killed in the entire 14th century; it killed more people in 25 weeks than AIDS killed in 25 years. By the time it was over, between 50 and 100 million people around the world had perished—more than the total number killed in World War I.

PANDEMIC
A worldwide epidemic; an exceptionally high number of cases occurring at the same time.

VIRUS
A noncellular infectious particle consisting of nucleic acid surrounded by a protein shell.

Why this particular strain of flu was deadlier than any pandemic that came before (or after), scientists of the time couldn't say. For decades after the pandemic had subsided, scientists thought they would never be able to explain the flu's severity, since the virus itself was lost to history. But recent discoveries have made it possible to peer into the viral past, and scientists have begun to uncover what made the 1918 virus so deadly.

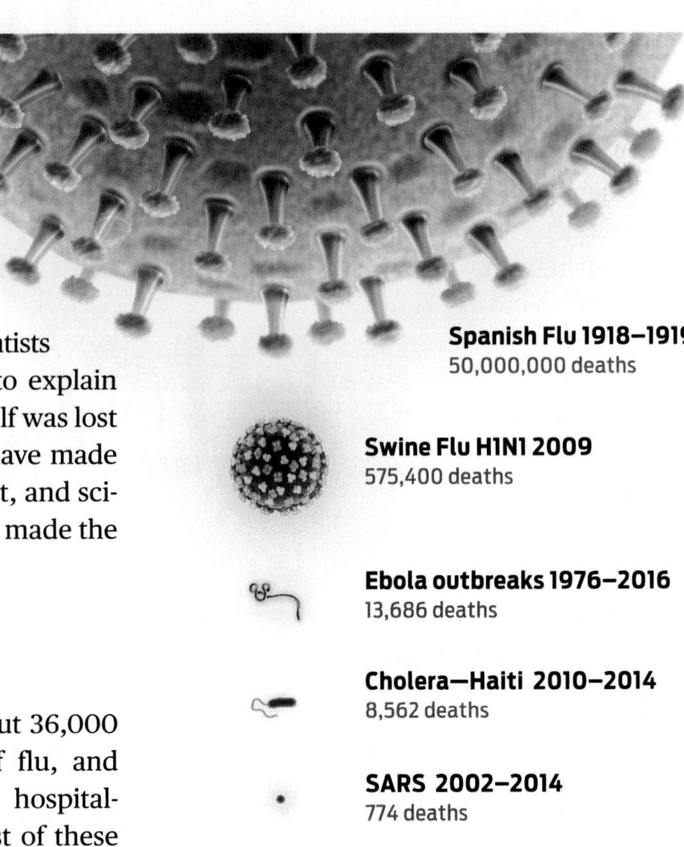

Silent Invader

Each year in the United States about 36,000 people die from complications of flu, and more than 200,000 people are hospitalized from flu-related illnesses. Most of these deaths strike young children and the elderly, whose bodies are less able to fight off the infection. The 1918 flu, by contrast, struck down people in the prime of life: people age 20 to 40 were the hardest hit. The death toll in this age group was so severe that it reduced the average U.S. life expectancy in 1918 by 12 years.

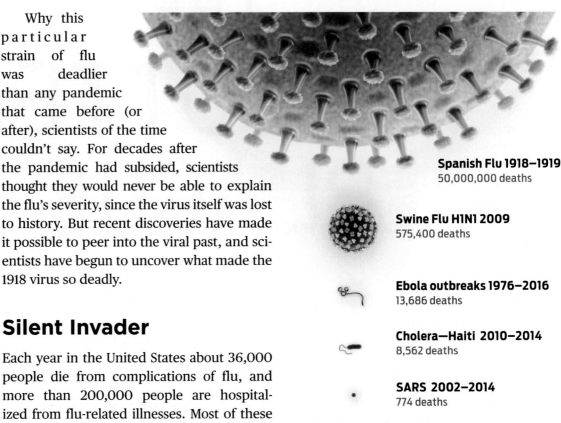

Police in Seattle, Washington, keeping the peace during the 1918 flu pandemic.

At the time, doctors did not know what caused influenza. Some suspected that it was caused by a bacterium, as were many other known diseases, such as cholera and tuberculosis. But research in 1933 showed that influenza is caused not by a bacterium but by a **virus.**

The influenza virus, like all viruses, is a tiny particle made up of nucleic acid surrounded by a protein shell. The nucleic acid serves as the virus's genetic material, while the protein shell (called a capsid) allows the virus to bind to and enter cells. Viruses are considered to be nonliving because even though all viruses have genes, viruses are not made of cells. Lacking the cellular machinery for replicating genetic material and synthesizing proteins, viruses cannot reproduce independently–they must infect a host cell in order to reproduce. Once inside, the virus then coopts the host cell's machinery to replicate its genetic material and make more viruses. Eventually the multiplied viruses burst from and destroy the host cell (**INFOGRAPHIC 32.1**).

Unlike cells, which rely on DNA as their hereditary material, viruses can store their genetic information in the form of either DNA or RNA (see Chapter 2). RNA viruses include the influenza virus and the viruses that cause colds, measles, and mumps. RNA viruses also cause more serious diseases, like AIDS and polio. Diseases caused by DNA viruses include hepatitis B, chicken pox, and herpes. Each type of virus has a distinctive protein shell with a characteristic shape.

Some viruses cause more damage than others. The severity of a viral illness typically

INFOGRAPHIC 32.1 Viruses Infect and Replicate in Host Cells

Viruses can replicate only within a host cell. Viral genes direct the host cell to synthesize new viral particles. Ultimately, viral replication kills the host cell either by causing it to burst or by depleting the cell's resources.

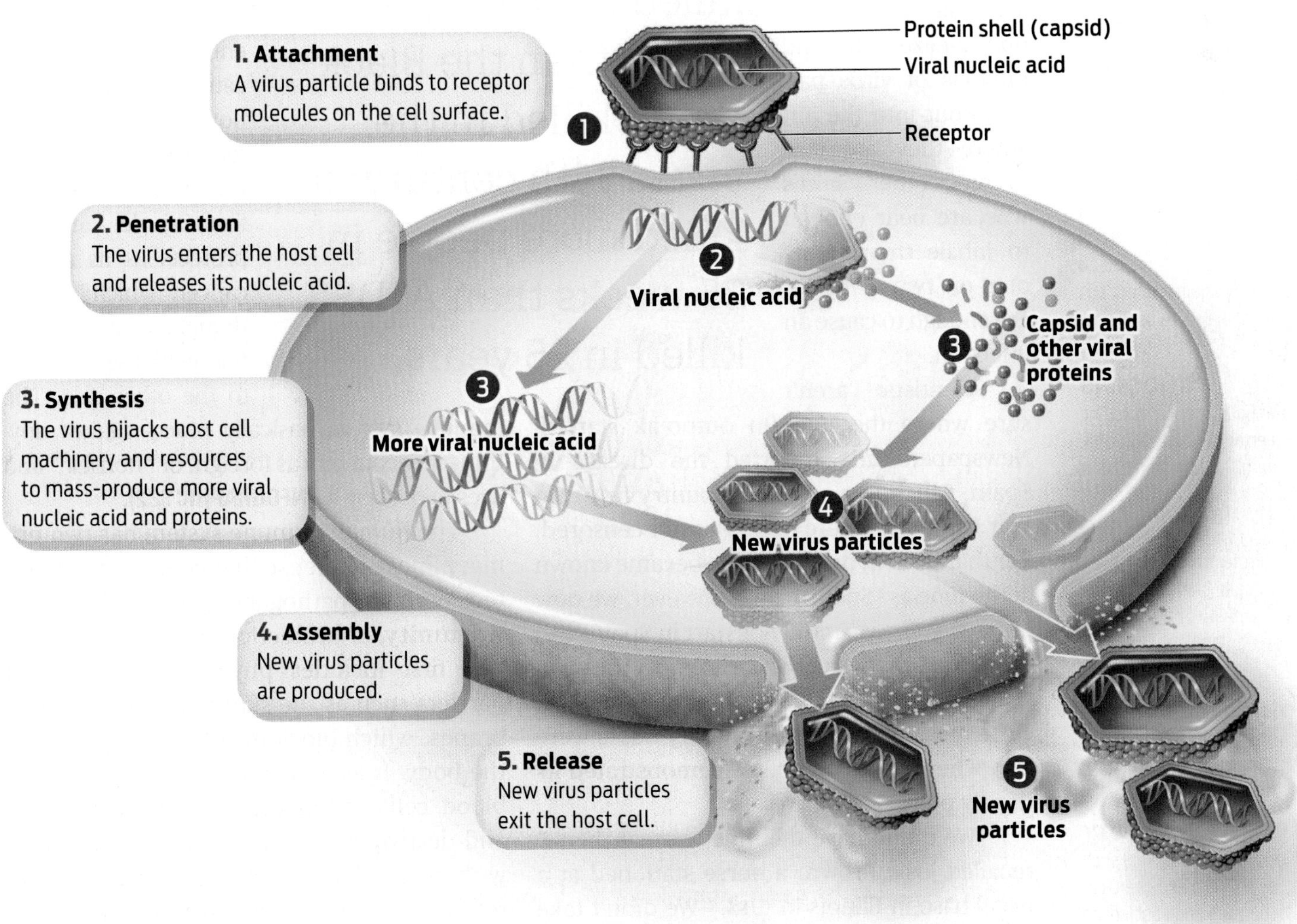

1. Attachment
A virus particle binds to receptor molecules on the cell surface.
Protein shell (capsid)
Viral nucleic acid
Receptor
1
2. Penetration
The virus enters the host cell and releases its nucleic acid.
2
Viral nucleic acid
3
Capsid and other viral proteins
3. Synthesis
The virus hijacks host cell machinery and resources to mass-produce more viral nucleic acid and proteins.
3
More viral nucleic acid
4
New virus particles
4. Assembly
New virus particles are produced.
5. Release
New virus particles exit the host cell.
5
New virus particles

? If a cell does not have receptor molecules to which a specific virus can bind, can that cell be a host for that virus?

depends on how quickly and successfully the body's defenses rally to kill the virus and how well the infected tissue can repair itself. Most people recover completely from the rhinovirus that causes the common cold because the respiratory tract contains rapidly dividing cells that quickly replace damaged ones. Poliovirus, however, attacks nerve cells that almost never divide, which is why polio infection may cause permanent paralysis. (Thanks to the polio vaccine, introduced in the 1950s, polio infection is no longer common in the United States.)

Like the virus that causes common colds, the flu virus spreads from person to person through coughs and sneezes. Viruses are extremely small and are easily aerosolized: one sneeze can send millions of virus particles out into the air, where they are free to infect other victims who are near enough to inhale them. Only one or two particles are enough to cause an infection.

The 1918 flu pandemic killed more people in 1 year than the Black Death killed in the entire 14th century; it killed more people in 25 weeks than AIDS killed in 25 years.

Scientists aren't sure where the 1918 flu outbreak started. Newspapers first reported the disease in Spain, which was a neutral country in World War I and one where news was not censored. For that reason, the pandemic became known at the time as "Spanish flu." However, we now know that the virus did not start in Spain, and that some of the earliest cases occurred in the United States, among soldiers housed in close quarters. Such U.S. military bases were also where Spanish flu first demonstrated its deadly power.

"We didn't have the time to treat them," recalled Josie Brown, a nurse stationed at a naval base in Illinois in 1918. "We didn't take temperatures; we didn't even have time to take blood pressure. We would give them a little hot whisky toddy; that's about all we had time to do."

A Microscopic Battleground

Some early clues to what made the 1918 flu infection so deadly came from examining victims' bodies. Autopsies performed at the time showed that the infection caused extensive damage to the lungs. In severe cases, the lungs looked torn apart, as if ravaged by shrapnel. The lungs, it seems, were the central battleground of the war between the influenza virus and the **immune system** of its human host.

Like an army, the immune system defends the body from different kinds of invaders—principally infectious agents, or **pathogens,** such as viruses, bacteria, and parasites that enter the body and cause disease. In addition, the immune system plays an important role in helping us heal from injuries, and even protects us as against cancer, which can be thought of as treasonous cells that go over to the other side. The immune system basically reacts against anything it encounters as foreign, or "nonself," and tries to destroy it **(INFOGRAPHIC 32.2).**

The human immune system has two primary lines of defense that coordinate to protect us from pathogens, providing us with **immunity,** or protection from these threats. The first includes physical and chemical barriers such as the skin and mucous membranes, which block invaders from entering the body. It also includes a variety of white blood cells called **phagocytes** that engulf and destroy invaders. Because we are born with these defense mechanisms, they are referred to collectively as **innate immunity.** Innate defenses are effective but nonspecific, in that they protect us against a wide variety of invaders but do not target any one in particular. The phagocytes of innate

IMMUNE SYSTEM
A network of cells and tissues that acts to defend the body against infectious agents and also helps to heal injuries.

PATHOGENS
Infectious agents, including certain viruses, bacteria, fungi, and parasites, that may cause disease.

IMMUNITY
Protection from a given pathogen conferred by the activity of the immune system.

PHAGOCYTE
A type of white blood cell that engulfs and digests pathogens and debris from dead cells.

INNATE IMMUNITY
Nonspecific defenses, such as physical and chemical barriers and specialized white blood cells, that are present from birth and require little or no time to become active.

INFOGRAPHIC 32.2 The Immune System Defends against a Variety of Pathogens

The immune system can protect the body against many specific pathogens. Viruses, bacteria, and parasites can all be detected as foreign—triggering a variety of defenses.

Viruses

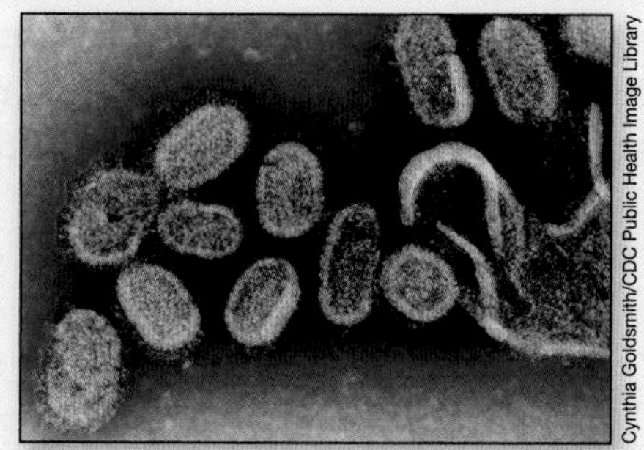
Cynthia Goldsmith/CDC Public Health Image Library

Influenza virus
Flu

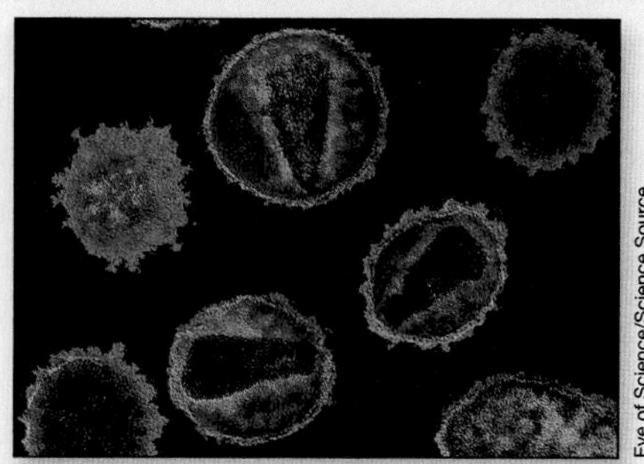
Eye of Science/Science Source

Human immunodeficiency virus
AIDS

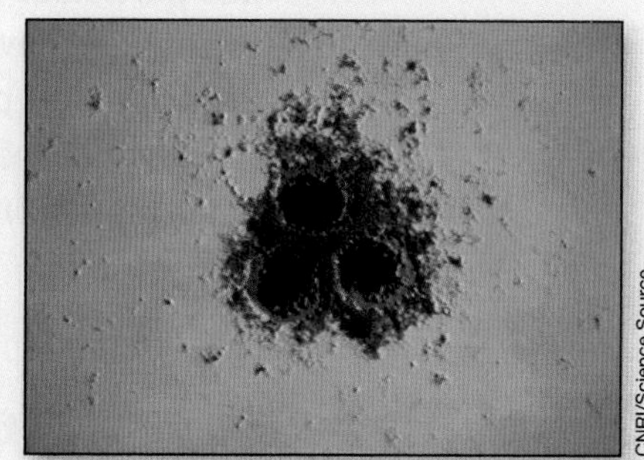
CNRI/Science Source

Varicella zoster virus
Chicken pox

Bacteria

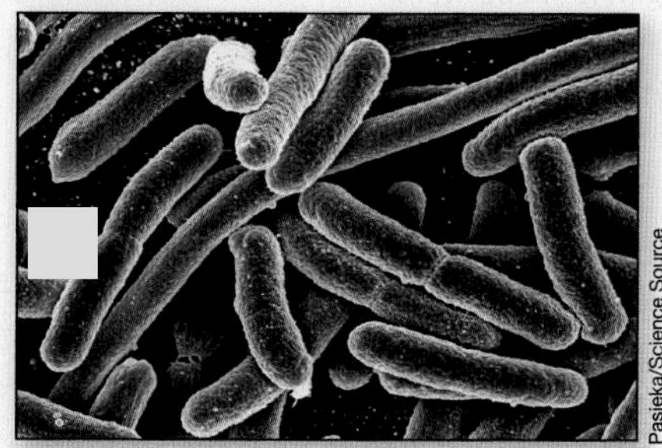
Pasieka/Science Source

Escherichia coli
Food poisoning

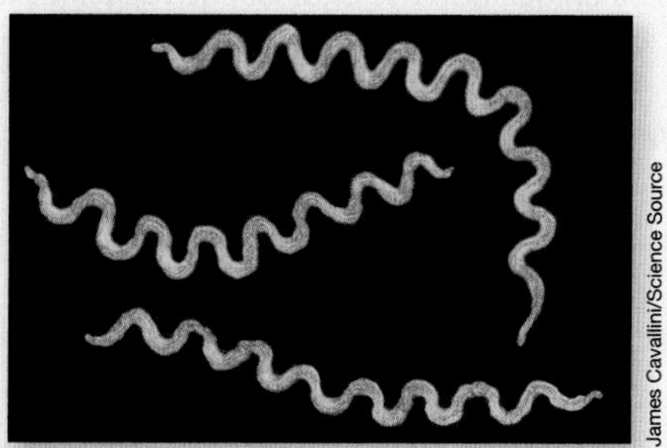
James Cavallini/Science Source

Treponema pallidum
Syphilis

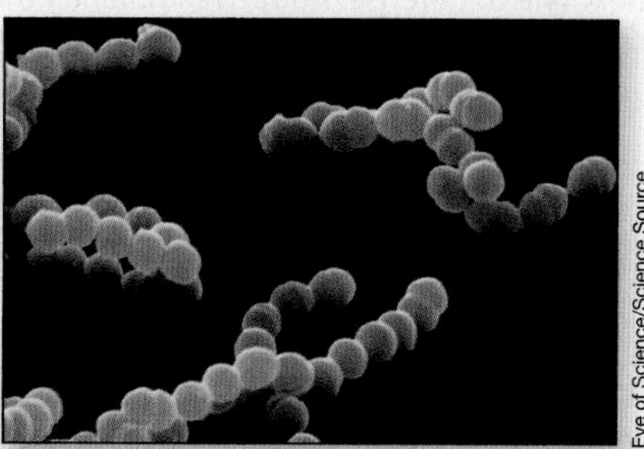
Eye of Science/Science Source

Streptococcus pyogenes
Strep throat, scarlet fever

Parasites

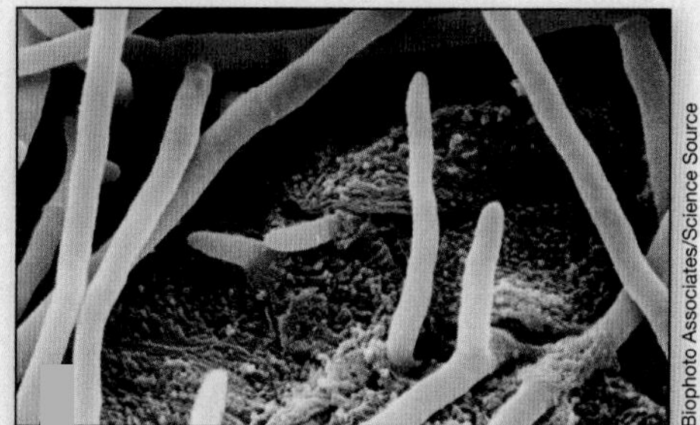
Biophoto Associates/Science Source

Trychophyton mentagrophytes (fungus)
Athlete's foot

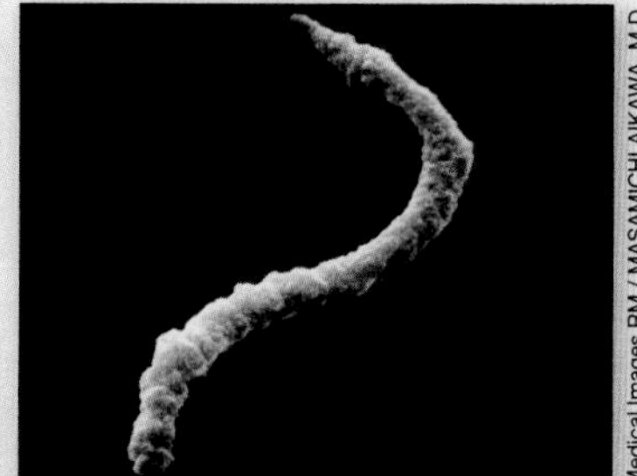
Medical Images RM / MASAMICHI AIKAWA, M.D.

Plasmodium falciparum (protist)
Malaria

Dr. Stan Erlandsen/CDC/Public Health Image Library

Giardia lamblia (protist)
Intestinal inflammation

? Can eukaryotic organisms cause disease in humans?

immunity, for example, respond to molecular patterns common to many types of bacteria, such as a particular carbohydrate found in many bacterial cell walls. These defenses are always present and require little or no time to become active.

The second line of defense is **adaptive immunity,** which includes the coordinated

ADAPTIVE IMMUNITY
A protective response, carried out by lymphocytes, that confers long-lasting immunity against specific pathogens.

LYMPHOCYTE
A specialized white blood cell of the immune system. Lymphocytes are important in adaptive immunity.

actions of specialized white blood cells called **lymphocytes** that take aim at specific pathogens. Adaptive immunity takes time to develop, but it has three main advantages over innate immunity. First, the adaptive immune system is *specific*: it is able to recognize particular pathogens–one type of virus, for example–even if it has never encountered that pathogen before. This feature is important because pathogens evolve quickly, and many have found ways to evade our innate defenses. Second, the adaptive immune system is *diverse*: it can mount an immune response to essentially any threat that comes its way. Finally, the adaptive immune system exhibits *memory*: once you have chicken pox, the activity of the adaptive immune system means you won't get it again. But even with an adaptive immune system, we still need our innate defenses to be in good working order; in fact, the two systems work together to protect us from illness (**INFOGRAPHIC 32.3**).

INFOGRAPHIC 32.3 Innate and Adaptive Immunity

Invading pathogens encounter a variety of defenses. Physical and chemical barriers are a first line of defense. Pathogens that breach these barriers encounter nonspecific cellular defenses. These early lines of defense are known as innate immunity, and are always present and ready to protect against a wide range of pathogens. Adaptive immunity, which relies on the activity of specialized white blood cells, is the next line of defense. Adaptive immunity mounts a unique defense against each specific pathogen it encounters.

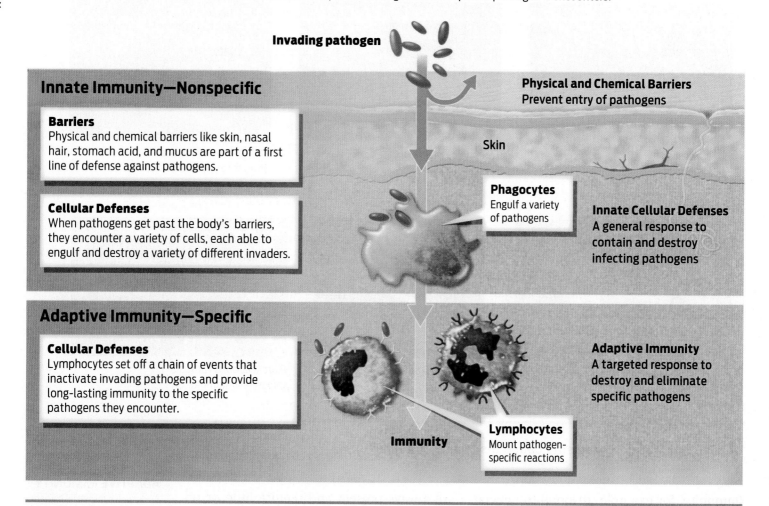

Invading pathogen

Innate Immunity—Nonspecific

Barriers
Physical and chemical barriers like skin, nasal hair, stomach acid, and mucus are part of a first line of defense against pathogens.

Cellular Defenses
When pathogens get past the body's barriers, they encounter a variety of cells, each able to engulf and destroy a variety of different invaders.

Physical and Chemical Barriers
Prevent entry of pathogens

Skin

Phagocytes
Engulf a variety of pathogens

Innate Cellular Defenses
A general response to contain and destroy infecting pathogens

Adaptive Immunity—Specific

Cellular Defenses
Lymphocytes set off a chain of events that inactivate invading pathogens and provide long-lasting immunity to the specific pathogens they encounter.

Adaptive Immunity
A targeted response to destroy and eliminate specific pathogens

Immunity

Lymphocytes
Mount pathogen-specific reactions

? What is a difference between innate and adaptive immunity?

Innate Defenses

The innate immune system starts defending at sites where the body is exposed to the outside world—your skin, eyes, mouth, and nasal passages. Physical barriers, like the skin itself, prevent the entry of many pathogens. Enzymes in saliva destroy some pathogens; nasal hair and the mucus that lines the throat trap others. Under the layer of mucus is a layer of cells with hairlike projections called cilia that sweep away pathogens to prevent them from taking hold in the throat. When pathogens do successfully breach these physical barriers, the body tries to flush them out with more fluid: runny noses, watery eyes, coughs, and sneezes are the body's way of expelling the invaders.

When pathogens overcome these physical and chemical defenses, they trigger additional innate responses. One of the most important functions of the innate immune system is to promote **inflammation** in response to tissue damage or infection. Inflammation is a suite of reactions that generates redness, heat, swelling, and pain at the affected site ("inflammation" means "setting on fire"). Inflammation is an important line of defense that kills invading pathogens and prevents them from spreading further in the body.

Several types of white blood cell contribute to inflammation. Most important are phagocytes, which surround and engulf the invaders in a process called phagocytosis. These cells, which include macrophages ("big eaters"), also help to clean up debris from dead cells and play an important role in promoting an adaptive immune responses (a topic to which we'll return).

"We didn't take temperatures; we didn't even have time to take blood pressure. We would give them a little hot whisky toddy; that's about all we had time to do."

— Josie Brown

What triggers inflammation? Tissues that have been damaged by injury or by infection release chemicals such as **histamine** that cause blood vessels to expand and leak fluid into surrounding tissues. These chemical signals also attract phagocytes from the blood and surrounding tissue to the region. The redness, heat, swelling, and pain associated with inflammation are the result of increased blood flow to the area (causing redness and heat), accumulation of fluid and immune cells from the leaky blood vessels (causing swelling), and release of chemical signals from damaged tissue (causing pain). The fluid at inflamed sites also contains clotting proteins that stop bleeding and prevent pathogens from spreading to neighboring tissues. Together, this collection of materials forms a milky white substance called pus.

The body has other nonspecific defenses. One of these is **fever.** In response to inflammatory chemical signals, the brain's hypothalamus raises the set point for body temperature by a few degrees. The higher temperature creates an environment hostile to bacterial growth and also signals immune cells to ramp up their reactions. (Aspirin, ibuprofen, and acetaminophen are anti-inflammatory drugs that help reduce fever and pain by interrupting the inflammatory signals received by the brain—which makes you feel better but doesn't treat the underlying cause of those signals.)

Other innate immune defenses include proteins that either attack pathogens directly or prevent them from reproducing. For example, defensive proteins in the blood called **complement proteins** coat the surface of pathogens, making them more easily destroyed by phagocytes, or directly destroy

INFLAMMATION
An innate immune defense that is activated by infection or tissue damage; characterized by redness, heat, swelling, and pain.

HISTAMINE
A molecule released by damaged tissue and during allergic reactions that promotes inflammation.

FEVER
An elevated body temperature.

COMPLEMENT PROTEINS
Proteins in blood that help destroy pathogens by coating or puncturing them.

INFOGRAPHIC 32.4 Some Important Features of Innate Immunity

Innate immunity is always "on" and ready to contribute to defenses. Innate immunity does not recognize specific pathogens, but relies on physical, cellular, and chemical defense strategies.

Physical barriers

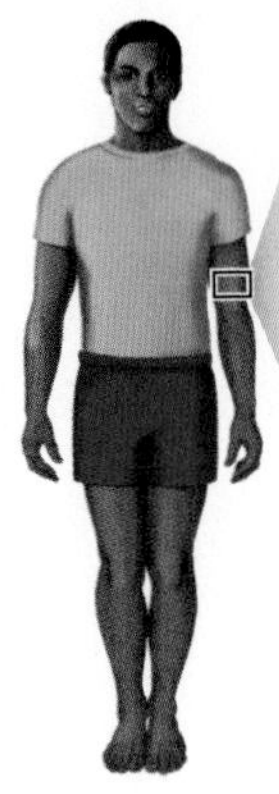

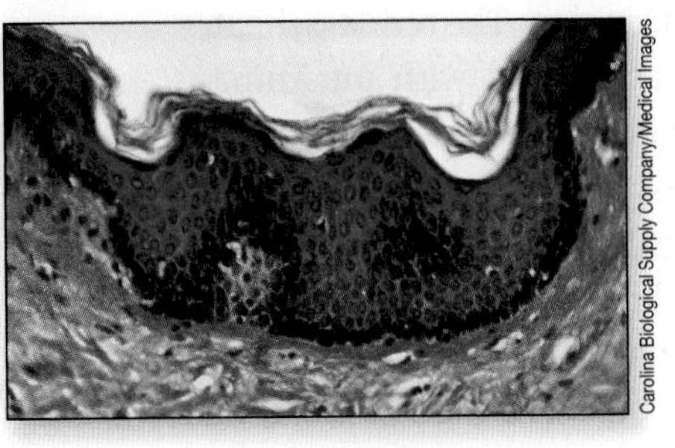

Carolina Biological Supply Company/Medical Images

Skin (on the outside of the body) and mucous membranes (lining the inside of the body) have layers of tightly packed cells that prevent pathogens from entering the body. The mucus that coats mucous membranes traps foreign substances.

Chemical defenses

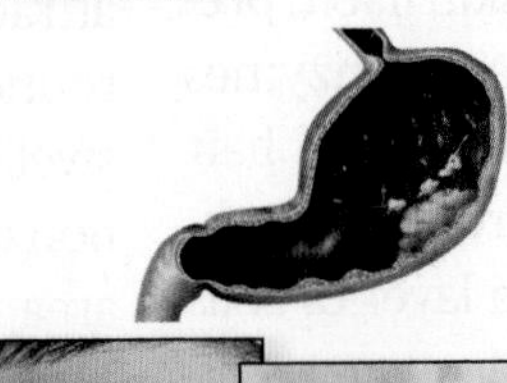

Acid in the stomach kills many of the microorganisms that we ingest.

Ocean/Corbis
fStop Images GmbH/Alamy

Tears and **saliva** contain an enzyme that breaks down the cell walls of bacteria, causing them to burst.

Complement proteins

Complement proteins in blood puncture holes in bacterial cells or coat the cell's surface, making them more easily destroyed by phagocytes.

Fever is induced by chemical signals acting in the brain in response to infection. Fever inhibits bacterial growth and stimulates the immune system.

Phagocytes

Phagocytes ingest and destroy pathogens. Phagocytes also trigger inflammation and adaptive immune responses.

Juergen Berger/Science Photo Library/Getty Images

Phagocyte (white) engulfs several pathogens (red).

Inflammation

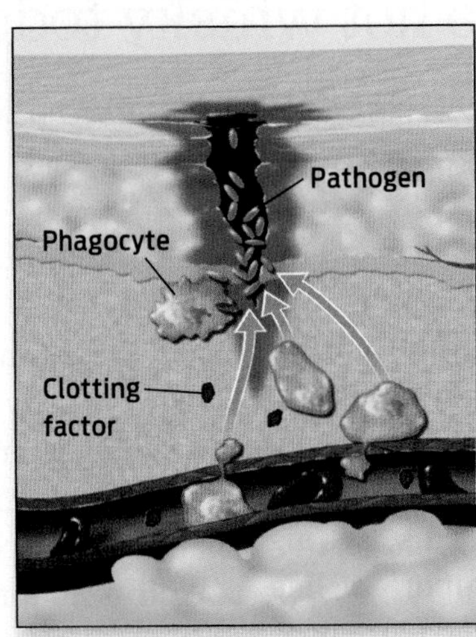

1. Pathogens get past physical and chemical barriers.
2. Pathogens and damaged cells release histamines and other molecules that increase blood flow and attract white blood cells to infected areas.
3. Blood vessels leak, causing surrounding tissue to swell with fluid that contains clotting factors and white blood cells.
4. Phagocytes ingest pathogens and trigger an adaptive response. Clotting reactions contain the infection.

? What innate defenses does an ingested pathogen encounter as it passes through the digestive tract?

some bacteria by punching holes in the cell membrane **(INFOGRAPHIC 32.4)**.

Though the innate immune system provides some protection against influenza, occasionally virus particles manage to evade the body's physical and chemical barriers and take up residence in the upper respiratory tract: the mouth, the nose, and the throat. Within minutes, these virus particles attach to epithelial cells that line the respiratory tract and slip inside the cells. The virus then hijacks the host cells' cellular machinery to replicate its own genetic material. Generally, about 10 hours after the virus invades a cell, the cell begins to release newly synthesized viral particles, sending out between 1,000 and 10,000 viruses capable of infecting other cells. Eventually the infected cells die, weakening the respiratory tract.

Normally our bodies are able to mount a sufficient immune response to contain the infection, so a typical course of flu lasts only about a week and is not generally fatal. Why the 1918 flu was so deadly, researchers at the time could not say. More perplexing still was why the virus tended to kill people in the prime of their lives—those with the healthiest immune systems. It would be more than 75 years before researchers could even begin to answer those questions.

James King-Holmes/Science Source

Sample blocks containing lung and brain tissue from 1918 Spanish flu victims.

Viral Time Capsule

In 1995, Jeffery Taubenberger, a virologist at the U.S. Armed Forces Institute of Pathology, became interested in the mystery of the 1918 flu, and wondered if advances in molecular biology might help him solve it. He realized that he could use the polymerase chain reaction (PCR; see Chapter 7) to reconstruct the viral genome—if only he had access to some original tissue samples from victims. He knew that the army had routinely saved tissue samples from soldiers killed during wartime and decided to check the institute's archives. Right away, he was able to locate tissues from several U.S. servicemen who had died of flu in 1918. Over the next several months, Taubenberger and his colleagues used PCR to recover viral RNA from these samples and then were able to sequence a few of the virus's genes. But the results were incomplete, and the team eventually ran out of tissue samples from which to obtain virus. Taubenberger published these preliminary results in 1997 in the journal *Science*.

Later that year, Johan Hultin, a Swedish pathologist and physician, came across Taubenberger's paper and realized he had a solution to the problem of finding more tissue samples. Back in the 1950s, while doing research for his dissertation, Hultin had discovered a mass grave of Spanish flu victims outside a remote village in Alaska. The village had been hit especially hard by the pandemic—85% of the villagers had died from flu. Because the victims were buried in permafrost, their tissues might still be intact, he reasoned. Hultin wrote to Taubenberger and offered to take him to the site.

With permission from village elders, the scientists dug into the permafrost and exhumed dozens of bodies. Most were not

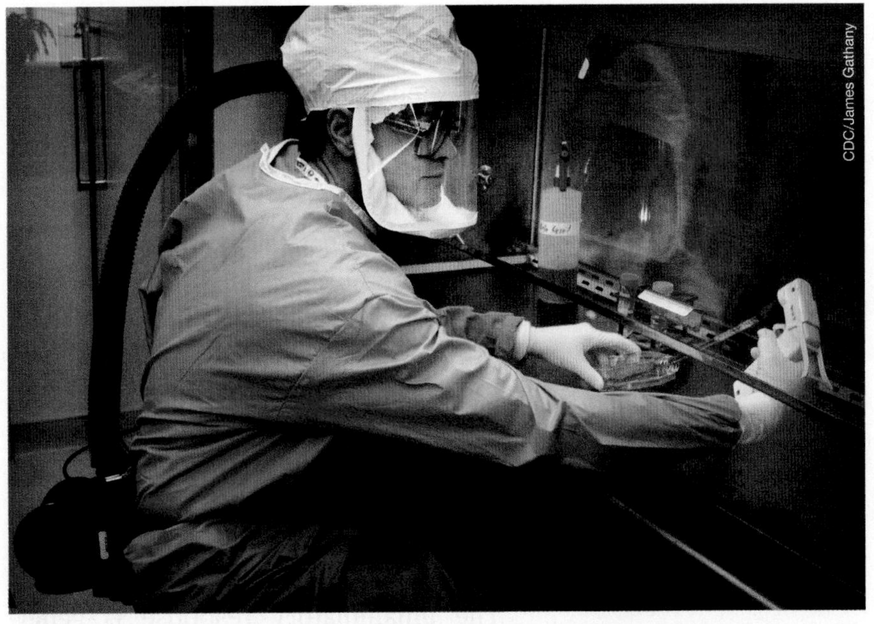

CDC/James Gathany

A CDC researcher working with samples of recreated 1918 flu virus.

Normal lung
A normal lung is clear on an x-ray, having plenty of space that can fill with air.

Pneumonia lung
An infected lung is cloudy on an x-ray. It is filled with pathogens, fluid, inflammatory cells, and debris, making it difficult to breathe. Bacterial lung infections were responsible for at least half of all deaths resulting from the 1918 flu pandemic.

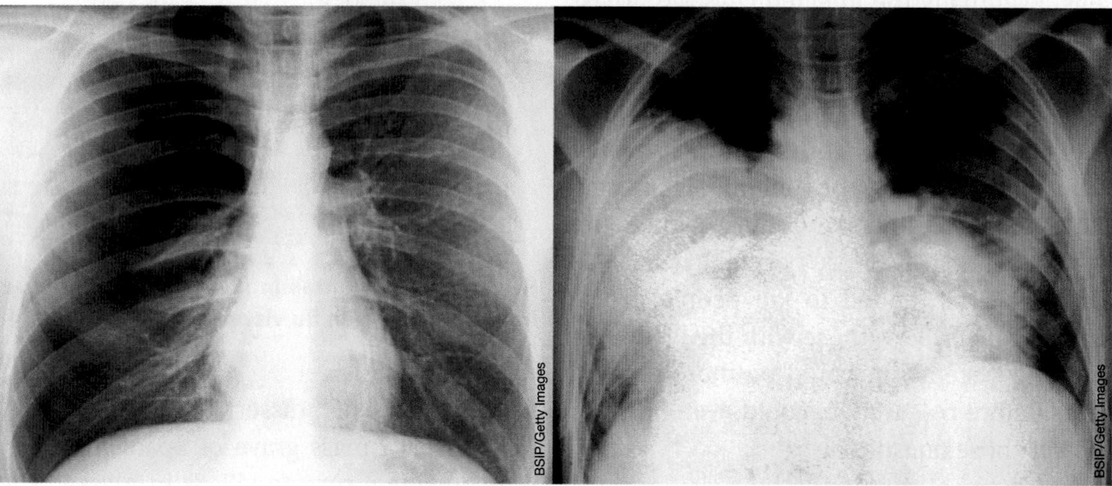
BSIP/Getty Images
BSIP/Getty Images

much more than bone, but the body of one obese woman was still relatively intact, preserved by layers of insulating fat. From her tissues, the scientists were able to obtain enough viral RNA to reconstruct the complete 1918 flu virus sequence—a landmark breakthrough. From this sequence, published in 2005, researchers were able to determine the identity of the virus and analyze its genetic relationship to other known strains of flu virus. But that's not all. Knowing the sequence of all eight of the virus's genes allowed researchers to determine which genes were responsible for the virus's extreme virulence.

It turns out that four genes with alleles specific to the 1918 virus allowed flu virus to penetrate deep into the lungs and cause infection there. While less lethal flu viruses replicate primarily in the upper respiratory tract, viruses with these four alleles cruise past the mouth, nose, and throat and invade the lungs. Replication of the virus in the epithelial cells lining the lungs killed these cells, and triggered a massive inflammatory response.

The inflammatory response is generally a good thing: it is a very effective way to contain and destroy invaders. But the balance between defense and destruction is delicate. In unusual cases, the inflammatory response can go into overdrive and destroy the very organ it is trying to save. This is what researchers now think happened with the 1918 flu. The inflammatory response of victims was so massive that it damaged lung tissue, ultimately causing people to suffocate from fluid buildup in the lungs. This aggressive immune response helps explain why victims with the strongest immune systems—people age 20 to 40—were the ones who suffered most.

Another important factor in explaining the large death toll is what came next. Because influenza destroyed the protective layer of cells lining the lungs and upper respiratory tract, bacteria and other pathogens had unimpeded entry into the lungs. Once inside, they grew like mad. Because antibiotics had not yet been discovered (see Chapter 3), doctors had no way of treating these infections. Bacterial pneumonia was responsible for at least half of all deaths during the 1918 pandemic, according to Taubenberger.

Immunological Memory

Not everyone who became infected with the 1918 flu died, despite its virulence. In fact, while there were approximately 50 to 100 million deaths, experts estimate that some 525 million people were infected—about 30% of the entire world population. How were so many people able to fight the infection while others died? Of those who became infected and then recovered, some may have had partial immunity from an earlier influenza infection. Such long-lasting immunity is conferred by the adaptive immune system.

Whereas the innate immune system is always ready to fight, the adaptive immune system must be primed over time. From birth into adulthood, our bodies are continually assaulted by pathogens. With repeated exposure, our adaptive immune system develops a memory of every pathogen we encounter that gets past our innate defenses. Should we confront the same pathogen again, immunological memory helps our bodies fight off infection before it can take hold.

To respond to and remember specific pathogens, our adaptive immune system must be able to recognize them, that is, to distinguish one from another. The particular identifying detail of a pathogen that lymphocytes recognize is called an **antigen**. An antigen could be a piece of a viral protein, a component of a bacterial cell wall, or a toxin made by the bacterium. Each antigen matches with molecular precision the exact shape of a receptor found on a lymphocyte. When a lymphocyte finds an antigen "match," it becomes activated—that is, it turns on specific immune response genes and begins to divide. You can think of an activated lymphocyte as being "armed" and ready to spring into action should it encounter an intruder with that antigen.

The phagocytes of innate immunity we encountered earlier play an important role in adaptive immunity: the antigens that activate lymphocytes are remnants of pathogens that phagocytes have swallowed and chewed up; when these cells are done chewing, they post little bits of the invaders—antigens—on their surface, where they present them to lymphocytes. This process is called antigen presentation, and the cells that do this vital work are called **antigen-presenting cells,** or APCs. Lymphocytes have receptors on their surface that can bind to antigens on the surface of APCs. A lymphocyte with a receptor that matches the antigen perfectly will bind to it, and that interaction will activate the lymphocyte. Each lymphocyte will recognize only one antigen, but since our bodies can produce billions of unique lymphocytes, each with a unique receptor, chances are good that at least one will match. This specificity is the basis of adaptive immunity **(INFOGRAPHIC 32.5)**.

Two types of lymphocyte play crucial roles in adaptive immunity: B cells and T cells. Like all blood cells, B and T cells develop from stem cells made in the bone marrow. Some immature lymphocytes in bone marrow become **B cells** (think "B" for "bone"). Other immature lymphocytes migrate from the bone marrow to the **thymus,** a gland in the chest, where they become **T cells** (think "T" for "thymus"). Both B cells and T cells eventually

Michigan Technological University Archives and Copper Country Historical Collections

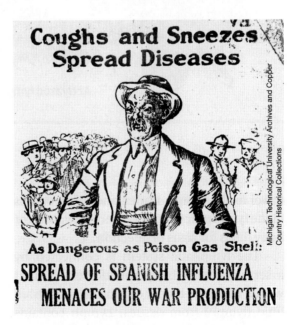

A U.S. Public Health Service ad that appeared in newspapers in 1918.

ANTIGEN
A specific molecule (or part of a molecule) to which immune receptors bind and against which an adaptive immune response is mounted.

ANTIGEN-PRESENTING CELL
A cell, such as a phagocyte, that digests pathogens and displays pieces of the pathogen on its surface, where they can be recognized by lymphocytes.

B CELLS
White blood cells that mature in the bone marrow and produce antibodies during an adaptive immune response.

THYMUS
The organ in which T cells mature.

T CELLS
White blood cells that mature in the thymus and play several roles in adaptive immunity.

INFOGRAPHIC 32.5 Lymphocytes Are Activated by Antigen-Presenting Cells

Phagocytes are antigen-presenting cells that play an important role in activating lymphocytes, arming them against specific invaders. Phagocytes ingest pathogens, digest them into small pieces, and then display these fragments on their surface. Once a phagocyte has displayed an antigen on a surface display molecule, a lymphocyte can recognize it and bind to it. Ultimately the lymphocyte is activated and can contribute to an adaptive response.

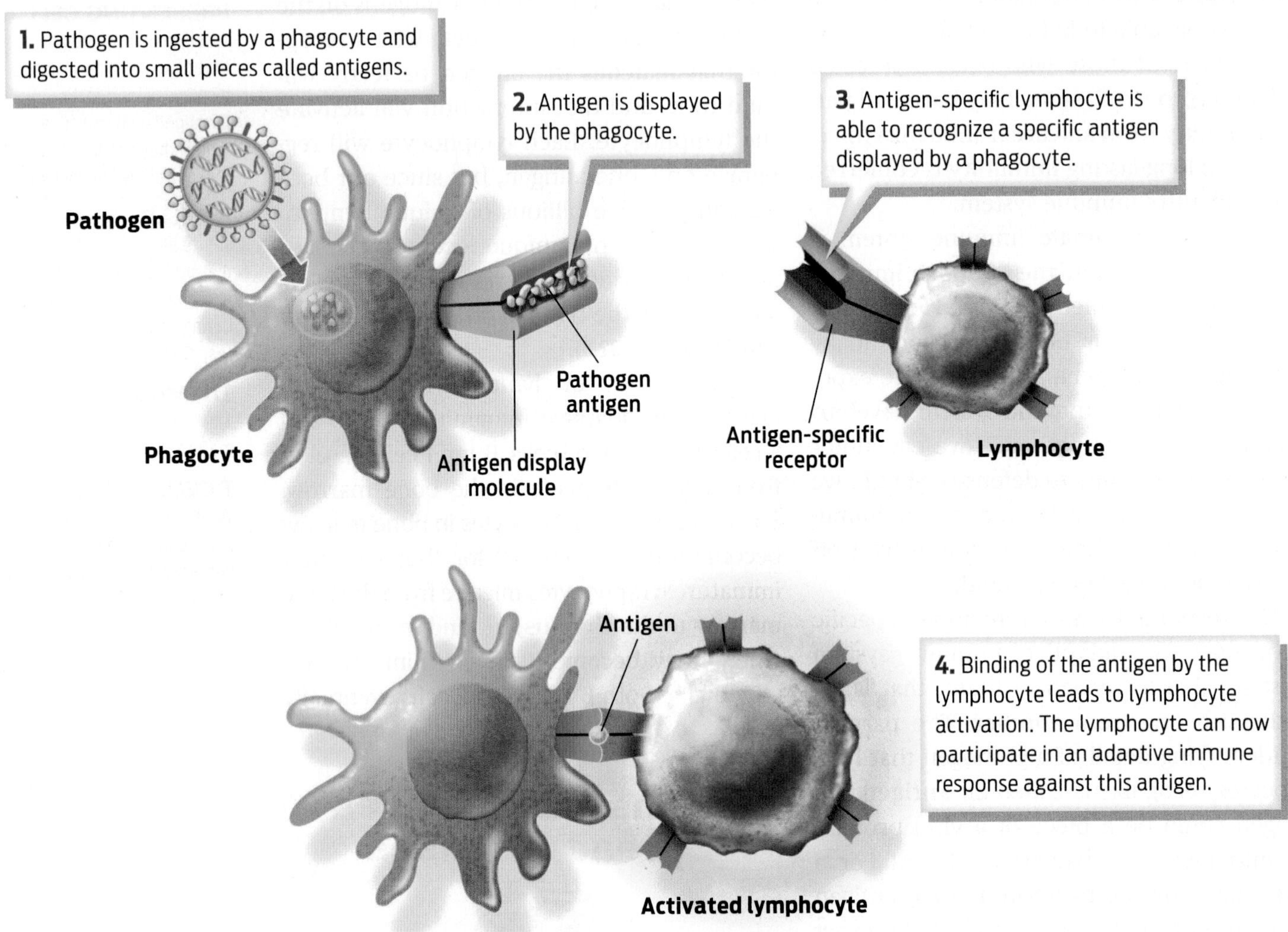

1. Pathogen is ingested by a phagocyte and digested into small pieces called antigens.
2. Antigen is displayed by the phagocyte.
3. Antigen-specific lymphocyte is able to recognize a specific antigen displayed by a phagocyte.
Pathogen
Pathogen antigen
Phagocyte
Antigen display molecule
Antigen-specific receptor
Lymphocyte
Antigen
4. Binding of the antigen by the lymphocyte leads to lymphocyte activation. The lymphocyte can now participate in an adaptive immune response against this antigen.
Activated lymphocyte

? Describe the role of phagocytes in recruiting lymphocytes to an adaptive response.

LYMPH NODES Small organs in the lymphatic system where B and T cells may encounter pathogens.

LYMPHATIC SYSTEM The system of vessels and organs that drains fluid (lymph) from the tissues and send it through lymph nodes on its way to the circulation.

travel to the **lymph nodes** and other organs of the **lymphatic system,** where they monitor body fluids for signs of infection.

The lymphatic system is connected by a set of vessels (the lymphatic vessels) that weave between blood capillaries and drain fluid (lymph) from tissues. Lymph contains white blood cells and sometimes pathogens picked up from tissues. Lymph travels from tissues to lymph nodes, where antigen presentation between phagocytes and lymphocytes occurs. From there, lymph is returned to the blood circulation through veins. In this way, the lymphatic system and the blood circulatory system are connected. Lymphocytes continually circulate through both the blood and the lymph as they patrol the body looking for invaders **(INFOGRAPHIC 32.6)**.

INFOGRAPHIC 32.6 The Lymphatic System: Where B and T Lymphocytes Develop and Act

The lymphatic system is a series of lymphatic vessels that move lymph and cells of the immune system around the body. Lymphocytes develop, mature, and act in a variety of tissues, including the bone marrow, thymus, spleen, and lymph nodes, which are part of the lymphatic system. There are two main types of lymphocyte (B and T), and each plays an important role in the adaptive response.

The Lymphatic System

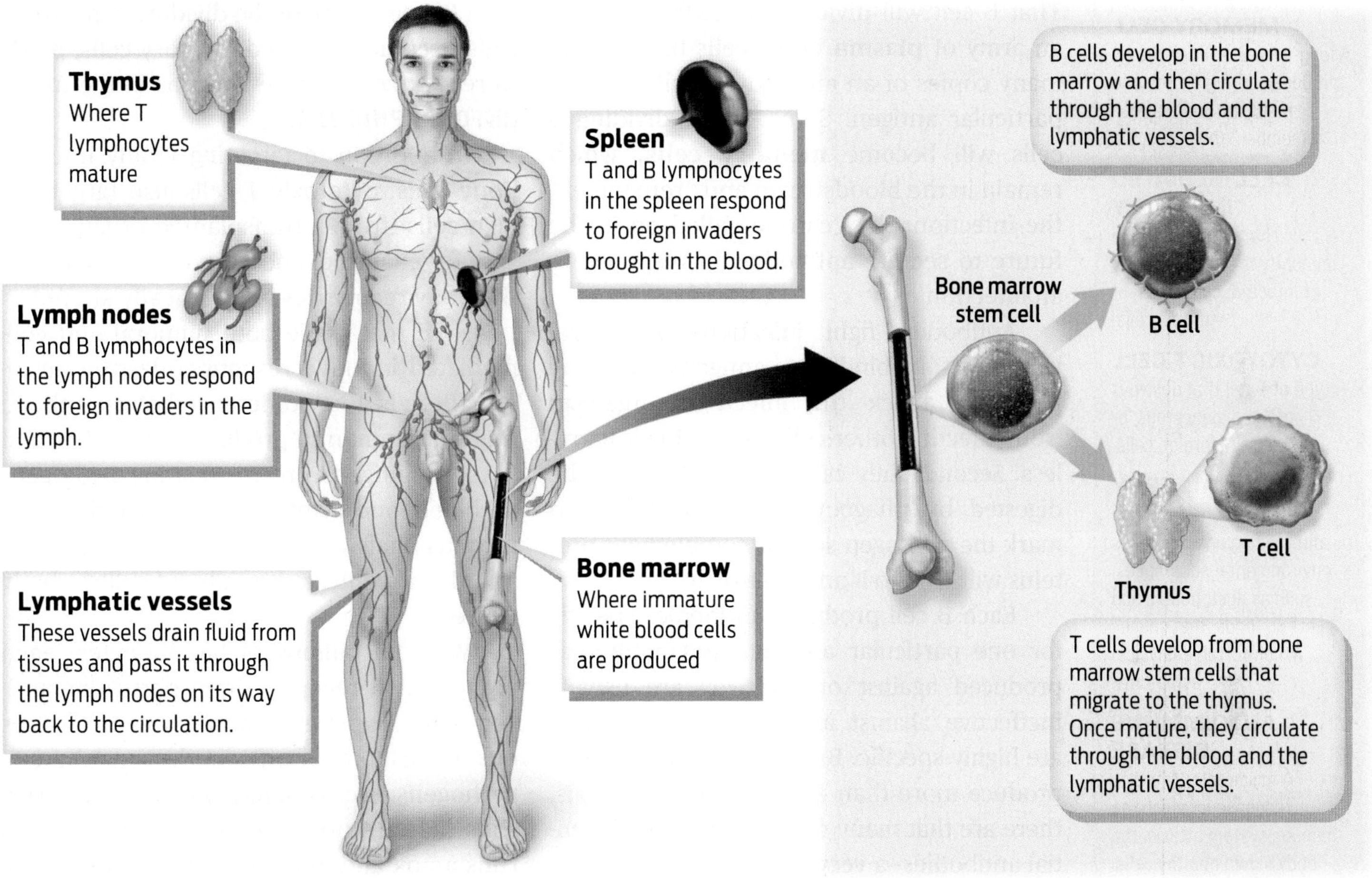

Thymus
Where T lymphocytes mature

Spleen
T and B lymphocytes in the spleen respond to foreign invaders brought in the blood.

Lymph nodes
T and B lymphocytes in the lymph nodes respond to foreign invaders in the lymph.

Lymphatic vessels
These vessels drain fluid from tissues and pass it through the lymph nodes on its way back to the circulation.

Bone marrow
Where immature white blood cells are produced

B cells develop in the bone marrow and then circulate through the blood and the lymphatic vessels.

Bone marrow stem cell

B cell

T cell

Thymus

T cells develop from bone marrow stem cells that migrate to the thymus. Once mature, they circulate through the blood and the lymphatic vessels.

? Where are B and T cells produced? Where do they mature?

B and T cells cooperate to produce two main types of adaptive immunity. One type, called **humoral immunity,** targets free-floating threats in the blood and lymph ("humoral" refers to the "humors," an old name for the liquids in the body). These threats may include virus particles, bacterial cells, and bacterial toxins.

Humoral immunity produces **antibodies**—proteins that circulate in body fluids and help to fight infections by binding to specific antigens on pathogens. Antibodies are produced by B cells. Like all lymphocytes, B cells have receptors on their surface that recognize specific antigens. When a B cell encounters an antigen on a pathogen, it binds to it with its receptor. The B cell then internalizes the pathogen, digests it, and presents these antigens on its surface. (B cells are a type of antigen-presenting cell.) To begin producing

HUMORAL IMMUNITY
The type of adaptive immunity that fights free-floating pathogens in the blood and lymph.

ANTIBODY
A protein produced by B cells that fights infection by binding to specific antigens on pathogens.

HELPER T CELL
A type of T cell that helps activate other lymphocytes including B cells and cytotoxic T cells.

PLASMA CELL
An activated B cell that divides rapidly and secretes an abundance of antibodies.

MEMORY CELL
A long-lived B or T cell that is produced during an immune response and that can "remember" the pathogen.

CELL-MEDIATED IMMUNITY
The type of adaptive immunity that rids the body of infected, cancerous, or foreign cells.

CYTOTOXIC T CELL
A type of T cell that destroys infected, cancerous, or foreign cells.

ALLERGY
An immune response misdirected against harmless environmental substances, such as dust, pollen, and certain foods, that causes uncomfortable physical symptoms.

AUTOIMMUNE DISEASE
A misdirected immune response in which the immune system attacks the body's own healthy cells.

antibodies that recognize and bind to that antigen, the B cell needs the assistance of specialized T cells called **helper T cells.** These cells also have receptors that can recognize antigens. When an activated helper T cell meets a B cell that has encountered the same pathogen, the helper T cell releases signaling molecules that trigger the B cell to divide. That B cell will divide repeatedly to create an army of **plasma cells**–cells that secrete many copies of an antibody specific to that particular antigen. Some of the dividing B cells will become **memory cells,** which remain in the bloodstream and "remember" the infection; they can be called on in the future to secrete antibodies in response to an infection.

Antibodies fight infections in several ways. First, by binding to antigens, they may physically block the infectious organism from infecting other cells, rendering it harmless. Second, they can flag a pathogen to be digested by phagocytes. Finally, they can mark the pathogen so that complement proteins will bind to it and destroy it.

Each B cell produces antibodies specific for one particular antigen. And antibodies produced against one antigen are usually ineffective against any other antigen–they are highly specific. But because the body can produce more than a billion unique B cells, there are that many different kinds of potential antibodies–a very large war chest indeed (**INFOGRAPHIC 32.7**).

The second type of adaptive immunity is **cell-mediated immunity,** which targets infected or altered body cells–for example, cells infected with the influenza virus. This response requires a type of T cell called a **cytotoxic T cell.** These important cells, also called killer T cells, are the immune system's elite fighting force; they are ones that actually kill body cells that are compromised by foreign invaders.

Cytotoxic T cells are activated by antigens they encounter on the surface of phagocytes that have engulfed pathogens. As with humoral immunity, helper T cells assist in the process: they provide a growth signal to killer T cells to multiply.

Activated cytotoxic T cells divide and patrol the body on the lookout for infected cells displaying the antigens that activated them. When such cells are found, the cytotoxic T cells bind to the antigens and release chemicals that kill the rogue cells. As with B cells, some of the dividing cytotoxic T cells become long-lived memory cells, ready to recognize the same antigens in the future (**INFOGRAPHIC 32.8**).

In addition to targeting virally infected body cells, cytotoxic T cells also target foreign cells (from a transplanted organ or tissue, for example) and even cancer cells that the body recognizes as genetically altered.

Note that while both humoral and cell-mediated immune responses destroy invaders, there is a key difference between them: humoral immunity produces antibodies that bind to antigens on free-floating pathogens in blood and lymph; cell-mediated immunity marshals cytotoxic T cells that bind to and destroy infected or altered cells in body tissues.

With its billions of lymphocytes, each equipped to recognize one particular antigen, the adaptive immune system enables our bodies to fight off countless numbers of pathogens. Occasionally, however, for reasons that are not entirely clear, the system runs amok and immune cells become active even against antigens that aren't harmful to us, or worse, they begin attacking healthy cells in the body.

When the immune system attacks harmless antigens from outside the body, like those in dust, pollen, or certain types of food, an **allergy** results. Allergies are quite common, often accompanied by a runny nose, watery eyes, and sneezing–all triggered by histamine, the same chemical involved in inflammation. When the immune system attacks the body's own healthy cells, a more serious **autoimmune disease** can occur. Multiple sclerosis, lupus, and rheumatoid arthritis are all autoimmune diseases, caused by a body at war with itself.

INFOGRAPHIC 32.7 Humoral Immunity: B Cells Produce Antibodies

The humoral immune response protects the body from free-floating pathogens found in the blood and lymph. When a B cell and a helper T cell recognize the same antigen, the T cell can activate the B cell to divide and become plasma cells. Plasma cells secrete antibodies that bind to the specific antigens present only on the target pathogens. The binding inactivates and destroys these pathogens. Specific memory B cells are also produced and spring into action if the same pathogen is encountered again in the future.

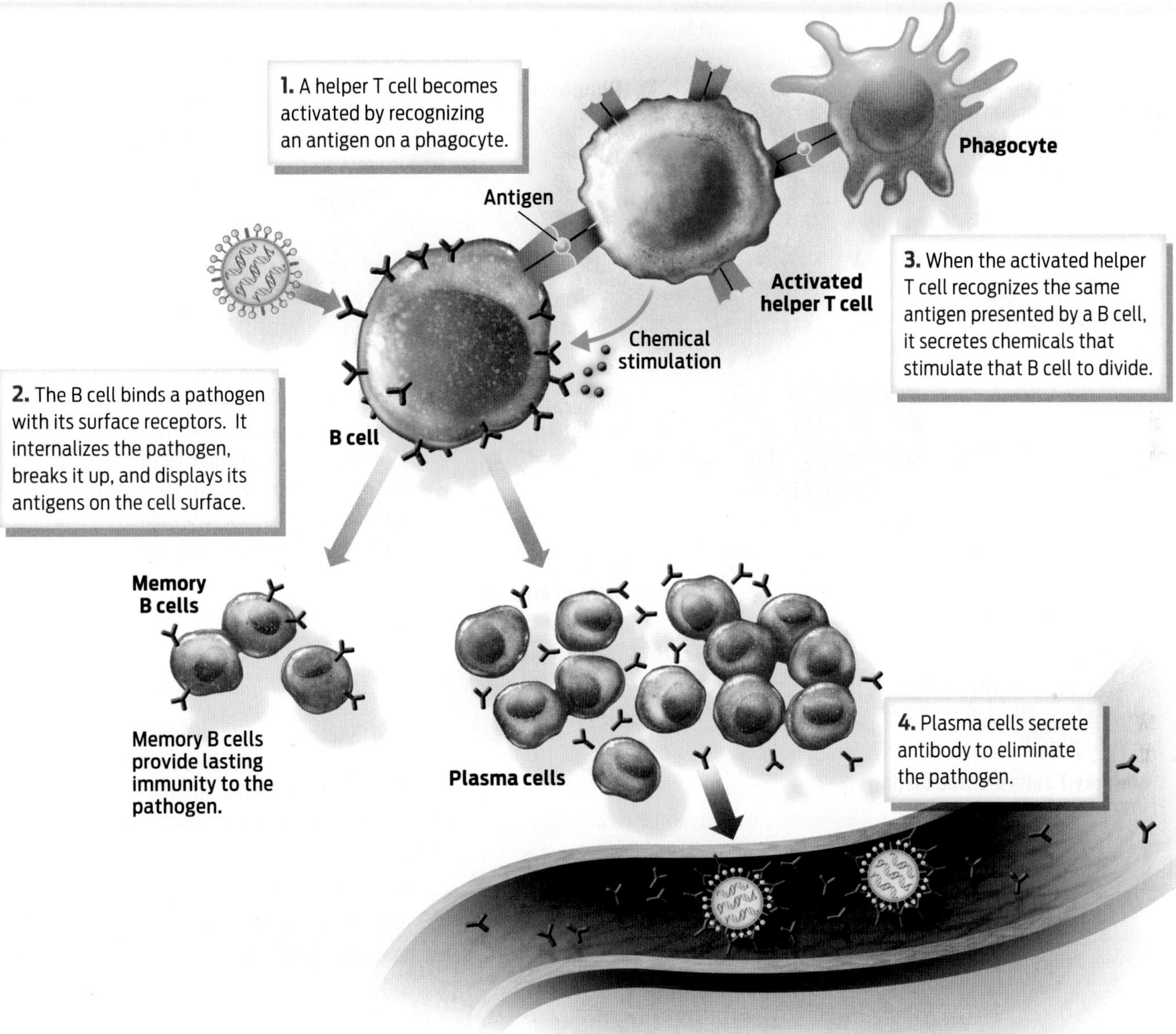

1. A helper T cell becomes activated by recognizing an antigen on a phagocyte.
Phagocyte
Antigen
Activated helper T cell
3. When the activated helper T cell recognizes the same antigen presented by a B cell, it secretes chemicals that stimulate that B cell to divide.
Chemical stimulation
2. The B cell binds a pathogen with its surface receptors. It internalizes the pathogen, breaks it up, and displays its antigens on the cell surface.
B cell
Memory B cells
Memory B cells provide lasting immunity to the pathogen.
Plasma cells
4. Plasma cells secrete antibody to eliminate the pathogen.

? What molecules are produced by plasma cells? How do these molecules protect the body during an adaptive response?

Building a Line of Defense

The advantage of adaptive immunity is that it "learns" to respond to new invaders and targets them specifically. The downside is that it takes a while for the response to kick in. First-time exposure to a pathogen that has breached our innate defenses will almost certainly cause illness, because the adaptive response takes 7 to 10 days to develop. Over time an exposed individual will recover as

INFOGRAPHIC 32.8 Cell-Mediated Immunity: Cytotoxic T Cells Kill Infected Cells

The cell-mediated immune response defends against infected or altered (e.g., cancerous) body cells. Once activated, cytotoxic T cells bind to antigens on target cells and release chemicals that destroy them. They also divide to produce memory cytotoxic T cells.

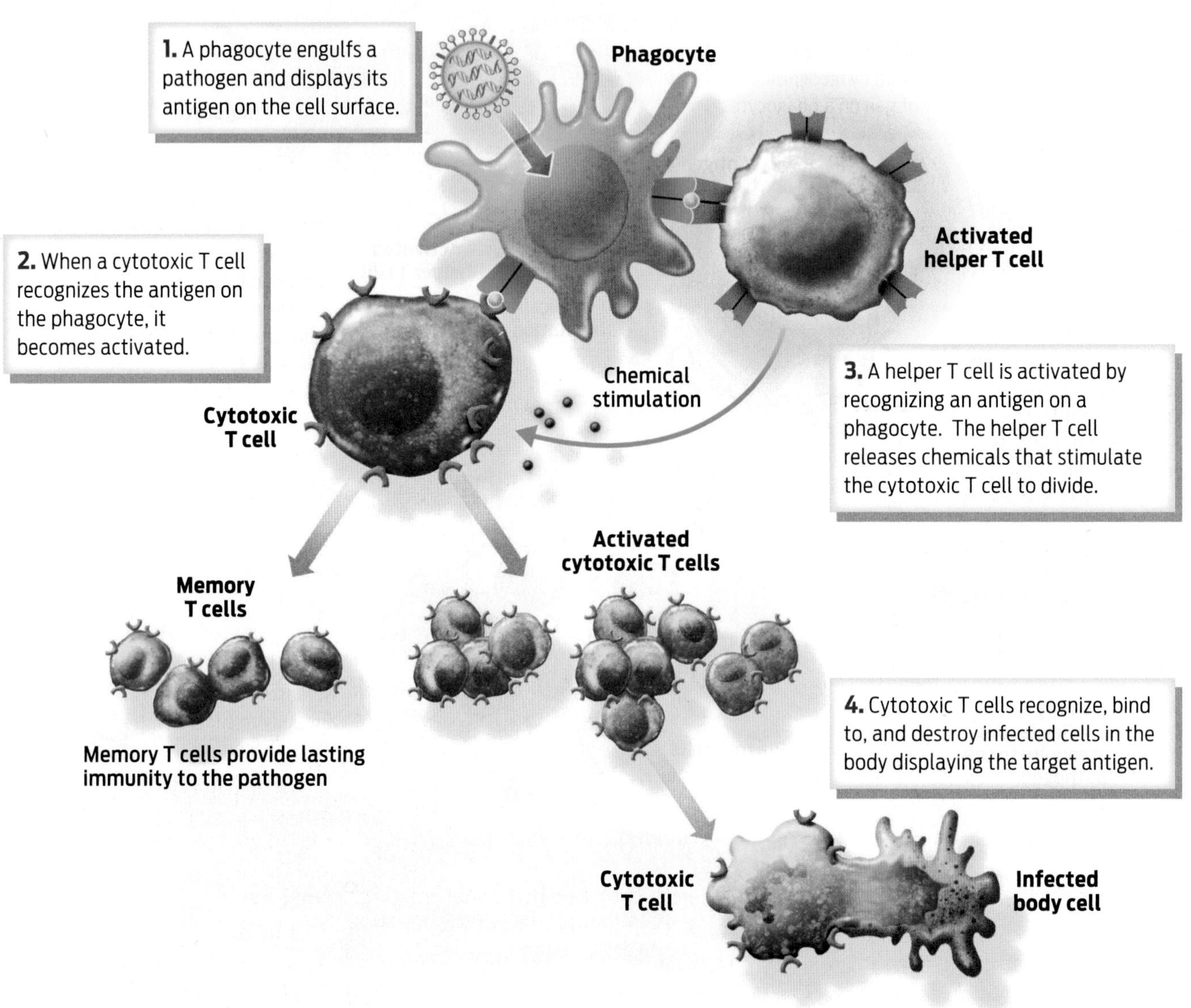

1. A phagocyte engulfs a pathogen and displays its antigen on the cell surface.
Phagocyte
Activated helper T cell
2. When a cytotoxic T cell recognizes the antigen on the phagocyte, it becomes activated.
Chemical stimulation
Cytotoxic T cell
3. A helper T cell is activated by recognizing an antigen on a phagocyte. The helper T cell releases chemicals that stimulate the cytotoxic T cell to divide.
Activated cytotoxic T cells
Memory T cells
Memory T cells provide lasting immunity to the pathogen
4. Cytotoxic T cells recognize, bind to, and destroy infected cells in the body displaying the target antigen.
Cytotoxic T cell
Infected body cell

? How does a cytotoxic T cell recognize a cell as a target for destruction?

PRIMARY IMMUNE RESPONSE
The adaptive immune response mounted the first time a particular antigen is encountered by the immune system.

T and B cells are activated and antibody levels increase. This initial slow response is the **primary immune response.** As B and T cells are churned out, some of them become memory cells. These memory cells remain in the body and "remember" the infection. The next time the same pathogen is encountered, memory B and T cells become active, dividing rapidly and leading to the production of very high levels of antibodies. They fight the specific pathogen so quickly that the illness usually doesn't occur a second time. This rapid reaction is the **secondary immune response** (**INFOGRAPHIC 32.9**).

INFOGRAPHIC 32.9 Memory Cells Mount an Aggressive Secondary Response

The adaptive immune system's primary humoral response is slow and produces low levels of antibodies. Upon subsequent exposures, memory B cells produced during the primary response respond quickly, rapidly dividing and developing into plasma cells that produce high concentrations of antibodies.

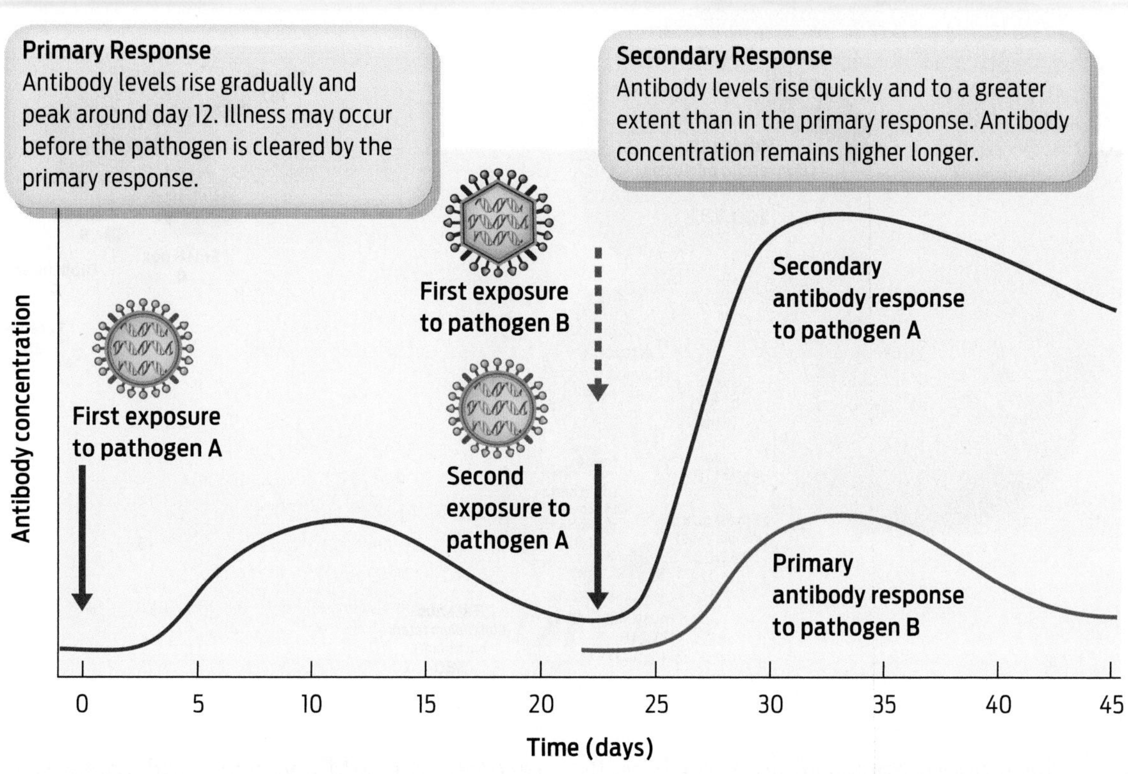

? If a third pathogen (pathogen C) is encountered at day 22, what do you predict about the response to that pathogen?

The secondary immune response is what confers immunity to a particular infection. It's also how **vaccines** work. Vaccines are essentially dead or weakened versions of a pathogen administered in order to generate immunity to that pathogen. The goal is to create a primary immune response in the body without causing disease. Thus, if the pathogen is subsequently encountered naturally, the secondary response is already primed. Vaccination is like being infected with a pathogen but not having the disease.

It is hard to overstate the impact that vaccines have had on public health. Many once-deadly and disfiguring diseases have been relegated to the historical dustbin thanks to these medicines. Many people today cannot remember the time when polio meant confinement in an iron lung in order to breathe, and often-fatal small pox caused blisters and then left scars all over the body.

In fact, you could say that vaccines have become victims of their own success. It's precisely because they have been so effective at reducing the burden of illness over the last hundred years that some people have begun to question whether it's really in their best interest to vaccinate their children. The movement against vaccination is sometimes called antivaccination.

SECONDARY IMMUNE RESPONSE
The rapid and strong immune response mounted when a particular antigen is encountered by the immune system subsequent to the first encounter.

VACCINE
A preparation of killed or weakened pathogen that is administered to people or animals to generate protective immunity to that pathogen.

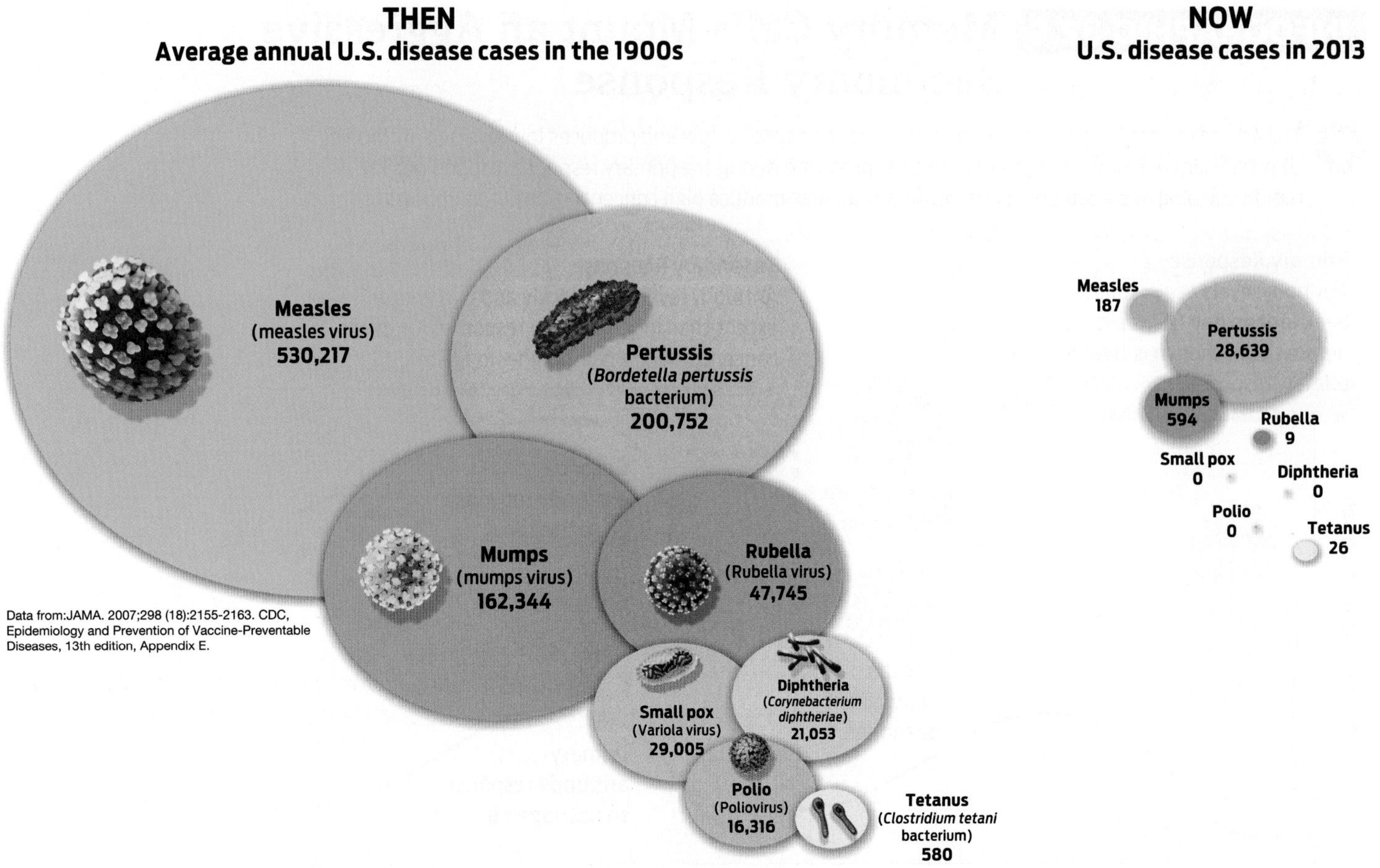

THEN
Average annual U.S. disease cases in the 1900s

Measles (measles virus) **530,217**

Pertussis (*Bordetella pertussis* bacterium) **200,752**

Mumps (mumps virus) **162,344**

Rubella (Rubella virus) **47,745**

Small pox (Variola virus) **29,005**

Diphtheria (*Corynebacterium diphtheriae*) **21,053**

Polio (Poliovirus) **16,316**

Tetanus (*Clostridium tetani* bacterium) **580**

NOW
U.S. disease cases in 2013

Measles 187

Pertussis 28,639

Mumps 594

Rubella 9

Small pox 0

Diphtheria 0

Polio 0

Tetanus 26

Data from:JAMA. 2007;298 (18):2155-2163. CDC, Epidemiology and Prevention of Vaccine-Preventable Diseases, 13th edition, Appendix E.

The modern antivaccination push really took off in the late 1990s, in the wake of a fraudulent research paper published by a doctor in England, Andrew Wakefield. That paper, which was published in the journal *The Lancet*, purported to show a link between the MMR (measles-mumps-rubella) vaccine and the development of autism in 8 children. However, numerous subsequent studies failed to confirm the finding, and Wakefield was ultimately accused of scientific misconduct and had his medical license revoked. The paper was retracted.

Since Wakefield's paper was published, many large studies have looked specifically at the relationship between vaccines and autism and found no link whatsoever. For example, a study published in 2002 in the *New England Journal of Medicine* looked at all children born in Denmark between 1991 and 1998—537,303 children—and compared the rate of autism among those children who received the MMR vaccine and those who did not. The authors' conclusion: "Overall, there was no increase in the risk of autistic disorder or other autistic-spectrum disorders among vaccinated children as compared with unvaccinated children."

Although the best medical science has disproved any link between vaccination and autism, a small but growing number of parents have begun to forgo vaccination of their children out of fear of possible side effects. The unfortunate result is a decrease in what's called **herd immunity.** In order for a population to be adequately protected from an infection, a certain percentage of its members must be immune, either through prior infection or vaccination. If the percentage falls below a certain number (different for different pathogens, but in the range of 80%-90%), then infections can take hold in the population and spread among those who are

HERD IMMUNITY
The protection of a population from an infection, based on a certain percentage of its members being immune.

not vaccinated—including those who are too young to be vaccinated or who cannot be vaccinated because they have weak immune systems (**INFOGRAPHIC 32.10**).

The decrease in herd immunity is responsible for the increased number of outbreaks of measles and mumps that have occurred in recent years in several U.S. states and Canada. These diseases are nothing to sneeze at: about 1 in 20 children with measles will contract pneumonia, which can lead to death. About 1 in 1,000 children with measles will develop encephalitis (swelling of the brain) that can lead to convulsions, deafness, and intellectual disability. And for every 1,000 children who get measles, 1 or 2

INFOGRAPHIC 32.10 Herd Immunity Protects against the Spread of Infections

When a sufficient proportion of people in a community is immune to a particular contagious pathogen (because they have been vaccinated, for example), then that infection cannot spread in that community. Low rates of vaccination lead to low rates of herd immunity and the spread of an infection in a community. High rates of vaccination protect those who cannot be vaccinated, for example, very young babies and people with cancer.

Large Outbreak:
No one is protected.

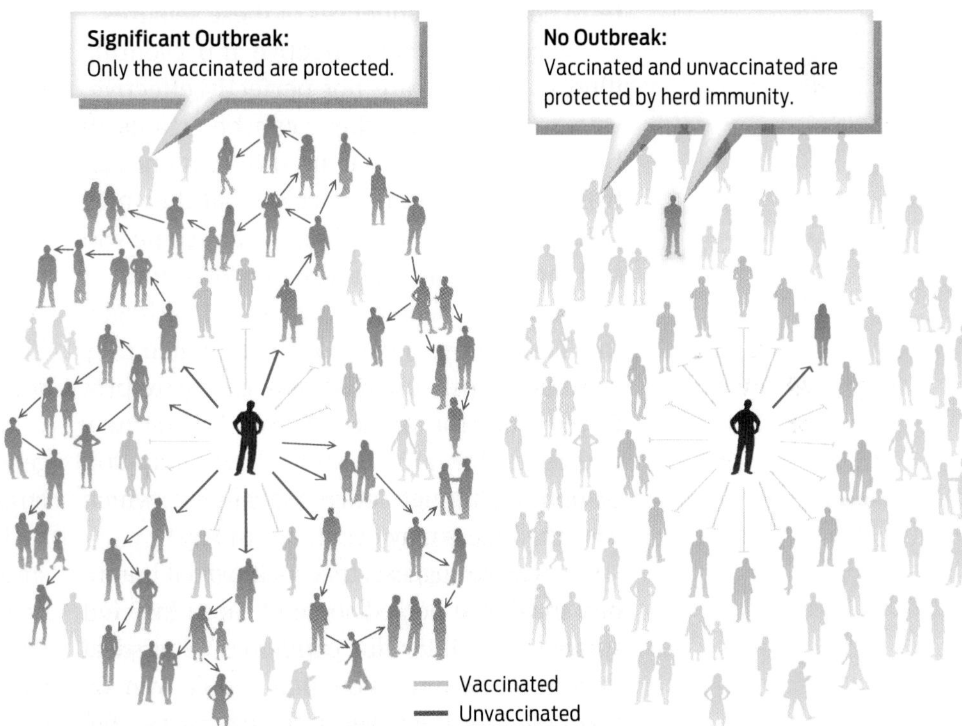
Significant Outbreak:
Only the vaccinated are protected.
No Outbreak:
Vaccinated and unvaccinated are protected by herd immunity.
Vaccinated
Unvaccinated

Unvaccinated Community

In a community where no one has been vaccinated, an outbreak can easily occur. One person can infect dozens of unvaccinated people, each of whom goes on to infect dozens more.

Partially Vaccinated Community

In a community where some people have been vaccinated, the infected person is able to infect multiple unvaccinated people, creating a significant outbreak of the disease. Some people (those who have been vaccinated) are protected.

Vaccinated Community

In a community where most people have been vaccinated, the infected person is surrounded by people who cannot become infected. An outbreak cannot occur. For this scenario to be effective, a high percentage of people must be immune to the disease.

? If a community is experiencing a measles outbreak, what can you infer about vaccination rates in that community?

ANTIGENIC DRIFT
Changes in viral antigens caused by genetic mutation during normal viral replication.

ANTIGENIC SHIFT
Changes in viral antigens that occur when viruses of one strain exchange genetic material with other strains.

will die. Yet this disease is largely preventable with routine vaccination.

How safe are vaccines? Remarkably safe. While any medicine comes with some amount of risk–and vaccines are no different–there is no comparison between the risks posed by vaccines and by the diseases they aim to prevent. For most people, vaccine side effects are nonexistent or very mild. In rare cases, vaccines may cause fever–a consequence of stimulating an immune response against the antigens in the vaccine. In even rarer cases, these fevers can cause seizures. For example, for every 10,000 children who get the MMR vaccine as infants, about 4 will have a fever-induced seizure during the week following vaccination; there are usually no long-term health consequences of these seizures.

Vaccines are very effective at preventing illness, but they are not perfectly effective. One reason is that there can be strains of bacteria or viruses that are different enough from that found in the vaccine that they may still cause illness. Even then, the vaccine will still provide some measure of protection.

Similarly, people exposed to a pathogen that is similar to a pathogen with which they were previously infected may be partly protected from the disease caused by the new pathogen. Memory B and T cells may still respond, although only partially–in which case the illness may occur, but mildly.

Evidence suggests that such partial immunity may have helped some of those infected survive the 1918 flu pandemic. Statistical data from the time show that people over 65 accounted for the fewest influenza cases, suggesting that they might over the years have acquired immunity or partial immunity from earlier infection. Partial immunity might have helped these people fight off the virus before it dug deep into their lungs.

An Evolving Enemy

When researchers discovered that the 1918 flu virus carried specific alleles that enabled the virus to replicate in the lungs, making it more deadly, a piece of the 75-year-old mystery was solved. But where did these alleles come from?

New influenza viruses are constantly being produced by two mechanisms: mutation and gene swapping. Because influenza viruses replicate their genetic material so rapidly and don't "proofread" the replicated copies, mistakes often occur, leading to mutations. **Antigenic drift** is the gradual accumulation of mutations that cause small changes in the antigens on the virus surface. Antigenic drift explains why there can be different types, or strains, of influenza circulating at the same time.

Two important antigens on the influenza virus are hemagglutinin and neuraminidase. Hemagglutinin is a viral protein that binds to receptors on host cells and enables the virus to enter host cells; the neuraminidase protein helps newly formed viruses exit host cells. A person's immune system mounts an adaptive response specifically to these two antigens. When there is a change in hemagglutinin or neuraminidase or both, his or her memory cells no longer recognize these viral proteins. These new antigens prompt a new and slow primary immune response.

An influenza virus strain can also swap genetic material with other strains of influenza, including strains that infect animals such as birds and pigs. While every influenza strain contains the same set of genes, the particular alleles of these genes that are present differ from strain to strain. The exchange of alleles between two strains that have infected the same cell does not simply create a small change in viral gene sequence–it introduces an entirely new allele, and therefore an entirely new antigenic protein. This process, called **antigenic shift,** is responsible for pandemic outbreaks of flu. Because we have no existing immunity to protect against infection by emerging strains of flu, these strains can spread rapidly throughout the human population. The severity of the 1918 flu is believed to have resulted from antigenic shift. Antigenic shift is also thought to have played a role in the emergence of avian flu in the 1990s and swine flu in 2009, both of which originated in animals.

Together, antigenic drift and antigenic shift create an increasing variety of strains over time until one of the variants is able to infect human cells so efficiently that it sweeps through the population and causes a pandemic. These strains are named by the type of H (hemagglutinin) and N (neuraminidase) proteins they carry, with some antigenic combinations more deadly than others (**INFOGRAPHIC 32.11**).

Tracking the strains as they move through the population helps public health officials predict how severe a coming flu season will be. But since viruses can mutate every time they

INFOGRAPHIC 32.11 Antigenic Drift and Shift Create New Influenza Strains

Mutation and gene exchange are two mechanisms by which influenza viruses can change. Mutations that accumulate gradually can cause variations in surface antigen. This process is called antigenic drift. Different strains can also swap genes and cause surface antigens to change more dramatically. This is called antigenic shift. Drift is responsible for annual seasonal variation in influenza; shift is responsible for dramatic pandemics.

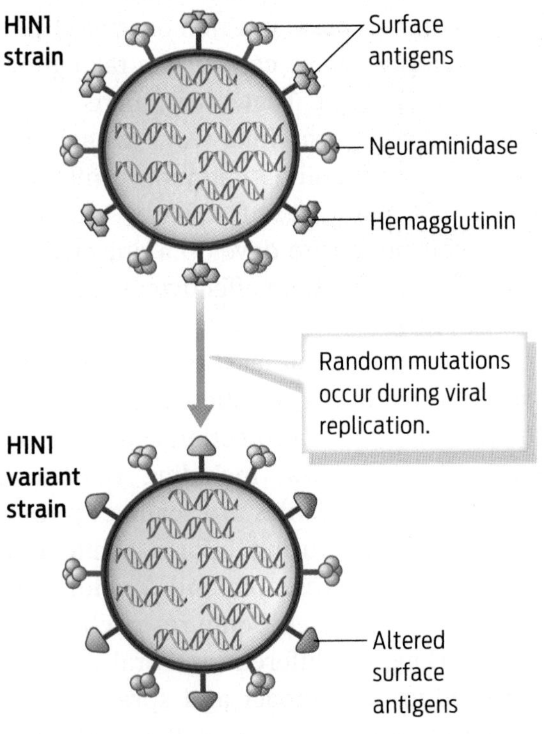

Antigenic Drift
Gradual change
Caused by point mutations that occur when the virus replicates
H1N1 strain
Surface antigens
Neuraminidase
Hemagglutinin
Random mutations occur during viral replication.
H1N1 variant strain
Altered surface antigens

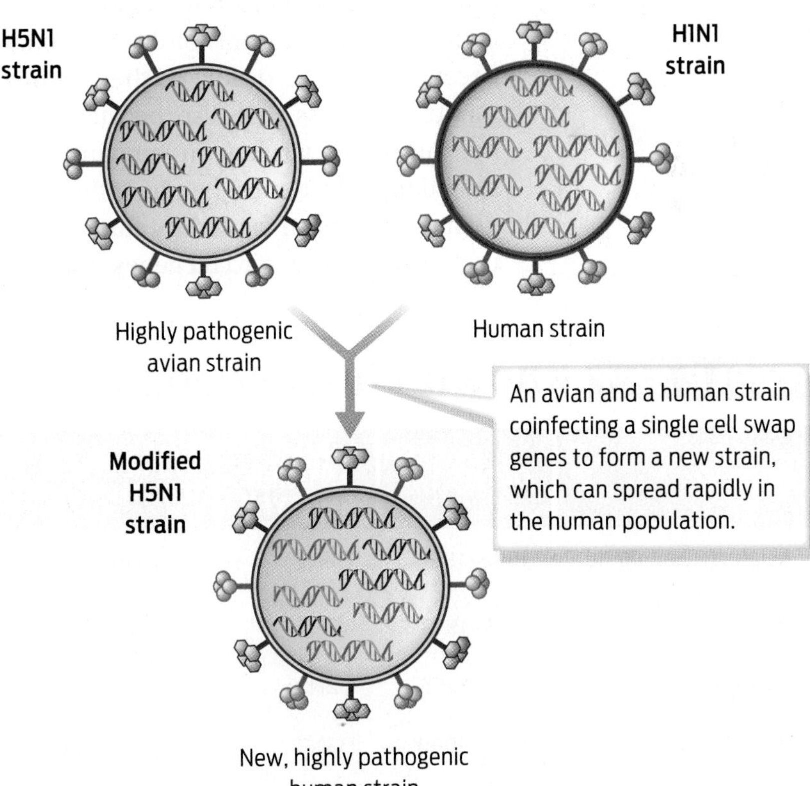

Antigenic Shift
Rapid change
Caused by gene exchange between two different viruses that simultaneously infect the same cell
H5N1 strain
H1N1 strain
Highly pathogenic avian strain
Human strain
An avian and a human strain coinfecting a single cell swap genes to form a new strain, which can spread rapidly in the human population.
Modified H5N1 strain
New, highly pathogenic human strain

? This year's flu virus has only slightly different H and N antigens relative to last year's. What mechanism generated this year's strain?

reproduce, they continue to evolve during epidemics. This unpredictability is what makes influenza so frightening: an apparently mild outbreak can suddenly become deadly.

Preventing Pandemics

Because viruses can swap genes so easily, public health officials closely monitor the types of virus that infect animals. Birds are a primary source of potentially pandemic viruses. Wild birds are routinely infected with viruses that live in their gastrointestinal tracts but cause little or no disease in the bird. Sometimes, however, bird flu viruses can jump from wild birds to poultry or other farm animals and become lethal. And since many human populations maintain close contact with poultry, pigs, and other domestic animals, the risk of an animal virus infecting humans is high.

In the 1990s, a lethal strain of the avian flu virus (H5N1) crossed over into humans from domesticated birds, but it was not easily transferred from person to person. Scientists continue to monitor this bird strain, not only because it can be deadly to the commercial poultry business, but also because it could become more virulent in humans through mutation. In fact, most human influenza pandemics have been caused by flu viruses that carry variations of bird flu genes (**TABLE 32.1**).

TABLE 32.1 Known Flu Pandemics

NAME	DATE	VIRUS ANTIGEN SUBTYPE	NUMBER OF DEATHS
Asiatic (Russian) flu	1889–1890	H2N2	1 million
1918 (Spanish) flu	1918–1919	H1N1	20–100 million
Asian flu	1957–1958	H2N2	1–1.5 million
Hong Kong flu	1968–1969	H3N2	0.75–1 million
Swine flu	2009–2010	H1N1	150,000–575,000

To help ward off flu infection, public health officials offer flu shots every year at the beginning of flu season, which in the northern hemisphere typically runs from October to February. A flu shot contains a vaccine: a weakened version or part of a flu virus is injected into the body in the hope of generating a primary immune response and memory cells against that strain of virus. However, since influenza viruses mutate so frequently, a flu shot may not protect us from getting the flu the following year.

Scientists create a new vaccine each year by tracking which strains of influenza are circulating worldwide and studying the antigens on their surfaces. But their predictions can be wrong. If public health officials decide to vaccinate against a strain of influenza with one variant of hemagglutinin (H) antigen but a strain with a different variant of the antigen strikes, those who are vaccinated may become ill anyway.

The 1918 flu pandemic taught scientists a lot about how viruses can become lethal in a matter of days and demonstrated that developing vaccines can be like taking aim at a moving target. That's why, increasingly, scientists are researching ways to make flu vaccines containing common influenza genes that mutate less frequently than others. Their hope is to develop a universal flu vaccine that would be effective over a number of years.

There are now at least two universal flu vaccines that have been shown in laboratory animals to provide protection against lethal strains of influenza, including H5N1 and H1N1. Some companies are scaling up their vaccine manufacturing efforts, while others are boosting research and development of universal vaccines. Some researchers are also working on different antiviral drugs to help reduce the impact and spread of the influenza virus. The hope is that such measures will suppress the next big flu outbreak, whenever and wherever it happens. ■

Defending Oneself Is a Fact of Life

The adaptive immune system that protects humans and other vertebrates from infection is a relatively recent adaptation. Evidence suggests that adaptive immunity first arose in fishes, some 500 million years ago. By contrast, the innate immune system is much older and is shared by creatures as diverse as sea sponges, fruit flies, and humans. In fact, all organisms have evolved ways to defend themselves from infectious threats.

Many bacteria are able to protect themselves from a class of viruses known as bacteriophage–"phage" for short. Phage infection causes a bacterial cell to burst, or lyse–a life-ending event for that bacterium. Bacteria defend themselves against phage by producing enzymes that act like scissors, cutting up the DNA that phage inject into the bacterial cell. A bacterium's own DNA is protected from these powerful molecular scissors by a mechanism that chemically modifies its DNA, ensuring that it won't be cut up by these enzymes. (This bacterial defense system is the basis for the genome-editing tool CRISPR discussed in Chapter 24.)

Many invertebrate animals have an innate immune system armed with specialized cells that attack or neutralize invaders. Sea stars are a good example. It has long been known that sea stars very rarely develop bacterial infections. It turns out that they have cells called amoebocytes that act much like the phagocytes of the vertebrate innate immune system. Amoebocytes circulate in the body cavity fluid of sea stars. When bacteria are injected

Bacterial enzymes defend against infection

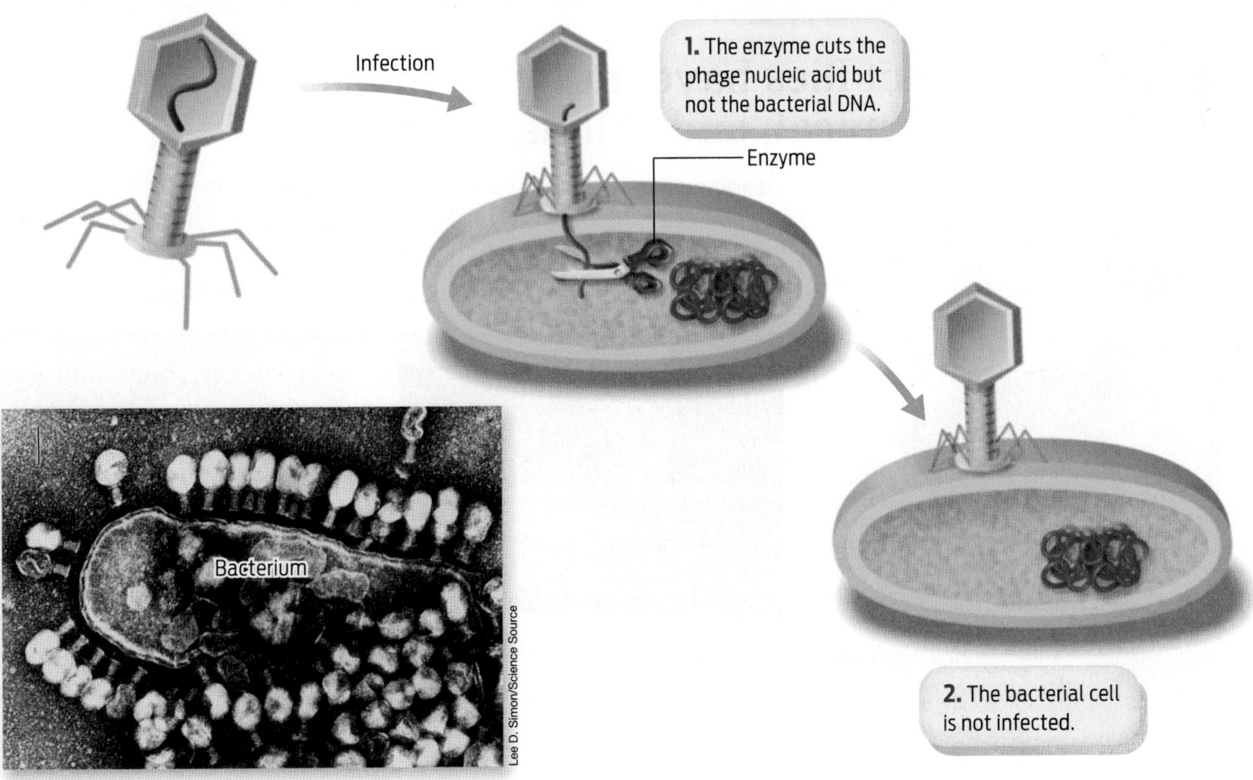
Infection
1. The enzyme cuts the phage nucleic acid but not the bacterial DNA.
Enzyme
Bacterium
Lee D. Simon/Science Source
2. The bacterial cell is not infected.

into the body cavity of sea stars, the amoebocytes engulf and destroy the bacteria within 10 minutes.

Fruit flies also have several innate defense mechanisms. Cells called plasmatocytes ingest and destroy foreign cells by phagocytosis. Fruit flies also have cells called lamellocytes that can coat and essentially wall off foreign objects that are too big to be phagocytosed, such as the eggs of parasitic wasps. This walling off, or encapsulation, helps destroy the foreign object. Humans carry out a similar process in response to lung infections, such as those caused by tuberculosis. In humans, immune cells surround and wall off the pathogen in a structure called a granuloma.

That many organisms from many different phyla share similar kinds of immune response is evidence that these defensive strategies evolved early in the history of life and were a crucial part of the organisms' survival and reproduction.

Sea stars have innate cellular defenses

Paul Kay/Getty Images

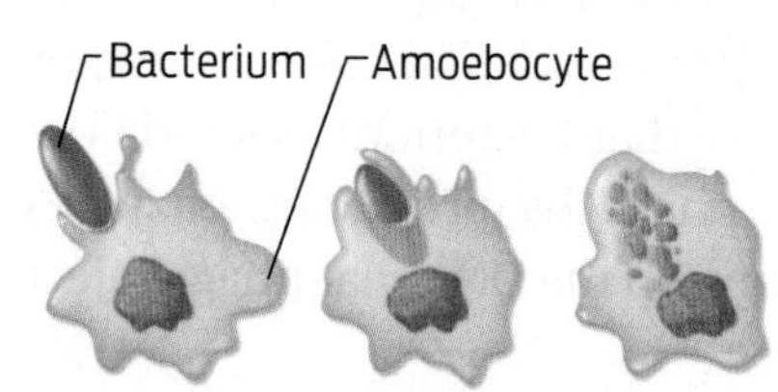

Phagocytosis
Amoebocytes in the body cavity fluid ingest and destroy invading bacteria.

Fruit flies have multiple physical, chemical, and cellular defenses

External Barriers
Fruit flies have a hard exoskeleton and specialized cells that line the airways and digestive system.

Encapsulation
Lamellocytes (green) surround and wall off infecting parasites (black) inside living *Drosophila* larvae.

Phagocytosis
Phagocytic cells known as plasmatocytes ingest and destroy invaders.

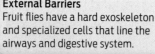

playTOME, FeaturePics.com

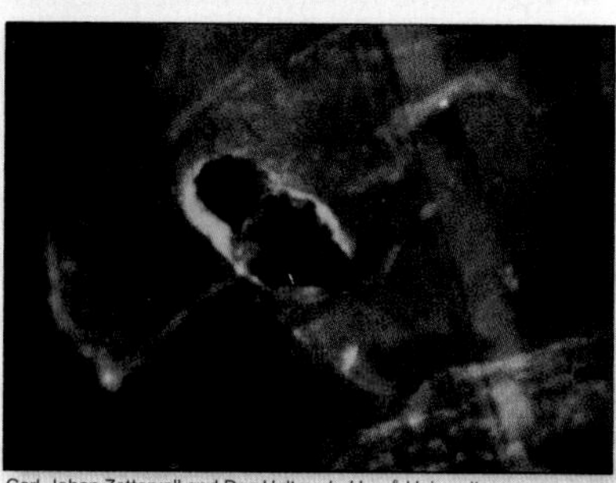

Carl-Johan Zettervall and Dan Hultmark, Umeå University.

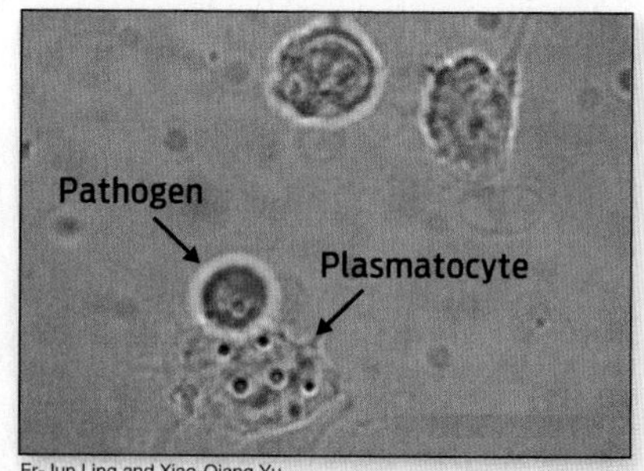

Er-Jun Ling and Xiao-Qiang Yu

CHAPTER 32 SUMMARY

- The immune system defends the body against infection by pathogens. It also helps to heal injuries and fight cancer.
- Pathogens are infectious agents that can cause disease; they include certain bacteria, viruses, fungi, and parasites.
- Viruses are non-cellular; they consist of a genome of nucleic acid (DNA or RNA) contained within a protein shell.
- The immune system has two main arms: innate immunity and adaptive immunity. Innate immunity is the first line of defense against invaders; adaptive immunity comes into play once innate defenses are breached.
- The innate immune system includes defenses with which we are born and which are always active: they include barriers such as skin and mucous membranes; antimicrobial chemicals in tears, saliva, and other secretions; and phagocytic cells that engulf and destroy pathogens.
- The inflammatory response, part of the innate immune system, is triggered by tissue infection or injury. During an inflammatory response, blood vessels swell and leak, marshaling phagocytic cells and protective molecules to the area to contain the infection.
- Adaptive immunity is conferred by specialized lymphocytes called B and T cells.
- B cells produce antibodies that recognize specific antigens unique to a pathogen and mark that pathogen for destruction.
- Cytotoxic T cells destroy infected, foreign, and cancer cells.
- The immune response can go awry, causing allergies and autoimmune conditions. Allergies are responses to intrinsically harmless substances (for example, pollen), and autoimmune conditions result when the immune system mounts a response against the body's own cells and tissues.
- At the first exposure to a particular pathogen, a primary immune response is generated that takes time to become fully effective; the primary response also produces memory cells.
- Memory cells remain in the body and become active at the time of a subsequent exposure to the same pathogen, producing a more rapid and vigorous secondary immune response that fights the pathogen and usually prevents the associated illness.
- Vaccination elicits a primary immune response. Memory cells produced during the primary response protect against illness following subsequent exposure to the actual pathogen.
- The adaptive immune response is highly specific for particular pathogens. If the pathogen changes, by antigenic shift or drift, the body must mount a new primary response for each strain of the pathogen.
- All organisms have ways of defending themselves against infection. The innate immune system is evolutionarily ancient.

MORE TO EXPLORE

- PBS (1998) *American Experience*, Influenza 1918 (film).
- Taubenberger, J. K., and Kash, J. C. (2011) Insights on influenza pathogenesis from the grave. *Virus Research* 162(1):2–7.
- Barry, J. M. (2005) *The Great Influenza: The Story of the Deadliest Pandemic in History*. New York: Penguin Books.
- Mnookin, S. (2011) *The Panic Virus: the True Story Behind the Vaccine-Autism Controversy*. New York: Simon & Schuster.
- Offit, P. (2015) *Deadly Choices: How the Anti-Vaccine Movement Threatens Us All*. New York: Basic Books.

CHAPTER 32 Test Your Knowledge

DRIVING QUESTION 1 What is the structure of a virus, and how do viruses cause disease?

By answering the questions below and studying Infographic 32.1, 32.2, and 32.11, you should be able to generate an answer for the broader Driving Question above.

KNOW IT

1 Which of the following is found in all viruses?

a. DNA
b. RNA
c. a membranous envelope
d. a protein shell
e. a cell membrane

2 Explain how viruses replicate within humans.

3 Why does poliovirus cause long-lasting damage, whereas those infected by influenza virus typically make a full recovery?

USE IT

4 Both viruses and bacteria can be human pathogens. Describe some key differences between them. (Hint: You may wish to refer to Chapter 3.)

5 Why do poliovirus, influenza virus, and HIV infections cause different symptoms?

DRIVING QUESTION 2 Why is nonspecific immunity important even though another system provides both specificity and memory?

By answering the questions below and studying Infographics 32.3 and 32.4, you should be able to generate an answer for the broader Driving Question above.

KNOW IT

6 Name three components of the innate immune system. For each, provide a brief description of how it offers protection.

7 What are phagocytes, and what do they do?

USE IT

8 From what you know about innate immunity, would you predict different or identical innate responses to infections from *E. coli* (a bacterium) and *S. aureus* (another bacterium)? Explain your answer.

9 Neutropenia is a deficiency in a type of phagocytic cell called neutrophils. Neutrophils are among the "first responders" to an injury or infection. Would you expect someone with neutropenia to be able to mount an effective inflammatory response? Explain your answer.

10 Why might someone taking anti-inflammatory drugs be more susceptible than others to bacterial infections?

DRIVING QUESTION 3 What is adaptive immunity, and how does vaccination rely on adaptive immunity?

By answering the questions below and studying Infographics 32.3, 32.5, 32.6, 32.7, 32.8, 32.9, 32.10, and 32.11, you should be able to generate an answer for the broader Driving Question above.

KNOW IT

11 Compare and contrast the features of innate and adaptive immunity.

12 B cells, plasma cells, and antibodies are all related. Describe this relationship, using words, a diagram, or both.

USE IT

13 Anti–hepatitis C antibodies present in a patient's blood indicate

a. that the patient is mounting an innate response.
b. that the patient has been exposed to HIV.
c. that the patient was exposed to hepatitis C within the last 24 hours.
d. that the patient was exposed to hepatitis C at least 2 weeks ago.
e. that the patient has hepatitis.

14 Vaccination against a particular pathogen stimulates what type of response?

a. innate
b. primary
c. secondary
d. autoimmune
e. b and c

15 Will someone who has been exposed to seasonal influenza in the past

a. have memory B cells?
b. still be at risk for seasonal influenza next year? Why?
c. still be at risk for pandemic swine flu? Why?

16 *Staphylococcus aureus* can cause a bacterial skin infection that can become very serious.

a. Why does the body exhibit innate and adaptive responses to *Staphylococcus aureus* but not to its own skin cells?
b. Will the innate response to *Staphylococcus aureus* be equally effective against *Streptococcus pyogenes,* another bacterium that can cause skin infections? Explain your answer.
c. Will the adaptive response to *Staphylococcus aureus* be equally effective against *Streptococcus pyogenes*? Explain your answer.

17 HIV is a virus that infects and eventually destroys helper T cells. Why do people with AIDS (that is, with advanced HIV infections) often die from infections by other pathogens?

apply YOUR KNOWLEDGE

MINI CASE

18 In 2008, there was an outbreak of measles in San Diego. The first patient was a 7-year-old unvaccinated boy who had returned home from a family trip to Switzerland. He began to develop a cough, sore throat, and fever, but continued to attend school. He was taken to his pediatrician when he developed a rash, and then was sent to the emergency room because of a very high fever. Blood tests revealed antimeasles antibodies. Eleven other children ended up developing measles; none had been vaccinated. The other cases were the 2 siblings of the first patient, 4 children in his school and 5 children who were in the pediatrician's office at the same time as the patient. Of the 11 additional cases, 3 children were less than 1 year old. (Data are from Outbreak of measles—San Diego, California, January—February 2008. (2008) *Morbidity and Mortality Weekly Report* 57(8): 203–206.)

a. Why was the presence of antimeasles antibodies in the first case an important finding?
b. What can you infer about how easily measles spreads?
c. What does this case suggest about the importance of measles vaccinations?
d. Were all the unvaccinated children necessarily behind on their vaccination schedule? (Hint: Look up the recommended measles vaccination schedule on the Centers for Disease Control and Prevention website, www.cdc.gov.)

apply YOUR KNOWLEDGE

BRING IT HOME

19 Almost 10% of the children in the school attended by the original patient described in the Mini Case were unvaccinated because their parents had filed Personal Belief Exemptions stating that they did not want to vaccinate their children. What is your local school district or state policy on vaccinations for enrolled students? This information is typically available at the school district website or the state's health department website. From the policies in place (and whether they permit any exemptions), do you think it is possible that a measles outbreak could occur in a local school?

DRIVING QUESTION 4 What are the features of the influenza virus that allow it to cause worldwide outbreaks?

By answering the questions below and studying Infographic 32.11, you should be able to generate an answer for the broader Driving Question above.

KNOW IT

20 What is the difference between antigenic shift and antigenic drift?

21 A strain of influenza can infect and replicate in birds without causing them disease. The same strain can be transmitted from birds to humans, causing severe illness in humans. What else would this strain need to be able to do in order to become pandemic?

USE IT

22 What processes are responsible for the emergence of pandemic influenza strains such as swine flu? Explain how these strains can spread so successfully through the human population.

23 Why are those with influenza infections susceptible to bacterial pneumonia?

apply YOUR KNOWLEDGE

INTERPRETING DATA

24 The table below gives the approximate cumulative rate of influenza hospitalizations in the United States in newborns to 4-year-olds and in people 65 years old and older for the 2009–2010 flu season and the 2014–2015 flu season. As the flu season runs from the fall of one year through the spring of the next, the data are reported by week (40th week of the first year to 12th week of the next).

a. Graph the cumulative incidence rates by week for each age group in each flu season.
b. Examine each season and compare and contrast the age-specific hospitalization rates in the two seasons.

Age Group and Season	Cumulative Incidence Rate (per 100,000)						
	Week 40	Week 44	Week 48	Week 52	Week 4	Week 8	Week 12
Birth to 4 years 2009–2010	40	45	50	60	60	60	60
65 years and older 2009–2010	10	18	20	20	20	20	20
Birth to 4 years 2014–2015	0	0	1	40	48	50	50
65 years and older 2014–2015	0	0	8	125	240	280	305

Data from Update: influenza activity—United States, 2009–10 season. (2010) *Morbidity and Mortality Weekly Report* 59(29):901–908; Influenza activity—United States, 2014–15 season and composition of the 2015–16 influenza vaccine. (2015) *Morbidity and Mortality Weekly Report* 64(21):583–590.

Answers to Infographic Questions

Chapter 1

IG1.1 Possible reasons include: Each study investigated different specific health conditions; different populations (differing by age, gender, or ethnicity, for example); different types and amounts of coffee.

IG1.2 Continue to refine and test the hypothesis using different experiments.

IG1.3 The dependent variable is memory, as assessed by a memory test. The independent variable is caffeine (decaffeinated coffee or caffeinated coffee). The independent variable (caffeine) is intentionally changed.

IG1.4 Larger sample sizes give more conclusive and reliable results because these results are less likely to be due to chance.

IG1.5 Theories are hypotheses that are supported by large bodies of evidence and have not been disproved. The reporter should have stated that researchers were testing the hypothesis that Zika virus causes birth defects.

IG1.6 There is a correlation between higher coffee intake and a lower risk of Parkinson's disease.

IG1.7 Information that may be missing includes the sample size, the population studied, and details of how data were collected and analyzed.

Chapter 2

IG2.1 It grows taller, it reproduces (by producing avocados with seeds), it maintains water balance by regulating water uptake and loss, its leaves can turn toward sunlight, and it uses the energy of sunlight to carry out photosynthesis.

IG2.2 The atomic number of magnesium is 12; it has 12 protons.

IG2.3 Every carbon in glucose has four covalent bonds. The carbon of CO_2 has four bonds.

IG2.4 Phospholipids and proteins

IG2.5 You can show the oxygen atom of the bottom right water molecule forming a hydrogen bond with a hydrogen atom of another water molecule.

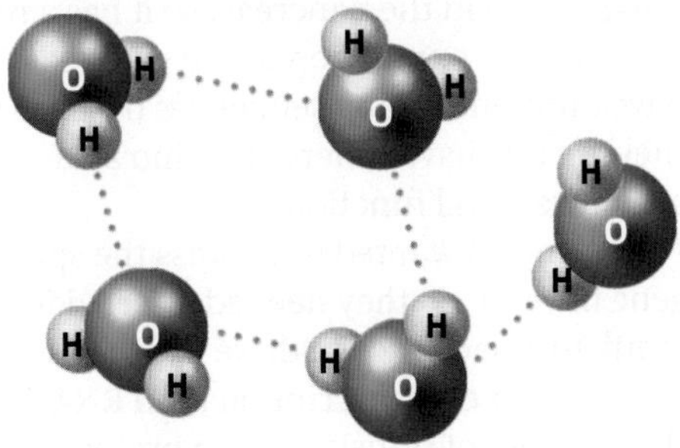

IG2.6 There are many possibilities, but in all cases, the oxygen atom of one water molecule will form a hydrogen bond with a hydrogen atom of another water molecule. Here is one possibility:

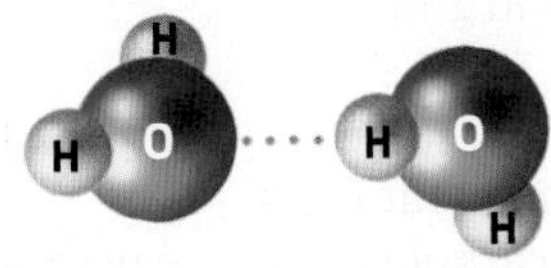

IG2.7 A sodium ion is positively charged because it has lost an electron. A chloride ion is negatively charged because it has gained an electron.

IG2.8 Because water is so crucial for life on Earth, the presence of liquid water is considered a proxy for life.

IG2.9 7.4

IG2.10 Both viruses and prions are infectious, and neither are made of cells. Viruses have genetic material enclosed in a protein shell and reproduce in infected cells. Prions do not have genetic material (they are made only of protein) and are also able to replicate in infected cells.

Chapter 3

IG3.1 There would not be any "clearing" of bacterial growth next to the fungus. The bacteria would grow well next to the fungus.

IG3.2 Diatoms, amoeba, molds, and *Elodea*

IG3.3 Eukaryotic cells have organelles, including a nucleus.

IG3.4 The cell will swell and may burst.

IG3.5 Eukaryotic ribosomes have a different structure from prokaryotic ribosomes. The antibiotic targets only prokaryotic ribosomes.

IG3.6 The hydrophobic tails

IG3.7 Diffusion does not require an input of energy; active transport does.

IG3.8 Three of the organelles present in both plant and animal cells and their functions are: the nucleus–it holds DNA; mitochondria–they convert energy from food into a useful form; the Golgi apparatus–it packages and processes proteins.

Milestone 1

IGM1.1 Circular

IGM1.2 Nonphotosynthetic bacteria

Chapter 4

IG4.1 Fats

IG4.2 Amino acids

IG4.3 Anabolic

IG4.4 An enzyme does not change the energy of the reactants or of the products. An enzyme reduces the activation energy of a reaction.

IG4.5 No. Although micronutrients are needed in only small amounts, they are critical for many functions in cells and organisms.

IG4.6 RUTF was more effective as a full dietary replacement than as a supplement.

IG4.7 These diet plans will be challenging to fit into a balanced diet because a balanced diet includes food from a variety of food groups.

Chapter 5

IG5.1 Petroleum, coal, and natural gas

IG5.2 Algal oil lipids and carbohydrates are used to produce biofuels. Algal cell wall biomass can be burned to generate electricity.

IG5.3 Kinetic energy of heat, muscle movement, and wheel movement

IG5.4 Potential energy (stored chemical energy)

IG5.5 Glucose can be used immediately as energy to power cellular functions, stored as potential energy (in molecules like starch), and used as energy to make cellular structures (such as the cell wall).

IG5.6 The carbon atoms in the glucose come from the carbon atoms in CO_2, and the oxygen atoms in the O_2 come from water (H_2O).

IG5.7 Violet

IG5.8 ATP

Chapter 6

IG6.1 Obese

IG6.2 Fat stores more than twice the number of Calories per gram as carbohydrates.

IG6.3 Approximately 5 hours and 15 minutes

IG6.4 Walking the dog and weeding the garden are NEAT activities. Studies show that increased amounts of NEAT activities can reduce fat gain in people who overeat and that lean people spend more time carrying out NEAT activities than do obese people.

IG6.5 They both require an energy source in the form of ATP.

IG6.6 Glucose from the digestion of carbohydrates is delivered from the small intestine via the circulation. Oxygen is delivered from the lungs, also via the circulation.

IG6.7 Glycolysis, the citric acid cycle, and electron transport. Glycolysis occurs in the cytoplasm; the citric acid cycle occurs in mitochondria; and the electron transport chain is in the mitochondria.

IG6.8 Pyruvate accepts electrons from NADH during fermentation. Pyruvate is converted to lactic acid or alcohol, depending on the organism and fermentation reaction.

IG6.9 A triglyceride stores more energy per gram than glycogen. Triglycerides are stored in fat cells.

IG6.10 In photosynthesis, CO_2 is the source of carbon used to generate glucose. In cellular respiration, CO_2 is a waste product.

IG6.11 There are many possible combinations. For example, using a headset to talk while making phone calls, walking 15 minutes back and forth to the convenience store and carrying milk home and spending an hour playing cards or working a puzzle.

Chapter 7

IG7.1 There are 46 chromosomes; each has a single DNA molecule. Therefore, there are 46 DNA molecules in the nucleus of each human cell.

IG7.2 TACGACT

IG7.3 DNA polymerase

IG7.4 DNA polymerase

IG7.5 He inherited six copies from one parent, and four copies from the other parent.

IG7.6 One half of the child's bands should match bands from his or her biological father.

IG7.7 $0.2 \times 0.2 \times 0.2 = 0.008$, or 1 in 125

Milestone 2

IGM2.1 A phosphate group, a deoxyribose sugar, and a nitrogenous base

IGM2.2 The phosphate groups on the outside of the helix can interact with water, as illustrated by the change in the structure of DNA in the presence of water.

IGM2.3 If a DNA molecules is 20% G, then it must also be 20% C. The remaining 60% is equally divided between A and T, giving the %A as 30%.

IGM2.4 A DNA molecule has two strands running in opposite directions, each made up of covalently bonded deoxyribonucleotides. The sugars and phosphates form the backbone of each strand, while paired bases form "rungs" in the middle of the molecule. The bases pair by hydrogen bonding.

Chapter 8

IG8.1 Alanine and glycine are hydrophobic, so they will tend to cluster together, away from water.

IG8.2 Because only the spidroin gene is being expressed ("turned on").

IG8.3 The regulatory sequence of each gene determines where it is expressed. A milk protein gene is expressed in mammary glands, so it has a regulatory sequence for mammary gland expression. Insulin is expressed in the pancreas, so it has a regulatory sequence for expression in the pancreas. The genes have different coding sequences because the proteins they encode have different amino acid sequences, structures, and functions.

IG8.4 The scientists wanted to express the spider spidroin gene in yeast, so they needed to provide a yeast regulatory sequence to drive expression in yeast.

IG8.5 The product of transcription is an RNA molecule; the product of translation is a protein molecule.

IG8.6 RNA polymerase is the enzyme responsible for transcription. It adds ribonucleotides to the growing RNA molecule.

IG8.7 tRNA reads the mRNA codons and brings amino acids to the growing protein. One end binds to a specific amino acid, and its anticodon end forms a complementary base pair with the mRNA codon.

IG8.8 The start codon has the sequence AUG; it specifies the amino acid methionine.

Milestone 3

IGM3.1 A T will be the next nucleotide to be added.

IGM3.2 Some nucleotides at the ends of fragments were repeated between fragments, generating overlaps during the assembly process.

IGM3.3 2001: ~$100,000,000 (100 million dollars); 2006: ~$10,000,000 (10 million dollars); 2015: ~$1,000 (1 thousand dollars). Answers will vary regarding the willingness to spend $1,000 to have one's genome sequenced.

IGM3.4 Coding; these sequences determine the amino acid sequences of proteins.

Chapter 9

IG9.1 Cell replacement is more likely to be on a schedule, because cells have defined life spans and need to be replaced at a constant rate. Damage is unpredictable and its repair requires the initiation of cell division.

IG9.2 Interphase prepares the cell for division. Interphase includes G_1, S, and G_2.

IG9.3 An engineered organ can be a perfect match to the patient, eliminating organ rejection.

IG9.4 A muscle cell in the arm, as both are a type of muscle (cardiac and skeletal).

IG9.5 All approaches require the use of chemicals to stimulate stem cells to divide and differentiate.

Chapter 10

IG10.1 *BRCA1*

IG10.2 So that mutations are not copied (during S phase) and passed on to daughter cells.

IG10.3 Avoid exposure to mutagens, including alcohol, charred meat, and tobacco smoke.

IG10.4 No. A missense mutation changes a single codon.

IG10.5 A tumor suppressor gene is most likely mutated, given that it is dividing despite the presence of damaged DNA. Tumor suppressors would normally detect this damage and pause the cell cycle to allow the damaged DNA to be repaired.

IG10.6 No. The *BRCA1* mutation would have contributed to the failure to repair other mutations, allowing cells to continue to divide even as they accumulate unrepaired mutations.

IG10.7 The cancer cell will now be recognized and attacked by specific immune cells.

IG10.8 Mammography

Chapter 11

IG11.1 Three nucleotides differ; there is a single amino acid difference.

IG11.2 Humans are diploid because they have two complete sets of chromosomes, one inherited from the mother and one from the father.

IG11.3 A sperm is haploid; a skin cell and a zygote are both diploid.

IG11.4 The first meiotic division (meiosis I)

IG11.5 Recombination mixes up combinations of alleles of genes on a single chromosome by the exchange of segments between homologous chromosomes.

IG11.6 The CFTR membrane protein allows chloride ions to move across the cell membrane. When ions leave the cell, water can follow, keeping the mucus thin and slippery. If the CFTR membrane protein does not work properly, chloride ions are retained in cells, as is water. The extracellular mucus is then thick and sticky.

IG11.7 The probability is 50% (2 in 4).

IG11.8 These parents cannot have a child with two copies of the dominant (*T*) allele (0% probability).

IG11.9 The probability is 1 in 16.

Milestone 4

IGM4.1 Both proposed that preformed humans were present in one of the gametes.

IGM4.2 Purple flower color is dominant.

IGM4.3 Only one allele for any given gene is segregated into a single gamete.

IGM4.4 Because the *R* alleles segregate away from each other at meiosis I, and the *Y* alleles also segregate away from each other at meiosis I.

Chapter 12

IG12.1 Their biological sex would be male, because of the Y chromosome.

IG12.2 All of their children (sons and daughters) will have red-green color blindness.

IG12.3 If Irene (a carrier) married Maurice (who has hemophilia), 50% of their daughters would be predicted to have hemophilia, and 50% of their sons would be predicted to have hemophilia.

IG12.4 Only his sons will inherit this allele.

IG12.5 Thomas, Edy, Beverly, Madison, and Eston would all be predicted to have the same Y chromosome as Thomas. If Harriet had a son, he would not share his Y chromosome with his grandfather Thomas (he would share his Y chromosome with his father, and his father's father).

IG12.6 All of their children will have the mild form of the disease.

IG12.7 The AB blood type exemplifies codominance. A person with AB blood could have parents with the following blood types: A and B; AB and A; AB and B; AB and AB.

IG12.8 Only people with AB positive blood can safely receive AB positive blood.

IG12.9 This person could have inherited three dominant alleles, or could have inherited four dominant alleles but have a nutritionally poor diet.

IG12.10 The results support a multifactorial inheritance model for depression (influenced by inherited alleles and by stressful life experiences).

IG12.11 The zygote will have 44 autosomes and a Y chromosome. The egg and the zygote are aneuploid in this case.

IG12.12 There would be a total of 45 chromosomes; 44 autosomes and an X chromosome.

Chapter 13

IG13.1 *S. aureus* ("staph") is most likely to be found on the skin. Other staphylococcal species (*S. epidermidis*) can be found in the urogenital tract.

IG13.2 The skin of other wrestlers; shared equipment or towels; contaminated surfaces (for example, mats or benches)

IG13.3 They would not rupture because there would be no influx of water to put pressure on the weakened cell wall.

IG13.4 The daughter cells are identical to each other and to the parent cell.

IG13.5 Mutation is more likely to introduce a new allele, by mutating an existing allele. Gene transfer can introduce an entirely new gene.

IG13.6 In the absence of antibiotics, there is no difference between the fitness of antibiotic-sensitive and antibiotic-resistant variants (both have average fitness).

IG13.7 Over time, the crabs will pose a selective pressure, and a higher proportion of the blue mussels will have the allele that permits the development of thicker shells.

IG13.8 Diversifying selection will select for both lighter and darker oysters in this environment.

Milestone 5

IGM5.1 Yes. The Lamarck model predicts that acquired traits can be passed on to offspring.

IGM5.2 Darwin did not visit North America or Asia; thus the development of his ideas was based on his observations in other parts of the globe.

IGM5.3 Darwin was influenced by substantial bodies of work ranging from 1798 to 1854.

IGM5.4 Darwin's key observations were made on his *Beagle* voyage (1831-1836), during which he visited South America (including the Galápagos Islands) and Africa. Wallace's key trip was to the Amazon (1848-1852). During their travels, they observed, collected specimens, and learned from the works of others.

Chapter 14

IG14.1 Evolution did not occur. The allele frequencies did not change over time.

IG14.2 The Highland Park and Willow Lake populations are likely to be isolated populations.

IG14.3 Both are types of genetic drift. In the bottleneck effect, a large population is reduced in size (and genetic diversity); in the founder effect, a small number of individuals (with limited genetic diversity) leave a larger and more diverse population.

IG14.4 A new red allele was introduced into population B.

IG14.5 A gene involved in detoxification of chemicals (like pesticides) would be most likely to vary between the two populations, given that agricultural chemicals are present in only one of the two populations.

IG14.6 Hybrid inviability, gametic isolation, and hybrid infertility

IG14.7 No. Differences in beak shape alone do not establish that species are distinct. Whether or not species are distinct is determined by their ability to reproduce and produce fertile offspring.

Chapter 15

IG15.1 Because flies do not have bones or other hard parts, they are most likely to be preserved if they are trapped in amber.

IG15.2 The middle toe is thicker and more prominent in *Merychippus*.

IG15.3 Approximately 511 million years

IG15.4 Weight-bearing forelimbs could be used to pull *Tiktaalik* onto land and crawl on land.

IG15.5 The humerus

IG15.6 All except the human

IG15.7 There are 2 differences in 30 nucleotides between the close and distant relative. 2/30 = 7% difference, or 93% similarity.

Chapter 16

IG16.1 The time it takes for half of the element present to decay.

IG16.2 The amount of uranium-238 decreases and the amount of lead-206 increases.

IG16.3 It was the first demonstration that organic molecules could form from inorganic molecules, and also showed that hypotheses regarding the origin of life could be tested.

IG16.4 Prokaryotes were dominant during the Archean time period; they were dominant in water.

IG16.5 If brown bears evolved in New Zealand, polar bears could have diverged from brown bears in New Zealand. Because penguins evolved in New Zealand, polar bears and penguins would coexist in New Zealand and Antarctica.

IG16.6 Approximately 65 mya

IG16.7 These animals are not closely related, so they differ in their evolutionary history. They differ in how they give birth to their young, and in where they evolved.

IG16.8 Animals

IG16.9 Supergroup

IG16.10 Bird

IG16.11 Archaea and Eukarya

Chapter 17

IG17.1 High temperature, low oxygen, high pressure, and extreme pH

IG17.2 Temperatures between 80°C and 100°C are those experienced by the microbes in their natural habitat.

IG17.3 Size (prokaryotic cells are typically smaller); organelles (absent in prokaryotes); DNA (contained in a nucleus only in eukaryotic cells)

IG17.4 There are 453.59 grams in a pound. If each gram contains 1 billion microbes, then each pound has 453.92 × 1 billion microbes = 453,590,000,000, or 453.9 billion, microbes.

IG17.5 They are classified into separate domains on the basis of distinct evolutionary histories.

IG17.6. Capsules and pili can be used to attach to host cells during the disease process. Flagella can be used to swim to host cells (for example, through a layer of mucus in the stomach).

IG17.7. High temperature, low temperature, high salt, extremely acidic pH

IG17.8 Abiotic methane: chemical reactions between carbon in the mantle rock and H_2 gas. Biotic methane: Archaea in chimneys make methane from H_2 and CO_2. Methane is an important carbon and energy source for bacteria and archaea living at Lost City.

Chapter 18

IG18.1 Animals

IG18.2 Cell walls made of cellulose; not mobile; autotrophic

IG18.3 Mosses have no vascular system, so they can't transport water long distances within the plant; their sperm require water to survive.

IG18.4 Vascular tissue

IG18.5 Cell walls made of chitin; external digestion; heterotrophic; not mobile

IG18.6 Molds and mushrooms

IG18.7 Various host-specific pests prevent new plants of the same species from growing near the parent plant.

IG18.8 They can evade predators, seek out food, and seek out mates.

IG18.9 Sand dollar–radial; parrot–bilateral; jellyfish–radial; earthworm–bilateral; octopus–bilateral

IG18.10 Asymmetrical–sponges; radial–jellyfish and starfish; bilateral–humans, birds, insects

IG18.11 Protists are not closely related to one another. They are found in many groups on the eukaryotic tree of life.

Milestone 6

IGM6.1 No. The parasite completes a portion of its life cycle in the mosquito.

IGM6.2 *Pneumocystis carinii* is more closely related to humans (both are opisthokonts).

IGM6.3 Malaria has been eradicated from Australia, both through insecticides to control the mosquito and antiparasite medications.

IGM6.4 No. The nonphotosynthetic host did not contain chloroplasts (or remnants of chloroplasts) before the endosymbiosis event. Because chloroplasts (and their remnants) are targets of the antimalarial herbicide, the host would not be susceptible to this drug.

Chapter 19

IG19.1 Because humans are 99.9% similar to one another, there would be only 0.1% difference between two people, or 1 nucleotide out of 1,000.

IG19.2 Melanocytes

IG19.3 Dark skin confers more of an advantage in a high-UV environment. Dark skin protects folate from degradation, which is more of a risk in a high-UV environment, and, where the UV level is high, still allows sufficient UV exposure for vitamin D production.

IG19.4 In a low-UV environment, the predominant pressure selecting for light skin is the need to make sufficient amounts of vitamin D.

IG19.5 Mitochondrial DNA is used to trace female lineages (mothers).

IG19.6 The female's mtDNA (shown as purple here) would be passed on to all the couple's offspring.

IG19.7 There would be higher levels of genetic diversity in Australian populations relative to African populations.

IG19.8 *Australopithecus africanus* (minimal overlap), *Paranthropus boisei, Paranthropus robustus, Homo rudolfensis, Homo erectus*

IG19.9 Dark-skin alleles would be predicted to experience negative selection in a low-UV environment.

IG19.10 Light skin is a disadvantage in a high-UV environment because it cannot protect folate from destruction.

Chapter 20

IG20.1 Ecosystem ecology

IG20.2 No. They estimate the population size on the basis of aerial counts of a portion of the island.

IG20.3 Clumped (clumps of plants near the watering holes)

IG20.4 Limiting amounts of food; disease; limiting amounts of suitable habitat

IG20.5 As the predator (wolf) population increases, more prey (moose) are hunted and killed, reducing the prey population.

IG20.6 Many large trees suggest that the moose population is low, presumably because of a large wolf population.

IG20.7 The composition of the vegetation in their diet and their overall nutritional status

IG20.8 Widespread flooding is abiotic and density independent.

IG20.9 Tick populations increase in warmer conditions. An abundance of ticks will lead to a reduction in the number of moose.

Chapter 21

IG21.1 Impact (not abundance) defines keystone species.

IG21.2 Approximately $11.2 billion

IG21.3 Bears can be characterized as consumers, herbivores, and carnivores.

IG21.4 10% of the energy stored in the hawk's body is stored in the top predator.

IG21.5 The difference is whether both species benefit or only one. In a commensalism, one species benefits and the other is not harmed. In a mutualism, both species benefit.

IG21.6 They can coexist by relying on nectar from different species of wildflowers.

IG21.7 Killer bees are out-competing other bee species in their niche, by aggressively defending their food resources.

IG21.8 They are keystone species, with a key role in pollination. Because they are a keystone species, the community would be jeopardized if they were lost.

Chapter 22

IG22.1 The temperatures in the southeastern United States are decreasing, while they are increasing in New England.

IG22.2 Increasing spring temperatures are causing earlier flowering. This means that plants flower well before deer are born, so there is limited food for young deer, and their survival is reduced.

IG22.3 There will be more oak, gum, and cypress forest and less longleaf and slash pine forest relative to the present.

IG22.4 Both temperature and precipitation are higher in July than in January.

IG22.5 Greenhouse gases trap in the atmosphere some of the heat energy that is reflected off Earth's surface.

IG22.6 The most recent data show the current temperature to be approximately 0.5°C higher than the 1901-2000 average. The temperature has been consistently higher than the 1901-2000 average since approximately 1980.

IG22.7 When birds migrate away, insects can proliferate. Many of these insects feed on and kill trees.

IG22.8 Northern regions have the greatest increase in temperatures, and the southern pole has colder than average temperatures.

IG22.9 If human activities had no effect on the carbon cycle, there would be 9 fewer gigatons of CO_2 added the atmosphere each year. This would result in greater amounts of CO_2 being taken out of the atmosphere than added to the atmosphere.

IG22.10 CO_2 remained less than 300 ppm for 800,000 years. It is now at 400 ppm. This is an increase of 133% ($400/300 \times 100$).

IG22.11 The proportion of emissions from transportation would decrease. If the cars are recharged from the electrical grid, the emissions from electricity may increase.

Milestone 7

IGM7.1 Typhus; malaria

IGM7.2 By eating earthworms

IGM7.3 Chemicals like DDT are retained in the bodies of organisms that ingest them (such as plankton in water). When many of these organisms are eaten by another (such as small fish), all the retained DDT ends up being retained in the consumer. At each level of the food chain, the concentration of the chemical increases (from small fish to large fish to birds of prey like eagles).

Chapter 23

IG23.1 Using a solar oven instead of an electric or gas stove will conserve stored carbon (that is, fossil fuels).

IG23.2 1970

IG23.3 China's footprint is so high because of its very large population size.

IG23.4 In 1950, the ranking (from highest consumption to lowest) was petroleum oil, natural gas, and coal. In 2015, the ranking was the same, with more of each fossil fuel being consumed compared to 1950.

IG23.5 Medical advances such as antibiotics and vaccines; clean water; intensive agriculture

IG23.6 Freshwater

IG23.7 1. nuclear; 2. biomass; 3. hydroelectric; 4. wind; 5. solar; 6. geothermal

IG23.8 Oceans and aquifers store water; groundwater returns water to lakes and oceans.

IG23.9 Because the population is increasing.

IG23.10 Answers will vary.

IG23.11 Answers will vary.

Chapter 24

IG24.1 Water–through roots; CO_2–through leaves; nitrogen–through roots; phosphorus–through roots

IG24.2 Human-produced fertilizers are sources of soil nitrogen and phosphorus. Nitrogen-fixing bacteria in soil produce usable soil ammonia. Weathering of rocks naturally introduces phosphorus into the soil.

IG24.3 An excess of these nutrients stimulates the growth of algae. When the algae die, decomposition removes oxygen, which harms the fish.

IG24.4 Pollination occurs at the stigma. Fertilization occurs in the ovules in the ovary.

IG24.5 One sperm fertilizes the egg to create the embryo. One sperm fuses with other cells in the ovule to generate the endosperm.

IG24.6 The root can still grow from the meristems at the tips of the side branches.

IG24.7 The presence of pesticide in the environment is a selective pressure for the emergence of pesticide-resistant pests.

IG24.8 Nonphotochemical quenching is important to protect plants from intense light. If NPQ were to be reduced, plants might be damaged when exposed to intense light. By engineering plants to recover quickly, photosynthesis can resume rapidly after NPQ.

IG24.9 CRISPR can target a specific gene of interest at a precise location in the genome, enabling targeted and planned changes that will contribute to the desired phenotype.

Chapter 25

IG25.1 Too much fertilizer will create a hypertonic environment in the soil, which will cause water to leave the roots by osmosis.

IG25.2 Xylem carries water from roots to shoots. Phloem transports dissolved sugars from shoots to roots or from roots to shoots.

IG25.3 Cherry trees are dicots. Some evidence includes blossoms with five petals and leaves with branched veins.

IG25.4 Nitrogen

IG25.5 Many plants close their stomata on very hot, sunny, dry days to prevent water loss. Closed stomata prevent CO_2 entry, limiting the amount of photosynthesis that can take place.

IG25.6 They help capture additional wavelengths of light for photosynthesis.

IG25.7 Sexual attraction

IG25.8 No. They can attach to the fur of any animal and do not need to be eaten to be dispersed.

IG25.9 The cells elongate.

IG25.10 Gibberellins

IG25.11 Both. Physical defense–tough exterior; chemical defense–toxins

Chapter 26

IG26.1 A tissue has different cell types, which work together to produce the function of the tissue. An organ has multiple tissue types, which work together to produce the function of the organ.

IG26.2 A body temperature of 36.5°C (97.7°F) is within normal range and will not trigger a response to restore homeostasis.

IG26.3 Shivering generates heat. We cannot generate "cool"–instead, we rely on heat dissipation through sweating and vasodilation to reduce body temperature.

IG26.4 Sweat glands and blood vessels. Sweat glands release sweat; the evaporation of sweat cools the body. Blood vessels dilate to release heat or constrict to conserve heat.

IG26.5 The pituitary

IG26.6 Insulin reduces the level of glucose (sugar) in the blood, whereas glucagon elevates the levels of glucose in the blood.

IG26.7 ADH is made in and released by the pituitary. Its target organ is the kidney. ADH causes the kidney to conserve water (by reabsorbing water).

Chapter 27

IG27.1 Salivary glands, liver, gallbladder, pancreas

IG27.2 Key events that occur in the stomach include mixing of food with stomach acid and digestive enzymes; digestion of proteins by pepsin; unfolding and loss of shape of proteins by the action of stomach acid; destruction of bacteria by stomach acid.

IG27.3 Both procedures reduce the size of the stomach. Sleeve gastrectomy does not re-route the passage of food; gastric bypass does.

IG27.4 They are all accessory organs that secrete enzymes and other substances into the small intestine.

IG27.5 A person lacking villi would not be able adequately to absorb nutrients.

IG27.6 No. Fats are fully digested before leaving the small intestine.

IG27.7 216 pounds. (In 10 years, 28% of body weight is lost; 28% of 300 pounds is 84 pounds; 300 − 84 = 216.)

Milestone 8

IGM8.1 The islet of Langerhans cells produce insulin; the duct cells produce digestive enzymes.

IGM8.2 When the pancreas was removed, the dog was unable to produce insulin and so was unable to control its blood-sugar levels.

IGM8.3 The primary difference relates to the production (or nonproduction) of insulin. In type 1 diabetes, insulin is not produced; in type 2 diabetes, insulin is produced but cells do not respond properly to the hormone.

Chapter 28

IG28.1 38.5%

IG28.2 An artery

IG28.3 Fatty streaks are still a cause for concern even if they don't presently obstruct blood flow because they may increase in size and harden, developing into plaques that can obstruct blood flow.

IG28.4 Oxygen diffuses into the tissue (from the blood) and carbon dioxide diffuses into the blood (from the tissue). The amount of oxygen in the blood increases, and the amount of carbon dioxide in the blood decreases.

IG28.5 Red blood cells transport oxygen, white blood cells contribute to immune defenses, and platelets form a blood clot at the site of an injury.

IG28.6 Blood in a vein in the pulmonary circuit is traveling toward the heart (from the lungs). It is oxygenated.

IG28.7 High blood pressure can cause tears in artery walls where plaque builds up and can cause plaques to rupture.

IG28.8 The extent of plaques on the lining of the aorta of those with 2 risk factors would be similar to their extent in people with 0 or 1 risk factors. The extent of plaques on the lining of the coronary arteries would be greater than in people with 0 or 1 risk factors, but not as extensive as in people with 3 or 4 risk factors.

IG28.9 Cholesterol is a lipid, so it is hydrophobic and not soluble in blood.

IG28.10 Texas and southern states have consistently had high proportions of overweight and obese people. Colorado has consistently had low proportions of overweight and obese people.

IG28.11 Texas and southern states have high rates of both overweight and obesity, as well as high rates of deaths due to heart disease.

Chapter 29

IG29.1 From the alveoli to the blood

IG29.2 In the lungs, O_2 enters the blood and CO_2 leaves the blood. In a tissue (such as muscle), O_2 leaves the blood and CO_2 enters the blood.

IG29.3 If the supplemental oxygen is sufficient to mimic the partial pressure of oxygen at the individual's home (lower) altitude, then that individual will not experience hypoxia and will not have an elevated RBC count.

IG29.4 Partial pressure at the peak of Mount Everest is 53 mmHg; at Colorado Springs, Colorado, it is 129 mmHg; at sea level it is 159 mmHg.

IG29.5 An athlete's RBC increases in hypoxic conditions. A higher RBC is advantageous when an athlete competes at sea level because it increases the O_2-carrying capacity of the blood.

IG29.6 Air flows in.

IG29.7 The change in oxygen uptake for males and for females is very close. The females appear to have a slightly larger increase (from 64 to 66 mg/kg/min) compared to males (from 75.7 to 77 mg/kg/min). The males had a slightly larger change in race time, with a 5.8-second improvement relative to the 5.5-second improvement by females.

IG29.8 A dietary iron deficiency would translate into an inability to make heme groups of hemoglobin, thereby reducing the ability of red blood cells to transport oxygen.

IG29.9 40%

IG29.10 The kidneys

Chapter 30

IG30.1 The nicotine in the chewing tobacco can still be absorbed into the bloodstream (through the mucous membranes that line the mouth). From the bloodstream it will reach the brain.

IG30.2 The peripheral nervous system senses and responds to the mosquito.

IG30.3 The dendrites receive signals; the axon terminals transmit signals.

IG30.4 Sodium ions (Na^+) flowing into the cell trigger an action potential at a particular spot along a neuron.

IG30.5 An injury to the cerebellum could impair movement, coordination, and balance.

IG30.6 An axon terminal

IG30.7 Drug-addicted people need to take increasing amounts of the given drug to compensate for the reduced number of dopamine receptors on receiving cells.

IG30.8 The participant who recalled the greatest number of words had more dopamine reuptake receptors than the participant who recalled the fewest number of words.

Chapter 31

IG31.1 Fertilization occurs in the oviduct; the embryo develops after implantation into the lining of the uterus.

IG31.2 Sperm travel from the seminiferous tubules through the epididymis, the vas deferens, and finally the urethra.

IG31.3 Sperm pass through the cervix from the vagina to the uterus. Once in the uterus, sperm must travel to the oviducts (fallopian tubes) in order to fertilize an egg.

IG31.4 Egg production occurs in the ovary; fertilization occurs outside the body, in a dish; embryo implantation occurs in the lining of the uterus.

IG31.5 FSH is produced by the anterior pituitary and acts on the ovary, causing a follicle to develop; LH is produced by the anterior pituitary and acts on the ovary, causing ovulation; estrogen is produced by a follicle in the ovary and acts on the endometrium, causing it to thicken; progesterone is produced by the corpus luteum in the ovary and acts on the endometrium, causing it to thicken.

IG31.6 The hormone hCG is made by the embryo. It is important in early pregnancy because it causes the corpus luteum in the ovary to continue secreting the hormones estrogen and progesterone to maintain the endometrium.

IG31.7 FSH and LH, which in males trigger spermatogenesis, are produced by the anterior pituitary.

IG31.8 There are many possible blockage sites, including the oviducts and the vas deferens.

IG31.9 In IVF, the physician can control the number of embryos that are implanted. In IUI, if many eggs are ovulated, there is a chance that all of them will be fertilized and implanted.

Chapter 32

IG32.1 No. A cell cannot be a host for a virus if the cell does not have the specific receptor molecules to which that virus binds.

IG32.2 Yes. Fungi and various other eukaryotic pathogens can cause disease in humans.

IG32.3 There are many differences between innate and adaptive immunity, including specificity, whether or not the defense is always present and active, and the types of cell involved.

IG32.4 Mucous membranes lining the digestive tract, saliva (with bacteria-destroying enzymes), and acid in the stomach

IG32.5 Specialized phagocytes process ingested pathogens, then display (that is, present) antigens to lymphocytes of the adaptive response. Lymphocytes bind to the displayed antigen, leading to lymphocyte activation.

IG32.6 B and T cells are both produced in bone marrow. B cells mature in bone marrow, and T cells mature in the thymus.

IG32.7 Plasma cells produce antibodies. Antibodies bind to specific antigens on pathogens, resulting in the elimination of the pathogen.

IG32.8 Cytotoxic T cells recognize target antigens by the pathogen displayed on the surface of the host cell.

IG32.9 The first response to pathogen C will be a primary response.

IG32.10 A community experiencing a measles outbreak likely has low rates of vaccination against measles.

IG32.11 Slight differences are the result of antigenic drift.

Glossary

abiotic Refers to the nonliving components of an environment, such as temperature and precipitation.

abscisic acid (ABA) A plant hormone that helps seeds remain dormant.

acclimatization The process of physiologically adjusting to an environmental change over a period of time. Acclimatization is generally reversible.

acid A substance that increases the hydrogen ion concentration of solutions, making them more acidic.

acidosis A dangerous condition in which blood is too acidic.

action potential An electrical signal within a neuron caused by ions moving across the cell membrane.

activation energy The energy required for a chemical reaction to proceed. Enzymes accelerate reactions by reducing their activation energy.

active site The part of an enzyme that binds to a substrate.

active transport The process by which solutes are pumped from an area of lower concentration to an area of higher concentration with the help of transport proteins; requires an input of energy.

adaptation The process by which populations become better suited to their environment as a result of natural selection.

adaptive immunity A protective response, carried out by lymphocytes, that confers long-lasting immunity against specific pathogens.

adaptive radiation The spreading and diversification of organisms that occur when the organisms colonize a new habitat.

adenosine triphosphate (ATP) The molecule in cells that powers energy-requiring functions.

adhesion The attraction between molecules (or other particles) and a surface.

aerobic respiration A series of reactions that occurs in the presence of oxygen and converts energy stored in food into ATP.

algae A diverse collection of aquatic, photosynthetic organisms, including both unicellular and multicellular species.

allele frequency The relative proportion of an allele in a population.

alleles Alternative versions of the same gene that have different nucleotide sequences.

allergy An immune response misdirected against environmental substances, such as dust, pollen, and foods, that causes uncomfortable physical symptoms.

altitude sickness An illness that can occur as a result of an abrupt move to an altitude with a reduced partial pressure of oxygen.

alveoli Air sacs in the lung across which gases diffuse between air and blood.

amino acid The building block, or monomer, of a protein.

amniocentesis A procedure that removes fluid surrounding the fetus to obtain and analyze the chromosomal makeup of fetal cells.

anabolic reaction Any chemical reaction that combines simple molecules to build more-complex molecules.

anatomy The study of the physical structures that make up an organism.

anchorage dependence The need for normal cells to be in physical contact with another layer of cells or a surface.

androgen A class of sex hormones, including testosterone, that are present in higher levels in men and cause male-associated traits like deep voice, growth of facial hair, and defined musculature.

anecdotal evidence An informal observation that has not been systematically tested.

aneuploidy An abnormal number of one or more chromosomes (either extra or missing copies).

angiogenesis The growth of new blood vessels.

angiosperm A seed-bearing flowering plant with seeds typically contained within a fruit.

animal A eukaryotic multicellular organism that can move and that obtains nutrients by ingesting other organisms.

annelid An invertebrate with a soft, segmented body; annelids are commonly referred to as worms.

anther The part of the stamen that produces pollen.

antibiotic A chemical that can slow or stop the growth of bacteria; many antibiotics are produced by living organisms.

antibody A protein produced by B cells that fights infection by binding to specific antigens on pathogens.

anticodon The part of a tRNA molecule that binds to a complementary mRNA codon.

antigen A specific molecule (or part of a molecule) to which immune receptors bind and against which an adaptive response is mounted.

antigenic drift Changes in viral antigens caused by genetic mutation during normal viral replication.

antigenic shift Changes in antigens that occur when viruses exchange genetic material with other strains.

antigen-presenting cell A cell, such as a phagocyte, that digests pathogens and displays pieces of the pathogen on its surface, where they can be recognized by lymphocytes.

aorta The large artery that receives oxygenated blood from the left ventricle.

apoptosis Programmed cell death; often referred to as cellular suicide.

aquifer An underground layer of porous rock from which water can be drawn for use.

Archaea One of the two domains of prokaryotic life; the other is Bacteria.

arteries Blood vessels that carry blood away from the heart.

arthropod An invertebrate having a segmented body, a hard exoskeleton, and jointed appendages.

atherosclerosis A condition in which fatty deposits build up in the lining of arteries, restricting blood flow; also known as hardening of the arteries.

atom The smallest unit of an element that still retains the property of the element.

atomic number The number of protons in an atom, which determines the atom's identity.

atria The chambers of the heart that receive blood. In humans, the right atrium receives low-oxygen blood from the body, and the left atrium receives high oxygen blood from the lungs.

autoimmune disease A misdirected immune response in which the immune system attacks the body's own healthy cells.

autosomes Paired chromosomes present in both males and females; all chromosomes except the X and Y chromosomes.

autotrophs Organisms such as plants, algae, and certain bacteria that can make their own food from inorganic starting materials (e.g., CO_2).

auxin A plant hormone that causes elongation of cells as one of its effects.

axon The long extension of a neuron that conducts action potentials away from the cell body toward the axon terminal.

axon terminals The tips of an axon, which communicate with the next cell or cells in the pathway.

B cells White blood cells that mature in the bone marrow and produce antibodies during an adaptive immune response.

Bacteria One of the two domains of prokaryotic life; the other is Archaea.

base A substance that reduces the hydrogen ion concentration of solutions, making them more basic.

benign tumor A noncancerous tumor whose cells will not spread throughout the body.

bilateral symmetry The pattern exhibited by a body plan with right and left halves that are mirror images of each other.

bile salts Chemicals produced by the liver and stored by the gallbladder that emulsify fats so that they can be chemically digested by enzymes.

binary fission A type of asexual reproduction in which one parental cell divides into two.

biocapacity The amount of Earth's biologically productive area–cropland, pasture, forest, and fisheries–that is available to provide resources and absorb wastes to support life.

biodiversity The number of different species and their relative abundances in a specific region or on the planet as a whole.

biofuels Renewable fuels made from living organisms (e.g., plants and algae).

biogeography The study of the distribution of organisms in geographical space.

biological species concept The definition of a species as a population whose members can interbreed to produce fertile offspring.

biome A large geographic area defined by its characteristic plant life, which in turn is determined by temperature and levels of moisture.

biotic Refers to the living components of an environment.

blood A circulating fluid that contains several types of cell and transports substances, including nutrients, gases, and hormones.

blood pressure The overall pressure in blood vessels, expressed as the systolic pressure over the diastolic pressure.

body mass index (BMI) An estimate of body fat based on height and weight.

bottleneck effect A type of genetic drift that occurs when a population is suddenly reduced to a small number of individuals, and as a result alleles are lost from the population.

brain An organ of the central nervous system that integrates and coordinates virtually all functions of the body.

brain stem The part of the brain that is closest to the spinal cord and which controls vital functions such as heart rate, breathing, and blood pressure.

bronchi Two airways that branch from the trachea; one bronchus leads into each lung.

bronchioles Small airways that branch from the bronchi.

bryophyte A nonvascular plant that does not produce seeds.

Calorie A Calorie (spelled with a capital "C") is 1,000 calories or 1 kilocalorie (kcal). The Calorie is the common unit of energy used in food nutrition labels.

calorie A calorie (spelled with a lower-case "c") is the amount of energy required to raise the temperature of 1 gram of water by 1°C.

cancer A disease of unregulated cell division: cells divide inappropriately and accumulate, in some instances forming a tumor.

capillaries The smallest blood vessels. Capillaries are the sites of gas, nutrient, and waste exchange between the blood and tissue cells.

capsule A sticky coating surrounding some bacterial cells that adheres to surfaces.

carbohydrate An organic molecule made up of one or more sugars.

carbon cycle The movement of carbon atoms as they cycle between organic molecules and inorganic CO_2.

carbon fixation The conversion of inorganic carbon (e.g., CO_2) into organic forms (e.g., sugars like glucose, $C_6H_{12}O_6$).

carbon footprint A measure of the total greenhouse gases produced by human activities.

carcinogen Any substance that causes cancer. Most carcinogens are mutagens.

cardiovascular disease (CVD) A disease of the heart or blood vessels or both.

cardiovascular system The system that transports nutrients, gases, and other critical molecules throughout the body. It consists of the heart, blood vessels, and blood.

carnivore An organism (typically an animal) that eats animals.

carrier An individual who is heterozygous for a recessive allele and can therefore pass it on to offspring without showing any of its effects.

carrying capacity The maximum population size that a given environment or habitat can support given its food supply and other natural resources.

catabolic reaction Any chemical reaction that breaks down complex molecules into simpler molecules.

catalysis The process of speeding up the rate of a chemical reaction (e.g., by an enzyme).

cell The basic structural unit of living organisms.

cell body The part of a neuron that contains the nucleus and most of the cell's other organelles.

cell cycle The ordered sequence of stages through which a cell progresses in order to divide; the stages include preparatory phases (G_1, S, G_2) and division phases (mitosis and cytokinesis).

cell cycle checkpoint A cellular mechanism that ensures that each stage of the cell cycle is completed accurately.

cell differentiation The process by which a cell becomes specialized to carry out a specific role by turning specific genes "on" and "off" and making different suites of proteins.

cell division The process by which a cell reproduces itself; cell division is important for normal growth, development, maintenance, and repair of an organism.

cell membrane A phospholipid bilayer with embedded proteins that forms the boundary of all cells.

cell theory The concept that all living organisms are made of cells and that cells are formed by the division of existing cells.

cell wall A rigid structure present in some cells that encloses the cell membrane and helps the cell maintain its integrity.

cell-mediated immunity The type of adaptive immunity that rids the body of infected, cancerous, or foreign cells.

central nervous system (CNS) The brain and the spinal cord.

central vacuole A fluid-filled compartment in a plant cell that contributes to cell rigidity by exerting turgor pressure against the cell wall.

centromere The specialized region of a chromosome where the sister chromatids are joined; it is critical for proper alignment and separation of sister chromatids during mitosis.

cerebellum The part of the brain that processes sensory information and is involved in movement, coordination, and balance.

cerebral cortex The outer layer of the cerebrum, the cerebral cortex is involved in many advanced brain functions.

cerebrum The region of the brain that controls intellect, learning, perception, and emotion.

cervix The opening or "neck" of the uterus, where sperm enter and babies exit.

chemical energy Potential energy stored in the bonds of biological molecules.

chemotherapy Treatment by toxic chemicals that kill cancer by interfering with cell division.

chlorophyll The pigment present in the green parts of plants that absorbs photons of light energy during the "photo" reactions of photosynthesis.

chloroplast An organelle in plant and algal cells that is the site of photosynthesis.

cholesterol A lipid that is an important component of cell structures; it is used to make important molecules and also plays a role in heart disease.

chromosome A single, large DNA molecule wrapped around proteins. Chromosomes are located in the nuclei of eukaryotic cells.

chyme The acidic "soup" of partially digested food that leaves the stomach and enters the small intestine.

citric acid cycle A set of reactions that takes place in mitochondria and helps extract energy (in the form of high-energy electrons) from food; the second stage of aerobic respiration.

climate The long-term average of atmospheric conditions.

climate change Any substantial change in climate that lasts for an extended period of time (decades or more).

closed circulatory system A circulatory system in which the blood remains in the blood vessels at all times.

coding sequence The part of a gene that specifies the amino acid sequence of a protein. Coding sequences determine the identity, shape, and function of proteins.

codominance A form of inheritance in which the effects of both alleles are displayed in the phenotype of a heterozygote.

codon A sequence of three mRNA nucleotides that specifies a particular amino acid.

coenzyme A small organic molecule, such as a vitamin, required to activate an enzyme.

cofactor An inorganic substance, such as a metal ion, required to activate an enzyme.

cohesion The attraction between molecules (or other particles).

commensalism A type of symbiotic relationship in which one member benefits and the other is unharmed.

community Interacting populations of different species in a defined habitat.

competition An interaction between two or more organisms that rely on a common resource that is not available in sufficient quantities.

competitive exclusion principle The concept that when two species compete for resources in an identical niche, one is inevitably driven to extinction.

complement proteins Proteins in blood that help destroy pathogens by coating or puncturing them.

complementary Fitting together; two strands of DNA are said to be complementary in that A in one strand always pairs with T in the other strand, and G always pairs with C.

conservation of energy The principle that energy cannot be created or destroyed, but can be transformed from one form to another.

consumers Heterotrophs that eat other organisms lower on the food chain to obtain energy.

contact inhibition A characteristic of normal cells that prevents them from dividing once they have filled a space and are in contact with their neighbors.

continental drift The movement of the continents relative to one another over time.

continuous variation Variation in a population showing an unbroken range of phenotypes rather than discrete categories.

contraception The prevention of pregnancy through physical, surgical, or hormonal methods.

control group The group in an experiment that experiences no experimental intervention or manipulation.

convergent evolution The process by which organisms that are not closely related evolve similar adaptations as a result of independent episodes of natural selection.

coronary arteries The blood vessels that deliver oxygenated blood to the heart muscle.

corpus luteum The structure in the ovary that remains after ovulation. It secretes progesterone.

correlation A consistent relationship between two variables.

cotyledon The part of a plant embryo that forms the first leaf of the seedling.

covalent bond A strong interaction resulting from the sharing of a pair of electrons between two atoms.

cuticle The waxy coating on leaves and stems that prevents water loss.

cytokinesis The physical division of a cell into two daughter cells.

cytoplasm The gelatinous, aqueous interior of all cells.

cytoskeleton A network of protein fibers in eukaryotic cells that provides structure and facilitates cell movement.

cytotoxic T cell A type of T cell that destroys infected, cancerous, or foreign cells.

dendrites Branched extensions from the cell body of a neuron, which receive incoming information.

density-dependent factor A factor whose influence on population size and growth depends on the number and crowding of individuals in the population (for example, predation).

density-independent factor A factor that can influence population size and growth regardless of the numbers and crowding within a population (for example, weather).

deoxyribonucleic acid (DNA) The molecule of heredity, common to all life forms, that is passed from parents to offspring.

dependent variable The measured result of an experiment, analyzed in both the experimental and control groups.

descent with modification Darwin's term for evolution, combining the ideas that all living things are related and that organisms have changed over time.

diabetes A disease characterized by chronically elevated levels of blood sugar.

diaphragm A sheet of muscle that contributes to breathing by contracting and relaxing.

diastolic pressure The pressure in arteries when the ventricle is relaxed.

dicots Flowering plants with two cotyledons (seed leaves) in their seeds.

diencephalon A brain region located between the brain stem and the cerebrum that regulates homeostatic functions like body temperature, hunger, thirst, and the sex drive.

dietary fiber Plant-based carbohydrates that contribute to proper digestion.

diffusion The movement of dissolved substances from an area of higher concentration to an area of lower concentrtion.

digestion The mechanical and chemical breakdown of food into subunits, enabling the absorption of nutrients.

digestive system The organ system that breaks down food molecules into smaller subunits, absorbs nutrients, and eliminates waste; it is composed of the digestive tract and accessory organs.

diploid Having two copies of every chromosome.

directional selection A type of natural selection in which organisms with phenotypes at one end of a spectrum are favored by the environment.

dispersion pattern The way that organisms are distributed in geographic space, which depends on resources and interactions with other members of the population.

diversifying selection A type of natural selection in which organisms with phenotypes at both extremes of the phenotypic range are favored by the environment.

DNA polymerase An enzyme that "reads" the nucleotide sequence of a DNA strand and incorporates complementary nucleotides into a new strand during DNA replication.

DNA profile A visual representation of a person's unique DNA sequence.

DNA replication The natural process by which cells make an identical copy of a DNA molecule.

domain The highest (most inclusive) category in the modern system of classification. There are three domains: Bacteria, Archaea, and Eukarya.

dominant allele An allele that can mask the presence of a recessive allele.

dopamine A neurotransmitter that is involved in conveying a sense of pleasure in the brain.

double fertilization The process in angiosperms in which one sperm fertilizes the egg and one sperm fuses with other cells in the ovule.

double helix The spiral structure formed by two strands of DNA nucleotides held together by hydrogen bonds.

duodenum The first portion of the small intestine, where mixing of chyme and digestive enzymes occurs.

ecological footprint A measure of how much land and water area is required to both supply the resources an individual or population consumes and also absorb the wastes it produces.

ecology The study of the interactions among organisms and between organisms and their nonliving environment.

ecosystem All the living organisms in an area and the nonliving components of the environment with which they interact.

ectotherm An animal that relies on environmental sources of heat, such as sunlight, to maintain its body temperature.

effector A cell or tissue that responds to information relayed from a sensor.

electron A negatively charged subatomic particle with negligible mass.

electron transport chain The transfer of electrons that takes place in mitochondria and produces the bulk of ATP during aerobic respiration; the third stage of aerobic respiration.

element A pure substance that cannot be chemically broken down; each element is made up of and defined by a single type of atom.

embryo An early stage of development reached when a zygote undergoes cell division to form a multicellular structure.

emigration The movement of individuals out of a population.

endocrine system The collection of hormone-secreting glands and organs with hormone-secreting cells.

endometrium The lining of the uterus.

endoplasmic reticulum (ER) A network of membranes in eukaryotic cells where proteins and lipids are synthesized.

endoskeleton An internal body skeleton, typically made of cartilage or bone.

endosperm A part of the seed containing nutrients that is the seed's source of energy, produced when a sperm fuses with cells in the ovule during double fertilization

endosymbiosis The scientific theory that free-living prokaryotic cells engulfed other free-living prokaryotic cells billions of years ago, forming eukaryotic organelles such as mitochondria and chloroplasts.

endotherm An animal that can generate body heat internally to maintain its body temperature.

energy The ability to do work. Living organisms obtain energy either directly from sunlight (through photosynthesis) or from food they consume.

enzyme A protein that speeds up the rate of a chemical reaction.

epidemiology The study of patterns of disease in populations, including risk factors.

epididymis A system of tubes in which sperm mature and are stored before ejaculation.

erythropoietin (EPO) A hormone produced by the kidneys that stimulates red blood cell production.

esophagus The section of the digestive tract between the mouth and the stomach.

essential amino acids Amino acids that can't be made by the body, and so must be obtained preassembled from the diet.

essential nutrients Nutrients that can't be made by the body, and so must be obtained from the diet.

estrogen A class of sex hormones, including estradiol, that are higher in women than in men and support female sexual development and function.

ethylene A gaseous plant hormone that promotes fruit ripening as one of its effects.

eukaryote Any organism of the domain Eukarya; eukaryotic cells are characterized by the presence of a membrane-enclosed nucleus and organelles.

eukaryotic cells Cells that contain membrane-bound organelles, including a central nucleus.

eutrophication An excessive enrichment of nutrients in bodies of water that can cause ecological problems.

evolution Change in allele frequencies in a population over time.

exoskeleton An external skeleton; in arthropods the exoskeleton is made up of proteins and chitin.

experiment A carefully designed test, the results of which will either support or rule out a hypothesis.

experimental group The group in an experiment that experiences the experimental intervention or manipulation.

exponential growth The unrestricted growth of a population increasing at a constant growth rate.

extinction The elimination of all individuals in a species; extinction may occur over time or in a sudden mass die-off.

facilitated diffusion The process by which large, hydrophilic, or charged solutes move across a membrane from an area of higher concentration to an area of lower concentration with the help of transport proteins; does not require an input of energy.

falsifiable Describes a hypothesis that can be ruled out by data that show that the hypothesis does not explain the observation.

feedback loop A pathway in which the output from an effector feeds back to a sensor and changes further output.

fermentation A series of chemical reactions beginning with glycolysis and taking place in the absence of oxygen. Fermentation produces far less ATP than does aerobic respiration.

ferns The first true vascular plants; ferns do not produce seeds.

fertilization The fusion of an egg and a sperm; the resulting cell is called a zygote.

fetus After the eighth week after fertilization, the embryo is referred to as a fetus.

fever An elevated body temperature.

fitness The relative ability of an organism to survive and reproduce in a particular environment.

flagella (singular: flagellum) Whiplike appendages extending from some cells, used in movement of the cell.

folate A B vitamin also known as folic acid, folate is an essential nutrient, necessary for basic cellular processes such as DNA replication and cell division.

follicle A structure in the ovary where eggs mature.

follicle-stimulating hormone (FSH) A hormone secreted by the anterior pituitary. In females, FSH triggers eggs to mature at the start of each monthly cycle.

food chain A linked series of feeding relationships in a community in which organisms further up the chain feed on ones below.

food web A complex interconnection of feeding relationships in a community.

fossil fuel A carbon-rich energy source, such as coal, petroleum, or natural gas, formed from the compressed, fossilized remains of once-living organisms.

fossil record An assemblage of fossils arranged in order of age, providing evidence of changes in species over time.

fossils The preserved remains or impressions of once-living organisms.

founder effect A type of genetic drift in which a small number of individuals leaves one population and establishes a new population, resulting in lower genetic diversity than in the original population.

fruiting body A fungal structure that is specialized for the release of spores.

fungus (plural: fungi) A unicellular or multicellular eukaryotic organism that obtains nutrients by secreting digestive enzymes onto organic matter and absorbing the digested product.

gallbladder An organ that stores bile salts and releases them as needed into the small intestine.

gametes Specialized reproductive cells that carry one copy of each chromosome (that is, they are haploid). Sperm are male gametes; eggs are female gametes.

gas exchange In the lungs, the process of taking up and releasing oxygen and carbon dioxide across capillaries.

gel electrophoresis A laboratory technique that separates fragments of DNA by size.

gene A sequence of DNA that contains the information to make at least one protein.

gene expression The process of using DNA instructions to make proteins.

gene flow The movement of alleles from one population to another, which may increase the genetic diversity of a population.

gene pool The total collection of alleles in a population.

gene therapy A treatment that aims to cure, treat, or prevent human disease by replacing defective genes with functional ones.

gene transfer The process by which bacteria can exchange segments of DNA between them.

genetic code The set of rules relating particular mRNA codons to particular amino acids.

genetic drift Random changes in the allele frequencies of a population between generations; genetic drift tends to have more dramatic effects in smaller populations than in larger ones.

genetic engineering Altering or manipulating the DNA of organisms by modern laboratory techniques.

genetically modified organism (GMO) An organism whose genome has been altered through genetic engineering techniques, often to contain a gene from another species.

genome One complete set of genetic instructions encoded in the DNA of an organism.

genotype The particular genetic makeup of an individual.

germination The process by which a new plant begins to grow from a seed.

gibberellins Plant hormones that cause cell division and stem elongation.

gills The gas exchange surface in fish.

glial cells Supporting cells of the nervous system, some of which produce myelin.

global hectare (gha) A unit of measurement representing the biological productivity (both resource-providing and waste-absorbing) of an average hectare of Earth.

global warming An increase in Earth's average temperature.

glucagon A hormone produced by the pancreas that causes an increase in blood sugar.

glycogen A complex animal carbohydrate, made up of linked chains of glucose molecules, that stores energy for short-term use.

glycolysis A series of reactions that breaks down sugar into smaller units; glycolysis takes place in the cytoplasm and is the first stage of both aerobic respiration and fermentation.

Golgi apparatus An organelle made up of stacked membrane-enclosed discs that packages proteins and prepares them for transport.

gonads Sex organs: ovaries in females, testes in males.

Gram-negative Describes bacteria with a cell wall that includes a thin layer of peptidoglycan surrounded by an outer lipid membrane that does not retain the Gram stain.

Gram-positive Describes bacteria with a cell wall that includes a thick layer of peptidoglycan that retains the Gram stain.

gravitropism The growth of plants in response to gravity. Roots grow downward, with gravity; shoots grow upward, against gravity.

greenhouse effect The natural process by which heat is radiated from Earth's surface and trapped by gases in the atmosphere, helping to maintain Earth at a temperature that can support life.

greenhouse gas Any of the gases in Earth's atmosphere that absorb heat radiated from Earth's surface and contribute to the greenhouse effect; for example carbon dioxide and methane.

growth rate The difference between the birth rate and the death rate of a given population; also known as the rate of natural increase.

gymnosperm A seed-bearing plant with exposed seeds typically held in cones.

habitat The physical environment where an organism lives and to which it is adapted.

half-life The time it takes for one-half of a sample of a radioactive isotope to decay.

haploid Having only one copy of every chromosome.

Hardy-Weinberg principle The principle that, in a nonevolving population, both allele and genotype frequencies remain constant from one generation to the next.

heart The muscular pump that generates force to move blood throughout the body.

heart attack Damage to the heart muscle resulting from the restriction of blood flow to heart tissue.

heat The kinetic energy generated by random movements of molecules or atoms.

helicase An enzyme that unwinds and unzips the DNA double helix during DNA replication.

helper T cell A type of T cell that helps activate other lymphocytes including B cells and cytotoxic cells.

heme groups Iron-containing structures on hemoglobin, the sites of oxygen binding.

hemoglobin A protein found in red blood cells specialized for transporting oxygen.

herbivore An organism that eats plants.

herd immunity The protection of a population from an infection, based on a certain percentage of its members being immune.

heterotrophs Organisms, such as humans and other animals, that obtain energy by consuming organic molecules that were produced by other organisms.

heterozygous Having two different alleles for a given gene.

high-density lipoprotein (HDL) A form of cholesterol and protein that is protective against CVD.

histamine A molecule released by damaged tissue and during allergic reactions that promotes inflammation.

homeostasis The maintenance of a relatively stable internal environment.

hominid Any living or extinct member of the family Hominidae, the great apes–humans, orangutans, gorillas, chimpanzees, and bonobos.

homologous chromosomes A pair of chromosomes that both contain the same genes. In a diploid cell, one chromosome in the pair is inherited from the mother, the other from the father.

homology Anatomical, genetic, or developmental similarity among organisms due to common ancestry.

homozygous Having two identical alleles for a given gene.

hormone A chemical signaling molecule that is released by a cell or gland and travels through the bloodstream to exert an effect on target cells.

human chorionic gonadotropin (hCG) A hormone produced by an early embryo that helps maintain the corpus luteum until the placenta develops.

humoral immunity The type of adaptive immunity that fights free-floating pathogens in the blood and lymph.

hydrogen bond A weak electrical attraction between a partially positive hydrogen atom and an atom with a partial negative charge.

hydrophilic "Water-loving"; hydrophilic molecules dissolve easily in water.

hydrophobic "Water-fearing"; hydrophobic molecules will not dissolve in water.

hypertension Elevated (high) blood pressure.

hypertonic Describes a solution surrounding a cell that has a higher concentration of solutes than the cell's cytoplasm.

hypha (plural: hyphae) A long, threadlike structure through which fungi absorb nutrients.

hypothalamus A master coordinator region of the brain responsible for a variety of physiological functions.

hypothermia A drop of body temperature below 35°C (95°F), which causes enzyme malfunction and eventually death.

hypothesis A tentative explanation for a scientific observation or question.

hypotonic Describes a solution surrounding a cell that has a lower concentration of solutes than the cell's cytoplasm.

hypoxia A state of low oxygen concentration in the blood.

igneous rock Rock formed from the cooling and hardening of molten lava.

immigration The movement of individuals into a population.

immune system A network of cells and tissues that acts to defend the body against infectious agents and also helps to heal injuries.

immunity Protection from a given pathogen conferred by the activity of the immune system.

immunotherapy A cancer therapy that uses the immune system to recognize and destroy cancer cells.

in vitro fertilization (IVF) A form of assisted reproduction in which eggs and sperm are brought together outside the body and the resulting embryos are inserted into a woman's uterus.

inbreeding Mating between closely related individuals. Inbreeding does not change the allele frequency within a population, but it does increase the proportion of homozygous individuals to heterozygotes.

inbreeding depression The negative reproductive consequences for a population associated with having a high frequency of homozygous individuals possessing harmful recessive alleles.

incomplete dominance A form of inheritance in which heterozygotes have a phenotype that is intermediate between the two homozygotes.

independent assortment The principle that alleles of different genes are distributed independently of one another during meiosis.

independent variable The variable, or factor, being deliberately changed in the experimental group relative to the control group.

inflammation An innate defense that is activated by infection or tissue damage; characterized by redness, heat, swelling, and pain.

innate immunity Nonspecific defenses, such as physical and chemical barriers and specialized white blood cells, that are present from birth and require little or no time to become active.

inorganic Describes a molecule that lacks a carbon-based backbone and C–H bonds.

insect An arthropod with three pairs of jointed legs and a body with three segments

insulin A hormone secreted by the pancreas that causes a decrease in blood sugar.

interphase The stage of the cell cycle in which dividing cells spend most of their time, preparing for cell division. There are three distinct subphases: G_1, S, and G_2.

intrauterine insemination (IUI) A form of assisted reproduction in which sperm are injected directly into a woman's uterus.

introduced species Species that are not native to a particular environment and which have arrived as a result of human activity.

invasive species Introduced species that do harm in their new environment.

invertebrate An animal without a backbone.

ion An electrically charged atom, the charge resulting from the loss or gain of electrons.

ionic bond A strong electrical attraction between oppositely charged ions formed by the transfer of one or more electrons from one atom to another.

isotonic Describes a solution surrounding a cell that has the same solute concentration as the cell's cytoplasm.

karyotype The chromosomal makeup of cells. Karyotype analysis can be used to detect chromosome disorders prenatally.

keystone species Species on which other species depend, and whose removal has a dramatic impact on the community.

kidney An organ involved in osmoregulation, filtration of blood to remove wastes, and production of several important hormones.

kinetic energy The energy of motion or movement.

kinetochore Proteins located at the centromere that provide an attachment point for microtubules of the mitotic spindle.

large intestine The last organ of the digestive tract, in which remaining water is absorbed and solid stool is formed.

larynx The opening to the lower respiratory tract; also known as the voicebox.

light energy A type of electromagnetic radiation that includes visible light.

lignin A stiff strengthening agent found in secondary cell walls of plants.

limbic system A set of brain structures that is stimulated during pleasurable activities and which is involved in addiction.

lipase A fat-digesting enzyme active in the small intestine.

lipids Organic molecules that generally repel water.

liver An organ that aids digestion by producing bile salts that emulsify fats.

logistic growth A pattern of growth that starts off fast and then levels off as the population reaches the carrying capacity of the environment.

low-density lipoprotein (LDL) A form of cholesterol and protein in the blood that contributes to CVD.

lung The major respiratory organ, the site of gas exchange between air and the blood.

luteinizing hormone (LH) A hormone secreted by the anterior pituitary. In females, a surge of LH triggers ovulation.

lymph nodes Small organs in the lymphatic system where B and T cells may encounter pathogens.

lymphatic system The system of vessels and organs that drains fluid (lymph) from the tissues and sends it through lymph nodes on it way to the circulation.

lymphocyte A specialized white blood cell of the immune system. Lymphocytes are important in adaptive immunity.

lysosome An organelle in eukaryotic cells that is filled with enzymes that can degrade worn-out cellular structures.

macromolecules Very large organic molecules that make up living organisms; they include carbohydrates, proteins, and nucleic acids.

macronutrients Nutrients, including carbohydrates, proteins, and fats, that organisms must ingest in large amounts to maintain health.

malignant tumor A cancerous tumor whose cells can spread throughout the body.

malnutrition The medical condition resulting from the lack of essential nutrients in the diet. Malnutrition is often, but not always, associated with starvation.

mammals Members of the class Mammalia; all members of this class have mammary glands and a body covered with hair.

mammogram An X-ray of the breast.

mass extinction An extinction of between 50% and 90% of all species that occurs relatively rapidly.

matter Anything that takes up space and has mass.

meiosis A type of cell division that generates genetically unique haploid gametes.

melanin A pigment produced by a specific type of skin cell that gives skin its color.

memory cell A long-lived B or T cell that is produced during an immune response and that can "remember" the pathogen.

menstruation The shedding of the uterine lining (the endometrium) that occurs when an embryo does not implant.

meristem A plant tissue containing cells that can divide and contribute to the growth of the plant.

messenger RNA (mRNA) The RNA copy of an original DNA sequence made during transcription.

metabolism All biochemical reactions occurring in an organism, including reactions that break down food molecules and reactions that build new cell structures.

metastasis The spread of cancer cells from one location in the body to another.

micronutrients Nutrients, including vitamins and minerals, that organisms must ingest in small amounts to maintain health.

mineral An inorganic chemical element required by organisms for normal growth, reproduction, and tissue maintenance; examples are calcium, iron, potassium, and zinc.

mitochondria (singular: mitochondrion) Membrane-bound organelles responsible for important energy-conversion reactions in eukaryotes.

mitochondrial DNA (mtDNA) The DNA within mitochondria; it is inherited solely from the mother.

mitosis The segregation and separation of replicated chromosomes during cell division.

mitotic spindle The microtubule-based structure that separates sister chromatids during mitosis.

molecule Atoms linked by covalent bonds.

mollusk An invertebrate with a soft, unsegmented body enclosed in a hard shell.

monocots Flowering plants with one cotyledon (seed leaf) in their seeds.

monocrop Single crop species grown in the same field over many seasons.

monomer One chemical subunit of a polymer.

monosaccharide The building block, or monomer, of a carbohydrate.

motor neurons Neurons that control the contraction of skeletal muscle.

multifactorial inheritance An interaction between genes and the environment that contributes to a phenotype or trait.

mutagen Any chemical or physical agent that can damage DNA by changing its nucleotide sequence.

mutation A change in the nucleotide sequence of DNA.

mutualism A type of symbiotic relationship in which both members benefit; a "win-win" relationship.

mycelium (plural: mycelia) A spreading mass of interwoven hyphae that forms the often subterranean body of multicellular fungi.

myelin A fatty substance that insulates the axons of neurons and facilitates rapid conduction of action potentials.

NAD^+ An electron carrier. NAD^+ can accept electrons, becoming NADH in the process.

natural resources Raw materials that are obtained from Earth and are considered valuable even in their relatively unmodified, natural form.

natural selection Differential survival and reproduction of individuals in response to environmental pressure that leads to change in allele frequencies in a population over time.

NEAT Non-exercise activity thermogenesis, the amount of energy expended in everyday activities.

nerve A bundle of specialized cells that transmit information.

nervous system The collection of organs that sense and respond to information, including the brain, spinal cord, and nerves.

neurons Specialized cells of the nervous system that generate electrical signals in the form of action potentials.

neurotransmitter A chemical signaling molecule released by a neuron to transmit a signal to a neighboring cell.

neutron An electrically uncharged subatomic particle in the nucleus of an atom.

niche The space, environmental conditions, and resources that a species needs in order to survive and reproduce.

nitrogen cycle The movement of nitrogen atoms as they cycle between different molecules in living organisms and the environment.

nitrogen-fixing bacteria Bacteria that convert gaseous nitrogen to ammonia, a form of nitrogen usable by plants.

nonadaptive evolution Any change in allele frequency that does not by itself lead a population to become more adapted to its environment; the mechanisms of nonadaptive evolution are mutation, genetic drift, and gene flow.

nondisjunction The failure of chromosomes to separate accurately during cell division; nondisjunction in meiosis leads to aneuploid gametes.

nonrenewable resources Natural resources that cannot easily be replaced; fossil fuels are an example.

nuclear envelope The double membrane surrounding the nucleus of a eukaryotic cell.

nucleic acids Organic molecules made up of linked nucleotide subunits; DNA and RNA are examples of nucleic acids.

nucleotides The building blocks of DNA. Each nucleotide consists of a sugar, a phosphate group, and a base. The sequence of nucleotides (As, Cs, Gs, Ts) along a DNA strand is unique to each person.

nucleus (atomic) The dense core of an atom.

nucleus (eukaryotic) The organelle in eukaryotic cells that contains the genetic material.

nutrients Components in food that the body needs to grow, develop, and repair itself.

obese Having 20% more body fat than is recommended for one's height, as measured by a body mass index equal to or greater than 30.

oncogene A mutated and overactive form of a proto-oncogene. Oncogenes drive cells to divide continually.

open circulatory system A circulatory system in which the blood is not entirely contained within blood vessels.

organ A structure made up of different tissue types working together to carry out a common function.

organ system A set of cooperating organs within the body.

organelles The membrane-bound compartments of eukaryotic cells that carry out specific functions.

organic (agriculture) Describes a way of growing crops that conforms to several regulations, among them that synthetic pesticides must not be used.

organic (chemistry) Describes a molecule with a carbon-based backbone and at least one C–H bond.

osmolarity The concentration of solutes in blood and other bodily fluids.

osmoregulation The maintenance of relatively stable volume, pressure, and solute concentration of bodily fluids, especially blood.

osmosis The diffusion of water across a membrane from an area of lower solute concentration to an area of higher solute concentration.

ovaries Paired female reproductive organs; the ovaries contain eggs and produce estrogen and progesterone.

ovary (plant) The structure at the base of the pistil that contains the ovules.

overweight Having a BMI of 25 or more but less than 30.

oviduct The tube connecting an ovary and the uterus in females. eggs are ovulated into and fertilized within the oviducts.

ovulation The release of an egg from an ovary into the oviduct.

ovule The part of a flower that develops into a seed after fertilization.

paleontologist A scientist who studies ancient life by examining the fossil record.

pancreas An organ that secretes the hormones insulin and glucagon as well as digestive enzymes.

pandemic A worldwide epidemic; an exceptionally high number of cases occurring at the same time.

parasitism A type of symbiotic relationship in which one member benefits at the expense of the other.

partial pressure The proportion of total air pressure contributed by a given gas.

pathogen A disease-causing agent or organism.

pedigree A visual representation of the occurrence of phenotypes across generations.

peer review A process in which independent scientific experts read scientific studies before they are published to ensure that the authors have appropriately designed and interpreted the study.

pepsin A protein-digesting enzyme that is active in the stomach.

peptidoglycan The macromolecule found in all bacterial cell walls that gives the cell wall rigidity.

peripheral nervous system (PNS) All the nervous tissue outside the central nervous system that transmits signals from the CNS to the rest of the body.

peristalsis Coordinated muscular contractions that force food down the digestive tract.

pesticide A substance that is toxic to pests, organisms that can damage crops or farm animals.

pH A measure of the concentration of H^+ in a solution.

phagocyte A type of white blood cell that engulfs and digests pathogens and debris from dead cells.

pharynx The throat.

phenotype The visible or measurable features of an individual.

phloem Plant vascular tissue that transports sugars throughout the plant.

phospholipid A type of lipid that forms the cell membrane.

phosphorus cycle The movement of phosphorus atoms as they cycle between different molecules in living organisms and the environment.

photons Packets of light energy, each with a specific wavelength and quantity of energy.

photosynthesis The process by which plants and algae harness the energy of sunlight to make sugar from carbon dioxide and water.

phototropism The growth of the stem of a plant toward light.

phylogenetic tree A branching diagram of relationships showing common ancestry.

phylogeny The evolutionary history of a group of organisms.

physiology The study of the way a living organism's physical parts function.

pili (singular: pilus) Short, hairlike appendages extending from the surface of some bacteria that enable them to adhere to surfaces.

pistil The female reproductive structure of a flower, made up of a stigma, style, and ovary.

pituitary gland An endocrine gland in the brain that secretes many important hormones.

placebo A fake treatment given to control groups to mimic the experience of the experimental groups.

placenta A structure made of fetal and maternal tissues that helps sustain and support the embryo and fetus.

plant A multicellular eukaryote that has cell walls, carries out photosynthesis, and is adapted to living on land.

plaque Deposits of cholesterol, other fatty substances, calcium, blood clotting proteins, and cellular waste that accumulate inside arteries, limiting the flow of blood.

plasma cell An activated B cell that divides rapidly and secretes an abundance of antibodies.

plate tectonics The movement of Earth's upper mantle and crust, which influences the geographical distribution of landmasses and organisms.

platelets Fragments of cells involved in blood clotting.

polar molecule A molecule in which electrons are not shared equally between atoms, causing a partial negative charge at one end and a partial positive charge at the other. Water is a polar molecule.

pollen Small, thick-walled plant structures that contain cells that develop into sperm.

pollen tube A hollow tube that grows from a pollen grain after pollination and transports the male gametes to the egg.

pollination The transfer of pollen from male to female plant structures so that fertilization can occur.

polygenic trait A trait whose phenotype is determined by the interaction among alleles of more than one gene.

polymer A molecule made up of individual subunits, called monomers, linked together in a chain.

polymerase chain reaction (PCR) A laboratory technique used to replicate, and thus amplify, a specific DNA segment.

population A group of organisms of the same species living together in the same geographic area.

population density The number of organisms per unit area.

population genetics The study of the genetic makeup of populations and how the genetic composition of a population changes.

potential energy Stored energy.

predation An interaction between two organisms in which one organism (the predator) feeds on the other (the prey).

primary immune response The adaptive immune response mounted the first time a particular antigen is encountered by the immune system.

prion A protein-only infectious agent.

producers Autotrophs (photosynthetic organisms) that obtain energy directly from the sun and form the base of every food chain.

progesterone A female sex hormone produced by the corpus luteum of the ovary that prepares and maintains the uterus for pregnancy.

prokaryote A (typically) unicellular organism whose cell lacks internal membrane-bound organelles and whose DNA is not contained within a nucleus.

prokaryotic cells Cells that lack internal membrane-bound organelles.

protein A macromolecule made up of repeating subunits called amino acids, which determine the shape and function of a protein. Proteins play many critical roles in living organisms.

protist A eukaryote that cannot be classified as a plant, animal, or fungus; usually unicellular.

proton A positively charged subatomic particle in the nucleus of an atom.

proto-oncogene A gene that codes for a protein that helps cells divide normally.

pulmonary circuit The circulation of blood between the heart and the lungs.

pulse The detectable force of blood entering arteries, which can be felt in the neck or wrist.

punctuated equilibrium Periodic bursts of species change as a result of sudden environmental change.

Punnett square A diagram used to determine probabilities of offspring having particular genotypes, given the genotypes of the parents.

radial symmetry The pattern exhibited by a body plan that is circular, with no defined left and right sides.

radiation therapy The use of ionizing (high-energy) radiation to treat cancer.

radioactive isotope An unstable form of an element that decays into another element by radiation, that is, by emitting energetic particles.

radiometric dating The use of radioactive isotopes as a measure for determining the age of a rock or fossil.

randomized clinical trial A controlled medical experiment in which subjects are randomly chosen to receive either an experimental treatment or a standard treatment (or a placebo).

recessive allele An allele that reveals itself in the phenotype only if a masking dominant allele is not present.

recombinant gene A genetically engineered gene that contains portions of genes not naturally found together.

recombination An event in meiosis during which maternal and paternal chromosomes pair and physically exchange DNA segments.

red blood cells Cells specialized for carrying oxygen.

regulatory sequence The part of a gene that determines the timing, amount, and location of protein production.

relative dating Determining the age of a fossil from its position relative to layers of rock or fossils of known age.

renewable resources Natural resources that are replenished after use as long as the rate of consumption does not exceed the rate of replacement.

reproductive isolation Mechanisms that prevent mating (and therefore gene flow) between members of different species.

respiratory surface A surface across which oxygen enters and carbon dioxide leaves.

respiratory system The organ system that allows us to take in oxygen and unload carbon dioxide.

ribosome A complex of RNA and proteins that carries out protein synthesis in all cells.

risk factor A behavior, exposure, or other factor that increases the probability of developing a disease.

RNA polymerase The enzyme that carries out transcription. RNA polymerase copies a strand of DNA into a complementary strand of mRNA.

root The belowground parts of a plant, which anchor it in the soil and absorb water and nutrients.

root nodule An enlargement on the root of a plant that contains nitrogen-fixing bacteria.

salivary glands Glands that secrete enzymes into the mouth to break down macromolecules in food. One such enzyme is salivary amylase, which digests carbohydrates.

sample size The number of experimental subjects or the number of times an experiment is repeated. In human studies, sample size is the number of participants.

saturated fat A fat, such as butter, that is solid at room temperature; often found in animal products.

science The process of using observations and experiments to draw conclusions based on evidence.

scientific theory An explanation of the natural world that is supported by a large body of evidence and has never been disproved.

scrotum The sac in which the testes are held.

secondary immune response The rapid and strong immune response mounted when a particular antigen is encountered by the immune system subsequent to the first encounter.

secondary metabolites Chemicals produced by plants that are not directly involved in growth or reproduction but that help protect the plant by their impacts on other organisms.

sedimentary rock Rock formed from the compression of layers of particles eroded from other rocks.

seed coat The hearty outer covering of a seed that protects the developing embryo.

semen The mixture of fluid and sperm that is ejaculated from the penis.

semiconservative DNA replication is said to be semiconservative because each newly made DNA molecule has one original and one new DNA strand.

seminiferous tubules Coiled structures that constitute the bulk of the testes and in which sperm develop.

sensor A specialized cell that detects specific sensory input like temperature, pressure, or solute concentration.

sensory neurons Cells that convey information from both inside and outside the body to the CNS.

sex chromosomes Paired chromosomes that differ between males and females. Females have XX, and males have XY.

shoot The aboveground parts of a plant: the stem and photosynthetic leaves.

short tandem repeats (STRs) Sections of a chromosome in which short DNA sequences are repeated.

simple diffusion The movement of small, uncharged solutes across a membrane from an area of higher concentration to an area of lower concentration without the aid of transport proteins; does not require an input of energy.

sister chromatids The two identical DNA molecules that result from the replication of a chromosome during S phase.

small intestine The organ in which the bulk of chemical digestion and absorption of food occurs.

solute A dissolved substance.

solution The mixture of solute and solvent.

solvent A substance in which other substances can dissolve. Water is a good solvent.

speciation The genetic divergence of populations, leading over time to reproductive isolation and the formation of new species.

spinal cord A bundle of nerve fibers, contained within the bony spinal column, that transmits information between the brain and the rest of the body.

spores (fungal) Fungal cells that are resistant to drying out and can be dispersed to new locations as part of sexual or asexual reproduction.

stabilizing selection A type of natural selection in which organisms near the middle of the phenotypic range of variation are favored by the environment.

stamen The male reproductive structure of a flower, made up of a filament and an anther.

staple crop A crop that is eaten in large quantities and provides most of the energy and nutrients in the human diet.

starch A complex plant carbohydrate made of linked chains of glucose molecules; a source of stored energy.

statistical significance A measure of confidence that the results obtained are "real" and not due to chance.

stem cells Immature cells that can divide and differentiate into specialized cell types.

stigma The sticky "landing pad" for pollen on the pistil.

stomach An expandable muscular organ that stores and mechanically breaks down food. Specific enzymes in the stomach digest proteins.

stomata (singular: stoma) Pores on leaves that permit the exchange of oxygen and carbon dioxide with the air and allow water loss.

stool Solid waste material eliminated from the digestive tract.

stroke A disruption in blood supply to the brain.

style The tubelike structure that leads from the stigma to the ovary.

substrate A molecule to which an enzyme binds and on which the enzyme acts.

sustainability The use of Earth's resources in a way that will not permanently destroy or deplete them.

symbiosis A relationship in which two different organisms live together, often interdependently.

synapse The site of transmission of a signal between a neuron and another cell; includes the axon terminal of the signaling neuron, the space between the cells, and receptors on the receiving cell.

synaptic cleft The physical space between a neuron and the cell with which it is communicating.

systemic circuit The circulation of blood between the heart and the rest of the body.

systolic pressure The pressure in arteries at the time the ventricle contracts.

T cells White blood cells that mature in the thymus and play several roles in adaptive immunity.

targeted therapy A cancer therapy that is specific for cancer cells and not harmful to normal cells.

taxonomy The identification, naming, and classification of organisms on the basis of shared traits.

testable Describes a hypothesis that can be supported or rejected by carefully designed experiments or observational studies.

testes (singular: testis) Paired male reproductive organs, which contain sperm and produce androgens (primarily testosterone).

testosterone The primary male sex hormone, which stimulates the development of masculine features and plays a key role in sperm development.

tetrapod A vertebrate animal with four true limbs, that is, jointed, bony appendages with digits. Mammals, amphibians, birds, and reptiles are tetrapods.

thermoregulation The maintenance of a relatively stable internal body temperature.

thigmotropism The response of plants to touch and wind.

thymus The organ in which T cells mature.

tissue An organized group of different cell types that work together to carry out a particular function.

trachea A large airway leading into the lower respiratory tract.

tracheal system A series of air-filled tubes in insects that deliver air to cells.

trans fat An unhealthful fat made from vegetable oil that has been chemically altered to make it solid at room temperature; often found in snack foods.

transcription The first stage of gene expression, during which cells produce molecules of messenger RNA (mRNA) from the instructions encoded within genes in DNA.

transfer RNA (tRNA) A type of RNA that transports amino acids to the ribosome during translation.

transgenic Refers to an organism that carries one or more genes from a different species.

translation The second stage of gene expression, during which mRNA sequences are used to assemble the corresponding amino acids to make a protein.

transpiration The loss of water from plants by evaporation, which powers the transport of water and nutrients through a plant's vascular system.

transport proteins Proteins involved in the movement of molecules and ions across the cell membrane.

triglycerides A type of lipid found in fat cells that stores excess energy for long-term use.

trisomy 21 Having an extra copy of chromosome 21; also known as Down syndrome.

trophic level The feeding level of an organism, based on its position in a food chain.

tumor A mass of cells resulting from uncontrolled cell division.

tumor suppressor gene A gene that codes for a protein that monitors and checks cell cycle progression. When these genes mutate, tumor suppressor proteins lose normal function.

turgor pressure The pressure exerted by the water-filled central vacuole against the plant cell wall, giving a stem its rigidity.

unsaturated fat A fat, such as olive oil, that is liquid at room temperature; often found in plants and fish.

urethra A tube that connects the bladder to the genitals and carries urine out of the body. In males, the urethra travels through the penis and is also used to carry sperm.

uterus The muscular organ in females in which a fetus develops.

vaccine A preparation of killed or weakened pathogen that is administered to people or animals to generate protective immunity to that pathogen.

vagina The first part of the female reproductive tract, extending to the cervix; also known as the birth canal.

vas deferens Paired tubes that carry sperm from the testes to the urethra.

vascular plant A plant with tissues that transport water and nutrients through the plant body.

vascular system A system of tube-shaped vessels and tissues that transport water and nutrients throughout an organism's body.

vasoconstriction The reduction in diameter of blood vessels, which helps to retain heat.

vasodilation The expansion in diameter of blood vessels, which helps to release heat.

vector A DNA molecule used to deliver a recombinant gene to a host cell.

veins Blood vessels that carry blood toward the heart.

ventilation The process of moving air in and out of the lungs.

ventricles The chambers of the heart that pump blood away from the heart. In humans, the right ventricle pumps blood to the lungs, and the left ventricle pumps blood to the body.

vertebrate An animal with a bony or cartilaginous backbone.

vestigal structure A structure inherited from an ancestor that no longer serves a clear function in the organism that possesses it.

virus An infectious agent made up of a protein shell that encloses genetic information.

vitamin An organic molecule required in small amounts for normal growth, reproduction, and tissue maintenance.

vitamin D A fat-soluble vitamin required to maintain a healthy immune system and to build healthy bones and teeth. The human body produces vitamin D when skin is exposed to UV light.

weather Local atmospheric conditions over a short period of time.

white blood cells (leukocytes) Cells involved in the body's immune response.

wood Hard, secondary xylem tissue found in the stem of a plant.

X-linked trait A phenotype determined by an allele on an X chromosome.

xylem Plant vascular tissue that transports water from the roots to the shoots.

Y-chromosome analysis The comparison of sequences on the Y chromosomes of different individuals to examine paternity and paternal ancestry.

zygote A diploid cell that is capable of developing into an adult organism. The zygote is formed when a haploid egg is fertilized by a haploid sperm.

thymus The organ in which T cells mature.

tissue An organized group of different cell types that work together to carry out a particular function.

trachea A large airway leading into the [illegible] respiratory tract.

tracheal system A series of air-filled tubes in insects that delivers air to cells.

trans fat An unhealthful fat made from vegetable oil that has been chemically altered to make it solid at room temperature; often found in snack foods.

transcription The first stage of gene expression, during which cells produce molecules of messenger RNA (mRNA) from the instructions encoded within genes in DNA.

transfer RNA (tRNA) A type of RNA that transports amino acids to the ribosome during translation.

transgenic Refers to an organism that carries one or more genes from a different species.

translation The second stage of gene expression, during which mRNA sequences are used to assemble the corresponding amino acids to make a protein.

transpiration The loss of water from plant leaves via evaporation, which powers the transport of water and nutrients through a plant's vascular system.

transport proteins Proteins involved in the movement of molecules and ions across the cell membrane.

triglycerides A type of lipid found in fat cells that stores excess energy for long-term use.

trisomy 21 Having an extra copy of chromosome 21; also known as Down syndrome.

trophic level The feeding level of an organism, based on its position in a food chain.

tumor A mass of cells resulting from uncontrolled cell division.

tumor suppressor gene A gene that codes for a protein that monitors and checks cell cycle progression. When they are mutated, tumor suppressor proteins lose normal function.

turgor pressure The pressure exerted by the water-filled central vacuole against the plant cell wall, giving plant stems rigidity.

unsaturated fat A fat, such as olive oil, that is liquid at room temperature; often found in plants and fish.

urethra A tube that connects the bladder to the genitals and carries urine out of the body. In males, the urethra passes through the penis and is also used to carry sperm.

uterus The muscular organ in females in which a fetus develops.

vaccine A preparation of killed or weakened pathogens that is administered to people or animals to generate protective [illegible] immunity to that pathogen.

vagina The first part of the female reproductive tract, extending to the cervix; also known as the birth canal.

vas deferens Paired tubes that carry sperm from the testes to the urethra.

vascular plant A plant with tissues that transport water and nutrients through the plant body.

vascular system A system of tube-shaped vessels and tissues that transport water and nutrients through a plant—primarily xylem and phloem.

vasoconstriction The reduction in diameter of blood vessels, which helps to retain heat.

vasodilation The expansion in diameter of blood vessels, which helps to release heat.

vector A DNA molecule used to deliver a recombinant gene to a host cell.

veins Blood vessels that carry blood toward the heart.

ventilation The process of moving air into and out of the lungs.

ventricles The chambers of the heart that pump blood away from the heart. In humans, the right ventricle pumps blood to the lungs, and the left ventricle pumps blood to the body.

vertebrate An animal with a bony or cartilaginous backbone.

vestigial structure A structure inherited from an ancestor that no longer serves a clear function in the organism that possesses it.

virus An infectious agent made up of a protein shell that encloses genetic information.

vitamin An organic molecule required in small amounts for normal growth, reproduction, and tissue maintenance.

vitamin D A fat-soluble vitamin that is required to maintain a healthy immune system and to build healthy bones and teeth. [illegible]

weather Local atmospheric conditions over a short period of time.

white blood cells (leukocytes) Cells involved in the body's immune response.

wood Hard, secondary xylem tissue found in the stems of a plant.

X-linked trait A phenotype determined by an allele on an X chromosome.

xylem Plant vascular tissue that transports water from the roots to the shoots.

Y chromosome analysis The comparison of sequences on the Y chromosomes of different individuals to examine paternity and paternal ancestry.

zygote A diploid cell that is capable of developing into an adult organism. The zygote is formed when a haploid egg is fertilized by a haploid sperm.

Index

Note: Page numbers followed by f indicate figures; t indicate tables.